ENERGY

Physical, Environmental, and Social Impact

Third Edition

Gordon J. Aubrecht, II

The Ohio State University at Marion

PEARSON

Prentice Hall

Upper Saddle River, NJ 07458

Library of Congress Cataloging-in-Publication Data

Aubrecht, Gordon J.
  Energy : physical, environmental, and social impact / Gordon J. Aubrecht II.
    p. cm.
  Includes bibliographical references and index.
  ISBN 0-13-093222-1
  1. Power resources. 2. Power (Mechanics) I. Title.

TJ163.2.A88 2006
333. 79--dc22   2005043168

Associate Editor: Christian Botting
Senior Editor: Erik Fahlgren
Editorial Assistant: Jessica Berta
Senior Marketing Manager: Shari Meffert
Buyer: Alan Fischer
Art Editor: Xiaohong Zhu
AV Art Director: Abby Bass
Production Supervision: Progressive Publishing Alternative
Composition/Art Studio: Laserwords
Art Director: Jayne Conte
Cover Designer: Bruce Kenselaar
Cover Image Credit: Getty Images, Inc.

Photo Credits for Part/Chapter Openers:
Part 1: NASA/Earth Observatory Team
Parts 2, 3, 4, 7: U.S. Department of Energy
Part 5: John Veil, Argonne National Laboratory
Part 6: Tennessee Valley Authority
Part 8: David Horsey, "Running on Empty". Copyright 2004, Seattle Post-Intelligencer. Used by permission.
Photos not otherwise credited were supplied by author.

Printed on Recycled Paper

For my children, and my children's children.
May our Earth abide.

BRIEF CONTENTS

CONTENTS

PREFACE

This book originally grew out of a course I taught at the University of Oregon in 1974 and 1975, just after the first of the "energy crises" wakened interest in the topic. The course was one quarter long and was one of a series of "mini-courses" instituted in the physics department because of student desire for relevance.

Preparation for the course was difficult because of the wildly diverse group of students who enrolled. I decided at the outset that I would not be apologetic about introducing the physics needed for the ensuing discussions. The ability to sift through data, winnow the salient features, and act upon that basis was something I tried to emphasize throughout the course. A few of the problems were discussed in some detail in response to specific student interest; for example, the material on insulation and lighting was originally incorporated because of interest expressed by a group of architecture students taking the course.

At the Ohio State University, I was involved in another physics course, Physics 100.02, which was generated in the early 1970s in response to the same pressures that led to Physics 114 at the University of Oregon. This course got me more involved in the experimental and phenomenological details of matters pertaining to energy. Many of these insights were incorporated in subsequent drafts of the book.

These courses focused in part on the moral dilemmas facing our technological society in its quest for some control of its own destiny. The lack of sensitivity to moral issues is embarrassingly present all around us. When moral issues are embedded in technological and economic decisions, even people of obvious good will are inclined to leave things to the "experts." These experts may encourage others in this; only they know the full facts of the matter, they say. The issues are too complex to allow free and open discussion, they say. We know best, they say. It is undeniable that experts do know a great deal about "current" technology. The problem is that many of these experts tend automatically to favor the status quo, and may foreclose options just because "it isn't done that way." Involved and informed outsiders can force people to reexamine their preconceptions. People are so much creatures of habit that this process can be extremely uncomfortable.

Responsible societal decisions cannot be made without broad involvement. If people remain willfully ignorant, politicians may decide these things, and politicians are as woefully uninformed as the rest of the citizenry. Knowledgeable citizens must be able to make their knowledge relevant and lead politicians to adopt responsible answers. This has happened in our past. The Clean Air and Clean Water acts of the 1970s came in response to concerned citizen involvement.

It is simply not true that laymen cannot grasp the issues or the technology involved in matters of social interest. In this book I attempt to inform people in a fashion that will allow them to use their good sense when they approach such questions as the application of technology to agriculture. We must have citizens who are not willing to "leave things to the experts."

"Then idiots talk," said Eugene, leaning back, folding his arms, smoking with his eyes shut, and speaking slightly through his nose, "of energy. If there is a word under any letter from A to Z that I abominate, it is energy. It is such a conventional superstition, such parrot gabble!"

CHARLES DICKENS
Our Mutual Friend (1865)

If our children were cold, we would wish to keep them warm. If we ever thought about it, we would wish the same for our grandchildren and great-grandchildren. The responsibility for their well-being and for the equitable distribution of resources between present and future—for example, for their access to the dowry of the last several hundred million years of stored solar energy (coal and oil underground)—rests clearly with all of us at the present time.

The purpose of this book is to give the reader some understanding of the important decisions which must be made soon; of the tradeoffs between risks and benefits as well as we are able to define them; of the consequent social and political choices; of the role the reader, as an informed citizen, is able to play. Choices will have to be made. There are no free lunches.

I have attempted to indicate in the text where my own value judgments have intruded. I have attempted to tell the story in a balanced way, and to present a lot of data in graphs and tables. I hope there is enough information given to allow the reader to decide which issues are important for the future. Hard decisions are looming and they will be made either in full knowledge of the issues involved or by default. My prayer is that we take the former course.

For the teacher: These issues must involve your value system; they cannot be addressed in a moral vacuum. Level with your students. Let them know where you stand.

For the student: **Question, question, question!** Question my values as presented here, your teacher's as given in class, and most of all, your own.

REFERENCES

Why the explicit references? The reader cannot fail to notice the explicit references to the popular and scientific literature in the book. There are about 5300 separate references in the book. I have included all these references for important reasons. First, they serve to indicate that there has been a discussion on the topic in the scientific community that bears on its credibility. Second, they provide resources for the interested teacher and student; they are a springboard to further knowledge. Third, they provide a teacher the opportunity to ask the students to research a particular topic with the knowledge that the student has good resources to turn to. In recent years, my students have turned to the internet to "research" their term papers, yet much "information" on the internet is incorrect. The cited references send them into the internet armed with some specific comparisons, and ensure their further work is based on knowledge of what the scientific community has done so far.

There are a very few references to pre-1960s literature to provide the reader access to the history of development of a few important ideas. I think it amazing how prescient these scientists were so many years ago. While most of the literature on energy is recent, there are really important ideas that have changed the worldview of physicists, and the general public, that benefits from the technology developed as a result of that advance.

The popular literature references are important because most public libraries (and all university libraries) have archived microfilm versions of the large-city newspapers, especially the *New York Times*, the *Washington Post*, the *Chicago Tribune*, and the *Los Angeles Times*, all national in scope. They also cover politics, which is inextricably mixed with questions of energy policy. Some students cannot read *Scientific American* or actual scientific journals with great comprehension because of lack of practice. Even if no journals are available to the reader, these library popular literature resources should be available locally.

The scientific literature is cited because of the tendency of many people to speak in generalities and misrepresent their knowledge of a particular subject. Contained within the vast scientific literature of energy is information that can lead the informed citizen to ask pointed questions. Having a broad base of knowledge support available is also a way to prevent oneself from becoming trapped into a mistaken point of view, and I have tried to provide the relevant documents so the reader will be able to draw on this support.

WHAT'S NEW IN THIS EDITION?

Currency

The current edition is as up-to-date as I can make it. In most cases, I chose to use year 2000 dollars as the monetary standard. I have studied the trends carefully and the political and social aspects of energy are also as current as I can make them. I have highlighted a few important aspects of the chapter topics with boxes that expand on them with the themes *A Closer Look, Applications, Case Study, Doing Your Part*, and *History of Energy*. The end-of-chapter exercises are virtually all new for this edition. We have made sure that there are some problems that can stretch a student at any level.

Reorganization

This third edition benefits from the input of colleagues and reviewers, who urged me to make a few organizational changes. In the previous edition, I faced the problem that many topics that needed to be mentioned could only be mentioned in one or two paragraphs. This often interfered with the flow of the story of the chapter and seldom was able to do justice to the subject. In this edition, I have been able to tighten the organization of the printed chapters by referring interested readers to Extensions (see below) that are able to treat the topics with the attention they deserve without interfering with the chapter's story. In the second edition, the sections were organized according to the Energy Supplies and the Consequences of Use. In this edition, I have grouped the chapters based on the physics involved. Thus, all the nuclear chapters are together, all the fossil-fuel energy chapters are together, all the solar energy chapters are together, etc.

Extensions

The Extensions are extra topics available for download in PDF format from this book's **Companion Web Site** (CW). The Extensions are referred to in short summaries, with the name of the Extension indicating the subject. These Extensions cover major topics that are either too technical to be in the printed text, or are not as important to the central story, and also allow more extensive discussion of important topics that had to be left out of previous editions or that were mentioned only in passing. Some Extension topics include: the costs and effects of the 1990 Clean Air

Act Amendments; the deregulation fiasco in California (but the success in Pennsylvania); the risks and possible benefits of oil development on the Arctic National Wildlife Refuge; the polluting effects of cars; the disposal of nuclear wastes; the linear no-threshold dose model; the near-accident at the Davis-Besse nuclear facility in 2002; the radioactivity we are exposed to from coal plants, and much, much more. These extra discussions add richness and vital information, but are not placed in the print text, where the detail would overwhelm the points being made.

Companion Web Site (*http://physics.prenhall.com/aubrecht*)

In addition to the Extension readings, the CW also contains Chapter-by-Chapter Learning Objectives, Online Web Destinations, Self-Check quizzes, Complete Chapter References (including all author names and complete title of the work to make identification easier), and Unit Conversion Spreadsheets.

Unit Conversion Spreadsheets

In most cases, no one learns much from doing conversions, especially if they are tedious. The Web Site has unit conversion spreadsheets I wrote that can convert virtually any unit of interest (energy-related or not) to any other—so if you want to know the energy in 12 tonnes of coal, equivalent, you can find out; if you want to know how many barrels of oil that represents, you can find that, too. Did you ever wonder what an acre-foot is? Most ordinary scientists and people do not use these units, and they are converted to International System units for ease in making comparisons to known units (I have tried to indicate conversions as appropriate in the text). A very compressed version of the spreadsheets is printed inside the book's covers.

Cross-Chapter Correlation

In this edition, I've tried to tie the similar topics more tightly together, emphasizing, for example how the relationship

$$\text{dose} = (\text{factor}) \times (\text{exposure})$$

is common for all sorts of cases—nuclear radiation, air pollution, etc., by writing the equations in parallel form in the appropriate chapters. The factor has different names depending on the pollutant, but the relationship is the same.

ACKNOWLEDGMENTS

First and most important, my wife Michelle and our daughters Katarina and Taryn put up with my foibles during the revision of the book and helped me in so many ways that it is impossible to express my abiding gratitude and love for their support. I want to thank my away-from-home daughters, Laurie Wessely-Baldwin and Dacia Jackson, for their long-distance support. I again thank my old friend Bob Friedman for continuing support and encouragement about the book over many years, which heartened me to the gargantuan task of revising and updating what you now read.

Thank you, former students! You know who you are. It is your interest that helped me be enthusiastic class after class, your questions that set me off in different directions and made me think, your patience with my jokes that helped us groan together as we grew together. I trust you are informed, responsible citizens.

On technical matters, I'd like to thank Seymour Alpert for arguing with me about the prudence or lack of it involved in burning petroleum in the late 1960s (he made me get my facts straighter); Amit Goswami for his advice and encouragement as I took on Physics 114 at Oregon; E. Leonard Jossem and the late James Harris for helping a klutzy theorist do class experiments that sometimes work; John Harte for interesting conversations on the Lotka-Volterra equation; Richard Bower for information on Canadian energy policy; and Bunny Clark, for making me revise many examination questions I'd written until they were reasonably comprehensible.

This book would not have been so complete without the services of our stalwart reviewers. They provided critical comments, found mistakes, and demanded focus. I was extremely lucky in having the insights of so many helpful reviewers. They are:

Justin Akujieze, Chicago State University; Zack Clayton, Ohio EPA; Doug Franklin, Western Illinois University; Jennifer Gitlitz, Container Recycling Institute; James Hansen, Goddard Institute for Space Science, Columbia University; Laurent Hodges, University of Iowa; Richard Holman, Carnegie Mellon University; Harris Kagan, Ohio State University; Jay Kunze, Idaho State University; Gene Lene, St. Mary's University; Gretchen Van Meer, Northern Illinois University; William Makofske, Ramapo College of New Jersey; Nebil Misconi, University of Central Florida; Michael O'Hadi, University of Maryland-College Park; Daryl Prigmore, University of Colorado-Colorado Springs; James Rabchuk, Western Illinois University; Ljubisa R. Radovic, Penn State University; Michael Ritzwoller, University of Colorado-Boulder; Alvin Saperstein, Wayne State University; Bill Simpkins, Iowa State University; Sunil Somalwar, Rutgers University; Marcy Vozzella, Northern Essex Community College; Richard Walton, Rocky Mountain College.

They have added much to the worth of this book. Any errors that remain are my own.

Librarians can help people doing scholarly research. I want to mention in particular the help afforded by OSU Marion librarian Betsy Blankenship in tracking down errant citations. I live in Delaware, and also used the resources of Beeghly Library at Ohio Wesleyan University, where science librarian Deb Peoples was extremely helpful.

Christian Botting, my editor at Prentice Hall, provided resources with patience. I really appreciate his help and support in making this third edition the best it could be.

During part of the time I worked on the first edition of this book I benefited from the support of the Aspen Center for Physics and the Alexander von Humboldt Foundation. I am deeply indebted to AvH for allowing me the opportunity to study energy conservation in Germany firsthand. My research on student understanding of radiation and radioactivity supported by a grant DUE-9950528 from the National Science Foundation informed the nuclear energy sections of this edition.

GORDON J. AUBRECHT
Delaware, Ohio
aubrecht.1@osu.edu

ABOUT THE AUTHOR

Gordon Aubrecht is Professor of Physics at The Ohio State University at Marion. He has been a university professor for more than 25 years and has taught well over 200 undergraduate and graduate courses during his career.

Gordon received his undergraduate degree in physics from Rutgers University. He did his doctoral work in theoretical particle physics at Princeton University, and did post docs at Ohio State and the University of Oregon. He was an Alexander von Humboldt Fellow at the University of Karlsruhe, Germany, and spent two years on leave at the University of Maryland and the American Association of Physics Teachers.

His research (outside of energy issues) focuses on the understanding of physics by students using the tools of physics education research. He has published over 50 articles on his research in refereed journals and has made many contributions to books and encyclopedias. In addition to *Energy*, he is author or co-author of about a dozen other books, including *Doing Physics with Spreadsheets* (Prentice Hall, 2000), and *The Charm of Strange Quarks* (New York: Springer-Verlag, 2001).

Gordon is a long-time member of many organizations, including the American Physical Society, the American Association of Physics Teachers, the National Science Teachers Association, the American Association of University Professors, the Institute of Electrical and Electronics Engineers, Phi Beta Kappa, and Sigma Xi.

Gordon served as the scribe of the Citizens Advisory Council of the Utility Radiological Safety Board of Ohio until 2001. He is secretary of the InterAmerican Council on Physics Education, an organization dedicated to improving communication among physics teachers in the Americas. He is secretary of IEEE Standards Coordinating Committee 14. He is a founding member and current chair of the Contemporary Physics Education Project (CPEP). He has run almost four dozen local, national, and international workshops for high school and college teachers.

Gordon is the recipient of numerous awards, including: the AAPT's Distinguished Service Citation (1996); Southern Ohio Section (AAPT)'s prestigious John B. Hart Award for distinguished service (2002); the Howard N. Maxwell Award for Distinguished Service by the Ohio Section, American Physical Society (2004); the Association for University Regional Campuses in Ohio (AURCO) Distinguished Service Award (2004); and the Ohio State Chapter of the American Association of University Professors Louis Nemzer Award for defending the principles of academic freedom against all challenges (2004).

For many years, until its demise, Gordon had been a board member of the Delaware Recyclers, a nonprofit recycling organization in Delaware. He served six terms as president of that organization. As a result of his devoted work with the Recyclers, he was given the Lifetime Commitment to Recycling award by *Keep Delaware County Beautiful* in 1998.

Gordon resides in Delaware (Ohio) with his wife Michelle and their daughters, Katarina and Taryn.

INTRODUCTION

Once upon a time, people thought that the supply of energy available to do useful work was inexhaustible. Once upon a time, fossil fuels—oil and coal—were so cheap that no one was concerned if they were wasted. Once upon a time, people were so few in number that they thought they could throw away anything, and it would never contaminate the air, the water, or the land, and there would always be room. More recently, we have come to realize that these are only fairy tale ideas. Fossil fuel resources on Earth are limited. The environment cannot be abused with impunity when the population in any area grows too large.

What should we, as citizens, do about energy? What are the long-term prospects for energy and resource supplies? Should we just enjoy the benefits of low gas prices? Gas was cheap in the 1950s, 1960s, the late 1980s, and the early 1990s, but it was very expensive in the 1970s and early 1980s, and prices are again rising in the 2000s. Was it responsible to buy gas-guzzling cars or SUVs when gas prices were lower? Should we concern ourselves about the health risks of carbon monoxide emissions from cars? Should government act on acid rain? Is climate changing, and if so, is the change due to human actions?

These questions are the concern of all citizens in a democracy. Many of them are discussed by our elected representatives in the political forum. We must consider what advice to give these representatives. Citizens and policy makers need to consider the long-term results of our actions and to construct long-term solutions. Unfortunately, the desire of officials to be reelected often inhibits the government's ability to deal well with these situations. Twenty years seems infinitely long to someone elected for two, or four, or even six years. There must be some way for the system to respond to pressing questions that have factual answers, or to set priorities after determining the best available answer when the facts are in dispute. Issues of energy and the environment are of concern to everyone. The world is, after all, finite.

Some of the questions concerning the future of our planet and our lives may be answered from the store of scientific knowledge available now. Politicians should solicit scientific advice and pay attention to it. For other questions, we need further scientific research to determine the best answers. In some cases, we may never know enough to be able to answer the questions in detail. We should realize that the scientific community disagrees about the results of research and may give conflicting advice.

What is known is that no method of generating energy is without risk. In obtaining and using energy, people are doing things that could result in events such as catastrophic climatic change; local pollution of air, water, and ground water, which could be hazardous to the health of large numbers of people; and long-distance pollution that causes harmful effects on agriculture, silviculture, and aquaculture.

Of course we cannot give up the use of energy. Without energy, we would return to the sort of life characterized by the philosopher Hobbes as "poor, nasty, brutish, and short." People do not want to give up their daily energy "fix." To have the energy to run businesses, hospitals, and schools, and to make life more satisfying, we have to accept the problems associated with generating energy.

In order to address these daunting problems, and problems of the Earth's finite resources, I believe that we, ordinary citizens of the world, must first acknowledge these problems if we are to succeed in overcoming them. There are many reasons for concern, but concern need not be synonymous with despair. It is by facing problems, not by denial, that we find solutions. I am optimistic that solutions will be found if the gravity of the problems of pollution, overpopulation, and irresponsible use of energy are acknowledged by the public and our politicians. Then we can work on solutions and find better ways to use our resources and preserve our environment.

As you read this book, I hope you will be sharpening your critical faculties. I hope that in the future, as you read articles in the newspapers and popular magazines, you will be able to distinguish the overstated and the untrue, the persiflage and the balderdash, the self-serving statements, from accurate information. Your responsibility is to make the future world one in which you wish to live, and one in which you would wish your descendants to live.

ORGANIZATION

The book is divided into several parts. In Part One: **General Introduction** (Chapters 1 to 3), we examine energy as the ability to do work, and power as the rate of doing work. In Part Two: **Generation and Transmission of Electricity** (Chapters 4 and 5), we consider the manner of generation of electric current, the distribution of electricity, exponential growth, the making of projections for demand, the logistic curve, and the pitfalls of projecting, and the role of population demand. In Part Three: **Thermal Aspects of Energy Generation** (Chapters 6 through 9) we consider constraints on heat engines from the laws of thermodynamics, the availability of water, and the effects on the water of the generation. In Part Four: **Material Resources** (Chapters 10 and 11), we look at the availability of mineral resources and prospects for reduction in the materials burden. In Part Five: **Fossil Fuel Energy Resources and Consequences** (Chapters 12 to 17) we consider the availability of fossil fuel supplies, and the consequent pollution and other problems that ensue from use of these fuels. In Part Six: **Nuclear Energy Resources and Consequences** (Chapters 18 to 20), we look at the physics of the nucleus and the generation of nuclear electricity, as well as some drawbacks of the current schemes. In Part Seven: **Solar Energy and Alternative Energy Resources and Consequences** (Chapters 21 to 25), we examine conventional hydroelectricity, new solar energy technology, biomass resources, and alternatives that may place less of a burden on public health, on society, and on the environment while preserving access to necessary energy. We finish with Part Eight: **Tocsin** (Chapter 26), which recapitulates many of the book's themes, offers a critique of current practice, and a voice of hope for the future.

Like Caesar's Gaul, there are three parts to the topic of energy: the **Physics**, the **Energy Supplies**, and the **Consequences of Use**. In this text, these three energy issues are addressed throughout.

OVERVIEW

THE PHYSICS OF ENERGY

Energy cannot be discussed without consideration of what it is physically. Sources of energy are many, and knowledge of physics is necessary to understand how most sources can be tapped. Chapters 1 and 2 discuss the history of energy demand and some constraints on energy policy and use. Chapter 3 defines energy physically and discusses its various forms. Chapter 4 introduces electricity; the electric utility system in North America is also discussed in Chapters 5 and 7 through 9: predictions of energy demand are found in Chapter 5, our introduction to the exponential function; Chapter 6 describes chemical energy, and Chapter 7 considers the heat engine and thermodynamics, the study of thermal energy in transit. Chapters 12 and 13 discuss physics aspects of the use of fossil fuel energy. Chapters 18 and 19 discuss nuclear energy. Chapter 21 sets solar energy in physical context. Chapter 25 involves the physics of alternative energy.

ENERGY SUPPLIES

Supplies of energy and related minerals form the basis of the world economy. In the following chapters, the various kinds of energy resources and resources in support of the energy economy are considered. Solar energy is the focus of this group of chapters. Forms of solar are considered in Chapter 12 (fossil fuel resources—stored solar energy), Chapter 21 (wind and photovoltaics), Chapter 22 (hydroelectricity, waves, and tides), Chapters 23 and 24 (biomass energy), and Chapter 25 (section on satellite capture of solar energy and geothermal energy). Resources mustered in support of energy are considered in Chapters 10 (mineral resources) and 11 (conservation). Energy resources themselves are discussed in Chapter 9 (conservation), Chapter 11 (recycling), Chapter 12 (coal, oil, gas), and Chapter 18 (nuclear).

CONSEQUENCES

The following chapters consider the consequences of current activities: electrical energy generation (Chapter 13); acid rain (Chapter 14); transportation sector of the economy (Chapter 15); climate (Chapter 16); human intervention in the Earth's weather system (Chapter 17); and nuclear energy (Chapter 20). Chapter 26 attempts to tie together all the issues involving energy in context.

INTRODUCTION TO ENERGY

PART I

CHAPTER 1

Energy and the Environment: Science, Technology, and Limits

Energy has been used by humanity ever since there have been humans. We give an indication of the way energy has been used. With use come consequences as well. Everything affects everything else. The study of energy involves scientific, economic, environmental, political, and social elements. Science works, because what we accept as a description of reality has been tested many, many times. The story of humanity's interaction with energy begins with examples of some unintended consequences of political and social decisions (both made and unmade).

KEY TERMS

energy / TANSTAAFL / falsifiability / ozone / catalyst / chlorofluorocarbon / atmospheric lifetime / "tragedy of the commons"

1.1   ENERGY IN HISTORY

Energy is what people use to exist and to grow. Energy is also what people use to make their lives easier. Energy makes our everyday lives of today possible. In this book, we will consider the benefits of energy in extending life, enhancing health, providing physical comfort, and making life more pleasant in many ways.

Energy measured in kilocalories (kcal), food "calories," may be familiar to dieters and family members. [Scientists seldom use the kcal; they usually use the joule (J) as the unit of energy. (See Chapter 3 for an explanation of energy and its units.)]

For our early ancestors, who lived by hunting, ate their kill raw, and were much smaller than we are, there was an energy budget of about 6000 to 8000 kilojoules per day. (Or, for those more familiar with food intake terms, this amounts to 1500 to 2000 kcal per day.)

When people learned how to use fire, the amount of energy used per day was probably about double or triple the former amount: about 12,500 to 17,000

TABLE 1.1

Daily energy use per person

	MJ per capita per day	kcal per capita per day
Hunters	8	2,000
Use of fire	17	4,000
Domestication of animals	40	10,000
Renaissance	100	25,000
1850,[a] U.S.	420	100,000
1900,[a] U.S.	460	110,000
1950,[b] U.S.	630	151,000
1960,[b] U.S.	705	168,000
1970,[b] U.S.	945	226,000
1980,[b] U.S.	970	231,000
1985,[b] U.S.	900	214,000
1990,[b] U.S.	945	226,000
2000,[b] U.S.	1,011	244,526

[a] Reference 1.
[b] Reference 2.
See the spreadsheet "energy converter" on the CW (companion website).

kilojoules per day. (I estimate about 20 MJ/kg heating value for wood, that is, a use of about 3 to 4 kg of wood per day.) Note that our simple definition of energy allows us to add all the contributions from all these disparate sources.

When human beings settled down to a more sedentary life of agriculture using oxen or horses, energy usage probably tripled per person; thus, a person used about 40,000 kJ per day. I obtain this by assuming that an animal uses about 100 terajoules (TJ) per year, based on an analysis[1] that estimates that 1 kg of feed must be supplied for each 50 kg of body mass and taking an average mass of 750 kg. At 16.75 MJ/kg feed, this is 250,000 kJ per day. Assuming one horse or ox to equal about 10 people, this is an additional 25,000 kJ per day per person.

In Europe during the early Renaissance, windmills and coal began to be used, and wood was used in large amounts for heating, cooking, metalworking, and

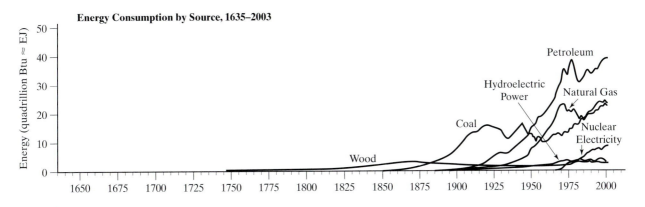

FIGURE 1.1

History of energy consumption in the United States by source, 1635–2003.
(Energy Information Agency, US Department of Energy, *Annual Energy Review 2003* (Washington, DC: GPO, 2003), DOE/EIA-0384(2003), August 2004, Fig. 5.)

other such tasks. Water power was also used to grind grain. Thus, energy use per capita probably doubled again, to 80,000 to 100,000 kJ per day.

In the mid-nineteenth century, the United States used about 400 MJ per capita per day, and today as an industrial power, it uses about 1000 MJ per capita per day (see Table 1.1). With only about a twentieth of the world population, we use about a fourth of the world's total manmade energy.[2] Some important energy equivalents and units are listed on the inside covers and in **Extension 3.7, *Various Units***.

In modern times, we have substituted other forms of energy for human labor, making life easier (Figure 1.1). The chemical energy of gasoline is transformed into kinetic and thermal energy. The chemical energy of wood (which is actually stored solar energy) powers boilers that generate electric energy. This electric energy can become mechanical energy (in machines), radiant energy (from light-bulbs), thermal energy (in ovens or water heaters), and so forth.

1.2 TANSTAAFL

John W. Campbell, editor of *Astounding Science Fiction Magazine*, formed a group of writers in the 1940s who wrote "hard science" science fiction. Some authors, such as Robert Heinlein and Eric Frank Russell, applied engineering principles to society. Many of these authors wrote stories of worlds that ran on the principle of *TANSTAAFL* (*tan-staff-el*): "There Ain't No Such Thing As A Free Lunch." The stories were antiutopian in that nothing in future societies was free, and everyone realized this and lived accordingly. In other words, if people want something, they have to pay for it.

We are certainly conversant with this principle in everyday life. We say: "You get what you pay for." This fact holds for physical phenomena as well. Life on Earth would not exist if the planet's energy deficit were not paid for by the Sun. (We will explore this dependence on solar energy later.)

Everything done by people has an effect on everything else. For example, a communications satellite (orbited with the laudable intention of allowing India to broadcast to the entire subcontinent with only one transmitter) interfered with radio astronomy. *SMS-1*, a synchronous meteorological satellite (synchronous means it does not move relative to one particular point on Earth), also caused problems for astronomers.[3]

Words of Science

We live in a world in a state of flux: things are always changing. Jokes change, politics change, the definitions of words we use to express ourselves change. Which word is this year's fashion?

Code words can be used to make complex issues seem simpler, but often they distort rather than simplify. Take "energy crisis": Experts, newscasters, and others said in the early 1970s that the days of cheap energy were over. It may look now as if they were right before their time; but between the 1980s and the new millennium, Americans thought they didn't need to worry. Cheap gasoline was available then, and so it always would be. We were led to that conclusion, because the 1970s energy crisis was partly to do with a temporary gasoline shortage caused by political decisions rather than a real shortage of resource. But, oil is not a resource that is increasing on Earth. This reality may take some time to be felt, but the code words allow us to make the real, underlying difficulties seem less of a bother.

One of the purposes of this book is to dampen the rhetorical extravagance of popular "energy language" in order to examine the underlying issues. First, however, we have to agree on definitions; that is, we have to speak the same language. Science uses words that have specific meanings. It is something like a foreign language, and some words will simply have to have their definitions committed to memory. Science words allow scientists to speak precisely to one another so that meanings do not get muddled.

What Science Is

Many people admire the fruits of science, but few take the time to think about what science really is—a process by which fallible humans attempt to pry Nature's secrets from her grasp for the common good (through increased knowledge). Scientists have a value system that rewards those who determine that a prediction was wrong, and therefore, allows checks to be built into the structure of science. Most people do not like to hear "no." The community of science values the "no" answer, when Nature says it does not work a certain way.

Many of you reading this book will have thought of scientists as having all the answers, but real scientists value the questions over the answers. It is true that the answers to some questions spawn other questions, and that is all to the good. So, why is science so good at prediction, at answers? It's no mystery. Real science bears some relationship to the scientific method you probably learned in elementary school. It is built on guesses (usually called hypotheses) of the way the world works. If that were all, nothing would have come of it.

The guesses, or models, of the way Nature works lead to predictions that can be tested and (this is crucial) be proved wrong if the model is wrong. The condition that the idea can be proven wrong is called **falsifiability**. If the prediction fails, the model is junked, and another is created. If the prediction succeeds, another prediction of the model is made and tested, again and again. In this way, only models that make predictions that *cannot be disproved* survive. False predictions doom models. After many thousands of tries at disproof by very clever people, a model that survives must have something right about it. It will be accepted provisionally as correct, until it fails to survive a test. Further, when the model is superseded by another, the part where conditions correspond to what was found to be correct before should give the same correct result as found before.

This does not mean that the experiments have proven the model. Science can never present proof that something is true, only proof that something is false. A model reigns only until dethroned. The process works to give us models that resemble reality—but they are still models, not the reality itself. And all models will fail somewhere.

Scientists have some prejudices about the way Nature works. They think that some experimental result would be equivalent to one done elsewhere or at another time (the result is independent of time or place). For example, what we learn about the Sun now should describe the Sun 1000 years ago. It should describe what would happen to an equivalent sun in another galaxy. The result is independent of the experimenter. (If it only works once, it isn't science.) Experimental scientists are also fanatic record-keepers, because they need to make sure of every effect that might affect their result.

Science is tentative, and scientists are willing to be led by the evidence to renounce what is false and use whatever hasn't so far been proven false. We think we know a pretty good description of the way the world works, because we've eliminated so many ways we've found that the world *doesn't* work. We treat all knowledge as provisional and accept the models that have not been proved wrong as tentatively correct.

What Technology Is

Technology, as expressed in engineering, is the process of developing scientific knowledge in the form of practical machines, ones that work. The wonders of modern technology include jetliners, DVDs, and personal computers. Scientists conceived of and investigated radio waves. Engineers designed and built radios and televisions. Scientists discovered x rays. Engineers designed x-ray tubes. Scientists discovered radioactivity. Engineers built machines that use radioactivity to help cure cancer. Scientists developed a theory of microwave radiation; engineers developed (microwave) radar and microwave ovens.

Without engineers, scientists would not have apparatus with which to ask questions of Nature. Without scientists, engineers would have to become scientists. Indeed, engineers sometimes act as scientists, and scientists sometimes act as engineers.

Energy and Pollution

The topic of energy and pollution is an inextricable mix of *physical* and *social* (or societal) questions. Energy is a physical entity; pollution, on the other hand, requires some value judgment to define, some recognition of the societal cost of an energy strategy. One person's pollutant may be another's raw material. In much of this book, we will explore the physical aspects of energy generation and the ensuing pollution. Inevitably, value judgments will be made. I will try to warn you when I make them. We will also explore how to use information without being overwhelmed by it. We must consider how to interpret data and how to judge what other people have done with these data.

To make these ideas more concrete, let us recognize two constraints that limit us in dealing with our problems. One constraint—TANSTAAFL, which we have already addressed—is basically physical; the other—the "tragedy of the commons"—is purely a social phenomenon.

The Ozone Layer

Normal oxygen (O_2) has two atoms in its molecule. **Ozone** is a form of oxygen in which three oxygen atoms (O_3) are bound together. As everyone probably knows, Earth possesses a layer of ozone in the stratosphere (upper atmosphere). Ozone is formed and destroyed by ultraviolet solar radiation in a cycle known as the Chapman cycle (see **Extension 1.1,** *Earth's Ozone Layer*). Under normal conditions, ozone is created and destroyed at the same rate, maintaining a constant concentration. Earth's ozone layer is unique, at least in our own solar system. The ozone layer, formed because plant life on Earth emitted oxygen into the atmosphere, has, in turn, molded life on Earth. Ozone concentrations change in the atmosphere throughout the year. In local winter and spring, ozone builds up at the poles through transport from lower latitudes. In summer, ozone levels decrease.

Even large-scale meteorological features of our planet, such as the protective ozone layer, which shields Earth from the Sun's ultraviolet radiation, can be altered by people. This provides an example of TANSTAAFL at work.

The ozone layer protects living things. If the ozone layer is damaged, our eyes and skin get more ultraviolet (UV) radiation, which causes sunburn and leads to increases in skin cancer and blindness. Skin cancer affects fair-skinned people the most. Even now, there are around 400,000 cases of skin cancer in the United States each year. A 1% ozone decrease causes roughly a 2% increase in UV radiation at Earth's surface. A 5% ozone decrease would lead to more than 8000 additional cases of cancer yearly among U.S. whites.[25] If the ozone concentration in the ozone layer decreased greatly, even people with the darkest of skins would suffer. Increased UV (of the type designated UV-B, wavelength 280 to 320 nm) has been shown to affect plants and animals adversely,[26] and the increase in UV-B in temperate zones has been documented.[27]

The Sun emits enormous amounts of radiant energy, which is spread over a very wide band (spectrum) of energies. Most radiant energy reaches Earth in the form of light. This energy can break apart oxygen molecules into atomic oxygen (O), which combines with normal oxygen molecule (O_2) to form ozone (O_3). Ozone absorbs certain bundles of energy from the Sun's UV radiation. This breaks ozone into an oxygen atom and a normal oxygen molecule. The oxygen atom then combines with another O_2 to reform ozone while re-emitting the energy in smaller bundles. The smaller bundles of light that get into the atmosphere below the ozone layer can no longer escape into space. The absorption of radiant energy by the ozone layer heats the atmosphere.

Naturally occurring chemicals can hasten the rate of ozone breakup. Three major atmospheric cycles can reduce ozone. These cycles involve nitrogen oxide (NO), bromine (Br), and chlorine (Cl). These chemicals interact with ozone to produce the normal molecular oxygen, destroying the ozone. In the absence of human intervention, small numbers of these molecules (as well as hydrogen) destroy ozone at the same rate it is created by radiant energy from molecular oxygen. If human actions add chlorine, nitrogen oxide, bromine, or hydrogen to the stratosphere, the equilibrium concentration of ozone decreases. See the section "Sources of methyl halides" in **Extension 1.1**.

Nitrogen Oxides and TANSTAAFL

When an NO molecule interacts with an ozone molecule, it forms nitrogen dioxide (NO_2) and a normal oxygen molecule (O_2). If NO_2 encounters a single oxygen atom (O), it forms NO and O_2. The whole cycle can then begin again. The NO acts as a **catalyst**—although it allows the reaction to occur, the NO remains unchanged at the end of a cycle and can repeat its performance. Nitrogen oxide gets into the atmosphere naturally when lightning breaks up molecular oxygen (O_2) and molecular nitrogen (N_2) into single atoms. Sometimes, a nitrogen atom and an oxygen atom get together. In fact, this process is responsible for the major part of all the nitrogen fertilizer applied each year to the world's soil. Very little of the lightning-formed NO gets into the ozone layer and so is little threat. Application of ever-increasing amounts of nitrogen fertilizer, however, is a threat.[28]

Supersonic transport (SST) airplanes, such as the *Concorde*, burn kerosene in the upper atmosphere, thereby producing carbon dioxide (CO_2), nitrogen oxides, and water. The nitrogen oxides are created in the combustion chamber, where some atmospheric nitrogen is broken up into atoms that combine with oxygen.[6,29] In addition, when the plane flies faster than the speed of sound (supersonic speeds), the shock wave causes most normal nitrogen molecules (N_2) and normal oxygen molecules (O_2) to break up; many do not recombine but form nitrogen oxides instead. Ozone is eaten up by nitrogen oxides from SSTs. Because SST exhaust is emitted in or near the ozone layer, SSTs pose an immediate threat to the integrity of the ozone layer.

There are few civilian supersonic planes now flying. In 1974, an estimate was made that a fleet of 500 continuously operating supersonic transports would result in a permanent reduction of ozone by 16% in the Northern Hemisphere and 8% in the Southern Hemisphere.[30] This estimate as well as concerns about sonic booms convinced the U.S. Congress that it was unwise to subsidize American civilian SSTs, thereby preventing major SST production. Another disincentive is that the *Concorde*, an SST subsidized by the British and French governments, has never been a commercial success, and stopped flying in 2003. Even the newest, fastest commercial planes planned are not supersonic.[31]

Nuclear weapons tests have provided information on this ozone-depleting effect, because explosion plumes from atmospheric tests raise huge amounts of nitrogen oxides into the stratosphere. The extensive testing of U.S. and Soviet bombs between 1948 and 1961 resulted in a 4% ozone depletion that took 2.5 years to regenerate.[30] The maximum effect of bomb tests on the ozone layer occurred several months after detonation. (It is possible that high-level nuclear explosions could cause nitrogen oxide to be fed into the stratosphere over a period as long as 10 years.)

Chlorofluorocarbons and TANSTAAFL

Just as with nitrogen oxide, if a free chlorine or bromine atom is loose in the ozone layer, it can act as a catalyst. Natural chlorine and bromine come into the atmosphere from sea salt. Chlorine and bromine atoms can destroy two ozone molecules and reconstitute themselves afterward, and do this thousands of times before being cleansed from the stratosphere.

The next illustration of TANSTAAFL concerns the effects of compounds containing carbon and fluorine, **chlorofluorocarbons** (mainly Freon, CF_2Cl_2, a common refrigerant once used widely in aerosol sprays; see Table 1.2), known as CFCs for short. Chlorofluorocarbons are a problem, because as they rise into or through the ozone layer, they break up, liberating atomic chlorine. Below the ozone layer, they are inert. (They were chosen as propellants in spray cans because they did not react with the contents.) Above the ozone layer, these gases encounter UV radiation, which results in their breakup. The chlorine is then free to destroy ozone.

By 1974, more than a million tonnes (metric tons) of CFCs per year were being released. That was the year that Molina and Rowland informed the world of the possible destructive effects of chlorofluorocarbons on the ozone layer.[33] There were fears at that time of an eventual 20% depletion in ozone, but little was known about the chemistry of the atmosphere. The atmospheric lifetimes (mean time before removal) of many CFCs are long, as shown in Table 1.2. Because it takes about 15 years for the material now entering the lower atmosphere

TABLE 1.2

Ozone-destroying chemicals

Chemical	Symbolic representation		In atm.	% Cl (yr)	Lifetime concentration ($\times 10^{-12}$)[a]
CFC-11	$CFCl_3$	Coolant, aerosol, foam	22%	60	220
CFC-12	CF_2Cl_2	Coolant, aerosol, foam	25%	130	375
CFC-113	$C_2F_3Cl_3$	Solvent	3%	90	30
CFC-114	$C_2F_4Cl_2$	Various	<1%	200	5
CFC-115	C_2F_5Cl	Various	<1%	400	4
HCFC-22	CHF_2Cl	Coolant	3%	15	100
Carbon tetrachloride	CCl_4	Solvent	13%	50	80
Methyl chloroform	CH_3CCl_3	Solvent	13%	7	130
Methyl chloride[b]	CH_3Cl		20%	1.5	600
Halon-1211	CF_2BrCl	Coolant, foam		15	1.5
Halon-1301	CF_3Br	Fire extinguisher		110	1.7
Methyl bromide[b]	CH_3Br	Pesticide		1.5	15

[a]Picomoles per mole, or parts per trillion.
[b]Naturally occurring compound.
Source: Office of Technology Assessment; 1985 values given; M. J. Prather and R. T. Watson, Reference 32.

to get into the ozone layer, CFCs stay around long enough to get their chances to wreak havoc with the ozone.

Chlorine chemistry in the atmosphere is now much better understood.[34] Ozone is increasing near the ground (some is generated by internal combustion at ground level, where—as a strong oxidant—it acts as a pollutant that causes plants to sicken and die)[35] but is decreasing in the stratosphere.[36–40] A decrease of 3% in stratospheric ozone has already been observed relative to the 1950s; the projected 20% decrease if no action on CFCs had been taken would have caused thousands of new skin cancer cases and as many as 30,000 extra deaths annually.[25] This is TANSTAAFL with a vengeance. Scientists and technologists involved in the development of spray cans never conceived that a danger to the entire human race could follow from their contributions to convenience and progress.

The Ozone Hole

Vivid evidence of ozone layer depletion was seen in measurements made by the *Nimbus 7* satellite,[41] as shown in Figure 1.2. The South Pole has a visible decrease in ozone concentration (popularly dubbed an ozone "hole"). The decrease is represented in "Dobson units." One Dobson unit represents a 10 μm thickness of the ozone at standard pressure and temperature in a column of air that runs between Earth's surface and the edge of the atmosphere (many kilometers up); a "normal" reading should be around 400 to 500 Dobson units, corresponding to an equivalent thickness of 4 to 5 millimeters of pure ozone.

This Antarctic ozone deficiency was first brought to public attention in 1986 by Farman, Gardiner, and Shanklin[42] and has been shown to occur at heights between 10 and 20 km.[43] After discovery of the Antarctic ozone depletion, scientists checked and reanalyzed the data from the solar backscatter ultraviolet detector aboard *Nimbus 7*.[44] Thinning of the ozone has occurred each year since 1979 from September to November, during the Antarctic late winter.[45] The data

FIGURE 1.2

Total Ozone Mapping Spectrometer (TOMS) satellite data for average ozone concentrations in the Arctic and Antarctic polar regions (63° to 90°). The Arctic has a smaller polar vortex (closed swirl of air circulation) than the Antarctic and a less extensive area of extreme cold because it is ice-covered water rather than an ice-covered continent. Thus, the possible occurrence of depletion is less in the Arctic than in the Antarctic, so the observed "hole" in the Arctic is less stark than in the Antarctic. In both hemispheres, the ozone depletion occurs in the spring, as the cold temperatures coexist with light from the Sun that can break chlorine-bearing molecules (CFCs) apart. Ozone is measured in Dobson units (DU; see text).
(NASA satellite data)

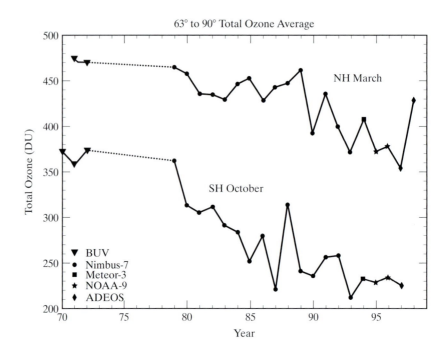

verified the decrease. It had been thought that there was a decline on the basis of the *Nimbus 7* data. It turned out on reanalysis to be much larger than the first analysis had shown,[46] which agree with model calculations.[47] The severity of the effect in the Antarctic has been increasing virtually every year since the early 1970s, when it began (see Figure 1.2 for Spring ozone levels from 1970 through 1998). By 1981, the hole was 1.5 million km², and in 2000, it registered a record area of over 28 million km².[22] Now, the Arctic also experiences a severe ozone hole, although over a smaller area.[41]

The Arctic has a smaller polar vortex (closed swirl of air circulation) than the Antarctic and a less extensive area of extreme cold, because it is ice-covered water rather than an ice-covered continent.[48] Thus, depletion is lower in the Arctic than in the Antarctic, so the observed "hole" in the Arctic, which hadn't existed until recently,[49] is less stark than that in the Antarctic. In both hemispheres, the ozone depletion occurs in spring, as the cold temperatures coexist with light from the Sun that can break the chlorine-bearing molecules apart. See the section "How ozone is destroyed at the poles" in **Extension 1.1**.

Stratospheric chlorine has gone from one part per billion in 1950 to about three parts per billion now, in eerie parallel with the ozone decrease.[21] Three pieces of evidence indicate clearly that CFCs, found in the stratosphere, are the cause of ozone depletion:

1. Chlorine oxide (ClO) concentrations at the poles are 50 times greater than "normal."
2. Ozone concentration decreases as ClO concentration increases.
3. Observed and predicted loss rates agree closely.

See **Extension 1.2, *Why Volcanic Eruptions can't be the Cause of Ozone Depletion.***

The Montreal Protocol

Fears of ozone loss led the United States to ban CFCs from spray cans unilaterally in 1978. Carbon dioxide is now used as the propellant in spray cans. Europe hesitated in 1978 and did nothing. By 1985, the relentless pace of new discoveries of the fatal consequences of CFC emission had finally attracted world attention. As a result, conditions for survival of the ozone layer have improved radically. An international agreement sponsored by the United Nations Environment Program, known as the Montreal Protocol, was signed by major producers in 1987.[53] The original Montreal Protocol would have reduced CFC emissions by 50% by the year 2000; the agreement has been amended twice, in London in 1990,[54,55] to eliminate CFCs altogether by 2000, and in Copenhagen in 1992,[56] to quicken the timetable for eliminating CFCs to 1996 and to ban halons starting in 1994 and freeze methyl bromide emissions at 1991 levels. In Vienna, in 1995, it was further agreed that methyl bromide use would end by 2010, and HCFCs would be phased out by 2040.[55] Both revisions took place because of increased knowledge of ozone destruction, including measurement of higher UV levels in both Northern and Southern Hemisphere mid-latitudes [4,57–59] as well as at the poles.[60]

It has been difficult to halt production entirely, because CFCs have been the predominant refrigerants used in air conditioners, the primary solvents used to clean circuitry in computer and appliance manufacture, and the foaming agent used to produce millions of rigid foam products each year. Manufacturers, however, have shown great ingenuity in designing replacements.[55,61] The amount of CFCs released by the United States, the European Union, and Japan has fallen from 725,000 tonnes pre-Protocol to 6800 tonnes currently.[62]

The Montreal Protocol succeeded. As of 2000, evidence shows that concentrations of some CFCs have peaked and have begun to decline (CFC-11 peaked in 1994–1995).[12,63] Despite this progress, the long lifetimes mean that the ozone layer will continue to degrade for some time (the recovery process is made even longer by global warming). The prognosis for various replacements is good,[64] although even with the replacements, models show continuing increase in ozone-destroying chemicals until about 2005, before eventual decline to pre-CFC levels by 2190.[32]

1.3   THE TRAGEDY OF THE COMMONS

Twentieth century socioeconomist Garrett Hardin has brought the concept of the **"tragedy of the commons"** to public attention through his investigations of the world population problem. In contrast to the views of the laissez-faire economist Adam Smith, who argued that an individual intent on personal gain would be led to benefit the public interest by an "invisible hand."[65] Hardin[66] holds that individuals behave in a selfish fashion that ultimately leads to societal destruction.

To illustrate his ideas, Hardin uses a village *commons*—a field open to all inhabitants of a village for use as pasture, park, or whatever they please. Each herdsman of the village is entitled to use the commons. Of course, each herdsman tries to keep as many cattle (or other domestic animals) on that commons as possible. This method might work well for a time, because war, famine, and disease hold both human and animal populations in check. Nevertheless, the day would eventually come when carrying capacity—the maximum number of cattle that could be supported on the commons indefinitely—would be reached.

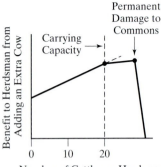

FIGURE 1.3

Benefit is perceived by the rational herdsman as he increases the number of cattle he runs on a commons. After carrying capacity has been reached, each cow will get less nourishment from its portion of the commons. Note that overuse, exceeding the carrying capacity, will cause the system to collapse eventually and no one will benefit from the commons.

Let us suppose that 10 villagers each had 20 cattle on the commons. In this case, there would be 200 cattle, which we take to be the carrying capacity. If another cow were added, each of the 201 cattle would get 200/201 of its requirements. The herdsman who owns the extra cow would see new assets: an additional cow; but he also would see new debits: each of his 21 cattle gets only 200/201 of its requirements, and consequently, each one is a little more bony, produces slightly less milk, and so on.

We might call his original assets +1, and his debits −21/201 ≈ 1/10, giving net assets of +9/10. Figure 1.3 shows how the situation appears to the herdsman. The rational herdsman, seeing he is getting 9/10 the benefit for only 1/10 the cost, would then add another animal, and another, and another. When carrying capacity is exceeded, the entire herd is more poorly fed. Eventually, the limit of permanent damage is attained, and all cattle (not just the added one) starve, become sickly, or die. The entire herd (and its economic benefit) is lost. "Each man is locked into a system that compels him to increase his herd without limit—in a world that is limited. ... Freedom in a commons brings ruin to all."[66] Because of its remorselessness, Hardin calls this process "the tragedy of the commons."

There is some possibility that humanity could ruin large parts of their environment through concentration on individual gain at societal cost, the tragedy of the commons. Ecologists Daily and Ehrlich believe that "the human enterprise has not only exceeded its current social carrying capacity, but it is actually reducing future potential biophysical carrying capacities by depleting essential agricultural capital stocks."[67] They point to the case of deer introduced to St. Matthew Island: their numbers grew from 29 to 6000. The great numbers destroyed vegetation; the carrying capacity was so damaged that the population of deer crashed to around 50.[68] Similarly, deer introduced on the Kaibab Plateau exploded in population and wreaked havoc on the vegetation.[67]

Air is known in economics as a "free good." It is breathed by everyone; everyone uses it. If an industry used air and polluted it without cost, all people would suffer, because the air would be polluted, but the industry's owner would reap economic gains. Such a case is also an example of the tragedy of the commons.

Hardin believes that tragedy of the commons will operate in every case in which personal gain is attained by distribution of losses to others, even if the losses may seem negligible to the losers. The wealth (or other advantage) of the gainers will lead them to destroy what all possess in common. If all possess something (the air, for example), the thing is perceived as having no individual economic value.

Earth is finite and thus can support only a finite population. Certain groups see advantages in increasing their numbers and so encourage people to have many children. Earth, in this case, is similar to an English commons, where people have use of common goods: air, water, land, and resources. If population grows without restraint, Earth may not be able to support that increased population. Hardin would coerce countries with exploding populations to limit the number of children born in each family, based on the reasoning that "injustice is preferable to total ruin."[66]

We are led to the conclusion that if we exploit resources in a laissez-faire manner, we favor people or institutions who focus narrowly on profits and distribute losses as widely as possible. We are practically forced to foul our own nest. The cost of dumping waste, as far as a profit-oriented company is concerned, is less than the cost of cleaning up. When the waste is discovered in the environment, the companies are no longer identifiable, and the public must pay for any cleanup.

See **Extension 1.3,** *Coal Mining as an Example of Tragedy of the Commons.*

We must know the ultimate gains and costs, as well as who wins or loses, to make informed decisions about energy strategy for the future. (When speaking of costs and benefits, we must quantify. Appendix 1 presents some of the background information needed for this book: powers of ten and scientific notation. We will use this material to practice the art of estimation in Chapter 2.) Our analysis will, of necessity, be incomplete. Society must make the best of a situation and make decisions from the best presently available knowledge. Decisions have to be made—even doing nothing involves making a decision.

Luckily, society can intervene to ameliorate the consequences of such local blindness—this is partly why humans have governments and social norms. Researchers have observed such effects in action through what they call "indirect reciprocity."[72] By this is meant that individuals who observe others "cheat" with no consequences will do so themselves, but when provided with the social opportunity, instead, those individuals will discipline the cheater and insure gain for society, even if the individuals must pay a personal price. In addition, there may be a common "social capital," a recognition of the local social interactions with the environment, that can help people manage the environment collectively and effectively.[73] It is clear that Hardin oversimplified the situation, and that humans can overcome the challenges of managing a commons, but only if they are aware of the problems.[73,74]

SUMMARY

Energy use involves gains in human comfort and ease. Energy is useful for many things and makes modern life possible. As a result, energy use has skyrocketed. Everything is connected to everything else, and everything we do exacts a price. One aim of this book is to explain both the gains and the physical and social costs associated with each strategy for generating energy. No choice connected with energy use is free of cost. Do without energy—there is a cost. Use energy—there is a cost. There is no such thing as a free lunch (TANSTAAFL).

When people exploit resources held in common, "free goods" in economic parlance, they are usually aware only of their own immediate gain or loss. This was the case for the production of CFCs, which led to ozone depletion. It seemed like a great idea to produce CFCs for propellants or coolants, because they were not reactive—they couldn't possibly become a problem. Often, the people using "free goods" are not aware of (or ignore) the costs to everyone (the discovery of the CFC ozone

problem occurred long after their development). The inexorability of the consequences of such actions constitutes the tragic character of the tragedy of the commons, illustrated in the extensions by the effects on the ozone layer and in coal mining. However, by their actions, human beings can overcome these natural tendencies. The Montreal Protocol, an international treaty that has greatly reduced use of CFCs, is an example.

Most choices involve benefits to some people and costs to others. Because of the costs associated with every action, and because the tragedy of the commons illustrates a way in which organizations or individuals avoid paying for the social costs inflicted by their actions (leaving others to bear that burden), society at large has a legitimate interest in regulation of these actions. Such regulation is best pursued when the citizenry is aware of the costs and benefits associated with some particular action, and a general framework equitable for all is set up.

PROBLEMS AND QUESTIONS

MULTIPLE CHOICE REVIEW

1. Which gas is *not* involved in ozone depletion?
    a. chlorine
    b. nitric oxide
    c. neon
    d. bromine
    e. hydrogen

2. The greatest cost per ton involved in buying coal for use in power plants is most likely for
    a. the mineral rights for the land on which the mine is built.
    b. transportation of the coal to the power plant from the mine.
    c. strip mining coal from the surface.

d.   mining coal from underground.

e.   reclamation of old strip mine sites.

3.   In which of the following cases does TANSTAAFL apply?

   a.   common use of land for grazing of sheep

   b.   use of air in industrial processes

   c.   use of water for power plant cooling

   d.   TANSTAAFL applies in all the above cases.

   e.   TANSTAAFL does not apply in any of the above cases.

4.   In which of the following cases would the tragedy of the commons apply?

   a.   common use of land for grazing of sheep

   b.   use of air in industrial processes

   c.   use of water for power plant cooling

   d.   Tragedy of the commons applies in all the above cases.

   e.   Tragedy of the commons does not apply in any of the above cases.

5.   A key difference between the tragedy of the commons and TANSTAAFL is that

   a.   TANSTAAFL applies to chemistry, and the tragedy of the commons applies to physics.

   b.   both apply to all situations; so there is no difference.

   c.   TANSTAAFL applies to sociology, the tragedy of the commons to economics.

   d.   TANSTAAFL applies to economics, the tragedy of the commons to physical systems.

   e.   TANSTAAFL applies to physical systems, the tragedy of the commons to social systems.

6.   Ozone is a pollutant

   a.   in the stratosphere.

   b.   near ground level.

   c.   in nitric-acid-rich polar stratospheric clouds.

   d.   in chlorine gas near Earth's surface.

   e.   that can be used as fertilizer on farm fields.

7.   The primary reason few SSTs were ever built was because

   a.   SSTs endangered the ozone layer.

   b.   SSTs were not economically viable.

   c.   the technology was risky and the planes likely to disintegrate in flight.

   d.   computers predicted that the bow wave would cause oxygen to recombine with nitrogen in the air.

   e.   SSTs travel through the stratosphere.

8.   In science, words always

   a.   have the same meaning as common English words.

   b.   mean vague things, so scientists can hide their ignorance.

   c.   have no reason for being; they must just be memorized.

   d.   are used faddishly by scientists and the public.

   e.   have specific meanings shared worldwide by all scientists.

9.   Science works efficiently by

   a.   testing models and keeping only those that work.

   b.   stating facts that need to be memorized.

   c.   rewarding those who prove things about how Nature works.

   d.   asking for answers to all pressing human problems.

   e.   valuing all models, no matter what happens in experiments.

10.   The statement that objects dropped from above the surface of Earth fall to the surface is

a.   probably true, but we'll have to observe it next time to be sure it's true.

b.   scientific truth, accepted as gospel.

c.   probably true on Earth and on other planets as well.

d.   only true if we look at the object while it falls.

e.   proven true by scientific experiments.

SHORT ANSWER

Explain whether the statement made is correct, and justify your answer.

11.   Scientific models should be falsifiable, that is, they should be able to be proven false.

12.   Ozone can be formed in the stratosphere by lightning.

13.   The materials within polar stratospheric clouds cause ozone to be destroyed through their presence.

14.   A scientific experiment that showed a favorite model to be wrong should be disregarded.

15.   The meaning of the word "energy" is immediately clear to anyone.

16.   Ozone is an unnatural form of oxygen.

17.   Materials from fire extinguishers can cause ozone depletion.

18.   The purchase by a power company of an area of ground on which to build a power plant is an example of the tragedy of the commons.

19.   Because all the chemicals that deplete ozone occur naturally in the ozone layer, there is no need to worry about releasing more of them into the air.

20.   Concentrations of ozone-eating atoms or molecules in the neighborhood of one in a thousand pose no threat to the ozone layer.

DISCUSSION QUESTIONS

21.   Why is falsifiability desirable in a scientific model?

22.   In an article in the *New York Times*,[75] A. Revkin pointed out that: "In the early 1970s, when scientists first reported that CFCs could destroy ozone, some theorized that the effect would be most discernible in the highest reaches of the atmosphere over the tropics, partly because of the abundance of sunlight." What happened when the experiments were performed? Explain how the conditions mandated this experimental result.

23.   How does ozone get into the stratosphere?

24.   If atomic hydrogen can destroy ozone, and most of the matter in the universe is hydrogen, why is hydrogen not a major cause of ozone depletion?

25.   Green cheese is unripe cheese, which has a yellowish color. Many hundreds of years ago, someone said that the Moon was made of green cheese. Was this model falsifiable in its time? Is it falsifiable now? Explain.

26.   What does the evidence that the Antarctic ozone hole has expanded over the past 20 yr mean? Explain.

27.   Only during the last decade of the twentieth century did the Arctic began to experience an ozone hole and formation of polar stratospheric clouds. Since its appearance, the hole expanded in a regular way. What does this indicate about Arctic stratospheric temperatures during this period? Explain.

28. How is it that the world's nations were able to forge the Montreal Protocol to reduce production of CFCs, when previously, many nations had disregarded scientists' warnings?

29. What does "carrying capacity" mean? Some have defined it as "the number that can be supported long term without damage to the environment." Is this a reasonable definition? Why or why not?

30. Does the success of the Montreal Protocol mean that Hardin's Tragedy of the Commons fears are groundless?

31. Excimer laser radiation from KrF 248 nm broad beam can produce ozone from air.[76] Discuss whether the lofting of low-level satellites containing excimer lasers could be used to restore the ozone layer.

32. Research has shown that Arctic plants are more injured by increased ultraviolet radiation than by global warming (see Chapter 17).[77] The authors of the study say "there is a growing belief that ozone depletion is of only minor environmental concern because the impacts of UV-B radiation on plant communities are often subtle. Here, we show that 5 yrs of exposure of a subarctic heath to enhanced UV-B radiation both alone and in combination with elevated CO_2 resulted in significant changes in the C:N ratio and in the bacterial community structure of the soil microbial biomass." What might this mean about the future Arctic ecology? Explain.

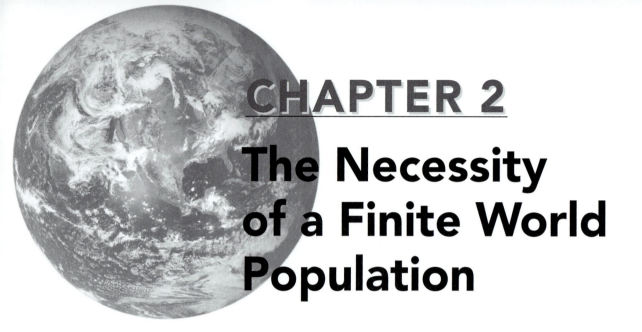

CHAPTER 2

The Necessity of a Finite World Population

The number of people who must be supported by energy or any resource determines to a large extent the amount that must be produced or extracted. In this chapter, we use some ideas from Appendix 1 to develop estimates of the human carrying capacity of Earth. The examples used illustrate the consequences of unimpeded geometric, or exponential, growth.

Because Earth is finite, there are limits to growth. We begin with an absurd assumption about external conditions limiting growth and end with a fairly realistic set of conditions limiting growth in the number of humans.

KEY TERMS

mass / doubling time / linear thinking / trophic pyramid / terraformation

"The whole world is watching," chanted antiwar demonstrators at the 1968 Democratic National Convention in Chicago. Even that long ago, what happened half a world away was as immediate as what happened on the other side of the city. The demonstrators were the vanguard in a shift of awareness, for, from time immemorial, humankind has thought in terms of an infinite world. Is it only accidental that concern about overpopulation, lack of resources, and the environment in general has followed hard upon the photographs of our blue–white planet (Figure 2.1) suspended atop the lifeless lunar plain?

2.1 MASS IS LIMITED ON EARTH

Each of us has intrinsic attributes; one such measurable one is our **mass**. Mass, roughly a measure of the amount of "stuff" making up something, is the agent

FIGURE 2.1

Earth looms over the lunar surface. Color photos of scenes such as this may have sparked people to become more aware of the fragility of their environment.
(NASA)

responsible for the force we feel pulling us toward the ground (see Chapter 3). A person who has twice the mass of another person is correspondingly pulled down toward Earth's surface with twice the force. In the metric system, the mass unit is the kilogram (a mass of 1 kilogram (kg) experiences a force of 2.2 lbs on Earth). Take a typical person to have a mass of about 80 kg (a weight of about 175 lb; men generally have a greater mass and women a smaller mass).

Scientists love to make estimates as a guide to thinking. The estimate may be about a real phenomenon. ("What is the number of piano tuners in Chicago?" was one of Enrico Fermi's favorite estimation examples.) To find the number of piano tuners, we would need to know the number of pianos, but we can get a reasonable number for that if we know how many people live in Chicago (about 3 million) and say that, perhaps, one-tenth of them own pianos (300,000 pianos). Then, we would want to know how long it takes a tuner to tune a piano. How many hours would a piano tuner work? We would have to take travel time into account, as pianos are not brought into the shop for tuning. Probably a tuner can tune four pianos a day, or about 1500 pianos a year. We'd estimate 200 piano tuners in Chicago. Most people, after a little thought, find that they are able to estimate reasonably well, say, the number of movie screens in their city. We may also make less realistic estimates to show, by example, that something is ridiculous. Here is such an estimate of the maximum population Earth can hold.

There is some kind of upper limit to the world population's mass. It is clear that Earth's mass, about 6×10^{24} kg, could never be turned entirely into people. Define the symbol N to represent the maximum number of typical people (of mass 80 kg). Then, the ultimate number of people making up Earth's mass would be given by

$$N \times (\text{mass of a typical person}) = \text{mass of Earth}$$

or

$$N = \frac{\text{mass of Earth}}{\text{mass of a typical person}} = \frac{6 \times 10^{24} \text{ kg}}{80 \text{ kg}} = 7.5 \times 10^{22} \text{ typical people.}$$

Even though this is a finite number, it is a mind-boggler. If we asked that many people to go jump in a lake, and 100 of them did just that each second, it would take over 20 trillion years (yr) for them all to do it. Of course, if a million people were to jump into the lake each second, it would take only a bit over 2 billion years.

This is a ridiculous number, it must be admitted. Something will intervene long before such a population could be reached (see the Verhulst function in Chapter 5). But what would Earth look like if such a population could exist on it? How long would it take us to get to that level?

How Fast Will the Limits of Population Growth Be Reached?

It's apparent that the number of babies born will depend on the number of possible mothers. The number of people who could be mothers, in turn, depends on how many people are alive, because a proportion of all people will be of the right age and sex. So, the growth in the number of people alive depends on how many people there are already. (More babies are born in a world with 6 billion people than in a world with 250 million people.) This sort of growth is characterized by the time it takes for their numbers to double. The present **doubling time** of Earth's population is about 35 years.[1] If we know the amount of time it takes the number of people to double, we are able to estimate how long it will take the population to grow to any size. For the purposes of our estimate, assume that the doubling time is not changing; that is, it will also take 35 years to double in 2500 A.D. This is not a good assumption if we look at the history of population increase over the past 400 years. The doubling time has been decreasing; that is, each succeeding generation has been doubling itself faster than the previous one, despite famine, pestilence, and war. So, if we assume the doubling time as it is now, we will find an upper limit on the time the population takes to reach a certain value.

There has been a crucial assumption made here that has not yet been made overt. The estimation procedure described has assumed that what we see now will be the same in the future. This assumption may or may not be true, but we must recognize that we are making the assumption. People exhibit a bias in their belief that the future will be more of the same, despite the experience of all of history to the contrary. Given our lack of information about the future, we have to estimate. However, we must not lull ourselves into thinking that we have described what must occur. This approach to the estimation problem is called **linear thinking**, and we shall see much more of it and its variants elsewhere in this book, where the same cautions will apply.

To recapitulate, we asked how long it would take the present population of Earth, 6 billion or 6×10^9 people, to reach our maximum estimate of population (N) at the present rate of increase. If we denote the present population by the symbol N_0, then in 35 years, there will be twice that many people ($2 N_0$); in 70 years, there will be $2(2 N_0) = 2^2 N_0$; in 105 years, there will be $2(2^2 N_0) = 2^3 N_0$; and so on. After m doubling times, the number of people is $N = 2^m N_0$. Solving with our numbers for the number of doublings, m, we have

$$2^m = \frac{N}{N_0} = \frac{7.5 \times 10^{22}}{6 \times 10^9} = 1.25 \times 10^{13}.$$

With this number and the approximation $2^{10} = 1024 \cong 1000 = 10^3$; $10^{12} \cong 2^{40}$, and so we obtain $2^{m-40} = 12.5$, an m between 43 and 44. Since each doubling time is 35 years, 44 doubling times is only 1540 years.

That is, if humanity insists on doubling itself every 35 yr, the mass of people alive in 3500 A.D. would have used up the entire mass of Earth. (Of course, Earthlings would have to bring mass from elsewhere if even a small fraction of such a number of people were living.) In fact, with this sort of proliferation, the mass of humanity would equal the mass of the known universe (10^{51} kg) in only about 4500 years. This is the magic (or downside) of this sort of growth.

To put this number, 7.5×10^{22} people, into context, note that the total world surface area (including oceans) is about 5.2×10^{14} m^2. The gross estimate of population density would be

$$\frac{7.5 \times 10^{22} \text{ people}}{5.2 \times 10^{14} \text{ m}^2} = \frac{150 \text{ million people}}{\text{m}^2}.$$

This would be something like squeezing the entire present population of Earth into a house trailer.

2.2   ENERGY LIMITS ON WORLD POPULATION

To make a more realistic estimate, it must be recognized that (at least crudely speaking) the ultimate factor limiting world population is the energy budget enforced by the Sun. Solar energy allows plants to grow; it is this energy that is stored up as deposits of natural gas, coal, and oil. This energy is also responsible for Earth's relatively high temperature. This discussion about energy anticipates our definition of energy and energy units in the next chapter. For our purposes here, it is enough to state that an important energy unit is the kilojoule (kJ); the average person on Earth uses about 8400 kJ of food energy per day to function normally (that's about 2000 kcal—called "calories"—per day to those who read diet books). The typical American diet contains about 14,000 kilojoules per day. Someone from Bengal, India, or from Bangladesh might use less than 6000 kilojoules per day, the bare minimum for survival.

Each minute, the surface of Earth receives about 10^{16} kilojoules[2] from the Sun. Because the Sun shines an average of 12 hours a day, Earth, on average, gets about 7.2×10^{18} kilojoules per day. If we imagine that people could make total use of this solar energy directly, Earth could support

$$\frac{7.2 \times 10^{18} \text{ kJ/d}}{8400 \text{ kJ/person/d}} = 8.5 \times 10^{14} \text{ people.}$$

This is about eight orders of magnitude (eight factors of 10, or 100 million) fewer than in our first estimate. Following the same reasoning as in the previous example, the number of doubling times is about 18; in about 600 years, at present population growth rates, the energy-sustainable limits of population would be reached. By this estimate, the ultimate population density of Earth is about 1.6 people/m^2—somewhat more plausible than before. This population density would be like crowding about 180 people into a typical middle-class American house.

Of course, this assumes that we could use the Sun's energy directly, over water and land surfaces, and that we would kill off all competing flora or fauna. Without the algae in the oceans and the plants on land, no oxygen would be produced. In fact, only 70% of the Sun's energy reaches the surface of Earth. The rest is reflected from the clouds, the atmosphere, and the surface. Moreover, we should not count the energy falling on the ocean (fish are only minor contributors to human nutrition) or on Antarctica. Let us ask what the human population could be if we eliminated all land animals and plants and got our energy requirements directly from the Sun.

In that case, using a ratio of land area to total area of 26%,[3,4] and taking into account the factor of one-half due to reflection of sunlight by Earth, the maximum number of people possible would be

$$\left(\frac{1}{2}\right)(0.26)(8.5 \times 10^{14} \text{ people}) = 1.1 \times 10^{14} \text{ people}.$$

Attainment of this population would take about 15 doubling times (until about 2500 A.D.). The population would have a land-area density of 0.2 people/m^2 in this case, which would be like fitting about 22 people into our typical house.

The Trophic Pyramid

The next problem is that we cannot directly use the Sun's energy for food. Plants use energy from the Sun to grow roots, pump water, and so on. Thus, plants can pass only about 10% of the energy they receive as food value to their consumers. Plants are called *producers*. Animals preying on the producers get only about 10% of the plant energy. They are called *primary consumers*. Animals use food energy to keep themselves warm and perform normal bodily functions; the food that is not used is excreted as waste. If we ate a cow, getting only about 10% of its energy, we would be called *secondary consumers*. If we were eating salmon that had fed on small fish, which, in turn, had fed on plants, we would be *tertiary consumers*. As you see, food consumption follows a series of steps resembling a pyramid of lowered available energy with each upward step (see Figure 2.2). This pyramid is called the **trophic pyramid**, or occasionally, the **trophic chain**. (The adjective *trophic* refers to nutrition.)

In America, people are about one-third primary consumers and two-thirds secondary consumers. Suppose, for the sake of our population estimate, that people are primary consumers (vegetarians), and they use sunlight with 10% = 0.10 = 10^{-1}

FIGURE 2.2

A trophic pyramid. The original level represents captured solar energy. Only about 10% of the energy available at any level in the pyramid can be turned into energy at the next higher level.

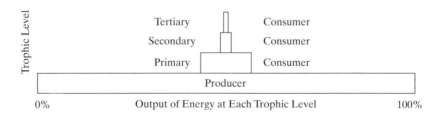

efficiency. Let us also take into account that photosynthesis (the process by which plants make sugars and oxygen out of sunlight and carbon dioxide) is about $6\% = 0.06 = 6 \times 10^{-2}$ efficient.[5] Therefore,

$$N_{max} = (6 \times 10^{-2})(10^{-1})(1.1 \times 10^{14} \text{ people})$$
$$= 6.7 \times 10^{11} \text{ people} = 670 \text{ billion people}$$
$$\cong 134 \times (\text{current world population})$$

so that the corresponding m is about 7; in about 345 years, we will have attained maximum population.

Arable Land

Finally, not all land is arable (capable of producing crops). If it were, our preceding calculation would be correct (assuming no other animals or insects competed for food, there were no weeds, and so forth). In fact, only about 25% of Earth's land is arable.[5-7] The cost of increasing the amount of arable land is about $500 to $1000 per hectare added. The hectare (ha) is a metric unit of area, 10,000 m^2, or a bit under 2.5 acres. Thus, only about a quarter of the 670 billion people can be accommodated unless we are able to increase arable land drastically, which seems unlikely—at least until large-scale weather control is possible. We probably do not want to kill off every other land animal and insect. We should take into account another factor of, say, one-half to share food with these other animals:

$$N_{max} = \left(\frac{1}{4}\right)\left(\frac{1}{2}\right)(6.7 \times 10^{11} \text{ people})$$
$$= 8.4 \times 10^{10} \text{ people}.$$

This number, 84 billion, is not that far above the maximum population estimate made by the Club of Rome study—15 to 20 billion,[7] or the maximum population estimate of the United Nations, 11.5 billion (2150). Using that number,

$$2^m = \frac{N_{max}}{N_0} = \frac{8.4 \times 10^{10} \text{ people}}{6 \times 10^9 \text{ people}} = 1.4 \times 10^1 = 14.$$

Because 2^4 equals 16, we see that m is about four (it is actually 3.8 by calculation). The human race will reach that limit (on Earth) in only 130 years at the present rate of increase. If we wish to avoid the consequences of unbridled growth, we need to act soon.

The density of population would be, in this case,

$$\frac{\text{total world population}}{\text{world actual land area}} = \frac{8.4 \times 10^{10} \text{ people}}{1.33 \times 10^8 \text{ km}^2} = 630 \text{ people/km}^2,$$

which is that of a suburban city (about 1600 people/mi^2). In comparison, the present world average population density is

$$\frac{6 \times 10^9 \text{ people}}{1.33 \times 10^8 \text{ km}^2} = 45 \text{ people/km}^2,$$

which is roughly the population density of Michigan or South Carolina (about 100 people/mi^2).

The foregoing discussion focused on outlining the barest essentials of survival for the maximum population. It has ignored the grim political, social, and psychological price all people would pay in order for Earth to support such a population.[11] Because the energy available to each person would shrink (by some near-future time, the stored energy in our planet's crust will have been totally squandered), the amenities available to each person would be fewer. People would be crowded together in nonproductive areas, such as deserts or mountainsides, in order to maximize productive land. Presently, good agricultural land is being paved over or built over to support increasing populations.

The basic question remains: Even if Earth could support 84 billion people, should it? What sort of price—physical, psychic, and spiritual—would be acceptable?[11] Thoughtful science fiction authors have addressed this problem, generally showing us chilling visions of the future.[12] Most demographers think that the actual maximum world population will be about twice the current population, attained around the year 2050. If so, we may avoid the unpleasant consequences of sustained exponential growth.

Agricultural Limits to World Population

As we shall discuss in Chapter 24, hand labor is the most energy-efficient means to harvest grain and vegetables. Yields of 4 million kJ per hectare (ha) per year are typical. We could reformulate our population question to ask how many people Earth could support at an adequate diet (say, 10,000 kJ per person per day, or about 4 million kJ per person per year). Using an estimate of 1.8×10^9 hectares for the current area under cultivation, the yield of 6.3×10^6 kJ/ha will feed about 5.4×10^9 people.[11,13,14] Earth's current population is just a bit higher than this, and most people are not vegetarians. Of course, more energy input would allow a greater number to be fed, but our question here concerns the ultimate situation, after all fossil fuels have been used. We could increase the area under cultivation, as previously noted, but the land in use now is the better half of all the world's arable land.

These poorer lands would need more nitrogen fertilizer than current farmland in order to grow crops for the increased population. This would again raise the specter of ozone-layer destruction,[14] because the nitrogen not only contaminates groundwater and surface water and wells, causing early aging of lakes (eutrophication) and making soil more acid, but it also sends more nitrous oxide (N_2O) into the atmosphere. The only way to avoid these deleterious impacts is for people to adopt vegetarian diets and find more efficient ways to fertilize agriculture,[14] or to hope that technology can be used to ensure a continuous supply of foodstuffs.[15,16] This seems somewhat unlikely, although some may adopt this lifestyle or have it forced on them. As an alternative, the population could be reduced, because the number of people alive directly affects how much food must be provided. Some claim that a population of 2 billion is the maximum possible if Earth is operated sustainably.[17] A meta-analysis of 69 studies yields a limit of 7.7×10^9 people (spanning the range between 0.65 and 98×10^9 people).[18]

The United Nations' estimate of the world population in the year 2050 is between 9.3 and 13×10^9 people.[19] Voluntary population reduction certainly seems less cruel than loosing the Four ancient Horsemen of the Apocolypse (war, pestilence, famine, and death) to ride unfettered once again.

Where Could the People Go?

We could send people to other planets (see **Extension 2.1,** *Humanity in Space*). It seems possible to make Mars and Venus habitable (a process called **terraformation**), and O'Neill pointed out that it is economically and physically feasible to construct giant space habitats at Lagrange point 5. None of these possibilities, however, could absorb a large population of immigrants from Earth.

SUMMARY

World population cannot grow indefinitely. Earth does not absorb all solar energy, and what energy is absorbed is absorbed by plants. People eat plants (are primary consumers) or eat animals that eat plants (are secondary consumers) or eat animals that eat animals that eat plants (are tertiary consumers). We estimated that Earth could support 10 times as many people as now live on it, but most of these people would probably not find life very pleasant by our standards, and no other animals of any size could exist. A more realistic limiting estimate is about two to four times the current population, a number that would be reached in a mere 35 to 70 years at the present doubling time of 35 years.

The growing human population could people surrounding planets in a short time if population growth rates do not decrease, even if few emigrants left Earth originally.

PROBLEMS AND QUESTIONS

MULTIPLE CHOICE REVIEW

1. Sustainability in a resource means
   a. ever-increasing amounts of the resource.
   b. doubling the amount needed means doubling the amount available.
   c. the amount of the resource grows as the population grows.
   d. production of the same amount of the resource forever.
   e. production of gradually decreasing amounts of the resource.

2. Suppose that we allowed each person 1000 m^3 of living space in a large building, with ceiling height 3 m. To the nearest order of magnitude, how many 1000 m high buildings would have to be built to house the entire population of 8.5×10^{14} people on the energy-limited Earth, if each building's base were 100 km on a side?
   a. 10^1
   b. 10^2
   c. 10^5
   d. 10^8
   e. 10^{13}

3. If the population of a country doubles every 25 years, a 1980 population of 10 million will reach 80 million in the year
   a. 2005.
   b. 2030.
   c. 2055.
   d. 2080.
   e. 2105.

4. Sustainable development of a finite resource is
   a. never possible.
   b. sometimes possible.
   c. possible half the time.
   d. usually possible.
   e. always possible.

5. If you eat a salmon that had fed on smaller fish, which, in turn, had fed on plants, you would be a _____ consumer.
   a. primary
   b. secondary
   c. tertiary
   d. trophic
   e. pyramid

6. Which of the following limits the ultimate world population?
   a. The size of Earth
   b. The energy coming to Earth from the Sun
   c. The amount of land mass
   d. The amount of food available
   e. All of the above are limits to world population.

7. The cockroach doubling time is about 1 week. After 6 weeks of reproducing, how many cockroaches will there be stemming from an original pair?
   a. 10
   b. 32
   c. 64
   d. 128
   e. 256

8. Rapidly expanding population effectively strangles most efforts to provide adequate
   a. education.
   b. nutrition.
   c. health care.
   d. shelter.
   e. amounts of all the amenities above.
9. Sustainable development can only be pursued if
   a. there is no "birth dearth," or declining population.
   b. extreme rates of population growth are limited.
   c. the utilization rate is the same as the production rate.
   d. we "grow" our way out of problems.
   e. population size and growth are in harmony with the changing productive potential of the ecosystem.
10. There is some possibility that humanity could ruin large parts of the environment through concentration on individual gain at societal cost. This is known as
    a. the tragedy of the commons.
    b. freedom.
    c. the free lunch.
    d. the outcome of economic progress.
    e. sustainable development.

SHORT ANSWER
Explain whether the statement made is correct, and justify your answer.

11. According to current estimates, rice agriculture will expand by up to 70% over the next 25 years to support the growing human population. This will not be any sort of massive problem for Earth's ecosystem to bear.
12. Sustainable development may be possible if materials are recycled to the maximum degree possible and if one does not have growth in the annual material throughput of the economy.
13. "People who take issue with control of population do not understand that if it is not done in a graceful way, nature will do it in a brutal fashion."
14. Growth is both good and inevitable.
15. Raising the level of economic opportunity in the underdeveloped countries and lowering it in the United States until economic opportunity is the same everywhere would decrease immigration to the United States.
16. It is not possible to sustain population growth or growth in the rates of consumption of resources.
17. Designing a "New Jersey Coastal Management Plan" that involves balancing the intense development pressures in the area with wetlands and wildlife protection, water quality, air quality, waste management, and other environmental considerations is a sensible approach.
18. There is an immediate need to develop strategies aimed at stopping world population growth.
19. Rapidly increasing demands for natural resources, employment, education, and social services make any attempts to protect natural resources and improve living standards very difficult.
20. One way to expand yields of crops is to increase the area under cultivation. This has few impacts other than expansion of crop supply.

DISCUSSION QUESTIONS

21. Bartlett enunciated a key question (Reference 11): Can you think of any problem, on any scale, from microscopic to global, with a long-term solution that is, in any demonstrable way, aided, assisted, or advanced by having continued population growth at the local level, the state level, the national level, or globally? Discuss whether this question can be answered, and, if so, how it could be answered.
22. Isaac Asimov, noted popularizer of science and author of science fiction, has addressed overpopulation: "Democracy cannot survive overpopulation. Human dignity cannot survive [overpopulation]. Convenience and decency cannot survive [overpopulation]. As you put more and more people onto the world, the value of life not only declines, it disappears. It doesn't matter if someone dies, the more people there are, the less one person matters." Respond to Asimov's statement by discussing any evidence available.
23. Under what circumstances is population growth "sustainable"? Discuss what sustainable means, and attempt to apply your definition to Earth's human population.
24. Bartlett (Reference 11) has observed that "sustained growth" is an oxymoron. Discuss his observation.
25. Applying the Tragedy of the Commons principle to the human population of Earth, we might phrase it as follows: "population growth and consumption benefit the few, while the costs are borne by all of society." What evidence is there to support this view? What evidence is there that contradicts this view?
26. Simon (Reference 24) said, "We have in our … libraries the technology to feed, clothe, and supply energy to an ever-growing population for the next 7 billion years." Is this a realistic view?
27. Simon (Reference 24) says "(W)e would be able to go on increasing our population forever." Could his statement have been made in earnest? Why or why not?
28. G. Easterbrook (Reference 25) has said: "The population growth for the next 2 generations is 100 percent cast in stone. Nothing short of world pestilence or a nuclear war can halt population growth for the next 2 generations. … And with the growth in the world's population that is inevitable through the next 2 generations at least, global resource consumption is going to have to rise. That's the only possible moral outcome, so that millions of people don't die in misery. We're going to have to manage a short-term drastic increase in global resource consumption. And the only way we're going to be able to get through the night with that, is with strong environmental regulations and optimism about the success of reform in technology." Agree or disagree with Easterbrook, citing specific reasons.
29. Referring once again to Easterbrook's quote (Problem 28), explain why he would assert that the next two generations of growth are inevitable.
30. In addition to destroying the ozone layer, nitrous oxide from fertilizer contributes to global warming by emission of methane, a greenhouse gas. What constraints can there be on Earth's population if we assume that about half the current population is (mostly) vegetarian? How does vegetarianism affect Earth's carrying capacity?

PROBLEMS

31. How many rooms would be needed to house the 7.5×10^{22} people who represent the mass of Earth? Assume 1000 m³ of living space with ceiling height 3 m. How does the volume needed to house these people compare to Earth's volume?

32. The energy flux from the Sun is about 1350 W/m². [The watt (W) is a unit of power, energy divided by time, a J/s.] If Earth's radius is 6.4×10^6 m, what is its apparent energy-gathering area? What amount of power is received by Earth from the Sun? In addition to the energy needed for life, a typical household uses a kilowatt of power. Assuming the average household consists of three persons, and recalling that the average person needs 8400 kJ of energy per day, how many people could be supported on Earth if all this power were used to support them?

33. Earth's oceans cover 3.5×10^8 km² and have an average depth of 4.0 km. How much ocean water is there on Earth? People in the United States, on average, use about 12 m³/d (including agricultural and industrial uses). Assuming that this limits Earth's population, what maximal population could be supported by all the water in the oceans? Is this a reasonable number to support at our current standard of living?

34. According to Einstein's energy relation $E = mc^2$, one kilogram of mass represents an energy content of 9×10^{16} J. How long could 1 kg of mass (assuming it was converted directly into usable energy) supply energy to a person using 8400 kJ of energy per day? If all Earth's mass were converted directly into usable energy, how long could it support the current world population of 6 billion people?

35. Assuming the population of Earth continued to double every 35 years, how long could Earth's mass last, and about how many people would there be when the last of the mass was consumed? (Refer to Problem 34.)

36. Given the growth rate of the human population on Earth, what effect would it have on Earth's population growth if 10,000 people each year were moved off the planet to the moon, Mars, or an O'Neill habitat? Explain how you arrived at your answer.

37. Given the growth rate of the human population on Earth, what effect would it have on Earth's population growth if 100,000 people each year were moved off the planet to the moon, Mars, or an O'Neill habitat? Explain how you arrived at your answer.

38. Given the growth rate of the human population on Earth, what effect would it have on Earth's population growth if 1,000,000 people each year were moved off the planet to the moon, Mars, or an O'Neill habitat? Explain how you arrived at your answer.

39. Let the beginning of the exodus off Earth be taken as year 0.
   a. Estimate the growth in off-planet population with the assumptions given in Problem 36.
   b. Estimate the growth in off-planet population with the assumptions given in Problem 37.
   c. Estimate the growth in off-planet population with the assumptions given in Problem 38.

40. We might make the estimates of growth in human population on Earth a bit more reasonable by assuming that no more than 1 m² of land per person is sustainable. Estimate Earth's maximum population on this basis.

41. Suppose that we allowed each person 1000 m³ of living space in a large building, with ceiling height 3 m. How many 500 m high buildings would have to be built to house the entire population, 8.5×10^{14} people, on the energy-limited Earth if each building's base were 1 km on a side? (The world's current tallest building is about 500 m high.)

CHAPTER 3

Work, Energy, and Power

So far, we have discussed matters relating to a vague concept of what is meant by energy. We need a more precise definition of energy. Energy has to do with work, and work has to do with the force and the movement of the object on which a force is exerted. These topics, as well as the definition of various types of energy and the definitions of efficiency and power, form the basis of this chapter.

KEY TERMS

work / displacement / vector / velocity / acceleration / force / inertia / mass / weight / gravitational field / energy / gravitational potential energy / thermal energy / kinetic energy / friction / conservation of energy / efficiency / power / basal metabolic rate

Leete's eyes flashed. "Set the divariable veeblefurtzer on that Seltz coil," he shouted to Igor, "or we shall all be blown to atoms!" As the snap of electric discharges continued, something dark began to swirl at the bottom of the vat....

This sounds like a 1930s pulp novel or horror flick. Yet even today, this sort of image of scientists and this language colors our attitudes. No horror movie is complete without a mad scientist mouthing pseudoscientific jargon. Even though true-to-life scientists introduce technical words, their purpose in doing so is to make their communications more precise. Often, a scientific word strays into general usage (usually with its meaning somewhat altered), and sometimes, an ordinary word is given a specific scientific meaning. An example of the former phenomenon is "radar," and examples of the latter are the use of "color" and "charm" as whimsical descriptions of the esoteric properties of the elementary particle building blocks of nature.

To enable us to communicate in this book (and to open the wider sphere of the scientific literature), we must agree on the use of certain words. These words are universally used in scientific prose, so no hint of misunderstanding can be allowed.

Many of the problems encountered in communications between scientists and the public can be attributed to their different definitions of the same words.

The word **work** is generally understood to mean toil, labor, or employment (it also means a military fortification). If you and I were speaking in a casual way, we could use one of these meanings, and each would know what the other meant. If, on the other hand, I spoke to you as a physicist about work, we might have some serious misunderstandings. Suppose we were watching an old movie in which a bride is carried over the threshold by the groom. Of course, many things happen when one person carries another: One person is picked up, then carried, then put down. Work is done in the picking up and putting down and in the starting and stopping of the walk. However, if we consider only the part during which the groom walks at a steady speed on a level floor carrying the bride, no work is done, in a technical sense. Does this distinction seem nonsensical? The problem arises because the word "work" has different meanings when used in casual conversation as opposed to when used in its technical sense.

3.1 BASICS OF MOTION: SPEED, VELOCITY, AND ACCELERATION

To define the scientific use of work, we must first consider the concepts of distance (and displacement), speed (and velocity), acceleration, and force.

Distance is a familiar quantity. We measure distances with tape measures, meter sticks, or other measuring devices. Distance measures how far away things are. **Displacement**, which is less familiar, merely adds information about direction to how far away something is: it tells how position has changed. For example, a newcomer to a neighborhood may be told the distance to the local school. Such information is incomplete unless the school is in sight. It would be far more useful to know the direction to the school in addition to the distance (for example, see Figure 3.1).

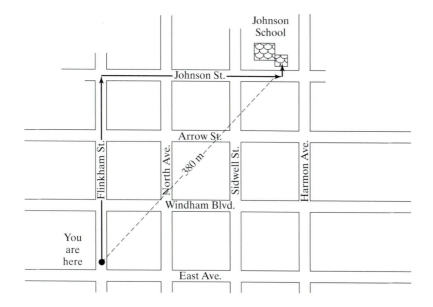

FIGURE 3.1

Which neighbor is more helpful? Arny: "Johnson School is 380 m away." Carol: "It's about 280 m north on Flinkham, 250 m east on Johnson, then 5 m north into the Johnson school door."

This additional information about direction, essential for everyday life, distinguishes quantities known as vectors from quantities specified only by numbers. **Vectors** are quantities that have a direction specified in addition to their numerical values. Position (location relative to some fixed point), displacement, velocity, acceleration, and force are all vectors. These physical quantities have a basis in our direct experience, and these common perceptions provide a basis for our mutual understanding.

Distance is just a number; when both the distance and the direction from one point to another are specified, it is displacement. To illustrate how distance differs from displacement, consider an object that travels in a circular path from some starting point and returns to that same point. The object will have traveled some distance, the circumference of the circle, but it will not appear to have been moved (or displaced) from its starting position, and so its displacement will be zero.

Speed

The speedometer in a car tells us how much territory we can cover in a given time. A reading of 100 kilometers per hour on the speedometer tells us that if the car kept moving at that speed, it would have gone 100 kilometers after 1 hour had passed, 200 kilometers after 2 hours, and so on. If we doubled the speed, we could double the distance covered in a given time (go 200 kilometers an hour) or cover the same distance in a shorter time (go 100 kilometers in half an hour). So, speed is found from

$$\text{speed} = \frac{\text{distance traveled}}{\text{time required to travel that distance}}.$$

People often use the terms "speed" and "**velocity**" interchangeably in common speech. Speed is different from velocity, however, because speed is a scalar—a number—and velocity is a vector. Specifying the velocity of an object imparts not only the information on speed but also the information about the direction in which the object moves. We might say, "I drove east at 55 miles per hour from Cheyenne to Omaha," or "I drove north at 80 miles per hour from San Francisco to Portland on I-5." In the former case, the speed is 55 mi/h, and the velocity is 55 mi/h, east. In the latter case, the speed is 80 mi/h, and the velocity is 80 mi/h, north. Velocity is a vector specifying both speed and the direction in which the object of interest moves. Of course, we would not make it to Portland without having paid several speeding tickets! We denote vectors by boldface, and indicate the direction associated with the vector by an arrow. The arrow length shows the size of the vector, and the direction of the arrow indicates the direction of the vector. So, velocity is found from

$$\text{velocity} = \frac{\text{change in position}}{\text{time required to make the change in position}}.$$

See **Extension 3.1,** *Calculating Average Speed*, to learn more about how to find these quantities, speed and velocity.

Acceleration

Acceleration is simply a measure of how rapidly the velocity changes. To say that an MG sports car has better acceleration than an antique Studebaker means that the MG can change its speed in a shorter time interval than can the Studebaker. An engine that can make the speed change from 0 to 100 kilometers per hour faster than another is said to produce greater acceleration. An "accelerator" is a device that allows you to change the speed of your car. The harder it is pushed, the greater the rate at which your speed changes (given the limitations of your car). The car's brake is also a sort of accelerator, because applying the brake changes the car's speed. Because acceleration happens when velocity changes, a car can accelerate even when the speed does not change. For example, if a car is driven along a circular track at a fixed speed, the direction of the velocity vector changes, and because there is a change in velocity, there is acceleration. In this sense, the steering wheel is also an accelerator of sorts. The acceleration is defined as

$$\text{acceleration} = \frac{\text{change in velocity}}{\text{time required to make the change in velocity}}.$$

See **Extensions 3.2** to **3.4,** *Calculating Average Acceleration, Circular Motion,* and *Newton's Laws of Motion* for more detail.

3.2   FORCE AND MOTION

Newton's First Law of Motion

We intuitively recognize that a force is a push or a pull. If you try to force someone to do something, it implies that you change a state of rest into a state of motion, or a state of motion into a state of rest. This intuitive understanding nearly describes the scientific use of the word as well. An outside agent that acts to change a body's motion is known as a **force**. There is always a direction associated with a force, so force is a vector quantity. Newton's First Law defines force by telling us that it is the agent responsible for a change in the motion of an object. Note that it does not say that force is responsible for motion, but rather a force is responsible for any *change* in the motion.

If you are a passenger in a car traveling in a circle, the acceleration acting on you causes you to remain on your seat in the car following the circular track. Your body is forced to follow the car by forces acting between you and the seat. Were those forces missing, your body would have traveled in a straight line. The passenger in this car feels the continuation of motion, the "urge to go as she was going," as an outward force, even though the actual force is provided by the car seat dragging the passenger along with the car.

The unit of force is named after Sir Isaac Newton, who did so much to explain motion: it is called the *newton* (symbol N). A typical apple weighs about a newton. [In the customary American system, the unit for force is the pound (symbol lb). A pint of water weighs about a pound.] Refer to **Extension 3.4,** *Newton's Laws of Motion* for an explicit statement of the three laws of motion.

Gravitational Acceleration

If I were to jump off a table, I would not expect to fly upward as a superhero might (especially because I am not a mild-mannered reporter, or even a mild-mannered physicist!), nor would I expect to wing to the right. I know I would plunge downward, my speed increasing as I fell. (I admit to having tested this time and time again during my earlier years.) Any object near Earth's surface experiences a downward force; the direction *down* (toward the center of Earth) characterizes the gravitational force. If the object is not supported by other objects, it will change its motion by increasing its downward velocity (or, equivalently, by reducing its upward velocity).

Because velocity changes as an object falls, it must be subject to a downward acceleration. Galileo first established that the gravitational acceleration near Earth is the same for all objects; it is denoted **g**. This acceleration "due to gravity" is approximately 10 newtons in magnitude for every kilogram of mass the object has ($\mathbf{g} = 10$ N/kg).

Forces may be exerted by objects such as walls and floors, or by rods or blocks, or by wires or ropes, or by "force fields" such as are exerted by the Sun on the planets (gravitational) or the nucleus of an atom on the atomic electrons (electromagnetic).

Consider a table sitting on the floor. How is it influenced by the forces acting on it? We know that the table would fall downward if the floor's support were removed because of the gravitational force on the object, its weight. It does not fall because of the countervailing effect of the floor. Imagine that you were under the table. The table legs would exert a force on you, but your body would be dented—the flesh would move inward, deform—because of the table's force. Your body would exert a force on the table legs, stopping the motion. The floor upon which the table rests appears immovable to us, but the force of the table acting on the floor causes the floor to bend, or deform. The floor underneath a heavy desk can often be seen to bear the permanent traces of such prolonged deformation. The deformation causes forces inside the floor that oppose further deformation. These forces ultimately act on the desk, exerting a force equal in magnitude to the weight of the desk. Such a force, because it is perpendicular to the supporting surface, is called a normal force. Because the floor exerts an upward force on the desk, and weight is a downward force, and both forces are the same size, the net additive effect of the two forces is zero. Here we see an important consequence of the vector nature of the force: adding two forces equal in magnitude may produce a net force anywhere from twice as great as either one to zero.

Bodies change the way they move when net nonzero forces act upon them. So, this idea of force as something that changes a body from a state of rest to a state of motion, or vice versa, must be connected to the idea of the acceleration experienced by that body. A scientist would want to do an experiment to check the hypothesis out, but it is important that in changing things to check it out, we change only one thing at a time. This is known as controlling the variables.

Imagine standing unmoving on rollerblades. Suppose that, during some short time interval, you were pushed by a certain force. The force acts to change your speed from zero to some nonzero speed *v* (Figure 3.2a).

In a second part to this experiment, suppose you now ran into an identical stationary cloned you. You would probably correctly guess that, after the collision, you would both be going at half the original speed, *v*/2 (Figure 3.2b) Both you and the clone's motions have changed; you both must have exerted forces on each other, because the only possible agents of change were you two. During collision with a clone, the clone exerts a force on the first rollerblader, slowing the rollerblader from *v* to *v*/2, while the rollerblader exerts a force on the clone, speeding the clone from 0 to $v/2$. The clone's speed increase came at your expense. You exerted a force on the clone equal in magnitude to that the clone exerted on you to cause the change in your motion. But, it must have been in the opposite direction from your force on the clone, because your changes were mirrored. In this experiment, we changed the number of people, but nothing else changed except conditions caused by the action of the collision itself.

Now, imagine starting all over again. This time, we will change only the force acting on you at the beginning—suppose you are pushed by twice the force that acted the first time, during an interval of the same length as before. The speed change you experience in this second experiment is twice as great as the speed change from the first, to $2v$ (Figure 3.2c). Because we controlled the variables, we can say that the greater the force, the greater the acceleration; the greater the acceleration, the greater the ultimate speed. In the same way, the greater the acceleration of a car, the greater the force the passengers would experience "pushing them into their seats." (Actually, the seat, as part of the car, is exerting a force on the passengers to accelerate them. They are being forced to change their states of motion along with the car, and they experience the force the seat exerts on them to change this motion as that of some agent pushing them into their seats.)

Return to the thought experiment with the rollerblades and the force. In a third experiment, suppose you were again teamed up with your clone. Both of you together begin motionless and are pushed with the same force as was used the first time, for the identical time interval. Only the fact that there are now two of you is different from that first part of the first experiment. The two of you would be skating after the force ceased at a slower speed than before (Figure 3.2b). The greater the number of clones holding together, the smaller the acceleration for a given amount of force and the slower the ultimate speed. The only difference is the greater amount of stuff present.

The third thought experiment reflects the idea that the more "stuff"—matter—there is, for application of a given force, the smaller the resultant acceleration. This quality of matter—that it resists changes in being put into motion—is called **inertia**. The quantitative measure of the inertia of matter is called **mass**: Mass is a measure of the amount of matter present. Two identical objects would have twice the mass of a single one. Five such objects would have five times the mass of one object. The mass of an object can be defined in some absolute way only by arbitrarily choosing a standard mass. This arbitrary standard mass is then duplicated and subdivided: the subdivisions and duplicates can be used to define other masses by comparison. The standard mass in the metric system is the kilogram. So, comparing the first part of the first experiment with the second, mass is held fixed, but force is varied. A greater push results in a greater change in motion. In comparing the first and third experiments, force is held fixed, but mass is varied. A greater mass results in a smaller change in motion.

Mass and Weight

Often, people confuse the idea of mass with that of **weight**. The weight of a mass is the force exerted on that mass by another nearby massive body. A nonliving body's mass is fixed once and for all the time it exists as that body. Its mass is

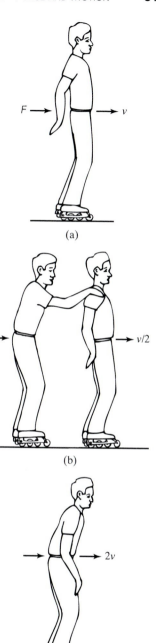

FIGURE 3.2

(a) A rollerblader, originally motionless, is pushed by a force for a certain time, and the motion is changed, ending at speed v. (b) The rollerblader and the clone move with speed $v/2$. (c) A rollerblader, originally motionless, is pushed by twice the force of (a) for the same time, and the motion is changed, ending at speed $2v$.

"stamped upon it." Its weight is the gravitational force exerted on it; this force depends on the position of the mass in the universe. If you were pushed on a cart along a concrete walkway on Earth and the Moon with the same total force, you would speed up the same way in both places. Your mass would be the same on the Moon as it is in your living room. Your bathroom scale reading on the Moon is not the same as on Earth, though. Your weight on Mars is different from your weight on the Moon and your weight on Earth, but you would still experience the same change in motion for a given net force acting on you.

According to Newton's First Law, an existing system continues as it was (particles moving apart from one another, or approaching one another, or whatever) unless an outside agent affects the bodies in the system. The Sun and planets exert forces on each other because of their masses. Even if a planet were not at some position in space, an influence would always exist there that would affect a mass (a planet) if it *were* placed there. We call the ratio of gravitational force to the mass at any point in space, which represents this influence, the **gravitational field** at that point. (We can think of Earth's gravitational field as the reason that bodies released near Earth's surface fall.)

Newton's Second Law of Motion

Both mass and the strength of a force influence the consequent acceleration. Acceleration increases if the net force is increased for a given mass. Acceleration decreases, for application of a given net force, if the mass is increased. The simplest hypothesis relating net force, mass, and acceleration that reflects the realities exhibited in the thought experiments with the skaters is

$$acceleration = force/mass, or$$

$$force = mass \times acceleration,$$

$$F = ma.$$

The force is in newtons when the mass is in kilograms, and the acceleration is in newtons per kilogram or meters per second squared ($N/kg = m/s^2$). This equation is a simplified way of stating Newton's Second Law of motion. Over hundreds of years, people have been unable to disprove it in any experiment. For example, if you are in an accelerating car, the force you experience pushing on you as the speed changes is simply your mass multiplied by the car's acceleration. Suppose the 2000 kg car is changing its velocity by 1.50 m/s every second in a certain direction during some time interval, that is, it is speeding up. The acceleration a is found as

$$a = \frac{change\ in\ velocity}{time\ required\ to\ make\ the\ change\ in\ velocity} = \frac{1.50\ m/s}{1\ s} = 1.50\ m/s^2.$$

The net force that must be acting is

$$F = (2000\ kg) \times (1.50\ m/s^2) = 3000\ N,$$

and it is acting in the direction in which the velocity is changing. Refer to **Extension 3.4, *Newton's Laws of Motion*** for an explicit statement of the three laws of motion and more examples.

3.3   WORK

We are now able to define work. The work done by any force is the product of the force and the distance moved in the direction of the force. If the force is exactly along the direction of motion, then the work done is just

$$\text{work} = \text{force} \times \text{distance moved.}$$

The work is in joules when the force is in newtons and the distance is in meters. If the force is exactly perpendicular to the direction of motion, there is no motion in the force's direction, and consequently, no work is done by the force (for example, see Figure 3.3).

To return to the example of the groom carrying the bride over the threshold, suppose he picks her straight up and then walks steadily into the house (Figure 3.4). In picking her straight up, he is moving her in the direction opposite to the gravitational force on her, which is her weight. He, therefore, does an amount of work on her equal to her weight (W) times the change in height (h):

$$\text{work by groom on bride} = \text{weight} \times \text{change in height}$$
$$= W \times h.$$

If we could ask the groom, he would agree that he worked to pick up the bride. He might also insist that he did a lot of work carrying her at constant speed across the threshold. In a technical sense, however, the only work he did resulted from the force he applied (parallel to the horizon) to accelerate the bride to the constant speed and the force he applied to stop. (These latter forces are of the same order of magnitude as the bride's weight but are applied over a very short distance.)

Assume that the bride's weight is directed downward, and the groom carries her horizontally across the threshold walking at a steady speed on a level. Because the groom is not accelerating at any time during the carry except at the very start and the very end, the work done in starting and stopping is probably, at most, a few percent of that done in lifting the bride. For most of the carrying time, there is no net force being exerted on the bride in the horizontal direction. Our conclusion is that no work is done during the part of the carry that took place at constant speed, because the distance moved *in the force's direction* is zero. The groom's perception seems to contradict the statement that no work is considered to have been done if the force is perpendicular to the direction moved.

We would also agree with a student who said she was working while she pulled an all-nighter to study for a physics exam. If someone pointed out that she sat in a chair and was not moving and claimed, therefore, that no work was being done, the student would probably become indignant.

In fact, work is being done by the principals in both these examples. In the case of the student, the body is constantly forcing blood to move through capillaries and veins, forcing the rib cage alternately to expand and shrink, moving lymphatic fluid, forcing blood through a filter, and so on. All these things constitute work, in both senses of the word. This is only a small part of the work that the body does continuously. For example, the muscles in the body are always moving slightly, preparing to be used, even when we think those muscles are perfectly

Direction of Motion

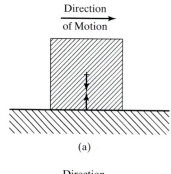

(a)

Direction of Motion

Rope being Pulled

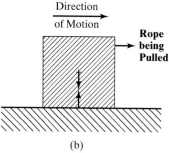

(b)

FIGURE 3.3

In both (a) and (b), the weight of the box acts downward from the center of the box, and there is an upward force on the box exerted by the table. (a) In this case, the box has forces on it that are perpendicular to the direction of motion, so no work is done. (b) In this case, the box is being pulled in the direction of motion, so work is being done.

FIGURE 3.4

A groom carries his bride over the threshhold. (The author and his wife reenact this here.)

still. Try holding your arm straight out from your body, still. After a short while, you will notice your arm moving perceptibly.

So, the groom is actually doing work, because the bride is being moved slightly up and down as the groom's muscles twitch, and he tries to compensate for it. He is doing work on her by raising her, just as the gravitational force is doing work on her as she is lowered. Assuming success in keeping her at about the same height, the *total amount of work done on her* in carrying her is zero. Because of the inevitable twitching of his muscles, the groom has done work, and we might understand his feeling tired. A robot doing the same job would not twitch and so would do no work on the bride.

3.4   ENERGY AND ENERGY TRANSFORMATIONS

In the discussion thus far, only work has been mentioned. Work is just one form of energy. This word **"energy"** was coined by Thomas Young in 1807.[1] It is from the Greek and means roughly "work within." As we shall use it, this is the key, that

energy is what allows useful work to be done,

or, as commonly put,

energy is the capacity to do work.

Work is central to the idea of energy. Now that we have some idea of what work is, we can consider without confusion the many forms of energy. As long as something can do work, it is able to use energy. Examples of several sorts of energy are listed in **Extension 3.5, *Different Forms of Energy***. As we shall see, any sort of energy can be transformed into any other sort.

Recall our groom. He did work against the gravitational force to lift the bride. Because of this input of work, the bride has extra energy associated with her new position above the surface of the ground. Any object in a gravitational field possesses energy by virtue of its position relative to the ground. We know this object has extra energy, because, if it falls, this energy has the potential to change into other forms of energy and to do work. For this reason, energy of position in the gravitational field is known as **gravitational potential energy**, potentially work. The gravitational potential energy is

gravitational potential energy $= mgh$.

The symbol m stands for the mass and g for the acceleration due to gravity, 9.8 m/s^2. What if the groom stumbled on the threshold? As he stumbles, the gravitational potential energy is converted into *energy of motion*—the bride falls, her speed increasing. As she hits the floor, her energy of motion is transformed into *energy of deformation*, her own and that of the floor. The energy of deformation of these two bodies is transformed (finally) into two forms of energy: *sound energy*, which propagates as a sound wave through the air (eventually warming the air), and *thermal* or internal *energy* spread throughout the two bodies (see Figure 3.5).

In our example, then, work becomes gravitational potential energy becomes energy of motion becomes deformation energy becomes sound energy and **thermal energy**. The final result is the heating up of the environment (and perhaps the relationship as well).

Kinetic energy is synonymous with *energy of motion*. The word "kinetic" comes from the Greek word *"kine,"* which means "motion." One of our words for movies, "cinema," comes from the same root. Kinetic energy, or energy of motion, would be zero for a body at rest. The kinetic energy of a moving body must depend on the speed. We define a body of mass m moving with speed v to

FIGURE 3.5
What might have been had the groom stumbled.

have kinetic energy defined (see **Extension 3.6, *Deriving Kinetic Energy from Work***) as

$$\text{kinetic energy} = \frac{1}{2}mv^2.$$

The distinction between forms of energy is sometimes obvious and sometimes a matter of convenience. For example, consider some solid material. It is made up of an array of atoms (or molecules) bound together by electric forces (see Chapter 4). Each atom remains in its place relative to the others. The atoms are vibrating in place. If we heat this solid material, and then could magnify the structure enormously, we would see that the distance moved by the atoms would, on average, be greater than that before the heating. So, an increase in temperature is reflected in an increase in the kinetic energy of the atoms in the solid, spread out among all of them. Temperature is simply a measure of the average kinetic energy of vibration of the material's atoms. Thermal energy is just this randomly distributed kinetic energy.[2]

Suppose a brick is carried up a mountain. Work is done in carrying it up the mountain. The increase in potential energy in the gravitational field is the amount of work done in lifting the brick to the top of the mountain by some path. One such path might be as shown in Figure 3.6. We can imagine breaking up any change in height into segments along the direction of W, the weight (that is, parallel or antiparallel to it), and segments having no length in the direction of W (that is, perpendicular to it). Thus, we have done work only along those segments parallel to W because of our definition of work. Those segments along W finally add up to h, the height above the ground. So,

$$\text{gravitational potential energy} = \text{work done} = W \times h,$$

which is the same result we found for the bride.

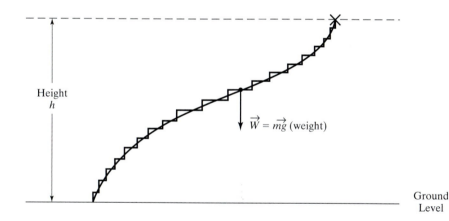

FIGURE 3.6

The vertical path segments are parallel to the weight (force of gravity) and contribute to the work. The horizontal path segments require no work to traverse. If you walk up a flight of steps, the work you do is done only when you go upward from one step to another, not while you walk on the level landings.

A pendulum can be made by hanging a mass from a ceiling by a string. When the mass is pushed aside and released, it undergoes a regular to-and-fro motion. Because the mass has to be pushed, so that a force is exerted on it, and it moves (changes position) as a result of the push, work is done, giving the pendulum gravitational potential energy. As the mass is released, it falls, losing some of this potential energy and increasing its kinetic energy. At the bottom of its swing, it has no potential energy (relative to a hanging mass), and all its energy is kinetic energy. Potential energy has been converted to kinetic energy. This kinetic energy converts itself into potential energy, the potential energy converts itself into kinetic energy, ad nauseam. Actually, in the operation of a pendulum, some energy is transferred from the pendulum to the air by collision with air molecules, and some energy heats the pivot due to friction there. Thus, if we considered only kinetic energy and potential energy, we would not have the whole story.

3.5   FRICTION AND THERMAL ENERGY

One of the difficulties we have in understanding motion is that objects moving near Earth's surface eventually stop. This phenomenon is called **friction**. Surfaces in contact grip one another, rather like Velcro, as they move relative to each other. We can observe motion at constant speed only when we apply a force to an object that matches the frictional force but acts in the opposite direction. The total force is then zero, but if we are unaware of friction, we would say that we needed to apply a constant force to achieve motion at constant speed. This is what led the ancient Greeks to the false idea that motion is only possible when a force is continually applied.

There are several kinds of friction: friction between bodies moving on one another, called *kinetic friction*; friction between motionless bodies in contact, preventing their motion, called *static friction*; and friction as a solid object moves through a fluid, called *drag*. Air resistance, a form of drag, is discussed in more detail in Chapter 15.

When objects in contact move, work is done by the frictional force between the bodies. This work is lost, as work, to the system, and heats the components, appearing as thermal energy. For this reason, mechanical systems produce less output work than the energy they take in.

What this really means is that some forms of energy, which can easily be transformed into useful mechanical work, are worth more (or are of higher quality) than forms of energy that are more difficult to transform, such as thermal energy. Equivalent amounts of energy do not necessarily have the same capacity to do work. This topic is discussed in more detail in Chapter 7.

3.6   CONSERVATION OF ENERGY

In real situations, then, energy appears to be lost. This loss is only chimerical, however, because what actually happens is that the energy is changed into a different form. Energy is neither lost nor gained in any process. We see examples of this in our everyday lives. In a car, the chemical energy of gasoline becomes thermal energy and work. (There is so much thermal energy generated in a gasoline engine that it must forcibly be cooled.) Water is caught in a dam to be transformed into mechanical kinetic energy in a turbine. And, because work can produce thermal energy, thermal energy can be made to do work.

The guiding principle in our approach to energy is that it can never be gained or lost. This tenet is called the *principle of* **conservation of energy**. The great physicist Hermann Helmholtz is credited with first formulating the principle in a useful way in the 1880s.

The amount of energy of the universe does not change. Whatever ways energy appears in the universe, the total amount is the same as it was 10 minutes ago, or 10^8 years ago, or 10^8 years in the future. This was essentially the principle enunciated when we defined TANSTAAFL. We may have energy in any of its forms, and the amount of any sort may change from instant to instant, perhaps, but the total energy is the same (see **Extension 3.7,** *Various Units of Energy*).

The Pendulum as an Illustration of Conservation of Energy

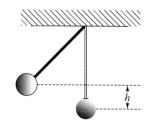

As an illustration of the principle of conservation of energy, let us consider the pendulum again (Figure 3.7). Recall that gravitational potential energy (PE) is given by $Wh(=mgh)$, and that kinetic energy (KE) is $\frac{1}{2}mv^2$. We have used the fact that W, the gravitational force (weight), is given by (mass) times (gravitational acceleration), which we write mg.

The maximum potential energy of an object that falls is transformed into kinetic and thermal energy as the object falls. We will consider an idealized pendulum (one in which we ignore thermal energy produced at the pivot and friction losses in collisions with air molecules), because energy is conserved; in this case,

$$KE + PE = \text{constant}.$$

Therefore, there is a relation between the maximum height, h, and the maximum speed, v; that is,

$$mgh = \frac{1}{2}mv^2, \quad \text{or} \quad v^2 = 2gh$$

no matter what the mass is. Suppose the maximum height, h, is 0.5 meters. Making the approximation that g, the acceleration of gravity at Earth's surface, 9.82 m/s², is 10 m/s² (or 32 ft/s²),

$$v^2 = 2(10 \text{ m/s}^2)(0.5 \text{ m}) = 10 \text{ m}^2/\text{s}^2.$$

To determine v, we must take the square root, giving

$$v = 3.1 \text{ m/s} \, [= 19.5 \text{ ft/s}].$$

We may thus use the principle of conservation of energy to find the speed of the pendulum mass at the bottom of its swing.

The notion that work and thermal energy are interchangeable took a long time to become accepted. Most scientists had thought that thermal energy was a fluid; this fluid was called the *caloric*. However, Benjamin Thompson, an American Tory who became the German Count Rumford, noticed in 1798 that a tremendous amount of heat was transferred in boring out cannon. Water was poured on the cannon to soak up the thermal energy being generated, and after a time, the water boiled. The more water put in, the more boiled off. Rumford was forced to conclude that the amount of "caloric" was infinite. Rumford recognized that the motion of the boring instrument became the motion of small pieces of borer, the cannon, and the water. In this conclusion, Rumford was far ahead of his time and was ignored. It was not until the 1840s that the so-called "mechanical equivalent of heat" was measured by James Prescott Joule. He found that a fixed amount of mechanical energy was converted to a fixed amount of thermal energy. This set the stage for the much later acceptance of the principle of conservation of energy.

Simple Machines

As another example of the conservation of energy, consider the car jack and the lever. Note that, with our definition of work, we could do the same amount of

FIGURE 3.7

A pendulum has its bob of mass m raised by h from its lowest point. It now has potential energy mgh. When the ball is released, it falls, and the potential energy is converted into kinetic energy. At the lowest point, where potential energy is zero, the kinetic energy, $\frac{1}{2}mv^2$, must be the same value. We may measure the exchange in an experiment and find that it is good.

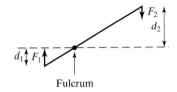

FIGURE 3.8

The lever is a simple machine. F_2 is the input force, which is applied through a distance d_2; F_1 is the consequent output force, acting through a distance d_1. The lever must be supported at one point (called the fulcrum) in order to work. Other simple machines include the wedge and the pulley.

work by applying a large force through a small distance as by applying a small force through a large distance. This is the principle of these simple machines: they multiply force, at the cost of increasing the distance through which the force must be applied, or change its direction. For the lever shown in Figure 3.8, an amount of work done is

$$\text{work} = F_1 d_1 = F_2 d_2.$$

Thus, if d_2 is large, F_2 can be small.

Pulleys are made of wheels, with ropes strung over them, as shown in Figure 3.9. Fixed pulleys, which are attached to walls or ceilings, act to change a force's direction. Movable pulleys are used to change the force's direction and to decrease the amount of force necessary to move an object. In Figure 3.9, which shows two pulleys, the force F needed to raise weight W at a constant speed is less than W (it is $\frac{1}{2}W$). For an ideal pulley, one with no friction in the pulleys and no inertia, the amount of work done to raise the weight a distance d (work $= Wd$) must equal the amount of work done by the force F. Conservation of energy guarantees that the amount of work coming out of a machine equals the amount of work put in. In the case of the pulley of Figure 3.9, the distance through which the force F must act is $2d$, because F is $\frac{1}{2}W$, and

$$(F)(\text{distance through which } F \text{ acts}) = Wd.$$

In this case, the ratio of the output force to the input force (known as the mechanical advantage) is

$$\frac{W}{F} = \frac{W}{\frac{1}{2}W} = 2.$$

Each additional set of pulleys added to such a system reduces the force required to move an object, at the price of increasing the distance through which the force is exerted. We cannot get something for nothing, but we may reduce the force necessary to do a job (or achieve a mechanical advantage) by increasing the amount of rope that needs to be pulled through the hands. The key point to remember is that although force can be multiplied by use of a machine, the amount of work that has to be done is the same (if an ideal frictionless machine is assumed), whether or not the machine is used.

Efficiency

Real machines have friction in them as well as the possibility of deformation. Ropes stretch, levers bend. In this real world, some input energy must either be used to battle friction or be stored in a stretched rope or bent rod. Consequently, the transformation of energy by a real machine involves a loss in available work to thermal energy (friction) or deformation energy (stretched string or bent rod). The ratio of work output to work input is one for an ideal machine; it is less than one for a real machine because of the losses mentioned. This ratio is called the **efficiency**:

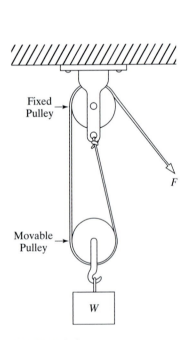

FIGURE 3.9

In a system containing one fixed and one movable pulley, a weight W may be lifted by application of a force F equal to $W/2$.

$$\text{efficiency} = \frac{\text{work output}}{\text{work input}}.$$

Consider a pulley system, as in Figure 3.9, that has negligible friction in the pulleys but in which the rope stretches by 10% when the weight is 500 N. Because the force F is determined solely by the fact that there are now two ropes holding up the weight W, F is still $\frac{1}{2}W$. Now, however, to raise the weight a distance d, we have to pull in a 10% longer rope, because of rope stretching, than we would have pulled in for a stretchless rope: 2.2 d. The work output is still Wd; the work input is now

$$(F) \times (2.2\,d) = \left(\frac{1}{2}W\right) \times (2.2d) = 1.1\,Wd.$$

The efficiency is, therefore,

$$\text{efficiency} = \frac{\text{work output}}{\text{work input}} = \frac{Wd}{1.1Wd} = 0.91,$$

or 91%. Similar calculations could be done for the other examples.

3.7 POWER

So far, we have discussed work and energy. In our personal lives, not only how much work was done, but also how fast it was done, are relevant. The rate at which work is done is called *power*. A hair dryer set to operate at 1200 W dries your hair twice as fast as one set to 600 W. A 1200 W toaster toasts twice as fast as a 600 W toaster.

We define **power** as work done divided by the time needed to do the work. This differs somewhat from the colloquial meaning of the word. In the international system, the unit of work is the joule, and the unit of time is the second. In that system, the unit of power is thus the joule per second (J/s). This unit is given the name watt (W), in honor of James Watt, the inventor of the modern steam engine. Because 1 W equals 1 J/s, the joule could also be written in terms of watt seconds. As 1 h equals 3600 s, a watthour (Wh) equals 3.6×10^3 J. The kilowatthour (kWh), a widely used unit of energy, is equal to 3.6×10^6 J. When you pay your electric bill, you pay for energy by the kWh. In this book, the J (and the kWh, as applicable) will be used as the energy unit as much as possible.

To demonstrate power developed, I can walk up a flight of stairs about 3 meters [~10 feet] high easily in about 20 seconds. In that case, I do an amount of work equal to the product of my weight and the change in the height:

$$80 \text{ kg} \times 9.8 \text{ m/s}^2 \times 3 \text{ m} = 2350 \text{ J}.$$

From the definition, the power developed is then

$$\text{my power} = \frac{2350 \text{ J}}{20 \text{ s}} \cong 120 \text{ J/s} = 120 \text{ W}.$$

[In the customary American system, this is about 180 lb × 10 ft/20 s = 90 ft-lb/s.] Suppose my wife Michelle, about 60 kg, runs up the same flight of stairs in 12 s. Her work is

$$60 \text{ kg} \times 9.8 \text{ m/s}^2 \times 3 \text{ m} = 1764 \text{ J}.$$

Her power is

$$\text{Michelle's power} = \frac{1764 \text{ J}}{12 \text{ s}} = 147 \text{ J/s} = 147 \text{ W}.$$

Many electrical devices are labeled by the amount of power that they draw. A 100 W light bulb can use electric energy only at that rate—100 J/s. A 1200 W electric heater uses energy at 1200 J/s and cannot use it any faster than that (if it does, it will burn up). Most electric devices list the number of watts the device should use.

Metabolic Power

I can do work at a rate of around 75 J/s. I get this estimate by imagining that I am digging. I can lift a 5 to 10 kilogram [10 to 20 pound] clod of dirt about a meter to throw it on a pile, and do it steadily. Thus, the potential energy changes by about 5 to 10 kg $\times$ 9.8 m/s^2 $\times$ 1 m, or 50 to 100 J. Because the process of lifting takes about one second, I would estimate that I could do work at the rate of 50 to 100 J/s. My estimate—averaging 75 J/s—is in the middle of this range.

I want to compare human energy use to that of machines. Because 1 calorie is 4.186 J (see the energy converter), 1 calorie = 4.186 J/(3.6 $\times$ 10^6 J/kWh) = 1.16 $\times$ 10^{-6} kWh, and so one kilocalorie is 1.16 $\times$ 10^{-3} kWh. An average human being must expend energy at a rate of about 2 $\times$ 10^3 kcal per day, or about 2.3 kWh per day. In the United States, the daily energy supply is somewhat higher, about 3.7 kWh per day. This amount of energy would burn a 100 W bulb for 37 hours, or a 150 W bulb for about a day. A typical monthly food bill for one person in the United States is in the neighborhood of $150. The cost of person-power at 3.7 kWh per day for 30 days is

$$\frac{\$150}{(30 \text{ d})(3.7 \text{ kWh/d})} = \frac{\$150}{111 \text{ kWh}} = \$1.35/\text{kWh}.$$

My electricity costs about 8 cents/kWh, still an order of magnitude cheaper than personpower. You are a lot more expensive to supply with energy than a 150 W bulb. That is because most of this food's energy goes into supporting body functions—heart, liver, internal temperature, and so on. As we found earlier, people are capable (over extended times) of an ordinary power expenditure of about 70 W (50 ft-lb/s).

Measures of metabolic power requirements of people at rest can indicate how much of this power is required for the body. This is a measure of the **basal metabolic rate** (BMR). Roughly, BMR is 45 to 70 kcal/h (200 to 300 kJ/h), depending on body mass. (It turns out that the relationship between the BMR and mass is about 0.6 kcal/kg body mass; if the body mass = 50 kg, the BMR is 30 kcal/h).[4] A person with a 70 kg mass (a weight of slightly more than 150 lb) has a BMR of 43 kcal/h = 180 kJ/h = 5 $\times$ 10^{-2} kWh/h = 50 Wh/h. Thus, the body of a 70 kg person uses energy at the rate of 50 W. With an average input of 150 W, we must either expend energy at an average rate of 100 W or get fat. I could probably run flat out up those stairs in 3 s, in which case I would be using the 2350 J calculated above in 3 s, for an average power of $\frac{2350 \text{ J}}{3 \text{ s}}$ = 780 W. If I could keep this up for a long time, I would have no trouble losing weight. (Ha!)

Of course, we are not totally efficient in using this excess power. Moreover, it is used in only about two-thirds of the day. During waking hours, about 150 W must be burned. Assuming energy usage in the body to be about 33% efficient, this means that a person can do ordinary work at a power output of 50 W, which agrees with the 70 W estimate previously obtained.

SUMMARY

In this chapter, we discussed the definitions crucial for understanding energy basics. Speed measures the rate at which distance is covered. Velocity measures the rate of change of displacement and is a vector quantity, by which we mean that we specify not only how fast, but also in what direction. The change of velocity with time is called acceleration, which is also a vector.

To make an object's acceleration increase, we have to push harder on it. We find that this force equals the product of the acceleration and the object's mass. The relationship $F = ma$ is known as Newton's Second Law. A body's mass is characteristic of the body. A body's weight is the gravitational force on the body, and so the weight of an apple on the Moon is different from the weight of that same apple on Earth or on Mars, although the mass of that apple remains the same regardless of its interplanetary travels.

For work to be done, there must be a force on an object, and the object must move: work = (force) × (distance moved in the direction of the force). The product of force and distance moved perpendicular to

the force is not work. It is not even relevant. Objects capable of doing work (or exerting a force) are said to have *energy*. Familiar sorts of energy include energy of motion (kinetic energy), energy of position (gravitational potential energy), energy of internal rearrangement of atoms in a material (deformation energy), energy stored in atomic compounds (chemical energy), energy stored in atomic nuclei (nuclear energy), energy carried by light (radiant energy), and energy stored in a car battery (electric energy). Energy is measured using many different units. We prefer to use the joule wherever possible. Other commonly used units are the kilowatthour, the kilocalorie, the British thermal unit, and the foot-pound.

The total amount of energy in the universe in all its various forms does not change. We speak of this as the principle of conservation of energy. Energy is not created or destroyed; it is transformed, or changed from one form to another. Real machines are less than 100% efficient. Power measures the rate at which work is done. Power is measured in watts.

PROBLEMS AND QUESTIONS

MULTIPLE CHOICE REVIEW

1. You ride in an automobile traveling at 100 km/h. It comes to a stop over a time interval of 0.5 s. You can be sure that
   a. the car's engine has stopped.
   b. a velocity acts.
   c. friction between the tires and the road stopped it.
   d. the car had a nonzero acceleration during that time.
   e. none of the other answers is correct.

2. The approximate daily energy consumption of primitive humans (before fire) was about
   a. 2000 kJ.
   b. 8000 kJ.
   c. 12,000 kJ.
   d. 48,000 kJ.
   e. 400,000 kJ.

3. Which of the following statements is false?
   a. Energy can be converted from one form to another.
   b. Energy comes in different forms.
   c. Energy can be stored to a greater or lesser extent.
   d. Energy can be moved or transferred.
   e. Energy is constantly being destroyed by humans.

4. The most basic and irrefutable proof that energy cannot be created out of nothing or completely destroyed is that
   a. Newton's laws of motion require it.
   b. no proof is necessary. Energy is energy.
   c. no experiments have ever proved the contrary.
   d. the Sun would have long since ceased to shine.
   e. it would contradict Einstein's relativity.

5. Which of the following conversions of energy is possible?
   a. kinetic energy to thermal energy
   b. solar energy to chemical energy

c. chemical energy to thermal energy
d. nuclear energy to electrical energy
e. All the other conversions are possible.

6. When a moving automobile stops for a stoplight, its kinetic energy
a. completely disappears.
b. is converted to thermal energy.
c. is transferred to the pavement.
d. is stored in the automobile's battery.
e. is stored in a flywheel.

7. An object moves 100 m along a frictionless table at a constant speed. The object weighs 50 N. The work done on the object by the gravitational force is
a. 0 J.
b. 2 J.
c. 50 J.
d. 100 J.
e. 5000 J.

8. The fact that a spacecraft does not have to fire its engines continuously as it moves in space is a consequence of
a. Snell's law.
b. Newton's Third Law of motion.
c. Einstein's theory of relativity.
d. Newton's First Law of motion.
e. Newton's Second Law of motion.

9. An 80 kg person walks up a flight of 3 m high stairs, and then walks down the same flight of stairs to the person's original position. What is the change in the person's gravitational potential energy?
a. 0 J
b. 240 J
c. 480 J
d. 2352 J
e. 4704 J

10. A rocket accelerates due to the force exerted on it by the
a. fuel.
b. oxidizer.
c. exhaust gases.
d. nozzle.
e. booster.

SHORT ANSWER

Explain whether the statement made is correct, and justify your answer.

11. Machines produce less work than that used to run the machines.

12. Some of the energy produced when machines are used is chemical energy, and some energy is lost.

13. The potential energy of an 80 kg male and a 60 kg female standing together on a promontory at some height above the surrounding plane is the same.

14. Heat energy is one of the many various forms of energy. It defines the amount of energy contained in a body at a certain temperature.

15. All the chemical energy in food is converted into running the body's metabolic processes.

16. A 1200 W hair dryer used for 10 min uses as much energy as a 1000 W hair dryer run 12 min.

17. A pulley system holding a rope that is pulled out a distance of 20 m is used for lifting a 2500 kg grand piano. The

piano can be raised 3 m by application of a 375 N force to the rope.

18. A pendulum bob on a long string is raised 1.1 m above its lowest point. When it is let go, we can determine the speed of the bob as it passes through its lowest point, even if we have no idea of its mass.

19. The weight of a body is the same, no matter where the body is located.

20. A 1200 W toaster does enough work in 10 s to raise an 80 kg man to a height of 15 m.

DISCUSSION QUESTIONS

21. Sharon cycles up a hill at 15 km/h. When she reaches the top, she immediately turns around and comes down the same road at 30 km/h, stopping where she started. Is her average speed greater than, less than, or equal to 22.5 km/h? Explain.

22. Willis thinks over the discussion of the second part of the first experiment described on pages 30–31. He is interested in the equality of the force he exerts on the clone and that the clone exerts on him. Later, he tells his girlfriend Sheena "It may be right in that case, because both people had the same mass. It's wrong if the masses are different. If I met the Hulk and he pushed me, it would be with a greater force than I could use to push him back." Discuss this reasoning.

23. A car drives down one hill and up another, maintaining constant speed (the road is a V-shape with a rounded bottom). What can be said of the acceleration of the car along the left side of the V halfway down? What can be said of the acceleration of the car along the right side of the V halfway up? Where on the track is the acceleration the greatest? Explain.

24. A force *F* is exerted on an object of mass *m*. A force twice as large (2*F*) will cause a mass 2*m* to change its velocity by a certain amount. Explain the relation between the change in velocity due to force *F* and that due to force 2*F*.

25. A force *F* is exerted on an object of mass *m* through a distance *d*. A force twice as large (2*F*) is exerted on a mass 2*m*, also through the distance *d*. Explain the relation between the amount of work done in the two cases.

26. A car's motor gives out just at the bottom of a big hill, but the car starts coasting up the hill. In what direction is the acceleration of the car just as it stops momentarily before rolling back down the hill?

27. Discuss whether it is possible to have a displacement with magnitude greater than the distance traveled.

28. Fred carried a 100 N basket to his apartment on the 15th floor of his building by climbing the stairs. June used a pulley hung from a 15th-floor window that amplified her force by a factor of three to raise a different 100 N basket up to the apartment. Fred claims he did more work. June disagreed and says she put out more power. Is either person correct? Explain.

29. A car coasts down a road into a valley and then coasts back up the next hill some distance. Where has its change in velocity been greatest—At the point at which it stops while going uphill? At the bottom of the hill? As it rolls back through the bottom to the other side? Explain.

30. If a new machine is more efficient than the old model, it uses a smaller amount of energy to do the same task. Why

hasn't the experience with all the electric machines over the 130 yr history of commercial electricity sales produced a machine that is 100% efficient?

PROBLEMS

31. You are playing Frisbee with a friend. She runs in a straight path after a Frisbee you threw to her, traveling 18.0 m in a time of 5.50 s from the time she started running. She then runs for an additional 4.00 s and comes back to a point 13.0 m from you before throwing the Frisbee back toward you. Determine her average speed and average velocity for each part of the trip and for the entire trip.

32. A pig waddles from its original position 3.00 m eastward in 5.00 s; it then waddles 4.00 m northward in 4.00 s; finally, the pig waddles south 5.00 m in 3.00 s. Determine the net distance traveled, the net displacement, the average speed and average velocity for each segment of the trip, and the net average speed and average velocity for the trip.

33. A bus leaves Pueblo, Colorado, at 10:00 and travels north to Denver, a distance of 180 km, arriving at 12:15. The bus leaves Denver for Cheyenne, Wyoming, at 1:00 and travels north, again 160 km, arriving in Cheyenne at 2:45. It leaves Cheyenne at 3:30 and travels without stopping 500 km to Rapid City, South Dakota, arriving at 10:30. Find the average speed on each segment and the overall average speed for the trip.

34. Consider the bus in Problem 33 again. If Pueblo, Denver, and Cheyenne lie along a line that is straight north–south, and Rapid City is 300 km north and 130 km east of Cheyenne, find the average velocity on each segment and the overall average velocity for the trip.

35. The moon travels around Earth in 27.32 d. The mean Earth–Moon distance is about 400,000 km. What is the average speed of the Moon's travel for one complete revolution about Earth? Explain how you get your answer.

36. Mark, 80 kg, walks up a flight of stairs 5.0 m high. What is the change in the person's gravitational potential energy? What is the average power used if it takes him 7.0 s?

37. Jerrod has walked to the top of a rock in Adirondack State Park that is 5.0 m above a small flat area. Jerrod falls off the edge of the rock. What would you expect his speed to be as he lands?

38. You wish to raise a 100 kg statue using an ideal lever. What distance will the mass be raised if you apply a 100 N force and lower your end 2.0 m?

39. Sarah, who has a mass of 55 kg, picks up Brittany, her daughter, who has a mass of 20 kg, and carries her upstairs for her nap. The second floor is 3.0 m above the first floor.
    a. How much work does Sarah do?
    b. By how much did Sarah's potential energy change?
    c. By how much did Brittany's potential energy change?
    d. If Sarah expended 215 W of power on this errand, how long must it have taken her to climb the stairs?

40. An apple hangs from the branch of an apple tree (the apple's mass is 0.65 kg). A 1.35 kg bird rests on the apple. Determine the apple's acceleration.

41. A diesel locomotive has a mass of 420 t, and there is a force applied to the locomotive by the track of 12,000 N. The force is applied for a distance of 2.00 km.
    a. What is the acceleration of the locomotive?
    b. How much work has been done?
    c. What is the kinetic energy of the locomotive at the end of 2.00 km? Neglect friction.
    d. What is the speed of the locomotive at the end of 2.00 km?

42. What is the mass of an object if a 50.0 N force produces an acceleration of the object of 0.50 m/s^2?

43. A glazed doughnut has a dietary value of 250 kilocalories, while a jelly doughnut has a dietary value of 330 kilocalories. Sarah, who has mass 55 kg, ate one of each. If all this energy were converted to change Sarah's potential energy, how far could the glazed doughnut raise her? How far for the jelly doughnut? How many jelly doughnuts would she need to raise her 30 m?

44. How much would it cost you to run your 400 W computer monitor for 2.0 h?

ELECTRICITY GENERATION AND TRANSMISSION

PART II

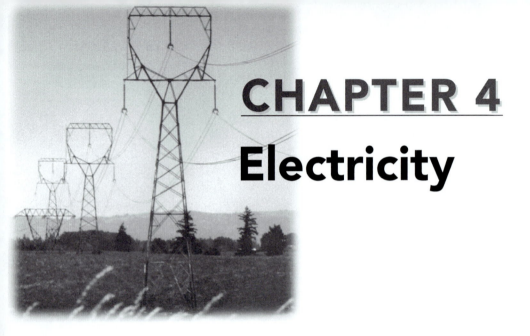

CHAPTER 4
Electricity

When the word energy is mentioned, most people think of electricity. Although this association overstates the case, electricity is an important component of energy use and generation worldwide. In this chapter, electricity and magnetism are described from the physicist's and engineer's points of view. The electric motor, generator, and electric transformer are explained; their respective roles in the electrical energy distribution are also described.

KEY TERMS

electric charge / atom / proton / electron / quantum / Coulomb's law / electrical potential energy / potential / voltage / current / circuit / resistance / Ohm's law / Joule heating / domain / magnetic field / electromagnetic induction / generator / induction motor / transformer

Our society uses many different fuel sources. Electricity comes mainly from burning coal, gas, or oil, or from nuclear energy. For some applications, however, electric energy is a preferred "fuel" source for various reasons. Its main advantages are its cleanliness at the point of use, convenience, safety, and ease of control. Its main disadvantage is its greater cost in comparison to coal, oil, or gas. Before we can understand the advantages and disadvantages of electric energy use, we must discuss electricity and how we get energy from it.

The electric plug and electric batteries in various forms dominate our experience with electricity nowadays, but people have coexisted with electricity and electric energy from that prehistoric time when humans first used lightning-kindled wood to keep watch fires and campfires burning. Lightning has fascinated people for centuries. Benjamin Franklin's well-known interest in lightning led to his invention of the lightning rod.

4.1   ELECTRICAL CHARGE

Franklin supposed that there was some electrical "fluid" that could be separated from (and distinguished from) some other electrical "fluid." Today, we call the "fluid" **electric charge** and follow Franklin's definitions of positive electric charge and negative electric charge. When two units of equal size but opposite charge combine, they form an electrically neutral material. When the charge is separated—accumulating in clouds, for example—it builds up until there is a breakdown in air as the charge tries to reach "the ground" to neutralize itself. Franklin, in his famous kite experiment, was able to drain off some of the cloud electricity into a Leyden jar (storage cell for charge) and show that it was the same as the so-called static charge produced by rubbing cat's fur on quartz, silk on rubber, or a shoe on carpet. Most of us have been jolted at one time or another when an accumulation of charge produced a spark from us to a door handle.

We are not accustomed to thinking about electric charge, because virtually everything in our experience is electrically neutral. Each positive charge has a corresponding negative charge bound to it. **Atoms** ultimately make up all material objects. An atom contains a nucleus of tiny particles ($\sim 10^{-15}$ m diameter) called protons and neutrons. The protons and neutrons have about the same mass and behave in similar ways, except that the **proton** carries a small unit of positive charge, whereas the neutron is electrically neutral. Outside the nucleus of the atom is a "sheath" of negatively charged **electrons** (which happen to be of exactly the same size as the proton's positive units of charge) at a radius of about 10^{-10} m. Electrons have only $\frac{1}{1830}$ the proton mass.

The electron has a size so tiny that it can be considered a point, yet in the atom, the electrons provide the outer "shape" by acting like a "solid" shell, the outer surface of the atom. We often refer to the atomic electrons as a cloud, because we cannot actually specify the position of any particular electron; in effect, they are everywhere on the surface. The total positive nuclear charge is offset by the equal amount of negative charge carried by the atom's electrons. The result is that most atoms of our acquaintance have no net charge. However, because the electrons are the outside of the atom (and the nucleus is so small), electrons may easily be removed from the atom. These electrons may even be removed by rubbing, as we experienced in the case of a spark between us and a door handle. Because electrons may be removed relatively easily from atoms, and they have smaller mass, by far, than protons, electrons move relatively more freely within and between materials than do protons.

It appears from our discussion that electric charge comes in bundles of specific size, because the size of the proton's charge is the same as the size of the electron's charge. The size of this bundle, or **quantum** (basic unit, from Latin, meaning "how much"), of charge was first measured by the American physicist Robert Millikan in the early 1900s. All subsequent experimentation has confirmed that all observable charge comes in integer multiples of this charge quantum. It is conventional to reserve the symbol e for the quantum of charge. Then, a proton has a charge of e, a neutron has a charge of $0\,e$, and an electron has a charge of $-e$. Charges can come in groups of $0\,e$, $-5\,e$, or $27{,}500{,}430\,e$, but never $\frac{1}{2}e$ (although particles called *quarks*, which exist only as constituents of the more familiar particles, have a quantum of charge of $\frac{1}{3}e$).

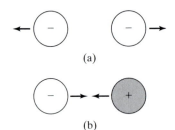

(a)

(b)

FIGURE 4.1

The effect of electric charge.
(a) Like sign charges repel each
other.
(b) Unlike sign charges attract
each other.

We measure charge in coulombs (C), honoring Charles Augustus Coulomb, the first physicist to state the law of electric force (**Coulomb's law**). The coulomb is very large compared to the quantum of charge, which is $e = 1.6 \times 10^{-19}$ C. This means that an object carrying a charge of -1 C has a net surplus of about 6×10^{18} electrons.

Electric Force

We have considered examples of charge separation leading to sound and motion (lightning) or doing work (starting a fire). Because energy is changed into other forms, it is clear that these phenomena result from forces exerted by electric charges. When the charge moves under the influence of the electric force, work is done (see Chapter 3).

Suppose you and a friend take, for example, a roll of transparent tape. Each of you should prepare two pieces of tape about 5 cm long, and bend one end on each piece over to form a lip that doesn't stick. Put the first piece (call it A) on a tabletop. Now, place the second piece (B) on top of the first, and rub (do work). After you pull B off A, hold it in one hand, and rip A up from the tabletop with the other. Your A tape will repel your friend's A tape, and your B tape will repel your friend's B tape. The As contain like charges and repel one another; the Bs contain like charges and also repel one another. We see what happens if positive and negative charges are separated; by experience, we have found that two positive charges or two negative charges repel one another, as illustrated in Figure 4.1a. Now bring the two different pieces of tape near to each other but not touching. They will pull toward one another. The charges are drawn toward or attracted to one another—the As attract the unlike Bs. This is illustrated in Figure 4.1b.

Another amusing experiment can be constructed to illustrate both forces. A charge that we may suppose to be positive can be built up by rubbing on an insulating rod, call it rod C, that is held in one hand. An insulator such as rubber or plastic prevents charges from moving. Ball D, made of a conducting material and uncharged, is hung as shown in Figure 4.2a. A conductor such as gold, silver, or

FIGURE 4.2a

A conducting ball, D, is attracted to
a charged rod, C, because some
charges from D are attracted toward
C.

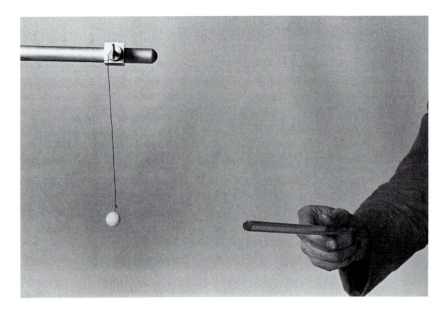

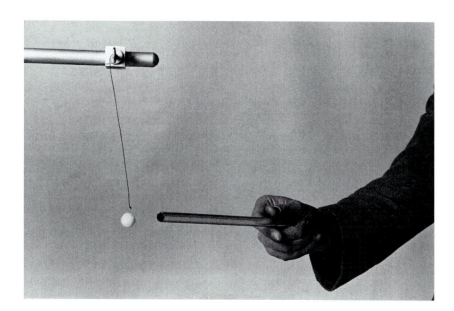

FIGURE 4.2b

In moving closer to *C*, ball *D*'s charges become more strongly separated, increasing the effect.

copper is a material that allows the carriers of electric charge, electrons, to move relatively freely. When the rod *C* is moved to a position near ball *D*, ball *D*, being free to move, is pulled toward the rod (Figure 4.2b) until the two touch (not shown). After a short time in contact, ball *D* is repelled from rod *C*.

We may explain our observation by noting that, because charge carriers are free to move in ball *D*, negative charges in the ball will be attracted toward the positive charge on rod *C*. This means that electrons will build up on the side of the ball that faces rod *C*. Ball *D* was originally neutral, so positive charge is left behind on one side as the electrons move away to the other side. Therefore, a positive charge builds up on the side of the ball facing away from rod *C*. The negative half of ball *D* is attracted to rod *C*, and the positive half is repelled. The negative half, because it is slightly closer to rod *C* than the positive half, wins, and the ball and rod move together. While the two are in contact, electrons flow from ball *D* to rod *C* in an attempt to neutralize rod *C*'s positive charge. Soon, however, enough charge is drained so that ball *D*, which now has a net positive charge, can push away from *C*. Then, rod *C* and ball *D* repel one another.

The reason that *D*'s negative half wins the tug-of-war with *D*'s positive half was discovered by Coulomb in his famous force experiments. Charges are usually assigned the symbol *q*. Coulomb found that the electric force, F_E, between two charges q_1 and q_2 was given by

$$F_E = (\text{constant})\, q_1 q_2 / (\text{distance of separation})^2.$$

This force law is now called Coulomb's law in honor of his work in determining the constant. In the international system of units,

$$F_E = (9 \times 10^9 \text{ N m}^2/\text{C}^2) \frac{q_1 q_2}{R^2},$$

where *R* is the distance between the centers of the two charges. See **Extension 4.1, *Electric Force Between Point Charges*** for two examples of the use of Coulomb's law.

We now apply Coulomb's law to the tug-of-war between D's positive and negative halves. Because D was originally neutral, the negative and positive charges on D are equal in magnitude. The numerator of the electric force equation is thus the same for both charges. The negative charge is slightly closer to C than is the positive charge, so the value of R corresponding to the negative charge is smaller than that corresponding to the positive charge. From the Coulomb's law equation, the force of attraction of the negative charge is thus larger than the force of repulsion of the positive charge on C. Consequently, D accelerates toward C.

Of course, Coulomb's law is of much broader scope than this particular example. It explains the electric force between any static charged bodies or bodies with distributions of charge.

Electric Energy

Now that we know what the electric force is, we can simply apply the formula for work as found in Chapter 3;

$$\text{work} = (\text{force}) \times (\text{distance moved along the force}).$$

Notice that, by Coulomb's law, the expression for the force changes as the distance between two charges changes. To determine the electrical work actually done by moving charges, we should calculate the tiny amount of work done at each point in moving a tiny distance and add all the tiny amounts of work. (This calculation is not difficult, but it requires the use of calculus; we will not treat that subject here.)

Consider most ordinary materials we know. They do not exhibit any electric characteristics, because they are neutral, made up of atoms in which the positive charges balance the negative charges. But, it is possible to separate charges, provided we are willing to do the work necessary to cause the separation. In separating the pieces of tape A and B, we achieved charge separation but had to do work (rub the tapes, then pull them apart) in order to do it.

We may recover the work we did by letting the A-type piece touch the B-type piece. The energy is higher when the charges are separated, because we had to do work to separate them, so we can say the system of A and B has **electrical potential energy**. Most devices that produce charges do so by separating them: one example is the Van de Graaff machine found in many physics labs. The electrical potential energy can be recovered as kinetic energy or as some other form of energy. (Of course, electrical potential energy is measured in joules, just as is any other form of energy.) For more information, see **Extension 4.2, *Electrical Potential Energy of Point Charges***.

Suppose you put styrofoam "peanuts" on a Van de Graaff machine and turn it on. The peanuts fly away, because they all picked up the same sign of charge. If you pushed one of the peanuts back toward the Van de Graaff machine, it would take work to do it. If you then let it go, that work would be converted back into kinetic energy. If a whole bunch of charged peanuts were held together, we could store the work we did to assemble them. This stored energy is the electrical potential energy of the system of peanuts. Recall that work may be expressed as the difference in potential energy. Doing work on the peanuts, pushing them together, made the electrical potential energy greater than it was when we began.

If two charged objects have an electrical potential energy of 100 J, 100 J of kinetic energy could be produced by allowing the charged objects to move until the electrical potential energy is zero. In determining the net final kinetic energy or the net work done, only the difference between the original value of the potential energy and the final value of the potential energy is important, because it is the difference that determines the kinetic energy according to the principle of conservation of energy: original kinetic energy plus original potential energy equals final kinetic energy plus final potential energy,

$$KE_F + PE_F = KE_0 + PE_0,$$

which implies that final kinetic energy minus original kinetic energy equals original potential energy minus final potential energy equals the negative of the difference in potential energy,

$$KE_F - KE_0 = PE_0 - PE_F = -(PE_F - PE_0).$$

In many applications, it is convenient to use the electrical potential energy difference per unit charge between two points in space. This is sometimes simply called the **potential**, symbolized V:

$$V = \text{potential} = (\text{electrical potential energy difference})/\text{charge}.$$

The unit of electrical potential is the *volt* (V), named for Alessandro Volta in honor of his pioneering electrical research in the eighteenth and nineteenth centuries. A volt is the potential difference between two points necessary to change the energy of 1 C of charge by 1 J in moving between the points (1 V = 1 J/C). Because the unit of potential difference is the volt, it is common to refer to the potential difference as **voltage**.

4.2   CURRENT AND CIRCUITS

The existence of differences in electrical potential at different points in space may result in the motion of charges. Suppose we connect each end of the Van de Graaff machine to a metal plate. A pith ball hanging in the space between the plates moves back and forth between the plates, shuttling charge between the plates. A net motion of a particular sign of charge in some direction is called an electric **current**. Current is measured by counting the number of charges going by in a given unit of time. You might imagine an elf entering in a great ledger the number of charge quanta going past him to right or left. If he counts more charge going one way than the other, a current is flowing. Because the quantum of charge is so small, the current in practice measures the number of coulombs passing each second. A current of one coulomb per second is called an *ampere*, in recognition of Ampère's contribution to the understanding of the interconnection of electricity and magnetism:

1 ampere = 1 coulomb per second (1 A = 1 C/s).

In most applications of electricity, it is the electron that moves. Thus, the electron is the energy carrier that supports the transfer of electrical energy in the medium. There are several reasons that the energy carriers are electrons. One reason is the small mass of the electron in comparison to the proton, or in comparison to an atomic nucleus that may contain many protons and neutrons. We saw the reason for this in Chapter 3—because force is given as mass times acceleration, for a given force, the larger the mass, the smaller the acceleration. The other reason is that most conductors of electricity are metals, and metals are used in electric wiring. In the metal, the atoms of the metal bind themselves together by letting one or two of their electrons free to wander randomly throughout the interior of the metal. Each of the myriad atoms frees an electron, and the atoms are left bound to one another into the metallic structure. If the atoms were to move, they would essentially have to move as the whole substance. In response to an electric force, the electrons accelerate. In many practical applications, this motion of electrons is achieved by making the metal into wire and then creating a potential difference between the ends of the wire.

These electrons responding to an applied electric force can neither be destroyed nor created. If I stuff electrons into one end of a wire, an equal number must exit from the other end of the wire, say, into the ground. But the electrons I originally stuffed into the wire had to come from somewhere; that somewhere is the ground, too. *Ground*, in electricity, represents the reservoir of electrons that are sent jaunting around doing the tasks we require of them in our transmission and use of electricity. It happens that Earth is a reasonable conductor, so that we can literally get electrons from the ground and dump electrons into the ground. It is from this fact that the terminology springs. Because no electrons are gained or lost, and because electrons are found in the ground and returned there, there is a wholeness to the process, a circulation. In a sense, the electrons make a closed loop.

In an electric distribution system, as well as in devices run from cells or batteries, there are also closed loops. Because it is possible to circle from the starting point through electrical elements back to the starting point, such a closed loop is called a **circuit**. In simple electric circuits, as well as in the electric distribution system, the closed loop is made completely of wire. In a circuit having a potential difference supplied by a battery (Figure 4.3), the lowest potential in the circuit is sometimes referred to as the ground. For example, if a single C- or D-cell battery is used to light a flashlight, the negative terminal could be ground, and the positive terminal is 1.5 V higher in potential and so would be at 1.5 V relative to ground.

In the electric distribution system, the two prongs in the outlet are for the hot wire (smaller opening) and the reference wire (larger opening), which functions as a ground for the circuit. It is possible for a potential difference to exist between this reference potential and true Earth ground; for this reason, to provide absolute safety to people operating them, three-prong devices have a connector between the metal parts and Earth ground built in.

In the electric generator or battery, charge is taken from the virtually infinite reservoir, and work is done on the charge. Because work is done on the charge, it gains energy. Because it has energy, when it travels in an electric circuit, it can give that energy up again as work, wherever we want the work done.

Picture the generating plant as a factory. Let the electrons be represented by little balls. These little balls roll out of the generator high in the air, in a gutter-like channel out into the distribution system. The gravitational potential energy of the balls in the gutters is analogous to the electrical potential energy of the electrons leaving a power plant. The gutters we imagine are sent out to houses and factories. The balls roll there, and at these places, they fall through downspouts, turning their gravitational potential energy into kinetic energy that can do work. This is analogous to the electrons falling in potential inside an electric device. Finally, the balls roll in low-lying gutters back to the generating plant. There, our elf (the one counting charges and entering them in a ledger) takes the balls from the low-lying gutters, climbs a ladder, and places the balls in the gutters high in the air, where the cycle can begin again. This is analogous to the return of the electrons to ground at low potential and to the raising of electrical potential in the generator.

The system is set up so that work is done on the electrons in the plant, where a lot can be done at once, and the electrons are sent out as messengers bearing

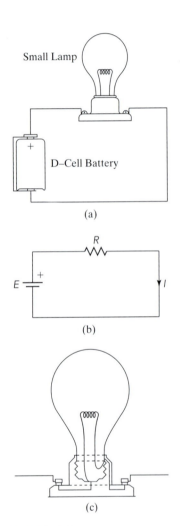

FIGURE 4.3

A simple electric circuit, with a source of potential difference (*V*), and a lamp, which provides resistance (*R*) to the flow of electric current (*I*). (a) A diagram of an actual apparatus (a flashlight, for example). (b) A more schematic representation. (c) A cutaway view of the wires' connections inside the socket.

their gift of work to be used wherever it is convenient to the user. This analogy best represents a direct-current generator, rather than the generators we use to supply electricity to our homes and schools. Even then, it is flawed as a description of what really occurs. Nevertheless, the analogy supplies a useful characterization of the process of energy transfer in the electric distribution system.

Resistance

Various materials resist the flow of these electrons by different amounts. Materials such as silver, copper, or aluminum offer little obstacle to the movement of electrons; consequently, they are said to exhibit low electrical **resistance**. Other materials, such as rubber or carbon, offer a great resistance. Actually, resistance depends on the size and shape of the material as well as on which material it is. *Resistivity* characterizes the resistive properties of materials in a way depending only on the material. All copper has the same resistivity, but a 2 m long copper wire has twice the resistance of a 1 m long copper wire of the same diameter.

Almost all materials exhibit some resistance to current flow. A few materials, called *superconductors*, show no measurable resistance whatsoever at very low temperatures (lower than −160 °C). These are discussed in more detail in Chapter 25.

We can easily understand why most materials have resistance to the flow of current. Electrons in a wire across which there is a potential difference are accelerated. Before they can move too far, however, they slam into one of the atoms making up the wire, transferring some kinetic energy to it, making it vibrate more. Then, the electron, which still feels the potential, starts accelerating again. Again, it collides with a wire atom. This will happen countless times as an electron wends its way through the wire. Thus, the wire's thermal energy increases; it heats up. Some materials stop the electrons efficiently and so offer high resistance. The coil of an ordinary toaster is made of such material: It stops the electrons so well that the wire glows red hot and toasts the bread. The current, a flow of charge, is proportional to the electrons' mean speed of progression (*drift speed*) through the wire. Incidentally, more complete analysis shows that the drift speed is slow, typically only tenths of millimeters per second.

Ohm's Law

Georg Ohm discovered that it was possible, in many cases, in electric circuits to define a constant resistance in terms of potential difference (voltage) and current. If we use the symbol I for current, and the symbol R for resistance, the relation (called **Ohm's law**) reads

$$V = IR.$$

This resistance to flow of current can be quantified by the following definition: the resistance offered a 1 A current between two points that differ in potential by 1 V is defined as 1 ohm (symbolized by Ω, capital Greek omega; $1\ \Omega = 1\ \text{V/A}$). For a circuit with a potential difference (voltage) of 5.00 V and a current 1.00 A, of the resistance must be 5.00 Ω, for example. A current of 10.0 A moving through a resistance of 100 Ω drops 1000 V in potential.

Recall from Chapter 3 that power is the work done per unit time. Because the potential is the work done per unit charge, and the current is the charge moving per unit time, the product of potential (work/charge) and current (charge/time) (VI) gives the amount of power (work/time) used in the circuit. That is, in a time t, an amount of energy VIt goes into the circuit. We write

$$P = VI.$$

The unit of power is the watt (1 W = 1 J/s), so volts times amperes must equal watts. Suppose an electric drill connected to a 120 V source draws 2.0 A of current when in use; then, the power being applied by the drill bit to the surface is

$$P = (120\,V) \times (2.0\text{ A}) = 240\text{ W}.$$

If Ohm's law holds, $V = IR$, and therefore,

$$P = I^2R.$$

For this case, the entire power supplied to the circuit is dissipated as thermal energy that is transferred from the resistor to the environment, and we speak of **Joule heating** (again, named for James Prescott Joule). Fuses (or circuit breakers) use this Joule heating to prevent circuits from drawing too much current for the wiring. If too great a current travels through the device, a wire melts from heating inside the fuse, breaking the circuit (see **Extension 4.3, *Using Ohm's Law***).

4.3   MAGNETISM

As children, we all played with little magnets. We found we could exert forces and pick up some sorts of metals. We also saw that the end labeled N (north) repelled

FIGURE 4.4

(a) The tiny domains (here shown magnified by a great factor) have magnetizations that point in random directions. (b) Here the magnetization in each domain is pointing in nearly the same direction, causing the magnetic field shown to the right.

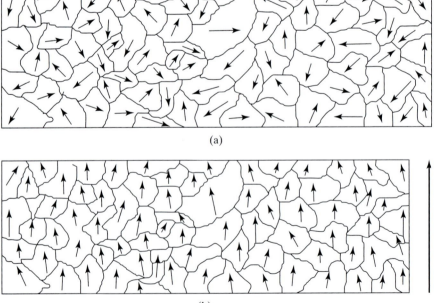

(a)

(b)

other *N*s and attracted the end labeled *S* (south), and likewise, that *S*s repel one another.

While magnets have been used for centuries as compasses, the modern uses of magnets and magnetic forces date to Christian Oersted's discovery in 1819 that a wire carrying an electric current can act as a magnet. The greater the current used, the stronger was the magnet created. This was the beginning of the recognition that electricity and magnetism are aspects of the same phenomenon. Since J. Clerk Maxwell's 1873 work on electrodynamics, we have known that electric and magnetic forces are just different aspects of the same phenomenon, electromagnetism. Magnets exert forces on moving charges. Moving charges (electric currents) produce magnetic fields.

At the microscopic level, circulating electrons in atoms can cause magnetic fields, as can circulating protons within the nucleus. In materials known as ferromagnets—iron or chromium, for example—these magnetic fields line up in the same direction. Many materials, such as aluminum, are not ferromagnetic.

We all know that ordinary iron is not a magnet. This is because the lining up of **magnetic fields** involves small volumes of the material (known as **domains**). A piece of iron of a size that we can see has many domains; in each domain, the magnetic fields line up. We can think of a domain as a little magnet. An ordinary piece of iron is made up of myriads of little magnets oriented at random (Figure 4.4a), so the effects of the little magnets cancel, and the iron is not a magnet. In the presence of an external magnetic field, the little magnets are forced to point in the same direction (Figure 4.4b). The piece of iron becomes a big magnet.

An electromagnet is constructed in this way: We wrap a wire around a piece of iron. The wires are connected to a battery, so that a current flows in the wires. The current produces a magnetic field. The magnetic field causes the domains to line up, and presto—we have a magnet. When the current is turned off, the piece of iron will return to its random domain orientation. The amplification of magnetic field possible in electromagnets is very large. Electromagnets are used to pick up entire cars in junkyards.

In permanent magnets, the domains are lined up. This can be done in a number of ways. The rocks of Magnesia in Asia Minor (from which the term *magnet* is derived) became magnets by being heated inside Earth and then cooled. Earth's magnetic field caused the domains in the iron-rich rock to line up. Another way to make a magnet is to subject a piece of iron to a strong magnetic field. When the field is turned off again, there is a remnant ordering of the domains, and the iron becomes a permanent magnet. If I could very carefully saw such an iron bar in half, perpendicular to the direction in which the magnet points, I would get two magnets, head to tail, each pointing in the same direction. If they remain in this orientation, they will attract one another and reconstitute the original magnet. If I saw the iron bar in half, parallel to the direction of the magnet, I get two magnets, side by side, each pointing in the same direction. These magnets will not attract one another. Will they repel one another?

Because an electric current is able to create something that acts like a magnet, it was inevitable that someone would try to see if a magnet can create a current. In 1831, Michael Faraday, the English physicist, found that a magnet moving through a wire coil will start a current flowing in the wire (see Figure 4.5b). Faraday called this **electromagnetic induction**. When the magnet is moved in one way, a current flows as shown in a galvanometer, similar to the one in Figure 4.5. The galvanometer is a device with a moving pointer; the pointer moves in response to current flow in such a way that a greater current causes a

(a)

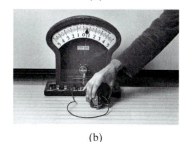

(b)

(c)

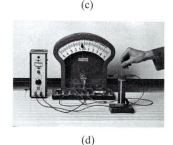

(d)

FIGURE 4.5

Magnetic induction. (a) A power supply causes a galvanometer to deflect. (b) A magnet moving through a loop of wire causes a galvanometer to deflect. (c) A moving loop of wire in the vicinity of a magnet causes a galvanometer to deflect. (d) Moving a piece of iron through a current loop causes a galvanometer to deflect.

FIGURE 4.6

Iron filings line up on a glass plate because of the magnet lying under the plate.

greater deflection (Figure 4.5a). If the magnet stops moving inside the coil, the current stops. When the magnet is withdrawn, a current flows in the opposite direction from that originally exhibited. Thus, one key to generating a current by using a magnet is to have the magnet move.

We can also generate a current by moving the coil while we keep the magnet stationary (Figure 4.5c). It is necessary only that the magnet and coil move with respect to each other to start the flow of current. (The orientation of the magnet and the coil are important as well, as will be discussed later.)

In the region about a magnet, there is some effect of the magnet. If we move a permanent magnet near a nail, the nail will be attracted to the magnet even if it does not touch it. If iron filings are placed on a piece of paper and a magnet is brought near it, the filings rearrange themselves in patterns such as that shown in Figure 4.6. It is as if there were lines in space connecting one end of a magnet to the other: north pole to south pole. Those lines would be there even without the filings to show us that they are there. They are results of the magnet. The abstract lines we imagine to run through space, we call *lines of magnetic field*. Outside a single magnet, they run from north pole to south pole. If we were to put two magnets south pole to north pole, we would expect magnetic field lines to go straight from the north pole to the south pole (with perhaps a few straying off if the magnets are not the same strength). For magnets of equal strength, there is a region of uniform magnetic field between the poles.

Now consider a coil put into such a region of uniform field. As we've mentioned, the orientation matters. If it is inserted as in Figure 4.7a, with the plane of the coil parallel to the faces of the magnets, magnetic field lines penetrate the coil. If the coil is as arranged in Figure 4.7b, with the plane of the coil perpendicular to the magnet faces, there is no magnetic field line to penetrate the coil. If we could turn the coil from parallel to perpendicular, we could change the amount of magnetic field cut by the plane of the coil. But, we know that when the magnetic field inside the coil changes, a current is produced. We can obtain a current simply by turning the coil.

This is a simplified description of an electrical generator. In an electrical generating facility, the coil between the magnets is turned by water falling through a turbine, or by steam from a boiler moving through a turbine. The device is then

FIGURE 4.7

The generator consists of a coil, which can be made to rotate, and magnets, which are fixed in place. (a) A coil in a magnetic field with the coil plane parallel to the magnet faces. (b) A coil in a magnetic field with the coil plane perpendicular to the magnet faces.

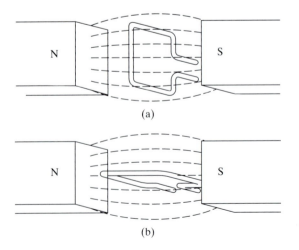

called a **generator**. As the coil rotates, the amount of magnetic field through the coil changes very little at first, as the coil moves from parallel configuration. The change grows faster as the coil approaches the perpendicular configuration. As the coil moves on past the perpendicular, the change grows slower again until the coil is in the parallel configuration again, when the change gradually stops. As it moves through the parallel configuration, the coil is moving in an opposite sense to that originally moved, so that the change continues through zero into negative values. The current generated in a complete cycle is shown in Figure 4.8. As the coil rotates, the charge moves back and forth, so this kind of current is called *alternating current* (AC); the kind of steady-flow current we considered previously is called *direct current* (DC).

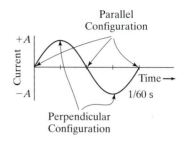

FIGURE 4.8

Alternating electric current resulting from coil rotation in a fixed magnetic field is plotted against time. One complete cycle takes 1/60 s, for North American generators; it takes 1/50 s for one complete cycle in Europe and much of the rest of the world.

Electric Motors

The same principle, worked in turnabout fashion, can produce rotation from the flow of electric current. Consider a wire in a magnetic field. As we have seen, as long as the wire is stationary, the charges in it are stationary, and the magnetic field does not change, so there is no magnetic force. Magnetic forces occur only when there is motion of charges in regions of magnetic field. A current flowing in the wire immersed in a magnetic field will cause a magnetic force to be exerted on the moving charges in the wire (that is, on the wire itself). Another way of saying this is that a current flow will create a magnetic field in the wire that interacts with the external magnetic field, and a force is exerted (see **Extension 4.4**, ***The Squirrel-Cage Motor***).

The most common AC induction motor achieves rotary motion by setting up two alternating magnetic fields at right angles to one another. An **induction motor** we all know is the watthour meter on each house that is read by the electric company every month. It consists of a disk that rotates; the disk is connected through gears to a counter mechanism similar to the odometer in a car, which tracks the total number of rotations of the disk. The motor is fed a small current proportional to the line voltage, and the magnetic field is set up by coils fed from the line current. As a result, the rotation speed of the disk is proportional to the product of current and voltage, that is, to the power. The total amount of rotation is the rotation speed times the time, which gives the total number of rotations of the disk. We, therefore, *pay* for power times time, or energy.

There are also direct-current electric motors. The principle of DC operation is that current-carrying wires in a region of magnetic field will feel a force. In the DC motor, there is a pair of magnets and a series of wires wound about the central cylinder connected to the voltage source in such a way that, as the cylinder rotates, the current can change direction (through a device called a commutator), as illustrated in Figure 4.9. Note that it is always necessary to have two or more magnetic fields to have an electric motor.

Transmission of Electric Energy

The output of generating plants is run into transmission lines for distribution to schools, homes, and businesses. In all these lines, energy is lost to resistive, or Joule, heating. Long-distance transmission lines are of high voltage so that Joule heating can be minimized. A generating plant produces a certain amount of power, given by IV, where I is the current generated in the plant, and V is the

FIGURE 4.9

Diagram of a direct current electric motor showing the two sides of the commutator, and the connections to the battery. The radial lines with the arrows are parts of vertically-oriented coils that "cut" the magnetic field.

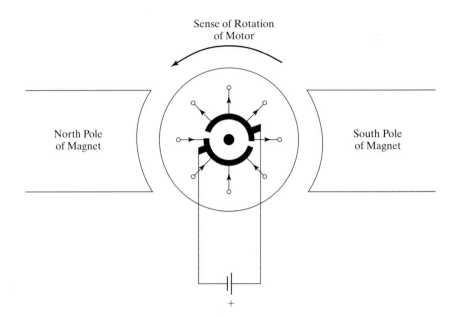

potential difference between the ends of the wire. (It is actually somewhat more complicated for AC than for DC, but our conclusions will be the same, so we will use the DC calculation.) Generator output at power plants is typically about 12,500 V. The power output of the plant must all be transmitted. The relation

$$P = IV$$

tells us that we may change the current and voltage of the transmitted electric energy together, as long as the product of current and voltage, the power generated at this plant, is a constant:

$$P_{\text{power plant}} = I_{\text{power plant}} V_{\text{power plant}}$$
$$= I_{\text{transmission line}} V_{\text{transmission line}}.$$

Because the transmission line has a certain fixed resistance, R, we may minimize Joule heating losses (I^2R) in the transmission wire by minimizing the current that the wire carries. This implies that the voltage of the transmission line should be very high, because the power transmitted is constant. Then, Joule heat dissipated in transmission

$$P_{\text{Joule heat}} = (I_{\text{transmission line}})^2 R$$
$$= (P_{\text{power plant}}/V_{\text{transmission line}})^2 R.$$

So, the higher the voltage, the lower the current, and the smaller the Joule heating losses. Typically, transmission lines carry electricity at voltages of 374 kV to 750 kV. When the destination is reached, the distribution takes place at around 22 kV to local substations, where it is typically lowered to 240 V for delivery from the local utility. The input is connected to give two 120 V circuits (except for ranges and dryers, usually run at 240 V). Overall, about 10% of generated electric energy is lost to Joule heat in the transmission process. Only about 2% is lost in long-distance (city-to-city) high-voltage transmission; the remaining 8%

of the energy is lost locally due to the much higher losses at the very low distribution voltage of transmission to homes and businesses.

Transformers

The voltage is raised from that generated at the power plant by a device called a step-up transformer (defined below). In your home, you use low-voltage power, because it is safer. The electric company transmits (locally) at a much lower voltage than that used in the long-distance transmission line. This is accomplished by use of a step-down transformer, which does the opposite of the step-up transformer.

Suppose we do an experiment in which we wrap a wire many times around a length of iron rod. If the ends of the wire are connected to the wall socket, an alternating current flows in the wire. Because the current changes with time, it causes a changing magnetic field in the rod. A wire wound once around the rod will enclose a changing magnetic field, inducing a current to flow in the wire and, therefore, inducing a voltage across the wire's ends. Suppose that voltage has a value of 0.10 V. If the wire is wound twice around the rod, each loop will have a voltage 0.10 V induced; because there are two loops, the net voltage across the wire's ends is 0.20 V. Likewise, wrapping the wire eight times around the rod will produce 0.10 V in each loop, or 0.80 V overall. A different wire wrapped around the same rod would also have a potential difference of 0.10 V for each loop wound around the rod.

The **transformer** consists of two separate unconnected wires wound around a common core, usually of iron. One of the wires is wound around the core many times; the other has fewer turns of wire. An AC current in one of the wires causes a magnetic field (which changes as the current changes) in the iron. The changing magnetic field in the iron, in turn, induces an AC current in the other coil of loops.

The voltage in any loop wrapped around an iron core will be the same, whether it is a loop on the input, or primary, coil or a loop on the output, or secondary, coil. This means that the voltage on the primary coil is divided among the loops. (If there are 100 loops on the primary coil, each loop will have a voltage $\frac{1}{100}$ of the input voltage.) If the number of loops on the primary and secondary coils is the same, then the device makes no change in the voltage across the device. If the number of loops on the primary is greater than the number on the secondary, the voltage on the secondary will be lower (that is, the voltage will be stepped down). If the number of loops on the primary is smaller than the number on the secondary, the voltage on the secondary will be higher than that on the primary (it will be stepped up). This phenomenon can occur only with AC. For DC, a magnetic field will be created, but the magnetic field does not change, and consequently, no current is induced in the secondary coil.

The transformer is a clever device (and is probably the reason that Westinghouse's advocacy of AC bested Edison's of DC). It provides another example of conservation of energy. In the ideal case, the amount of energy going into a transformer in a given amount of time is equal to the amount coming from the transformer in the same time. Of course, in real life, some energy is lost to Joule heating in the transformer's iron core, but we can say that, to a good approximation,

$$(VI)_{\text{output}} = (VI)_{\text{input}}.$$

Later development of technology (the inverter, for example, which electronically converts DC to AC) allowed the use of direct current transmission lines. They are less lossy than AC transmission lines, but the large capital expense involved in the inverters and rectifiers means that only the longest transmission lines can be competitive economically with AC.[1] The longest U.S. DC transmission line is 1361 km long and runs between The Dalles, Oregon, and Los Angeles.[1]

Power transmission is often accomplished by constructing large pylons to carry the wires in the air (Figure 4.10). The pylons are built in this manner because air acts as an insulator. Thus, the bare wires are protected from breakdown by the air, because air is a good insulator (poor conductor). If the wires were to be buried underground, we would need to wrap them in large amounts of glass and plastic to protect them from breakdown. This alternative is obviously much more expensive because of the cost of excavation and insulation. There are problems with aging of transmission lines and the high amount of current brought about by wheeling and deregulation. The August 2003 blackout that hit the East and Midwest was a sign of these troubles (refer to **Extension 4.5, *The 2003 Blackout*** for more details.).

When the transmission lines enter your house, they go through a watthour meter, the induction motor described earlier. The meter is geared; the turns of the disk are recorded on dials, one for each power of 10 recorded. Because the stored information is the number of turns of the disk, what has been recorded is *energy* use. The disk speed is proportional to power consumption, but you do not pay for the rate of energy use. You pay for the *total amount of energy used*.

In your home, the transmitted energy is finally transformed into radiant energy, thermal energy, mechanical energy, and so on. Motors run from electricity by running induction in reverse. In Chapters 7 through 9, we discuss where this energy goes.

FIGURE 4.10
Transmission lines in rural Ohio.

SUMMARY

Charged objects can exert forces on other charged objects. Charge comes in two forms: positive and negative. Two positive charges repel each other; two negative charges repel each other; positive charges attract and are attracted by negative charges.

Because charges exert forces, and objects move, work can be done by them. The amount of electrical work done per unit charge is a measure of the electrical potential energy per unit charge, otherwise known as the potential, or the voltage. A voltage can cause charges to flow—that is, cause a current. Ohm's law states that current is proportional to voltage. The proportionality constant connecting them is called the resistance: $V = IR$.

The electric power produced by a source having potential V and delivering current I is $P = IV$. This is also the power delivered to something drawing current I at voltage V. In the latter case, if the object is a simple resistance (not a motor), then $V = IR$, so that $P = I^2R = V^2/R$. This produces heating in the resistor, which is called Joule heat.

Magnets also exert forces on other magnets or on materials such as iron. The magnets arise at the nuclear and atomic levels from the circulation of currents. A current produces a magnetic field. Contrariwise, a current can be made by moving a magnet near a coil of wire, or by moving a coil of wire around a stationary magnet. That is, the changing magnetic fields cause charges to move.

These principles allow for the construction of electric generators, in which a coil rotating in a magnetic field produces a current by induction, converting kinetic energy of rotation into the energy of moving electric charges; electromagnets, in which a current loop surrounds a piece of iron and makes the domains in the iron line up to produce a strong magnet; and electric motors, converting the electric energy into kinetic energy and useful work.

Transmission lines operate at high voltage to minimize Joule heating. Voltages may be increased by using a device called a step-up transformer and may be decreased by using a step-down transformer. Transformers work only with AC; this is another example of the utility of magnetic induction.

PROBLEMS AND QUESTIONS

MULTIPLE CHOICE REVIEW

1. Charge comes in pieces of size
    a. 6×10^{18} C.
    b. $\frac{1}{3}$ C.
    c. $\frac{1}{3} \times 1.6 \times 10^{-19}$ C.
    d. 1.6×10^{-19} C.
    e. not given by any other answer.

2. Most atoms are
    a. electrically charged.
    b. uncharged but made up of electrically charged pieces.
    c. unitary things that cannot be broken up in any way.
    d. sources of easily removed protons.
    e. sources of easily removed neutrons.

3. Electric forces are
    a. the same as electric charges.
    b. exerted between all objects.
    c. exerted between any charged objects.
    d. exerted between pieces of tape only.
    e. not exerted except between manmade objects.

4. Point A in space is at potential 20.0 V, while point B is at potential 100 V. How much work will have to be done to move a charge of 0.0050 C from A to B?
    a. -80 J
    b. -0.40 J
    c. 0 J

    d. 0.40 J
    e. 80 J

5. A constant electric force of magnitude 5.00 N acts on a charge of 0.250 C, while the charge is moved a distance of 2.50 m along the direction of the force. How much electrical work is done?
    a. 0.0 J
    b. 0.500 J
    c. 1.25 J
    d. 12.5 J
    e. 50.0 J

6. A 100 W light bulb is plugged into a 120 V DC outlet and turned on to produce light and heat. What is the resistance of the filament in the bulb?
    a. 0.007 Ω
    b. 0.833 Ω
    c. 1.20 Ω
    d. 100 Ω
    e. 144 Ω

7. A 100 W light bulb draws 0.833 A of current when plugged into an outlet. What is the voltage of the outlet?
    a. 0.0083 V
    b. 83.3 V
    c. 100 V
    d. 120 V
    e. 240 V

8. Magnets may be found only
   a. in natural domains in the environment where there is magnetite.
   b. in pieces of all metals.
   c. in domains.
   d. where there are electric outlets.
   e. where there are aligned domains in a material.
9. A DC current of size 10 A flows through one coil of a transformer. What is the current flowing through the other coil of the transformer if that second coil has 10 times as many loops as the first?
   a. 0 A
   b. 1.0 A
   c. 10 A
   d. 100 A
   e. 1000 A
10. A 250 MW power plant sends its power out on a high-voltage line at 50,000 V. If the total resistance of the line is 1.00 Ω, then what is the power lost in transmission due to Joule heating?
   a. 125×10^6 W
   b. 25×10^6 W
   c. 5×10^4 W
   d. 5×10^3 W
   e. 0.0025 W

SHORT ANSWER

Explain whether the statement made is correct, and justify your answer.

11. Charge must come in specific-sized bundles. Only integer multiples of this smallest amount of charge can be found in Nature.
12. The resistance of a longer piece of wire is greater than the resistance of a shorter piece of wire.
13. A circuit breaker connects two parts of a circuit if the current flowing in the circuit is too large.
14. Only negative charge travels as electric current in our electric system.
15. Ohm's law is one of the fundamental principles of physics.
16. Electromagnets work by aligning little regions of magnetic field in iron so they mostly point in the same direction. This produces a greatly strengthened magnetic field.
17. If a normal D-cell battery is used in a device, for every 1.0 C of charge that moves through the battery, 1.5 J of work is done on the charge.
18. Claude the Magician carefully cuts a bar magnet in half. Afterwards, Claude will have two magnets, whether he cuts the magnets lengthwise or crosswise.
19. A generator works by having a spinning turbine impart a spinning motion to the charge carriers in the electrical system, thereby giving them additional energy.
20. The equation $P = IV$ is true only if Ohm's law is valid.

DISCUSSION QUESTIONS

21. Why is the wire connecting your dryer to the circuit breaker in your house thicker (of greater diameter) than the wire connecting a table lamp to the electric outlet?
22. A plastic hose stuffed more or less densely with steel wool and a source of water can act as a model for resistance and electric current in wires. Suppose two 1m lengths of such a hose were stuffed identically with steel wool. Discuss how one could use this model to determine which of the following would have greater resistance: one length; two lengths connected one after another; or two lengths connected side by side, both connected to one source of water. Speculate on the answers as best you are able.
23. In the classical Drude model of electricity, resistance occurs because the charge carriers, which are accelerated due to electric forces, collide with the atoms of the metal wire, causing them to bounce backward. What should you expect the Drude model to predict for the relative resistance of two materials, one of which has twice as many atoms in each cubic meter of material?
24. A visual physical model for Ohm's law uses a wood board with nails in it and metal balls. The nails in a board represent a fixed amount of electric resistance in a wire. As the metal balls roll through the nail obstacle course, they run into nails and bounce in various directions. The balls represent the charge carriers in a current; the greater the speed of the balls running through the nails, the greater the current represented. The electric potential energy difference between the ends of the wire is represented by the tilt of the board, because a greater tilt represents a greater difference in gravitational potential energy. Explain how this analogy works.
25. Recently, superconducting wire was made using magnesium diboride [S. Jin, H. Mavoori, C. Bower, and R. B. Van Dover, "High critical currents in iron-clad superconducting MgB$_2$ wires," *Nature* **411**, 563 (2001)]. This is assumed to hold the promise of cheap, efficient energy transmission. Explain how this might be true.
26. How can we know that energy is conserved in a transformer? Explain.
27. Your friend Alf tells you that the magnetic force on a stationary 10 C charge near the north pole of a magnet having a magnetic field of 0.10 tesla is twice as great as that on a stationary 5.0 C charge at exactly the same position. Discuss this with your friend, and explain how he is correct or incorrect.
28. If a watt is a unit of power, how can a kilowatthour be a measure of energy? Explain.
29. What would happen if a transformer had the same number of loops on each side?
30. Consider a step-up transformer (the output voltage is greater than the input voltage). What happens to the current in the output side compared to the input side?

PROBLEMS

31. Two charged pith balls are hanging from a common point. Both have the same sign charge, but one is 2.0 mC, while the other is 8.0 mC. Draw a labeled picture showing the positions of the charges. How does the force on the 8.0 mC charge compare to that on the 2.0 mC charge? How would this differ from the case for which both charges are equal at 4.0 mC? Explain.
32. In the tape experiment described in the section on electric force, let us suppose that you and a friend measure a hang angle of 30° when your fingers hold the tops of two *A* pieces 3.0 cm apart. How will the hang angle change when your fingers are moved to 6.0 cm separation? Will

it be half or twice the original hang angle or some different value?

33. In the tape experiment described in the section on electric force, let us suppose that you and a friend measure a hang angle of 30° when your fingers hold the tops of one *A* and one *B* piece 3.0 cm apart. [The hang angle is the sum of the angles between the A piece and the vertical and the B piece and the vertical.] How will the hang angle change when your fingers are moved to 1.5 cm separation? Will it be half or twice the original hang angle or some different value?

34. In a certain region of space, the electric force is 35.0 N to the right along the *x*-axis. The charge is originally at a position given by $x = 20.0$ cm, $y = 20.0$ cm. The charge is moved to a new position, position 1, at $x_1 = 2.20$ m, $y_1 = 20.0$ cm. How much work has been done? What is the electric potential energy at point 1 if it was 25.0 V at the original position? Suppose instead that the charge had been moved to point 2, at $x_1 = 20.0$ cm, $y_1 = 2.20$ m. How much work has been done? What is the electric potential energy at point 1 if it was 25.0 V at the original position? Explain.

35. In a certain region of space, point *A* is 100 V higher in potential than point *B*. If we move a +2 C charge from point *A* to point *B*, then how much work will we do?

36. A flashlight bulb operates at 3.0 V and has a power rating of 4.5 W. How much current does this bulb draw from the battery?

37. There are 17 million electric heaters in use in the United States. Electric heaters each use a yearly average of 176 kWh of energy. If the cost of electricity is 10 cents per kWh, what is the total yearly amount of money spent to operate electric heaters in the United States? If the average heater is 1200 W, how many hours, on average, are the heaters used every year? How high could the average energy use for a heater lift a person of mass 80.0 kg if it were used to do that instead? What is the resistance of the average heater? (You may assume that the wall outlet operates at 120 V.) How much current does this average heater draw?

38. There are approximately 600 million televisions presently in use in the United States. About half are the energy-saver models that stay on at a lower power level after they are turned off. Each of these TVs draws 450 W of power when on and 60 W when in the "off" state. If the cost of electricity is 10 cents per kWh, and the TV is typically watched 4 h a day, what is the total yearly amount of money spent to keep all the TVs "ready for use" in the United States? How does that compare with the cost of watching them? What is the total cost of TV watching in the United States? What is the current used when the TV is "on" and when it is "off"? How do the total energy costs of the always-on models compare to the costs of the older models?

39. A 750 MW power plant sends its power out on a 365,000 V high-voltage line. If the total resistance of the line is 0.60 Ω, what percentage of the power is lost due to Joule heating?

40. Consider a step-up transformer that takes 120 V to 960 V. If the input current is 3.0 A, what is the current on the output side?

41. A 250 MW power plant sends its power out on a high-voltage line at 50,000 V. What total current is produced by the plant's generators? How would this compare to sending the power out on a 500,000 V high-tension line? What if it were at 5000 V instead? How would the Joule heating at 50 kV compare to that at 500 kV? 5 kV? Explain.

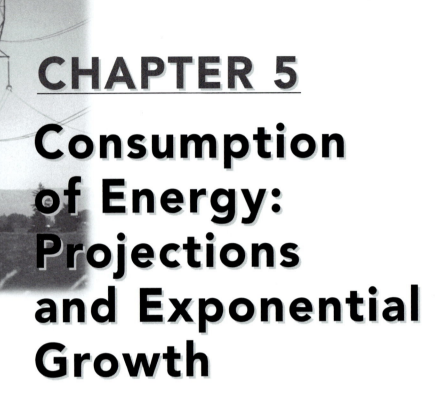

CHAPTER 5

Consumption of Energy: Projections and Exponential Growth

This chapter and the next focus on specific details of energy generation and distribution systems. Electricity is one of the most common forms of energy we encounter in our everyday lives. In this chapter, we consider some means that people have developed to predict the future: future population, future energy use, and future electricity demand. The most important modes of thinking about the future are linear extrapolation and exponential extrapolation. Various extrapolation methods and the consequences of using each method are discussed. Each extrapolation method has drawbacks.

KEY TERMS

linear projection / exponential growth / growth rate / doubling time / geometric progression / logarithmic scale / price elasticity / logistic curve

Immense expansion in the use of energy, especially electricity, and huge increases in the size of generating facilities characterize our century. The utilities loom as the fastest-growing energy users.[1,2] Electric power consumption in the United States grew at a rate of 7% annually from 1961 to 1965, at 5.6% annually from 1965 to 1969, and at 9.25% in 1970,[1] while the population grew by only about 1.6% each year. This trend did not continue through the 1970s or 1980s; two successive energy crises caused a falloff in consumption, first after 1973 and again in 1979 (Figures 5.1 and 5.2).[3] Though energy use rose again in the 1990s because energy was so cheap, it did not rise as spectacularly as it did in the 1960s.

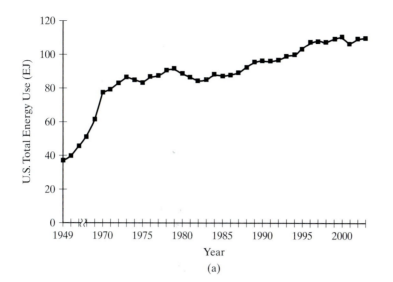

(a)

FIGURE 5.1

(a) Electricity generation from 1949 to 2003. (b) Component growth in electricity generation, 1949 to 2000. (U.S. Department of Energy, *Annual Energy Review 2003* (Washington, DC: GPO, 2004) August 2004)

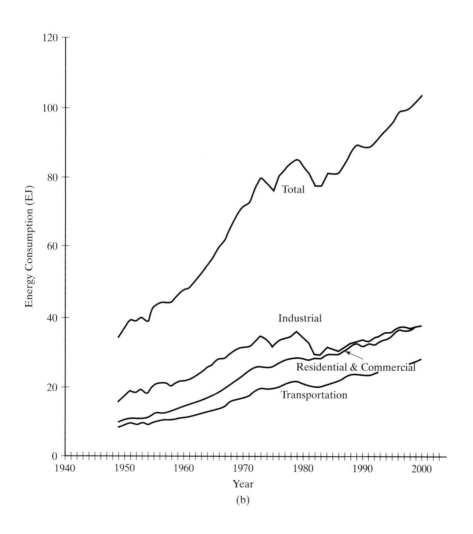

(b)

FIGURE 5.2

Energy use by source.
(U.S. Department of Energy, *Annual Energy Review 2000* (Washington, DC: GPO, 2001), August 2001)

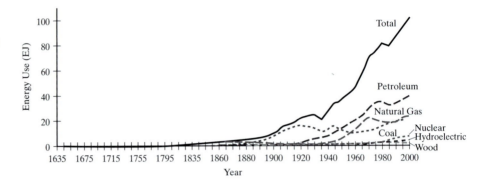

5.1   MAKING PROJECTIONS

A legitimate question to ask when presented with the current world energy picture is: What will happen to energy demand in the near future? There are many answers. By the last decade of the last century, many pundits were telling us what would happen (or perhaps more modestly, what they predicted would happen) by its end. See **Extension 5.1, *Predicting 2000*** for more detail. The U.S. Department of Energy has predicted annual energy (Table 5.1) use out to the year 2030 (at between 102 and 147 EJ).[4]

There were a rash of predictions, with outcomes between 54 EJ to 468 EJ (alternatively, between 15 and 130 trillion kWh, or 57 and 494 quadrillion Btu, or quads). With actual energy use in the year 2000 of 104 EJ (28.9 trillion kWh = 110 quads), we can see that some of the predictions were more successful than others. It is useful and important to know how people go about making

TABLE 5.1

U.S. total energy use in exajoules and trillion kWh
(1 trillion kWh = 10^{12} kWh = 3.6×10^{18} J = 3.6 EJ)*

Year	Actual use (EJ)	Actual use (10^{12} kWh)
1950	34.2	9.5
1960	45.36	12.6
1970	68.76	19.1
1980	78.84	21.9
1990	84.24	23.4
2000	104.04	28.9
2010[b]	118.44	32.9
2020[b]	135.36	37.6
2030[b]	101.88–146.88	28.3–40.8

[a] Because the kilowatthour is treated as a unit in its own right in most industry and government publications, we apply the prefix, in many cases, to kWh. Note that a strict interpretation of scientific notation would be PWh instead of trillion kilowatthours. The "quad," short for quadrillion Btu, is often used in the United States when presenting energy data; it is just about 1 EJ (1 quad = 1.055 EJ).

[b] Projections, not actual use, are given for 2010, 2020, 2030.

Source: Energy Information Agency, U.S. Department of Energy.

such predictions about energy. Before we can evaluate the plausibility of others' predictions, we need to know how to make our own.

Linear Projections

If a given quantity is fixed in value, that means it will not change. Most quantities of interest to us—money in the bank, for example, or the amount of beer sold—change. Suppose we observe a change in a quantity during some given time interval. For instance, suppose that in 1997, 500 million pizzas were sold nationwide, and that in 1998, 506 million pizzas were sold nationwide. We might assume that in 1999 (and even onward), the increase in the number of pizzas sold would be 6 million. In 2010, we would predict, 578 million pizzas would be sold. This is an example of **linear projection**. Figure 5.3 shows such a projection. This linear projection is conceptually the simplest way to predict the future. People who make such extrapolations are said to be using "linear thinking."

Projections, however, cannot be made in a vacuum. Outside information is often relevant in evaluating a projection. In Table 5.2, for example, we see the Federal Power Commission's 1972 predictions of utilities' energy requirements.[5] Table 5.2 looks absurd in hindsight, as may be seen from the comparison with actual 2000 values (only coal was roughly correct). It was probably not realistic to expect half the energy generated in 1990 to come from nuclear installations, even from the perspective of 1972. By that time, there had been a consistent trickle of antinuclear referenda on state ballots, which started in the late 1960s, long before the incident at Three Mile Island (near Harrisburg, Pennsylvania) in 1979 galvanized opposition to nuclear energy, or the disaster at Chernobyl in 1986 convinced even more people of the hazards of nuclear energy. Concern about greenhouse warming (Chapter 17) may lead to a renaissance for nuclear energy. In this chapter, we consider how to address some of these concerns and try to identify ways to do better projections in the future (because we need to project demand for many things).

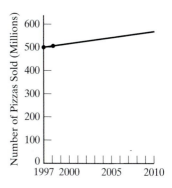

FIGURE 5.3

The number of pizzas sold illustrates linear growth in this hypothetical example.

TABLE 5.2

FPC[a] prediction from 1972 of utilities' energy requirements/production (in trillion kWh)

	1970	1975	1980	1985	1990	2000	2005	2010	2015	2020
Fossil										
Coal	0.75	0.99	1.16	1.38	1.61	1.94[b]				
Gas	0.40	0.42	0.39	0.40	0.43	0.52[b]				
Oil	0.21	0.35	0.40	0.45	0.50	0.11[b]				
Subtotal	1.36	1.76	1.95	2.23	2.54	2.57[b]				
Hydro	0.28	0.29	0.32	0.35	0.37	0.27[b]				
Nuclear	0.02	0.24	0.96	1.73	3.10	0.75[b]				
Total	1.66	2.29	3.23	4.31	6.01	3.64[b]				
Actual[b,c]		1.918[b]	2.286[b]	2.470[b]	3.038[b]	3.802[b]	3.793[c]	4.055[d]	4.429[d]	4.811[d]

[a] As of 1977, the Federal Power Commission became part of the Department of Energy.

[b] Source for actual energy production is Department of Energy, *Annual Energy Review 2003* (Washington DC: GPO, 2004), DOE/EIA-0384(2003), Tables 8.2 and 8.3. Actual production in 2003, the latest available, was 3.848 billion kilowatthours.

[c] Source for projected energy production for 2005 is Office of Integrated Analysis and Forecasting, Energy Information Administration, U.S. Department of Energy, *International Energy Outlook 2002* (Washington, DC: GPO, 2002), DOE/EIA-0484(2002), Table 20.

[d] Source for projected energy production for 2010–2020 is Office of Integrated Analysis and Forecasting, Energy Information Administration, U.S. Department of Energy, *International Energy Outlook 2004* (Washington, DC: GPO, 2004), DOE/EIA-0484(2004), Table 14.

Exponential Projections

Linear projections often fail. One reason they fail is that many quantities grow or decay in a way that depends on their initial size. For example, when people catch a cold, they have been exposed to a virus, and the virus has found conditions in the body favoring its rapid growth. Viruses increase in number more rapidly, as there are more of them in the bloodstream. The number of viruses "explodes" and affects the sufferer with typical cold symptoms—sneezes, a runny nose, and so on. If it takes 1 d for the number of viruses in our body to double, then after a week, one invading virus becomes 128; after 2 weeks, 16,384; after 3 weeks, 2.1 million; after 4 weeks, 256 million; and so on. This sort of growth is known as **exponential** or **geometric growth**.

Many quantities in nature exhibit exponential growth or decay. In Chapter 2, we discussed the exponential growth of population (we use the population example again, shortly), and in Chapter 18 we will discuss radioactive decay. These are just two of many examples of the ubiquity of exponential effects: the number of things at any time determines the rate of growth or decay in that thing.

Projection of the demand for electricity (our main concern in this chapter) certainly depends on the growth of population, because the more people who can use electricity, the more is used. We would expect this projection to involve exponential growth.

Growth rate and **doubling time** are intimately related. Table 5.3 presents the doubling times for given growth rates. Bankers often use the "rule of 70"—the product of growth rate and doubling time is just about 70 for rates up to 10%. Note that an investment of money at 3% per year interest would double its original value in only 23.4 years. This table allows us to discover how long something will take to double if we know its current rate of increase (or we can use the rule if we're willing to accept an approximate answer). Thus, we can discover how long, at a certain compound interest rate, a bank will take to double the money we have deposited in it, or how long it will take a mosquito-infested lake to become doubly unbearable. As an example, if the mosquito population grows at a rate of 7% per week, the number of mosquitoes will double in a little over 10 weeks.

The growth rate is a very important aspect of growth, but how precise can we be about how long growth takes? We do not know the best way to judge what will

TABLE 5.3

Growth rates and doubling times (compounded annually)

Growth rate (%)	Doubling time	Growth rate (%)	Doubling time
0	∞		
1	69.7	14	5.29
2	35.0	16	4.67
3	23.4	18	4.19
4	17.7	20	3.80
5	14.2	30	2.64
6	11.9	40	2.06
7	10.2	50	1.71
8	9.01	75	1.24
9	8.04	100	1.00
10	7.27	200	0.631
11	6.64	500	0.387
12	6.12	1000	0.289

occur in the future. As we will see, the doubling time (or growth rate) in real-life situations can change. How do we decide what growth will occur?

Exponential change can be extremely surprising to people, because the final stages of growth seem to occur so rapidly. A jar that is a quarter full of fruit flies (doubling time, say, 3 days) and looks reasonably empty will be totally full 6 days from now. The first changes, though at the same growth rate, are perceived as slower (and thus nonthreatening to the environment). The final filling of the jar seems to happen entirely too rapidly, because the growth rate is applied to so much larger an amount.

Consider the example of growth in human population. For something that progresses geometrically, the new amount depends multiplicatively on the old amount. For a population, the number of children born (and, hence, the added population) depends on the number of women of childbearing age. The number of women of childbearing age is some proportion of the total population. The number of people dying also depends on the population size. We can, therefore, say that the change in population (call this ΔN) is proportional to the population (call this N). Any proportionality can be written as an equation. The equation resulting from the proportionality for population is

$$\Delta N = aN\Delta t.$$

Here, a is the net rate of addition to the population, assumed constant. This is the birth rate (number of children born per capita per unit time) minus the death rate (number of people dying per capita per unit time). The time interval involved in going from a population N to a population $N+\Delta N$ is Δt.

Let us assume temporarily that the birth and death rates are constant. For this case, the equation just given for ΔN has a solution for the population N as a function of time [$N(t)$]. If the population at some starting time ($t = 0$) is N_0, then

$$N(t) = N_0 e^{at}$$

or

$$N(t) = N_0 10^{ct},$$

where the constant c is related to a (see Appendix 1). Here, we consider a situation in which the power of 10 is not an integer but is a real number, ct, that changes with time. The properties of powers we discuss in Appendix 1 continue to hold. It is just that the power (of e or 10, respectively) is continuously changing (the exponential factor at changes linearly with time).

If we were to try to graph the function $N(t) = N_0 10^{ct}(N_0 e^{at})$ versus t on graph paper with one logarithmic scale and one linear scale (semilogarithmic graph paper), it would appear as a *straight line*. This follows, because the logarithm of 10^{ct}, which is ct, is a straight line when plotted against t. It is easier to extend a straight line with a ruler laid to a piece of paper than it is to extend an exponential curve.

People are relatively comfortable making projections once geometric growth is presented as a straight line on semilogarithmic graph paper, because it is possible to visualize the increase as a straight line. This feels comfortable, because it is linear (which is why linear thinking is so attractive). Several examples of geometric, or exponential, growth are given in **Extension 5.2,** *Exponential Growth,* and in the box, *A Closer Look:* Fable of the Boy and the King.

A CLOSER LOOK

Fable of the Boy and the King

There is an old story of a boy who played chess with the king and beat him. The king offered him his choice of payment; the boy chose to be paid on the chessboard in the following way. On the first day, the king would put one grain of wheat on the first square of the chessboard's 64 squares. On the second day, he would double the number of grains on the first square and place them on the second square. On the third day, the king would double the number of grains on the second square and place them on the third square. And so it was to go.

The king was pleased at the boy's generosity, for he imagined that he was getting away easily. As we saw in Chapter 2, however, if we double something 10 times, we multiply it by about a factor of 1000. This means that on the 64th day, the boy would be entitled to approximately 1.6×10^{19} grains of wheat. This is more than has been harvested in all human history.

Consider the example of the growth of the U.S. population and its projections. This will reinforce some important concepts about exponential growth. The population of the United States is measured by the Census Bureau every decade.[6] Figure 5.4 shows this historical growth. In the discussion of population doubling in Chapter 2, it was shown that the first measured doubling took about 20 years. This doubling time characterized population growth up until the time of the Civil War (1860s). By the turn of the century, the doubling time had reached 30 years. The last doubling has occurred from about 1925 to 1980—that is, 55 years. For the United States, as opposed to the world as a whole, the population growth rate has been dropping (with one notable "bump down," the depression, and one notable "bump up," the baby boom between about 1945 and 1962). For more information, see **Extension 5.3,** *U.S. Population*.

FIGURE 5.4

The U.S. population as recorded by the Census Bureau.

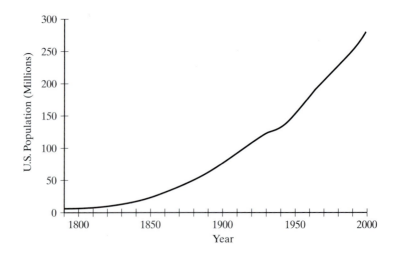

If we wish to project the U.S. population into the future, the simplest way is to continue the line in the graph of Figure 5.3. The trouble is that we really cannot judge very well which way the line should be extended. This is because of the characteristic steep rise that is occurring, what economist Thomas Malthus referred to as "**geometric progression**." We need a better way to make judgments about growth.

Using Semilog Graph Paper to Make Projections

In Figure 5.5, we replot the curve of Figure 5.3 on semilogarithmic graph paper. Note first that it is not a straight line. (If we were to extend the original growth of the 1790–1850 era by a straight line, the population in 1990 would have been predicted at about 1 billion.) This means that the net rate of population increase is declining for the United States. For any small segment of the curve, though, a straight line gives an adequate description. (In **Extension 5.3,** note also the projections of population increase indicated.) The projections follow from hypothesizing various straight-line increases from the graph. Projections are usually made, therefore, by graphing the past behavior of some quantity (say, demand for electricity) on semilog paper, then drawing the best possible straight line through the past behavior on into future behavior.

This is a plausible way to project short-term growth. The Edison Electric Institute in 1960 made a 10-year projection of electricity demand, predicting 1.31 trillion kWh demand;[8] the actual 1970 demand was 1.39 trillion kWh. This agreement of projection with reality is good; here, the straight-line method worked well.

Suppose that on the basis of the 1959–1964 figures for electricity demand (7.35% sales growth per year), you predicted the 1969 and 1970 sales by extending a straight line on semilog graph paper.[8] The straight line represents an annual increment of 7.35%. Six eventful years pass from 1964, with military buildup in Vietnam, a U.S. invasion of the Dominican Republic, a six day war in the Middle East, riots in the centers of many American cities, and so on. In 1970, you recall your predictions:

1969   1.28 trillion kWh
1970   1.37 trillion kWh

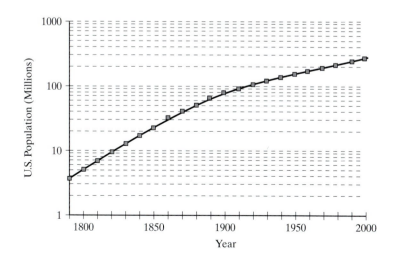

FIGURE 5.5

The U.S. population as recorded by the Census Bureau, as plotted on semilogarithmic paper.

Actual energy sales were 1.31 trillion kWh in 1969 and 1.39 trillion kWh in 1970. In 1972, you look at your prediction for 1971: 1.47 trillion kWh. This was the actual sales figure for 1971. It is this kind of experience, in all sorts of applications at all levels, that has reinforced the predominance of this method of prediction.

Lacking other input, this is an acceptable way to make projections, and we need to make them. However, we must be aware of the pitfalls involved in making projections when we are in the process of interpreting them. The greatest pitfall is the belief that the future will bring the same as the past, only more so. This is a reasonable enough belief for projections of a few years, but the premise can fail miserably when extended indefinitely (remember our already mentioned projection of a U.S. population of a billion for 1990). Circumstances that are assumed to remain steady change all too soon, vitiating even the best-informed predictions. This is true in longer-term projections, whether our predictions are linear or exponential.

If the **logarithmic scale** is used, the resulting graphs may be misleading. The "agreement" of the data with the prediction is emphasized in a plot on full logarithmic graph paper as compared to a linear scale. Individual points would appear closer to the line than they would with a linear plot because of the compression resulting from logarithmic rescaling. (A straight line on a graph on full logarithmic graph paper is a straight line on linear paper only in special circumstances.)

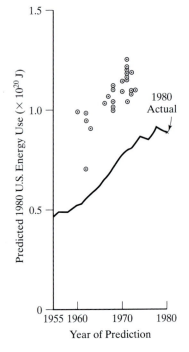

FIGURE 5.6

Predictions of energy use in 1980 as compared to actual production of energy are plotted in terms of the year the projection was made. (Adapted from Reference 5 by permission, Duxbury Press)

5.2 MAKING THE ACTUAL ELECTRIC CONSUMPTION PROJECTIONS

There is much uncertainty in long-term population growth projections, even for the United States, as we have just seen. As an example of the difficulty in making even short-term predictions, consider the sterling record of short-term predictions in the "dismal science," economics. An article[9] written in October 1993 pointed out that the economists' predictions of economic growth for the year 1993 (average prediction for the year: 3%), made starting in November 1992 (at ~2%), had varied by 0.8%, even during the year for which the prediction was being made. The greatest predicted growth rate was 3.3% in April 1993; the last rate quoted was 2.6% in September 1993. Who would want to bet the store on their predictions of economic growth in the year 2000 on the basis of how successful economists' predictions have been?

The conservation ethic has become more pervasive in the developed countries as a result of the energy crises of the 1970s. Sales of appliances that use large amounts of energy are not rising much, as we shall see in Chapter 9. For such reasons, projections that consist in drawing straight lines on graphs can lead to estimates that may be far too high. An illustration of how expert opinion can differ is given in Figure 5.6,[10] which shows a wide range of energy demand projections for 1980 made during the 1970s. (Few such predictions were made during the 1980s, as concern over energy use declined.)

The amount of energy sold by utilities and the rate of growth are historically large. One reason for the tremendous growth is that electricity, from World War II onward, had been a better buy (real price decreasing), as shown in Figure 5.7.[11] The trend of reduced real cost, which is shown on the graph to extend until about 1970, could not continue indefinitely; in fact, the real price rose slightly and then

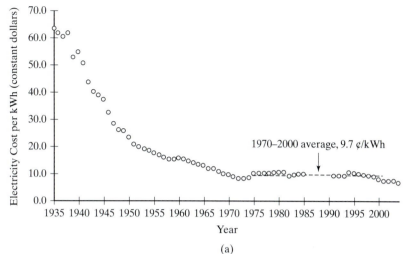

FIGURE 5.7

(a) The real (deflated) average cost of electricity in central Ohio, 1935–1985, using year 2000 dollars. (Data courtesy of Southern Ohio Electric Company through 1985, Aubrecht family average annual electricity cost thereafter)
(b) The real average national cost of electrical energy, 1960–2003. (U.S. Department of Energy, *Annual Energy Review 2003* (Washington, DC: GPO, 2004), August 2004, Fig. 8.10)

has held reasonably steady since about 1971. The costs of pollution control are mainly responsible for the price rise between 1970 and 1973 and are the reason the price has not fallen further. (U.S. utilities spent only 3.8% of their total capital budget for pollution control, as compared with the 10.3% budgeted by the iron and steel industry.)

Price Elasticity

The main cause of utility price increases is the swift rise in utilities' fuel prices after decades of little change. The historical picture of real fuel costs, shown in Figure 5.8,[10] verifies the extreme stability of fuel prices until 1973. The meteoric rise of adjusted fuel prices is astounding, and it did get the public's attention (at least for a while).

Real price is the most important determinant of demand;[8] there is substantial **price elasticity** (that is, price rises cause reduced demand, and price decreases cause increased demand) in electricity use. An elasticity of 1 would mean that a 1% price increase would cause a 1% consumption increase. A Canadian study[12] showed that energy price elasticity in Canada in 1984 was −0.3 to −0.6 (and about −0.7 for petroleum products). These negative numbers mean that a price increase causes a consumption decrease. Elasticities in the United States should be similar to those in Canada. Thus, raising the price of electricity should promote reduced use of energy.

Various studies of the effect of cost on use have led to different predictions for the amount of electricity to be used in the future. A "think tank" analysis by RAND Corporation[13] of energy demand in California concluded that the use growth rate would be between 3.4% and 6.3% for all reasonable assumptions (a doubling of energy cost would lower the growth rate from 6.3% to 4.7%). A study of national energy requirements by another group attempted to provide perspective by varying the possible population growth rates and prices assumed.[8] Possibilities for electricity use in the year 2000 ranged from a low of 2 trillion kWh, if the price of electricity doubled, to a high of 9.9 trillion kWh, if costs decreased approximately as they did prior to 1970 (the actual use was 3.01 trillion kWh). An Environmental Defense study in Wisconsin indicated that industrial demand would slacken if industrial users paid a true share of the cost. (As we shall see in the next chapter, the utilities signed up industrial users by giving them the lion's share of the savings realized by increasing the base load.) This reflects one of the major recommendations of the Ford Foundation Energy Policy Project:[14] namely, that promotional rates and subsidies be eliminated.

As discussed earlier, the U.S. population growth rate is dropping spectacularly. The U.S. market should approach the saturation point soon in major appliances.

FIGURE 5.8

Real (deflated) fuel costs, 1950–2003, using 2000 dollars. (U.S. Department of Energy, *Annual Energy Review 2003* (Washington, D.C.: GPO, 2004) August 2004)

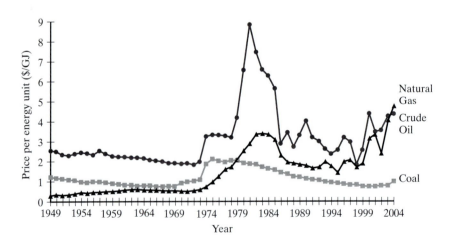

(Of course, new appliances are being sold all the time—the home computer, the DVD player, and so on. However, there have been no new *major* appliances developed.) Fisher,[15] in 1973, analyzed U.S. energy use for the year 2000, taking into account revised population growth estimates, saturation, projected increases in industrial efficiency, and different growth rates of different sectors of the economy. His conclusions were shown earlier in this chapter, in **Extension 5.1**. Each successive lowering of the estimate followed on the use of more care in the choice of assumptions, recognition of changes in energy-use habits of individuals and industry,[16] and more experience in seeing the effects of energy conservation measures in everyday life.

In this sense, the 2001 report by the Vice President's Task Force[17] suggested a rise by 2020, as shown in Figure 5.9. There seem to be several things wrong with this picture. Given the energy use shown in Figure 5.1, and the growth rate of 1.7% per year, we would predict energy use in 2010 of 123 EJ (117 quads) and in 2020 of 147 EJ (140 quads), almost a 20% increase in each decade. This is hardly going to be a straight line as shown in Figure 5.8. That line would be curved upward (the

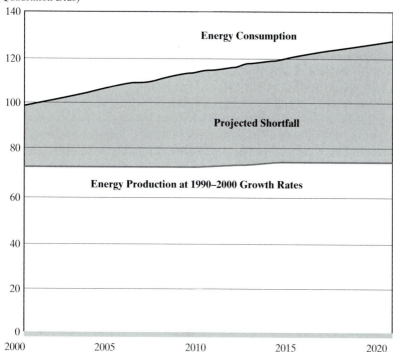

Growth in U.S. Energy Consumption Is Outpacing Production

(Quadrillion Btus)

Energy Consumption

Projected Shortfall

Energy Production at 1990–2000 Growth Rates

Note: Over the next 20 years, growth in U.S. energy consumption will increasingly outpace U. S. energy production, if production only grows at the rate of the last 10 years.

Sources: Sandia National Laboratories and U.S. Department of Energy, Energy Information Administration.

FIGURE 5.9

The projection of U.S. energy use to 2020 from Reference 17. Note the lower line that shows U.S. production as roughly constant; the remainder is imported energy, which is expected here to grow rapidly.

use in 2010 shown is about what should be expected in 2010). Actual predicted 2020 use at today's rates would be 140 quads in 2020, significantly higher than shown. The report says[17] "[t]o meet projected demand over the next two decades, America must have in place between 1,300 and 1,900 new electric plants." This statement reflects the rise of electricity as the preferred source of energy for Americans (Figure 5.10a). If we take the "real business-as-usual prediction" as above, this would imply an even more drastic need for construction of generating facilities. However, the report underplays the effect of conservation brought about by higher energy prices, as the experience after 1973 showed (electricity per dollar of GDP peaked in 1976, Figure 5.10b). It is likely that energy growth will be significantly smaller than the report predicts.

Predicting into the Far Future

Thoughtful scenarists of future energy demand are worried. In 1989, the average annual per capita world energy use was 57 GJ (1500 kWh) per person. However, the world's use was not evenly distributed: Americans used about

FIGURE 5.10

(a) Per capita electrical energy use is rising approximately linearly (at 214 kWh/yr) because of the convenience of electricity. (b) Electricity use per unit of economic output is declining after having reached a peak in 1976, just after the first energy crisis. (U.S. Department of Energy, *Annual Energy Review 2000*, Washington, D.C.: GPO, 2001), August 2001, Table 8.2)

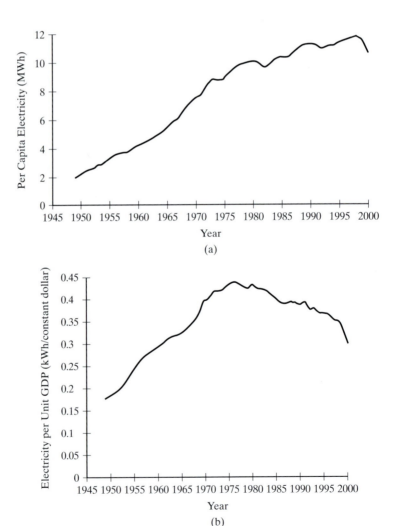

300 GJ per person per year; Europeans used about 127 GJ per person per year; and Third World countries used less than 10 GJ per person per year (many as little as 1 GJ per person per year).[18] Shall we assume a future world of wastrel Americans, or a modest rise in per capita energy use? Even if the per capita energy use does not rise, the number of people using energy will rise exponentially. Global energy production of almost incomprehensible proportions and investment of hundreds of billions of dollars to supply this energy[18,20] will be needed, as the world continues running (as does the Red Queen in *Through the Looking Glass*)[19] just to stay in the same place.

Sassin[18] and Häfele[20] see the need for drastic changes in the world's economic structure. As we shall see in Chapter 12, fossil fuels are inexorably running out. As this happens, there must be a shift to some other resource, and such changes in the United States have historically taken about 60 years (in the rest of the world, about 100 years) to achieve a 50% penetration for the new fuel.[20] We do not even know now what the new fuel will be. If it is solar energy (discussed in Chapters 21 through 24), the only realistic scenario at present involves a tremendous civil engineering undertaking, using the bulk of cement now available in world production.[20] If it is nuclear energy (discussed in Chapters 18 through 20), the experience of the past decade indicates that a "transition from fossil-fuel energy to nuclear energy will probably be an uphill fight in terms of cost and effort."[18]

Any transition to a sustainable energy future will become more difficult the longer it is put off. People would like a world with a stable price structure, full employment, and increasing wealth for the people in it. These goals are often mutually exclusive, except for short time-spans such as the three decades after World War II in the developed world.[21] The decline in the ratio of American energy use to gross national product (Figure 5.10b), which started in the 1970s and which many see as a hopeful sign for the future, is almost entirely attributable to four factors:[21] decrease in the proportion of petroleum used, increase in the proportion of hydroelectric and nuclear energy generation, increase in the proportion of direct fuel use in the ratio of household to manufacturing energy use, and the efficiency of energy use.

Much of the decline is due to a shift to fuels having greater energy return on investment (EROI).[21] Not all fuels are equal; some take much more energy than others to find, exploit, and refine to a handy form. Per joule, oil is worth 1.3 to 2.45 times as much as coal.[21] Any transition has to involve a shift to fuels with a smaller EROI, because the cheaper options will already have been exploited. People are usually content with the status quo and want to stave off installing a new energy infrastructure. This is problematic, because the longer we put off the change, the less rewarding the change will appear to be, and so the speed of the change will be even slower.[18,21]

The economic recovery from the recession of 1980 to 1983 was due, in part, to declining oil prices set by the Organization of Petroleum Exporting Countries (OPEC), which dropped as a result of decreased oil demand caused by the recession. Many believe the economic growth of the 1990s was fueled by cheap oil. As the experience of very low oil costs shows, very cheap oil may be disadvantageous, because it leads to increased dependence on imported oil and encourages wastefulness on the part of consumers. In any case, cheaper oil will again become expensive oil, as price rises and falls continue to show. Soon, oil will be gone forever (see Chapter 12). The cost of extraction of energy resources, which hovered around 3% for most of this century, rose in 1982 to 10%.[21] The switch to the sustainable energy future should not be put off until we have squandered our one-time access to inexpensive, useful fuel.

Fisher[15] looked even farther into the future than the turn of the century: He assumed that U.S. population growth would drop steadily to zero in the 2200s, with a stable population of 1 billion, and that a stable total energy consumption of about 11 trillion kWh would ultimately be reached. But such predictions, based on our earlier analysis, should serve to remind us that no matter how thoughtful the predictor, any but short-term predictions are fraught with difficulties. We must look on *any* two-century prediction with extreme skepticism.

The Logistic Curve

The exponential method of prediction fails to recognize that all systems of our acquaintance are finite. Consider the example of a new species exploiting a hitherto untapped ecological niche. At first, with superabundant resources, there are no checks on the growth of the new species. Individuals of the species are too few to have attracted predators, for example. As a result, the growth in numbers of this new species is exponential. As the numbers increase, the species becomes more interesting to predators and exploits more of the finite resources available to it. At some stage, the growth in numbers must cease. The population thereafter remains essentially stable, unless the environment is somehow altered. The phenomenon of untrammeled growth in the beginning, approaching a steady-state situation at the end, occurs in many facets of our experience. (See the saturation of appliances curve in Chapter 6 and the yield versus energy-input curve in Chapter 23, for example.) The curve describing this behavior is known as **logistic curve** (or sigmoidal curve); a comparison of the logistic curve and its corresponding exponential (for small *t*) is shown in Figure 5.11.

In a finite world, the logistic curve describes many situations in which there is, for a time, what appears to be exponential growth. Farther out on the logistic curve, we could approximate exponential growth by a slightly different exponential curve. The logistic curve, as opposed to the exponential curve, begins after a certain point to turn over, and the rate of change gradually decreases until the value becomes constant. While the logistic curve is below its turnover, one cannot be certain whether it is exponential or logistic. It is only after the turnover has occurred that the logistic curve clearly reveals its logistic character.

The exploitation of mineral resources often follows a logistic function. In many cases, however, economic activity suppresses production when it would have risen without economic interference. This makes the time past the turnover (fall time) longer than the time it takes to rise to turnover (rise time). The amount produced per unit time in such a case may be described by what is called a *Verhulst function*. The production of mineral resources is described by Verhulst functions that seem to have ratios of fall time to rise time clustered at 1 (logistic curve), 5, and 10 (Figure 5.12).[22]

The reason that care is needed in interpreting projections is that the straight-line semilogarithmic projections are often approximations to the logistic curve; the turnover is ignored. If the data projected are far below turnover, the projection may be valid over broad swaths of time. If the data to be projected are near turnover, the "straight-line" prediction is grossly wrong.

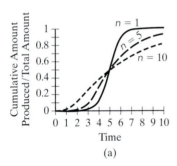

FIGURE 5.11

The logistic curve and exponential curve compared.

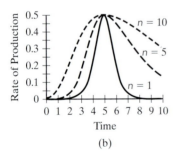

FIGURE 5.12

The Verhulst function for $n = 1, 5, 10$ (arbitrary time units). (a) Cumulative amount produced/total amount that can be produced. (b) Rate of production (arbitrary units).

SUMMARY

Exponential growth occurs if the growth in a quantity during a particular period depends on how much of that quantity there was at the beginning of the period. The rate of change in such a quantity is the growth rate times the quantity. Exponential, or geometrical, growth occurs much faster than linear growth, in which a fixed amount is added to a quantity in every unit of time.

Exponential growth is characterized by its doubling time; exponential decay, by its halving time (half-life). Specification of either of these times is equivalent to specification of the growth or decay rate, respectively. Exponential growth catches most people unaware, because by the time the problems associated with this type of growth become obvious, the time available to do anything about those problems is very short. If it takes 1 week for water hyacinth to double the area of a lake it covers, it may take until the lake is a quarter (or more) covered before the observer realizes that there is a problem with the plants; only 2 weeks remain in which to prevent the complete choking of the lake.

Special graph paper (semilogarithmic graph paper) can be used to show exponential growth as a straight line. Most projections of future energy use (or consumption of any sort) consist in extending such a straight line to future times. Such projections work well if trends are extrapolated for short intervals (one or two doubling times) into the future. No such projection can be trustworthy for predictions many doubling times hence.

PROBLEMS AND QUESTIONS

MULTIPLE CHOICE REVIEW

1. Use a linear model to predict the amount of energy that should have been used in the United States in the year 2000 given the data below.

Year	Actual energy use
1980	82.7 trillion EJ
1990	89.0 trillion EJ

   a. 85.9 trillion EJ
   b. 89.0 trillion EJ
   c. 92.2 trillion EJ
   d. 95.3 trillion EJ
   e. 171.7 trillion EJ

2. Use a linear model to predict the amount of energy used in a particular country in the year 2010 given the data below.

Year	Actual energy use
1980	21.9 trillion kWh
1990	23.4 trillion kWh

   a. 22.7 trillion kWh
   b. 23.4 trillion kWh
   c. 26.4 trillion kWh
   d. 27.9 trillion kWh
   e. 43.8 trillion kWh

3. An amount of $100 is deposited in a bank at a fixed annual interest rate compounded annually. At the end of a 2 yr period, the amount of money has doubled. About what was the annual interest rate?

   a. 10%
   b. 41%
   c. 50%
   d. 100%
   e. 200%

4. Which of the following statements about exponential growth is/are TRUE?

   a. Exponential growth is characterized by its doubling time.
   b. Exponential growth occurs when growth in a quantity depends on how much of the quantity there was at the beginning.
   c. Exponential growth occurs much faster than linear growth.
   d. All of the above statements are true.
   e. Only statements *a* and *b* are true.

5. Suppose the amount of fish caught worldwide, assumed to be a small proportion of the total fish population, grows at an annual rate of 3%. About how long will it take for the fish catch to double?

   a. 10 yr
   b. 13 yr
   c. 18 yr
   d. 23 yr
   e. 33 yr

6. If a lake is about an eighth covered by water hyacinths when you first look at it, and completely covered 3 weeks later, what is the water hyacinth doubling time?

   a. 0.25 week
   b. 0.5 week
   c. 1 week
   d. 2 weeks
   e. 3 weeks

7. Elephants are hunted illegally in Africa for ivory. If the current poaching experience is any guide, the number of elephants will decrease by half in about 25 yr.[23] If the elephant population in 1995 was 16,000, by what date would you expect the number to have decreased to 1000?
   a. 2011
   b. 2045
   c. 2070
   d. 2095
   e. 2120

8. A certain piece of land can support up to 100 cattle. If the number of cattle doubles every 3 yr, how many years will it take for the original cattle population of 25 to reach the land's carrying capacity?
   a. 3
   b. 6
   c. 9
   d. 12
   e. 15

9. You deposit $500 in a special investment. It doubles to $1000 after 7 yr. What was the approximate interest rate paid on the investment?
   a. 2%
   b. 3.5%
   c. 7%
   d. 10%
   e. 15%

10. Suppose the data set below is shown to you. Pick the sort of function of the independent variable $y(x)$ that could describe these data.

Independent variable, x	Dependent variable, y
0	0.5
5	5
10	15
15	30
20	45
25	55
30	60

   a. constant
   b. linear
   c. exponential
   d. logistic
   e. None of the above choices could describe these data.

SHORT ANSWER

Explain whether the statement made is correct, and justify your answer.

11. It makes little difference in making a prediction whether a linear or an exponential growth in some quantity is assumed.

12. HDTV is now being introduced in the United States. It is likely that sales of HDTVs will follow an exponential growth curve during the earliest years of sales.

13. The prediction of the value of the dollar in 2050 in terms of today's dollar is easy to get right.

14. Exponential growth has been assumed in Figure 5.8, which shows the energy use prediction utilized by the Vice President's 2001 report on future energy use.[17]

15. Coal exploitation gives a better return on investment than exploitation of other fossil fuels.

16. We would expect the curve for number of can openers sold versus time to be a logistic curve.

17. We are certain that fossil fuels will be in use for thousands of years to come at current use rates.

18. For use rates that depend on the number of people, the rates must increase exponentially as the number of people increases.

19. The United States has one of the highest energy use rates per person in the world.

20. It makes little difference in making a prediction over a very short period of time whether a linear or an exponential growth in some quantity is assumed.

DISCUSSION QUESTIONS

21. Is there a connection between population growth and land use? Explain.

22. Is there a connection between population growth and energy use? Explain.

23. Describe some of the difficulties people have in making good long-term predictions.

24. Why does exponential growth seem so surprising in its outcome to most people?

25. Iomega Corp. makes Zip drives. How would investors in the early 1990s have perceived the market growth for these removable drives? Would knowledgeable investors have drawn straight lines on regular or on semilogarithmic graph paper? Why? Would they do so now? Explain.

26. What is the greatest difference between quantities described by linear and by exponential growth?

27. The 1920s represent the dawn of commercial radio. How would those investors interested in the new technology have decided what growth possibilities there were? Would knowledgeable investors have drawn straight lines on regular or on semilogarithmic graph paper? Why?

28. If the number of disease-causing bacteria can grow exponentially in the body, why don't all people succumb to them at an early age?

29. Try to identify reasons that the population growth rate, both in the United States and worldwide, has declined in the last century rather than remaining constant.

30. What are the major reasons that the energy use rate has decreased over the last quarter century as compared to the preceding quarter century?

PROBLEMS

31. Determine the amount of widgets expected to be manufactured in 2010 if the growth rate is projected to be 2.5% per year, and in 2000, there were 250,000 widgets made.

32. If you put $1000 into a 10 yr CD at an APR of 7% per year, how much money should you have at the end of the 10 yr period?

33. Toaster oven manufacturers sold 750,000 1200 W units last year on an installed base of approximately 3.5 million. If the growth rate in sales is projected to be 4% per year, how much more energy will the United States use 10 yr hence solely because of the sales of toaster ovens? If the growth remained the same until every other American owned a toaster oven, how long would

it require? How much more time would it take until each American owned one?

34. Predict the American population in 2050 assuming that the 1990–2000 growth rate remains constant and the 2000 population figure is accurate. What would it have been had the growth rate not decreased from its 1950–1960 value, assuming the growth had continued from the 1960 population figure until 2050? What would it have been if the Depression growth rate (1930–1940) had continued from 1940 on to 2050? How large a difference does this represent?

35. You put $600 into a bank account at a 4% annual interest rate. At the end of the 10 yr period, how much money is in your account?

36. Determine the actual sort of growth and describe the increase assumption used in Vice President Cheney's task force projection, shown in Figure 5.8.

37. Using the 1990 energy use figure and the known U.S. population growth between 1990 and 2000, explain what energy use figure would have been expected for 2000 solely on the basis of population growth. How does this compare to the experience making the same assumption between 1980 to 1990?

38. Suppose that 100,000 frammises have been sold up to now, and 1200 are sold in the succeeding year.
   a. Compute the projections of sales over the next 30 yr assuming that this increase is linear.
   b. Compute the projections of sales over the next 30 yr assuming that this is characteristic of the rate of increase for exponential growth.
   c. Compare these results, and discuss them.

39. Suppose that 100,000 frammises have been sold up to now, and 10,000 are sold in the succeeding year.
   a. Compute the projections of sales over the next 30 yr assuming that this increase is linear.
   b. Compute the projections of sales over the next 30 yr assuming that this is characteristic of the rate of increase for exponential growth.
   c. Compare these results, and discuss them.

40. How much more energy would be required to bring the entire world up to the average per capita U.S. energy use? Recall that world average energy use is 57 GJ per capita, and the U.S. average energy use is 300 GJ per capita.

41. Western Europe has a standard of living similar to the that of the United States, while per capita energy use is only about half that of the United States. If the United States could be about as energy efficient as Western Europe, how much energy could be saved in the American economy? Would this help us meet the Kyoto Climate Treaty reductions (a return to 1990 emissions levels)? Explain. What would be the projected energy necessary to raise the entire world to the European standard?

THERMAL ASPECTS OF ENERGY GENERATION

PART III

CHAPTER 6

Atoms and Chemical Energy

In preceding chapters, we defined work, energy, and power. We also discussed some of the various forms of energy and how they may be converted into other forms of energy. The electric energy we use comes in large part from chemical processes. (This is discussed in more detail in Chapters 12, 13, and 14.) Burning—a chemical reaction—supplies energy and allows energy to be used to change materials and to produce goods. Solar energy often causes chemical changes. Thus, a basic understanding of the process of chemical change is essential to any understanding of energy and energy production.

KEY TERMS

chemical energy / state / Pauli exclusion principle / energy levels / ionize / periodic table / inert / exothermic / oxidation / activation energy / heat of combustion / endothermic / catalyst / catalysis / enzyme / photosynthesis / battery / electrode / electrolyte / cell / fuel cell

Many of the chemical compounds used to produce energy are burned and involve several chemical recombinations. Gasoline powers automobiles and trucks; kerosene serves as a jet fuel. Natural gas may heat our homes directly or generate electricity. Coal and fuel oil are burned to generate electricity as well. The root source of the energy used for heating, transportation, and industry in most of the world is **chemical energy**. Of course, that chemical energy originally came from the Sun, which operates on nuclear fusion energy. (We discuss photosynthesis in this chapter as well as in Chapter 23.)

In any chemical process, there is a change in the condition of the constituents of the atoms involved. In some processes, the result is more stable than before the process occurred; in others, the result is less stable. To find out what happens, we have to look into the atom. If we were able to see inside a material and shrink our field of vision again and again, we would eventually be able to identify what seems like an impenetrable shell. We would be looking at the outer surface of an atom or molecule. This surface is provided by the atom's electrons. A typical atomic "size" (diameter) is 0.1 nm (10^{-10} m).

Chemistry is the basis for processes of life. Cells are made up of molecules, and molecules are made of atoms. Molecules and atoms and their properties are the study of chemists, physical chemists, chemical physicists, and biochemists. Atoms consist of the central atomic nucleus with electrons occupying, in some fashion, the region surrounding it. The nucleus, made of protons and neutrons, is the subject of Chapters 18 to 20.

6.1   PROPERTIES OF ATOMS

The chemical properties of atoms are determined by the configuration of the outermost electrons in the atoms. Atoms are held together by electric forces. The truly tiny atomic nucleus (diameter $\sim 10^{-15}$ m, 100,000 times smaller than that of the atom) carries a positive charge, and the atom's electrons, one for each proton, carry negative charges (of the opposite sign to the protons) in the space of the atom, so the atom is electrically neutral, as noted in Chapter 4.

Although the electrons in the atoms are wavelike and are not really localized in space, they do have a high probability of being near positions they would have occupied if they were particles (the electrons form a cloud, see Figure 6.1). The electrons repel each other electrically, just as they are attracted electrically to the nucleus. Furthermore, they cannot all get close to the nucleus. We find experimentally that no two electrons may exist in the same **state** (defined by their mean distances, energies, and so forth) in an atom. This phenomenon is referred to as the **Pauli exclusion principle**. Because of this fact of nature, electrons from one atom do not interpenetrate those of another atom. They provide what we consider to be the outer surface of the atom, the "size."

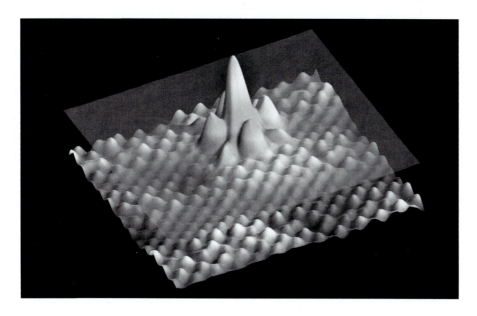

FIGURE 6.1

An image of a two-dimensional layer of copper oxide doped with a single zinc impurity atom. The electron cloud is centered above the zinc impurity. The image was made with a scanning tunneling microscope. (Séamus Davis, Lawrence Berkeley National Laboratory)

The electrons thus take certain "positions" (orbitals) around the atomic nucleus. The first electrons occupy the lowest, most stable, of the possible "positions." The closest-in electrons have given up the most energy to become part of the atom. Subsequent electrons added to the atom give up less energy and are then less stable than the deepest electron and more easily removed from the atom.

Figure 6.2 shows what we might see if we looked at a glow discharge from a tube containing a pure vapor that is excited by a high voltage. To observe more than a general colored glow, we use a device called a diffraction grating, which is a piece of glass or plastic on which tiny parallel ridges separated by a distance of ~1 μm cause different colors of light to be bent by different amounts. Only discrete lines are seen. Electrons joining an atom can give up only certain discrete amounts ("quanta") of energy. The energy is quantized (discrete), because the

FIGURE 6.2

A look at a glow discharge tube through a diffraction grating.

angular momentum of the electron in its orbit (a measure of how rapidly it "goes around" the atom) is quantized in units of Planck's constant, $\hbar$. That is, angular momentum is an integer multiple $(0, 1, 2, \ldots)$ of $\hbar$.

The Bohr Model Picture

Each filled "position" corresponds to a certain amount of energy that the electron gave up to join the atom; we speak of the allowed energy values as **energy levels**. Figure 6.3 shows the first few allowed electron energy states for hydrogen labeled by their quantum numbers (numbers that taken together define the state of the electron in the atom). Hydrogen has only one proton in its nucleus and one electron. An electron may transfer from one energy level to another by emission or absorption of light that has the exact energy corresponding to the specific energy difference between the two levels. Quantum numbers characterize the discrete states of the electrons. The energy is mostly determined by the principal quantum number, denoted n. In ordinary hydrogen, the single electron is in the state $n = 1$. Higher values of n denote excited states, which are less stable. Normally, an excited electron emits light as it goes to lower energy orbits until the lowest state, the state $n = 1$, is reached.

In helium, there are two protons in the nucleus of the atom. There are, consequently, also two electrons. Both electrons can fit in the $n = 1$ level. In more complex atoms, states of higher n are filled. (Each level characterized by n has sublevels defined by other quantum numbers known as the orbital angular momentum—l—and its projection in a fixed direction—m.) In each state, two electrons can fit in each sublevel. There can be two electrons in each sublevel, because electrons can take on two separate guises: as "up" electrons or as "down" electrons. No sublevel can have two "up" or two "down" electrons, but if one electron is "up" and one is "down," the two electrons are in different states for that energy level, which *is* allowed by nature (by the Pauli exclusion principle).

As an analogy, consider a eucalyptus tree with pairs of thin leafy branches going up the tree trunk and koalas trying to eat the leaves. Because the koalas have some weight, we imagine that they would break the thin branches if more than one koala were to sit on each branch. Because each branch can hold only one koala, a new koala climbing the tree would have to climb to the next available free branch to have a safe perch on which to eat eucalyptus leaves. The electrons in atoms are forced by the Pauli exclusion principle to take "positions" higher in energy than those already there. In this way, each atom has its own configuration of "branches" on the "tree" (available electron energy levels), and if one sees this configuration, it is possible to identify the type of atom. Groups of levels sharing a certain characteristic (called *orbital angular momentum*) are known as shells or subshells.

Chemistry and the Periodic Table of Elements

In fact, most of the chemical properties of atoms depend on the configurations of the *outer electrons*. If the outer electron configurations of two atoms are the same, they will combine with still other atoms (which basically "see" only the outsides of these atoms) in approximately the same way. Therefore, they will form similar chemical compounds, have similar physical characteristics (boiling or melting points), and so on. It will take similar amounts of energy to remove an

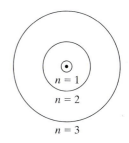

(a) Relative Radii of First Three Hydrogen Orbits

(b) Energy Levels in the Hydrogen Atom

FIGURE 6.3

(a) The first few allowed electron orbits in the Bohr model (1912) for hydrogen. R, the Bohr radius, is 72.9 pm. Allowed orbits have radii R, $4R$, $9R$, and so on. Each discrete electron state has a corresponding quantized angular momentum $l\hbar$ and projection $m\hbar$. (b) Diagram of energy levels for hydrogen. The Bohr model ultimately led to the development of quantum mechanics, which, in principle, can describe all atoms. The eV is a unit of energy equal to 0.16 aJ.

outer electron from, or **ionize**, atoms with similar outer shell configurations. This leads to recurring similarities among atoms. The atoms appearing in the columns of the **periodic table** of the elements exhibit similar melting points and have similar chemical behaviors (Figure 6.4).

Atoms possessing certain configurations of electrons are resistant to indulging in chemical interactions. The electrons in these atoms are not easily removed, nor can an extra electron be added to these atoms easily. Such atoms in nature are gases at room temperature and are called **inert** (or noble) gases. Such atoms have a full outer shell. They make up the last column in Figure 6.4. If an atom, in order to balance the positive charge of the protons in the nucleus, has such a configuration plus one additional electron, it is easy to get it to give up that extra electron and become a positively charged, or positively ionized, atom. These atoms are in group I A in Figure 6.4. If an atom is lacking one electron of such a full outer shell, it is easy to get it to accept an extra electron and become a negatively charged, or negatively ionized, atom. These atoms are in group VII A of Figure 6.4.

Many chemical combinations occur because an atom of the first type (such as sodium) gives up one of its electrons to an atom of the second type (such as chlorine). Because the ionized atoms have opposite charges, they attract one another

The Periodic Table of Elements

I A																	VIII A
1 H 1.0079	I A	1→Atomic Number H→H—Symbol 1.00794→Atomic Mass										III A	IV A	V A	VI A	VII A	2 He 4.0026
3 Li 6.941	4 Be 9.012											5 B 10.81	6 C 12.01	7 N 14.01	8 O 16.00	9 F 19.00	10 Ne 20.18
11 Na 22.99	12 Mg 24.31	III B	IV B	V B	VI B	VII B		VIII B			I B	II B	13 Al 26.98	14 Si 28.09	15 P 30.97	16 S 32.07	17 Cl 35.45	18 Ar 39.95
19 K 39.10	20 Ca 40.08	21 Sc 44.96	22 Ti 47.88	23 V 50.94	24 Cr 52.00	25 Mn 54.94	26 Fe 55.85	27 Co 58.93	28 Ni 58.69	29 Cu 63.54	30 Zn 65.39	31 Ga 69.72	32 Ge 72.61	33 As 74.92	34 Se 78.96	35 Br 79.90	36 Kr 83.80
37 Rb 85.47	38 Sr 87.62	39 Y 88.91	40 Zr 91.22	41 Nb 92.91	42 Mo 95.94	43 Tc 98.91	44 Ru 101.1	45 Rh 102.9	46 Pd 106.4	47 Ag 107.9	48 Cd 112.4	49 In 114.8	50 Sn 118.7	51 Sb 121.8	52 Te 127.6	53 I 126.9	54 Xe 131.3
55 Cs 132.9	56 Ba 137.3	57 La to Lu 71	72 Hf 178.5	73 Ta 180.9	74 W 183.9	75 Re 186.2	76 Os 190.2	77 Ir 192.2	78 Pt 195.1	79 Au 197.0	80 Hg 200.6	81 Tl 204.4	82 Pb 207.2	83 Bi 209.0	84 Po 209	85 At 210	86 Rn 222
87 Fr 223	88 Ra 226.0	89 Ac to Lr 103	104 Rf 261.1	105 Db 262.1	106 Sg 263.1	107 Bh 264.1	108 Hs 265.1	109 Mt 268	110 Ds 269	111 272	112 277		114(?) 289		116(?) 289		

Lanthanide Series

57 La 138.9	58 Ce 140.1	59 Pr 140.9	60 Nd 144.2	61 Pm 144.9	62 Sm 150.4	63 Eu 152.0	64 Gd 157.3	65 Tb 158.9	66 Dy 162.5	67 Ho 164.9	68 Er 167.3	69 Tm 168.9	70 Yb 173.0	71 Lu 175.0

Actinide Series

89 Ac 227.0	90 Th 232.0	91 Pa 231.0	92 U 238.0	93 Np 237.0	94 Pu 244.1	95 Am 243.1	96 Cm 247.1	97 Bk 247.1	98 Cf 251.1	99 Es 252.1	100 Fm 257.1	101 Md 258.1	102 No 259.1	103 Lr 262.1

FIGURE 6.4

The periodic table of the elements. Elements falling in the columns have similar physical and chemical properties.

(forming compounds such as sodium chloride—table salt). In other sorts of compounds, electrons are shared among the constituents in such a way as to make the environment of each atom more like the "stable" electron configuration.

The other atoms shown in Figure 6.4 have two "extra" electrons (group II A) or three "extra" electrons (group III A), or they lack two electrons of a full shell (group VI A) or lack three electrons of a full shell (group V A). The atoms of group IV A can be thought of as either possessing four extra electrons or lacking four electrons. The first element in group IV A is carbon. Because of the many ways carbon can combine with other elements, molecules containing carbon can exist in countless forms. There is a branch of chemistry that deals only with carbon compounds; it is called organic chemistry. This name recognizes the importance of carbon for life. The most common other form of life imagined by science fiction writers is one modeled after the second element in group IV A, silicon. Silicon binds in ways similar to carbon. However, a silicon life form is difficult to imagine, and the proposed chemistry seems a little far-fetched. Silicon and the rest of the group IV A elements (except lead) have found use in semiconductor technology (see Chapter 21).

6.2   CHEMICAL COMBINATION OF ATOMS

Exothermic and Endothermic Reactions

Formation of compounds from atoms occurs, generally, when the compound is more stable than the constituents were when they were alone. The constituents become more stable by giving up some of their energy to become a compound by making a bond. Such chemical reactions, which give off energy as they proceed, are called **exothermic**. The word "exothermic" really means "heat out" (*ex* in Latin means "out" or "from," and *therm* is from the Greek for "heat"). Chemical reactions involving burning (classed as **oxidation** reactions) are exothermic. As an analogy for such a reaction, consider billiard balls on a pool table. It is easy for the balls to roll into holes and fall below the surface, where they can mix with the other balls already down there. To unmix the billiard balls, we have to give them back the gravitational potential energy they gave up in falling down into the hole, by doing work as we lift them back up to the table. Similarly, atoms give up some of their energy when they share or exchange electrons to form a compound; the constituents may be re-formed only if energy is added to the system. Because the electrons are in something analogous to a hole after they attain the more stable state, we characterize the atom as an "energy well." (See Figure 6.5.)

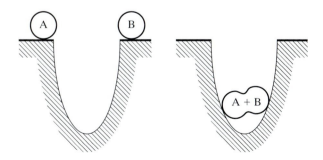

FIGURE 6.5

The analogy between a ball falling into a hole and chemicals bonding chemically by falling into an energy "well."

APPLICATIONS

Using the Periodic Table

The periodic table can be used to predict the way compounds form from constituent atoms. Here we concentrate on the groups identified by an A in Figure 6.4. The column headings I through IV can be taken to represent the number of extra electrons over the closed shell; column headings IV through VII can be subtracted from eight to determine the number of electrons an element is lacking from a closed shell. The resultant number is known as the chemical valence of the atom.

Thus, atoms from group I A combine one to one with atoms from group VII A (for example, LiF); an atom from group II A will combine with two atoms from group VII A ($MgCl_2$, where the subscript indicates that two chlorine atoms combine with one magnesium), or with one atom from group VI A (CaO). All atoms from such groups can be combined in this way.

Activation Energy

In most cases, some energy is necessary to get the atoms started to give up or share their electrons. Of course, they then give up a great deal more energy as they form the compound. The small amount of energy necessary to get the formation of compounds started is called the **activation energy**. Something as simple as one atom's bumping into another can supply the activation energy necessary to start a reaction. For example, water is formed from hydrogen gas and oxygen gas. If we put the gases into a container and waited, there would be a very slow reaction. Occasionally, there will be a random collision between molecules of such high kinetic energy that they are able to supply the activation energy. After thousands of years, some molecules of hydrogen and oxygen would break into atoms, which would occasionally hit appropriate other atoms as they jostle about with their thermal energy, and the compound water would form. Activation energy is supplied for macroscopic (large) amounts of chemicals by methods as varied as rubbing them together with a pestle in a mortar, by heating them with a match, by exposing them to light (as in photographic film), or by heating them in a crucible.

Once the activation energy starts a reaction that is exothermic, or energy producing, more energy is produced in the chemical process than was put in as activation energy. This energy surplus can be a source of activation energy to spread the reaction. The energy released (as constituent kinetic energy) is spread out through the entire gas mixture by the random collision process. This energy supplied breaks up molecules farther away, they combine, and so on, until no further combination is possible. In the case of water, supplying a small amount of activation energy (with a match, say) starts off the reaction of hydrogen and oxygen. A large number of molecules break into atoms near the match head. The atoms combine to form water, which liberates a large amount of energy. At that point, the constituents' kinetic energy represents a concentrated supply of gas molecule activation energy so that molecules adjacent to the first molecules also combine, producing more energy, causing more water to be produced, causing more energy to be released, and so forth. When the energy liberated is much greater than the activation energy, the reaction could take place so rapidly that an explosion may occur. Water formation is exothermic, and the reaction rate is high, because the energy produced in forming the bond between hydrogen and oxygen to make water so far exceeds the activation energy.

Much of the time, chemicals combining in exothermic reactions give off their energy in just this way—by increasing their kinetic energy (and thereby, their speed). Thus, energy is transferred *to* the substance. When molecules speed up, the temperature of the compound increases (as we shall see in Chapter 7). The energy evolved from a reaction need not necessarily cause a temperature increase, so we speak of exothermic reactions in general as energy-producing chemical reactions. The energy may be drawn off as electrical energy instead of thermal energy, as in the battery (in which chemical energy is transformed to electrical energy). Or, it may be produced in the form of radiant energy. The light sticks and light necklaces you may have seen for sale in amusement parks produce light—but do not heat up in doing so—from the chemical reaction taking place between the two liquids inside the tube.

Making and Breaking Chemical Bonds

When chemicals burn, that is, combine with oxygen in the air, they form bonds and release energy. Combustion involves liberation of energy (usually as thermal energy) and so is an exothermic process. The **heat of combustion** is the amount of energy given off per kilogram of the substance when it is combined with oxygen (burned). Table 6.1 shows the energy available from combustion for several

TABLE 6.1

Heat of combustion values for various substances

Substance	In MJ/kg	In kWh/kg
Fuels		
Crude oil	45.0	12.5
Fuel oil (mid-continent)	45.1	12.5
Gasoline	46.9	13.0
Kerosene	46.7	13.0
Average coal (bituminous and lignite)	26.0	7.2
Ohio coal (bituminous)	29.5	8.2
North Dakota coal (lignite)	13.9	3.9
Fuel alcohol	27.5	7.5
Hydrogen, to liquid water (to steam)	141.9 (122.0)	39.4 (33.9)
Methane (chief constituent of natural gas)	55.2	15.3
Foods		
Butter	38.5	10.7
Mean animal fats	37.8	11.0
Egg white	23.9	6.6
Egg yolk	33.9	9.4
Linseed oil	39.5	11.0
Olive oil	39.3	10.9
Other		
Bagasse (sugarcane refuse, 12% water)	16.9	4.7
Oak (13% water) wood	16.7	4.6
Pine (12% water) wood	18.5	5.1
Dynamite (75%)	5.4	1.5
Iron	6.6	1.8

Source: Reprinted with permission from *Handbook of Chemistry and Physics*, 32nd ed. Copyright 1959, 1974 by CRC Press, Inc., Boca Raton, Florida. Note that the energy produced is thermal, so there is a penalty factor of 1/3 in converting to electricity.

important sources. See **Extension 6.1, *Finding Total Energy Released in Combustion***.

Because the combination of two chemicals can often release energy, we might imagine that we could subsequently break the chemical bonds if we paid a high enough energy price. Such a reaction sucks in energy from its surroundings, perhaps even lowering the temperature, in order to break the chemical bonds. The word **endothermic** (*endo* is from the Greek for "within") characterizes such reactions. When ammonium chloride or ammonium nitrate is dissolved in water, the temperature of the mixture decreases. The mixing of ammonium chloride with water absorbs so much energy from the surroundings that water touching the vessel may actually freeze. Water can be broken up into its constituents, hydrogen and oxygen, when an external energy source is available. If electrical energy is supplied, the process of water breakup is called electrolysis.

Energy Conservation and Chemical Change

So far, we have discussed simple compound formation or simple compound breakup. Many chemical reactions, however, involve a multitude of steps, some of which may be exothermic and some of which may be endothermic. The overall reaction is classified as exothermic if the *net* result of all steps in the reaction is energy released and bonds made overall and as endothermic if the *net* result of all steps in the reaction is energy absorbed and bonds broken overall. If a reaction takes place in which 30 joules of energy are absorbed in one step, and 50 joules of energy are produced in another step, the net result is that

$$50 \text{ J} - 30 \text{ J} = 20 \text{ J}$$

of energy is produced; this is an exothermic reaction. The reverse reaction (if it is possible) would be an endothermic reaction, and in the two parts, 30 J would be produced, while 50 J would be absorbed because of conservation of energy; thus, a net of

$$30 \text{ J} - 50 \text{ J} = -20 \text{ J}$$

would be produced.

To summarize, we can say that exothermic reactions result when the overall result of a chemical reaction is the making of bonds. The reactants fall into their most stable available state by giving up energy. Conversely, endothermic processes involve the breaking of bonds. Energy from outside is required to break the bonds and make the endothermic reaction occur. All chemical bonds involve only exchange or sharing of the atoms' electrons.

Catalysis

It is possible to change the rate at which reactions occur. Electron exchanges can occur when electron shells are in the vicinity of one another. Suppose we add to a reaction materials that allow the appropriate surfaces in the constituents to come into contact with one another. The materials would enhance the speed with which the chemical reaction occurs, but the added materials would not be used or even affected in the process. Materials that have this property are known as **catalysts**; the process of using catalysts is called **catalysis**. Examples of catalysts include platinum powder, used to increase the rate of combination of hydrogen and oxygen to form water; finely ground ashes, used to allow a cube of sugar to burn (without the ash, the sugar simply melts and does not burn); and, more important from our stand-point, catalysts used with the chlorophyll in plants to enhance the rate at which carbon dioxide and water are converted into sugars and oxygen.

All of the reactions just listed as examples would occur without catalysts. They would simply take a longer time to complete. The catalysts are unchanged, because after the reaction occurs, the constituents leave the surface as a compound and allow their places to be taken by new sets of constituents.

A CLOSER LOOK

Chlorine Catalysts in Ozone Depletion

In Chapter 1, ozone depletion by molecules containing chlorine and bromine was discussed. The two species of molecules go through the chemical transformations shown here, in both cases changing two ozone molecules into three ordinary oxygen molecules. Chlorine molecules allow ozone to be broken into ordinary oxygen molecules; the chlorine molecules are reconstituted at the end, and so act as catalysts. The reactions also go in reverse (run backwards).

Chlorine cycle:

$$ClO + ClO + \text{neutral air molecule} \longrightarrow (ClO)_2 + \text{neutral air molecule}$$
$$(ClO)_2 + \text{solar energy} \longrightarrow Cl + ClO_2$$
$$ClO_2 + \text{neutral air molecule} \longrightarrow Cl + O_2 + \text{neutral air molecule}$$
$$2(Cl + O_3) \longrightarrow 2(ClO + O_2)$$

Bromine–chlorine cycle:

$$ClO + BrO \longrightarrow Cl + Br + O_2$$
$$Cl + O_3 \longrightarrow ClO + O_2$$
$$Br + O_3 \longrightarrow BrO + O_2$$

I have implied that catalysts are essential to life. In living things, catalysts called enzymes make life possible. **Enzymes** are proteins in living things that act to speed up reactions. Enzymes act as catalysts in photosynthesis.

Many chemical reactions occur quite slowly. Iron rusts (combines with oxygen) very slowly at normal temperatures. In the presence of water, especially salt water, the reaction occurs more rapidly. People who live in areas in which salt is spread on roadways for ice control can observe their cars corroding after only a few years' exposure. Nevertheless, rusting and corroding do not happen as rapidly as a wood fire reduces a log to ashes. Paper is made from wood; it is burning (oxidizing) just by existing. The burning is slow (acid-free paper oxidizes much more slowly). In several hundred years, the paper in this book will have disintegrated. You may have seen old books with their brittle, yellowed, half-burnt paper. We can hasten the process by supplying the activation energy for a chain reaction and burn the book in minutes.

Many reactions in the human body are of this "burning" kind (called oxidation) and usually proceed slowly. The enzymes may catalyze specific reactions, allowing them to occur rapidly. As with any catalysis, reactions cannot be made to happen that could not otherwise have occurred; the catalyst merely hastens them.

Photosynthesis

The body of an animal mines the lode of energy deposited in plant cells by the Sun. Energy-carrying sunlight is absorbed by chlorophyll in plants and allows an endothermic reaction of water and carbon dioxide to take place. This reaction, called **photosynthesis** (*photo* means light), is not yet fully understood. Carbon dioxide and water form glucose, a sugar, which stores much of the energy. The chemical equation that describes the overall transformation is

$$6CO_2 + 6H_2O + \gamma \text{ (radiant energy)} \longrightarrow C_6H_{12}O_6 + 6O_2.$$

Thus, six units of carbon dioxide are cycled into sugar by photosynthesis. The simplest organisms that photosynthesize are the cyanobacteria (blue-green algae). Other algae, and the higher plants, also photosynthesize. A green cell can produce up to 30 times its volume of oxygen each hour.[1] Recent work has shown that manganese is part of the catalysis cycle in photosynthesis.[2]

The reverse of photosynthesis, respiration, is used by animals to utilize the food they eat:

$$C_6H_{12}O_6 + 6O_2 \longrightarrow 6CO_2 + 6H_2O.$$

Note that the six carbon dioxide molecules locked up by photosynthesis have been released by the animals in combination of the sugar with oxygen.

Photosynthesis proceeds by using enzymes to make oxygen out of carbon dioxide, then transferring hydrogen atoms using radiant energy (sunlight) to effect the transfer, and then finally combining carbon dioxide with the hydrogen to form the sugars. In this process, energy of about 6.2 kJ/kg (1474 kcal/kg) is stored as chemical energy—the sugars—in plants.[1]

Many millions of years ago, plant life in swamps used photosynthesis to produce stored energy. Some of the plant life fell into the muck, making the swamp into a peat bog. Geologic processes may have led to the laying of sediment over the peaty material. Then, biological action, the heat from inside Earth, and the pressure of the overlying sediment (turned to rock) resulted in the formation of a coal deposit. Therefore, coal is merely stored solar energy. A similar process, involving other kinds of organic matter, leads to formation of oil and gas deposits; they, too, are stored solar energy.

Coal and oil resources for 50 to 100 million years from now are currently being formed here on Earth. These *are* renewable resources, but on a time scale somewhat long for humanity. Wood and plants and ocean algae form a much shorter-term reservoir of stored solar energy, and it is this reservoir that animals tap to utilize the energy necessary to sustain their daily existence.

Carbon Compounds

As mentioned previously, carbon compounds play a vital role in the existence of life on Earth. Plants "inhale" carbon dioxide and "exhale" oxygen. Animals reverse this process: they combine oxygen with the carbon compounds in their food to produce the energy for life and produce carbon dioxide as a waste product. Foods consumed by animals are often carbohydrates (compounds containing carbon, oxygen, and hydrogen). Life also plays a role in the formation of the fossil fuels (see Chapter 12), which are deposits of hydrocarbon compounds (compounds containing carbon and hydrogen). These hydrocarbon compounds can produce energy as they burn in the presence of oxygen (see **Extensions 6.1,** *Finding Total Energy Released in Combustion* and **6.2,** *Formation of TNT*).

Carbon can bond easily with itself and other compounds, and its symmetrical configuration of electrons allows it to bond in many different ways. Many carbon compounds form closed rings and other interesting three-dimensional shapes. Natural gas is mostly methane, CH_4. The other fossil fuels have a more complicated structure than methane, with carbons bonding to other carbons. Petroleum's chemical composition is approximately CH_2 because of this carbon–carbon bonding, and so burning it produces relatively more carbon dioxide and less water than burning methane. Coal has a chemical composition with even more carbon–carbon bonds. Its composition is approximately $CH_{0.8}$; its burning produces relatively more carbon dioxide and less water than petroleum.

When methane burns completely in oxygen, only carbon dioxide (CO_2) and water are produced. Carbon monoxide is partially burned carbon; it is dangerous to animals, because it binds with red blood cells (as does oxygen) but cannot be removed (in contrast to oxygen, which is easily removed). The red blood cells transfer oxygen to the rest of the body from the lungs; hence, carbon monoxide can prevent transfer of oxygen from the lungs (see Chapter 14).

When methane burns in air, a mixture of nitrogen and oxygen, it can produce nitrogen compounds as well as carbon compounds. In a real combustion situation, methane and air can produce carbon monoxide (CO), carbon dioxide (CO_2), and various compounds of nitrogen and oxygen: N_2O, NO, NO_2, N_2O_2, and so on (these compounds are known collectively as NO_x). The effects of these and other combustion products are discussed in Chapter 14.

6.3   LEMON POWER, OR THE BATTERY

When I was a child, I found out about lemon power. We would take pieces of two different metals (say, copper and nickel or zinc—we used coins) and stick them into opposite ends of a lemon. Then we would stick our tongues onto both metal pieces at the same time. Presto! We experienced a distinct tingling in the tongue due to a flow of current—lemon power.

We did not really know it, but we were discovering the principle of the **battery**. Batteries need two **electrodes** made of different materials and a fluid, the **electrolyte**, to conduct electricity between the electrodes. The lemon could work as a battery, because lemon juice is an acid and conducts current relatively easily between the coin electrodes. Any fruit acid will do as an electrolyte—acetic acid (cider vinegar), wine vinegar, and so on.

The battery is so named because any one **cell** (composed of the two electrodes and the electrolyte) does not deliver much voltage; a battery, or series, of cells connected together, as in Figure 6.6, could deliver an appreciable voltage between the ends. Such a battery is not very useful, because the acid will spill if one attempts to move it. Several methods have been used to overcome these

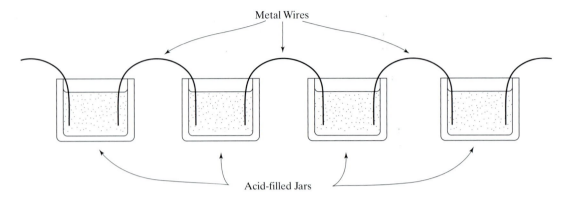

Metal Wires

Acid-filled Jars

FIGURE 6.6

A battery of simple cells.

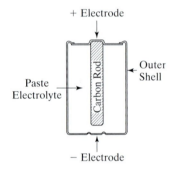

+ Electrode

Paste Electrolyte

Carbon Rod

Outer Shell

− Electrode

FIGURE 6.7

A dry cell. The outer shell is the negative electrode. The carbon rod is the positive electrode.

problems. For example, a container could be built so that a series of bimetal strips is held closely in the container, acid is put in between the strips, and the top is closed. This describes the essentials of an auto battery. A dry-cell flashlight "battery" (Figure 6.7) is made so that the outer container acts as one electrode, the other is replaced by a carbon rod in contact with the metal top, and a paste electrolyte fills the rest of the space. A dry cell is not really dry, but moist; however, the acid contained in it will not spill.

There are two general types of batteries: primary and secondary cells. A primary cell is a throwaway (nonrechargeable) battery; a secondary cell is designed to be recharged.

6.4   FUEL CELLS

Another source of electrical energy from chemical energy is found in **fuel cells**. Fuel cells were first developed in 1839 by Sir William Grove. In the hydrogen–oxygen fuel cell, hydrogen is fed to one porous electrode and oxygen to the other. The electrodes are separated by an electrolytic material. The hydrogen in the electrode is converted catalytically to hydrogen ions, releasing electrons to run through an external electrical circuit. The electrons then combine with oxygen at the other electrode, producing oxygen ions. As the ions travel through the electrolyte, they meet and react to produce water. The only outputs of fuel cells are water and electrical energy, and the output continues as long as hydrogen (or other fuel) and oxygen are supplied to the cell. This process is electrolysis run backwards.

A practical fuel cell first became possible in the 1950s with the development of suitable ceramics. Rapid development followed in the 1960s, as fuel cells were designed for use in the space program. Most current models use natural gas to supply the hydrogen. Moderate-scale fuel cell power plants for electricity have been tested in New York City ($4.8 \, MW_e$) and Tokyo ($11 \, MW_e$), although both plants experienced problems.[3] Solving problems in prototypes leads to advances in technology, allowing wider uses. Smaller-scale fuel cells are undergoing testing in various places around the country. For example, a 40 kW prototype was installed by Southern California Edison on an old landfill in City of Industry, California, which is now the site of a hotel complex and golf course. Anaerobic digestion of the garbage in the landfill produces methane. The gas from the landfill supplies hot water for the complex and fuel for the fuel cell. Fuel cells are attractive, because they have proved in prototype to be up to 80% efficient (see Chapter 7), and their installation price is about the same as that for coal or gas power plants. Fuel cells are being used in many ways commercially, on scales both large and small.

SUMMARY

Atoms are the building blocks of matter. They are composed of protons and neutrons in the nucleus and a cloud of electrons surrounding the nucleus. Atoms cannot interpenetrate because of the Pauli exclusion principle. Electrons in atoms may occupy many possible states, but one configuration is stable. Atoms are normally in this stable, or ground, state. The ground state has the lowest energy possible for the atom by itself.

Chemical energy arises from the exchange or sharing of electrons among atoms. When chemical bonds are made in a reaction, energy is given off, and the process is called exothermic. Such reactions occur, because the compounds are more stable afterward (they lower their energies). When chemical bonds are broken in a reaction, energy must be absorbed from the surroundings, and the process is called endothermic.

Most reactions involve both the making and the breaking of chemical bonds. If the overall effect is the making of bonds, the reaction is exothermic; if the overall effect is the breaking of bonds, the reaction is endothermic.

Activation energy is the energy necessary to begin the process of the formation of compounds. Catalysts are materials that can change the rates of reactions without being consumed in the reaction. Enzymes are biological catalysts.

Some exothermic chemical reactions can be arranged to produce electric current. To make an electric cell, we need two dissimilar metals and an electrolyte. A battery consists of a series of cells. A fuel cell produces electricity from the combination of hydrogen and oxygen to form water.

PROBLEMS AND QUESTIONS

MULTIPLE CHOICE REVIEW

1. The approximate energy needed to ionize the $n = 1$ (the deepest) state in hydrogen is
   a. 10 eV.
   b. 10 keV.
   c. 10 MeV.
   d. 10 GeV.
   e. 10 TeV.

2. Most of the chemical properties of atoms depend on
   a. the neutrons in the nucleus.
   b. the electrons closest to the nucleus.
   c. the configuration of outer electrons.
   d. the configuration of protons in the nucleus.
   e. the attraction of protons to neutrons.

3. The approximate size of the nucleus in an atom is
   a. 10^{-8} m.
   b. 10^{-10} m.
   c. 10^{-12} m.
   d. 10^{-15} m.
   e. 10^{-18} m.

4. According to the periodic table, which of the following reactions should occur?
   a. H + He $\longrightarrow$ HHe
   b. He + C $\longrightarrow$ HeC
   c. Li + F $\longrightarrow$ LiF
   d. Be + B $\longrightarrow$ BeB
   e. All of the above should occur.

5. Which of the statements about hydrogen–oxygen fuel cells is/are *true*?
   a. Electricity may be produced when the hydrogen and oxygen react.
   b. The only outputs are water and electrical energy.

c. The output continues as long as hydrogen and oxygen are supplied.

d. Prototype fuel cells have proven to be up to 80% efficient.

e. All of the above statements are correct.

6. According to the periodic table, which of the following reactions should *not* occur?

  a. $C + 2H_2 \longrightarrow CH_4$

  b. $B + N \longrightarrow BN$

  c. $2Li + O \longrightarrow Li_2O$

  d. $H + He \longrightarrow HHe$

  e. $H + H \longrightarrow H_2$

7. In a certain chemical reaction, chemical bonds are broken. The reaction is

  a. endothermic and liberates energy.

  b. endothermic and absorbs energy.

  c. exothermic and liberates energy.

  d. exothermic and absorbs energy.

  e. described by none of the above choices.

8. The approximate size of an atom is

  a. 10^{-6} m.

  b. 10^{-8} m.

  c. 10^{-10} m.

  d. 10^{-12} m.

  e. 10^{-14} m.

9. If two atoms appear in the same column of the periodic table, they will

  a. form similar compounds.

  b. have similar physical characteristics.

  c. have similar melting points.

  d. do all of the above.

  e. do only (a) and (c) above.

10. Consider the chemical process of photosynthesis. In this process,

  a. energy is absorbed by the chlorophyll in plants.

  b. water and carbon dioxide are turned into sugar.

  c. oxygen is given off.

  d. all of the above phenomena occur.

  e. only (a) and (c) may occur.

SHORT ANSWER

Explain whether the statement made is correct, and justify your answer.

11. Atoms are mostly empty space.

12. The Bohr quantization principle is responsible for the relative indivisibility of atoms.

13. In a neutral atom, the number of electrons is equal to the number of protons.

14. Ionized atoms are quite resistant to joining in chemical interactions.

15. The energy levels in the atom are continuous.

16. The greater the number of carbon atoms in a compound, the greater its heat of combustion.

17. There is direct evidence for the discreteness of energy levels in atoms.

18. In an exothermic reaction, bonds may be broken.

19. It is not possible that electricity may be produced when the hydrogen and oxygen react.

20. It is quantization of electron angular momentum that is the heart of the Bohr model.

DISCUSSION QUESTIONS

21. What does it mean to say that electrons are not point particles, but rather are wavelike?

22. Why are enzymes important to life?

23. What is the importance of the fact that methane has a chemical formula CH_4, while coal has a chemical formula approximately $CH_{0.8}$?

24. Methane combustion is basically carbon combining with oxygen to form carbon dioxide. Why don't we represent the chemical equation for methane burning as $C + O_2 \longrightarrow CO_2$?

25. Heptane has the chemical formula C_7H_{16}. How would one represent the burning of heptane accurately?

26. Is it possible for any compound to produce *less* carbon dioxide per unit mass than that for methane? Explain.

27. Menthene has the chemical formula $C_{10}H_{18}$. How would one represent the burning of menthene accurately?

28. Explain what occurs in the combustion of methyl formate (HCO_2CH_3).

29. Explain what occurs in the combustion of methylhexyl ketone ($CH_3CO_2C_6H_{13}$).

30. Consider certain reactions

$$A + B \longrightarrow C$$

and

$$A + B \xrightarrow[\text{catalyst}]{} C.$$

How does the energy output of these two reactions compare?

PROBLEMS

31. The heat of combustion of heptane, C_7H_{16}, is 48 MJ/kg.

  a. How many kilograms of CO_2 are emitted for each kilogram of heptane burned? (HINT: You will need to refer to Table 6.1 and Figure 6.4.)

  b. How much CO_2 is emitted per joule of heptane burned? How does this compare to methane?

32. Consider combustion of TNT.

  a. How many kilograms of CO_2 are emitted for each kilogram of TNT burned? (HINT: You will need to refer to Figure 6.4, Table 6.1, and **Extension 6.2.**)

  b. How much CO_2 is emitted per joule of TNT burned? How does this compare to methane?

33. Methyl formate (HCO_2CH_3) has a heat of combustion of 61 MJ/kg.

  a. Write out the chemical relationship for the combustion of methyl formate in air.

  b. How much of each kilogram of methyl formate is carbon?

  c. Suppose you could burn methyl formate with 100% efficiency (impossible) to run a 1000 MW power plant. At what rate must you supply the power plant with methyl formate?

34. A chemical reaction takes place in which at first, 50 kJ of energy is absorbed, then 200 kJ of energy is given off. What sort of reaction is this, and how much energy is absorbed or given off overall?

35. Given the size of the nucleus and the size of the atom, about how much of the volume of the atom does the nucleus occupy?

36. Pure carbon has a heat of combustion of 33 MJ/kg. How much energy would you get from burning 15 t of coal, assuming it to be pure carbon (it is not, as we have noted)?

37. Suppose reaction A involves the making of bonds and 225 J of energy, reaction B involves the breaking of bonds and 400 J of energy, and reaction C involves the making of bonds and 125 J of energy. What is the overall $A + B + C$ reaction energy? Is this an exothermic or an endothermic reaction? Explain.

38. Suppose reaction A involves the making of bonds and 22.5 J of energy, reaction B involves the breaking of bonds and 40.0 J of energy, and reaction C involves the making of bonds and 125.0 J of energy. What is the overall $A + B + C$ reaction energy? Is this an exothermic or an endothermic reaction? Explain.

39. Suppose reaction A involves the breaking of bonds and 75 J of energy, reaction B involves the breaking of bonds and 40 J of energy, and reaction C involves the making of bonds and 125 J of energy. What is the overall $A + B + C$ reaction energy? Is this an exothermic or an endothermic reaction? Explain.

40. Suppose reaction A involves the making of bonds and 44.5 J of energy, reaction B involves the making of bonds and 17.3 J of energy, and reaction C involves the breaking of bonds and 12.5 J of energy. What is the overall $A + B + C$ reaction energy? Is this an exothermic or an endothermic reaction? Explain.

CHAPTER 7

Efficiency of Energy Generation and Thermodynamics

In our world, machines are not 100% efficient. Most machines evolve thermal energy through friction. The study of thermodynamics allows for an understanding of the limitations of efficiency for machines. Thermodynamics deals with the study of thermal energy and the transfer of energy to and from particular systems. An understanding of energy conservation and entropy is essential in choosing suitable strategies for producing energy.

KEY TERMS

thermodynamics / temperature / Celsius scale of temperature / heat / thermal energy / internal energy / Kelvin scale of temperature / first law of thermodynamics / specific heat / latent heat of fusion / latent heat of vaporization / conduction / convection / radiation / statistics / Brownian motion / diffusion / evaporation / temperature reservoir / second law of thermodynamics / entropy / disorder / order / thermodynamic efficiency / Carnot cycle / second-law efficiency / waste heat

---

7.1   INTRODUCTION TO THERMODYNAMICS

Chemical change by burning is involved in most of the methods we use to generate energy. Figure 7.1 shows the flow of energy in the U.S. electricity system in 1999. We utilize the heats of combustion of materials by burning coal, oil, and gas in our power plants. In addition, we burn oil and gas to heat our homes, gasoline in our cars, kerosene for our jets, and diesel fuel for our trucks. Most of our energy-generating methods are inherently inefficient, as the figure shows that twice as much energy is lost as is used. Some ways of transforming energy are less inefficient than others. To understand the limitations on efficiency in such processes, we must study the effects of temperature changes in energy flow—**thermodynamics**.

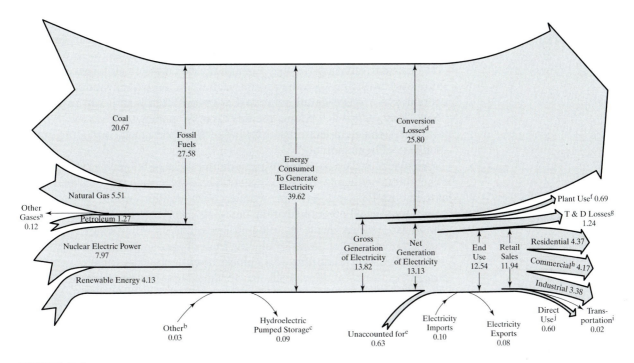

FIGURE 7.1

The flow of energy through the U.S. electrical system in 2003. Units are quads ($\approx$EJ)
(U.S.Department of Energy, *Annual Energy Review 2003* (Washington, D.C.: GPO, 2004), DOE/EIA-0384(2004), Diagram 5)

Temperature

Temperature is a useful concept because it can be defined macroscopically (in the large): if two systems are at the same temperature as a third system (say, a thermometer), then they are at the same temperature as each other and are in thermal equilibrium with each other. We therefore have a way to measure temperature. (Measurement is near and dear to the scientist's heart!) Bring a thermometer to a system, allow it to reach a state of equilibrium with the system, and then read off the temperature from the markings.

Temperature is normally measured in **degrees Celsius**, which are of such a size that the interval between the temperatures at which water freezes and boils at sea level is exactly 100 degrees. The freezing temperature is set to be 0 °C in this scale. [In the United States, temperature is also measured in Fahrenheit degrees; water freezes at 32 °F and boils at 212 °F at sea level on this scale. The spreadsheet Unit Converter on the CW shows how to transform from one unit to the other: $T(°F) = \frac{5}{9}T(°C) + 32$.]

Hot and Cold

We think of temperature as related to how hot or cold a body feels to us. If the other body is at a temperature different from ours, we would say it is warmer or colder than we are, heating us up or cooling us down. Our senses are reacting to energy that is transferred to us or away from us. (For more detail see **Extension 7.1,** *Explaining Hot and Cold.*) This energy transfer can do work (for example, a metal at high temperature can boil water). By **heat,** we refer to energy being transferred from one object to another because of a difference in temperature between them. According to this definition, an object cannot have a "heat content." (For more detail, refer to **Extension 7.2,** *There is no Such Thing as "Heat Energy."*) When people use this expression, they are probably referring to how hot—what temperature—a body is. What does the temperature measure, then, if not "heat content"?

An object at the South Pole is most definitely a cold object: but by comparison to an object on the planet Pluto, it is hot. The difference between a wooden block at the South Pole and that same wooden block on Pluto would not be obvious to an observer who only looked at the block. Its color and shape would be the same. The difference lies inside the block. In wood, as in other solids, the atoms are fixed relative to other atoms by electric forces between them. However, if there is energy available inside the wood block, it will appear as kinetic energy of the atoms; the atoms can move back and forth as if connected to springs. Each atom in the wood is moving and has kinetic energy. The atoms have different speeds, with probabilities described by the Maxwell-Boltzmann distribution.

Thermal and Internal Energy

The total energy inside the block is distributed among all its atoms. We call this randomized kinetic energy or **thermal energy.** Temperature is a direct measure of this thermal energy. (Temperature is essentially the average kinetic energy per molecule; for a monatomic ideal gas, $T = m<v^2>/3k$, where k is Boltzmann's constant, 1.38×10^{-23} J/K (the brackets $<>$ mean to take the average of what is inside)). A wood block at the South Pole has more thermal energy, and thus a higher temperature, than a similar block on Pluto. The **internal energy** is the sum of the thermal (kinetic) energy and the potential energy of the individual molecules in the object (for example, because the object is a solid and the individual atoms are "locked" in place by interatomic electric forces).

An object at 0 °C has thermal energy, because we know that colder temperatures exist. We shall use a scale with degrees of the same size as Celsius degrees, but beginning from the place that, in theory, should have zero internal energy. (Actually, this zero temperature is not attainable, because substances change their forms—the scientist would say change their phases—when sufficiently cold.) This temperature scale we have chosen to use is called the *Kelvin,* or absolute, temperature scale: its zero is at −273 °C. On the Kelvin scale, ice freezes at 273 K (0 °C), room temperature is near 300 K (27 °C), and water boils at 373 K (100 °C). The relation between temperature scales is T(K) = T(°C) + 273 K.

7.2   FIRST LAW OF THERMODYNAMICS

We saw that heat is energy in transit from one body to another at a different temperature. If we have a container of gas on a thermally insulated stand, we may add heat to it by, for example, bringing energy from an object at a higher temperature ("add heat"). Clearly, the container will have a greater amount of energy after the transfer. If we push on a piston connected to the container to compress a gas, we do work on it. By adding the work to the system, we have increased the energy in the container.

Contrariwise, we could take energy from the container and transfer it to another container at a lower temperature ("reject heat"). Or, we could allow the gas in the container to expand. A gas that expands can push the piston out, and in so doing, do work on its surroundings. In either of these cases, the amount of energy in the container has decreased.

If we transfer heat to the system but do no work on it, we increase the amount of energy in the system—its internal energy. If we do work on the system but do not add heat, we also increase the internal energy. If we let the system do work and do not add heat, or if we take out heat and do no work, or if we take out heat and have the system do work, the internal energy of that system must decrease. This statement of energy conservation can be summarized as the **first law of thermodynamics**, as follows:

(heat absorbed *by* the system) + (work done *on the* system)

= change in internal energy of the system.

The signs shown are important. They are chosen assuming that heat is absorbed *by* and that work is done *on* the system. If heat is given off *by* the system, the first term in parentheses is negative; if work is done *by* the system, the second term in parentheses is negative. Denoting heat absorbed by the system by ΔQ, work done on the system by W, and the internal energy change by ΔU, the relation just expressed in words may be written

$$\Delta Q + W = \Delta U.$$

This particular statement of energy conservation is known as the first law of thermodynamics.

We use the first law in many contexts, for example, when considering what happens when objects at different temperatures are brought in contact. Objects in contact will eventually come to thermal equilibrium. What will the final temperature be? To answer that question, we must know a bit more about the connection between energy transfer and temperature.

Specific Heat Capacity

The heat capacity of an object is the ratio of the amount of heat transferred to an object to the rise in the temperature of the object. (The heat capacity is $\Delta Q/\Delta T$, where ΔT is the temperature change.) For example, a pot of water on a stove, a bathtub full of water, and Lake Erie are all water. Naturally, Lake Erie has a greater heat capacity than the bathtub, which in turn, has more than the pot of water, simply because it contains more water. Said another way, an amount of heat that could boil the pot of water would hardly raise the bathtub water's temperature and would disappear into Lake Erie with barely a trace. Each body has its own heat capacity. But as scientists, we believe that fresh water should basically be the same everywhere (ignoring minor variations in dissolved mineral content), just as iron should be the same the world over. Differences arise from differences in the amount of matter, which as we saw in Chapter 3 is measured by the mass. The greater the mass, the greater the heat capacity.

If we divide the heat capacity of any object by its mass, we should have something that is characteristic of the composition of the object, specific to it. The heat

TABLE 7.1

Mean specific heats of water and various elements (at constant pressure)

Element or compound	Specific heat, kJ/(kg °C)
Water	4.186
Aluminum	0.900
Carbon (graphite)	0.712
Carbon (diamond)	0.519
Copper	0.268
Helium	5.191
Hydrogen	14.274
Iron	0.444
Lead	0.159
Mercury	0.139
Oxygen	0.917
Platinum	0.133
Silver	0.237
Tin	0.226
Zinc	0.388

Source: Reprinted with permission from *Handbook of Chemistry and Physics*, 32nd Ed. Copyright 1959 by CRC Press, Inc., Boca Raton, FL.

capacity of Lake Erie divided by the mass of water in Lake Erie should be the same as the heat capacity of the water in the pot divided by its mass. We call this quantity the specific heat capacity (or **specific heat** for short, $s = [\Delta Q/\Delta T]/m$). The specific heat of water is $4186 \text{ J/(kg °C)} = 1.16 \times 10^{-3} \text{ kWh/(kg °C)} = 1 \text{ kcal/(kg °C)}$. Water has an amazing property—most substances have specific heat capacities much smaller than that of water. Table 7.1 presents several specific heats of substances. (These are specific heats taken for fixed pressure.)

Let's see how knowledge of specific heat can be used to find final temperatures of mixtures. Suppose 1 kg of water at 95 °C is poured into 5 kg water that was originally at room temperature, 20 °C, in an insulated container. What will the final temperature be? We use the first law—any energy lost by the hot water must be gained by the cold water. So, if we knew the final temperature (call it T_F), we would know

$$\text{energy lost by the hot water} = [4186 \text{ J/(kg °C)}] \times (1 \text{ kg}) \times (95 °C - T_F)$$

and

$$\text{energy gained by the cold water} = [4186 \text{ J/(kg °C)}] \times (5 \text{ kg}) \times (T_F - 20 °C).$$

These energies must be equal by conservation of energy (the first law), so

$$[4186 \text{ J/(kg °C)}] \times (1 \text{ kg}) \times (95 °C - T_F) =$$
$$[4186 \text{ J/(kg °C)}] \times (5 \text{ kg}) \times (T_F - 20 °C).$$

In this case, the specific heats of the two objects are the same, and this reduces to

$$(1 \text{ kg}) \times (95 °C - T_F) = (5 \text{ kg}) \times (T_F - 20 °C), \text{ or}$$
$$95 °C - T_F = 5(T_F - 20 °C).$$

Solving for T_F, we obtain $6T_F = 195 °C$, so

$$T_F = 195 °C/6 = 32.5 °C.$$

We may find final temperatures by this method as long as there is no change that involves interatomic potential energies (such as would be the case if we mixed 1 kg of ice with 5 kg of water at 20 °C, discussed below).

Latent Heat

Matter usually appears to us in one of three states or phases: solid, liquid, or gas. Atoms in a solid are fixed in place relative to other atoms. Atoms in a liquid, while they are pulled toward adjacent atoms, are not locked into place. Atoms in gases are essentially free from the influence of one another.

In a glass of ice water, the water and the ice (solid water) are at the same temperature, 0 °C, because they are in contact. The two phases differ but can coexist at the same temperature. We know that the water will remain very cold until all the ice is gone. The reason is that it takes energy to "unlock" the solid, to pull the molecules far enough apart so that they are somewhat free to move. We must pay an energy price to free the molecules: this price, measured in energy per kilogram required to change the phase from solid to liquid, is known as the **latent heat of fusion**. The latent heat of fusion of water is 333 kJ/kg. (Fusion is the inverse process of melting; both involve the same energy.)

In a similar fashion, steam (water vapor) and liquid water may coexist at 100 °C at sea level. The energy price paid per kilogram to unbind the liquid molecules and make them free gas molecules is called the **latent heat of vaporization**. Water's latent heat of vaporization is 2.25 MJ/kg. (Condensation is the inverse process to vaporization.)

Latent heats and specific heats are important to know in dealing with the measurement or calculation of heat transferred when objects of different temperatures are brought together and finally reach some equilibrium temperature. As with specific heat, water has anomalously large latent heats compared to other materials.

How would this information and the first law allow us to find the final temperature of the mixture of 1 kg of ice and 5 kg of water? The first law is the same — any energy lost by the hot water must be gained by the cold water. The energy lost by the hot (20 °C) water is

$$\text{energy lost by the hot water} = [4186 \text{ J/(kg °C)}] \times (5 \text{ kg}) \times (20 \text{ °C} - T_F),$$

while energy is gained by ice at 0 °C until it is all melted. The amount of energy required to melt all the ice is (1 kg) × (333 kJ/kg) = 333 kJ. So, assuming all the ice melts, the energy gained by the ice is

$$\text{energy gained by the ice} = 333 \text{ kJ} + [4186 \text{ J/(kg °C)}] \times (1 \text{ kg}) \times (T_F - 0 \text{ °C}).$$

The maximum amount available if all the 5 kg of water ended up at 0 °C would be [4186 J/(kg °C)] × (5 kg) × (20 °C] = 418,600 J, so we are sure that the ice will all be melted and will increase in temperature. The first law now reads

$$333 \text{ kJ} + [4186 \text{ J/(kg °C)}] \times (1 \text{ kg}) \times (T_F - 0 \text{ °C})$$
$$= [4186 \text{ J/(kg °C)}] \times (5 \text{ kg}) \times (20 \text{ °C} - T_F).$$

This may be rewritten

$$333 \text{ kJ} + [4186 \text{ J/°C}] \times (T_F) = 418,600 \text{ J} - [20,930 \text{ J/°C}] \times (T_F),$$

and the like terms gathered to read

$$[4186 \text{ J/°C}] \times (T_F) + [20,930 \text{ J/°C}] \times (T_F) = [25,116 \text{ J/°C}] \times (T_F)$$
$$= 418,600 \text{ J} - 333,000 \text{ J} = 85,600 \text{ J}.$$

So $T_F = 85,600 \text{ J}/[25,116 \text{ J/°C}] = 3.41 \text{ °C}.$

7.3   HEAT TRANSFER

If two glasses are stuck, one inside the other, we can get them apart by running hot water on the glass on the outside. Why? When we increase the temperature of the glass, the atoms travel out farther from their position of equilibrium. Because each atom takes up a little more space, the glass expands when heated. Thus, the atoms in the glass are forced to move through a certain distance, and work is done. Because there is a difference in temperature, heat can be converted into work, or internal energy, or both. This is the central characteristic of heat transfer: heat will be transferred between two systems only if there is a temperature difference between the systems.

Types of Heat Transfer

Conduction

The transfer of energy by **conduction** occurs in situations in which there is a temperature difference and there is a solid material between the two temperatures. A window during winter is roughly at room temperature on the inside and outdoor temperature on the outside. [There is a "boundary layer" of cooler air just inside the window and warmer air just outside the window. The boundary layer acts as an insulator and cuts down the rate of heat transfer to a certain extent. The actual conduction depends on the size of the boundary layer, which is determined by the amount of wind outdoors and air circulation indoors (see Chapter 24).] Even if there are no air leaks, energy would be conducted through the window from the inside to the outside of the room because of the temperature difference. Energy may also be conducted through walls, through the bottom of a pan from a heating element or flame, through a coffee cup, and so on.

Convection

Energy may also be transferred by hot material as it moves; in this case, we say it is transferred by **convection**. In forced-air furnaces, gas or oil is burned to heat air, and the hot air is pushed by a fan out through the heat registers into the room. When this hot air touches other air or material objects, thermal energy is transferred from hot air to colder air or colder objects. If a person wears a strong perfume and walks through a room, the person moving through the air moves the air as well. In this case, the moving air carries the smell with it. The smell is mixed by convection. Convection and conduction are discussed in more detail below, in Chapter 9, and in **Extension 7.1**.

Radiation

The third (and final) way to transfer energy between bodies at different temperatures is by **radiation**. In contrast to the other two methods, transfer of energy by radiation takes place even in the absence of material, often by electromagnetic waves (including light). The Sun transfers heat to Earth by radiation. In Chapter 21, the first chapter on solar energy, we discuss radiation in more detail.

The Statistics of Heat Transfer

Energy conservation still does not tell us all we must know about thermal systems. To understand the crucial concept of time variation in the system, consider the following example: You have prepared your supper and just put it on the table. The phone rings, and a long-lost friend launches a reunion with you over the phone. Half an hour later, you sit down at the table. What does your knowledge of thermodynamics so far tell you?

1. Objects have a measurable temperature.
2. Energy is conserved.

As far as these precepts are concerned, you might expect to find your potatoes, coffee, and roast beef hot, and your ice cream cold, when you return. Energy is certainly conserved this way.

The dismal truth is that the potatoes, coffee, and roast beef are now rather cold, and the ice cream is a sort of cool puddle in the dish. If we left it long enough, everything on the table would be at the same temperature. See **Extension 7.3,** *Statistics and the Dinner* for more detail of how statistics assures these changes.

In the transfer of energy by conduction, the original energy gets randomly spread throughout a substance. Each atom gets its bit of additional vibrational energy (on average) or loses its bit of extra vibrational energy (on average). There is no way to distinguish or idiosyncratically label a constituent atom or molecule. The beef cools, the ice cream melts, and the plate and the air become slightly warmer.

Without this large number of atoms, or the large number of collisions, we could not be certain of the outcome. That is why our discussion focuses on statistics. In fact, it is this statistical probability phenomenon that allows us to distinguish future from past—to label "time's arrow."

Brownian Motion and Diffusion—Evidence for Statistical Collisions of Molecules

Several phenomena prove beyond any reasonable doubt that collisions occur between atoms (at least in gases and liquids). In 1821, Robert Brown, a botanist, observed under a microscope that pollen particles in water jiggled. The explanation of this **Brownian motion** was provided by Albert Einstein (see **Extension 7.4,** *The Maxwell–Boltzmann Distribution*). The jiggling of the pollen particles, which are visible, is due to their collisions with water molecules, which are invisible. Smoke particles in still air also exhibit Brownian motion, reflecting smoke particle collision with air molecules.

The phenomenon of *diffusion*, in which small amounts of impurities mix thoroughly in gases or liquids, takes place because the molecules are in constant motion. One example of diffusion is provided by the release of a smelly substance (for example, cheap perfume) in a still jar. Even if there is no gross air motion—caused, say, by someone moving through the room dragging the smell along—the smell will slowly but surely penetrate every corner of the room. Another example of diffusion is the mixing of a drop of food coloring put into still water. The color will spread out from the drop and, after some time, produce water of uniform color. In both of these examples, collision with the gas or liquid molecules is responsible for the mixing. The drop will not spontaneously unmix, nor will the smelly substance become concentrated in one part of a room. Again, we can see "time's arrow."

7.4   THE SECOND LAW OF THERMODYNAMICS

Heat transfer from a warmer to a colder body resembles transfer of water from a reservoir to a lower level. Work can be obtained from water dammed in a reservoir by allowing it to fall the height of the dam and run turbines in powerhouses, thereby producing electricity. On top of the dam (in the reservoir) all the water is at one level. There is no way to have water in the reservoir do work while remaining at its single level. There must be a difference created in height to allow gravitational potential energy to be converted into work (illustrated in Figure 7.2).

FIGURE 7.2

The analogy between water falling from the top of a dam, doing work, with the water falling into the river, and heat taken from a high-temperature reservoir, doing work, and the remaining heat rejected to the low-temperature reservoir. The barrier in the left picture is a dam; that in the right is a thermal insulator.

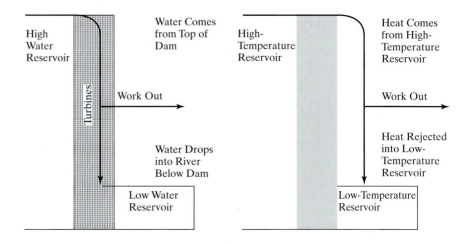

We can imagine that a large body at a given temperature is a **temperature reservoir**. This temperature reservoir contains gigantic quantities of thermal energy at one temperature. The reservoir is so large that transfer of thermal energy out or in does not affect its temperature. Work is produced by heat flowing "down" from a reservoir at a higher temperature to a reservoir at a lower temperature (just as water flows from a higher to a lower elevation).

It is not possible to use the water in the higher reservoir to do work, even though the water contains gravitational potential energy in immense quantities; only when that water flows to a reservoir at a lower elevation can work be done. Likewise, it is not possible to get work from the thermal energy in a temperature reservoir at a fixed temperature. Only by allowing a transfer of some of that thermal energy to a reservoir at a lower temperature can work be done.

Air molecules form an immense reservoir of energy; however, it is not possible to run cars on this energy. We must transfer energy from a higher to a lower temperature to do work. We, therefore, burn gasoline to produce a high enough temperature. We have seen that temperature is a measure of internal energy or of randomized vibrational energy. Recall that energy is conserved: if work is done, then heat has been transferred, or internal energy has been changed, or both (first law of thermodynamics). Our discussion of the irreversible changes of the roast beef dinner (see **Extension 7.3**) has prepared us for one way of stating the **second law of thermodynamics**—that systems achieve equilibrium. When there is equilibrium, there can be no temperature difference. If there is no temperature difference, there is no transfer of heat, so no work can be done. This is another way to state the second law (there are many more ways).

The Second Law of Thermodynamics is as follows: Systems isolated from the rest of the universe will move toward equilibrium with their surroundings.

It is impossible for a machine to take heat from a reservoir at a certain temperature, to produce work, and to exhaust heat to a reservoir at the *same* temperature.

Entropy

By defining order, we have also defined disorder. It is possible to identify "disorder" quantitatively with something known as **entropy**. We can then say that *the entropy in the universe always increases* in any process. (The only exception occurs when a system is already at equilibrium. In this case, any process taking place within the system will leave it in equilibrium and thus leave its entropy unchanged.) Just because entropy increases globally does not mean that there can never be a localized entropy decrease. Such local decreases are possible as long as the entropy of the universe as a whole increases in the process. The inexorability of the increase in entropy in the universe is still another way to express the second law of thermodynamics. See **Extension 7.5,** *Why Conserve Helium?*

A CLOSER LOOK

Order and Disorder

Still another way to phrase the second law of thermodynamics is to say that processes go in such a way that **disorder** increases. Before this statement can make sense, we must decide what we mean by "order." **Order** is set by rules. The 52 playing cards in a deck, for example, look similar to one another to an untrained eye. As you learn more, you are able to distinguish two sets of equal size: red cards and black cards. You could order by color: black, red, black, red or black, black, red, red. Closer inspection reveals that there are two sorts of red cards—diamonds and hearts—and two sorts of black cards—clubs and spades. There could be an arbitrary decision that, say, spades come before hearts, hearts before diamonds, diamonds before clubs. This is an ordering of the suits you would use in playing the game of bridge. You could now order by the "value" of the suits: clubs, diamonds, hearts, spades, clubs, diamonds, hearts, spades, and so forth.

Someone could also notice that most cards have numbers. The cards may be ordered according to number. As we know, all these orderings are important if we are to be able to play cards well. The orderings are completely arbitrary, and the rules are based on the arbitrary orderings; once we accept them, we can play bridge, poker, or whatever.

A newly opened deck of cards has the cards in a standard order. By moving just one card, we have disordered the deck. If we then move another card, chosen at random, the deck will become even more disordered. Only if we chose the card first moved and happened to restore it to its original position would the deck become more ordered (this would happen by chance only about once every 2700 tries). As more cards are moved, the probability for restoration of the original order gets even smaller. This is another example of the statistical basis of the second law of thermodynamics.

In talking about the second law of thermodynamics and order or disorder, we must define our (arbitrary) order. Only when this is done can we discuss changes in order. There is an overwhelming probability that order will be lost, whether we are talking about the moving of cards or the collisions of molecules. In every real-world case—in which the number of ways to increase disorder is so much greater than the number of ways to increase order—disorder will increase.

For example, consider a typical kitchen (such as the one in my house). Periodically, we clean it up. A few days later, the opened and unopened letters, cookies, and the butter knife are spread all around. There is only one ordered state of the kitchen. There are many, many other states, all disordered in some way. It is much more probable that the kitchen will get more disordered—at least until we do work to restore the order. A kitchen seemingly left to itself will become disordered with no apparent effort.

> Remember that roast beef dinner? In the beginning, there was an order, a hierarchy, of objects by temperature. The roast beef was hot, the plate and air were at room temperature, and the ice cream was cold. At the end, all of the vibrational energy was randomly spread out, with the ice cream being about the same temperature as the roast beef and the air and the plate. There was no longer a temperature order or hierarchy. That original *order* was now lost; in its place was *disorder*. And so it is with other, potentially more complex, systems.

How can we have cold ice cream if all processes involve an increase in entropy? A freezer surely has a lower entropy than its surroundings (it is more ordered at a lower temperature). But, the freezer has to be cooled by a motor that exhausts heat into the air. It takes energy input to decrease entropy locally. Generation of this energy causes an entropy increase in the universe, which more than compensates for the entropy decrease in the freezer and increases the total entropy of the entire universe.

Entropy may be decreased in a system by input of energy from outside the system: a system in equilibrium may be put into a state of disequilibrium by doing work on it. Likewise, as systems return from a state of disequilibrium to one of equilibrium with their surroundings, useful work may be done. For example, when we compress air into a tank, we create a pressure disequilibrium by doing work on the air. We now have a pressure difference where there was none before, at an energy cost. This pressure disequilibrium could now provide useful work by running a power tool, raising a hydraulic lift, or whatever.

Examples of large entropy increases are present in Humpty Dumpty's fall, in the breaking of an egg yolk, or in the mixing of cookie batter. Note that all these processes are irreversible. All real-world processes lead to increasing disorder, unless energy is used to decrease the disorder in some locality. As mentioned, however, production of the energy involved a larger entropy increase than the local decrease achieved by its use.

Another familiar example of entropy increase as a transition from order to disorder is the sponge. A sponge is an object riddled with tiny holes. When it is slightly wet, water can travel inside the sponge by a phenomenon known as capillary action. In capillary action, water in thin tubes rises due to water surface tension, and therefore, water can be drawn up in the sponge. Suppose there is a big spill and a slightly damp sponge. Although this might seem to be a disordered system, in fact, it is *ordered* with respect to the degree of wetness. As water flows into the sponge, the wet–dry order vanishes, and everything becomes uniformly wet. At this point, no more water (or milk or whatever) is absorbed by the sponge. Nothing more happens until energy is used to restore the wet–dry order (by manually wringing out the sponge, for example). Then, the sponge can be used again in the same manner.

7.5   MAXIMUM THERMODYNAMIC EFFICIENCY OF A SYSTEM

For doing work with a machine, we defined efficiency as the ratio of the work done by a machine to the total energy used to run the machine. In heat engines, energy is put into the system, and work is done. In all cases, the ratio of the work done in some process to the total energy input is the *efficiency* of that process (sometimes called the first-law efficiency):

$$\text{efficiency} = \frac{\text{work output}}{\text{total energy input}}.$$

It is also possible to speak about the theoretical efficiency of a thermal process. At a given temperature, there is a total amount of energy involved in the vibrations of atoms in a material. If all that energy could be used to do work, we would have the *maximum* possible efficiency. Here, the total energy input would be the total amount of internal (vibrational kinetic and potential) energy in the material above what there is at absolute zero (the so-called zero-point energy).

Exact methods exist for proving the relation of efficiency to temperature; these were delineated by the founder of the science of thermodynamics, the French engineer Sadi Carnot. Recall that the internal energy of a substance is proportional to temperature. The constant of proportionality varies from substance to substance. (It is called the heat capacity at constant volume, as previously discussed.) We usually assume that the heat capacity for a certain substance is constant for all temperatures. The internal energy is thus proportional to the Kelvin (absolute) temperature, T:

$$\text{internal energy} = (\text{heat capacity})\, T.$$

It is then plausible that efficiency may be expressed as a ratio of temperatures for thermal systems transferring energy—it is a **thermodynamic efficiency**.

Using the absolute temperature scale, we can consider a special process (called a Carnot cycle; see the box *History of Energy:* The Carnot Cycle) in which work is done in such a way that all parts of the process can be reversed and all intervening steps retraced. A prototypical heat engine is shown in Figure 7.3. The **Carnot cycle** is an idealization and is never encountered in the "real world." In particular, no real engine is reversible. This is still another way to express the second law of thermodynamics.

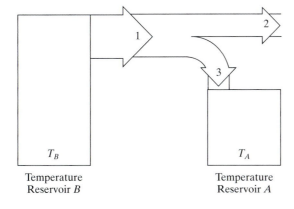

T_B

Temperature
Reservoir B

T_A

Temperature
Reservoir A

FIGURE 7.3

Thermodynamic representation of a process by which thermal energy is absorbed from the high-temperature reservoir, some thermal energy is rejected to the low-temperature reservoir, and work is done. The device illustrated works as a heat engine.

1 Energy Absorbed from Reservoir at High Temperature

2 Output of Useful Work

3 Energy Rejected to Reservoir at Low Temperature

The cycle (see box *History of Energy:* The Carnot Cycle for a complete description) consists of absorbing heat from a high-temperature reservoir, B, at absolute temperature T_B, doing work, and disposing of the remaining heat at the low-temperature reservoir, A, at absolute temperature T_A. This very special cycle has a thermodynamic efficiency of

$$\text{Carnot efficiency} = \frac{\text{work output}}{\text{total energy input}}$$

$$= \frac{\text{total energy input (at } T_B) - \text{energy exhausted (at } T_A)}{\text{total available energy}}$$

$$= \frac{T_B - T_A}{T_B}.$$

Even though it is an *ideal* cycle, the thermodynamic efficiency is not 100%, because some energy must be rejected at the lower temperature. Only if all the original internal energy that is available is used can the Carnot cycle be 100% efficient; in this case, the low temperature T_A must be 0 K. As an example, consider a coal-fired plant with a boiler temperature of 400 °C (673 K) that uses a local river at 20 °C (293 K) to cool the spent steam. The Carnot (maximal) efficiency is (673 K − 293 K)/673 K = 0.564. (Note that we do *not* use the high temperature as 400 °C in the denominator. This is an important point.) The plant's maximum possible first-law efficiency is 56.4%.

In terms of machines, the second law of thermodynamics says that it is impossible to devise a machine that takes heat from a reservoir and produce only useful work with it. In a similar way, a boat could not run on the gravitational potential energy stored behind a dam in a reservoir.

HISTORY OF ENERGY

The Carnot Cycle

The Carnot cycle consists of four successive steps:

1. From its initial state, the working fluid expands at fixed temperature T_B, absorbing heat from its surroundings.
2. The working fluid is removed from temperature reservoir B and insulated from all outside influence. No heat can be absorbed or given off; such a process is termed *adiabatic*. The working fluid expands adiabatically, and the temperature decreases to T_A.
3. The working fluid is placed on temperature reservoir A. It is compressed at fixed temperature T_A, exhausting heat to its surroundings.
4. The working fluid is removed from temperature reservoir A. It expands adiabatically until it returns to the initial state.

The Carnot cycle is a closed cycle. In the process, energy is absorbed at the high-temperature reservoir (B), and a smaller amount of energy is given off at the low-temperature reservoir (A). The difference between the energy absorbed and the energy rejected is the amount of work produced by the Carnot engine. If we run the Carnot engine in reverse, the result is a Carnot refrigerator. Real engines are less efficient than Carnot engines. Real refrigerators or air conditioners require more energy input for a given amount of cooling than does a Carnot refrigerator.

The Carnot engine is a special heat engine, the one that has the greatest efficiency when it operates between two temperature reservoirs. In all heat engines, work is done as the material expands and then is compressed again. In the Otto engine (used in a car, Figure 7.4), the gasoline burned in one particular cylinder causes the volume of the cylinder to expand and contributes part of its resultant energy to compressing the air in other cylinders. A little later in the cycle, another cylinder will fire and cause the air in the first cylinder to be compressed in preparation for the gasoline to be burned again. In the Otto engine, the low-temperature reservoir is the outside air, and the high-temperature reservoir is the combusted hot air and gasoline mixture just after the spark ignites it.

The Carnot efficiency is the maximum theoretical efficiency of a heat engine operating between temperature reservoirs at temperatures T_B and T_A. No engine may exceed this thermodynamic efficiency; however, there is no guarantee that it can be attained. The degree of attainment depends on the particular engineering of the heat engine. The automobile engine is about half as efficient as it theoretically could be. The second law of thermodynamics guarantees that the thermodynamic efficiency $(T_B - T_A)/T_B$ is the absolute maximum efficiency attainable by any possible contrivable heat engine. This constraint has only to do with the second law of thermodynamics and could be said to preclude the existence of perpetual motion machines otherwise allowed by energy conservation. We shall encounter many examples of Carnot efficiency in later chapters.

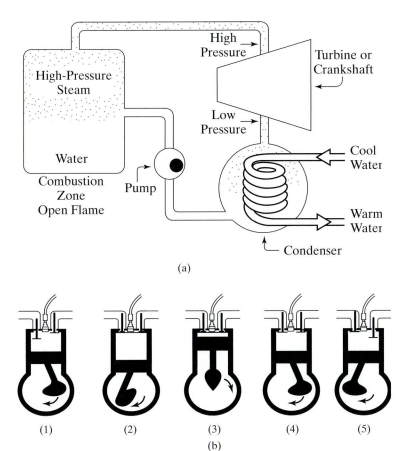

(a)

(b)

FIGURE 7.4

(a) The closed cycle external combustion engine. The "external" refers to the fact that the combustion takes place in air outside the steam chamber. The presence of the condenser maintains the pressure difference that allows the steam to turn the turbine. The closed cycle reuses the water instead of exhausting it (in the latter case, it would be referred to as an open cycle). (b) The internal combustion engine. The Otto cycle (car internal combustion engine). In the internal combustion engine, the fuel is burned inside the chamber. The flywheel keeps the crankshaft turning; it is connected to all cylinders of the engine. (1) The air–fuel mix is introduced into the cylinder from the carbeuretor through the intake valve. (2) The intake valve closes, and compression begins. (3) The air–fuel mix is compressed, and the spark plug fires, igniting the mixture. (4) The expanding gas pushes the piston down. (5) The exhaust valve opens, allowing the exhaust gases out.

FIGURE 7.5

A schematic thermodynamic representation of a refrigerator, in which thermal energy is absorbed at the low-temperature reservoir, work is done on the system, and the total (thermal energy plus work) is rejected at the high-temperature reservoir.

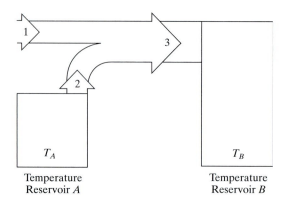

Temperature Reservoir *A*

Temperature Reservoir *B*

1 Work Input

2 Energy Extracted from Reservoir at Low Temperature

3 Energy Rejected to Reservoir at High Temperature

If we reverse the cycle for the heat engine just discussed, we have a process that takes in heat from a low-temperature reservoir, does work on it, and exhausts the original heat plus the additional work to the high-temperature reservoir. Such a process can be thought of as cooling the low-temperature reservoir or as heating the high-temperature reservoir. A refrigerator is a device for cooling a low-temperature reservoir; a prototypical refrigerator (or air conditioner) is shown in Figure 7.5. When such a device is used to heat the high temperature reservoir, we call it a heat pump. (See **Extension 7.6,** ***Real Power Plants and Refrigerators.***)

An electric heat pump is more efficient than resistive electric heating. In resistive heating, the thermal energy transferred is Joule heat (see Chapter 4). In the heat pump, thermal energy is taken from outside air or water (remember, because they are far above absolute zero in temperature, they have a lot of thermal energy). Electric energy is put in to run the pump and compressor, and the sum total of the outside thermal energy plus the work done is delivered to heat the room, the high-temperature reservoir. Hence, the electricity provides only a small part of this thermal energy, and we pay only for that small part. Thermal energy from the great outdoors does not cost money directly. (See also Chapter 9.)

It is clear from these explanations that the greater the temperature difference, the more work that can be done by a given amount of thermal energy. (Here, we are comparing high and low temperatures—a greater amount of material at a low temperature can contain as much thermal energy as a smaller amount at a higher temperature—see box *History of Energy:* The Carnot Cycle).

7.6   SECOND-LAW EFFICIENCIES

Most definitions of efficiency, in contrast to the one we just developed, do not take the dissipative effects of entropy into account. Such definitions can be misleading; they can severely misstate the efficiency of a given process. This problem led a group of physicists involved in a project under the auspices of the American Physical Society to propose a different definition of efficiency, which they call **second-law efficiency**.[6] This definition is equivalent to the efficiency as calculated for thermal systems (such as power plants), and it takes the "quality" of the energy into account in other applications. The idea of "quality" of energy is meant to clarify how a given amount of energy differs in its ability to do useful work. Electric energy is of high quality, because it may be used easily to do work of many kinds. Room-temperature thermal energy is of low quality; it may be used to heat a room, but it is difficult to use it to raise a weight or run a stereo.

The point of the second-law efficiency idea is that some conversions of energy are more rational than others, in terms of performing a certain amount of work. The authors of the American Physical Society study defined a quantity they call "available work": the "maximum work that can be provided by a system (or by fuel) as it proceeds (by any path) to a specified final state in thermodynamic equilibrium with the atmosphere."[6] Available work then defines the minimum amount of energy needed to do a certain amount of work (at maximal efficiency); it is consumed (rendered unavailable) during a process. For real systems, the actual energy used in a device for some specified purpose is greater than the available work. The second-law efficiency is defined as the available work divided by the actual work. This definition of available work allows us to make quantitative the distinction between high-quality energy (which has a great deal of available work) and low-quality energy (which has a much smaller amount of available work). Work is of highest quality, as is electric energy. Thermal energy is of low quality.

There are great differences between conventional measures of efficiency (which we may call first-law efficiency) and the second-law efficiencies. The definition of efficiency as the ratio of the work output to the total available energy is a definition of a first-law efficiency. Table 7.2 presents a comparison of first- and second-law efficiencies from Reference 3. An alternative statement of the second-law efficiency is that

$$\text{second-law efficiency} = \frac{\text{first-law efficiency of a chosen process}}{\text{Carnot efficiency of the best possible process for the task}}.$$

The calculation of second-law efficiencies is an important component of any national policy of energy conservation, because it shows how much room is left for

TABLE 7.2

Comparison of efficiencies of household heating systems at typical ambient and heating or cooling temperatures

Task	First-law efficiency	Second-law efficiency
Water heat		
Electric	0.75 (0.25)	0.015
Gas	0.50	0.029
Space heat		
At room	0.60	0.028
At register	0.60	0.074
At furnace	0.75	0.145
Air conditioning	2.00 (0.67)	0.045

NOTE: Numbers in parentheses refer to efficiencies, taking into account the loss of a factor of three in converting from thermal to electric energy production (see text for explanation of the factor of three in efficiency).

improvement.[7] A maximization of the second-law efficiency provides a standard against which proposed improvements can be measured. This standard was used in an action by the Federal Power Commission,[8] in which only part of a proposal by an oil company consortium to use offshore gas to run refineries was granted (on the basis of a second-law argument).

For a furnace in a power plant, the available work is the heat of combustion, so that our thermodynamic efficiency and the second-law efficiency are identical. For a solar hot water heater, in contrast, the first-law efficiency is given as the ratio of the heat Q_2 added to the warm reservoir at temperature T_2 to the heat Q_1 taken from the hot reservoir at T_1: Q_2/Q_1. In this case,

$$\text{second-law efficiency} = \frac{Q_2}{Q_1} \times \frac{1/T - 1/T_2}{1/T - 1/T_1},$$

where T is the atmospheric absolute temperature.

Let us calculate the second-law efficiency of a power plant operating at an upper temperature of 500 °C. A typical river temperature is about 20 °C, and river water will probably be used as a coolant in our imaginary generating facility, so we will take 20 °C as the temperature of the lower-temperature reservoir. The thermodynamic efficiency is calculated with the absolute temperature, so first we must change Celsius to absolute temperatures: 500 °C is

$$T_B = (500 + 273) \text{ K} = 773 \text{ K};$$

20 °C is

$$T_A = (20 + 273) \text{ K} = 293 \text{ K}.$$

We may then write that the thermodynamic efficiency equals

$$\text{Carnot efficiency} = \frac{T_B - T_A}{T_B}$$
$$= \frac{773 \text{ K} - 293 \text{ K}}{773 \text{ K}} = \frac{480 \text{ K}}{773 \text{ K}} = 0.65.$$

The idealized, or Carnot, engine operates with a maximum efficiency of 65% between the typical generating facility temperatures. All actual engines are far less efficient than the Carnot engine. The actual efficiency of all U.S. fossil-fuel generating facilities rose from 24.3% in 1950 to 31.7% in 1960, 32.5% in 1970, 32.9% in 1980, and 33.0% from 1992 onward (see also **Extension 8.1**).[9] As a good approximation, we may take 33.3% (that is, one third) as the actual power generation efficiency of American fossil-fuel plants; this is only half the theoretical maximum efficiency (refer to Figure 7.1). The second-law efficiency of a nuclear-generating facility is also about one-third (32.1% in 2000).[9]

The first-law efficiency is defined differently for different processes, which is one of the reasons that the definition of second-law efficiency is superior. For example, for an electric motor, the first-law efficiency is defined as work out/work in.

A power plant burning fuel energy at a rate of 3000 MW to produce electricity for sale transmits only $\frac{1}{3} \times (3000 \text{ MW}) = 1000 \text{ MW}$ to its customers. The other 2000 MW is exhausted to the environment and turns up as **waste heat**. This is heat of low thermodynamic quality; that is, it will not do work. We will discuss this waste heat in detail in Chapter 13.

We have given energies in kilowatthours (kWh) in many places in the text. However, as we have just seen, there is a distinction between electric energy and fuel (or thermal) energy of a factor of three. We will avoid any confusion in the rest of the book by distinguishing between electric kilowatthours (kWh$_e$) and thermal kilowatthours (kWh$_t$) where that distinction matters:

$$1 \text{ kWh}_t = \frac{1}{3} \text{ kWh}_e; \text{ or } 1 \text{ kWh}_e = 3 \text{ kWh}_t.$$

SUMMARY

Heat is energy in transit from higher to lower temperature. Heat may be transferred by conduction, convection, and radiation. In conduction, heat moves through a material that does not move itself (a metal knife held in boiling water will eventually cause burning pain in the fingertips of the knife holder). In convection, hot material carries heat along with it; for example, hot air from a jet engine can make people in the backwash hot. In radiation (such as by light), no material medium is required to transfer heat.

No energy is gained or lost by the universe by any process; energy is merely changed into other forms. Heat is work, and work is heat: We call this restatement of the principle of energy conservation the first law of thermodynamics. According to the first law, the thermal energy gain of a system is the sum of the heat transferred into the system and the work done on the system.

Heat is transferred only if there is a temperature difference. Although we cannot measure the "heat content" of a body, we can measure the internal energy, which is the sum of the kinetic energy of all atoms and the potential (binding) energy of the atoms in a material. Absolute temperature is a measure of the internal energy, or "degree of hotness." Atoms in materials may have any energy. The average energy is measured by the absolute temperature, but the individual atoms have energies that fall on a characteristic distribution, the Maxwell–Boltzmann distribution, which describes the number of atoms at a particular energy level as a function of its energy.

Entropy is a measure of disorder. To define disorder, we must first agree on what we will call order. There are many ways to define order. Once the order is defined, there is only one way for something to be ordered, and many possible ways for it to be disordered; any random change is likely to decrease order. This tendency toward disorder, or entropy, is called the second law of thermodynamics. The heart of the second law is its statistical character. The second law may be stated in many different ways:

- A system tends toward equilibrium with its surroundings; the entropy of the universe increases or remains the same in any process.
- It is possible to decrease entropy locally only if work is brought in from outside the system.
- It is impossible to build a machine that takes heat from a temperature reservoir and produces only useful work.
- It is impossible for any machine to absorb heat from a temperature reservoir, do work, and exhaust heat to a reservoir at the same temperature.
- All real processes are irreversible.

(This list by no means exhausts the possible alternative statements of the second law.)

Even the Carnot engine, theoretically the most efficient heat engine, must exhaust waste heat to the low-temperature reservoir. The efficiency, which is the ratio of the work done (equal to the difference between heat absorbed at the high-temperature reservoir and heat rejected at the low-temperature reservoir) to the heat absorbed at the high-temperature reservoir is therefore always less than one. In fact, it is possible to show that the efficiency of the Carnot cycle is given by $(T_H - T_L)/T_H$, where T_H is the absolute temperature of the high-temperature reservoir, and T_L is the absolute temperature of the low-temperature reservoir. The second law of thermodynamics guarantees that any real engine has an efficiency less than that of a Carnot engine operating between the same high and low temperatures.

Electric energy can do many things; it is an energy source of high second-law efficiency, and, therefore, we say its quality is high. Thermal energy is much more difficult to use; due to its low second-law efficiency, we say it is of low quality.

PROBLEMS AND QUESTIONS

MULTIPLE CHOICE REVIEW

1. The spreading of heat by conduction
   a. is statistical in nature.
   b. flows from tall objects to short objects.
   c. can occur instantaneously across large distances.
   d. cannot be quantified.
   e. none of the above.

2. The amount of energy required to melt 10 kg of ice to water and then raise its temperature from 0 °C to 20 °C is
   a. 83.7 kJ.
   b. 837 kJ.
   c. 3330 kJ.
   d. 4167 kJ.
   e. 5004 kJ.

3. What is the ratio of average kinetic energy per molecule in a monatomic gas at 1200 °C to that same gas at 400 °C?
   a. 3.0
   b. 2.2
   c. 1.0
   d. 0.46
   e. 0.33

4. Which of the following statements is *false*? Friction
   a. causes the efficiency of real machines to be less than 100%.
   b. operates in all real machines.
   c. involves conversion of mechanical energy to random motion of the molecules in a substance.
   d. can be eliminated with proper cleaning of surfaces.
   e. would not operate were it not for electric forces between adjoining surfaces.

5. An ideal thermal engine operates by drawing 2500 J of thermal energy from its environment and exhausting 200 J of thermal energy to a low-temperature reservoir on each cycle. How much work does the engine do in one cycle?
   a. 200 J
   b. 400 J
   c. 2100 J
   d. 2300 J
   e. 2700 J

6. In an ideal monatomic gas, the absolute temperature of the gas
   a. is constant.
   b. depends on the average of the square of the speed of the atoms in the gas.
   c. is proportional to the radius of the atoms in the gas.
   d. slowly changes.
   e. is described by none of the above choices.

7. The amount of energy absorbed when 10 kg of water is heated from 80 °C to 100 °C and then is turned into steam is
   a. 837 kJ.
   b. 3330 kJ.
   c. 4167 kJ.
   d. 22500 kJ.
   e. 23337 kJ.

8. Which is *not* a statement of the second law of thermodynamics?
   a. Heat flows spontaneously from a hot object to a cold one.
   b. All real processes are reversible.

   c. Isolated systems tend toward equilibrium with their surroundings.
   d. It is possible to decrease entropy locally only if work is brought in from outside the system.
   e. None of the above is a statement of the Second Law.

9. A heat pump is a device that takes heat from a ____ temperature reservoir and heats a ____ temperature reservoir.
   a. low-, low-
   b. low-, high-
   c. high-, low-
   d. high-, high-

10. An engine operates between two temperatures with a maximum efficiency of 0.80. If the lower operating temperature is 100 °C, what is the upper operating temperature?
    a. 80 °C
    b. 180 °C
    c. 500 °C
    d. 1592 °C
    e. 1865 °C

SHORT ANSWER

Explain whether the statement made is correct, and justify your answer.

11. All forms of energy are equivalent in terms of doing useful work.

12. Substance A has twice the heat capacity of substance B. If equal masses of both substances are raised 10 °C, and substance A gains 30 kJ of energy, substance B gains 15 kJ of energy.

13. The average kinetic energy per molecule in an ideal gas at 0 °C is zero.

14. It is possible to decrease the entropy of a system locally without doing work on the system.

15. In an ideal monatomic gas, the average of the square of the speed of the molecules in the gas increases as the number of molecules is increased.

16. Heat flows spontaneously from a hot object to a cold one.

17. Substantial heat transfer by conduction can occur over great distances.

18. In friction, the kinetic energy of an object may be turned into the random thermal motion of the molecules making up the object.

19. It is possible for a mixture of hot water (90 °C) and ice to have a final equilibrium temperature of 0 °C.

20. It is easy to look up the heat capacity of most objects in a reference book.

DISCUSSION QUESTIONS

21. Why does the high specific heat of water make it useful as a coolant?

22. A school principal tells a new teacher that "the orderly classroom is quiet." Using this definition of order, and assuming a class of 20 pupils, list several ways that the class could be disordered.

23. What distinction is there between a thermal kilowatt-hour and an electric kilowatthour?

24. If no one has ever seen a molecule, how do we know that there are molecules in fluids?

25. Which costs more energy: changing 5 kg of ice into water at 0 °C, or changing 5 of water at 100 °C into steam at 100 °C?

26. Where do the "conversion losses" shown in Figure 7.1 come from?

27. An inventor wants you to finance construction of a machine that will do work by running a continuous chain through a pulley, letting weights attached every so often fall down do the work. The machine should run forever, because the weights will fall again and again as the chain goes round the pulley. You both can sell energy so cheaply, you would make millions. Should you invest $100,000? Why or why not?

28. An inventor wants you to finance construction of a machine that will run at room temperature by using the free thermal energy available in the air. Should you invest $100,000? Why or why not?

29. If two machines have the same first-law efficiency, they must also have the same second-law efficiency.

30. Hydrogen should be a more useful coolant than water, because its specific heat is greater. Explain why this is not the case in practice.

PROBLEMS

31. Determine the temperature on the absolute scale of the following Celsius temperatures:
    a. 20 °C
    b. 37 °C
    c. −100 °C
    d. +100 °C
    e. −200 °C
    f. 6000 °C

32. Determine the Celsius temperature for the following absolute temperatures:
    a. 0 K
    b. −10 K
    c. 100 K
    d. 273 K
    e. 300 K

33. A school principal tells a new teacher that "the orderly classroom is quiet." Using this definition of order, and assuming a class of 20 pupils, enumerate possible states of the system (a state specifies each one of the 20 and whether the student is quiet or talking). Why is it so hard to have a classroom completely in order?

34. Consider ice water that is in a beaker on a hot plate and is being stirred. There is originally 0.50 kg of water and 1.00 kg of shredded ice in the beaker. What is the temperature of the ice water if the plate is at $T = 85$ °C. Why is it at that temperature? If the hot plate is a 1200 W model, how long would it take to raise the temperature of the beaker contents to 20 °C?

35. An ideal thermal engine operates by drawing 2500 J of thermal energy from its environment and exhausting 200 J of thermal energy to a low-temperature reservoir on each cycle. How much work does the engine do in one cycle? What is the efficiency of this engine? If the low-temperature reservoir is at 27 °C, what is the temperature of the high-temperature reservoir?

36. An ideal thermal engine does 2500 J of work on each cycle and exhausts 1000 J of energy to its low-temperature reservoir in each cycle. How much work does the engine do in one cycle? What is the efficiency of this engine? If the low-temperature reservoir is at 27 °C, what is the temperature of the high-temperature reservoir?

37. If the average kinetic energy per molecule of a certain monatomic gas is 4.63 times that of the same gas at room temperature (20 °C), what is the temperature of the gas?

38. A glazed doughnut contains 239 kcal. A power output of 100 MW is equivalent to the consumption of how many glazed doughnuts per second?

39. A small meteorite comprised of iron with a mass 1.32×10^3 kg enters Earth's atmosphere from space and crashes into Lake Champlain (a body of fresh water, which has a total mass of 8.56×10^8 kg). The entry into Earth's atmosphere increased the temperature of the meteorite to 3240 °C just before it came apart in small pieces and impacted the water at a speed of 3500 m/s above the deepest part of the lake. The temperature of the lake is a chilly 10 °C just before the impact. What is the final temperature of the system of lake and meteorite? For this purpose, assume the lake bottom is a good thermal insulator.

40. Suppose an equatorial lake is 10,000 m^2 in surface area, and that the Sun deposits 0.500 kW/m^2 into the lake. What is the maximum conceivable amount of electric power such a power plant on the lake could generate? You then build a solar-powered electric plant on the lake, which operates by using the temperature difference between the top (30 °C) and bottom (15 °C) waters of this lake. Taking into account the efficiency of this process, what is the actual maximum power you could get from the plant?

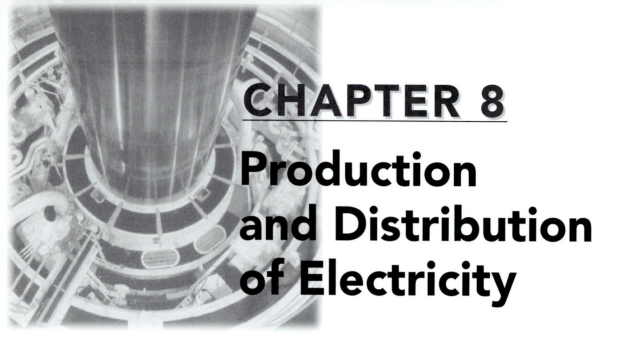

CHAPTER 8

Production and Distribution of Electricity

Electricity is central to national energy policy. In this chapter, we study the history of the utilities, and show how economy of scale encouraged monopoly. The discussion in Chapter 5 of how utilities have projected future demand provides the basis for a consideration of base load, peak load, and the choice of generating facility. The electric grid and transmission of electricity also play an important role. Cogeneration and conservation (see Chapter 11) are discussed briefly.

KEY TERMS

base load / intermediate load / cycling load / peak load / reserve capacity / economy of scale / screening curve / hard energy path / soft energy path / cogeneration / district heating / total energy system / saturation / electric grid / superconductivity

8.1   ENERGY END USE AND ENERGY SECTORS

People use energy for many different things: to heat themselves, to cool themselves, to cook their foods, to move from one place to another. Natural gas, fuel oil, or electricity may be used to heat a home; natural gas or electricity to heat foods; and gasoline, kerosene, and diesel fuels to power transportation systems. The major sources of electricity for the U.S. economy are shown for 2003 in Figure 8.1.[1]

Energy may be employed in a variety of ways. It may, for example, heat a home or a place of business. Although the end use is for the purpose of heating in both cases, in each case, the energy serves a different economic purpose. Heat in business is used as part of the cost of production of goods or services.

It is conventional to allocate energy use to four sectors of the economy: industrial, transportation, residential, and commercial. In the transportation sector, of course, all energy is fuel energy. In the residential sector, coal, oil, and natural gas are used directly to generate space heating. They are also used indirectly; that is, they are used to generate electricity that is used in various ways in the home. At present, about a quarter of residential energy use is electrical energy. Plants built for generating this electricity are among the most concentrated sources of air and water pollution in the world, as will be seen. In this chapter, we study the electric generation and distribution system. This will lead us in succeeding chapters into discussion of the importance of water and other forms of resources.

Figure 8.2[1] presents the 2003 energy picture from the end-use perspective. Notice that the total energy use of 108 EJ does not match the 104 EJ of end use energy. Energy use continues to increase from year to year. See also **Extension 8.1,** *Correcting the Figures*.

Space heat looms as the largest share of both the residential and commercial sectors, as we see from the details of Reference 1. In Chapter 24, we shall discuss some useful economy measures in the space heating of residences, as well as other conservation measures applicable to residences, commercial establishments, and industrial use.

The residential sector is probably the most familiar to most of us. Aside from space heating, water heating demands the most energy, followed by cooking, refrigeration, and air conditioning. Other uses, such as for washing machines and TV, are minor constituents of the total demand.[2]

8.2 EVOLUTION OF THE UTILITIES

The electric industry has grown mightily in this country since Thomas Edison began formation of generating companies in 1881. Edison, as we have mentioned, had embraced direct current (DC) generation of electricity. Because the voltage was only 110 V (about that used today), and because transformers could not be used with DC, Joule heating losses in the system were high. Electricity could be transmitted only a few miles; DC generating facilities had to be widely scattered. Edison began his DC system before George Westinghouse initiated his alternating-current (AC) system, so that in the beginning, many sources of electricity were scattered around the country-side. It was the usual practice for industry to generate its own energy as needed rather than buy more expensive energy from a generating company.

As AC came to be accepted, it became feasible to have larger facilities to generate energy. This led the many scattered electric companies to the path they have followed ever since: their growth as utilities rather than as competing businesses. Samuel Insull, of Chicago Edison, led the utilities onto this path.[3] The utilities had begun life stringing duplicate sets of wires to compete, and this made no economic sense. By obtaining monopoly rights as public utilities, they were able to construct a single distribution system and to trade competition for regulation by state public utilities commissions (PUCs).

Because utilities are usually allowed profits as a fixed percentage of costs, this removed some incentives for innovation. The utility makes its profit no matter what. An additional advantage for the utility in its monopoly position is that it has a steady supply of captive customers to provide the capital necessary for expansion. Expansion increases capital costs; because PUCs give profits based on costs, expansion means increased profit. The extra energy must be sold to customers who are encouraged to use it in ever more frivolous ways. This provides more expansion capital, and so on.

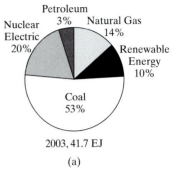

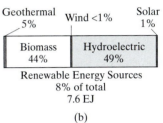

FIGURE 8.1

(a) U.S. Electricity sources totaled 39.6 quads, or 41.7 EJ, in 2003. This is almost 40% of total U.S. energy use in 2003, 107.9 EJ (102.3 quads). (Department of Energy, *Annual Energy Review 2003* (Washington, D.C.: GPO, 2004), DOE/EIA-0384(2004), Tables 1.1 and 8.1)
(b) Renewable sources (exclusive of hydroelectricity) have grown from under 4% in 1992. (Department of Energy, *Annual Energy Review 2003* (Washington, D.C.: GPO, 2004), DOE/EIA-0384(2004), Table 8.2b).

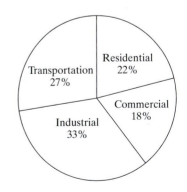

FIGURE 8.2

U.S. Energy end uses, 2003. Total use was 104 EJ. (U.S. Department of Energy, *Annual Energy Review 2003* (Washington, D.C.: GPO, 2004), DOE/EIA-0384(2004), Table 2.1a)

The deregulation fervor of U.S. politicians led in the late 1990s to measures that would dismantle the monopoly position of the utilities. See **Extension 8.2, Deregulation in the Twenty-First Century**. The results so far have been decidedly mixed, with Californians suffering from rolling blackouts and "sticker shock" prices that nearly bankrupted the state in 2001,[7,12,20,25,26] thought to have resulted from the "gaming" of the system by energy suppliers.[20,21] On the other hand, deregulation seems to have benefited Pennsylvanians.[39]

The historical amalgamation of the small utilities into giant energy conglomerates proceeded apace since the last decade of the nineteenth century for essentially two reasons: the difficulty of storing electricity and economy of scale. These characteristics still remain, and always will.

Because storage of electric energy is difficult, the generating facility must have on hand enough capacity to supply all its customers when they want electricity. When an electric switch is flicked, the generator puts out a little more current, which is used as it is generated. If the electric company has only residential users, it has to supply almost all of them in the early morning and then again in the evening, when most are at home. The capacity is virtually idle the rest of the time. Business abhors tying up capital in capacity that is needed sporadically. Of course, even with long spans of low use, less capacity is needed than would be the case if each residence had its own generator. A family's peak use may be 10 kW, and the family next door may also have a peak use of 10 kW; however, the families probably operate on different schedules. One family may be sitting down to dinner just as the other is getting up from the table. Thus, the peak use from the two families together would be considerably less than 20 kW. This was an illustration used by Samuel Insull in his speeches on behalf of utility monopolies:[3] An apartment house that would need 68.5 kW capacity if each user were to be able to draw at will had an actual maximum demand of 20 kW. The more people there are using electricity, the more the utility can smooth out its demand. The larger the utility, the more customers it can serve, and the smaller the minute-to-minute fluctuations in demand.

As a rule of thumb, we may say that in the first decade of the twenty-first century, every megawatt is enough power to supply 1000 households. This is an average of 1 kW of power per American household. Because the U.S. is a high per capita energy user, the power per household would be lower elsewhere. In Europe, for example, average power use is only about 750 W per household.

To keep their expensive generating equipment in use, the electric companies began an aggressive selling campaign to woo industrial users away from generating their own energy. An industrial user will continue to generate energy unless a utility can sell its energy cheaply. Because industrial users tend to have a steadier demand, utilities can further smooth out demand by having many industrial users. For these reasons, the utilities set industrial rates lower than residential rates.

The daily energy use profile of a typical large-city utility is shown in Figure 8.3. Only about half the capacity is in constant use; this constant demand is called the **base load**. Another 40% or so of the demand cycles on and off slowly; this is known as **intermediate**, or **cycling, load**. About 10% of capacity is used only a few hours a day; this is called **peak load**. The peak load comes around late afternoon for most utilities.

Base load actually changes from month to month. In the southern states, for example, energy demand peaks in the summer months because of air conditioner use. Air conditioning may be the most important contributor to the explosive growth of population in the Sunbelt. Who would have considered living in Houston or Phoenix without it? Farther north, in Canada, a typical annual load curve peaks in

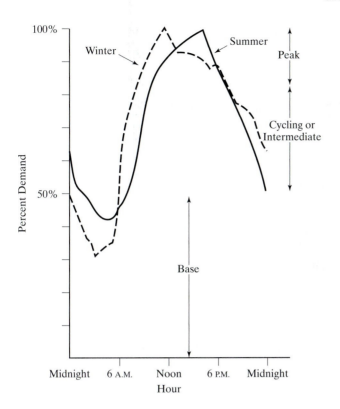

FIGURE 8.3

Winter and summer load curves for a city in the northeast are indicated schematically.

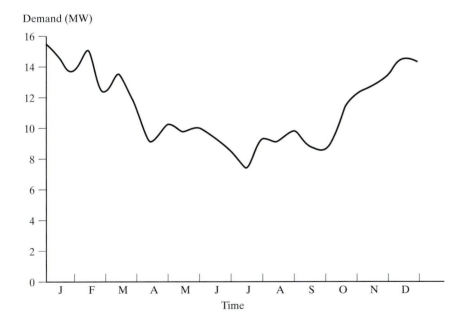

FIGURE 8.4

This annual load curve is for a Canadian utility, and it shows peak annual demand in winter. A curve from Texas would show peak annual demand in midsummer.
(Reproduced with permission of the Ministry of Supply and Services Canada)

the winter months, as shown in Figure 8.4.[48] Such variation is also of concern to the utilities, and it is another important reason for soliciting industrial business.

Base load is most economical for the utility to supply. Base load is generated all day and is always running; in addition, maintenance can proceed continuously. The utilities' largest generating facilities are dedicated to baseload production. Because it is difficult to start up and shut down nuclear plants, they are used to

FIGURE 8.5

A load duration curve for calculating the minimum cost of generation for a utility.
(Reproduced with permission of the Ministry of Supply and Services Canada)

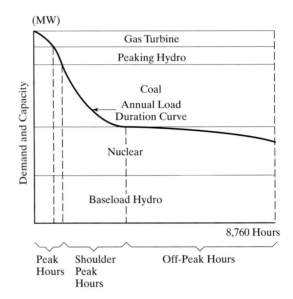

provide baseload energy. The utility usually has special additional generators that are turned on for only a few hours a day for the peak load; this is the most expensive energy generated by the utility. Figure 8.5 [48] presents a different view of the load curve—an annual perspective. The figure is indicative of the situation in Canada, where hydroelectricity is more important than in the United States. For this reason, hydroelectricity in the United States would be baseload rather than baseload and peak, as in Canada. Nuclear electricity generation is a somewhat smaller proportion of total electricity generation in Canada than in the United States.

Table 8.1[1,48] shows the mix of sources of electric energy in the two countries; note the heavy Canadian dependence on oil and hydroelectricity and the heavy U.S. emphasis on coal as the sources of electricity. Although it is important to be able to meet peak demand, it is also important not to have too much **reserve capacity**. If the reserve capacity is too large, equipment sits idle, not earning a return on the capital invested in it. Figure 8.6[48] shows that Canadian utilities had a peak demand over capacity averaging 20% over the past three decades. This is about where the rule of thumb in the industry sets reserve margins. The high reserve capacity of the late 1970s in Canada is probably due to predictions in the early years of the decade of an impending shortage of electricity. Such faulty prediction has also been a problem

TABLE 8.1

Electricity production (%) by source

	United States[a]	Canada[b]
Hydroelectricity	7.2	57.9
Coal	51.2	16.0
Oil	6.5	6.7
Gas turbine	16.6	4.8
Nuclear	19.9	12.9
Internal combustion, wood, waste, geothermal	2.3	1.3

[a]U.S., 2003, Table 8.2b, Reference 1.

[b]Canada, 1997, Natural Resources Canada, http://www2.nrcan.gc.ca/es/ener2000/online/html/chap3g_e.cfm, Table K (coal, oil, gas); 2001, CIA World Factbook (hydroelectricity, nuclear, other).

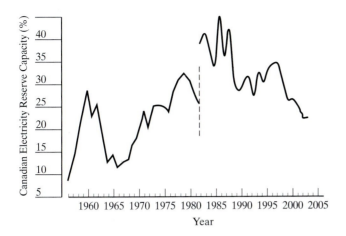

FIGURE 8.6
Canadian reserve capacity as a
proportion of peak demand for the
years 1956 through 2002.
(Data from 1956 to 1982 reproduced
with permission of the Ministry of
Supply and Services Canada, data
from 1980 to 2002 from Statistics
Canada)

in the United States. Of course, without the energy crises of the 1970s, the predictions might have been correct.

Economies of Scale

Efficiency problems occur with the use of electricity as compared with other "fuels." Inappropriate use of electricity has been responsible for some diseconomy in certain industrial applications, where direct heat use would have been more appropriate.[49] To an extent, economy of size (or scale) may counterbalance some of these inefficiencies. By **economy of scale** for a generating facility, we mean that the larger the capacity of a facility, the lower the cost per kilowatt of capacity. For example, doubling a boiler's size does not double the amount of steel needed or the amount of fabrication necessary. Tripling the volume of a building does not triple the architect's fees or the cost of erecting it. Erection cost is roughly proportional to the outside surface area of a building. Suppose the building is a cube. If I tripled the length of each side, the surface area would increase by a factor of nine, and the volume would increase by a factor of 27. This holds true for other components in a generating facility, as well. Table 8.2 shows how power plant size is affected by economy of scale. Note that the number of plants that must be built to supply a given amount of energy decreases with size, as it must, and costs are lowered as a result. (Another example, that of the economies of scale in the transport of crude oil by tanker, is

TABLE 8.2

How economies of scale affect power plant size in Germany

Capacity (MW)	Number	Total production capacity (MW)
<1	409	119
1–10	263	982
10–50	106	2663
50–200	89	9926
200–500	52	16,883
500–1000	29	21,002
>1000	13	23,113

Source: die Zeit, 21 September 1984.

FIGURE 8.7

Economies of scale in production of ammonia.
(*Industrialization and Productivity*, U.N. Bulletin 10, 1966; Fig. VII, p. 16. Reproduced by permission.)

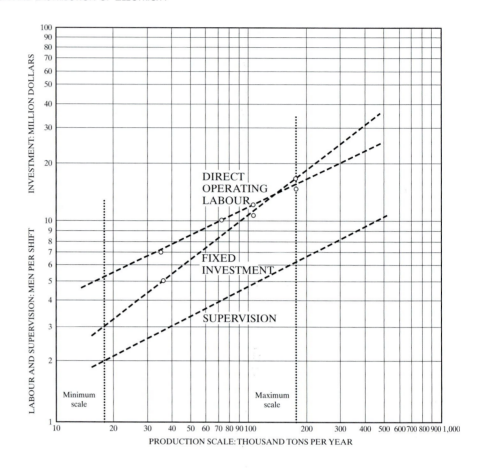

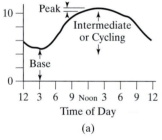

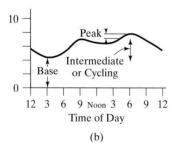

FIGURE 8.8

(a) Summer demand in southern California (30 August 1976).
(b) Winter demand in southern California (6 January 1976).
(Data courtesy of Southern California Edison)

presented in **Extension 8.3,** *Economies of Scale in Tanker Transport of Crude Oil*.) Figure 8.7 illustrates economy of scale in industrial production of ammonia. The lines would be diagonal if the scale were linear, but they are well below this.

8.3   THE MIX OF GENERATING FACILITIES

Most utilities use a mixture of generating facilities from fossil fuel and from nuclear fission; that is, some energy is supplied by using turbines fired on natural gas, some by oil or coal, and some by nuclear reactor. If the utility is burning fossil fuel, it might use an advanced design boiler (discussed in Chapter 12) instead of the standard grate furnace. Why would a utility want to have such a diversified mix of generating facilities?

As pointed out earlier in this chapter, the utilities are faced with a fluctuating demand that depends on the hour of the day. Only about 50% of the total demand is constant, all-day (baseload) demand. Figure 8.8 is a schematic showing the daily winter and summer demands for energy in southern California.[51] Note that, in either case, peak demand lasts only several hours a day. It would be wasteful for the utility to be capable of generating peak demand at all times during the day, but the capacity to handle peak demand must, of course, be there in order to allow the generators to work harder, when necessary, as the demand arises.

It is reasonable for the utility to have some generators that will be on call for only a few hours a day. But of course, different types of generators cost different amounts to build and to run, which brings us back to the tradeoff between capital costs and operating costs. Owners of small businesses often do not think about operating costs and consider only capital costs of construction; in contrast, the utility, using vast amounts of energy resources, must address the ongoing costs of operation. In the utility business, machinery can do more work at lower cost than human labor; this machinery requires investment, and large companies such as utilities are able to finance the investments from operating revenues or financing, while smaller companies cannot.

These cost tradeoffs can best be represented for the utility by a **screening curve**. Figure 8.9 shows screening curves based on the lowest cost experience for

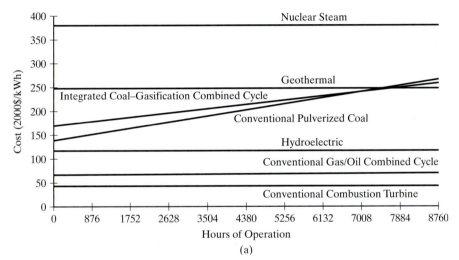

(a)

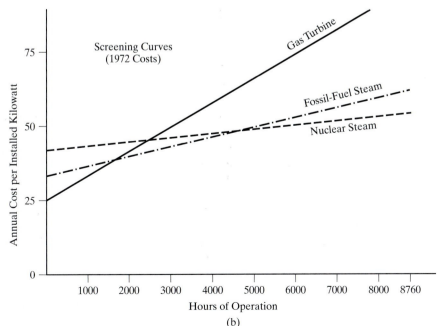

(b)

FIGURE 8.9

(a) Screening curves for production of energy from fossil-fuel steam, gas turbine, hydroelectric, and nuclear steam plants are shown using 2001 costs.
(U.S. Department of Energy, *Assumptions to the Annual Energy Outlook 2002* DOE/EIA-0554(2002), Table 38, and *Electric Power Annual 1999*, Vol. II. Some data (conventional nuclear) are from 1991, Department of Energy, *Electric Plant Cost and Power Production Expense 1991* (Washington, D.C.: Government Printing Office, 1991), DOE/EIA-0545(91)]
(b) Screening curves for production of energy from fossil steam, gas turbine, and nuclear steam plants are shown using 1972 costs.

each type of facility as of 2001 and 1972. Utilities use such curves to decide on the source that is most cost-effective for each purpose. Technological developments have changed the situation greatly since the 1970s. Then, gas turbines provided the cheapest peak energy, and nuclear facilities provided the cheapest baseload (actually, hydroelectric energy, not shown, was cheapest). A utility would not have constructed a fossil-fuel or nuclear facility for peaking power or used gas turbines all the time. Some hydroelectricity has been subject to limitations in water flow due to drought and other conditions; in these cases, it will be used as peaking supply. Of course, costs will change for both construction and for fuel, but the tradeoffs indicated will remain relatively the same. Turbine engines adapted from jet engines have changed the dynamics in the 1990s and 2000s, and these are typically 160 MW, much smaller than, say, conventional coal plants. Conventional nuclear energy is much too expensive. Advanced nuclear energy is much less expensive than conventional nuclear.

TABLE 8.3

Capital cost ($) per kilowatt of nuclear energy versus other sources

	Coal 1000 MW	Nuclear 1000 MW	Hydroelectric 1000 MW	Gas turbine 160 MW
1971[a]	1411	1491		
1978[a]	1520	2335		
1991 (low)	2073[b]	3056[c]	1091[b]	332[b]
1991 (mean)	2545[b]	4844[c]		404[b]
2001	2798	3573[d]	1117	456

[a] Adapted by permission from Reference 52, corrected to 2000 dollars. Copyright © 1981 by the New York Times Corporation.

[b] Department of Energy, *Electric Plant Cost and Power Production Expense 1991* (Washington, D.C.: Government Printing Office, 1991), DOE/EIA-0545(91), corrected to 2000 dollars.

[c] Department of Energy, *Commercial Nuclear Power 1991* (Washington, D.C.: Government Printing Office, 1991), DOE/EIA-0438(91), corrected to 2000 dollars.

[d] No conventional nuclear plants will be constructed; the "advanced" nuclear option is shown.

TABLE 8.4

Comparative cost of nuclear energy in various countries (1984$/kW installed)

Country	Cost
France	680
Italy	812
Belgium	876
The Netherlands	1148
West Germany	1213
United Kingdom	1298
United States	1434
Japan	1438

Source: Reference 10. Copyright © 1984, 1985 by The New York Times Company. Reprinted by permission.

Table 8.3 illustrates how costs for nuclear energy changed over the decades of the 1970s and 1980s.[52] Nuclear energy is still the most capital-expensive conventional alternative and is getting more so (as the table shows).[53] The United States is just about the most expensive place in the world to build a reactor (the relative cost comparisons in Table 8.4 are still valid today).[54] As a result, no orders for nuclear plants since 1974 resulted in completed plants until 1990; five reactors have opened since then. Nuclear energy has borne a heavy burden of regulation and public disenchantment because of perceived safety issues and because of large cost overruns in the construction of nuclear plants. The costs of fossil-fuel plants will be increasing even faster than nuclear in the future because of the 1990 Clean Air Act Amendments.[55] Nuclear plants emit no greenhouse gases, so there may again come a time, soon, when nuclear is the cheapest alternative for base load. Costs for construction of various facilities in the United States are presented in Table 8.5 in 2000 dollars.[56,57]

Note that the installation cost (Figure 8.9a) of a conventional nuclear power station is the greatest of the alternatives shown ($377/installed kW), but its operating and fuel cost is lowest (0.70 ¢/kWh); except for advanced nuclear (0.47¢/kWh).[58] The installation cost is the capital cost amortized over the life of

TABLE 8.5

Capital costs of various generating units (2000$/kW installed)

Technology	Unit size (MW)	Capital cost per kW	Lead time (yr)
Current conventional			
Conventional pulverized coal	400	1046	4
Integrated coal–gasification combined cycle	428	1250	4
Conventional gas/oil combined cycle	250	435	3
Conventional combustion turbine	160	323	2–4
Nuclear steam	1000	3130	10–14
Geothermal	50	1663	4
Hydroelectric	1000	1117	5
Current less conventional			
Biomass	100	1536	4
Fuel cells	10	1810	3
MSW—Landfill gas	30	1336	3
Solar photovoltaic	5	3317	2
Solar thermal	100	2157	3
Wind	50	918	3
Advanced			
Advanced combustion turbine	120	451	2
Advanced gas/oil combined cycle	400	546	3
Advanced nuclear	600	1772	4

Sources: Department of Energy, *Assumptions to the Annual Energy Outlook 2002* DOE/EIA-0554(2002), Table 38, all except nuclear steam and hydroelectric, from Department of Energy, *Electric Plant Cost and Power Production Expense 1991* (Washington, D.C.: Government Printing Office, 1991), DOE/EIA-0545(91), costs adjusted to 2000 dollars.

the installation (assumed to be 40 years at 10% interest). With the escalated cost of conventional nuclear, the figures would indicate that the nuclear facility is never cheaper than fossil fuel as of 2000 (it once was, see Figure 8.9b). The 1970s utilities would tend to construct any nuclear power stations for the base load. One of the biggest problems facing the nuclear industry then was its poor reliability, but reliability has increased,[59] so much so that nuclear energy is now the most reliable base load.

The installation cost of the gas turbine is the lowest ($41/installed kW), and turbines can be turned on and off easily, but the fuel and operating expense is the most expensive (2.2 cents/kWh).[58] This is the cheapest way to generate the energy using steam for any period of time (because of the efficiency increases in turbine technology). Experience in California, where gas has been favored as a pollution reduction measure, has shown the severe financial effect of gas price fluctuations (see **Extension 8.4, *Gouging?***). Coal-fired generators are almost as costly as nuclear. It is clear why utilities mixed their modes for generating energy in the 1970s. In the 2000s, mixing is done because the plants still function and because it guards against huge price increases such as those experienced in California in 2000.

Other schemes are available for meeting peak demand, such as energy trading (made possible as Enron developed the methods, then crashed and burned; see **Extension 8.5, *Enron and Deregulation*** for more information on the rise and fall of Enron). In Chapter 25, we discuss many alternatives for storage of energy to be used at peak demand.

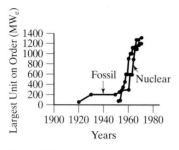

FIGURE 8.10

The trend in size of U.S. steam-generating facilities between 1920 and 1969 is indicated in data supplied by the U.S. Atomic Energy Commission.

Generating Units Become Larger, then Smaller

Many types of generating facilities built today are large. The history of the largest-size unit on order is shown in Figure 8.10.[152] The largest nuclear units built are now at about 1300 MW, so the logistic curve is coming into play yet again. The historical increase in size is a reflection of both the effect of technology in keeping electricity prices low and the incentive to realize economies of scale.

In recent times, units have become smaller (modular). As the producers of energy learn more from their experience in running a facility, they are better able to innovate the next time they build a plant to produce energy. (This phenomenon is known as a "learning curve.") The same holds true in other endeavors; for an example from another industry, the price of hand calculators has declined as the total number produced has increased. This experience has been summarized by Fisher:[153]

> Each time there is a doubling of the total cumulative number of units manufactured, or transported, or processed in some economic activity, there results a fixed percentage decline in cost.

The electricity industry has found that the learning experience with modular units gives them an economic advantage despite loss of economy of scale. And modular units are produced off-site, in a factory (see the section on plant construction in Chapter 9).

In 1973, the Atomic Energy Commission predicted an increased demand of 16 trillion kWh in the year 2000 (the actual value was 3.01). We would have had to have built more than one new 1000 MW plant every week to meet that goal; an eerie echo of this recommendation reverberates in the Vice Presidential task force recommendation to build a power plant a week until 2020.[152] We have already noted that far less construction is likely to be needed (Chapter 5). Still, a lot of new facilities will have to be built in the next two decades; they will most probably be smaller gas-fired modules (60 to 300 MW) because of their low capital cost (Figure 8.9a).

"Hard" and "Soft" Energy Paths and Cogeneration

In addition to economies of scale, a larger power plant can recover efficiency by operating at higher temperatures than are possible for a smaller facility. These factors have led to ever-increasing centralization of electric generating facilities. In this case, however, there is a price to be paid—increased pollution. More nitrogen oxides are emitted when the combustion temperature is higher.[154] Lovins[155] has called the continuation of the trend toward more centralized energy generation the **hard energy path**. Lovins favors development of renewable, decentralized energy sources, his so-called **soft energy path**, as the rational and fruitful way to approach the future. He views the strategy of "more of the same" with great disaffection and sees disincentives to change as structural; he believes that market forces can induce changes that will lead to increased conservation efforts. The price rise in gasoline throughout the 1970s and early 1980s has proved this to be true for energy efficiency in cars.

Lovins sees "soft" technologies as able to provide all energy use after about 2025. These soft paths are characterized by

- Reliance on renewable energy sources
- Diversity of sources of supply, with total supply aggregated from many modest inputs

- Flexibility and use of equipment of low technological level
- Match in scale and geographic distribution of supply to end use
- Match in energy quality

As we emphasized earlier, we lose two-thirds of the energy value in producing electricity from fossil fuels. If we want to heat a room, it is usually better to burn the fuel oil or gas directly. This is what Lovins referred to when he spoke of energy quality. Smaller installations would minimize transmission losses; roughly half the amounts of our electric bills in the United States go to pay transmission and distribution costs, not generation costs.[155] Lovins favored **cogeneration**, a time-honored method by which, for example, a paper mill could generate steam for making paper and incidentally get electricity as a byproduct. The Public Utilities Regulatory Policies Act (PURPA), passed by Congress in 1978, allowed cogenerators or producers to sell their extra electricity (up to 80 MW) to utilities at the marginal (peak, or "avoided") cost of generating energy. Avoided cost is the most expensive energy that the utility buys or generates. The PURPA has encouraged many producers and cogenerators (Figure 8.11),[156] even including the Bronx Zoo,[156] in a sense turning the clock back to the situation a century ago, when generators were widely scattered. Utilities claim that the law costs their customers more for energy because utilities are forced to pay producers at the utilities' own highest cost of production under PURPA and have carried their point about reliability in the courts. Nevertheless, PURPA has been successful in supporting alternatives to coal- and oil-fired facilities.

District Heating and Other Alternatives

In **district heating**, a generating facility supplies waste heat for space heat in large districts of cities. The Europeans have used the district heating concept extensively,[157] and conditions favoring district heating exist in some American cities. Baltimore Steam Company in Baltimore, Consolidated Edison of New York, Detroit Edison in Detroit, and Eugene Water and Electric Board in Eugene, Oregon, among others, have significant district heating programs; and

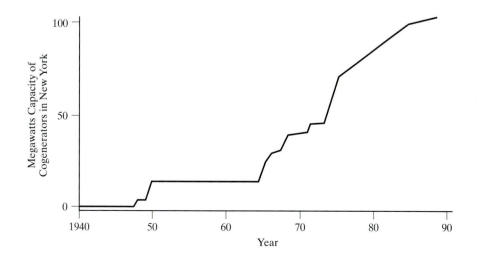

FIGURE 8.11

The number of kilowatts of cogeneration capacity available in the New York Consolidated Edison service area have increased substantially, especially since the passage of the Public Utilities Regulatory Policies Act.
(Data courtesy of Consolidated Edison)

the Trenton, New Jersey,[158] and Cedar Rapids, Iowa, business districts are heated by cogeneration.

Similar ideas have been suggested that would lead to smaller generating facilities. For example, in **total energy systems**, small generators run on the burning of trash from a small housing or apartment complex, thereby producing electricity. Transmission losses are small, and waste heat from the generator is used to supply space heat in the homes. This is a promising idea that should certainly be pursued. At present, it suffers from problems in maintenance and efficiency (trash combustion temperature is low). Maintenance costs at least 10 times as much per kWh for a small power installation as for a utility. It would be difficult for such a small operation to afford more than the cost of a full-time operator to run the facility; it would not be enough to hire a full-time technician. The utilities spend less than 0.1¢/kWh on maintenance and have qualified personnel on hand to deal with emergencies,[51] again demonstrating economies of scale.

It is probably more feasible to set up such a total energy system for a large town or part of a city. This would assure a larger supply of trash and waste to be burned, and it would increase the number of customers for the electricity generated and the waste heat to be used for space heat. If the usage base is large enough, it should be possible to afford proper maintenance.

The idea of coordinating electricity generation with the supply of low-grade waste heat for space heating or for process steam (used in industrial applications) is not new. As previously mentioned, Consolidated Edison of New York sells much of its reject heat. Yet another example is that of paper companies that must maintain boilers to obtain steam for processing of paper. Before the state governments stopped them from competing with the electric companies by setting up electric utility monopolies, paper companies sold much of the electricity they were able to generate in this incidental way. PURPA has restored this possibility.

Some utilities are following the suggestion of researchers like Lovins and are getting involved with renewable energy. California utilities, in particular, have embraced various renewable energy sources. Pacific Gas and Electric buys energy from wind generators in the Altamont Pass. Southern California Edison buys energy from wind generators near Palm Springs, and built "Solar Two," a large-scale solar-generating facility developed in cooperation with the Department of Energy. These alternative sources are discussed in more detail in Chapter 21. A 1984 survey of 65 utilities conducted by the Electric Power Research Institute[159] revealed that 24 of them see future hydroelectric generation as important, 20 are doing research on biomass alternatives (see Chapter 22), and 14 are working on wind energy.

Other factors may be pushing the utilities in the direction suggested by Lovins. This trend was reflected in the industry as a whole. Orders for new plant capacity fell from 94 GW in 1974 to an average of 3.5 GW in 1979–1981[160,161] to an average of 0.9 GW for 1982–1984 because of the success of post-energy-crisis conservation measures and increased capital costs.[161] No new nuclear plants have been ordered since 1979; and utilities have canceled most plants on order before that date.[161] By the end of 1993, only one plant remained under construction. This downward trend caused industry analysts some concern.

Most generation facilities take roughly a decade from order to first use. If more energy is needed a decade hence, the plants must be started now. Peak demand increased from 465 GW in 1985 to 546 GW in 1994.[161] The 1985 projected capacity in 1994, 704 GW, would have brought the reserve margin below 20%, and as Californians learned in 2000, such a low margin causes problems. The capacity as of 2000 is 812 GW.[162] A healthy reserve margin is necessary to deal with weather-induced demand, unanticipated shutdowns, and necessary maintenance. The utilities met the demand in the mid-1990s and late 1990s by purchasing hydroelectricity and nuclear electricity from Canada, not by building power plants. Net electricity imports peaked in 1987 and again in 1994 at about 45 billion kWh, 50% above the 1980–2000 average.[1] Demand, which rose to an annual rate of 5.7% from about 2% as the recovery from the 1981–1983 recession began, fell back to an annual rate of 3.4% in 1984 and then began an uninterrupted rise.[163]

With all the uncertainty, it appears that the best way for the utilities to proceed in the present poor investment climate[160] is to continue to encourage conservation and to plan to build modular units that can be erected quickly and can be added to as necessary (see, for example, the discussion of the Cool Water Coal Gasification Plant in Chapter 12). In this instance, we see again, as we saw in the previous chapter, how important reliable projections are and how difficult they are to make. Electricity is vital to the functioning of the country, but we have little idea of what even the near future will bring. This uncertainty may well be the most persuasive cause of the change to a softer energy path if the projections by Lovins prove true.

The director of the Italian Nuclear and Alternative Energy Agency has argued that electricity could be useful for development in the Third World:[164] It is still coupled to growth in gross domestic product (see Chapter 7); it is convenient; rural electrification could stem a rush to the cities; and it could be small-scale and based locally on renewable resources. These arguments follow Lovins' arguments for the soft approach. Europeans have been somewhat more receptive to these ideas than Americans.[165]

8.4 ENERGY USE BY APPLIANCES AND THE PHENOMENON OF SATURATION

Of particular interest is the distribution of energy use by small appliances. These appliances, as shown in Table 8.6,[2] consumed over 47 billion kWh in 1969, but this amounted only to a 3% share of total electricity use. Although it is useful to practice economies by halting gratuitous use of these appliances, their effect on total national consumption is small indeed. Table 8.7 presents a more balanced picture of household appliance energy use.

By this time, over a century past the first large-scale commercial use of electricity, we may reasonably hope that we have reached the end of the road in these small ingenious devices. In any case, they all take such a small share of demand that they are of no practical importance. While the U.S. population was growing at a rate of only 1.3%/yr between 1960 and 1968, basic energy consumption increased an average of 4.1%/yr.[2] It is clear that faddish devices such as electric toothbrushes and electric carving knives could not be responsible for this great growth; the increase is attributable mostly to space heating and water heating, frost-free refrigeration, and growth in air conditioning. The distribution of energy to the consumer changed very little between 1960 and

TABLE 8.6

Electrical consumption of selected small appliances

	Annual kWh$_e$ per item[a]	Number of items (millions)[b]	Annual consumption (billion kWh$_e$)
Bed coverings	147	27.0	3.97
Blenders	1	16.0	0.02
Broilers	85	14.0	1.19
Clocks	17	55.0	0.94
Coffeemakers, automatic	140	50.0	7.00
Dehumidifiers	377	3.8	1.43
Fans (circulating)	43	75.0	3.23
Food disposers	30	13.5	0.41
Hair dryers	25	22.5	0.56
Humidifiers	163	4.0	0.65
Fry pan, skillets	100	33.0	3.30
Heaters (portable)	176	17.0	2.99
Hot plates	90	15.0	1.35
Irons	60	57.0	3.42
Knives (carving)	8	13.0	0.10
Mixers	2	49.0	0.10
Radios	86	57.0	4.90
Shavers	0.5	24.0	0.01
Toasters	39	54.0	2.11
Toothbrushes	1	15.0	0.02
Vacuum cleaners	46	53.0	2.44
Total			40.12

[a] *Source*: Edison Electric Institute, 1987.

[b] Estimated number in households as of mid-1969.

Sources: *Merchandising Week*, 23 February 1970, and Stanford Research Institute estimates.

1968, except for heating. This was probably a result of an overselling of totally electric homes during the 1960s; the miscalculation became unhappily apparent when the price of electricity skyrocketed in 1973. Older uses of energy, such as for clothes washing and cooking, have not really increased much since the 1950s.

The relative stagnation of the older uses of energy is a reflection of the fact that there is a limit to how much hot water, how many stoves, and how many washing machines we need. When every family has a stove, then the rate of sale of stoves will simply reflect the growth rate of the population and the retirement of old stoves. This phenomenon is called **saturation**, which is defined as the proportion of households having a particular appliance to the total number of households. Refrigerators and stoves are in just about all American homes. Washing machines are in three-quarters of homes (apparently, this is about the limit, because some apartment dwellers must use, and others prefer to use, laundromats). Black-and-white TVs are a supersaturated commodity, but the market for color sets is still alive and well. Clothes dryers and air conditioners still have some growth potential, but water heaters are in practically every home already. At some future time, the market for newer uses of energy will become just as saturated as that for stoves or water heaters. Figure 8.12 shows a typical saturation curve from its beginning to total saturation. We have seen such a logistic curve before, in Chapters 4 and 5. Figure 8.12 gives us a basis for estimating how long it will take an appliance to reach saturation. The growth

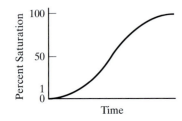

FIGURE 8.12

Saturation of appliances with time follows a logistic curve.

TABLE 8.7

Electricity consumption of appliances

	Power (W)	Average hours used per year	Annual use (kWh$_e$)[a]
Food			
Dishwasher	1200	30	36
Drip coffeemaker	1500	60	90
Freezer			
15 ft^3 manual	320	8760	1512
15 ft^3 automatic	440	8760	1824
Hot plate	2350	24	56
Knife	100	6	0.6
Microwave oven	1500	360	540
Range			
Small surface unit	1300	163	212
Large surface unit	2400	82	197
Oven	3200	240	768
Self-cleaning oven	3200	240	614[b]
Refrigerator			
20 ft^3	150	8760	1344
20 ft^3 best, post-1993	59	8760	515
Toaster oven	1400	24	34
Trash compactor	400	3	1.2
Clothes			
Dryer (Regular)	3000	240	792
Dryer (Permanent Press)	5000	72	360
Iron	1000	60	60
Washer	500	145	72
Water heating	2500	1600	4000
Household			
Air conditioner (whole house, 3.5 t)	410	924–4080	4540–20,000
Air conditioner (1 room)	1100	2920	3210
Blanket	175	1200	210
Clock	2	8760	18
Computer (10 h/d)	160	3650	576
Computer (always on)	160	8760	1400
Lamp	75	200	15
Pool pump	1970	1825	3600
Shaver	14	60	0.8
Stereo	100	120	12
TV, black and white	55	2160	119
TV, color	200	2160	432
Toothbrush	7	60	0.4
Vacuum cleaner	650	24	16
Standby power loss (appliances that appear to be off)			420–600

Sources: Ohio Edison, 1987. N. R. Brooks, "Crying foul over energy baselines," *The Los Angeles Times*, 8 July 2001. Department of Energy, *Energy Consumption & Conservation Potential: Supporting Analysis for the National Energy Strategy* (Washington, D.C.: GPO, 1990), SR/NES 90–02, Tables 2–11 and 2–12.

[a] Estimated.

[b] The self-cleaning oven does not use power as much of the cooking time as does the conventional oven.

rate of all large appliances will eventually decrease to the growth rate of the American population.

Refer to Appendix 3 for some additional background material as well as a discussion of how graphs can be misleading. Included in this material are explanations of how the tables and charts of this section follow from the tables of Reference 2.

8.5   TRANSPORTATION OF ELECTRICITY

In Chapter 4, we discussed the reasons that high-voltage transmission lines are used to transport electricity. Recall that power generated by a power plant is the product of the current and voltage produced, and that the greater the current, the greater the power lost to Joule heating. It follows that the power plant's output is stepped up to high voltage so that the energy loss is minimized.

Long-distance transmission has changed in character over the years. Originally, "long distance" referred to interlinkages between adjoining utilities. The depression of the 1930s brought massive federal involvement in energy in the United States: in the South and Midwest, the TVA; in the West, the massive Columbia River and Colorado River projects. With these projects far away from urban concentrations, it became necessary to transport energy over long distances. Those first transmission lines were 138 kV; today, the lines are 765 kV or greater.[166] There is still disagreement as to whether there are significant dangers to people living near high-voltage transmission lines,[166,167] though most experts doubt any effect because of the small size of the fields involved compared to naturally occurring ones. See **Extension 8.6, *Low-Frequency Fields*** for more detail.

Despite possible risks, transmission lines interconnect utilities, and since 1976, the entire country has been joined into one interconnected **electric grid**, which also reaches into Canada.[167] These interconnections are pictured in Figure 8.13.[184] They help utilities smooth out hour-to-hour variations among users (for example, near adjoining time zones, where peak usages occur an hour apart), and they allow the United States to import much surplus hydroelectricity from Canada.[48] The main use of transmission lines has been changing from smoothing to bulk electricity transfers.[185] The huge hydroelectric project at James Bay in Canada is partly based on the assumption that the excess electricity may be sold in the United States.[186]

FIGURE 8.13

The interconnected system of electrical distribution in the United States.
(Courtesy of the National Electric Reliability Council)

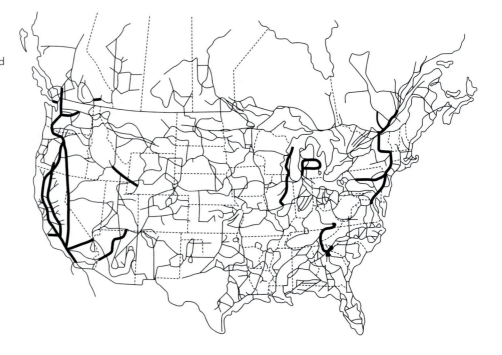

There have been some decided problems as a result of interconnections. The blackout in most of the northeast states on November 9, 1965, was traced to a problem in the transfer point at Niagara Falls. A blackout in the New York metropolitan area on July 13, 1977, was traced to three lightning strikes on transmission lines in upstate New York.[187] Surge protectors and automatic cutoffs prevented this blackout from spreading as far as the one in 1965, although the 1977 blackout lasted more than a day. The massive August, 2003, blackout that spread to most of the northeast, Ontario, and a swath of the midwest was due to system failures coming from unneeded relay trips and lack of communication (see **Extension 4.5**). Such large-scale blackouts have greatly decreased in frequency as transmission technology has developed, but complex systems will remain vulnerable, and the effects will be severe.

Most of the cost of bringing electricity to a location is the local transmission cost (see the switchyard of Fig. 8.14). Also, these local lines remain problematic, causing occasional death from downed power lines or ungrounded connections (especially when electricity is being stolen).[188] Long-distance transmission of electricity costs about 8% of what local transmission costs per kilometer.[189] Nevertheless, the total transmission cost can loom large—up to more than 50% of the cost of energy.[153] As a result of this high cost, research is proceeding on methods of electric transmission for which Joule heating is much less of a problem than in the current methods. Ironically, for the Edison–Westinghouse wars discussed in Chapter 5, one way to save money in transmission is to use solid-state voltage conversion and DC transmission lines (which can carry a lower voltage for a given power transfer and avoid the electromagnetic radiation losses associated with high-voltage AC).[190] Another promising method involves use of superconducting transmission lines (see Chapter 25 and the accompanying box).

FIGURE 8.14

A switchyard used in long-distance transmission of electricity.
(U.S. Department of Energy)

HISTORY OF ENERGY

Superconductivity

Kammerlingh Onnes discovered in 1911 that very cold lead ($\approx$11 K, or -262 °C below 0) altogether lost its resistance to the flow of electric current. He soon discovered other materials for which this happened at low temperatures. This phenomenon is called **superconductivity**. In a superconducting current loop, the current circulates without changing for a very long time. (An experiment carried out for over a year verified this aspect of the phenomenon.)

Many materials are superconductors. Each material exhibits a transition to the superconducting state at a different temperature. Most transition temperatures are very close to absolute zero—liquid helium at 4.2 K is used to cool most samples. Since the discovery of superconductivity at very low temperatures, there has been research aimed at finding superconductors that work at higher temperatures.[191] As time passed, materials were discovered for which the transition occurred at higher temperatures. Although no room-temperature superconductors have yet been found, compounds made of yttrium, barium, copper, and oxygen have exhibited superconductivity in the region up to about 100 K, with hints of superconducting behavior at higher temperatures.

Materials with high transition temperatures, if they are found, would be ideal to use for electricity transmission. Even so, the building of superconducting transmission lines at present is problematic. Materials for transmission lines remain to be fabricated. Estimates are that it will be economical to install such superconducting transmission facilities only for transmitted powers in the gigawatt range, unless true room-temperature superconducting materials can be discovered. Even though in theory there are no losses, eddy currents (induced currents circulating in metals, see Chapter 4) will still cause losses. The eddy current Joule heating losses must be kept to below 0.1 μW/mm^2.[191] There are unresolved problems with current surges (three to 10 times the rated current in AC lines) and doubts about the ability of the relatively high-temperature superconductive materials to doubts about the repeated ability to withstand large temperature changes.

Research continues on superconductor transmission, because it will make possible the cutting of losses by over 50%. Even if the highest transition temperatures remain at about 125 K, this is above the temperature at which nitrogen liquefies, and it is cheaper to use a nitrogen liquefier to supply refrigerant than to continue to accept Joule heating losses in conventional transmission lines.

SUMMARY

Transportation and space heating uses account for about half of the yearly energy use in the United States. In both these uses, liquid fuel or natural gas provides the major energy source. A true accounting of energy involves distinguishing thermal from electric energy, as discussed in **Extension 8.1**.

Electric utilities, originally small local facilities competing for business, became publicly regulated monopolies around the turn of the twentieth century. The major reason for this change is the difficulty of storing electricity. Demand varies from hour to hour and month to month, and without monopoly, the utilities would not have had the financial resources to build and maintain power plants that are idle (or have large unused capacity) except at peak demand. Another reason has to do with economies of scale, which make energy generation cheaper as the facilities are made larger.

For a utility to decide what sort of facility it wishes to use to generate energy, the capital cost of the facility must be known, and the operating cost of the facility must be projected. The utility's screening curve is then used in conjunction with its load curve to determine the lowest cost alternative.

The law known as PURPA has caused the monopoly grip of the energy utilities to loosen a bit. Small installations can supply energy by cogeneration or alternative energy sources and sell the excess, up to a maximum, to the utilities. The utilities must buy it at the "avoided cost" from reliable producers.

The push toward electricity deregulation, led by California, gathered steam in the late 1990s, but the California energy crisis and gouging experienced by state residents cooled deregulation in other states, as discussed earlier in **Extensions 8.2, 8.4** and **8.5**. Some states have had good experiences with deregulation, but most regulators are chastened by the California experience.

Because of advances in the technology of electricity transmission over long distances, almost all of North America is interconnected in a grid. This allows for the shuffling of cheap hydroelectricity and nuclear electricity from Canada into the United States and among domestic utilities, and it helps utilities smooth out peaks and valleys in demand. Energy trading companies such as Enron and Dynergy were involved in supplying this energy using the national grid. Concern about health effects of transmission lines is discussed previously in **Extension 8.6**. At some time in the future, superconducting transmission lines may further lessen transmission losses.

PROBLEMS AND QUESTIONS

MULTIPLE CHOICE REVIEW

1. Which is NOT a source of energy in the United States?
   a. Coal
   b. Electricity
   c. Natural gas
   d. Nuclear
   e. Petroleum

2. The base load of daily energy use represents
   a. the constant load of consumers.
   b. the load that cycles every day.
   c. the early morning load.
   d. the load supplied by nuclear fuel.
   e. none of the above.

3. The main reasons for the amalgamation of small utilities into giant conglomerates in the last 100 years is economy of scale and
   a. the difficulty in storing electricity.
   b. the lack of transmission lines.
   c. the decrease of residential customers.
   d. the development of TV.
   e. government intervention.

4. In the daily energy use profile of a typical large city, about 10% of capacity is used for only a few hours a day. This is called
   a. the constant load.
   b. the intermediate load.
   c. the peak load.
   d. the early morning load.
   e. the summer load.

5. Electric deregulation may encounter difficulties with implementation, because
   a. it is difficult to store electricity.
   b. the utilities may attempt to gouge their customers.
   c. state legislatures have never had experience with writing rules for these former monopolies.
   d. there will be less oversight by state regulators.
   e. of all the above reasons.

6. Which of the following is an economy of scale? Savings due to
   a. doubling a boiler's size.
   b. larger-capacity power plants.
   c. more capacious oil tankers.
   d. building bigger buildings.
   e. all of the above-listed measures are economies of scale.

7. For a nuclear power station, the installation costs are _____, while the operating costs are _____.
   a. high, high
   b. high, low
   c. low, high
   d. low, low
   e. There is not enough information available to answer this question.

8. All 160 million households in the United States have a refrigerator. (Ignore the fact that some households have more than one). The newest automatic frost-free models use 515 kWh per year and can be found in about 20% of households. Using the data of Table 8.7, determine how much energy would be saved overall by replacing the old models.
   a. 16.5 billion kWh
   b. 41.9 billion kWh
   c. 58.4 billion kWh
   d. 65.9 billion kWh
   e. 167.5 billion kWh

9. About half the 160 million American households have an electric knife. How much energy is used to run electric knives annually? HINT: Use data from Table 8.7.
   a. 0.6 kWh
   b. 96 kWh
   c. 600,000 kWh
   d. 96 million kWh
   e. 267 million kWh

10. Suppose it costs a utility $0.04/kWh to generate electricity at the generating facility, that long-distance energy transport from the generating facility to your local substation costs the utility $0.0002/kWh/km, and that local transport costs $0.0016/kWh/km. If the plant is 100 km from the substation, and the substation is 12.5 km from your home, what is the total cost of 1 kWh of electricity delivered to your home?
    a. $0.06
    b. $0.08
    c. $0.10
    d. $0.12
    e. $0.14

SHORT ANSWER

Explain whether the statement made is correct, and justify your answer.

11. Peak load is the most expensive to supply.
12. It would be most efficient if each household supplied its own electricity.
13. A screening curve exhibiting utility prices is only useful as long as the fuel price assumptions going into the curve are reflective of actual cost.
14. The United States is one of the least expensive places to build a power facility.
15. It would be cheapest if power plants came prefabricated as much as is possible.
16. Deregulation is a certain solution to energy shortages and blackouts.
17. The energy use in a household is different at different times of the year.
18. Experts are absolutely sure that there is danger to the health of people living underneath power transmission lines.
19. The energy used for "minor" uses adds up to truly gigantic amounts.
20. Superconducting transmission lines may revolutionize the energy distribution system.

DISCUSSION QUESTIONS

21. Explain why it is so much more expensive to transmit electricity locally than over long distances. Give as many reasons for this as you can.
22. Why do some people think that cogeneration is a useful strategy for dealing with energy generation? Explain.
23. Why should the U.S. electricity transmission system be so much better connected to Canada than to Mexico?
24. What were the major reasons for the great increase in size of the typical power plant, as seen in Figure 8.10?
25. How is it that Insull's apartment building allows more efficiency in the distribution and supply system?

26. Why do the smaller household gadgets have such a small impact on the system overall?
27. A typical computer system draws about 400 W. What effects do you think the rise of the home personal computer has had on the supply of electrical energy?
28. How does the summer load curve for a utility differ from one that might have characterized the 1930s? Ignore the overall difference in scale in your answer.
29. Why have refrigerators been the focus of so much research in lowering energy use?
30. Lovins' soft-energy path would lead to decentralization of energy production and distribution. List at least two reasons each why this could be a good, or a bad, thing.

PROBLEMS

31. Suppose it costs a utility $0.04/kWh to generate electricity at the generating facility, that long-distance energy transport from the generating facility to your local substation costs the utility $0.0002/kWh/km, and that local transport costs $0.0016/kWh/km.
    a. If the plant is 250 km from the substation, and the substation is 25 km from your home, what is the total cost of 1 kWh of electricity in your home?
    b. How much of the cost is for local, as opposed to long-distance, transmission?
    c. Can the utility sell it for $0.06/kWh and make a profit? for $0.10/kWh? Explain.

32. In 1972 (see Figure 8.9b), the screening curve differs from that from 1991 (the last year this information was readily available).
    a. In 1972, what was the cheapest option for producing energy needed for 4000 h/yr? Explain.
    b. Ignoring hydroelectricity, what was the cheapest option for producing energy needed for 4000 h/yr in 1991? Explain.
    c. In 1972, what was the cheapest option for producing energy needed for 7000 h/yr? Explain.
    d. Ignoring hydroelectricity, what was the cheapest option for producing energy needed for 7000 h/yr in 1991? Explain.

33. In 1972, a utility owned a nuclear plant that was online 70% of the year, the oil-burning plant was only out for unscheduled maintenance 10% of the time, and the gas-fired turbines were online 15% of the time. The nuclear plant is rated at 1000 MW, the oil plant at 940 MW, and the turbine at 200 MW.
    a. What is the total cost of running these three plants for the utility?
    b. How much would the utility have to charge per kWh to make a profit?

34. In 1991, a utility owned a nuclear plant that was online 95% of the year, the oil-burning plant was only out for unscheduled maintenance on the air pollution system 15% of the time, and the gas-fired turbines were online 15% of the time. The nuclear plant is rated at 1000 MW, the oil plant at 940 MW, and the turbine at 200 MW.
    a. What is the total cost of running these three plants for the utility?
    b. How much would the utility have to charge per kWh to make a profit?

35. How much does it cost a household to run its coffee machine if it is used 2 h/d and draws 1500 W? Assume the cost of electricity is $0.08/kWh.

36. A stay-at-home mom watches soap operas for 1.5 h each day, and her husband veges out in front of the tube for around 2 h a night. He also watches the football games on Saturday and Sunday during football season (two games each day, 6 h of watching). How much does this family pay to watch TV every year? Assume the cost of electricity is $0.08/kWh.

37. Determine the cost of operating an electric range for a year. Assume that the cost of electricity is $0.08/kWh.

38. Joe's toaster broke. He bought a toaster oven to replace it, telling his wife Annie that it is cheaper to make toast in the toaster oven than in the range's oven. Toast takes 5 min to cook in the toaster oven, while it takes only 3 min in the range's oven. Is Joe right, or is Annie, who says they've wasted their money? Explain.

39. In California in 2001, the wholesale price of natural gas jumped from about $4 per thousand cubic feet to $11. The former price translates into a cost of $0.04/kWh for the utility, while the latter translates into $0.11/kWh. California utilities can only charge around $0.08/kWh. Given this price and assuming the other costs are the same as those in 1991, determine about where the fossil-fuel steam generation will be the same in cost as gas.

40. Repeat the problem posed in Problem 39, but compare to nuclear generation of electricity.

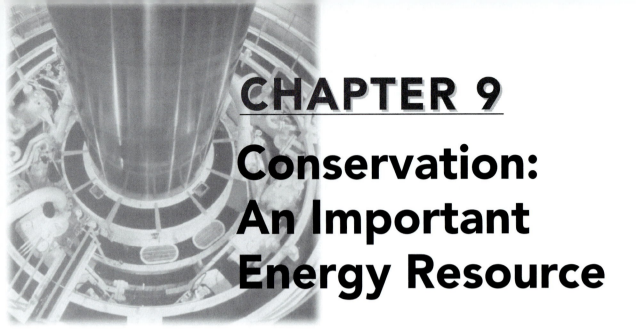

CHAPTER 9

Conservation: An Important Energy Resource

In considering the supply of energy available to heat or cool houses or produce electricity, it is important to understand that major savings have come from using less to do the same amount with equal comfort—a practice known as conservation. *Uses of energy that make do on less reduce the need to build new generating facilities, thus saving everyone money. In this chapter, we consider different ways of achieving this goal, including lighting, appliances, home heating and cooling, and active and passive solar energy uses in the home. Some aspects of conservation in the commercial and industrial sectors are considered. Chapters 9 through 12 and 18 through 20 are devoted to the examination of conventional resources (Chapter 25 to alternate resources).*

KEY TERMS

energy conservation / incandescent / fluorescent / visual acuity / demand water heater / demand-side management / R value / thermal conductivity / superinsulated house / infiltration / payback time / active solar system / passive solar system / direct gain / Trombe wall / sunspace / radiant heat / pulsed combustion furnace / condensing furnace / heat pump / Rolladen / life-cycle costing / variable-speed AC motor

---

9.1 HOW "CONSERVATION" DIFFERS FROM "CONSERVATION OF ENERGY"

In Chapter 3, we discussed the definition of energy and the principle of conservation of energy: that energy can be neither created nor destroyed, but rather that the forms of energy can be interchanged, one transformed to another. Since the first modern energy crisis in 1973, people have been speaking of energy conservation in another sense. What is meant by **energy conservation** in this chapter is the provision of the same level of goods or services with a smaller drain than the drain at present on high-quality energy resources (ultimately fossil or nuclear fuels). Energy conservation is here, in other words, concerned with wise use of energy resources, not with the physical principle (which always holds).

Similar or Greater Benefit at Smaller Energy Cost

The science of thermodynamics has shown us that the efficiency of any heat engine will be less than that of the Carnot engine operating between the same high and low temperatures. Even with the Carnot engine, there must be waste heat rejected. With real engines, more energy is wasted than in a Carnot engine. If we could devise a more efficient real engine, we could reduce the amount of waste heat, thereby, increasing useful work for the same amount of heat taken from the high-temperature reservoir. Thus, more of the fossil-fuel energy would have done useful work. Similarly, a device that produced the same amount of work for a smaller electricity input would have saved fossil fuel, which would not have to have been burned. These measures to increase useful work for the same energy input or to decrease the amount of fuel for the same energy output are conservation measures. The 55 mi/h speed limit on the interstate highways for many years after the 1970s energy crisis was a conservation measure: it required a smaller amount of fuel to go the same distance (albeit at a slower speed).

Most people view conservation measures as "doing without." These people would rather not have to think about such an unattractive topic. They are not interested in energy, per se, but only in having life made as comfortable and convenient as possible, and moreover, getting the comfort at the lowest cost. In many cases, however, rather than making people have to "do without," energy conservation provides *similar or greater comfort* to people *at reduced cost*. People's preconceptions may prevent them from being able to get what they really want. Vice President Dick Cheney said in 2001 that conservation is perhaps "a sign of personal virtue, but it is not a sufficient basis—all by itself—for a sound, comprehensive energy policy."[1] Conservation is not just virtuous behavior—it is sensible behavior! See **Extension 9.1, *What Government Can Do About Conservation.***

The Holistic View of Energy Systems

Often, people do not have a unified concept of the entire energy delivery system. Consider home heating by gas as an example. Natural gas is brought into the home by pipeline and delivered to the furnace. To use the furnace for heating, we need a pilot light, a thermostat, a valve in the furnace, a plenum, a forced-air system (fan), and furnace ducts and vents. This delivery system must be considered as a whole in design, not as a series of separate components, if it is ultimately to function best as a unit. People who think in terms of systems are more likely to realize the positive aspects of conservation. A similar example is in the construction of a building using subcontractors. Subcontractors have little stake in the continuing costs of a system. By minimizing their part of the costs, they may look more responsible to the purchaser while delivering a less efficient and more poorly functioning system.[48]

Efficiency increases between 1973 and 1988 cut $45 billion off the U.S. energy bill. Conservation is already saving at least $150 billion/yr in energy costs.[49] In this chapter, we discuss conservation in lighting, in space heating, and in air conditioning, as well as passive solar techniques, conservation in action in Germany, and conservation in industry.

9.2   SAVINGS IN ILLUMINATION

According to a study by General Electric (GE), about 5% of all the energy generated in the United States is spent on lighting.[50] (This estimate is lower than the ~10% of electricity use, which means ~7.5% of total energy use, reported by the Department of Energy.)[51] Because lighting takes such a large share of electricity use, any savings here would be welcome.

As a first step, people can turn out lights as they leave a room. When a 100 W incandescent bulb is turned on, some energy is used to warm it up. In fact, it takes 2.13×10^{-5} kWh$_e$ to warm the 100 W light bulb.[52] As 1 kWh equals 3.6×10^6 J, 2.13×10^{-5} kWh is about 76.7 J. This is the amount of energy it would take to lift a 1 N weight 76.7 m [or a 1 lb weight 56.3 ft]—quite a lot. To determine how long you should be out of a room before you plan to turn out the 100 W bulb, recall that if power consumption is constant, the total amount of energy used in a time t is

$$\text{energy} = Pt,$$

where P is the constant power. The energy used in turning on the light bulb is 76.7 J; at a power of 100 W, this amount of energy is used in a time of only 0.77 seconds. That is, if you were to leave the room for 1 second, you would still save energy by turning off the light bulb. This is not quite true, because there is a surge that occurs every time the bulb is turned on. This surge fatigues the filament, which shortens bulb life. A good compromise here is probably a minute: If you plan to be out of a room more than a minute or two, turn off the bulb! The situation for fluorescent bulbs is different because of the energy needed to start the ballast up. In this case, the fluorescent light should be left on if you are away less than about 15 minutes.[53]

Another area of energy savings involves the decision about which type of light bulb to use. An **incandescent** bulb contains a tungsten coil filament in a vacuum. A current running through the filament heats it red hot, and it emits light. A normal 100 W incandescent bulb puts about 95% of its power into Joule heat and only 5% into light. In **fluorescent** lights, a mercury gas is excited by an electric discharge and emits ultraviolet (UV) light. The phosphor coating of the tube absorbs this UV light and reemits it as visible light. A fluorescent bulb uses about 80% of its power for Joule heating and about 20% of its power as light.[54,55] Table 9.1 indicates the efficacy of the various types of light bulbs.[50]

TABLE 9.1

Efficacy of illumination measured in lumens per watt

Type	Power	Efficacy
Incandescent lamp	40 W	12
	100 W	18
Fluorescent lamp	40 W (+13.5 W)	59
	75 W (+11 W)	73
Sodium vapor lamp	180 W (+30 W)	154
Mercury vapor lamp	400 W (+26 W)	53.5

Note: The lumen measures effective illumination light per unit time. The ballast power demand is also listed, where applicable.
Source: Reference 50.

The ballast in a fluorescent light provides high voltage and limits current to the tube. Electromagnetic ballasts have replaced standard iron-core ballasts, which has saved a great deal of energy as well as provided better function. Electronic ballasts are more efficient and reduce the annoying fluorescent flicker and audible hum. High-frequency ballast reduces eyestrain and headaches.[56]

Fluorescent lighting uses less than a quarter as much energy as incandescent lighting. In the commercial sector of the economy, lighting is 30% to 50% of the total energy load. Lighting is where most energy can be saved in the commercial sector.[57] Overall, around 20% of our lighting is incandescent, and a bit more than 75% is fluorescent.[50,51,58]

New incandescent bulbs introduced originally by GE are three times as efficient and last four times as long as ordinary light bulbs but sell for around $10.[59,60] Such bulbs are cost-effective relative to ordinary incandescent bulbs, even if used only half an hour per day; at an hour's use per day, the return on investment is 13%, even at an electric cost of 6 cents per kWh.[61]

Compact Fluorescent Lights

New designs increase electrical energy-to-light efficiency by a factor as high as four.[62] The same price considerations apply to compact fluorescents (Figure 9.1) as to more efficient incandescents. The new breed of compact fluorescents being marketed touts the value of saving on energy costs, but the price tag still causes a bit of "sticker shock" for many buyers. (The prices of compact fluorescents have dropped substantially since the mid-1990s.) Nevertheless, as the manufacturers point out, a bulb that lasts longer and uses much less energy will end up saving the buyer money in the long run. (The trouble is that it is really difficult to get people to think in the long run.)

The bulb coatings have been developed to give the bulbs spectral qualities more resembling incandescents than "cool light" fluorescents. I am writing this book in a room lighted by two compact fluorescents that provide as much light as a 60 W and a 75 W incandescent bulb but draw a total of only 29 W. This is over a fourfold savings in electricity use for the same amount of light. Compact fluorescents are not quite as comfortable as the incandescents they replace.[14] Tri-chrome lamps, with rare earth elements emitting in the regions humans see, will give more "natural" lighting; they are expected to become dominant, eventually.[14]

European energy generally costs more than energy in the United States, so Europeans have experimented more with compact fluorescent replacement when these became available. Nine buildings in five European countries were studied, and pre- and postretrofitting data were gathered.[63] Savings of 36% to 86% were realized with payback times of 1 to 8.5 years.[63] An experiment in the German state of Schlewig-Holstein involved the loan of 100 million marks by a utility to replace incandescent lamps by compact fluorescents. Of a possible 600,000 lamps that could have been replaced, 77,000 have been (saving 3.5 MW and 3.4 million marks annually).[64] A Danish utility offered two compact fluorescents to each customer; 90% accepted. Consumers substituted these in the lamps that burn most, in the living room and kitchen, and saved 15% of their lighting costs.[65] The average societal cost of the saved energy was 2.1 cents/kWh, far less than the cost of new power plants.[66]

The United States finally caught on, too. The "Green Lights" program, which was started by the Environmental Protection Agency (EPA) to convince business that it could reduce expenses and pollution by buying compact fluorescents, has

(a)

(b)

FIGURE 9.1

Energy-saving compact fluorescent light bulbs.
(a) helical compact fluorescent bulbs.
(U.S. Department of Energy)
(b) Other styles of compact fluorescent bulb.
(National Renewable Energy Laboratory, U.S. Department of Energy)

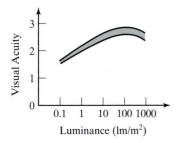

FIGURE 9.2

Visual acuity in people depends on the luminance, measured in lm/m². Acuity becomes saturated, following a logistic-shaped curve, so that, beyond a certain level of illumination, extra light does not contribute much to acuity.

been very successful. Compact fluorescents alone could save 10% of U.S. electricity use.[15,67] Industrial lighting has options not available to the ordinary consumer; for example, light-piping of microwave-excited sulfur bulbs can light huge spaces cheaply.[68]

Daylighting

Residences and business establishments should use sunlight for illumination as much as possible ("daylighting"). Several studies estimate an energy saving of around 70% when daylighting is used.[69] A business office, for example, could use photocells to turn out the lights if the illumination from outside is bright enough. Prismatic panels, light pipes, holograms, and other devices can bring the light from outside deep into a building.[70,71] Such devices have the added advantage that the cooling load of the office is reduced.

Levels of illumination recommended to architects in the United States have been repeatedly raised over the years until they are now about twice the European levels of illumination.[72] New York City's recommended level of illumination increased from 20 lm/ft² in 1952 to 60 lm/ft² by 1971 for schools (from about 2 to about 6 lm/m²).[73] Modern office buildings typically have illumination levels of 8 to 10 lm/m².[50] Such higher illumination levels are not necessary. Norwegian light levels are generally lower than U.S. levels for comparable tasks.[74] **Visual acuity** levels do increase somewhat with increases in light, but as Figure 9.2 makes apparent, acuity increases only a few percent in doubling lighting levels (elderly eyes benefit more from greater illumination than do young ones). There is certainly room for energy savings here. In addition to energy conservation, we must consider the comfort of people using the lighting and the space in determining proper illumination levels. People apparently feel better about working where the light is brighter and feel best working in natural light.

9.3 SAVINGS IN APPLIANCES

By the end of the 1970s, many appliances were much more efficient than they were at the start of the decade: Refrigerators were 45% more efficient; freezers, 48%; dryers, 8%; room air conditioners, 12%; and dishwashers, 50%.[60,75] This improvement continued through the 1980s into the 1990s. Appliances, particularly refrigerators and freezers, still have great potential for energy savings.[75,76] Europe is ahead in energy-efficient washers, or was, anyway, until Whirlpool introduced their energy-efficient European model into America in 1994.[77] Refrigerators and freezers are on all the time, and they typically last 20 years. Even small savings per refrigerator loom large when the total number of households is so large. However, there does not seem to be as much consumer demand as would be expected for efficient ones. In fact, even on current models, the connection between price and energy efficiency is essentially nonexistent.[48,49,76] (More efficient models of all sorts of appliances may be found by paying attention to Energy Star labeling.)

The role of certain states, such as California, in setting energy standards has had an exemplary effect on national efficiencies. By 1993, all refrigerators sold in the United States had to meet California standards. These refrigerators were over 150% more efficient than those sold in 1972. By as early as 1986, refrigerators overall had been improved by over 70% relative to 1972.[78] In 1992, the Super Efficient

TABLE 9.2

U.S. residential appliance lifetime and efficiency, 1987

Appliance	Lifetime (yr)	Efficiency (Average)	(New)	(Best)
Refrigerator[a]	19	1134	979	515
Room air conditioner[b]	—	7.20	8.10	NA
Central air conditioner[c]	14	7.90	9.00	16.9
Heat pump[c]	14	8.40	8.70	16.4
Electric water heater[d]	13	0.82	0.84	3.5
Gas water heater[d]	13	0.49	0.51	0.76
Gas warm-air furnace[e]	23	0.66	0.75	0.97
Oil warm-air furnace[e]	23	0.76	0.80	0.95

[a] Energy use in kWh/yr; lower is better.

[b] EER: cooling Btu/h divided by rate of energy consumption in W.

[c] SEER: cooling for a season in Btu divided by total energy input in Wh.

[d] EF: thermal energy output in hot water divided by energy consumption.

[e] AFUE: dimensionless ratio characterizing furnace operation.

Source: Department of Energy, Reference 57, Tables 2-11 and 2-12.

Refrigerator Project (a group of utilities) offered a $30 million prize for the first CFC-free refrigerator to beat 1993 standards by 25% in energy consumption.[79] Whirlpool Corporation won the prize for best refrigerator design.[80] Table 9.2 shows the situation for several appliances as of 1987. **Extension 9.2,** *Savings on Appliances* provides additional information on more recent developments, such as **demand water heaters**, which make the water hot at the point of use.

Even more energy can be saved in future refrigerator models. Sunpower, Inc., an Ohio firm, had EPA funding to develop a technique for making energy-saving compressor motors, but manufacturers have not shown interest.[126] The EPA says that the motors are reliable, will last 20 yr, and save will 15% in electricity. The technique involved uses a coil to move the piston up and down, and it eliminates the crankshaft. The piston stroke is adjustable, as well.[126] New insulating materials such as aerogel (Figure 9.3) will eliminate CFCs in refrigerator foam and save energy as well.[70]

Refrigerators that are driven by sound waves are under development at Los Alamos National Laboratory.[127] The first such system was designed in the 1920s. The sound sets up a standing wave, and the wave carries thermal energy from the thinner regions to the denser regions of the wave, where a plate accepts it and conducts it away from the refrigerator. The device would have to be more bulky than a present refrigerator and might work best in a large installation.[127]

Magnetic cooling is currently under investigation. So-called giant magnetocaloric materials might be used instead of pumps, saving huge amounts of energy and wear and tear.[128] The first magnetic refrigerator was developed in Iowa in 2001.[129]

See **Extension 9.3,** *Demand-Side Management* for a look at how the regulated utilities encouraged conservation.

FIGURE 9.3

Aerogel materials, developed at Lawrence Berkeley Laboratory for use in particle physics experiments, are transparent and make spectacular insulators. This piece of aerogel can insulate the crayons from the effects of the propane torch. Thin, light aerogels can replace thick layers of insulation in refrigerators.
(NASA/JPL-Caltech)

9.4 SAVINGS IN SPACE HEATING AND AIR CONDITIONING

As you can see from Table 9.3,[50] space heating and air conditioning account for about 20% of U.S. energy use. Space heating is the largest home consumer

TABLE 9.3

Perspective on efficiency of energy use

Use	Percent of U.S. fuel consumption	Second-law efficiency
Space heat	18	0.06
Process steam (industrial)	17	~0.25
Auto transport	13	0.1
Truck transport	5	0.1
Water heat	4	0.03
Air conditioning	2.5	0.05
Refrigeration	2	0.04

of energy. In fact, space heating, water heating, refrigeration, and cooking combined use 95% of commercial energy and 85% of residential energy.[61] Space heating now accounts for about a third of our energy use, whereas in 1900, it accounted for about half, and in 1850, it comprised the huge majority.[55,142,143] Clearly, energy savings, even on an individual level, can be extremely important. In Chapter 8, we saw that second-law efficiencies are a better measure of energy efficiency than the usual first-law efficiency. The very small second-law efficiencies in the U.S. energy economy, shown in Table 9.3, indicate vast room for improvement.

To illustrate the possible effect of price increases, an early study[144] claimed that sharp rises in gas prices were ineffective in causing changes in household energy use. Because natural gas is still so cheap, price increases are unlikely to cause changes until the price has risen enough to be really painful. Nevertheless, gas furnaces available on the market are now up to 95% (first-law) efficient, while pre-1980 models are only 50% to 60% efficient.[145] Often, a time lag occurs between the cause and the response in many cases (here, between the early 1970s energy crisis and the means of saving energy).

In countries with hot climates (and usually not with strong economies), buying energy to use for space conditioning might be difficult. Latent heat absorbed by water evaporating from the roof could allow apartment houses and commercial buildings to be cooled at minimal cost.[146] Common construction techniques in Mediterranean countries (and California) and simulations all attest to the remarkable comfort achievable with the right building orientation and shade conditions,[147] and the use of natural ventilation.[148]

Insulation

Insulation is one of the most important factors in reducing leakage of heat from the home. About 12% of conductive heat loss occurs through the ceiling, and 17% occurs through walls.[149] As a result of greater recognition of the increasing expense of energy, accepted insulating standards have been tightened. The 1965 FHA standards allowed $0.6 \text{ kWh}/10^3 \text{ ft}^3/\text{d}$ heat loss in the installation of home insulation. By 1970, this had been reduced to $0.45 \text{ kWh}/10^3 \text{ ft}^3/\text{d}$, and by 1972, to $0.3 \text{ kWh}/10^3 \text{ ft}^3/\text{d}$. Buildings now consume about 40% more than they would if the 1972 standards had been enforced earlier;[150] if all buildings met these standards, there would have been a 13.5% national decrease in energy consumption.

Anyone who has visited a home improvement store, or insulated a house, knows the insulation is sold with an ***R*-value** rating the efficacy of the insulation. The higher the *R*-value, the more effective the insulation, that is, the *smaller* the flow of thermal energy between reservoirs at different temperatures. This means that the rate of heat transfer varies inversely with R. We may write

$$rate\ of\ heat\ transfer = A\ \Delta T / R.$$

An R-value of 1 is equivalent to about 1 inch (2 cm) of still air in insulating value. The R-value of insulation is defined to be additive, so that if one puts two layers, each made of R-12 material, the effect would be of R-24. See **Extension 9.4, R-Values,** for more detail about housing insulation standards and the way insulation works. (In every region, ceilings are recommended to have more insulation than floors, and wall insulation ratings are less than or equal to those for floors.) The units used to measure R-values, as presented in the extension, are $m^2\ °C/W$. In common English units, the unit would be the $ft^2\ °F/(Btu/h)$. Figure 9.4 shows the number of heating and cooling degree days for the United States and the recommended insulation thicknesses.

In Canada, which generally experiences extremely cold winter temperatures, there was an 11% decrease in energy per household between 1976 and 1982.[165] Had *no additional insulation* (compared to previous practice) been installed in Canada after 1978, the 1985 demand would have been the equivalent of 60 to 80 million barrels of oil higher.[165] The installation of additional insulation alone accounted for half the decrease in energy use. Canadian homes are now much better insulated than was the case in bygone years. Some houses are constructed with R-50 in the walls, R-70 in the ceiling, and R-10 in the floors.[166] Such homes are said to be **superinsulated** (see **Extension 9.4**). The additional costs of such housing range from 5% to 10% of the purchase price; the mean cost of such extras in Minnesota is $44/m^2$ of living space (typical homes have $138\ m^2$ of living space).[61]

Housing construction carries energy costs of around $2.7\ GJ/m^2$ of living space in New Zealand (probably it is about the same elsewhere).[167] Commercial building energy costs are $1.8\ GJ/m^2$ (timber construction) to $3.2\ GJ/m^2$ (steel construction) in New Zealand;[167a] it is estimated at 8 to $12\ GJ/m^2$ in Japan.[168]

Government programs to increase conservation have shown little long-lasting effect in the residential sector.[169] Changes in building codes helped in some places. California has some of the highest thermal resistance (insulation) standards in the country. In 1975, California legislators mandated Title 24, to go into effect fully in 1995, which said that homes should use 75% less energy as compared to 1975.[170] Builders seem well on the way to meeting the requirements. For homes, the standards are R-30 for ceilings, R-19 for walls, and use of R-1.5 double-glazed windows.[171] An early study's [149] "optimal" values are much less than have already been accomplished, and this reflects the lingering effects of the higher fuel prices after the 1973 and 1979 hikes in crude oil prices as well as California's continued efforts to save energy. There is still a lot of room for improvement in California industrial-sector building energy use.[172]

More heating is necessary on cloudless, still nights than on cloudy, still nights. This indicates that radiation is an important energy loss. Radiation loss is large through windows; for double- or triple-glazed windows, it is larger than any other source of energy loss. In fact, about 3% to 5% of total national energy consumption has been attributed to windows.[153,173,174]

Indoor Air Quality

Infiltration of outside air is one of the important problems in space heating and air conditioning. Air exchange accounts for about 40% of heat loss in the home:[50]

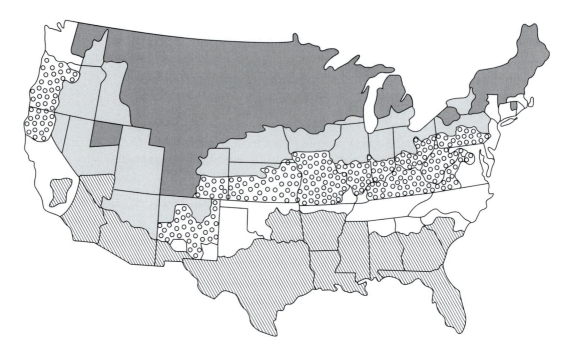

Zone 1		Less than 2000 CDD and greater than 7000 HDD
Zone 2		Less than 2000 CDD and 5500–7000 HDD
Zone 3		Less than 2000 CDD and 4000–5499 HDD
Zone 4		Less than 2000 CDD and less than 4000 HDD
Zone 5		Greater than 2000 CDD and less than 4000 HDD

(a)

FIGURE 9.4a

Weather zone map of heating degree days (HDD) and cooling degree days (CDD). A degree day is referred to 65 °F. HDDs are given as the number of degrees the mean daily temperature falls below 65 °F on each day. CDDs are given as the number of degrees the mean daily temperature rises above 65 °F on each day. HDDs and CDDs are cumulative. (U.S. Census Bureau)

estimates range from 25% to 50%.[72] One problem is control of those areas getting outside air. The kitchen and the bathroom have high ventilation requirements. If infiltration could be reduced elsewhere and directed to high-use areas, large energy savings could result.

The American Physical Society study on efficient uses of energy [50] found that total energy supplied to heat and humidify the incoming air is almost as large as (75% of) the energy supplied to compensate for the conductive losses through the shell. The study targets 0.2 exchanges of air per hour as sufficient for ventilation requirements while minimizing energy waste. Existing infiltration rates in U.S. homes are about 1 to 1.5 changes per hour.[154] The superinsulated Swedish home has only one-fifth of an air change per hour.[161] Because houses are so leaky, it is important to try to minimize the leaks. A technique called pressurization appears to be very useful in finding leaks.[175] With the house closed up, a powerful fan over- or underpressurizes the house to a much greater extent than expected for normal weather. The leaks can be found easily from influx or efflux of air by inspection and then plugged.

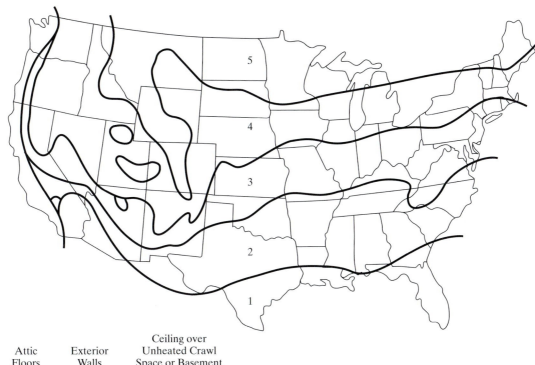

Heating Zone	Attic Floors	Exterior Walls	Ceiling over Unheated Crawl Space or Basement
1	R-26	Full-wall	R-11
2	R-26	insulation	R-13
3	R-30	Which is	R-19
4	R-33	3 1/2" thick	R-22
5	R-38	R-14	R-22

(b)

FIGURE 9.4b

Climatic zones map for determining winter insulation needs. Zones 1 and 2, in the Sunbelt, need more insulation than indicated to fight summer heat.
(U.S. Department of Energy)

It has been estimated that capital expenditures on thermal upgrading of existent structures can pay for themselves in 5 years (or less time, if energy costs resume their upward spiral). An analysis of the **payback time** for simple efficiency changes in appliances, which have not yet been made, show that even with short payback times, the changes are not made.[158] Consumers and producers seem to demand unrealistically short payback times (1 to 2 years) in making energy-conserving choices.[61,158] Time and again, savings from efficiency investments were underestimated.[176] Consumers underinvest in energy efficiency.

Superinsulated houses need to provide for sufficient fresh air. It is reasonably easy to put in a heat exchanger to minimize heat loss.[166] But with most people at home for 12 to 16 h a day, any deterioration in air quality at home can be worrisome.[177–179] For example, urea formaldehyde foam insulation has been abandoned in the United States and Canada because of high formaldehyde levels in homes.[179] Kerosene heaters can cause degradation of air quality and emit toxic and carcinogenic gases.[177,179] Gas stoves can produce carbon monoxide levels of 200 to 700 μg/m^3 (the U.S. National Ambient Air Quality Standard is

$<100\ \mu g/m^3$). Brickwork can emit radon, a radioactive gas with decay products that can easily become lodged in the lungs.[177–180] (Refer to Chapter 20 for a more detailed discussion.) It has been estimated that as many as 10,000 people die each year of cancer brought on by exposure to radon in the home.

Levels of 11 chemicals have been found to be higher inside homes than in the outside air; these are generally from solvents used in the house and stored indoors and from use of chemicals in building construction and furnishings.[177,180–183] Even benzene (produced outside) turns up at higher levels indoors, at two or three times the outside air concentration.[184,185] Chloroform is also present in high concentrations, mainly from exposure to chlorinated water in taking hot showers and from the water vapor from tap water in household air.[185] Because of new materials being introduced as building materials, even more chemicals will become important and need to be investigated.[186] The exposure to in-home pollutants for people who live in clean-air environments is about the same as for people who live in dirtier areas.[182,185]

More and more, measurements are being made where the people are, in their homes and at work, and more problems are being identified.[185,187] There is strong evidence that building characteristics influence incidences of asthma, respiratory illness, and allergies.[183,187] Reduction of exchange of polluted air within buildings is important.[186,187] Even if the inhabitants do not smoke, all these considerations imply the necessity of sufficient air exchange. The biggest change in building engineering is the focus on the health of the building occupants. Cooking fumes are second to indoor exposure to tobacco smoke as risk factors for children's health and that of the elderly (both groups spend much more time indoors than normal adults).[188] That U. K. report suggested that several childhood health conditions could be exacerbated by indoor air pollutants, especially common childhood illnesses. This suggests strongly that ventilation rates should be increased.[186,187] ASHRAE Standard 62 and the Uniform Building Code Section 1202.2.1 both designate a minimum of 15 cubic feet per minute (0.4 m³/min) of fresh air per person occupying the area. (ASHRAE Standard 62.2 [2004] deals specifically with residences in this regard.) There is a possibility of direct savings of $17 to $48 billion annually, from illness avoided or reduced severity, from improvements to indoor air; and there are indirect savings possibilities of $20 to $160 billion. Some of this gain in indoor air quality comes from changes in lighting as well as introduction of enhanced ventilation.[187,189]

House plants can be part of the solution to indoor air pollution. The Gerbera daisy and the mum are effective at reducing levels of benzene. Aloe vera removes formaldehyde at small concentrations. English ivy, peace lilies, golden pothos, and mother-in-law's tongue—all common houseplants—remove large amounts of pollutants.[190]

9.5 ACTIVE AND PASSIVE SOLAR ENERGY AND CONSERVATION

Active Solar Energy Systems

Solar energy has been used for decades to produce cheap hot water for home use in Israel, Japan, Florida, and Australia. Many such systems were being marketed across the United States because of the tax credit available for investment in solar energy. Some space heating systems are competitive even without tax incentives, but only in the Sunbelt.[191] Mixed systems of solar plus conventional energy supplements are feasible everywhere in the United States. Some argue that the technological match of solar systems with the utilities is poor, and that most of the savings from solar energy come from energy storage rather than the solar component.[192] Others argue that the match is appropriate, especially when conservation is taken into account.[193]

Because solar energy can be intermittent, it is always necessary to have backup systems as well as storage or to disperse widely the components of the system.[193,194] Flat-plate collectors, as shown in Figure 9.5, oriented toward the Sun can be up to 70% efficient in warm weather but are only about 10% efficient in colder weather because of conductive losses.[195] In recognition that we are mechanically assisting the circulation, such systems are known as **active solar systems**.

There are two kinds of active systems, liquid and air. Air systems store and distribute heat produced by air collectors. Liquid systems store and distribute thermal energy produced by the liquid collectors, such as Freon, water, or antifreeze. Liquid systems produce fewer drafts and use less energy to run; they are also easier to retrofit.[196] However, indirect systems require a heat exchanger. Storage is provided by 40 to 80 liters per square meter (1 to 2 gallons of water per square foot) of collector. Very few direct systems, which circulate city water through a line for heating, are being used. Air collectors produce heat both earlier and later in the day than do liquid systems, and cannot freeze.[196] Typically, the storage is a rock storage bin in the basement. Warm air heats the rocks; this warmth may be used later. Rock bins should be 0.5 to 1 ft^3 per ft^2 of collector. The rock bins occupy a larger volume than the liquid storage containers.[196]

These systems cost energy to build, so it is reasonable to ask if a widespread switch to solar energy will involve so much new manufacture that there may be a cumulative energy debt, and if so, how long it will take to break even. Indications are that the payback period for a system bought could extend more than 5 years.[197,198] Of course, new homes should last much longer than that. On the other hand, consumers and businesses are remarkably resistant to the charms of any method of savings that has a payback of longer than 2 years.[170,199] Governments accept an 8 to 10 year payback period.[170] Solar hot water heating is discussed in **Extension 9.2**.

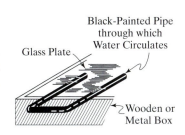

FIGURE 9.5

Flat plate solar collectors.

Passive Solar Energy Systems

A system that uses the Sun's energy but uses no external energy to collect or distribute it is called a **passive solar system**. The distinction between passive solar and "conservation measures" is fading away as both become more common. The use of passive solar energy in new homes complements "conservation measures" in construction. Modern building practice works with both conservation and passive solar on the same basis. It makes sense to treat them simultaneously here, so that information on both will be available to those who want to design a new home or repair or rebuild an old house.

The Romans, in response to deforestation of the land around Rome, were forced to bring wood from farther distances (up to 1000 km) for heating. In response to this pressure, the Romans began to design and build solar homes using glass.[143,183] Since that first use of passive solar heat, the technology was developed and lost twice. The losses occurred in the Dark Ages, as much knowledge was extinguished, and again in the late nineteenth century to the late twentieth century, in the age of cheap energy.

There are three basic types of passive heating systems: direct gain systems, storage wall systems, and sunspaces. In passive solar systems, it is essential to have a net heat gain from the windows or other openings during the heating season and helpful to have a net loss over the course of the cooling season. Passive systems are generally less expensive to build and operate than active ones, and some passive designs have even been found to outperform more active systems.[201,202] As simple a measure as installing a plastic covering on a window in winter can be effective as well as economical.

Direct gain in solar space heat is possible through windows, which presently account for 30% of all heat losses in the average house.[150] The newer generations of windows, "superwindows," have been designed to have R-8 to R-10 and cost 50% more than ordinary glazing ($<R$-1).[70,203,204] These windows have several layers, sometimes with a vacuum or insulating gas between layers (Figure 9.6). The new superwindows decrease the amount of UV allowed through. New windows that are low emissivity, or low-e, are coated with a metal oxide, a technology developed in Europe.[70,203,204] Low-e windows cost 20% more but double the insulating ability. The problem of heat leak through the frame becomes more prominent as the windows become more insulating. The coatings can be designed to reduce heat gain without cutting incoming light; this is an important consideration for commercial buildings that spend most of their energy cooling and lighting.

Active areas of research focus on windows that can darken in response to light, heat, or electric control—photochromic, thermochromic, and electrochromic windows. A cloud gel is a mixture of a polymer and water. At low temperatures, the gel admits 90% of the light. Above its transition temperature, it admits only 10% of the light. The transition can be engineered to be at any temperature between 16 °C and 65 °C.[205] The transition can also be controlled by

FIGURE 9.6

Glass beads are used for separation in vacuum window glazing in an R-9 window.
(U.S. Department of Energy)

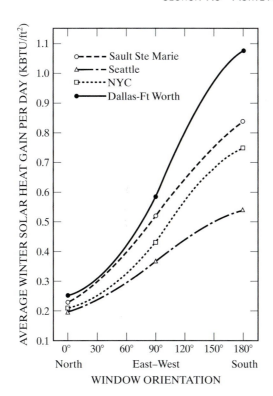

FIGURE 9.7

The daily average solar heat gain for four U.S. cities.
(From A.I.P. Conference Proceedings #25, *Efficient Uses of Energy*, © 1975)

use of electric currents. The inventors have set up a company called Suntek in Albuquerque, New Mexico, to market the cloud gel.[205] Electrochromic windows are the most exciting technology because of the possibility that people can control them (automatic changes can be quite annoying). The technology should be commercialized by the time you read this by Sage Electrochemics of Faribault, Minnesota.[71]

On a sunny day in winter, the peak radiation energy flux from the Sun can be about 1 kW/m² on the ground. Figure 9.7 shows the average energy gain possible through windows facing in various directions.[50] Another factor influencing the amount of sunlight received is the overhang on the window; it is possible to design windows so that summer sunlight, when the Sun is high in the sky, is not incident on the window, while winter sunlight, when the Sun is low in the sky, is captured by the window. Passive solar homes with limited eastern and western exposures function well during the summer.[201,202] The main rooms should not be exposed to direct sunshine.

Deciduous trees can also shield a house from summer Sun while in leaf and let winter Sun through bare branches, although care must be taken to prevent shading. In passive cooling systems, plants cool the area around a house by evapotranspiration. Trees at the northeast/southeast and northwest/southwest sides of the house are most useful. Trees should not be placed to the south, because even bare branches can block winter Sun, but rather on the west and east.[201] In addition to landscaping, night ventilation, roof radiation to the night sky, and earth tubes are passive cooling measures.[202]

Masonry walls exposed to the Sun through a window during the day (**Trombe walls**) radiate the heat later, as the temperature drops, cutting costs for space heating. Trombe walls work well, because people feel warmer at a lower temperature if they are in an environment in which the surroundings radiate heat. Some studies in Germany have led to the development of heating systems that heat interior masonry walls or concrete floors during the heating season.[206] While it may be somewhat expensive to raise the temperature of the walls at the start of the heating season (stone has a substantial heat capacity), the temperature of the interior walls could be maintained at a modest cost. Thick-walled structures could absorb substantial amounts of heat during the daytime that would be released after sundown. Perhaps the medieval castles were more comfortable, when drafts were closed off by draperies, than we have traditionally supposed. Please read **Extension 9.5,** *Comfort on the Cheap* for more details on Trombe walls, radiant heating, and shade.

The **sunspace**, a 1970s idea, is actually a revival of the sun porch. Sun porches, first developed by the Greeks, were brought to a high stage of art in America in the eighteenth and nineteenth centuries, only to vanish in the early twentieth century.[200] Sunspaces are south-facing vertical walls with brick backing, installed to take advantage of incident solar energy. Such spaces can be very pleasant, in addition to helping to decrease home energy consumption. It is important not to have too much glazed area, or sloped glass with little or no effective shading in sunspaces, to prevent overheating in summer.[225]

Much research on solar energy in less-developed countries focuses on its use in agriculture. The use of passive solar energy in construction of poultry sheds[229] and for drying of grains or fruits[230] has been very popular.

Passive Solar Techniques of Earlier Times

Many natural heating and cooling techniques were developed in the Seistan region (on the Iran–Afghanistan border), where the windmill was invented in the eighth or ninth century (windmills reached Europe by the twelfth century, as mentioned in Chapter 19). Some of these techniques can work elsewhere as well. Thick adobe walls act as insulators and mitigate temperature swings (as already mentioned). Adobe wind towers act to cool in all sorts of summer weather. If there is no wind at night, the warm air rises and draws in cool air.[231,232] If a wind blows, the circulation is reversed.[232] In daytime, the cool tower causes air to sink, causing a flow of cooling air. If the air passes over water, it cools still more as it picks up water vapor[232] and loses latent heat of vaporization. Air vents or roofs serve a similar function, causing hot air near the ceiling to be pulled out, cooling the building.[232] Wetted bushes placed outside windows can cool a room by 8 °C if there is a breeze.[231] The Seistanis also designed natural refrigerators and freezers, although they used stagnant water, which often led to health problems.

The open areas in buildings in Saudi Arabia's semidesert areas have high walls with openings reminiscent of Iranian towers. The air may be excluded from the rooms or brought inside by sliding wood doors. The streets are winding, and the houses have courtyards just as in desert areas, for the same reasons.[199]

In hot and humid regions of Saudi Arabia, similar needs and solutions were found. The louvered "Rowshan," which is similar to an ariel- or bay window, allows air to flow freely into the room while not collecting heat the way a sunspace would and providing protection against glare.[205] The conical huts seen in hot and humid areas in Africa have roofs of dried thatch that serve much the same purpose—the hot air rises to the peak of the roof, thereby allowing the cooler air to condition the air below naturally.[205]

9.6   SAVING ENERGY IN HOME HEATING AND COOLING

Radiant Heating

The time delay inherent in massive masonry or stone structures is well known. Cobblestone houses take 2 months to warm up or cool down.[233] Meanwhile, they provide a comfortable environment for people, because radiated heat makes people feel more comfortable. Warm walls allow the air temperature to be reduced with maintenance of comfort.[234,235] The temperature used to determine comfort is the mean radiant temperature, that is, the average of room air temperature and wall temperatures. This would seem to indicate that the folks who are insulating their cobblestone houses on the inside[200] are ignoring the better return on investment they would get if they were to install heating onto the walls to maintain their temperature and take advantage of insulation on the stone, even in winter. The latter practice is becoming more common in Germany.[206]

A technique applied both here and in Europe involves *radiant heating* through the floors with plastic pipe.[206,234] The pipes are polypropylene or polybutane, and medium-temperature water (30 °C to 50 °C) is circulated through the pipes, which are embedded in the concrete floor.[233] The system scheme is shown in Figure 9.8.[233] For the most affluent Americans, the system

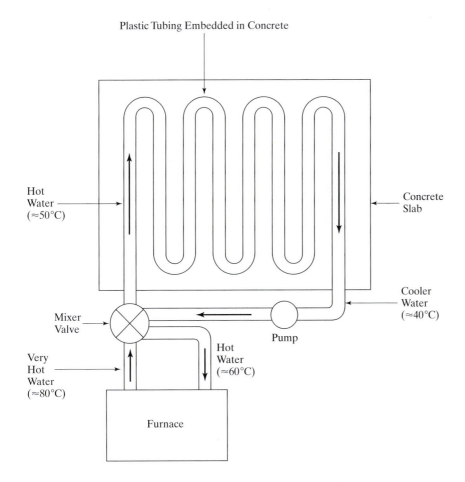

Plastic Tubing Embedded in Concrete

Hot Water (≈50°C)

Concrete Slab

Cooler Water (≈40°C)

Mixer Valve

Pump

Hot Water (≈60°C)

Very Hot Water (≈80°C)

Furnace

FIGURE 9.8

Slabs may be heated by pumping warm water through embedded plastic pipes to provide radiant heat to room inhabitants.

of choice today is hydronic radiant floor heating.[235] This system is expensive to install but cheap to run. About 7% of all homes have this system.[235]

Quartz electric heaters increase comfort levels by providing radiant heating. A more speculative proposal along these lines was the suggestion that homes use microwave radiant energy to provide warmth directly.[232] This type of energy is said to be more effective than infrared, and it works by providing a radiant temperature component to raise comfort levels. However, there may be hazards to such use of microwaves.[233]

The number of people in the house can also influence heating requirements. People are approximately 100 W heat sources, as we saw in Chapter 3. It is for this reason that we see open windows and doors in winter at crowded parties.

Furnaces

Another simple method for reducing heat use is the timing of the furnace. For example, the only reason for heating a house above about 15 °C (60 °F) while people are sleeping is so that they are not uncomfortable when they get out of bed. By setting a timer to turn the furnace on half an hour before awakening, energy wastage can be cut. (There is no truth to the myth that it is less energy-expensive to keep the house at a constant temperature by running the furnace continuously. Energy loss is proportional to the inside–outside temperature difference, as discussed earlier in the chapter.) If no one is at home during the day, the timer can also be used to turn the furnace on shortly before anyone is expected home.

If the home is heated by radiators circulating hot water, you might want to see if the radiators are near outside walls. If so, you will benefit from getting a piece of corrugated paper, painting one side matte black, gluing aluminum foil to the other side, and setting the result upright halfway between the wall and radiator, black side toward the radiator.[236] The black side absorbs and reradiates well, the foil side does not. The gap between radiator and black side is efficient in setting up convection currents, so with a small shelf mounted above the radiator and the corrugated paper, warm air is deflected into the room and somewhat away from the windows (normally the radiator is placed by a window to help set up convection currents), so you get more heating for the same price. This may reduce heating bills by as much as 10%.[236]

Recall that the average efficiency at a conventional power plant is about 33%, and the efficiency of local electric transmission is about 91%. Resistive house heating is 100% efficient with floorboard or baseboard heat but is only about 50% efficient for ceiling heat. Thus, the overall efficiency of electric heating is 30% or less, except if heat pumps are used.

The use of gas- or oil-burning in-house facilities can be 40% to 95% efficient; the average is about 60%.[145,149] The upper figure represents full-capacity use; typical usage is about half that. Soot buildup can soon lead to substantial loss of efficiency, typically to 50% of that expected.[72] Some units are as low as 35% efficient.[237] People should have their gas furnaces checked to get maximum heating from each cubic meter of gas. (Also see the discussion of ducts' contributions to energy efficiency in **Extension 9.2**.)

The newest gas furnaces beginning in the 1980s became much more efficient than the older ones (Annual Fuel Utilization Efficiency [AFUE] >90% vs. <70%). The major technological change has been the development of pulsed furnaces and of condensing gas furnaces (separately or together).[61,75,238]

In the **pulsed combustion furnace**,[238] a cylindrical chamber with one open end and one valved end is supplied with air and gas. A spark starts ignition. The burning of the gas causes the pressure of the air in the chamber to rise, which closes the valve. Combustion continues until the air is exhausted, when the effluent rises. This reopens the valve, and the chamber refills with air. Since it is hot, there is no need for another spark. The furnace cycles on and off in this model. It operates at resonance, pulling in air as the valve opens, burning the gas–air mixture, and exhausting the effluent to begin again.[238] First-law efficiencies of 95% are possible with a pulsed furnace.

The pulsed furnace is simple to supply and of simple construction. The furnace is somewhat noisy, because it opens and closes half a billion times a heating season, and the valve (which may be inaccessible) is a maintenance problem.[238]

Much energy goes up the stack as water vapor in an ordinary furnace. In some new furnaces, called **condensing furnaces**, this additional thermal energy is extracted from the flue gases, which increases the efficiency.[61,75,238] There are several ways of achieving this: stack damping (temperature activated); vent damping; use of a power burner (a blower forces combustion gases through the furnace); and heat exchanging.[75] Some condensing furnaces are over 92% efficient. When condensing flue furnaces are used, the exhaust gas is so cool (about 40 °C),[61] that polyvinyl chloride pipe can be used to eliminate the gas through a sidewall of the house.[61] A model that combines pulsed combustion and condensation is shown in Figure 9.9.[75] The savings possible in use of newer furnaces is

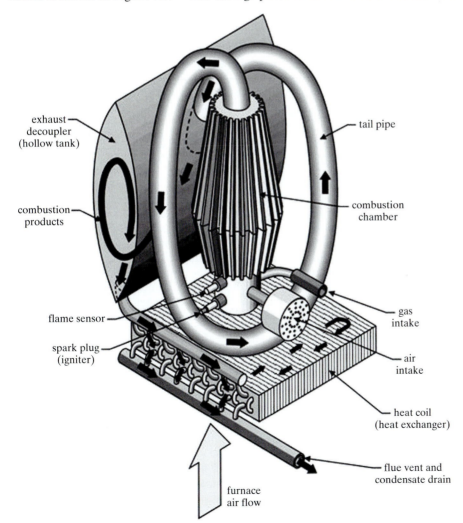

FIGURE 9.9

A pulsed combustion condensing furnace. These furnaces provide first-law efficiencies of ~95%.
(Courtesy of Lennox Industries, Inc.)

TABLE 9.4

Savings in heating systems between new and old German homes

	Older homes	Newer homes
Exhaust losses	18%	7%
Radiation losses from heating vessel	5%	1%
Losses from keeping the system running even when not directly in use	24%	4%
Distribution losses	2%	1%
Losses from overheating due to poor regulation of temperature	13%	0%
Useful heating value of fuel	38%	87%

Source: Reference 239.

illustrated in a German study that predicted a savings of 60% to 70% in heating energy costs.[239] The comparison is shown in Table 9.4.

Wood Heating

Because more people are using wood for energy nowadays (up to a third of all households in New England use wood for some space heating),[240] it is important to burn wood more efficiently. Normal fireplaces are only about 10% efficient, and there is a net loss in energy from a wood fire when the outside temperature is less than ~5 °C because of the escape of warm air up the chimney.[241] Logs in the fireplace (or even home furnaces) take fuel energy, turn it into high-temperature thermal energy (in the vicinity of the flame), and then mix it inefficiently with interior air to warm us. However, new ideas, such as use of outside air instead of inside air in the fire, and use of more efficient grates—the "Texas fireframe" is about 30% efficient[213]—can also save energy.

Most people who do substantial heating with wood have enclosed woodstoves modeled after the sort first developed by Benjamin Franklin. Franklin touted the stove's energy conservation potential:[243]

> Wood, our common fuel, which within these 100 years might be had at every man's door, must now be fetched nearly 100 miles to some towns, and make a considerable article in the expense of families. We leave it to the political arithmetician to compute, how much money will be saved to a country, by its spending two-thirds less on fuel; how much labor saved in cutting and carriage of it; how much more land may be cleared for cultivation; how great the profit by the additional quantity of work done, and to physicians to say, how much healthier thick-built towns and cities will be, now half suffocated with sulfury smoke, when so much less of that smoke shall be made, and the air breathed by the inhabitants so much purer.

Modern stoves, with airtight enclosures lined with firebrick, can provide warmth for hours on relatively little fuel. People, especially in rural and suburban areas, have installed woodstoves for partial heating of their homes. In the richly forested Pacific Northwest, many city-dwellers have woodstoves.

This success of stove technology has caused a problem for many cities and towns in the western United States, such as Vail and Denver in Colorado. In Portland, Oregon, on winter days, nearly 50% of ambient particulates come from residential wood burning.[240] Emissions from woodstoves have essentially wiped out air quality gains from control of industrial emissions.[244] Residential stoves emit much more polycyclic organic material than commercial burning,[244] and these agents are known carcinogens. Oregon tackled the problem by first comparing performance of current stove models, then mandating a 50% reduction in particulate emission by July 1986 and a 75% reduction by July 1988[244] in stoves sold in retail stores. This single state regulation has resulted in sales of much cleaner stoves nationwide since 1988.

Heat Pumps

Figure 9.10[245] reminds us, as we learned in Chapter 7, that the freezing point of water and room temperature are very close in absolute temperature, which is a measure of their thermal energy. This means that the amount of thermal energy of an object at room temperature is only marginally greater (~7%) than the thermal energy of that object at freezing. It would be wonderful to be able to "piggyback" on all that energy around us, even when the temperature is low.

The **heat pump** uses low-grade ambient thermal energy to increase the efficiency of providing thermal energy at room temperature. A heat pump's operation is similar to that of a refrigerator. The refrigerator cools itself by heating the room it is in. The heat pump warms the inside of a house by cooling outside air (or water), and cools inside air by heating outside air (or water).

Thus, the heat pump allows us to use the thermal energy in the outside (cold) environment, do work on it, and exhaust the thermal energy plus the work at higher temperature, warming the inside of the house. Because there is so much thermal energy in the outside environment relative to absolute zero temperature, the electric energy input (in theory) can be only 7% of the net energy delivered to the interior of the house. Of course, there is a factor of three loss in generating the electricity in the first place, and no heat pump is even close to ideal. Many commercial home heat pumps sold operate on the temperature difference between outside and inside air. These work reasonably well when the temperature difference is rather small, but not so efficiently on extremely cold or hot days.

The groundwater heat pump is a more expensive, but better, choice for comfort management. See **Extension 9.6, *Heat Pump Systems*** to learn more about these devices and how the systems work. A large-scale heat pump using ambient air and having exhaust heat recovery can be an efficient heating or cooling device for large buildings or groups of houses.

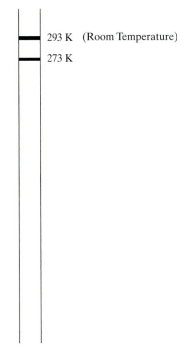

FIGURE 9.10

Room temperature is only about 7% greater than outdoor freezing temperature. The thermal energy in the environment at temperatures between 273 K and 293 K can contribute to heat release in a warm (293 K) environment when a heat pump is used.

Air Conditioning

Insulation put in to reduce heating loss reduces energy loss in air conditioning as well. The 1972 insulation standards brought about an 18% savings for electricity and a 26% savings for gas air conditioners in comparison to the 1970 standards,[78] and later standards have saved far more energy. (See also **Extension 9.2** for more information on advances in air conditioner efficiency.) Most popular models of window air conditioners use about 4 kWh$_e$ while running. (Commercial consumption for air cooling far exceeds residential consumption.)

The air-cooling component of energy consumption has been doubling about every 5 years. The saturation index had reached about 40% for air conditioners as early as 1968[251] and is now substantially higher. Air conditioners have spurred the growth of the U.S. "Sunbelt" by making it possible for northerners to live comfortably in southern regions.

Many models of air conditioners have been tested to determine efficiency. The models ranged from 4.7 to 12.2 Btu/kWh$_e$. Of 90 models tested at 10,000 Btu/h capacity in one case,[149] the lowest energy use was 880 W; the highest was

2100 W.[149] This would entail an energy cost of 976 kWh$_e$ higher for the highest energy user as compared to the lowest. At the 2004 national average cost of 7.4 cents/kWh$_e$, this is an added yearly running expense of $72.50. If we assume a 10 year life and constant energy cost, the lowest energy use model would have cost $725 less to run than the highest. We can justify spending a considerably larger amount to buy the more efficient unit because of the electricity cost savings. An added factor is that the lower-priced models have no thermostat. Thus, the air conditioner tends to operate continuously, cooling a room to a temperature lower than desired. This decreases the lifetime of the air conditioner. As with furnaces, average efficiencies of air conditioners have been increasing in recent years (about 3%/year).[75]

I learned many things about energy conservation from my time living in Europe such as the use of **Rolladen** to control sunlight entering a home. I share some of what I learned in **Extension 9.7**, *Lessons from Europe on Energy Conservation*. Join me to see how we could do more with the same or less effort.

9.7   COMMERCIAL AND INDUSTRIAL ENERGY CONSERVATION

Buildings

Many office buildings are energy guzzlers. In the wake of the energy crises of the 1970s and 1980s, however, businesses began to consider **life-cycle costing** (that is, using continuing as well as acquisition costs to determine which alternative was cheaper overall). One of the first such buildings was Citicorp Tower in New York City. Most buildings in Manhattan require an energy input of 2800 MJ/m^2/yr; the Citicorp building uses less than 40% of that amount.[259] Glass panes are double thick, with a reflective coating on the glass. The National Audubon Society built in energy conservation in its renovation of its New York headquarters. The building is daylighted, which means it is lighted to a large extent by natural light. The building is insulated much more than is standard practice. Air circulates through filters six times per hour, twice the normal rate in commercial buildings. Normal buildings use 26 W/m^2, whereas this building uses 11.[260] Green construction on average saves 30% of annual energy costs at the cost of an additional 2% in construction costs (see **Extension 9.5**).[214]

In a building with a floor area over about 90,000 m^2, enough waste heat is available that the building must be cooled all year round.[259,261] The phenomenon of large buildings needing to be cooled year-round is not hard to understand. In the large building, energy inputs are determined by the Sun, by the people inside the building, and by the amount of lighting (and, of course, heating) done. Recall that each person is radiating heat equivalent to a 100 W bulb, and every electric apparatus dissipates heat. Because these buildings are well lighted, and much energy is emitted as heat, lighting produces much indoor heating.

In large buildings, people on a floor provide a heat flow of about 10 W/m^2, and lighting provides a heat flow of over 30 W/m^2.[262] The heat loss is due to conduction through the walls, ceiling, and floor, radiation through the windows, and infiltration and exfiltration of air (which must be warmed as it invades the building and carries heat away convectively as it leaves the building).

If losses are less than gains, the building cools to some new equilibrium temperature if not heated. If losses equal gains, nothing occurs. If gains exceed losses, the building warms to some new equilibrium temperature if not cooled. The former situation is analogous to the ease of evaporation from a material with a large ratio of surface area to volume (discussed in Chapter 7): Evaporation occurs rapidly. The latter situation is analogous to the case of evaporation from a material of small ratio of surface area to volume: The volume is so large that evaporation proceeds quite slowly. The larger the building, the smaller its ratio of surface area to volume, and the less important the surface is to any process. Thus, above some volume (90,000 m^2 × 3 m, the interfloor height), a building will always produce more waste heat than it is able to dissipate, even in winter. It must then be cooled at all times. Another problem is the maldistribution of thermal energy in such a building: It is very hot at the center and cold at the periphery.

The low cost of energy in the early 1970s caused construction of many "dinosaurs"; the worst buildings required an energy input as large as 1930 MJ/m^2/yr.[49] Numark and Bartlett documented some of the diseconomies of a building constructed in 1973 for student recreation at the University of Colorado.[263] They demonstrated that use of the heat rejected in freezing water in the ice rink could be used to heat shower water in the same building at an estimated annual savings of $40,000.

The energy crisis in California and resulting price rise affected commercial establishments adversely, as seen in **Extension 8.5**. As the effects lingered into 2002, California businesses suffered.[264] This has resulted in more acceptance of conservation measures by small businesses, some of which have saved tremendous amounts of money.[265] Such conditions have been successful at bringing increased conservation in California, a cut of 10% for peak use in 2001.[140]

Computers in the Citicorp Tower reallocate thermal energy from people, lights, and computers to the periphery of the building, where it is needed in winter, thus saving energy. The Citicorp building (~900 MJ/m^2/yr) and the Audubon headquarters (340 MJ/m^2/yr) are still worse than a typical Swedish office building and pale in comparison to the Folksam building in Stockholm, which uses only 150 MJ/m^2/yr.[160] Buildings can be built that need only 110 MJ/m^2/yr or less.[49] It costs no more to construct the energy-efficient building. Even use of ASHRAE Standard 90.1P-1989 would provide savings; buildings constructed according to this standard are estimated to use 550 to 900 MJ/m^2/yr, about half that of existing buildings.[18,174,266]

The Swedes use a system called "Thermodeck" in which warm or cool air is supplied through hollowed-out regions in concrete floor slabs. Even though every Swedish office must have a window, the triple glazing used in the windows assures a solar gain without corresponding radiation loss, and the building can be occupied virtually without heating in winter.[171] Although the building must be cooled in summer, the cooling can occur in the evening when the rates for electricity are lower, due to the large heat capacity of the Thermodeck slabs.[171] Thermodeck hollow-core concrete slabs have a time constant of 100 h.[174] See also the **Extension 9.5** for more information on "green" construction, which can deliver savings of 10 times the additional cost of the construction.[214]

Overall, the efficiency of many components of the home and office have led to substantial savings, in part because of new standards (see **Extension 9.1** and **Extension 9.2**). It is estimated that each dollar spent on implementing such standards means a net savings of $165 (in terms of discounted, or present, value) for the U. S. economy during the period between 1990 and 2010, and benefits of the efficiency added over that period are estimated to be 3.5 times larger than costs.[267] The standards are predicted to have saved 30 EJ cumulatively by 2015 and over 60 EJ by 2030.[268]

Cogeneration and District Heating

Some cities and towns have restricted areas within which the buildings are heated using a central utility. This is called district heating (see also Chapter 6). For example, the Ohio State University attaches its buildings to a central system, and steam pipes carry the hot water and steam around campus. This allows some economy of scale in the production of the heated material and is part of the electricity generating system already in place. It allows for a valuable way to use the "waste heat" produced.

While some American district heating occurs, the method has a longer history, and is of more importance, in Europe. Europe has a different land-use pattern, and for the most part, people gather into villages, even in farming areas, rather than have stand-alone houses. The relative compactness of cities and towns lends itself to exploiting district heating schemes. Again see **Extension 9.7** for more information on this aspect of European practice.

The utility and the university in Saarbrücken, Saarland, Germany, are involved in an experiment begun in 1980 to reorient energy use.[253] The region is in the middle of vast "dirty" coal deposits. The air quality was awful in the Saarland, and people were contemplating changing their heating from coal to something else. The time was ripe for change. Following a German federal government report, this was the only region in Germany to attempt to set a 15 year plan to rationalize energy use and turn to renewable and alternative energy.[253]

In 1980, as the project began, 1900 MW was the demand for heating; the plan was to reduce this to 1400 MW by 1995. District heating (cogeneration, using the hot water generated in producing electricity to heat space) is being used in one part of the city (the inner 25 km² or so), while heating in the rest is by natural gas. Cogeneration and district heat were already in some use in 1980 as the project began. Competition between the two was forbidden; service zones were partitioned out by the utility. Already by 1988, the total use had decreased to 1500 MW. The bonus was lower pollution with higher employment.[253] District heating causes *negative* emissions (in comparison to what would have been the case).[254] The people of Saarbrücken seem satisfied by what they have collectively wrought.[135,255]

Saarbrücken is financing the project in a clever way. The available money for energy-saving investments is limited. Saarbrücken considers measures that save money in the long run. An example is the gorilla house in the zoo: it costs about 23,000 euros to heat each year using resistance heating.[254] If the gorilla house is connected to the district heating system, it costs 10,000 euros to heat. The city pays the money to connect it to the district heating system, 85,000 euros, and continues to allocate 23,000 euros to pay for heat for the gorillas. The difference between the real experience and the budgeted allocation is used to pay back the investment. Once it is paid back, the smaller real expense is budgeted, and the "surplus" money is freed to make other energy-saving investments.[254]

Conservation in Industry

Many industrial processes are still very inefficient,[166] but compared to the 1970s, manufacturers have reduced their energy intensity by 50%.[269] The net reduced growth in the 1970s came from slower growth in the economy (1.4%/yr decrease), efficiency improvements (1.2%/yr decrease), and changes in the composition of the output.[270] Overall, productivity increased about 18% for all the forms, and between 1973 and 1982, about a year's worth of oil consumption was saved.[270] Compared to 1985, by 1987, steel energy intensity fell 11.5%; in paper, it decreased by 2.1% between 1987 and 1988; in chemicals, it fell by 5% from 1985 to 1988.[271] Compared to 1973, primary energy use was reduced 14%.[272]

The 1988 energy intensity by industry for the G-7 nations is shown in Figure 9.11.[273] The energy-intensive industries (primary metals, chemicals and petroleum, and paper) use about two-thirds of all industrial energy in the United States, while in the United Kingdom, they use only 40% of all industrial energy.[273]

Little money has been put into increased energy efficiency.[144] Also, the mix of energy use in industry is changing. In 1980, manufacturing accounted for 85% of U.S. industrial energy use;[272] in 1984, it was 78%.[274] In 1980, mining accounted for 7% of U.S. industrial energy use;[266] in 1984, it was 10%.[274] In 1980, construction accounted for 4% of U.S. industrial energy use;[266] in 1984, it was 6%.[274] The remainder, 4% in 1980[266] and 6% in 1984,[274] is used in agriculture. Most of this energy use is for process heat. Redesign using heat recovery and other forms of conservation will probably decrease the amount of low-temperature process heat generated directly.[272,274]

Industry in the United States is becoming less dependent on materials processing, which should indicate a shift to more value added per unit material processed.[272] However, comparisons with Japan and Europe imply that the United States still has a long way to go (a factor of roughly two) to equal foreign efficiencies.[272] Streamlining of operations is still possible. New, more efficient equipment is being ordered.

There is always room for improvement, and one promising place is in AC motor drives.[15,70,172,234,275–277] About half the electricity in the world runs motors.[15,70] Electronic speed control systems, new materials and designs, new drive trains, and new bearings altogether could save about 60% of the cost of electricity. A new motor would pay for itself in 16 months, according to A. Lovins.[15] Almost all AC motors are currently constant speed (3600 revolutions per minute), and because so much of electricity consumption is for motors, any small improvement in efficiency will pay large dividends. Constant-speed motors are not well suited to a majority of applications. For example, in a pump with a constant-speed motor, a valve is partly shut to cut down the flow rate by increasing the resistance, dissipating energy. The **variable-speed AC motor** is easily controlled and can be made to match the demand much more closely. Although it is more costly to buy initially, it saves energy in the long run. More such innovations are needed. The Consortium for Energy Efficiency has worked with manufacturers to increase the efficiency of their motors by many small steps that lead to several percent efficiency increases (a big thing for an old technology): putting more copper in the windings, using thinner laminated steel, optimizing the gap between stator and rotor, using a more efficient rotor bar design, reducing stray losses, reducing fan losses, and reducing resistance losses. The costs are 28% to 51% more, but recovery of costs is within a year at the average electricity cost of 8 cents per kWh.[278]

Potentially, the greatest savings occur if one can design new, more efficient processes to replace older ones. The chemical industry does this routinely. Large savings in fabrication of steel can be made by forming it, while hot, to near its final desired shape rather than reheating many times.[279] Large-scale savings in petroleum refining are also possible, in principle.[274]

Energy-intensive industries in the United States do not face a bright future. Expect to see relocation of these industries to the energy sources (just as iron ore was once brought to coal-rich Pittsburgh). If U.S. industry researches and adopts new technology, the future is relatively bright. If not, the United States will become even more a debtor nation, with sorry consequences for us all.

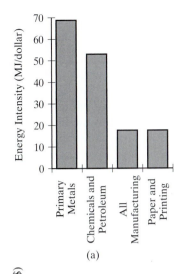

(a)

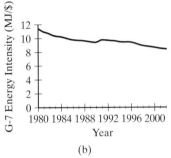

(b)

FIGURE 9.11

(a) Energy intensity of the G-7 nations in 1988, in MJ per dollar. Since then, G-7 energy intensity has fallen by over 10% on average. (U.S. Department of Energy)
(b) G-7 energy intensity, 1980–2002. Since 1988, G-7 energy intensity has fallen by over 10% on average. (U.S. Department of Energy; Economic Commission for Europe, Economic Survey Of Europe 2003 No. 2, Geneva: United Nations, 2003).

SUMMARY

Energy conservation means getting more for the same energy price or getting the same at a lower energy cost. Reduction of illumination levels and the purchase of longer-lasting, energy-efficient bulbs seem advisable. Certain government officials and reports say that conservation is moral but ineffective. **Extensions 9.1** and **9.2** show how incorrect this characterization is. Demand-side management has shown some utility (see also **Extension 9.3**).

Buildings should be properly insulated. Recommended *R*-values increase generally as we go from south to north. Increased insulation generally saves energy and money (see opportunities in **Extension 9.4**). Passive solar techniques show great promise for residential energy savings (see some described in **Extension 9.5**). Heating of interior masonry walls or concrete floors can increase comfort at modest cost, as can construction of Trombe walls in homes, strategic placement of windows with appropriate overhangs, and the use of furnace timers. In some areas, natural ventilation can save energy. Energy stored in the ground can reduce heating and cooling costs if groundwater heat pumps are used (as expanded on in **Extension 9.6**).

The experiences of other countries that have higher energy costs than the United States indicate that it is possible to live reasonably in residential communities with a lower per capita energy expenditure (see especially **Extension 9.7**).

Business and industry learned about life-cycle costing in the 1973 and 1979 energy crises. As a result, the newer generation of buildings is much more energy efficient than were the previous generations.

Innovations in energy-use devices can also help conserve valuable energy. Industry will become less dependent on materials processing than it is now, with more emphasis on value added. The long-term health of the economy demands rationalization, replacement of inefficiencies, and innovation of technology.

PROBLEMS AND QUESTIONS

MULTIPLE CHOICE REVIEW

1. *Conservation* is
   a. doing without.
   b. being cold in winter and hot in summer.
   c. paying more for less.
   d. described correctly by only (a) and (b).
   e. not described correctly by any other choice given.
2. In terms of supplying lighting to an area, the most efficacious form of lighting is
   a. incandescent lighting.
   b. compact fluorescent lighting.
   c. mercury vapor lighting.
   d. sodium vapor lighting.
   e. halogen lighting.
3. Daylighting is
   a. having every room have windows.
   b. setting illumination levels as if the room were in daylight.
   c. making it as bright as day in a room.
   d. bringing light from outside to deep inside a building by various means.
   e. lighting the room only during the day.
4. The greatest amount of heat transfer from a heated building can be traced to
   a. escaping air.
   b. radiation from windows.
   c. radiation from the building structure.
   d. convection along the structure.
   e. conduction through the building's walls.
5. What area of window having an *R*-value of 2.4 m² °C/W allows conductive heat flow through the window to be 10 J/s, when the temperature difference across the windowpane is 12 °C?
   a. 0.5 m²
   b. 1 m²
   c. 2 m²
   d. 4 m²
   e. 5 m²
6. Which of the following measures is passive solar in nature?
   a. Gas forced air
   b. Heat pump
   c. Trombe wall
   d. Weatherstripping
   e. Solar-pumped hot water
7. If a pulsed combustion furnace is used for home heating, what is the greatest first-law efficiency possible?
   a. 80%
   b. 85%
   c. 90%
   d. 95%
   e. 100%
8. Which of the following would NOT help reduce energy use?
   a. Lowering the thermostat at night during winter months.
   b. Switching from tungsten filament to fluorescent bulbs.
   c. Purchasing the highest efficiency refrigerator.
   d. Changing attic insulation from *R*-19 to *R*-2.
   e. Using double-pane glass windows.
9. What is the conductive heat flow rate across a window 4.5 m² in area that has an R-value of 2.5 m² °C/W when

the outside air is 50 °C lower in temperature than the inside air?
   a. 0.036 W
   b. 0.225 W
   c. 27.8 W
   d. 90 W
   e. 563 W
10. Which of the following would help conserve energy use?
   a. driving at 55 mi/h rather than at 65 mi/h.
   b. keeping a house at a constant temperature all the time rather than letting the temperature fall at night.
   c. using double-pane glass windows.
   d. All of the above measures help conserve energy.
   e. Only measures (a) and (c) above help conserve energy.

SHORT ANSWER
Explain whether the statement made is correct, and justify your answer.
11. New incandescent bulbs are more efficient than older types of incandescent bulbs.
12. Refrigerators have become much more efficient in the last decade.
13. Indoor air is much cleaner than outdoor air.
14. Payback time is important in a choice of any home improvement.
15. Heating that uses masonry at higher temperatures to warm indoors is more effective than circulating water in a system of hot water/steam radiators.
16. Passive heating measures are among the most expensive ways of heating spaces in buildings.
17. The amount of heat transferred between two reservoirs depends on the temperature difference between them.
18. *R*-09 insulation provides more resistance to the flow of heat than does *R*-12 insulation.
19. Industry has been faster to adopt conservation measures than private homeowners.
20. In lighting practices, lots of energy can be saved.

DISCUSSION QUESTIONS
21. Is the conductive heat flow rate the greatest, about the middle value, or the least amount of total heat flow through the window when both convection and conduction are important? Explain.
22. Explain the advantages and disadvantages of "superinsulated" homes.
23. What do you think are the reasons that European buildings are, in general, more energy efficient than American buildings?
24. Explain why a heating system made by circulating hot water through a concrete floor could be desirable. Would it be better to circulate the hot water through the ceiling? Why or why not?
25. When using solar energy systems, why are backup systems necessary to have? Explain.
26. A recent announcement indicated that magnetic motors can be used to cool refrigerators.[129] In the method, a gadolinium-based motor warms and cools off in response to application of a magnetic field. The gadolinium rotating wheel is the only moving part. Why might this be a breakthrough?

27. Which window would give the best solar gain in winter? One facing north, south, east, or west? Why?
28. Would switching your home from incandescent to compact fluorescent bulbs be a measure you would take? Why or why not? Explain your answer.
29. Are you reading the Energy Guide material on the appliances you purchase? Do you look for the Energy Star label? Why or why not? Do you think you should if you don't? Do you think it's helped if you do? Explain your answer.
30. Evaluate the success of demand-side management techniques in decreasing the demand for power plants. Could the method have worked better?

PROBLEMS
31. Compact fluorescent bulbs cost $10 per bulb and last 10 yr. Tungsten bulbs cost $0.50 per bulb and last 1 yr. For a given amount of light, a tungsten bulb draws 75 W, and a compact fluorescent bulb draws 10 W. If the cost of electricity is $0.10/kWh and the bulbs are operated for 2000 h/yr, what amount of money is saved by use of a compact fluorescent instead of a tungsten bulb over a 20-yr period?
32. Compact fluorescent bulbs typically last 10 yr, while incandescent bulbs last about 1 yr. For a given amount of light, a tungsten bulb draws 100 W, and a compact fluorescent bulb draws 15 W. If the cost of electricity is $0.10/kWh, how long would you have to operate a compact fluorescent bulb to save $5.00 compared to the costs of operating a tungsten bulb?
33. You are looking for a new electrical appliance that will last a long time. You are considering buying a high-efficiency model that uses an average of 60 W of power and costs $800 or a lower-efficiency model that uses an average of 90 W and costs $600. If the cost of electricity is $0.10/kWh and the appliances are operated continuously, how much money will you have saved by purchasing and operating the high-efficiency model over the low-efficiency model after 15 yr?
34. The cooling efficiency of an average new heat pump [in Btu/(Wh)] is 8.70. The most efficient unit has a cooling efficiency of 16.4. A heat pump, typically lasts 14 yr. How much more energy will the less-efficient heat pump use over its lifetime? Describe the assumptions you must make to answer this question.
35. The average American refrigerator uses over 1140 kWh/yr. In 1992, should one have bought an average refrigerator (1992 model) for $635 or a more energy-efficient refrigerator costing $838? The average refrigerator then on sale used 979 kWh/yr, while the more efficient one used 515 kWh/yr. (Use your local electricity price; if not known, use $0.10/kWh.) The refrigerator life span is 19 yr.
36. In 2005, there are 20 ft³ refrigerators on sale that use 440 kWh/yr. If you scrap your old refrigerator that uses 1140 kWh/yr and buy the new one for $500, how long will it take before you start saving money?
37. In 1987, the best refrigerator being sold used 515 kWh/yr and the average used 979 kWh/yr. How much money would have been saved on your electric bill between then and 2000 by buying the more efficient one?
38. In 2006, you have a choice between keeping your 13-year-old model that uses 515 kWh/yr or buying one that uses

440 kWh/yr. Given the average refrigerator lifetime of 19 yr, would you save money or spend money overall by 2012 if you bought the efficient model for $500? Suppose you know that the best refrigerator available in 2012 when the old one really bites the dust will use 380 kWh/yr. Should that affect your decision to buy now? Explain.

39. Large-capacity ($\sim$22 ft^3) refrigerators that have the freezer on top use about 600 kWh/yr, while side-by-side refrigerator–freezers of the same capacity use about 660 kWh/yr. Each costs about $1000. What will be the total cost after 19 yr? Which one should you buy? What considerations enter into your decision?

40. What is the temperature difference across a window if the conductive heat flow rate across it is 20 J/s? The window is 4.5 m^2 in an area that has an R-value of 2.5 m^2 °C/W.

MATERIAL RESOURCES

PART IV

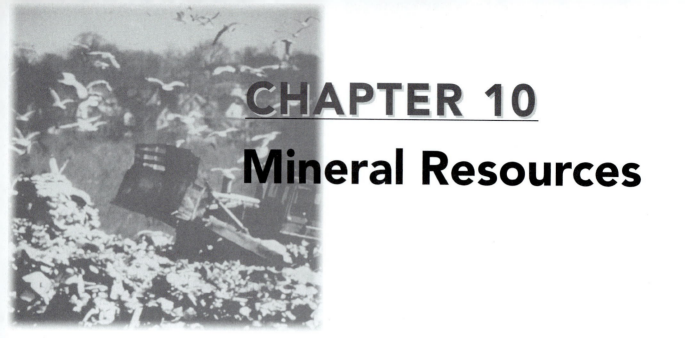

CHAPTER 10
Mineral Resources

*Energy and resource exploitation are closely related. Many fuels are mineral re-
sources—coal, oil, gas, oil shale—and this chapter explains the nature of such
resources, how they are produced, and the ultimate limitations and costs of their
production. In this chapter, we discuss the distribution of minerals in Earth's
crust and the interconnection of the availability of materials, standard of living,
energy use, and growth of world population.*

KEY TERMS

*core / plate / mantle / crust / magma / seafloor spreading / subduction zone /
plate tectonics / ferromagnetic / Curie temperature / triage / clarke / reserves /
resources / recoverable / paramarginal / submarginal / manganese nodules*

The distribution of minerals is intimately connected to energy flow. Energy is
necessary to obtain the mineral. Energy is required to refine the raw material. An
estimated 1% of world energy use is the energy price the world pays for its yearly
production of aluminum, for example. Without energy, the final products could
not be made. Coal and petroleum are minerals as well as energy sources and will
be discussed in the next chapter. Fertilizer minerals (see Chapter 24) are necessary for crop production.

Without raw materials to work with, we could not support the economy. In some
products, the raw material is an insignificant part of the cost, but it is absolutely essential to create product value (for example, consider the small amount of germanium necessary to be added to a semiconductor to make a transistor element in an
integrated circuit). The phenomenon of great value added for small amounts of materials used will probably be a characteristic of modern economies.[1]

Production of goods and services is the basis of value added. That is why reuse
and recycling can be attractive (Chapter 11). Without any material production,

we could not afford to produce energy. To discuss energy without considering raw materials is like discussing cooking without mentioning the ingredients. Conversely, without the necessary materials, we would not be able to exploit energy resources. It is becoming more difficult to extract energy, as the most easily obtainable resources are used first (see Chapter 12).

10.1   THE SOURCE OF MINERALS

Earth and Geological Change

Earth's surface is constantly being reshaped by sedimentation, weathering of exposed rock, movement of the continents, volcanism, and the activities of its living inhabitants. Some 200 million years ago, there was just one continent on Earth. It existed for an eyeblink in geologic time. In 1912, the irregularities of the continents' outlines, which seemed almost to match one another, led the German meteorologist Alfred Wegener to postulate that some of the continents were once adjacent and had subsequently moved apart somehow. In his time, he was met with skepticism, but by the end of the 1960s, his ideas had been vindicated. The continents and ocean basins rest on rock platforms called **plates**, and the plates move relative to one another. At one point, the motion of the plates brought the continents together, and then the motion of the underlying plates broke them apart again.

Earth has a liquid **core**, surrounded by a **mantle**, and overlain by a **crust** (Figure 10.1a). The crustal plates are patches of light rock that cover Earth like a patchwork quilt (Figure 10.1b). The internal thermal energy of Earth drives up molten material, **magma**, between the places where adjacent plates abut, causing them to be pushed apart. This generally occurs on the seabed and is referred to as **seafloor spreading**. At their other edges, the plates collide, and one is pulled downward under the other plate, in regions called **subduction zones**, by the cooler, descending parts of the plates. Many geologists have spent the last four decades (since Wegener's vindication) studying the plates and their motions as they move about over the ductile (semiliquid) mantle. They call their area of study **plate tectonics**; the word "tectonics" comes from the Greek word for "builder." In architecture, tectonic refers to the construction of buildings; in geology, it refers to the study of the structural deformation of Earth's crust.

Evidence of Change

Through subduction at plate boundaries and direct volcanic action, material is fed into the mantle rock and back to Earth's crust in the largest recycling project ever conceived. The Himalayas rise out of the collision between the plate bearing India and that bearing the Asian landmass. The Pacific plate had traces of its movement written on it by a hot spot in the underlying mantle; the movement produced chains of parallel islands and seamounts. The Hawaiian Islands–Emperor Seamounts are one visible remnant of this movement. The Andes and the Rocky Mountains were pushed up by the collision of the South and North American plates with plates under the Pacific Ocean, causing subduction of these plates under the mountains. The rocks are in plastic flow under isostatic equilibrium (the masses above and below the surface are balanced, a principle discussed in Chapter 3), so that wherever a mountain rises from the surface, there is an upside-down mountain underneath it.

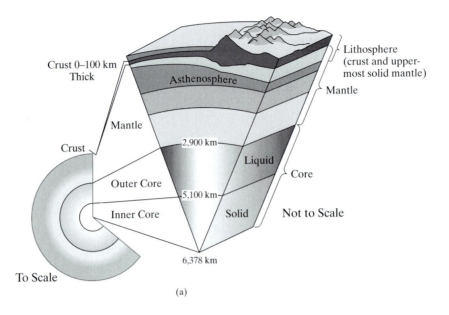

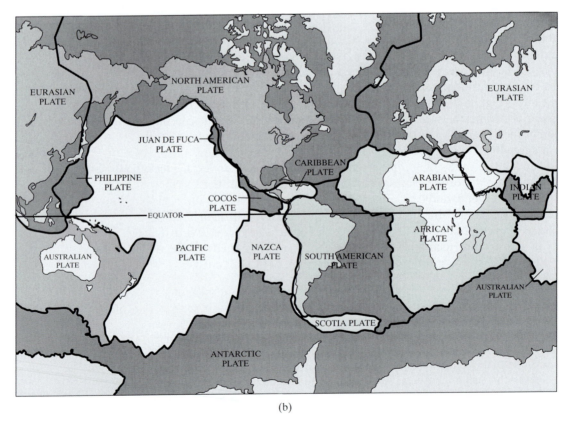

FIGURE 10.1

(a) Cutaway view of Earth (not to scale). (b) Plates covering Earth's surface.
(W. J. Kious and R. I. Tilling, *This dynamic Earth: The story of plate tectonics* [GPO: Washington, DC, 1996], illustrations by Jane Russell, pp. 5, 8)

Many elements are found in the magma, or liquid material, that oozes from the plate boundaries or is ejected from volcanoes as lava. Iron is one of these elements. Iron and other materials that can form magnets are called **ferromagnetic**. Magnets form in nature when hot rock containing iron cools in the presence of an external magnetic field. Above a temperature known as the **Curie temperature**, the magnetic domains (see Chapter 4) are free to move about and line up pointing along the direction of the external field. As the material cools below the Curie temperature, the domains become frozen, unable to jiggle about. If there is an external magnetic field on the material as it cools below the Curie temperature, the result is the formation of magnetic rock.

Geologists have been able to verify seafloor spreading because of this physical principle and because Earth's magnetic field occasionally flips direction from north to south. Seafloor spreading from a midocean ridge will have a record of Earth's magnetic field at the time of cooling locked into the solidified lava and magma. The flip-flops of Earth's magnetic field cause symmetrical patterns of "magnets" in the rock of the seafloor—a sort of geological "tape recording." Such symmetrical patterns have been observed near the Strait of Juan de Fuca in the Pacific Northwest, on the seafloor along the Atlantic rift, and elsewhere.

All these motions and collisions give rise to geologic processes that generate ores. Western American ores seem to have been deposited in spurts (most ores of one kind are found in restricted geographic provinces created at roughly the same time).[2,3] The spurts are correlated with tectonic and magmatic activity in the American cordillera,[3] and ores are formed by a combination of volcanic processes, erosion, presence of water, and bacterial action (see also **Extension 19.4, *Uranium Ore Formation***).[2]

Mineral Distribution

The United States, once a supplier of raw materials, has been a net importer of minerals since the 1920s. Of the key metals, chromium, aluminum, nickel, and zinc, the United States gets more than half of its supply from abroad. A survey of stockpiles of minerals that would be necessary in case of a national emergency (a war of 3 years' duration) showed that for 24 of the 42 most critical minerals, the nation depends on overseas sources.[4] The five most critical minerals are cobalt (93% imported from the Congo and Zambia), chromium (91% imported from South Africa and Zimbabwe), manganese (all imported), platinum metals (87% imported from South Africa and the former Soviet Union), and titanium.[4] Titanium is mostly imported from Australia, which produces three times as much as the United States; Canada produces twice as much titanium as the United States.[5] Our reliance on foreign supply is illustrated in Figures 10.2[6] and 10.3.[7] With 5% of the world population, the United States imports 27% of the world's raw materials exports.[5] The estimated value of U.S. minerals production was $44 billion in 2004, about 13% more than in 1999, but only about 0.5% of U.S. gross domestic product.[6]

We are not unique in finding it necessary to import raw materials. Virtually every nation must import some materials, because no part of Earth is self-sufficient in all minerals. North America is rich in molybdenum but poor in tin, tungsten, and manganese; the situation is the reverse in Asia. South Africa and the former Soviet Union have much of the world's gold and platinum. Cuba and New Caledonia supply half the world's nickel.[5,8] Our supplier nations are distributed worldwide; about a third of the world's countries are major suppliers for some minerals.

2004 U.S. NET IMPORT RELIANCE FOR SELECTED NONFUEL MINERAL MATERIALS

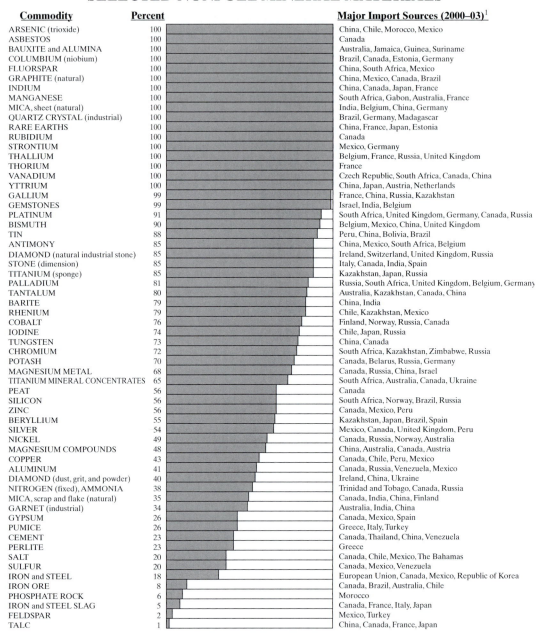

Commodity	Percent	Major Import Sources (2000–03)[1]
ARSENIC (trioxide)	100	China, Chile, Morocco, Mexico
ASBESTOS	100	Canada
BAUXITE and ALUMINA	100	Australia, Jamaica, Guinea, Suriname
COLUMBIUM (niobium)	100	Brazil, Canada, Estonia, Germany
FLUORSPAR	100	China, South Africa, Mexico
GRAPHITE (natural)	100	China, Mexico, Canada, Brazil
INDIUM	100	China, Canada, Japan, France
MANGANESE	100	South Africa, Gabon, Australia, France
MICA, sheet (natural)	100	India, Belgium, China, Germany
QUARTZ CRYSTAL (industrial)	100	Brazil, Germany, Madagascar
RARE EARTHS	100	China, France, Japan, Estonia
RUBIDIUM	100	Canada
STRONTIUM	100	Mexico, Germany
THALLIUM	100	Belgium, France, Russia, United Kingdom
THORIUM	100	France
VANADIUM	100	Czech Republic, South Africa, Canada, China
YTTRIUM	100	China, Japan, Austria, Netherlands
GALLIUM	99	France, China, Russia, Kazakhstan
GEMSTONES	99	Israel, India, Belgium
PLATINUM	91	South Africa, United Kingdom, Germany, Canada, Russia
BISMUTH	90	Belgium, Mexico, China, United Kingdom
TIN	88	Peru, China, Bolivia, Brazil
ANTIMONY	85	China, Mexico, South Africa, Belgium
DIAMOND (natural industrial stone)	85	Ireland, Switzerland, United Kingdom, Russia
STONE (dimension)	85	Italy, Canada, India, Spain
TITANIUM (sponge)	85	Kazakhstan, Japan, Russia
PALLADIUM	81	Russia, South Africa, United Kingdom, Belgium, Germany
TANTALUM	80	Australia, Kazakhstan, Canada, China
BARITE	79	China, India
RHENIUM	79	Chile, Kazakhstan, Mexico
COBALT	76	Finland, Norway, Russia, Canada
IODINE	74	Chile, Japan, Russia
TUNGSTEN	73	China, Canada
CHROMIUM	72	South Africa, Kazakhstan, Zimbabwe, Russia
POTASH	70	Canada, Belarus, Russia, Germany
MAGNESIUM METAL	68	Canada, Russia, China, Israel
TITANIUM MINERAL CONCENTRATES	65	South Africa, Australia, Canada, Ukraine
PEAT	56	Canada
SILICON	56	South Africa, Norway, Brazil, Russia
ZINC	56	Canada, Mexico, Peru
BERYLLIUM	55	Kazakhstan, Japan, Brazil, Spain
SILVER	54	Mexico, Canada, United Kingdom, Peru
NICKEL	49	Canada, Russia, Norway, Australia
MAGNESIUM COMPOUNDS	48	China, Australia, Canada, Austria
COPPER	43	Canada, Chile, Peru, Mexico
ALUMINUM	41	Canada, Russia, Venezuela, Mexico
DIAMOND (dust, grit, and powder)	40	Ireland, China, Ukraine
NITROGEN (fixed), AMMONIA	38	Trinidad and Tobago, Canada, Russia
MICA, scrap and flake (natural)	35	Canada, India, China, Finland
GARNET (industrial)	34	Australia, India, China
GYPSUM	26	Canada, Mexico, Spain
PUMICE	26	Greece, Italy, Turkey
CEMENT	23	Canada, Thailand, China, Venezuela
PERLITE	23	Greece
SALT	20	Canada, Chile, Mexico, The Bahamas
SULFUR	20	Canada, Mexico, Venezuela
IRON and STEEL	18	European Union, Canada, Mexico, Republic of Korea
IRON ORE	8	Canada, Brazil, Australia, Chile
PHOSPHATE ROCK	6	Morocco
IRON and STEEL SLAG	5	Canada, France, Italy, Japan
FELDSPAR	2	Mexico, Turkey
TALC	1	China, Canada, France, Japan

[1] In descending order of import share

FIGURE 10.2

Selected mineral imports as a proportion of U.S. supply are shown. A dark horizontal bar all the way across means that all of that resource is imported. A dark horizontal bar one-third of the way across indicates that one-third is imported.
(U.S. Geological Survey)

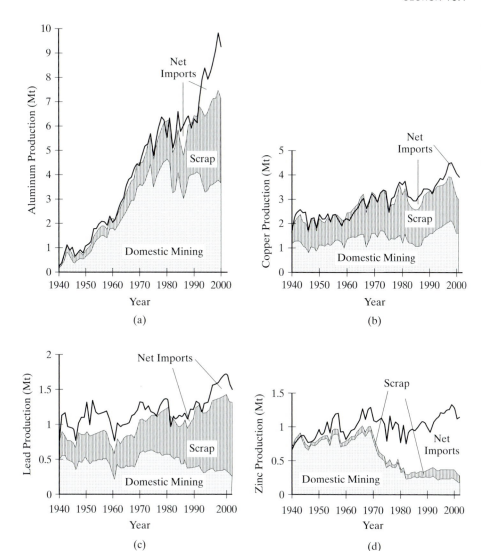

FIGURE 10.3

Supplies of aluminum, copper, lead, and zinc, are shown for the U.S. for the period between 1940 and 2002. (T. Kelly, D. Buckingham, C. DiFrancesco, K. Porter, T. Goonan, J. Sznopek, C. Berry, and M. Crane, "Historical Statistics for Mineral and Material Commodities in the United States," U.S. Geological Survey Open-File Report 01-006, Version 8.9)

Thus, there is a politics of materials. Reasons for our diplomatic involvement with so many countries and our worldwide strategic interests include our vulnerability to interruptions in the immense flow of materials into our country. We will need even more minerals in the future, because our rate of consumption of raw materials has tended to increase at *twice* the rate of population growth (and those who want raw materials tend to get them; see **Extension 10.1, *Minerals, Conquest, and War***)

This is the point at which population growth, standard of living, and questions of international morality intersect (see **Extension 10.2, *What Drives Exploitation?***). Each additional American born or immigrating exerts a disproportionate leverage on use of materials. Increasing our share of use of the world's materials will become more difficult as the rest of the world tries to catch up to our standard of living. The pressure on resources will be intense—recall that per capita materials use and standard of living are related (sometimes inversely; many people in other advanced countries have a standard of

FIGURE 10.4

World births are plotted by country in terms of the countries' GNP. (Chancellor and Goss, "Balancing Energy and Food Production, 1975–2000," *Science* **192**, 213 (1974))

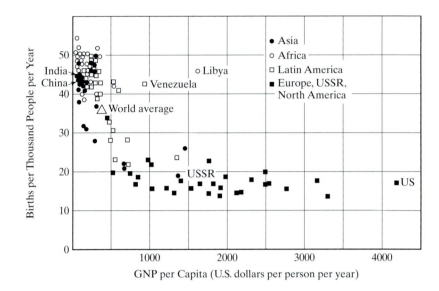

living equal to or better than Americans, while smaller amounts of raw materials and lower amounts of energy are used per person, and more amounts are recycled and reused; see Chapter 11 and **Extension 10.3, *Minerals and the Environment***). The dilemma is that the high population growth rates of the undeveloped world ensure that the standard of living in those countries cannot become "American" (Figure 10.4); yet the high population growth rate is, in a sense, a response to the low standard of living (children provide security in illness or old age). The world must keep running the Red Queen's race simply to stay in the same place.

In addition to pondering the morality of continuing to amass riches in contrast to most of the world, we must also consider the ramifications of our minerals acquisitions policy on the supplier countries. There is some truth to the claim that supplier countries benefit monetarily, but resource development seldom initiates other industrial development. Bolivian tin mines have not boosted Bolivian industrialization. The copper supplies of Chile and Zambia have not spread benefits of industry broadly or caused the growth of industrial development. Small nations also leave themselves open to manipulation by powerful nations, which generally find ways to get resources when they need them.[64] The question of what is to be done is still open.

10.2 MINERAL SUPPLY ADEQUACY

Our discussion has made it obvious that our domestic supplies of many materials are insufficient. We must turn to world supplies. Are world supplies *adequate* for sustained world production? *Adequacy* is a slippery concept. It might mean that the required amounts of the particular material should be available at present prices; or that the rise in prices should be no steeper than some specified rate; or that with a certain percentage of scrap being recycled, materials can meet domestic demand. What is to be done with nations that have no resources? See the box, *A Closer Look*: Of Triage and Lifeboats, for one person's suggestion.

A CLOSER LOOK

Of Triage and Lifeboats

In World War I, the French were terribly short of medical supplies and doctors to treat their troops. They decided that they had to use the medical resources they had in the most efficient ways. Some of the wounded soldiers would survive without treatment of any kind; some of them would die no matter what the treatment; and some would die *unless* they were treated. They decided that the best way to use their meager resources was to separate the wounded into these three categories and attempt to treat immediately those who would die without treatment or medicine. Because of the division into three groups, the practice became known as **triage**.[65]

The late socioeconomist Garrett Hardin has made the controversial suggestion that we practice a form of triage among nations[66] that he characterizes as the "lifeboat" analogy. Hardin pictures the world as an ocean and the various countries as survivors of a downed ship. The world's food and resources constitute the lifeboat. If all nations share resources equally, the lifeboat will be overloaded, and all will perish. He suggests that the weakest should be jettisoned in the interest of saving those who can be saved.

It is important to put mineral use into perspective. As Matos and Wagner commented about U.S. consumption:[67]

> During this century, the quantity of materials consumed has grown, from 161 million metric tons in 1900 to 2.8 billion metric tons by 1995, an equivalent of 10 metric tons per person per year.

Will world use grow similarly and become more like the U.S. use rate? If so, with a current world population of 6 billion, we would infer a future world materials use rate of 60 Gt/yr, far higher than today's (not even taking into account the growth in population). What effects would such a demand cause? Would it be possible to return to relying on renewable materials for one-half the demand, as was the case in the United States at the start of the last century, or will we maintain the one-twelfth proportion of renewables that characterizes the end of that century?[67] Matos and Wagner continued:

> Of all the materials consumed during this century, more than half were consumed in the last 25 years.

As we will see below, this is one consequence of the exponential nature of growth (recall the discussions of Chapters 2 and 5).

Determining the Lifetime of a Resource

Let us make an attempt to estimate the adequacy of presently known supplies by considering the connection between supply and rate of use. The use rate is historical information, expressed as number of tons of a material used in a year (or any other convenient measure of time). Known supplies, given in tons of material, divided by the use rate, gives an estimate of supply lifetime:

$$\text{lifetime of supply of a material} = \frac{\text{known supply}}{\text{rate of supply use}}.$$

Table 10.1 was prepared using data from the Bureau of Mines of the U.S. Department of the Interior.[4] There are three ways in which this estimate is deficient:

1. The actual rate of use of a supply of a certain material changes with time; the general trend in the use of materials is that the rate rises. The Great Depression of the 1930s is an exception to this rule, as was the recession of the early 1980s.
2. The amount of known supply increases as new resources are discovered. For example, tremendous amounts of iron ore and bauxite were discovered in Australia in the 1960s.
3. The market for a material could become saturated.

The table illustrates several things. For one, it is clear that there is still abundant fluorine from fluorspar available a decade past the point it was supposed to have been unavailable. New supplies were discovered. For another, there is a far larger supply of copper now than in 1968, because fiber optic lines have replaced paired copper cables for telephone applications. Note that most materials exhibit a decreased consumption in 2000 compared to 1968. The American economy is less dependent on materials today. Discovery of new supplies and technological replacement play significant roles in the supply picture.

Note that the table's predicted lifetime is a linear lifetime; the future use rate has been assumed to be *the same as* the present rate. If the number of tonnes used each year grows, the resource will not last near as long. So, this table illustrates one basic point: it is difficult to sustain exponential growth in anything, because even at current use rates, some resources are predicted to be exhausted soon. (For definitions of resource terms, see **Extension 10.4,** ***Definition of Resource***.)

TABLE 10.1

Mineral supplies, use rates, and projected lifetimes

Material	Supplies (Mt)	Rate of use in 1968 (kt/yr)	Lifetime at 1968 rate (yr)	2000 U.S. consumption rate (kt/yr)
Cobalt	2.4	22.1	109	10.9
Iron	96,720	423,000	229	76,000
Manganese	736.5	3250	96.7	795
Molybdenum	5.4	33.575	161	21.5
Nickel	75	233.5	322	158
Aluminum	1170	10,288	114	7900
Magnesium	2600	5274	493	175
Titanium	137.4	1424	96.5	1190
Copper	307.9	7290	42.3	3120
Fluorine	38.8	1800	22	612
Phosphorus	21,800	11,475	1,900	39,500
Potassium	109,800	14,250	7,600	5000

Source: U.S. Bureau of Mines, U.S. Geological Survey

Some would argue for unrestrained use of those resources that are likely to last a long time *at current use rates*. However, this leads us back to consideration of the explosive nature of exponential growth. Emrick[68] pointed out that we have a hard time *not* being linear thinkers (Chapter 8). But, even a small growth rate has surprising results. Consider the case of potassium, which has a lifetime of 7600 years at 1968 rates (Table 10.1). Suppose that exploitation grows at a mere 1% a year. In this case, the doubling time is 70 years (refer to Table 5.3). We think of the resource as infinite, because 7600 years is 109 times as long as a doubling time. Following Emrick's ideas,[68] each box in Figure 10.5 shows the amount of potassium exploited in 70 years at the current use rate. To see the quality of exponential growth, look at what happens in the next 70 years: two boxes are used. During the next 70 years, four boxes are used, and so on. Note how fast the resource disappears. The 109th box is gone long before 500 years is up! Instead of lasting 7600 years, the material is gone in 450 years if the growth is a mere 1%. If it is 2% instead, the material will last only 225 years.

There is some indication that a decline in per capita steel use sets in after a per capita income of $2000 is attained.[69] If all materials are subject to such a use peak, perhaps there is yet hope that we do not need an infinite resource supply to support a reasonable living standard.

Mining Average Rock?

Total consumption of any mineral resource is all but impossible. One cubic kilometer of "average" crustal rock contains 200 million tonnes (Mt) of aluminum, 100 Mt of iron, 0.8 Mt of zinc, and 0.2 Mt of copper. Even though some see advantages in mining "average" crustal rock rather than ore (the "mine" can be adjusted to the level of technology available rather than vice versa),[70] dilute minerals in rock will probably never be mined.[71] Some minerals may be extractable from the seawater of which they are minor constituents.

To get some measure of how much of an element exists, and to see what we mean by an ore, we define the **clarke**. The clarke of an element is its average crustal abundance (the proportion of the element in the outer crust). Table 10.2[72,73] compares ores to the respective clarkes of several elements. We presently mine high-grade ores, those for which the abundance is much higher (by two to four

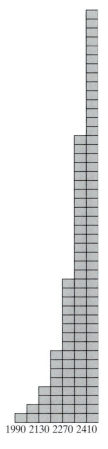

1990 2130 2270 2410

FIGURE 10.5

If potassium is exploited at the current use rate, it would last 2900 yr. Because growth in exploitation is exponential, the effect of a 1% growth rate is to double use every 70 yr. If each box represents 70 yr of use at current rates, the resource will last a much shorter time than 2900 yr. The resource will have disappeared by about the year 2450, only 450 yr from now. If the use grew at 2%, the resource would be gone in a mere 230 yr.

TABLE 10.2

Comparison of clarke and ore for selected materials

Element	Clarke (% abundance)	Typical ore (%)	Cutoff grade (%)	Cutoff/ clarke
Abundant				
Aluminum	8.3×10^{-2}	2×10^{-1}	1.85×10^{-1}	2.2
Iron	5.8×10^{-2}	3×10^{-1}	2×10^{-1}	3.4
Manganese	1.3×10^{-3}		2.5×10^{-1}	190
Intermediate				
Nickel	8.9×10^{-5}		9×10^{-3}	100
Copper	8.7×10^{-5}	6×10^{-3}	3.5×10^{-3}	40
Chromium	1.1×10^{-4}		2.3×10^{-2}	210
Rare				
Mercury	9.0×10^{-8}	3×10^{-3}	1×10^{-3}	11,000
Tungsten	1.1×10^{-6}		4.5×10^{-3}	4100
Gold	3.5×10^{-9}		3.5×10^{-6}	1000

orders of magnitude) than the clarke. The grade of an ore is its concentration in the ore body.

For some materials, the relatively high concentrations of ore gradually change to the clarke as we move outward from the center of the ore body. An analysis by Lasky[74] is applicable to these ores. Lasky found that there was a relation between ore grade and amount mined:

$$\text{average grade of ore} = c - c'\log_{10}(\text{cumulative tonnage mined}).$$

Here, c and c' are constants for a given body of ore. The same principle may be stated alternatively as

$$\text{volume of ore of a particular grade} \approx 10^{-(\text{grade})}.$$

This relation means, roughly, that resources increase in quantity as the grade mined declines. Lasky's work has been interpreted to mean that raising the price increases the volume of mined rock and, hence, the amount of material available. In such an interpretation, the world will not run out of its inexhaustible resources; those resources will simply increase in a monetary and energy cost. Because the dollar value of minerals used has averaged only about 2% of the gross national product (GNP) over the past five decades, it is probably possible to increase mineral prices severalfold[8] if we are willing to ignore the cost in energy efficiency of extraction.

For most materials, however, the ore bodies are in geochemically isolated outcrops. Different sorts of rocks are separated by sharp boundaries. Limestone carrying metal of grade 2 to 3×10^{-5} is extremely close to a body of ore of grade 10^4 times the clarke. Lowering the grade from 20% metal to 10% metal in this case does not greatly increase reserves.[72,73] Thus, the solution of simply mining greater volumes of rock for our materials is probably not practicable, except for materials such as iron and aluminum, for which the Lasky analysis holds.[8] (See **Extension 10.5, *Why Lowering Grade Sometimes does not Increase Reserves by Much.***)

We could always return to the idea of mining average crustal rock. This alternative shares the same problem just described in the mining of lower-grade ores in immense quantities: an increase in energy consumption per unit of output. Cheaper energy, in fact, would do little to reduce total costs (chiefly capital and labor) required for mining and processing rock. The enormous quantities of unusable waste produced for each unit of metal in ordinary granite (in a ratio of at least 2000 to 1) are more easily disposed of on a blueprint than in the field.[69]

Consequences of Untrammeled Exploitation

The difficulty in obtaining energy for exploitation of resources was part of the resource catastrophe evidenced in the "limits to growth" analysis.[77] The history of the energy cost of mineral production is shown in Figure 10.6.[72]

In addition, there are mountains of rubble produced in mining large volumes of rock. An area the approximate size of Puerto Rico (or one and a half Delawares or three Rhode Islands) is covered with mine wastes in the American West. The groundwater in these areas will probably leach out contaminants and become polluted. As many as 10,000 miles of streams in the West (see **Extension 10.3**) have already been made acidic by the leaching of materials from the pulverized rock.[78] The decrease in effluent per unit of production must more than overcome an increase in production. The inevitable accident is made more likely the

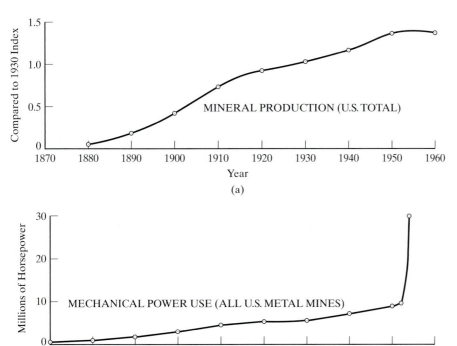

FIGURE 10.6

(a) U.S. mineral production from 1880 to 1960. (b) Mechanical power is used in ever-greater amounts to extract minerals, even though production has not grown commensurately. Note that this rapid rise is caused by two factors—the fact that the average grade of minerals mined in the U.S. has decreased, and the development of machines that replaced human labor with fossil energy (machines are cheaper than people, despite the capital cost of the machine and the energy to run them).
(Figure from article by P. Cloud, *Resources and Man.* Copyright by W. H. Freeman and Company. Used with permission)

longer the production, and its impact increases in step. Most pollutants cause problems even with chronic low-level exposure. There is the possibility of pollution on a truly global scale. Mines have other consequences as well: About 30 people die in abandoned mines every year in the United States.[78]

10.3 ESTIMATING RESOURCES

Most mineral resources are hidden under Earth's surface. **Reserves** (economically recoverable material in identified deposits) can be distinguished from **resources** (which include reserves, identified deposits not presently recoverable, and deposits not yet discovered).[69] It is even difficult to estimate reserves with a high degree of accuracy until they are largely mined out. Estimates made in advance of production are generally 25% too low. The error in estimating incompletely explored deposits is much greater.[69]

We may classify resources as proved, probable, possible, and undiscovered (see **Extension 10.4**). The distinction between some resources and reserves is economic (Figure 10.7). This distinction can be made by applying one of three descriptors:[69] **recoverable** (at less than or equal to present cost), **paramarginal** (recoverable at a cost less than 1.5 times present cost), and **submarginal** (recoverable at some price above 1.5 times present cost). History shows that there is a movement from submarginal to paramarginal to recoverable, mostly attributable to technological advances. The cutoff grade of copper ore has been reduced by a factor of 10 since 1900, and by a factor of 250 over the history of mining.[69] As of this writing, copper is in oversupply, because its main

FIGURE 10.7

How resources are distinguished
from reserves.

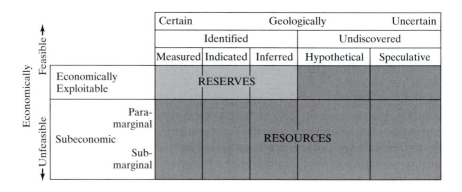

use—electric cable for telephony—has been superseded by the use of fiber-
optic lines.

With these definitions, it is possible to relate reserves to the abundance of minerals in the geologic environment. The tonnage of minable reserves of well-ex-
plored elements in the United States is approximately related to the crustal
abundance (reserves are generally 10^9 or 10^{10} times the clarke—the natural
abundance)[69] (Figure 10.8). This provides us with a tool for estimating, with a
given level of technology, the total amount of resources.

FIGURE 10.8

A comparison of domestic reserves
to their abundance in Earth's crust.
The dots represent tonnage of
minable ore at present; tonnage
exploitable at higher prices is
indicated by the bars. The dotted
lines are multiples of the average
crustal abundance (10 billion, 1
billion, and 1 million, decreasing
toward the bottom right).
(V. E. McKelvey, "Mineral resource
estimates and public policy," *Am.
Sci.* **60**, 32 (1972). Copyright 1972
by the Society of the Sigma Xi.
Reprinted by permission of
American Scientist.)

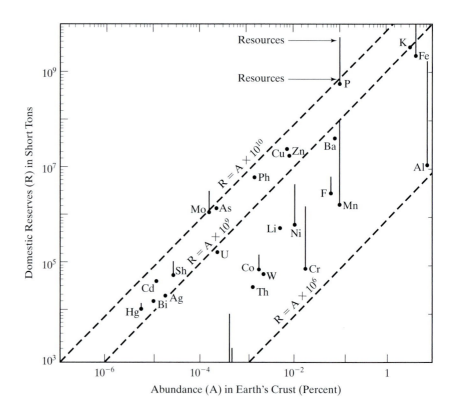

TABLE 10.3

Estimated lifetimes[a] of selected resources at current world use rates

Resource	Meadows et al.,[b] 1972 (Static)	Meadows et al.,[b] 1972 (Exponential)	German Bureau for Geological Sciences and Raw Materials, 1984 (Static)	MMSD[c], 2002 USGS, 2005[d] (Static)
Aluminum	100	31	253	202[c], 16[d]
Chrome	420	95	364	48[d]
Cobalt	110	60	112	149[d]
Copper	36	21	70	28[c], 32[d]
Iron	240	93	184	132[c], 64[d]
Lead	26	21	43	21[c,d]
Mercury	13	13	28	69[d]
Nickel	150	53	110	41[c], 44[d]
Tin	17	15	41	37[c], 24[d]
Tungsten	40	28	49	48[d]
Zinc	23	18	39	25[c], 24[d]

[a]There are two ways to interpret the lifetime: lifetime at current use rate, with use of the current year's demand, for which there is no future growth in use assumed (static assumption); and lifetime at current use rate, assuming that the demand for the quantity has been growing and will continue to grow exponentially (exponential). We use both in this table. Note that the economy was near the top of a business cycle in 1972 and near the bottom of a business cycle in 1984.

[b]D. H. Meadows, D. L. Meadows, J. Randers, and W. W. Behrens, III, *The Limits to Growth: A Report for the Club of Rome's Project on the Predicament of Mankind* (New York: Universe Books, 1972), Table 4.

[c]Mining, Minerals and Sustainable Development (MMSD) Project, *Breaking New Ground: Mining, Minerals, and Sustainable Development* (London and Sterling, VA: Earthscan Publications Ltd., 2002), Table 4-1.

[d]U.S. Geological Survey, *Mineral Commodity Summaries 2005* (Washington, DC: GPO, 2005)

Static and Exponential Lifetimes

Table 10.3[79] presents contrasting use "lifetimes" of various metals as given in Meadows et al.[77] and as estimated by the German government.[79] The difference is partly due to replacement of metals by plastics and by the reduction in weight of cars since the early 1970s, as well as the effect of the general world economic malaise in the 1980s, which caused a huge drop in demand.[79] Table 10.4[79] may hint that the peak use rate of some common metals may have passed (with caveats as indicated, a recognition that ore grade may be lowered and due recognition that exponentially growing populations will eventually force use rates to rise).

Other Resources: Resource Substitution and Resources from the Ocean

Many argue that resources can be extended by substitution of one material for another.[1,80] For example, transistors replaced vacuum tubes, and germanium, selenium, and other once-useless materials became resources.[81] Old, abundant materials may be used in new ways (see, for example, Hillig's discussion, in Reference 82). Magnesium or aluminum may be used instead of steel. The development of "metal-ceramic" materials that can withstand a thermal shock of 1500 °C may lead to their replacement of metals in airplane engines.[83]

Another source of materials is common seawater and the seabed. Table 10.5 lists some common elements found in seawater and their values.[80,84] **Manganese nodules** are found in certain regions of the ocean in amounts up to 10 kg/m^2, assaying up to 27% to 30% manganese, 1.1% to 1.4% nickel, 1.0% to 1.3% copper,

TABLE 10.4

Metals use (in kt/yr) in 1973, 1983, and 2005

Metal	1973[a]	1983[a]	2005[b]
Aluminum	11,187	11,760	28,900
Copper	6927	6765	14,500
Lead	4052	3795	3150
Nickel	514	472	1400
Tin	204	154	250
Zinc	4822	4317	9100

[a]Top of business cycle.
[b]Bottom of business cycle.
(U.S. Geological Survey)

TABLE 10.5

Concentration and 1992 values of selected elements in seawater (not including water)

Element	Concentration (kg/m³)	Value per 10^6 gallons (1992$)
Chlorine	0.4859	1183
Sodium	0.4180	484
Magnesium	0.485	5290
Sulfur	0.0255	130
Calcium	0.0091	190
Potassium	0.0088	117
Carbon	0.0021	0.0001
Bromine	0.0007	243

and 0.2% to 0.4% cobalt.[85] These nodules grow at a speed of from 1 mm to 10 mm per million years by several different mechanisms.[86,87] Some areas of the seafloor, such as in the Red Sea, contain new deposits of minerals of virtually inestimable value. As of the early 2000s, only India is contemplating large-scale seabed mining operations to recover cobalt and manganese.[88] The possible worth of these seabed resources has made agreement on rules for exploitation of the ocean basins difficult and has exacerbated arguments about territorial limits that were once centered on disagreements over fishing rights. It is good to remember Rona's cautionary words, "Sea-floor resources are immense and of great scientific interest and increasing economic value. It is important to remember that mining marine minerals is only economic when the costs of bringing them to market under prevailing conditions are factored in and environmental impacts are evaluated."[89]

Bartlett proposes a different sort of solution to the problem of extracting depletable resources,[90] which he calls "sustained availability." He proposes that resource exploiters voluntarily arrange that the rate of extraction will decline exponentially with time. Such a program would assure that a resource would be available forever (in declining amounts, to be sure). He admits that the plan would seem too simple and naive to those advocating immediate plunder for financial gain, but its simplicity is a powerful asset. He believes that we should adopt a more responsible attitude toward exploitation of resources and leave to our descendants some modicum of resources. And, of course, reusing and recycling (Chapter 11) lengthen the availability of materials as well.

SUMMARY

Minerals are extracted using energy, and some minerals supply energy. Minerals are deposited in place by chemical and geological processes, especially processes involving molten rock and the subsequent interaction of cool rock and water. Minerals are concentrated at various grades in various parts of the world. Seldom does one country or even continent contain all or most minerals. The United States depends on the rest of the world for most of its minerals. Many minerals appear to be in short supply; the supply of most materials is measured in decades.

Minerals and energy resources have been cause for contention among countries because of the uneven distribution of them worldwide. See **Extensions 10.1** and **10.2** for more information. Exploitation of minerals has consequences. (See **Extension 10.3** for more detail on "heavy metal" contamination and other consequences.)

Reserves are economically recoverable materials in identified deposits. (See **Extension 10.4** for the definitions.) Resources include reserves, unrecoverable deposits, and deposits not yet discovered. If the grade is lowered, the amount of resources available increases (see **Extension 10.5**). Recycling also increases resources and lengthens the amount of time before exploitable grade is lowered.

PROBLEMS AND QUESTIONS

MULTIPLE CHOICE REVIEW

1. How does one estimate the volume of ore of a particular grade?
   a. the Cloud relation
   b. the Lasky analysis
   c. the Hillig prediction
   d. the Hardin choice
   e. the Club of Rome

2. Which material in 1 m^3 of seawater has the greatest economic value?
   a. manganese
   b. carbon
   c. salt
   d. magnesium
   e. calcium

3. When comparing lifetimes within categories (exponential, static), which is true?
   a. The relative rank of the exponential lifetimes is the same as for the static lifetimes.
   b. Static lifetimes agree in their relative rankings, but exponential rankings do not.
   c. Exponential lifetimes agree in their relative rankings, but static rankings do not.
   d. There is no correlation between relative lifetimes in terms of static or exponential growth.
   e. None of the other answers is correct.

4. Which method is most likely to supply materials at reasonable prices?
   a. Substitution of materials
   b. Exploitation of "average" crustal rock
   c. Exploitation of the next lower grade of materials
   d. Extraction of the material from seawater
   e. Decrease of the rate of growth of materials use

5. Use of materials grows exponentially because
   a. resource use makes the economy grow.
   b. more materials are available, as more money is available to the average consumer.
   c. all growth is exponential.
   d. resources are continually being formed.
   e. population grows exponentially.

6. Abandoned mines are a problem, because
   a. people can fall into mineshafts and die.
   b. groundwater quality can be injured by leakage from the mine.
   c. rainwater can leach out toxins that get into local streams.
   d. discarded tailings can leach "heavy metals" into the environment.
   e. of all the above reasons.

7. Reserves are
   a. always accurate.
   b. impossible to know until after the resource has been mined out.
   c. measured by geologists using hole borers over the ore field.
   d. generally 20% greater than resources.
   e. not characterized by any of the other choices.

8. Of the following countries, which would most likely be *triaged* out of the world economy if Hardin's lifeboat analogy were actual policy?
   a. Chile
   b. Australia
   c. Indonesia
   d. Sudan
   e. Argentina

9. Mining for resources
   a. always endangers groundwater.
   b. is always going to supply enough of the resources needed to keep the economy going.
   c. will provide many jobs in a local area, because the work is body-labor intensive.
   d. releases arsenic.
   e. may be done responsibly but costs more.

10. The grade of an ore changes
    a. linearly lower from the center of an ore body.
    b. exponentially lower from the center of an ore body.
    c. gradually, but not necessarily linearly lower from the center of an ore body.
    d. abruptly at the edge of the ore body.
    e. in different ways for different ore bodies.

SHORT ANSWER

Explain whether the statement made is correct, and justify your answer.

11. Lifetimes of resources of a specific type (static, exponential, etc.) are fixed and may never be changed.

12. The estimated lifetime of a resource is longer for exponential than for static growth.

13. The ocean floor is currently the main supply source for manganese.

14. Exploitation of a resource only occasionally leads to greater development.

15. Tectonic plates on Earth are moving rapidly geologically at some number of meters per year.

16. Some materials in seawater are economic to extract at the present.

17. Material replacement is one solution to mineral resource depletion.

18. Pollution is always a consequence of exploitation of resources.

19. Material use rates rose substantially in the twentieth century.

20. U.S. industry uses mostly renewable resources for its high economic rate of return.

DISCUSSION QUESTIONS

21. Critics in 1972 decried the report of the Club of Rome known as the "limits to growth." Why did this Club of Rome study, which predicted that materials would become economically too expensive to exploit in the future, cause such a stir?

22. Is it good or bad that many materials must be imported to run the economy of any country?

23. What sort of politics flow around materials? Explain and give any example(s) you can.

24. Anne S. Moffat, in a paper in Science, "Engineering plants to cope with metals" (Reference 41) said: "Many plants

cope with … metals by binding them in complexes with a class of peptides called phytochelatins and sequestering the complexes inside their cells. Now three groups have isolated genes for the enzymes, called phytochelatin synthases, that make the metal-binding peptides when the cell is exposed to toxic metals.… After searching genome databases, they also found counterparts of the plant genes in the roundworm *Caenorhabditis elegans*." What are the implications of finding the same genes to inactivate metals in both plants and animals?

25. In Moffat's article (Reference 41), she quoted U.S. Department of Agriculture agronomist Rufus Chaney as saying that it would cost "less than one-tenth the price tag for either digging up and trucking the soil to a hazardous waste landfill or making it into concrete" to use plants for removing metal from contaminated soil. Suggest reasons why this approach should be so much less costly.

26. Research suggests that lead exposure leads to irremediable results, so that the best way to reduce lead levels is to remove lead-based materials. How realistic is such a tactic? Explain.

27. Suppose you were near a site with mercury pollution and were told that a plant could take mercury up, but that it would emit mercury vapor. Would it be worth planting these plants? Why or why not? (This is an expected phenomenon.)

28. In Reference 59, it was found that lead deposition decreased after 1979. What could be the reason? Suggest a possible answer. Justify your suggestion.

29. Why should resource use in the United States have increased by a factor of 17 in the past century? Suggest as many different reasons backed by some evidence as you can.

30. What difference to resource lifetimes does the state of the economy make? Explain.

PROBLEMS

31. In a record from 12,370 carbon-14 years before the present to the present time, in Reference 35, it was found that there was a decreasing lead-206/lead-207 ratio in deposited lead starting 3000 carbon-14 years ago. This is an indicator of lead mining.
    a. If the pre-anthropogenic deposition rate is 0.01 mg/m^2/yr, and the highest rate found was 15.7 mg/m^2/yr in 1979, determine the relative effect on humans in 1979.
    b. How much more lead is being used in 1979 relative to the year 100, if the deposition in the year 100 was 0.12 mg/m^2/yr?

32. How long would each of the materials listed in Table 10.1 last at the 2000 use rate, assuming the reserves given in 1968 were still correct? Compare your answers to those given in Table 10.3.

33. Table 10.1 shows the U.S. use and world reserves. Adjust to determine the *actual* lifetime. Assume U.S. use constitutes the following percentages of world use: cobalt, 25%; iron, 30%; manganese, 34%; molybdenum, 40%; nickel, 22%; aluminum, 33%; magnesium, 45%; titanium, 50%; copper, 24%; fluorine, 35%; phosphorus, 42%; and potassium, 20%. Compare your answers to those given in Table 10.3.

34. Suppose molybdenum use grows at 7% per decade. How long would the 1968 resources last (assuming the figure is still accurate)?

35. Show how the static and exponential lifetimes for cobalt in Table 10.3 are consistent.

CHAPTER 11
Recycling and Reuse

We extend the discussion of the last chapter, of the effect of conservation on energy supplies, to the discussion of the economy of recycling and reuse. Urban wastes of today will probably be mines for resources in the future. Reuse is clearly preferable even to recycling, if the political will to institute reuse can be summoned.

KEY TERMS

MSW / reduce / reuse / recycle / source reduction / electronics recycling / bauxite / beneficiation / cullet / PET / HDPE / PVC / flow control / refillables / bottle bills / throwaways

11.1 THE THROWAWAY SOCIETY

The amount of material thrown away is staggering. About half a megatonne a day is generated (about 2 kg per person per day by American consumers on average, of which 0.6 kg is recycled; it is as high as 2.5 kg per person per day in New York City[1]). In this country alone, then, a municipal solid waste (**MSW**) discard volume equivalent to piling about 50 m of garbage onto an area 100 m by 100 m (about the same volume occupied by 350 middle-class American homes) must be added to landfills or otherwise disposed of each day. This means that another *cubic kilometer* of garbage must be disposed of every four years or so. The United States is truly the throwaway society (see Tables 11.1 and 11.2 and Figure 11.1) and getting more so; it leads the world, but other countries are catching up. The typical German throws away about half of what the typical American does, about 1 kg per person per day.[2–4] The numbers in Table 11.1 show the effect of increasing recycling between 1986 and 2001, discussed in more detail later in this chapter and in the extensions.

TABLE 11.1

Composition of typical landfilled refuse by percentage by weight and by volume

Material	Weight (%, 1986)	Volume (%, 1990)	Weight (%, 2001)
Paper, paperboard	41	32	35.7
Yard waste	18	10	12.0
Metal	9	11	7.9
Glass	8	2	5.5
Rubber, leather	—	6	2.8
Textiles	—	6	4.3
Wood	—	7	5.7
Rubber, leather, textiles, wood	8	—	12.8
Food	8	—	11.4
Plastics	6	21	11.1
Other	2	5	1.8

Source: Environmental Protection Agency.

TABLE 11.2

Gross discards by weight (million tonnes) and total gross waste, 1960–2000

Material	1960	1965	1970	1975	1980	1985	1990	1995	2000
Paper, paperboard	27.5	34.5	41.4	39.1	47.9	55.9	57.9	54.0	55.2
Glass	7.3	7.9	13.9	12.3	15.9	12.0	11.5	10.7	11.0
Ferrous metals	11.3	9.2	13.5	11.2	13.5	9.9	11.5	8.3	9.8
Aluminum	0.4	0.5	0.9	1.0	1.6	2.1	2.0	2.2	2.5
Other metals	0.2	0.5	0.4	0.8	0.7	0.9	0.4	0.5	0.5
Plastics	0.4	1.3	3.2	4.1	7.5	10.5	18.5	19.7	25.8
Food scraps	13.4	—	14.1	—	14.3	—	22.9	23.3	27.8
Yard waste	22.0	19.6	25.6	22.9	30.3	27.3	34.0	22.8	13.2
Other waste	8.4	—	11.8	—	19.5	—	30.9	32.6	35.3
Total (MSW discarded)	91.0	—	124.6	—	151.1	—	189.6	174.1	181.1

Source: 1965, 1975, 1985: Department of Commerce, *Statistical Abstract of the United States, 1992* (Washington, D.C.: Government Printing Office, 1992), Table 360 and Dept. of Commerce, *Statistical Abstracts of the United States, 2000* (Washington, D.C.: GPO, 2000), Table 396; U.S. EPA, Office of Solid Waste and Emergency Response, *Municipal Solid Waste in The United States: 2001 Facts and Figures* (Washington, DC: GPO, 2003), EPA530-R-03-011.

FIGURE 11.1

Municipal solid waste generated in the United States before recycling or composting, 1960–2001.
(U.S. EPA, Office of Solid Waste and Emergency Response, *Municipal Solid Waste in The United States: 2001 Facts and Figures* (Washington, DC: GPO, 2003), EPA530-R-03-011)

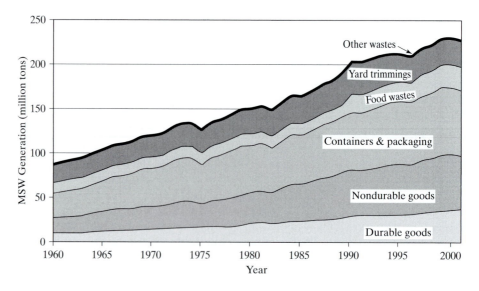

Once you think of the huge amounts of trash being generated, it must occur to you that some of the materials being thrown away could be valuable. Indeed, they are. Previous generations of Americans threw away far less. My mother and father can remember the scrap-metal collector with his horse-drawn wagon. He *bought* anything metal and then recycled it. The ragpicker would come around periodically to buy old clothing and paper. Such people made a good living, and performed a useful service.[5] These sorts of service industries disappeared after the end of World War II; many postwar industries adopted "planned obsolescence" as a way to retain wartime manufacturing capacity, and sold consumers on the inevitability (and desirability) of throwaways through advertising. Resources and energy were so cheap that it seemed to make sense to people to squander them.

Not everyone was content with this new American way of doing things. The counterculture youth of the late 1960s questioned their parents' ways with peace and war; they saw in their parents' attitudes the effect of grasping materialism, which permeated the entire society. As a result, many people turned to a simpler life and to the values of their grandparents. The *Whole Earth Catalog*[6] gave advice on returning to the land and on reuse of materials. These attitudes, in turn, affected ordinary citizens, making them more aware of the price they were paying for their material comforts: air pollution, water pollution, and destruction of the environment. In 1970, groups of people gathered across the country to celebrate the first "Earth Day" and helped raise consciousness even more. The time was ripe for the return of recycling, and in small towns across the United States, recycling groups were formed. (One such group of recyclers worked for many years in Delaware, Ohio, where I live. I joined the board of Waste Watchers/Delaware Recyclers when I moved to Delaware in 1976 and was involved until the organization was killed for political reasons in 2000.)

11.2 ENERGY IN INDUSTRY AND INDUSTRIAL WASTES

While residential wastes are large in amount, industrial wastes loom large as well. The U.S. EPA estimated the 1970 total waste from residential, commercial, and institutional establishments as 228 million tonnes (228 Mt); industrial wastes totaled 100 Mt; mineral processing wastes totaled 1545 Mt; and agricultural wastes totaled 2073 Mt.[7] By 1990, municipal wastes alone totaled 178 Mt, and by 1998, 200 Mt.[8]

Industry consumes tremendous amounts of energy. It has been pointed out that energy by itself is not only part of the process, but also that it "is not only a natural measure of the resource inputs to an economic system. It is also a measure of the material outputs."[9] The authors of Reference 9 argue that energy in a time series is the best proxy measurement for supplying useful environmental data for assessing efficiency and waste generation for industrial processes.

Figure 11.2 shows the distribution of energy among the economic sectors of the economy and how it has changed.[10] The electrical and chemical industries are among the largest users of energy and have traditionally been most conscious of its costs. For the industrial sector (not including the electrical utilities), energy use in the energy-profligate 1960s grew at an annual rate of 3.5%, slower than

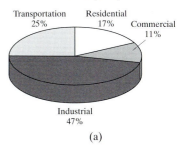

Net Energy Consumption by Sector (1950)

Transportation 25%
Residential 17%
Commercial 11%
Industrial 47%

(a)

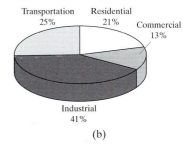

Net Energy Consumption by Sector (1975)

Transportation 25%
Residential 21%
Commercial 13%
Industrial 41%

(b)

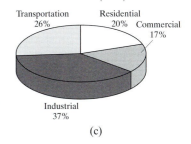

Net Energy Consumption by Sector (2000)

Transportation 26%
Residential 20%
Commercial 17%
Industrial 37%

(c)

FIGURE 11.2

The breakdown of energy use among the sectors of the economy. (a) 1950; (b) 1975; (c) 2000. Note that the commercial sector has grown relative to the industrial sector
(U.S. Department of Energy, *Annual Energy Review 2000* (Washington, D.C.: GPO, 2001), DOE/EIA-0384(2000), Table 2.1a)

any other sector, while energy use increased from 46.2 EJ in 1960 to 70.1 EJ in 1970 (an annual rate of 4.3%).[11] Energy use remained roughly constant after the 1970s energy crises (it was 78.0 EJ in 1985), then rose steeply to 85.7 EJ in 1990; industrial energy use was actually lower in 1990 than it was in 1980.[12] Overall, during the period from 1960 to 1990, the industrial sector's energy use grew at an annual rate of just under 4%, while overall energy use grew at a rate of 6.4%.[11] Industry is much more energy efficient than the other economic sectors.

The industries offering the most energy-saving possibilities are those with the largest energy-to-labor ratios: the chemical, automobile, paper, and construction industries.[13] As Table 11.3 shows, these are among the largest energy consumers. The labor-intensive industries, such as the leather and leather products industry, are the places in industry for which the energy growth rate is smaller than the growth rate of value added. Industry, in general, is becoming more economically efficient in energy use. The sectors becoming less energy-efficient comprise less than 2% of the total energy consumption.[14]

TABLE 11.3

Industrial fuel consumption by industry group (trillion Btu)

	Coal	Natural gas	Petroleum products	Electricity	Total energy
Primary metal industries	2838	836	306	1291	5298
Chemicals and allied products	666	1219	1426	1626	4937
Petroleum refining and related industries		1012	1589	225	2826
Food and kindred products	263	593	134	338	1328
Paper and allied products	467	341	211	280	1299
Stone, clay, glass, and concrete products	406	449	87	280	1222
Subtotal	4640	4477	3753	4040	16,910
All other industries	976	4781	721	1572	8050
Total	5615	9258	4474	5612	24,960

Source: U.S. Department of Energy.

We know that it is possible to save substantial amounts of energy. Japanese steel firms used an average of 30% less fuel energy per tonne of shipped steel than did the United States in 1978.[15] The chemical industry's organic chemical plants used 40% less fuel per tonne of output in 1981 compared to 1971.[15] The paper industry sees an overall 20% to 40% reduction in energy use for new plants coming on line.[15]

Much industrial waste of energy occurs because industries often find it cheaper to leak energy than to modify or replace outmoded equipment. Industry is expending some capital funds on pollution control at present; Figure 11.3 gives the amounts spent for pollution control,[10] and Table 11.4 indicates total investment compared to pollution abatement by sector. The highest expenditures are occurring in industrial applications, where retrofit is the only option. Where *new* machinery is being installed, pollution control is often incorporated at modest additional expense. In some cases, money as well as energy is saved. For example, gas-fired vacuum furnaces use about 25% of the fuel used in past models as a result of new research. With the introduction of vacuum furnaces during the 1960s, the steel industry's energy use per ton of coal decreased one-fifth from 1959 to 1969, from 9.5 to 7.7 MWh_t.[11] The decrease continues. A German firm developed a new steel process, thin-slab casting, in which the output is much thinner than that from current practice, and which

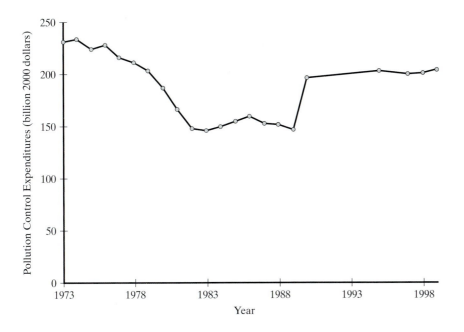

FIGURE 11.3

Total pollution control expenditures, 1973–1999, in billions of constant (2000) dollars.

(Adapted from Department of Commerce, *Statistical Abstract of the United States, 1984* (Washington, D.C.: Government Printing Office, 1984), Table 365; Department of Commerce, *Statistical Abstract of the United States, 1987* (Washington, D.C.: Government Printing Office, 1987), Table 340; Department of Commerce, *Statistical Abstract of the United States, 1992* (Washington, D.C.: Government Printing Office, 1992), Table 362; and Department of Commerce, *Statistical Abstract of the United States, 2000* (Washington, D.C.: Government Printing Office, 2000), Table 403)

TABLE 11.4

Investment for pollution control, 1982, billions of 2000 dollars

	Total expenditures	Pollution abatement expenditure
Total nonfarm business	561.52	15.07
Manufacturing	212.38	8.38
Durable goods	100.16	3.12
Primary metals	13.24	1.35
Blast furnaces, steel works	6.16	0.73
Nonferrous metals	4.81	0.53
Fabricated metals	4.60	0.07
Electrical machinery	18.85	0.27
Machinery, except electric	22.87	0.32
Transportation equipment	26.90	0.71
Motor vehicles	14.05	0.57
Aircraft	10.72	0.05
Stone, clay, and glass	4.63	0.14
Other durables	9.10	0.27
Nondurable goods	112.20	5.25
Food, including beverage	13.73	0.67
Textiles	2.36	0.05
Paper	10.59	0.53
Chemicals	23.55	1.19
Petroleum	47.36	2.66
Rubber	3.03	0.07
Other nondurables	11.57	0.07
Nonmanufacturing	349.14	6.69
Mining	27.42	0.92
Transportation	21.21	0.25
Railroad	7.77	0.14
Air	6.97	0.02
Other	6.46	0.09
Public utilities	74.44	5.32
Electric	59.27	5.13
Gas and other	15.17	0.20
Trade and services	154.30	0.16
Communications and other	71.80	0.04

Source: Department of Commerce, Bureau of Economic Analysis.

saves energy, because the slabs can be worked more rapidly.[16] Substantial increases in the cost of energy here in the 1970s and early 1980s have caused installation of new equipment for small steel mills while forcing closing of many obsolescent mills.[17]

The cost of transportation has the potential for great energy savings. More than 20% of the energy cost of industry spent is in bringing the energy to the materials processor.[18] The shipment of raw ore, and its associated mining costs, are not included in estimates of the energy cost of ore refining. Because ore includes nonuseful material, the shipping of ore can be very expensive; the lower the grade of ore, the more substantial the expense.

Why Recycle?

The mantra of all concerned for the environment is this hierarchy: **reduce, reuse, recycle**; only after these actions are taken should other waste disposal options be explored. An important consideration is that when products are made, energy must be committed to produce them. If the product is eventually discarded (or even if it is recycled), the energy committed to the product is lost for good.[19] That is the reason we recommend that we first reduce, then reuse, then finally recycle. In approaching recycling, we need to focus first on materials for which the greatest gain is possible.

Source reduction is the best way to minimize the effect on the environment. Packaging constitutes roughly one-third of the waste stream in developed countries. Most paper products, aside from newspapers, that end up in landfills are originally packaging. That is, of course, true for glass and plastic bottles and metal cans that contained some product. Simplicity in our lives results from reduction, but it is hard to do this by oneself without refusing to buy any products at all. Even though costs of raw materials give businesses some incentive to reduce waste, the cost of raw materials is artificially low (due to subsidies and uncounted environmental costs),[20] industry does not pay the cost of disposal, and maximization of profits does not always mean minimization of costs (a lot of money is spent by businesses on glitzy packaging to sell products). The Germans have taken a clever approach to solve this problem: put the responsibility on the manufacturer to dispose of the product (see **Extension 11.1, *Waste in Germany***), and the European Union is following suit.

Reuse of materials happens all the time. Many products are still being sold in reusable containers, even in the United States. The glass jar that stores the summer's harvest can store it year after year. A glass bottle that now stores screws is a reuse. Anything that gives a value to the material qualifies as reuse. Reuse involves materials already on hand and so does not require additional energy input. For example, obsolete computers are being disassembled to obtain small amounts of metal and chips.[58] A reuse would be to put the working but obsolete computers in the hands of people who could use them, perhaps in less-developed countries. What is commonly called recycling of used auto parts is actually reuse. (Ford Motor Co. is buying many auto parts "recyclers.")[50]

Recycling is the breaking down of a material into its components or into its purest form so that it can be used as a raw material again. A can melted down and made into another can is an example of recycling of a resource. Some resources are not degraded in recycling; for example, can and bottle recycling produces a product identical to the original. Some resources are degraded; for example, fibers in paper are shortened each time the paper is recycled.

CASE STUDY

Computer and Computer Part Recycling

Electronics recycling includes recycling of computers and computer equipment as well as audio and video equipment. Hewlett-Packard and Apple Computer both take their laser toner cartridges back. Apple even includes prepaid mailing labels for 13 countries in its cartridge package. Hewlett-Packard accepts returns of its own inkjet cartridges.

All inkjet cartridges are recycled in Franklin, Tennessee, home to SM Engineering USA, arm of a Swiss company that pioneered manufacturing inkjet cartridge remanufacturing machines. Prepaid mailers are available at universities and computer stores. Write to the Recycling Center, PO Box 683000, Franklin, TN 37068-9911 for mailers. Inkjet recyclers scrap some of the takebacks, but reuse and recycling of materials certainly saves materials and energy.[59]

Both Hewlett-Packard and IBM have pay-for-service old computer recycling, costing about $30 per computer (less in bulk). There are many agencies that support computer reuse, for example, in shipping old but working computers abroad to help people in poorer countries. See also **Extension 11.3**.

A list of computer equipment recyclers by state is available on the National Safety Council website at *http://www.nsc.org/ehc/epr2/cntctlst.htm*, a copy of which is available on the *Energy* CW. The EPA has an electronics recycling website, www.epa.gov/epaoswer/osw/conserve/plugin/index.htm.

For many of the recyclable components of household waste, the market value is not going to cover costs for a long time. There *is* money to be saved by recycling scrap metal, and this is where the commercial recyclers are involved. Recycling did increase some in the 1970s (from 6.4% in 1960 to 6.6% in 1970). By the late 1980s, the United States was starting to recycle more (Table 11.5 shows the situation in 2004). From 1980 to 2004, the recycling of lead increased from 35% to 77% and that of aluminum from 4% to 19% (though copper decreased from 24% to 9%).[70] Recycling rates have also increased in other industrial countries during this period[71] the recycling of lead increased to 49%,

TABLE 11.5

Per ton energy requirements to refine selected metals, and scrap use

Material	Energy use (Thermal MWh per ton)	Scrap as percent of 2004 U.S. Consumption
Mg (magnesium)	27	25
Al (aluminum)	17.8	19
Zn (zinc)	13.4	7
Mn (manganese)	13.4	0
Sn (tin)	5.9	12
Fe (iron)	7.6	60[a]
Cu (copper)	8.2	9
Glass	5.1	22[b]
Pb (lead)	3.3	77 (batteries, 93.5)[b]

[a]Includes imports and exports; 14% from residential discards; 46% from purchased scrap.
[b]2001 figure, Reference 1, U.S. EPA, Office of Solid Waste and Emergency Response, *Municipal Solid Waste in The United States: 2001 Facts and Figures* (Washington, DC: GPO, 2003), EPA530-R-03-011, Table 5 (glass), p. 96 (lead).
Source: Energy use, Reference 7; Scrap, Reference 75, U.S. Department of the Interior, U.S. Geological Survey, *Mineral Commodity Summaries 2005* (Washington, DC: GPO, 2005).

TABLE 11.6

Energy costs in making aluminum

Facet of production	Energy used (MJ/kg finished aluminum)
Ore extraction (mining, drying, shipping)	
Caribbean bauxite	6
South American bauxite	9
Average	7
Production of alumina from ore	
Caribbean bauxite	42–57
South American bauxite	40–56
Average	48
Production of aluminum from alumina	
Electrode ⎰ Prebaked	208–272
⎱ Soderberg	246–277
Cost of fluorine compounds and calcining	4–10
Average	260
Overall energy cost from ore	314
Overall energy cost from scrap	
Pure	6
30% to 40% contaminated	25

Source: U.S. Bureau of Mines

that of copper to 40%, and that of aluminum to 25%.[60] Worldwide, as of 1984,[61] 40% of lead, 33% of aluminum, 33% of copper, 10% of chrome, and 6% of zinc were recycled. Now, much of the scrap goes to China.[72] The huge demand led to looting in Iraq and thievery in Argentina.[73] Of the materials listed in Table 11.6, lead and zinc do not save much energy per se in recycling. However, because lead and zinc are difficult to find in appreciable concentrations, a reasonable fraction of production represents recycled material. With the advent of curbside municipal recycling programs in the late 1980s and 1990s, however, the rates have increased substantially for steel and aluminum. By 1999, the overall American national recycling rate was 27.8%.[63]

The success of recycling in America has been spotty. America recycles about one-quarter of the waste stream, as mentioned above.[63] By many measures, recycling is decreasing.[64] In Europe, where the population is denser, recycling is much more successful (see **Extension 11.1**, and note that European countries are also struggling with the "throwaway society").[65] Only Nova Scotia in North America has achieved a really respectable recycling rate—80% of beverage containers are returned, takebacks are mandated for some goods, and composting has proved to be a moneymaker.[66]

Aluminum

Consider aluminum beer or soda cans. At one time, the aluminum was in a chemical compound in the ground, in **bauxite** deposits. The ore had to be dug up, crushed, washed and dried, then shipped to another site to be refined into alumina, then shipped to yet another site to be electrolytically smelted (reduced) into aluminum ingot. The ingot is then sent to a rolling mill to be formed in sheets, and sent to the can manufacturer—all this before the brewer bought cans to put beer or soda into them. If aluminum is recycled, it has to be transported from the consumer to the recycling processor and then to the smelter for remelting and rerolling into can sheet, and so on.

From this description, it does not seem that recycling of aluminum should be a great savings. So, why am I an avid recycler of aluminum cans? The reason becomes clear when the first few steps in aluminum production are examined. Much of our bauxite (aluminum ore, Al_2O_3) comes from Jamaica, Suriname, Canada, and Australia.[7,62] In bauxite deposits, the ore is about 20% aluminum. This ore is usually transported to coasts and loaded onto ships. It is shipped to a U.S. port and brought by rail to the smelter. Thus, each tonne of aluminum transported has an additional 4 tonnes of waste material transported along with it. Even though the transportation is not expensive, one must pay five times more than necessary just to get the aluminum to the smelter.

The refining of aluminum is very energy expensive, because it uses electricity (that is why aluminum is refined in regions with cheap electricity, such as the Pacific Northwest). Since 1 kWh of electric energy would cost about 3 kWh of thermal energy (at typical generating plant efficiencies, about 33%), we have to correct the energy budget to arrive at a true energy cost. We can compare energy costs of various procedures only if we choose a consistent basis for comparison. I will quote thermal energies here. This means that where thermal energies are used, for example, if fuel oil is used, the energy quoted is used as given in the table entry (or is converted into joules). Where electric energy is used, the energy given in the source must be multiplied by a factor of approximately three before entry under its appropriate heading. For aluminum, the cost of refining is about 300 MJ/kg, while the cost from scrap is about 7 MJ/kg (see Table 11.6).[67,68] In fact, when one takes the energy costs of transportation into account, it will be clear that the production of aluminum from scrap must cost only about 2% of that from ore, as well as saving the disposal of the four-fifths of the ore material that is useless slag (and I have not counted such industrial wastes in the volume of waste to be disposed of as trash every day). Although a process developed by Alcoa reduces energy input in smelting by one-third,[15] it is still advantageous to recycle as opposed to refining virgin ore.

Much of the cost of transport and refining of the ore arises from the dispersal of those useful aluminum atoms among all those other sorts of atoms. This configuration has greater entropy than one with the aluminum concentrated. In scrap aluminum, the atoms of aluminum are already concentrated. As we all remember from basic thermodynamics, it is easy to decrease entropy locally by paying the appropriate energy price. If aluminum is just thrown away, entropy is again increased. The energy cost of processing contaminated scrap is also much greater: ~30 MJ/kg at ~30% contamination. It would be incredibly expensive to retrieve cans from the landfills, where they would be mixed in among many other sorts of materials. The least energy-expensive way to gather aluminum is for it to be saved separately at the point of use. This means that the individual household, at the point of use, should be saving its metal separately. About 53% of aluminum cans produced in the United States, 600 kilotonnes worth, were recycled in 1986,[82] and this had decreased to 44% by 2003.[77] The same logic applies to metals other than aluminum as well as to glass and to waste paper.

Aluminum consumes 71% of the energy used by all nonferrous metals.[68] The total cost of aluminum from ore is about 43,700 kWh$_t$/t (180 GJ per tonne), whereas the cost of aluminum from scrap is about 2440 kWh$_t$/t (9.6 GJ per tonne).[68] Therefore, recycling aluminum saves 94.4% of the total energy used in

FIGURE 11.4

(a) Recycling rates for beverage containers in the United States by container type, 1970–2003.
(b) Beverage container recycling rates in selected deposit states and national average, 1990–2003.
(Container Recycling Institute)

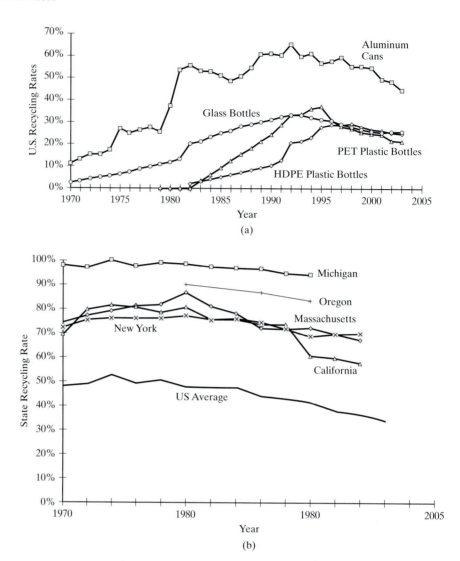

(a)

(b)

aluminum smelting (can fabrication requires a significant amount of energy, no matter how the ingot is produced; thus, the overall energy savings involved in recycling aluminum cans to new aluminum cans is only about 65%). The steep rise in cost of electricity in the 1970s was partly responsible for the increase in aluminum recycling from 17% to 32% in the United States[83] (Earth Day consciousness and bottle bills—Section 11.5—also contributed). By 1990, aluminum can recycling had reached about 65%, but it dropped through the 1990s to under 50% by 2003, due to increased away-from-home consumption, inadequate recycling opportunities in public spaces, and consumer, media, government, and businesses apathy.[77] In 2000, about 101 million aluminum beverage cans were used.[64] See the situation for soft drink containers and beer containers in Figure 11.4.

Steel

The story for scrap steel is similar to that of aluminum. The energy cost of steel from ore in 1979 (see Table 11.7) was 28.5 MJ/kg, while that from scrap was only 1.26 MJ/kg (less than 5% that from ore). These costs do not include the energy costs of mining, beneficiation, or transport. In contrast to the case for aluminum,

TABLE 11.7

Energy costs in making steel from iron ore (MJ/kg)

	From ore	From scrap
Coke (0.4 kg)	11.5	
Coal (0.02 kg)	0.7	
Electricity (0.46 kWh)	5.6	
Fuel oil (4.0 L)	2.5	
Tar, pitch (0.3 L)	0.2	
Liquid petroleum gas (0.5 L)	0.02	
Natural gas (0.13 m^3)	5.0	
Coke gas (0.16 m^3)	3.0	
Total from ore	28.5	
Total from scrap		1.26

Source: U.S. Bureau of Mines.

much of the ore used for steel is produced domestically, mostly in Wisconsin and Minnesota. Taconite is low-grade iron ore mixed with minerals (it does not have a high concentration of iron), so about 3 kg of crude ore is mined for each kilogram of usable ore transported.[62] The ore is in deep formations, and large quantities of overburden must be stripped away—an average 1 kg of waste to each kilogram of crude ore (the range for the United States is one-third to 3).[62] In addition to the production of some 660 Mt of waste in mining, a gigantic amount of water is used in the pelletization (beneficiation) of taconite ore: over 3 L/kg of pellet ore. This water is then dumped, mostly into Lake Superior. (Several years ago, concerns were raised about the pollution of the lake, and lawsuits were initiated over the presence of asbestos in the tailings.) Energy consumed in mining ranges from 36 kJ/kg for high-quality ore (~50% iron) to 1.4 MJ/kg for taconite made into magnetite or hematite concentrate Fe_3O_4 (pellets).[70]

Clearly, scrap steel is very good for conserving energy; for this reason, nearly 60% of the total weight of steel produced comes from scrap (about half from in-plant scrap and half from recycled steel). Scrap is very important in the newer, smaller specialty mills.[17]

Recycling of scrap steel costs only 4.4% of the energy cost of refining the steel from ore. This rerefined steel is used in place of the foundry pig iron previously used. There are still many opportunities to save energy in steel manufacture, most associated with the multiple heating and cooling steps used to process steel. The most efficient mills can produce steel at 24 GJ per tonne, while it is estimated that it would eventually be possible to manufacture steel for 12.5 GJ per tonne.[71]

It has been estimated that there is 700 million tonnes of scrap steel accumulated in the United States, growing at a rate of about 11 million tonnes a year.[72]

Copper

There are two stages involved in the smelting of copper: production of matte from ore and refining of the matte to the required purity. The first process requires about 32,000 ft^3 of gas and 375 kWh of electricity per ton; because 1 ft^3 of gas provides 1.035 Btu, or about 0.32 kWh, this means that 10,825 kWh/ton is required to produce matte. Refining the matte requires 615 kWh$_e$ and 4700 ft^3 of gas; thus, the total energy cost of refining matte is 3265 kWh$_t$/t.[68] Overall, this is roughly 160 GJ per tonne (at 0.3% ore), which compares well with ~180 GJ per tonne from Reference 70. Recycling scrap eliminates the first process, thereby achieving a 77% savings over the energy cost of smelting from ore.

The two processes discussed for copper generally occur at two separate locations. The continuous smelting of the ore can conserve thermal energy for an overall energy savings. This also reduces the emission of noxious sulfur dioxide in the process.[73]

Glass

In contrast to both steel and aluminum, there is not much energy to be saved in recycling glass. It takes almost as much energy to melt glass as it takes to melt the silicon dioxide–soda ash mixture at about 1400 °C. Glass manufacturers have gradually been able to raise the amount of **cullet** (recycled glass) in their mixes. By 1984, a 40% cullet mix was not uncommon in the United States. In Switzerland, a cullet mix of 90% is apparently the highest in the world.[74] For each 10% of the batch that is cullet, an energy savings of 2.5% results. There are advantages for the manufacturer to using recycled glass; like butter in a pan, glass melts in the furnaces before the raw materials (1200 °C) and allows the mixture of sand and soda ash to melt at a somewhat lower temperature. Using the lower melt temperature prolongs furnace life. Soda ash is getting more expensive (about 25% of the raw material is soda ash), and glass cullet allows less to be used, saving a bit of money.

The major reason for encouraging glass recycling, rather than reuse, is practical. Many people simply throw glass away wherever it is convenient for them to do so. Delaware, Ohio, has a population of about 25,000; it is a city with a good school system, a university, and a large middle-class population. Even so, I pick up one to two bottles or cans a week that people have simply dropped on my lawn. It is even worse in mostly rural Delaware County. The roads are badly littered. Recycling centers can provide an alternative to rural dumping. Bottle bills (see the discussion later in this chapter) can address part of such problems.

Delaware County has a problem with rural dumping, especially because our landfills were closed, and all trash is transferred to a landfill in Logan County, some distance away. Rates for trash taken to the transfer station have increased greatly over rates at our former landfill, and some residents now scatter their garbage along the county roads. This has happened elsewhere. The county or municipality must then pay for the cleanup, and cleanup along roads costs about 10 times what it would cost if the material were concentrated in one place (the entropy problem again).

Broken littered glass can constitute a transportation hazard, can cause lacerations if stepped on, and can injure grazing livestock and damage farm equipment. Another problem with glass, as opposed to steel and aluminum, is its long degradation time. A steel can along a highway will rust away within 5 to 50 years; an aluminum can may disappear after a century or so; a glass or plastic bottle may not have disappeared even after a millennium. Recycling of glass may prevent some of these problems, leaving walks or drives more pleasant for all, and saving energy.

The rationale of local recycling organizations is that people will recognize that it is possible to save materials at home and bring them in all at once to a recycling station that takes in all items. At one point two decades ago, the Delaware Recyclers paid over twice what it received from selling glass in order to get the glass to the point at which it could be reused. We wished to remain the "universal acceptor" for recyclables, and the economics of the situation did not deter us from doing what we thought right. With the advent of massive recycling programs in the eastern United States, the value of glass that is recycled has dropped substantially. About 3 million tonnes of glass was recycled in America in 1997, for a recycling rate of 31%.[64] This compares with a rate of 89% in Switzerland, at the top of the list, 81% in the Netherlands, 79% in Germany, 75% in Norway, and 72% in Sweden. American recycling rates for glass did best the United Kingdom's shameful 22% recycling rate.[64]

Paper

As Table 11.1 showed, paper and paper products may make up nearly half the weight and volume of waste deposited in landfills. Paper ends up elsewhere as well. Recycling paper allows this volume to be used more effectively and saves some trees (and a bit of energy). Every ton of newsprint recycled saves 3.3 yd^3 (each tonne saves 2.3 m^3) of landfill space.[75]

Here, too, entropy considerations are important. Mixed paper commands a much lower price than pure de-ink newsprint (used newspaper), pure computer printout, or pure rag paper if it can be sold at all. There has almost always been a market for unmixed paper. This is reflective of both time and energy necessary to segregate the mixture.

The massive recycling programs in the northeastern United States caused the paper market to crash in 1989, and some communities paid to have their paper recycled.[76] Delaware Recyclers was hit particularly hard by the drop in value of newsprint, which went from about $100/ton to $5/ton after the glut registered in the used paper markets (as new recycling plants opened in the mid-1990s, the price rebounded somewhat, to ~$15/ton). The country was getting rid of its massive paper glut by exporting the used paper, which caused the market to crash elsewhere.[77] The situation overseas had long been better than in the United States. In Germany, about 52% of all paper (over 70% of newsprint) is recycled.[78] In 1992, Germans saved 8 Mt of used paper. Much of it ended up in incinerators or was exported to the Far East because of the continuing glut.[78] Still, Germany planned to recycle 70% of all paper by 1997, and made that handily. Some paper is sold for burning, but its value for combustion is only about $20 per tonne; its value as recyclable paper is greater in much of the world.[70]

One good result of the oversupply of de-ink newsprint is the decision of many North American paper companies to expand recycling. In 1989, there were 43 paper mills in North America, of which only nine could use recycled material. Now there are 26 mills that can use recycled paper.[76] This helps increase the market for our newsprint and boosts recycling. Another help is that the "premium price" charged for recycled paper is a thing of the past.[79] Florida put a tax on nonrecycled newsprint, giving the recycled product a competitive advantage.[80] Perhaps the most important change in helping to support recycling is the decision of the U.S. government to buy large amounts of recycled paper.[81]

DOING YOUR PART

Recycling Newspapers

The de-inking facility used to make old newspapers into new paper follows a sophisticated sequence of steps.[82] It begins with the hydrapulper (which makes the paper into a slush). A coarse screen then removes large impurities (staples, paper clips, and so forth), and a cyclone removes dust and stones. This is followed by fine screen (which removes very small impurities). At this stage, the ink is finally removed in two steps, first the de-inking flotation, which removes large ink particles, and then the de-inking wash, which removes small ink particles. The paper mixture is bleached and sent on to paper machines, as would ordinary bleached pulp.

The bleaching stage is polluting in all paper making because of the large amounts of chlorine used. Because of the imperfect removal of ink, old newspaper must be bleached more than virgin pulp. The Union Camp Paper Co. is now using ozone to bleach paper instead of chlorine.[83] The emitted ozone is relatively harmless in comparison to the chlorine.

Plastic

Six classes of plastics are being used; the major four are **PET** (polyethylene terephthalate, known as number 1 plastic, 2-liter soda bottles); **HDPE** (high-density polyethylene, known as number 2 plastic, beverage bottles and soap bottles); **PVC** (vinyl chloride and polyvinyl chloride, known as number 3 plastic, water bottles); and **LDPE** (low-density polyethylene, known as number 4 plastic, the plastic bags used in the grocery store). Actual recycling of plastic in 1992 was 2.9%.[84] Only PET and HDPE are being recycled in any numbers nationally. The PET beverage container recycling rate was 37.3% in 1995; but by 2003 it had dropped to 21.4%. The HDPE beverage recycling rate was 30% in 1998; in 2003 it was down to 25%.[77] It takes a lot of labor to separate and bale the plastic. Contamination is a problem as well. One PVC clear bottle among a batch of PET bottles can ruin the whole batch. (The PVC bottle burns at the PET melting temperature.)

In 1990, in the early days of plastic recycling, Du Pont and Waste Management (a national waste disposal firm) entered a joint venture to make money recycling plastic. They found they were paying $1500 per ton to collect and process plastic worth $80 to $100 per ton.[44] (Delaware Recyclers could have told them that!) The National Solid Waste Management Association has produced figures that agree with this assessment: Table 11.8 shows the costs per ton to handle various materials that are being recycled. Clearly, all materials cost a substantial amount to handle, but plastic is generally the worst of the lot. For this reason, many plastic waste materials are sent to China, where low labor costs allow such handling to be done,[85] and where PET is desirable because the Chinese lack the chemical refining infrastructure to make PET resin. The Chinese demand for reclaimed PET has made it difficult for American reclaimers to procure sufficient supply; many may close. In addition, the recycling rate for plastics continues to drop.[99]

Companies using recycled plastics are prospering because of the low price the plastic fetches. Wellman, Inc., for example, makes polyester, nylon, and other

TABLE 11.8

Handling costs per ton for various recycled materials

Material	Cost ($/ton)
HDPE	188
PET	184
Aluminum	143
Amber glass	112
Green glass	87
Clear glass	73
Steel cans	68
Mixed glass	50
Corrugated paper	43
Mixed paper	37
Newspaper	34

Source: National Solid Waste Management Association.

materials from waste plastic. It has been successful in the marketplace over the past decade.[86] Ground-up PET bottles are used for insulating material. A man's ski jacket uses two PET bottles worth of plastic for insulation. A sleeping bag uses 36 PET bottles worth of plastic.[87] Several manufacturers are making polyester fiber out of recycled material. One promising avenue for recycling plastic is "plastic wood," used in fencing and park benches. It does not rot like wood would, and has even been used to build a bridge over the Mullica River in the pine barrens in New Jersey.[103]

Perhaps the most notorious plastic is number 6, polystyrene, used in styrofoam drinking cups and plates. The plastic was said to be recyclable, but as only five plants were set up (mostly with McDonald's money before McDonald's eliminated the so-called styrofoam clamshell), this was rather an overstatement. (It is true with all plastics that they are only recyclable if there is a market to which the recycler can sell.) The "environmentalist" choice between styrofoam and paper cups is not as clear as some had thought (reusable ceramic cups are the best choice, in any case). Paper cup production takes 36 times as much electricity and 580 times as much water as styrofoam cups.[104] Similarly, the "environmentalist's" choice between paper sacks and plastic bags in the grocery store is not clear-cut.[103] An elegant solution here is a cloth bag that can be reused for years, or reuse of last week's plastic or paper bag. Plastic's failure to biodegrade is a problem on land, with the unpleasant impact of windborne bags and styrofoam cups; it is a worse problem in water, where plastic debris harms marine life.[105]

Plastic recycling rates have been dropping over time. Part of this is due to the rise of the 20 oz convenience store soda, which usually ends up in a wastebasket or other trash container rather than in a recycling bin.[78] In 1994, the beverage plastic bottle recycling rate was 48%, but it had dropped to just 24% by 1999.[78]

One response to the plastic waste problem was exploration of degradable plastics; however, few have lived up to such a name.[106] The more durable the plastic, the less biodegradable it is. Attempts have included blends of polymers with starch, synthetic polymers with weak links, bacterially produced polymers, polymers with pre-oxidants, and cellulose-based materials.[106] Cellulose, a plant-based material, may be making a comeback.

Waste and Entropy

The physical quantity entropy is intimately connected to these questions of ore concentration and the throwing away of recyclable materials. Ores are used because the materials are found in concentrations significantly above the clarke (the average concentration in the crust). Ore bodies thus have a lower entropy than "average rock." The process of refining the ore introduces a further drop in its entropy. The material is then used for some purpose in products. The products are widely distributed, causing entropy to be increased. After use, the material is thrown away. This increases its entropy still more. If, instead, the material were separated, the entropy would not be increased any further.

Entropy is characteristic of the natural systems that make up our ecology. It is also characteristic of the waste stream.[107] One innovative way to approach the waste stream is from the point of view of a mimicking of the natural ecology by industrial systems. The study of systems according to this view is called industrial ecology. Industrial ecology's successes and limitations are documented in **Extension 11.2,** ***Industrial Ecology***.

11.3 HANDLING WASTE

Curbside Recycling

At this point, materials can be recycled at a local recycling station or be collected by the municipality, separated by the user so that it can be recycled. This is known as curbside collection. Money from municipal recycling sales can be used to prevent rises in the cost of refuse collection, mostly because the tipping fee is avoided, and some income comes from the sale of material. Many municipalities already do this to a certain extent, for example, Urbana, Illinois; Kitchener, Ontario; and Mississauga, Ontario. In the past few years, many more cities have begun collections of recyclables. Curbside recycling is practiced in more than 5000 communities serving 85 million people.[125] Recycling now diverts 30% of MSW from landfills (twice as large as the 15% diverted by combustion).[126] Such recycling is not really new. Many years ago, when I lived in Princeton, New Jersey, we had to separate the metal trash from the rest.

Municipal recycling programs seem to present a sensible approach to the problems of landfills, to enhance recycling of materials, and to reduce cost increases for refuse collection. The cities are interested in what is known as **flow control**, control of all refuse collected within their geographical areas, because otherwise, they cannot plan for recycling mandated by laws nationwide and assure that their landfills meet the stringent requirements set for governmental bodies. The funding for a recycling facility or the landfill may be predicated on collection of a dumping fee charged to all haulers. The U.S. Supreme Court allowed a lower court flow control decision to stand: the effect is that flow control is legal.[127] So, one community's successful recycling programs "infect" nearby communities, in a happy feedback loop.[128]

If such programs can be implemented nationwide, they will be an appropriate solution to the problem. (Increasing costs cause difficulties for municipal recycling programs, leading to serious consideration of the European "producer pays" approach to packaging reduction. See **Extensions 11.1** and **11.2** for more discussion of this option.) If not, more recycling facilities should be opened, and recycling should be pursued as vigorously as possible. Of course, if everyone recycled, there would have to be additional measures taken to assure that the materials are reused. A careful analysis of the economics of recycling and landfills that takes into account differences between use of virgin and recycled materials showed clearly that for every way of performing an analysis, recycling is superior in terms of reducing waste, reducing emissions, and reducing energy use.[74] Only if material recovery or waste management activities were analyzed by themselves, without consideration of the larger system, did it appear that use of virgin materials was superior.[74]

Large appliances are being successfully recycled in Germany and in Japan, which enacted the toughest appliance recycling law in Asia in 2001.[129] The greatest push is coming in recycling computers. The European Union has a directive mandating takebacks, as do some individual non-EU countries. Japan has a law mandating recycling of both home and commercial computers.[129] Computer and other appliance recycling and reuse is discussed in **Extension 11.3, *Recycling Computers***.

Landfills

Household recycling probably never directly pays for itself from sales of recyclables. The point is not to lament this—this is *waste disposal*, not a productive

TABLE 11.9

Amount of glass recovered (tons)

Percent recovery	Population				
	10,000	25,000	50,000	100,000	500,000
5	31	76	153	305	1525
10	61	153	305	610	3050
20	122	305	610	1220	6100
50	305	763	1525	3050	15,250
70	427	1068	2135	4270	21,350
100	610	1525	3050	6100	30,500

business, and we should not expect it to pay for itself. Note especially the glass figure; in 1993, while still operational, Delaware Recyclers recycled 382 tons of glass. A town of 25,000 will produce, on average, 1525 tons of glass waste per year (Table 11.9). On this basis, we were reaching only 25% of the population of the city (and only about 8% of the county).

In Delaware, Ohio, alone, the volume of unbroken glass produced each year as trash is about 3900 m^3. Even if the glass were crushed, it would occupy over 2400 m^3. The space problem in landfills is even worse for large cities. A city of 500,000 would generate between 47,000 and 78,000 m^3 of glass waste volume each year. No wonder some landfills are running out of space.

The amount of trash is growing exponentially (Table 11.2). People are becoming aware too late to make anything less than a complete change. Landfills have been closed in many regions of the country because of inadequate pollution control or simply because they were full. New Jersey, for example, had 350 landfills in 1977 and accepted out-of-state trash. By 1987, New Jersey had only 80 landfills, of which fewer than 10 took municipal garbage; as a result of landfill closings, costs of disposal jumped by factors of two to five.[161] By 1988, only 13 landfills were left in New Jersey.[162] New Jersey now exports over half its waste.[162,163] Hempstead, Long Island, ships trash to Youngstown, Ohio, for disposal, because its landfill closed.[164] Landfills accepted only 63% of U.S. MSW in 1990, down from around 90% in the 1970s.[126] While there are still a lot of landfills, they are only the biggest ones. The smaller landfills have closed. The receiving states are not necessarily overjoyed by the bounty. In Ohio, "garbage imports rankle."[165] Imports in Ohio peaked in 1989 at about 20% of the total landfilled and are currently running a bit under 10%.[165]

Not many of us in Delaware, Ohio, want a landfill next door. The Delaware city council met solid opposition from the potential neighbors of a proposed landfill. In some landfills, there have been leaks of leached materials into groundwater. The EPA is now requiring 30 years of monitoring after the closing of a landfill.[166,167] In 1979, there were almost 20,000 U.S. landfills; by 1991, there were 5800, and now there are about 5300. Over 20% of Superfund (see Chapter 26) cleanup sites are former landfills. New landfills must be lined properly (clay and plastic) and have pumps to collect leachate, which must be collected and treated.[166]

Experts used to think that the sanitary landfill degraded everything in it rather quickly. Biodegradation does not occur in all landfills. William Rathje of the Garbage Project found 40-year-old newspapers that look as fresh as today's,[4] and 15-year-old meat with fat still hanging on the bone, and a frank that looked like a frank.[168] A landfill just doesn't look like a good neighbor. This lack of trust is happening everywhere. No matter how obvious it is that facilities are needed for landfills, treatment plants, and the like, people want them somewhere else: NIMBY! (not in my back yard).

Because of this attitude, many old landfills are used past the ends of their projected lives. Several are even being mined for incineration or for resources. In Lancaster, Pennsylvania, the local landfill is supplying material to keep the new incinerator running.[167] In Collier County, Florida, the landfill is one in which decay occurred, and the composted material is being dug up for new fill (earth must be trucked in at great expense, otherwise).[169] At both facilities, the recyclables are being recycled, and the landfill life is expanded. Waste expert Alan Hershkowitz of the Natural Resources Defense Council was quoted [167] as saying that the Lancaster effort "underscores the absurdity of entombing everything we throw out."

One of the clearest illustrations of the NIMBY attitude became news in early 1987. A barge, the *Mobro*, was filled with garbage from Islip, Long Island, to be sent to North Carolina. This began a 3-month odyssey for the unwanted garbage barge, which was refused entry everywhere it tried to dock. The *Mobro* had to return to its point of origin. The garbage was finally unloaded, incinerated, and disposed of 5 months after its peregrination began.[170] A similar thing, lasting much longer, happened to the *Khian Sea*, which moved 15,000 tons of incinerated trash from Philadelphia from place to place around the world from 1987 until 2001, when the materials were returned to Florida, then finally to Pennsylvania; the waste was incinerated once again and trucked to a landfill.[171]

11.4   CAPITAL COSTS VERSUS OPERATING EXPENSES

Though there is great potential in our present-day midden heaps, even the most optimistic estimates before widespread municipal recycling programs projected only a quarter of all discarded materials recovered (some states have bettered this goal by significant amounts). The difficulty with recycling now is that "at the point of utilization, economic justification of energy consuming equipment tends to be governed by initial costs."[15] This means that operating costs are not considered in buying equipment, only the capital costs are. The operating expenses are carried on company books as overhead, and savings in operating costs possible with different equipment, thus, are not considered (recall the air conditioner discussed in Chapter 9 and in **Extension 9.2** and the discussion in **Extension 11.2**).

The problem of operating vs. capital costs occurs in waste management. Should landfills be small and local, or larger and involve considerable trucking of waste? Economies of scale favor larger landfills—the number of active landfills fell by more than 75% between 1988 and 2001 (from 7924 to 1858).[126] Despite worries about declining space available to dispose of wastes, which led to short-term pressure for increased recycling, the landfills grew fewer—and larger. Larger landfills have more monetary leeway (per ton of landfilled waste) to deal with problems such as leaching of toxic materials from the landfill, collecting the gases generated, and so on. However, these larger landfills are generally much farther from the point of collection and make the cost of disposal much more dependent on the ongoing cost of fuel. The pressure for increased recycling is lowered as well.

The Diseconomy of Trash Recycling

A similar situation for trash recycling occurs because of inertia and lack of understanding of the entire process of trash collection. In terms of generation, in the

United States, we overpackage things (see the discussion of the German *Grüne Punkt* program in **Extension 11.1**). Approximately 10% of total food costs are due to packaging.[172]

Consider the "disposal" of trash. A New Yorker generates over twice as much trash per day as someone from Hamburg, Germany, and 3.6 times as much as someone from Manila, the Philippines.[4,172] In other countries, such as Egypt, Mexico, and Thailand, many people make their livings from trash recycling. In the United States, we see only the disposal problem. The increasing costs of landfilling, however, are causing a rethinking of the situation. New Jersey, which already had a significant number of municipalities begin recycling, passed a *mandatory* recycling law in 1987.[173] Without recycling, New Jersey would have been totally without landfill space by 1991.[174] Oregon has made recycling mandatory for all communities of more than 4000 people,[69] and many other cities and states have done the same. In New York and other places, "mongo hunters" reduce waste by reusing what others throw away.[175]

The new perception comes from a "holistic" approach to trash disposal. It is being recognized that even though municipal recycling loses money, it loses *less* money than conventional disposal of trash. According to one expert,[82] per-tonne burning costs in the northeastern United States are $65 to $110, and landfill per-tonne disposal costs begin at $40 (costs for Long Island trash are $140 or more); it costs only $20 to $30 per tonne to recycle from a weekly trash collection, depending on whether or not materials are presorted.[82] See the sad story of the rise and fall of a local recycling organization in **Extension 11.4, *Delaware Recyclers***. A large initial investment (capital cost) is required to get the equipment necessary to operate effectively. As a result of this investment, however, operating expenses are lower in perpetuity. Multinational companies such as BFI and Waste Management, Inc. have access to capital that local businesses do not have. (Consolidation of the waste management and recycling industries is a complex and controversial subject beyond the scope of this book.)

11.5 ADDRESSING BOTTLE AND CAN LITTER—RECYCLING, REFILLABLES, AND BOTTLE BILLS

Incredible numbers of beverage containers are made each year. In 1975, about 40 billion glass containers (7 Mt of glass), about 10 billion steel containers (2 Mt of steel), and about 10 billion aluminum containers (0.5 Mt of aluminum) were made.[194] By 2003, beverages filled 100 billion aluminum cans, 47 billion PET plastic bottles, 35 billion glass bottles, and 10 billion HDPE plastic bottles.[195] Soft drink containers cost manufacturers more than the contents.[196] In the foregoing figures, we have considered only the fabrication cost of throwaways. The proportion of beverage containers in waste is growing at a rate over twice that for all refuse.

There are other costs, such as the esthetic costs of litter (Figure 11.5) and the actual dollar cost of cleaning up the litter. Litter on 1 mile of Kansas highway in 1975 contained 590 beer cans, 130 soda bottles, 120 beer bottles, and 110 whiskey bottles as well as 10 tires and 2 sets of bedsprings.[197] In fact, containers for alcoholic beverages are still common on highways: Eighteen percent of Ohio's litter was such containers in 2004 (9% were nonalcoholic);[198] in California, half the containers had held alcoholic beverages,[199] but in Texas, the alcohol proportion was only 8% (nonalcoholic was 16%).[200] The cost of disposal of highway litter

FIGURE 11.5

Litter along a California highway (actual unstaged photo). The cost of cleaning up such highway litter is enormous; it has been found to be substantially less in states with "bottle bills."

was four times as expensive as residential refuse disposal in 1969.[7] It remains much more expensive to clean litter than to prevent it. Bottle and can litter has been reduced by 35% to 40% by volume (~80% in numbers) in states with mandatory deposits.[83,201]

Refillables

Recall the environmental anthem: reduce, reuse, recycle. Beverages have grown in popularity across the world as noted above. Almost one trillion beverage containers are produced every year—in the absence of reduction in this number, we must turn to reuse or recycling. Reuse in this case means refilling beverage containers. In the case of **refillable** containers, it is desirable for the returnable containers to go through as many trips as possible. In the United States today, refillables make up under 1% of the market. In Canada (Prince Edward Island, Ontario, and Quebec), and more especially in Europe, refilling is still prevalent.[202] Evidence from the United States shows that as the use of refillables declines, a concomitant decline in the number of trips made by refillables is observed. In the 1950s, the average number of trips per bottle was about 40 in the United States; by 1972, it had dropped to 15, and by 1978, it had dropped to 8.[203] Conversely, when refilling is enforced, the average number of trips increases.[203,204]

Evidence from many life-cycle studies shows that refilling is desirable-it saves energy, reduces emissions, and is simply good stewardship of the environment. It is no surprise that Europe, more devoted to the Precautionary Principle, leading to Extended Producer Responsibility ("producer pays"), is still hospitable to refillable containers. In many European countries, refillables still make up the majority of the market.[202] My hope is that we will enact legislation to mandate a return to refillable containers in the United States to save energy, to create jobs, and to preserve scarce resources for more important uses (if it can be done in Canada, it can happen here). Incidentally, the return to refillable bottles favors local bottlers over national bottlers, so that this may be one way to prevent the concentration of all beer brewing into the hands of a few national breweries.

Bottle Bills

Eleven U.S. states (led by Oregon in 1972) have enacted **bottle bills**—laws that require a deposit be placed on beverage containers (whether glass, plastic, or metal). These laws recognize that collectively consumers can accomplish conservation, even if they cannot individually. A mechanism must be set up to repay the deposit and recycle the containers. In most cases, beverage distributors and bottlers pay the retailers and redemption centers taking the container to cover storage, staffing, and hauling costs. In states with the bottle bill, there appears to be overwhelming support for the laws. States with bottle bills recycle an average of 72% of the beverage containers sold. In Michigan, the only state with a dime instead of a nickel deposit, the redemption rate exceeds 90% of beverage containers sold.[83,183,202] In Canada, the deposit (bottle bill) provinces have over twice the recycling compliance as curbside recycling provinces (86% vs. 42%).[205]

There is also a belated realization of the huge amount of energy invested in beverage containers. In fact, as we shall see, the energy cost of the various **throwaways** is six to nine times the energy value of the beverage (see *A Closer Look*: Energy in a Beverage and **Extension 11.5, *Energy Analysis of Refilling*** for more detail).

A CLOSER LOOK

Energy in a Beverage

We can look up the heat of combustion of alcohol (C_2H_5OH) in a table in a book, such as the *Handbook of Chemistry and Physics*, and find that it is about 320 kilocalories per gram molecular mass (the gram molecular mass is known as the mol). We can also find that the mol occupies a volume of about 46 milliliters—about 1.5 oz. Therefore, for pure alcohol, we can convert kcal to kWh_t to obtain a heat of combustion of (0.28) kWh_t/oz.

A 12 oz bottle of high beer (about 4% alcohol by volume) will have an energy content of 0.14 kWh_t. Similarly, a shot of whiskey, about 1.6 oz, which is 86 proof (43% alcohol) on ice has an energy content of

$$(0.43)(1.6 \text{ oz})(0.38 \text{ kWh}_t/\text{oz}) = 0.17 \text{ kWh}_t.$$

That is, a shot of whiskey has about the same energy content as a bottle of high beer.

To compare to soft drinks, we must look up the calorie content in something like the U.S. Department of Agriculture tables: cola, 145 kcal; fruit flavor, 170 kcal; ginger ale, 44 kcal; and root beer, 150 kcal. This translates into energy contents of 0.17, 0.20, 0.13, and 0.17 kWh_t, respectively.

Typically, then, the energy content of a beverage is about 0.2 kWh_t. As seen in **Extension 11.5,** the energy content of the bottle is around or greater than 0.4 kWht.

I was a resident of Oregon when the bottle bill went into effect, and I observed the drop in litter that followed. Oregon highways became obviously cleaner than those in the surrounding states, at a reduced cost to the taxpayer. On the occasion of the thirtieth anniversary of the bottle bill, an editorial in the *Portland Oregonian* aptly summarized the situation: "After three decades, the bottle bill has done much more than help clean up litter. It has helped foster a recycling ethic that's become part of the fabric of Oregon life."[208]

Another pleasing aspect of bottle bills is the fact that enactment leads to an increase in jobs:[32,83,171,201–209] A nationwide bottle bill would produce 80,000 jobs by one estimate. It would also save our resources. Nationwide, an equivalent of 30 to 45 million barrels of oil would be saved each year, by recycling about 80% of the aluminum, glass, and plastic.[201,209]

The New York bottle bill has caused the creation of "redeemers," people who are self-employed and collect bottles and cans.[210] Redeemers may make $60 a day during the year[211] and as much as $10,000 in a summer recycling cans from Central Park.[175] Their work and the effect of the bottle bill in general saves an estimated 500 tonnes of trash per day in New York City alone (of 20 kilotonnes production). For this reason, the reaction to Mayor Michael Bloomberg's decision to try to save $57 million in the budget of the City by suspending recycling was met with general protest, and not only from the redeemers, especially when he characterized all recycling as "costly and inefficient."[212] A modified bill went into effect in July, 2002 that retained recycling of metals and suspended recycling of glass, plastic bottles, and beverage cartons.[213,214] It seems that the result was less recycling, loss of interest, and more material headed for the landfill.[215] In 2003, the City again began recycling plastic,[216] and in 2004, it began collecting glass for recycling (reinstituting the full City curbside recycling program after finding that the cancellation had cost the City more in trash hauling costs than it had saved in lower recycling costs).[217]

The bottle bill seems like motherhood and apple pie to most average citizens in states with bottle bills in operation. The public see it as "normal." Although the states do redemption differently, this seems to have little impact on public satisfaction. Some states have state facilities that buy back the deposit items. Others pay retailers to take the containers back. There are difficulties in execution in some states. For example, in Connecticut you can redeem bottles in stores only if the store sells the product even if the container has the recycling label. In Massachusetts, retailers claim that the state's handling fee is too small and return less than the stated deposit amount to customers.[218] Of course, bottle bills are no panacea, but they are a help.

The counterargument by opponents generally focuses on four issues: inconvenience, health, job quality, and energy costs of recycling. Consumers are inconvenienced because they must save bottles; local stores must provide space and staff to handle the bottles. Some merchants have problems with vermin in their bottle storage areas. Some are concerned that the bottles could constitute a health hazard, although no state health department has ever documented a health code violation due to bottle bill redemption.[183] The number of jobs overall increases with a bottle bill, but employment of skilled workers decreases; the jobs gained are not as good as the jobs lost. Some also claim that the energy cost of transporting bottles and cans for refilling or recycling is prohibitive. This last argument is analyzed in **Extension 11.5,** and the result is favorable to the case for returnables. In addition, numerous studies have shown net energy savings benefits from recycling.[183,202] The health argument seems not to be a serious problem in states with bottle bills. The inconvenience and job quality arguments have merit. In the final analysis, one must weigh the value of esthetics and reduced road hazard against inconvenience and job redistribution.

It is undeniable that the states with bottle bills have higher recycling (and refilling) rates than states without such laws. States and provinces with bottle bills were also more successful in use of refillables (**Extension 11.5**).[74,206] Among bottle bill states, Iowa is particularly committed to recycling and waste reduction.[219,220] By 1995, Iowa has reduced its waste stream by one-third from 1988 levels, and the bottle bill was a large part of that success.[188] California bottle laws are partly responsible for the state's high recycling quotient, but some is due to municipal curbside recycling.[221] California's recyclables reached a new recycling record: 5.8 billion containers in the first half of 2004 out of about 9 billion sold.[222] Overall, *lack* of bottle bills in other states led to a sad record: in 2004, 129 billion beverage containers—out of about 230 billion manufactured—ended their journey in trash cans.[223] Similarly, Saskatchewan, which has the highest deposits in Canada, also has one of the highest return rates, 94%.[224] Only Prince Edward Island, with a "producer pays" approach is higher (98% for soft drink containers, 95% for beer containers, and even 60% for liquor containers).[206]

According to Jennifer Gitlitz of the Container Recycling Institute, "Deposit systems produce bang for the buck."[225] She cites a 2002 multi-stakeholder report by the group Businesses and Environmentalists Allied for Recycling (BEAR),[226] which found that a combination of recycling methods operating in the nation's 10 deposits states recycled 490 containers per capita per year, at an average unit cost of 1.53 cents, while the nation's 40 non-deposit states (which rely on curbsides and drop-offs to do the whole job) recycled only 191 containers per capita per year, at an average unit cost of 1.25 cents. "In other words," she says, "at an additional cost of just over one and half cents per six-pack, the recovery rates in bottle bill states are more than two and a half times higher than in states without bottle bills."[225] In the absence of the refilling alternative (**Extension 11.5**), recycling makes a great deal of economic and ecological sense.

SUMMARY

The throwaway society creates immense mounds of trash. Trash is not only created by throwing away things; it is also created in the process of extracting the raw materials themselves, and at every step in the transformation of ore into finished product. The best way to reduce trash is to reduce use of packaging materials and increase reuse. This course has been followed in Germany (see **Extension 11.1**).

Reuse is much more efficient than recycling (see **Extension 11.2** for ideas on rethinking industrial resource use). The cost of aluminum and steel from recycled metal is less than 5% of the cost for ore, and for aluminum, it is considerably less. It is not so energy efficient to recycle glass, but glass recycling saves some energy and removes problematic material that can last for millennia. Many computers that are thrown away are still valuable (**Extension 11.3** gives alternatives for disposal), and recycling efforts could save energy and money. Computer supplies, laser cartridges, inkjet cartridges, etc., can usually be recycled, too, and consumers should be alert and recycle these items where possible.

Universal adoption of recycling would greatly reduce the volume of waste sent to local landfills. Recycling is very labor-intensive (see **Extension 11.4** for an account of my personal experience with recycling over many years and **Extension 11.5** for a discussion of refillable containers). At present, recycling is not usually a money source, but it does allow an offsetting of the cost of disposal in some cases, it reduces litter, and it prolongs the life of landfill by reducing the volume of waste. It also reduces the use of raw materials.

PROBLEMS AND QUESTIONS

MULTIPLE CHOICE REVIEW

1. What is the predominant material found in landfills?
   a. kitchen waste
   b. glass containers
   c. paper
   d. yard waste
   e. steel or aluminum cans
2. Which material takes the most labor to process for recycling?
   a. aluminum
   b. cardboard
   c. ferrous scrap
   d. newspaper
   e. PET
3. The energy content of a beverage is equivalent to 0.2 kWh of thermal energy, and a deposit bottle costs 0.9 kWh of electricity to make and 0.2 kWh to clean. After how many uses will the energy content of this bottle per use be equal to the energy content of the beverage?
   a. 1
   b. 2
   c. 50
   d. 500
   e. never
4. Which of the following arguments might an opponent of recycling use?
   a. Recyclables are more inconvenient than throwaways.
   b. There are health problems associated with the recycling of dirty bottles.
   c. A recycling job is of poorer quality than a bottling job.

d. Recycling has substantial energy costs.
   e. An opponent could use all of the above arguments.
5. A recycling center handles 2400 lb of steel scrap, 400 lb of aluminum cans, and 3000 lb of clear glass a certain week. It receives $0.01/lb for steel, $0.29/lb for aluminum, and $0.015/lb for clear glass. If its handling costs are $0.01/lb for steel, $0.29/lb for aluminum, and $0.015/lb for clear glass, how much money did the recycling center make that week?
   a. $-$57.60
   b. $-$32.40
   c. $0.00
   d. $31.40
   e. $91.20
6. A recycling center receives $0.01/lb for steel, $0.29/lb for aluminum, and $0.015/lb for clear glass. Its handling costs are $0.034/lb for steel, $0.062/lb for aluminum, and $0.037/lb for glass. If the recycling center handles 500 lb of aluminum and 2000 lb of steel a week, how much glass can it handle per week before it begins to lose money?
   a. 400 lb
   b. 1783 lb
   c. 3000 lb
   d. 6136 lb
   e. 7364 lb
7. Recycling steel, aluminum, and glass saves energy compared to making objects out of raw materials. Order the energy savings from most to least.
   a. glass, steel, aluminum
   b. aluminum, steel, glass
   c. steel, aluminum, glass

d. glass, aluminum, steel
e. aluminum, glass, steel
8. What material is found in municipal waste in the greatest amounts?
   a. kitchen waste
   b. glass containers
   c. paper
   d. yard waste
   e. steel or aluminum cans
9. In comparing populations and percentage of glass recovered in various cities,
   a. a city of 100,000 that recycles 10% of its glass recovers less glass than a city of 10,000 that recycles 70% of its glass.
   b. a city of 500,000 that recycles 15% of its glass recovers less glass than a city of 100,000 that recycles 50% of its glass.
   c. a city of 50,000 that recycles 50% of its glass recovers less glass than a city of 25,000 that recycles 90% of its glass.
   d. a city of 100,000 that recycles 20% of its glass recovers less glass than a city of 50,000 that recycles 50% of its glass.
   e. a city of 50,000 that recycles 25% of its glass recovers less glass than a city of 10,000 that recycles 70% of its glass.
10. Which of the following statements is TRUE?
    a. Recycling is a money source.
    b. Recycling is not labor intensive.
    c. It is more efficient to recycle aluminum than glass.
    d. Can and bottle recycling produces a degraded product.
    e. All of the above statements are true.

SHORT ANSWER
Explain whether the statement made is correct, and justify your answer.
11. Recyclables are more inconvenient for the consumer to deal with than throwaways.
12. Recycling computers is a moneymaking operation.
13. Everyone is very happy with returnable containers in states with container recycling laws.
14. In former times, most beverage containers were refillable.
15. Recycling steel is more energy efficient than recycling aluminum.
16. Recycling costs energy as well as time.
17. Municipal recycling programs are generally successful.
18. Jobs associated with recycling involve handling garbage, and so are dirtier than most comparably paid jobs.
19. All objects made from metals have much smaller cost when the metal comes from recycled material rather than ore.
20. There are health problems associated with the recycling of dirty bottles.

DISCUSSION QUESTIONS
21. Explain why aluminum cans won out in the beverage wars, even though steel cans cost less energy per use to make.
22. What is the reason that glass is less of an energy saver than steel or aluminum if it is recycled? Explain your reasons for your answer.

23. Explain why plastic is currently so uneconomic to recycle. Give specifics.
24. Magazine paper is shiny because it is coated with clay. Why would it be wise to separate magazines from newspapers?
25. Many people think that it's no big deal to separate their recyclables, while others think the opposite. Imagine being in both these persons' minds, and discuss how recyclable separation could seem easy or hard to that person.
26. Why is it possible for markets for recyclables to fluctuate wildly? Explain how this could occur. Would we expect the swings to be bigger for recyclables than for ores or smaller? Explain your reasoning.
27. Why do the Germans find it so much more expensive to produce an aluminum beer can as opposed to a glass bottle?
28. Explain reasons why the environmentalist's mantra is "reduce, reuse, recycle." Is this a proper order? Is there a downside to following this motto?
29. How can a focus on the original cost of a product distort economic decisions?
30. Examine flow control from your own perspective. Is it justified, or should the free market reign? Explain your reasons clearly.

PROBLEMS
31. How much glass does an average person use every year?
32. Scrap steel in the United States is accumulating at 11 million tonnes per year. How much is this per capita? What proportion does this make up of the average amount of material each person throws away?
33. After how many uses would the amount of energy in the beverage of a returnable aluminum can (if such a thing could exist) be equal to the energy content per use of the can?
34. Using Table 11.2, determine the growth rates (in percent) in the mass of the various components of garbage listed.
35. Using Table 11.6, determine which material saves the greatest amount of energy.
36. The heat of combustion of alcohol (C_2H_5OH) is 1.01 MJ/oz. How much energy is there in a 32 oz bottle of beer that is 5% alcohol?
37. The total amount of waste generated in the United States doubled between 1950 and 1978. What is the average annual rate of increase of waste generated?
38. The energy content of a certain beverage in a bottle is 1.15 MJ, and a deposit bottle costs a total of 5.4 MJ to make and 1.08 MJ to clean. After how many uses will the energy content of the bottle per use be equal to the energy content of the beverage?
39. The energy content of a certain beverage in a bottle is 1.02 MJ, and a deposit bottle costs a total of 5.4 MJ to make and 1.08 MJ to clean. After how many uses will the energy content of the bottle per use be equal to the energy content of the beverage?
40. The energy content of a certain alcoholic beverage in a bottle is 15 MJ, and a deposit bottle costs a total of 5.4 MJ to make and 1.08 MJ to clean. After how many uses will the energy content of the bottle per use be equal to the energy content of the beverage?

FOSSIL FUEL ENERGY
RESOURCES AND
CONSEQUENCES

PART V

CHAPTER 12
Fossil-Fuel Resources

In this chapter, we continue the examination of resources available for production of energy by considering the long-term possibilities for fossil fuels: petroleum, coal, and natural gas, and conventional alternatives such as shale oil. All fossil fuels are present in finite amounts. How do we predict how long a supply will last? When does it make sense to use coal in preference to oil? This chapter attempts to address some of these issues.

KEY TERMS

fossil fuel / hydrocarbons / hopanoids / production rate / logistic curve / normal curve / primary recovery / secondary recovery / tertiary recovery / finding factor / kerogen / block carving / longwall mining / fluidized-bed combustion / power gas / synthetic natural gas

12.1   HISTORY

Surface oil and asphalt deposits have been used since ancient times for medicinal purposes. In the late Middle Ages, the British began to use their coal resources for energy—the first large-scale use of fossil-fuel energy. This began a march of ever more intense use of stored energy to make human life easier.[1] They were forced to do this, because the profligate use of wood as fuel for heating and for the cottage industry had reduced the once magnificent forests of the British Isles to pitiful remnants. Nevertheless, it is safe to say that the total use of fossil-fuel resources was negligible prior to 1800.

Figure 12.1 shows how use of coal and oil has grown over the past 200 years. Compare this world use history to the use in the United States, shown in Figure 12.2. Coal use grew at an average of 4.4%/yr in the period from 1860 to 1913, and at a lower average, 0.75%/yr growth rate, in the period from 1913 to

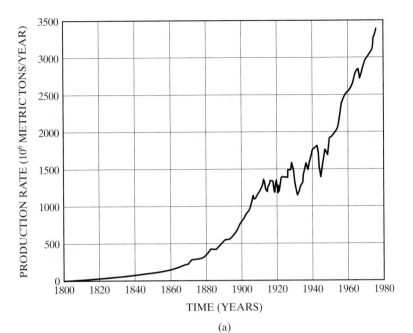

(a)

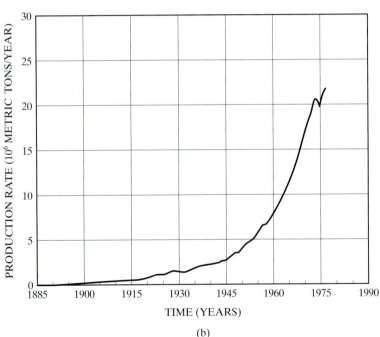

(b)

FIGURE 12.1

(a) World production of coal and lignite as it varied between 1800 and 1982
(M. K. Hubbert, Figure 3, in *U.S. Energy Resources as of 1972*, U.S. Senate Committee on Interior and Insular Affairs, document 93-40 (Washington, D.C.: Government Printing Office, 1974) and U.S. Department of Energy, *Coal Data: A Reference* (Washington, D.C.: Government Printing Office, 1985)]
(b) World production of crude oil as it varied between 1880 and 1977 [M. K. Hubbert, "The world's evolving energy system," *Am. J. Phys.* **49**, 1007 (1981), Figure 3. Reprinted with permission)

1945—two major wars and a depression occurred during this time—and has grown at a 3.6%/yr growth rate from 1945 onward (a doubling time of 20 yrs).[2] Note that U.S. coal use lagged somewhat behind the total world use; this happened because the wood economy was able to persist longer than in developing countries elsewhere. U.S. oil use led the world in development, because, by the late nineteenth century, the United States was already an industrial power, and industry was developing in the coal-energy-poor northeast part of the country. World crude oil use (crude oil is often measured in barrels, bbl; the barrel has an energy content of about 6 GJ) has doubled every decade (a 7%/yr growth rate).[3]

FIGURE 12.2

(a) U.S. production of coal and lignite as it varied between 1800 and 1971
(M. K. Hubbert, Figure 5, in *U.S. Energy Resources as of 1972*, U.S. Senate Committee on Interior and Insular Affairs, document 93-40 (Washington, D.C.: Government Printing Office, 1974)]
(b) U.S. production of crude oil as it varied between 1880 and 1971
[M. K. Hubbert, Figure 6, in *U.S. Energy Resources as of 1972*, U.S. Senate Committee on Interior and Insular Affairs, document 93-40 (Washington, D.C.: Government Printing Office, 1974). Reprinted with permission)

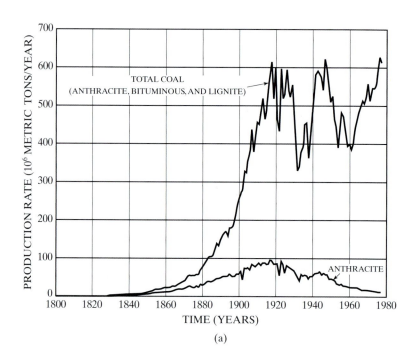

(a)

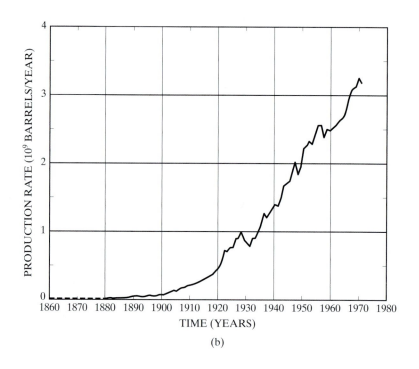

(b)

These **fossil fuels** are used in small vehicles for transportation of goods and people (Chapter 15), and in large power facilities to supply industrial steam and electrical energy. The burning of fossil fuel in a boiler is described in Chapters 7 and 13. In the automobile, **hydrocarbon** fuels refined from fossil-fuel resources provide energy for transportation. A fuel (gasoline or diesel fuel) is mixed with air, introduced into a cylinder, and then ignited by a spark device in conventional vehicles or its own heating by compression for diesels. The resultant explosion increases the pressure of the gas in the cylinder as the chemical energy in the fuel is transformed into kinetic energy of the gas molecules (see **Extension 7.6,** *Real Power Plants*). Work can be done, and the piston inside the cylinder is driven, pushing on an eccentric crankshaft device (cam) that ultimately couples to the wheels.

HISTORY OF ENERGY

The Origin of Coal, Oil, and Natural Gas

No matter which of the fossil fuels we concentrate on, we find we are using it at a rate so that it will last mere centuries at most, while it takes on the order of a hundred million years to develop new fossil-fuel resources. Most Eastern U.S. coal was formed 280 to 320 million years ago. Most Western U.S. coal was formed about 60 to 140 million years ago. Much of the world's oil was formed about 500 million years ago. The future is clear: The supply of fossil fuels is small and will soon disappear.

Coal results from geological processes that involve the burying of vegetation under anaerobic conditions in swamps. There, it becomes peat and is overlain gradually by rock, raising the temperature and pressure of the partially-oxidized organic matter. Petroleum forms from more dispersed organic matter, such as organic sediment on a continental shelf that is buried by geological processes and, as with coal, subjected to a high-pressure cooking deep underground. After millions of years, in either case, the materials are what we call coal and petroleum. Coal is a rock; petroleum (literally, rock oil) inhabits tiny pores in underground rock, such as sandstone. Under certain conditions, oil can migrate through cracks and form large underground pools.

All nongaseous fossil fuels are made up of complicated molecules with backbones of many carbon atoms. Over 500 different chemicals have been identified in crude oils.[4] Analysis of samples from many different parts of the world led researchers to a surprise—all samples contained similar profiles of complicated carbon molecules containing from 27 to 35 carbon atoms each.[4] These compounds, characteristic of all fossil fuels, seem to have been ubiquitous among precursors of modern fossil fuels. Compounds of mass ≈191 atomic mass units [u, see Chapter 18] and 35 or 36 carbons, called **hopanoids**, were implicated somehow in the formation process. Further research isolated such compounds in primitive bacteria and blue-green algae living today.[4,5] The compounds are probably early ancestors of the cholesterol produced in animals (which has 27 carbons), and their presence means that all fossil fuels owe their existence to bacterial and algal biological processes.

Oil tends to have come primarily from marine sources of carbon (86%), with smaller amounts from deltas (11%) and lakes (3%).[6] These precursors of oil do not come from the abysses, but rather from ancient shallow waters.

Natural gas, or methane, CH_4, is formed in association with coal and oil by similar processes (though they may be much more rapid). The difference is that coal remains in place, and most oil migration is halted by caps impermeable to liquids. See **Extension 12.1,** *Finding Oil and Gas* for more information. Sometimes, these caps are permeable to gas. In any case, the gas migrates through pores in the rock containing coal or oil and either gathers in large volumes above the oil (because it is less dense) or migrates such great distances that it is no longer associated with the oil or coal deposit that produced it, and the gas trapped in a formation may have come from many sources. Gas is found alone in large reservoirs underground, in association with oil (it is produced as a byproduct) or in association with coal beds (where it may migrate through fractures or remain trapped within the coal). Canaries were once used in coal mines to warn miners of danger. The canaries would die of the gas before it could kill miners; miners could then leave before they were killed.

Dukes calculated (with an admittedly large uncertainty) the amount of organic carbon that was necessary to form the fossil energy reserves we exploit.[6] He finds that the amount of organic material that the world used as fossil energy in 1997 totaled 4.4×10^{19} g C, over 400 times Earth's net primary productivity (the total production of photosynthesis). Only 0.81% of the original store of carbon at a coal deposit eventually becomes coal, 0.094% at an oil deposit eventually becomes oil, and 0.085% eventually becomes gas.[6] Most of the rest is lost to the atmosphere through decomposition. That means that one liter of gasoline represents the end point of 23 tonnes of ancient buried organic matter (one gallon, 89 tonnes). Dukes further calculated that "replacing the energy humans derive from fossil fuels with energy from modern biomass would require 22% of terrestrial NPP, increasing the human appropriation of this resource by ~50%."[6]

The Aftermath of the Energy Crises of the 1970s

The United States (and the world at large) had gotten used to cheap crude oil. Such low cost led to the massive importation of foreign oil, even in the oil-rich United States. After the first energy crisis, the oil proportion dropped, and coal and nuclear rose (Figure 12.3a).[11] The price increases of the 1970s were triggered by a relative oil scarcity in the United States. The increases caused lowered demand,[12–14] which, in turn, impelled conservation measures (Figure 12.3b).[14] (See Chapter 17 and **Extension 12.2, *Politics and Fossil Fuels*.**)

FIGURE 12.3

(a) U.S. energy consumption by source, 1949–2000. The nuclear component has been multiplied by 10 to make the trend clearer. [Taken from or based on various tables in U.S. Department of Energy, Energy Information Agency, *Energy Annual 2000* (Washington, D.C.: GPO, 2001)]

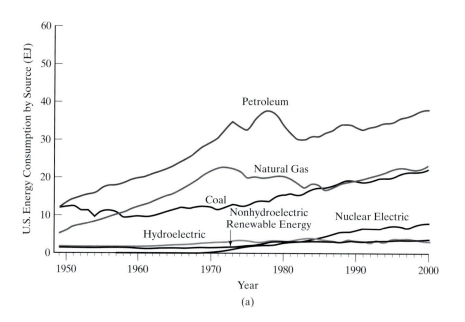

(a)

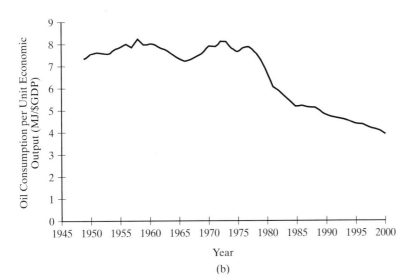

(b)

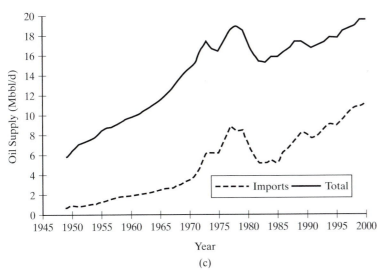

(c)

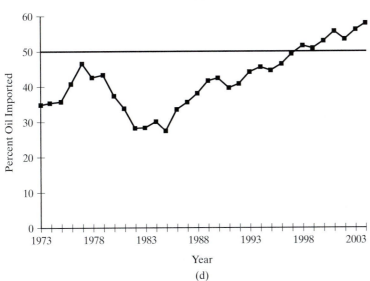

(d)

FIGURE 12.3 (Continued)
(b) Oil consumption per GNP dollar is given from 1949 to 2000.
(c) Oil supply, 1949–2000 including imported oil.
(d) Percentage of oil imported into the United States.
[Taken from or based on various tables in U.S. Department of Energy, Energy Information Agency, *Energy Annual 2000* (Washington, D.C.: GPO, 2001)]

By 2000, the total consumption of petroleum products was 19.7 million barrels per day (Mbbl/d), primarily used in the transportation sector of the economy (Figure 12.3c).[11] The United States imported about 11 Mbbl/d that year, over half the total use (Figure 12.3d shows the percentage of imports), not that much more than the 7.95 Mbbl/d used for gasoline (a bit over 1.6 Mbbl/d is used for jet fuel).[11] Gasoline use alone dwarfs the 5.83 Mbbl/d of crude oil produced by the United States, and gasoline and jet fuel together about equal total U.S. production of all petroleum products.[11] Other oil-importing countries have recognized their vulnerability to a cutoff of supply and have taxed imported oil to dampen consumption. The U.S. government has taken a more *laissez-faire* attitude, and consumers (especially contractors) have taken this as an invitation to waste oil.

According to the economic view, the oil price had to fall again as it did in 1986, because a glut had built up due to increased conservation measures in response to the price rises following the oil crises[14] (see Figure 12.4 for droll comments on

FIGURE 12.4

(a) A 1985 cartoon comments on the political demise of OPEC in the mid-1980s.
(Dana Summers, © 1986, Washington Post Writers Group, reprinted with permission)
(b) and (c) Cartoons from 2000 comment on the oil price rise of 2000.
(Jim Borgman, © 2000, Cincinnati Enquirer)

(a)

(b)

FIGURE 12.4 (Continued)

(c)

these changes). For more detail on these effects, see **Extension 12.3,** *Consequences of Cheap Oil*.

Energy Projections

Energy use for the last half century was shown in Figure 12.3a. Note that all four major sources are increasing steadily (although the nuclear component will level off, because no new plant has been ordered since the late 1970s). Everyone agrees that, barring some spectacular happening such as the 1970s energy crisis, petroleum and gas use will continue to increase. Table 12.1 shows projections of U.S. oil consumption in 2010. These may be compared to the data shown in Figure 12.3c. Note how much lower the Wharton's and Department of Energy's 1991 projections are than the others. (They were made when the rates of increase in consumption were lower than those of today.) Figure 12.5 shows how much lower the 1984–2000 trend is than in earlier stages of growth, and also how use of a steeper trend would affect predicted consumption.

Uncounted Energy Costs

There are many hidden costs of energy. For example, oil costs did not include the costs of defending the oil states or the Persian Gulf war with Iraq. The cost to support Kuwait was estimated to add $23.50 per barrel above the market price (around $18 at the time) to the real price of a barrel of oil in 1991.[69] At many wells, gas produced with oil is flared (burned) to eliminate it. Also not counted in the energy cost the consumer pays are tax credits given to oil producers, the costs of fossil-fuel research, health consequences of combustion (see Chapters 14 and 15), and so on.[69]

Old oil and gas wells are recognized as sources of pollution. Wastes seep from drilling pads, drums leak, and wastes are disposed of in unlined pits.[70] Brine pumped out of wells contains radioactive materials and lots of salt.[71] Pumped-out uncapped wells can cause salinity problems for the soil. Texas alone has about

TABLE 12.1

U.S. oil consumption projections for 2010

Source	Consumption estimate (Mbbl/d)
1950 Actual use	6.46
1990 Actual use	16.99
2000 Actual use	19.48
1984 to 2000 trend continued to 2010	22.0
1949 to 2000 trend continued to 2010	24.3
Wharton Econometric Forecasting Assoc. (1991)	19.97
Department of Energy (AOOG1991)	20.27
Data Resources, Inc.	21.25
Gas Research Institute	23.24
National Energy Policy Development Group (2001)	22.5

Source: References 11, 17, 68.

FIGURE 12.5

Oil consumption, 1949 to 2000, showing how the trends from 1949 to 2000 and 1984 to 2000 affect predicted consumption in 2010. (Actual: U.S. Department of Energy)

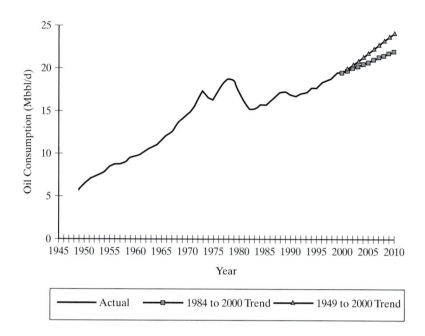

50,000 unplugged wells; Oklahoma has 15,000 to 20,000 unplugged wells. Starting in 1990, Texas and Oklahoma set up a fund to cap old wells, paid by taxpayers, not by the users of the oil.[71]

There are consequences in transportation of oil as well. Large oil spills such as from the *Torrey Canyon*, the *Amoco Cadiz*, and the *Exxon Valdez* foul coasts (see Chapter 15). If the total real costs of oil production and use were paid, the price of oil would certainly be much higher.

12.2 OIL AND THE PREDICTION OF DEPLETABLE RESOURCES

The Hubbert "Blip"

At the beginning of the chapter, we presented geologist M. K. Hubbert's data on coal and oil production. You might have noticed in Figure 12.2a that anthracite coal exploitation followed a curve that rose from zero to a maximum and fell rather symmetrically toward zero again. Hubbert realized that this should be true generally when a finite resource is exploited.[2,3] The amount of production must be zero at two times: *before* exploitation of a resource and *after* exhaustion of the resource. In between, a certain amount is produced, and the time over which it is produced depends on the **production rate** (the amount that is produced per unit time). Of course, this rate of production depends on human decisions, politics, economics, and the like.

Suppose for a moment that the rate of production is constant, say, 1 million barrels of oil per day. Then in one day we'd produce 1 million barrels (1 Mbbl/d × 1 d = 1 Mbbl), in two days, 2 million barrels, and so on. So the amount produced, ΔP, in a certain time interval Δt is (see Figure 12.6)

$$(\text{rate of production}) \times \Delta t.$$

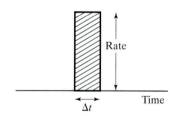

FIGURE 12.6

Production equals rate times time interval. The shaded area is equal to ΔP, the amount produced.

The shaded rectangular area in Figure 12.7 is just this amount, ΔP. For a time twice as long, we would get twice the area. Of course, real rates of production are not constant (Figures 12.1, 12.2, 12.5, and 12.6), so the rectangles will not all be the same size. Even so, we can add up areas just like those in Figure 12.6 for all the changing rates and their corresponding time intervals Δt to find the total area beneath the rate of production curve. The total area under the rate of production curve is the total amount of resource produced.

The reasoning described in **Extension 12.4, Hubbert and World Oil Production,** allows us to predict how much energy it will ultimately be possible to extract; that is, how much fossil fuel there is in some area. The method used in these predictions was originally derived by M. K. Hubbert, who worked at various times for Shell Oil and the U.S. Geological Survey. Hubbert began with this (perhaps obvious) point that what is used today had to have been discovered prior to today. Hubbert recognized that in the early years of production, utilization is proportional to the amount produced, and discovery is proportional to the amount hitherto discovered (which leads to an exponential rise in both production and utilization). If the time lag between first discovery and resource development is known, we can find the production from the rate of resource discovery. When the discovery rate drops, because more of the original resource has been used, production will follow (after the time lag). Now the rate of discovery must decrease, because it can only find what has not already been discovered. The cumulative amount produced starts exponentially and then turns over and approaches the final amount produced slowly—it is the **logistic curve** introduced in Chapter 5.

Now, amount = rate × time interval; if this is extrapolated, or projected ahead, observations related to the discovery rate of energy resources lead to the production rate of the resource, which leads to a prediction for the total area under the curve—the total energy resource that will have been produced by the time the resource is exhausted. (This method could also apply to rates of commodities use, or rates of industrial activity, or other such kinds of activity.)

Using his method, Hubbert predicted that U.S. oil production in the Lower 48 would peak in 1969 or 1970. In fact, 1969 *was* the year of maximum production from domestic wells;[72–74] Alaskan production peaked in 1988. Hubbert's reasoning turned out to be correct. Hubbert also predicted that the ultimate amount that would be produced in the Lower 48 would be $Q_\infty = 171$ billion barrels (Gbbl) of crude oil.

Let us denote the rate of production of a resource by P and the amount produced by Q. Thus, we can say

$$P = Q/\Delta t$$

or

$$Q = P\Delta t,$$

as we found earlier. The ultimate amount produced we will call Q_∞; it will be greater than (or equal to) the estimate of Q from geological information. The production curve may assume any number of shapes. However, our latitude of choice is reduced, because the technology of production mandates an exponential growth of production at the start of exploitation. One curve with these properties is the bell-shaped, or **normal curve**, illustrated in Figure 12.7. We will assume here that this normal curve is the actual production curve.

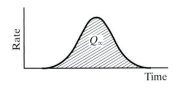

FIGURE 12.7

A normal curve.

Figure 12.8a shows the actual U.S. production data for crude oil (compare to Figure 12.2b) along with the corresponding normal curve. Domestic output, exclusive of Alaska, has been dropping since 1969. Production curves are shown in Figure 12.8b for the Lower 48 and Alaska. Note that we can say with pretty fair certainty from Figure 12.8a that about three-fourths of all oil that will *ever* be

FIGURE 12.8

(a) Data on U.S. oil production through 2000 show that the normal curve was an appropriate choice to describe domestic oil production. My best fit has Q_∞ of 216.5 Gbbl, a maximum production occurring in April 1975, and a standard deviation of 26.65 yr (the standard deviation is a measure of the width of the curve, which is here 63 yr). These numbers agree well with Hubbert's original predictions as well as with those of others who worked on this problem.
(b) Data for total U.S. oil production for 1953 through 2000 showing detail of the supply. Alaska oil is shown separately as the dotted lower curve.
(Energy Information Agency, Department of Energy)
(c) Data compared to the normal curve if it is assumed that Q_∞ = 285 Gbbl.

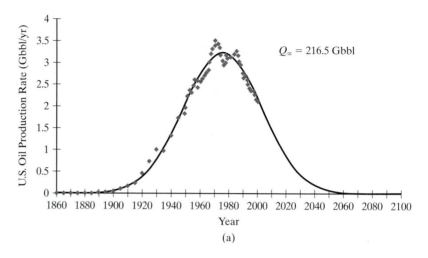

(a)

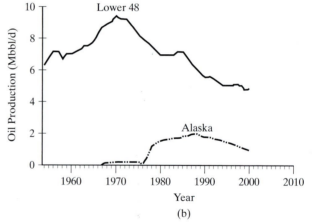

(b)

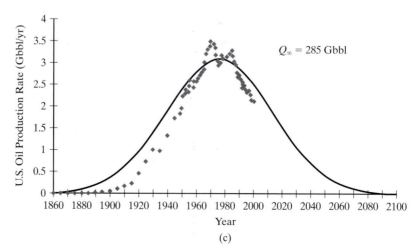

(c)

produced by the United States *has already been produced*. Note that the value of $Q_\infty = 216.5$ Gbbl of crude oil is the current prediction for all the oil that will ultimately be produced.

The U.S. Department of Energy estimates the remaining "Technically Recoverable Petroleum Resource" for the United States at 124.736 Gbbl.[11] When this is added to the cumulative production of approximately 160 Gbbl, one may see that the total is around 280 to 290 Gbbl, substantially greater than Hubbert's analysis would suggest (Figure 12.8c). Clearly, it does not describe the data as well as the value $Q_\infty = 216.5$ Gbbl, which suggests that only around 50 Gbbl are yet to be recovered from the United States. Three other common estimates, those of *Oil & Gas Journal* (21.8 Gbbl),[75] *World Oil* (21.8 Gbbl),[75] and BP Amoco (28.3 Gbbl)[76] agree better with the Hubbert analysis but are lower.

Expert opinion is that the ultimate amount of world oil still available to be recovered, 1 Tbbl, is just about the amount that has already been produced.[74–76] It seems likely that world oil production will hit its peak around or before 2010, and the ultimate amount produced will be in the neighborhood of 2 to 3 Tbbl. See **Extension 12.4** for more detail on Hubbert's methods as applied to world oil resources. Of course, there is a huge volume of fossil-fuel resources left, and there still will be when oil runs out.

The Hubbert method's weakness is that the phenomena it analyzes strongly reflect human activity, which is influenced by economic and political factors. McKelvey,[72] in particular, has criticized this method and prefers to extrapolate observations relating to the abundance of a mineral in its geologic environment. One such method used in dealing with oil exploration (the Zapp hypothesis) is discussed in the **Extension 12.4**. These methods are purportedly better capable of allowing for major breakthroughs in technology that would lead to increased recovery. See **Extension 12.5,** *Increasing Recovery,* which discusses the various levels of recovery (**primary recovery**, the simplest, **secondary**, and **tertiary**, the most complex and expensive). However, Hubbert's methods have stood the test of time and continue to be used.

Hubbert's method obviously may be generalized to other fossil-fuel resources, and it has been accepted by the government as its method of estimation. Using a modified Hubbert-style method, the estimate of remaining undiscovered (technically recoverable) crude oil in the United States is 51 ± 12 Gbbl (44 ± 21 Gbbl, excluding areas of restricted access, such as wildlife refuges and other such areas; see **Extension 12.4** and Reference 68).

A CLOSER LOOK

Other Estimates of U.S. Reserves

Depending on whose estimates are believed, the United States could run out of oil sometime between now (1998 was the lowest estimate) and 2075 (highest estimate) at 1975 rates of consumption.[112] The Workshop on Alternative Energy Strategies predicted in the early 1970s that world production would fail to meet demand sometime between 1990 and 1997.[113] The group was premature in this prediction; the result will still occur, but in the more distant future (most oil will probably be gone by about 2070 unless great changes occur).

A National Academy of Sciences (NAS) study[112] predicted remaining reserves as of 1975 to be 113 Gbbl of oil and 15 Tm^3 [530 Tft^3] of gas.

FIGURE 12.9

The geographic distribution of North American oil reserves.

(Gas volume is measured in cubic meters, about 38 MJ worth of energy.) The U.S. Geological Survey predicted 200–400 Gbbl of oil and 28–56 Tm^3 [1000–2000 Tft^3] of gas.[114] To put this in some perspective, British probable reserves as of this same time are about 13.6 Gbbl from its North Sea oil finds.[115] The U.S. Geological Survey has, in recent times, given estimates more in line with Hubbert's projections.

The major disagreement involves the estimate of the **finding factor**—the amount of oil ultimately available as a result of drilling in barrels per foot drilled. (The Zapp hypothesis concludes that it is 1; Hubbert concludes that it is about 0.1.) The argument against the Hubbert–NAS view is that they are essentially *economic* rather than *geologic*. In fact, some critics have claimed that as much as 485 Gbbl is yet recoverable.[116] A sampling of estimates is shown in Table 12.2.[84,101,117–119] Figure 12.9 shows the areas containing oil resources in North America. Predictions of total American reserves include oil and gas from the continental shelf. In 1981, the U.S. Department of the Interior estimated that about 27 Gbbl of oil would be found offshore (about 40% of all undiscovered oil).[120] By 1985, 100 dry wells had been sunk offshore, and the Department of the Interior had reduced its estimate to 12.2 Gbbl of oil.[99] R. Nehring of RAND Corporation estimated that offshore wells will produce only 3.5 Gbbl of oil.[84] Alaska has between 14 and 19 Gbbl of oil.[84]

For comparison, Masters and Root[121] suggested identified world reserves of 922.1 Gbbl, undiscovered reserves of 542 Gbbl, futures (reserves and mean undiscovered oil) of 1464 Gbbl, and ultimate world reserves of 2074 Gbbl. Another recent prediction of total world oil remaining is just under 1 trillion bbl.[122]

TABLE 12.2

Estimates of ultimate oil recovery in the United States

Reference	Oil in place (billion barrels)	Ultimate reserves (billion barrels)	Undiscovered reserves (billion barrels)
Hubbert (Lower 48)[2,3]		190	20
National Petroleum Council	727	228	48
American Association of Petroleum Geologists	824	257	78
U.S. Geological Survey[117]		1718	508
Doescher[101]	460	148	
Halbouty and Moody[65]		987	
Department of Energy, 1987[a] [119]	28.4		40.6
BP Amoco[76]		29.7	
Department of Energy, 1990[a] [123]	26.8		71.3
Fainberg consensus[a] [123]	84		46
Department of Energy, 2000[a] [11]	64		78

[a] Undiscovered recoverable reserves have been estimated by subtracting the present cumulative production of 142×10^9 bbl and the expected reserves of 38×10^9 bbl from ultimate recoverable reserves if unavailable from the source.

Because of America's (profligate) increases in demand from 2.6 to 4.3 Gbbl/yr between 1955 and 1975,[83] there has been unprecedented exploration after the advent of the first oil crisis (Figure 12.10). Even so, the number of drilling rigs has dropped dramatically since 1980, when about 4000 were operating, to under 1000 by 2000.[11]

Because the United States is the best-explored region in the world, and the region with the most experience in extrapolating production, there is less certainty

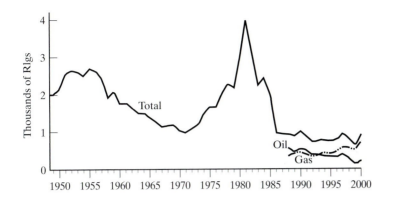

in applying this experience to estimation of world energy resources. Therefore, estimates of world oil resources are even less trustworthy than estimates of American discoveries. Even though the ground in the United States has been studied and drilled extensively, there are great disparities among U.S. estimates, as we have seen in Table 12.2. (Refer to **Extension 12.4**.)

More clues can be found in the history of oil in the United States. The peak in discoveries in the 1930s coincides with the great number of giant fields discovered. Reserves peaked in the 1960s and began to decline exponentially following the exponential decline in discoveries[81,86] by the delay time (see **Extension 12.4**). Peak oil production occurred in 1969. Oil production had fallen so far by 1991 that *total* U.S. energy production began to fall.[11]

World reserves have stopped growing, and it is unlikely that the finding rate of the early 1970s (15 Gbbl/yr) can ever be attained again.[80] One intriguing possibility is that cooler oil reservoirs, which had been thought to have been made worthless by bacterial action, might have been sterilized at earlier times before uplift and formed an overlooked large source of oil.[103] Worldwide reserves are beginning to fall; the U.S. Department of Energy estimates 697 billion barrels of proved reserves worldwide.[66] The former Soviet Union, another large oil producer, saw its production of oil peak in 1985.[124] The lack of growth is a problem, because it is not possible to recover more than 10% of reserves in any year without reducing the amount ultimately recoverable.[59]

The Hubbert analysis, applied to the exploitation of petroleum over written history, 2500 B.C. to 2500, is shown in Figure 12.11. Note how small a time is represented

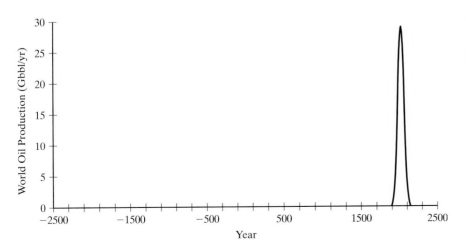

FIGURE 12.11

The "Hubbert blip." The world production of oil in historical perspective. We have assumed $Q_\infty = 3$ Tbbl for showing the "blip."

by the age of oil. While the age of coal may be a bit wider, as we shall see, it too represents just a small part of human history. This is the Hubbert "blip."

The U.S. Gasoline Shortage in the 2000s

The gasoline shortage in the 2000s appeared suddenly, when the price jumped from around $1.40/gal to well over $2.30 in some areas.[125,126] Many in the public attributed the jump to greedy gouging by the oil companies. That may have played a role, but the ensuing studies were inconclusive.

The shortage apparently developed because of a combination of factors. The number of refineries has decreased, because the margins on production were so small in the 1990s. In 1981, the number of refineries was 324, while by 2000, the number was 158.[11] Fewer refineries mean more concentration, and as with many industries, the merger mania led to consolidation and plant closures. Also, because refineries produce all grades of product, if there is a need for, say, home heating oil, the amount of heating oil is increased, and the amount of gasoline is decreased. That was, in fact, the case; the winters of 2000 and 2001 were colder than during the preceding decade. In addition, because there are fewer refineries, the ones operating are more likely to be operating all the time, meaning that needed repairs are sometimes put off as long as possible. When the refinery has a problem, it might have to shut down for an extended period for repairs.[126]

In addition, because of pollution control regulations, there are several formulations of gasoline that must be sold in certain markets. The more different grades of gasoline produced, the more complex the refining and distribution. To learn about more details of the refining process, see **Extension 12.6, *Refineries***.

12.3   IS THERE SUCH A THING AS ENERGY INDEPENDENCE?

Before the energy crises of the 1970s, developing countries as well as developed countries were hooked on cheap energy. They were, however, much less able to adjust to the price rise than the more developed countries, and we had quite a time, even in the United States, in coping with the gigantic reallocation in capital from oil consumers to oil producers. The oil price rises were responsible for a 10-fold jump in the trade deficit of the developing countries in the decade 1973–1982, as well as the sixfold increase in national debt.[127] Because the systems of all countries, developed and undeveloped, came to depend on oil as an essential commodity, it has been extraordinarily difficult to switch people's unconscious expectations. People want the status of car ownership, and automobiles need to have gasoline for fuel.

All these considerations taken together indicate that President Richard M. Nixon's Project Independence, the immediate postcrisis effort to insulate American consumers' costs from the rest of the world, would not have been able to end American dependence on imported oil and (eventually) gas. As was pointed out by President Jimmy Carter (1977–1981), only conservation of our scarce reserves and resources has a chance of protecting the United States and Canada from total dependence on imports at some future time. Carter's plan would have emphasized coal production, as we will see later in this chapter, because of the vast quantities of coal that we still have, and renewable energy (discussed in Chapters 19 to 25).

No country of any reasonable size in the world is insulated from oil addiction, and it is just as impossible to go "cold turkey" from oil as it is from an addictive drug. All countries must find their own alternative energy sources, be they coal, or geothermal, or energy from biomass, or alternate means of transportation. Whatever one's idea of estimates of total oil reserves, it is absolutely clear that oil will run out, for all practical purposes, by 2050 at the latest, unless civilization itself collapses. The alternatives to oil, whatever they are, will have to be found.

The United States established a Strategic Petroleum Reserve in salt caverns hollowed out in Louisiana, which is designed to hold between 750 Mbbl and 1 Gbbl of oil, eventually. The reserve has continued to be filled, in spite of opposition from oil-producing countries, at a rather slow rate since its beginning in 1977 as shown in Figure 12.12.[128] This reserve should be of some help in delaying interference in

(a)

FIGURE 12.12

(a) The salt cavern at Weeks Island, Louisiana, part of the Strategic Petroleum Reserve storage site, which was established after the first energy crisis.
(b) The Strategic Petroleum Reserve has been filled gradually.
(U.S. Department of Energy)

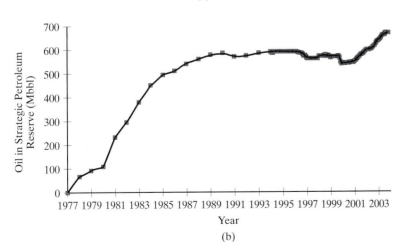

(b)

internal American politics by the oil producers. A test sale from the reserve was held during the period of the Persian Gulf war and again when prices rose steeply in the summer of 2000.

Some observers, such as the politician Ross Perot, have advocated a national tax on gasoline or on oil consumption (a gas tax, Btu tax, and a carbon tax were considered by President Clinton before the Btu tax was proposed to Congress, only to be voted down) similar to what has been done in other countries. The purpose of the tax is to encourage substitution of other fuels where possible, to prevent an undesirable shift back toward gas-guzzling cars, and to make an attempt to have consumers pay the actual full cost of fuels.[129] Such a realistic tax is unlikely to be instituted, despite the prudence of such a course, because of a commitment to the "free market" concept in political thought. This was proven in the budget debate of 1993, in which President Clinton's proposed Btu tax was transformed into a tiny gas tax increase (less than 5 cents per gallon). The decline in consumption as a result of the price increases of the 1970s, however, proves that eventually, as the price moves high enough, demand drops. We may hope that politicians will develop their ability to "think outside the box" and see past immediate consequences if they are to gain the conviction that leads to the courage to act responsibly.

Shale Oil, Tar Sands and Oil Mining, and Increasing Recovery

Shale Oil

Another alternative for increasing the reserves of oil is the utilization of the hydrocarbon compounds (called **kerogen**) in oil shale. Kerogen (a solid hydrocarbon) is found in the process of oil formation; it is partially cooked petroleum. The oil shale in Colorado was reportedly discovered to contain fuel resources by a local settler who used the shale to construct a fireplace and chimney. This rock's oil caught fire; the settler's first fireplace fire was his last in *that* house.

Some of the consequences attendant on exploitation of oil shale are studied in **Extension 12.7, *Shale Oil Exploitation***. Water is needed in staggering quantities, and waste dumps have a greater volume than the originals. But, the amounts of energy in the shale oils are also staggering. The question is whether the oil can be extracted at competitive costs; so far, the answer has been negative.

Tar Sands and Oil Mining

There are also heavy oils in oil sands in various locations around the world. The amount of oil locked up as bitumen in oil sands is staggering. Estimated reserves in Venezuela are somewhere in the neighborhood of up to 1200 to 3000 Gbbl; in Canada and the former Soviet Union, over 1000 Gbbl; and in the United States, about 175 Gbbl.[136] The tar sands of Utah contain 53 Gbbl of oil.[102] For the sake of comparison, the Prudhoe Bay oil find contained about 15 Gbbl and ANWR about 3 Gbbl. In Venezuela, "Orimulsion" a heavy oil from the Orinoco tar sands, is being marketed.

Oil mining commonly refers to lifting the oil-bearing sands out of the ground (much as coal is mined in strip mines) and bringing it to a giant distillery that separates the oil from the sand. The Athabasca oil sands in Canada contain 869 Gbbl of oil, of which under 10%, 74 Mbbl, is readily available in an area of surface strip-mining.[137] Of this, about 300 Gbbl of the total can be mined with current technology.[135,138] At present, about one barrel of oil per tonne of sand is being produced.[138] The biggest producers as of 2004 are Syncrude, at 230 kbbl/d and Suncor, at 233 kbbl/d.[88] The Surmont project south of Fort McMurray, Alberta, is estimated to have 5 billion barrels of oil recoverable from the sand, a size similar to the spectacular oil finds in Texas in the last century.[139]

The process of separation used builds on the process invented by Karl Clark of the Alberta Research Council in 1920: add bitumen, caustic soda, and hot water, and the oil floats to the top. Obviously, the oil mining process costs a lot of water and energy and produces tailings (all that rock has to be disposed of).[135] The wash water also contains "fines" that don't easily settle out and must be held in ponds for periods of decades. In addition, a lot of carbon dioxide and sulfur dioxide are released during production, so the environmental CO_2 burden of mined oil sands is more similar to coal.

The main mining problems encountered in Canada are abrasiveness of the sand (mostly quartz) and the ferocity of the Canadian winters. Large chunks of steel can shatter as machines try to loosen and transport oil sands from frozen windrows to the refinery.

During the era of extremely low oil prices, oil sand extractors worked effectively to reduce prices, so oil extraction from the oil sands is economically feasible even when the oil price is as low as $20/bbl.[88] The oil price rise of the late 1990s and early 2000s caused a boom in Alberta's oil sands industry. Suncor plans are to produce nearly half a million barrels per day by 2016 with reduced energy input and greatly reduced sulfur dioxide and carbon dioxide emissions.[88]

Increased Recovery

An alternative technology called steam-assisted gravity drainage involves steam injection. The oil becomes thinner and gathers near the injection site.[135] This technique would reach deposits too deep to mine conventionally.

Still another way to increase fossil-fuel supplies is to mine oil. We have seen that oil wells typically leave two-thirds of all oil in place, whereas coal mines can recover up to 95% of the coal in a formation by stripping. Exploration of this connection led to construction of several working oil mines. The oldest appears to have been a gravity flow of oil to collection sites through holes drilled into an oil formation in Ventura, California, in 1866.[132] During World War I, Pechelbronn in Alsace was the site of an oil mine of the same sort as in Ventura. By the end of World War II, 5.4 billion bbl had been recovered.[132] The Germans also mined oil in Wietze, near Hannover, for a cumulative production by 1950 of 5 million barrels.

The Russians mine oil in Yarega, 185 m down. The Yarega mine holds viscous heavy oil, so in addition to the holes drilled into the rock (at ≈30 m intervals), other holes are drilled to inject steam into the formation. The oil that drips out is collected and then pumped to the surface. The Russians have achieved about 50% to 60% recovery.[132] See **Extension 12.5**.

Another method called **block carving** undermines weak oil-bearing rock, which fractures and can be removed to the surface for processing. Finally, there is open pit surface mining. This method is the least expensive of the alternatives discussed. Costs at the terrace mine in Kern River, California, are approximately $11/bbl, and at the terrace mine in Sunnyside, Utah, about $21/bbl.[132] An open pit mine in Edna, California, produces oil at $18/bbl. A strip mine in Santa Cruz, California, produces oil at a cost of $18/bbl.[132] Because this is at or below the world oil price, these oil resources are economically exploitable. The other methods are more capital-intensive, and the price of oil will have to rise again and maintain consistency before it would be economically worthwhile to use them to produce oil.

12.4 NATURAL GAS

Natural gas, which is perceived as a "clean" fuel, is almost pure methane (CH_4). It can be made to burn to produce water, small amounts of nitrogen oxides, and carbon dioxide (much less CO_2 than is produced by burning coal or oil). Because it is so clean compared to other fossil fuels, use increased by over 230% between 1970 and 2000, and gas is expected to have the highest growth rate among all fossil fuels for the foreseeable future.[78] By 2020, gas utilization is expected to double again to 450% of 1970 levels.[78]

We estimate, for comparison, the Q_∞ for natural gas using historical experience as a guide: for each barrel of oil discovered, 1700 cubic meters (m^3) of gas (60,000 ft^3) is discovered. This gives a rough estimate of 28.3 Tm^3 for the U.S. Q_∞. The Potential Gas Committee estimated that Q_∞ is 36.5 Tm^3, while a 1989 estimate gave 26.95 Tm^3,[140] rather good agreement among all these estimates.

The emissions-consciousness of the 1980s and 1990s led to greater use of natural gas (each cubic meter of gas has an energy content of about 38.3 MJ) as a replacement for other fossil fuels. As Figure 12.3a shows, natural gas use has risen to levels of coal use. The current American use rate for natural gas is 614 Gm^3/yr (21.7 trillion ft^3/yr).[141] The U.S. Department of Energy puts U.S. reserves at 31 Tm^3 ($\sim$1000 Tft^3), including proven recoverable reserves at 10.2 Tm^3,[142] and undiscovered reserves at 25 Tm^3.[142]

Figure 12.13a shows the major U.S. oil and gas fields, and Figure 12.13b shows how the current production is distributed among the world regions. Probable and proven world reserves total 150 Tm^3,[76] although MacDonald believes that world proven reserves total 102.9 Tm^3 and recoverable resources total 246.5 Tm^3,[143] and Masters and Root believe the total to be 305 Tm^3.[121] Whereas crude oil will likely last about 48 years at current use rates, natural gas will last about 69 years.[144] Consumption of natural gas fell after the beginning of government price deregulation but climbed slowly and steadily after 1983 (Figure 12.3a). Deregulation has increased gas supplies by stimulating exploration, but the end of these supplies is also in sight. If one applies the Hubbert analysis done for oil earlier in this chapter to U.S. resources of natural gas, the middle 80% of natural gas should be produced between 1950 and 2015, with peak production in the 1980s.

Gas production is growing in areas of the United States and Canada to feed the pipeline network and heat homes and produce electricity. Canadian pipelines bring 98 Mm^3/d of gas into the United States,[145] 20 Mm^3/d of which goes to

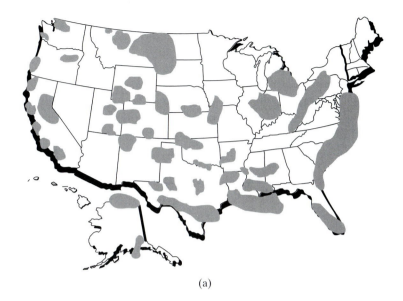

(a)

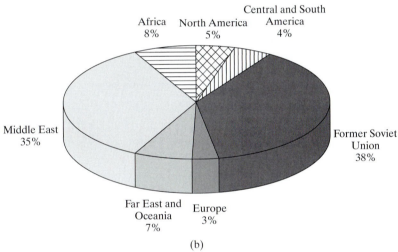

(b)

FIGURE 12.13

(a) Major oil and gas fields in the United States. (b) Estimates of gas production by world region. (Assembled from BP Amoco, 2004; World Oil, Oil & Gas Journal, 2002)

California.[38] Mexico supplies the United States with $58 \, \text{Mm}^3/\text{d}$. Domestic pipelines are capable of delivering $2.66 \, \text{Gm}^3/\text{d}$ throughout the Lower 48.[145] Figure 12.14 shows the U.S. gas pipeline connections as of 2000.

Wyoming production is booming, and there is controversy elsewhere as to whether the environmental costs are worth the gains of development.[146] Alaska has at least $1 \, \text{Tm}^3$ of gas on the North Slope alone,[147] while the National Energy Policy Development Group uses the figure $2.8 \, \text{Tm}^3$.[17] The George W. Bush administration has pushed hard against the protests of ranchers and environmentalists to develop Western gas.[148] The North Slope gas, $250,000 \, \text{m}^3/\text{d}$, is currently reinjected. The Mackenzie River Delta has $1.8 \, \text{Tm}^3$ of gas that is exploitable; there is no pipeline to connect either Canada or the North Slope to the Lower 48, though planning is under way for a pipeline by 2010.[37,149] Pipelines are generally safe, though there have been deaths, and major pipeline accidents have been increasing at a rate of 4%/yr.[150]

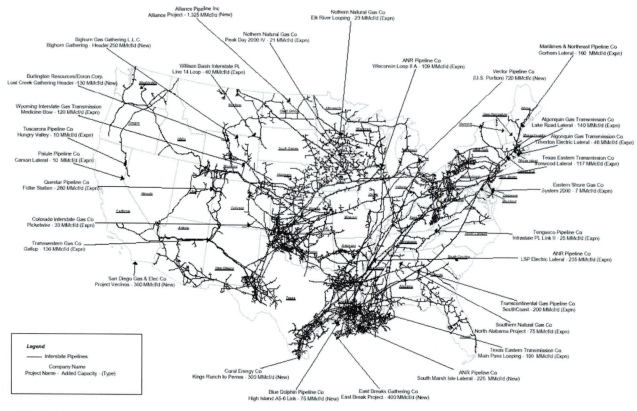

FIGURE 12.14

The U.S. natural gas pipeline system. (Energy Information Agency, "Status of natural gas pipeline system capacity entering the 2000–2001 heating season," *Natural Gas Monthly*, October 2000, pp. vii–xviii, Fig. SR 3)

Clathrate Resources

Clathrates are hydrated gases, mostly methane (waterized natural gas), frozen into a crystal structure. The clathrates are often referred to as gas hydrates. The water forms a rigid structure that contains the gas molecule under certain conditions. If conditions change, the methane can be released.[151,152] Researchers on a fishing ship have even seen such releases from clathrate deposits on the seafloor; if they'd been there underwater, they might have been able to see the 500 m plume of bubbles heading upward. Instead, they saw it using fish-finder sonar.[153]

While the oil industry knew of clathrates since the 1930s, it was not realized until the last several decades that a significant amount of methane worldwide was tied up in the form of clathrate. A total methane volume of 21,000 Tm^3 is the consensus estimate (the range of estimates is 1000–140,000 Tm^3).[153–155] This is far larger than the resources of conventional natural gas, and the resource is near the solid surface in permafrost, or at relatively shallow depths (500–2000 m). However, the seafloor methane is trapped in clays at low density and does not move easily to a collection point.[153–155] The difficulties seemed so enormous that industry lost interest, and the Department of Energy disbanded its gas hydrate team in 1993.[156]

However, changes in phase of the water allow the release of methane—the plumes observed. One fear is that global warming will allow methane to "blow out" of its formations and make the global warming that much more of a problem. (This was the premise of a science fiction novel *Mother of Storms*[157] by John Barnes.)[158] There is some indication that a hot climate followed release of methane from the seafloor 130 million years ago.[158,159] However, experts believe that most methane would be changed to carbon dioxide before it emerged from the sea ($CH_4 + 2\,H_2O \rightarrow CO_2 + 4\,H_2$). The greatest danger of a massive "blow out" could be to human structures attached to the seafloor and coastal settlements (because of the expected tsunami).[154]

Coal Mine Gas

A much neglected source of natural gas is coal mines. Methane (natural gas) has long been a great danger to coal miners because of the possibility of explosion. It is estimated that there are from 230[122] to 2000 Tm^3 [160] of natural gas recoverable from U.S. mines. The first coal-bed gas recovery well was drilled by Amoco in 1977.[161] Progress in the use of coal-seam methane is palpable, with a significant amount being used. Wyoming is a major source of coal-bed methane.[145]

The newest method for recovering coal-bed methane is called the stress-cavity technique.[161] A hole is drilled down into the top of the coal seam and filled by concrete and a steel pipe. Then water is pumped down to carve out a hole in the coal formation. A filter is inserted, and natural gas flows. It is much more productive than the old method, in which a concrete pipe was laid into the coal seam and bullets were fired through into the seam to generate gas flow.[161] It has generated controversy where it has been used because of the large quantities of water released and its effects on livestock.[162] In already-opened mines, a horizontal drilling technique is used.[163] The mine is much safer for miners after the gas has been removed. Furthermore, methane is implicated in the greenhouse effect (see Chapter 15), and it is better to burn it than to allow it to be released unburned, because then, at least some energy can be extracted first.

12.5   COAL

Coal replaced wood in the nineteenth century, and until the 1980s, coal use was pretty stable (each ton of coal has an energy content of about 22.5 GJ and each tonne an energy content of about 24.8 GJ). Coal comes in various grades: anthracite, bituminous, subbituminous, and lignite, in descending order of energy content and hardness. Anthracite, almost pure carbon, is found in very small quantities. An average bituminous coal might have an energy content of 30 MJ/kg, while lignite may have an energy content of 16 MJ/kg. As with oil, some coal is left in place when it is mined.

The growth in coal use is a recent development. Prior to that time, the other forms of carbon were cheaper (see Figure 5.7), and because much of domestic coal is "dirty" (has a high sulfur content) and must have sophisticated equipment available to control emissions, growth in coal use lagged. Further, because most coal is used to make electricity and the U.S. economy became more efficient in the years after 1973, the demand for coal was decreasing.

TABLE 12.3

Estimates of world coal resources

Estimate	Amount (gigatonnes)	Reference
Averitt	7623	2,165
British Petroleum	987	76
Department of Energy	988	11
Workshop on Alternative Energy Sources	11,500 (740 Gt recoverable)	166
World Energy Council	1036	168

The U.S. production rate for coal is about 1 Gt/yr, and world production is 4.3 Gt/yr.[11] An estimate of domestic coal reserves and resources is about 250 Gt,[76] or perhaps as much as 450 Gt,[164] and an estimate of world resources is about 988 Gt [about 1 Tt, or 10^{12} tonnes].[11,76] Averitt (Table 12.3) has estimated world coal reserves[2,165] to total the much larger 7623 Gt, of which the former Soviet Union has total estimated reserves of 4310 Gt and the United States 1486 Gt; together the two countries account for three-quarters of total world coal resources. The Workshop on Alternative Energy Sources estimates a total of 11.5 Tt of coal available, of which only 1300 Gt is *known* reserves and only 740 Gt of that is economically recoverable (agreeing with the British Petroleum and World Energy Council estimates, Table 12.3).[166] It would be difficult to place too much reliance on predictions for coal production, because we are only *beginning* the cycle of coal use. Recall that the Hubbert analysis can determine Q_∞ only after the peak has been reached.

North American reserves are found in large areas (Figure 12.15)—a band stretching along the Rockies from Albuquerque, New Mexico, to north of

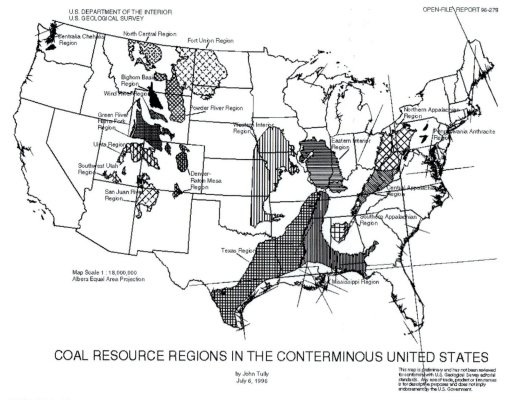

COAL RESOURCE REGIONS IN THE CONTERMINOUS UNITED STATES

by John Tully
July 6, 1996

FIGURE 12.15

The geographic distribution of North American coal reserves.
(U.S. Department of the Interior, USGS, John Tully, Open-File Report 96-279)

Edmonton, Canada; a band running from an Omaha–St. Louis axis south toward Dallas; a large area south of Chicago; and a band stretching along the Appalachians, as well as smaller fields dotted around. Coal-bearing rock underlies 13% of the total U.S. land area. Each meter of coal results from compression of a thickness of 3 to 10 m of compacted plant matter. The greatest thickness of a coal seam is 7.4 m, the thinnest 5 cm. Average thickness of Eastern coal is a bit more than a meter, Midwestern coal about 2 m, and Western coal about 10 m.[167]

In the United States, there are 938 underground and 910 surface mines (which give two-thirds of production).[164] The largest underground mine is Enlow Fork/Consol PA Coal Co. in Pennsylvania, which produced 9.84 Mt in 1999.[164] The top eight surface mines are in Wyoming. The Rochelle Mine Complex/Powder River Coal Co. is the largest (69 Mt in 1999) and the Black Thunder Mine is the second largest (49 Mt in 1999).[164] Surface, or strip, mines generally remove 85% to 95% of the coal present.[168] Clearly, economics is forcing production to switch from underground to strip mining, and from East to West.

In deep mines, anywhere from 20% to 75% of the coal may be left to support the overlaying rock. A new technique, adopted during the 1980s, reduces the amount left behind. **Longwall mining** is an automated technique to obtain coal using mining robots. It can now extract 80% of the coal in a formation, more than the typical 50% of normal mining. Two miners dig a hole the height of the seam, attach brackets to hold the ceiling in place, and set up a robot that runs 230 m (750 ft) across the seam. The brackets move forward as the robot mines more coal.[169,170] Longwall mining is not without its drawbacks. House walls crack above ground and water supplies mysteriously vanish or change. Sediment leaches into streams.[171] About 100 longwall systems are in use in the United States.

Underground mines are much less safe for miners than are strip mines, and mine fires are common. There are even fossil records of fires in coal seams millions of years ago.[172] Mine fires can last years, decades, even centuries. The Centralia, Pennsylvania, fire is more than 40 years old.[173] It is expected to burn another century. The Liu Huangou fire in Xinjiang, China, is about the same age as the Centralia fire.[174] Devil's Oven, near New Straitsville, Ohio, is above a fire that has been burning in the Middle Kittanning coal seam for 120 yr. It was set during a labor dispute.[175] There are over 3000 coal mine fires in China, India, and the United States, and perhaps 1000 in Indonesia.[176] Mine fires in China are estimated to burn 200 Mt of coal per year, 20% of what is burned to supply electricity in the United States; these fires may be producing 3% of the world's carbon dioxide.[176] The fires are only a part of what makes China such a bad place to work as a miner (an estimated 10,000 Chinese miners die every year).[177]

Coal has, roughly speaking, a chemical formula $CH_{0.8}$ as compared to diesel fuel with a formula of roughly CH_2 or to natural gas, methane, which has a chemical formula CH_4,[171] as discussed in Chapter 7. A fossil fuel described by the chemical formula CH_x has an atomic mass of $(12 + x)$ u (atomic mass units), so the proportion of carbon in the fuel is $12/(12 + x)$; one can see that there is more carbon in coal (94%) than in methane (75%), and the heat of combustion of methane is greater than that of coal, so more carbon dioxide is emitted per unit of energy released for coal (2.1 mol CO_2/MJ) than for alternative fossil fuels (for methane it is 1.16 mol CO_2/MJ, or about half that of coal; see also Chapter 15).

The wanton wastage of the environment by strip miners led Congress to pass the Surface Mining Act of 1977, which mandates environmental safeguards. The Act has added $1 to $5 per ton extra cost for surface coal to pay for reclamation.[178] This reclamation cost may or may not be a factor in actual sales, depending on the cost of transportation of the coal to its point of combustion. Transport cost may be two to three times the actual cost of the coal. The Clean Water Act also contains safeguards against strip mines contaminating the environment. Slurry pipelines, pipelines carrying ground-up coal suspended in water, are in competition with unit coal trains to transport coal. Large ships carry coal around the world; the United States averaged about a fourth of world coal trade between 1973 and 1989, 55% from Appalachia.[165] The most important problems hampering construction of more pipelines are the lack of rights-of-way and the difficulty of obtaining water in the dry West.[178,179] There are sometimes also environmental objections, as occurred, for instance, in a plan to have pipeline facilities within Grand Canyon National Park.[180]

Cleaning Coal?

Burning of coal produces many pollutants: waste heat (see Chapter 13); nitrogen and sulfur oxides, which produce adverse effects on health; acid rain; materials corrosion, and other consequences (see Chapter 14); and carbon dioxide (see Chapter 17). Local emission problems have been tackled by the addition of tall smokestacks, which aid dispersal, and by application of the best available control technology (known as BACT). The United States and Canada are cooperating on a Clean Coal Technology Program that began in 1985.[181–183] Special envoys on acid rain for the United States and Canada, Drew Lewis and William Davis, proposed a $5 billion program. President Reagan proposed $2.5 billion in his budget, and Congress went along.[181,182] Coal-fired plants cause 70% of the SO_x, 30% of the NO_x, and 35% of the CO_2 produced in the United States.[184] The Clean Air Act Amendments of 1990 impose a nationwide cap of 8.9 Mt of sulfur dioxide and 11 Mt of NO_x per year after 2000.[67,68,181,184]

So far, scrubbers are the technology of choice, and 146 have been installed on 82 of the 370 coal-fired power plants operating. In a scrubber, a limestone slurry spray is mixed with exhaust gases to trap sulfur-bearing molecules in the flue gas. The designation "scrubber" has become more of a generic term that now includes the precipitators that sweep particulates from the smoke electrostatically or baghouses that act like vacuum cleaner bags, as well as the original meaning of ridding (scrubbing) the flue gas of sulfur dioxide (see Chapter 13). The scrubbers have had many problems, not the least of which is the generation of great volumes of throwaway material—sludge from the limestone–gas mixture.[178,179,185] This material is still undesirable waste, but it is in a concentrated form. It is usually stored in "ponds" and allowed to weather indefinitely. Typical emissions from coal-fired generating facilities are presented in Table 12.4 and are discussed in more detail in Chapter 14.

Coal-Fired Power Plants

The fossil-fuel combustion process described in Chapter 6 applies to coal as well as to oil. In earlier days, generating facilities burned lump coal. In more modern times, the industry standard has been the pulverized coal boiler. The next step appears to be atmospheric or pressurized **fluidized-bed combustion**;[181–183,186] many units are already in service worldwide.[181,182,187]

There are several reasons for this development: many different fuels, of different characteristics, may be burned in fluidized beds; fuels of varying heats of combustion may be burned equally satisfactorily; and many designs incorporating

TABLE 12.4

Average yearly emission estimates from an average coal-fired 1000 MW$_e$ plant

Emissions	Amount (kilotonnes)
Air	
Carbon monoxide	2.5
Hydrocarbons	2.6
Nitrogen oxides	40.0
Particulates	6.3
Sulfur oxides	32.0
Water	
Organics	0.4
Dissolved solids	95.0
Suspended solids	1.6
Solid Wastes	3500.0

two-stage combustion allow lowering of emissions of nitrogen oxides (by keeping combustion temperature under ~830 °C).[181,182,186] A reactor that has a bed of combustion through which the fuel circulates may well become the newest standard for larger-scale (>60 MW$_t$) combustion.[186] These reactors began to be sold commercially in the early 1980s. As we will see in the next section, fluidized-bed reactors are also very important in the gasification of coal, another promising new technology that reduces effluent emission.

Coal is a dirty fuel compared to oil and gas (see Chapter 13). More than 600 coal plants still operating are exempt from the regulations of even the first Clean Air Act (1970).[188] They were grandfathered with the expectation that they would be retired on the normal schedule. That did not happen. The situation in Ohio illustrates this phenomenon. According to the EPA, four Ohio utilities rank among the top 10 air polluters nationwide (AEP, second; Cinergy, fifth; FirstEnergy, sixth; and Dayton Power and Light, 10th).[189] Investigative reporters for *The Akron Beacon Journal* found a web of deceit spun by Ohio utilities. As indicated in one of the series' articles,[190] the utilities had a "dirty little secret":

> Even utilities' closest allies were deceived. "We'd be urging them to pay the money and put in scrubbers," said former coal lobbyist Neal Tostenson. "And they'd say the plants were too old, that they'd be retired before the scrubbers got paid for. They'd say the old ones don't have much useful life left. They'd do it on the new plants. I haven't seen too many new power plants, have you?"
>
> More than half of Ohio's coal-fired power plants were built in the 1940s or 1950s.... What Ohio utilities did instead was keep their old plants alive. The federal lawsuits claim they did that by illegally modifying and rebuilding the old plants, without applying for the permits and installing the pollution controls the modifications required.

Many utilities do act in good faith and work to reduce emissions.

Coal Gasification

A widely broached possibility is gasification or liquefaction of coal or high-sulfur residual oil. The main advantage of these techniques is that it is relatively cheap to remove sulfur and other impurities from gas. Squires[191] estimated that the H_2S from "power gas" costs only $20/kW to remove, whereas the equivalent cost of removal from stack gas is $70/kW (in 1984–1985 dollars). Fluids are also easily moved by pipeline (see Chapter 15), so there may be an economic reason to gasify coal to transport it.

These alternatives look more attractive in light of persistent delays in construction of nuclear plants and concomitant cost increases. Newer oil plants (those built after 1974) need a daily ration of 1.25 Mbbl of oil.[191] When oil costs are low, this is attractive. When oil prices rise, however, coal may become more economic. Most energy experts look to coal to fill the energy needs of the near future, because so much electricity is produced by coal. The current mix of fuel sources for electricity generation was shown in Figure 12.3a. Although the coal has been there all along, it has not been gasified for several reasons, according to Osborn:[192] the low cost, convenience, and availability of natural gas; the convenience and availability of low-sulfur oil; the inability to obtain long-term low-sulfur coal supplies; the uncertainty as to practicality of sulfur removal from stack gases; the more stringent environmental, health, and safety regulations in coal mining; and recurrent transportation problems. Additionally, for many years, coal did not enjoy a significant price advantage over oil or gas.

FIGURE 12.16

The advanced coal conversion project at the Rosebud coal mine in Montana shows that gasification is *not* particularly high-tech. (U.S. Department of Energy)

The first-known coal gasification took place in 1792 when a Scotsman, William Murdock, used gas to light his home.[193] Two types of gas are usually produced. **Power gas** is a mixture of carbon monoxide, hydrogen, and nitrogen. The simple technology for producing power gas dates to the 1880s.[194] It has about one-sixth the heating value of natural gas, about 5.6 MJ/m³, which is too low a heating value to be worth transporting any distance. **Synthetic natural gas** (SNG) is made from power gas by shifting the ratio of hydrogen to carbon monoxide (to 3 : 1), cleaning, and then transforming to methane.[192] It can then become ammonia, methanol, or synthetic gasoline, or it can be fed into a natural gas pipeline as methane, if desired.

The proven Lurgi (medium pressure) and Koppers–Totzek (low pressure) processes convert only about 55% of the coal to synthetic natural gas. These processes are also expensive; the gas costs two to four times as much as natural gas from the wellhead,[193–198] but the cost is falling steadily. (See **Extension 12.8, Newer Gasification and Liquefaction Technology**.)

The Cool Water Coal Gasification Plant

A relatively new technology is the integrated gasification combined cycle (IGCC) plant, in which energy efficiency is gained through use of both a topping (Brayton) cycle and a bottoming (Rankine) cycle to gain energy that would normally have dissipated as waste heat. At the experimental Cool Water IGCC plant, the initial gasification took place in oxygen instead of air so that no nitrogen oxides were formed in the gasification stage.[207,210–213] A form of power gas was generated that is low in methane. Methane is not desired, as the gas is used in situ, and thus, the extra energy cost required to produce methane would have been wasted. The plant, a 100 MW$_e$ modular demonstration facility, was completed in under 2 years and under budget (a feat not often equaled today).[213] In its first year of operation, 24 June 1984 to 23 June 1985, the plant generated 413 GWh of electricity with an online factor of ≈60%.[212–214] After the successful 5-yr test period, Southern California Edison sold the plant to Texaco. Texaco plans to burn a mixture of sewage sludge and coal.[183] The experiment was considered an engineering success but not an economic one. One large problem was the need to reduce the fuel gas temperature to clean it.[215] General Electric has a new project that uses a cyclone to remove the particles that can ruin turbine blades and cause contamination of water vapor.[215] The experience gained with Cool Water is making it more likely that more IGCC plants will be built in the near future. Because of the fact that energy is generated in several parts of the process, not just in one stage, it holds the promise of 60% efficient power plants soon.

In the test, a slurry of ground Utah coal and water was fed to the gasifier. The gas from the gasifier, which was preheated to 1260 °C and brought to 40 times atmospheric pressure, consisted of 42.5% carbon monoxide (CO), 38.2% hydrogen (H$_2$), 18.6% carbon dioxide (CO$_2$), 0.3% methane (CH$_4$), 0.4% argon (A) and nitrogen (N$_2$), and about 50 parts per million (ppm) of hydrogen sulfide (H$_2$S) and carboxylsulfate (COS).[213] The gas had an energy content of 10.7 MJ/m³. The sulfur was scrubbed out from the synthesis gas using commercial processes developed for oil refining (Selexal, SCOT, and Claus processes).[213]

At such high temperatures, over 1200 °C, no tars or phenols are produced. Heat produced in the gasifier is used to generate steam that is mixed with the steam generated in the gas combustion so as to run the turbine that produces electricity (this gives the facility a "bottoming cycle"). Therefore, the facility runs two

TABLE 12.5

Cool water power plant pollutant levels

Pollutant	Plant data	EPA new source performance standard
NO_x	0.061 lb/MBtu	0.6 lb/MBtu
SO_2	97% removal	90% removal
	(0.034 lb/MBtu)	(1.2 lb/MBtu)
Particulates	0.0013 lb/MBtu	0.03 lb/MBtu

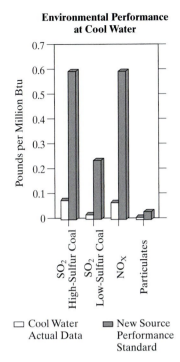

FIGURE 12.17

Cool Water effluent compared to the New Source Performance Standards. (U.S. Department of Energy)

generators—one operating on the "normal" generator fired on the coal gas (the generator is modified to take larger airflow rates), and the other operating on steam from both combustion and gasification. Steam is injected with the coal gas in the main generator to reduce combustion temperature and thereby cut nitrogen oxide emissions. Essentially all the carbon in the coal is burned. The levels of pollution are compared to the EPA standards in Table 12.5.[213] Such a combined cycle plant is a good bet for clean combustion of coal.

The Cool Water plant was the only coal-fired plant ever to be licensed in California, and it has exceeded the requirements of the California clean air standards during its operation. It used less land and 60% less water than conventional plants, as well as consumed 10% less coal to generate the same amount of energy. No scrubbers are necessary on such combined-cycle plants. The effluents are, as Table 12.5 and Figure 12.17 show, CO_2, coal clinker, and sulfur. The clinker is nonhazardous and was sold for use in road fill. The sulfur was sold. The plant demonstrated that coal plants can be constructed to meet stringent effluent standards. The Cool Water plant, built in 1984, cost $2380/kW as compared to a 1984 cost of $1000/kW for a conventional plant.

The Cool Water plant was experimental, and as such, cost more than the proposed production modular units. An advantage of the modular units is that they can be installed in phases. It has been suggested that the first purchase for the utility be gas turbines, at first burning natural gas or oil. As prices rise or load requirements change, the turbines could be integrated into a coal gasification facility.[216]

The Cool Water plant saved importation of 4300 bbl of oil per day.[213,217] Nevertheless, the project was completed only because of $130 million in guarantees from the government-owned Synthetic Fuels Corporation (the plant cost $238 million to build [217,218]). The owners were guaranteed $12.50/MBtu for the first 255 Gm^3 and $9.75/MBtu thereafter.[218]

This subsidy may be considered justifiable if the plants developed based on Cool Water can replace dirty, older plants as they are retired. The 430 MW Fort Lonesome plant of Tampa Electric is a descendant of the Cool Water plant.[219] Fort Lonesome was expensive to build but is cheap to operate, and it sells products made from the material it scavenges from the coal it burns in its two syngas-fired turbines. It sells sulfuric acid, it turns coal ash into concrete, and it might sell carbon dioxide as well.[219] Another descendent of Cool Water has been built in Terre Haute, Indiana.[206]

Alternative Strategies

Additional opportunities exist for savings at the burning end. Other methods and strategies could be technologically competitive right now, especially in the area of increasing overall combustion efficiency. One such strategy involves construction of a centralized station, a coal refinery or "fuelplex."[191] The fuelplex would address energy generation from a combination of methods, using and/or producing low-sulfur coke, synthetic natural gas, clean liquid fuel, electrochemical generation, power gas, electricity, and so on. This concentration could increase efficiency and allow a wider range of options.

The electric utilities' thirst for coal in nearly all developed countries means that coal will continue to be in worldwide demand. Because strip mining allows so much more coal to be extracted and is so much more labor efficient and safer for miners (see **Extension 20.12, _Nuclear versus Coal_**) than deep (even longwall) mining, more coal will inevitably be coming from strip mines. Up to the 1990s, that meant that coal moved to the west, abandoning Appalachia, home of nineteenth and twentieth century "coal towns." Appalachia still has significant coal reserves, but the topography—mountains and valleys—meant that deep mines were the historical norm. A new technique introduced by the remaining Appalachian coal companies, called mountaintop mining or mountaintopping,[220] allows strip mining in the eastern coalfields. In this method of coal mining, the top 200 m of the mountain is consumed in 130 ton chunks and dumped into the valleys (for more detail, see **Extension 12.9, _Mountaintopping_**). Oil will soon become too expensive to compete with coal. Already, even excluding costs of Middle East wars that never would have occurred without oil, the true cost of a gallon of gasoline is between $4 and $5.[245]

SUMMARY

Within the last three decades, the United States has experienced two energy crises. Each was caused by forces outside government control. In **Extension 12.2**, the political ramifications are examined more deeply. There are many consequences of cheap oil, such as waste, a tendency to build gas guzzlers, and so on, as discussed in **Extension 12.3** in more detail.

The vulnerability we have experienced has led to increased domestic oil exploration at the same time reserves are decreasing (see **Extension 12.1**). This has occurred because U.S. oil is fast on its way to exhaustion—and world oil is sure to follow soon. The model of M. K. Hubbert, which asserts that production follows a bell-shaped ("normal") curve, seems to be verified by domestic experience (see also **Extension 12.4**). The largest oil fields are the most quickly discovered, and it has become progressively more difficult to find oil.

Research has increased the amount of oil ultimately recoverable from an oil-bearing formation from about 20% to around 40% of the oil in place (see **Extension 12.5**). The world seems to be "hooked" on liquid fossil fuel. However, experience with the price increases of the 1970s verifies that once the price rise is high enough, the demand decreases.

Shale oil and oil sands, which are found worldwide, contain staggering amounts of oil. Exploitation of most of these resources involves large-scale environmental disruption (see **Extension 12.7**). The oil must be processed, and in **Extension 12.6** the way it is broken into smaller fractions is discussed.

Natural gas will last somewhat longer than oil, but coal reserves are so huge that coal may be able to supply energy for the next few centuries, while oil and natural gas will be gone within a century (at _current_ use rates—but of course, current use rates change with time). All fossil energy has environmental problems associated with its exploitation, as attested to in **Extension 12.7** and **Extension 12.9**. Coal burning by traditional techniques produces the most pollution. Research has proceeded on ways to reduce pollutants from coal, and it appears that some gasification processes are inherently cleaner and more efficient than usual combustion modes now in use. Alternatives involving synthetic gases are being developed around the world.

PROBLEMS AND QUESTIONS

MULTIPLE CHOICE REVIEW

1. In the United States, which has been thoroughly explored for oil,
   a. large oil fields are commonly being discovered.
   b. new oil fields are being discovered that are generally smaller than those discovered in the past.
   c. no oil fields have been discovered in the Lower 48 in the last few decades.
   d. the rate (amount per time) at which new fields are being pumped surpasses the rate at which the old fields are being pumped.
   e. the number of wells discovered per meter of exploratory hole drilled is increasing because of technological advances.

2. In an oil field, oil is being produced at a rate of 10 kbbl/d. The amount of oil produced in a week is
   a. 70 kbbl.
   b. 70 kbbl/d.
   c. 1.42 kbbl.
   d. 1.42 kbbl/d.
   e. given by none of the other answers.

3. The amount of oil imported by the United States in the future will most likely be _____ the amount produced in the United States.
   a. far above
   b. somewhat above
   c. approximately equal to
   d. somewhat below
   e. far below

4. In the Hubbert model, the amount of oil ultimately produced by oil fields in the United States (Q_∞) is known, because
   a. the rate of oil pumping averaged over the country tells it to us.
   b. it is simply the sum of the past production and the reserves.
   c. it is simply the product of the past production and the reserves.
   d. geologists can tell it from their seismic data on the reservoirs.
   e. the shape of the curve determines the value of Q_∞.

5. The price of crude oil
   a. affects industry choices in buying more energy-efficient energy-intensive equipment.
   b. does not vary by much over very long periods of time.
   c. does nothing to change the consumption rate, because people make long-term commitments to their energy sources.
   d. allowed SUVs to take a large role in conserving energy.
   e. forces other energy prices to swing higher and lower in tandem.

6. In the Hubbert model, the amount of natural gas ultimately produced by gas fields in the United States (Q_∞) is
   a. determined to be 28.3 Tm^3.
   b. determined to be 36.5 Tm^3.
   c. determined to be 614 Tm^3.

   d. cannot yet be determined.
   e. given by none of the other answers.

7. Gasification is a useful option, because
   a. gas is cheaper than coal.
   b. gas takes up less space than coal.
   c. energy is gained through the change.
   d. it is easier to remove sulfur from a gas than a solid such as coal.
   e. gas may be released to the atmosphere without problems.

8. The coal found around the world
   a. constitutes the largest resource of fossil fuel.
   b. is evenly distributed around the globe.
   c. was created when Earth formed.
   d. is uniform in composition everywhere, with only minor differences.
   e. was once oil, but changed as a result of the pressure of the overlying rock.

9. America is dependent on the rest of the world for
   a. oil.
   b. gas.
   c. coal.
   d. methane.
   e. clathrates.

10. Emissions from coal-to-gas electric plants are lower than those from
    a. nuclear reactors.
    b. natural gas electric plants.
    c. oil electric plants.
    d. fluidized bed coal plants.
    e. none of the other possibilities.

SHORT ANSWER

Explain whether the statement made is correct, and justify your answer.

11. Of the fossil fuels, oil produces the most pollutants per kilowatt of energy produced.

12. It is possible to predict with certainty how much oil will have been extracted worldwide when the "oil era" is over.

13. We do not yet know when (in what year) the maximum amount of crude oil will be produced per year worldwide.

14. The lifetime of a fossil-fuel resource, formed by dividing the total reserves by the current use rate, is an accurate measure of the number of years of that resource remaining to be exploited.

15. Fossil fuels are what they are, the amount of effluent cannot be changed in any way.

16. We do not yet know when (in what year) the maximum amount of crude oil will be produced per year in the United States.

17. The amount of reserves of oil in a given formation is determined at the beginning of expolitation of a resource and never changes much thereafter.

18. Gasoline price jumps almost always occur because of price gouging by the "big oil" companies.

19. The utilities always follow the practice of retiring fossil-fuel electric-generating plants at the end of their original lifetimes.
20. Underground mines produce smaller amounts of coal per year than strip mines by their very nature.

DISCUSSION QUESTIONS

21. "Sting me once, shame on you. Sting me twice, shame on me." Apply the moral of this folk saying to the energy resource price fluctuations over the past 30 yr.
22. Why is gasification being pursued as an option for coal use?
23. What reason did Hubbert have to think that the normal curve would describe the production of oil? Would the Hubbert description apply to a natural resource such as coal in addition to oil? Why or why not?
24. Why did some Ohio utilities extend the life of their coal-fired plants? What were the consequences of this extension?
25. One method of mining coal, applied in Tennessee, Kentucky, and West Virginia, is called "mountaintop mining."[220] What possible consequences of such mining can you propose? Explain.
26. What evidence is there that oil has its origins in algae? Why is this of importance?
27. The Fort Lonesome and Terre Haute gasification plants were subsidized with public money. According to one observer quoted in Reference 219, the company would have built the plant in any case. Was this a good or poor use of public money? Try to argue both sides.
28. In Reference 210, it was suggested that natural gas should be used to produce motor fuels. Discuss the reasonableness of this proposal, given that this is feasible.
29. According to two different estimates, coal reserves can be 250 Gt or 450 Gt, almost twice as much. Why is it so difficult to pin down the reserves' total?
30. A punster in the utility industry has referred to the burning of coal in the Cool Water plant of Southern California Edison as "immaculate combustion." Explain how serious he could have been by reference to Table 12.4 and Figure 12.17.

PROBLEMS

31. Determine the ratio of emissions of an ordinary coal plant in terms of those from the Cool Water plant.
32. How many days' use of oil is stored in the Strategic Petroleum Reserve?
33. Assuming Squires is correct (and that the prices he quotes still hold: removal of H_2S from "power gas" costs $20/kW, whereas removal from stack gas is $70/kW),[191] how much money would a 1 MW plant save per year by gasifying? What maximum could the plant afford to spend at an interest rate of 8% to install the equipment to gasify? (Typical payoff is 30–40 yr for a power plant.)
34. Under the Clean Air Act Amendments of 1990, how much sulfur dioxide and nitrogen oxides could be emitted by power plants? The Act's limits are 8.9 Mt of sulfur dioxide and 11 Mt of NO_x. Coal-fired plants cause 70% of the SO_2 and 30% of the NO_x released in the United States. Justify any assumptions you make.
35. About what is the volume (or mass, as appropriate) of each fossil fuel (natural gas, oil, coal) that represents 1 GJ of energy?
36. About how long would U.S. coal last (assuming none is exported) at the current consumption rate? How long would the world's known coal reserves last at the current world use rate?
37. Assuming that Hubbert is correct about U.S. oil, how long would domestic oil last at the current U.S. use rate if all crude oil were able to be supplied from domestic sources?
38. Assuming that clathrate methane can reasonably be tapped and 25% utilized as a fuel resource, how long would the resource last at current use rates?
39. World use rates for coal have been dropping at approximately 0.2%/yr.[76] How long would the world's known coal reserves last at the current world use rate if that decrease continues unabated? Explain.
40. Oil use is increasing worldwide at about 1.5%/yr.[76] Assuming that 2.5 Tbbl is the value of Q_∞, and given that the current use rate is 75 Mbbl/d, how long will the oil last? How would your answer change if Q_∞ were 2.0 Tbbl, 3.0 Tbbl, or 4.0 Tbbl? Explain.

CHAPTER 13

Environmental Effects of Utility Generating Facilities

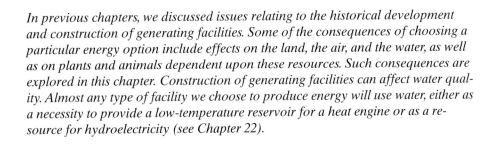

In previous chapters, we discussed issues relating to the historical development and construction of generating facilities. Some of the consequences of choosing a particular energy option include effects on the land, the air, and the water, as well as on plants and animals dependent upon these resources. Such consequences are explored in this chapter. Construction of generating facilities can affect water quality. Almost any type of facility we choose to produce energy will use water, either as a necessity to provide a low-temperature reservoir for a heat engine or as a resource for hydroelectricity (see Chapter 22).

KEY TERMS

steam vessel / turbine / condenser / reactor / boiling water reactor / pressurized water reactor / waste heat / biochemical oxygen demand / hypolimnion / epilimnion / stratification / eutrophication / wet cooling / dry cooling / precipitation / transpiration / evapotranspiration / gain

13.1   GENERATING ENERGY

Conventional electricity sources in the developed world are fossil-fuel combustion and nuclear energy. Both methods generate waste heat, and both use water to transfer energy between low- and high-temperature reservoirs. In this chapter, we consider effects common to both fossil-fuel and nuclear energy generation. These include the problem of waste heat production, the various negative effects of constructing a generating facility, the ultimate gain in energy of a generating facility, and effects on water quality. Issues peculiar to one particular mode of power production are considered later in separate chapters.

The Fossil-Fuel Process

A simplified schematic of a fossil-fuel facility is shown in Figure 13.1. Fuel is introduced and burned in the boiler. There are various ways to burn fuel in the boiler, but the most common are similar in operation to those used in home furnaces. The gas or fuel oil burns in an open flame fed by pipes. Lump coal is carried to a grate by a conveyer, dumped onto the grate, and burned with the ash falling through the grate. Coal is sometimes piped in as slurry or as small-sized lumps of coal in water. Furnace heat evaporates the water, the coal then burns, and the ash falls as before. Most often, coal is pulverized into dust-like powder and blown into the furnace.

The open flame heats the feed water in the **steam vessel**, which produces steam under pressure in a closed container. This increases the overall efficiency of the process, because it raises the temperature at which water boils, thereby enhancing the temperature difference between high- and low-temperature reservoirs. The high pressure is also useful in driving the steam along its path. The steam is then piped to the **turbine**. In the turbine are blades that intercept the steam as it flows. The blades are canted at an angle (like a propeller), and the steam bounces off the blades, pushing at them. This causes the turbine blade assembly to rotate about its center. Because the blades are connected to a shaft at the center, the shaft can be connected to a coil outside the turbine. The coil then rotates in a region where there is a magnetic field, as described in Chapter 4. This produces the AC current we ordinarily use.

The steam, which has continued through the turbine, continues on to a **condenser**, where the spent (cooling) steam condenses again to water. This is the lowest temperature point in the cycle. To keep the condenser at ambient temperature, cooling water from a lake, river, or cooling tower is employed. This cooling water merely circulates through the system and is later exhausted (slightly warmer). The steam turns to water in the condenser and is generally recycled back into the steam vessel as feed water to close the cycle.

A recent development, called the Kalina cycle, holds the promise of raising the efficiency of conversion to electricity from ≈33% to 45%.[1] Instead of using water

FIGURE 13.1

Schematic of a steam electric generating facility.

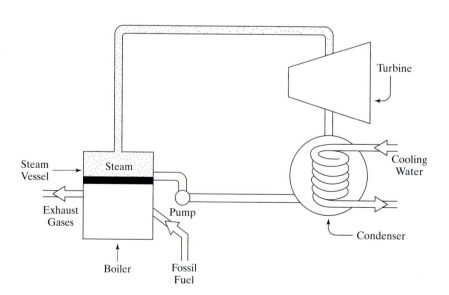

alone in the closed cycle as the working fluid, in the Kalina cycle, water is mixed with ammonia to lower the boiling point and, consequently, liberate more steam than water alone at the high temperature. If the ammonia is diluted further with water to raise the dew point, condensation occurs at a reasonable temperature, and the water is distilled out to reconstitute the original ammonia–water mixture.

The Nuclear Process

The term **reactor** is used to denote a closed vessel in which nuclear processes produce thermal energy (see Chapter 19). There are several different types of nuclear reactors used in generating electricity on a commercial basis. The most common types of reactors in use in the United States (and the world), the **boiling water reactor** (BWR) and the **pressurized water reactor** (PWR), use water in the region of the reactor containing the fuel (the core).

In the BWR, the fuel elements are loaded in the reactor core, surrounded by water. As fissions occur, fragments are produced that move through the elements and the water, transferring their energy to other materials. This kinetic energy thus gets "spread out" in many collisions to become thermal energy. The thermal energy boils the water in the core, producing steam at about 70 times atmospheric pressure (6.75 MPa, or 1000 lb/in^2). The steam drives the turbine and arrives in the condenser, where it collects as liquid feed water and is pumped back to the core. The maximum thermodynamic (Carnot) efficiency for this process is about 45%, and the actual efficiency is about 33%, substantially less than that of the newest fossil-fuel boilers. The BWR is shown schematically in Figure 13.2. Thirty-eight BWRs are operating in the United States.

In the PWR, the fuel elements are in a closed, highly pressurized container. The high-temperature water ($\approx$320 °C) at about 150 times atmospheric pressure (15.2 MPa, or 2250 lb/in^2) is circulated through a "steam generator" (a heat exchanger for the high-temperature steam). The water in the heat-exchanger vessel boils as a result of the heat provided by the reactor, and this steam runs through a turbine as in a conventional fossil-fuel plant. The maximum thermodynamic efficiency for this process is 49%, but the actual efficiency is again about 34%. The

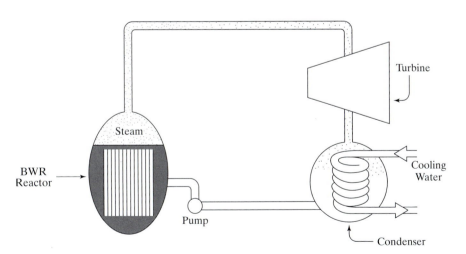

FIGURE 13.2

A schematic representation of a boiling water reactor (BWR).

FIGURE 13.3

A schematic representation of a pressurized water reactor (PWR).

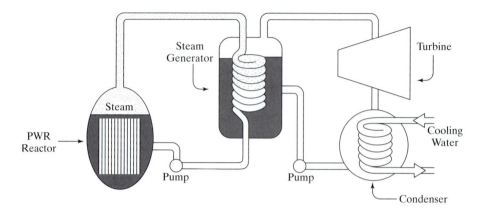

PWR is shown schematically in Figure 13.3. Sixty-seven PWRs are operating in the United States.

The common factor in both types of energy generation is the thermal energy produced. In accord with thermodynamics, this process must entail thermal energy discharged at low temperature. Consider a power plant burning fuel energy at a rate of 3000 MW to produce electricity for sale. Of the 3000 million joules per second produced, 1000 million constitute electrical energy transmitted to the consumer, and the other 2000 million joules (the "conversion losses" of Figure 8.1) are exhausted into the environment as **waste heat**. In practice, this means that an amount of water is heated in the power plant and then returned at a higher temperature to a river or lake, where it gradually cools by heat exchange with other water and the air, or the waste heat is exhausted directly into the atmosphere by cooling towers.

13.2   EFFECTS OF WASTE HEAT ON AQUATIC SYSTEMS

Waste heat exhausted into water is a particularly serious problem for two reasons: its effect on the dissolved oxygen content of the water and the effect of warmer water on aquatic systems. Both fossil-fuel and nuclear plants generate waste heat and use large quantities of water.

Almost any pollution in a river or lake demands oxygen in its neutralization. If the pollutants are organic, the amount of oxygen necessary for bacteria and protozoa to break them down is called the *biological oxygen demand*. If the pollutants are chemical, the amount of oxygen necessary to neutralize them is called the *chemical oxygen demand*.[2] We lump these together and simply use the term **biochemical oxygen demand** (BOD for short). Cold water contains more dissolved oxygen than warm water. The chance of a collision ejecting the entrapped gas is lower at a lower temperature, because the mean energy of the water molecules is lower (Chapter 7). Consequently, waste heat imposes additional stress on the "cleaning power" of a river or lake and reduces the amount of oxygen available. This has a debilitating effect on aquatic life.

The other effect of waste heat on aquatic life has to do with the rates of chemical reactions. It is a rule of thumb that rates of reaction in plants and animals double for each 10 °C rise in temperature (see **Extension 13.1, *Rate***).[3] It is probably to control reaction rates within their bodies that mammals evolved a fixed internal body temperature. Among other effects, waste heat causes increased plant and algal growth.[4] For fish, which take on the temperature of their environment, the problem is most serious. A slight temperature change can result in the replacement of previously dominant species by others (trout and salmon, for example, prefer relatively cool, well-aerated water and are especially sensitive). There are adverse thermal effects on most physiological processes, including reproduction, at temperatures several degrees below lethal high temperatures;[5] fish under constant heat stress are more susceptible to disease. Sex ratios can become skewed.[6]

Many electric facilities are situated on lakes (or bays) rather than on rivers. A temperate-zone lake generally exhibits thermal stratification in the late spring and summer, and the addition of waste heat increases and prolongs the stratification. The lower, denser layer of the lake is called the **hypolimnion**; this is much colder than the **epilimnion**, or upper, less dense, layer of the lake. During the summer, the epilimnion gradually becomes thicker, and the amount of dissolved oxygen in the lake—in the cooler hypolimnion—gradually decreases as a result of natural BOD. In the fall, the cooling of the epilimnion, which increases its density, allows the layers to mix (or turn over), replenishing hypolimnion oxygen levels. Valuable fish, such as lake trout, live in the hypolimnion and need this oxygen to stay alive and healthy. Thus, the longer the stratification exists, the less oxygen there is for the animals living there.

13.3   DISPOSING OF WASTE HEAT

Many power facilities use "once-through" cooling to eliminate waste heat. If a lake is used as a source of cooling water (a low-temperature reservoir) for a generating facility, water is taken from the lake and pumped through condensers. The steam or other working fluid is cooled in the condensers after flowing through the turbines. It is then recycled to the boilers to be made into steam again. The cooling water is exhausted to the lake 5 °C to 10 °C warmer than it was when it was pumped from the lake. Most likely, water will be taken from the cool hypolimnion and exhausted to the warmer epilimnion, increasing the difference in density. The exact temperature change of the water depends on the rate at which water is pumped through the condensers. To produce a given amount of energy requires a plant to exhaust a given amount of waste heat. This amount is fixed by the energy requirements of the plant, not the pumping rate of the cooling water. If we pump water through faster, we simply take away a smaller amount of waste heat for each liter of water pumped through.

An Environmental Defense Fund study for a proposed nuclear facility on Cayuga Lake in New York[7] found effects that *could* seriously affect the lake and its inhabitants:

- Stratification would begin earlier in spring and extend later into fall.
- The length of the growing season would be increased in the epilimnion.
- Water from the hypolimnion that would be brought to the surface would contain nutrients that were previously unavailable in the epilimnion.
- There would be a greater capacity for biological production in the epilimnion.

Such considerations apply to all plants that produce waste heat.

Prolonged **stratification** extends the period of oxygen depletion in the hypolimnion. The effect of the waste heat burden and the environmental stress on the lake could cause breeding problems for the fish in the lake. The increased biological production is on a primary level—algal growth is encouraged; then, when the algae decay, the process uses up oxygen, to the detriment of fish and invertebrate life. This process, called **eutrophication**, often occurs gradually in the natural aging of lakes, but it is accelerated by waste heat exhaust. Waste heat exhausted into rivers is generally less of a problem than it is in lakes, but river-dwelling fish can still suffer from the effects of higher temperatures.

Once-through cooling has become a rarely chosen strategy to dissipate waste heat,[8] because most prime sites, those with sufficient quantities of water available, have already been taken, and because the public is concerned about the effects of temperature increases. Use of artificial cooling ponds or canals for once-through cooling is possible, as the thermal energy is subsequently dissipated by evaporation of some of the hot water. The pond option is not often chosen, because sufficient land is not available for ponds in wetter climates, and sufficient water is lacking in drier climates. Occasionally, it is possible to build a facility on the site of an abandoned water-filled quarry.

There is another strategy for dissipation of waste heat: cooling towers, either of the natural-draft or the forced-draft type (Figure 13.4). Currently, this is the most popular method for dissipating waste heat. In **wet cooling** towers, the heat is dissipated by evaporation. This evaporation can cause fog to form in areas around the generating plants and has even been responsible for increased local rainfall and snowfall levels.[9] These "wet" methods are less effective when the relative humidity of the area is high.

In **dry cooling** towers, the heat is taken away directly by the atmosphere. Air passes across pipes containing hot water and carries the heat away by convection (that is, the actual mass motion of the air does the cooling). The warm air will rise, expand, and cool until its temperature is the same as the air around it. It expands because the air is normally at lower pressure as the altitude increases. The rate at which dry air cools with increase in height (decreasing pressure) is called the dry adiabatic lapse rate (0.01 °C/m, see Chapter 14), because the process takes place adiabatically (so rapidly that there is no exchange of energy to the surroundings). Dry towers are less expensive to maintain than evaporative towers, and they can be used even in areas with little water. However, dry towers are less thermally efficient than wet towers because of the energy used in running the fans.[8]

These alternative cooling plants are expensive to build and run compared to those that use once-through cooling. In addition, once-through cooling can usually provide lower water temperatures in the condenser than can be provided by the water that is run through cooling towers.[9] A comparison of the continuing expenses of competing strategies is shown in Table 13.1.[10] Only the mechanical draft takes an appreciable amount of energy to run (and so dry cooling is very expensive). Despite the costs, the cooling pond or tower method has been adopted because of the growing public perception expressed succinctly by Walter Heller in public testimony:[11] "The public is subsidizing these industries at least twice—once by rich tax bounties and once by cost-free or below-cost discharge of waste and heat." The EPA's regulations have made once-through cooling almost impossible.

In the current economic climate, upgrading older towers is more important than new construction, because there is very little new construction.[8] Natural-draft hyperbolic towers (the outer shell is in the shape of a segment of a hyperbola), more

FIGURE 13.4

Cooling towers such as these have been installed at both nuclear and fossil-fuel electric installations. (U.S. Department of Energy)

TABLE 13.1

Power requirements for cooling systems (percentage of output of 800 MW$_e$ plant)

Cooling system	Type of fuel	
	Fossil	Nuclear
Once-through: discharge to lake, river, ocean, or cooling pond	0.4	0.6
Wet cooling tower		
Natural draft	0.9	1.3
Mechanical draft	1.0	1.7
Dry cooling tower		
Natural draft[a]	0.9	1.5
Mechanical draft	3.0	4.8

[a] Construction of natural-draft dry cooling towers is quite expensive.

Source: Compiled from data in M. Eisenbud and G. Gleason, eds., *Electric Power and Thermal Discharges* (New York: Gordon and Breach, 1969), 372.

energy-efficient than forced-draft units, were a more popular option in the 1970s. Such units have performed well in North America and Europe. New units being designed are 130 m in diameter and 200 m high.[8] Europeans have favored smaller towers, with mechanical-draft assist units, because of lack of space. In America, most plants in operation have cooling towers of the forced-draft variety. More towers are being built of concrete rather than wood because of substantially lower maintenance costs.[8]

13.4   WATER AND POLITICS

A student of water politics has said in regard to water and politics: "Water is energy, and in arid lands it rearranges humans and human ways and human appetites around its flow."[12] Groundwater is a nonrenewable (or at best slowly renewable) source of such energy in many places.

Water is necessary for cooling most types of power plants. Water is a necessity for making energy from heat. The heat engine would not work if there were just one temperature reservoir. The water is the low-temperature reservoir, whether it is used as a once-through system or circulates to a cooling tower where it is evaporated.

Water can also make energy directly: Hydroelectricity is energy won from water as it falls. Water from oceans and lakes absorbs energy from the Sun to supply latent heat of vaporization (see Chapter 7), thereby changing liquid water to water vapor. The water vapor forms ice crystals and drops, which make clouds. Clouds eventually release their water as rain. The rain falls on land, runs off into streams or rivers, and may be caught behind a dam. The gravitational potential energy of the water behind the dam is, simply, stored solar energy.

Water withdrawal for use in the year 2000 was estimated to be $3.9 \times 10^{12} \text{ m}^3$, while water used was estimated to be $2.3 \times 10^{12} \text{ m}^3$ (the difference is wasted in the system).[13] Per person, Americans use 100 times more water than people living in an undeveloped country,[14] about 1.4 m^3 per day (1400 L/d), though only a tiny

FIGURE 13.5

Fresh water consumption in the United States 1940–2000. (S. S. Hutson, N. L. Barber, J. F. Kenny, K. S. Linsey, D. S. Lumia, and M. A. Maupin, *Estimated Use of Water in the United States in 2000*, Circular 1268, U.S. Geological Service, U.S. Department of the Interior, 2004; Statistical Abstracts of the United States, (Washington, DC: GPO, 2001))

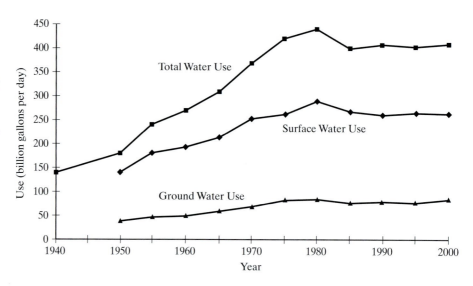

amount of total consumption, about 8%,[15] is used by public water utilities for drinking and sanitation. The United States withdrew 1.52 billion m^3 of water per day in 1995[15,16] (see Figure 13.5). Steam electric utilities account for 47% of total withdrawals, but only 2% of that amount was used for generating energy and was not returned to bodies of water (most withdrawn water was returned to the original source, albeit warmer).[15] Irrigation accounted for about one-third of withdrawals,[15] and another 5% was for manufacturing and mineral extraction.[15] Unfortunately, much of the water not used for drinking is wasted. The United Nations estimates that in 2000, industry wasted 4×10^{11} m^3 of water, while agriculture wasted 8×10^{11} m^3.[17]

Water exists on Earth in staggering quantities. To get a feeling for how much water there is, imagine Earth to be a smooth sphere coated with all Earth's water.[18] The estimated 1.5×10^9 cubic kilometers is distributed as described in Table 13.2.[15,16,19] Note that only two water molecules in every 10,000 are fresh surface water molecules.

TABLE 13.2

Water inventory on a smooth Earth

	Percent of all water	Height (m) above smooth spherical earth
Salt water	97	2700
Ice, snow	2.25	120
Antarctic ice cap	1.93	
Greenland ice sheet	0.20	
All other glaciers	0.12	
Fresh groundwater	0.73	45
Surface (water in soil, rest below soil)	0.006	
Above ≈1 km (below ground level)	0.36	
Below ≈1 km (below ground level)	0.36	
Fresh water	0.022	1.00
Water vapor	—	0.03

Source: References 16, 19, and 20.

Precipitation, averaged over the world, amounts to about 1 m/yr. Over the United States, the precipitation averages about 750 mm/yr,[21] which is marginally better than the world average of 710 mm/yr for land precipitation. For normal years, the precipitation in the continental United States ranges between 2.06 m/yr in the Pacific Northwest to less than 100 mm/yr in parts of the Pacific Southwest.[21] Figure 13.6 shows annual precipitation in the United States; regions with rainfall levels under 60 cm (24 in) are short of water in most years. The variability of wet and dry years can make the interregional disparity even greater. And this will get worse as populations grow in the world's drier areas.[22]

Agricultural irrigation is the major water consumer, particularly in the United States, India, China, and Russia.[15,23] (Agriculture uses 80% to 90% of water in the western United States.) Irrigation is useful, because it increases the cultivated area (raising yields); it increases crop yield; it allows for multiple cropping; and it provides farmers with some independence from the vagaries of weather.[24] It costs energy to irrigate; the major portion of energy used in irrigated agriculture is for supplying the water. Roughly two-thirds of all fresh water used is utilized to irrigate crops.[23] By 2025, almost half the world population may face water scarcity.[14,23,25–28]

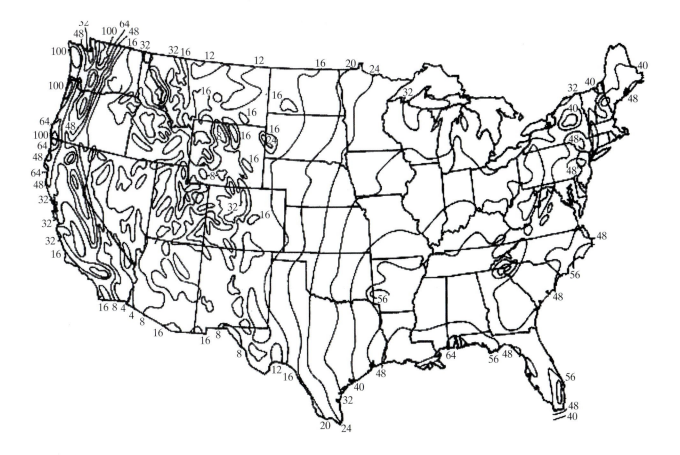

FIGURE 13.6

Average Precipitation in the U.S. in inches per year.
(National Weather Service)

Sometimes there are unexpected consequences of damming and irrigation. Vegetation loses water through its surfaces, a process called **transpiration**. Because most precipitation eventually evaporates from open-water surfaces, building dams or irrigating dry land increases evaporative losses. Over a large part of the North American landmass, evaporative loss exceeds precipitation. **Evapotranspiration** in the United States averages 630 mm/yr (leaving about 120 mm/yr for river flow from the 750 mm/yr of precipitation). Both of these effects are present in the salt loading of the Colorado River, which has caused friction between the United States and Mexico. The loss from open water in the 17 western states amounts to about 8% of capacity.[29]

Irrigation can have political ramifications as well. For example, American farmers in the Wellton–Mohawk district using the Colorado River for irrigation have made the lower Colorado so saline that the United States has built a 400 million L/d desalination plant in order to fulfill its water treaty obligations to Mexico.[30–32] This deficit of water has led to proposals such as the "Parsons plan," which would have pumped water from Canada and the American Pacific Northwest to Colorado, New Mexico, and Arizona.[32,33] California cities have bid irrigation water away from farmers.[34] In Arizona, where water rights go with land, cities buy up land to get water.[35] In California, the Metropolitan Water District paid $128 per acre-foot for conservation and took the conserved water.[35] There is more discussion of what to do to get water where it is dry and some alternative ideas for increasing fresh water supplies in **Extension 13.2, Getting More Fresh Water**.

13.5 CONSIDERATIONS IN BUILDING A POWER PLANT

In this section, specific real-life examples illustrate issues that were of concern to the utilities and the public during the preparatory environmental impact study for the Kaiparowits coal-fired plant. Kaiparowits was proposed by California utilities and was to be built in remote southern Utah. Partly as a result of public response to the impact statement, this plant was abandoned in 1976.

Construction

Many large nuclear power plant installations as well as conventional installations are built on the site of the plant rather than in a factory. Factory fabrication is advantageous, if it is feasible, because it is cheaper to build units in a factory and transport them to the site. This follows from our discussion of economy of scale in Chapter 5. Building a power plant at the site essentially means that one builds a factory with output that is a power plant; by using prefabricated units, the factory cost is amortized over all units produced.[97] Often, such onsite construction is necessary, because the facilities are one of a kind or because shipping already-constructed modules to the site is difficult.

To illustrate the savings associated with factory production, suppose the cost per installed kilowatt drops by a quarter each time production is doubled (in accordance with expectations developed in our earlier discussion of the learning curve). We may assume a large factory to produce 1000 MW$_e$ of installed capacity per year, and we suppose that we wish to build a 1000 MW$_e$ generating facility. Typical construction time is 5 to 10 yr; therefore, we shall choose a construction time of 8 yr (that is, 125 MW$_e$ of electric capacity installed per year). A comparison of the construction costs for differing amounts of on-site construction is shown in Figure 13.7. It is clear that the more construction that can be done in a factory, the better. The start-up of on-site construction introduces the greatest increase in construction cost. The cost has increased over 50% if only 10% of construction is in the field; full on-site construction is only about 60% more expensive than the first 10%. The last 84% of construction introduces only as much extra cost as the first 16% of construction on-site.

Construction in remote, underpopulated areas can result in even greater cost if worker housing must be supplied and labor brought in from outside. In addition, field labor is often of lower quality than that employed in factories, which can cause expensive maintenance or repair problems after the plant is in operation.

Boomtown Syndrome

Undeveloped areas that have a supply of natural resources may be suddenly transformed by plans for rapid use of the resource (the coal lands of Wyoming and Montana are examples of this phenomenon). The problem here is that a small indigenous population is pressed to provide greatly expanded local service to large numbers of new residents. Because of the lack of school space, recreational facilities, and other infrastructure necessities, it is difficult to attract the service personnel needed, and high turnover is thus much more likely. This leads to a drop in productivity, and thence profits, and further, to a drop in public services as tax revenues drop. This, in turn, causes even greater strain on the abilities of public servants, and coping becomes harder, leading the cycle farther downhill.[98]

While local government is stalled in its effort to minimize the impact of development because of widespread local sentiment against any interference with property rights, state and federal government can help by assisting the locality in the institutional changes necessary to cope with development.

The "boomtown" problem is now part of the overall environmental impact considered in siting new energy facilities. The now-abandoned plan for a large coal-fired power facility near Utah's Kaiparowits plateau required preparation of an environmental impact statement that dealt, in part, with where the incoming population would be housed and serviced.[99] We might regard this part as dealing with "social pollution."

Other social problems include a high rate of divorce (at least a third higher than in comparable counties, as reported in a study of Wyoming), greatly increased use of alcohol and consequent drunkenness, and a higher rate of school dropouts than in comparable counties.[100]

The Kaiparowits plan envisioned a scheme in which 25% of the plant's support staff would be housed 80 km away in Page, Arizona, where facilities were available to decrease the boomtown pressure. (Of course, this would greatly increase the net fossil-fuel cost of plant operation, because that 25% of the workforce would commute each day by private car or contractor bus.)

The plan was ultimately dropped because of lower-than-expected increases in electric consumption in Los Angeles,[101] a casualty of the first energy crisis.

Environmental Effects

Like other proposed generating facilities, the Kaiparowits plant was to be located in an area of great natural beauty. Kaiparowits had been sited close to Glen Canyon,

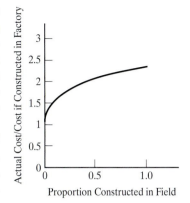

FIGURE 13.7

Due to the learning curve, the unit cost of power plant capacity decreases as the amount produced increases (here, we assume a 25% decrease when production doubles). For a 1000-MW$_e$ plant built over 8 years, the minimum cost of field construction as compared to that in a factory producing 1000 MW$_e$ of installed capacity per year is 2.3 times as great.

Capitol Reef, Bryce Canyon, Zion Canyon, Lake Powell, and several national forests. A smoke plume would be out of place in this natural setting. The plant would have been visible over great distances, and the transmission lines would have impaired the splendid natural vistas.

There would have been "unavoidable deterioration of air quality"[99] were the Kaiparowits plant to be built, including "periodic yellow-brown atmospheric discoloration"[99] due to nitrogen oxide emissions (perhaps as much as 10.4 tons per hour). The air quality would also have been lowered by the emission of particulates (0.58 tons per hour) and sulfur dioxide (2.2 tons per hour), and by the minor estimated emissions of 980 t/yr of particulate matter and 700 t/yr of sulfur dioxide added by the new residents brought in to work on the plant. (Air pollution is discussed in detail in Chapter 14.)

The proposed removal of 420 megatonnes (Mt) of coal from a nearby strip mine over the lifetime of the plant would have affected 63 mi^2, with concomitant noise and contamination of aquifers (porous rock features containing groundwater) from mined-out coal beds. Also, another 420 Mt of coal would have been rendered unrecoverable. The proposal would also have involved excavation of 1.6 million yd^3 of sand and gravel for construction and mining of 6.8 Mt of limestone from the region for concrete.

Other damage from implementing the Kaiparowits project would have included pollution of Lake Powell from the final estimated 50 million yd^3 of fly ash scrubber residue (containing arsenic, barium, fluorine, lead, mercury, and other materials); increased sediment yield; and increased salinity in local streams as well as in Lake Powell. Erosion losses would have been serious, because the local soils have such a poor capacity for recovery (almost three-fourths of the local soils would not recover during the 20-year project span). Salinity would have increased for several reasons: pumping-induced rock fracture between salt and fresh water aquifers, leading to mixing; evaporative salinization of the evaporation ponds, liquid from which would have then leaked into groundwater; and salt drift from the cooling towers onto the nearby vegetation, thence into the waters. The Colorado River would have increased in salinity by 2 mg/L.

Thus, to build Kaiparowits would have meant a loss of vegetation, impairment of habitat for wildlife, a decrease in amount and aesthetic value of recreational land, and general degradation of the environment. Although some of the particular effects cited are unique to the Kaiparowits plan, most of the problems would be encountered in the attempt to build a large coal-fired facility anywhere.

Energy Gain

Given the environmental impacts of Kaiparowits, it is interesting to consider the net gain the plant would have contributed in the energy supply system. By **gain**, we refer to the high-quality energy available to people (mostly as electricity) as a result of construction of a facility. In a universal sense, we merely transform some energy into other forms—energy is conserved. It is high-quality energy, however, that is of direct use to us. Earlier, we referred to one of the "invisible" energy costs of Kaiparowits: the amount of fuel that would have been used by workers in commuting. Other such "invisible" costs occur in this sort of project, including the energy cost of construction materials and machines and the energy value of the fuel used in the construction and operation of the facility.

It is not possible to evaluate these hidden costs of building an energy-generating facility comprehensively. The losses that can be measured are staggering enough. Using the projections in the Kaiparowits study,[102] only 50% of the coal in the underground formation would actually have been mined (the remainder, left in place to support the excavation, would have been lost forever), and 25% of the coal mined would have been wasted. Each year, on average, mining would have cost 710 GWh_e, and transporting the coal to the plant would have required 730 GWh_e. Assuming the plant to be 38% efficient in converting kWh_t to kWh_e, its overall efficiency in terms of the energy in the coal mined would have been only 25%; if we include in the energy available the other half of the coal resources that are rendered permanently useless, the overall efficiency would have been 12.5%. (Reference 100, which does not properly allocate kWh_e, obtains an overall efficiency of 15.5%.) This efficiency is further reduced if one takes into account the energy cost of producing the cement and steel used in the plant, the energy cost of transporting the material, the requisite earthmoving, and so forth. And this still does not count the energy cost of moving people to the site or the energy cost of daily commuting.

To be fair, this cost is shared by any such large undertaking and is not restricted to coal-burning facilities. Analyses of the energy cost of a 1000 MW_e nuclear installation used for 25 years[103] show that it would provide a net energy loss during the first 8 to 9 years after it is placed on line for reasons similar to those cited. Only after this time does the high-quality energy produced by the plant exceed the energy cost of its construction. (Using a thermal energy criterion, the break-even time is reduced to about 7 years.) Ultimately, of course, there is a substantial net energy gain, as indicated in Table 13.3.[103] This is also true of solar energy (see Chapter 21). Solar facilities built in the 1990s, for example, may not become net producers until after 2010.[104]

The point of these remarks is that we must be careful, if we build any sort of generating facility, to ensure that the pace of construction results in a rather slow increase in generating capacity. If we were to try to build all these facilities at once, we could cause a net energy *consumption* in the short run.

TABLE 13.3

Ratio of energy-out/energy-in reactors

	0.3% Ore	0.007% Ore
Pressurized water reactor	8–19	0.5–4.8
CANDU[a]	5–13	5–7.4
Heavy Water Reactor[b]	9–13	2.1–2.9

[a] Canadian-designed PWR.
[b] Canadian-designed BWR.
See Chapter 19, "Conventional Reactor Types."

13.6 WATER QUALITY

One often hears that in future years, power plants in the United States will require the total national water supply for cooling purposes. Such statements are somewhat exaggerated, because water used for cooling upstream can, at a later time, be used for cooling downstream. Nevertheless, any use degrades the water to some extent and renders some unfit for consumption. Tests of municipal water supply systems [2,105,106] have shown substantial quantities of carcinogenic material and "heavy metals," including arsenic. Arsenic exposure causes various nasty diseases, including cancer.[55] Arsenic is also a water problem in Britain,[107] where information on pollutants has not, in general, been available to the public. The inadvertent poisoning of the Hudson River (New York) by General Electric with the noxious polychlorinated biphenyls (PCBs; see **Extension 13.3, *The Hudson Story***) and of the James River (Virginia) by a subsidiary of Allied Chemical with kepone also stand out as water-degrading uses. These examples are indicative that the "pollution dilution solution" is simply not feasible over broad areas of our country.[131] Germany has also found that there is no solution in dilution, especially with the heavily polluted Elbe River.[132]

Water is required for many uses. In many parts of the United States, water use is already perilously close to supply. Thirsty Las Vegas is considering tapping an aquifer that underlies Death Valley National Park.[133] Four states already use more groundwater than surface water.[134] In most cases, aquifer recharge is less than what is withdrawn (see also **Extension 13.4, *Coastal Aquifers***). The Tragedy of the Commons applies here, as groundwater use is almost always unregulated, so the people who pump the most, win at the expense of others.[134] Making separate projections by region, only New England, Ohio, and the southern Atlantic and Gulf states will have an assured supply that exceeds commitment.[139]

The loss of wetlands has become a political issue in the United States; the major culprit in the Midwest appears to be the farmer.[140–142] The concern voiced about our water supplies has generated new ideas for water purification. One of the most interesting proposals is the use of bulrushes (*Scriptus lacustris*) in artificial marshes as filters for human sewage.[137] It is possible to spray wastes onto sandy soil. Water percolates through the sand filter to arrive purified at groundwater level, while solid wastes remain at the surface and eventually become humus.[106] This is impractical if the wastes contain metals, and the site would have to be isolated from people because of bacteria in the wastes.

13.7   WATER AND ENERGY IN SEMI-ARID REGIONS

Another area of new perception involves water use in obtaining energy in the semi-arid western United States. Montana, Wyoming, and the Dakotas are problem areas. The push for energy will have a mammoth effect there, because there are very thick coal beds near the surface (in contrast to experience in the eastern part of the country). With the coal beds near the surface, extraction will be by strip mining—removing overburden and piling it, thereby despoiling many hectares of land. In addition, these states do not get much rain, and thus, the weathering of the piles is very slow. There are 52 million ha (128 million acres) of this coal and lignite.[143]

The North Central Plan, a joint endeavor of 25 utilities, proposed development of water and coal resources in over 250,000 mi^2 centering on the Gillette-Coalstrip (Wyoming–Montana) area. This plan, now temporarily shelved, proposed 42 sites in five states for coal-fired steam-generating plants to produce an additional 50 GW$_e$ by the year 2000. (The 39 existing coal plants in the western states now generate 9.3 GW$_e$.) Thousands of miles of 765 kV transmission lines would have linked Medicine Bow, Wyoming, to St. Louis, Missouri. This may yet happen if the plan is revived.

Three new pumped storage facilities (see Chapter 22) would provide another 3 GW$_e$. The project would consume 855,000 acre-feet per year (half the New York City consumption; an acre-foot is the amount of water required to cover an acre of land a foot deep in water). If gasification plants are added, the demands rise to 2.6 million acre-feet per year.[141,143] This is one-third the average flow of the Yellowstone River.

In a wide swath of the semiarid western states, surface and near-surface rock is permeated with an oil-like mineral. Such rock is called oil shale (see Chapter 12). Some attempts have been made to exploit the energy locked up in the oil shale, and prototype treatment facilities have been designed. Oil shale treatment would require large amounts of water: three barrels of water for each barrel of oil produced.[142] In addition, there is the problem of disposing the spent shale in the arid climate. Runoff from a shale oil project could increase the salinity of the Hoover Dam reservoir by 50%.[144]

Other water problems follow from exploitation of such energy minerals as coal and oil shale. A study by the National Academy of Science[145] has issued several warnings:

1. Only areas with rainfall greater than 250 mm/yr seem to have high potential for rehabilitation (given good management). These make up about 60% of strippable western coal lands.
2. *Restoration* is "not possible anywhere."
3. Coal seams are also aquifers, serving livestock and domestic wells. Mining could "dewater" a large number of wells.
4. There is the problem of destruction of ephemeral streams (dry gullies and arroyos) that carry off thunderstorm or snowmelt. These are vital features of the arid bioscape.

With the water shortages already mentioned in the western region, it behooves us to decide rationally whether we wish to use local water to create another Four Corners (an enormous coal-fired plant near the point where Utah, Colorado, New Mexico, and Arizona meet; its smoke plume is visible from orbit) in the beautiful landscape of Wyoming, Montana, or North Dakota.

SUMMARY

Utilities build generating facilities, thereby effecting local land, air, and inhabitants. In most processes, thermal energy is converted to electricity, with actual thermodynamic efficiencies around one-third. In the most common means by which electricity is generated, fossil fuel and nuclear, water is used to provide a low-temperature reservoir. The warmed water is expelled, carrying waste heat to the environment. Waste heat can have deleterious effects on life in the body of water supplying the facility. For this reason, many newer facilities use wet or dry cooling towers to release the waste heat to the atmosphere instead of to local bodies of water. (Dry cooling is very expensive.)

Other uses of water can affect the environment on a continental scale through evapotranspiration from dam reservoirs, salinization, and so on. Water is already in short supply in the southwestern United States and in arid regions the world over. It may become so in other areas of the country and the world in the future.

In making a decision about which type of generating facility to build, a utility must consider several factors to the extent required by law: (1) it must assure that sufficient water is available for cooling, so that the effects of waste heat rejection on the local biota are minimized, including suppression of biochemical oxygen demand; (2) it must decide how much of the facility should be prefabricated as opposed to constructed onsite (it is much cheaper to build using prefabs); (3) it must consider the effect of building a facility on the social structure of the community servicing it; and (4) it must deal with any emissions of pollutants into the atmosphere and water. Utilities must also consider the rate at which they build facilities in order to spread out demands on capital, as well as to assure that some net output of energy is available. It takes over 5 years before there is net energy output from a plant (considering the energy required to build it).

PROBLEMS AND QUESTIONS

MULTIPLE CHOICE REVIEW

1. Water quality depends most on the presence of
   a. dissolved oxygen.
   b. PCBs.
   c. plastic.
   d. high BOD.
   e. carcinogens.

2. Waste heat exhausted into water
   a. affects the dissolved oxygen content of the water.
   b. affects aquatic systems.
   c. helps stabilize the temperature of the water.
   d. does all of the above.
   e. does only a. and b. above.

3. The actual efficiency of a boiling water reactor generating facility is approximately
   a. 75%.
   b. 50%.
   c. 30%.
   d. 10%.
   e. 5%.

4. In a steam electric generating facility,
   a. an open flame heats the feed water in the steam vessel.
   b. steam drives a turbine to produce electricity.
   c. a condenser is used to cool the steam to water.
   d. the typical efficiency is 33%.
   e. all of the above statements correctly characterize the situation.

5. The actual efficiency of a pressurized water reactor generating facility is approximately
   a. 75%.
   b. 50%.
   c. 30%.
   d. 10%.
   e. 5%.

6. The biological oxygen demand is
   a. the minimum amount of oxygen in a river or lake.
   b. the amount of oxygen necessary for fish to survive.
   c. the amount of oxygen necessary to break down chemical pollutants.
   d. the amount of oxygen necessary to break down organic pollutants.
   e. none of the above choices.

7. During the early years of a power facility's operations,
   a. the facility produces its cheapest energy.
   b. the facility represents a net loss in energy because of the energy cost of its construction.
   c. the facility has practically no environmental impact.
   d. the facility uses very little cooling water.
   e. all of the above conditions correctly characterize the situation.

8. Dry cooling towers are
   a. more efficient at cooling than wet cooling towers.
   b. more efficient for nuclear plant cooling than for fossil-fuel plant cooling.
   c. an environmentally acceptable way to dissipate waste heat from a generating facility.

d. more expensive to operate than wet cooling towers.
   e. likely to cause increased local snowfall.

9. Water is a cause of friction between people and governments because
   a. it is important for economic activity.
   b. without it, crops wither and die.
   c. it is necessary for life.
   d. it is essential to proper sanitation.
   e. of all the other reasons.

10. In making a decision about which type of generating facility to build, a utility must
    a. be sure that there is sufficient water available for cooling.
    b. consider the effects of building the facility on the community.
    c. deal with any emission of pollutants into the atmosphere.
    d. decide how much of the facility should be prefabricated.
    e. do all of the above measures.

SHORT ANSWER

Explain whether the statement made is correct, and justify your answer.

11. Cooling water is not needed for the pressurized water reactor.
12. Aquatic life is affected by waste heat.
13. When temperature goes up, BOD goes up.
14. It is not possible to increase the efficiency of thermal power plants because of thermodynamics.
15. All thermal power plants run basically the same way.
16. Wet cooling towers are the most efficient way to eliminate waste heat.
17. Irrigation is one of the largest uses of water by volume.
18. Coal power plants release particulates as well as other materials.
19. There are social as well as physical effects to power plant construction.
20. Strip mining can be environmentally benign.

DISCUSSION QUESTIONS

21. Because evaporation of water from reservoirs is an important water loss mechanism, especially in dry climates, it is important to reduce losses. What measures could you suggest for reducing such evaporation? Are any of these actually feasible?
22. The EPA suggested that General Electric should pay to clean up PCBs that were allowed to get into the Hudson River. General Electric has objected. The EPA suggests that wildlife is being affected more than can be explained by leaks in the factories; the company counters that dredging will release even more. Discuss this controversy with as much objectivity as you can.
23. What is the reason cities buy water rights from farmers? Is this a good or a poor way to address the problems?
24. In what ways could a plant such as Kaiparowits be built with less environmental impact?

25. Discuss the similarities and differences between the various sorts of thermal energy production systems.

26. Why are some regions of the United States, such as Ohio, unlikely to run short of water, while others are very likely to run out of water?

27. Explain why the temperature change of coolant water in once-through cooling depends on the rate at which water is pumped through the condensers.

28. Why is stratification sometimes dangerous to the fish that live in a lake?

29. How likely is oil shale to be a component of the U.S. energy policy? Explain.

30. It is often stated that natural systems are capable of cleaning pollution (at least when the contamination is not large). How can plants and animals in the environment clean water?

PROBLEMS

31. A certain chemical reaction combines 2.55 kg/s of reactant when the temperature is 10 °C. At what rate do the chemicals combine if the temperature is
    a. 20 °C?
    b. 30 °C?
    c. 60 °C?

32. Suppose a power plant, which produces a total of 1000 MW of electric power, is 35% efficient. Water (of density = 1000 kg/m^3) passes through a heat exchanger at a rate of 30 m^3/s. What temperature increase, in °C, is experienced by the cooling water?

33. If a 1000 MW power plant were able to reach the thermodynamic greatest efficiency, it would be approximately 66% efficient. Water (of density = 1000 kg/m^3) passes through a heat exchanger in this hypothetical plant at a rate of 20 m^3/s. What temperature increase, in °C, would be experienced by the cooling water in this plant?

34. A 1000 MW electric power plant is 33% efficient. If the cooling water used in the plant is raised in temperature by 7.5 °C, how much water is being pumped through the heat exchanger?

35. A 600 MW nuclear facility with an efficiency of 30% operates at two-thirds the maximum thermodynamic efficiency possible. What is the temperature of the plant's pressure vessel (the high-temperature reservoir) if the plant is being cooled by river water of temperature 17 °C?

36. In a certain region of the United States having an area of 3.5 × 10^4 km^2, evapotranspiration is 68 cm/yr, and rainfall is 80 cm/yr. How much water is carried away by the region's streams and rivers? Give the answer as both water mass and water volume.

37. Given that a chemical reaction takes 5 min to complete at a temperature of 20 °C, how much time should the reaction take at a doubled temperature (40 °C)?

38. A 1000 MW PWR with a lifetime of 40 yr and an expected capacity factor of 96% is constructed in such a way that the lifetime ratio of energy out to energy in is expected to be 16. How much energy was used to construct the reactor?

39. A 1000 MW PWR with a lifetime of 40 yr and an expected capacity factor of 96% is constructed in such a way that the lifetime ratio of energy out to energy in is expected to be 4.2. How much energy was used to construct the reactor?

40. What would happen if the entire United States got the rainfall characteristic of the Pacific Southwest? What if instead it were the rainfall characteristic of the Pacific Northwest?
    a. How much extra water would be available in the latter case nationwide?
    b. What would the water deficit be compared to the average rainfall of 0.75 m/yr if the entire country experienced Pacific Southwest rainfall?
    c. How much extra water would there be compared to the average rainfall of 0.75 m/yr if the entire country experienced Pacific Northwest rainfall?

CHAPTER 14

Pollution from Fossil Fuels

Earlier chapters of this book described how energy is generated for electricity and other purposes. In this chapter, we extend the discussion of the consequences of burning fossil fuels. This chapter emphasizes effects of what are called fixed sources (for example, industrial and utility installations). We consider how pollution is generated, what the health consequences of combustion can be, and how it might be possible to ameliorate the effects of pollutants in the fuel. We look at how acid rain is generated and what its consequences can be. In Chapter 15, we discuss some of the effects that the transportation system has on the air around us that contribute to pollution problems.

KEY TERMS

pollution / heat island / electrostatic precipitator / scrubbers / adiabatic lapse rate / temperature inversion / PAN smog / linear nonthreshold dose hypothesis / collective dose hypothesis / heavy metal tracers / acid rain / pH / buffering / wet deposition / dry deposition

14.1 THE ATMOSPHERE

There is an ocean surrounding us—an ocean of air. The volume of the atmosphere is, in fact, over twice that of the world's water oceans (see box *A Closer Look*: Mass and Volume of the Atmosphere). It is useful to think of the air around us as an ocean, because the interdependencies of the parts of the system are then emphasized. Experience with Lake Erie in the early 1970s, when it was written off as dead, shows that any *enclosed* body that is overused may suffer a loss of utility for all. Lake Erie has recovered somewhat, but only through great effort on the parts of the bordering states and Ontario, Canada.

What is **pollution**? Human beings by their very presence change the local situation where they congregate. Things will get into the air and water that were not there before. We could designate any such change as pollution. However, the atmosphere and flowing and still water themselves constitute a natural change control system. Rain washes pollutants out of the air. Particulates eventually fall to Earth under their own weight, the buoyant force that acts on them, and air resistance. Natural sources contribute "natural changes." In this book, we will follow the obvious definition of pollution—pollution is *recognizable degradation* in conditions. A few people who live in one place would probably not notice any changes in their environment, because the impact is so small. In many cases, whether there is *recognizable* pollution will depend on how many people are crowded together. If people gather densely enough to overcome natural cleansing, pollution is the result.

People have been polluting the atmosphere for quite some time. Such pollutants come in two classes:[1]

> those which take part in natural cycles and can be assimilated if their levels are trivial compared with the background levels of the natural cycles,

and

> those which do not take part in natural cycles and so accumulate and cause injury.

Pollution is an unavoidable consequence of energy and materials production. Our response to pollution must take this fact into account. The prudent response is to determine the level of a given pollutant that will not accumulate injuriously and aim to produce no more than this amount. Care must be taken to recognize that materials that are not injurious pollutants in a *global* sense may still be injurious pollutants in a *local* sense.

In this chapter, we shall consider in greater detail the problems of urban environments, air pollution, and acid rain (see Figure 14.1). In the following chapter, we consider effects on gross properties of the entire atmosphere—the climate— of the uses to which we are putting our atmospheric ocean through the burning of fossil fuels and wood.

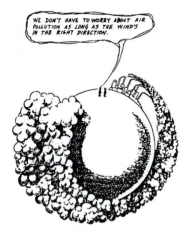

FIGURE 14.1

"We don't have to worry about air pollution as long as the wind's in the right direction."
(*Die Zeit*, 26 October 1984. Used by permission, Zeitverlag Gerd Bucerius KG.)

14.2 AIR IN URBAN ENVIRONMENTS

A smaller body of water is affected more severely than a large body of water when pollution is introduced into the water system. It is easy to see how small bodies of water can exist in isolation from other bodies of water. How, though, can we be allowed to make an analogy between small bodies of water and local environments of air?

The most important reason for the analogy is the knowledge that weather may be local. Los Angeles' smog does not affect San Francisco—San Francisco has its own. Each city has its own conditions, but we can generalize somewhat, and note that cities are typically 2 or 3 °C higher in temperature than rural areas in the afternoon (and can be as much as 7 °C higher), especially in winter, because the stone, brick, and asphalt absorb heat better than Earth, and because there is no transpiration;[2–7] see Table 14.1.[2]

TABLE 14.1

Comparison of urban and rural areas

Condition	Urban to rural comparison
Temperature	Annual mean 0.5 °C to 1 °C higher
	Winter minimum 1 °C to 2 °C higher
Relative humidity	Annual mean 6% lower
	Winter 2% lower
	Summer 8% lower
Dust particles	Ten times more
Clouds	5% to 10% more
Winter fog	100% more
Summer fog	30% more
Total horizontal surface	15% to 20% less
Ultraviolet radiation	5% to 30% less
Wind speed	Annual mean 20% to 30% lower
	Extreme gusts 10% to 20% lower
	Calms 5% to 20% more
Rainfall	5% to 10% more

Source: References 2, 3.

A CLOSER LOOK

Mass and Volume of the Atmosphere

The density (amount of mass in a unit volume) of air near Earth's surface is 1.29 kg/m^3. We may estimate the mass of the atmosphere atop us in the following way. Consider an imaginary cube of air in still atmosphere. A cube of side l has area l^2 and volume l^3. The cube does not move, because the air is still, despite the fact that the cube of air has weight. There must be an upward force on the bottom face of the cube just equal to the weight of the cube plus the net downward force on the top face of the cube (see Figure 14.2). This upward force is known as the buoyant force, a force that acts in the upward direction; the atmosphere buoys objects up, just as salt water buoys a swimmer's body up. The difference is that the atmosphere is much less dense than salt water.

The weight of the cube is its mass times g, or

$$(\text{density}) \times (l^3) \times (g),$$

because the factor (density) (l^3) is just the cube mass. The force on any cube face is the product (pressure)(area), so,

$$P_{\text{top}} \times (l^2) + (\text{density}) \times (l^3) \times (g) = P_{\text{bottom}} \times (l^2)$$

or

$$P_{\text{top}} + (\text{density}) \times (l) \times (g) = P_{\text{bottom}}.$$

The pressure at the bottom of the atmosphere, then, 1.013×10^5 Pa, is due to the weight of all the atmosphere pressing down from on top. Hence, imagining the atmosphere to be uniform in density, we get (for $P_{\text{top}} = 0$) an estimate of the thickness, l, of a uniform atmosphere:

$$\text{thickness} = P_{\text{bottom}}/(\text{density} \times g)$$

$$= (1.013 \times 10^5 \text{ N/m}^2)/(1.29 \text{ kg/m}^3 \times 9.8 \text{ m/s}^2) = 8 \text{ km}.$$

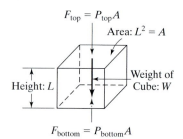

$$F_{\text{top}} = P_{\text{top}}A$$

Area: $L^2 = A$

Height: L

Weight of Cube: W

$$F_{\text{bottom}} = P_{\text{bottom}}A$$

FIGURE 14.2

The vertical forces on a cube of air are due to the air pressure at the top and bottom of the cube and the cube's weight. The net upward force, $F_{\text{bottom}} - F_{\text{top}}$, is numerically equal to the weight of the cube. It is called the buoyant force on the cube.

This answer (for a uniform density atmosphere) is much smaller than the actual atmospheric thickness. It actually extends to heights over 100 km. In fact, satellites in low Earth orbit (at 200 km, for instance) experience some atmospheric drag that leads to eventual return to the surface.

The volume of the real atmosphere can be found, because it is the same as for the uniform atmosphere, by

(surface area of Earth) $\times$ (thickness of the uniform atmosphere),

where the surface area of Earth is $4\pi R_E^2$, and R_E is Earth's radius, about 6400 km. As a result, our uniform atmosphere has volume 4.09×10^{18} m^3. The total atmospheric mass is

mass of the atmosphere = (air density) $\times$ (volume),

or

$$\text{mass of the atmosphere} = \frac{(\text{air density}) \times (4\pi R_E^2) \times (P_{\text{bottom}})}{(\text{air density}) \times (g)}$$

$$= (4\pi R_E^2) \times (P_{\text{bottom}}) = 5.28 \times 10^{18} \text{ kg}.$$

We do not know the oceans' exact volume, but it may be estimated. Ocean covers about three-quarters of Earth's surface. The average depth is about 2.7 km (Table 13.2). The ocean's volume, then, is

$$(0.75) \times (4\pi R_E^2) \times (d_{\text{average}}) = (0.75) \times (5.14 \times 10^{14} \text{ m}^2) \times (2700 \text{ m})$$

$$= 1.04 \times 10^{18} \text{ m}^3.$$

The uniform atmosphere has about four times the volume of the oceans. The mass of salt water in the oceans is

mass of salt water = (salt water density)(volume)

$$= (1025 \text{ kg/m}^3) \times (1.04 \times 10^{18} \text{ m}^3) = 1.07 \times 10^{21} \text{ kg}.$$

The oceans have over 200 times the mass of the atmosphere.

Heat Islands

Cities are classified as "**heat islands**." Luke Howard, an English amateur meteorologist, originated the idea in the nineteenth century. He compared temperatures from several places in London with those measured away from the city on the same day at the same times. In his 1818 book *The Climate of London*,[8,9] he found "London always warmer than the country, the average excess of its temperature being 1.579 degrees." At night, a modern city may be 5 °C to 7 °C warmer.[3] That warmth is costly: Electrical cooling cost increases 0.5% to 1% for each degree temperature increase. About 2% to 8% of all electricity used is spent to compensate for the heat island effect (about one-sixth of Los Angeles' energy is used to cool buildings).[3,10] Not only are cities hot, they are getting hotter by about 0.5 °C per decade![11] Tokyo's nighttime average temperature rose by 4 °C over the last century, and the average temperature is up 4.5 °C since 1990.[12] The situation with thermal air pollution can have serious consequences; mortality rises when hot air comes to stay.[13,14] The air pollutants that cause smog are more likely to cause it to form when temperatures are high.

Recently, it was found that there are even regional heat island effects—the entire East Coast might be one huge heat island.[8] Presently, urban areas cover about 3% of the U.S. land area, and that continues to increase. Worldwide, there is flight of people from the rural areas to cities, so this may become a more worldwide problem in the near future. Different cities' populations react to high temperature levels in different ways; a University of Delaware program used by the National Weather Service uses historic data and weather forecasts to allow warning of anticipated problems when the weather gets hot.[15]

There may even be "cool islands," as the National Aeronautics and Space Administration (NASA) found that Salt Lake City may have reduced temperatures after residents planted water-guzzling trees from the East.[8] This makes it clear that people can do something about the problem. Arthur Rosenfeld and his group at Lawrence Berkeley Laboratory (LBL) have determined many actions that can help.[10,16] Cities can reverse the effect of the heat island by planting a lot of trees and taking other actions. The U.S. Department of Energy estimates a savings of $1 billion/yr if 100 million trees were planted. Trees raise water from the ground level to the canopy and allow it to evaporate, cooling the surroundings. Planting three trees per house can save up to 50% of the energy used for air conditioning.[3] For example, Nanking, China, has planted 34 million trees to beautify the city and has seen its average temperature fall 3 °C.[3] The cost of taking action against the heating of American cities is about one cent for each kilowatthour of electricity saved, much lower than the cost of the electricity.[11]

A primary cause of the effect in cities is dark roofs; not only do these cause heat absorption, but they also increase air-conditioning costs.[10,16] In the computer models the LBL group used, the Los Angeles afternoon temperature decreased by 3 °C, and as a result, ozone decreased by 10% when trees were planted, pavements were changed, and lighter roofing materials were used.[16] Table 14.2 summarizes the savings determined for Los Angeles if these changes were made.

Buildings with light-colored roofs reflect more sunlight. A new rating system called the solar reflective index (SRI) is being developed for roofing materials so that the homeowner may choose the most appropriate roofing material.[8] A homeowner trying to cool a house could paint it to reflect 90% of the incident sunlight or increase the vegetative cover around the house by 30% to have the same effect.[17] One computer simulation showed that whitewashing buildings could save Sacramento, California, 14% of its cooling peak power and 19% of

TABLE 14.2

Savings to Los Angeles if trees were planted, pavement were reflective, and roofs were white

		Direct savings	Indirect savings	Smog benefit	Totals
Cooler roofs	Average peak power	400 MW	200 MW	104 M$	600 MW
	A/C cost savings	46 M$/yr	21 M$/yr		171 M$/yr
Trees	Average peak power	600 MW	300 MW	180 M$/yr	900 MW
	A/C cost savings	58 M$/yr	35 M$/yr		273 M$/yr
Cooler pavement	Average peak power	0	100 MW	76 M$/yr	100 MW
	A/C cost savings		15 M$/yr		91 M$/yr
Total	Average peak power	1000 MW	600 MW	360 M$	1600 MW
	A/C cost savings	104 M$/yr	71 M$/yr	360 M$/yr	535 M$/yr

Source: Reference 10.

total electricity used for cooling.[17] It is more effective to whitewash buildings that have little insulation. The nice thing about this finding is that roofs are generally replaced on a regular basis; changing to a light roof when the change is made does not cost anything extra and provides monetary savings. The LBL group is working to find materials that increase infrared reflectivity that hold the promise of saving even more in air-conditioning costs.[9]

The change in temperature and other effects of pollution affect weather patterns. A study of the St. Louis area's precipitation patterns showed that, while nearby farms lost income because of air pollution, they gained back more income than they had lost from the increased rainfall.[7] Such increased precipitation has also been found at sites as varied as Long Island, Australia, and Washington, D.C.[18] The National Weather Service found that Atlanta's heat island caused increased thunderstorms downwind.[8,9] Storms even bypass the city, and the rain falls on the suburbs.[8] There are "pollution tracks" blown by the wind from pollution sources. In these tracks, there was very little rain, because the water droplets were much smaller than normal. This apparently occurs because there are so many small particles on which the water vapor in air can condense.[19]

The "heat island" effect makes heat waves more deadly. From the deadly European heat wave of 2003 (over 2000 in the United Kingdom and about 1400 excess deaths in the Netherlands alone, and probably 35,000 premature deaths overall),[20] the Chicago heat wave of 1995 (739 excess deaths), the New York heat wave of 1972 (891 excess deaths), to the Los Angeles heat wave of 1955 (946 excess deaths), these deaths occur mostly in cities.[21]

Factors in Local and National Pollution

Local pollution comes from cars, trucks, and other moving sources; from evaporation of paint and solvents; from industrial processing; and from power plants burning coal, oil, or natural gas. The average auto a decade ago that traveled 10,000 miles emitted much more of every pollutant than the 1998 model, as we see in Table 14.3, over five times as much on average. Drivers of the Volkswagen Lupo emit under one-tenth as much as the average 1990 car. While each car or truck, at least on average, emits less, there are more of them around driving greater distances. Total emissions are falling more slowly. And localities' emissions may have increased if a lot of people with vehicles are moving in, even if nothing else happens.

Local pollution can also come from sources that are industrial or utility and do not move; they are known as stationary sources. See **Extension 14.1,** *Pollution*

TABLE 14.3

Pollutants emitted in driving 16,000 kilometers (10,000 miles), in kilograms

	CO	HC	NO$_x$	PM$_{10}$
Average 1990 car[a]	295	48	23	5.45
Average 1998 car[b]	159	17	17	0.33
Average 1998 pickup/SUV[b]	182	0.1	19	0.06
2001 Volkswagen Lupo[c]	—	4.5	2.8	0.26

Note: CO, carbon monoxide; HC, unburned hydrocarbons; NO$_x$, nitrogen oxides; and PM$_{10}$, particles of size under 10 μm.

[a]Reference 22.

[b]Reference 23.

[c]Reference 24.

Control Systems for a discussion of ways to decrease pollutants, such as particulates, sulfur dioxide (SO_2), and nitrogen oxides (NO_x) using **electrostatic precipitators** that catch particles using high voltage, **scrubbers** that strip pollutants out of flue gases by chemical reactions, and strategies that reduce the proportion of oxygen in the combustion chamber to prevent NO_x from forming in the first place. Both oil and coal have some sulfur. Coal from the eastern United States contains large amounts of sulfur, while western U.S. coal is lower in sulfur content (but is also lower in heat of combustion). (Some oils from the Middle East are high in sulfur as well. Libya, for example, has done very well selling crude oil in world markets because of its very low sulfur content, even though the Libyan leader Muammar Ghaddafi made many enemies among the customer countries by his former support of international terrorism.) Utilities emit additional pollutants in large amounts, including so-called "heavy metals." The effects of "heavy metal" emission are detailed in **Extension 14.2**, *Heavy Metals*.

Even with scrubbers to remove sulfur, sulfur emissions can be substantial. The sulfur encounters water to become sulfate, a haze. The local pollution added together becomes regional pollution. As an example, visibility in the western United States is 150 km, as it should be everywhere; visibility in the eastern part of the country is never more than 20 km or so because of the permanent haze hanging over the entire east coast.[56] A day after the 14 August 2003 blackout, an aircraft was flown to test air pollution changes. The researchers found that SO_2 decreased by about 90%, ozone was down by about half, and visibility increased by more than 40 km.[57]

Local pollution episodes can be bad indeed; it is more dangerous to health to live in cities than in rural areas, even at "normal" times. Local differences have been apparent for some time in the case of cities surrounded by heights. Pittsburgh, Pennsylvania, for example, was notorious for its pollution as early as the turn of the century; it is in a valley surrounded by the Alleghenies. Los Angeles is in a bowl-shaped plain surrounded to landward by mountains. What has occurred more recently is noticeable local pollution, even in areas not hemmed in by such topography. This results from weather-induced density differences in the air.

Thermal Inversion

In Chapter 13, we referred to the **adiabatic lapse rate**, the rate at which temperature decreases as altitude increases (for dry air)—about 10 °C/km or 0.01 °C/m. In normal conditions, the temperature decreases with height for a few kilometers. Figure 14.3 shows the situation. If the air temperature drops more steeply than the lapse rate, air is unstable. If the air temperature drops less steeply than the lapse rate, the air is stable.

If the air is in an unstable condition, air from near the surface will rise and be replaced by cooler air sinking from above. This is what happens in home heating; you will see that all cold air returns are at floor level. Unstable air is good for an area in that the local pollution gets carried away and diluted by the rising air. Stable air conditions in flat areas are responsible for the buildup of pollutants in the air. Pollution alerts are often called when these conditions prevail.

In a valley, the air may become stagnant as the wind blows across the top of the hills or mountains surrounding the valley. For areas that are relatively flat, this cannot be the mechanism of stagnation. In the case of flat areas, stagnation has to do with the dynamics of the atmosphere. In hot weather, the ground (and buildings) heats up more quickly than the air during the day and cools more quickly than the air at night. At night, the ground cools, and air near the surface may be cooler than the overlying air layers, setting up a condition in which the higher the air

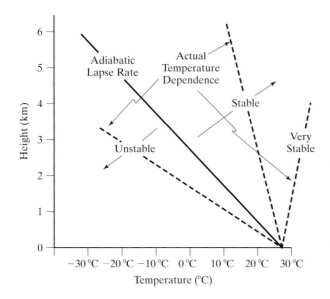

FIGURE 14.3
The temperature profile expected for a surface temperature of 27 °C involves a decrease in temperature of 10 °C for every kilometer rise above the surface. This line is the adiabatic lapse rate. If the actual profile is to the left of the lapse rate, the air is unstable. If to the right, the air is stable.

A CLOSER LOOK

How Air Rises and Sinks

Suppose that the actual lapse rate is greater than the adiabatic lapse rate for an imaginary cube of dry air. If we move the cube upwards, it would cool at the adiabatic lapse rate and become warmer than the surrounding air. Air cools as it rises, because the atmospheric pressure decreases with height. As air rises, it expands; the work for the expansion comes from the air's thermal energy, causing its temperature to decrease. (Recall from Chapter 7 that temperature is the measure of thermal energy.) Because the imaginary cube of air is warmer than the surrounding air, it is less dense and would keep on rising. This means that air under such conditions is not stable. Conversely, if the actual lapse rate is smaller than the adiabatic lapse rate, upward movement of the imaginary cube will result in the cube becoming cooler, and thus more dense, than the surrounding air. Dense air will tend to sink; it will return to its original position if it is moved from it. Air is, therefore, stable when the actual lapse rate is less than the adiabatic lapse rate.

The lapse rate changes if the air is wet. Cold air holds less water than warm air (which is why we use humidifiers on forced-air furnaces, as discussed in Chapter 9), so cooling air raises the relative humidity of the air. After the air has cooled (at the adiabatic lapse rate) to the point at which the air holds as much water as it possibly can at that temperature, so that the relative humidity is 100%, any additional cooling will cause the water to condense as drops (as we can observe from the condensation of drops on cold glasses in summer). As the drops form, thermal energy is released, so that the wet air will not cool as fast as would dry air.

is, the warmer it is—a **temperature inversion**, which means the temperature changes opposite to the way predicted by its lapse rate. So, the air temperature will actually rise with altitude for a few hundred meters before falling again with increasing height. In Figure 14.3, this profile would show a line sloping to the right (as the stable air profile shows) before curving back to slope to the left (as the adiabatic lapse rate does). Such an inversion disappears soon after the Sun comes up.

Air in high-pressure regions is forced downward, thus becoming warmer, and is denser as the air molecules are forced together. This sinking takes place on top of the usual air layers, so that temperature decreases as the air rises from the ground and meets the warmer air from the high-pressure region at an altitude of a few hundred meters. In the region where the temperature increases with height, the air is, of course, stable, and the stability is so strong that what is in the air becomes trapped in place and very high pollution levels are likely. This condition is another type of temperature inversion, caused by local air pressure conditions: the air temperature first decreases with height (as usual), then increases with height, and then finally decreases with height. (In Figure 14.3, such a profile would slope left from the ground, then turn right, and then turn left again.)

A well-studied pollution episode, Episode 104 (extreme cases are given numbers), which affected much of the eastern part of the United States for an extended period in 1969, was of this latter sort.[6] This episode occurred in late summer, a time during which inversions are common. During the day, the hotter ground warms the air near it and usually causes the inversion to break up, as indicated above. Episode 104 was so severe because a large high-pressure system existed over the entire eastern half of the country. Because the temperature did not decrease with altitude, local air was trapped underneath the layer of warm air. This isolated the air in the locality. Because there were sources of pollution there, the air became more polluted the longer the high-pressure system prevented mixing with clean air. Clearly, because we cannot control weather, the one way to prevent such occurrences as Episode 104 is to decrease local pollution levels.

As with the heat islands, the flow of people to the cities has led to high exposures to particulates (in almost every megacity in the world, levels are above World Health Organization (WHO) guidelines; see **Extension 14.3**, **WHO Air Quality Guidelines**),[72] ozone (in Los Angeles, Mexico City, São Paulo, and Tokyo, especially),[72] SO_2 (in Beijing, Mexico City, and Seoul, with observable effects on health),[72] and lead (Cairo, Karachi, and Dhaka).[45,72] Mexico City experiences peaks of 900 $\mu g/m^3$ in ozone.[72]

Again, larger forces often control what the pollution concentrations will be in any given location. For example, El Niños (see Chapter 16) have a large role in determining whether Los Angeles and Santiago, Chile, will experience severe air pollution episodes.[73] The number of Stage 1 smog alerts in Los Angeles has declined on average since the 1970s (148 in 1970, 122 in 1978, 1 in 1997, none in 1999 and 2000),[73,74] and the number of days carbon monoxide exceeded federal standards also declined (38 days in 1987, 20 in 1996, and 15 in 1997),[73] while some years have much lower numbers than expected and some higher, the overall trend is clear. This trend is the result of actions taken; the smaller fluctuations are the result of longterm weather patterns.

Will the Atmosphere Always Be Able to Clean Itself?

As mentioned above, the atmosphere has always been able to clean up the effects of pollution eventually, and there are natural sources of sulfur and nitrogen oxides. Carbonyl sulfate is given off by soils and the oceans, dimethyl sulfide by marine organisms, volatile organic compounds (VOCs) by plants, and so on.[75–77] Predation on plants increases VOC release. For example, animals that chew pine needles multiply needle monoterpene releases by a factor of 15.[78] (Also, the increase of carbon dioxide discussed in Chapter 17 will lead to increased VOC emissions from plants.)[78]

About a gigatonne of carbon is present in the atmosphere at any time in the form of VOCs.[77] These compounds participate in chemical cycles in the atmosphere and are part of the natural cleansing cycle. Their presence affects how much rain falls and where it falls. However, the most important chemical involved in cleansing is the hydroxyl radical (written OH^-).[75–77,79]

The hydroxyl radical is made by ultraviolet (UV) light, which can remove a hydrogen atom from water in the atmosphere in the presence of ozone. It is effective at eliminating VOCs and sulfur dioxide. (It has been called "the detergent of the atmosphere,"[75,80] and it destroys most air pollutants and many gases involved in ozone depletion and the greenhouse effect as well as VOCs and SO_2.) There appears to be one little problem. The hydroxyl radical has been on the decline in atmospheric concentration since 1988.[77,81] Prinn et al.[81] found that the greatest effect in the northern, most heavily industrialized, hemisphere.[81,82] The answer to the question at the start of this section is still unknown, although increasing effects of Asian winter haze might make it impossible for the atmosphere to cleanse itself.[83,84]

14.3   HEALTH EFFECTS OF POLLUTION

Air Pollution Catastrophes

Emissions and weather conditions such as those described above led to severe pollution events in the 1960s and 1970s. Those events served to shatter complacency, to awaken the public to the dangers of air pollution. The public recognized the health problems that followed a severe episode and realized that there may still be a health problem, even when the concentration of pollutants is not as great as in a severe episode.

But, warning flags had been set out earlier. The first of the extreme episodes to be clearly documented occurred in the Meuse Valley of Belgium in 1930 and caused about 63 deaths; another took place in Donora, Pennsylvania (a Pittsburgh suburb) in 1948 and caused about 20 deaths. Undoubtedly the most acute air pollution disaster (and the outstanding example of the power of such incidents to stimulate governmental action) was the December 1952 London smog that resulted in an originally estimated extra 4000 deaths,[85,86] although a reanalysis estimates about 12,000 excess deaths.[87] (There was controversy about whether the flu was responsible for many of the deaths; that was not the case.)[88] Pollution levels during the London smog were from about five to 20 times above current regulatory standards, and both soot and "heavy metal" particulates were found in archived tissue from decedents.[88] In this case, mortality was concentrated among the very young and the very old, the two most susceptible groups. The large number of deaths led to the appointment of a Royal Commission and to the close monitoring of mortality and air pollution data (see Figure 14.4). The less severe episodes of 1956 (about 1000 excess deaths), 1957 (about 900), and 1962 (about 700), as well as the close monitoring of death rates, made a compelling case that sudden increases in the death rate followed each abrupt change in pollution levels. The subsequent studies of both mortality (death) and morbidity (sickness) seem to have been

FIGURE 14.4

Air pollutant levels and excess mortality in London, December, 1952.
(D. Bates, Reference 63, by permission McGills-Queens University Press. Over 4000 excess deaths occurred.)

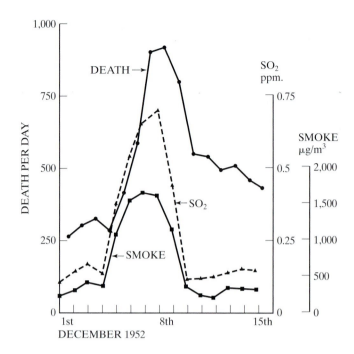

able, to some extent, to quantify the deleterious effects of urban (polluted) life on health. The 1952 disaster prompted changes in British laws and practice that have been effective in preventing a recurrence.

Perhaps Americans should not have been surprised that more of these would happen; but happen they did, to continuing surprise. Some 180 New Yorkers died prematurely in a nasty 1953 smog episode.[89] On one infamous Thanksgiving weekend in 1966 in New York, an estimated 270 extra deaths occurred (Episode 273).[6] This may have been the most severe incident in the east in the 1960s, but Los Angeles in the west became synonymous with smog and hacking coughs. The Los Angeles smog, caused by VOCs and nitrogen oxides and sunlight, and featuring thick haze and high ozone levels, is known as **PAN smog** (see **Extension 14.4**, *PAN Smog*).

National experiences with poor air quality led citizens to demand that government do something about it, first in California, then nationwide. The Environmental Protection Agency (EPA) was created by President Richard Nixon in 1970, and the first Clean Air Act (CAA) cleared Congress. Much progress has been made since, with consequences that are described in **Extension 14.5**, *The 1990 Clean Air Act Amendments (CAAA) NO_x, particulates, and the EPA* and below.

In the pollution incident in Clairton, Pennsylvania (near Pittsburgh), in November 1975, 14 people may have died as a result of high particulate levels:[18] 929 $\mu g/m^3$, compared to national standards of 50 $\mu g/m^3$ average, and the maximum allowed concentration of 75 $\mu g/m^3$ in one 24-h period per year (see Table 14.4). The level of pollutants in inversions can get even higher. In January 1985, the Ruhr region in Germany suffered a 5-day pollution episode. In Dortmund, measured sulfur dioxide levels were reported as 1118 $\mu g/m^3$ for a long period.[258] In the Western Ruhr, all traffic was banned and all factories closed. In the rest of the Ruhr, traffic was banned until 10:30 in the morning. (As the Sun heats the ground during the day, the inversion disappears, only to

TABLE 14.4

California standards (1999) and National Ambient Air Quality Standards (NAAQS)

Air Pollutant	Period[a]	California		Federal		
				Primary[f] (ppm)		Secondary[f] (ppm)
Ozone[b]	1 h	0.09	180 µg/m³	0.12	235 µg/m³	Same as primary
	8 h			0.08	157 µg/m³	
Carbon monoxide	8 h	9.0	10 mg/m³	9.0	10 mg/m³	None
	1 h	20	23 mg/m³	35	40 mg/m³	None
(Lake Tahoe)	8 h	6.0	7 mg/m³		None	
Nitrogen dioxide	1 h	0.25	470 µg/m³			Same as primary
	Annual mean	None		0.053	100 µg/m³	
Sulfur dioxide	1 h	0.25	655 µg/m³	None		
	3 h	None	None	0.50	1300 µg/m³	
	24 h[b]	0.04	105 µg/m³	0.14	365 µg/m³	
	Annual mean	None		0.03	80 µg/m³	
Respirable particulate matter (PM$_{10}$)	24 h			30 µg/m³	50 µg/m³	Same as primary
	Annual mean		30 µg/m³[d]	50 µg/m³[e]	50 µg/m³[e]	
Fine particulate matter (PM$_{2.5}$)	24 h				65 µg/m³	Same as primary
	Annual mean				15 µg/m³	
Sulfates	24 h		25 µg/m³			
Lead	30 d		1.5 µg/m³			
	quarter				1.5 µg/m³	1.5 µg/m³
Hydrogen sulfide	1 h	0.03	42 µg/m³			
Vinyl chloride[c]	24 h	0.010				No federal standards
Visibility-reducing-particles	8 h (10 A.M. to 6 P.M., PST)	Range to less than 10 mi (30 mi at Lake Tahoe) at relative humidity <70%				

[a] The air quality to be acceptable under the standards must not exceed the value given during the length of the period more than once a year. If annual is indicated, it may not exceed the level given.

[b] With ozone ≥0.10 ppm, 1 h average, or total suspended particulates (TSP) ≥100 µg/m³, 24 h average.

[c] Vinyl chloride is identified as a Toxic Air Contaminant by the ARB, which allows action if the 10 ppb 1978 standard is exceeded.

[d] Geometric mean.

[e] Arithmetic mean.

[f] ppm, parts per million, µmol/mol, or 1×10^{-6}.

be reestablished in the evening.) The developed world recognizes these as serious problems. The irony is that the 1952 London smog pollution levels, far above that in these examples, approximate current pollution levels in some areas of the developing world.[87]

Quantifying Health Effects

The pollutants most directly linked to increased sickness and death include ozone, particulate matter, carbon monoxide (CO), sulfur dioxide (SO_2), volatile organic compounds (VOCs), and oxides of nitrogen (NO_x).[193,259,260] Bronchitis incidence is much higher for people living in regions of high particulate and sulfur dioxide levels.[13,261] Heart patients are adversely affected by elevated CO levels.[177,262] In Poland, where pollution is much greater than in the United States, the health effects are more stark. Dr. H. Kowalski is quoted[263] as saying "Whenever the smog is heavy, the wards fill up." Low birth weights, premature births, and stillbirths are correlated with air pollution levels.[14,264] In Poland, doctors send some afflicted people into uranium mines for the good air![263] It has been found that children have difficulty in breathing if the annual average SO_2 concentration is >0.037 μmol/mol.[265] Nitric acid in polluted areas has been implicated in increasing incidence of nephritis (kidney disease) as well as in increases in the levels of fats in the blood.[266] For NO_2, there are health effects at concentrations as low as 0.11 μmol/mol. At 0.45–1.5 μmol/mol, there is impaired pulmonary function, increased incidence of acute respiratory disease, and breathing difficulties, even in healthy individuals.[265,267] It has been suggested that if pollution were to decrease to minimal levels, the death rate in the United States would be reduced by 38%,[268,269] and medical expenditures for chronic illness would be reduced as well. Even in relatively clean Norway, researchers found an 8% increase in premature deaths due to an increase of 10 μg/m^3 in average exposure to nitrogen oxides.[270] Nitric acid also causes cirrus clouds to thin, which may affect climate (see Chapter 17).[271]

Smoking is an air pollution menace by itself, known to cause cancer in innocent bystanders as well as smokers. Smoke contains carbon monoxide and small particles in the smoke. Particulates and tars can lead to sickness, whether the smoke is from tobacco or another source. This is why cities have higher rates of lung disease than suburbs, and suburbs higher than farmland areas. Lung cancer death rates drop (when smokers are compared to smokers or nonsmokers to nonsmokers) from 500 per 100,000 people in urban areas to 39 per 100,000 people in rural areas;[261] and stomach cancer death rates in high-pollution regions of New York are double those for low-pollution regions.[261]

Epidemiology of Pollution

In working with pollution's risks to health, a model must be adopted. (A similar model is discussed in Chapter 20 for risks from nuclear decay.) Exposure is defined as the pollutant concentration. The dose is the amount of pollutant that gets into a body, which depends on the exposure. The general model for health risks is[272]

$$\text{Dose} = \text{Potency} \times \text{Exposure},$$

where potency measures the relative strength of the effect. The total risk to the population is assumed to be

$$\text{Health risk} = \text{Dose} \times \text{Exposed population}$$
$$= \text{Potency} \times \text{Exposure} \times \text{Exposed population}.$$

The best way to reduce health problems when the exposed population cannot be separated from pollution altogether is to reduce exposure. This has driven attempts to find the causes of various maladies and to find effective ways to eliminate those causes. More detail on the epidemiology for NO_x and particulates is available in **Extension 14.5**, and for SO_x in **Extension 14.8**, *Acid Lakes, Acid Rain, and Nitrogen and Sulfur Oxides*. One point that deserves mention is that exposure to pollution is correlated inversely with income level, so minorities, who are poorer on average, are more exposed to pollutants.[273]

Throughout this book, conclusions about effects are derived from what is traditionally designated as the **linear nonthreshold dose hypothesis** or the **collective dose hypothesis** (see also **Extension 20.9**). The detrimental effects of low doses are difficult, if not impossible, to obtain. However, high doses often have perceptible effects. The reasonable hypothesis is that data of the effects of high dose or concentration on a relatively small group (whether a test on humans, animals, or plants, or an unexpectedly affected group of humans) can be extrapolated collectively and linearly to the effects of much lower doses on a very large group. This process would tend to overestimate the effect at small dose and so is conservative (i.e., estimated death rates and costs are probably higher than actual ones). There can be situations for very low doses where the conclusions may actually have no (i.e., a zero) effect or even an oppositional (i.e., a beneficial) effect. A well-known natural example is sunlight: too much can cause cancer, but some direct exposure is beneficial to health.

As an example of the linear nonthreshold dose idea, suppose an incident occurred in which 10 of the people died out of 1000 who were exposed, and the average exposure came from breathing a 500 mg dose of a certain pollutant. The death rate would then be 1% per 500 mg dose. According to the collective dose hypothesis, if 1 million people were exposed to a 100 mg dose, the death rate would be (100 mg/500 mg) $\times$ 1% = 0.2%, which means (because 1 million people were exposed) 0.002 $\times$ 1,000,000 people = 2000 people would die. The conservativism comes because it is possible that a 100 mg dose could readily be tolerated by the body, and no one would die (a threshold for producing deaths could exist somewhere between a 100 and a 500 mg dose). On the other hand, trying to extrapolate to doses higher than 500 mg using this linear extrapolation might lead to an underestimate of the death rate, because the toxic effect may be worse than a mere linear extrapolation to higher doses.

Pollution's Costs to People

It is clear that pollution costs people in both money and health. Health damage to humans can lead to costly medical bills. Health damage to plants can be substantial and cause monetary losses for farmers.

Atmospheric particulates encourage corrosion, soiling of textiles, and weakening of paint and other protective coverings. By as early as 1957, spinach crop losses of growers in Los Angeles from ozone totaled half a million (1957) dollars; New Jersey pollution damage in 1966 was estimated at $5 billion.

A 1913 Pittsburgh study indicated a pollution cost of $20 per person per year (excluding health costs); the 1954 Beaver report estimated a cost of $35 per person per year in urban areas; and studies in Steubenville, Ohio, and Uniontown, Pennsylvania, indicated an increased average cost of $84 per family in increased house maintenance, laundry, and personal grooming care.[274] A similar Canadian study in 1965 indicated an average cost of over $52 per person per year (rising to $94 per person per year in Toronto) due to air pollution.[268] More recent studies[275] based on more complete information about the costs of air pollution damage have found much greater costs today, even accounting for the effects of inflation (Table 14.5). China's costs due to air pollution are, at 2% of the Chinese gross domestic product (at least $13.25 billion), quite a drag on development and health.[274]

In response to such evidence, the U.S. Congress authorized the EPA to propose federal air quality standards (see **Extension 14.5**). The standards for air

TABLE 14.5

Comparison of per capita pollution costs (excluding health) in various studies, given in constant (2000) dollars

Study	Per capita yearly cost
Pittsburgh, 1913	384
Beaver Report, U.K., 1954	242
Steubenville/ Uniontown, 1966	456
Canada, 1965	
Average	288
Ontario	397
Toronto	522
New Jersey, 1966	392
California, 1992	923

Source: Adapted from W. Bach, Reference 274, except California figure from Reference 275.

TABLE 14.6

Tolerance levels and relative toxicity compared to NAAQS and German air quality standards (in µg/m³)

Pollutant	Tolerance level (µg/m³)	Relative toxicity	NAAQS (µg/m³)	German (µg/m³)
Carbon monoxide	40,000	1	44	50
Hydrocarbons	19,300	2.1	160	
Sulfur oxides	1430	28	0.09	300
3 h maximum			0.40	1000
Nitrogen oxides	514	78	100	500
Particulates	375	107	50	
24 h maximum			150	

Source: Selected from References 113, 114, and 277.

quality are designed to protect the health of the most sensitive groups in the population affected. Other developed countries are following the lead. In Table 14.6, the U.S. standards[265] (see Table 14.2), the less strict German standards,[276] tolerance, and toxicity levels are given.[278] "Tolerance level" implies that below such a level, the pollutant is tolerable for some time for members of sensitive groups. Toxic substances can cause illness or death; the table indicates that particulates are a hundred times as toxic as carbon monoxide.

Damage to the Lungs

In order for atmospheric pollutants to affect us, they must enter our bodies. The body has two main defenses against pollutants: in the airways leading to the lungs, mucus and cilia entangle large particles; in the air sacs (alveoli), macrophages, or large cells that engulf foreign matter, can sop them up. Those pollutants that make it into the lungs usually transfer through the approximately 70 m² of surface available there (the lung tree has 15 million branches). The bronchial system is illustrated in Figure 14.5.

Particulates

Particles of certain sizes are much more likely to reach and remain in our lungs. Larger particles are trapped in the nose and upper respiratory tract by impact (being carried by momentum into tissue where airflow direction changes, mainly for particle diameter >0.5 µm) and sedimentation (particles falling under influence of gravity, also mainly for diameter >0.5 µm; this mode increases as the residence time increases).[279] Smaller particles, those of size 1/100 to 1/2 µm, will be preferentially deposited in the lungs.[13,280] Diffusion deposition (particles being pushed about as a result of impact of other molecules) is the most important trapping mode for diameter <0.2 µm.[279]

The pernicious role of particulates was not fully recognized until the 1990s. In an experiment in Utah, C. Arden Pope checked hospital admissions before, during, and after a steel mill strike in Provo, Utah, and compared them to a measure of particulates, PM$_{10}$ (the concentration of particles having diameters less than 10 µm). Admissions to hospitals jumped 50% to 90% when PM$_{10}$ was above 50 µg/m³,[197] which is below the legal limit (Table 14.2). Pope found that the death rate increased by 16% per 100 µg/m³ of PM$_{10}$. Given this "smoking gun," research went on to identify even smaller particulates, those with diameters less

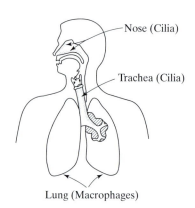

Nose (Cilia)

Trachea (Cilia)

Lung (Macrophages)

FIGURE 14.5

The human bronchial system and its defense mechanisms.

than 2.5 μm, $PM_{2.5}$, as even more important to control. The EPA has responded by regulating $PM_{2.5}$. (For more detail on particulates and EPA regulatory decisions, see **Extension 14.5**).

The atmospheric lifetime is greatest for particles with diameters 0.2–2.0 μm.[281] Many natural particles can irritate the lungs: Dust from dust storms and salt from spray evaporation are larger than 0.3 μm; particles from Sun-induced gas reactions and smoke are smaller than 0.2 μm.[282] In the past, the measure of particulate pollution quoted was "total suspended particulates," or TSP. However, only the smaller particles (those of diameter smaller than roughly 10 μm) are dangerous to health; for this reason, the concentration of particles of diameter less than 10 μm, PM_{10}, and particles of diameter less than 2.5 μm, $PM_{2.5}$, are now the parameters of choice with which to measure particulate air pollution.

The particles penetrating the lung's defenses can cause great trouble, especially if toxic gases or liquids are present on the surface of the inhaled particle. About 30% of particles smaller than 1 μm remain in the lung once they are there.[94] Once these particles get into the lung, the soluble particles can be rapidly transferred across the air sacs in the lungs and circulated in the blood. Small amounts of gases that dissolve in water (that is, are soluble) will be trapped in the nose unless they are carried through on small particles.

Ozone, nitric acid, and particulates are particularly prevalent in urban areas. Ozone causes breathing problems in asthmatics, children, and the elderly. Inhalation of ozone leads to pulmonary edema (for short-duration exposure) and to pulmonary fibrosis, or premature aging of the lungs (for chronic exposure).[283,284] Exposures allow recovery, but the lung heals to, as B. C. Jordan of the EPA was quoted as saying,[285] "something like scar tissue." Tissues are replaced by thicker ones. The lungs become stiffer.[285] Studies in beagles show a delayed response to air pollution, especially to NO_x. The disease is at least irreversible, possibly progressive.[283,286] In humans, NO_2 exposure leads to emphysema, but O_3 exposure does not.[281] Many materials emitted in exhaust are organic compounds that are mutagenic or carcinogenic.[287] Overall, it has been found that there are approximately 0.02–0.60×10^{-4} lung cancers/$(\mu g/m^3)$, suggesting 100 to 3500 excess cancers per year.[288]

The smaller particles put directly into the atmosphere are produced almost exclusively in engines and furnaces,[94,289,290] and it is just these that are most dangerous, as we have seen.[279,290,291] Diesel exhaust particles are typically 0.2–0.3 μm in diameter, with a dense core.[19,279] Metals from exhaust are typically 2–3 μm in diameter; lung absorption efficiency is about 50%.[279] Small particles are also produced in the daytime by sunlight and last for about half a day before they coagulate into particles large enough to fall under gravity. (Particles larger than about 10 μm fall to the ground rapidly when the air is still.) The atmospheric distribution of particles is shown in Figure 14.6.[103,279,292,293] It has two peaks: one for particles smaller than 1 μm peaking at about 0.1 μm, and the other for large particles, peaking at about 10 μm.[103,279,292,293] There is increasing evidence that poisonous elements tend to concentrate in the smaller particles, but even dust particles can cause silicosis[94] and impair lung function. An autopsy on an Egyptian mummy showed scarred lungs from the omnipresent desert sand.

These smaller particles we are discussing, which most easily evade body defenses to penetrate deep into the lungs and are the most toxic, are most heavily concentrated in urban areas. This is so for two reasons: because of the high concentration of cars and trucks in the urban areas (60% to 90% of particulates from autos are fine particulates [290]) and the other sources of coal and oil combustion

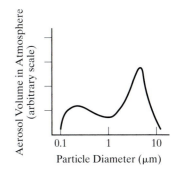

FIGURE 14.6

The distribution of total numbers of particles in the atmosphere in terms of their sizes.

TABLE 14.7

Comparison of health effects of selected pollutants

Chemical	Concentration for health hazard	Infant mortality	Cardiovascular disease	Viral diseases, respiratory tract	Chronic bronchitis, asthma	Lower respiratory, emphysema
Sulfur oxides, sulfuric acid	25 µg/m^3	X	X	X	X	X
Nitrogen oxides	8–11 µg/m^3		X	X	X	X
Lead						
Cadmium			X		X	X
Nickel						X
Beryllium				X	X	X
Mercury				X		X
Arsenic					X	
Vanadium						X
Chromium						
Asbestos				X	X	X
Polycyclic organic matter						

there; and because of condensation of particles formed by chemical reactions stimulated by the Sun (peracetyl nitrate smog—PAN smog; see **Extension 14.4**).

Research on coal combustion has found that there are two distinct peaks in the amount of mass released as fly ash. One peak occurs for large particle diameter, as expected. The surprise was the very sharp peak for particle diameter near 0.1 µm.[292] This peak occurred for all types of coal burned. Although the peak represents only a small mass release, it is responsible for most of the number of particles (and surface area) in the fly ash. Reducing the combustion temperature to control the release of nitrogen oxide (NO_x) has the desirable side-effect of reducing generation of fine particles (see **Extension 14.1**).[26,292] A compilation of some environmental chemicals and their effects on health is presented in Table 14.7.

The southern California region,[106,265] the New York City region,[294] and Texas are all pollution problems.[122–124] However, California and New York have seen great reductions in pollution levels, while Texas has not improved—in fact, now Houston is the most polluted city in the United States.[122] In 1991, the South Coast Basin exceeded at least one federal standard 184 days.[109] The PM$_{10}$ highest annual reading is in the South Coast Basin; the highest 24-h value is found in the adjacent Southeastern Desert Air Basin.[265] The pollution results because about 10 kt/d of pollutants (1375 tons HC, 1208 tons of NO_x, 4987 tons CO, 134 tons SO_x, and 1075 tons of particulates)[109,265] are released in greater Los Angeles. Four of six federal standards (for O_3, CO, NO_2, and PM$_{10}$; see Figure 15.9) were exceeded in the South Coast Basin.[295] They had dropped significantly over the period from 1976 to 1990, except for PM$_{10}$. Between 1955 and 1991, the ozone peak declined from 680 nmol/mol to 300 nmol/mol.[109] Smog had been reduced. In the decade since, pollution has declined even as more people moved into California and the Los Angeles area.

Sulfur Compounds

Sulfur dioxide is a pernicious health hazard, as Table 14.8 shows.[291] Closer to home, haze over the Great Smoky Mountains in Tennessee appears to be due to sulfate from coal burned in the Ohio Valley, and American urban and rural populations are exposed to roughly the same sulfate concentrations.[293] The surface deposition of

TABLE 14.7

(Continued)

Diseases of the central nervous system	Kidney damage	Anemia fatigue	Bone changes	Cancer	Hypertension	Skin ulcers, dermatitis	Visual disorders	Gastrointestinal disorders
X	X	X						
	X	X	X	Prostate, renal system	X		X	
				Lung		X		
X						X		
X						X	X	X
				Lung, Skin		X		
						X		
X		X		Mesothelomia		X		X
				Scrotum			X	

sulfur in the states in the upper Midwest and Northeast is over 8 t/km², while the U.S. average is only 1.4 t/km².[296] One report estimated that some 20,000 people east of the Mississippi die each year as a consequence of sulfur dioxide emission.[297]

Overall Costs of Pollution from Utilities in Mortality and Health

We have indicated throughout this chapter that pollution from utilities is responsible for sickness, death, and other deleterious health impacts. The American Lung Association found that more than half of Americans are breathing polluted air that will cause such problems. The median number of lives lost to air-pollution-related causes is estimated at 27,000 to 58,000 lives per year, at a median economic cost of $178 billion per year.[298] By 2007, many regulations will have taken effect (including the EPA Summer Smog rule) that should reduce the number of deaths substantially.

A study by the consulting firm Abt Associates, Inc. for the Clean Air Task Force of Boston, Massachusetts, predicted that a 75% reduction in particulate emissions from 1997 levels would enhance air quality considerably over expected 2007 levels and result in a median savings of $111 billion per year, mostly attributable to reduced mortality (mortality reduction estimated at 13,000 to 20,000 lives spared per year [median values quoted]).[298] Another Abt study funded by

TABLE 14.8

Annual health effects associated with sulfur oxide emissions, based on postulated conversion rates of sulfur dioxide and sulfur tetroxide and EPA epidemiological data for representative power plants in the Northeastern United States

	Remote location	Urban location
Cases of chronic respiratory disease	25,600	75,000
Person-days of aggravated heart–lung disease symptoms	265,000	755,000
Asthma attacks	53,000	156,000
Cases of children's respiratory disease	6,200	18,400
Premature deaths	14	42

Source: C. L. Comar and L. A. Sagan, "Health effects of energy production and conversion," *Annu. Rev. Energy* **1**, 581 (1976). Reproduced, with permission, from the *Annual Review of Energy*, Volume 1, © 1976 by Annual Reviews, Inc.

the Rockefeller Brothers Foundation and based on the epidemiological literature models found that, overall, the eight largest utilities will be responsible for about 5900 premature deaths in 2007, a rate considerably lower than that now (but still large), as well as a large number of asthma attacks and bronchitis cases.[299] According to the study, American Electric Power (AEP) alone is to be responsible for 1400 deaths per year, followed closely by the Southern (Electric Company) utility system at 1200 deaths per year. These two largest utilities are responsible for almost half the deaths predicted.

Is Pollution Local?

San Francisco does not seem to suffer from Los Angeles smog. Isn't this sort of pollution just local? Los Angeles doesn't normally export its pollution because it's hemmed in by mountains and open to the Pacific (and Japan is 7000 km away). Urban pollution in other parts of the country may travel long distances. One strategy for reducing *local* pollution is to construct high chimneys. This has the effect of more thoroughly mixing sulfur dioxide in the atmosphere. Tall chimneys in Britain cause sulfur dioxide pollution in Norway,[259] as the pollutant is released at such a high altitude. Recently, strides have been made in improving our ability to track down sources of pollutants.[300]

Ozone levels of air entering the New York metropolitan area are often already above federal standards. New York adds its own pollutants, which travel up to 300 km through northeast Massachusetts.[301] Plumes from single power plants can cause excess ozone concentrations to distances greater than 56 km.[302] A study in the New York metropolitan area indicated that ozone concentrations are regional in character, as are concentrations of PAN smog, acrolein (CH_2CHCHO), nitric acid, and other such pollutants. Control of metropolitan NO_x appears to be the most effective strategy for overall reduction in air pollution in the New York metropolitan area.[303]

New techniques using **heavy metals tracers** (arsenic, selenium, zinc, antimony, indium, and noncrustal vanadium and manganese) are able to distinguish the area emitting the pollution by its distinctive chemical signature.[304–306] Most coal has high selenium levels.[293,304] However, manganese and vanadium may be removed from the atmosphere faster than sulfur.[36] In addition, the soot picks up special signatures known as "sootprints" from the way it is burned (there are roughly 20 classes of soot).[306] It is not only the eastern United States that is covered by haze; Arctic haze, first noticed in the 1960s, can cover up to 9% of Earth's surface,[304,307] and at its peak is 2% of New York's peak pollution values.[308]

Because of the tracer techniques, we now know that most of the sulfur measured in northern Canada ($\approx 80\%$) comes from Russia and Europe.[307] The Arctic suffers from European and North American releases.[308] The spring Arctic haze decreased 50% between 1982 and 1993.[309] The decrease is a result of more pollution control in Europe and the switch of Russia from coal and oil to gas.

These tracer techniques allow researchers to point back to the sources of pollution. Canada has pointed its figurative finger at the U.S. Midwest as the origin of much Canadian air pollution, and New York State has long implicated the same source as the major contributor to Adirondack lake acidification. Contrariwise, much Northeast pollution is attributable to Canadian sources.[310] The methods were tested in two New England cities, one in Rhode Island, one in Vermont, and worked well. It was found that local sources in the Northeast contribute about as much pollution as the Midwest in New England (even though

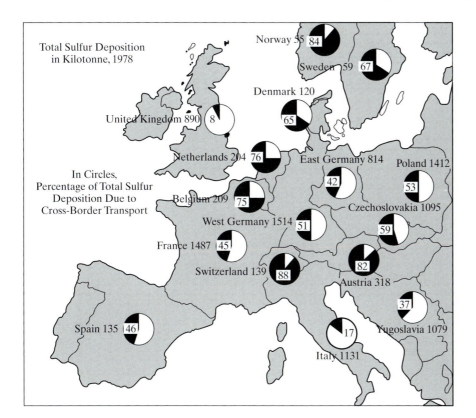

Total Sulfur Deposition
in Kilotonne, 1978

In Circles,
Percentage of Total Sulfur
Deposition Due to
Cross-Border Transport

Norway 55 | 84 |
Sweden 59 | 67 |
Denmark 120 | 65 |
United Kingdom 890 | 8 |
Netherlands 204 | 76 |
East Germany 814 | 42 |
Poland 1412 | 53 |
Belgium 209 | 75 |
Czechoslovakia 1095 | 59 |
West Germany 1514 | 51 |
France 1487 | 45 |
Switzerland 139 | 88 |
Austria 318 | 82 |
| 37 |
Spain 135 | 46 |
Italy 1131 | 17 |
Yugoslavia 1079

FIGURE 14.7

The boundary of a country is no defense against sulfur deposition. The countries in Europe are shown with the measured deposition of sulfur in kilotonnes. The circles show how large a percentage of this total came from outside (black) or inside (white) the country. (Adapted by permission of the Spiegel-Verlag from a figure in *Der Spiegel*, 1981.)

the Midwest emits 10 times as much SO_2),[304] although the Midwest contribution dominated in Vermont. The scientists were even able to determine that a 5-day pollution episode was actually a 3-day local episode followed immediately by a 2-day Midwest episode.[304]

Thus, what we call "local" pollution may be anything but local. Acid deposition is a regional phenomenon. For example, Ohio emits 2.4 Mt of SO_2 and 1.0 Mt of NO_2 per year.[311] Emission exceeds deposition standards by factors of 10 to 20 for sulfur and three to five for nitrogen.[311] If pollution were local, we would need to reduce emissions to the level of allowed deposition to meet the standards. Figure 14.7[312] shows this for the countries of Western Europe. As a result, the APHEIS group found, in the 26 European cities studied, that PM_{10} levels range from 14–73 mg/m^3 and smoke soot from 8–66 mg/m^3. Such levels are known to be health risks.[313] Worldwide, people contribute about 20% of the total atmospheric burden of fine particulates.[314] As we have noted, urban areas suffer heavier precipitation than rural ones, partly because of the amounts of particles available as the nuclei for formation of raindrops.[3,7,18,315] An extended discussion of transport across national boundaries is found in **Extension 14.5**. Many other places around the world are much more polluted than the United States.[316]

With this health burden, which must be borne, it is not surprising that a study[317] indicated that a coal-fired power plant is up to 18,000 times more costly to health than a nuclear pressurized water reactor, as indicated in Table 14.9.[290,318] In fact, coal-fired plants emit more radioactivity than nuclear plants are allowed to emit,[318] as we shall discuss in Chapter 20 (see also **Extension 14.2**). And it doesn't count the health effects consequent on mining of, for example, coal by "mountaintopping," or slicing off mountain tops and dumping them into valleys, with

TABLE 14.9

Significant health risks associated with electricity generation

Stage of fuel cycle	Coal	Oil	Gas	Uranium
Exploration extraction	Mine drainage Subsidence Scarred land Pneumoconiosis of miners Accident hazards	Spills Subsidence	Risk of explosion	Subsidence Scarred land Radioactive tailings Silicosis of miners
Transport processing	Accident risk Pneumoconiosis of workers	Spills Residuals in air (SO_2, NO_x, hydrocarbons) Residuals in water (degradable and nondegradable)	Risk of explosion	Accident risk Silicosis of workers
Generation	Air pollution (SO, NO, hydrocarbons, trace elements) Water pollution (heat, nondegradable residuals) Sulfur Nitrogen			Accident risk Waste heat Radioactive waste Radionuclides in air
Waste disposal	Sludge			Reprocessing or disposal of fuel
Ranking[a]	4	3	1	2

[a]Based on health and environmental effects per kilowatthour.

Source: L. B. Lave and L. P. Silverman, "Economic costs of energy-related environmental pollution," *Annu. Rev. Energy* **1**, 601 (1976). Reproduced, with permission, from the *Annual Review of Energy*, Volume 1, © 1976 by Annual Reviews, Inc.

consequent effects on local waters (see Chapter 12), or clouds of sulfuric acid from pollution-control equipment.[235] In addition, coal mining can lead to underground fires that are almost impossible to put out. Hundreds of such mine fires are burning in the United States alone; China and India have similar problems.[319]

The Costs of Cleaning Up Pollution

Evidence mounts that control of acid rain (and sulfur emission) will provide *gains* as well as losses, even for the state of Ohio, America's major sulfur emitter. A cost-benefit study done for the U.S. Congress showed opportunities for benefits to Ohio far in excess of the costs necessary to control sulfur emissions. The study pondered many scenarios, including a national switch to mandatory scrubbers and to fuel switching as an alternative to installation of some scrubbers.[320]

Phase I of the CAAA was a bargain for everyone, despite claims that it would be prohibitively expensive. Additional information is found in **Extension 14.6, *Costs of Attainment of Phase I of the CAAA***.

Clearing the Air?

Many proposals have been made to try to clean up the air. It has been pointed out that healthy forests and grasslands can remove ambient pollutants with minimal harm as long as concentrations are sufficiently low.[115] A different sort of proposal for reducing pollution has been more controversial:[354] J. Heicklin of the Pennsylvania State University has suggested that addition of tiny amounts of a compound diethylhydroxylamine—DEHA: $(C_2H_5)_2NOH$—to polluted air on days having a high probability of smog formation will hinder the pollution process. Tests in laboratories have shown that reactions causing smog proceed much slower in the presence of DEHA. However, the idea of adding an additional pollutant to air to cut down smog has little romance. In addition, there is some evidence that DEHA may cause cell mutations.[345]

14.4   ACID RAIN

Carbon dioxide, sulfur dioxide, and nitrogen oxides in the air gradually react with water vapor and become acids adsorbed onto particulates, which then accrete more water and can become rain. Alternatively, small acid droplets can form the nucleus of a water droplet that accretes more water and will become rain. Rain becomes acidic by the mixing of wet air with these gases in the air. Figure 14.8 shows areas in the United States subject to high acid deposition.

Sources and Cross-Border Transport

All three gases have natural as well as human sources. Table 14.5.1 in **Extension 14.5** shows sources of atmospheric NO_x to be automobiles, industry, ships, biomass burning, lightning, agriculture, the stratosphere, and aircraft (in that order). Sulfur oxides are produced by volcanoes, industry, automobiles, plants, and phytoplankton in the ocean that emit substantial amounts of dimethyl sulfide.[76,355]

Pollution can travel great distances through the atmosphere, as we have seen here and in **Extension 14.5**. Midwestern pollution affects New England;[303] European air pollution has penetrated the Arctic; and Canada continually argues with the United States over cross-border pollution. Figure 14.7 showed how

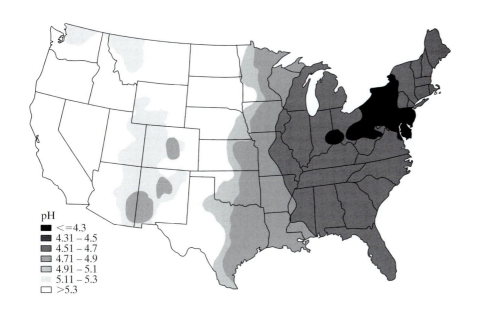

pH
- ■ $\leq=4.3$
- ■ $4.31 - 4.5$
- ■ $4.51 - 4.7$
- ■ $4.71 - 4.9$
- □ $4.91 - 5.1$
- □ $5.11 - 5.3$
- □ >5.3

FIGURE 14.8

The United States has substantial areas of highly acidic precipitation. pH is explained in the text. (J. Carlin, "Environmental Externalities in Electric Power Markets: Acid Rain, Urban Ozone, and Climate Change," Environmental Protection Agency, 1995.)

FIGURE 14.9

This 1982 cartoon pokes fun at the
Reagan Administration's cynical
laissez faire attitude toward the
environment.
(Wiley, © 1982, Copley News
Service, *San Francisco Examiner*.)

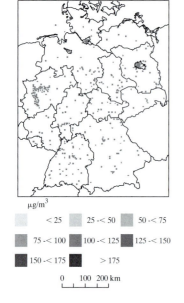

μg/m³

< 25	25 -< 50	50 -< 75
75 -< 100	100 -< 125	125 -< 150
150 -< 175	> 175	

0   100   200 km

FIGURE 14.10

Sulfur deposition in Germany comes
mainly from the heavily industrialized
area of the country and big cities.
(Umweltbundesamt, *Daten zur
Umwelt*, Erich Schmidt Verlag, 1998,
Fig. III.3.)

pollution crosses European borders. Much of this phenomenon is due to the tall
stacks installed to minimize the amount of local pollution; instead of falling near
the stack, the pollutants are carried long distances through the air, giving sufficient time for interaction with water vapor to form acids.[356,357] A droll comment
on transborder acid rain is shown in Figure 14.9.

The pollution from Europe in the Arctic causes acid rain there. Even areas farther away from industrial development than the Arctic have acid rain: Hawaii[358]
and Samoa[359] have experienced significant excess sulfur deposition. Some of this
sulfur comes from biological processes in the upper ocean; some may have a
volcanic source.

Germany, containing five times as many people as Ohio on three times the area,
emits only 3 Mt of sulfur each year (see the emissions locations in Figure 14.10).[360]
Both are turning to sulfur removal technology, but Germany allowed construction
of a coal facility at Buschhaus that emits 12 g of SO_2 per cubic meter of smoke—30
times the official standard (new pollution control equipment was installed in
Buschhaus).[361] German emissions, low on a per capita basis (Table 14.10), are high

TABLE 14.10

Comparison of 1990s per capita emissions, Canada, European Union, United States

	(kg per person)		
	Canada (1995)	**European Union (1996)**	**United States (1998)**
Carbon monoxide	654	130	301
Nitrogen oxides	94	36	82
Total particulate matter	599	NA	117
Sulfur dioxide	101	43	66
Volatile organic compounds	136	38	60

Source: Constructed from Tables ES-2 and ES-3 of Reference 218 and Table 1352 from
Department of Commerce, *Statistical Abstract of the United States 2000* (Washington, D.C.: GPO, 2001).

partly because the former East Germany had few pollution controls, but Germany is still producing more SO_2 than Ohio. The European situation will continue to improve because of the First and Second Sulfur Protocol agreements, which commit the European states to emissions reductions.[353] Canada is also pursuing reductions in emissions to fight acid rain.[362] Asia, already a significant contributor to worldwide pollution and becoming more so as the economies develop, is also beginning to take some collective action.[363]

14.5   THE pH SCALE AND ACID RAIN

The term **acid rain** refers to the deposition of sulfur and nitric oxides by rain or other processes such as condensation on plants, buildings, and the ground. We use the **pH** value to measure acidity or alkalinity in terms of the concentration of hydrogen ions (see **Extension 14.7** for a more detailed explanation of pH). Distilled water, with a concentration of 10^{-7} moles of hydrogen ions per liter, therefore, has a pH of

$$\text{pH}_{\text{water}} = -(\log_{10} 10^{-7}) = 7.$$

Because the negative logarithm is used, *more* acidity means a *lower* pH.

Rain is actually somewhat acid in its normal state, not neutral; it has a pH less than 7 because natural sources of NO_x and SO_x exist, as well as the air's natural complement of CO_2, which forms carbonic acid. Any precipitation with a pH value of less than 5.6 is classified as acid rain. The average pH of rain in the eastern United States is 4.5,[295] though it has been increasing somewhat as sulfur and nitrogen oxide emissions decrease. Acid rain comes from the gases NO_x and SO_2, as we mentioned above; the sulfur oxides cause about two-thirds of the total acidity and the nitrogen oxides, one-third.[295] These pollutants react with oxygen and water vapor in the air to produce nitric and sulfuric acid, respectively:

$$NO_x + H_2O \longrightarrow HNO_3 \ (+ \text{ oxygen}),$$
$$SO_2 + O_3 \longrightarrow SO_3 + O_2.$$
$$SO_3 + H_2O \longrightarrow H_2SO_4.$$

There are many possible ways for these reactions to occur in the air. (They are intermediate steps in chemical reactions that take place normally in the air.)

Causes of Acid Rain

Approximately 70% to 90% of acid rain appears to be human generated (the remainder due to carbonates, nitrates, and sulfates put into the air through natural causes), and the acid in acid rain varies with the amount of sulfur in the air.[364] This makes control of sulfur a burning issue (see **Extension 14.1**), because one tonne of 6% sulfur coal contains 60 kg of sulfur, which produces 120 kg of SO_2 when burned. Furthermore, increases in acid deposition parallel local increases in SO_2, and the highest concentrations of acid rain are found where SO_2 emissions are highest.[36]

There now appears to be a danger that acid rain can be a problem in the American West as well as in the East. Figure 14.11[365] shows how sulfur concentration depends on sulfur emission—it simply follows the same curve. Although

FIGURE 14.11

Correlation between sulfur emissions from smelters in the American west and sulfur in rain, 1980–1983. (M. Oppenheimer, C. B. Epstein, and R. E. Yuhnke, "Acid deposition, smelter emissions, and the linearity issue in the Western United States," *Science* **229,** 859, 1985.)

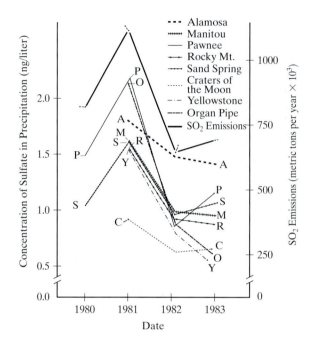

models may disagree in details, all agree that decreases in emission are going to lead to decreases in deposition.[366] Sulfur oxide emission fell by approximately 90% in Arizona and New Mexico during a copper smelter strike in 1980, and consequently, measured sulfate levels fell by over half from the previous summer.[36] The acid rain in the West comes from nitrogen oxides from cars (implicated in the German forest disaster as well) and from sulfur dioxide from copper smelters and utilities. There is great concern for thousands of lakes in the Rockies, especially because of the opening of the Nocozari copper smelter in Mexico just south of the border (see **Extension 14.8**). The plant emitted about a half-million tonnes of SO_2 each year (50% more than any U.S. source then).[371,405] Mexico and the United States agreed in July 1985 to control pollution from the three giant smelters located in the border area, but the results of the agreement remain to be determined.[406]

A National Research Council Study claims that a 50% reduction in sulfur and nitrogen gases leads to a 50% reduction in acid rain.[407] The only long-term study, from Hubbard Brook, New Hampshire, indicates that sulfur in the northeast United States is at two to 16 times greater concentration than in remote areas, and that the water acidity is due mostly to sulfur (about 85%) and to nitrogen oxides (about 15%).[354,359,370,408] Average sulfur oxide concentrations have declined 39% in the United States between 1989 and 1998, and nitrogen oxide concentrations declined by 14% during that time.[346,371] The same sort of decrease is occurring in Europe.[360] Figure 14.12 and Figure 14.13 shows both the American and German pollutant decreases throughout the 1990s.

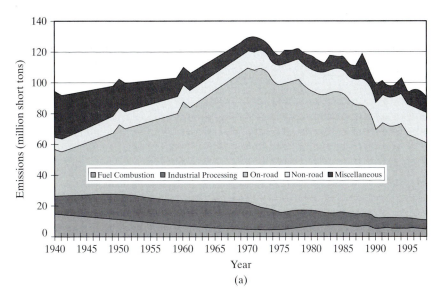

FIGURE 14.12a

American CO, NO$_x$, and SO$_2$ pollutant levels all declined in the 1990s.
(a) Megatons of Carbon monoxide (CO) emitted in the U.S., 1940–1998. (Reference 218, Fig. 3-2)

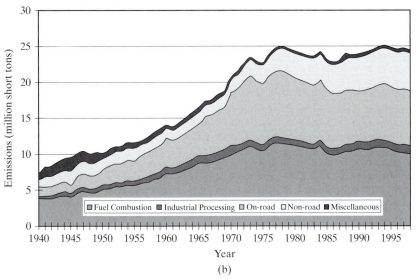

FIGURE 14.12b

Megatons of nitrogen oxides (NO$_x$) emitted in the U.S., 1940–1998. (Reference 218, Fig. 3-3)

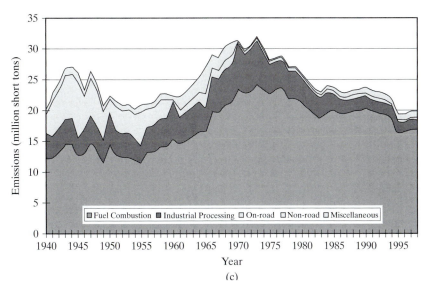

FIGURE 14.12c

Megatons of sulfur dioxide, (SO$_2$) emitted in the U.S., 1940–1998. (Reference 218, Fig. 3-5)

FIGURE 14.13

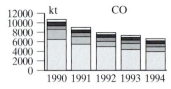

(a)

Kilotonnes of carbon monoxide (CO) emitted in Germany, 1990–1994. (Reference 360, Fig. III-2)

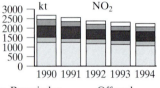

(b)

Kilotonnes of nitrogen oxides (NO$_x$) emitted in Germany, 1990–1994. (Reference 360, Fig. III-3)

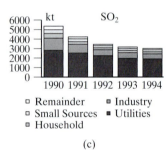

(c)

Kilotonnes of sulfur dioxide (SO$_2$) emitted in Germany, 1990–1994. (Reference 360, Fig. III-5)

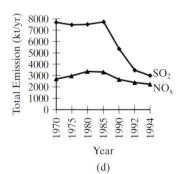

(d)

A longer-term time series view of German SO$_2$ and NO$_x$ emissions, 1970–1994. (Reference 411)

Decreases in NO$_x$ and SO$_2$ translate into increases in pH. Figure 14.14 shows how the average pH measured at various locations in Germany has increased in response to the decrease in amounts emitted. Figure 14.15 shows the wet sulfate deposition in the east and midwest. Note, for comparison, how well deposition correlates with emissions (Figure 14.8). A study of rainwater at 26 of 33 collection sites showed increases in pH and decreases of sulfates between 1980 and 1991.[409] Nitrates have not generally decreased much and are at around 50 μg/m^3 (Figure 14.12b; note from Figure 14.13b that the German experience is similar). In the remote ocean, the nitrate load is only 0.11 μg/m^3.[410] The decline in sulfur oxides in the United States is at least partly due to the 1990 CAAA (see the steep decline in the mid-1990s in Figure 14.12c).

Sulfur Release

Canada and the United States each have reason to complain of and to one another. The Sudbury smelter of INCO, Ltd. in Canada produces about 20% of Canadian SO$_2$—about 1% of total sulfur emissions from all human activity.[411]

Much of the acid rain in the northeastern United States is due to sulfur from the Midwest. In the late 1970s and early 1980s, utilities dumped some 16 Mt/yr of SO$_2$ into the atmosphere east of the Mississippi;[412] Ohio alone released 5.1 Mt of SO$_2$ in 1978.[413] By 1980, Ohio's emissions had fallen to 3.8 Mt/yr, but Ohio emissions were still higher than those of any other state,[36] about 12% of entire national emissions. The other Ohio River Valley states contributed large amounts as well: Illinois, 2.1 Mt/yr; Indiana, 2.8 Mt/yr; Pennsylvania, 3.0 Mt/yr; and West Virginia, 1.5 Mt/yr.[36] The emissions from these six states were *half* the total utility emissions of the entire United States. By 1998, all the states' emissions were lower. Ohio's emissions were still in first place with 1.74 Mt of SO$_2$, and the same five states combined totaled about 3.6 Mt, about 20% of the 128 Mt emitted nationwide. (Ohio alone still produces 10% of national SO$_2$ emissions!) Meanwhile, Texas has leaped into the number two position, responsible for about 5% of national SO$_2$ emissions.[218]

Figure 14.16 shows the American trend over the last century in four major pollutant concentrations: NO$_x$, VOCs, SO$_2$, and PM$_{10}$. In the early part of the century, there was a steep rise in all concentrations, and most pollutants reached maximal concentrations in the early 1970s (here the effect of the CAA can be seen). The only exception to the trend appears to be in nitrate concentrations, which are essentially flat (only small changes occurred after the CAAA controls, as we saw in Figure 14.12). The highest nitrate concentrations in America are to be found in Ohio, Indiana, and Illinois, while the highest sulfate concentrations are found adjacent to the Ohio Valley and in northern Alabama, which corresponds to the locations of large electric utilities. The amount of sulfur released by Ohio is very large.

Effects of Acid Deposition

Sulfur pollution has been around since the dawn of the industrial revolution (as recorded in the Greenland ice cap), and acid rain was described as early as 1852[391] and 1872.[414] The sulfuric and nitric acid load from combustion has health effects. The pollution episodes in the Meuse Valley, in London, and in Donora were so deadly because of acidic pollutants. Inadvertent experiments in

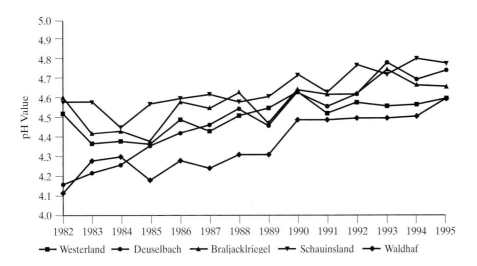

FIGURE 14.14

The pH value in precipitation in Germany shows an upward trend for all five locations between 1982 and 1995.
(Umweltbundesamt, Daten zur Umwelt (Berlin: Erich Schmidt Verlag, 1998), Fig. III-19)

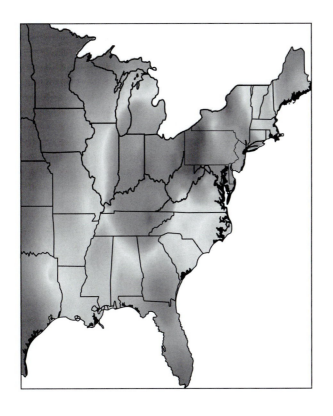

FIGURE 14.15

Trends in wet sulfate deposition (kg/ha); 1995–1997.
(Reference 346, Fig. 7-3a)

FIGURE 14.16

The century-long history of emissions of NO$_x$, VOCs, SO$_2$, and PM$_{10}$ in the U.S. (Reference 218, Fig. ES-1)

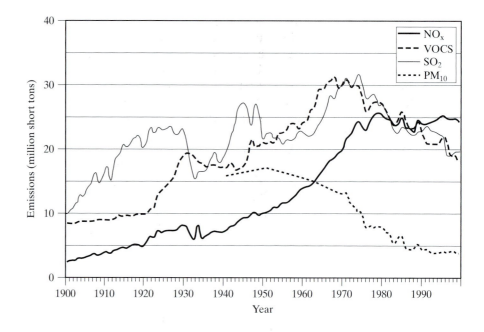

Yokkaichi, Japan, and in Ontario, Canada, have established that lung disease and hospital admissions are correlated with acidic concentrations and decrease as distance from the source increases.[392] Trees are affected as well: 60 µg of SO$_2$/m^3 leads to leaf function harm, and at 80 µg/m^3, the tree becomes deeply sick.[415] The German forests are suffering profoundly; as many as 70% of the trees are affected.[416] What happened to the German forests, which was long brewing, but not really noticed until 1981, was a rapid deterioration. There are indications that eastern forests in the United States are in decline, too. Softwoods at high elevation are dying or failing to reproduce in the Southeast as well as in the Northeast.[417,418] About 10% of annual forest growth in the United States takes place in areas of high sulfur deposition.[416] The U.S. Forest Service has documented a large-scale, rapid, simultaneous 20% to 30% drop in growth rate for at least six species of coniferous trees. This pattern was observed in former West Germany 20 years ago (Table E15.6.1).[417,418]

A parallel problem is that increased damage to trees seems to be accompanied by increased concentrations of heavy metal in the soil.[391,415,417–419] Europe is leading the United States in damage to trees, because there has been a chronic problem in Europe for years (the central European pH of approximately 3.4 had been stable from the 1930s to the 1980s).[420] The "heavy metal" problem, mixed with the acidity of the precipitation, can cause problems for soils, especially poor soils. There is some indication, however, that soils can produce acidity independent of acid rain.[421]

As seen in **Extension 14.8** and in Figure 14.15, sulfur emissions are dropping (as are nitrogen oxides, though more slowly), but the problem of acid rain does not seem to be decreasing everywhere. With European sulfur emissions decreasing (except for Great Britain) and nitrate emissions steady, as is the case in the United States, and with Canadian decreases in sulfur emissions in the 30% to 40% range,[399] the situation in the Northern Hemisphere has started to improve somewhat.[353,372,399] Indeed, surface waters in Europe and North America have begun to improve, except for some regions in Germany,[422] southcentral Ontario, the Adirondack and Catskill Mountain region, and the midwest.[372,399] This is apparently due to regional declines in base cation concentrations. Chemical weathering can restore soils, but on a long time scale. In addition, evidence has emerged that ecosystems are becoming less able to retain nitrogen.[353,367,368,370,372,376] This is important, because the uptake of carbon is governed by the nitrogen cycle.[369]

The pH of rain depends upon the concentration of pollutant in the atmosphere and the amount of water in which it is dissolved. Acid deposition began 200 years ago, but it was first brought to general attention by Swedish scientist Svante Odén in the 1960s. Part of the reason for the long lag is the neutralizing or **buffering** effect that arises naturally. Only after enough time will the acidification be pronounced enough to notice.[380] Clayey soils buffer by ion exchange. When most calcium, magnesium, and potassium is leached out, the pH goes down to 4.2, then aluminum takes over. Thus, because so much soil is aluminum, the acidification does not get any worse.[380]

Wet and Dry Deposition

There is disagreement about how much of the problem is due to **wet deposition** (that is, in actual rainfall or snowfall) and how much is due to **dry deposition**. The problem is that measurement of dry deposition is not possible, because physical collectors cannot duplicate natural surfaces, such as leaves.[405,423] Conventional collectors in Ontario measure only about 60% of the sulfate, because some particles are too fine to precipitate out.[423] At Hubbard Brook, about a third of the deposition is in dry form, and in many lakes, more sulfur flows out of them than flows into them in feeder streams. This indicates that sulfur is being absorbed from the air on the surfaces. Dry deposition may even be half the total sulfur deposition. The acidification of the Elbe River basin in central Europe appears to be caused by dry deposition of sulfur dioxide and runoff from chemical fertilizers.[424] Further research in progress on this question should resolve it.

Of course, other damage occurs to metal and to stone monuments.[391,418,425] Damage was shown to occur to German monuments and old buildings built of stone, even at concentrations below 25 $\mu g/m^3$. In the United States, more than 35,000 buildings and 10,000 monuments are endangered by acid rain.[426] It is not only old stone that suffers; a tablet erected in Alexandria, Virginia, in the 1920s shows extreme damage where it has been exposed to rain (Figure 14.17). Acid rain falling on Pennsylvania is obliterating the monuments to the Civil War in Gettysburg.[427] Stone is eroded, bronze eaten away.[427] Carbon-containing particles cause acceleration of the destructive work of acid rain.[347,428] Acid rain falling on southern Mexico is destroying Mayan artifacts. Only vandalism is damaging Mayan relics more than acid rain.[429]

Costs of repairing the damages are awesome. A study by the Organization for Economic Cooperation and Development in 1981 estimated these costs as $805 million for Belgium and Luxembourg, $462 million for Denmark, $316 million for France, $873 million for the Netherlands, $4.331 billion for the United Kingdom, and $7.042 billion for West Germany. Research has identified "biomortars" that can be sprayed on or put on a wrapping.[430] The bacteria can penetrate limestone (impossible to protect using resins, which clog the pores). Limestone must "breathe" and cannot withstand weather. When the pores are clogged, the stone can crack and crumble.

FIGURE 14.17

This stone tablet in Alexandria, Virginia, that describes the history of the area's settlement has been damaged where it's been exposed to acid rain. The top has been protected by the enclosure and is still readable from afar, while where the overhang no longer protected the tablet, the text has become hard to read.

Acid Lakes

Acid rain can lead to acid lakes. A study of acidified lakes showed that acid deposition from the atmosphere was the dominant cause of acidification in 755 of the 1180 lakes studied and 47% of the 4670 acidic streams studied.[431] Acidic mine drainage was the dominant acid source for 26% of the streams.[431]

Lakes have the capability of neutralizing some acid. However, about one-fifth of lakes in the Northeast and one-third of Florida lakes were extremely vulnerable to acid attack.[364] Acidification has been less than feared. Six areas of the United States (Adirondacks, New England, Appalachia, Atlantic Coastal Plain, Florida, and Eastern Upper Midwest) account for almost all of the acidic lakes and streams.[401] (More detail is available in **Extension 14.8.**)

The final word on acid rain and acid lakes has not yet been spoken. There is enough circumstantial evidence to warrant joint American–Canadian action to reduce sulfur and nitric oxides and to warrant the European Union's taking action to try to save European forests before it is too late. Evidence suggests that reduction in sulfur dioxide levels gives at least as great a decrease in the acidity of rain.[432] We would need to reduce sulfur emissions by 25% to 50% to bring back already acidified lakes.[393]

SUMMARY

Urban areas suffer pollution from cars, trucks, industrial processing, and electric utilities. Urban areas are generally warmer as well as more polluted than rural areas. Many great conurbations are in valleys or are surrounded on one side by hills. There, air may become trapped by temperature inversion. In the normal case, air temperature decreases as one moves upward in the first few kilometers of air. Occasionally, conditions arise in which air temperature increases with height, the inverse of the usual condition; in other words, there is a temperature inversion. In this condition, the air is so stable that it is quiescent, trapping all the pollutants within the urban area. Such extreme events may cause sickness or even death.

Health damage has been documented for people living in urban areas. Plants are also affected adversely by pollution—especially ozone and acid rain. In addition, pollution causes property damage in the billions of dollars.

Large particles cause exhaust gases to look awful, but these particles are trapped effectively by the lungs'

defenses. Small particles may not appear to be in the exhaust gases, but they can cause more severe injury than the larger particles and set the stage for serious lung disease.

Local air pollution is bad, but pollution respects no boundaries. Midwestern acid rain falls in Canada and New England. Britain's air pollution invades Norway, also as acid rain. The pH scale is used to measure acidity (see **Extension 14.7**). A pH of 7 is neutral; when the pH is less than 7, the liquid is acidic; when the pH is greater than 7, the liquid is alkaline. Acid rain has damaged German forests, and U.S. forests are showing symptoms similar to those observed in Germany one or two decades ago.

Lakes with a pH less than 5.4 have no reproducing fish. Many lakes in the eastern United States have such pH values, but lately, acidity has been observed in the West as well. This is also discussed in **Extension 14.8**. With the pollution problems identified, and the health consequences of pollution known reasonably well, the time has come to address the issue.

PROBLEMS AND QUESTIONS

MULTIPLE CHOICE REVIEW

1. For precipitation to be classified as acid rain, it must have a pH of
   a. less than 5.6.
   b. more than 5.6 but less than 6.
   c. less than 7 but more than 6.
   d. more than 7 but less than 9.6.
   e. less than 9.6.
2. Acid rain has poisoned lakes and destroyed monuments. Which chemical(s) below is (are) implicated in acid rain?
   a. Carbon monoxide
   b. Carbon dioxide
   c. VOCs
   d. Sulfur oxides
   e. All of the above chemicals are implicated in acid rain.

3. Photochemical smog is caused by
   a. the reaction of unburned hydrocarbons with nitrogen oxides and sunlight.
   b. the reaction of ozone and sulfur oxides with sunlight.
   c. the reaction of carbon monoxide with sunlight.
   d. the reaction of nitrogen oxides and sulfur oxides with sunlight.
   e. none of the above.
4. In general, hydrocarbons, when burned, produce
   a. carbon dioxide and water.
   b. carbon dioxide, water, and sulfur dioxide.
   c. methane and water.
   d. sulfur dioxide and nitrogen dioxide.
   e. none of the compounds given in the other answers.

5. Per capita costs to health and property due to pollution are most likely to be between
   a. $0 and $100.
   b. $100 and $1000.
   c. $1000 and $10,000.
   d. $10,000 and $100,000.
   e. $100,000 and $1,000,000.

6. In a temperature inversion,
   a. the warmer air is underneath the cooler air.
   b. there is no temperature difference between air regions, but there is between the ground and the air.
   c. the air temperature changes at the same rate as the lapse rate.
   d. the air temperature changes inversely to the lapse rate.
   e. polluted air can become trapped.

7. Transport of pollution has been known to
   a. bring "heavy metals" into the Arctic regions.
   b. carry sulfur-laden air from Ohio to Massachusetts.
   c. bring Southern Hemisphere smoke from biomass burning into the Northern Hemisphere.
   d. cause the air quality desert regions of California far from Los Angeles to degrade.
   e. cause all of the above-listed things to happen.

8. The best description for pollution is
   a. bad air.
   b. poor water quality.
   c. emissions from utilities.
   d. recognizable degradation in the environment.
   e. not entirely given by any other statement.

9. The best way of those listed below to eliminate particulates in smoke from utility coal-fired power plants is
   a. scrubbers.
   b. cooling towers.
   c. lower-temperature combustion.
   d. fuel switching.
   e. electrostatic precipitators.

10. The main contributor(s) to carbon monoxide pollution is (are)
   a. cars and trucks.
   b. operation of heavy industry.
   c. electric trains.
   d. power plants.
   e. home heating.

SHORT ANSWER

Explain whether the statement made is correct, and justify your answer.

11. Acid deposition is decreasing due to the decrease in sulfur emissions, so acid rain problems should disappear soon.

12. London smog has a deserved reputation as a killer—the smog killed thousands of people.

13. Temperature inversions involve a higher temperature at ground level than in the air over the ground.

14. A heat island is annoying but not dangerous to health.

15. VOCs can cause cancer, even if present in small amounts.

16. Air is denser than water, which has implications for underwater divers.

17. If the lapse rate is smaller than the adiabatic lapse rate, the air is stable.

18. Cities have higher rates of lung disease than rural areas.

19. The EPA primary standards guarantee that no person can be harmed by air pollution.

20. The lungs occupy a small volume but have a very large surface area.

DISCUSSION QUESTIONS

21. Should the world be concerned that China, with enormous coal deposits, is undergoing explosive economic development? Explain why or why not, with reference to the subject of this chapter.

22. Explain the history of emissions shown in Figures 14.12 and 14.15, explaining how the emissions have changed over this time.

23. Explain how air that is colder than predicted by the adiabatic lapse rate at a given altitude will be moved.

24. Driscoll et al.[365] wrote that:

    Without strong acid anions, cation leaching in forest soils of the Northeast is driven largely by naturally occurring organic acids derived from the decomposition of organic matter, which takes place primarily in the forest floor.... [O]rganic acids tend to mobilize Al through formation of organic–Al complexes, most of which are deposited lower in the soil profile ... This process ... results in surface waters with low concentrations of Al.... Acidic deposition to forest soils with base saturation values less than 20% increases Al mobilization and shifts chemical speciation of Al from organic to inorganic forms that are toxic to terrestrial and aquatic biota.

    Exactly what are they saying? Explain.

25. Some lakes are "naturally acidic." How can that be possible?

26. "North America and Europe are in the midst of a large-scale experiment. Sulfuric and nitric acids have acidified soils, lakes, and streams, thereby stressing or killing terrestrial and aquatic biota."[365] Assess this statement. What is meant by the word "experiment" here?

27. What evidence supports or undercuts stiffer actions to protect the air? Explain.

28. The Great Smokies got their name because natural VOCs created haze. Does this mean we do not need to be concerned about the effects of air pollution in the Park?

29. What is likely if the particulate NAAQS is exceeded by a factor of 10? Explain.

30. Why should the smallest particulates cause the greatest health problems? Explain.

PROBLEMS

31. What is the pH of water having a concentration of hydrogen ions of 10^{-4} (one ion in 10,000 molecules)?

32. What is the pH of water having a concentration of hydrogen ions of 2.5×10^{-5}?

33. If the tolerance level for carbon monoxide is 40,000 $\mu g/m^3$, and the tolerance level for nitrogen oxides is 500 $\mu g/m^3$, what is the relative toxicity of nitrogen oxides compared to carbon monoxide?

34. If the tolerance level for carbon monoxide is 40,000 $\mu g/m^3$, and the relative toxicity of particulates to carbon monoxide is 50, then what is the tolerance level for particulates?

35. If the tolerance level for carbon monoxide is 40,000 $\mu g/m^3$, and the tolerance level for particulates is 400 $\mu g/m^3$, what is the relative toxicity of particulates to carbon monoxide?

36. If the tolerance level for carbon monoxide is 40,000 $\mu g/m^3$, and the relative toxicity of hydrocarbons to carbon monoxide is 2, what is the tolerance level for hydrocarbons?

37. If the temperature of dry air at the surface of Earth is 20 °C, in normal conditions, what is the temperature at an altitude of 1000 m above the surface? 2500 m? 5000 m?

38. If the temperature of dry air at the surface of Earth is 15 °C, in normal conditions, what is the temperature at an altitude of 1000 m above the surface? 2500 m? 5000 m?

39. On a hot day, the temperature on the ground is 35 °C. Under normal conditions, what is the temperature at 7 km expected to be?

40. The actual lapse rate on a certain day is 90% of the adiabatic lapse rate. Predict the temperature at 100 m, 1000 m, 2000 m, 2500 m, 3000 m, 4000 m, and 5000 m altitude.

41. The actual lapse rate on a certain day is 1.2 times the adiabatic lapse rate. Predict the temperature at 100 m, 1000 m, 2000 m, 2500 m, 3000 m, 4000 m, and 5000 m altitude.

42. The actual lapse rate on a certain day is 125% of the adiabatic lapse rate. Predict the temperature at 100 m, 1000 m, 2000 m, 2500 m, 3000 m, 4000 m, and 5000 m altitude.

43. Rowe has calculated that 1 t of coal flue gas particulates translates into 100 $\mu g/m^3$ of exposure.[340]

   a. What exposure results as a result of a power plant burning 1000 t of coal per day?

   b. What number of cancers could result?

   c. If the pollutants are 55% particulates with diameters smaller than 10 μm, by how much should the local death rate increase?

44. At what concentrations of the following pollutants will the toxicity relative to the toxicity level of carbon monoxide be 1.0?

   a. Carbon monoxide

   b. Hydrocarbons

   c. Sulfur oxides

   d. Nitrogen oxides

   e. Particulates

CHAPTER 15

Moving Down the Road

Transportation is the lifeblood of economic and social exchange. The transportation system is run mainly on fossil fuel, almost 30 EJ worth, over 20% of the total energy use in the United States. This chapter considers the context of the transportation system—use of fossil-fuel energy and pollution. While the main subject of the chapter is the automobile, mass transit, load factors, and other modes of transportation are considered as well. Following the discussion in Chapter 14, we see that pollutants from mobile sources can have deleterious effects on the health of people as well as on the health of forests in the industrialized countries.

KEY TERMS

load factor / circuitry / drag / drag coefficient / catalytic converter / stratified charge engine / photochemical smog / Waldschäden / specific energy

15.1   THE TRANSPORTATION SYSTEM

There are various ways to transport people and goods. People can travel from place to place on foot, or by bicycle, horse, motorcycle, bus, car, tram, train, ship, or plane. In a logical transport system, modes of transport would be rationally apportioned to minimize energy expenditure, while maintaining ease of connections among populated locations. This is not now the case. Most people in the American transportation system travel by private car, and they travel alone rather than together. For this reason, the emphasis of this chapter will be on the automobile.

Countries elsewhere have made more of a commitment to mass transportation than in North America, perhaps partly because of the vast areas that are lightly peopled. Europe's population density is generally much higher than that of the

United States and Canada, except for compact regions within these countries (e.g., the Boston–Washington corridor, Chicago, and southern California). Public transportation before the end of World War II was much more robust in North America. Local transit systems existed, especially in the northeast, Chicago, and southern California, and the railroads had knitted the countries firmly together (for example, one of independent Canada's first acts was authorization of a Canadian transcontinental railroad).

As with people, freight can travel from source to destination by different means: wagon, truck, train, barge, ship, pipeline, or plane. Each mode of transport has its own niche, and its own efficiency. But the world's developed countries (and the United States and Canada, in particular) depend mainly on the truck for movement of bulk freight.

The amount of energy used for fuel for transportation is about one-quarter of all the energy used in the United States (see Figure 8.2), about 28 EJ. The energy cost of transportation depends mainly on the energy use by automobiles and trucks. Autos presently account for 21% of *all* U.S. energy consumption (accounting for fabrication, fuel, and maintenance), and that is mainly oil. Figure 15.1 shows that two-thirds of all petroleum used in the United States is for transportation.[1] Figure 15.2 shows that the average car uses less fuel annually than average vans, sport-utility vehicles (SUVs), or pickups. A truck uses much more fuel on a per-vehicle basis.

The petroleum distillate consumption is about 24 billion liters, illustrated in Figure 15.3.[2] In the total transportation sector, 27×10^{18} J (27 EJ) was used in 2000.[2] Over 55% of this energy was used by pickups and cars; 17% by freight trucks; and 18% for air, rail, and marine. Only one-third of a percent of the energy is used for mass transit (buses, subways, and trolleys).[3] Before focusing on the automobile, this chapter's main topic, we discuss some other parts of the transportation system.

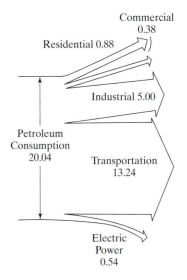

FIGURE 15.1

Most petroleum used in the United States in 2003 was burned to run the transportation system. The numbers denote use in million barrels per day. (Dept. of Energy, *Annual Energy Review 2003* (Washington, D.C.: GPO, 2004), DOE/EIA-0384(03), p. 159.)

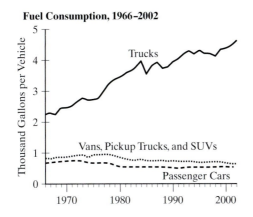

FIGURE 15.2

Automobiles use somewhat less fuel per vehicle than other passenger vehicles, and all passenger vehicles use much less fuel annually than trucks, because trucks are run a greater portion of the time. (Dept. of Energy, *Annual Energy Review 2003* (Washington, D.C.: GPO, 2004), DOE/EIA-0384(03), p. 94.)

Long-Distance Mass Transit

Table 15.1 compares the energy efficiencies of various transportation modes. One study showed that in the 1960s, the number of railroad passengers declined by half, the number of airline passengers tripled, and the number of auto passengers increased by a factor of 1.5.[4] Ways of saving energy are discussed more fully below.

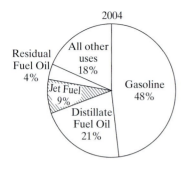

FIGURE 15.3

The relative proportion of gasoline, jet fuel, and residual and distillate fuel oils is shown. "All other uses" includes petroleum coke, still gas, asphalt, LPG, road oil, and other products. The proportion of gasoline is relatively steady despite burgeoning jet air travel. The total annual use increased from 8.32 billion liters in 1950 to 23.88 billion liters in 2000.
(U.S. Department of Energy, *Monthly Energy Review, March 2005.*)

TABLE 15.1

Comparison of energy costs of various passenger transportation modes

Mode	Energy cost (kJ/passenger km)			Energy cost (kJ/vehicle km)	
	Lincoln[a] 1972	ORNL[c] 2000	DOT[b] 1987	ORNL[c] 2000	DOT[b] 1987
Local					
Bicycle	130				
Walking	200				
Motorcycles		1368	2276	1642	
Urban auto	5300	2407		3850	
Personal truck		2936		4698	
School bus				12,914	10,837
Transit bus		2778		24,982	25,416
Mass transit railroad	2500	2108		46,397	
Commuter railroad		2300		73,655	
Intercity					
Automobile			1957		
Intercity bus	1050	467	771	10,847	
Intercity railroad		1600	1679	26,970	
General aviation		7266	7256		
Certified air carriers		2622	3175	242,022	

[a] G. A. Lincoln, "Energy conservation," *Science* **180**, 155 (1973).

[b] Department of Transportation, *National Transportation Statistics* (Washington, D.C.: Government Printing Office, 1987).

[c] S. C. Davis, *Transportation Energy Data Book: Edition 20*, Oak Ridge National Laboratory, ORNL-6959 (November 2000), Tables 2.11, 2.14, and 2.15.

Airplanes

Air transport is responsible for 1.2% of all the energy used in the United States, which is only about 5% of the energy used for transportation.[5] Table 15.1 shows that flying the airlines is much more energy expensive than most other modes of moving people (urban buses are especially costly, see the discussion in urban mass transit, below). The cost per person for air transport is great, although the advent of jumbo jets a generation ago reduced this cost to almost the level for cars.[6] Many possibilities exist for reducing per passenger cost, from more intelligent computer control systems to more efficient jet engines, to reductions in drag using new materials and coatings.[7]

Railroads

We already mentioned mass transit as a good way to save energy and money. Long-distance buses cost about 6 cents per passenger-mile to run in 1980 and 4 cents per passenger-mile in 1997. Local buses required $1.69 per vehicle-mile to run in 1980 and 0.76 per vehicle-mile in 1997.[8] The fact that mass transit is so much more popular in Europe and Japan than it is in the United States may be due to the greater population density and the more than twofold greater cost of gasoline in other countries as compared to prices in the United States, as well as to more attention paid to creature comforts. (See **Extension 15.1, *Mass Transit or Highway Construction?*)

The Acela train, which tilts on turns, is bringing Boston and Washington closer together at speeds over 200 km/h. One new technique has been developed for reducing side friction between wheel and rail. Lasers carried on some cars can glaze the inner rail surface, reducing friction by 40%. This process also apparently reduces the incidence of stress cracks in rails, lengthening their life significantly.[5]

As mentioned, in Europe and Japan, railroads are more important people movers than in the United States. Many Americans have heard about the Japanese "bullet trains," the French TGV, and the German *Schnellbahn*. The national railroad system outside the United States is seen as a necessary component of a national transportation plan; there is a danger that, by the time you read this, Amtrak will have been scuttled. Introduction of high-speed rail and magnetically levitated trains will cost a lot of money in subsidies, but the other components of the transportation system actually get gigantic subsidies from governments in the form of public construction and maintenance of highways and airports. It is entirely possible to rebuild the rail transport system in America given sufficient political recognition of the important role an alternative transportation system could play (especially when other system components labor under terrorist threat).[11]

High-temperature superconductors (Chapter 25) may lead to magnetically levitated trains that could be run more energy efficiently than current trains. The roadbed will not need much maintenance (because the cars do not come into contact with the rails except in station areas).

Urban Mass Transit

To examine the various transportation options carefully from an informed perspective, we must know not only the energy consumption per passenger-mile, but also the **load factor**, or proportion of passengers carried. What makes mass transit efficient (where it is) is its large load factor, and conversely, what makes mass transit inefficient (where it is) is its small load factor. As Table 15.1 shows, on average, urban mass transit has a low load factor. The cheapest way to decrease transport system cost is to increase the utilization of already extant transport. The report of the Office of Emergency Preparedness estimated that improved communication facilities, urban clustering, and construction of walkways and bike paths could lead to an overall 15% to 20% energy saving for transportation.[12]

The best way to transport people in urban areas is by mass transit with high load factor. Unfortunately, the load factors for urban mass transit are typically 20% to 25%. This happens because transit systems generally have to run during hours when there is little passenger demand, which is done to encourage usage, build ridership, and eventually raise the load factor.

Critics often argue that off-hour service should be eliminated, and that buses should run only to transport people to and from work. They appear to feel that the low load factor encountered in off-hour service is *prima facie* evidence that auto transport is more efficient. I argue with this view. If off-hour service were not available, some riders would be forced to buy cars. There is a huge energy cost hidden in the auto itself—it costs about as much energy to build the car as it will ultimately use in its lifetime. Hence, it is often energy efficient to run off-hour mass transit despite the low load factor, even though it may not at first glance appear to be so. Also, critics assume that large buses will be used for service during off hours; a rational system apportions the size of the transportation to the expected clientele. Another alternative is call-up door-to-door delivery during off-hours, but this has a high labor cost.

My argument is based, to some extent, on intangibles—I cannot quantify the number of people who would buy cars if service were curtailed. It is well known, however, that decreases in mass transit services go hand-in-hand with increases in auto ownership. Other aspects that should be considered here include the necessity of increased street work as auto density increases; the increased pollution concomitant with running car engines cold; the crowding of cars in streets, causing increased pollution and frustration levels; and elimination of important transport services to the poor and the elderly, who often use mass transit in off-peak hours.

We may use load factors to find the average energy cost in the hypothetical American city, Midtown. Urban transit in Midtown is the local bus line that we will suppose has a load factor of 25%, and the car that we will suppose has a load factor of 30%. (This would be an average of 1.5 passengers per car, fairly typical, and just about the number given in the Oak Ridge figure in Table 15.1.) We will use the graph, Figure 15.3, to take the loading into account. According to the graph, for the bus at 25% load factor, the energy use is 2778 kilojoules per passenger km, while for the car, it is 2407 kilojoules per passenger km (straight from Table 15.1). Buses account for only 10% of Midtown's in-city passenger kilometers; cars account for the remainder. Because buses account for 10% of the passenger kilometers in Midtown, the average energy cost for transport for all modes is

$$(0.1)(2778 \text{ kJ/passenger km}) + (0.9)(2407 \text{ kJ/passenger km})$$
$$= 2444 \text{ kJ/passenger km}.$$

If Midtown could persuade people to use buses, we could imagine a doubling of the load factor to 50%, in which case buses would account for 20% of the passenger kilometers. The bus transport average energy cost is only 1488 kJ/passenger km (Figure 15.4), and the average cost of transport for all modes is

$$(0.2)(1488 \text{ kJ/passenger km})$$
$$+ (0.8)(2407 \text{ kJ/passenger km})$$
$$= 2223 \text{ kJ passenger km}.$$

FIGURE 15.4

The energy cost per passenger decreases as the number of passengers increases. The load factor is the proportion of the total number of passengers on the bus. The inset shows the cost on an expanded scale for load factors between 25% and 100%. The graph was created using the Oak Ridge figure for energy use from Table 15.1, assuming each passenger had an average mass of 80 kg, and that energy use depends linearly on mass. The Oak Ridge bus carries 9 passengers. For this graph, a fully loaded bus carries 36 passengers.

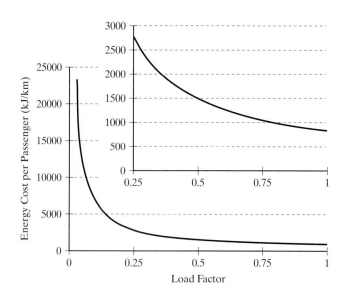

If Midtown could use its buses at 100% load factor, buses would account for 40% of the passenger kilometers, the cost per passenger would be 845 kJ/km (Figure 15.4), and the average energy cost for Midtown would be reduced to 1782 kJ/passenger km. It is clear that raising a low load factor results in great energy savings.

The city would get a reduction of about a third in its energy use for transportation simply by *filling* its buses. If more buses are added for increased ridership, the savings are not so great, but any savings are welcome. Public transportation must compete with the car's ability to take people where they want to go when they want to go. This makes it difficult to keep buses filled if they do not run often enough, and this cannot happen where the number of riders is too small to justify the number of buses. The major successful mass transit systems in North America run in the largest cities—New York, Toronto, Chicago, Washington, Atlanta, San Francisco, Mexico City. Salt Lake City, Houston, Sacramento, Denver, Portland, and Seattle have added light rail systems.

Subways in New York have successfully carried hundreds of thousands of people per day for almost a century (although, it must be admitted, in conditions often approaching squalor during the last quarter-century). Despite the expense of subway systems (averaging $5.38 per vehicle revenue mile in 1984),[13] the most successful modern American mass transit systems of the present are Washington's Metro and San Francisco's BART system. Both of these systems are moving large numbers of people, causing economic growth near stations, and helping to relieve traffic congestion. Los Angeles has a new subway, too.

In smaller cities, such as Columbus, Ohio, or Eugene, Oregon, and in extended suburban areas, bus systems are more rational than subways. Bus systems in this sort of area have met with mixed success, and costs were an average of $3.86 per revenue mile by 1984.[13] Suburban New Jersey has lost its bus service over the past quarter-century. Columbus' system is holding its own but is not a thriving concern. Eugene's system was a clear success.

Movement of Goods

Many alternatives exist for the transportation of goods to market, or for shipping of components to assembly factories, or to transport fuels. Energy cost (shown in Table 15.2) is one indication of the attractiveness of various commercial

TABLE 15.2

Energy efficiencies of various freight modes (kilojoules per tonne kilometer)

	DOR[a]	ORNL[b]	(ORNL)[d]
Truck	10,216		2810
Class I freight railroad	1081	264	
Water commerce		315	
Air freight[c]		8998	
Pipeline	790		197
Coal slurry pipeline			2003

[a] Department of Transportation, *National Transportation Statistics* (Washington, D.C.: Government Printing Office, 1987).

[b] S. C. Davis, *Transportation Energy Data Book: Edition 20*, Oak Ridge National Laboratory, ORNL-6959.

[c] This number was constructed using Table 12.01 of (b), assuming an average passenger weight of 150 lb, and allocating the freight proportionally.

[d] Pipeline figures from S. C. Davis and S. G. Strang, *Transportation Energy Data Book: Edition 13* (Oak Ridge, TN: Oak Ridge National Laboratory, 1993). The number for the truck came from the use of Table 2.07 of (b) and Table 1008 of Department of Commerce, *Statistical Abstract of the United States*, 2000 (Washington, D.C.: GPO, 2001).

transportation modes, but not the only one. Time and convenience are often important considerations. Freight takes about 10% of the energy supply in the developed countries and 9% to 13% in developing countries.[14] Safety is important for truck drivers, who must spend so much time in the cab (an average of 66 h/wk).[15] A new product, Howard Power Center Steering, developed by River City Products, helps drivers reduce road wander and fatigue at the same time. The Howard system supplies power directly to steering to keep the rig on-center (such a system would be unnecessary if the steer-by-wire systems are adopted). The system allows the driver to choose how much help is desired.[15]

Trucks and Railroads

The **circuitry**, or deviation from the shortest path, for rail transport is comparable to that for truck transport (1.24 and 1.21).[9] This cannot account for the switch from rail to truck transport (refer to **Extension 15.1**.)[8] Part of the reason may be that the railroad must pay its full share of facility construction and maintenance costs. The truck pays only 76% of its incurred costs; semis and full trailers pay only 56% of the actual cost for their use of interstate highways.[8] Although railroads are more efficient than trucks, trucks carry a lot of traffic, because they are more flexible than the railroads in that they provide door-to-door service. In addition, the new generation of diesel locomotives are more efficient and less polluting than previous generations.[16]

There are opportunities to have the best of both worlds—containers can be shipped by sea and rail, then delivered to local destinations by truck after offloading. This is clearly an example of efficient use of resources.[17] Table 15.3 shows how energy efficiency of commerce is improving over time, especially that of the railroads (and also shows that the railroads are less energy expensive than water transport).

TABLE 15.3

How efficiency of freight modes has improved, 1970–1998 (Common and SI Units)

Year	Trucks (mpg)	Class I freight railroads (Btu per ton mile)	Domestic waterborne commerce (ton miles per barrel)
1970	5.5	645	4820
1980	5.4	590	3680
1990	6.0	420	3370
1995	6.2	370	3580
1996	6.2	365	3580
1997	6.4	370	3770
1998	6.1	360	3660

Year	Trucks (MJ/km)	Class I freight railroads (megajoules per tonne kilometer)	Domestic waterborne commerce (megajoules per tonne kilometer)
1970	15.22	0.47	0.87
1980	15.50	0.43	1.14
1990	13.95	0.30	1.24
1995	13.50	0.27	1.17
1996	13.50	0.26	1.17
1997	13.08	0.27	1.11
1998	13.72	0.26	1.15

Source: U.S. Department of State, *U.S. Climate Action Report—2002* (Washington, D.C.: GPO, 2002), Figure 2.9.

Air Traffic

The air carriers constitute the second-fastest-growing segment of passenger transportation in the United States (refer to **Extension 15.1**) after the automobile. The lower crude oil prices of the mid-1980s did not help the airlines, which were reeling economically from the competition attendant on deregulation. The airlines knew that the prices must eventually rise again. For this reason, any new development that increases fuel efficiency is of great interest. The current average aircraft efficiency is 48 seat mi/gal.[18] The Boeing 777 is twice as efficient as the 727: 94.5 versus 50 seat mi/gal.[19] In fact, the energy intensity (energy used per passenger kilometer) of the 1999 aircraft fleet is 60% below that of the 1971 fleet. Of this improvement, about 60% is due to improvements in engines, and 20% each from improved aerodynamics and the more efficient use of capacity.[18] Improved aerodynamics might offer ways to reduce aircraft noise as well.[20]

There are still many savings to be found in aircraft design. The GE unducted turbofan jet engine uses 40% less fuel.[19] The prop-fan engine, which looks something like the old turboprops[21] but allows the plane to fly at jet speeds, may increase fuel efficiency by 30%. These new engines should make air travel comparable to or cheaper than energy use in private cars. With ultrahigh bypass engines, a 10% to 20% increase in efficiency could occur.[18] The prop-fan engine, developed in the early 1980s but dropped when energy prices remained low, could be revived.[22] Winglets, upward-pointing extensions at the end of a wing, save 5% in fuel consumption.[22] The scramjet engine is a combination rocket and ramjet, and it promises to give a huge boost in the race for speed.[23]

Covering an airplane wing with a porous metal skin punctured by tiny holes can save energy. Total drag can be reduced by 10%, according to a partial experiment on a 757.[24] Airbus Industrie designed and installed a plastic grooved skin on planes that gives a 4% savings.[22] Boeing is developing the 7E7, made largely of composites; it would get an estimated 20% better fuel economy than Boeing's 737 and 747;[25] trading economy for speed, Boeing scrapped development of its "Sonic Cruiser," a superfast plane.[25,26]

Delays in the skies are often more the rule than the exception nowadays. The Federal Aviation Administration (FAA) is proposing a "Free Flight" system,[27] opening the skies and ending the practice of enforcing air routes. Software is being upgraded locally after the huge procurement system faltered after "eating" many millions of dollars. This is now feasible because of the information from GPS systems that each plane can use to determine a flight route.

Sea Traffic

Ships still represent an economical way to send large volumes of goods great distances. However, technology has lagged, as other methods of transport have seemed more interesting. For a hundred years, shipbuilders have recognized that ships' captive waves (the waves the ships make as they travel through water) reduce the maximum speed a ship can make by creating vortices and draining energy.[28] This can be partially overcome by making the ship longer and thinner, but this can make the ship more unstable in rough seas. Ships also have problems with their propellers when they turn too fast—cavitation, or boiling, occurs, and this can lead to vibrations that can crack ship hulls.[28]

Replacing propellers with gas turbines or water jets can circumvent this problem. Jetships are being built for the transatlantic trade.[29] Many fish shed water from their skins, and the discovery of polypropylene-based superhydrophobic substances could lead to naval and aircraft coatings that make travel through these media more efficient.[30] This could be a big help, because many antifouling paints contain materials such as TBT and Irgarol 1051, which are environmentally dangerous for aquatic life.[31]

The semiplaning monohull (called Fastship) has a deep-V-hull forward and a shallow stern. It appears that this type of hull will be able to move twice as fast as current ships, albeit at a higher energy price.[28] Much sea traffic is made up of tankers, the ships that supply the fuel we use. See **Extension 15.2, Transport of Fuel** for information on the ways fuel is transported and the consequences. Use of superconducting technology (see Chapter 25) can cut ship power plant size by 70%, and the U. S. Navy is testing a 5 MW motor and developing a 36.5 MW motor.[39]

15.2   THE AUTOMOBILE

The American public has carried on a love affair with the automobile from its earliest days, in spite of the weather, economic depression, and war or peace. The car has had an almost unbelievable effect on the world's way of life: changing the duties of family members; providing a new method of courtship; and giving every Walter Mitty his own chariot. One of America's great cities, Los Angeles, has developed around dependence on the private car as the mode of transportation. The flight to the suburbs gained momentum as the car came of age, so that today, the U.S. population is about 80% urban or suburban. An astounding 18% of the population moves each year.[40]

The breakdown of fuel economy of various transportation modes shows that the efficiency of most other modes is lower than that of cars (see Figure 15.2).[13] About 55% of travel involves trips of less than 16 km [10 mi];[4,41] 56% of all commuting trips involve cars with only one occupant;[4] 26% of all commuting trips were on public transportation, and 4% were by other means.[41]

Most of the chapter discussion will focus on the automobile, because it makes up such a huge proportion of passenger transportation in the world, especially in the United States.

The Cost of Owning a Car

"Romance of the road" has occurred despite the tremendous economic burden of car ownership. Approximately 20% of the total energy use in the United States goes into operating cars. About 8% of the gross national product (GNP) is connected to automobile expenditures, and approximately 25% of retail sales are car related. Some 20% to 25% of all American deaths by accident occur in car-related accidents. These figures are staggering. And the rest of the world would like to emulate America, with worldwide consequences.

As we have long known, American cars are gas guzzlers, and after a reduction in fuel consumption in the 1980s, they have been becoming more so again. The average (constant dollar) cost of a car per mile driven over the last several decades averaged about 53 cents, including purchase, maintenance, repairs, and fuel. The 1999 average cost per mile was higher, 56.74 cents.[42,43]

Between 1950 and 1973, American average car mileage declined from 18.27 mi/gal to a low of 11.85 mi/gal,[44] basically because cars got heavier during that time, and because the size of the engines increased. Mileage rose fairly steadily for cars from 1978 to 1990 (Figure 15.5a), though with cheaper gas (Figure 15.5b),

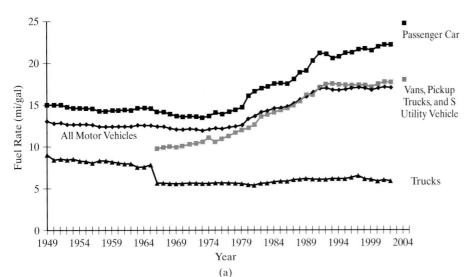

(a)

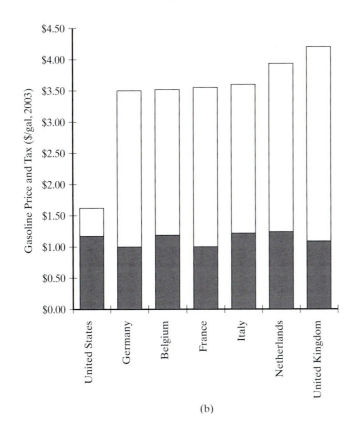

(b)

FIGURE 15.5

(a) Mileage of passenger cars, 1949-2004. Note that the solid line demonstrates that the fuel economy of each year's models does not translate into the fleet immediately. (U.S. Department of Energy, Annual Energy Review 2003 (Washington, DC: GPO, 2004), DOE/EIA- 0384 (2003), Table 2.8)

(b) Cost of gasoline in various countries. Gasoline in the United States is quite cheap compared to other developed countries, mostly because the fuel tax (open box) is so low.

(U.S. Department of Energy.)

the increase in the number of pickup trucks for family use, the rise of SUVs (as Figure 15.2 showed; see **Extension 15.3, *SUVs and CAFE Standards***), and the public infatuation with horsepower, mileage has stagnated since then.

The energy we use in a car depends on physics. The total force acting on a car, the sum of all the individual forces, must be found. We know that going uphill, we increase potential energy and so must transform fuel to work. The work done is the force times the distance driven, and that force depends on the amount of weight we are moving upward. (Of course, work done may also be expressed as power $\times$ time, which is why increasing the horsepower of a car motor allows faster passing and more "zip" climbing hills.)

Going downhill, we decrease potential energy, but this does not give the car additional energy; fuel is used anyway. Much of the energy is transformed into thermal energy when going downhill—we use the brakes. That energy is gone for good, as friction in the brakes has dissipated it. Frictional forces acting on cars are of various sorts: friction in the brakes, friction between tires and road (called rolling resistance), and friction between air and car (air resistance).

Overall, we have four separate pieces of the total force acting on a car: those speeding up or slowing down the car (accelerating and braking); those acting uphill or downhill; rolling resistance; and air resistance. There are then four separate pieces to the work done, each of which contributes separately to energy consumption by the car.

How Weight Matters

The weight of the vehicle is the most important predictor of mileage. This is because the work done going up and down hills depends on the weight being moved up and down, and because the amount of rolling friction depends on the weight. Figure 15.6 shows the relation between weight and fuel consumption that seems to hold for 2000 model cars (presumably similar to most recent models).[56] From the graph, we can find that the specific fuel consumption, the number of gallons per 100 miles per ton weight. It is approximately given by the constant value 2.29 for 2000 model year cars. This graph translates into a relation between fuel consumption in gallons and weight in pounds of

Fuel consumption (in gallons per thousand miles)

$$= 0.01116 \text{ (weight in pounds)} + 2.871$$
$$= 0.02456 \text{ (mass in kilograms)} + 2.871,$$

or

Fuel consumption (in liters per hundred kilometers)
$$= 0.0001044 \text{ (mass in kilograms)} + 0.012207.$$

How Speed Matters

As the speed of a car increases, the force of the air holding it back (**drag**) increases. The retarding force of the air grows faster for a given speed increase (as the square of the speed) as speed increases. Because the amount of work done by the car in overcoming the retarding force is equal to the retarding force times distance driven,

$$W_R = f_R \times d,$$

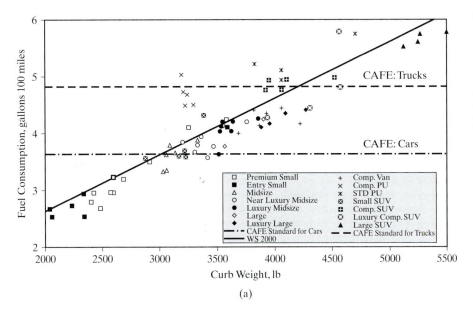

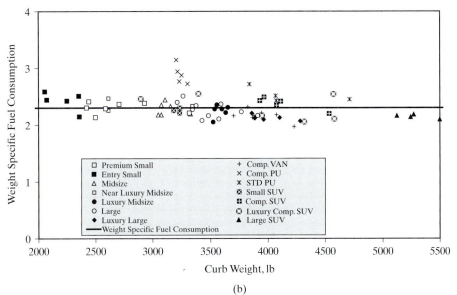

(b)

FIGURE 15.6

(a) Data show that the fuel consumption goes up when the weight goes up. (b) The weight specific fuel consumption is the number of gallons per 100 miles per ton weight. It is approximately given by the constant value 2.29. (National Academy of Sciences, Reference 46, Ch. 5-A, Figs. 5A-4 and 5A-5.)

this work must also increase faster as the speed increases. The total retarding force acting on a car is the drag force plus the force of rolling resistance. The energy to overcome the retarding force is supplied by gasoline—the greater the retarding force, the more gasoline must be burned just to overcome that retarding force—and thus, gasoline mileage must decrease.

The problem with increasing speed is illustrated in Figure 15.7. The best 1997 model fuel economy is experienced at a speed of 50–55 mi/h; in previous studies in 1973 and 1984, the best fuel economy had been at speeds of 35–40 mi/h.[91] The aerodynamics of cars has improved since then.

Drag from air resistance is a major source of energy loss; it increases as the square of the speed and is responsible for the rapid rise in Figure 15.7 as speed increases. Rolling resistance also depends on the speed, but it increases linearly with speed (as the straight line part of the graph). Rolling resistance is a much

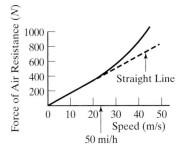

FIGURE 15.7

The drag force increases faster than speed increases as the speed increases above about 20 m/s. For larger speeds, drag force increases as the square of the speed.

smaller contributor to the resistance at high speeds. The **drag coefficient** measures the ratio of drag force to area. For a flat plate, the drag coefficient is 1.17.[92] By the early 1980s, drag coefficients of cars ranged from 0.4 (for the Volkswagen Golf) to 0.55;[92,93] by 1985, U.S. models had drag coefficients between 0.3 and 0.4.[94] The current level of ≈0.25 still leaves room for improvement.

In the years before the first energy crisis, the average speed of cars on rural highways rose from 48.7 mi/h (1950) to 60.6 mi/h (1970).[95] Overall, the average speed rose through 1973, dropped as a result of the 55 mi/h speed limit enacted by Congress in 1974, then has edged upward again (Figure 15.8a).[95] Figure 15.8b shows a somewhat different perspective: the proportion of post-1974 speeders, which also continues to edge upward.[95,96] Congress officially allowed the speed limit to be raised to 65 mi/h for rural interstate highways in 1987. The upward drift was probably partly responsible for small increases in highway fatalities in the mid-1980s after years of decrease. Statistics and Figure 15.8c show the effect of speed on the death rate; the numbers of deaths have continued to fall slightly

FIGURE 15.8a

The average speed of highway vehicles, which dropped precipitously following passage of the 55 mile-per-hour speed limit, is shown here from 1960 to 1993. (U.S. Department of Commerce, *Statistical Abstracts of the United States*, 2000, Table 1040.)

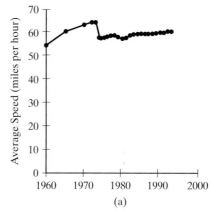

(a)

FIGURE 15.8b

The percentage of cars traveling at speeds greater than that indicated is shown from 1960 to 1993. This curve also shows a sharp drop after the national 55 mile-per-hour speed limit was passed in 1973. (U.S. Department of Commerce, *Statistical Abstracts of the United States*, 2000, Table 1050.)

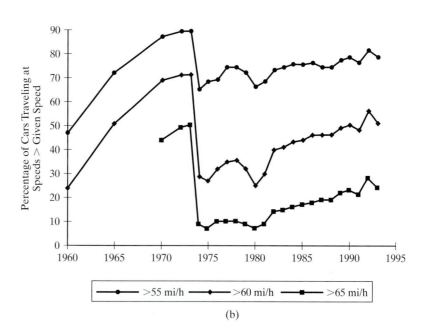

(b)

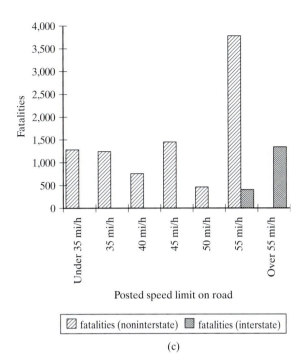

FIGURE 15.8c

Fatalities on roads in terms of posted speed limit.
(U.S. Department of Commerce, *Statistical Abstracts of the United States 2000*, Table 1040.)

(c)

overall because of automakers' incorporation of new safety systems and stricter enforcement of drunk driving laws.[97] The speed surveys of Figures 15.8a and 15.8b were discontinued in 1993, but there is evidence that "speed creep" may be accelerating. For instance, Connecticut found in 1998 that along its interstates (posted at 65 mi/h), the smooth flow speed was 74 mi/h, and in Utah, drivers routinely exceed the posted 70 mi/h speed limit by 10–20 mi/h.[98]

The increasing weight and increasing speed of cars during the two decades shown led to the mileage stagnation observed in the 1990s. In addition, the 55 mi/h national speed limit was removed, and, for example, in Illinois, led to increases in fatalities in rural areas.[99] The Insurance Institute for Highway Safety reached the same conclusion: they calculated 1900 extra deaths in 22 states between 1996 and 1999.[100]

15.3   POLLUTION FROM CARS

Pollutants from cars include unburned hydrocarbons and other volatile organic compounds (VOCs), which can escape while filling the gas tank and from leaks and from the engine under certain conditions; nitrogen oxides that are formed endothermically at high combustion temperatures; sulfur oxides from any residual sulfur in the fuel (especially a concern in diesels, but a problem for any converters); and carbon monoxide from incomplete combustion of the fuel. In earlier times, lead from the antiknock component tetraethyl lead was an environmental problem.[101] The problem is long-lasting; even now, lead emitted by American cars many years ago is being flushed into San Francisco Bay.[102] The costs of car-emitted pollutants have been estimated to be between 2 and 41 cents per km for older (pre-1997-model) cars and 1 to 9 cents per km for newer cars.[103] The damage cost is comparable (in Europe) to the price of the fuel. The U.S. Environmental

Protection Agency (EPA) uses six "criteria pollutants" as indicators of air quality: ozone (which is a measure of volatile organic compounds), carbon monoxide, nitrogen dioxide, sulfur dioxide, particulate matter, and lead.

One way to gauge the effect of transportation on emissions is to consider the proportion of various emissions from highway vehicles as shown in Table 15.4.[104–106] The fear of auto pollution is clearly well justified, despite the relative harmlessness of most auto particulates as compared to those from power plants and the low levels of sulfur dioxide in the exhaust.

The advent of pollution control devices is one reason for the poorer fuel economy of cars built in the late 1970s. Engines performed less efficiently when a **catalytic converter** was added to the exhaust system. In catalytic converters, the carbon monoxide (from incomplete combustion) and unburned hydrocarbons are completely burned, and nitrogen oxides react at high temperature with platinum catalyst pellets stacked in the path of the exhaust gases. The high temperatures arise because a series of baffles slows the exhaust gases to allow sufficient time for the catalyst to cause the pollutants to react and become carbon dioxide, normal nitrogen, and normal oxygen. This slowing causes a back pressure, increasing the pressure at the valve through which the combustion gases exit the cylinder, increasing gasoline consumption. Later (post-1985) model cars use computer chips to help the engine run in its most efficient condition despite this loss, thereby contributing to the increase in mileage observed between then and 1988 (see Figure 15.1)

Catalytic converters reduce nitrogen oxides in the exhaust. The catalyst provides a surface on which exothermic reactions can take place to reconstitute ordinary nitrogen and oxygen molecules when the converter temperature exceeds 300 °C. It also supplies the activation energy to allow the reactions to begin. Cars with converters can still emit a lot of pollutant as the converter warms up, but they typically remove 95% or more of the pollutants once they are hot. Several types of converter warmers that would heat the converter in seconds after the car started (battery-powered,[110] salt thermal storage,[111] and swirl-control valves[112]) are under development. Catalytic converters can be a pollution problem if disposed of improperly, because the platinum, palladium, and rhodium used are "heavy metals" and can harm living things. (Typical converters use 3–6 g of expensive platinum group metals, costing perhaps $100.[113,114]) In addition, overheating can lead to vaporization of these "heavy metals," which can be a problem in city stop-and-go

TABLE 15.4

Emission from highway vehicles compared with total emissions (megatonnes per year)

	EPA[a]	South Coast Basin[b]	United States[c,d]	United States[c,d]	United States[d]
	1975	1985	1980	1990	1998
CO	100.8/135.1	1.32/1.80	70.8/106.5	52.5/89.4	45.7/81.2
Reactive organic gases	17.7/22.6	0.17/0.41	8.2/23.8	5.7/19.0	4.8/16.2
NO_x (nitrogen oxides)	10.6/20.2	0.14/0.35	7.8/22.1	6.4/21.9	7.1/22.2
Particulates (PM_{10})	0.7/23.2	0.015/0.55	0.4/6.4	0.3/27.2	0.3/31.5
SO_2	0.9/26.5	0.007/0.04	0.4/23.4	0.6/21.2	0.3/17.9

[a] Reference 106.

[b] Reference 107.

[c] Reference 108 for sulfur dioxide.

[d] Reference 109.

driving.[113] An additional saving from installation of converters is the elimination of lead from gasoline (see **Extension 15.4, *The 1990 CAAA, California Standards, and Zero Emission***). Lead causes mental disorders, even in small quantities, but it was added in the 1920s to make burning more complete at low temperatures. Because lead incapacitates the platinum catalyst, it cannot be used in a car equipped with converters. Analyses of health benefits in America and elsewhere find that they far outweigh the costs of requiring converters. U.K. researchers say "Projections through 2005, when full catalyst penetration is expected, suggest net benefits of more than £2 billion."[169]

A new generation of converters utilizing mixed-metal oxides are being adopted because of the very high prices commanded by the platinum group metals.[114] Converters must be replaced, because the finely divided platinum tends to clump when heated, and over time, the loss of surface area renders the converter ineffective. There are indications that the perovskite-based catalyst $LaFe_{0.57}Co_{0.38}Pd_{0.05}O_3$ does not clump,[170] which could extend converter life substantially. (Palladium is roughly as expensive as platinum.) The cars should be very-low-emission vehicles and should cost less.

Honda and Nissan have introduced models that meet California's super-ultra-low-emission vehicle standard, 90% cleaner than current regulations for most cars require. In California, they can replace part of the mandate to build electric vehicles.[112] The Nissan Sentra CA boasts three catalytic converters.[112] An even more impressive idea is the "plasmatron" that zaps the exhaust with electric arcs and reduces emissions while burning leaner.[171] As the foregoing might indicate, the Japanese automobile companies build the cleanest, greenest cars.[172] As a result, one journalist wrote, "the Big Three no longer exist as such in California, which trails only Connecticut and Washington D.C. in the percentage of import sales."[173]

Not all strategies for pollution control cause increased gasoline consumption, as is the case for the catalytic converter. In fact, the experience since the early 1970s is that pollutant reductions have actually gone hand in hand with mileage increases. The common positive crankcase ventilation (PCV) valve merely allows gas leaking from the cylinders to be reused. The **stratified charge engine** has a modified cylinder, which has a small chamber with fuel-rich air fired by the spark plug connected to the main cylinder. The explosion causes a large chamber with fuel-poor air to ignite (as in the Honda Civic). This design provides reduced emissions without a mileage penalty. Honda also designed the VTEC-E engine for the Civic. This engine uses lean burn; the "swirl effect" directs gas to the hottest part of the chamber, insuring more complete combustion. These Civics get 55 mi/gal.[174] Combustion Electromagnetics, Inc., a small company, combines a lean-burn engine dubbed "Eco-Fire," which runs well with an air–fuel ratio of 26 to 1 (normal engines run at about 15 to 1), with a novel spark ignition system to achieve 5% to 20% reductions in emissions at a price estimated at $20 per car.[175]

Cars are the leading cause of photochemical smog (see Chapter 14, and especially **Extension 14.4**) and are important contributors to total emissions, as shown by the Los Angeles data in Table 15.5[104] and by the national data of Figure 15.9.[103] The amounts of pollutants emitted increased year by year during the 1970s and early 1980s, but finally began to come down as the effect of succeeding Clean Air Acts took hold. In Table 15.3, we also see the "baseline predictions" for the South Coast Air Basin for 2007 if no new controls on emissions are instituted (many new controls have been proposed). Controls had to be instituted; otherwise, ozone levels would have continued to exceed federal standards by a large factor,

TABLE 15.5

Air pollutant levels in the South Coast Air Basin [Los Angeles basin] (tonnes per day)

| | 1985 levels | | | | |
	Reactive organic gases	NO$_x$	Fine particulates	SO$_x$	CO
Transportation	590	585	645	75	4070
Oil products, marketing	120	90	10	30	50
Commercial	310	185	30	20	150
Solvents, home heating	80	35	15	5	270
Total	1100	895	700	130	4540

| | 2007 "Baseline" levels | | | | |
	Reactive organic gases	NO$_x$	Fine particulates	SO$_x$	CO
Transportation	310	580	975	80	2375
Oil products, marketing	80	45	5	40	20
Commercial	325	155	45	45	160
Solvents, home heating	210	50	5	5	430
Total	920	830	1030	170	2980

and levels of carbon monoxide and fine particulates would have exceeded them by a factor of almost two. The new 8-hour ozone standard that replaces the 1-hour standard (which was subject to strong effects of meteorological conditions) will

FIGURE 15.9

The 1998 distribution of air pollution emissions by source. Total emissions were 186.2 million tons (168.9 Mt). The proportion of emissions from stationary sources has been increasing. Air pollutants included are particulates, sulfur dioxide, nitrogen oxides, volatile organic compounds, carbon monoxide, and lead; their weights cannot strictly be added to reflect the net effect of air pollution as we have done here. Most of the miscellaneous emissions are particulates.
(Department of Commerce, *Statistical Abstracts of the United States 2000*, Table 394.)

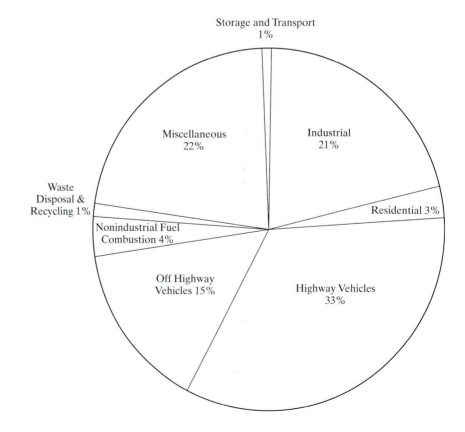

be difficult for some localities to meet.[120] However, given that almost 70 million Americans, 14 million Canadians, and 20 million Mexicans are exposed daily to heightened ozone concentrations,[120] the effort is worthwhile (even if it has not been entirely successful in the past). Many strategies for reducing ozone are currently available, and straightforward advances in current technology should ultimately help reduce ozone levels below the standards.[117,120]

As an illustration of the overwhelming presence of the auto in overall air pollution, Figure 15.10[176] illustrates the effect of the imposition of the first stringent pollution controls in the South Coast Air Basin of Los Angeles in 1965. According to the U.S. Office of Technology Assessment, autos emit 40% of the hydrocarbons, 70% of the carbon monoxide, and 45% of the nitric oxides in the United States.[177] If catalytic converters had been installed by 1965, the reduction in pollution would have been spectacular. Note that there is a "hidden saving" made obvious in the diagram. This occurs because pollution levels would have grown about as they had from 1955 on without pollution control; the actual gain is the difference between what *would* have happened and the drop that occurred. Figure 15.11,[178] from Los Angeles County, gives additional evidence of how emissions from autos pollute urban areas.

Since 1970, tailpipe emissions from new cars have fallen by 96% because of the imposition of pollution controls.[145] California auto standards have reduced the allowed emissions by 95% since 1966, but total emissions are not down that much (the number of people, and cars, keeps growing).[115] Nevertheless, the number of days of ozone alert was down 22% from 1976 to 1991.[117] The controls do not always reflect real-world occurrences, though. More miles are driven per car, and there are more cars. Poor maintenance and the lengthening lifetime of used

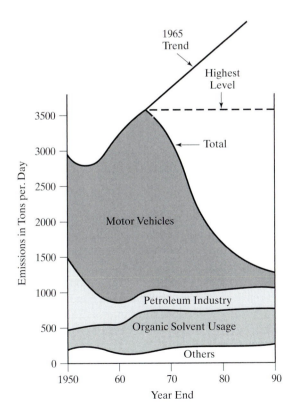

FIGURE 15.10

Actual (to 1965) and estimated emissions in the South Coast (Los Angeles) Basin.
(A. J. Haagen-Schmidt, "Air resources," in *Energy Needs and the Environment*, eds. R. L. Seale and R. A. Sierka, Tucson: University of Arizona Press, 1973, Figure 3.13, 80.)

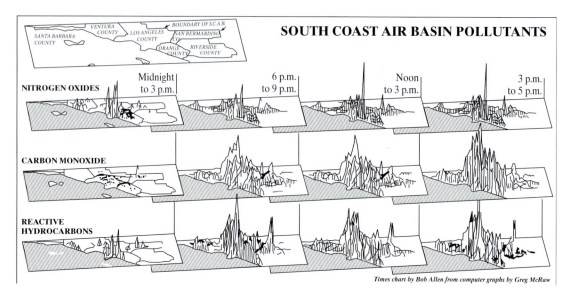

FIGURE 15.11

Profile of emissions of nitrogen oxides, carbon monoxide, and reactive hydrocarbons in the Los Angeles metropolitan area, showing the effect of commuting.
(Data from G. J. McRae, cited in an article by S. Blakeslee, © 1989, *Los Angeles Times*. Reprinted by permission.)

cars also contributes. An experiment in 1987 showed that mean emissions were 2.7 times higher for carbon monoxide (CO) and 3.8 times higher for hydrocarbons (HC) than expected.[117] California's "Smog Check" program is doing little or nothing to reduce in-use emissions. In fact, evidence from a test of 4421 vehicles implies that the most efficient way to reduce emissions is to measure the vehicles on the road and to fix them.[117] Ten percent of the California fleet is responsible for 60% of the exhaust hydrocarbons, and 10% of the fleet is responsible for 60% of the carbon monoxide emitted.[115,135] Similar results have been found in other states.[117] High emitters were scattered among all model years. The EPA tests cars' emissions, but the tailpipe tests are so expensive that the cost-benefit ratio is questionable.[179] If there were a cheap way to identify on-road polluters, and there is a device that uses the effect of exhaust on infrared light to measure pollutants as they are emitted on the road,[121] identification and repair should reduce pollution substantially.

Cars also cause pollution indirectly. An estimated 3 kilotonnes per day of hydrocarbons is emitted nationwide as vapor from filling station pumps; this is an indirect cause of car-related pollution and a waste of natural resource, as well as an economic loss. Now, California filling stations and those in some other states have a vapor recycling system mandated by law. The gas pump has a rubber collar that fits on the gas tank intake of the car. As gas is pumped in, the displaced vapor is pushed into the collar and collected. The EPA has mandated controls nationwide to prevent evaporation from "breathing" using activated carbon canisters in each vehicle.[117,180]

We look at some of the EPA criteria pollutants (Table 15.3) below.

Ozone

Ozone (O_3) is a major pollutant in all cities, and it is the one implicated in smog formation (see Chapter 14). The Clean Air Act Amendments (see **Extension 15.4**) mandate a reformulation of fuel to reduce ozone. One of the steps being taken is to use natural gas, ethanol, or methanol as well as ethanol/methanol blend fuels (see **Extension 15.5, *Methanol and Ethanol***). Methanol-fueled cars are already sold in Japan and California.[145] General Motors, at one time, shipped 2000 multi-fuel cars to California for use by state agencies.[196] The mixture of 85% methanol and 15% gasoline (called M85) is the usual choice, because cars fueled with pure methanol are hard to start. The GM multifuel cars worked on any mix. Computers check fuel flow and set the engine for the fuel mix available.[196]

Carbon Monoxide

Carbon monoxide (CO) is a colorless, odorless, and poisonous gas coming from incomplete burning of carbon in fuels. Carbon monoxide levels arise primarily from cars and trucks. Carbon monoxide is a deadly pollutant because it can re-place oxygen in the hemoglobin in the blood, and once it has done so, it is very difficult to pry out. In 1970, the U.S. government set a standard specifying a max-imum "safe" concentration; an area is deemed unhealthy at levels of over 9 μmol/mol (or 9 *parts per million* [ppm]) during an 8 hour period at least twice a year, or if the level exceeded 35 μmol/mol for two or more 1 hour periods per year.[197]

Most city-dwellers suffer from exposure to elevated levels of CO. A schematic representation of a round trip from Cal Tech to downtown Los Angeles is shown in Figure 15.12. Note how high the concentration of CO becomes. In downtown Los Angeles, the concentration exceeded 100 μmol/mol (100 ppm) on this trip. Los Angeles is not the only city with a carbon monoxide problem. Workers in New York City's garment district apparently experience a concentration of CO greater than 100 μmol/mol every day.[197]

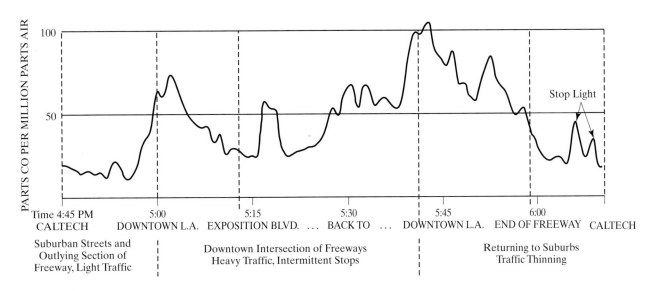

FIGURE 15.12

Carbon monoxide levels in a trip from Pasadena to downtown Los Angeles.
(A. J. Haagen-Smit, *Arch. Env. Health* **12**, 548, 1966. Reprinted with permission of the Helen Dwight Reed Educational Foundation. Published by Heldref Publications, 4000 Albemarle St. N.W., Washington, DC 20016. © 1966.)

One day of six, Denverites used to live through some of the highest levels of CO in the country: 24 ppm.[198] However, Denver, despite its impressive growth, has reduced pollution enough to become the first city to achieve compliance with the Clean Air Act; it reduced its carbon monoxide level by reformulating gasoline, and it forced a local power plant to switch from coal to natural gas.[199]

Much of the 80 million tonnes of CO emitted each year is from cars.[106] Because people sitting in nearby cars are affected, it is particularly hazardous to sit in traffic jams, when about 10% of the gas emitted from the exhaust is CO.[200] This problem could be reduced by placing car exhausts at roof level, as diesel truck exhaust is placed. This placement improves dispersal of CO by a factor of 10.[200] One major problem with CO is the burning sensation in the eyes associated with its presence; another is that it interferes with judgment by hampering brain function. Because we are less able to deal with difficult situations when affected by CO, being caught in a traffic jam can also be dangerous to health because of the increased risk of accidents. One lucky effect of making car exhausts cleaner is that the number of suicides by carbon monoxide inhalation has decreased dramatically. (One person was rescued, slightly dazed, after more than 8 hours in a garage with the car running.)[201]

One puzzle about CO is its disappearance from the environment. Estimated worldwide CO emissions should be raising the atmospheric level by about 40 nmol/mol (parts per billion) each year. (It takes about a month for half the CO to oxidize in the atmosphere.) The CO concentration does not appear to be increasing.[197] No one knows where it is all going (also see Chapter 16).

Volatile Organic Compounds

Carbon monoxide is a major pollutant in all cities, but other pollutants are present as well. (Figures 15.13a through d show why Los Angeles was once considered the most polluted city in the country, an honor now belonging to Houston.) **Photochemical smog** is created by the reaction of unburned hydrocarbons with sunlight. The photochemical smog composition for Los Angeles is shown in

FIGURE 15.13

Frequency of NAAQS violations by county for (a) carbon monoxide, 8 h. (S. Cohanim et al., *Summary of Air Quality in California's South Coast and Southeast Desert Air Basins*, 1991.)

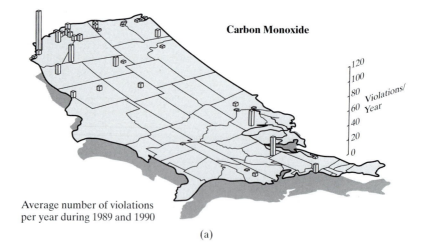

Carbon Monoxide

120
100
80   Violations/
60   Year
40
20
0

Average number of violations per year during 1989 and 1990

(a)

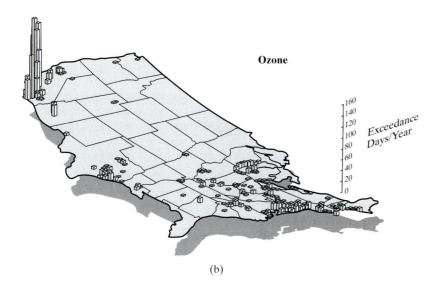

Ozone

(b)

FIGURE 15.13 (Continued)
Frequency of NAAQS
violations by county for
(b) ozone; (c) nitrogen dioxide;
(d) particulates (PM$_{10}$).
(S. Cohanim et al., *Summary of Air
Quality in California's South Coast
and Southeast Desert Air Basins*,
1991.)

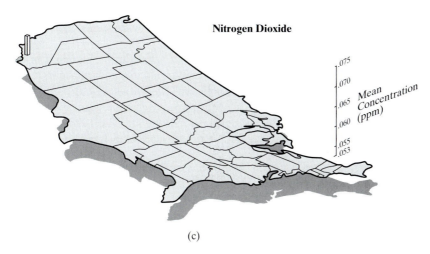

Nitrogen Dioxide

(c)

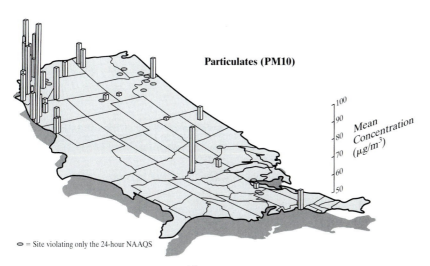

Particulates (PM10)

⊂⊃ = Site violating only the 24-hour NAAQS

(d)

TABLE 15.6

Typical concentrations in Los Angeles smog and clean air

Material	Smog concentration	Clean air concentration
Water vapor	2×10^{-2}	
CO_2	4×10^{-4}	4×10^{-4}
NO_x	2.5×10^{-7}	7.5×10^{-10}
CO	2×10^{-5}	2×10^{-7}
O_3	2×10^{-7}	4×10^{-8}
Reactive organic gases[a]	5×10^{-6}	1×10^{-8}
Nonreactive organic gases[b]	2.5×10^{-6}	
SO_4	2×10^{-9}	
NO_3	5×10^{-9}	

[a] Reactive organic gases exclude carbon monoxide, carbon dioxide, carbonate, and nonreactive organic gases.
[b] Nonreactive organic gases are CH_4, CH_3Cl, CH_2Cl_2, and the chlorofluorocarbons.
Source: Smog, Reference 202; clean air, J. Cassmassi, T. Chico, H. Hogo, J. Lester, and S, Mitsuomi *Ozone Modeling—Performance Evaluation* (unpublished) (Los Angeles: South Coast Air Quality Management District, 1990).

Table 15.6.[202] Note that it is still difficult to determine pollution levels accurately: Los Angeles readings on smog levels disagreed by 30% with readings in nearby counties obtained using the state-mandated method.

California, with what was in the 1950s the worst air pollution problem in the nation, created a board to set automobile standards for California cars. With (then) 20% of the nation's population, the board's decisions would have clout. Standards slowly got more restrictive. California was the first state to mandate the use of the catalytic converter, now standard nationwide. The 1998 California standards actually mandate zero emission vehicles (see **Extension 15.4**).

Combustion Products and Waldschäden

The major reason for German enthusiasm for unleaded gas was not the 4.5 megatonnes of lead emitted each year into European air;[203] rather, it stemmed from their perception of the impending destruction of the German (and Swiss) forests (*Waldsterben*, forest death) from the effects of acid rain.[204–208] (Forests generate archetypes in the German psyche—think Hänsel and Gretel and other Grimm brothers tales—and are important beyond their economic use.) The situation remains serious, but researchers have recognized that it is not likely to end in the death of all trees, so the problem is now designated **Waldschäden**, or forest damage, instead of forest death. Waldschäden is defined as loss of 25% or more of leaves or needles from the trees. The connection between unleaded gas and the nitrogen oxides is, of course, the catalytic converter.

A CLOSER LOOK

Pollutants and Politics in Germany

European countries, with the exception of Germany, have speed limits between 100 and 130 km/h. Cars traveling 130 km/h produce emissions at twice the rate of those traveling 100 km/h,[209] but because the faster car does not travel for as great a time, the total emissions are only 54% higher than for the car traveling 100 km/h. Part of the reason that other European

Union countries were not inclined to rush installation of converters, as discussed earlier, is the lack of speed limits on the German autobahns. Adoption of a speed limit by Germany would cut pollution levels, as these other countries point out. Germany has adopted many zones with speed limits, rather than cut the speed limit directly, and enforced these limits by having cameras that snap pictures of speeders' license plates along with the speed; the speeder must pay the ticket that arrives in the mail. On the other hand, if cars are fitted with converters, they must be driven correctly to prevent overheating.[203] Hot converters on cars parked near fallen leaves have caused fires, for example.[210] The Germans who buy cars with converters have to pay a performance price in any case. BMW engineers say that a converter-equipped car loses 15% in horsepower.[211] One way or another, German speeds on the highway have to be limited if the NO_2 emissions are to be reduced.

Because converters for German cars cost 300 euros (€300 = about $350),[203] and because the Germans eliminated leaded gas in 1989,[212] manufacturers and drivers had expected substantial subsidies from the government. Both groups were disappointed. The government simply waived the engine capacity tax that car owners pay.[213] The payback period for the improvement in emissions seems too long to most observers. The normal wisdom is that buyers will pay only for fuel economy savings that can realize a net return within 2–3 years.[214] This is probably the attitude most people have toward payments for emission control. Remember that this air is similar to a commons. Individuals will have to be coerced or cajoled into doing something that will cost them money personally to benefit everyone by preserving the forests.

Although the damage to German forests can be seen in hindsight to have begun as long as 30 yr ago (the diameter increase in trees in forests fell from about 20 mm/yr in the mid-1960s to a few mm/yr in the mid-1980s), it was not obvious to the Germans until 1981. For more details, see **Extension 15.6, *Sulfur Oxide, Nitrogen Oxides, and* Waldschäden**. Similar changes have occurred in the United States. An abrupt change in growth rates simultaneously observed in New England and the Southern Appalachians may bode the same fate for American forests.[222]

15.4   POSSIBLE ECONOMIES OF VARIOUS MODES OF TRANSPORTATION

The steady increase in speed and power have, as we have seen, caused a decline in efficiency, and, as discussed, emission control by catalytic converter causes even more. Artificially low energy costs, sustained over decades, and urban sprawl have also contributed to the inefficiencies of the system. With energy cheap, people choose power instead of efficiency.[223] In Europe, where high gas taxes (Figure 15.1b) make people use energy more rationally, the number of person kilometers traveled is 40% lower.[224] Fuel costs are a major variable cost but not a large part of a person's total travel costs (personal time is valued considerably more highly). Gas taxes would need to be substantial before they would discourage travel.[225]

The most important measure of economy in the United States is the number of passenger-miles per gallon. Suppose we compare a Honda Civic with an EPA rating of 41 mi/gal with one passenger, to a Chevrolet station wagon with an EPA rating of 20 mi/gal carrying four passengers. The Honda in this example gives 41 passenger-mi/gal on a trip; the Chevy delivers 80 passenger-mi/gal. The people travel more efficiently in the larger car because there are more of them—the load factor is higher.

Raising Fuel Economy

The American economy, as we indicated, depends heavily on the automobile sector. In 2000, the United States spent $90 billion for imported oil, which produced roughly the amount of gas used by cars.[2] If we do not want to scrap the current system, the coming exhaustion of fluid fossil fuels (Chapter 12) should impel us to increase fuel economy. Unfortunately, there are shysters who try to sell to unwary consumers mysterious devices "guaranteed" to increase mileage.[226] Remember the phrase *caveat emptor* (buyer beware) when reading ads promising a 20% increase in fuel economy; the only guarantee is that the consumer will lose any money invested in such devices.

Americans slowly switched to smaller cars in the wake of the first energy crisis because of concern about fossil-fuel supply, but the switch had profound effects for the American auto industry. Massive layoffs by the automakers in the 1974–1975 recession (in the aftershock of the first energy crisis) contributed to that recession's prolongation, as did the layoffs of the 1981–1983 and 1990–1993 recessions.

As gasoline prices fell in the early 1980s, demand for bigger cars remained stronger than the automakers anticipated. This situation led automakers to attempt to water down Title V of the Energy Policy and Conservation Act of 1975, the Automotive Fuel Economy Program. The bill mandated that by 1985, the fleet average of the large automakers be at least 27.5 mi/gal (the so-called corporate average fuel economy, or CAFE standard, see **Extension 15.3**). Of the big three American automakers, only Chrysler had met the standards by 1985. The Reagan administration succeeded in its efforts to allow the required fleet average to be lowered to 26 mi/gal. President Bill Clinton campaigned on a platform of higher CAFE standards, and the Clinton administration implemented this change early on; further gains were mired in an unreceptive Congress. Also, automakers are selling more light trucks and SUVs, which did not have to meet the same mileage standards (they did after 2000); this kept the rise smaller than it would have been.

Any possible dollar savings we make must be balanced against losses of time, comfort, and flexibility that would be encountered in the shift to a more energy-efficient mode. The old saying "time is money" has some truth to it. It is also true that each 1 mi/gal improvement in the fleet average beyond 28 mi/gal reduces hydrocarbon emissions by 0.0229 g/mi and other emissions accordingly.[227]

Turbocharging, in which the exhaust gases run a compressor to increase air pressure in the cylinder and allow more gasoline to be mixed in, increases efficiency by 10% to 15%.[181] Because kerosene contains more energy per volume than gasoline, use of diesels is good for fuel economy, although probably not so good for health because of increased emissions of NO_x and small particulate matter.[123] Continuous torque transmissions can also contribute to energy conservation.[94] A major help for the U.S. fleet fuel economy has been the shift back to manual transmissions from conventional automatics during the 1980s.[94] Porsche developed the Tiptronic, an expensive automatic transmission using a micro-processor. The computer considers 40 sets of data in deciding on shifting. Shifting back and forth between two gears and shifting at the wrong moment are avoided. This automatic equals manual transmissions for emissions and fuel economy.[228]

All in all, currently available technology could raise automobile fuel economy for modified versions of cars such as the VW Golf (a midsize car) above 70–80 mi/gal.[94] However, each mile per gallon improvement costs $2–$4 billion (if the re-tooling cost is attributed solely to this factor), so one must consider the tradeoff in-volved in achieving such fuel economy levels. Table 15.7 shows various efficiency measures that could raise the efficiency of all cars. According to L. Schippers,[229] existing devices scattered among car models would save 30% more efficiency if they were all in one car. Advanced technologies could save 50%;[229] applying 17 ex-isting technologies more broadly would raise average mileage to 43.8 mi/gal from 28.3 mi/gal.[230] **Extension 15.7, *Examples of Efficiency Possibilities*** has many ex-amples of ways that new technology can impact fuel economy—weight reduction, novel engines, and the like.

TABLE 15.7

Improving efficiency measures (many already adopted by the auto industry)

Measure	Improvement
Decrease drag	10% reduction implies 2.3% efficiency gain
Reduce weight	10% reduction implies 6.6% efficiency gain
Rear-wheel to front-wheel drive	5% to 10%
Use overhead cams	3% to 6%
Add two valves per cylinder	5%
Variable valve timing	15% to 25%
Torque converter lockup	3%
Accessory improvements	0.5% to 1.0%
Install advanced transmissions	4.5% to 8.0%
Electronic transmission control	0.5%
Fuel injection systems	3% (throttle body); 6% (multipoint)
Improved tires	0.5%
Improved lubricants	1.0%
Engine friction reduction	≈6%

Source: References 94, 230 (quoting a study by Ledbetter and Ross), and 249.

Diesel Engines

The diesel engine is already efficient. In the United States, the fact that the diesel engine is "dirtier" than the gasoline engine has meant that diesels have represented only a small portion of the personal vehicle market. The Volkswagen Lupo "3 liter" uses only 2.99 L per 100 km (equivalent to 79 mi/gal), the best mileage of any car manufactured.[156,250] Meanwhile, President George W. Bush's administration has declared the 80 mi/gal Clinton administration proposal an unrealistic goal.[250] According to a reporter who drove a Lupo around the New York area, "Several people swore they would buy one if it were offered for sale. Many used the word 'cute.'" That has apparently been the reaction in Europe. The Opel Corsa, a larger diesel car, uses just under 5 L per 100 km (50 mi/gal).[251] The Audi A2, also a larger car, gets 78 mi/gal.[161] The average diesel car gets around 40 mi/gal.[157]

Because of the high price of fuel in Europe, and because diesel is significantly lower in cost, diesels have captured a third of the European market (diesels ac-count for half of French sales).[157] American consumers had bad experiences with GM diesels in the 1970s, and they have never been as popular (see the discussion of diesel pollution in **Extension 15.4**). Further, older diesels were rather sluggish, but the diesels manufactured during the last decade have been quite peppy. The Lupo goes from 0 to 100 km/h in about 15 seconds (0 to 60 in about the same time). While no drag racer, this is respectable.[156] The old characteristic diesel knock is gone. The National Academy report on the CAFE standards suggests that a greater emphasis on diesels will be warranted if they can meet American emissions standards.[56] There is hope that plasma devices can be used in place of converters on diesel cars and trucks and can clean the exhaust sufficiently to meet pollution standards, even in states adopting California standards.[252]

Hybrids and Pure Electric Cars

There has been a lot of concern about the poor mileage of modern American cars, and one way of making it better might be to produce cars that use electricity, wholly or in part. In addition, eliminating tailpipe pollution is a worthy goal. This might also be improved by a switch to all or part electric cars.

Pure electrics have been produced (for example, the GM Impact, now the EV-1) but have not been successful in the marketplace. Some problems are due to lack of good alternatives to the common lead-acid battery used in cars. Other problems have to do with the ability of the battery to supply sufficient power and to be able to store enough energy but not weigh too much (a drawback of the lead-acid battery). The energy content of batteries is normally characterized by the **specific energy**, the number of watthours per kilogram of battery mass. A typical lead-acid battery is in the neighborhood of 30 Wh/kg (for comparison, gasoline, at 44 MJ/kg, has a specific energy of 12 kWh/kg, 400 times as much) and certainly under 40 Wh/kg.[253] Pure electric cars demand a much higher specific energy. The specific power, a measure of the amount of power that can be delivered per kilogram, also limits possibilities; if there is too low a value, the battery will weigh too much. Lead-acid batteries have specific powers around 100 W/kg.[253]

Hybrids also need more specific energy and specific power than common lead-acid batteries can produce, but with different characteristics from those needed for pure electric cars. Many manufacturers have turned to nickel-metal hydride batteries as the solution.[254] Many hybrids have been sold, and many more will be produced, both by Japanese and American carmakers. For more detail about cars that run on electricity, see **Extension 15.8, *Hybrids and Electric Cars***).

Economy and Pollution

It *is* possible to reduce emissions and clean the air as well. The conventional fuel economy of a car is measured, according to the EPA, through its emissions in grams per mile of carbon monoxide (CO), carbon dioxide (CO_2), and unburned hydrocarbons (HC):[44]

$$\text{fuel economy} = \frac{2423 \text{ mi/gal}}{[0.866(\text{HC}) + 0.429(\text{CO}) + 0.273(CO_2)]}.$$

Typical emissions and the effects of various proposed and adopted standards[308] were shown previously in Table 15.4.

Fuel economy should not be pushed without regard to the other aspects affected by increased fuel economy. For example, for a car tuned for best fuel economy, the engine burns very hot so that NO_x emissions are a thousandfold larger than the best attainable levels, and hydrocarbon emissions are 19% higher than achievable.[210] Similarly, for the minimum level of NO_x emission, the temperature of combustion is so low that fuel economy is 30% lower than the best value, and hydrocarbon emissions are 29% higher. Finally, to minimize hydrocarbon emission, the engine must run hot which supplies the energy for endothermic NO_x-creating processes, increasing NO_x emissions to 433% above the lowest levels possible,[210] and fuel economy for a hot engine is 10% lower than the best possible. The ratio of fuel to air is the most important variable determining emissions. When CO and hydrocarbons are minimal, A/F ~ 16–17, NO_x is maximal, and fuel economy is maximal.[309] If A/F > 17.5, the "lean burn" region, there can be misfires.[309] Simultaneous extremization (maximization of mileage, minimization of HC, CO, NO_x) occurs for A/F = 14.7 ± 0.05. Computer engine control has been essential for simultaneous pollutant minimization and mileage maximization.[309] Clearly, the tradeoff must be made in which all these costs are balanced and an acceptable tuning point found.

One sort of "pollution" not considered in the standards is mosquitoes breeding in discarded tires; neither is the emission of particles of rubber, carbon black, and other fillers in tires. Wilson[200] estimated the rate of these emissions at 0.43 tonnes of particles per million miles from road wear and 0.39 tonnes per million miles from burning of old tires. Over 1.6 Mt of tire particles go into American air annually.[200]

Tires can be used to make asphalt. Asphalt costs more with rubber but lasts three times as long. Each year some six million tires, 45 kt of rubber, are used in road construction.[310] Reclamation of tires used to be routine (there were some 30 plants in the 1960s), but the tires have become more difficult to reclaim, and there are only two such firms left.[310] American Tire Reclamation shreds tires and purifies them by pyrolysis to get oil products and carbon black (but not carbon black of original quality). The oil is separated and refined.[311] Oxford Energy Corporation, a tire burner located in Modesto, California, operated at a loss as late as 1989, despite being paid 8.5 cents/kWh.[310] A devastating fire closed the Modesto facility permanently in September 1999.[312]

The safety hazard from lead, much from cars, was discussed in **Extension 14.2**.

Economy and Safety

Another consideration is safety. The alternatives to cars are much safer. Dying in an automobile is the most common accidental death: 20% to 25% of the people who die in accidents each year in the United States die because of car accidents (of course, accidents account for only about 10% of all deaths, so car accidents produce about 2% of all deaths). The next most common cause of accidental death, falls, is responsible for only about 10% of all deaths, fire and explosion about 4%, and drowning about 3%. Railways and planes each claim less than a life per 1,000,000 people carried and cause much less than 1% of all accidental deaths.[313]

There is an additional safety factor to be considered when the weight of a vehicle is reduced. Amazing advances in adhesion have led directly to applications in painting plastics for cars.[314] Bigger cars are generally safer (although smaller cars can be made safer by clever use of materials). Reducing the car mass from 2000 to 1500 kg results in an increase of 4% to 6% in the chance of death or injury.[210] The imputed cost of increased risk of fatalities due to weight reduction is $0.16/lb reduced weight.[249] In 1977, the car fleet on American highways had a

42% small-car share. Had the fleet been all large cars, the 27,400 fatalities would have been reduced to 22,600; if it had been all small cars, fatalities would have risen to 34,000.[210] There is a trade-off at work here, too. Most car buyers ignore safety; as Lave said:[210]

> When given a choice, the vast majority of new car buyers choose an automobile with style, power, and comfort, and these preferences have been persistent in the United States and the rest of the world.

The most cost-effective method of preventing death and injury is to use an ordinary seat belt. Air bags cost three to four times as much per death or injury averted as installation of seat belts.[210] Most states have seat belt laws, but enforcement varies. The State of Washington, which had a high compliance (~80%), ran a "Click It or Ticket" campaign that raised compliance to 95%.[315] Enforcement works, when that heavier enforcement lasts a year or more. Fatalities dropped by 13%, and, as the authors of Reference 315 stated, "The 5% of non-users contribute half of the motor-vehicle occupant deaths!" Because Washington had such a high compliance to start, the study refutes the argument that increased enforcement is not worthwhile in such a case. Antilock brakes, also effective, are on about half the cars being sold. Improvement of highway construction and maintenance practices is probably the next most cost-effective.

New Technologies Used in the Car

Many cars already come ready for Internet and fax connections. The owner simply pays a monthly fee.[316] At present, the idea is to install most connections in the back seat to minimize driver distraction. Virtually all fleet trucks have GPS positioning systems, which enhance driver safety and give more control to the fleet owner to allow adjustments. Many cars already have GPS connectivity and can be tracked and located. "Telematics" systems will allow better navigation assistance, tracking of stolen vehicles, and automatic notification to the police of accidents.[5,316]

Car accidents kill or injure over 10 million people a year and cost over 1% of the world's gross domestic product,[317] so increasing car safety has a high priority. But now, the methods must move beyond air bags, seat belts, and antilock braking systems.

Already, there is the equivalent of a PC under the hood of most cars. One use of this computer power is to give advice on how best to drive to maximize fuel economy under a system developed in Sweden and the Netherlands. Drivers with advice used 16% less fuel than drivers without and did even better, 23%, in urban areas.[318]

One new technology is the drive-by-wire system, which would substantially upgrade that computer and replace a car's hydraulic system with microcontrollers. This system is used in advanced aircraft and may help cars adjust automatically to road conditions, making for a safer car trip. It is possible that future systems will automatically adjust speed and steer the car.[319,320] At present, there is no industry-wide standard that can be used, though work is in progress. Perhaps the most difficult thing to replace is the "feel" for the road that the mechanical couplings used in cars give. Heads-up displays have already been implemented in "high-end" cars.[320]

One advantage of such systems is a savings in energy, as each system will be activated only as it is needed and will not run all the time. The hydraulic systems currently in use require relatively large amounts of energy (for pumps, etc.).[319]

Throttle-by-wire has been in use since 1996 in diesel-fueled cars and in a few high-performance cars.[319] Pressing the accelerator sends a message to the throttle through the onboard computer. This computer control allows increased fuel economy, because the throttle is always adjusted to its optimal setting automatically.[319]

In steer-by-wire, turning the steering wheel (or joystick) changes a sensor readout, which sends a signal to the motor controlling the front (or rear) wheels. The system increases vehicle stability. Brake-by-wire offers more traction control than today's antilock braking systems, because each wheel is controlled individually.[319] Electromechanical braking systems, with a motor-controlled caliper at each wheel, require the 42 V electrical systems being adopted gradually. These systems would replace the current yaw systems available on some cars, such as StabiliTrak on Cadillac.[321] Luxury carmakers had some brake-by-wire available in the 2004 model year. This braking system is already very popular in Europe.[81,322]

The by-wire system has safety implications in other aspects. The absence of the mechanical steering linkage allows different crumple zone possibilities for driver protection in the event of a crash. An alternative is electric power steering (the current power-steering systems use hydraulics).[81] In addition, carmakers are adopting collision-avoidance ideas, such as adaptive cruise control, which use lasers (lidar) or radar to judge the speed and distance of other vehicles on the road.[5] These systems have been installed on big-rig trucks since 1994. By 2001, only foreign automakers such as Toyota and Mercedes-Benz have adopted this technology, but domestic carmakers began to offer these systems on high-end cars in 2002.

Current systems are expensive, typically adding $1500 to the price of the car.[81,317] Laser-based systems are cheaper but perform poorly in bad weather conditions, because they are too sensitive. Radar systems allow a "view" of 150 m regardless of outside weather conditions. The systems occupy a volume of only 14 cm $\times$ 10 cm $\times$ 7 cm.[317] A collision avoidance system presents a difficult problem. The big question, not yet decided: Should the systems merely warn, or should they interfere actively? DaimlerChrysler believes the answer is the latter; it developed systems to prevent driver distraction actively.[323]

The next generation of collision avoidance is known as cooperative adaptive cruise control. In this system, cars could communicate to avoid collisions. When all cars are capable of this, the density of cars on the road can rise substantially, with intervals of a mere half-second between cars (two car lengths at 100 km/h).[317] Fujitsu Ten, Ltd. is working on a stop-and-go adaptive cruise control. The system would stop the car when the car in front stops using a "fusion sensor" camera that detects only vertical edges of obstacles. Toyota mounted such a motion sensor in its 2005 model Lexus.[324] Nissan is the first car company to embed a camera in the rearview mirror in the U.S., while in Japan, both Nissan and Honda sell such cameras as part of systems that prevent lane drift.[324] German researchers have also tackled this as part of a system that prevents drivers from falling asleep at the wheel.[325]

"Smart" cars that will operate more energy-efficiently are coming; they will even read information from local and distant sensors and make automatic adjustments.[326] In Los Angeles, power companies are cooperating to build a road that can transfer electricity by magnetic induction to cars driving on it.[327] This could work with battery-powered cars. The freeways could be rebuilt to carry the electricity to run the cars and recharge their batteries.[327] By means of sophisticated software, a car could locate itself by dead reckoning and a built-in map. Or, the navigation could rely on radio signals from navigation satellites.[328]

The satellite system detectors are expensive, and the Global Positioning System (GPS) could be shut down by the Pentagon at any time. Smart cars could be packed closer without collisions, allowing roads to carry more traffic. Lanes could be narrower.

The Europeans are working on complementary ideas. The goal of the Prometheus project (PROgraM for a European Traffic with Highest Efficiency Unprecedented Safety), begun in 1986, is to improve traffic flow by making vehicles more "intelligent." It is subsidized by European governments and involved over a dozen European automakers and a score of universities.[329] Infrared cameras mounted by the rearview mirror can sense heat from objects 100 m away, for example. One important use of such a system is to prevent drivers from backing up into pedestrians, especially children.[330] The cameras can produce clear images to be displayed on a dashboard TV or projected onto the lower windshield so the driver's eyes can stay aimed at the road.[328] Many other technologies were worked on, and some have been introduced into the cars being sold both in Europe and the United States.[331]

New Technologies for Road Surfaces

The mean life of concrete roads, the 20% of the nation's roads that must carry the heaviest traffic, is around 30 years.[332] A new process, called dowel-bar retrofitting, is cheaper than repaving the road and may extend the life of the pavement by more than 10 years. Though this is old technology, it used to be prohibitively expensive to install the bars into the highway, but a slotter machine has been developed that makes the job much easier and, therefore, more affordable, less than half the price of repaving.[332] The success of the technique has led to the new highways being installed with dowel bars. The bars bind the slabs of concrete together, reducing battering of its edges.[332] In the wake of terrorist attacks on the World Trade Center, research into use of fiber-reinforced concrete has proceeded for buildings, but this may also end up helping us to build concrete roads better.[333]

The other 80% of the roads that are asphalt have also benefited from new techniques. Asphalt roads typically go 10 years between repaving.[334] Superpave is a method developed to measure the properties of the hot mix asphalt and to design the high- and low-temperature behavior of the material to extend the lifetime of the road. Superpave can extend the road life to 15 years.[334] The Federal government invested $50 million in research in the Strategic Highway Research Program and $150 million in implementation to achieve these gains. Energy and monetary savings can be substantial, over half a billion dollars per year.[334] Of course, less paving activity will eventually translate into fewer delays for drivers, saving them time and money. The technique essentially provides a method to standardize road surfaces, and they seem to hold up much better than conventional asphalt pavements.

SUMMARY

Most forms of transport of goods and people are more efficient than cars and trucks as used. However, they are not as flexible or as convenient. Most people choose to own cars rather than use public transportation, and many businesses choose to use trucks to ship materials long distances instead of railroads, barges, or ships. Ships and pipelines are used to transport fuel, because they are intrinsically more energy efficient. However, fuel is so essential to the American economy that the cost of oil is subsidized in various indirect ways (see **Extension 15.2**).

One form of transport that is highly energy inefficient (but extremely time efficient) is aircraft. There are promising ways of increasing aircraft efficiency, and there is hope that new jet engines and composites will help.

Load factor is important to use in assessing the efficiency of various transportation modes. A full station wagon may transport seven people for a lower energy cost per person than a subcompact carrying one person. Mass transit holds the promise of more efficient transportation of large numbers of people than at present, if we can figure out how to set up such a system. Were the mass transit infrastructure to receive subsidies similar to those given car and truck transportation (as discussed in **Extension 15.1**), it might be able to attract more riders.

Car mileage depends most strongly on car weight and also on the profile it presents to the air (drag). For this reason, the change in buying habits that has resulted in more SUVs and pickups on the road has adversely affected the average economy of the American transportation system (see **Extension 15.3**).

In addition to emitting pollutants as they are driven, cars emit pollutants as they are refueled. The worst pollutants are carbon monoxide, which replaces oxygen in the blood, and lead, which can cause mental disorders. Nitrogen oxides and ozone can cause smog, which exacerbates lung ailments. There are many opportunities for increasing car mileage. California has led the way in forcing automakers to reduce pollution (as described in **Extension 15.4**). Pollutants from petroleum-based motors have contributed to sickening of forests (noticeably in Germany, see **Extension 15.6**).

There are many opportunities for increasing overall efficiency, and efficiency of future cars (see **Extension 15.7**). Methanol and ethanol blends are more fuel efficient than pure gasoline (see also **Extension 15.5**), and they also reduce emission of pollutants. Use of diesel engines (if the fuel can be cleaned up) and of hybrid cars, running on gasoline but using various techniques to reduce fuel use, already available in the Toyota Prius and the Insight may capture a greater market share, especially in California. This is discussed in more detail in **Extension 15.8**.

PROBLEMS AND QUESTIONS

MULTIPLE CHOICE REVIEW

1. Automobiles use about what proportion of all the energy produced in the United States?
   a. 1/100
   b. 1/20
   c. 1/10
   d. 1/5
   e. 1/3

2. Which pollution control device listed below does NOT increase gas consumption?
   a. Positive crankcase ventilation valve
   b. Stratified charge engine
   c. Catalytic converter
   d. All of the above increase gas consumption.
   e. Only (a) and (b) increase gas consumption.

3. The mileage a car that weighs 1500 lb should get is
   a. 0.051 mi/gal.
   b. 19.6 mi/gal.
   c. 25.2 mi/gal.
   d. 39.7 mi/gal.
   e. 51.0 mi/gal.

4. The purpose of a catalytic converter in a car is to
   a. increase the back pressure in the engine.
   b. turn carbon monoxide into carbon dioxide.
   c. turn nitrogen oxides into nitrogen and oxygen.
   d. do all of the above.
   e. do only (b) and (c) above.

5. In Midtown, the average energy cost to operate an empty bus is 23,170 kJ/km, and the load factor for bus operation is 30% (Figure 15.3). A fully loaded car (five people) costs 4183 kJ/km to operate, while with just the driver, the energy cost is 3791 kJ/km. If the operating costs of the bus per passenger kilometer are half that of the average car per passenger mile, about what is the load factor of the car?
   a. 0.17
   b. 0.24
   c. 0.33
   d. 0.40
   e. 0.84

6. Which factor contributes to the cost per mile of owning a car?
   a. Price of fuel
   b. Insurance
   c. Maintenance
   d. Purchase price of the car
   e. All of the above factors contribute to the cost.

7. Which statement below is FALSE?
   a. Lead in gasoline allows complete burning at a lower temperature.
   b. Lead levels on the Greenland ice cap increased after about 1930.
   c. Ingesting lead can cause mental disorders.
   d. Lead in gasoline "poisons" the platinum in a catalytic converter.
   e. None of the above is false.

8. In Columbus, the COTA bus line has a load factor of 40%, while cars have a load factor of 20% (just the driver). If the

loaded bus costs 1800 kJ/passenger-km to operate, buses account for 40% of the downtown passenger kilometers, and cars account for the rest (60%), what is the average energy cost for downtown transport in all modes?
a. 1800 kJ/passenger-km
b. 3030 kJ/passenger-km
c. 2825 kJ/passenger-km
d. 3850 kJ/passenger-km
e. 5650 kJ/passenger-km

9. Replacement of up to 15% of gasoline by ethanol or methanol has which effect(s)?
a. Loss of performance
b. Less engine knocking
c. Lower emissions
d. All of the above are effects.
e. Only (b) and (c) above are effects.

10. The fuel consumption of a car weighing 4000 pounds is
a. 0.021 gal/mi.
b. 9.90 gal/mi.
c. 31.0 gal/mi.
d. 47.5 gal/mi.
e. 101 gal/mi.

SHORT ANSWER

Explain whether the statement made is correct, and justify your answer.

11. Replacing manual transmissions by automatic transmissions will increase gas mileage.
12. Drive-by-wire has the potential to decrease gas mileage.
13. While air freight is energy inefficient, aircraft carry the greatest amount of freight.
14. The cost of oil as sold does not include all the concomitant costs, such as those for pollution, the cost of maintaining a defense establishment to protect oil supply, etc.
15. A typical SUV uses more gas in a year than a typical family sedan.
16. Generation of photochemical smog involves the reaction of unburned hydrocarbons with sunlight in a closed basin.
17. Both rolling resistance and drag increase with increasing speed as the square of the speed.
18. Transport companies like the use of GPS transponders on their trucks because it allows them more flexibility in moving their rigs to where they are needed.
19. Carbon monoxide is necessary to lung function in healthy adults.
20. Drag forces in cars depend on the profile the car offers to the air.

DISCUSSION QUESTIONS

21. Would increasing the load factor for air travel reduce per passenger energy use? Explain.
22. Discuss why research into transport by ship would be less emphasized than research into other modes of transportation.
23. Explain why California would mandate the installation of catalytic converters in cars at a time when their use cost energy and degraded performance.

24. Explain why cars are responsible for so much of the carbon monoxide emitted every year (see Table 15.4).
25. As we know that the 55 mi/h speed limit saves lives and gasoline, try to find reasons the speed limit on interstate highways has been set to 65 mi/h and above in most states.
26. Why doesn't a car driven at 200 km/h emit twice as much pollution in a 400 km trip as a car traveling at 100 km/h?
27. Long-distance bus service is cheaper per passenger than urban bus transit. Explain as many as possible of the factors that make this so.
28. Consider methods to identify the major polluters from among the many cars on the road (a few are responsible for most of the pollution). List as many as you can, and explain how they can help reduce the problem. How many of these do you know are being used in your state?
29. Why does a truck transportation company have an advantage over a railroad in government subsidy? Explain.
30. List as many ideas as you can that will result in increased efficiency in the transport system.

PROBLEMS

31. What mileage should the DaimlerChrysler Corp. Unimog, which weighs 12,500 lb, get?
32. The drag force increases as the square of the speed. At what speed is the drag force four times as great as at 50 mi/h? At what speed is the drag force 16 times as great as at 50 mi/h?
33. For the Midtown urban transit example given in the text, graph the transit energy use per passenger mile for bus load factors from 0.20 to 1.00. Assume the car load factor is a constant 30%.
34. For the Midtown urban transit example given in the text, graph the transit energy use per passenger mile for car load factors from 0.30 to 1.00. Assume the bus load factor is a constant 40%.
35. How much more pollution in a 400 km trip is emitted by a car driven at 200 km/h than one driven at 100 km/h? Explain your reasoning.
36. The drag force increases as the square of the speed. By what factor must the drag force at a highway speed of 65 mi/h be greater than at 55 mi/h? On some highways, the speed limit is 70 mi/h. By what factor is the drag force on a car going at this speed greater than that at 55 mi/h?
37. On German highways, the speed limit on Bundesstrasse is 100 km/h. While the Autobahns have many regions with no speed limit, in many areas, a speed limit of 130 km/h applies. How much greater a drag force would a car going 130 km/h have than one going 100 km/h? I've been passed by cars I estimate are going 200 km/h. By what factors would the drag be greater at 200 km/h than at 100 km/h? By what factors would the drag be greater at 200 km/h than at 130 km/h?
38. Suppose you drive a Volkswagen Lupo, having a fuel consumption of 3 liters per hundred kilometers, at 130 km/h. If you are passed by a small Mercedes at 200 km/h, estimate

the fuel consumption of the Mercedes. What factors enter into this calculation? Explain.

39. A car emits 2.2 g/km of carbon monoxide, 0.33 g/km of NO_x, and 0.17 g/km of unburned hydrocarbons. What should you expect its fuel economy to be?

40. Leaded gasoline contained 0.10 g/L of lead. How much lead would have been emitted in 2000 if the United States still allowed leaded gas?

41. What is the energy cost of a trip of length 20 km for the following modes of transportation? Indicate any assumptions you have to make.
    a. A single person bicycling.
    b. Two people on a tandem bike.
    c. Two people on a motorcycle.
    d. Three people in a pickup truck.
    e. Twenty passengers in a local bus.

CHAPTER 16
Weather and Climate

An atmosphere envelops Earth, interacting with its surface. Local and regional atmospheric pollution affects the atmosphere as a whole, as seen in the preceding chapter. Now we turn our attention to the effects of energy circulation in the atmosphere. This involves weather and climate. Weather behaves unpredictably from day to day, but predictably from season to season. Climate, the predictable seasonal pattern of the weather, changes more slowly. What causes climate to change? How is energy transferred in the global system? We examine the interconnection of air and ocean and look at possible cosmological, geological, and biological mechanisms of climate change.

KEY TERMS

troposphere / stratosphere / weather / teleconnections / climate / Hadley cell / Coriolis effect / ocean conveyor / thermohaline circulation / albedo / Southern Oscillation / El Niño / interglacial / Little Ice Age / isotope ratio / CLIMAP / Suess effect / nuclear winter / Milankovitch cycles / eccentricity / cosmological / solar constant / sunspots

16.1   EARTH'S ORIGINAL ATMOSPHERE

Originally, the major gases on Earth, Mars, and Venus were probably CO_2, H_2O, and N_2.[1] Today these atmospheres are very different. Earth's atmosphere is a unique construct of its geological and biological history. Life probably began 3.8 billion years ago (Gyr BP; BP stands for "before the present").[2,3] Oxygen, a waste product of the original anaerobic life, became a substantial part of the air before 2.0 to 2.5 Gyr BP.[1–4] Land plants and fungi arose between 700 and 1000 million years BP.[5]

The **troposphere** is the part of the atmosphere near Earth's surface. It contains air at breathable density, water, and clouds, and its temperature decreases with height. The **stratosphere**, in which temperature rises with height, lies above the troposphere; this is the region above about 8 km at the poles and 18 km at the equator. The composition of most of the atmosphere is uniform in the troposphere and stratosphere (water vapor and ozone are major exceptions). Above the stratosphere are the mesosphere and the thermosphere.

16.2   WEATHER

The interaction of the atmosphere with Earth's surface determines **weather**. Weather is a short-term *local* phenomenon in the atmosphere, occurring over a small time span in a circumscribed area. It results from the combination of atmospheric characteristics and events, such as the ambient temperature and its changes, precipitation or the lack of it, the amount of water vapor, or cloudiness. The world's weather is nothing more than the sum total of all these brief occurrences at each place, each one causing another effect somewhere else later on. Weather *is* "predictable" in a sense, because one of the best ways to forecast weather for any day is to look at the weather experienced on that date in the past. Figure 16.1[6] illustrates this principle. The temperature for 1998 and 2004 in New York City fluctuated mostly within the range of "normal highs" and "normal lows" (these are long-term means of the high and low temperatures for each of the days). The seasons are shown clearly in Figure 16.1.

Weather forecasts have, historically, a poor track record. The reason is that everything in the atmosphere is connected to everything else. If I want to predict the weather in New York City for tomorrow, and I wish to model it using the physics of atmospheric motion, I would use as input weather conditions in the Atlantic through the midline of the continent (including the Gulf of Mexico). Predicting daily weather in western Europe requires knowledge of conditions in Greenland, West Africa, and the Urals; weekly weather forecasts involve the weather conditions at the North Pole, Honolulu, Rio de Janeiro, Jakarta, and Teheran.[7] Without modern supercomputer facilities, there would be no hope of success: The weather is predicted by setting "initial" weather conditions in a grid pattern on the globe at several height levels. Weather patterns among the grid points are traced through time. Better weather predictions result from use of more grid points, but increasing the number of grid points requires many more calculations. The new North American supercomputer model, called RAFS (Regional Analysis Forecast Systems) uses points 80 km apart on land and 320 km apart over the oceans, and it has 16 vertical layers. Running the weather on the old model, LFM (Limited-area Fine Mesh), with grid spacing 190 km apart and seven layers, was eight times faster but cruder.[8] Seven days is the current limit for accurate weather prediction.[9,10] Two-week forecasts are reasonably descriptive of the weather to come.

Although weather forecasts of longer than about a week cannot yet be taken totally seriously,[8] we can argue that weather history is a good guide to what will happen. Weather patterns may exist in two different, but equally stable, flows—a sort of equilibrium, which may bring similar weather day after day, then suddenly switch.[11] While weather fluctuates from day to day, patterns persist and determine

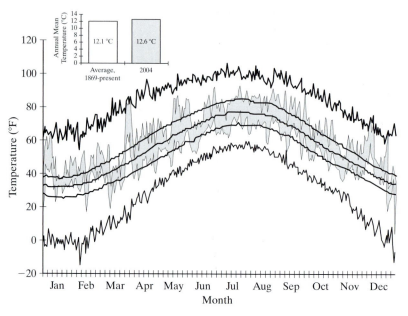

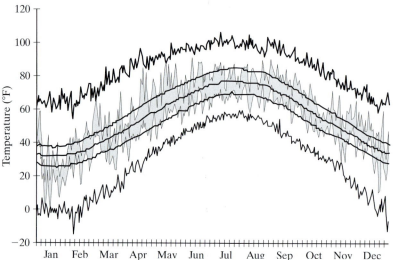

mean values that do not change greatly. The best long-term forecast accuracy comes from using a statistical technique called a "superensemble" that examines past weather situations, looking for analogies with the current weather situation.[12]

Meteorologists can now warn people when their prediction of weather rests on shaky ground. The 90-day forecast is only 8% better than pure chance at predicting the weather; but in winter, it is 18% accurate, while in fall, it is 0%.[9] The predictions are also different in different geographical locations—bad in the Rockies, but 40% accurate in the southeastern United States. Occasional stable weather patterns (for example, El Niño and La Niña) help make predictions more accurate. For forecasts made when the Pacific–North America pattern is present (a chain of highs and lows arcing over the ocean), forecasts are reasonably accurate for a longer period, around 2 weeks.[9] We here in Ohio are, of course, interested in weather forecasts, because we tend to get the flu and colds more in winter, and the chilliest days seem to make it easier to get sick. In summer, we can get heat prostration. In addition to anecdotal evidence, an epidemiological study shows that cardiovascular death rates depend on the weather. In "cold" cities, both temperature extremes are associated with excess deaths.[13]

Did you know that events occurring in the Pacific can affect weather in the U.S. Midwest? Long-distance connections between distant spots abound: These are called **teleconnections**. In addition to the North American teleconnection to the Indian monsoons, Greenland and Europe are teleconnected through the North Atlantic Oscillation. There is a high correlation between Eurasian snow cover in October and November and winter air temperature the following winter in Siberia, as well as the intensity of the summer monsoon in south Asia.[14] There appears to be an association between hurricanes in the Americas and the amount of west African rainfall.[15] Even stranger correlations seem to exist: Weather forecasts for the British Atlantic coast err by a significantly greater amount when the number of thermal (slow) neutrons in the lower stratosphere (presumably produced by cosmic rays) and troposphere decrease substantially.[16]

Weather can change rapidly—it is inconstant from day to day. Weather history emphasizes year-to-year comparisons rather than day-to-day ones. As Figure 16.1 showed, the temperature varies by the day as well as by the season. The seasons characterize weather over a longer time. If we look over a year, though, we would expect the mean temperature to be fairly stable from year to year. The inset in the upper left-hand corner of Figure 16.1 shows that this is so—the mean annual temperature is about 11 °C. It is clear that deviations from this mean occur every day in New York City. If we look at the weather over other places and average the data over a reasonable number of years (10, 20, 30, or 50, say), we would get a general idea of the weather pattern over that place. Similar deviations or variations will occur in other localities, and the patterns of these variations are important to consider. For example, while the mean annual temperatures of San Francisco and Cincinnati are about the same[17]—13.8 °C (57 °F) versus 13.6 °C (55 °F)—the average deviation of the temperature from the mean is much greater for Cincinnati (9.3 °C) than for San Francisco (3.1 °C). This generalized averaged weather pattern in some restricted geographical region is called **climate**. We use the average temperature, the extremes of temperature, the amount of precipitation, and other such indicators to describe the climate, which should not change rapidly.

Of course, one of the important things about climate is that it *is* always changing. The numbers that describe the climate do not have to change much to be important.[18-20] A study of the mean annual temperature in Iceland over the past millennium shows that a 1 °C decrease reduces the length of the growing season by about 2 weeks and the number of growing-degree-days (the sum of the average daily number of degrees above the minimum growing temperature over the growing season) by about 25%.[18]

What is ultimately responsible for the weather? It is the Sun. Radiation (Chapters 7 and 21) transfers energy from a hotter body (the Sun) to a colder body (Earth). Earth absorbs solar radiation from the Sun, which serves to heat it and drive the winds and the weather. Wind patterns arise because the poles intercept very small amounts of light per square meter, while the equatorial regions

FIGURE 16.2

The disequilibrium between the amount of solar radiation intercepted per square meter at the poles and the equator drives the world's weather. Rays of light from the Sun (far background) are all parallel at Earth's distance from the Sun (the perspective distorts this here).

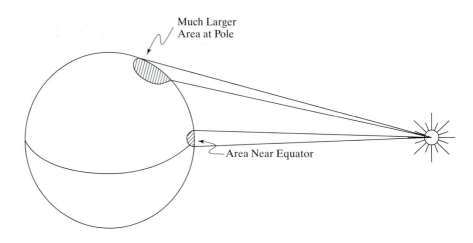

capture large amounts per square meter (Figure 16.2). The net result is an influx of ≈ 600 W/m^2 in the equatorial regions and an influx of ≈ 100 W/m^2 at the poles.[21] The existence of large land areas between the tropics of Cancer and Capricorn and relative lack of land at the poles enhances this effect.

Radiation travels in straight lines, so it cannot equalize the spherical Earth's temperature differences. Conduction requires physical contact between temperature extremes and so is a minor source of equalization. It is convection, involving both gaseous air and liquid water, that plays the major role in minimization of temperature differences. The heated air at the equator rises and cools as it moves to higher latitudes, slowly sinking back to ground level, where it recirculates back to the equator. There it is reheated and recycled. The air that sank north or south of the equator dragged more northerly or southerly air with it, setting up another circulation pattern. This air rises at still higher latitudes, dragging local air along to set up still another circulation pattern. These segregated regions of air are called **Hadley cells**. This circulation redistributes solar energy; combined with the effects of Earth's rotation (the Coriolis effect, which causes air to move eastward in the Northern and oppositely in the Southern Hemisphere), which causes the winds to swirl within the cells, the circulation sets up familiar wind patterns, such as trade winds and the "furious forties" (Figure 16.3). Of course, this is a much simplified description of the heat circulation in the atmosphere.

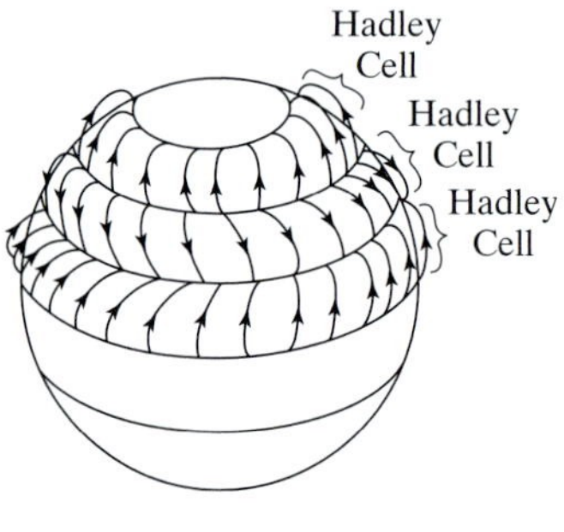

Hadley
Cell

Hadley
Cell

Hadley
Cell

FIGURE 16.3

So-called *Hadley cells* partition the atmosphere between the equator and the poles; three Hadley cells are shown in the northern hemisphere in the diagram. (Technically speaking, only the cell adjacent to the equator is a Hadley cell.) The boundaries of the cells are known as *convergence zones*.

The Ocean Conveyor

Almost three-quarters of the globe is water surface. Oceans are important in absorbing solar energy at one place and releasing it somewhere else. Convection is an important element here, too. Ocean water moves in large volumes, much as the air does, but the ocean currents move more slowly than the winds.

The North Atlantic is saltier than the rest of the oceans. Warm saline surface waters from the tropics and subtropics flow into the North Atlantic, where some water evaporates, lowering the temperature and increasing the salinity. Vapor loss alone from the Atlantic is equivalent to about two Amazon Rivers.[22] As the water is cooled by air from Canada, it becomes even more salty and denser. South of Iceland, some 200 million m^3/s of water,[22] over 20 times the water in all the world's rivers, sinks from the upper layer of ocean. This water, called

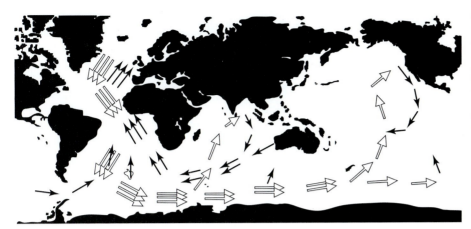

FIGURE 16.4
The ocean conveyor transfers water around the globe. Heavy arrows: warm surface waters. Open arrows: deep cold nutrient-laden waters. Note downwelling in the North Atlantic Ocean south of Iceland and upwellings in the Indian Ocean and in the Northern Pacific Ocean south of Alaska.

North Atlantic Deep Water (NADW), is saline and low in nutrients. NADW flows southward through the deep Atlantic. Around Antarctica, NADW mixes with deep Antarctic water, becomes less saline, and flows on into the Pacific.[22] The water (with its gift of Antarctic nutrients) eventually rises in the Pacific and Indian oceans, where it can be warmed by the Sun and complete the surface flow back to the North Atlantic. W. S. Broecker:[22] has referred to this circulation (Figure 16.4) as the **ocean conveyor**. The conveyor makes a 5 °C to 8 °C temperature difference to northern Europe. It is the oceanic salt gradient (maintained by the Sun) that drives the conveyor. The circulation of salty water within the ocean is known as **thermohaline circulation**.

Another mechanism of heat transfer works in the ocean. Energy from the Sun can supply latent heat of vaporization (Chapter 7) in low latitudes, where the solar flux is highest. The water that is evaporated becomes part of the atmosphere as water vapor and moves with the air to higher latitudes. For example, about 35% of the Sun's energy falling onto the Atlantic is released by evaporation in creating NADW. Eventually, atmospheric water vapor becomes rain that falls where the temperature is lower, releasing the latent heat absorbed in the equatorial regions.

Radiation Balance

Earth absorbs solar radiation with a spectrum of radiation corresponding to that from a body at a temperature of about 5800 K, the Sun's surface temperature (as will be discussed in Chapter 21 and Figure 21.4). This radiation warms Earth, increasing its temperature. The temperature does not rise indefinitely, as is obvious to all of us residing here on Earth. Earth's average temperature has been constant to within a range of about 10 °C over millions of years. The Earth must, therefore, emit radiation to space as the Sun does in order to maintain this long-term balance. Because the temperature of Earth is so much lower than that of the Sun, it emits radiation with a spectrum appropriate to a body of temperature about 255 K (infrared radiation). A quick calculation shows that the radiation balance alone would predict Earth's temperature to be about 35 K lower than the observed surface temperature (below the freezing point of water).

A CLOSER LOOK

Radiation Balance and Earth Temperature

Earth appears from the Sun to be a disk; that is, it looks flat, just as the full Moon does from Earth. The area of the disk is πR_E^2, where R_E is Earth's radius. The total energy absorbed for a solar constant, or available power per unit area at Earth's orbit, of 1.35 kW/m² is, assuming an albedo (reflectivity) of 30%,

$$P_{absorbed} = (1 - 0.3)(1.35 \text{ kW/m}^2)(\pi R_E^2).$$

A body at absolute temperature T radiates according to the Stefan–Boltzmann law (see Chapter 21) as

$$P_{radiated} = (5.67 \times 10^{-8} \text{ W/m}^2 K^4) T^4 (4\pi R_E^2),$$

where Earth's total spherical surface area is $4\pi R_E^2$. Thus, since in the long term the power absorbed must again be radiated for Earth to be in equilibrium, $P_{absorbed} = P_{radiated}$, and πR_E^2 is a common factor, so

$$(1 - 0.3)(1.35 \text{ kW/m}^2) = 4(5.67 \times 10^{-8} \text{ W/m}^2 K^4) T^4.$$

Therefore,

$$T^4 = (0.7)(1350 \text{ W/m}^2)/4(5.67 \times 10^{-8} \text{ W/m}^2 K^4) = 4.17 \times 10^9 \text{ K}^4,$$

or

$$T = 254 \text{ K}.$$

We can roughly estimate the entropy increase (Chapter 7) as solar energy runs into, and Earth's radiation out of, Earth's system. By definition, the increase or decrease in entropy consequent in transferring a certain amount of heat is that amount of heat divided by the absolute temperature of the fluid. We make the approximation that Earth's temperature is constant. Because the Sun's temperature is roughly 6000 K and Earth's is roughly 250 K, an amount of heat as it is radiated by the Sun will cause an entropy increase about 25 times as great as the entropy decrease consequent on the reception of that radiation by Earth (heat/250 K versus heat/6000 K).

As emphasized in Chapter 1 (and as will be discussed in Chapters 21 and 23), not all the Sun's energy reaches the surface of Earth. Much of the infrared and ultraviolet radiation is absorbed (see Figure 21.4)—the ultraviolet by atmospheric ozone and the infrared principally by carbon dioxide and water vapor in the air.

Let us consider solar radiation of wavelengths in the range from 300 nm to 3000 nm (0.3 μm–3.0 μm, the principal gap in Figure 16.5). The intensity of solar radiation at Earth's distance from the Sun is about 1350 W/m². Not all solar radiation that reaches Earth makes it through the atmosphere. About 30% of this radiation is reflected (the proportion of light reflected is called the **albedo**). Suppose we call the available intensity 100 units; then 30 units are reflected. This takes place from clouds (20 units), from the air (six units), and from the surface (four units), as illustrated in Figure 16.6.[21,23] Because Earth's temperature is not zero, it emits infrared radiation in the wavelength range

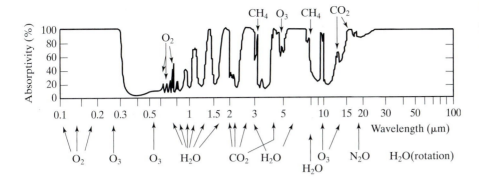

FIGURE 16.5

Atmospheric absorptance at sea level during clear weather. The parts due to each molecule are shown.

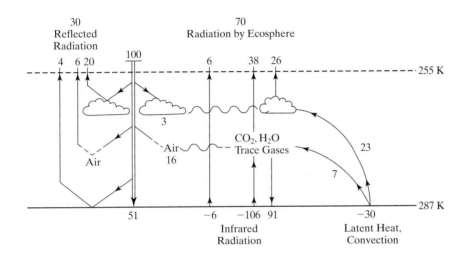

FIGURE 16.6

The mean annual radiation and heat balance of the atmosphere, based on satellite measurements and conventional observations. The total incoming radiation is given a value of 100, and all fluxes are measured relative to this.

6–60 μm (about 5100 TW).[21,23,24] As the figure shows, some of this radiation is absorbed by the carbon dioxide, water, and other trace gases in the air, and some of this is reradiated to Earth. Altogether, 38 units are radiated to space from the gases, and six units go directly to space from the surface. Because most of the surface is ocean, the ocean absorbs most of the incoming radiation. Some of this is transferred to the atmosphere (seven units) in the form of turbulent diffusion from land and ocean.[23,24] Finally, water is evaporated, absorbing latent heat of vaporization and transferring this latent heat to the clouds as it recondenses to water (23 units, or ~4400 TW).[21,23,24] The clouds then radiate 26 units into space. Note that the temperature of the upper atmosphere is 255 K, just about what we calculated when we took albedo into account. The surface temperature is higher than that by 33 °C because of reradiation from the carbon dioxide, water vapor, methane, and other atmospheric trace gases.[23] The net effect is to send a total of 240 W/m² into the lower atmosphere.

The Oceans' Weather Connection

Modern studies of the oceans involve determination of sea surface temperature and atmospheric pressure. Sea surface temperatures are highest in the western Pacific of any of the large ocean basins, and the central equatorial Pacific is a key region in ocean–atmosphere coupling.[25]

We mentioned that there is a teleconnection between the South Pacific's weather and that in midwestern North America. The Quasi-Biennial Oscillation is a 26-month cyclic windshift pattern in the lower stratosphere over North America. The **Southern Oscillation** is a seesawing of the coupled South Pacific ocean and atmosphere system first described by Sir G. T. Walker in the 1920s. Tahiti, in the central Pacific, and Darwin, Australia, are at opposite ends of the Southern Oscillation.[25] As put by Walker:[26]

> When pressure is high in the Pacific Ocean it tends to be low in the Indian Ocean from Africa to Australia; these conditions are associated with low temperatures in both these areas and rainfall varies in the opposite direction.

Actually, Walker found three oscillations, one in the North Atlantic, one in the North Pacific, and the Southern Oscillation. The Southern Oscillation and the Quasi-Biennial Oscillation are connected.

Under normal conditions, the ocean off Peru exhibits cool temperatures (due to upwelling from the Humboldt current, a cold current in the upper ocean that flows toward the equator along South America). Sea level is elevated in the western Pacific, because the trade winds blow water westward. Off the coast, there is a shallow surface mixed layer, and a slow temperature decrease with increasing depth.[25,27] The flow of current causes an upwelling of cold water that brings nutrients up from below and leads to high biological productivity of various fish species.[27]

An interesting ocean temperature phenomenon known as **El Niño** occurs every three to seven years, usually around Christmas time, off the Peruvian coast (*El Niño* means "the child" in Spanish and is a reference to the birth of Christ). Fishermen notice it because of the failure of their anchovy catches during these occurrences. The appearances of El Niño are connected to anomalies in the Southern Oscillation and the seasonal migration of the intertropical convergence zone,[25,27–30] so the events are now called ENSO, for El Niño–Southern Oscillation.

The ENSO is the most obviously dramatic weather event involving the ocean (see **Extension 16.1**, *ENSO Events*). During the 1988 drought in the United States, it was first realized that the drought was connected to the concurrent ENSO.[76] The ENSO controls droughts and floods in India.[77] Hurricanes and typhoons are driven by similar mechanisms of uplift and deposition of energy (latent heat transfer). The circulation of thermal energy through the ocean is important for mitigating large temperature variations in the midlatitudes. (See **Extension 16.2**, *Physics of the Atmosphere*.)

16.3   OUR CLIMATE

The climate is described by the mean and the variability of measured variables of the atmosphere, oceans, and land masses, such as temperature, rainfall, or snowfall, or extent of snow cover. Most studies of climate focus on the air and water (both ocean water and water vapor) and their effects. The reaction time for air is relatively short (we have all noticed hot days and cool nights—the air temperature can change a great deal within hours). The atmospheric heat engine is driven by the mismatch in energy received at the equator and the poles. Most of the energy taken to the polar regions is carried by water's latent heat of vaporization. Because the specific heat of water is much greater than that of air, the huge mass of water in the oceans acts as a buffer. Overall, it can store roughly 20 times as much thermal energy as the air and moderate the diurnal and seasonal temperature changes.[8,21,94] It takes a much longer time than one season for any long-term heating to affect climate. The distribution of land masses and the amount of ice locked up in ice sheets and glaciers also affect climate, but as they change slowly over long periods, their effects on climate are usually ignored.

Climate depends on the region considered and the time interval chosen. We think of our current climate as normal. But climate is not permanent, and our "normal" everyday climate is not normal for Earth.[95] We are in a time of rather benign climate, one that's existed only about 10% of the past 2 million years.[96,97] We live in what is called the Holocene **interglacial**, a time between two glacial ages.

The Little Ice Age

To illustrate how precarious our position is, how variable our climate is, consider the cases of the "Middle Ages optimum" and the "**Little Ice Age**." In the Middle Ages optimum (900–1050 A.D., also known as the Medieval Warm Period),[23] the Vikings settled on an Iceland that was about 60% forested, ice free, and could grow its own grain. They even sent a colony from the Iceland colony to Greenland. A complete tree ring reconstruction of 14 sites in the Northern Hemisphere shows evidence of this warm period across the Northern hemisphere.[98] The Caspian Sea was 32 m lower than it is today.[23] At this time, the world's mean temperature may have been about 1 °C greater than now.[23] In the Little Ice Age (~1550 to ~1850), northern Europe's mean temperature was about 1 °C less than now[23,99] (tree ring studies in Sweden show a much shorter cold period, 1570–1650).[100] The Icelandic colony in Greenland disappeared,[19,23,95,97] and Iceland was locked in pack ice for 6 to 9 months a year.[23] There were general famines in Iceland, Norway, and Finland,[23,97,99,101,102] and crop failures and political unrest in much of Europe. Breughel and other Dutch artists painted icy winter landscapes in Holland (Figure 16.7). Glaciers in the Alps, Scandinavia, and Alaska were at their greatest extents in thousands of years.[23]

As part of the effect of a previous cold period, Great Plains Indians abandoned their villages about 1200 A.D., and Mesa Verde Pueblos left their homes in about 1300 A.D. due to a long-term drought in the Great Plains.[103] The tree line in Canada, which was north of today's position, moved to south of today's position during this period.[103] As late as the 1850s, Taos, New Mexico, had the rainfall that Madison, Wisconsin, has nowadays.[103]

Climate is slowly changing elsewhere in the world. There is evidence for a steady progressive desiccation in the sub-Saharan zone of Africa from at least 1800.[104] The droughts of the 1970s and 1980s may be part of the continuation of a long trend that may get worse.

FIGURE 16.7
Breughel painted scenes of Holland winters, indicating that the climate was more servere in the sixteenth century.
(P. Breugel, *The Hunters in the Snow*; Kunsthistorisches Museum, Vienna.)

The Temperature Record

How can we know the temperature record without time travel? We have several methods: some are physical; some are chemical; some involve the study of plankton and pollen; and some are models that can be checked against data generated in the other methods. As an example of the physical method, the top layer of soil records the mean external temperature. Fluctuations die out as the depth increases:

seasonal records cannot be measured below a depth of 15 m; the century-long cycle penetrates to 150 m, a millennial cycle to 500 m.[105] The chemical method rests on the fact that many kinds of atoms come with different masses. Oxygen has two stable forms (isotopes, see Chapter 18), oxygen-16 (^{16}O) and oxygen-18 (^{18}O). (A third stable form, ^{17}O, makes up only 0.04% of all oxygen.) The oxygen **isotope ratio**, the ratio of ^{18}O to ^{16}O in a sample, known as $\delta^{18}O$, is an indicator of the global total of glacial ice (see **Extension 16.3**, *Proxy Measurements* for more information about measuring temperature and ice volume).

The pattern is the same for all major evidence.[57] Loess in China gives a 2.4 Myr record of climate that agrees with the $\delta^{18}O$ timescale from seafloor cores.[119] Plant pollen reflects temperature conditions at some particular location and results of analyses agree with other evidence.[120] Alkenones are long-chain carbon molecules produced by sea life; they can give sea surface temperatures. The results of a 170 thousand year (kyr) record agree with published results.[121] Corals can be analyzed by uranium-thorium dating techniques (see **Extension 16.3**); mostly they agree with the sea core record,[122] except for a frequency doubling that is due to coral recording signals of warmings in both the Northern and Southern hemispheres (the equatorial region is connected to both hemispheres).[123] Calcite formation in caves gives a climate record inland,[124,125] as do fossil fish ear bones. All these records agree on the timing of changes.[126]

The data developed by the **CLIMAP** (Climate: Long-range Investigation, Mapping, and Prediction) Project can then be compared to the models of glacial climate, to see how realistic the data are. Figure 16.8[127] shows the results of the calculations for 18,000 years ago at the end of the last glacial period. The resemblance between calculated and reconstructed sea surface temperature is striking. It means that the models of the atmosphere used to do the calculations are reasonable. The temperature in the Greenland ice sheet was -32 ± 2 °C during the last ice age, about 12 °C colder than now.[128]

It is very important to study climate in the past in order to learn about past changes. This can help us by allowing us to see how changes in past climates worked. This point was made strongly by Bard.[129] Knowing what sorts of events occurred in the past is really our only guide to the future. That is because the model calculations we can do are so difficult to be completely certain of. The importance of thermohaline circulation was elucidated by discoveries about the changes in systems in the past. In finding out about the past, we discovered the existence of massive pulses of fresh water as glaciers melted (see Chapter 17).[130,131] If we want to know what could happen if the ice sheets at the poles melt, we have examples in the past.[130]

16.4   CAUSES OF LONG-TERM CLIMATE CHANGE

Now that the record shows obvious climate changes, we want to find the causes for these changes. The causes may be classified generally as biological, anthropogenic (or human-caused, a subset of biological causes), geological, and cosmological. Biological changes include the effects of living things on Earth. Geological changes include the effects of continental drift on land distribution. Cosmological changes occur when Earth is influenced by extraterrestrial agencies. In Chapter 17, we shall examine the evidence for climate change caused by our activities and those of our ancestors. Despite this classification, it is not entirely possible to segregate the causes; there is always cross-category feedback.

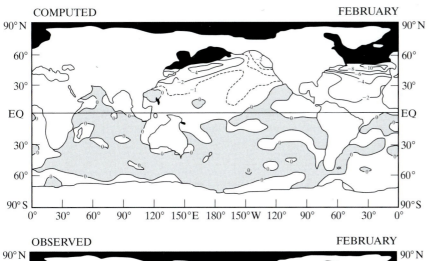

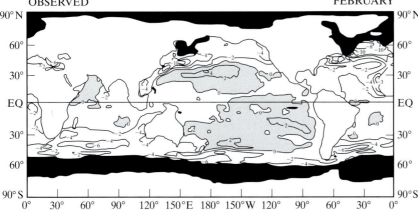

FIGURE 16.8
February monthly mean sea surface temperature difference in kelvin. TOP: observed difference between ice age and the present. The text and Extension 16.3 explain how we can "observe" past temperatures through proxies. BOTTOM: reconstruction by CLIMAP. (S. Manabe and A. J. Broccoli, "Ice age climate and continental ice sheets: some experiments with a general circulation model," *Ann. Glaciology* **5**, 100, 1984, reproduced by courtesy of the International Glaciological Society.)

Biological Change—The Carbon Cycle

The carbon cycle is made up of two sources (three, if the human source is included; see Chapter 17) and two sinks. (For more details of the carbon cycle, see **Extension 16.4, *The Carbon Cycle*.**)

The oceans and the biota clearly must be involved in the carbon cycle, because the CO_2 content of the atmosphere has risen only about half as much as it would have if all CO_2 from fossil-fuel production had been released to the air.[105,160] Also, there is less carbon monoxide in the air than would be expected. Carbon monoxide had increased relative to 1950 by the 1980s, but it has lately been decreasing.[161] The ocean deeps, which hold 35,000 billion tonnes of carbon dioxide at present, have a virtually unlimited capacity for absorbing carbon.[162] The problem is that the transfer of carbon between the surface and the abysses is slow (transfer takes hundreds of years in the Atlantic and thousands of years in the Pacific) and thus cannot account for the relative dearth of carbon dioxide in the atmosphere. The rates of transfer to the abyssal regions are well known from studies with tritium from oceanic nuclear weapons tests and carbon-14 (radioactive carbon). The dilution of carbon-14 in the atmosphere, called the **Suess effect**, has been studied extensively.

The global carbon cycle is, at the moment, still shrouded in mystery. Ground reservoir estimation is difficult. "Reconstruction" tries to estimate changes in vegetation and soil. Ecological models such as those in References 103, 148, and 163 have the biota as a net source of CO_2. For example, Woodwell et al.[100] estimated the biota release to be between 1.8 and 4.7 billion tonnes of CO_2 per year. Typical results of the calculations are shown in Figure 16.9.[164] These models produce a larger amount than do geochemical models that find the amount by "deconvolution"— the amount needed to bring carbon inputs and outputs into balance.[151] In these, the biota constitute a sink of carbon.[164] The geochemical models use the biota "box" to balance their flux of carbon, as the ecological models use the oceans. The geochemical models say the ocean takes up 2.0 ± 0.6 billion tonnes of carbon per year.[164] Tans, Fung, and Takahashi claimed that this was an overestimate;[165] their work led to better identification of carbon transfer and supported the earlier estimates.[166] The best guess is that 1.8 ± 1.3 billion tonnes of carbon per year is unaccounted for,[167] a "missing sink." Meanwhile, the geochemists can use the known rates of ocean carbon transfer to suggest that the ecologists' models are wrong, while the ecologists point to the huge releases from burning wood (1.5 billion tonnes CO_2 per year, or ~30% of fossil-fuel release)[148] and from deforestation, and ask how the geochemists can see the biota as a carbon sink. Some think the explanation of the missing sink of carbon lies in increased sedimentation

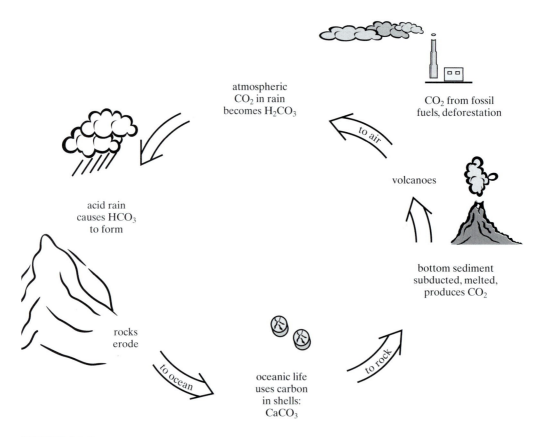

atmospheric CO_2 in rain becomes H_2CO_3

CO_2 from fossil fuels, deforestation

to air

acid rain causes HCO_3 to form

volcanoes

rocks erode

bottom sediment subducted, melted, produces CO_2

to ocean

oceanic life uses carbon in shells: $CaCO_3$

to rock

FIGURE 16.9

The carbon cycle: atmospheric carbon dioxide erodes rocks, carbon is carried to the oceans and incorporated in the shells of marine organisms, the carbon sinks to the ocean floor as they die, the floor is covered and the pressure creates rock; the rock is carried under a plate and melted, and carbon dioxide is released to the atmosphere.

rates, 1.8 billion t/yr by one estimate,[103] but many experts think that the amount of biomass burning has been overestimated and that regrowth of forests (mainly in the north) has absorbed a huge amount of carbon in recent years.[167]

Geological Change

Geological changes include isostatic uplift (the upward movement that results when the weight of an ice sheet is removed), mountain building and continental uplift, changes in the latitudes of continents, changes in the characteristics of the ocean, volcanism, and alterations in the circulation of material in the core. Some 200 million years ago, there was only one continent on Earth. It fractured, and the pieces are still moving about on the plates making up Earth's outer surface (see Chapter 10). It is clear that continental masses at high latitudes are heavily glaciated (Antarctica, Greenland). The continents that drifted poleward became more susceptible to glaciation.

From the Cenozoic era to the present, Earth has been in (essentially) an ice age climate state. The cooling in the Cenozoic era may have resulted from the uplift of the Himalayas and from the 5-km-high ice domes of the Laurentide ice sheet. Uplift resulted in stronger deflections of the jet stream, a more intense monsoon, increased rainfall on the front slope of the Himalayas, greater chemical weathering, and so decreased CO_2.[168] The Tibetan Plateau uplift is consistent with the predictions of general circulation models (GCMs) of the atmosphere (see Extension 16.2).[169]

Earth reflects, on average, about 30% of the incident sunlight back into space; we say it has an albedo of 30%. Pack ice and snow have an albedo of 80%, whereas the ground the ice covers has an albedo of only about 20%. The mean annual coverage of the poles by ice has been about 30 million km^2 in recent times.[170] The polar snow cover seems to have been decreasing through the 1980s and 1990s; a comparison of submarine measurements of polar ice thickness 20 years apart showed a significant thinning of the polar ice cap.[171] Similar changes are seen in other records as well.[172] In the last full glacial, mean annual coverage was 60–70 million km^2.[170,173] Satellite monitoring of the ice cover has revealed that the mean annual coverage increased from 32.9 million km^2 in 1970 to 36.9 million km^2 in 1971. This area remained enlarged for several years.

Just 7 years more of such an increase might have reestablished the glacial surface albedo and could have initiated or accelerated the change to a glacial era[99,170] in our "marginally interglacial" climate. The snow and ice not only reflect incoming visible radiation, but they also consume substantial latent heat as they melt, and they reduce the energy and moisture exchanged between the atmosphere and the surface insulated from it by the overlying ice.[14,174] The energy deficit from snow is in the lowest 2 km of the atmosphere.[14] Figure 16.10[175] shows that the increase was not sustained; nevertheless, it is possible that, without the human-caused effects discussed in the next chapter, we might have been entering another glacial age. The data shown were obtained from satellite monitoring using microwaves (because those wavelengths are not affected by the omnipresent cloud cover). Clouds are also very important in cooling, because they contribute over half Earth's albedo.[176] We already mentioned how important water vapor is in warming, so the tradeoffs must be examined closely.

Data from the past 100,000 years show[18,177] that temperatures over the Greenland and Antarctic ice sheets decrease as the amount of volcanic dust falling on them increases. Volcanic outbursts in the North Pacific coincided with the descent into the glacial Pleistocene epoch ~2.6 Myr ago.[178] The measured strength

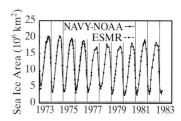

FIGURE 16.10

Seasonal cycle of the extent of Antarctic sea ice.
(H. J. Zwally, "Observing polar-ice variability," *Ann. Glaciology* **5**, 191, 1984, reproduced by courtesy of the International Glaciological Society.)

of sunlight shows the effects of great volcanic eruptions.[177,179] The most stunning historical example is of the year without a summer: 1816, known to those who lived through it[180] as "eighteen hundred and froze to death." Prices of grain rose substantially in both America and Europe in the wake of crop failures (made worse in Europe because of the Napoleonic Wars that had just ended). There is also some evidence[180] that the world's first cholera pandemic was partly due to what happened in 1887 when the volcano Tambora on Sumbawa injected approximately 100 km^3 of dust into the atmosphere and caused a worldwide temperature drop of 1 °C.[97] (For more details, see **Extension 16.5,** *Effects of Dust and Aerosols*.)

The big particles in the air tend to settle out quickly; while they are present, they cause warming, but they soon disappear. Small particles in the stratosphere have a mean residence time of 0.5–1 yr (which means that about one-third will have fallen out by that time, and all but 5% will have gone by 3 yr).[185] Because big particles of dust quickly leave the atmosphere, and small particles remain for long times, the dust particles injected by volcanic action or air circulation (such as a hurricane or nuclear explosion in the atmosphere) cause Earth's surface to cool and the stratosphere to become warmer.[157,185]

These models can be tested in simulations of Mars' atmosphere, because the weather on Mars is less complex than that on Earth due to the absence of clouds and exposed bodies of water. Comparisons of model calculations and the results of lengthy Martian dust storms, in which surface temperatures plunge, led to the development of the idea of **nuclear winter** (see **Extension 16.6,** *Mars and Nuclear Winter*).

Volcanic Eruptions

The lesson is clear: The more dust and particulate matter in the atmosphere, the greater the drop in temperature. Thus, a sustained siege of volcanic activity could trigger a glacial episode. Evidence that this is the case comes from the Deep Sea Drilling Project[217] from which it was found that the rapidly oscillating climatic conditions of the past 2 million years are synchronized with eras of greatly increased volcanic activity.

Other evidence abounds from the Roman era to today. Greenland ice cores show that there was three years of acidic fallout about 50 B.C., at a time when histories of Rome report a dimming of the Sun after Caesar's assassination.[218] The explosion of Thera, which led to the death of the Minoan civilization on Crete, was dated in the same manner to 1645 B.C.[219] The sulfur released by volcanoes can cause increases in levels of sulfuric acid in the atmosphere. The sulfur and other volcanic aerosols injected into the stratosphere during a volcanic eruption can cause global cooling.[185] Ice cores from 1601 and 1602, during the depths of the Little Ice Age, show sharp peaks of sulfuric acid, too,[218] indicating increased volcanic activity.

In our century, there were many more eruptions from 1900 to 1925 and 1955 to 1980 than from 1925 to 1955.[220] Between 1945 and 1970, the average number of eruptions per year doubled.[220] If the optical depth of the atmosphere increased enough to reduce the direct beam at Earth's surface by 5%, the resulting increase in diffuse radiation would cause a cooling of 0.85 °C. The increasing carbon dioxide levels (see Chapter 17) should have caused an increase of ~0.35°C; the combined result of −0.5°C temperature change compares well with the measured change of −0.3°C observed between 1945 and 1975 as the measured value of the solar constant at Earth's surface decreased 5% because of particulate emission by volcanoes (smaller cause) and human activity (larger cause).[217]

Radiocarbon dating results suggest increased volcanic activity 600 years ago (Little Ice Age), 1200 years ago, 3700 years ago (coincident with a long-term failure of the Indian monsoon, which is tied to the Southern Oscillation), and 8000 years ago (a time of ice stillstand).[220]

Orbital Change—*Milankovitch Cycles*

It was an old idea, resuscitated by Miltun Milankovitch in 1930,[221] that variations in Earth's orbit may cause changes in insolation—**Milankovitch Cycles**. The changes seem too small (there is a variation due to **eccentricity**, for example, which characterizes Earth's oval orbit, of only 0.3% over the past million years[222]) to actually cause ice ages. Milankovitch met these objections by suggesting that the insolation at high latitudes, which changes by 20%, could cause great changes in climate by affecting the amount of ice and snow cover in the Northern Hemisphere. (More details of the theory and its triumphs and shortcomings are given in **Extension 16.7, *Milankovitch cycles*.**) Minor questions remain, but the idea has been a great success. Ice ages can be the initial condition feeding into the Milankovitch theory.

Recent indications are that carbon dioxide levels are intimately involved in the Milankovitch approach to climate.[215,232,246,260,261] Changes in carbon dioxide levels lag behind changes in orbital geometry by a few thousand years and precede changes in ice volume.[217,261] This seems to be in accord with models of the atmosphere, which give 3000-year delay times.[261] Adding the effects of carbon dioxide to the "pure" Milankovitch predictions tremendously improves the agreement between prediction and actual data.[216] Ice sheet conditions interact with the orbital forcing to make the response nonlinear.[262]

Polar ice core records indicate that orbital changes precede changes in CO_2 concentrations, which precede ice sheet growth or melting. The 100,000-year cycle appears to be enhanced by this mechanism in ways that are still not clear. It may be that there are two (or more) stable modes for the climate; an "ice age" one, with reduced deep water production in the North Atlantic, and an interglacial one, in which water is pumped from North Atlantic to North Pacific. The climate could "snap" from one to the other[120] the way a bicycle wheel, tied to the ceiling by rubber bands fastened at one particular place on the rim, is stable in two positions (Figure 16.11). Sudden changes in the flow of warm Atlantic water into the Norwegian Sea and changes in the flow through the Bering Strait happened often during the last major glacial episode.[263] A two-basin model showed the North Atlantic with three circulation modes, each corresponding to a particular climate.[264]

Cosmological Change

Possible **cosmological** causes are many. The solar system could be passing through stationary dust clouds,[260,265] or the **solar constant** (see Chapter 2) may be *in*constant.[266–268]

Out of This World? Changes in the Solar Constant

The number of **sunspots** (magnetic storms on the solar surface) changes roughly regularly in 11-year cycles. Droughts in the midwestern United States seem to be correlated with the number of sunspots.[19,100] When the winds are blowing from

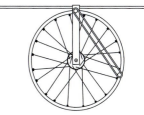

FIGURE 16.11

Climate can "snap" from one stable equilibrium to another, much as a bicycle wheel hung from the ceiling and also attached to a point on the ceiling by a rubber band can be stable in two positions.

the west and the sunspot activity is near peak, the winter in the United States is colder than normal.[269] London annual mean temperature may be determined by sunspot number; and mean rainfall seems to be correlated as well.[263,269] It has been suggested that the Maya collapse stemmed from solar changes leading to drought.[270] Figure 16.12[260] shows the sunspot number as it has changed during the course of over 200 years.

Upper troposphere and stratosphere temperatures have been found to vary in phase with the changes in length of the 10- to 12-year cycle of the Sun.[268] The northern hemisphere land air temperature is almost perfectly correlated with long-term variation of solar activity.[268] The sunspot link may well be the effect of solar radiation on stratospheric ozone.[271]

Even though there is no absolutely convincing physical mechanism for short-term changes in the Sun to influence weather (although the ozone layer may provide that), and despite the changes on this scale being small relative to total output, the influence appears to have some observational bases.[260,272] The monsoon in Oman appears to be correlated with solar-radiation variations.[273] A difficulty seems to be that there is not much evidence for large solar constant changes; measurements show no more than 0.15% change over an 11-year sunspot cycle,[274] in fact, the results from the Active Cavity Radiometer Irradiance Monitor, perhaps the best measurements, show only a 0.036% per decade change.[274]

A careful analysis of seafloor cores by Bond et al. showed that the ice drifts could be correlated with the production of carbon-14 and beryllium-10 (the production of which depends on the Sun's output) in Greenland ice cores.[275] As proxies for the amount of ice cover as measured from the sea cores, they use hematite-stained grains, Icelandic glass, and detrital carbonate. These have been shown in the past to be correlated with ice cover.[275] The two plots appear to march in synchrony, suggesting that the Sun had an effect on the last 12,000 years of climate, despite the relative weakness of the signal.[275] However, it also seems clear that solar variations are not nearly large enough to offer an explanation for the late-twentieth-century warming that has occurred.[276,277] Most researchers agree with Lean that some have "exaggerated the role of the Sun in climate change."[277]

The Maunder Minimum

The number of sunspots, however, seems to be connected to secular increases and decreases in the solar constant of as much as 1% to 2%.[19,260,278,279] Stars of the (low metal content) HK group observed by astronomers have changing solar constants.[280] The peaks are most intense during the shortest cycles. The changes are correlated with vigorous magnetism and the largest number of sunspots.[280] It is probable that a 1% change in the solar constant would cause the polar caps to melt; a 2% change in the other direction would cause glaciation.[21] Caldiera and Kasting suggested that this is extreme, because the early solar flux is estimated to have been 25% to 30% lower than now (this is known as the faint early Sun problem).[281] However, scientists disagree on this point.[282]

Climate was cooler during a period called the Maunder minimum (1645–1715), when sunspots virtually ceased to be observed, at least in Europe (there is some evidence that Chinese astronomers continued to observe sunspots during this period). The Maunder minimum may be an important contributor to climatic change. (Note that the Maunder minimum occurred during the 1550–1850 coolest part of the Little Ice Age.) The predicted temperature variations are too small to explain the whole of the surface temperature anomaly.[279,283] The record of carbon-14 production as

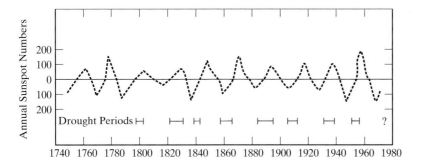

FIGURE 16.12

Correlation between midwestern droughts and Zürich sunspot number. The zero line represents the long term average number of sunspots. When the number is below the line the number of sunspots is lower than average; when above, higher than average.
(S. Schneider, *The Genesis Strategy*, © 1975 by Plenum Publishing Corporation. Reprinted by permission.)

seen in tree rings is reflective of the number of sunspots. The frequency of droughts is correlated with sunspot activity (see Figure 16.12). The 860-year record shows that there may be at least two other minima in sunspot activity—the Spörer minimum (1416–1534) and the Wolf minimum (1282–1342).[196]

Use of computer models of the effects of a reduced solar constant during the Maunder minimum show that the irradiance affects the AO/NAO (Arctic Oscillation/North Atlantic Oscillation, see **Extension 16.1**) through the ozone layer,[275,283] causing the index to lower and causing cooling in Europe.[284]

There are indications that the price of wheat in seventeenth-century England was correlated with solar activity.[285] Pustilnik and Yom Din use beryllium-10 isotope abundance in Greenland ice cores to infer solar activity between 1600 and 1700. They find that prices at minimum activity are higher every time than the price at maximum activity and claimed that "[t]hese results imply a causal connection between solar activity and wheat prices in medieval England."[285]

However, studies show that large variation in the solar constant is unlikely.[286] If the effects occur, then there must be an amplification that somehow occurs to make the effects strong enough, as in the connection through the ozone layer. Perhaps solar forcing amplifies weather noise in the system.[287] Cosmological changes can certainly have serious consequences, as we explore in **Extension 16.8, *Where Did All the Dinosaurs Go?***

Out of This Solar System?

Another cosmological possibility was raised by Shaviv and Veizer in 2003. They proposed that passage of our solar system through the spiral arms of the Milky Way could be responsible for some of the climatic variation seen over the past 545 million years. They calculate the cosmic ray flux at various times in the past, taking supernova and increased interstellar dust into account, and find surprising consonance between their predicted flux of cosmic rays and the paleotemperature of tropical seawater. They claim that as much as two-thirds of the variance in climate could be due to extrasolar-system effects.[302]

SUMMARY

Weather describes the local atmospheric state over short periods of time. Climate describes the state of the annual and seasonal average weather; these averages are stable for long times. Weather is driven by the Sun's energy input and the difference between the solar power per unit area of the poles and the equator. The energy flux of Earth is in long-term balance—as much is radiated away by Earth as is absorbed, or the mean temperature would have to increase or decrease steadily (and, of course, this is not observed).

Some solar energy is absorbed directly by the oceans and some is transferred by the winds to the ocean. The interplay between atmosphere and ocean over long distances is evidenced by El Niño–Southern Oscillation events, which involve half the globe (see **Extension 16.1**), and many other ones, the North Atlantic Oscillation being the one that affects us most after El Niño (refer to **Extension 16.2**). The large heat capacity of the oceans helps to damp rapid fluctuations in weather and climate.

Climate has changed over periods of hundreds of years. How scientists can determine and document change is discussed in some detail in **Extension 16.3**. The mean temperature in the Little Ice Age was no more than 1 °C cooler, but there were large effects, especially near the poles. Warm climates correlate with high CO_2 levels. Glacial ages have lowered CO_2 levels, as seen in ice cores and ocean sediment cores.

Although the carbon cycle of weathering, sedimentation, and volcanic release seems to be reasonably well understood in its overall characteristics (for more of the details, see **Extension 16.4**), some details are still not understood. There is a sink of carbon somewhere—but where?

Volcanic activity injects aerosols into the upper atmosphere. These aerosols as well as natural and human-caused aerosols may lead to a lower surface temperature (investigations described in **Extensions 16.5, 16.6,** and **16.8** indicate the consequences of too many aerosols in the atmosphere). Hence, climate is affected by agencies outside the atmosphere. In the next chapter, we consider the consequences of human intervention in the atmosphere-climate system.

The movement of continents (which is driven by the motion of the plates) from equator to pole or vice versa can cause climate change. The Milankovitch cycles (see **Extension 16.7**) have clearly caused climate changes, even though the exact mechanism enhancing the 100,000-year cycles is still a subject of scientific debate.

The solar constant, when it changes, has weather effects. The solar constant was smaller in the past, and a change of much less than 1% could cause a great deal of climate change.

PROBLEMS AND QUESTIONS

MULTIPLE CHOICE REVIEW

1. If Earth's albedo were 0, what would its radiation balance temperature be?
   a. 0 K
   b. 255 K
   c. 278 K
   d. 398 K
   e. 77,152 K

2. Proxy measures of temperatures can be made using
   a. $\delta^{18}O$ in benthic foraminifera interred on the sea floor.
   b. alkenones in benthic foraminifera interred on the sea floor.
   c. calcite formation in stalagtites in caves.
   d. otoliths (ear bones) of fossil fish.
   e. all the other means listed.

3. Climate may be described by indicating the mean and the variability of
   a. temperature.
   b. snowfall.
   c. rainfall.
   d. extent of snow cover.
   e. all the other variables listed.

4. Ocean circulation could cause an abrupt surprise in the future, because
   a. it has "flipped" to other circulation states in the past.
   b. the top and deeper layers thoroughly mix periodically.
   c. the ocean stores so much more thermal energy than the land.

   d. sea ice leads to fluctuations in icebergs in the sea lanes.
   e. the Arctic Oscillation/North Atlantic Oscillation controls European temperatures.

5. Which changes could affect climate?
   a. Geological changes
   b. Orbital changes of Earth
   c. Human interference
   d. Changes in the output of the Sun
   e. All of the above can affect climate.

6. The Little Ice Age
   a. was a figment of collective imagination, investigators being misled by local events.
   b. did not affect the behavior of people in Europe.
   c. was very much colder than now, ice-age temperatures lasted several centuries.
   d. was a period of glacial retreat.
   e. was a period with temperatures in Europe about 1 °C cooler than in the twentieth century.

7. Milankovitch cycles
   a. have one main period, 23 kyr.
   b. have several important periods corresponding to angular momentum changes in the sea–atmosphere system.
   c. have several important periods that reflect the position and orientation of the poles and the land to the Sun.
   d. have been shown by many experiments to have been mere statistical association rather than a real effect.
   e. are seen in both $\delta^{13}C$ and $\delta^{18}O$, but nowhere else.

8. Records of glacial ages in past times may be found in
   a. the amount of calcium carbonate deposited on ocean floors.
   b. the ratio of ^{18}O to ^{16}O in plankton in the Pacific Ocean.
   c. accumulation of wind-blown soil containing cold-resistant species.
   d. pollen found in ancient lake bottoms.
   e. all of the above alternatives.
9. Weather is caused by
   a. solar energy input.
   b. geological processes.
   c. orbital changes.
   d. the difference between insolation at the polar and equatorial regions.
   e. the Coriolis effect.
10. The temperature difference between the Equator and the South Pole is due to
   a. energy transport in Hadley cells.
   b. geological processes.
   c. the Coriolis effect.
   d. the difference between insolation at the polar and equatorial regions.
   e. the joint action of (a) and (d).

SHORT ANSWER
(Explain whether the statement made is correct, and justify your answer.)
11. The Pacific Ocean plays a key role in climate.
12. Oceans absorb much of the incident solar radiation.
13. Local climate patterns in a cooler world may involve more rainfall.
14. The difference in temperatures between the ice age and non-ice age climate is less than 20 °C.
15. There is no reason to believe that there have been any changes in the Sun that could affect climate.
16. We know that orbital changes cause changes in ice cover, because they occur simultaneously.

17. Volcanic eruptions can affect weather in the short term.
18. Terrestrial plants can absorb many billions of tonnes of carbon from the atmosphere.
19. Planktonic and benthic foraminifera can absorb many billions of tonnes of carbon from the atmosphere.
20. Gases in equilibrium in the atmosphere are not subject to rapid (year-to-year) fluctuations.

DISCUSSION QUESTIONS
21. Because the planet receives a net of about 100 W/m^2, would a change of 4 W/m^2 be considered substantial? Explain. (Chapter 17 shows that this is the size of the greenhouse effect.)
22. The oceanic conveyor is often referred to as thermohaline circulation. Explain why this is an appropriate name.
23. Local climate patterns may involve more rainfall in a cooler world and also in a warmer world. Why is this not contradictory? Explain.
24. Volcanic eruptions may affect climate because volcanoes release sulfur compounds and small particles into the stratosphere.
25. What is the connection between an ENSO and the Peruvian anchovy catch? Explain.
26. Many terrestrial weather patterns have been observed and named, for example, the Southern Oscillation, the Arctic Oscillation/North Atlantic Oscillation, the Pacific Decadal Oscillation, and so on. How are these indices useful in the discussion of weather and climate?
27. What is the utility of studying weather on Mars or Venus, our planetary companions? Explain.
28. Suppose you heard that 22-yr sunspot cycles are correlated with periodic droughts in Nebraska. Does this mean that sunspots have caused the droughts? Explain.
29. Could the resonance between the Moon's orbit and Earth affect Earth's weather? Why or why not?
30. What effect could Wegener's continental drift (Chapter 10) have on climate? Explain.

CHAPTER 17

Climate Change and Human Activity

Climate is less steady than we used to think. Any change could tip the world into a different climatic state. Knowing the basic nonhuman-generated causes of climate change from the previous chapter, we consider here the effects of the "human volcano." People have historically affected local conditions by contributing to desertification in some regions and to long-term fertility of soils in other regions and establishing and perpetuating prairielands in temperate regions. The industrial revolution led to the burning of fossil fuels in growing yearly volumes. In Chapter 14, we discussed the effects of "pollutants" in the effluent from burning processes. Another component of the effluent—considered harmless by many people until only a few years ago—is carbon dioxide. We extend the discussion of Chapter 16 to consider human input to the climate system. Humanity has embarked unknowingly upon a long-term experiment in climatic stability. We examine some possible consequences of that experiment in this chapter.

KEY TERMS

flickering climate / anthropogenic / deforestation / savanna / desiccation / greenhouse effect / resonance / forcing / general circulation model / climate sensitivity / CO$_2$ scrubbing

17.1   THE TENUOUS NATURE OF THE PRESENT CLIMATE

Not long ago, the globe experienced a mild ice age that had profound consequences for people living at the higher latitudes. The current mild weather is decidedly not the norm. Earth is in an ice age configuration[1,2] of 80,000- to 100,000-year glacials alternating with 10,000- to 15,000-year interglacials.[2] Even though we all tend to take our interglacial climate as the way things should be, we are enjoying climatic beneficence, and the stability is extremely fragile (as we saw in considering pack-ice cover in the preceding chapter). The major evidence in the climate record is provided by ice cores from Greenland and Antarctica, oceanic sediment cores showing the fossil record, and growth rates of corals in the tropic oceans and calcite (stalagtites and stalagmites) in caves on islands and continents. It is sobering that the world's ice age climate differed so little from that of today,[3,4] while the effect was so striking.

17.2   HEINRICH EVENTS AND DANSGAARD–OESCHGER CYCLES

Even more sobering is the evidence for climate instability and rapid changes compiled from the record of ice cores drilled in Greenland in the early 1990s. Two groups, known as GISP2 (Greenland Ice Sheet Project 2) and Greenland Ice-Core Project (GRIP), found evidence for **"flickering" climate** changes from glacial to interglacial times.[5] The annual layers of snow are compressed under the weight of the overlaying snow, and the changes in snow characteristic of the different seasons show up as visible layers in the ice that forms. The evidence indicates that two "cold snaps" known as the Younger Dryas/Preboreal and Older Dryas/Allerod transitions happened over fewer than 20 years (perhaps as few as 3 years to 5 years).[5,6] During the last glacial, there were swings of 10 °C over a century (or perhaps even a decade or less).[5,6,7] The rapid warmings followed by slower coolings are now called Dansgaard–Oeschger cycles. They appear on a cycle of about 1500 years.

The story took a dramatic turn during the late 1980s. A. Heinrich discovered that during at least six time intervals of 1 to 2 thousand years over the past 100,000 years, Canadian glacial debris was dropped down to the ocean bottom, forming layers as far south as the latitude of Portugal.[8] The debris was carried by icebergs and dropped to the ocean floor when they melted. These episodes of armadas of icebergs calved off the Laurentide ice sheet are now called Heinrich events. They were caused by the interaction of the Laurentide ice sheet, the underlying Earth, and the ocean conveyor. Earth leaks thermal energy through its surface, even when covered by glaciers. But when the ice over Hudson Bay became 3 km thick, it blocked the escape of this internal geothermal flux.[8] The bottom layer melted—if there's enough melting, southward go the armadas. The melting of icebergs disrupted the ocean salinity and chilled Northern waters; the Gulf Stream was rerouted, causing the climate to chill.[9] The events were felt over the entire North Atlantic Ocean[10] and were also seen in fluctuations in the moraines of Sierra Nevada glaciers.[11] During this time, the Mississippi was rerouted to drain through the Hudson or St. Lawrence.[12]

In addition to the Greenland ice cores[13] and the Arctic Coring Expedition (ACEX) cores from under the Arctic Ocean,[14] there are cores from Antarctica, for example, the 420,000 Vostok ice cores,[15] the 750,000-year-old (3 km) European Project for Ice Coring in Antarctica (EPICA) ice cores,[16] and others.[17] Though there was some controversy about the conditions across ancient Antarctica as inferred from the cores, it seems to have been resolved.[17] It is clear that interglacials similar to the present one occurred only over the last 430,000 yr; before that, there seems to have been much less temperature difference between glacials and interglacials.[16]

Correlations exist between Greenland ice cores, Vostok ice cores, and oscillations of glaciers recorded elsewhere,[15,18] Heinrich events, and Dansgaard–Oeschger cycles. These signals are also seen in stromatolite data.[17] Andean glacier debris were formed coincident with Heinrich events;[19] the release of methane from continents (indicating wet swamp formation) coincided with Dansgaard–Oeschger events. The major difference is that the Heinrich events are less cyclic, occurring erratically.

The Dansgaard–Oeschger events are thus correlated with the Heinrich events. The "flickering switch" is triggered by the melting and (at least partial) shutdown of the ocean conveyor.[9] This sort of change represents the "catastrophic shift" that was mentioned in Chapter 16. The eerie thing is that the changes were so sudden and intense, and that our Holocene interglacial is so stable. The catastrophe may be ready to strike with no warning, caused by some subtle change human beings may be making, returning the world to a much warmer or a much colder climate, as has occurred in model calculations.[20] There was a "systematic freshening"[21] in the North Atlantic Ocean between 1950 and 1990.[21,22] Is this a foretaste of such a change?

17.3 HISTORICAL EVIDENCE FOR ANTHROPOGENIC CLIMATE CHANGE

We examine here the relationship between climate change and human activity (human-caused change is called **anthropogenic** change). Relatively minor changes could spark the advent of new ice ages. Human activity seems quite capable of supplying these minor changes; indeed, we are now shaping the future climate. Perhaps the most important ways we can cause climatic change are by changing the albedo (or amount of light reflected), by changing the amount of dust in the atmosphere, by causing a redistribution of thermal energy, and by increasing or decreasing the amounts of certain gases in the atmosphere.[3] Crutzen and Stoermer proposed calling modern times (starting around 1800) an "Anthropocene," or recent human-caused, Era.[23] The current geological interglacial era that began around 12,000 years ago was officially known as the Holocene, or whole recent, Age; with Crutzen and Stoermer's acknowledgment that the present age is due to human intervention, the Holocene would be split in two.

Ruddiman has proposed a time around 8000 years ago as the divide after which humans began influencing climate by felling forests (producing CO_2).[24] Humans amplified their effects around 5000 years ago by starting to cultivate rice (producing CH_4). Ruddiman asserts that "evidence from palynology, archeology, geology, history, and cultural anthropology shows that human alterations of Eurasian landscapes began at a small scale during the late stone age," so he would say the Anthropocene Age began around 6000 B.C.[24]

Desertification and Deforestation

Albedo, or reflectivity, can readily be changed by human action. At relatively minor cost, dust or soot could be dumped on the North Pole ice cap, which would cause it to absorb more sunlight and melt.[25] But even normal activity can result in albedo change. Many peoples practice swidden (slash and burn) agriculture. The ratio of carbon black to total carbon in the smoke gives a signal for the type of burning, whether forest or savanna.[26] More than 80% of biomass burning takes place in the tropics.[27–29] Substantial areas can be put to the torch (Figure 17.1).[30] Tropical **deforestation** alone releases 1.6 ± 1.0 Gt, and perhaps as much as 6.3 ± 0.4 Gt of carbon per year is released by worldwide deforestation.[31] This tropical release figure is somewhat higher than that found by interpretation of satellite imagery, 0.5 Gt to 1.4 Gt.[32] However, Amazonian wetlands and rivers may be adding another 0.5 Gt/yr.[33] There was a recent discovery in the tropics of a feedback loop that means at higher temperatures, partly caused by deforestation, trees take up less carbon dioxide.[34] Evidence from Alaska's Kenai Peninsula shows that warmer weather led to a spruce bark beetle (*Dendroctonus rufipennis* Kirby) population explosion that killed millions of trees, reinforcing the effects of the warming.[35] Such insect pest outbreaks are projected to grow as the climate warms.[36]

Evidence that burning alters the climate was also found in Australia.[37] Researchers in Alaska see links between forest fires there and climate change.[38] Kurtz et al. proposed that widespread peat fires that burned 55 million years ago for thousands of years caused the Paleocene–Eocene thermal maximum (a time of high global temperature).[39]

The world total of forests is about 3.4 billion hectares (Gha),[40] and about 2 Gha has been lost to forest through human activity.[31] One hectare of forest contains an average of 100 to 200 t carbon, so the world's forests contain 500 to 800 Gt of carbon, perhaps even 1 Tt.[40,41] The Amazon basin is undergoing especially rapid deforestation from fires and cutting.[42,43] From 1990 to 1997, 5.8 ± 1.4 million hectares of tropical forest were lost, and another 2.3 ± 0.7 million hectares of forest were significantly degraded.[44] An international study has been organized to investigate the Amazon and its future effects on climate.[45] Deforestation has turned 40% of the African equatorial forest into savanna over the past several thousand years.[46]

It is well known that much of the Great American Prairie was maintained by fires set by American Indians.[46,47] The prairie reflects much more sunlight than forest does. The British energy crisis in the mid-sixteenth century was caused by a shortage of wood, and, in turning the British to coal, may have sparked the industrial revolution.[48] Deforestation has dramatic effects on winter albedo in snow areas: Snow-covered farmland has double the albedo of snow-covered trees;[46] over deciduous forests, visible and near-infrared albedos are almost equal; over coniferous forests, near-infrared albedo is greater than that in the visible; and over scrubland and snow-covered fields, the visible albedo is greater than that in the near infrared.[49]

Overgrazing also alters albedo. Desert expansion, such as has been seen in the recent advance of the Sahara into the Sahel, can be stimulated by increasing the albedo.[25,50,51] An example of the effect of overgrazing in the spread of deserts was seen in 1974 satellite photos of the Sahel, which showed desert surrounding a fenced-in green area.[52] The green area was a ranch with natural local grass; it was operated to produce cattle by a commercial establishment that allowed the land to regenerate after grazing. In neighboring regions, cattle belonging to individuals had eaten every blade, which altered the microclimate enough to prevent grass from regrowing. The interaction of humans and climate is hard to disentangle, but we do know that small, individual human actions over large geographic scales can cause important changes.[53]

Sagan, Toon, and Pollack[46] argued that people have influenced the climate by altering abundance and distribution of vegetation beginning as early as 20,000

FIGURE 17.1

Seasonal burns, 1972 and 1973 dry season, in equatorial Africa. (J. Hidore, "Population explosion in Africa: further implications," *Journal of Geography* **77**, 214 (1978). Reprinted by permission of The University of Chicago Press)

years ago. The name "Tierra del Fuego" (land of fire) was given by Magellan to the southernmost tip of South America because of the many fires set by the natives that burned off all the local bushes and trees. The degraded volcanic Mexican highlands were formed by native Mexican agricultural practices ~2500 years ago.[54] The Carthaginian Hanno referred to annual burning south of the Sahara, which was followed by **desiccation**, drying. Deforestation has destroyed 60% of the central European forests since the eleventh century. Traces of Rome's public road building from 171 B.C. caused changes in pollen species seen in lake sediments in Italy, as Italy lost forest cover and started to suffer from floods.[55] Italy suffered "an *ecological catastrophe* during ... Antiquity" evidenced in sediments and caused by people.[56]

Albedo

Table 17.1 lists changes in albedo due to desertification, salinization, deforestation, and urbanization.[46] The Rājasthān Desert, now about 1 million km^2 in area, was once the site of the Indus valley civilization. The desert appears to have been caused by overgrazing, as grass grows rapidly when it is protected from overgrazing. Ancient farming villages are buried in the Sahara,[46] and North Africa was the breadbasket of the Roman Empire; ecological change was hastened by human abuse and importation of goats by Arab invaders. Central Anatolia, now semidesert, was once the breadbasket of Byzantium before goat husbandry destroyed the delicate ecological balances and helped cause desertification. All major deserts are subject to a radiation deficit; they reflect so much that they radiate more thermal energy than they absorb.[57–59] The Sahara Desert loses about 35 W/m^2 in radiation; the subsequent sinking of air and its heating supplies this energy.[57]

Iraq, which was once Mesopotamia, home of several mighty civilizations, has had 20% to 30% of its farmland destroyed by salt pulled up from the underlying

TABLE 17.1

Anthropogenic albedo changes

Albedo process	Surface albedo	Global change — Last few million years	Global change — Past 25 years
Desertification (savanna → desert)	0.16–0.36	0.004	0.0006
Salinization (open field → salt flat)	0.1–(0.25–0.71)	0.00015–0.0001	0.00025–0.0004
Temperate deforestation (forest → grassland field)			
Summer	0.12–0.15	0.00025	Small
Winter	0.25–0.60	0.0004	
Tropical deforestation (forest → field, savanna)	0.07–0.16	0.001	0.00035
Urbanization	0.17–0.15	−0.0025	−0.001
Totals		0.006	0.001

Source: C. Sagan, O. B. Toon, and J. B. Pollack, "Anthropogenic albedo changes and the Earth's climate," *Science* **206**, 1363 (1979).

soil.[60] The Aswan Dam, by preventing the annual flooding of Egyptian fields, has led to increased salinization there.[52] In the American West, also a dry region, irrigation water leaches salts and brings them to the surface. The salty irrigation water causes American rivers to become salt loaded; dams built to regulate water flow for irrigation and to supply electricity have also contributed to the salt problem. Salty water from upstream can harm farmers downstream who try to use the water to irrigate. The United States and Mexico signed a treaty in 1944 to try to regulate the salinity of the Colorado River; a desalting plant has been built under the treaty terms.[61]

Albedo is extremely important in assessing climate change, because, as implied by climate models discussed subsequently, a change in the 0.30 albedo by 0.01 can cause a change in mean annual temperature of 2 °C.[46] Sagan, Toon, and Pollack concluded that it is "likely that the human species has made a substantial and continuing impact on climate since the invention of fire."[46]

Despite the longterm increase, a decrease in Earth's reflectance began in 1984, with a significant drop after the mid-1990s.[62] Presumably, this fluctuation is related to changes in Earth as well as to climate changes being brought about by human beings. The reflectance is increasing again in the 2000s.[62]

The burden of human-generated dust now in the atmosphere is comparable to that emitted by volcanic action.[63,64] Reid Bryson has spoken of the "human volcano." An article by *New York Times* science writer Walter Sullivan that appeared four decades ago discussed human-caused desertification and salinization; its sardonic title was "Is There Intelligent Life on Earth?"

17.4   CARBON DIOXIDE AND THE GREENHOUSE EFFECT

Perhaps the three most important trace gases in the atmosphere are ozone, water vapor, and carbon dioxide. Ozone is important because it absorbs ultraviolet radiation from sunlight. High levels of ultraviolet light could lead to widespread eye damage and increases in incidence of skin cancer (see Chapter 1). Both water vapor (concentrated near Earth's surface) and carbon dioxide (distributed throughout the atmosphere) act to absorb infrared radiation, which raises Earth's temperature. Both carbon dioxide and water vapor absorb the long wavelength (infrared) radiation emitted by Earth and reemit it in all directions. As a result, the ambient average temperature is not 254 K but 288 K, over 30 °C higher. Water vapor is extremely important in this process, because it typically appears in the lower atmosphere at a concentration of 10,000 to 50,000 µmol/mol (this unit is often referred to as parts per million, ppm). Of course, water vapor also forms clouds that help cool Earth by reflecting sunlight. Carbon dioxide is relatively uniformly distributed with height, with an average concentration of about 350 µmol/mol, but it varies from a low in the Antarctic to a high in the Arctic.[65]

The Greenhouse Effect

Water vapor and carbon dioxide act as two layers of a multilayer window to surround Earth and warm it. A somewhat similar thing happens in a closed-up car during the spring and summer. The glass allows light in, and the radiant energy is

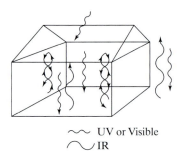

~~~ UV or Visible
⌒ IR

FIGURE 17.2

In a greenhouse, the glass transmits visible electromagnetic radiation but reflects infrared. As a result, the temperature of the quiescent air rises.

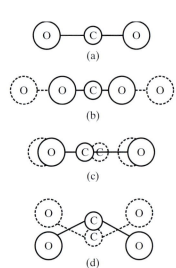

FIGURE 17.3

Possible oscillations of carbon dioxide. (a) The carbon dioxide molecule has the atoms strung out along a straight line. (b) One mode of oscillation has both oxygen molecules simultaneously approaching and fleeing the stationary carbon. (c) One mode has the oxygens moving right while the carbon moves left and vice versa. (d) One mode has the carbon bending upward as the oxygens simultaneously bend downward and vice versa.

absorbed by the interior material of the car, raising its temperature. The car interior begins to radiate in the infrared, but the glass reflects the infrared radiation coming from the car interior and traps the hot air inside, ultimately heating up the car. A greenhouse works in just this way (Figure 17.2). Of course, having the car or greenhouse closed so that the air is quiescent is important for them but irrelevant for the atmosphere. This trapping is known as the **greenhouse effect** because of the functional similarity between the glass in the greenhouse and gases in the atmosphere in trapping radiation. The greenhouse effect has always been present during the time human beings have been alive, and it has made it possible for our form of life to occur. In fact, it is known that there must have been an extreme greenhouse effect earlier in Earth's history, because the Sun was weaker then than it is now. Evidence was found in microfossils from the Proterozoic, 1.4 Gyr ago, that CO_2 levels then were 10 to 200 times as large as now.[66]

The "atmospheric window" was shown in Figure 14.5. The atoms in the atmosphere's molecules can vibrate or rotate in various ways at certain natural frequencies. (Figure 17.3 shows how the atoms in CO_2 can move, leading to the structure shown in Figure 14.5.) Radiation near these frequencies causes a huge effect on the molecules, and we refer to the vibrational response as **resonance**. The resonance frequencies for vibration of carbon dioxide are in the infrared. Some strong carbon dioxide resonances in the atmosphere are near 3.6, 3.85, 4, 7.8, 8.1, 9.3, 10.4, and 10.8 μm in wavelength.[58,67] If infrared radiation from Earth with a frequency near the resonant frequency impinges on a molecule, it will be absorbed, exciting the molecule's oscillations. The radiation is then reemitted in all directions, so half of the radiation emitted by an average molecule is directed back to Earth. Of the half radiated outward, most encounter another CO_2 molecule, and half that is radiated back toward Earth. There are so many molecules in the air that in the end, most radiation is trapped. Water vapor has concentrations between 1% and 5% in the air and has many resonances in the infrared, so it plays a large role in the atmosphere's infrared absorption. Carbon dioxide and ozone also have many resonances. The window between 300 and 700 nm is due to the *lack* of resonances there. The gaps, extending to around 12 μm, are important, because they allow some of the emitted infrared radiation to escape Earth.

Atmospheric scientists refer to an imposed change in Earth's energy balance as **forcing**. Forcing is measured in W/m^2. Forcing may be due to the changes in the Sun's output, to volcanic activity, to orbital changes, or to human causes (see Chapter 16). The upper "edge" of the atmosphere is at a temperature of 255 K, as determined by radiation balance. Most radiation is absorbed and emitted there, but 240 W/m^2 enters the lower atmosphere, and the same amount is emitted from the upper atmosphere. The solar forcing of the lower atmosphere thus would be 240 W/m^2. The forcing resulting from a 1% change in solar irradiance (Chapter 16) would be 2.4 W/m^2. Climate forcing by CO_2 and other greenhouse gases is considered later.

Venus, with a high carbon dioxide concentration (95%) and a surface temperature of 470 °C, is an example of a runaway greenhouse effect[68] despite the presence of sulfur dioxide clouds that prevent most solar radiation from being absorbed by the atmosphere. Mars is also warmer than it would have been without its substantial component of carbon dioxide. The greenhouse effect on all three planets comes from the presence of carbon dioxide and, on Earth, from "trace gases" present in the atmosphere. Where do these gases come from?

Establishing the Increase in Carbon Dioxide: The Carbon Budget

Over the long term (Myr), excess carbon dioxide on Earth will eventually be stored, and there will be no carbon left but the relatively small emissions from volcanic activity. Over a shorter, but still long term (tens of kyr), the concentration of CO_2 will be in rough equilibrium between the atmosphere and the biosphere. This was the case between the time the Younger Dryas cooling ended (11,600 years BP) until the industrial age began.

Carbon dioxide concentrations are increasing partly because of the burning of 3 Gt of coal, 3 Gt of oil, and 1.6 Gm^3 of natural gas each year.[49,69] As a result, some 6 Gt of carbon is released each year from fossil-fuel burning alone, over four times the rate of release in 1950, and over 10 times the rate in 1900.[69–71] These almost unimaginable figures correspond to a release of about 1 tonne of CO_2 for each person on Earth,[72] and the release rate is increasing. The cumulative release since 1850 is in the neighborhood of 270,000 Mt (270 Gt).[73] Figure 17.4 indicates how this has risen since 1751.

Figure 17.5[74] shows per capita world energy use, total world energy use, and growth in energy use by region. Clearly, the per capita use (as well as total use) is greatest in the developed world. The growth in use is marked in the less-developed regions of the globe; growth in North America paused in the 1980s in delayed response to the 1970s energy crisis, then rose again, while growth in western Europe has stopped, and serious governmental efforts are in place to reduce emissions. The major source of this energy is fossil fuel. Energy production from fossil fuel results in carbon emissions. Figure 17.6[75] shows the results of the trends seen in Figure 17.5: since 1982, the less-developed countries have contributed more of the carbon burden than the developed countries (as represented by those countries belonging to the Organization for Economic Cooperation and Development [OECD]).

Other contributions than fossil fuels have added to the carbon dioxide burden of the air. Deforestation contributes between 0.5 and 5 Gt to the carbon balance each year,[29,31,76] with a best guess of 1.6 ± 1.0 Gt.[70,71] People burning trees, brush,[30,47,48,69] and fossil fuels have managed to increase the carbon dioxide concentration in the air from about 260 µmol/mol in the Arctic[77] and the Antarctic[78] during a long period prior to the mid-nineteenth century to around

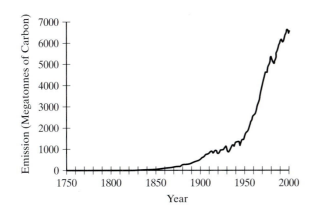

FIGURE 17.4

Carbon emissions, 1751, to 2000. (G. Marland, T. A. Boden, and R. J. Andres, "Global CO_2 emissions from fossil-fuel burning, cement manufacture, and gas flaring: 1751–2000," in *Trends: A Compendium of Data on Global Change*, Carbon Dioxide Information Analysis Center, Oak Ridge National Laboratory, Department of Energy, 2003)

FIGURE 17.5

(a) Per capita world energy use.
(b) Total world energy use. (c) Growth in energy use by region. (d) Per capita CO_2 emission versus per capita GDP. The curve is a best-fit regression. Only a few countries are shown by name (193 countries represented).
((a,b,c) U.S. Department of Energy, *International Energy Annual 2002* (Washington, DC: GPO, 2004).
(d); CIA, *World Factbook* and G. Marland, T. A. Boden, and R. J. Andres, "Ranking of the world's countries by 1998 CO_2 per capita emission rates," in *Trends: A Compendium of Data on Global Change*, Carbon Dioxide Information Analysis Center, Oak Ridge National Laboratory, Department of Energy, 2001)

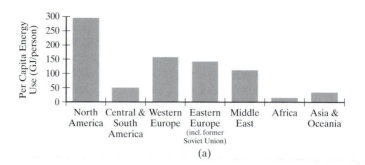

(a)

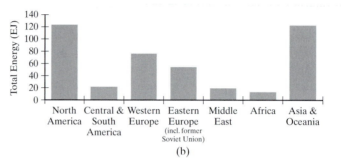

(b)

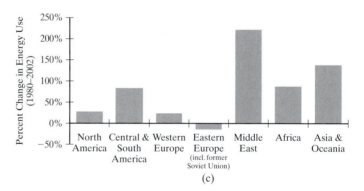

(c)

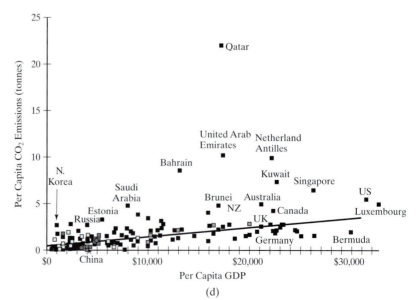

(d)

370 μmol/mol now.[25,69–71,77–79] (The topic of carbon uptake by the biosphere will be discussed in more detail later in this chapter, especially in the extensions.) Known sources and sinks are shown in Table 17.2. The net imbalance between sources and sinks (Chapter 16) is close to zero (but was substantial in earlier times).

The recent trends—including seasonal variations, with less carbon dioxide during summer—are shown in Figure 17.7,[81] which shows measurements made at Mauna Loa, Hawaii: A continuous increase of about 2 μmol/mol per year is measured.[81,82] Mauna Loa is a very high mountain, and Hawaiian air is just about as clean as air can be. Hence, the CO_2 levels from Mauna Loa must reflect the average atmospheric CO_2 concentration in the Northern Hemisphere. The current concentration, 370 μmol/mol, is much higher than the preindustrial level of 280 μmol/mol. Similar plots from other regions of Earth made by many groups show that the annual cycle is opposite in phase in the Southern Hemisphere, as expected, because the decrease in atmospheric CO_2 is due to vegetation using it for growth, while the rise is due to the decay of plant litter; and that the cycle amplitude is much smaller than in the Northern Hemisphere, which follows because the land area in the Northern Hemisphere is much greater than that in the Southern Hemisphere.

TABLE 17.2

Known sources and sinks of carbon (1990)

	Carbon per year (Gt)
Sources	
Fossil-fuel Burning	6.0 ± 0.5
Land-use Changes	1.6 ± 0.8
Sinks	
Oceans	2.0 ± 0.8
Atmosphere	3.4 ± 0.2
Land	2.3 ± 1.3

Source: IPCC, References 70, 71, 80.

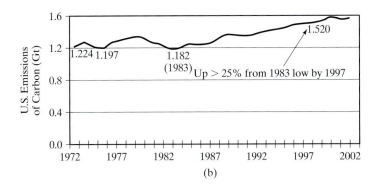

FIGURE 17.6

(a) Carbon emissions compared between developed and developing countries.
(U.S. Department of Energy; International Energy Agency, *International Energy Outlook 2000* (Paris: OECD/IEA, 2001), Table 2.4)
(b) U.S. emissions.
(Adapted from U.S. Department of Energy, Energy Information Administration, *Emissions of Greenhouses Gases in the United States* 1997, DOE/EIA-0573 (1997) and *Emissions of Greenhouse Gases in the United States* 2002, DOE/EIA-0573 (2002))

FIGURE 17.7

The concentration (in micromoles per mole, otherwise known as parts per million) of atmospheric carbon dioxide measured at Mauna Loa Observatory, Hawaii, which has very clean air. The gas analyzer samples Northern Hemisphere CO_2 values. Note that, in the so-called Keeling plot with the year labeled at January, summer values are lower and winter values are higher (as one might expect). The dots indicate monthly average concentrations. A smooth line through the middle would represent an annual cycle (increasing roughly linearly with time; however, the best fit is to a sum of a parabola and an exponential).
(C. D. Keeling and T. P. Whorf, "Atmospheric CO_2 records from sites in the SIO air sampling network," in *Trends: A Compendium of Data on Global Change*, Carbon Dioxide Information Analysis Center, Oak Ridge National Laboratory, U.S. Department of Energy, Oak Ridge, Tenn., U.S.A., 2004)

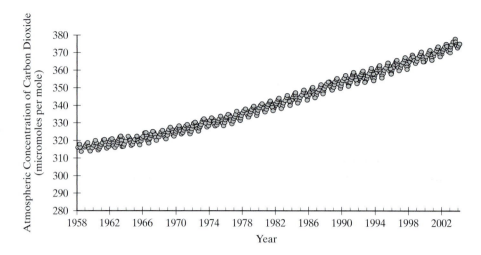

Ancient CO_2 Concentrations

How can we know what the carbon dioxide concentration was a century ago? It makes a nice scientific detective story. It is known so far back from analyses of air bubbles trapped in Arctic and Antarctic ice.[77,78] The ice is crushed in an evacuated vessel, and the CO_2 concentration is measured spectroscopically. The strength of the response is proportional to the amount of CO_2 present. This CO_2 concentration is in agreement with that of the 260 to 276 µmol/mol measurement from analysis of $^{13}C/^{12}C$ in tree rings.[72] Extrapolation of the data of Figure 17.7 backward in time gives a best estimate of the preindustrial level of carbon dioxide of 290 µmol/mol.[68,69,83,84]

As we pointed out, over rather long time periods, concentrations will be in equilibrium between atmosphere and biosphere. Were all fossil-fuel and land-use change emissions of CO_2 to cease as of today, in 1000 years the oceans would have absorbed about 85% of the cumulative emissions (total emissions up to the present).[85] Over the course of many millennia, this oceanic CO_2 would decrease again as the carbon is cycled to limestone. So, a large portion of the carbon we are putting into the air will be around "permanently," as least as far as we now living are concerned. Either we and our descendants will live in a warmer world, or we will have to become very active managers of carbon stocks.[85]

There is evidence from at least the last 500,000 years that climate and atmospheric carbon dioxide are linked. Periodic asteroid impacts, volcanism, and mass extinctions led to high carbon dioxide levels, followed by weathering and decreases in carbon dioxide with global temperatures in lockstep.[86]

Earth Climate during the Last Millennium

The ability to be able to learn about climate conditions in the past involves the use of proxy indicators of temperatures and other important parameters, simply because the instrumented record stretches back only to about 1860 (some instrumental records go back 400 years). See **Extension 16.3,** Proxy Measurements, in particular Table 16.3.1. which indicates a few of the most important proxy measurements available. There are many other important proxy methods not listed in the table, for example, the use of alkenones, methyl ketones in foraminifera that are normally single-carbon-bonded but form multiple bonds at a rate that depends on temperature.[87] As another example, the sea surface temperatures (SSTs) can be found for the South Pacific from the differential uptake of strontium and calcium in coral. The instrumental record for the last two decades was compared to the coral record and indicated excellent agreement, which led to a good knowledge of SSTs for the past several hundred years.[88]

Given that the instrumental record is but a century and a half in duration, and that good written records exist perhaps only through the last half millennium, we can say that we have good climate records for the last 150 years, we have a good sense of the past 500 years' climate, a somewhat hazier picture of the past millennium, and a very hazy view farther into the past than that. As was discussed in Chapter 16, the Little Ice Age was a cold period that followed the Medieval Warm Period and led directly to our own climatic period. Ice cores from Greenland and from glaciers worldwide provide powerful evidence of the climate during this period. The Dye 3 Greenland record is about 7000 years long and the GRIP borehole gives a 250,000 year record, clearly showing the warm and cold eras.[5,89]

Glaciers are in both hemispheres, and glacial ice cores show features over very long periods of time. Tibetan ice cores may contain records over 500,000 years, and the millennial cooling and warming is seen, though with poor resolution.[90] The South American glaciers show local droughts but not the Medieval Warm Period or the Little Ice Age.[19,91]

Much of the other evidence for the transitions on land must be nontraditional (at least for science).[91] Nevertheless, experts believe they can find the temperatures from the proxy evidence to within about 0.1 °C over a 30-year time interval over the last 300 years, but with increasing uncertainty before that time up to about 0.25 °C.[91] The eleventh and twelfth centuries were probably warm (Figure 17.8),[91-93] although not quite as warm as the last half century.[91] It is still

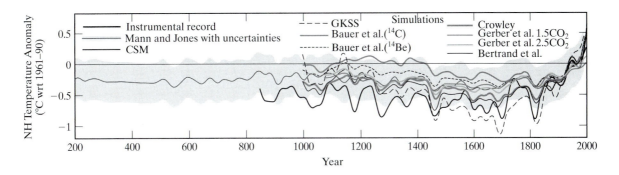

FIGURE 17.8

Overview of temperature reconstructions for the past two millennia. The zero line is set for the interval from 1961 to 1990. The shaded area represents the range of uncertainty in the reconstructions.
(P. D. Jones and M. E. Mann, "Climate over past millennia," *Rev. Geophys.* **42**, RG2002
(2004), Fig. 8. © 2004 American Geophysical Union. Used with permission.)

not clear, for example, whether the Medieval Warm Period was a worldwide event[94] or localized to the Northern Hemisphere.[95]

Glaciers are known to have advanced, from facts such as the destruction in 1292 of an aqueduct built in 1200 from the Aletsch glacier, which indicates that the Little Ice Age succeeded a warmer period.[94] In addition, records were found by Kiefer of dates of transfer of a statue of the Madonna across the Bodensee (Lake Constance) ice from Germany to Switzerland churches, if the lake froze over.[96] This evidence shows that the number of crossings increased substantially after the twelfth century, that is, that cooling had occurred from a warmer period. Data from the Sargasso Sea show both the Medieval Warm Period and the Little Ice Age.[97] The Little Ice Age is also seen in the Antarctic,[98] as well as in drought cycles in African rainfall.[99]

Earth Climate during the Last Half Millennium

We can say a bit more about the last 500 years, because the proxy records include more evidence. For example, borehole evidence and mountain snowlines are reasonable measures of temperature,[94] better than tree rings in the estimation of Broecker,[94] and mountain snowlines show a drop worldwide until around 1860, when they begin rising.[94,95] A study of 358 boreholes scattered around the world showed clear evidence of generalized warming throughout the last 500 years.[100] Clear evidence for the Little Ice Age is seen. (The evidence is still there in individual boreholes, but it is less clear.)[101] A study of paleoclimate records from the Arctic showed the Little Ice Age and the warming from the mid-1900s to the present.[102]

While tree-ring studies provide limited evidence (pointing to summer temperature averages), trees at the tree line are also affected by the yearly temperature fluctuations and can give evidence about past climate. A study of tree rings from treeline trees in the Tarvagatay Mountains of Mongolia from 1550 to the present shows the colder period in earlier times and substantial recent warming.[103]

Both the Mann et al.[92] and the Crowley[93] millennial reconstructions show considerably better records of fluctuations during this time as compared to the preceding 500 years. Both sets of records correlate well with known volcanic, solar forcings until the last century, and Crowley's reconstruction, which also includes land-use changes for the nineteenth century,[104] shows that the pre-1900 record is basically explicable in terms of just these few forcings.[93]

Earth Climate during the Last Century and a Half

During the last century and a half, the world has warmed, cooled, then warmed again. Figure 17.9 shows global temperatures through the twentieth century. The cooling and warming are clearly seen. Overall, Earth's temperature has increased by 0.6 ± 0.2 °C during this period.

This is the time during which instrumental records are available pretty much worldwide (with increasing coverage with the passage of time). Much can be learned. For example, between 1846 and 1995, in a study of 39 lakes with definitive records, freeze times of Northern Hemisphere lakes averaged 8.7 days later, and ice breakup times averaged 9.7 days earlier.[96] The date of Tenana River ice breakup for Alaska's Nenana Ice Classic has advanced and receded with climatic cooling and warming. The general cooling of the early twentieth century (see Figure 17.9) and subsequent warming are clearly visible.[105]

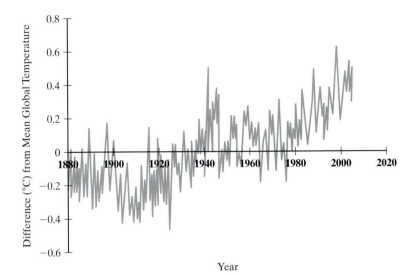

FIGURE 17.9

Monthly global surface mean
temperature anomalies 1880–2004
(baseline 1961–1990).
(National Climatic Data
Center/NESDIS/NOAA)

General circulation models (GCMs) predict that the Arctic regions will experience much greater warming than the temperate regions, which in turn, experience greater warming than the tropics. Adélie penguins have abandoned rookeries in the Antarctic that had been occupied for over 600 years, as the necessary winter pack ice disappeared.[106,107] Seven ice shelves have vanished from the Antarctic during the past half century.[106]

The human imprint becomes clear in this part of the record, as Mann pointed out that human land-use changes are needed to square known physical forcings with the nineteenth century temperature record.[104] A simulation of early twentieth century warming found that human changes plus an unusually strong multidecadal oscillation account for that warming.[108] A study of the entire twentieth century found both natural and anthropogenic causes for the observed warming (Figure 17.9).[109]

Earth's Climate during the Last Decade

While it is clear that numeration does not give a complete story, consider the results in Table 17.3. Over half of the top 20 U.S. temperatures registered occurred

TABLE 17.3

Ten warmest years, United States and world, through 2004

Rank	U.S.	Temperature (°C)	World	Temperature (°C)
1	1998	12.74	1998	14.53
2	1934	12.73	2002	14.46
3	1999	12.52	2003	14.46
4	1921	12.50	2004	14.44
5	1931	12.41	2001	14.41
6	1990	12.36	1997	14.37
7	2001	12.35	1990	14.30
8	1953	12.33	1995	14.30
9	1954	12.30	1999	14.28
10	1987, 1986, 1939	12.27	2000	14.27

Source: U.S.: NOAA/NCDC, "Climate at a glance"; World: NOAA/NCDC, "Global Surface Temperature Anomalies."

in the 1980s through 2004. The world global (land plus ocean) temperature top 10 come exclusively from the 1990s to 2004 time period (see the discussion in Reference 110).

Even clearer indications of change surfaced. "Iceberg Alley" finished 1998–1999 without a report of an iceberg.[111] Ice cover over the Arctic Ocean decreased by 43% between 1976 and 1996,[112] and it thinned by almost 2 m.[112,113] Arctic rivers are sending more fresh water toward the Arctic Ocean, and this will have some effect on the polar climate.[114] While the Greenland ice sheet (~10% of glacial ice) is maintaining a balance, according to some researchers, the southwest having thickened 21 cm per year as 30 cm is being lost annually in the southeast,[115] some 50 km^3 of ice volume is being lost each year.[116] Others assert that Greenland's glaciers are thinning by only 10–15 cm/yr.[117] Melting from Alaskan glaciers is larger than has been thought, a loss in thickness of about 0.5 m/yr, which means that Alaskan glaciers' losses are roughly twice as great as Greenland's.[118] The situation is similar in the Antarctic; the Pine Island glacier is thinning by 1.6 m/yr,[119] and the glacier retreated over a kilometer per year in the 1990s.[120] The Ross Sea is less salty than in the past, at least partly because of fresh glacier meltwater,[121] and the Ross Ice Shelf is growing.[122]

A sensitive indicator of whether such warming continues will be the growth or continued melting of the lower part of valley glaciers.[123] Glaciers are in retreat in mountains around the world. The ice core records from a Tibetan glacier show that the current era is the warmest in the last millennium.[124] The mountain glaciers of Africa,[125–127] South America,[127–129] and Asia[127,130] are vanishing—all will be gone in the next half century, according to renowned glaciologist L. Thompson.[127,128]

Given the improvements in GCMs and the massive instrumentation mounted on the planet in recent years, one would expect a near-perfect description. Even with this, satellite studies of tropospheric temperatures and measured surface temperatures appeared in the late 1990s to differ by up to 0.1 °C per decade. While much smaller than the difference (incorrectly) asserted in the early 1990s,[131] and while anthropogenic and volcanic forcings could explain most of the differences, the remainder was of concern.[132] However, there were reconsiderations in the early 2000s that led to reconciliation of the satellite records with surface warming.[133,134] A typical satellite result is now 0.22 °C to 0.26 °C per decade, in agreement with ground measurements.[134] It is very important to make certain of calibrations among different datasets and to use multiple datasets to characterize uncertainty.[135]

One study refuted the possibility that problems with missing coverage could be inflating climate numbers; the study found the reverse to be true, which means that if anything, calculations *under*estimate warming.[136]

The biggest story of the 1990s was the identification of the climate "smoking gun," which was the determination that the oceans were being warmed (by 0.06 °C between 1948 and 1999).[137] As long as there was no evidence for this warming, it might be argued that the warming was evanescent. Because GCMs predict ocean warming, the absence of warming would have cast great doubt on those models.

The evidence showed that the top 3 km of ocean water had warmed (half the warming was in the top 300 m).[137–139] Additionally, warming was found in the Southern Ocean of 0.17 °C between the 1950s and the 1980s at depths between 700 m and 1100 m.[140] In order for the models to reproduce the observed warming, they had to include anthropogenic forcing effects.[139,141] The models used by the two groups were different (Levitus et al. used the Geophysical Fluid Dynamics Laboratory (GFDL) model and Barnett et al. used the National Center

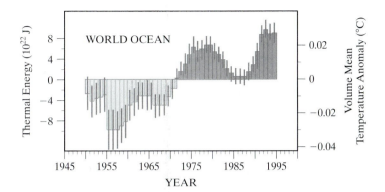

FIGURE 17.10
Thermal energy stored in Earth's oceans, in the top 3 km, 5-year running averages. The zero line is the 1945 to 1995 average. (S. Levitus, J. I. Antonov, T. P. Boyer, and C. Stephens, "Warming of the world ocean," *Science* **287**, 2225 (2000), Supplementary Material, URL: *http://www.sciencemag.org/ feature/data/1046907.shl*, adapted from Web Figure 3d © AAAS, 2000)

for Atmospheric Research (NCAR) Parallel Climate Model), but the same results were found, and a third model (using the Max Planck Institute (MPI) climate model) also gave identical results.[141] While the temperature increase might seem modest, the amount of thermal energy stored in the oceans is an order of magnitude greater than that in any other element of the climate system, an incredible 2×10^{23} J (it took about 8.1×10^{21} J to melt continental glacier ice and about 1.1×10^{21} J to melt freshwater ice, see Figure 17.10). The amount of energy stored in Earth's oceans since the 1950s is equivalent to about 2000 years' energy supply for the United States at the year 2000 energy use rate.

The Effect of Trace Gases

Trace gases, such as nitrous oxide, methane, and chlorofluorocarbons, seem to be increasing at an extremely rapid rate. It was not until 1976 that it was first realized that these so-called trace gases could cause warming.[142] The trace gases absorb infrared radiation in regions of the spectrum that are clear to other atmospheric constituents, and they are extremely strong absorbers.[79,143,144] Data lead to estimates of increased forcing from all the trace gases combined of 2.2 to 7.2 W/m^2.[143] Nitrous oxide emissions come mainly from agriculture (fertilizer, animal dung, and indirect emissions); agricultural emissions total around 6.3 Mt/yr.[145] Methane releases are especially closely tied to human activities. A spate of wildfires and booming economies in the 1990s led to a huge methane emissions spike (maximum emissions rate was 12.7 ± 0.6 nmol/mol in 1998),[146] while the world economic slowdown in the early 2000s led to a leveling off at an atmospheric methane concentration of 1.75 μmol/mol.[147]

According to the German Parliamentary Inquiry Commission,[148] a unique combination of legislators and scientists, in preindustrial times, carbon dioxide was responsible for 7.2 °C of the 32 °C rise over the 254 K expected had Earth no atmosphere; ozone caused 2.4 °C; nitrous oxide, 1.4 °C; and methane, 0.8 °C. The largest warming contribution was from water vapor, at 20.6 °C. A World Meterological Organization panel concluded that the combined effects of changes in the trace gases are as large as that due to CO_2.[149] For this reason, one early U.S. Environmental Protection Agency (EPA) study on prevention of greenhouse warming found that little could be done to reduce the impact of the warming by changes in energy policy now.[150]

There is little doubt that the chlorofluorocarbon refrigerants CFC-11 and CFC-12 will have a pronounced heating effect.[70,71,151] Absorption of CFC-11 in the infrared is, for example, eight times as strong as that of one of CO_2's principal bands.[143] Also, because the gases are *extremely* dilute (unlike CO_2), the increase in absorption is directly proportional to the increase in concentration, and vice versa. The Montreal Protocol (Chapter 1) mandates that levels of CFCs fall, and they have begun to do so as a result of the success of united world action. While in the mid-1980s, CFCs were increasing at about 6%/yr,[79] currently, the CFCs are actually decreasing, and the decrease is expected to accelerate.[152]

The trace gas N_2O, now 310 nmol/mol (parts per billion),[149] is increasing at 0.2% to 0.4% per year (Table 17.4). Some gets into the atmosphere, even in the absence of human intervention, from lightning and volcanic gases.[153] The recent rise in concentration attests to the importance of the human source—fossil-fuel combustion, application of nitrogen fertilizer,[143,154,155] discharge of sewage,[143,154,155] clearing of land,[156] and nylon manufacture.[157] These have a significant effect in causing the rise in the world average temperature.[143]

The concentration of methane has fluctuated rapidly over many millennia. The concentration's fluctuations are in synch with proxy measures of Earth temperature (including CO_2).[158,159] It has been variously guessed that the very recent increases in the amount of methane have been coming from flatulence in cattle (because their numbers are increasing), from the digestive tracts of termites (termite numbers increase due to increases in forest clearing), or from rice paddies (with an area worldwide that is on the increase).[71,143,154] Cows in barns emit about 550 liters per day of methane, a substantial number when you realize that there are 1.3 billion cattle alive now, and that free-ranging cows produce about 650 liters per day.[160] The Australian Commonwealth Scientific and Industrial Research Organisation developed a vaccine undergoing tests that would reduce methane emissions from sheep and cattle by about 20%.[161] If it is approved, it should be a win-win situation—methane will be reduced, and the energy that went into methane generation will go instead into meat and milk.

Landfills emit one-sixth of all human-generated methane.[162] Reservoirs are significant sources of methane emissions and are becoming more so.[163]

Methane has increased substantially; analysis of air bubbles in glacial ice indicates that the atmospheric concentration was a steady ~0.7 μmol/mol from 27,000 years ago up to 500 years ago.[159] By about 1985, it had risen to

FIGURE 17.11

Methane concentrations from 1860 to 1994.
(D. I. Stern and R. K. Kaufmann, "Annual estimates of global anthropogenic methane emissions: 1860–1994," in *Trends Online: A Compendium of Data on Global Change* (1998). Carbon Dioxide Information Analysis Center, Oak Ridge National Laboratory, U.S. Department of Energy, Oak Ridge, Tenn., U.S.A)

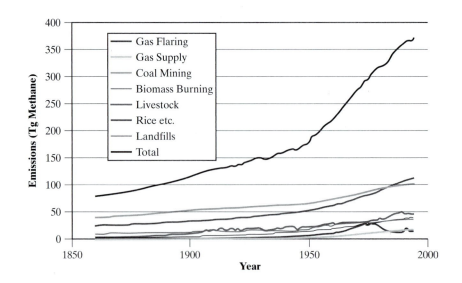

1.25 µmol/mol,[143,154,155,164,165] and by 2001, the concentration of methane was 1.75 µmol/mol.[166] The methane concentration increased by $1.1 \pm 0.2\%$/yr from the 1940s to the 1980s;[164] but the rate of rise has slowed since 1983 (from ~15 nmol/mol/yr to the present value of ~5 nmol/mol/yr).[166,167] Still, a doubling has occurred between 1900 and the present (the approximate atmospheric concentration is 1.75 µmol/mol). Methane may cause 15% to 20% of global warming in the next 50 years (see Table 17.4 and Figure 17.11).[168] However, it has been suggested that it is possible that the rise in methane could be halted, showing how global warming can be slowed.[152,169]

The total mass of methane in the atmosphere is 4.9 Gt.[149] Permafrost in the Arctic has about 10 Tt of methane hydrates stored in the form of hydrates.[170] During glacial times, the methane could have been trapped by the snow cover; when the ground was uncovered, the large concentrations of methane could have explosively increased the rate of warming.[171] Other evidence shows that methane emissions increase (if not explosively) under warming,[172] while another finds increased CO_2 release (a loss of 0.3 tonnes of carbon per hectare per year).[173] The fact that methane levels decreased during the 1982–1983 El Niño by some mechanism not yet understood, as well as the as yet unexplained 1990s slowdown in the methane growth rate, illustrates our ignorance of how methane enters and leaves the atmosphere.[174]

Many additional trace gases have been identified, as can be seen in Table 17.4. The fact of identification (that is, awareness) has led, in some cases, to rapid decreases in emissions, as occurred for methyl chloroform and the CFCs.[175] The net CFC forcing, now high at 0.85 W/m^2 even though there is only about one CFC molecule for every 100,000 molecules of CO_2 in the atmosphere, should have decreased by the year 2050.

The Intergovernmental Panel on Climate Change (IPCC) was set up under the auspices of the United Nations to address the threat of additional greenhouse warming (see **Extension 17.1**, *The IPCC*). The IPCC consensus is that the CO_2 increase is responsible for about half the additional warming.[82] They have gathered many strands of evidence, some of which is given in Table 17.4, which describes the growth rates of the greenhouse gases and their effect relative to the strength of the infrared absorption by carbon dioxide (their "global warming potential").[200,217] More detailed information on CO_2 absorption is gathered in Reference 67.

The greatest net acceleration in climate forcing occurred around 1980 at 5 W/m^2/century. By the turn of the millennium, this increase in growth had been reduced to 3 W/m^2/century.[152,220] There is a possibility of decreasing the rate still farther, and even at some point, of beginning to roll back the increases.

The effects of the warming observed are providing a "fingerprint" of the culprit—humanity. In simulations using many different models for North America, observed trends are tagged as due to human influence.[221] Ocean sea level pressure has responded to human emissions of greenhouse gases and sulfate aerosols.[222] Root et al.[223] and Parmesan et al.[224] showed the effects of those fingerprints on other species sharing Earth with us.

GCMs and Past and Future Warming

We have no crystal ball to see the future. The only way we can tell what will happen is to try to predict it. This is impossible, because we cannot take an Earth and

TABLE 17.4

Global warming potential of greenhouse gases

	Preindustrial concentration	Present concentration	Concentration units	Relative forcing per molecule	Lifetime (years)	Global warming potential (per molecule) (time horizon in years)		
						20	**100**	**500**
Carbon dioxide (CO_2)	280	370	$\times 10^{-6}$	1	~250	1	1	—
Methane (CH_4)	700	1760	$\times 10^{-9}$	21	12b	62	23	7
Nitrous oxide (N_2O)	285	312	$\times 10^{-9}$	206	114b	275	296	156
Hydrofluorocarbons								
HFC-23 (CHF_3)					260	9400	12,000	10,000
HFC-32 (CH_2F_2)	—	1.6	$\times 10^{-12}$		5	1800	550	170
HFC-41 (CH_3F)					2.6	330	97	30
HFC-125 (CHF_2CF_3)					29	5900	3400	1100
HFC-134 (CHF_2CHF_2)					9.6	3200	1100	330
HFC-13138 (CH_2FCF_3)	—	—	$\times 10^{-12}$	9550	13.8	3300	1300	400
HFC-143 (CHF_2CH_2F)					3.4	1100	330	100
HFC-65 (CF_3CH_3)					52	5500	4300	1600
HFC-152 (CH_2FCH_2F)					0.5	140	43	13
HFC-78 (CH_3CHF_2)					1.4	410	120	37
HFC-161 (CH_3CH_2F)					0.3	40	12	4
HFC-227ea (CF_3CHFCF_3)					33	5600	3500	1100
HFC-236cb ($CH_2FCF_2CF_3$)					13.2	3300	1300	390
HFC-236ea (CHF_2CHFCF_3)					10	3600	1200	390
HFC-236fa ($CF_3CH_2CF_3$)					220	7500	9400	7100
HFC-245ca ($CH_2FCF_2CHF_2$)					5.9	2100	640	200
HFC-245fa ($CHF_2CH_2CF_3$)					7.2	3000	950	300
HFC-365mfc ($CF_3CH_2CF_2CH_3$)					9.9	2600	890	280
HFC-43-10mee ($CF_3CHFCHFCF_2CF_3$)					15	3700	1500	470
CFCs								
CFC-11 ($CFCl_3$)	0	275	$\times 10^{-12}$	12,400	50	5000	4000	1400
CFC-12 (CF_2Cl_2)	0	515	$\times 10^{-12}$	15,800	100	7900	8500	4200
CFC-113 ($C_2F_3Cl_3$)		82	$\times 10^{-12}$		85			
HCFCs								
HCFC-22 (CF_2HCl)	0	117	$\times 10^{-12}$	10,700	13	4300	1700	520,000
HCFC-141b ($C_2FH_3Cl_2$)	—	3.5	$\times 10^{-12}$		9	1800	630	200
HCFC-142b ($C_2F_2H_3Cl$)	—	6.8	$\times 10^{-12}$		20	4200		2000
HCFC-123 ($C_2F_3HCl_2$)				9920		300	93	29
HCFC-124 (C_2F_4HCl)				10,790		1500	480	150
Fully fluorinated species (includes perfluorocarbons)								
SF_6	0	4.2	$\times 10^{-12}$	38,400	3200	15,100	22,200	32,400
CF_4	40	80	$\times 10^{-12}$	5460	50,000	3900	5700	8900
C_2F_6	0	2.6	$\times 10^{-12}$		10,000	8000	11,900	18,000
C_3F_8					2600	5900	8600	12,400
C_4F_{10}					2600	5900	8600	12,400
$c\text{-}C_4F_8$					3200	6800	10,000	14,500
C_5F_{12}					4100	6000	8900	13,200
C_6F_{14}					3200	6100	9000	13,200
SF_5CF_3	0	0.12	$\times 10^{-12}$	42100	~1000		~17,500	

(Continued)

TABLE 17.4 (*Continued*)

	Preindustrial concentration	Present concentration	Concentration units	Relative forcing per molecule	Lifetime (years)	Global warming potential (time horizon in years)		
						20	100	500
Ethers and halogenated ethers								
CH_3OCH_3					0.015	1	1	≪ 1
HFE-125 (CF_3OCHF_2)					150	12,900	14,900	9200
HFE-134 (CHF_2OCHF_2)					26.2	10500	6100	2000
HFE-65 (CH_3OCF_3)					4.4	2500	750	230
HCFE-235da2 ($CF_3CHClOCHF_2$)					2.6	1100	340	110
HFE-245fa2 ($CF_3CH_2OCHF_2$)					4.4	1900	570	180
HFE-254cb2 ($CHF_2CF_2OCH_3$)					0.22	99	30	9
HFE-7100 (C_4F9OCH_3)					5	1300	390	120
HFE-7200 ($C_4F9OC_2H_5$)					0.77	190	55	17
H-Galden 1040x ($CHF_2OCF_2OC_2$)(F_4OCHF_2)					6.3	5900	1800	560
HG-10 ($CHF_2OCF_2OCHF_2$)					12.1	7500	2700	850
HG-01 ($CHF_2OCF_2CF_2OCHF_2$)					6.2	4700	1500	450
Other								
Chloroform ($CHCl_3$)						15	5	1
Methyl chloroform (CH_3CCl_3)	0	160	$\times 10^{-12}$		5.4	360	110	35
Methyl bromide (CH_3Br)		12	$\times 10^{-12}$		1.3			
H-1211 (CF_2ClBr)		2.5	$\times 10^{-12}$		20			
H-1301 (CF_3Br)		12	$\times 10^{-12}$		65	6200	5600	2200
Carbon tetrachloride (CCl_4)	0	98	$\times 10^{-12}$		42		1400	
Sulfate	25	24–29	$\times 10^{-9}$		hours		20	

Note: Concentrations are absolute portions (mole per mole) and are given as numbers (in columns 1 and 2) times the "concentration units" in the third column; thus, the preindustrial CO_2 concentration is $280 \times 10^{-6} = 2.8 \times 10^{-4}$; that is, 2.8 out of every 10,000 molecules in the atmosphere were CO_2.

Source: Reference 158, Table 1; Reference 185; Reference 204; Reference 212, Table 3 or Reference 211, Table 6.7; Reference 218, Reference 219, Tables 1, 3

see what happens in speeded-up time. We have to make models—mathematical models. These are discussed in more detail in the **Extension 17.2,** *Modeling in Science.* Without the models, there would be no way to peer into the cloudy future. But all models have drawbacks that must be recognized and factored in when talking about what is being predicted.

Warming is incontrovertibly historically connected with increases in carbon dioxide concentration. Figure 17.12 shows carbon dioxide concentrations over the past 150,000 years, as inferred from the Vostok ice cores from the Antarctic and the correlation with temperature.[185]

So-called **general circulation models** (GCMs) of the atmosphere, which also incorporate the ocean, were first developed by Manabe and Wetherald in 1967.[265] They indicated that a doubling of carbon dioxide would cause an average temperature increase of 2 °C to 3 °C overall, with a temperature rise of 8 °C to 10 °C near the poles.[266] There have been more detailed analyses carried out since 1967 using a variety of GCMs,[231–233] some are one-dimensional and exclude the oceans, some are fully three-dimensional and include oceans and clouds more realistically. All obtain similar results: the doubling of the preindustrial value of the CO_2 concentration to 600 μmol/mol will cause a rise in global mean temperature of 3°C ± 1.5°C. An atmospheric profile typical of such models is shown in

FIGURE 17.12

The Vostok ice cores from Antartica show a correlation between methane and carbon dioxide trapped in air in Antarctic ice and the local temperature over the last 160,000 years.
(*IPCC Summary Report*, © 1990 Cambridge University Press)

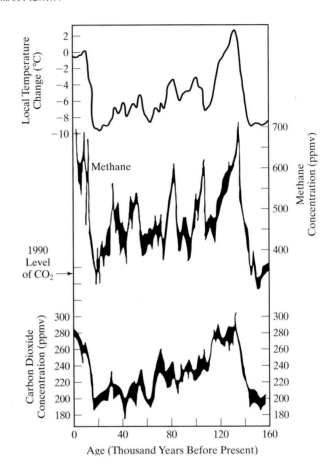

Figure 17.13 [265] and Figure 17.14.[266] The current consensus is that 3 °C is the best value of that temperature rise.[267]

IPCC Conclusions

The IPCC has worked with models to come to a consensus, which is explained in **Extension 17.1**. Continued research has not altered the IPCC conclusion that warming has occurred since the late nineteenth century,[268,269] though the size has increased from 0.45 ± 0.15°C to 0.6 ± 0.2°C.[212] The current rate of temperature rise is 0.17 °C per decade.[134] Their major statement in their Second Assessment Report is that the "balance of evidence suggests a discernible human influence on global climate."[204]

The IPCC Third Assessment Report says "new and stronger evidence" exists "that most of the warming observed over the last 50 years is attributable to human activities" and "is likely (66% to 90% chance) to have been due to the increase in greenhouse gas concentrations. ... The current rate of increase [of CO_2] is unprecedented during at least the past 20,000 years."[211] It also warns that "Anthropogenic climate change will persist for many centuries."[211] The warming range quoted for 1990 to 2100 is 1.4 °C to 5.8 °C, a dramatic upping of the high temperature from the Second Assessment Report's 0.8 °C to 3.5 °C.[211] This change represents a better assessment of scientific uncertainties.[270] The IPCC median predicted warming rate is fivefold what has occurred in the past (~0.6°C/decade versus 0.1 °C/decade).[213]

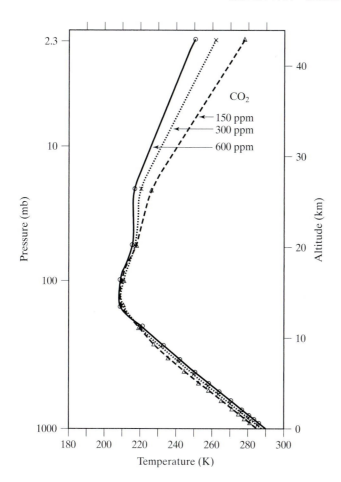

FIGURE 17.13

The temperature of the atmosphere varies with altitude. The effects of carbon dioxide increases may be calculated in general circulation models of the atmosphere. Effects depend on altitude.
(S. Manabe and R. T. Wetherald, "Analytical study of the evolution of an amplifying baroclinic wave. Part II: Vertical motions and transport properties," *J. Atmos. Sci.* **24**, 241 (1967), © 1967 by the American Meteorological Society)

The effect on the global surface temperature for a certain change in forcing (due to, say, a doubling of carbon dioxide from preindustrial times, which would lead to a forcing of 3.5 to 4.0 W/m²)[241] may be written

$$\lambda \Delta F = \Delta T_s,$$

where λ is the "**climate sensitivity**" and ΔF the change in forcing. Values of λ are the products of models (and of comparisons to earlier climates). Climate sensitivities from earlier times were estimated by the CLIMAP paleoclimate reconstruction. The evidence is of a 3 °C to 5 °C cooling in the tropics, for an implied climate sensitivity of 2.5 °C to 4 °C for the ice age.[271] Paleoclimate sensitivity in the last glacial maximum (21.5 kyr BP) was estimated by Hoffert and Covey and leads to 2.0 ± 0.5 °C, while for the Middle Cretaceous maximum (~100 Myr BP) they found 2.5 ± 1.2 °C.[272] These values are not far from the current prediction of 3.0 ± 1.5 °C.

In 2001, the Bush administration asked the National Academy of Sciences to answer a series of questions about climate change. Their response was that "Despite the uncertainties, there is general agreement that the observed warming is real and particularly strong within the past twenty years.... The committee finds that the full IPCC Working Group I (WG I) report is an admirable summary of research activities in climate science, and [that] the full report is adequately summarized in the Technical Summary."[216]

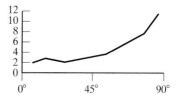

FIGURE 17.14

Estimated effect, by latitude, on surface temperatures consequent upon a doubling of atmospheric carbon dioxide levels.
(S. Manabe and R. T. Wetherald, "The effects of doubling the CO₂ concentration on the climate of a general circulation model," *J. Atmos. Sci.* **32**, 3 (1975), © 1975 by the American Meteorological Society)

The Role of Clouds, Water, and Aerosols

The role of clouds is not completely understood and is mixed up with the attempt to understand aerosols as well. Water vapor is the most common greenhouse gas in the atmosphere, and it must be better understood before the models can give better results. A good feature is the growing stream of data on atmospheric properties taken over decades at one point on Earth's surface. These data can be used to winnow models that cannot reproduce their richness. **Extension 17.3**, *GCMs' Problems* addresses the difficulties outlined here in more detail, as well as other difficulties. It is still an open question whether aerosols cause overall warming or overall cooling. In addition, the extension deals with other uncertainties in the models. Held and Soden believe that the climate sensitivity problem is the most important one that needs to be solved,[289] while Hansen et al. put the finger squarely on climate forcings as the most important unknown.[220] A problem for GCMs is the scanty data on oceanic uptake of greenhouse gases; there are just three sites giving long-term data, only one in the Northern Hemisphere (in the Atlantic by Bermuda).[274] Without comparison data, it is hard to decide which models better handle oceanic predictions.

Radiation Balance Revisited

Consider again radiation balance. Radiation balance means that, in the long term, all energy absorbed is radiated. Any forcing of the system, ΔF, leads to a change in temperature, ΔT, such that the balance is preserved. The connection is the climate sensitivity, λ:

$$\Delta T = \frac{1}{\lambda} \Delta F.$$

A change in the forcing F may arise from many sources: changes in emitted flux, secondary changes in albedo or cloudiness, anthropogenic change, and so on. Early satellite measurements gave 2 $W/m^2/K$,[277] while looking at paleoclimate leads to a value around 1.33 $W/m^2/K$, which is about the sensitivity found in GCMs ($\Delta T = 3 \pm 1.5\,°C$ for the forcing of doubled CO_2, 4 W/m^2).

Clearly, more fine-tuning of the models is necessary. Recent comparisons of GCMs by Cess et al.[232,233] found general agreement among all the different models except for the effect of clouds and greenhouse forcing. In another comparison, the doubling of CO_2 led to a threefold discrepancy among the models' climate sensitivity predictions.[235] This situation has improved (see **Extension 17.2**).

17.5   EFFECTS OF TEMPERATURE CHANGES FROM INCREASED CO_2 LEVELS

If there is additional radiation trapped in the future, we should expect that the mean Earth temperature should increase. The effect should already be apparent, according to the models. There is also other circumstantial evidence of greenhouse warming. It is even possible that, without the greenhouse warming caused by human-generated CO_2 and trace gas releases, Earth now could have been undergoing cooling comparable to that of the Little Ice Age.[191] Before the scientific evidence of warming had grown clear, there had been concern about the possibility of Earth leaving the current interglacial and entering another ice age climate, as discussed in Chapter 16.

Changes Expected from Increased CO_2 Levels

Weather Changes

In the warm Altithermal era, from about 5000 to 10,000 years ago, rain decreased by 10% to 25% over what is now the U.S. Midwest[343] while temperature rose 0.5 °C to 2.0 °C. Africa was much wetter; Lake Chad had 10 times its present area.[344] Similar patterns could occur in our warmer world.

Some facts about warming already observed are known. Most of the warming that has taken place over North America, the former Soviet Union, and China took place at night.[268,345,346] That occurred, it is thought, because of the daytime cooling effect of aerosols and their effect on clouds; most aerosols are found in the Northern Hemisphere. The Southern Hemisphere has so far been more affected by warming than the Northern.[347] The GCMs that do not take aerosols into account predict, contrary to observation, that the Southern Hemisphere warms more slowly and the Northern Hemisphere warms more rapidly.[348] Models that include aerosols predict decreasing ground-level temperatures and increasing stratospheric temperatures, as has been observed in Europe and the eastern United States.[349] Clouds over North America increased by 10% during the twentieth century, either due to concentrations of aerosols or human-enhanced greenhouse warming.[191,350]

Climate models[63,232,233] also predict a larger increase in rain and runoff at high latitudes. Rainfall could increase in summer rainfall areas and decrease in winter rainfall regions.[351] Current growth of greenhouse gases is modeled to cause an increase in frequency and severity of droughts, with greatest effects in the subtropics and midlatitudes. Between latitudes of 35° and 45°, there is predicted to be less rain and more evaporation.[40,352] There will also be a greater frequency of extreme wetness—more thunderstorms, for example.[353]

Europe has experienced many more windstorms since the early 1990s; these may be a harbinger of a warmer greenhouse world's weather.[354] Flooding after storms seems to have been increasing.[355] There will also be more destructive tropical cyclones.[353] By the end of the century, global warming effects could increase hurricanes' intensity by about 20%. Near-storm rainfall would also increase substantially.[356] Other estimates based on August mean conditions over tropical oceans predicted by a GCM with doubled CO_2 give a 40% to 50% increase in destructive hurricanes.[357]

Lightning activity increases strongly with the temperature of boundary-level air.[358] Lightning should increase by 32% by 2060, with increased risk of forest fire as forests migrate north, leaving drying forests behind.[359]

Agricultural Changes

When the historical record shows decreases of 1 °C in temperature, food production drops because of a decrease of two weeks in the growing season.[360] The response to temperature increases of 1 °C is not so clear, but the record shows that the climate was wetter in North Africa and India in such times.[347] It also shows that a 1 °C warming caused a 10% decrease in precipitation and reduced the Colorado River flow by 25%.[361] A schematic representation of the changes is shown in Figure 17.15.[347] Russia might do somewhat better and the United States somewhat worse under the inevitable changes (see **Extension 17.4**, ***Effects of a Warmer Climate on Societies and Ecosystems***). If CO_2 doubles, what are

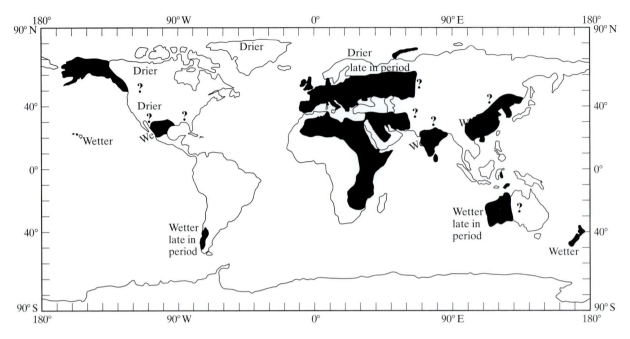

FIGURE 17.15

Estimated rainfall patterns, 4000 to 8000 years before the present, when the mean annual temperature was *several degrees warmer* than now. Such a pattern could well develop in the near future as the world mean temperature rises by several degrees.
(W. W. Kellogg, "Predicting the climate," in J. Gribben, ed., *Climate Change* (New York: Cambridge University Press, 1978), © 1978 by Cambridge University Press; reprinted with permission of Cambridge University Press)

now the fruitful areas of the Mediterranean, the United States, and Russia will probably be more desiccated; northern Canada, Argentina and Chile, and Siberia will warm up and become wetter.[69] The boundary between permafrost (or tundra) and normal soil moves north by 100 km to 200 km for each degree Celsius of temperature increase.[532] Much more of Canada should be able to grow wheat, and wheat production in the southern Great Plains will decrease.[533]

In a simulation covering ~75% of current world production, Rosenzweig and Perry used two different GCMs to predict changes.[534] The crops studied were limited to the most important grain crops. They also modeled world food trade patterns and crop growth in 18 countries. Crop yields in the study are reduced 11% to 20%,[361,532] but with CO_2 effects (closing stomata, fertilization), the reduction is only 1% to 8%; the largest negative changes occur in developing countries. Though the effect was worse in the north, the effect on crops was more severe in the south, because the basic subsistence crops did not fare as well. Adaptation in developing countries will do little to redress their economic disparity relative to developed countries. Historically, changes in grain available for world trade of about 1% have sparked price increases (see Chapter 24).[425] Nevertheless, on the basis of the study, no disaster for global food production is in sight solely from warming.[534] However, another study implies that pollution combined with warming could have a devastating impact on the world food supply.[535] The expanded production of NO_x by power plants and industry will lead to increased ozone pollution at ground level and adversely affect the industrial countries that currently produce most of the world's grain. The outcome could be a 10% to 35% drop in grain production by 2025.[535]

The U.S. corn belt will move by about 175 km south–southwest or north–north-east for each 1 °C change in the mean daily temperature during the growing season.[536] Of course, although the warming would permit growing in more northern latitudes, these soils are generally poorer than the ones that would no longer be usable. Farmers would change their crops and methods, one would hope, or, more likely, just move north. See **Extension 17.4**.

The species of plants growing would change as well. Plant communities persist long after the climate change has wiped out favorable habitats; about two millennia separate climate changes from the consequent changes in vegetation.[537] After a glacial episode when climate changes, vegetation shifts. A meta-analysis showed a shift to earlier spring and consequent effects on flora and fauna.[223,538] Dispersal rates are crucial to a species' ability to colonize new habitat.[537] The average migration during the ice age was about 30 km/century.[420] The biota in some areas remained behind during ice ages in refuges (refugia), from which they spread out again after the end of the ice ages.[539] Beech trees will be worst hit by rapid warming. In the last ice age, beech advanced 10–45 km/century, while spruce, the fastest disperser, moved 200 km/century.[540] Each 1 °C temperature change causes a change in range of 100–150 km.[420,540] Around the Great Lakes, the contours of constant temperature would be displaced north by 300 km in 100 years, an order of magnitude faster than trees move.[223,420,492,538] There are new barriers to migrations that did not exist in former times.[420] An increase of 1 °C also increases the rate of forest respiration (the decay of biomass, with release of carbon dioxide), as extrapolated from measurements made on Long Island, by 3% to 25%.[76] Data from tree inventories on the tropical continents show a large increase in the annual turnover of trees. Carbon dioxide hastens both growth and death of trees. Massive trees are being replaced by lianas, vines, which are less massive and take less CO_2 out of the air.[541,542] The new forests will take up less CO_2, and shade-tolerant species will decline.[543] This is still another feedback loop accelerating the warning.

Effects on Living Things

The ranges of many animals are linked to vegetation. Generally, mammals respond quickly to changes, whereas other parts of the ecosystem do not.[544] How will plants and animals respond to greenhouse temperatures? In the Rockies, fossil beetle evidence shows that up-mountain migration occurred in times of warming, and down-mountain migration occurred in cooling times.[545] Species from high latitudes are more adaptable and colonize new areas more easily than do tropical species,[546] as is especially easily seen in mountainous areas.[547] Even so, many extinctions occurred.[545] See **Extension 17.4** for more information about the responses of plants and animals to warming.

Sea Level Rise

The ocean off California has warmed by 0.8 °C in the top 100 m, and temperature rise has penetrated to 300 m.[548] The world ocean has warmed by 0.06 °C over the recent past. As explained in **Extension 17.5**, *What Does Sea Level Rise Mean?*—this means that the ocean volume has expanded, so the sea level will have to rise as a result, even if glacial melting were neglected.

Rising sea level will increase surface runoff and decrease groundwater discharge due to a rising water table.[605] According to a model calculation, a 20 cm sea level rise will double runoff and reduce groundwater discharge by half.[605] Water is stored in reservoirs and channeled into aquifers, removing water that could raise sea level. On the other hand, human activities have increased water diversion from underground reservoirs. Humans account for at least one-third of the observed sea level rise.[589] The average rise due to human influence since 1960 is 0.54 mm/yr.[589]

Rapid sea level change is associated with climatic change. Sea level has changed rapidly at the beginnings and endings of glacial episodes.[595] About 95,000 years ago, sea levels rose by 15–20 m inside a century.[590] If the heating is prolonged, the Greenland and Antarctic ice sheets (about 8% and 90%, respectively, of the total ice in the world) would melt over a period of centuries, bringing the sea level up another 60 m or so.[68] This would cause an even larger shift, but because the change would take place slowly, it would probably not be socially wrenching. The shorelines of the continents would be *very* different if all the ice melted.

Effects on Societies in the Past

We do not know exactly what changes will occur in society in response to warming. However, some information is available on past responses to past changes. Some millions of years ago, changes in the climate in Africa caused ecosystem changes that spurred hominid evolution, so we humans would not be here without that response.[606] Societies around the world have been affected by past climate changes. Civilizations have risen and fallen based on the weather patterns. More detail on such effects can be found in **Extension 17.4**.

Economic and Political Effects

Caldiera and Kasting[607] note that "climate forcing by CO_2 is more difficult to reverse on human timescales than is forcing by other greenhouse gases. Continued high rates of CO_2 forcing will commit us to many centuries of warmer climates."

As with any human activity, there will be winners and losers. Northern cities would be cheaper to heat, for example, and southern cities more expensive to cool. The EPA has estimated annual *additional* electricity costs of $33 to $73 billion.[496] Some island states may vanish.[608] If the sea rises, social action will have to be taken, which will cause conflict. Already, economists are calculating that the human-enhanced greenhouse will not be any particular problem, without realizing that they cannot quantify everything with a dollar value. A shocking report by a prominent economist revealed that only the natural scientists on a panel were concerned by the human tragedy that might occur; the economists were blasé.[609] The scientists had "profound concerns" about severe economic consequences. The economists, whose only concern was with costs, did not share the scientists' concern, nor did they have much concern about the ability of ecosystems to adapt to warmer times. The scientists were very worried about the possibilities for irreversible deleterious changes.[609] Perhaps the old saying, "ignorance is bliss," is relevant.

Human migration may arise as a result of greenhouse warming effects. The developing countries will be especially hard hit. Egypt's Nile delta will suffer from encroaching salt water. As much as 5 million km^2 could be threatened by a rise of 1 m.[608] China, India, and Bangladesh will all have problems.[608] Of course, the developed world will have problems, too. These nations will command more resources to combat the effects than the poorer countries.

It is ironic that, if such extreme climate changes do come to pass, the United States will have been mostly responsible for its own predicament. The United States burns a quarter of all the coal burned in the world, and the former Soviet Union and Canada together burn another quarter.[610] In view of the possible advantages to be gained, perhaps the former Soviet Union and Canada would want to encourage continued reliance on fossil fuels, but the same could not be said for the United States.

17.6   MINIMIZING THE IMPACT OF GREENHOUSE WARMING

The cost of the coming warming is expected to be substantial.[611] The United Nations has estimated the cost at about $300 billion per year to pay for losses from the greater number and increased intensity of storms, loss of land to the sea, problems with fisheries, and so on.[612] Who will bear the cost of the liability? Some industries will be hit harder than others.[613,614] Choices seem limited to government (society as a whole pays), the insurance industry, or the courts (find and sue the cause).[615] Insurance companies are pondering how to deal with the increase in claims expected in the warmer world.[613,616,617]

Natural disasters are increasing in scope and frequency. This has happened because of increasing population and infrastructure in exposed areas.[617] A true disaster threatens four vital societal functions: biological survival, social order, meaning of a person's life, and people's motivations.[618] Some, if not all, of these are threatened by greenhouse warming. To deal with any hazard, communities can[618]

1. Modify the hazard,
2. Prevent or limit impacts,
3. Move or avoid losses,
4. Share losses, or
5. Do nothing and bear any losses

The future responses of communities are conditioned by their experiences of past hazards.

What is the world doing about the greenhouse warming that most scientists believe is coming? Until recently, choice 5 seemed to be the preferred option. However, with worldwide changes in political climate after the Rio Environmental Summit of 1991 and with the weight of the scientific opinion of IPCC, some of the other options, especially options 1, 2, and 3, are beginning to be pursued. We discuss some of these next, in **Extension 17.4**, and in the remainder of the book.

Consider option 2. One primary issue is coastal defense (or nondefense). Planning is required decades in advance for construction of seawalls, and so forth.[619] The city of Charleston, South Carolina, is thinking about the rise in sea level in planning construction of new sewers.[620] The Netherlands, at war with the sea already, is also planning for the future.[619]

Option 1 is now receiving a great deal of thought. Wider use of nuclear energy (Chapters 18–20) or development of viable solar energy (Chapters 21–24) would slow greenhouse warming by decreasing emissions of CO_2.[150,621] Conservation, alternative fuels, and prevention of waste of energy are necessities (Chapters 9,

11, and 25). More recycling, better urban planning, and use of automobiles with greater efficiency all can help.[622] It has been estimated that emissions of CO_2 can be reduced 35% below that of 1987 levels within 25 years without need for any technological breakthrough.[623] Reduction in energy intensity in industry and end use could show spectacular real savings.[623–625]

Perhaps the hardest change will be a wrenching of our minds to consider other possibilities. The world is hooked on fossil fuels. Continued use will lead to more intense greenhouse warming. Even a drastic reduction to half of today's use of fossil fuel by a century from now is not likely to prevent the CO_2 concentration from climbing to about 450 μmol/mol without some major change.[163,164] As Kasting and Walker[626] point out, though, "(m)odest decreases in fossil fuel consumption cannot stabilize atmospheric CO_2," but they do slow onset by warming.

Even in this case, new ways of thinking are possible. Caldeira et al. suggest we must move massively to new carbonless technologies if we are to be able to manage climate change.[627] Hoffert et al.[628] identified many likely candidates. Not all fossil fuels contribute the same amount of CO_2. Coal, of which the world supply is most abundant, is most productive of CO_2. This occurs because the chemical formula for natural gas is CH_4; for oil, it is about CH_2; for coal, it is about $CH_{0.8}$. The long chains of carbon–carbon bonds reduce the number of non-carbon atoms in the molecules in coal and oil. Fuel switching (to natural gas) or techniques for burning fuel more cleanly (Chapter 14) can help reduce emissions. New technology rather than extension of lifetime is an important method of decreasing all emissions.[629]

The National Academy of Sciences Panel on Policy Implications of Greenhouse Warming pointed out that there are many measures that could be taken that would make sense environmentally and economically.[176] These are many "no regrets" measures society can take, some that even lead to overall savings of money and should be taken first, because they are easy to carry out.[176] Indeed, Pacala and Socolow identified many "stabilization wedge strategies" that allow humanity to delay major disruption for the next half-century and temporarily "solve" the carbon dioxide problem. All involve readily available technology.[630] The "no regrets" policy will not solve the greenhouse problem, but it is a start. Table 17.5 shows some of the suggested measures. Efficiency is the watchword, and it is happening; a fairly conservative expectation is that light-duty vehicles will be 40% more efficient by 2016.[631]

TABLE 17.5

Effects of reductions in carbon dioxide

	CO_2 equivalent saved	Cost ($/Mt)
Industrial end use	24	−60
Transportation	115	−22 → +530
Electric supply technology	1780	50 ± 20
Halocarbon use	1409	$1.4^{+1.6}_{-0.5}$
Domestic agriculture	223	2^{+3}_{-1}
Reforestation	242	7^{+3}_{-4}

Note: A negative cost is a benefit.

Source: National Academy of Sciences study on Policy Implications of Greenhouse Warming, Reference 143.

CO$_2$ Scrubbing and Other Removal Options

Another possible approach is **CO$_2$ scrubbing,** in which the gas is removed from the effluent before release. There are several possibilities discussed in **Extension 17.6**, *Methods of Scrubbing CO$_2$*.

If the world finally finds it must live with the enhanced greenhouse effect, it would be nice to have time during which to adjust. The best currently known prospect for supplying such a pause is the growing of billions of trees. The planting of trees is a *certain* way to delay the onset of the greenhouse era by decades. This idea was first proposed in 1976 by F. J. Dyson and G. Marland.[719] Young trees can absorb large amounts of CO$_2$ as they grow ($\sim$7 t/ha/yr). Older trees store the CO$_2$ for long periods, and many species live for up to 100 or more years. The problems with the idea are *political* not practical. More details about carbon sinks and sources from the biosphere are found in **Extension 17.7**, *Planting Trees*.

Additional options are technical fixes. Lightweight mirrors shading a fraction of Earth could undo greenhouse warming.[787] The "shades" would be made of a very thin material. The result would be a human-caused decrease in solar input (see Chapter 25). Jets can produce particulates to counter warming. Turning 1% of fuel into 1.5 Mt of soot in the stratosphere would change the albedo by 1% and offset 1.5 °C of greenhouse warming.[636,788] Residence times should be about 10 years for these particles. Sulfur aerosols could also be used.

The Kyoto Protocol

Hansen et al. have suggested that it is much harder to reduce carbon dioxide emissions than all other "trace gas" emissions. Their goal is to stop—and even reverse—the growth of trace air pollutants, especially soot and tropospheric ozone (and precursors, including methane). All these gases increase in every IPCC scenario. Therefore, a good strategy, they say, is to attack these gases vigorously, while keeping CO$_2$ emissions from growing.[135] They anticipate that advances in technology will allow CO$_2$ emissions to begin being reduced before 2050.[789] Others agree that adding trace gases reduces costs of compliance.[790,791] However, De Leo et al. argue that the costs of Kyoto Protocol compliance are not so high if *all* costs are accounted for.[792] Another approach would be somehow to reward the "first adopters," who test out systems for reducing emissions.[793]

There are unrecognized health benefits as well. Cifuentes et al.[794] noted that deaths from air pollution rank among the world's top 10, and that much evidence exists that hospital admissions decline when air pollution decreases (see Chapter 14). They propose that using existing methods for reducing greenhouse gases would lead to reductions in particulate matter, which as we saw in Chapter 14 is a very nasty material. They estimate that reduction in greenhouse gases could prevent 64,000 premature deaths, 65,000 chronic bronchitis cases, and 37 million person-days of reduced activity or need to remain home from work.[794] This is an incentive to attack pollutants in a way that allows the "co-benefit" of CO$_2$ reduction. Adoption of such an unarguably reasonable approach is worthy of itself, and the developing countries will be the biggest beneficiaries.

Some economic measures can help, too. There have been proposals to make a free market in emissions trading, which would let the market determine the most cost-efficient reduction solution.[795] For example, New England Electric is paying Innoprise, a Malaysian forest production company, to leave trees in place.[796] An economic model of a transition to a system to slow global warming indicates that a modest carbon tax could be optimal, whereas rigid emissions limits would cause severe economic dislocations. The optimal path, however, would not reduce warming significantly.[797] An EPA report suggested that a cold-turkey phaseout and a carbon tax would make the greatest decrease in warming.[798] Perhaps a mixture of regulation and tax would work best. Whatever is developed, it needs to be equitable (such as a per capita emissions cap).[799] **Extension 17.8,** *Why We Should Support the Kyoto Protocol* presents four reasons to support the Kyoto protocol—low cost "no regrets" measures, the added cost of noncompliance, equity, and stewardship.

In fact, global warming could have important consequences for U.S., and global, security. If a catastrophic abrupt change in thermohaline circulation[210,335,843,844] or sea-ice cover [845] were to occur and humanity were not prepared, many millions could starve, and there could be massive uncontrollable migrations from "loser" regions to "winner" regions. This could be a political and military nightmare, even leading to war.[846] A rerun of the "Medieval Warm Period" in the modern age "with more than 10 times as many people on Earth as in High Medieval time, could be catastrophic."[847] In light of this, it is important for the United States to become effective in reducing emissions and preparing for adaptation.[848]

Greenhouse warming may be inevitable in the long run. Fossil fuels may continue to be used until they are used up. It may be too late to prevent a significant rise in average world temperature. Stott et al. of the British Met (Meterological) Office found in their study of the deadly 2003 European heat wave that "there is a greater than 90% chance that over half the risk of European summer temperatures exceeding a threshold of 1.6 K is attributable to human influence on climate. ... On this basis, human influence is to blame for 75% of the increased risk of such a heatwave."[849] This is the first scientific identification of a climate change fingerprint on a specific weather event.

Stott et al. go on to say that the "probability of European mean summer temperatures exceeding those of 2003 increases rapidly under ... [one of the IPCC scenarios], with more than half of years warmer than 2003 by the 2040s. By the end of this century, Figure 1 shows that 2003 would be classed as an anomalously cold summer relative to the new climate, for the scenario and model under consideration."[849] They conclude it is "likely that past human influence has more than doubled the risk of European mean summer temperatures as hot as 2003, and with the likelihood of such events projected to increase 100-fold over the next four decades, it is difficult to avoid the conclusion that potentially dangerous anthropogenic interference in the climate system is already underway."[849] The adjustments to a warmer world will have to be made, since prevention of the consequences of warming is impossible.

W. W. Kellogg,[344] in looking at the future "warmer Earth," sees a balance: Some countries will be better off, others worse. Earth, he believes, will be better able to support its increased population. Despite such optimism, Kellogg still sees the grim reaper of starvation as part of a change to a warmer Earth. We can only hope that we are farsighted enough to prepare to meet the consequences that we were apparently not farsighted enough to prevent. If we finally manage to prepare, Kellogg's nightmare need not become reality.

SUMMARY

Climate has changed over the course of millennia. Climate changes for various physical reasons, but also because human beings can promote desertification by cutting too many trees, or cause overgrazing, burn fossil fuels, or change the albedo in other ways. And once the change is made, it is difficult to reverse.

Increased volcanic activity and the action of the "human volcano" in producing aerosols in the late twentieth century have decreased the impact of the carbon dioxide increase to an extent. The least expected agents of climate change have been found in the gases that are normally present in small amounts in the atmosphere. Carbon dioxide and other "trace gases" can cause Earth's mean temperature to rise through the greenhouse effect. The CO_2 that stays in the atmosphere will raise Earth's mean temperature, according to models developed by the IPCC and individual researchers (see **Extensions 17.1, 17.2 and 17.3**). The effects on humanity are still the subject of speculation, although the effects of CO_2 increase on climate are relatively certain — a lot of warming at the poles and a very small amount of warming at the Equator.

The United States may be about the same or a bit worse off overall by ending up with a smaller area available for agriculture. Some countries, now in the Third World, may make gains and others suffer losses (the predictions of loss far outweigh those of gains). Warming will be accompanied by more storms. World hunger will probably not increase, but more food will be sold by temperate countries and bought by more equatorial countries. Sea level will rise 1 m during the next century (see **Extensions 17.4 and 17.5**), causing some dislocation.

Mitigation strategies range from preventing the CO_2 from reaching the atmosphere, to squirreling CO_2 under the ocean (see more options in **Extensions 17.6**), to planting trees to soak up the CO_2 in the air (see **Extension 17.7**). This suggests we should support the Kyoto Protocol to minimize such disruptions and inequities (see **Extension 17.8**). We should hope that we will do better at managing the change than we have done in preventing it.

PROBLEMS AND QUESTIONS

MULTIPLE CHOICE REVIEW

1. If a warmer world is to live at a temperature 3.5 °C higher than at present, and the forcing is determined to be 4 W/m^2, what is the climate sensitivity (in appropriate units)?
   a. 14
   b. 3.5
   c. 2
   d. 1.14
   e. 0.875
2. What is(are) possible way(s) humans encourage desertification?
   a. Uprooting plants
   b. Herding goats
   c. Irrigating fields
   d. All of the above encourage desertification.
   e. Only (b) and (c) encourage desertification.
3. How is Earth's greenhouse effect *different* from what happens in a greenhouse?
   a. Earth does not trap infrared radiation as a greenhouse does.
   b. Plants transpire in a greenhouse but not on Earth outside a greenhouse.
   c. There is no water vapor in the air in a greenhouse, while there is on Earth.
   d. Air on Earth can move much more than air in a greenhouse, the air is not trapped.
   e. There are no differences—the situations are identical.
4. It is necessary to use models and scenarios to predict global warming because
   a. Earth has never experienced such warming before.
   b. if we pay attention to data, we can rule out the warming.
   c. laws of physics and knowledge of chemistry determine how much warming will occur.

   d. they provide the only way to "see the future."
   e. we cannot know exactly what will happen without living through it, but we must prepare.
5. The warming experienced by Earth has been recognizable, looking back, since
   a. ~1000 A.D.
   b. ~1500 A.D.
   c. ~1940 A.D.
   d. 1985 A.D.
   e. 1995 A.D.
6. Mountain glaciers are
   a. uniquely sensitive to warming.
   b. in retreat worldwide.
   c. prone to collapse.
   d. correctly described by all the above.
   e. correctly described only by (a) and (b).
7. Which of the following is *not* a greenhouse gas?
   a. Water vapor
   b. Tropospheric ozone
   c. Hydroxyl
   d. CFC-11
   e. Methane
8. If a warming rate of 0.6 °C/decade persists for another century, the mean temperature of Earth in 2100 will be about
   a. 6 °C.
   b. 21 °C.
   c. 65 °C.
   d. 75 °C.
   e. 90 °C.
9. Heinrich events occur because
   a. huge ice sheets crash into one another.
   b. a warming cycle produces so much meltwater that the ice floats out to sea.

c. the cold phase of the Dansgaard–Oeschger cycle is under way.
d. huge bursts of hydrates are released by sediments off continental shelves.
e. the climate is basically glacial.
10. What measure(s) would be most useful in reducing carbon emissions?
a. Carbon tax
b. Voluntary compliance
c. Increased SUV safety
d. Increased rice cultivation for food
e. All the measures are useful in reducing carbon emissions.

SHORT ANSWER
Explain whether the statement made is correct, and justify your answer.
11. There is no irrefutable evidence of global warming.
12. There are many no-cost measures that can be adopted that would reduce warming.
13. Nations such as Nauru and Tonga are threatened by continued warming.
14. Glaciers are thinning in high mountain regions, but there is no evidence of thinning on the Antarctic, Arctic, and Greenland ice sheets.
15. Clouds have a profound effect on climate.
16. There is no real evidence that greenhouse warming could cause abrupt catastrophic changes in climate on Earth.
17. Cultures of the past have generally been unaffected by climate changes.
18. Given the expected magnitude of coming warming, it is a good idea to invest in beachfront property in Miami.
19. All past proxy evidence of climate change is anecdotal.
20. Soot has a clear warming effect.

DISCUSSION QUESTIONS
21. What is the role, if any, of aerosols in climate? Explain.
22. What does equity have to do with the global warming issue? Explain.
23. Explain why it is plausible that adding all of the trace gases to carbon dioxide as the mix of gases to be controlled should reduce the cost of compliance with the Kyoto Protocol.
24. What "technical fixes" might be used to reduce or eliminate the effects of warming?
25. What information is available that can allow people to give quantitative measures of climate in earlier times? Give examples.
26. Explain how increasing the efficiency of cars could help contain greenhouse warming.

27. How descriptive is Bryson's "human volcano" designation for human-caused climate effects? Explain your reasoning.
28. If a tree's range can migrate 30 km in a century, and warming is expected to cause isotherms to move north- and southward 300 km in a century, what can be expected to occur? Explain. Would Scotch pine (200 km per century) be similarly affected?
29. How is it that absorption and reemission of radiation by carbon dioxide molecules can cause trapping of infrared radiation in Earth's atmosphere? Explain.
30. What evidence can you find that GCMs are getting better? For example, could the change in the IPCC prediction from $3.0 \pm 1.5\,°C$ to $1.4\,°C$ to $5.8\,°C$ be considered an improvement? Explain.

PROBLEMS
31. What is the average decadal global warming experienced by Earth in the last 150 years? How does this compare with the average decadal warming during the last 50 years?
32. What is the percentage change in the atmospheric carbon dioxide concentration from preindustrial times?
33. Determine the typical proportion of a time that is interglacial in glacial ages.
34. Find the approximate rate of increase in carbon emissions since 1950 from the data of Figure 17.4. How does this compare to the approximate rate between 1900 and 1950? What could be responsible for the differences, if any?
35. What are the values of climate sensitivity λ corresponding to the findings of Hoffert and Covey about past climates?
36. By what amount would the Earth albedo alone have to change to cause a warming of $5.8\,°C$? Is this a large or a small relative change? Explain.
37. How do carbon sources and sinks compare? Is some carbon unaccounted for?
38. Given that trees can sequester carbon at 7.5 t/ha/yr, what area of trees would have to be planted to sequester all the carbon emitted by humanity?
39. Values in Table 17.3 may have resulted from a statistical fluctuation. If the temperature distribution were random, rather than a result of warming, the warmest years should be distributed with equal probability among the decades. Suppose N of the 10 warmest years on record in the past century were in the last decade. The probability in a completely random distribution of any decade containing a warm year is 10%. Find the probability that N of the 10 warmest years cluster in one decade for $N = 1, 3, 5,$ and 7.
40. One may continue less conservatively to note that the top eight warmest years are all in the last decade of the record.
a. Determine the random probability of any one being in a decade in the past century.
b. Comment on your result.

NUCLEAR ENERGY RESOURCES AND CONSEQUENCES

PART VI

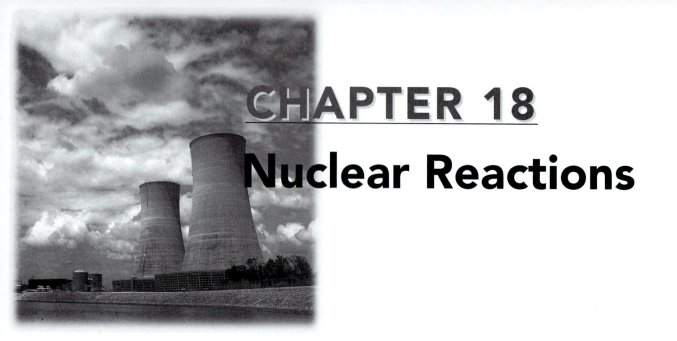

CHAPTER 18
Nuclear Reactions

In Chapters 7 and 13, we examined chemical processes that can be used to generate energy and some of the attendant consequences. Another thermal process that is important for generating electricity is nuclear fission (nuclear breakup). This occurs when the nucleus can reach a more stable state by breaking in two. Nuclear fission reactors supply about 20% of electricity demand in the United States and over 80% of France's electricity demand. This chapter provides the introduction necessary to understand how reactors work. Fusion occurs when the combination of two nuclei is more stable than the original nuclei were by themselves. There are no fusion reactors generating electricity, but fusion in stars is responsible for all the elements beyond helium. Radioactive decay, the breakup of particles by alpha, beta, or gamma emission, is exponential in character and is described in terms of mean life or half-life. The following two chapters discuss the present state of the art and the environmental concerns associated with nuclear energy.

KEY TERMS

nucleus / interaction / strong interaction / mean life / half-life / nucleon / electronvolt / mass number / isotopes / atomic number / binding energy / fissile nucleus / nuclear activation energy / chain reaction / moderator / radioactive decay / alpha particle / beta particle / gamma radiation / big bang / nucleosynthesis

---

18.1   THE NUCLEUS

Atomic Characteristics

If we examine a material on a much smaller scale than the atom, $\sim 10^{-10}$ m, we find that the atom is composed mostly of empty space. At the atom's core lies the massive atomic **nucleus**, which has a diameter measured in femtometers (10^{-15} m). If we imagine a nucleus to have a diameter of about 10 mm, the atom would have a diameter of about 1 km—100,000 times greater. If a college football stadium were an atom, a large pea at the center of the playing field would represent the nucleus.

An atomic nucleus is built from protons and neutrons. These particles can also exist outside a nucleus. A proton is the nucleus of a hydrogen atom; because the universe is mostly hydrogen, many protons exist in the universe. Neutrons are generally stable when they are inside atomic nuclei. If a neutron remains outside the nucleus of an atom, it will eventually decay. Inside a nucleus, it generally does not decay. Despite their propensity to break apart, there are many neutrons around us coming from the Sun and from the breakup of radioactive materials.

Free Neutron Decay, Mean Life, and Half-life

A free neutron is one found outside the nucleus of an atom. The neutron is unstable when it exists outside an atomic nucleus; that is, free neutrons decay. This process is caused by the so-called weak interaction, which is one of the four fundamental interactions known to exist at our energy scale (see box, *A Closer Look*: The Four Fundamental Interactions).

A CLOSER LOOK

The Four Fundamental Interactions

Particle physicists prefer to use the word **interaction** to describe what happens to subatomic particles that affect one another. The word *force*, while it has nearly the same meaning, implies to many people an "off-stage" bystander (in practice, the words "force" and "interaction" are used almost interchangeably). Experimentally, two particles do things to one another, and the result of this mutual dance is observable in changes in both particles' motions. It is not the case of one causing the effect on the other and being unaffected itself.

If dimensionless numbers are defined to characterize the respective sizes of these four basic interactions, at our energy scale, they are in the ratio

strong interaction : electromagnetic interaction : weak interaction :

gravitational interaction : : $1 : 10^{-2} : 10^{-5} : 10^{-39}$.

The relative size of the interactions depends on the energy scale and the separation of the particles (the strong and weak interactions' ranges are very short, while electromagnetism and gravitation act at arbitrarily large distances as well as small distances). It is believed that in the first moments after the universe began with the big bang, all the forces were the same size—that is, indistinguishable. At the present time, work is continuing on the task of unifying the four interactions so that that original unity can be understood, and great progress has been made in tying the first three (quantum chromodynamics, quantum electrodynamics, and the electroweak theory) together into a theory known as the Standard Model. Sheldon Glashow, Steven Weinberg, and Abdus Salam shared the 1979 Nobel Prize for the electroweak theory, which ties the weak and electromagnetic interactions together into one theory.

A further Nobel Prize was awarded in 1984 to Carlo Rubbia and Simon van der Meer for their experimental discovery of two particles (W and Z) predicted by this theory.

While the **strong interaction** is greater than the electrical force of repulsion, it acts only for separation distances less than $\sim 10^{-15}$ m. Because of the strong attraction between nucleons (protons or neutrons), nucleons can give up energy by binding together to form a nucleus. They become more stable by doing so, just as two balls rolling into a trough end up sitting at the bottom of the trough, or two chemicals making bonds release chemical energy and make themselves more stable. Nucleons can then be released from a nucleus, or balls be taken out of the trough, or the original chemicals reconstituted, only if a substantial energy price is paid. By conservation of energy, this price is exactly the energy given off in the original process.

Given that the free neutron will decay, no one can know whether some particular neutron will decay at any given time. However, we may say that some number of an assemblage of many neutrons will decay in a time interval characteristic of neutrons. (Because the description is statistical in nature, we must have a lot of nuclei to make this prediction, not just one or two.) The random decays combine for large numbers of decaying particles to produce a predictable curve called the decay curve (Figure 18.1a). The curve of the number of neutrons at any given time, $N(t)$, is predicted to be given by the relation

$$N(t) = N_0\, e^{-t/\tau},$$

where the time τ is called the **mean life** (lifetime) for the decay. Exponential decay is the inverse of exponential growth (Chapter 5). After one mean life, about 38% of the original nuclei remain; the others have become other nuclei—in the case of a free neutron, the decay products are a proton, an electron, and an electron antineutrino. The neutron mean life is about 15 minutes. The free proton has so far not been observed to decay; if it decays as predicted in some models, it has a mean life perhaps 10^{18} as long as the universe's current age.

More opaquely, the number $N(t)$ may be expressed as

$$N(t) = N_0\left(\frac{1}{2}\right)^{t/T},$$

where T is the **half-life** of the neutron. The half-life is defined as the time it takes for the number of original nuclei to fall to half the original number. The neutron half-life is about 10 minutes. Mean lives and half-lives, in the statistical sense described here, may be defined for all unstable elementary particles and nuclei.

Just as exponential decay is the inverse of exponential growth, half-lives are the inverses of doubling times. In exponential growth, it takes the same time—the doubling time—to increase from 2 million to 4 million as from 1 million to 2 million. In exponential decay, it takes the same time—the half-life—to decay from 4 million to 2 million as from 2 million to 1 million. Recall that in exponential growth, the growth rate depends on the number of things or particles at the beginning. Doubling times may be defined for all exponentially growing quantities. The number of decays of particles depends on the number of particles—the greater the number of particles, the greater the number of decays. The decay curve in Figure 18.1b shows that this is

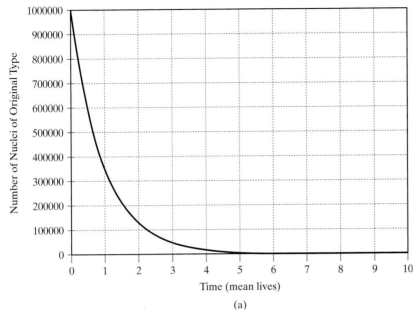

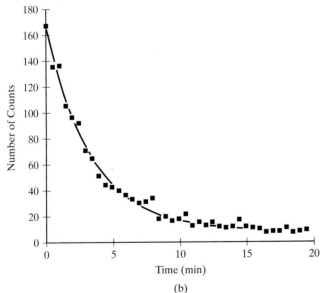

FIGURE 18.1

(a) Predicted number of radioactive nuclei of the original type remaining versus time measured in mean lives. (b) The actual decay curve for decay of ^{182m}Ba as measured by a Geiger counter. It essentially follows an exponential decay curve; the best-fit exponential curve is shown in the diagram.

true for exponential decay. Because the number of decays is exponentially decreasing, the number of particles is exponentially decreasing as well.

Protons and neutrons are almost indistinguishable. The fact that the proton has one unit of electric charge and the neutron has none is responsible for the difference between them. Because of this similarity, we often speak of them as **nucleons** in cases where the charge is not an important difference. A nucleon is a proton or a neutron; nucleons are either or both. Both have essentially the same size and mass ($m_{\text{neutron}} = 939.6 \text{ MeV/c}^2$ or 1.008664904 u, and $m_{\text{proton}} = 938.3 \text{ MeV/c}^2$ or 1.007276470 u, where u is the unified atomic mass unit, $1.66053873 \times 10^{-27}$ kg; see **Extension 18.1, *Mass in Nuclear Physics*** and the box *History of Energy*: The Electronvolt).

HISTORY OF ENERGY

The Electronvolt

Before we continue, we should define an energy unit that is convenient for discussing nuclear processes: the MeV (see the preceding discussion of free neutron decay). The size of the charge on the electron and proton, denoted e, was found to be 1.6×10^{-19} C. A charge e falling through an electrical potential difference (energy per unit charge) of 1 V (1 J/C) gains 1.6×10^{-19} J in the process. This amount of energy is defined to be 1 **electronvolt** (1 eV). It is convenient, because there are sources of potential difference (voltage) that an experimenter can use to "dial" a desired energy by setting the dial to some value. To make a 100 eV proton, for example, we can just have the proton fall through a potential difference of 100 V.

Most atomic processes take place on the electronvolt scale. The ionization energy of the electron (the energy needed to rip off the outermost electron) in a hydrogen atom is 13.6 eV (2.18 attojoules).

An MeV is a million electronvolts (160 femtojoules). Most nuclear processes take place on the MeV scale. In these energy units, the total energy release in an average fission is about 200 MeV or 30 picojoules. The energy equivalent of a proton is about $E_{proton,rest} = m_{proton}c^2 = 940$ MeV, almost a billion electronvolts (GeV). The newest particle accelerators produce particle energies a thousand times greater, over a trillion electronvolts (TeV) (about one-sixth microjoule).

Mass Number and Atomic Number

Because protons and neutrons have practically the same mass, the mass of a nucleus containing a number A of nucleons has a mass roughly A times the average proton–neutron mass. The number of nucleons is called the **mass number** and is given the symbol A. Of course, the number of protons is important for an atom, because it determines the number of electrons necessary to make the atomic charge zero. These electrons are all that we "see" of ordinary atoms, and so the number of electrons determines how a particular atom acts chemically. That is, the number of protons in an atom determines what kind of an atom it will be. Thus, atoms of the same element (same proton number) can have different nucleon numbers: helium-3 and helium-4; carbon-12 and carbon-14; and iron-56 and iron-58. Such nuclei are called **isotopes**. The number of protons is called the **atomic number** and is given the symbol Z. The number of neutrons is given the symbol N; clearly

$$A = Z + N.$$

Even though Z and A totally specify an atom (and its nucleus), it is common to give the elements names and symbols. Hydrogen has $Z = 1$, $A = 1$, and has the symbol H; helium has $Z = 2$, $A = 4$, and has the symbol He; and so on. It is customary to specify an element by using all of this information: ^A_Z(element name). Thus, ordinary hydrogen is also known as $^1_1$H; helium as $^4_2$He; and so on. See **Extension 18.2, *Isotopes***.

One interesting fact, illustrated in Figure 18.2, is that the higher Z is, the greater the number of neutrons N. Why might that be? Think about a bunch of free protons—we know protons have the same sign charge, so the protons repel

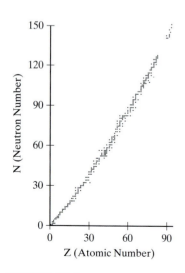

FIGURE 18.2

The neutron number, N, is plotted in terms of atomic number, Z. Note that the number of neutrons in a nucleus increases faster than the number of protons.

one another electrically; all these protons would leave each other's company. But in an atom, the protons are all squeezed together inside a nucleus, at distances less than 1 fm $= 10^{-15}$ m apart. The electrical repulsion forces between any two protons in a nucleus may be 10^{30} times the electrical force between protons that are 1 m apart.

The nucleons in a nucleus are bound together by the strong interaction, which is greater than any electrical force of repulsion when the nucleons are less than about 1 fm apart. The strong interaction between any two nucleons is attractive, while the electric force between protons is repulsive. If neutrons get between protons, separating them a bit from one another, the protons experience a weaker electric repulsion than before. This makes the nucleus more stable by lowering its overall energy. If there are two protons, it takes just one neutron to get between them. If there are three protons, it takes three neutrons. But if there are, say, seven protons, it will take more than seven neutrons to be able to fill the region between each pair of protons. So, the reason there are more neutrons than protons in a nucleus is so that there are enough neutrons to get in between the protons.

Binding Energy

When two nucleons interact, they generally "bounce" off one another because of electrical repulsion or because there is a hard core of "protection" surrounding the nucleon. When two nucleons are close, however, they can manage to be attracted to one another. By emitting some energy as light (gamma rays), they can remain together. They act as if they have fallen into a hole or a "well." That is, by giving up energy, they can become bound together (Figure 18.3). The amount of energy they give up to gather together is called the **binding energy**. The binding energy grows as more nucleons become bound inside a nucleus. Each incoming nucleon gives up energy to become bound. A very massive nucleus will have a large binding energy, while a low-A nucleus will have a small binding energy.

If you were to see balls filling a cylindrical hole in the ground in your mind's eye, you would see that the first balls into the hole fall the deepest, while later balls do not fall so far in (because the first balls are filling the hole). In a similar way, the first nucleons in a nucleus give up more energy than later ones. Because no one can see the nucleons filling the nucleus, it is best to concentrate on the *average* binding energy for the nucleons in a nucleus. As the number of nucleons increases, there is a rapid increase in the average binding energy per nucleon. This comes about, because, as more neutrons are taken in, they get between the

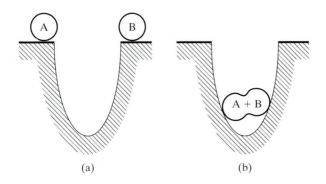

(a)   (b)

FIGURE 18.3

Nuclear constituents "fall" into a well and give up binding ("nuclear potential") energy, just as a ball falling into a hole gives up gravitational potential energy. (a) Before forming the nucleus. (b) After forming the nucleus.

protons to shield them from one another. The protons then do not need as much energy as before to repel other protons electrically, so they can give up more energy as they gather together to form a nucleus. At high mass number (high A), the neutron's shielding effect cannot save any more energy; after this, we would expect the binding energy per nucleon to be approximately constant. The curve of the actual average binding energy per nucleon as a function of A, the nucleon number, is shown in Figure 18.4. Note that the general features are *approximately* as we discussed. However, for A greater than that of iron-56 (^{56}Fe), the average binding energy per nucleon slightly decreases. It is precisely *this* characteristic that permits fission in nature to occur.

(Actually, we shall see later that slow "thermal" neutrons can penetrate a nucleus and interact while there.)

The binding energy can be found only if the mass of the nucleus is known. The nuclear masses have been carefully measured experimentally. These experiments gave rise to the data shown in Figure 18.4. The difference between the nuclear mass and the mass of the constituents of the nucleus (the nucleons) must represent the binding energy given up to bind the nucleus together. The mass of the bound nucleus is lower than the mass of the constituents, $Zm_p + Nm_n$ by

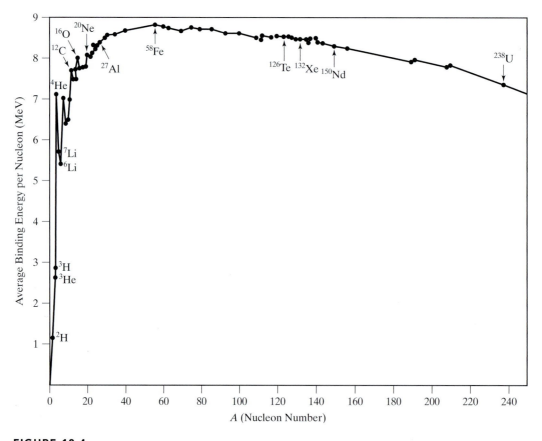

FIGURE 18.4

The average binding energy per nucleon is plotted in terms of the mass number or nucleon number, $A = N + Z$.

the binding energy/c^2. Suppose I know that the mass of a nucleus is 3.01605 u. (I will use hydrogen-3, tritium—hydrogen with a proton and two neutrons, as—an example.) The neutron has a mass 1.008664904 u, while the proton has a mass 1.007276470 u. Tritium has one proton and two neutrons, for a total constituent mass of 2(1.008664904 u) + (1.007276470 u) = 3.024606278 u. This is greater than the actual nuclear mass because of binding energy. The mass difference, (3.024606278 u − 3.01605 u) = 0.00856 u, multiplied by c^2 gives the binding energy:

$$\text{binding energy} = 0.00856\ uc^2$$
$$= (0.00856)(1.66053873 \times 10^{-27}\ \text{kg})(3 \times 10^8\ \text{m/s})^2$$
$$= 1.28 \times 10^{-12}\ \text{kg m}^2/\text{s}^2 = 1.28 \times 10^{-12}\ \text{J}$$
$$= (1.28 \times 10^{-12}\ \text{J})/(1.602176 \times 10^{-19}\ \text{J/eV}) = 7.98 \times 10^6\ \text{eV}$$
$$= 7.98\ \text{MeV}.$$

To find the binding energy from Figure 18.4, multiply the average binding energy per nucleon by the mass number (the number of nucleons). Consider the binding energy of oxygen-16. The graph gives an average binding energy per nucleon of just about 8 MeV; because there are 16 nucleons, the total binding energy of the nucleus is 16(8 MeV) = 128 MeV.

18.2   FISSION

Fission is the breakup of something into two (or possibly more) pieces. **Fissile nuclei**—that is, nuclei that can undergo fission—are generally stable until something happens to give them a little extra energy (for example, when they absorb another neutron) that destabilizes them and leads to the breakup. This energy is called **nuclear activation energy** and is analogous to the chemical activation energy discussed in Chapter 6. At this stage, the nucleus seems to act something like a drop of liquid (see **Extension 18.3, *The Liquid Drop Model***) that oscillates between a spherical shape and a dumbbell shape (something like an accordion being opened and closed rhythmically). The nucleus separates into two parts, which become nuclei of lower-A elements. In stable nuclei, the ratio of neutrons to protons decreases as Z decreases (see Figure 18.2), so that, when fission into two pieces occurs, there are extra neutrons in the product nuclei. These are rapidly emitted to allow these nuclei to become stable. Most fissions end up producing two or three extra neutrons. These neutrons can then interact with other fissile nuclei, causing additional fissions. This is the essence of the idea of the **chain reaction** (note that each neutron going in produces two or three new neutrons in causing a fission).

No way exists to predict exactly which nuclei will be produced in a fission reaction; it depends on the details of how the "drop of liquid" oscillated. There is a distribution of fission product masses. That distribution of product mass numbers for the fission of a uranium-235 ($^{235}_{92}$U) nucleus that absorbs a neutron is shown in Figure 18.5.[1] The mass of each of the fission products is roughly half the original mass. The diagram makes it obvious that what typically occurs is production of one nucleus with $A \sim 90$ and one nucleus with $A \sim 140$.

FIGURE 18.5

The yield of fission products is plotted in terms of the mass number for fission products resulting from the fission of an atom of ^{235}U that has absorbed a thermal neutron. (F. J. Shore, "Commercial nuclear steam-electric power plants, part I," *Phys. Teach.* **12**, 327 (1974), reprinted by permission from *The Physics Teacher*)

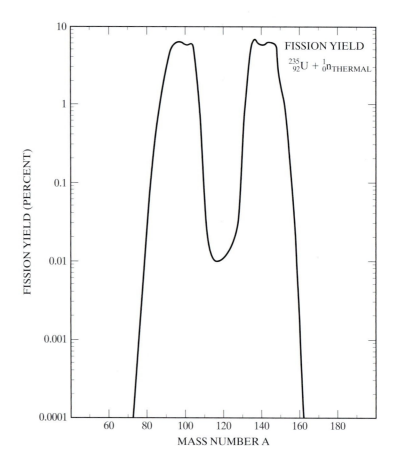

TABLE 18.1

Distribution of energy release in typical fission

Source	Energy (MeV)
Kinetic energy of fission fragments	167
Kinetic energy of fission neutrons	5
Instantaneous gamma rays	7
Emissions from fission fragments	
Gamma rays	7
Beta particles	7
Neutrinos	11
Total	204

Source: Reference 2.

Typical fissile nuclei such as ^{235}U, ^{239}Pu (plutonium), and ^{232}Th (thorium) react with slow incident neutrons to produce isotopes of greater mass number,

$$n + {}^{235}_{92}\text{U} \rightarrow {}^{236}_{92}\text{U}$$
$$n + {}^{239}_{94}\text{Pu} \rightarrow {}^{240}_{94}\text{Pu}$$
$$n + {}^{232}_{90}\text{Th} \rightarrow {}^{233}_{90}\text{Th},$$

that then decay into two fission fragments, neutrons, and radiant energy.

If there is a very massive nucleus, the nucleons in it could give up energy—that is, be more tightly bound into the nucleus containing them—if the massive nucleus were to break up into two less massive fragments. This additional energy can then appear as radiant energy (gamma rays) from decay of highly excited states, as well as kinetic energy of the fission fragments (see Figure 18.4). The energy distribution from nuclear fission is shown in Table 18.1.

All processes in which mass is transformed into energy and energy into mass obey Einstein's mass-energy relation $E = mc^2$, where c is the speed of light, 3×10^8 m/s. Let's use Einstein's mass-energy relation to see how much mass of the uranium-235 nucleus was transformed into kinetic energy. (See **Extension 18.1** for more examples of calculations). In this case, E is 204 MeV, the difference in binding energies of the two nuclei from that of the original nucleus. We can find out how much energy this is in joules by using the conversion factor 1 eV equals 1.6×10^{-19} J:

$$204 \text{ MeV} = (204 \text{ MeV}) \times (1.6 \times 10^{-19} \text{ J/eV}) = 3.26 \times 10^{-11} \text{ J}.$$

Rewriting the Einstein mass–energy relation as $m = E/c^2$, we find

$$m = E/c^2 = (3.26 \times 10^{-11} \text{ J})/(3 \times 10^8 \text{ m/s})^2 = 3.62 \times 10^{-28} \text{ kg},$$

or, alternatively

$$m = E/c^2 = 204 \text{ MeV} \times (\text{u}/931.5 \text{ MeV}/c^2) = 0.218 \text{ u},$$

and we can see that a very small amount of nuclear mass, 0.218 u, has been transformed into kinetic energy in this typical fission.

A **chain reaction** is a reaction that continues once it is begun without outside interference until all the fuel is used. The extra neutrons produced in fission allow for controlled chain reactions (fission reactors in nuclear power plants) and uncontrolled chain reactions (the "atomic" bomb). It takes a slow ("thermal") neutron to initiate the sequence, and, on average, for each neutron going into uranium-235, 2.43 come out. If these cause other fissions and enough fissile uranium is present (bombs use highly enriched uranium; see **Extension 19.1**), a chain reaction can ensue, producing a lot of energy at once as a bomb.

Slow Neutrons and Fission

If uranium is instead put into water, and carbon rods are intermingled with the uranium, the neutrons produced in fission are slowed by collisions with the atoms in the water and the carbon rods. The water molecules (Figure 18.6) can absorb neutrons; the carbon cannot, but it acts to slow the neutrons. Materials that slow neutrons so they can cause another fission are called **moderators**, so water is a moderator (the main one in most reactors). Some neutrons escape the reactor, and some are captured by the nonfissile uranium atoms, such as $^{238}_{92}\text{U}$. It is thus possible to make certain that there are always the same numbers of neutrons crossing a unit area in a given time. This system can then persist for lengthy periods of time in a state of dynamic equilibrium.

Energy from Fission

We shall investigate two representative decays of uranium-235 plus a slow neutron, $^{235}_{92}\text{U} + n$, which is $^{236}_{92}\text{U}$, using the binding energy curve (Figure 18.4).

$$^{236}_{92}\text{U} \rightarrow ^{90}_{36}\text{Kr} + ^{143}_{56}\text{Ba} + 3n. \tag{1}$$

(This is the decay into krypton-90 and barium-143, Figure 18.7.) From Figure 18.4, we read that the average binding energy per nucleon is about 0 MeV for n, 7.6 MeV for ^{236}U, 8.8 MeV for ^{90}Kr, and 8.4 MeV for ^{143}Ba. The total binding energies for the respective nuclei are, therefore,

$$236(7.6 \text{ MeV}) = 1794 \text{ MeV, for } ^{236}\text{U},$$
$$90(8.8 \text{ MeV}) = 792 \text{ MeV, for } ^{90}\text{Kr, and}$$
$$143(8.4 \text{ MeV}) = 1201 \text{ MeV, for } ^{143}\text{Ba}.$$

FIGURE 18.6

A reactor is shown being refueled. The water surrounding the core is visible, and the glow comes from radiation released as particles travel through the water at a speed faster than that at which light can travel in water. These faster-than-light particles excite electrons as they interact with atoms, and the electrons give off blue light as they return to the ground state. (U.S. Department of Energy)

FIGURE 18.7

A thermal (slow) neutron hits a ^{235}U nucleus, and after a time, ^{90}Kr and ^{143}Ba plus three neutrons emerge, producing energy from some of the mass of the original nucleus.

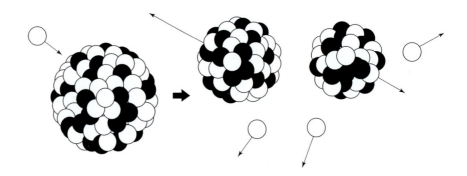

The net energy emitted in the fission is the difference between the binding energies of the fission products and the binding energy of the original nucleus:

$$BE(^{90}\text{Kr}) + BE(^{143}\text{Ba}) - BE(^{236}\text{U})$$
$$= 792 \text{ MeV} + 1201 \text{ MeV} - 1794 \text{ MeV}$$
$$= 199 \text{ MeV}.$$

$$^{236}_{92}\text{U} \rightarrow {}^{97}_{38}\text{Sr} + {}^{137}_{54}\text{Xe} + 2n. \tag{2}$$

(This is the decay into strontium-97 and xenon-137.) Again, reading from Figure 18.4, the binding energy per nucleon is about 0 MeV for n, 7.6 MeV for ^{236}U, 8.6 MeV for ^{97}Sr, and 8.4 MeV for ^{137}Xe. The total respective binding energies are

$$236(7.6 \text{ MeV}) = 1794 \text{ MeV, for } {}^{236}\text{U},$$
$$97(8.6 \text{ MeV}) = 834 \text{ MeV, for } {}^{97}\text{Sr, and}$$
$$137(8.4 \text{ MeV}) = 1151 \text{ MeV, for } {}^{137}\text{Xe}.$$

This time, the net energy emitted in the fission is

$$BE(^{97}\text{Sr}) + BE(^{137}\text{Xe}) - BE(^{236}\text{U})$$
$$= 834 \text{ MeV} + 1151 \text{ MeV} - 1794 \text{ MeV}$$
$$= 191 \text{ MeV}.$$

These numbers are in general agreement with the sum of all the energies listed in Table 18.1—204 MeV—for the "typical" fission. Note, in these examples, how the binding energies increase for more massive nuclei in comparison to less massive nuclei, as was predicted above.

Radioactive Decay

Fission—breakup into two roughly equivalent pieces—is not the only way to change one atom into another sort of atom. Many nuclei exhibit radioactive decay (see the discussion of free neutron decay and the explanation of the exponential character of radioactive decay). By **radioactive decay**, we mean the spontaneous change of a nucleus into a different nucleus, accompanied by the release of other particles and energy. There are three common ways for nuclei (or other particles) to decay, called, respectively, *alpha-emission*, *beta-emission*, and *gamma-emission*. Free neutron decay is an example of beta decay. See **Extension 18.4, *Alpha, Beta, and Gamma*** for more examples of these decays.

Alpha Decay

Alpha particles, which are helium nuclei ($^4_2$He) and so consist of two protons and two neutrons, are rather massive and travel relatively slowly. Alpha decay is the process by which a nucleus emits an alpha particle spontaneously and becomes a nucleus of an atom lower in atomic number by two and lower in mass number by four than the "parent" nucleus. The nuclei must be transmuted into other sorts of nuclei, because the alpha particle is a helium nucleus (with two protons and two neutrons, that is, $Z = 2$ and $A = 4$). The energy released in alpha decay is typically a few MeV. The two protons in an alpha particle can ionize air fairly effectively. Alpha particles cannot penetrate very much matter, because they do interact so effectively, and so they can be stopped even by a thin sheet of cardboard, a sheet of bond paper, or a person's skin.

An example of an alpha decay is the decay of radium-226 to radon-222 (Figure 18.8a). As we discuss in Chapter 20, radon is responsible for much of the health impact of radioactivity on Earth. Another example is the decay of thorium-232: $^{232}_{90}$Th decays by α emission to $^{228}_{88}$Ra, or

$$^{232}_{90}\text{Th} \rightarrow \, ^4_2\text{He} + \, ^{228}_{88}\text{Ra}.$$

Table 18.2 presents several interesting half-lives for α emitters.

TABLE 18.2

Selected half-lives for radioactive decays

Element	Half-life	Decay mode
$^3_1$H	12.32 yr	β^-
$^{15}_8$O	122.2 s	β^+
$^{14}_6$C	5715 yr	β^-
$^{29}_{13}$Al	6.5 min	β^-
$^{24}_{14}$Si	0.10 s	β^+, proton emission
$^{32}_{14}$Si	160 yr	β^-
$^{40}_{19}$K	1.26 Gyr	β^-, β^+, electron capture
$^{44}_{23}$V	0.09 s	α, β^+
$^{52}_{26}$Fe	8.28 h	β^+, electron capture, internal conversion
$^{38}_{90}$Sr	29.1 yr	β^-
$^{81}_{39}$Y	1.21 min	β^+
$^{137}_{55}$Cs	30.3 yr	β^-
$^{106}_{52}$Te	60 μs	α
$^{197}_{83}$Bi	10 min	β^+, electron capture
$^{226}_{88}$Ra	1599 y	α
$^{232}_{90}$Th	13.9 Gyr	α
$^{232}_{92}$U	68.9 yr	α
$^{238}_{92}$U	4.46 Gyr	α

Beta Decay

Beta particles are electrons (or positrons, antielectrons) and, thus, are much less massive than alpha particles. Beta-minus decay occurs when a neutron inside a

FIGURE 18.8

In each type of decay, the energy of the nucleus has decreased, and the nucleus has become more stable. (a) The process of alpha decay. A parent nucleus emits a helium nucleus (an alpha particle), and the daughter nucleus moves in the opposite direction. (b) The process of beta-minus decay. One of the neutrons has been converted into a proton inside the nucleus. An electron and an antineutrino, a particle with no charge or mass, are also emitted. (c) The process of beta-plus decay. One of the protons has been converted into a neutron inside the nucleus. An antielectron (positron) and a neutrino, a particle with no charge or mass, is also emitted. (d) The process of gamma decay. The photon, a particle with no mass or charge, is emitted.

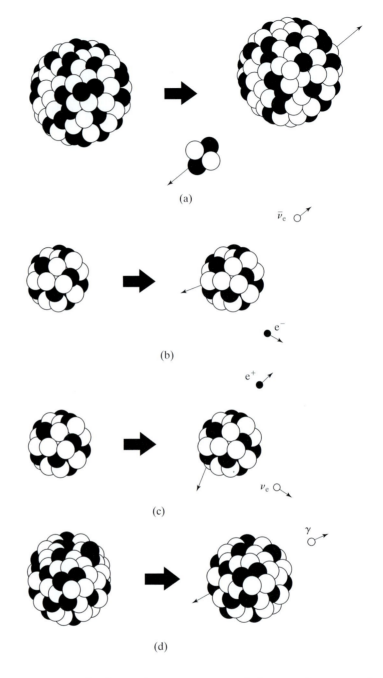

nucleus spontaneously decays into a proton, emitting an electron and an antineutrino $(\bar{\nu})$; the atomic number Z increases by one in nuclear beta decay (Figure 18.8b). A nucleus that emits a beta particle does not change its mass number A, as in the decay

$$^3_1\text{H} \rightarrow \text{e}^- + ^3_2\text{He} + \text{antineutrino}.$$

The atomic number Z has gone from one to two, but A has remained three. In this case, conservation of charge and several more exotic conservation rules require that one of the neutrons in the tritium nucleus decay into a proton, an electron, and an antineutrino. Another example is the decay $^{165}_{66}\text{Dy} \rightarrow ^{165}_{67}\text{Ho} + \beta^- + \bar{\nu}$. Here, the atomic number Z also increases by one from 66 to 67, and the mass number A is unchanged. Typical energies released in beta decays are around 1 eV.

A decay producing a positron is also called beta decay, but is a beta-plus decay; in this case, a proton changes into a neutron, with a positron (antielectron) and a neutrino emitted. Therefore, the atomic number Z would decrease by one (Figure 18.8c), but again, the mass number A would remain unchanged. For example, in the decay $^{207}_{86}\text{Rn} \rightarrow ^{207}_{85}\text{At} + \beta^+ + \nu$, the atomic number decreases from 86 to 85, while the mass number remains 207.

As electrons and positrons are singly charged, they are not as effective at ionizing air as the alphas. Betas can penetrate matter more easily than the alphas and need a thickness of perhaps 5–10 cm of water or plastic or a sheet of a metal such as aluminum to stop them. Alpha and beta particles, as well as more exotic particles, can be detected by use of Geiger counters, or by observation in cloud chambers, bubble chambers, or more sophisticated particle-detection devices (see **Extension 18.5**, *Detecting Decay Particles*).

Gamma Decay

Gamma decay occurs when a nucleus in an excited state falls to a lower state (which could be the ground, or lowest, state) and emits electromagnetic radiation equal to the energy difference in the process (an example is shown in Figure 18.8d). Many alpha and beta decays are followed by gamma decay, because they result in an excited nucleus. Another example is the decay of the metastable state of barium into the ground state,

$$^{137\text{m}}_{56}\text{Ba} \rightarrow ^{137}_{56}\text{Ba} + \gamma.$$

The energy of the gamma ray in this decay is 662 keV.

Still another example of gamma decay is a beta decay from an excited state of cobalt-60 to an excited state of nickel-60. Three gamma transitions of the nickel-60 are observed, with energies 826 keV, 1.352 MeV, and 2.158 MeV representing transitions from two excited states to the nuclear ground state (note that 826 keV + 1.332 MeV = 2.158 MeV because of conservation of energy). This decay is discussed in more detail in **Extension 18.4**.

Gamma rays are very energetic forms of electromagnetic radiation or light, and most gammas emitted are in the interval of 10 keV to 10 MeV. They do not ionize air very well. In order for **gamma radiation** to be detected, it must interact with the matter in its path. For example, matter in its path can cause pair production by γs, electron–positron pairs can form out of pure energy (another example of the Einstein mass–energy relation, $E = mc^2$), the γ can scatter from a charged particle, and so forth. The electrons are then easily detected in a Geiger counter, for example. Gamma radiation is the most penetrating, so γs require great thicknesses of lead or another dense substance to stop them.

Several other processes can result in a change in nuclear identity. These are electron capture and spontaneous fission. They are discussed in **Extension 18.4**.

18.3   NUCLEAR FUSION

To understand how nuclear fusion differs from nuclear fission, it is useful to return to Figure 18.4, the curve of the average binding energy per nucleon versus A.

Note the steep rise in average binding energy per nucleon for small A. This implies that if we push low-A nuclei together to form higher-A nuclei, we can obtain energy release. However, to get that energy out, we must put a lot of energy in. If the colliding nuclei do not have sufficient energy, they will recoil from one another due to the internuclear proton–proton electric repulsion (because the nucleons cannot get close enough together for the strong interaction to cause attraction). On Earth, we can give the colliding nuclei enough energy by raising the temperature to very high levels (thermonuclear fusion on Earth requires temperatures of 50–100 MK); stars achieve fusion because of the high interior pressure and temperatures of around 10 MK. The need for this fusion activation energy introduces a considerable complication in the implementation of any scheme for fusion power.

Fusion reactions include the following:

$$^2_1H + {}^2_1H \rightarrow p + {}^3_1H + 3.25 \text{ MeV}$$
$$(104 \text{ TJ/kg} = 29 \text{ GWh/kg})$$
$$^2_1H + {}^2_1H \rightarrow n + {}^3_2He + 4.0 \text{ MeV}$$
$$(96 \text{ TJ/kg} = 27 \text{ GWh/kg})$$
$$^2_1H + {}^3_1H \rightarrow n + {}^4_2He + 17.6 \text{ MeV}$$
$$(337 \text{ TJ/kg} = 94 \text{ GWh/kg})$$
$$^2_1H + {}^3_2He \rightarrow p + {}^4_2He + 18.3 \text{ MeV}$$
$$(351 \text{ TJ/kg} = 97 \text{ GWh/kg})$$
$$^6_3Li + {}^2_1H \rightarrow n + 2\,{}^4_2He + 22.4 \text{ MeV}$$
$$(268 \text{ TJ/kg} = 75 \text{ GWh/kg})$$

Here, ^2_1H is hydrogen with both a proton and a neutron in its nucleus, the sort known as deuterium; ^3_1H is tritium. Deuterium (^2_1H) is stable and thus can be found at some small concentration, one of every 6500 hydrogen atoms,[3] in ordinary hydrogen (^1_1H); tritium decays rapidly (Table 18.2) and is not found in nature. Deuterium–tritium fusion is the focus of most research efforts. The collisions, given sufficient energy to initiate fusion, first produce a momentary intermediate state that decays; for instance, deuterium and tritium collisions produce helium-5, which then becomes an α particle and a neutron (Figure 18.9).

FIGURE 18.9

Deuterium (hydrogen-2) and tritium (hydrogen-3) collide to produce a helium-5 nucleus that then emits a neutron to become an alpha particle (a helium-4 nucleus). In this fusion process, 17.6 MeV of mass energy is transformed into kinetic energy (the neutron has 14.1 MeV; the α has 3.5 MeV).

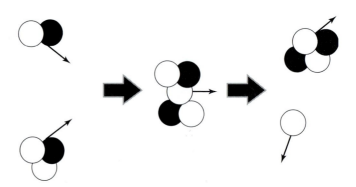

A CLOSER LOOK

Nucleon Constituents

The incredible variety of our experience with material objects can be explained in terms of only about 100 building blocks, the atomic elements. After experiments in Lord Rutherford's laboratory early in the twentieth century, it was found that these hundred atoms are composite—made of nucleons (protons, neutrons) and electrons. Could the nucleon be composite? The Standard Model of Particles and Interactions, the widely accepted model of particles, implies the existence of constituents known as quarks (Figure 18.10). Nucleons are made up of three quarks: the proton is made of two up quarks and a down quark; the neutron is made up of two down quarks and an up quark. Because the up and down quarks have different electric charge, proton and neutrons have different charges.[4,5]

Quarks are never observed alone. They are always in the company of other quarks or antiquarks. It is as if the quarks were at opposite ends of a continuous piece of an elastic material that stretches, then finally tears and knits both ends to make two complete pieces. The result is two elastic pieces, each of which has quarks at its end. Only when two nucleons hit one another really hard—at high energy—can the two nucleons' quarks interact with each other as if the others were not present.

Because the quarks are so deeply bound, the interquark forces play a subtle role in nuclear interactions inside a nucleus. Some chemical bonds are made because the pieces are uncharged, but polarized. For example, solid water is made because the little water molecules, though neutral, have their charges separated—more positive charge is found on the protons and more negative charge on the oxygen. The polarized molecules attract and can form strong enough bonds from this "leftover," or "residual," electric force to make water a solid. In a similar fashion, the strong force between protons and protons, between neutrons and protons, and between neutrons and neutrons inside the nucleus is a "leftover," or "residual," result of the forces among the quarks making up the protons and neutrons.

Quarks	spin = 1/2	
Flavor	**Approx. Mass GeV/c²**	**Electric charge**
u up	0.003	2/3
d down	0.006	−1/3
c charm	1.3	2/3
s strange	0.1	−1/3
t top	175	2/3
b bottom	4.3	−1/3

Leptons	spin = 1/2	
Flavor	**Mass GeV/c²**	**Electric charge**
ν_e electron neutrino	<1 × 10⁻⁸	0
e electron	0.000511	−1
ν_μ muon neutrino	<0.0002	0
μ muon	0.106	−1
ν_τ tau neutrino	<0.02	0
τ tau	1.7771	−1

FIGURE 18.10

The six fundamental quarks and six fundamental leptons.
(Courtesy of Contemporary Physics Education Project © 1999)

This curve of the binding energy is useful for explaining how the elements we know got here. In the "**big bang**," the universe formed in an explosion, and there was only radiant energy in it. As the universe expanded, it cooled. When it cooled far enough, matter could exist.

The Early Universe

The matter in the early universe was mostly hydrogen with a small admixture of helium; in fact, big bang model calculations predict the correct ratio of helium to hydrogen in the universe. As matter collected, stars grew. Inside stars, all of which are natural fusion reactors, the temperature can climb to tens of millions of kelvins (our Sun's core is at a temperature of over 10 MK). At these temperatures, nuclei can get to within 10^{-15} m of one another, and fusion reactions can occur. The curve shows that particles fused to other particles can be more deeply bound until the product mass reaches 56, so fusion in the massive stars can proceed until a layer of iron is in the core, surrounded by layers of less-massive elements.

If this were all, our universe would be uninteresting and humanless. However, as the fusion fires are banked (that is, more iron is formed in the stellar core), the heat output goes down. If the temperature goes down, the outer layers of the star can fall inward (because the pressure is related to the temperature, as we saw in Chapter 7). In certain circumstances, this can cause a supernova, which is a catastrophic explosion in which the collapsing outer layers of a star pump incredible energy into the central core.

With so much energy available, endothermic fusion reactions can occur. Such reactions are responsible for the synthesis of all elements having an A greater than 56. The supernova explosions spread the elements around the universe. When new stars form from interstellar gas, some higher-mass elements are incorporated. Our planet is made mostly of such recycled stardust. This is possible, because our solar system formed some 10 billion years after the big bang, leaving plenty of time for many supernova events.

Much progress has been made in our understanding of the formation of nuclei—**nucleosynthesis**—in stars during the past several decades. William Fowler was awarded the 1983 Nobel Prize for his work on the problem of formation of higher-mass elements inside stars. We think we understand the general way elements are formed, but many exciting problems remain to be solved.

Once the elements are formed, many decay into less-massive elements or undergo fission into less-massive elements. Some of the original energy available in the supernova can be recovered by allowing a large-A nucleus to fission. We already know how to recover this energy. Nuclear energy generation is the subject of the next chapter.

SUMMARY

Nucleons are protons or neutrons. Nuclei having the same number of protons but different numbers of neutrons are isotopes; see **Extension 18.2**.

Nuclear energy arises from the binding of nucleons into a nucleus. In going into the nuclear energy well more deeply, each nucleon gives up some energy. As a result, the bound nuclei have masses smaller than the sum of the masses of their constituent nucleons. The difference is the binding energy of the nucleons in the nucleus. The energy scale of nuclear physics is the MeV, and nuclear masses are given in terms of MeV/c^2 or of the unified atomic mass unit u (as described in **Extension 18.1**).

Because larger nuclei have more nucleons, they have a greater binding energy simply by virtue of the greater number of nucleons they contain. We remove this spurious dependence by looking at the average binding energy per nucleon. When this quantity is plotted against the total nucleon number, the curve reveals regions where fusion (combining nuclei) allows nucleons to become more stable ($A < 56$), and regions in which fission (nuclear breakup) allows nucleons to become more stable (very large A).

When a large-A nucleus fissions, it oscillates and breaks into two smaller pieces (see **Extension 18.3**). Because the number of neutrons per proton increases with increasing A, a large-A nucleus that breaks into two smaller-A nuclei produces a few extra neutrons. This allows further fissions of fissile nuclei to occur, producing more neutrons, producing more fissions, producing more neutrons, producing more fissions … all in an eyeblink. This is an uncontrolled fission reaction, a nuclear explosion. In reactors, moderators and control rods absorb extra neutrons, consequently allowing a nuclear reaction to be controlled.

Confined fusion in the stars provides light and energy to the universe (and to incidental planets). Uncontrolled fusion in supernovas leads to nucleosynthesis. Uncontrolled nuclear fusion on Earth occurs in thermonuclear bombs. No fusion reactor currently produces electricity.

Radioactive decay is a random process that proceeds exponentially. The exponential curve is characterized by a decay time known as the mean life. Alpha emission is ejection of a helium-4 nucleus from a larger nucleus.

Beta emission involves an electron or positron leaving the decaying nucleus. Gamma radiation is emitted when an excited nuclear state gives off energy in the form of photons, or light quanta. More details on the properties of the radioactive decays are found in **Extension 18.4**. Decays are detected using the ionization caused by passage of a charged particle through matter, as described in **Extension 18.5**.

PROBLEMS AND QUESTIONS

MULTIPLE CHOICE REVIEW

1. Nuclei with which values of A are most likely to be involved in a fusion reaction?
   a. $A < 25$
   b. $25 < A < 50$
   c. $100 < A < 150$
   d. $150 < A < 200$
   e. $200 < A$
2. Some nuclei decay by alpha emission. Which of the following correctly characterizes the atomic number of the new decay product nucleus with respect to the original nucleus?
   a. It increases by four.
   b. It increases by two.
   c. It stays the same.
   d. It decreases by two.
   e. It decreases by four.
3. Nuclei with a value of $A >$ _____ could be involved in a fission reaction.
   a. 4
   b. 25
   c. 30
   d. 54
   e. 211
4. Some nuclei decay by beta-plus emission. Which of the following correctly characterizes the mass number of the new decay product nucleus with respect to the original nucleus?
   a. It increases by two.
   b. It increases by one.
   c. It stays the same.
   d. It decreases by one.
   e. It decreases by two.
5. In causing a fissile nucleus to break up, which of the following is most effective? A collision of the fissile nucleus with a
   a. fast neutron.
   b. fast proton.
   c. decay product of another fissile nucleus.
   d. fast neutron.
   e. slow neutron.
6. In the decay of a collection of radioactive nuclei, the number of original nuclei left after two mean lives is
   a. the same as the original number.
   b. half as many as the original number, because there are *two* mean lives.
   c. one-quarter the original number, because there are *two* mean lives.
   d. less than the original number but greater than half the original number.
   e. not correctly given by any other alternative listed here.

7. For the nucleus $^{15}_{6}C$, we may say that in that particular nucleus,
   a. six is the number of nucleons, and 15 is their mass.
   b. six is the number of protons, and 15 is the number of nucleons.
   c. six is the number of neutrons, and 15 is the number of protons.
   d. six is the number of protons, and 15 is the number of neutrons.
   e. six is the number of electrons, and 15 is the number of protons.
8. Some nuclei decay by beta-plus emission. Which of the following correctly characterizes the atomic number of the new decay product nucleus with respect to the original nucleus?
   a. It increases by two.
   b. It increases by one.
   c. It stays the same.
   d. It decreases by one.
   e. It decreases by two.
9. The strengths of the interactions in nature, from strongest to weakest, are in the order
   a. strong, weak, electromagnetic, gravitational.
   b. strong, electromagnetic, weak, gravitational.
   c. strong, gravitational, electromagnetic, weak.
   d. strong, electromagnetic, gravitational, weak.
   e. gravitational, electromagnetic, weak, strong.
10. Some nuclei are changed by a reaction called electron capture, in which the nucleus absorbs one of the atomic electrons. Which of the following correctly characterizes the atomic number of the new nucleus with respect to that of the original nucleus?
    a. It increases by two.
    b. It increases by one.
    c. It stays the same.
    d. It decreases by one.
    e. It decreases by two.

SHORT ANSWER
Explain whether the statement made is correct, and justify your answer.
11. Carbon-15 and nitrogen-15 are isotopes of the same element.
12. The average binding energy per nucleon is the best parameter to use in determining whether a given nucleus is likely to undergo fission.
13. There are more neutrons than protons in most nuclei, because neutrons have a smaller mass than protons.
14. Because nucleons are constituents of atoms, and quarks are constituents of nucleons, quarks must have constituents.

15. Just as some chemicals must be given activation energy to start an exothermic chemical reaction, some nuclei must be given nuclear activation energy in order to undergo an exothermic fission.
16. Binding energy is roughly constant for all higher-A nuclei.
17. The mass numbers of fission decay products appear to be random, because the excitation of the nucleons in the nucleus randomly oscillates as a liquid drop does.
18. The electronvolt is the unit of energy well suited to describing energies of nucleons and electrons at the atomic level.
19. Gamma radiation is characterized by greater ability to penetrate materials than alpha or beta.
20. There are more neutrons than protons in most nuclei, because neutrons are physically smaller than protons.

DISCUSSION QUESTIONS

21. What is the unknown product nucleus X in the reaction $^{235}_{92}U + ^{5}_{2}He \rightarrow 4\,n + ^{112}_{48}Cd + X$? Explain how it may be determined.
22. Consider the fission reaction $n + ^{238}_{94}Pu \rightarrow 8n + ^{159}_{65}Tb + X$. Explain how the unknown product nucleus X may be found.
23. The proton and neutron are clearly different. How is it that we can designate them collectively as nucleons?
24. How would a nucleus $^{137}_{55}Cs$ change to become a nucleus of $^{137m}_{56}Ba$? Explain. The "m" stands for metastable. The metastable nucleus is in an excited state and will later undergo a decay itself.
25. How would the excited nucleus $^{137m}_{56}Ba$ probably decay? Explain.
26. How many neutrons should be released in the reaction $^{235}_{92}U + ^{5}_{2}He \rightarrow ^{117}_{48}Cd + ^{121}_{50}Sn + $ neutrons? Explain.
27. In the reaction $n + ^{238}_{92}U \rightarrow ^{114}_{48}Cd + ^{124}_{50}Sn + $ neutrons, how many neutrons will be released? Explain.
28. The americium nucleus $^{242}_{95}Am$ undergoes α decay. What will the daughter nucleus be? How do you know?
29. The curium nucleus, $^{241}_{96}Cm$, undergoes electron capture as one of its possible decay modes. What is the daughter nucleus? Explain.
30. The californium nucleus $^{242}_{98}Cf$ decays via α decay with a half-life of 210 s. If we have two californium nuclei, how many will remain as californium after one half-life? Explain.

PROBLEMS

Note: The mass of a proton is 1.007825 u; that of a neutron is 1.008665 u. An electron has a mass 6.4858×10^{-4} u. The conversion between u and MeV/c^2 is 1 u = 931.5 MeV/c^2.

31. What is the total binding energy in the nucleus $^{2}_{1}H$? Explain how you obtain your answer.
32. What is the total binding energy in the nucleus $^{4}_{2}He$? Explain how you obtain your answer.
33. What is the total binding energy in the americium nucleus $^{242}_{95}Am$? Explain how you obtain your answer.
34. What is the average binding energy per nucleon in the nucleus $^{4}_{2}He$? The mass of this nucleus is 6.018886 u. Explain how you obtain your answer.
35. What is the average binding energy per nucleon in the nucleus $^{112}_{48}Cd$? The mass of this nucleus is 111.9028 u. Explain how you obtain your answer.
36. What is the average binding energy per nucleon in the nucleus $^{238}_{94}Pu$? The nuclear mass is 238.0496 u. Explain how you obtain your answer.
37. How much energy would be required to remove a neutron from the nucleus $^{69}_{31}Ga$, which has a mass of 68.925581 u?
38. What amount of energy would be required to dismantle the nuclei of 1.00 g of aluminum (^{27}Al, nuclear mass = 26.981541 u) into separate protons and neutrons?
39. If the nucleus of a certain atom is the diameter of an orange, about 10 cm, approximately how large is the atom?
40. If we visualize nucleons as little balls of diameter 1 fm (1.0×10^{-15} m), we can imagine the nucleus to be made up of A such balls. How would the volume of the nucleus $^{238}_{94}Pu$ compare to the volume of one nucleon? How would the volume of the nucleus $^{112}_{48}Cd$ compare to that of one nucleon? How does that of cadmium compare to that of plutonium?
41. In the reaction $n + ^{239}_{94}Pu \rightarrow ^{240}_{94}Pu \rightarrow ^{100}_{44}Ru + ^{134}_{50}Sn + 6\,n + energy$, what amount of energy is released?
42. In the reaction $n + ^{239}_{94}Pu \rightarrow ^{240}_{94}Pu \rightarrow ^{112}_{44}Ru + ^{124}_{50}Sn + 4\,n + energy$, what amount of energy is released?
43. The californium-242 nucleus $^{242}_{98}Cf$ decays via α decay with a half-life of 210 s.
    a. What is its daughter nucleus?
    b. If we originally have 10,000 californium-242 nuclei, how many will remain as californium after one half-life? Explain.
    c. If we originally have 10,000 californium-242 nuclei, how many will remain as californium after five half-lives? Explain.
44. The berkelium-247 nucleus $^{247}_{97}Bk$ decays via α decay with a mean life of 2000 yr.
    a. What is its daughter nucleus?
    b. If we originally have 100,000 berkelium-247 nuclei, how many will remain as berkelium after one mean life? Explain.
    c. If we originally have 100,000 berkelium-247 nuclei, how many will remain as berkelium after three mean lives? Explain.

CHAPTER 19
Energy from Nuclear Reactors

In Chapters 13 through 17, we saw that energy from fossil fuels can be hazardous to human health. If solar energy, conservation, and recycling cannot replace necessary electricity in the short run, and if the hazards of fossil fuels are recognized, where shall we get our electricity? The focus of this chapter is generation of energy through nuclear processes. In the preceding chapter, we considered the basic nuclear physics involved in fission and fusion of nuclei. This physics is utilized here to explain in some detail the operation of reactors. We try to dispel some of the myths surrounding the operation of boiling water reactor (BWR) and pressurized water reactor (PWR) nuclear fission reactors and the fuel cycles of these reactors. We consider the breeder reactor; compare American and foreign experience with nuclear energy; and discuss the possibility of controlled nuclear fusion energy (which has already become surrounded by myths of its own, even though no such power plants can be operating until at least 20 or 30 years from now).

KEY TERMS

gaseous diffusion / moderator / prompt neutrons / delayed neutrons / control rod / SCRAM / loss-of-coolant accident (LOCA) / core melt / emergency core cooling system (ECCS) / CANDU / high-temperature gas-cooled reactor / graphite-water reactor / yellowcake / capacity factor / radioactive / activity / becquerel / curie / converter / conversion coefficient / burner / breeder / critical mass / plasma / heating / containment / Lawson criterion / tokamak / inertial confinement fusion

The nuclear age, properly speaking, dates to the late 1930s, when a series of experiments, culminating in the observation of fission by Otto Hahn and Lise Meitner, demonstrated the potential for controlled nuclear fission to those physicists working in the exciting new field. It was but a short time later, 1942, that

Enrico Fermi and his coworkers demonstrated controlled fission in a squash court under Stagg Field at the University of Chicago. Even as these newly christened "nuclear scientists" worked on nuclear weapons for the Manhattan Project during World War II, many were already thinking of the blessings they expected to be conferred on society by the cheap nuclear energy they would make available, energy too cheap to bother metering. They worked hard afterward to build *peaceful* uses of nuclear energy. Despite their work, the most easily remembered symbol of this nuclear age is the mushroom-shaped cloud caused by the release of nuclear energy in a bomb. Some believe that the specter of such a destructive force unleashed from the nucleus has poisoned nuclear energy in the mind of the public.

All current and proposed forms of nuclear energy involve production of steam to drive turbines. The delivery system, from turbine on, is identical for all facilities producing electricity (see Chapter 13), so we concentrate on the parts of the facilities that are different for nuclear systems as opposed to fossil-fuel systems. Consult References 1 and 2 for more detail than it is possible to give in this chapter.

19.1   THE FISSION PROCESS IN REACTORS

Fission is the breakup of a nucleus into smaller parts. The thermal energy provided by a fission reactor comes from the kinetic energy of the fragments of the broken-up nucleus. Because protons and neutrons have practically the same mass, the mass of a nucleus containing a number A of nucleons has a mass roughly A times the average proton–neutron mass. It is slightly less than the sum of all proton and neutron masses involved, because the nucleons give up binding energy in order to gather together; that is, some mass energy is converted into kinetic energy during the binding process, resulting in a more stable bound system.

As the number of nucleons increases, there is a rapid increase in the average binding energy per nucleon. At high mass number (high A), the average binding energy per nucleon is approximately constant. Refer to the curve of the actual average binding energy per nucleon as a function of A, the mass number, Figure 18.2.

Fission reactions involve the breakup of a high-A nucleus into two smaller A nuclei, in each of which the "average" nucleon is more deeply bound than in the more massive nucleus. The process produces this mass energy difference as kinetic energy of the products as well as that of the two or three extra neutrons (recall that higher-A nuclei have a greater neutron-to-proton ratio than lower-A nuclei). These neutrons are ejected during formation of the fission products; they can then move away and interact with other fissile nuclei, causing additional fissions. This is the essence of the idea of the chain reaction (note that each neutron going in produces two or three new neutrons in causing a fission).

A tabletop model of an uncontrolled chain reaction has been made using mousetraps and Ping-Pong balls. If all the mousetraps are set, and several Ping-Pong balls are placed on the front of the trap, one Ping-Pong ball can cause a mousetrap to snap, sending three balls flying. Each of these springs a trap, and so on. Balls fill the air with explosive suddenness.

Reactors

Commercial boiling water reactors (BWRs) and pressurized water reactors (PWRs) were discussed briefly in "The Nuclear Process" in Chapter 13. The uranium fuel used in reactors is enriched uranium (reactors use ~4% uranium-235, while ore is 0.7% uranium-235). Several methods of enrichment have been used: **gaseous diffusion** (^{235}U and ^{238}U go through a successive series of membranes, with ^{235}U slightly more likely to get through), gas centrifuges (the less dense gas is more likely near the center of motion), and laser-selective isotope enrichment (energies of electrons are slightly different in the two isotopes; the laser energy can be tuned to either). More information on enrichment is available in **Extension 19.1, *Enrichment***. When uranium fuel rods are put into water in reactors, and carbon rods are intermingled with the uranium fuel elements, the neutrons produced in fission are absorbed in water molecules and slowed by collisions with the atoms in the water and the carbon rods.

It is essential that the neutrons be slowed, because slow neutrons are most effective at being captured by the ^{235}U nucleus and causing it to fission. **Moderators** are materials that are in reactors to slow fission neutrons. The carbon rods and water in reactors act as moderators. In a sense, slow neutrons dawdle long enough inside the nucleus to cause the internal nuclear excitation leading to fission, whereas fast neutrons just zip by without having time to interact with the protons and neutrons in the ^{235}U nucleus. The nonfissile ^{238}U can actually be made to fission by fast neutrons, but its cross section ("effective size") is much smaller than that of the ^{235}U for slow neutrons.

Some of the neutrons released in fission emerge immediately upon fission (**prompt neutrons**). In some fissions, however, neutrons are held in the nucleus of the fission product for a time before release (**delayed neutrons**). The presence of some delayed neutrons among the average net 2.43 neutrons per fission allows for greater operational safety in reactors by giving the operators leeway in mechanically initiating shutdown if even a hint of danger is detected.

The reactor functions in a steady state when exactly one neutron from each fission is allowed to cause a subsequent fission. In use in the reactor, enough material is present so that there is a net of less than one prompt neutron per fission. The delayed neutrons make up the difference and allow the reactor to continue operation. This means that there is an extra time interval during which safety devices can work. About 0.65% of the neutrons from the decay of fissile uranium are delayed; only 0.3% of those from fissile plutonium are delayed.[8] This gives an extra margin of safety to the uranium reactor as compared to the plutonium reactor. Delay times are so short that shutdown must depend on mechanical devices. **Control rods**, which absorb neutrons and can cause the reactor to cease operation, can be inserted mechanically at any time on seconds' notice. (Figure 19.1 shows a schematic fuel assembly and control rod.) The mousetrap and Ping-Pong ball model can be extended to controlled reactions by using flypaper hanging above the table as a model for control rods. Some Ping-Pong balls can stick to the strips of flypaper; the number of strips can be adjusted until only one ball from each "fission" escapes to produce a subsequent "fission."

It is important to be able to shut down the reactor in case of emergency. See **Extension 19.2, *Reactor Safety Systems*** for a description of **SCRAM** (emergency shutdown, originally "safety control rod ax man") and **loss-of-coolant accident** (LOCA). A severe LOCA combined with other system failures can lead to a **core melt**. There is an **emergency core cooling system** (ECCS) built into reactors to

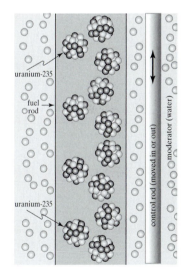

FIGURE 19.1

Inside a nuclear reactor, fuel pellets are held in fuel rods that are surrounded by water, which acts as a moderator and slows neutrons produced in fissions. Control rods, containing neutron absorbers, may be inserted or removed. The control rods can turn down or turn off the reactor if need be.

remove residual thermal energy from the reactor fuel rods if the normal core cooling system fails.

19.2   CONVENTIONAL REACTOR TYPES

Several different types of nuclear reactor are used in generating electricity on a commercial basis. The two most common types, the BWR and the PWR, were briefly discussed in Chapter 9. Three types of reactor use water in their cores: the BWR, the PWR, and the Canadian-designed pressurized water reactor (**CANDU**). In addition, some energy is generated in **high-temperature gas-cooled reactors** (HTGRs), and in the former Soviet Union, in **graphite-water reactors**. As of 2001, there were 35 BWRs and 69 PWRs operating in the United States—104 altogether, the greatest number of any country in the world. In the United States, the active nuclear reactors produced 753.9 billion kWh (754 TWh) of electricity in 2000.[13–15] Figure 19.2 shows the history of U.S. power capacity (available power) and energy generation.

FIGURE 19.2

(a) The nuclear electric power capacity has grown to about 100 GW since the first commercial reactor was opened. (b) Net nuclear electric generation has grown to about 800 TWh, about 20% of U.S. production.
(U.S. Department of Energy)

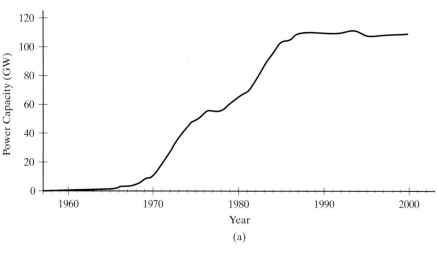

(a)

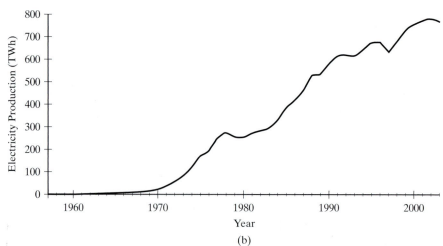

(b)

TABLE 19.1

World's operating reactors and capacity, 2002

Country	Number of units	Total capacity (MW)
United States of America	104	98,230
France	59	63,073
Japan	54	44,287
United Kingdom	31	12,252
Russian Federation	30	20,793
Germany	19	21,283
Republic of Korea	18	14,890
Canada	14	10,018
India	14	2503
Ukraine	13	11,207
Sweden	11	9432
Spain	9	7524
Belgium	7	5760
China	7	5318
Slovak Republic	6	4884
Czech Republic	6	3468
Slovak Republic	6	2408
Switzerland	5	3200
Bulgaria	4	2722
Finland	4	2656
Hungary	4	1755
Argentina, Brazil, Lithuania, Mexico, Pakistan, South Africa	2	Average, 1465
Armenia, Netherlands, Romania, Slovenia	1	Average, 540
Total	441	358,661

Source: International Atomic Energy Agency, *Country Nuclear Power Profiles*, Annex II, Table II-1, February, 2005

There are 438 nuclear electricity reactors operating worldwide. Another 32 were under construction as of 2002. Table 19.1 provides information on the world's operating reactors. In 1996, 77% of French electricity was generated in nuclear reactors; only Lithuania gets a higher percentage from nuclear, 83%.[16] The United States gets about 20% of its electricity from nuclear energy (Figure 19.2). For the world in 1996, the capacity was 351 GW, and the total electricity generated was 1649 TWh from 442 operational reactors. See **Extension 19.3, *Uranium Resources*** for an overview of world nuclear reactors.

Other Current Reactor Types

The CANDU (CANadian Deuterium–Uranium) reactor uses "heavy water" (deuterium oxide, D_2O) as a moderator (and, in a separate loop as a coolant) and can, as a result, use natural uranium in its core, rather than enriched uranium (which has a higher concentration of fissile uranium). In CANDU, the moderator is enriched rather than the fuel. Because of its simplicity and ease of construction, it is used in eight different countries around the globe, but its efficiency is a bit lower than the BWR or the PWR, and its energy release per tonne of fuel is much lower.[28]

The British use so-called MAGNOX magnesium alloy reactors, which are cooled using carbon dioxide gas, but they do not operate at the elevated temperatures of the HTGR reactors. The HTGR has been used mainly in Europe, but it may now be more attractive in the United States, partly because of recent concern over reactor safety and because it operates at a higher temperature than

FIGURE 19.3

Fuel elements for the Jülich type HTGR are shown. They are graphite balls 6 cm in diameter (about the size of tennis balls) containing uranium in 20 000 coated particles. (See a description of their fabrication in E. Brandau, "Microspheres of UO_2, ThO_2 and PuO_2 for the high temperature reactor," in IAEA, *Proceedings of the Conference on High Temperature Reactors*, Petten, NL, April 22–24, 2002 (Vienna: International Atomic Energy Agency, 2002). Photo courtesy Kernforschungszentrum Jülich)

conventional BWRs and PWRs. Because of the higher operating temperature, it should have a greater thermodynamic efficiency. The thermal energy is transferred through helium gas. The fuel used in HTGRs is highly enriched uranium wrapped into 6 cm diameter balls of graphite and ceramic. Because this material has such a high melting point, it is possible that a total loss of helium, which would raise the temperature to 1500 °C or higher, could be withstood without long-term reactor loss. However, there would still be problems in case of such an accident.

The reactor fuel is made out of uranium ore mined and made into a uranium oxide called **yellowcake**. See **Extensions 19.1, 19.3,** and **19.4,** *Uranium Ore Formation*.

The graphite-water reactors (known as RBMK from the name's Russian acronym) in use in the former Soviet Union are known as pressure-tube reactors.[31] They operate on scantily enriched uranium fuel and differ from BWRs and PWRs in that it is possible to refuel RBMKs as they run. The typical 1000 MW_e RBMK reactor is made of 2488 columns of graphite blocks, through which pass 1661 pressure tubes containing zirconium-niobium fuel elements and water (Figure 19.4). The water system for this reactor has been characterized as a "plumber's nightmare."

The rate of reaction is controlled by 210 boron carbide control rods that are cooled separately from the fuel rods. The reactor is heavily monitored and computer controlled because of the fact that it has a "positive reactivity coefficient." This means that if the temperature rises (as it would if power output increased), the core becomes *more* reactive, because steam bubbles absorb fewer neutrons than the water they replace. BWRs and PWRs have negative reactivity coefficients and

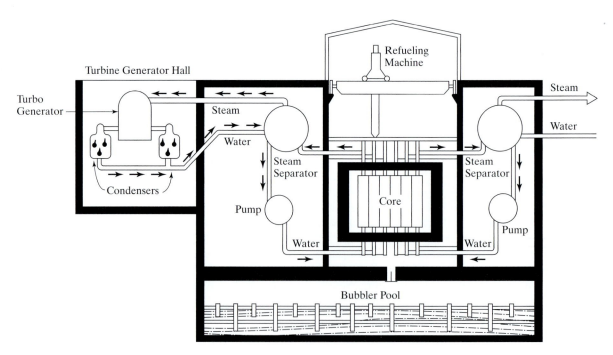

FIGURE 19.4

The RBMK reactor of the type used in Chernobyl.

so are less prone to and less dangerous during accidental failure. This type of reactor, the RBMK-1000, was involved in the 1986 disaster at Chernobyl in the Ukraine (discussed in Chapter 20).

New Reactor Construction

Reactor construction worldwide has slowed because of public perceptions of risk in the wake of reactor accidents, and no more so than in the United States. Utilities do not want to build nuclear plants at all. No plant ordered between 1974 and 1978 is still under construction,[32] and no new plants have been ordered in the United States since 1978 (see Figures 19.2a and b).[33,34] Partly, this was due to construction cost overruns that never led to gains for others, because most reactors were of differing designs. The final cost of the reactors constructed between 1966 and 1977, projected to be $45 billion, was $145 billion.[35]

This was also partly because the plants were originally not as reliable as many had hoped; early supporters of nuclear energy had thought that the reactors would operate 80% to 90% of the time, which would help lower the cost of nuclear as opposed to fossil-fuel generation. Critics pointed out that nuclear reactors were operating far below capacity, thus raising costs.[36–39] The **capacity factor** (percentage of time operational) had long been disappointing, but as Figure 19.5 shows, it has been improving dramatically. In 1990, there were 1.6 unplanned shutdowns per plant; in 1980, it had been 7.4,[40] and the 2001 capacity factor was 89.1%. Most analyses now find that capacity factors for coal and nuclear energy are similar (nuclear is now a bit better).[39] The busbar costs (that is, the costs of the electricity as it leaves the facility) for older nuclear plants are consistently lower for nuclear than for other sources of energy.[25,24] One incentive that raised the capacity factor is that some states allow the utilities to charge higher rates when they maintain higher reliability.[41]

Each U.S. plant has been constructed in the field (very expensive), and very few are like any other. The experience has not really helped the learning curve. Another problem, at least in the United States, is the incredibly high cost of construction (tied to the cost of borrowing money for such a long construction time). The average time for construction of reactors in the United States in 1990 was 14 years, while it was 7 years in Japan,[42] though a pair of plants were built in Japan in about 4 years in the late 1990s,[43] and a mere 5 years in France.[44] In the early 1970s, construction in the United States took only 6.3 years per plant; later in the 1970s, it rose to 11 years per plant, and it rose rapidly again after the Three Mile Island accident (see Chapter 20).[45] Long construction times have driven up the cost of reactors substantially (see the screening curves in Chapter 8).

Although France is continuing its massive reactor construction program, and South Korea continues to build reactors, most of the rest of the world has acted as the United States. In Europe, concern about the environment led Belgium, Germany, the Netherlands, Sweden, and Switzerland to plan to shut down their nuclear power industries over a period of years.[46] However, Sweden appears to be having second thoughts.[18]

Innovative Reactor Designs

In a world concerned about the greenhouse effect, any energy production option that does not put carbon into the atmosphere must be considered seriously. For

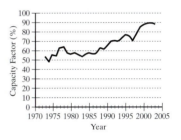

FIGURE 19.5

Capacity factor of nuclear plants. (U.S. Department of Energy)

this reason, some thought has been put into how to construct reactors that are less frightening than those currently being used. Some progress has been made in conceiving "inherently safe" reactors. Of course, there is no absolute safety (see Chapter 18), but there can be designs that are simpler than those in use now. The U.S. Department of Energy granted General Electric, General Atomics, and Westinghouse money to help with the design of commercially viable, safer, more reliable reactors, which have been licensed by the Nuclear Regulatory Commission (NRC).[47,48] The new reactor should be inexpensive (probably modular), cheap to maintain and operate, last a long time, operate reliably, and shut down safely. See **Extension 19.5,** *New Reactor Types* for more detailed information about different possible designs and descriptions of passive safety features of new types of reactor. Passive safety means that no voluntary effort is necessary to prevent disaster if an accident occurs. Measures could include no need for external energy, such as having the emergency core cooling water reservoir above the reactor level, so gravitational potential energy supplies the energy needed to deliver the cooling water.

HISTORY OF ENERGY

The Oklo Reactor

Reference 29 tells a fascinating story of scientific detective work. A French scientist analyzing ore from Oklo, Gabon, in 1972, stumbled on an anomaly. While in normal uranium ore 0.7202% is $^{235}_{92}U$, in his sample, the concentration was only 0.7171%. Further investigation turned up a sample containing only 0.44% $^{235}_{92}U$.

The only acceptable hypothesis is that pieces of the ore body had been a reactor at some point in the distant past. This is possible because, in the past, not so many of the $^{235}_{92}U$ atoms present when Earth formed had decayed. With higher concentrations of $^{235}_{92}U$, a natural reactor could have formed (the concentration in a modern PWR or BWR is about 4%). Billions of years ago, before so much uranium had decayed, mineral accumulations of uranium concentrated by natural geological processes could have been "naturally enriched."

Further studies have revealed that there were six lens-shaped natural reactors in the ore that released about 500 PJ of energy, approximately the energy release of a 1000 MW reactor in 15 years. The reactor ran for at least 150,000 yr (most probably it ran four to five times as long), about 2 billion years ago.

Recent advances in separation technology have led to discovery of several processes using much less energy per kilogram of enriched uranium than gaseous diffusion: gas centrifuge enrichment (Figure 19.6 shows gas centrifuges) and laser enrichment (see **Extension 19.1**).

Irradiation in Reactors

In nuclear fission, the decay products are **radioactive**, which means that they can decay and give off ionizing radiation. There is always radiation from nuclear disintegrations in operating reactors. All the concrete and steel used to contain the reactor is irradiated, and that radiation can damage some materials

FIGURE 19.6
Centrifuges in the gas centrifuge enrichment plant near Portsmouth, Ohio. Construction of the plant was suspended, and the facility has been shut down. The centrifuge hall is shown.
(U.S. Department of Energy)

(see **Extension 19.6**, *Effects of Irradiation*).[77] Neutron bombardment of the re- actor materials may change the reactor's physical properties (making it brittle, for example) and allow leaks to develop or lead to structural failures.

An amount of radioactive nuclei or particles can be characterized by the number of nuclear disintegrations per unit of time. The rate of nuclear disintegrations is known as the **activity**. Activity is measured using the **becquerel** (Bq), which is one disintegration per second. A gram of radium, for example, produces 3.7×10^{10} dis- integrations per second, or 3.7×10^{10} Bq = 37 GBq of activity. An older unit of activity that is still sometimes used is the **curie** (Ci), which represents the activity of 1 g of radium.

Soon after the discovery of radioactivity, it was noticed that at least three dif- ferent kinds of radioactivity were present: the so-called alpha "rays," beta "rays," and gamma (or x) rays (see Chapter 18 and **Extension 18.4**). The alpha particles (helium nuclei, the bound states of two protons plus two neutrons) travel only a few tens of millimeters in air and may be stopped by cloth or cardboard; the beta particles (electrons from the decay of a neutron into a proton in the nucleus) travel up to a few meters in air and are stopped by varying thicknesses of aluminum, iron, and lead; and the gamma rays (electromagnetic energy released from the nucleus) are stopped only by substantial thicknesses of lead. While alpha-particles are eas- ily stopped, if they do penetrate tissue, they are over 10 times as dangerous to health as gamma rays (see Chapter 20). The activity of any radioactive material is the total number of disintegrations per second, no matter which sort of product is produced (alpha, beta, gamma, or other decay products).

Why Reactors Cannot Explode

There are internal safeguards in nuclear reactors in addition to moderators and control rods. As temperature increases, the ^{238}U nuclei absorb more neutrons.

Because the moderator expands, there are more open spaces, and it does not slow down as many neutrons as at a lower temperature. Because only the slow neutrons can initiate the fission reaction, the reactor must then cool down. There is really no possibility of a reactor undergoing a nuclear explosion, because the density of fissile material in the reactor's core is much too low, as shown in Table 19.2.[8] We shall discuss some actual dangers of reactor operation in Chapter 20.

TABLE 19.2

Dilution of fissile materials in reactors

	Commercial[a] reactors	Research[b] or test reactors	Weapons
Fuel enrichment, percent	1–4	90–95	90–100
Fissile concentration in fuel (kg fissile/mfissile/m³ of fuel)	100–400	1000–1500	18,000
Assembled density[c] (kg fissile/mfissile/m³ of solid, including clad and structure)	100–300	200–400	18,000
Fissile concentration in core (kg fissile/mfissile/m³ of core volume)	50–150	100–200	18,000

[a] Thermal neutron power reactors, excluding those fueled with natural uranium.
[b] Thermal neutron type.
[c] Without transient compaction.

However, there are many other things that can happen during the fuel cycle, in the reactor, or even as part of the research that has been done on nuclear weapons that have unfortunate effects. See **Extension 19.7, *Carelessness and the Nuclear Industry*** for some of them, including the most serious recent American reactor problem, the corrosion that ate almost completely through the reactor vessel head at FirstEnergy's Davis-Besse facility in Ohio. The NRC classified this as the fifth most serious incident at a utility reactor in the United States (Davis-Besse is also the location of the second most serious reactor incident).[121]

In addition, since September 11, 2001, there are more serious concerns among politicians about terrorists attacking nuclear installations and using the fuel inside to contaminate large areas.[141] Many experts pointed out that current containment vessels would not be penetrated by a fully-loaded plane strike.[142,143]

19.3 BREEDER REACTORS

Reactors such as the BWR and the PWR, are called **converters**. They produce some ^{233}Th and ^{239}Pu, which could, in turn, be used to run reactors from the reaction of fast neutrons with ^{232}Th and ^{238}U, but they extract only 1% to 2% of the energy theoretically available in uranium.[8] The typical BWR or PWR produces 0.55 atoms of ^{239}Pu[144] per atom of uranium spent. This number, called the **conversion coefficient**, is used to classify reactors. If the conversion coefficient is zero, the reactor is called a **burner**; if it is between zero and one, it is called a **converter**; and if the conversion coefficient is greater than one, the reactor is classified as a **breeder**.

In a breeder, extra ^{238}U is put into the reactor core in such a way as to encourage fast neutrons to make ^{239}Pu. The ^{239}Pu is highly fissile and can be used as fuel for a converter. Because the conversion coefficient is greater than one, the breeder produces more fuel than it burns. The higher the conversion coefficient, the more fuel bred. Breeders can multiply the nuclear fuel supply. The Russian RBMK reactor can be used to breed plutonium (it was originally designed for the Soviet weapons program).

The liquid metal fast breeder reactor (LMFBR) is a breeder that uses liquid sodium metal as the working fluid to achieve higher temperatures than those possible with conventional reactors that use water as the working fluid and because water would moderate the reactor (make it into a converter). The breeder utilizes reactions such as

$$^{238}_{92}\text{U} + n \rightarrow ^{239}_{92}\text{U} \rightarrow ^{239}_{93}\text{Np} \text{ (plus electron and antineutrino)},$$

which beta decays to $^{239}_{94}\text{Pu}$ (the reaction $^{238}_{92}\text{U} + n \rightarrow ^{239}_{92}\text{U} \rightarrow ^{239}_{94}\text{Pu} + 2\,e^- + 2$ antineutrinos can also occur in the breeder). Thus, the uranium atoms capture fast neutrons, and the resultant nucleus undergoes one or several neutron beta decays to produce the more stable, but fissile, isotopes of neptunium and plutonium. Germany and the United States considered building breeders, but environmental concerns halted the projects in the early 1980s. The French built two prototypes—the Phénix and Superphénix, referring to the mythical bird that is reborn from ashes—to set the stage for further research.

A **critical mass** is the minimum amount of fissile material that can sustain a chain reaction internally (and therefore explode). Plutonium has a much smaller critical mass than uranium, so it is easier for it to achieve a state of self-criticality (become a "bomb"). For this reason, plutonium is a proliferation concern for those who worry that rogue states could build a bomb, and that breeders could fuel such bombs or become a weapon in the hands of terrorists. Such a bomb would not have to make much of an explosion; it is hard to build a bomb to explode—the more likely result is a "fizzle." Even a "fizzle" would be bad; it could spread contamination (such a bomb is called a "dirty bomb").[145]

19.4   NUCLEAR FUSION AND PLASMAS

The average binding energy curve of Figure 18.2 indicates that fusion reactions release energy as long as the product has a mass number less than 56. Fusion reactions are responsible for the production of all elements besides hydrogen. Our entire world, exclusive of the primordial hydrogen and helium from the big bang, was built of material forged in the centers of stars. The fusion reactions can take place above 10^7 K, and at this high temperature, all atoms are ionized—all particles are charged. The gas of charged particles is called **plasma**. Most of the universe is plasma, despite our experience here on Earth's surface.

Of the many possible reactions shown in Chapter 18, consider the first,

$$^2_1\text{H} + ^2_1\text{H} \rightarrow p + ^3_1\text{H}.$$

(Note that in these reactions, the atomic number and mass number must add correctly, as discussed in Chapter 18: Here, $2 + 2 = 1 + 3$ for A and $1 + 1 = 1 + 1$ for Z.) We may estimate the extent of the energy resource using this fusion reaction by finding out how much energy is available in common seawater if all the deuterium were to be used in fusion devices. Because 1/6500 of hydrogen atoms are deuterium atoms,[26] 1 m^3 of water contains 34.4 g of deuterium (1.03×10^{25} atoms). Taking the volume of all the oceans to be about 1.5×10^9 km^3, the oceans contain

$$34.4 \text{ g/m}^3 \times 1.5 \times 10^{18} \text{ m}^3 = 5.16 \times 10^{19} \text{ g} = 5.16 \times 10^{16} \text{ kg of deuterium.}$$

Using the first reaction listed, the oceans can supply

$$(104 \times 10^{12} \text{ J/kg}) \times (5.16 \times 10^{16} \text{ kg}) = 5.37 \times 10^{30} \text{ J.}$$

At the 2000 U.S. use rate, the deuterium would last

$$(5.37 \times 10^{30} \text{ J})/(1.04 \times 10^{20} \text{ J/yr}) = 5.17 \times 10^{10} \text{ yr.}$$

The age of our universe is only about 13.7 billion years, so we could supply the United States at its current consumption for about five ages of the universe. Even if we used only 1% of the ocean's deuterium, the energy would last the United States for over half a billion years. If we further assume that the rest of the world would use energy at the 2000 U.S. rate (with present populations), the resource would last, at 1% utilization, almost 30 million years.

If the second reaction, fusion of two deuteriums ($^2_1$H) to produce helium ($^3_2$He)

$$^2_1\text{H} + {}^2_1\text{H} \rightarrow {}^3_2\text{He} + n,$$

is considered, the entire United States could be supplied with energy at the present rate for input of 10 kg of deuterium per hour.[146] As $^2_1$H costs 3×10^{-6}/kWh,[147] and demand is about 350 GW (350 million kW), we find a cost of $1050/h, if deuterium supplied all U.S. power needs.

Fusion looks as if it is a cornucopia—unlimited power for all. The catch is that small-scale controlled fusion has not yet been developed. We do know that it is possible, because all normal stars in our universe run confined fusion reactions. The hydrogen bomb is an uncontrolled fusion device. The question is whether controlled fusion will be able to operate economically on a scale much smaller than that found in a star.

The Lawson Criterion

The two key problems for fusion on Earth are **heating** and **containment**. The reactants must be at sufficiently high temperature ($\sim 10^8$ K) so that they collide with enough energy to fuse. This heat must not be allowed to dissipate before more energy emerges than we had to put in. If 10^8 K particles were to hit any material obstacles, they would quickly give up their energy, and all possibility of fusion occurring would be lost. How long the nuclei must be confined depends on the density. The greater the density, the more collisions, and the more energy is given out in a given time. This is expressed by the **Lawson criterion**:[148]

$$(\text{number density}) \times (\text{energy confinement time}) > 6 \times 10^{19} \text{ m}^{-3} \text{ s.}$$

The Lawson criterion must be satisfied if sustained fusion is to occur. Many research machines have now surpassed it. The first case of a device exceeding the Lawson criterion occurred in late 1982 at the Alcator, a doughnut-shaped fusion reactor, at the Massachusetts Institute of Technology. The (density) times (energy confinement time) was about 6 to 8×10^{19} m^{-3} s.[149] Alcator operated at too low a temperature, however, to be able to produce energy.

In stars, in contrast to reactors, both the high temperature and the confinement are produced by the gravitational compression of the stellar material. Because of the high densities in our sun, its interior temperature can be rather low (~10 MK). In the Sun, the density is about 1.6×10^5 kg/m^3, so the number density is

$$\text{number density} = (1.6 \times 10^5 \text{ kg/m}^3)/(1.66 \times 10^{-27} \text{ kg}) \sim 10^{32} \text{ hydrogens/m}^3.$$

According to the Lawson criterion, if the confinement time is greater than 6×10^{-13} s, fusion can occur, and it does occur in stars, including the Sun.

Approaches to Fusion on Earth

Fusion on Earth is being pursued by three main strategies: magnetic confinement (nonmaterial container), laser implosion (reaction before the possibility of escape to the container walls), and particle beam fusion. Inertial confinement is the term applied to both laser implosion and particle beam fusion. These latter approaches are more speculative, but they arose in response to a desire to achieve fusion while avoiding the many problems associated with plasmas.

Magnetic Confinement

The magnetic confinement idea arose because only a magnetic field (if it is intense enough) can keep the charged particles in a plasma confined inside for long times—by forming a "magnetic bottle" or "magnetic mirror"—and prevent the plasma from touching material objects, which would cool it and convert it back to "ordinary" matter. Some nuclei in the plasma will eventually collide and fuse. However, the plasma, being made up of charged particles having high speeds, is *extremely* unstable. A reliable fusion reactor would need to be able to control the many possible instabilities automatically, which is the focus of current plasma research.

The most promising (and best-funded) approaches thus far are the **tokamak** toroid, the doughnut toroid geometry, and the tandem mirror (Figure 19.7). In a figure-eight configuration, such as was used in Stellerator research fusion reactors, the plasma is trapped inside a figure-eight magnetic field. Researchers have chosen to use tokamaks to do the bulk of the magnetic confinement fusion research.

Recent developments in tokamak research have brought the plasma above the Lawson criterion at temperatures in excess of 75 MK. One source of progress was changing the shape of the plasma trapped from a circular to an oval cross section.[150] The new mirror reactors differ substantially from older ideas that assumed magnetic mirroring through which plasma tended to leak. Electrostatic plugs are now expected to take care of trapping the plasma. However, minimizing power losses from flow of plasma to the plugs is still a focus of research.[150] In addition, most tokamaks operate in a pulsed mode. The most recent development of radiofrequency waves for driving currents holds out the possibility of continuous operation.[150]

The simple toroid geometry, a poloidal magnetic field (running around the periphery of the doughnut) is used to keep the particles in the doughnut, and a

FIGURE 19.7

(a) The Princeton TFTR (tokamak fusion test facility) design.
(b) National Spherical Torus Experiment (NSTX) tests the physics principles of spherically-shaped plasmas.
(U.S. Department of Energy)

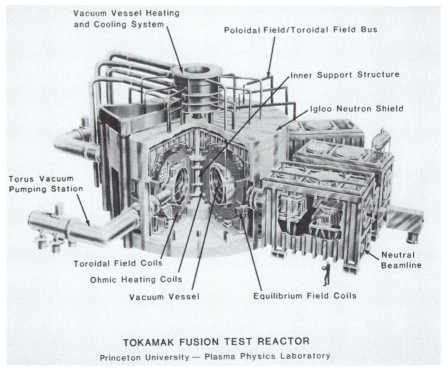

(a)

(b)

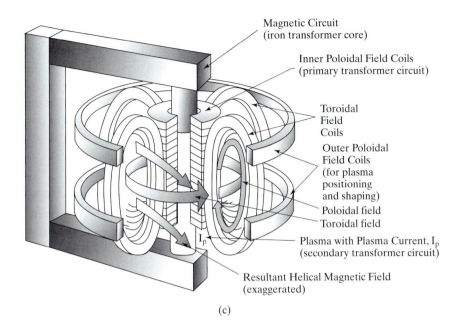

Magnetic Circuit
(iron transformer core)

Inner Poloidal Field Coils
(primary transformer circuit)

Toroidal
Field
Coils

Outer Poloidal
Field Coils
(for plasma
positioning
and shaping)

Poloidal field

Toroidal field

Plasma with Plasma Current, I_p
(secondary transformer circuit)

Resultant Helical Magnetic Field
(exaggerated)

(c)

FIGURE 19.7   (*Continued*)
(c) The Joint European Torus (JET) vessel. Thirty-two toroidal field coils surround the vacuum vessel. Six outer poloidal field coils are used for positioning, shaping and stabilizing the plasma inside the vessel.
(d) Inside the JET torus (1998). (European Fusion Development Agreement)

(d)

vertical field is used to push plasma to the doughnut's center.[150] The Tokamak Fusion Test Reactor (TFTR) at Princeton, a tokamak operating on these principles (Figure 19.7a shows the TFTR installation), clearly exceeded the Lawson criterion and produced a burst of 6 MW of fusion power in December 1993.[151,152] The maximum deuterium–tritium fusion energy (see Chapter 18) output from TFTR was 7.6 MJ.[153] TFTR work was mainly focused on transport, plasma stability, and energetic particle confinement in the reactor core.[153] While TFTR has been shut down, Princeton is still doing fusion research: it is home to the National Spherical Torus Experiment (NSTX, seen in Figure 19.7b), which is investigating properties of spherically-shaped plasmas.

Research outside the United States has involved greater emphasis on torus geometry machines other than tokamaks.[150] The European collaboration, JET (Joint European Torus), at Oxford, England, has had success in generating fusion at 300 MK using the deuterium–tritium reaction.[154] In 1997, JET achieved 13 MW of deuterium–tritium fusion (peak) output power (14 MJ of energy total),[155] and later reached 16 MW (with input of 25 MW).[156] JET is about 15 m around and 12 m tall and contains a doughnut-shaped vacuum vessel with a major radius of about 3 m; the "doughnut" is 4.2 m by 2.5 m (Figure 19.7c shows the main components and Figure 19.7d shows the actual torus).

The largest collaboration in fusion research is that on the International Thermonuclear Test Reactor (ITER). Canada, China, the European Union, Japan, Russia, and the United States will jointly build a huge tokamak, 25 m by 4.3 m by 8.4 m, which is hoped to be begun by 2005 in either France or Japan and completed within 15 years.[156,157] (As of this writing, Canada may withdraw, the consortium is deadlocked between the two sites, and Europe is considering financing the project by itself.)[158] The United States was originally part of the ITER consortium, withdrew over funding issues in 1998, then rejoined in 2003.[159]

Inertial Laser Fusion

Inertial confinement fusion avoids plasma instabilities but not plasmas. If the deuterium–tritium fuel mixture can be held in place, a bath of incoming radiation can raise its temperature and make it into plasma, which then allows fusion to occur on a very short time scale. This is done by packaging the fuel in glass or plastic beads ("micropellets") 50 μm in diameter at pressures 50 to 100 times atmospheric pressure and freezing it. The radiation vaporizes the outer skin, and as it moves away at high speed, it forms a shock wave pushing the fuel in the ball's interior into a dense mass that fuses hydrogen into helium. Because the shock wave keeps pushing the fusing hydrogen together by inertia (continuation of its prior motion), this is known as **inertial confinement fusion**.

Inertial confinement fusion was first demonstrated in 1986. A nuclear weapon was detonated underground to produce high-energy x rays (it powered an x ray laser). The x rays bombarded a small deuterium–tritium filled sphere and succeeded in causing fusion.

The National Ignition Facility (NIF), being built at Lawrence Livermore Laboratory, is a powerful x-ray laser with energy that will "ignite" micropellets (see how micropellets are "ignited" and "burned" in the sequence of Figure 19.8a). The NIF (Figure 19.8b shows a Livermore prototype) will generate successively amplified 6 TW x rays in 5 ns pulses that will be split (see below), amplified, and then focused by lenses and mirrors onto a small spot occupied by the fuel pellet to vaporize the outer part of the pellet and cause implosion of the fuel at temperatures higher than 10^7 K. The resulting microexplosion should give out more energy than was put in.

It is important to compress the micropellets uniformly (in all directions) to make sure the reactants cannot escape. The original laser beam in the inertial confinement scheme is therefore divided into many beams. These beams are reflected and aimed so as to hit the pellet from many separate directions to allow more uniform irradiation of the fuel pellets. (The usual number of beams is 16 or 32.)

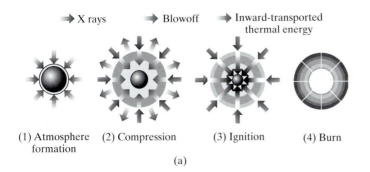

(a)

FIGURE 19.8

(a) The National Ignition Facility (NIF) at Lawrence Livermore National Laboratory will use x-ray laser beams to achieve fusion. The micropellet containing the fuel is irradiated by an x-ray laser (1) and compressed enough (2) to allow fusion to occur (3), which transforms mass energy to kinetic energy (4). (b) The Beamlet, a prototype for NIF. (Sandia National Laboratories)

Sixteen (or even 32) separate beams would not seem enough to provide uniform compression, but the Z-pinch experiment at Sandia Laboratory has demonstrated that such uniform pellet compression can occur.[160] X rays from fusion have been observed, representing 15% of the energy input.[161]

The advent of frequency-converted lasers that raise the energy of the laser light has been essential to recent advances.[162] An inertial fusion device would run rather like an internal combustion engine; rapid bursts of x rays would ignite the fuel, it would "burn," and the kinetic energy would ultimately become thermal energy, which would be tapped in the conventional way. In a 1000 MW$_e$ fusion plant using inertial fusion, there would be about 10 pulses per second, with the lasers or particle beams having to deliver 2–10 MJ during a time of under 1 ns. This is not yet achievable, because currently, laser cycling times are about half an hour.[155] Inertial fusion work is proceeding mostly because of its potential for weapons research applications (making certain that nuclear fuel is not deteriorating) rather than for its use in a practical energy device.

Inertial Particle Fusion

There is some hope that accelerators designed for high-energy physics research could be used to focus beams of particles onto one point and cause the micropellets to implode. This is also a form of inertial confinement fusion. Some researchers are optimistic that these focused particle beams can bring workable fusion technology more quickly and cheaply than can other methods.[163]

Fusion Reactor Engineering

It is hard to arrange net energy output from fusion reactions, because it is so hard to bring together the things that are supposed to be fused. It takes huge magnetic fields or intense laser light or intense particle beams to make a fusion reaction go. This is heartening, because the reaction would never lead to runaway reactors! Fusion is safe in that sense. It is also a challenge for the engineers who will design real fusion reactors as well as for physicists who will conceive them. Much reactor engineering research has focused on tokamaks and, to a lesser extent, tandem mirrors.[150]

Methods of energy conversion are still under discussion. Some argue for direct conversion of the current of charged particles from the reaction. This may more easily be achieved with the linear tandem (magnetic) mirror than with toroids. Most designs advocate deuterium–tritium fusion (which produces α particles and neutrons), capturing the neutrons liberated from the fusion reaction as heat in a liquid lithium blanket surrounding the reactor, which slows or stops neutrons. The blanket would contain ^{7}Li, so that the reaction $n + {}^7\text{Li} \rightarrow 2\,{}^3\text{H} + {}^2\text{H}$ or $\rightarrow {}^3\text{H} + {}^4\text{He} + n$ will replace the tritium burned. It will also keep the high temperatures confined and external radiation deposition below unacceptable levels. Thermal energy is then converted to electricity in the conventional way. The lithium is highly reactive chemically, and so this raises the possibility of fires as a cause for some concern. Use of a lithium-lead alloy instead of pure lithium would reduce the fire hazard.[164]

Capital costs for fusion reactors are thought to be within a factor of two of being competitive in today's market,[150] and the cost of fuel is very cheap. The designs to be used must be for a reliable, steadily available machine that will not produce large volumes of pollutants. The major contaminants to be released to the environment from the fusion reaction are tritium, radioactive hydrogen, ^3_1H, and neutrons, if the deuterium–tritium reaction is used, that will make everything in the vicinity radioactive.

A typical 1200 MW$_e$ deuterium–tritium reactor would contain radioactive materials undergoing 2.4×10^{19} to 1.5×10^{20} nuclear disintegrations per second when shut down. The disintegration rate would drop by a factor of three to 10 a day later. Wastes from fusion-generating facilities are projected to be amenable to near-surface burial.[150] Assuming that some other reaction is chosen, such as protons on boron-11,[165] compared to a fission plant of the same energy output, activity emissions are expected to be 5000 to 100,000 times lower.[164] Other environmental costs of fusion reactors are similar to those of any thermal system.

Holdren[164] believes that fusion energy may prove more acceptable to the public than fission, because

- A fusion severe accident is, of itself, less severe than a fission accident.
- There is lower activation of structural materials than fission.
- No fissile material that could be diverted by terrorists is produced.
- There would be no off-site radiation deaths, even from a severe accident.
- Tight emissions standards could be met.

The timetable for fusion of any sort is not certain, but most authorities do not believe it is feasible to develop fusion for energy before near midcentury. Part of the fusion problem may be that fusion research for weapons is easier and has a greater priority than fusion research for energy. As to the status of fusion as a possible energy source, one scientist was quoted in 1976 as saying, "We see smoke but not fire."[166] This remains true three decades later.

SUMMARY

There are four major types of reactor currently in use: the boiling water reactor; the pressurized water reactor; the Canadian deuterium reactor; and the breeder reactor (still experimental, and only in Europe). The high-temperature gas-cooled reactor has several working prototypes, but it is still not being used for everyday generation of electricity. Pressure-tube (RBMK) reactors have been built only in the former Soviet Union.

Fission reactors are identical to other sorts of reactors in converting steam to electricity in a turbine. The difference is found in the reactor vessel, where nuclei split and give the energy released in the fission to the products of the decay, which further transfer energy to the material in the reactor vessel. Eventually, the temperature gets high enough to produce steam for the turbine. All reactors but CANDU use enriched fuel (see **Extension 19.1**). Uranium ore is often found near geologically active areas (see **Extensions 19.3** and **19.4**).

All working reactors are fission reactors. In fission, the many nucleons in the large nucleus can gain stability by dropping deeper into each of the daughter nuclei. In going into the nuclear energy well more deeply, each nucleon gives up some energy. As a result, the larger nuclei have masses smaller than the sum of the masses of its constituent nucleons. The difference is the binding energy of the nucleons in the nucleus.

BWRs and PWRs cannot undergo a nuclear explosion, because the proportion of ^{235}U is too low, only around 4%. Bombs contain over 90% ^{235}U. In addition, reactors are controlled by relying on delayed neutrons to produce the average of one neutron per fission needed for normal

reactor operation and have special emergency safety systems (see **Extension 19.2**). CANDU and RBMK reactors can be refueled during normal operation, while BWRs and PWRs cannot. Breeder reactors, such as the French-built Phénix and Superphénix, are reactors blanketed in ^{238}U that produce ^{239}Pu fuel as a "waste" product. Breeder reactors are more unstable than conventional reactors, but they hold the promise of an almost infinite supply of reactor fuel.

Neutrons can cause physical deterioration of reactors, which could cause safety concerns. Some effects are described in **Extension 19.6**. Attention to safety in the wake of accidents at Three Mile Island and Chernobyl led the nuclear industry to propose "inherently safe" reactors. These are passive, they do not depend on external power to work, and are described in greater detail in **Extension 19.5**. The industry made problems for itself, as described in **Extension 19.7**.

Nuclear fusion research in the 1990s and 2000s demonstrated that fusion is feasible, but working fusion reactors are probably more than 30 years in the future. Magnetic confinement has provided energy from fusion, but plasma instabilities make the prospect of commercial energy seem remote. Inertial confinement has also demonstrated fusion, but the need for many x-ray lasers also makes this route to commercial fusion seem difficult. Fuel for fusion reactors is virtually inexhaustible. Best estimates of pollution emission for fusion are that the levels should be low, but this must be the subject of engineering research as the fusion reactor is developed.

PROBLEMS AND QUESTIONS

MULTIPLE CHOICE REVIEW

1. Which country has the greatest number of nuclear reactors?
   a. France
   b. Germany
   c. The United States
   d. Russia
   e. Japan

2. The conversion coefficient of a breeder reactor is
   a. less than 0.
   b. equal to 0.
   c. between 0 and 1.
   d. greater than 1.
   e. not able to be defined, because it does not convert.

3. Which reaction below is NOT a possible fusion reaction?
   a. $^1_1H + ^2_1H \rightarrow ^3_2He + \gamma$
   b. $^2_1H + ^2_1H \rightarrow p + ^3_1H$
   c. $^3_2He + ^3_2He \rightarrow ^4_2He + ^1_1H + ^1_1H$
   d. $^{12}_6C + ^1_1H \rightarrow ^{14}_7N + \gamma$
   e. None of the above reactions is NOT a fusion reaction.

4. In nuclear bombs, the concentration of ^{235}U is over 90%. In light water reactors, the concentration of ^{235}U is around 4%. If natural uranium as mined, the concentration of ^{235}U is about 0.7%. Scientists have found that uranium deposits in Gabon functioned as reactors about 2 billion yr ago. What does this imply?
   a. The deposit in Gabon was made of artificially enriched uranium.
   b. Aliens visited Earth about 2 billion yr ago.
   c. Some ^{235}U has spontaneously changed to ^{238}U.
   d. The uranium had much more plutonium in it 2 billion yr ago.
   e. Uranium-235 has a half-life of about 3/4 billion yr.

5. The two key problems for making controlled fusion work on Earth are
   a. obtaining fuel and containing the plasma.
   b. cooling the plasma and radioactive by-products.
   c. heating the plasma and containing the plasma.
   d. obtaining fuel and cooling the plasma.
   e. heating the plasma and radioactive by-products.

6. Which statement below about a fission reactor is FALSE?
   a. The heat from a fission reactor comes from the kinetic energy of the fragments of the broken-up nucleus.
   b. The presence of delayed neutrons allows greater operational safety.
   c. A reactor functions in the steady state when exactly one neutron from each fission is allowed to cause a subsequent fission.
   d. Moderators, which absorb neutrons, can cause a reactor to cease operation.
   e. In use, enough material is present so that there is a net of less than one prompt neutron per fission.

7. Which of the following is NOT a method of uranium enrichment?
   a. Use of a laser to ionize only the ^{235}U in a mixture of isotopes.
   b. Use of a centrifuge to separate isotopes.
   c. Progressive separation by diffusion through membranes of uranium hexafluoride gas.
   d. Use of gravity to separate isotopes of uranium hexafluoride gas.
   e. None of the above is not a method used to separate uranium.

8. The conversion coefficient of a burner reactor is
   a. less than 0.
   b. equal to 0.
   c. between 0 and 1.
   d. equal to 1.
   e. greater than 1.

9. Which country gets the greatest proportion of its electricity from nuclear energy?
   a. France
   b. Germany

   c. The United States
   d. Russia
   e. Japan

10. Which reaction below is a possible fusion reaction?
    a. $^1_1H + ^2_1H \rightarrow ^4_2He + \gamma$
    b. $^2_1H + ^2_1H \rightarrow p + ^3_1H$
    c. $^3_2He + ^4_2He \rightarrow ^4_2He + ^1_1H + ^1_1H$
    d. $^{12}_6C + ^1_1H \rightarrow ^{12}_7N + \gamma$
    e. None of the above reactions is a possible fusion reaction.

SHORT ANSWER

Explain whether the statement made is correct, and justify your answer.

11. Nuclear energy could be part of the answer to the danger of the world's greenhouse warming.
12. A 1 m thick lead plate can provide protection from α, β, and γ radiation.
13. Control rod material must absorb neutrons.
14. The fact that some neutrons are released from a fission after some delay helps make fission plants more stable.
15. Nuclear reactors are bombs waiting to go off.
16. The HTGR is attractive, because its higher operating temperature indicates a greater thermal efficiency in operation.
17. Lithium is a good material to use for a possible fusion reactor, because it is so easy to handle chemically.
18. RBMK reactors are inherently safe, because their graphite moderator gets smaller with increasing temperature.
19. A reactor with a conversion coefficient of 0.87 is a burner.
20. Moderators absorb neutrons.

DISCUSSION QUESTIONS

21. What is so difficult about making a working fusion reactor? Explain.
22. How does allowing the utilities to charge higher rates help increase the capacity factor of nuclear plants? Explain.
23. What is the reason that nuclear utility plants are so expensive? Cite evidence in your answer.
24. Explain why the laser beams used in inertial confinement fusion are even in number (such as 16 or 32).
25. Why do experts think that nuclear fusion energy might be more acceptable to the public than nuclear fission energy?
26. Why are nuclear engineers concerned about neutron embrittlement in fission power plants?
27. What are the advantages of making generating facilities modular? Explain.
28. Why has fusion been described as a "nearly inexhaustible" energy source?
29. The lithium reaction with neutrons $n + ^6_3Li \rightarrow ^4_2He + ^3_1H$ produces 4.8 MeV plus tritium that is used in the deuterium–tritium fusion reaction, $^2_1H + ^3_1H \rightarrow n + ^4_2He$. What is the overall reaction for this deuterium–tritium fusion process followed by neutron capture by lithium?
30. Why is the effect of neutrons on materials such as steel of concern in the design of fusion reactors using the deuterium–tritium reaction?

PROBLEMS

31. Given that we know that the average fission produces 200 MeV in energy, and that nuclear plants have a thermo-dynamic efficiency of about 33%, how many fissions are there taking place each second in a nuclear plant producing 1000 MW of electricity? How much uranium-235 is being used per day?

32. The deuterium–tritium reaction favored by fusion re-searchers, $^2_1H + ^3_1H \rightarrow n + ^4_2He$, produces 17.6 MeV. Given that the electricity-generating parts of fusion plants are going to be conventional thermal plants, how many fusions would be taking place every second during operation of a fusion electrical energy plant producing 1000 MW?

33. The heat of combustion of TNT is about 15 MJ/kg. How much energy is released in complete combustion of 1 t of TNT? The "kiloton" is often used to express the explosive energy of a nuclear weapon. It refers to the energy released in burning 1 kiloton of TNT. A kilotonne is about 1.1 kiloton. How much energy is released in a kiloton explosion? How many uranium-235 fissions would it take to supply this much energy?

34. Average coal has a heat of combustion of 23.4 MJ/kg. At what rate must coal be burned to supply a $1000\ MW_e$ plant? How much coal will be burned in a day? What mass of uranium-235 could supply this same plant?

35. Because the average uranium fission produces 200 MeV, how much uranium-235 would have to be fissioned to produce an amount of explosive energy equal to that of the Hiroshima "atomic bomb"? That bomb delivered 20 kilotons; 1 kiloton of energy is about 15 TJ.

36. The lithium reaction with neutrons $n + ^6_3Li \rightarrow ^4_2He + ^3_1H$ produces energy plus tritium that is used in the deuterium–tritium fusion reaction. How much energy is produced? As the deuterium–tritium reaction itself produces 17.6 MeV, what is the total energy produced per fusion in a fusion reactor with a blanket made of pure lithium-6? The average binding energies per nucleon for lithium-6 and lithium-7 are 4.877272 MeV and 4.116765 MeV,

respectively; for tritium, the average binding energy per nucleon is 2.825811 MeV; and for helium-4, it is 7.072832 MeV.

37. Lithium-6 makes up only 7.5% of lithium. Lithium-7 makes up the rest. Examine the reaction $n + ^7_3Li \rightarrow$ $^4_2He + ^3_1H + n$, and determine whether it produces or absorbs energy. What is the overall energy produced in a single deuterium–tritium reaction when the neutron hits a lithium-7? See Problem 36 for the relevant binding energies. Will there be a tritium supply problem for such a fusion plant?

38. In normal seawater, deuterium makes up one hydrogen in 6500. Suppose a sample of seawater contains one molecule of water that is made up of two deuteriums and an oxygen atom. How many ordinary hydrogen–deuterium–oxygen molecules are there in this same volume? How many ordinary water molecules are there?

39. A plasma has a density of $1.2 \times 10^{20}/m^3$ (this is the actual peak density experienced at TFTR).[154] How long would this plasma have to be confined to make sure fusion occurs?

40. What density would a plasma that is confined for 1 ms have to have?

41. The ratio of prompt (approximately immediate, within about 10^{-14} s) neutrons from a fission from this generation to the previous is known as the reactor constant, denoted k. If $k = 1$, the reactor is critical for prompt neutrons. If $k > 1$, the number of neutrons will grow exponentially with a time constant ("response time") $\tau/(k - 1)$, where τ is the neutron mean life in the reactor, the average of the 10^{-14} s for prompt neutrons and about 14 s for the delayed neutrons. Control rods are set to make sure that $k < 1 + f$, where f is the fraction of delayed neutrons. In this case, the reactor has a response time of greater than τ/f. In uranium-235, the delayed neutrons make up a fraction 0.007 (99.3% are prompt). What is τ? What is the reactor response time if $k = 1 + f$? For what reactor constant k is the response time 100 s? Two minutes? (Inserting control rods can decrease k to any desired value; reactors are typically run with response times of 100 s.)

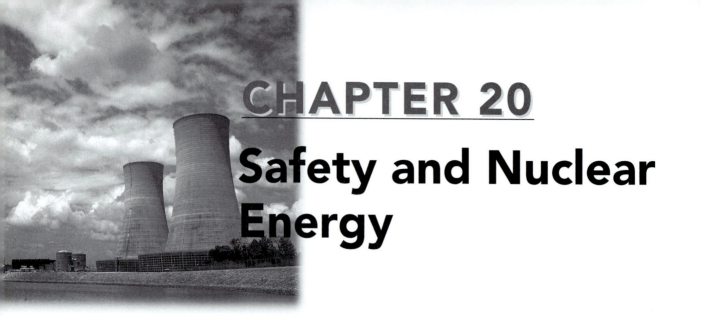

CHAPTER 20
Safety and Nuclear Energy

In this chapter, we examine the consequences of choosing a nuclear option. Two events of recent times have made the public increasingly wary of nuclear energy: the Three Mile Island accident in 1979 and the Chernobyl disaster in 1986. We examine the risks—what could go wrong in the reactors involved in nuclear energy generation; in addition, we consider health effects of exposure to ionizing radiation. The nuclear industry will have to convince the populace that nuclear wastes may be stored safely and that disasters such as Chernobyl are not possible before more people will again accept the nuclear alternative. The chapter compares realistic risks among various methods of generating energy with results that may be surprising to some readers.

KEY TERMS

worst-case assumptions / event tree / fault tree / voluntary risk / involuntary risk / source terms / exposure / dose / linear energy transfer (LET) / gray / sievert / tailing / linear dose theory / threshold / deep burial

Nuclear scientists who worked on nuclear weapons during World War II and wanted instead to turn to peaceful uses after the war recognized early on that safety would be an issue, and they were convinced that they could make nuclear energy absolutely safe. Their insistence that nuclear energy would be absolutely safe has haunted the nuclear industry up to the present day, and it is partly responsible for public disenchantment with nuclear energy.

The prospect of the peaceful uses of nuclear energy led to the formation of the Atomic Energy Commission (AEC) after World War II; that agency had the responsibility to promote and to regulate nuclear energy in the United States. History of this time, which describes the interplay of personality and politics, is available in readable form in Jungck's book, *Brighter than a Thousand Suns*.[1]

20.1 PUBLIC MISTRUST OF NUCLEAR ENERGY

The 1979 accident at the Three Mile Island (TMI) reactor (discussed later in this chapter), propelled the question of the safety of nuclear reactors from a concern of a minority into an issue of concern to the majority of Americans. In such a state of uncertainty as to the hazards of nuclear energy, it is no wonder that citizens have signed petitions, demonstrated at construction sites, or voted to prohibit state licensing of nuclear reactors in some states. Since the TMI accident, no new nuclear plant has been ordered in the United States. California enacted a law that effectively put a stop to further plant construction there. The people of Sweden voted to retire all reactors as they became obsolete and to replace them with energy from other sources. Voters in Austria have kept a completed reactor from opening. The serious accident at Chernobyl in Ukraine hardened opposition to nuclear energy.

There are many issues to address in any examination of the safety of nuclear plants: proliferation of nuclear weapons, catastrophic accidents at nuclear reactors, chronic health-endangering emissions from reactors, health-related concerns from other parts of the nuclear fuel cycle (see Chapter 17), and the ultimate disposal of nuclear wastes. Many people believe that without human intervention, there would be no radioactivity.[2] All these issues are discussed here, but I shall examine in the most detail questions of catastrophic accidents and of relative and absolute health dangers of nuclear energy as compared to other forms of energy generation. **Extension 20.1**, *Background Radiation* which points out that we are always exposed to radioactive decay products, provides some context for the discussion in the book.

20.2 CATASTROPHIC ACCIDENT

The incident at Three Mile Island and the disaster at Chernobyl were severe, but there still has never been a truly catastrophic accident with a nuclear reactor, one that released a large proportion of the core's activity. Of course, the lack of catastrophe is desirable, but it means that it is hardly possible to know the consequences of catastrophes. Our lack of experience means that other approaches must be adopted if anything is to be learned about the possibility of accidents. These alternative approaches are only good insofar as the assumptions made are reasonable. Several quantitative studies of BWRs and PWRs have undertaken[3-6] to assess hypothesized consequences of large reactor accidents.

The earliest guess, in an Atomic Energy Commission (AEC) report known as WASH-740,[3] assumed that an accident would release 50% of all fission products in the reactor core to the environment; it also assumed the worst possible weather conditions. Such **worst-case assumptions** had to be made at this time because of lack of experience and because of the desire to make estimates conservative. The result of these pessimistic assumptions was that this accident in a 167 MW_e nuclear plant could cost 3400 deaths, cause 43,000 acute illnesses, contaminate up to 150,000 mi^2, and do $50 billion damage (2000 dollars). The 1964 revision to WASH-740 was suppressed by the AEC until 1973, when portions were released under the Freedom of Information Act.[7] The results of the revised study were even more pessimistic than WASH-740: fatalities could be as high as 45,000 and an area "the size of the State of Pennsylvania" could be contaminated.

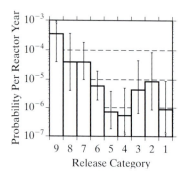

FIGURE 20.1

Histogram of PWR radioactive release probabilities. The categories are classes of accidents; category 9, which has the greatest probability of occurrence (one accident every 3300 reactor years), is least severe. The vertical bars give an indication of the uncertainty in the numbers (one standard deviation).
(U.S. Nuclear Regulatory Commission, [Rasmussen report])

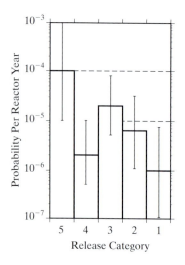

FIGURE 20.2

Histogram of BWR radioactive release probabilities. The categories are classes of accidents; category 5, which has the greatest probability of occurrence (one accident every 10,000 reactor years), is least severe. The vertical bars give an indication of the uncertainty in the numbers (one standard deviation).
(U.S. Nuclear Regulatory Commission, [Rasmussen report])

The probability of such a large number of fatalities occurring and such a large area being contaminated is so extremely low that it is not possible to produce reasonable numerical estimates. Of course, the effect of contamination decreases with time because of the decay of radioactive nuclei as $N(t) = N_0 e^{-t/\tau}$, where τ, the mean life, is a parameter that describes how long, on average, it takes for a nucleus to decay. The activity, which is the number of disintegrations per second, also decreases in the same way:

$$A(t) = \lambda N(t) = A_0 e^{-t/\tau}.$$

See **Extension 20.2, *Nuclear Units and Applications***. The relation between the mean life (τ) of a radioactive particle and the half-life (T), $\tau = T/0.693$, is explained in this extension. See also **Extension 20.3, *Irradiation and Contamination*** for more detail about the distinction.

These reports, flaws and all, were useful, because they forced regulators to formulate a safety philosophy and caused the AEC and the reactor operators to take steps that had not been, but could have been, taken to improve safety. As one example, it was recommended that an evacuation plan be developed for each nuclear facility in case of a release.

The Rasmussen Report

The Rasmussen report analyzed risks for plants such as those now in operation in the United States. To do this, two designs were selected, one each for BWRs and PWRs. Because of the wide variation in designs, some features of the modeled reactors are not found in all reactors in use. The study relied heavily on the use of "fault tree" and "event tree" analyses. An **event tree** begins with an event and then traces all paths stemming from an event (within reason) to see if an accident results. A **fault tree** begins with an accident and attempts to trace backward in time to discover the circumstances leading to the accident in question. Illogical sequences are eliminated, as are those characterized by small consequences.

The analyses described resulted in identification of 78 accident sequences of interest (out of a possible 130,000) that were categorized into 14 groups. These were studied in great detail. The results of the analyses are summarized in Table 20.1, and the health effects—immediate and delayed consequences—as presented in the Rasmussen report, are given in Tables 20.2, 20.3, and 20.4. The categories of release of Table 20.1 correspond roughly (for each reactor type) to increasing probability of occurrence (see Figures 20.1 and 20.2) as well as to decreasing severity. Roughly, the worse the consequences, the less probable the accident.

The study was supposed to compare the risks of nuclear energy to other risks. Some of the most controversial results are presented in Figures 20.3, 20.4, and 20.5. The comparison shown in Figure 20.4, in particular, between the danger of being struck by meteorites and accidents at 100 nuclear power plants, was overstressed in press reports. However, the graphs do express the main conclusion of the Rasmussen report, namely, that the danger of nuclear energy pales in comparison to most other hazards. As a result, the study has been hailed by the nuclear industry as a vindication of nuclear power as is. Its recommendations were reasonable. We learned after the Three Mile Island accident discussed below, however, that despite the Rasmussen report's recommendation that evacuation plans be prepared, none had been.

TABLE 20.1

Summary of accidents involving core

Release category	Probability per reactor-yr	Time of release (h)	Duration of release (h)	Warning time for evacuation (h)	Elevation of release (meters)	Containment energy release (10^6 Btu/h)	Fraction of core inventory released[a]							
							Xe-Kr	Org. I	I	Cs-Rb	Te-Sb	Ba-Sr	Ru[b]	La[c]
PWR 1	9×10^{-7}	2.5	0.5	1.0	25	520[d]	0.9	6×10^{-3}	0.7	0.4	0.4	0.05	0.4	3×10^{-3}
PWR 2	8×10^{-6}	0.5	0.5	1.0	0	170	0.9	7×10^{-3}	0.7	0.5	0.3	0.06	0.02	4×10^{-3}
PWR 3	4×10^{-6}	5.0	1.5	2.0	0	6	0.8	6×10^{-3}	0.2	0.2	0.3	0.02	0.03	3×10^{-3}
PWR 4	5×10^{-7}	2.0	3.0	2.0	0	1	0.6	2×10^{-3}	0.09	0.04	0.03	5×10^{-3}	3×10^{-3}	4×10^{-4}
PWR 5	7×10^{-7}	2.0	4.0	1.0	0	0.3	0.3	2×10^{-3}	0.03	9×10^{-3}	1×10^{-3}	1×10^{-3}	6×10^{-4}	7×10^{-5}
PWR 6	6×10^{-6}	12.0	10.0	1.0	0	N/A	0.3	2×10^{-3}	8×10^{-4}	8×10^{-4}	1×10^{-3}	9×10^{-5}	7×10^{-5}	1×10^{-5}
PWR 7	4×10^{-5}	10.0	10.0	1.0	0	N/A	6×10^{-3}	2×10^{-5}	2×10^{-5}	1×10^{-5}	2×10^{-5}	1×10^{-6}	1×10^{-6}	2×10^{-7}
PWR 8	4×10^{-5}	0.5	0.5	N/A	0	N/A	2×10^{-3}	5×10^{-6}	1×10^{-4}	5×10^{-4}	1×10^{-6}	1×10^{-8}	0	0
PWR 9	4×10^{-4}	0.5	0.5	N/A	0	N/A	3×10^{-6}	7×10^{-9}	1×10^{-7}	6×10^{-7}	1×10^{-9}	1×10^{-11}	0	0
BWR 1	1×10^{-6}	2.0	2.0	1.5	25	130	1.0	7×10^{-3}	0.40	0.40	0.70	0.05	0.5	5×10^{-3}
BWR 2	6×10^{-6}	30.0	3.0	2.0	0	30	1.0	7×10^{-3}	0.90	0.50	0.30	0.10	0.03	4×10^{-3}
BWR 3	2×10^{-5}	30.0	3.0	2.0	25	20	1.0	7×10^{-3}	0.10	0.10	0.30	0.01	0.02	3×10^{-3}
BWR 4	2×10^{-6}	5.0	2.0	2.0	25	N/A	0.6	7×10^{-4}	8×10^{-4}	5×10^{-3}	4×10^{-3}	6×10^{-4}	6×10^{-4}	1×10^{-4}
BWR 5	1×10^{-4}	3.5	5.0	N/A	150	N/A	5×10^{-4}	2×10^{-9}	6×10^{-11}	4×10^{-9}	8×10^{-12}	8×10^{-14}	0	0

[a] A discussion of the isotopes used in the study is found in Appendix VI of Reference 13. Background on the isotope groups and release mechanisms is found in Appendix VII of Reference 13.

[b] Includes Mo, Rh, Tc, Co.

[c] Includes Nd, Y, Ce, Pr, La, Nb, Am, Cm, Pu, Np, Zr.

[d] A lower energy release rate than this value applies to part of the period over which the radioactivity is being released. The effect of lower energy release rates on consequences is found in Appendix VI of Reference 13.

TABLE 20.2

Consequences of reactor accidents for various probabilities for one reactor

Chance per reactor-yr	Early fatalities	Total early illness	Property damage (10^9)	Decontamination area (square miles)	Relocation area (square miles)
One in 20,000[a]	<1.0	<1.0	<0.1	<0.1	<0.1
One in 1,000,000	<1.0	300	0.9	2000	130
One in 10,000,000	110	3000	3	3200	250
One in 100,000,000	900	14,000	8	—	290
One in 1,000,000,000	3300	45,000	14	—	—

[a] This is the predicted chance of core melt per reactor-year

TABLE 20.3

Consequences of reactor accidents for various probabilities for one reactor

Chance per reactor-yr	Latent cancer fatalities (per yr)	Thyroid nodules[b] (per yr)	Genetic effects[c] (per yr)
One in 20,000[a]	<1.0	<1.0	<1.0
One in 1,000,000	170	1400	25
One in 10,000,000	460	3500	60
One in 100,000,000	860	6000	110
One in 1,000,000,000	1500	8000	170
Normal incidence	17,000	8000	8000

[a] This is the predicted chance of core melt per reactor-year.
[b] This rate would occur approximately in the 10- to 40-year period following a potential accident.
[c] This rate would apply to the first generation born after a potential accident. Subsequent generations would experience effects at a lower rate.

TABLE 20.4

Consequences of reactor accidents for various probabilities for 100 reactors

Chance per yr	Early fatalities	Early illness	Total property damage (10^9)	Decontamination area (square miles)	Relocation area (square miles)
One in 200[a]	<1.0	<1.0	<1.0	<1.0	<1.0
One in 10,000	<1.0	300	0.9	2000	130
One in 100,000	110	300	3	3200	250
One in 1,000,000	900	14,000	8	[b]	290
One in 10,000,000	3300	45,000	14	[b]	[b]

[a] This is the predicted chance per year of core melt considering 100 reactors.
[b] No change from previously listed values.

Risk-Benefit Analyses

Let us suppose for the sake of argument that the Rasmussen report's conclusions are correct insofar as the probability of catastrophic hazard is small. This might not be as reassuring as it sounds at first. Large accidents cause more impact (even if rare) than many smaller accidents of similar or even larger total mortality. Many

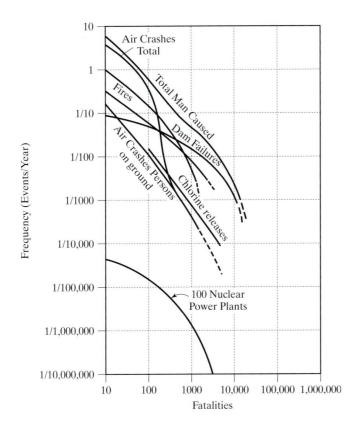

FIGURE 20.3

Frequency of human-caused events involving fatalities compared to that predicted from nuclear reactor accidents. Note that in all cases, the greater the number of fatalities, the lower the probability that the accident will occur.
(U.S. Nuclear Regulatory Commission, [Rasmussen report])

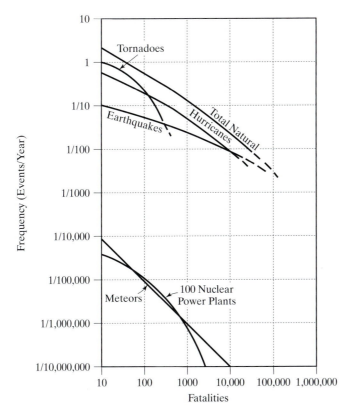

FIGURE 20.4

Frequency of natural events involving fatalities compared to that predicted from nuclear reactor accidents. The scale is the same as that of Figure 20.3.
(U.S. Nuclear Regulatory Commission, [Rasmussen report])

FIGURE 20.5

Frequency of accidents involving property damage compared to that predicted from nuclear reactor accidents.
(U.S. Nuclear Regulatory Commission, (Rasmussen report])

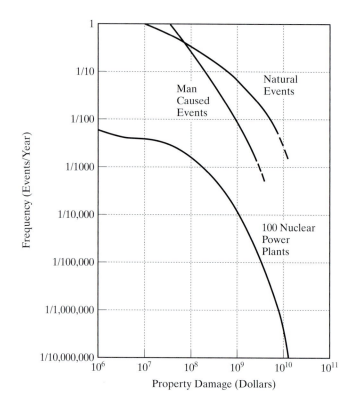

people who have no qualms about driving a car (and in the United States alone there are about 45,000 deaths from auto accidents each year), are unwilling to travel by air (typically 2000 deaths each year in the entire world). Deaths from auto accidents occur by ones and twos, while those from air disasters occur by the hundreds (see Table 20.5). Some of these considerations have been investigated by Chauncey Starr and coworkers.[42] Starr observed that risk (R) and benefit (B) seem to be related both for voluntary activities and for involuntary activities (see Figure 20.6).

FIGURE 20.6

Model of risk versus perceived benefit for various activities. The higher curve shows the limit for activities in which people consciously engage. The lower curve describes the situation for risks outside a person's immediate control.
(C. Starr, "Social benefit versus technological risk," *Science* **165**, 1234 (1969). To convert from 1969 dollars, multiply by 4.5)

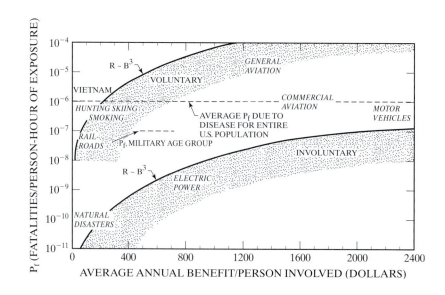

TABLE 20.5

**Individual risk of early fatality by various causes
(U.S. population average, 2002)**

Accident type	Total number for 2002	Approximate individual risk early fatality probability per yr[a]
Motor vehicle	42,643[b]	3×10^{-4}
Falls	NA	9×10^{-5}
Fires and hot substances	3,024	4×10^{-5}
Drowning	3,399	3×10^{-5}
Poison	14,670	2×10^{-5}
Firearms	28,428	1×10^{-5}
Injury at Work	5,252	1×10^{-5}
Water transport	750[c]	9×10^{-6}
Air travel	609	9×10^{-6}
Falling objects	NA	6×10^{-6}
Electrocution	4000[c]	6×10^{-6}
Railway	596	4×10^{-6}
Lightning	3,696	5×10^{-7}
Tornadoes	55	4×10^{-7}
Hurricanes	34[d]	4×10^{-7}
All others	NA	4×10^{-5}
All accidents	85,964	6×10^{-4}
Nuclear accidents (100 reactors)	—	$2 \times 10^{-10[d]}$

Source: National Center for Health Statistics, Centers for Disease Control and Prevention, U.S. Department of Health and Human Services, *Health, United States, 2004* (Washington, DC: GPO, 2004)

[a] Based on total U.S. population, except as noted.

[b] U.S. Department of Transportation

[c] For year 2000

[d] 1970–2004 average

A CLOSER LOOK

Risk Versus Benefit

The risk one is willing to take seems to increase faster than the ensuing benefit ($R \sim B^3$). Note, however, that people seem to tolerate risks that are **voluntary** at a level about 10,000 times greater than risks that are **involuntary**, for the same benefit. The disease risk (about 10^{-6} in Figure 20.6) seems to provide the benchmark for voluntary toleration of risk.[42] There are also indications that acceptable risk decreases in response to involvement of larger numbers of people, for example, as indicated above in the comparison of risk of auto and air accidents. Accidents threatening society (very large mortality over a very short time) are not acceptable until the risk is *much* lower than indicated in Figure 20.6. Other characteristics influencing risk assessment include familiarity, level of knowledge about risk, and considerations of equity.[43]

To illustrate these ideas, Starr and associates considered the case of a fire in the 2 million barrels of oil stored for a 1000 MW_e oil-fired plant as compared to an equivalent catastrophic release from a nuclear reactor.[42] They found the fire to be substantially greater as a real hazard than the nuclear accident, although it seems that the nuclear accident was more frightening. This result is similar to the conclusions reached by the Rasmussen report and serves to demonstrate that the report is not *totally* unrealistic.

The U.S. public is very undereducated on risk issues.[44] One clear example is in the seat belt laws. Ohio's law does not allow police to ticket a motorist for not wearing a seat belt unless the motorist is stopped for some different reason. If the seat belt proportion rose to 85% from 66%, 4200 fewer people would die each year.[45] We all would save—$7 billion in medical costs as well as those lives! And, it would not cost a thing, because seat belts are standard equipment on every car. Contrast this to the Federal Aviation Administration rule that airlines install smoke detectors and fire suppression systems in cargo holds in the wake of the Valujet crash of 11 May 1996. Each of 3700 jets would have to be altered, at a cost of $75,000 each—a total of $277.5 million. The chance of fire is 1 in 40 million flights.[45] The chance of getting in an accident in a car is 1 in 340,000 km (210,000 mi), about 16 years of normal driving for Americans.

This lack of reasoning leads to difficulties with sensible regulation. How should funds be allocated when there are limited resources? The public would have to have some ideas about how to compare two events, have an idea about the magnitudes of probabilities, be able to make mental models, and have some idea of what risk means.[44] None of these is likely. Perhaps the most difficult issue is the meaning of probability. A probability of 5×10^{-5} corresponds to events unlikely to exist; a probability of less than 2.5×10^{-10} is essentially meaningless; and a 10^{-11} probability event would not have been observed in the lifetime of Earth.[46]

Many believe that the report's flaws do not overwhelm the reasonableness of its conclusions. The Birkhofer report,[6] which was prepared for the West German government, reached conclusions similar to those of the Rasmussen report: 104,000 eventual fatalities over 30 years for the worst case, which had a probability of 5×10^{-7} per reactor-year.

The Rasmussen report was reconsidered by a committee of the American Physical Society, and it was seen as representing a substantial advance over previous attempts, particularly in its study of the probabilistic assessment of risks of accidents, though criticism was warranted (see **Extension 20.4, *The Lewis Report*** for more detail).[47] According to these responsible comments, the Rasmussen report should not be accepted uncritically. In particular, its use of numbers for probabilities is tantamount to misrepresentation. We should be extremely skeptical of results presented in the manner of Figures 20.3, 20.4, and 20.5. Perhaps the most compelling reason that the Rasmussen report should be treated with caution is the occurrence of the accident at Three Mile Island unit 2 near Harrisburg, Pennsylvania, which did not follow one of the possible failure modes discussed in the Rasmussen report (the committee had considered its failure scenarios exhaustive). There have been no modeling studies of graphite-steam reactors such as that at Chernobyl, so despite the accident there, one may make no conclusions about the value of modeling from it.

20.3   THE THREE MILE ISLAND ACCIDENT

On 28 March 1979, a serious accident involving a partial core melt took place at the 3-month-old Three Mile Island PWR owned by the Metropolitan Edison Company, a subsidiary of General Public Utilities Corporation. Middletown, Pennsylvania, became the focus of unprecedented public attention; public opinion polls after the accident showed that only 33% were willing to have a nuclear plant in their communities, down from 56% in the previous poll.[49] Subsequent events at Chernobyl (see the following section) had an even stronger negative public impact.

Causes of the Accident

It now appears that valve malfunction, pump malfunction, faulty design, and human error were responsible for the accident at Three Mile Island. Ironically, it was just such causes that were implicated as major factors by the Rasmussen report.[5] The course of events in the accident[9,50] actually began 2 weeks before, when two valves were mistakenly left closed after a routine maintenance check. The accident began when a pump stopped. One of the operators had been working near the pump just before it failed. The pump that failed should have been recirculating water from the condenser back to the reactor vessel. Three backup feedwater pumps should have begun to operate when the pump failed—to take off the excess heat generated—but did not, because the closing of the valves 2 weeks previous had deprived them of their water supply. Because heat removal was inadequate, the temperature rose, thereby increasing the pressure inside the reactor vessel. An automatic reactor shutdown accompanied by a shutting off of the turbine occurred because of the high pressure.

When the pressure rose unnoticed through about 153 times atmospheric pressure (2250 psi), a relief valve opened as it had been designed to do. This reduced the pressure. As the pressure dropped back through normal operating pressure, 150 times atmospheric pressure (2200 psi), the valve failed to close as it should have, and the control board incorrectly showed that it had. The pressure, therefore, continued to drop. When the pressure hit 109 times atmospheric (10^7 Pa 1600 psi), the high-pressure coolant injection system, part of the emergency core cooling system, automatically turned on.[50]

The operators turned off one pump for this system 2 minutes after the system turned on and turned off a second pump 10 minutes thereafter. This was done because a water indicator gave a faulty reading that indicated the water level was high enough. (It was built for steady operation and was unreliable during the rapid changes that were occurring.) This act guaranteed that the core would be uncovered. The operators also turned off one of the primary circulation pumps an hour and a quarter into the accident and turned off the other pump after 1.8 hours, because they were vibrating (again causing heat buildup and exposing even more of the core).[9]

Effects of the Accident

At the worst point in the accident, 15% to 30% of the core was uncovered and apparently remained so for 13.5 hours.[50] The uncovered part of the core became very hot, because there were steam voids, and so there was no water to conduct the heat away as would have happened if water had been there. The damage was recognized as severe just after the accident because of the presence of a "hot" spot in the core of the cooled reactor.

When the pumps were finally restarted, it was found that a large gas bubble, which contained about 2% hydrogen, had developed in the core. At the time, this development raised fears of an explosion. The presence of hydrogen actually helps oxygen to reform water, so these fears appear baseless. Incidentally, the production of hydrogen gas by a hot reactor core was anticipated and discussed in the Rasmussen report (see Reference 51, Section 3.3, main report).

The accident became more severe when the liquid released from the reactor vessel by the stuck valve caused a seal to rupture in a holding tank, ultimately spilling some 400,000 gallons of radioactive water onto the floor of the containment building.

As a result of the accident, 52% of the core melted, and about 70% was damaged. Fuel assemblies broke open and dumped about 20 tonnes of core material into the lower part of the reactor.[52,53] This was a much more severe consequence than expected and raised the question of how the vessel was able to maintain its structural integrity during the accident.[52] Overall, 150 Mt of debris was removed from the containment building during the cleanup.[53]

We have learned from this accident that it is easier to ruin a core than had been believed. The reactor vessel showed little sign of damage, so it was also harder to penetrate the core than had been believed. Decontamination of the reactor was very difficult and took 13 years. The cost of cleanup came to $1 billion.[54]

The Human Factor

The human factor proved important. One of the problems contributing to the accident was lack of skill on the part of the operators and their casual attitude to their work. Another problem had to do with the engineering design of the control room, which was an intimidating place, with alarms going off occasionally. The instrumentation was designed for ease of fabrication, not for communication of information.[55] Human error cannot be eliminated totally, but faulty design and lack of training are reparable. This realization, and the engineering response of instrument simplification and training, has been a long-term good outcome of the accident.

The major release from the plant was radioactive iodine, which was produced when the fuel rods cracked; much less escaped than would have been anticipated. No radioactive krypton or xenon, released at the same time, got out of the plant. The total activity released was only 630 GBq (17 Ci), so the dose to the population adjacent to the plant is thought to be low,[50,56] but it appears to be impossible to determine it exactly.

Another cause for concern brought to light by the accident is the misrepresentation of the facts by the utility, Metropolitan Edison, before the NRC became involved. One observer characterized the utility's description of the reactor as "stable" as roughly equivalent to the description of an out-of-control train as stable, because its speed isn't changing. This sort of misrepresentation is unfortunate and contributes to public mistrust of nuclear energy and of the utilities. It also weakens the voice of the responsible commentators in the nuclear energy debate. Furthermore, it lends credence to the suspicion that all utility executives are venal, as they are characterized in the film *China Syndrome*, a fictional account of a nuclear accident.

The accident at Three Mile Island was almost repeated at the Davis-Besse nuclear plant in Ohio in June 1985.[57] The same set of pumps failed; the valve did not reclose; the operator pushed the wrong buttons in response to system failure; and 14 separate pieces of equipment failed. This time, however, the problem was recognized in time to prevent damage.

What Was Learned from TMI

One of the most useful results of the Three Mile Island accident was the re-analysis it spurred of the release of radioactivity (so-called **source terms**). It had been expected that a lot of iodine-131 would be released. There were 2.4×10^{18} Bq (see **Extension 20.2**) of radioactive iodine in the containment vessel, but only 6.3×10^{11} Bq (17 Ci) was released, one four-millionth of the total. Three reviews were done: the American Nuclear Society (ANS); the industry degraded-core rule-making program (IDCOR); and the American Physical Society (APS). All studies agreed that source terms are lower than calculated in WASH-1400, so that many of those accident sequences have lower releases.[58–60]

The APS study, called the Wilson report,[60] was more cautious than the others. It suggested that it is impossible to say that the "calculated source term for any accident sequence involving any reactor plant would always be a small fraction of the fission product inventory at reactor shutdown."[60] The studies agree that the iodine forms cesium iodide (less volatile than iodine), and that tellurium, another radioactive element, forms compounds with zirconium and stainless steel. The Wilson report presented its caution with a scenario for release of material from containment: With the power off and primary coolant lost, the ECCS is activated. The core eventually melts through the containment vessel and gets to the floor of the containment building. The core boils off the water surrounding it, melts again, and interacts with the concrete of the structure to generate hydrogen and carbon monoxide.[58–60] The pressure could cause confinement to be breached, emitting aerosols to the atmosphere. Such a sequence of events could cause contamination of soil and bodily harm.

20.4   CHERNOBYL

The Accident

On 25 April 1986, the worst known nuclear accident of the nuclear age came to pass at the Ukrainian generating facility at Chernobyl. A fire combined with a breach of the reactor's shell "containment structure" to spray radioactive contamination over the local area as well as over much of eastern and western Europe. According to the report written by the USSR State Committee on Utilization of Atomic Energy for the International Atomic Energy Agency (IAEA),[61] the Chernobyl accident occurred mainly because of human error. The operators committed at least six serious violations of the operations protocol, including disabling all technical protection systems. "The designers of the reactor facility did not provide for protective safety systems capable of preventing (such) an accident … since they considered such a conjunction of events to be impossible."[61]

The conditions that led to the accident arose because of the operators' concern about what would happen if there were a failure in the offsite electrical supply (see also Chapter 19). All nuclear generating stations draw their operating electricity from offsite, and all have local backup generators to provide onsite energy in case of failure of offsite supply. The concern was justified, because an offsite electricity loss was actually experienced at the Kursk nuclear station in 1980.[62,63] For the Russian reactors of the type RBMK, it is especially important to protect against loss of offsite electricity for two reasons: because of the necessity for sufficient water circulation[62] and because of the need for computer control of the reactor at all times, as its design entailed a "positive void coefficient" that made it liable to runaway reactions (see Chapter 19 and **Extension 20.5, *Void Coefficient***).

The nuclear engineers in the former Soviet Union decided to use the kinetic energy stored in the rotating turbogenerators at the stations as they ran down to supply electricity to the pumps, computers, and so forth, during the time required to engage the diesel backups (15 to 60 seconds). (If passive safety features as discussed in Chapter 19 were universal, this would not have been of concern.) This method had already been subject to testing at various stations, including Chernobyl.[61,65,66] During the day of 25 April 1986, the operators of the graphite-steam nuclear reactor unit 4 at Chernobyl were engaged in such a "turbine inertia" test.

The test began at 1 o'clock, as operators reduced the power output to half (1600 MW_t) over a 12 hour period. At 13:05 o'clock, one of the turbines was switched off (in accordance with the test protocol), and at 14 o'clock, the emergency cooling system was disconnected. At this point, the shutdown was stopped, because the electricity from the reactor was needed in the distribution system.[61,62] Stopping a shutdown is a violation of experimental and operating protocol. At 23:10 o'clock, the shutdown resumed, and the test (to take place in the vicinity of 700 to 1000 MW_t) continued. However, as the reactor was being slowly shut down, xenon had been building up. Xenon is in equilibrium in normal operation because it captures neutrons easily, which eliminate the xenon as the new nucleus is formed. The neutron flux stops when the reactor stops, and the xenon builds up for about 10 hours before decaying.[65] As a result, when the operators shut off the local automatic regulating system according to the operating rules, the xenon absorbed the neutrons, and the power output plunged to 30 MW_t.

The operators pulled all the manual control rods to raise the power output, and attained 200 MW_t by 1 o'clock on 26 April.[62–66] This placed the reactor in a precariously stable condition; nevertheless, it was decided to conduct the experiment. Two additional pumps were started in order to join the six operational pumps, so that four could be shut down during the test. This caused the flow rate to jump and the reactor steam pressure to drop toward the emergency trip level. This was a violation of operating instructions; when too much water flows through the pipes, regions of voids, or bubbles, build up and cause vibrations.[61,62]

The operators cut off information to prevent the automatic trip and ignored a printout requiring immediate shutdown. Because of the drop in steam production, all the automatic control rods withdrew.[62–66] At 1:23 o'clock, the operators blocked the closing of the emergency regulating valves so that the test could be repeated if necessary, again in violation of all operating procedure and test protocol.

At 1:23:40 o'clock, the shift foreman realized that something had gone dreadfully wrong and ordered emergency SCRAM. The control rods began to engage but then halted with a shock. Computer analysis indicated that within 3 seconds of SCRAM, the power rose above 530 MW_t for some seconds. The increase in heat output probably caused some pressure tubes to rupture.[61,67,68] As the pressure tubes ruptured, water reacted with zirconium from the fuel rod cladding to produce hydrogen, and steam and graphite reacted endothermically to produce hydrogen and carbon monoxide.[61,62] Presumably, the high pressure breached the containment around the many pipes that penetrate it, allowing oxygen inside the containment. It is probably at this point that the 1000 tonne cover plate lifted.[69] This led to immediate ignition of the hydrogen at the high temperature inside the core.

At 1:24 o'clock on 26 April 1986, there was a loud bang, followed seconds later by a fireball and two explosions. The explosions pulverized fuel rods, and the rising plume from the explosion carried the debris upward about 500 m,[70–72] which protected the region nearby from immediate contamination. Flames rose over 30 m into the air as the graphite caught fire.

The high flames carried most radiation upward so that relatively few deaths occurred (all 31 of the dead were plant workers or firefighters),[68,69] and the evidence of severe irradiation of firefighters and workers did not appear until the next day in most cases. An additional firefighter and many workers developed acute radiation sickness. The fire itself was put out by about 5 o'clock that morning, although smoldering continued until the core was buried by the lead and boron was dropped by helicopter.

The Human Factor

Local authorities apparently thought they had a "normal" fire, not a nuclear accident, according to a report by *New York Times* correspondent Harrison Salisbury, an expert on the Soviet Union.[73] As a result, no one in authority knew what the Swedes were talking about when they detected the radioactive cloud and demanded an explanation on 28 April.[73,74] This misconception on the part of the local authorities may explain why the unit 3 reactor was not shut off until 3.5 hours after the start of the accident; why units 1 and 2 were not shut down until a *day after* the accident; and why local residents were not evacuated until over 36 hours after the start of the accident. The Russian delegates at the IAEA meeting in Vienna in August 1986 were reportedly still incredulous at the circumstances surrounding the accident.[62] The Russian report[61] said that

> the prime cause of the accident was an extremely improbable combination of violations of instructions and operating rules committed by the staff of the unit. The accident assumed catastrophic proportions because the reactor was taken by the staff into a non-regulation state in which the positive void coefficient of reactivity was able substantially to enhance the power excursion.

J. M. Hendrie, former NRC chairman, agreed with the Russian analysis of the actions of the operators at Chernobyl:[61] "What they did sounds like a very poorly considered invitation to disaster."

The command system should not be exonerated of responsibility. The RBMK reactor was developed from a military reactor designed for plutonium production (and converted to civilian use). For this reason, the reactor is designed for refueling while running. This leads to the presence of the thousands of pipes cutting through the upper surface of the containment shell, which is a rectangular cavity more liable to breach than the curved containment vessels used elsewhere. It is known from the isotopes observed in the fallout that the Chernobyl reactor was not used for plutonium production.[68]

The reactor design was faulty because of the positive void coefficient (see **Extension 20.5**), and the other instability problems plaguing the RBMK reactor design. The bureaucratic mentality was involved as well; nuclear energy expert R. Wilson[69] said that "(t)hey specified a set of operating rules to be rigidly followed. But they forgot that rules that are not understood are often not complied with."

The accident released about 2.5% of the radioactivity in the core into the environment.[74] By far, the accident was the largest nonweapon release of radiation publicly known. The radiation, in addition to being detected in Europe,[71,74,75] was

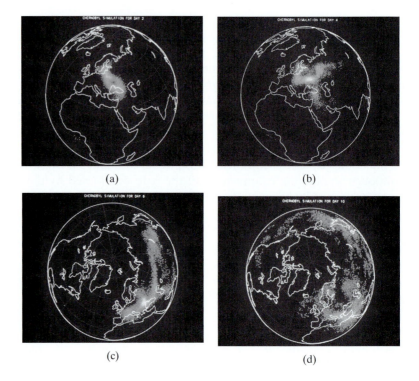

(a)　　　　　　　　(b)

(c)　　　　　　　　(d)

seen clearly on Japanese monitoring devices.[76] Deposition of radioactivity was extremely nonuniform; where rain fell, much was deposited. The debris of the accident was carried so far (Figure 20.7), because it was carried upward by the hot gas plume from the graphite fire. Levels of radioactivity were higher for countries bordering the former Soviet Union than in many parts of that country.

Windscale

There had been an accidental release from a graphite reactor in 1957 at Windscale (now Sellafield) in England. The reactor produced plutonium for weapons. Because of operator error, it caught fire during a routine procedure to release "Wigner energy," an energy of dislocation caused by collisions of neutrons with the carbon atoms. The fire burned for about 4 days and released a lot of radioactivity locally.[67,77] That accident released only one-150th of what was released at Chernobyl. Health consequences of releases will be discussed after we define some terms.

20.5　HEALTH EFFECTS OF IONIZING RADIATION

To be able to discuss the health effects of radioactivity, it is necessary to discuss radioactive **exposure** and **dose**. Radioactive decay involves emission of one to several particles from a nucleus as it changes its identity. These particles may have high or low energy. The biological effects ensue due to energy loss along the path of the particle through the body. For example, a particle in the soft tissues of the body can ionize water, which reacts destructively with other cell units. For this reason, this radiation is known as ionizing radiation.

Background Radiation

As discussed more fully in **Extensions 20.1** and **20.2**, we are at all times bathed in radiation. Some is from our surroundings and some comes from inside our own bodies. The net effect of all this radiation in our environment is known as background. While it is somewhat difficult to measure the dose from internal body radiation, if a detector such as a Geiger counter is available, it is easy to measure the radiation from our surroundings. In my classroom, we average about one count every 5 seconds on my counters (the actual count found will vary according to location and the area of the detector window—mine have windows about 1 cm^2).

Exposure and Dose

Exposure is a measure of the energy deposited in matter by radiation. The effects of exposure to radiation depend both on **linear energy transfer** (LET) and on the energy of the particle. Generally, the amount of charge produced in ionization and the energy deposited in a sample are reasonably easy to measure and are useful measures of exposure. The **gray**, which is a measure of energy absorbed traveling through matter, is the unit of exposure (officially, "absorbed dose"); gamma radiation that loses 1 J/kg of material (such as tissue) would result in an exposure of 1 gray.

There are many different decay products in nuclear disintegrations, each having a different effect on health because of differing biological effectiveness. (See **Extension 20.2**) While all exposures could now be measured in grays, the biological effectiveness is different for the different types of ionizing radiation—protons, neutrons, and alpha particles are many times more dangerous than electrons or gamma rays. These effects are characterized by the LET (see **Extension E20.2**). The LET of protons and neutrons is about 10 times that of γ rays, while for α particles, it is about 20 times larger than for γ rays.

To discuss the relative hazards of radiation, then, requires a unit that incorporates both energy loss and biological effectiveness. The biological effectiveness of alpha particles increases as their energy increases in contrast to the case for gamma rays and electrons. It is thus convenient to define a unit based on the gray or the rad, which also takes into account this difference. **Dose** is the biological effect of exposure. We write

$$\text{Dose} = (\text{Quality Factor}) \times (\text{Exposure}).$$

The quality factor is "the value of absorbed dose of any radiation that engenders the same risk as a given absorbed dose of reference radiation," which means the risk compared to that from gamma radiation. The unit of dose (officially, "dose equivalent") is the **sievert** (Sv), which is equivalent in biological effect of 1 gray of gamma rays.

One effect we have not considered here is that doses from some sources concentrate in specific body organs. For example, iodine goes directly to the thyroid. The effects of various doses on people are given in Table 20.6. This table refers only to the effects of ionizing radiation (see **Extension 20.6, *Nonionizing Radiation*** for a discussion of other types of radiation).

Everyone in the world is exposed to radiation. It comes from cosmic rays, from bricks, from weapons fallout, from diagnostic x rays, and so on. In Table 20.7, we present the exposures in units of mSv (10^{-3} Sv) per person[17] and in total dose

TABLE 20.6

Somatic effects of radiation dose (soma refers to the body)

Dose (mSv)	Short-term effect
0–250	None detectable
250–500	Slight; temporary blood change
1000–2000	Nausea, fatigue
3000–7000	Half of exposed people die (LD$_{50}$)
25,000	Most tissue destroyed

TABLE 20.7

Radiation doses

	U.S. average personal dose (mSv)	Total dose (10^3 person Sv/yr)
Radon	2.00	500
Cosmic rays	0.27	68
Terrestrial (rocks and soil)	0.28	70
Internal body	0.40	100
X-rays	0.39	97
Nuclear medicine	0.14	35
Consumer products	0.10	25
Occupational	<0.01	
Fallout	<0.01 (0.04 [1963])	10–16
Nuclear fuel cycle	<0.01	0.56
Weapons development	<0.01	1.65×10^{-3}

per year to the United States population.[17,84] The largest component of dose to individuals on average comes from exposure to radon. This is discussed in more detail in **Extension 20.7, *Radon*.**

For 1984, the estimate for medical uses in the U.S. averaged 0.75 mSv/yr[90] and in West Germany, 0.50 mSv/yr.[99] (The mean bone marrow dose from x rays varies from 0.52 mSv/yr for young adults and adolescents to 1.51 mSv/yr for persons over 65.[90] A chest x ray gives a 0.10 mSv dose when properly administered, and a barium enema gives 8.75 mSv.[90]) For comparison, the world average internal body dose is about 0.2 mSv. Also, new consumer products such as smoke detectors using americium-241 contribute a dose estimated at 0.01 to 0.04 mSv/yr.[90] A pictorial representation of the data of Table 20.7 is shown in Figure 20.8.

FIGURE 20.8

Exposure of the U.S. population to ionizing radiation.

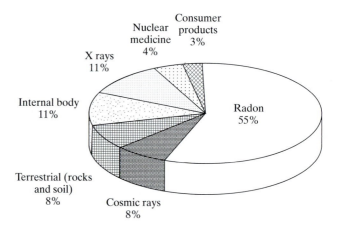

The population was exposed to relatively high levels of ^{90}Sr and ^{137}Cs from fallout from weapons tests in the 1950s and 1960s. The atmospheric nuclear test ban treaty in 1963 had clear effects on the amount of radioactivity in milk (Figure 20.9).[99] The dose, even in 1963—the year of maximum exposure—is about 3% of the natural background exposure. People engage voluntarily in many activities, such as travel by airplane, that cause an increase in yearly dose by amounts larger than this 3%.[13]

The total dose a person gets depends strongly on location. For example, the annual background dose in Denver from cosmic rays is some 0.60 mSv higher

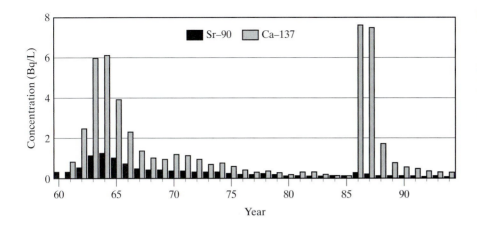

FIGURE 20.9

Activity in milk in Germany. Increases due to fallout from weapons testing and Chernobyl are apparent. Natural levels may be assumed to be the same as between 1980 and 1985. (Umweltbundesamt, *Daten zur Umwelt* [Erich Schmidt Verlag, 1998], Fig. 11.21)

than at sea level. At some locations in the world, radioactivity from subsurface uranium deposits is an order of magnitude higher than at Denver. Airline passengers and crew get doses of about 2 μSv/h at 8 km altitude. Crew members can get 1 mSv/yr, about 10 times that of a chest x ray (0.1 mSv), and more than expected for nonnuclear workers.[11,13,14,100] There have also been large exposures to some individuals because of inappropriate use of mine refuse (see **Extension 20.8, *Civilian Wastes, Defense Wastes, Tailings, and Exposure***). Given the existence of human exposure to radiation, whether or not there are nuclear weapons tests or nuclear power plants, advocates of nuclear energy must estimate the health cost (remember TANSTAAFL) and attempt to compare these costs to the costs of competing energy sources.

Linear Dose Theory

Comparison of health costs of competing energy sources is very difficult indeed, as well as controversial. We do have *some* evidence of radiation effect at very high doses, but what happens at low doses is difficult to determine (refer to the discussion of the Three Mile Island accident). Nevertheless, there must be ways to estimate the effects, so that something can be said about low doses. The problem is that of making a *responsible* estimate. This problem is discussed in detail in **Extension 20.9, *Linear Nonthreshold Dose Relationship*. A threshold** is a greatest exposure for which there is no effect. The result is that most health physicists continue to feel that the linear nonthreshold dose model, while perhaps not accurate, is the most conservative approach.

In the **linear dose theory**, we make these assumptions:

- There is no recovery from any dose.
- Doses add up, no matter how much time has elapsed between them.
- The graph of hazard versus dose is a straight line.

Figure 20.10[86] shows some indication of such straight-line behavior. A study of cancer risk to children as a result of x rays given to the mother during pregnancy gives an apparent straight line (Figure 20.11[86]). More support comes from experiments in induction of mutations in fruit flies; see Figure 20.12.[197] Note that the mutation rate is not zero, even when no dose is applied; this is the mutation rate due to background radiation (and in fruit flies accounts for only about 1%

FIGURE 20.10

Effect of whole-body exposure to ionizing radiation.
(U.S. Atomic Energy Commission)

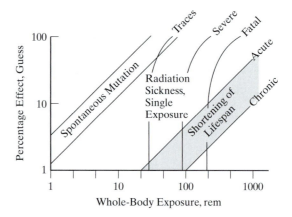

of spontaneous mutations). Other support for linearity comes from studies of the leukemia rate in Hiroshima and Nagasaki after World War II.

One of the problems with the acceptance of the Rasmussen report (discussed earlier) was the use of a dose-reduction factor for low dose. That is, the report assumed that low dose entails less risk than would be present if the linear hypothesis were accepted. The early BEIR (biological effects of ionizing radiation) reports[196,198] set no precedent for this, and the linear standard is the generally accepted one. In the opinions of the experts, the Rasmussen dose-reduction factor seems unjustified. The linear dose theory does have experimental support.

Estimating Health Effects

The estimate of the cancer death rate from the latest BEIR revision for which a consensus existed[199] is 7 to 35.3 fatal cancers per 1000 person-sievert per year. This is about 30 to 140 person-sievert per fatal cancer. Upton[11] quoted a best value of 8.5×10^{-2} cancer deaths per person-sievert. J. Gofman,[198] a persistent critic of the status quo, would prefer 0.266 fatal cancers per person-sievert, about a factor of three greater than Upton; compared to BEIR V,[17] 8.6×10^{-5} to 4.28×10^{-4} fatal cancers per person-sievert, both values are much greater. Current workers disagree on risk estimates but generally accept higher values (lower risk) than in BEIR, as shown in Table 20.8. To be most conservative, we should probably choose a value around 4.0×10^{-2} cancers/Sv.

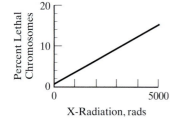

FIGURE 20.11

The risk of cancer grows with the number of x rays to which one is exposed. The data are described by a straight line.
(U.S. Atomic Energy Commission)

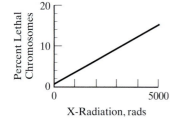

FIGURE 20.12

The genetic effect of exposure to x radiation.

TABLE 20.8

Various estimates of cancer risk from radiation

Source	Risk (cancers/Sv)	Dose effect (Sv/cancer)	Reference
BEIR VI	$0.7–3.5 \times 10^{-2}$	30–140	199
Upton	8.5×10^{-2}	12	11
DOE	2.3×10^{-2}	43	200
EPA	2.0×10^{-2}	50	200
Wilson and Crouch	1.25×10^{-2}	80	201
Garwin	4.0×10^{-2}	25	151
UN benchmark	1.0×10^{-2}	100	200

If the average allowable dose to a person from a particular nuclear plant is 5 mrem, 50 μSv, we may use the linear theory (and BEIR values) and assume that *at worst*, the entire U.S. population gets the average dose (although this is not likely, it would be allowed). Then, with a population of 250 million, the total yearly dose to the population would be 12,500 person-sievert (or 1.25 million person rem). The plant could cause about 500 extra fatal cancers per year under present standards.

Even beneficial uses of radiation have corresponding costs. X rays used for diagnosis are thought to cause 5600 cancers per year in the United States and 700 in the United Kingdom.[202]

TMI

The Three Mile Island accident resulted in the exposure of nearby residents to some radioactivity. The accident released 630 GBq (17 Ci)[60] of activity into the environment. Estimates of dose range from 16 to 53 person-sievert within 80 km of the plant (with about 90% of the dose concentrated within 15 km); an average release estimate of 33 person-sievert has been adopted. The highest conceivable dose—that to a person who would have stood at the plant boundary throughout the course of the accident—is 1 mSv (100 mrem) compared to yearly background of 3.6 mSv. People who were within 1.5 km of the plant the entire time probably received a dose of about 0.5 mSv, and those continuously within 3 to 5 km of the plant probably received about 0.3 mSv.[26,56]

Estimates of the effects on health using the standard linear dose model indicate less than one extra death among the 2 million people exposed. The severest critic, using a model enhancing the effect of low doses, made a prediction that about 50 extra deaths from cancer would result over the course of 30 years (compared to the 110,000 normal cancer deaths over this time). You will see that it is going to be extremely difficult to find any effect of the accident on health. The statistical fluctuation on the estimate of 110,000 is the square root of 110,000, about 330 deaths, over six times the number predicted to die in this worst case.

According to the BEIR guidelines, the dose from Three Mile Island will cause 0.1 to 1.9 extra cancers per 1000 years. In the same way, the Windscale accident, which released 11 PBq (300 kCi) of ^{133}Xe, 0.7 PBq (20 kCi) of ^{131}I, and 0.4 PBq (12 kCi) of ^{132}Te[71,77] is estimated to have caused an extra 39 fatalities.[203] Most investigations of the health effects of the Three Mile Island accident found no cancer or other health effect,[204] even one involving cancers among the 159,684 inhabitants of the area over 15 years;[205] only one study (to my knowledge) has claimed any such effects at all.[206] The most thorough retrospective 20-year study showed no discernable effects (though a connection to a small possible rise in breast cancer could not be ruled out).[207]

Chernobyl

The Chernobyl accident released about 1.85 EBq (50 MCi) into the environment, as totaled on 6 May 1986.[61] Because of the decay in activity [recall the decay relation $N(t) = N_0 e^{-\lambda t} = N_0 e^{-t/\tau}$] of many of the products released during the time between the start of the accident and 6 May, the actual *integrated* exposure is about 2.5 EBq (70 MCi). Because the plume carried the activity up so high, the exposed population at risk totals about 400 million. The total amount of radioactivity within the zone of 30 km radius about Chernobyl was estimated to be 0.7 EBq.[208] Levels of radioactivity measured outside the former Soviet Union ranged from 130 to 750 times background levels in Poland; 1 to 40 times background in Sweden; 1 to 30 times background in Finland, Switzerland, and West Germany; 1 to 10 times background in Austria; and 1 to 2 times background in Britain.[71,74–76] In West Germany, activity of 8 to 12 kBq/m^3 of air was measured after Chernobyl; for comparison, weapons test fallout over West Germany in the 1960s produced measurements of 0.8 kBq/m^3.[71] Thus, peak exposures in western Europe from Chernobyl were 10 to 15 times that from fallout. Of course, the exposure from fallout lasted much longer than that from Chernobyl.

The heaviest exposures at Chernobyl were suffered by the 444 technicians and firefighters on the plant site; 30 of these people died quickly of radiation, and one is probably buried within the plant itself. Out of 600 Chernobyl workers, 134 received high exposures ranging from 0.7 to 13.4 Gy.[209] There are rumors of 5000 to 7000 more fatalities among decontamination workers, the "liquidators,"[100,210] but they have not been substantiated. The 226,000 workers did get exposures ranging from 0.01 Gy to 0.5 Gy.[209]

The 45,000 people of the town of Pripyat were evacuated on 27 April, 1.5 days after the accident. By early morning of that day, the exposure in the streets nearest the power plant was 16 to 53 milligray per hour (mGy/h),[211] and it was increasing. The IAEA report estimates their probable dose from γ rays at 15 to 50 sieverts and the exposure from β radiation to the skin at 0.10 to 0.20 grays.[211] Somewhat later, an additional 60,000 people within a 30 km radius were evacuated. Residents of Pripyat received a total average dose of 30 mSv, and the 24,000 people living in the zone between 3 and 15 km radius received a total average dose of 430 mSv.[69] The report estimated a total collective dose of 16,000 person-sievert[211] for all evacuees. The projected exposure and health effects are shown in Table 20.9. [It should be noted that for large—greater than about 3 Gy—individual doses, the dose in sieverts is thought to overstate the effects, so doses are often given in grays.]

The Chernobyl disaster resulted in two fatalities during the fire as well as 29 deaths from among the total of 203 people diagnosed as suffering from acute radiation sickness. The long-term followup data from the Chernobyl survivors is of immense value to the rest of the world in determination of the effects of low doses of radiation. Elaborate plans for the followup should assure its success.[211] Using the BEIR numbers and the linear dose theory, the 135,000 evacuated people are expected to have about 17,000 spontaneous cancer deaths. Their natural death rate from cancer will be increased less than 2% as a result of Chernobyl for external exposures, that is, less than 2700 extra deaths.[211] The rest of the former

TABLE 20.9

Projected health effects from the Chernobyl nuclear accident

	Population affected (millions)	Collective exposure (person-Gy)	Fatal cancers	
			Natural (thousands)	From Chernobyl (thousands)
Former Soviet Union	279	326,000	35,000	6.5
Europe	490	580,000	88,000	10.4
Asia	1900	27,000	342,000	0.5
North America	250	1200	48,000	0.02

Soviet Union is expected to suffer a comparable number of deaths. For more information, see **Extension 20.10, *More on Chernobyl*.**

Reactor Safety Standards

Now that we have some idea of the risk of radiation dose, it is possible to discuss reactor safety standards. The history of these standards shows a continual lowering of occupational and allowable public exposures, except in the area of medical technology.[86,231] The permissible limits for population exposure in the United States have most recently been lowered from 1.70 to 0.05 mSv/yr (from 170 to 5 mrem/yr) for emissions from nuclear power plants, and they may be lowered again. The limits were lowered in response to the BEIR reports[196,231] and because of the previous use of too low a cancer-to-leukemia ratio by the International Committee on Radiation Practices. The studies on which the standards had been based had been conducted over too short a term to measure the rates adequately, because of the long delay time for the occurrence of some types of cancer.[231,232] A related problem has to do with biological concentration mechanisms.[233] Concentration travels up the food chain; it can be dangerous to eat fish because of their high radiation levels, even though they are taken from water that meets the safety standards for emission from nuclear power plants. (Biological concentration was discussed in Chapters 13 and 15.)

The normal American cancer death rate is about 360,000 per year, of which 80% to 90% are due to environmental effects.[234] To make the estimate another way, leukemia occurs naturally at 0.10 to 0.20 cases of leukemia per million person-sievert per year. For the allowed nuclear plant dose, 12,000 person-sievert, we could have as many as 0.012 to 0.024 extra leukemia deaths. Using an estimate of 20 eventual cancers per leukemia produced[125] leads to 0.24 to 0.48 extra cancers per year from nuclear plants on top of the 360,000 natural deaths. Notice that this is in rough agreement with the numbers of the BEIR report. It also gives a flavor of the sort of accuracy with which the estimates are made.

Over a 30-year plateau period, the worst allowable extra burden of this nuclear power plant is 0.4 to 0.7 extra leukemia cases and seven to 14 extra cancer deaths out of a total exposed population of 10 million. This assumes that the plant operators are careful and the techniques of nuclear energy are perfected. Perhaps the assumption is somewhat unrealistic, as we suggested in the discussion of the accidents at Chernobyl and Three Mile Island.

Occupational and Voluntary Risk

We can also estimate the health cost to people in various professions. In the United States, radiologists die 5.2 months sooner than other doctors. If we assume a 10 sievert lifetime dose, taking delayed effects into account by dividing by five, the linear dose theory tells us that a one sievert dose shortens life by about a day. Thus, we can see the effects on medical x-ray technicians, who get an average whole-body dose of 3.0 to 3.5 Sv/yr; dental x-ray technicians, who get an average dose of 0.5 to 1.25 Sv/yr; medical technicians handling radionuclides, who get a dose of 3.6 to 5.4 Sv/yr; and workers in the civilian nuclear power industry, getting 6 to 8 Sv/yr.[196] Over a 40-year working life, people in these professions should lose about 14, 5, 22, and 32 days of life, respectively, for practicing their work.

For comparison, we give up, on average, half a year of life because of our desire to drive cars given the probability of fatal accidents.[45,199] The dose equivalent to smoking a pack of cigarettes is 0.07 Sv (7 mrem),[199] so each pack shortens life by about 2 hours. A one-and-one-half-pack-a-day smoker would lose over 5 years life expectancy. And this does not even take into account that cigarettes have a relatively high level of polonium-210, a radioactive isotope. The isotope is deposited in the lungs from the smoke that carried it into the smoker's lungs. The one-and-one-half-pack-a-day smoker is said to get a yearly dose equivalent to 1500 chest x rays from ^{210}Po.[235]

Genetic Hazards

In addition to direct hazards of ionizing radiation, there is a *genetic* hazard for humanity. Figure 20.12[197] showed how lethal chromosomes increase in fruit flies with dose; so it must also be with human beings. Radiation alters genes and chromosomes. The nonlethal mutations are of greatest long-term worry, because the offspring will survive and perhaps pass on damaged germ cells in turn. Also, the increase in the mutation rate eventually means a corresponding increase in the genetic death rate.

Everyone has deleterious genetic material (to some extent), and this is passed on to the next generation. What is deleterious to some may be advantageous to others in other circumstances. This helps give humans the *genetic* ability to adjust to change. Nobel Prize winner Joshua Lederberg estimated that at least 25% of our health care burden is of genetic origin.[236] By acting in a humanitarian way, we can add to our species' genetic load of mutations. There is no excuse for gratuitous additions to the genetic burden; we should be as careful as we possibly can. This is not, however, always the course that is taken. In Western society, men wear pants, which heats the gonads, causing an increased mutation rate.[125] If men were to wear kilts, the rate of mutations would be reduced, but kilts are not the latest fashion.

CASE STUDY

Fallout and Nuclear Accidents

Fallout from nuclear testing in the atmosphere increased in the 1950s and decreased through the 1960s as a result of the nuclear test ban treaty. The U.S. government has been accused by "atomic veterans" of being less than candid about the risks of exposure to nuclear blasts when it sent troops into the Nevada desert during nuclear weapons tests. The tests exposed unwitting civilians as well. Utah was downwind from the test site. As a result, residents of sparsely settled Washington County, Utah, received 10 times the global average dose.[237] Residents of Salt Lake City were exposed on those occasions when the cloud moved farther north than expected. They consequently received four times the global average dose.[237] The 133,000 people in high fallout counties were exposed to a total of 0.86 ± 0.14 R (75 ± 12 mGy), while the 556,000 people in lower fallout areas were exposed to 1.3 ± 0.3 R (114 ± 26 mGy).[237]

Fallout is important for accidents as well. The fallout is, of course, much less than that from a nuclear weapon. The effects of a worst-case reactor accident, in which the core melts and the containment vessel ruptures, and a thermonuclear detonation[238] were compared. The comparison reveals that the lethal zone, defined as the area that would expose people to a dose in excess of 4 kSv/day, is less than 1 mi^2 for the reactor and about 400 mi^2 for the 1 Mt bomb. The Hiroshima and Nagasaki bombs had a lethal zone of about 29 mi^2. The area of contamination for which the dose would exceed 20 Sv/yr for over a month was 1800 mi^2 for the reactor accident and 20,000 mi^2 for the bomb.[238] Of course, dropping a bomb on a nuclear plant is a dreadfully effective way to lay waste to large areas. An attack on a nuclear reactor at the confluence of the Rhine and Neckar Rivers in Germany would render a third of the country uninhabitable.[238]

20.6   DEALING WITH WASTE FROM NUCLEAR ENERGY

No discussion of commercial nuclear energy can be complete without consideration of the problem of disposal of the waste from power plants. There are several categories of wastes from nuclear plants; most is what is called "low-level" waste, such as workers' clothing that may have accidentally come in contact with something radioactive, while the spent fuel is "high-level" waste. Low-level waste is not a long-term problem, but high-level wastes must be removed from the environment as completely as possible for as long as possible. **Extension 20.8** discusses these and other classes of waste.

There are two viable sequestration options for high-level waste: disposal in the ocean deeps and disposal in geological formations underground. There is little chance that the wastes will be disturbed by human efforts in either location. The deep oceans are not teeming with life and are not economic to exploit. Many geological formations are very stable and are not affected by groundwater intrusion, both of which imply that wastes isolated in these formations can rest undisturbed for millennia as the activity decays toward background levels. At present, the geological storage option is the only one in use. **Extension 20.8** explains the status of the Waste Isolation Pilot Project (WIPP) in New Mexico and the Yucca Mountain Waste Storage Facility in Nevada.

Temporary Storage Measures

Our nuclear wastes are being temporarily stored until a decision is made on permanent disposal (see **Extension 20.11,** ***Trust Funds for Wastes*** to understand how the federal trust funds are allocated among various energy-producing methods, one of which is to fund permanent storage of the wastes of reactors for electricity). Congress tried to encourage states to join compacts to set up nuclear waste disposal facilities for the 140,000 m^3 of low-level wastes being generated in the country each year.[239] After January 1986, the regional compacts were supposed to be able to exclude wastes from nonmember states, but movement in this direction is glacial. It looks as if the high-level waste will ultimately, through a political decision and despite stiff opposition, be stored in Nevada (the Yucca Mountain site is described in **Extension 20.8**). Meanwhile, several Indian tribes were investigating whether they could earn revenue by storing nuclear wastes, and the Goshute tribe in Utah has actually gained NRC approval.[240]

Salt versus Rock Storage

Salt formations were originally favored because the heat and pressure would cause the salt to melt and fill up any cracks or cavities. However, the salt is impermeable only when no water is present. The investigations of the AEC in Lyons, Kansas, and the AEC/NRC in southeast New Mexico found that salt formations are not as isolated from water as had previously been thought.[110,124,125,241] A pocket of brine in the salt formation is only about 200 m from the proposed storage site at WIPP.[111,124] The site of the repository was moved to another part of the formation. Water migration toward the waste through the salt occurs because of the thermal energy release, causing severe corrosion problems for any container.[242] This migration of water takes place because solubility increases as temperature increases. The part of a brine pocket near a hot waste container would take in salt from the formation, while the salt in the cooler brine farther away would precipitate out. The salt water in contact with the metal will eat it away in short order. The temperature will be high, initially—a typical canister of one-year-old waste generates 25 kW of thermal energy, but after about 10 years, the flux drops to 3.5 kW.[127,243] Vitrification of waste (making waste into glass) provides some safeguard against spillage and degradation (Figure 20.13). Even glass, however, is susceptible to high leaching rates at the high temperatures expected in the brine.[244] Ceramic containers provide the best prospect of attaining long-term impermeability.

Tests with heated canisters in holes drilled in rock were run for over a year and showed that the theory of rock movement was not yet completely understood. The rock moved less than predicted.[127] In the same experiment, the hydraulic conductivity (ease of water passage through rock) was measured to be about 10^{-11} m/s.[127,239] Water migration through the formations adjacent to the WIPP facility is also slow.[243] Rock appears a better medium for disposal than salt if water might intrude because of the low speed of water migration and because salt dissolves easily in water.

FIGURE 20.13

Nuclear wastes being turned into a glass-like substance in a furnace (vitrification). Note the glassy surface in the foreground. The Department of Energy is investigating this method for isolating some nuclear wastes from the environment. (U.S. Department of Energy)

Reprocessing and Proliferation

The current waste treatment system leaves much to be desired. Washington has Hanford Reservation horror stories—contaminated groundwater, leaking storage tanks. Most people in Nevada do not want a Yucca Mountain waste repository in their state; the same could be said of every state with a nuclear waste site. For a slight cost increase estimated at 0.9% of the total, all useful material could be reused rather than throwing much of it away. With reprocessing, a million-year high-level waste problem would become a 700-year problem.[245] The difficulty with reprocessing is nuclear proliferation: that the plutonium recovered in the reprocessing can be used by nations having the reactors to build bombs, even with imposition of inspections (as with India, Pakistan, and North Korea). It takes only a small amount of even dilute plutonium to make a bomb, a critical mass as low as 10 kg—in contrast to the case for a uranium bomb, with a critical mass of 100 kg.[246] One additional difficulty is that with expensive oil and the recent U.S. tendency to reduce its role as uranium supplier, more nations will decide to "go nuclear" and reprocess their own waste.[247]

Various proposals to avoid nuclear weapons proliferation include construction of breeder reactors, each of which would supply fuel for three to four BWRs or PWRs[248] (recall that the United States has suspended its breeder research); development of so-called advanced converter reactors (ACRs) using the uranium–thorium cycle with quick removal of spent fuel;[249] spread of the newer CIVEX ion-exchange process in place of the more risky old PUREX ion-exchange process could help thwart diversion.[250] There is practically no U.S. research on ACRs, but breeder and ACR research are continuing in France and Russia. The breeder has the burden of producing a lot of plutonium and is more prone to core disassembly (a euphemism for nuclear runaway).[251] Also, if one considers nuclear energy to be a short-term solution to the energy supply problem, the economic argument for the breeder is questionable.

20.7   RISKS RELATIVE TO OTHER METHODS OF GENERATING ENERGY

While nuclear energy has a bad reputation, a dispassionate look at the evidence, as in **Extension 20.12, *Comparing Nuclear and Fossil-Fuel Energy Risks***, shows that nuclear energy is, relatively speaking, safe. By extrapolating the data on dose from coal plants in Reference 234, I find that a 1000 MW_e plant can produce a dose of 360 mSv/yr. The coal effluents are especially dangerous for bones,[199] and emissions are higher in toto than allowed for nuclear reactors. However, there are risks to reactor aging that have not been fully understood, discussed in **Extension 19.7**.

In addition, as Gaia originator James Lovelock, among others, pointed out, nuclear energy does not produce greenhouse gases, and so may be considered more environmentally benign than fossil fuel. Lovelock wrote: "I am a Green, and I entreat my friends in the movement to drop their wrongheaded objection to nuclear energy."[268] Lovelock also said "Perhaps the strangest thing about the Earth is that it formed from lumps of fall-out from a star-sized nuclear bomb. This is why, even today, the Earth's crust has enough uranium left to reconstitute the original event on a minute scale."[269] He thus agreed with the Royal Academy of

Engineering.[143] His plea caused discussion among environmental groups but seemed to change few minds.[270]

In terms of chronic emissions, a nuclear utility is probably safer than a coal-fired plant. The radiation from natural gas is 1000 times higher than that allowed to be emitted from nuclear plants. There are estimates of 15 deaths per year resulting from exposure to radiation in natural gas. Both oil and natural gas also emit much more radiation than would be allowed from a nuclear plant.[115,271] As already discussed in Chapter 13, coal effluents are produced in substantially greater volume than nuclear effluents, and the scrubber residue, in particular, presents a long-term disposal hazard. Nuclear electric energy appears to be a safer form of generating energy than coal electric energy.

SUMMARY

The nuclear industry was expected by its advocates to be safe and to generate energy so cheaply that it would not have to be metered. An extensive research program quantified the hazards of radioactivity, and nuclear plants were designed with safeguards in mind. Because so much more was known about nuclear hazards, and because of assurances that nuclear energy would be absolutely safe, people set standards for nuclear generation that are much more stringent than those for fossil-fuel generation. In addition, expensive retrofit changes have been required in already operating reactors. These costs, together with the extra costs caused because most U.S. nuclear plants are one-of-a-kind, have made nuclear energy very expensive.

Reactors are designed to foreclose the possibility of catastrophic accident, although it is impossible to ensure that an accident will not occur. Reactors cannot cause a nuclear explosion, because the fuel is not concentrated enough (too dilute). They are housed in reinforced concrete containment vessels. Thermal energy is transferred by heat exchangers to isolate the radioactivity.

The Rasmussen report analyzed the probability of catastrophic accident and concluded that chances are quite small. The German Birkhofer report reached a similar conclusion that chances of serious accident are on the order of one per million reactor years. See also **Extension 20.4,** which describes a critical study of the Rasmusssen report.

The accident at Three Mile Island crystallized opposition to nuclear energy in the United States. No new reactors have been ordered since, and many previously on order were canceled. Because most of the problems at Three Mile Island involved the operators, the NRC made changes in licensing requirements. The small release of activity led advisory groups to the conclusion that risks, as quantified before the TMI accident, had probably been overstated. **Extension 19.7, _Carelessness and the Nuclear Industry_**, recounts some new problems of aging reactors.

The accident at Chernobyl was a nightmare come true. People everywhere felt threatened by nuclear energy in a way they never had before. The ubiquity of human error reinforced conclusions reached in the wake of Three Mile Island. The long-term consequences of the Chernobyl accident include 50,000 to 330,000 premature deaths, as well as a greater awareness of the fragility of the technology of nuclear energy. More detail is given in **Extensions 20.5 and 20.10.**

Ionizing radiation can cause health problems; nonionizing radiation cannot (see **Extension 20.6**). The unit of exposure to ionizing radiation is the gray (the rad and the röntgen are also used; see **Extension 20.2**). The unit of dose is the sievert, though at present, the rem is more commonly used. Natural background (see **Extension 20.1**) is about 0.3 sievert. Radon (see **Extension 20.7**) is a large component of natural exposure to radiation. Exposure from nuclear energy is less than 10 mSv. The linear dose theory is used to extrapolate effects from high dose, where the effects can be measured, to low dose, where they cannot (more detail is given in **Extension 20.9**). The health consequences of bomb blasts for people living in Hiroshima and Nagasaki are extremely important in determining risk; at present, there is concern that the exposure data in use are incorrect and that people are at higher risk than estimated.

The problem of wastes from nuclear energy has to be solved. Current thinking is that waste could be buried in salt deposits or deep in rock. Wastes from the civilian nuclear industry are much smaller in volume than those from the military uses. No completely satisfactory solution is at hand. See **Extensions 20.8 and 20.11** for an extended discussion of waste issues.

In comparison to other forms of energy generation, nuclear energy is relatively safe. The public should probably be made more aware of the risks inherent in other common choices for energy generation.

It is appropriate to have stringent standards for generation of nuclear energy. These protection standards should be applied to nonnuclear energy generation as well. Remember TANSTAAFL. Energy benefit has its cost. Those enjoying the benefits should be prepared to pay the costs; these costs should not be borne disproportionately. **Extension 20.12** spells out comparative costs and benefits.

PROBLEMS AND QUESTIONS

MULTIPLE CHOICE REVIEW

1. What was (were) the basic cause(s) of the accident at Three Mile Island?
   a. Human error
   b. Pump malfunction
   c. Valve malfunction
   d. All of the above were causes.
   e. None of the above was a cause.

2. The Rasmussen report analyzed the risk for nuclear plants in the United States. Which of the following is a conclusion of their work?
   a. The danger of nuclear energy pales in comparison to most other hazards.
   b. Three Mile Island was inevitable.
   c. Nuclear fission energy is unsafe.
   d. Create a Nuclear Regulatory Commission.
   e. Dose-reduction factors from the linear dose model are unjustified.

3. You know someone who was exposed to 5 Gy of beta radiation. What can you expect to happen to that person?
   a. Nothing
   b. Temporary blood change
   c. Nausea
   d. 50% chance of death
   e. Most tissue destroyed

4. You are exposed to 0.5 sieverts of protons with a quality factor of 10. Given the effects given in Table 20.6, what do you expect will happen to you?
   a. Nothing
   b. Temporary blood change
   c. Nausea
   d. 50% chance of death
   e. Most tissue destroyed

5. For release of radioactivity outside the body, which of the emissions below has the highest ionizing ability or LET (linear energy transfer)?
   a. Neutron emission
   b. Beta emission
   c. Gamma emission
   d. Neutrino emission
   e. All of the above particles have equal LET.

6. If the conclusions of the Rasmussen report are correct for the probability of a nuclear accident, then the individual risk of early fatality due to a nuclear accident in the United States is
   a. smaller than that of being killed by lightning.
   b. larger than that of being killed by lightning.
   c. about the same as being killed by lightning.
   d. impossible to determine, because the Rasmussen report is not about nuclear accidents.
   e. impossible to determine, because no analysis was done in the report.

7. What was the cause of the accident at Chernobyl?
   a. Human error
   b. Pump malfunction
   c. Valve malfunction
   d. All of the above are causes.
   e. None of the above is a cause.

8. You are exposed to 0.04 Gy of protons with a quality factor of 10. Given the effects given in Table 20.6, what do you expect will happen to you?
   a. Nothing
   b. Temporary blood change
   c. Nausea
   d. 50% chance of death
   e. Most tissue destroyed

9. The main danger in a reactor core melt is
   a. that power would be lost.
   b. the core would melt its way through Earth to China.
   c. the formation of a blow hole through which vaporized core material would escape.
   d. an explosion.
   e. none of the above dangers.

10. You are exposed to 5.10 mGy of neutrons with a quality factor of 21.1. Given the effects given in Table 20.6, what do you expect will happen to you?
    a. Nothing
    b. Temporary blood change
    c. Nausea
    d. 50% chance of death
    e. Most tissue destroyed

SHORT ANSWER
Explain whether the statement made is correct, and justify your answer.

11. Nuclear fission plants generate no pollution.
12. Explosions could be common in reactors were they not computer controlled.
13. A Chernobyl-type accident would not be possible for a Western-style reactor.
14. It was because the health effects of the Three Mile Island accident in 1979 were so severe that it caused general public alarm.
15. Compared to other systems for generating electrical energy, nuclear energy is relatively unsafe.

16. Fewer nuclear reactors have been built than might have been expected because of the high costs of satisfying unreasonable demands for safety compared to other sorts of energy.
17. The radioactivity in wastes from nuclear reactors decays away quickly.
18. If 100 people are exposed to a total (collective) dose of 1 Sv, there will probably be four cancers experienced among the people in this group.
19. A collection of 1 million radioactive particles will become fewer than 1 million particles after some of the radioactive particles decay.
20. Chernobyl demonstrated that nuclear energy can never be a secure form of generating energy.

DISCUSSION QUESTIONS

21. Why was the presence of a steam void so dangerous in the Three Mile Island accident? What could have happened?
22. Why is it so important to model the possible mechanisms of failure in nuclear reactors? Explain why.
23. Why is it important to realize that there is background radiation? Explain.
24. In the monazite sands region of Kerala, India, adults, children, and unborn children receive relatively high doses of radioactivity. The cancer rate is no different there from the rest of Kerala, where people receive much lower doses.[92] How could this be possible? Give your reasoning in your answer.
25. Is it possible for the decay product of radioactive decay to be radioactive? Explain.
26. One problem for civilian nuclear energy is the creation of long-lived radioactive wastes. Explain why they are created and why they are a problem.
27. Three Mile Island and Chernobyl were not the only nuclear accidents in the last six decades. Were there any worse accidents (answer for both Three Mile Island and Chernobyl)?
28. Is the Chernobyl accident an advertisement for passive reactor safety features? Explain.
29. Why can we say that 1963's fallout value was the highest? Explain why this could be so.
30. Why is it impossible for the mass of radioactive particles to decrease by half in radioactive decay after a half-life? Explain.

PROBLEMS

31. A 70 kg man is fatally exposed to 8 Gy of radiation. How much energy was transferred to this person? Would this amount of thermal energy cause death also? What about this amount of gravitational potential energy?
32. A 60 kg woman is given a dose of 40 mSv from an alpha particle of quality factor 20. How much energy was deposited in her body by this dose? Explain.
33. Assuming a life expectancy for females of 79.4 yr, what is the average dose absorbed during a lifetime? What is the chance someone with that dose would die of cancer? Explain.
34. Some residents of Pripyat (the town adjacent to the Chernobyl reactor) were exposed to 2.5 Sv/h for 15 h before evacuation. What effects are to be expected?
35. Cobalt-60 can emit 310 keV beta particles. In the nuclear accident in Goiânia, Brazil, cobalt-60 was thrown away. Suppose a curious raccoon had ingested material containing 2 g of cobalt-60, which has a half-life of 5.3 yr. The raccoon excretes the cobalt after 32 h.
    a. How much of a dose has the raccoon gotten from the beta particles?
    b. What is the likely consequence of such a dose?
    c. How much power would have been transferred by radioactive decay inside the raccoon?
36. Cobalt-60 can emit γ rays of energy 1.16 MeV and 1.30 MeV. In the nuclear accident in Goiânia, Brazil, cobalt-60 was thrown away. Suppose a curious raccoon had ingested material containing 2 g of cobalt-60, which has a half-life of 5.3 yr. The raccoon excretes the cobalt after 32 h.
    a. How much of a dose has the raccoon gotten from the γ rays? Assume the γ rays are emitted in equal numbers.
    b. What is the likely consequence of such a dose?
    c. How much power would have been transferred by radioactive decay inside the raccoon?
37. Plutonium-236 α-decays with a mean life of 3.9 yr. The α-particles have an energy of 5.75 MeV. If 4×10^{-9} kg of ^{236}Pu is inhaled and lodges in the lungs of a 75 kg man for 50 yr, how much energy has been transferred? What dose has been delivered?

SOLAR ENERGY AND ALTERNATIVE ENERGY RESOURCES AND CONSEQUENCES

PART VII

CHAPTER 21

Solar Energy: Wind, Photovoltaics, and Large-Scale Installations

Continuing the examination of various energy resources, this and the next several chapters examine several forms of solar energy. This chapter begins with an examination of the Sun (fueled by nuclear fusion—see Chapter 18) and its output of energy. Wind energy, solar cells, and large-scale solar energy projects are discussed in detail. We examine how likely these sources are to satisfy energy needs, when this might occur, and their limitations and costs. Chapter 22 considers the role of solar energy in hydroelectricity. In Chapters 23 and 24, solar energy resources stored in biomass are examined.

KEY TERMS

electromagnetic radiation / photovoltaic (PV) cells / frequency / period / wavelength / visible spectrum / infrared radiation / ultraviolet / photosphere / photoelectrochemical reaction / blackbody / photodissociation / semiconductor (n-type, p-type) / bands / band gap / valence band / conduction band / central receiver / heliostat / load-shifting / parabolic trough / parabolic dish / solar farm / heat islands / heat balance

21.1   USES OF SOLAR ENERGY REACHING EARTH

The surface of the Sun, which is what we see if we look at the Sun, is hot enough to cause the electrons in atoms to be excited and to emit energy in the form of what is called **electromagnetic radiation**. The Sun delivers all of the energy that makes objects visible, keeps Earth warm, and drives the winds and the ocean currents.

There are many different ways to use the solar energy that reaches Earth. Active and passive solar systems were already described as part of the discussion on energy "conservation" in Chapter 9. In this chapter, we discuss more direct ways to use this energy: wind power, photolysis for use in fuel cells, **photovoltaic (PV)** solar **cells**, and large-scale direct solar projects, including "farms."

21.2   ELECTROMAGNETIC WAVES

Electromagnetic radiation can be produced in many ways; one way is to wiggle a charged particle (such as an electron). For example, our electric generating facilities make electrons wiggle back and forth 60 times per second. As they wiggle, they cause electric and magnetic fields to be generated that travel off in all directions. The electric and magnetic fields, taken together, are the entirety we refer to as electromagnetic radiation. Familiar forms of electromagnetic radiation include radio waves, microwaves, infrared radiation, visible light, ultraviolet radiation, x rays, and gamma rays. We distinguish forms of electromagnetic radiation by their **frequencies** (number of repetitions per second, measured in hertz, abbreviated Hz). Hence, electric power produces electromagnetic radiation at 60 Hz.

If we were on a boat in a lake in which there is a continuous periodic water wave, we could measure the period of time it takes our boat to go through a complete cycle of motion up and down (see Figure 21.1). This is called the **period** of the wave. If we know the period, we also know how many repetitions there are in a second; that is, we know the frequency, the inverse of the period. A water wave with a period of 2 seconds will complete half its cycle in 1 second $\left(\frac{1}{2}\text{ Hz}\right)$; a wave with a period of $\frac{1}{2}$ second will complete two cycles in a second (2 Hz); and so on. This relation is summarized in the equation

$$\text{frequency} = \frac{1}{\text{period}}.$$

If someone were to take a snapshot of the water surface as in Figure 21.1, we would see that there is a repetitive behavior of the wave, one that repeats itself after a certain distance. The distance necessary for the pattern to repeat is called the **wavelength**. If the person in the boat timed the wave for 1 period, it is easy to see from the diagram of Figure 21.2 that the wave would have traveled

FIGURE 21.1

Water waves illustrate the concept of wavelength.

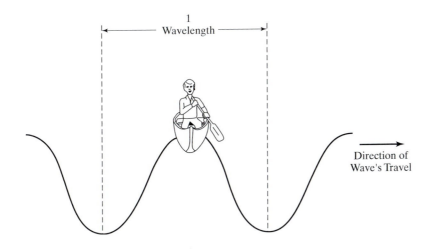

FIGURE 21.2

Water waves move past a boat:
(a) start, (b) 1/4 wavelength after
start, (c) 1/2 wavelength after start,
(d) 3/4 wavelength after start, and
(e) 1 wavelength after start.

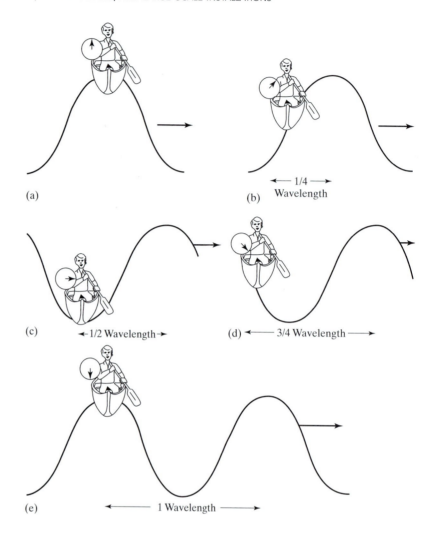

1 wavelength. Thus, its wave speed, which we denote c, is (see Chapter 3) wave-length/period, or alternatively,

$$c = (\text{wavelength})(\text{frequency}).$$

Knowing wavelength, we also know frequency—and vice versa, *if* we know the wave speed c. This means that the frequency and wavelength stand in an inverse relation: a smaller wavelength means a larger frequency, and a smaller frequency means a larger wavelength, because the product must always be the fixed wave speed. For example, if we consider a sound wave that moves through air at 330 m/s, a 165 Hz tone has a wavelength $= \frac{330 \text{ m/s}}{165 \text{ Hz}} = 2.00$ m, while a 1000 Hz tone has a wavelength $= \frac{330 \text{ m/s}}{1000 \text{ Hz}} = 0.330$ m $= 33.0$ cm, and a 2500 Hz tone has a wavelength $= \frac{330 \text{ m/s}}{2500 \text{ Hz}} = 0.132$ m $= 13.2$ cm.

Electromagnetic radiation, which includes light, travels through empty space at a speed of 3×10^8 m/s (300,000 km/s). This speed is usually called the speed of light. The equation relating wavelength, frequency, and wave speed holds for electromagnetic radiation as well. Our 60 Hz radiation from power

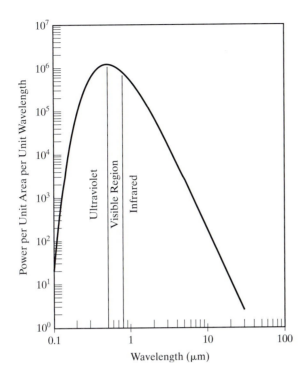

FIGURE 21.3

The blackbody radiation spectrum from the Sun peaks in the visible region of the spectrum. Note the sharp falloff in power on each side. (This is a logarithmic scale, and each labeled tick mark is 10 times larger or smaller than the adjacent one.)

plants corresponds to a wavelength of

$$\text{wavelength} = \frac{c}{\text{frequency}} = \frac{3 \times 10^8 \text{ m/s}}{60 \text{ Hz}} = 5 \times 10^6 \text{ m} = 5000 \text{ km.}$$

Much of the radiation from the Sun is in the visible region of the spectrum (Figure 21.3). The visible white light from the Sun reaching Earth can be made to pass through a slit and then onto the face of a prism. The light is spread out by the prism into a continuum of color that changes uniformly from red (lower frequency) to deep blue (higher frequency), as shown in Figure 21.4. This spread-out light is called the **visible spectrum**. The wavelengths corresponding to red and deep blue are about 650 nm and 450 nm, respectively. In fact, we receive other wavelengths of electromagnetic radiation from the Sun, but these are not visible to us. The "heat" we feel from the Sun comes from wavelengths longer than that of visible red and is known as **infrared radiation**. The radiation causing sunburn comes from regions of wavelengths smaller than visible blue and is

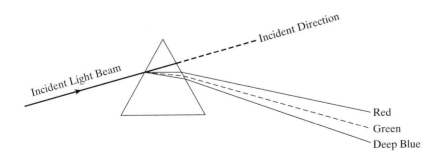

FIGURE 21.4

An incident light beam is refracted (bent) and dispersed (separated by color) by a prism.

FIGURE 21.5

The electromagnetic spectrum.

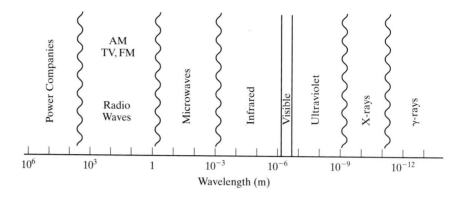

called **ultraviolet**. We extend the meaning of the word spectrum, originally applied only to visible light, to the entire set of electromagnetic radiation. Figure 21.5 presents the spectrum on a logarithmic scale. Note that visible light constitutes only a tiny region of the electromagnetic spectrum. The boundaries between regions on the spectrum are rather vaguely defined (indicated in Figure 21.5 by wavy lines).

FIGURE 21.6

A radiometer, a device with vanes that turn under the influence of light.

HISTORY OF ENERGY

Detecting Invisible Electromagnetic Radiation

Even though we cannot see it, infrared and ultraviolet radiation cause a radiometer (Figure 21.6) to rotate just as visible light does. The fact that infrared radiation is invisible, but acts in the way that light does, allows instruments to be designed that send out infrared light and detect and amplify the reflected infrared light. Such a device, called a *sniperscope*, was developed in the 1960s. The scope was attached to a rifle and used by U.S. soldiers in Vietnam to detect warm objects (people) in the dark. It uses an image intensifier to make infrared light visible to the observer. A more modern version now in use is the night vision goggle, which amplifies light intensity by a factor of about 50,000 and is small and easy to use. Law enforcement and search and rescue teams use the technology routinely.

Light, or electromagnetic radiation, is a form of energy. When energy is transferred, its presence can be detected, as is the case with the sniperscope. The energy associated with an electromagnetic wave of frequency f is $E = hf$, where $h = 6.626 \times 10^{-34}$ J s. The constant h is known as Planck's constant. Einstein first explained in 1905 that radiation is delivered in particle-like "packets," now called photons. The sniperscope's light "packets" carry energy of about 4×10^{-20} J (0.25 eV). This energy is enough to allow them to be detected quite easily.

The color of light is detected in our eyes by special structures called cones that respond to three different (overlapping) ranges of radiation frequencies—red, green, and blue. The cones have been shown to respond to as few as three photons. Our eyes can see both red and blue lines in the hydrogen spectrum. This red has a frequency of 4.6×10^{14} Hz, and the red photons have energy $(6.626 \times 10^{-34}$ J s$) \times (4.6 \times 10^{14}$ Hz$) = 3.0 \times 10^{-19}$ J, or 1.9 eV, much higher than the energy of the sniperscope photons. The blue has a higher frequency than red of 6.2×10^{14} Hz, and so blue photons have an energy $(6.626 \times 10^{-34}$ J s$) \times (6.2 \times 10^{14}$ Hz$) = 4.1 \times 10^{-19}$ J, or 2.6 eV, higher than red photons. Ultraviolet photons carry higher energy than blue photons. The common red laser produces light of frequency 4.6×10^{14} Hz and energy 1.96 eV. A typical 5 mW laser in 1 s transforms 5 mJ into electromagnetic radiation; after transformation, that 5 mJ is being carried by about 16 quadrillion photons {5 mJ/1.96 eV = 5 mJ/[(1.96 eV) $\times$ $(1.6 \times 10^{-19}$ J/eV$)] = 1.59 \times 10^{16}$}.

The Sun

The surface of the Sun that we can see is called the **photosphere**. The temperature of the photosphere is about 6000 K. The gases at that temperature emit a characteristic set of light intensities at different wavelengths, known as the blackbody radiation spectrum (see **Extension 21.1, Blackbodies**). A **blackbody** is a perfect absorber (and emitter) of radiation.

The amount of power radiated by the Sun is 3.9×10^{26} W. This awesome amount of power can scarcely be comprehended. The total energy produced in the world each year by all the world's energy sources is equivalent to the amount the Sun produces in about 1 billionth of a second.

The Sun's energy is derived from fusion reactions taking place inside the Sun, where the temperatures can be measured in megakelvins (MK). The minimum temperature at which fusion reactions can occur is about 10 MK (10^7 K), and the Sun's core is slightly hotter. Using the curve of the average binding energy per nucleon versus nucleon number (Figure 18.2) allows us to calculate that this reaction liberates about 97 TJ/kg. This large energy release reflects the incredible stability of the helium nucleus. The net result of fusion is to transform hydrogen into helium.

Chapter 7 discussed conduction, convection, and radiation as means to transfer thermal energy between temperature reservoirs. The thermal energy produced travels through the gases making up the Sun (mostly hydrogen and helium) by conduction and convection to the limits of the solar atmosphere. The various layers of material in the Sun are somewhat insulated from one another, allowing the Sun's surface (the photosphere) to be at its much lower temperature than the core (6000 K versus 10 MK).

Because of the temperature difference between the Sun's surface and that of interplanetary space (at about 3 K) or Earth (about 300 K), thermal energy may be transferred. Conduction and convection require a material medium to transfer energy, but there is virtually no material medium in interplanetary space. The Sun's electromagnetic radiation, needing no medium, can travel through vacuum between the Sun and Earth.

Solar energy streams off from the Sun into space in all directions at the rate of 64 MW/m^2. In most directions, there is nothing much to intercept the light. As the energy gets farther and farther away from the Sun, it is spread out uniformly over a larger and larger spherical area. A very small fraction of the Sun's radiated power falls on Earth, because each square meter of surface at Earth's orbital distance, about 150 million km from the center of the Sun, receives 1382 W.[1] Put a different way, the power emitted by just 21.5 mm^2 at the Sun's surface falls on 1 m^2 at Earth's orbit.

Electromagnetic Radiation Reaching Earth

The Sun emits radiation mainly in the visible wavelengths, with lower but still substantial intensities of ultraviolet and infrared radiation that decrease as we move away from the visible on either side of the distribution. As we saw in Chapter 17, the atmosphere prevents some of the solar radiation that reaches Earth's orbit from reaching its surface. Most of the infrared and ultraviolet radiation is reflected or absorbed in the upper atmosphere by water vapor, ozone, methane, and other trace gases. A vastly greater proportion of the Sun's visible light makes it through, which is why Earth animals have light detectors (eyes) sensitive to these frequencies (colors) of light and why plants absorb radiation at these frequencies (see the discussion in Chapter 24). The power that reaches the surface, about 1000 W for every square meter, is absorbed by vegetation and is responsible for life on Earth, and for all the coal, oil, and gas deposits found on Earth (see Chapter 12).

Because the Sun's radiation output becomes smaller on either side of the visible region, few x rays and gamma rays (the short wavelength, high-energy region of the spectrum) and few microwaves and radio waves (the long wavelength, low-energy region of the spectrum) reach us from the Sun.

21.3   UTILIZING RENEWABLE ENERGY

A study conducted in 1976 by the National Science Foundation (NSF) and the National Aeronautics and Space Administration (NASA) predicted that by 2000, solar energy could account for 35% of heating and cooling, 30% of U.S. gaseous fuels, 10% of U.S. liquid fuels, and 20% of the nation's electric needs.[2] The study's projection was clearly wildly optimistic. Later realities have dampened enthusiasm for solar power somewhat after substantial growth, especially in California.[3] (The United States even reduced the proportion of renewables in 2001.[4]) Table 21.1 shows the status as of 1999, when there was a renewable capacity of 117 GW.[5] Still, well over 1 million homes in the United States now use solar water heating,[6,7] and the Million Solar Roofs project involves government incentives for far greater use

TABLE 21.1

Capacity and operating status by technology (kW)

Fuel source	Operating	Planned	Retired	Out of service	Standby	Testing	Unknown status
Ag waste	357,773	7,800	358,212	90	723,875		
Biogas	1,063,949	195,896	67,825	4,618	7,200	4,425	1,343,913
Waste-to-energy	2,563,038	476,898	1,200	72,550	3,113,686		
Wood residues	6,584,827	68,800	477,211	8,625	64,700	7,204,164	
Total biomass	10,569,588	272,496	1,380,146	9,818	15,825	141,765	12,385,638
Geothermal	2,697,150	225,499	398,120	55,000	272,000	3,647,769	
Hydro[a]	94,789,367	579,910	414,362	109,624	500	156,760	406,243
Photovoltaics	15,432	66,773	7,562	3	117	89,886	
Solar/thermal	353,925	2,000	15,575	8	371,508		
Wind	2,601,694	1,174,060	302,100	25,214	170,578	4,273,647	
Total	111,027,155	2,320,738	2,517,866	195,659	16,325	156,760	990,710

Source: Reference 15.

[a]Includes pumped storage hydro.

TABLE 21.2

Power available in renewable resources, world, and in use, world and United States

Source	World power resources (GW)	World power (GW)	U.S. power (GW)
Wind power	130,000	13.3	1.54
Tidal power	158	0.26	
Geothermal power	130	24.3	12.5
Hydropower	40,700	1640	117
Biomass		1800	107

Sources: M. R. Gustavson, "Limits to wind power utilization," *Science* **204**, 13 (1979), World Energy Council, *Survey of Energy Resources 2001* (WEC: London, 2002), Office of Coal, Nuclear, Electric and Alternate Fuels U.S. Department of Energy, *Renewable Energy Annual 2000* (GPO: Washington, DC, 2002), DOE/EIA-0603(2000); for world biomass, we use 14% of total power.

of solar energy.[8,9] California has pushed renewables, and has its own million solar roofs project,[10] while the West as a whole,[11] the Midwest,[12] Virginia,[13] and New Jersey also have seen gains by renewables.[14]

Meanwhile other countries are "going renewable" much faster than the United States.[16,17] Germany, for example, had a solar roofs program (originally 10,000, then 100,000 roofs) in place in the early 1990s.[18] The United Kingdom pays half the cost of installing rooftop solar panels. (Costs run from £6700 to more than £20,000.)[19] And, as we will discuss below, solar is economically feasible elsewhere. Europe is much more committed to wind energy than the United States, after a strong U.S. start—or, perhaps one should say, *California* start. As early as 1983, about 5% of California's energy was from renewable or geothermal sources; officials predicted then that this would be up to 10% by 1990.[20] In fact, it was close to 20% (13% geothermal, 4% biomass, 1.3% wind, and 0.4% solar).[21] California has continued to stress these forms of energy for obvious reasons. (Chapter 14 detailed California's air pollution plight.)

Table 21.2[22] shows the worldwide potential for use of renewable resources, which is far in excess of current power use. Table 21.3[22] shows the breakdown of 2003 U.S. energy use, showing that one-sixth was from renewables (much more could be exploited).[23] The United States currently has 90 MW of renewable non-hydroelectric (biomass, geothermal, wind, solar photovoltaic) capacity.[5,24]

TABLE 21.3

Renewables and U.S. energy use in 2003

	Use (quads)	Corrected use[a] (EJ)	Percent of total[a]
Coal	22.31	23.54	29.2
Natural gas	19.64	20.72	25.7
Petroleum, gas liquids	12.14	15.29	19.0
Nuclear	7.97	8.41	10.4
Renewables		12.71	15.8
Hydroelectric	2.78	8.80	10.9
Biomass	2.88	3.04	3.80
Geothermal	0.31	0.31	0.39
Solar	0.06	0.25	0.25
Wind	0.11	0.42	0.42
Total energy use	76.37	80.58	100

Source: Reference 22.

[a]Altered to reflect the difference between electrical and thermal kilowatthours.

FIGURE 21.7

A farmer tends his windmill. Wind energy has been used in the United States for almost a century on the farm. (NASA)

Wind Energy

The windmill is a proven technology. Windmills have been used for more than 1000 years, mostly to pump water and grind grain. Windmills were brought to Europe by merchants and returning crusaders. In the fourteenth century, the Dutch were making great strides in creating the basis for the modern windmill. By the eighteenth century in the Netherlands, there were 10,000 windmills reclaiming the land from the sea,[25] and today, the Netherlands has well over 1000 modern wind turbines.[26] In the United States, the old farm windmill was used for decades to pump water and generate electricity (Figure 21.7). It is still used for electricity in areas too remote for rural electrification. Table 21.2 shows how wind power compares to other renewable sources of power, while Table 21.3 documents how little it currently contributes to the American energy picture.

By 1985, wind energy alone was providing California with 600 million kWh (2 EJ) annually;[27] more windmills have been installed since then. California wind energy in the 1980s cost 5 to 7 cents/kWh[28] but now is not returning as much of a profit to investors because the tax credits expired in 1990;[27] the cost is now at the high end of the cost of energy from conventional sources. By the late 1980s, there were about 15,000 wind turbines in California, generating 1 million kWh,[21,29] which is significant in local terms. In 2000, the cost of wind energy was under about 4 cents/kWh, and the cost of construction had dropped from $1100/kW to $900/kW;[21,30] it may drop as low as $100 to $600/kW.[31] A federal subsidy of 1.5 cents/kWh for renewable energy became available after 1994 as part of the Clinton energy policy.

About 2% of solar energy is fed into the winds, in such a way that, at any one time, about one day's worth of solar energy is stored in the kinetic energy of the atmosphere. It has been estimated that about 10 EWh/yr could be captured from the winds (130 TW from Table 21.2 times 8760 hours in a year, at a low 1% efficiency), about 20 times the current world electricity use.[32] Over the long term (that is, seasons), wind is quite predictable. Over the short term, though, it is very unpredictable. Currently, it is thought desirable to limit wind to 5% to 15% of the total energy demand[31] (it is close in Denmark, 14%, and the next closest is Germany, at 2.5%). More detail on wind energy is provided in **Extension 21.2, *How a Windmill Works***.

This great untapped source has fired the imaginations of some engineers. One proposal calls for a chain of 333,000 windmills 240 meters tall, across the great plains from Texas to Canada to take advantage of the strong night winds over the Great Plains. Kansas, Nebraska, and the Dakotas could be "the Saudi Arabia of wind energy."[52] Some believe that such a system could be built in less than 30 years and could ultimately generate 1.5 TWh/yr (13 EJ/yr, more than the electric utilities supply now).[53] In 1970, it was estimated that such huge windmills could have been constructed at a cost comparable to that of coal-fired power plants. The economics of construction is much better today, as are the size, reliability, and efficiency of modern wind turbines. Fewer turbines would be needed, for more electricity supplied.

The largest current market for windmills is in the United States because of the large population and huge demand for nonpolluting electricity. (Europe as a whole is a larger market but is fragmented among countries.) Just four countries account for over three-quarters of the wind energy produced: the United States, Germany, Denmark, and India.[41] Hawaiian Electric bought thirty 2.5 MW machines in 1980 to operate on the local trade winds (about 25 km/h), and about 10% of Hawaii's electrical energy comes from the wind. California is the nation's leader in the scale of its renewable energy use, and is the leader in use of wind energy, but other states are gaining.

Wind is famously intermittent, so a single windmill might sit idle some of the time. There are geographic regions with steady winds, and with wind speeds at above 7.5 m/s at 50 m height; this translates into wind power densities over 500 W/m². These areas constitute prime wind energy resources (see **Extension 21.3,** *Power From the Wind* for more detail on prime wind areas within the United States). Even so, some windmills will be idle some of the time. Someone using a single windmill needs a backup (such as connection to the grid or batteries) to keep electricity flowing all the time. If many windmills are in different locations, the fluctuations from individual machines are evened out and the supply is reasonably predictable (very important for a utility buying it).

The temperature difference between the cool ocean and the warm interior valley of California causes an air exchange, with cool air coming in from the coast and warm air rising. Passes in the mountains funnel the winds, which blow steadily through them. The average power density in the passes is 380 W/m² at 10 m elevation, and around twice that at 50 m, which classifies as outstanding (so-called wind power class 6, on a scale of 1 to 7).[25] Southern California Edison buys energy from "wind farms" in the San Gorgonio Pass and in the Tehachapi Mountains (Figure 21.8). Pacific Gas and Electric buys wind energy from Altamont Pass wind farms.

FIGURE 21.8

Windmills on a windmill farm in the Tehachapi Mountains of California. (U.S. Department of Energy)

The International Picture: Europe Is Ahead in Wind

European countries had over 9 GW of wind energy installed by the end of 2000 compared to about 6 GW for the United States.[41] The European Wind Energy Association predicts that Europe will have 100 GW of wind turbine capacity installed by 2020.[96] Collectively, the European countries are ahead of the United States, and several beat the United States in proportion of national energy supplied by wind.

Denmark, which has rejected nuclear energy and has wind in abundance, leads the world in windmill manufacture (it builds about half of all wind turbines in the world).[97] The Danes, with 2800 installed windmills, currently get over 14% of their energy from the wind. The village of Vester Hjermitslev, Denmark, with 500 inhabitants, is supplied fully from biogas (another renewable from solar energy, see Chapter 23), and uses a windmill backup.[35] The Danes planned to be getting 20% of their energy from wind by 2005. Denmark experiences an operational and maintenance (O&M) cost of about $0.01/kWh, comparable to O&M for fossil-fuel plants; of course, the capital cost must be paid off, and that amortized payoff is considerably greater than the O&M cost.

Local opposition to wind farm siting is common worldwide. Denmark developments include coastal and offshore development. Wind machines have been erected along windy stretches of the coast. Many locals have a NIMBY attitude, just as occurs for other modes of electricity generation, and want the farms to be as far away as possible, preferably in deep water well off the coast. The offshore sites used are in shallow water near shore despite this local opposition and not in deeper water because construction costs increase as water depth increases.

Germany has gained on the United States in wind capacity. It now has 2 GW capacity, up from an anemic 100 MW in 1990.[41] Part of the reason is the high subsidy in the German market. Germany pays generators a premium subsidy so that instead of the production cost of about 3 euro cents per kWh, renewables earn about

8.9 euro cents (roughly 12 U.S. cents).[18] Even utilities have taken part, as I found from a visit to SWK (Stadtwerke Karlsruhe), the Karlsruhe electricity municipal utility, where I spoke to the engineer in charge.[98] Wind would not be economic compared to other forms of energy (assuming a 20 year life for the windmill) without the subsidy, which runs about 15 euro cents per kWh (about 6 cents per kilowatthour, comparable to American utilities). Officials at EnBW (Energie Baden-Württemberg AG), the state electricity utility, agreed that the subsidy was essential to amazing growth of wind energy utilization in Germany.

German churches have organized groups of investors to support erection of wind turbines to take advantage of the subsidy. They view this as part of their mission of stewardship. Part of the success of renewables in Germany and Denmark is this religious dimension, much more obvious there than in America. As with Germany, part of wind's success in Denmark lies with Danish energy laws: The Danish utilities must buy from private wind turbine owners at 85% of the consumer price of electricity.[41,72] The utilities, in turn, get a subsidy from the government.

India is the renewable energy leader in the Third World. As of 1998, it had 850 MW of wind energy available.[41] China is planning to use windmills where it lacks other energy sources, and now has 166 MW installed.[41,99] One site in Egypt could supply virtually the entire current Egyptian demand for electricity.[40] Large areas of the world are suitable for wind energy installations. For example, South America's Cape Horn is the windiest place on Earth, and the rest of the west (Andean) coast of South America has a lot of potential. At many places in the world, the flow of wind through a square meter (at 25 m height) can produce 500 W, currently a criterion for viable windmill development.[37]

A pair of energy experts has suggested that building 40,000 wind turbines in the United States using government money, then selling the resulting electricity, could be done at no net cost.[77] The federal government would sell electricity over the estimated lifetime of the turbines to pay back the original cost, which would absorb about 4% of the national budget. The wind energy would substantially reduce carbon dioxide emissions, allowing the United States to meet Kyoto Protocol emissions reduction levels.

How Wind Energy Has Grown

Private companies are building machines to tap the winds, then selling electricity to power companies. In large part, this is due to the Public Utilities Regulatory Policies Act (PURPA) passed by Congress in 1978. Under the terms of PURPA, companies may sell energy to utilities at the utility's marginal or avoided cost, not their average cost. (Marginal, or avoided, cost is the cost of adding capacity; under PURPA, this has been interpreted to mean the cost of the most expensive additional energy, usually from gas turbine generation.) Southern California Edison in the 1980s had contracts to purchase about 5 GW from renewable energy sources, paying 5.7 cents/kWh as its avoided cost.[100] Now this is much larger,[101] especially since California deregulation forced Edison out of the generating business. However, many renewable generators went unpaid for months during the 2000–2001 California energy crisis.[102]

The utilities' monopolies are "protected," because the private companies are prohibited from supplying more than 80 MW to the utilities. This has led to the unintended consequence of windmills being forced to shut down, while California utilities burned fossil fuel in their existing plants to produce electricity that could have come from renewable sources (without the pollution). Many Californians mistakenly believe that windmills sit idle because they are broken, rather than because they are forced to shut down by the law as written. The public similarly believes, incorrectly, that windmills are unreliable. (This may be because many of the early 1980s machines in the original California wind farms *did* break down.) Windmills built in the 1990s are quite rugged and extremely reliable compared to other energy-generation methods. They also operate more quietly than earlier models.

Utilities in California have been oversubscribed at times under PURPA, and the California Public Utility Commission has suspended sign-ups at these times. California utilities have been working against the bid auctions through which renewable energy is sold, and they are fighting the California Public Utility Commission's support of renewable energy in all forms so as to maintain their monopolies.[100] Even the most forward-thinking California utility (and most California utilities are national leaders in the shift to renewable energy sources) is not immune to monopoly self-interest. Nevertheless, without PURPA, the explosion in energy alternatives of the 1980s and 1990s could not have occurred.

Photoelectricity
Light-Caused Chemical Reactions

In the methods discussed so far, solar energy is converted to electricity through intermediate steps or is converted to "lower-quality" thermal energy. One reaction that is direct is the **photoelectrochemical reaction**. Such reactions are chemical reactions in which light causes production of a fuel, such as hydrogen, that may subsequently be burned. The most attractive such reaction is **photodissociation** (breakup directly caused by light) of water into hydrogen and oxygen. Unfortunately, direct photolysis in the laboratory is not presently possible.[103,104] To this point, there has not been much success in satisfying all the constraints of such a fuel cycle. One group claims to have photoelectrochemical cadmium–selenium ferrocyanide solar cells made that show about 13% lab efficiency.[105]

There are other glimmerings of possibility:[106] Small particles suspended in electrolytes liberate H_2 and O_2 under certain circumstances,[106] and do so with high efficiency if the "catalyst islands" cover only a small fraction of the surface.[107] It also appears possible to use catalysts for the production of methanol from carbon dioxide. If any one of these possibilities is realized, it would mean that fuel could be produced by the Sun and either piped to a central location or used locally to supply energy as needed.

Photovoltaic, or Solar, Cells

Direct conversion of sunlight to electricity is accomplished in photovoltaic (pv) cells. The photovoltaic effect was first discovered by the French physicist Edmund Becquerel in 1839.[8,108] In Chapter 4, we discussed materials that are conductors (allow charge to flow freely) and insulators (do not allow charge to flow). Some materials are neither one nor the other—charge flows, but not as freely as in a conductor. In photovoltaic cells, the material that makes up the cell is a material that allows electric charge to flow but in which the properties of electrons make the amount of conductivity dependent on temperature. At higher temperature, more charge can flow. Such a material is known as a **semiconductor**.

Recall from Chapter 7 that electrons in atoms can exist only with certain allowed energies. Groups of atoms gathered together also have only certain allowed energies, but there are groups of energies instead of just one energy allowed. These groups of possible energies are known as **bands**. In conductors, the bands overlap. In insulators and semiconductors, the band energies for electrons are separated from other allowed band energies. There is at least one region of energies that electrons are not allowed to have. In insulators, the difference between allowed energies, the **band gap**, is large. In semiconductors, it is small.

Thermal energy can cause some electrons in semiconductors to be promoted from a filled band (the **valence band**) to an unfilled band (the **conduction band**). In the valence band, the electrons cannot move; in the conduction band, they can move freely and so conduct. Light from outside coming in can also cause electrons to be promoted and move. If the electron gives up its energy immediately, it cannot move, but if it does not recombine immediately, the electron can be shunted to an external circuit to do something before returning and recombining. This is the principle of a photovoltaic—solar—cell. For more details about semiconductors and the photovoltaic effect, see **Extension 21.4, *The Physics of Solar Cells***.

Physicists at Bell Labs produced the first modern photovoltaic cell (solar cell) in 1954.[116] It was made with silicon, a component of sand. However, two different types of silicon were used, in order to prevent recombination. In one type, the charge carriers were positive (**p-type semiconductor**), while in the other type, the charge carriers were negative (**n-type semiconductor**). When a p-type semiconductor is laid next to a piece of n-type semiconductor, it produces a p–n junction. Random thermal motion makes some of the charge carriers from each region cross the boundary into the other, where they combine with opposite charges and produce a region from which charge is depleted (one that has no charges that can move). An electric field exists across the region. It can prevent charges from moving in one direction and allow movement in the other. When light produces more charge carriers in a solar cell, the charge carriers must move through an external circuit to get to the other side, and so we can get electricity from the cell.

Research on increasing cell efficiency continued, especially after satellites began to orbit Earth. As a result, the cost per watt dropped from $1000 before the space program began in 1960 (almost $6000 in 2000 dollars) to about $5 today, which is still too high to compete with conventional energy sources where they are available.

A typical single-crystal silicon 10 cm square photovoltaic cell (having area of 100 cm^2) can give 1.5 W of power at a potential difference of 0.5 V DC and 3 A under full summer sunlight (the light intensity is about 1 kW/m^2). The electricity is direct current, because the charges can flow only one way, as we mentioned. An electronic device known as an inverter can make grid-compatible 60 Hz (or 50 Hz) AC electricity out of the DC supplied by the solar array. The potential difference (voltage) is determined by the band gap, not the physical dimensions of the cell. Cells can be connected in series to increase the potential difference, or in parallel, to increase the current.

Uses of Solar Cells

Industry has invested billions in photovoltaic research since 1973.[114] A cost of under about $2.50/W (in 2000 dollars) appears necessary for there to be real competition with established sources.[127] The cost should be about $1500/kW for central-station photovoltaics to be clearly economical.[128] This is quickly approaching reality.[129] Actual operations and maintenance experience for photovoltaic cells varies from 0.4 to 7 cents/kWh (the average is 2.3 cents/kWh), with small installations at the high end and large installations at the low end. Already, 30 utilities have small grid-connected photovoltaic cell facilities.

For the 55% of the world population not connected to an electric grid in remote areas, the price is already competitive.[130] It may cost $7.50 to $10 per meter to extend grid service ($7500 to $10,000/km). Installation of photovoltaic cells at a site is currently cheaper than connection to the electricity grid when the line must be extended more than half a kilometer to reach the site.[110] At current prices, if the load is smaller than 8 kWh per month, solar cells are cheaper than extending the utility line 65 m. If the load is 3000 kWh per month, solar cells would be cheaper than a 3 km extension of the utility line.[119] Tunisia has equipped 200 schools with photovoltaic systems of 250 W each. They are used for light in classrooms, for TVs, and for radio.[131]

Germany, as with wind energy, is one of the leading countries in backing other renewable energy sources. Germany has a 100,000 solar roofs program under way, giving subsidies to homeowners who use solar water heat or photovoltaics.[18] The U.S. Department of Energy launched a bigger "Million Solar Roofs Initiative."[9] As part of the program, the federal government pledged to put 20,000 solar energy systems on federal buildings. The program's aim is to have a million solar energy systems on public and private U.S. rooftops by 2010. These programs obviously build on a public desire for clean energy. The Sacramento Municipal Utility District (SMUD) used to have a program that allowed homeowners to pay $4 each month to host solar cells on a home's roof to send energy to SMUD, not the homeowner. Even so, the program was oversubscribed.[132] Now SMUD even pays for the energy if it is in excess of that used by the customer (this is referred to as " net metering"). Refer to **Extension 21.5, *Photovoltaic Systems*** for more information about connected systems.

An important question is how the energy costs of production of an energy converter compare to the energy ultimately supplied by the converter. Investigations of "payback," the time at which cumulative energy production is equal to fabrication and installation energy costs, for solar photovoltaics has revealed that crystalline silicon modules achieve energy payback in 3 to 4 years, while for thin-film copper indium diselenide modules, payback is 9 to 12 years.[172] Despite this long payback, the energy eventually produced is about an order of magnitude greater than for fabrication because of the long lifetime of solar panels.[172]

Production costs for solar cells dropped about 80% during the 1980s.[110] Japanese companies are mounting a challenge to U.S. manufacturers with plans to produce amorphous silicon solar cells at $3 to $4/W. In response, many U.S. companies have abandoned the field.[108,173] ARCO Solar was sold to Siemens (a German company) in 1989.[174] As the experience of the past decade leads to ever-cheaper solar cells, we can expect to see mass market penetration by 2010. Solar energy strikes a resonant chord in many people, because it promises independence from the power companies and the government. Most American production of solar cells is being exported at present. Demand is currently greater outside the United States. When cells do become cheap, there may be an explosion of demand, even if tax incentives are lacking. Unfortunately, U.S. industry may not reap the benefits of its investment because of foreign penetration of the U.S. market[175] and construction of state-of-the-art factories elsewhere, similar to that of 10.5 MW capacity-per-year Flabeg Solar International in Gelsenkirchen, Germany.[176]

TABLE 21.4

Maximum efficiencies at light level of 1 kW/m² measured at the National Renewable Energy Laboratory

Type of cell	Efficiency (%)
Crystalline silicon	23
Gallium arsenide	25
Polycrystalline silicon	18
Edge-defined film	14
Dendritic web growth	15.5
Thin films	
Cadmium telluride	15.8
Silicon	15.7
Copper indium diselenide	12–13
Amorphous silicon	12

Source: Reference 110.

Various cells have different efficiencies (efficiency defined as $\frac{\text{electrical energy out}}{\text{energy input}}$), as shown in Table 21.4. The greatest measured actual efficiency of a cell thus far is 24.7%.[177] This compares to a "natural" efficiency of under 1% for corn (maize), measured as $\frac{\text{grain energy content}}{\text{energy input}}$.

One distinction between wind energy and photovoltaic energy is that geographic distribution does not help the intermittancy problem. A solar cell cannot convert radiant solar energy to electricity at night. Storage in batteries or in thermal systems must be arranged or grid connection must be made if the user wants to have electricity during the dark. Battery or thermal storage introduces other losses and additional capital costs into the system. Grid connection, if available, solves this problem easily, and cheaply, if net metering is possible.

21.4 LARGE-SCALE SOLAR ENERGY PROJECTS

There are three main choices for a large-scale solar power station: central receivers, parabolic dishes, and parabolic troughs (including solar farms). Such projects are clearly not meant to appeal to an individual user, but rather to mesh with the present character of the publicly regulated electric utility industry. These technologies could affect the world's heat balance if widely adopted.

Central Receivers

Central receivers use large fields of tracking mirrors (**heliostats**) to focus sunlight onto central receiving towers. A pilot plant named "Solar One" was built at Daggett, near Barstow, California, by Southern California Edison with the support of Sandia Labs, the U.S. Department of Energy, the Los Angeles Department of Water and Power, and the California Energy Commission.[178–189] It operated between 1982 and 1988. Other large installations have operated worldwide (see Table 21.5).[188] "Solar Two" operated between 1996 and 1999.[189]

The tower's absorber panels, although they appear to be white in Figure 21.9, are painted black and absorb 88% (96% when newly painted) of the incident

FIGURE 21.9

The tower at Solar One in Daggett, California.

TABLE 21.5

Other large-scale central receiver solar installations

Name	Location	Power	Operation	Comments
Eurelios	Adrano, Spain	1 MW$_e$	1981–1984	Poor design
Solar One	Daggett, California	10 MW$_e$	1982–1988	See text
Solar Two	Daggett, California	10 MW$_e$	1996–1999	See text
CESA-1	Tabernas, Spain	1 MW$_e$	1983–1984	Long startup times, cloudiness a problem
Themis	Targasonne, France	2.5 MW$_e$	1983–1986	Lower output than predicted
MSEE	Albuquerque, New Mexico	0.75 MW$_e$	1984–1985	Too small, negative output
PHOEBUS	Aqaba, Jordan	30 MW$_e$	?	Dropped in aftermath of Persian Gulf War

Source: Reference 188.

FIGURE 21.10

A view of Solar One from the air.
(U.S. Department of Energy)

light.[100] The heliostat field (Figure 21.10) is oriented mostly toward the north of the central tower to capture sunlight from the south. The heliostats to the south (to the right on the photo) focused sunlight on six panels to preheat the water that went to the 18 superheat panels. The superheat panels received a much larger amount of sunlight from the northern heliostats. Water left the superheat panels at 510 °C at 100 times atmospheric pressure.[100,175,179]

Load Shifting

The thermal storage system was an important part of the design, and it is cost-effective. Off-peak electricity was worth only 5.7 cents/kWh, while peak electricity was worth 12 cents/kWh in winter and 14 cents/kWh in summer to Southern California Edison. This temporary storage of energy until the price is higher is referred to as **load-shifting** and could have made Solar One profits possible. Solar One did prove that the water/steam cycle is reliable, that the system can meet expectations, but also found that thermal storage was not really as cost-effective as had been believed (although that was the one aspect that could easily be improved). The efficiency experienced, 8.2%, fell short of design.[188] The real problems with Solar One were its unexpectedly high cost and its limited storage capacity.[183,189] The computer control system, partly developed in use on Solar One, is now used in other utility control rooms.[179]

Solar Two, utilizing the lessons of Solar One, was more successful by any measure. It was highly reliable, capable of running without maintenance for extended periods of time; the energy storage system was so improved that Solar Two ran continuously for 153 hours; its measured efficiency for thermal storage was 88%; and it delivered a peak output of nearly 12 MW.[189] See **Extension 21.6, *Solar One and Solar Two*** for more details.

FIGURE 21.11

The Luz solar facility (SEGS-I) at Daggett, California.

Parabolic Troughs

Adjacent to the Solar One site in Daggett is a **parabolic trough** facility, the Solar Electric Generating System (SEGS), consisting of modules SEGS I through SEGS IX,[185,188] built by Luz International. The parabolic shape focuses light on a central receiver tube that contains oil that is heated by the sunlight.

The Luz facility generates 350 MW$_e$ of solar electricity altogether. SEGS is the largest commercial solar energy project in the world; SEGS I produces 13.8 MW$_e$, SEGS II through SEGS VII, 30 MW$_e$, and SEGS VIII and IX, 80 MW$_e$.[185,188] The Luz experience is that it takes about 2 years to bring their facilities to full operation.[188] The SEGS I system has 72,000 m^2 and SEGS II 165,000 m^2 of the parabolic trough collectors (Figure 21.11). The first pilot projects cost $5979/kW$_e$, and their energy cost was $0.265/kW$_e$h. The intermediate units were standardized at an area of 202,000 m^2, a power generation of 30 MW$_e$, covering 65 hectares and built at a cost of $2900/kW$_e$.[183] Over the years of experience, the solar-to-electric efficiency has risen from around 10% to around 15%.[188]

The parabolic shape focuses sunlight on a pipe, delivering concentrated solar thermal energy. The black collector lines (they appear bright in the photos) are along an east–west axis. The east–west orientation was chosen to give a more constant output throughout the year. The black chrome lines at the focus of the parabola were replaced by a ceramic/metallic selective surface in SEGS VIII and IX. The lines carry oil that is heated to 245 °C and runs into storage tanks (Figure 21.12). The hot oil is used later to produce steam, which is raised to a temperature of 415 °C with a gas superheater and then used to drive a turbine generator.

Luz depended on a 10% federal tax credit that expired in 1990. The investors filed for bankruptcy in 1991 because of the loss of that tax credit.[174,186] The bankruptcy calls further expansion plans into question. Despite the bankruptcy, Luz's system still represents almost all the solar thermal capacity in the United States.[5]

FIGURE 21.12

Storage tanks for hot oil at the Luz solar energy facility in Daggett, California.

Parabolic Dishes

Parabolic dishes are based on the same optical principles as parabolic troughs but focus sunlight at a single point.[192] The dishes are mounted on individual stands

TABLE 21.6

Large-scale parabolic dish solar installations

Location	Operation	Comments
Shenandoah, Georgia	1982–	114 collectors using hot oil
White Cliffs, NSW, Australia	1982–	14, 5 m diameter collectors, 25 kW$_e$ overall; judged too costly
Sulaibyah, Kuwait	1982–	Test facility using hot oil; less efficient than designed
Rancho Mirage, California	1984–1985	23% efficient 25 kW$_e$ SE modules[a]
King Abdul Aziz City, Saudi Arabia	1982–1985	20% efficient 50 kW$_e$ SE modules[a]
Warner Springs, California	1984–	700 concentrators using water/steam; 4.92 MW$_e$ overall; many problems

Source: Reference 188.

[a]SE stands for Stirling engine.

and can be arranged to track the Sun. They can be equipped with hot-oil lines (as in the SEGS), or run Stirling engines, a type of gas cycle engine that runs between the high-temperature reservoir and the atmosphere. Over the years, collectors have gotten larger and cheaper and have been moderately successful in stand-alone applications. Some parabolic dish projects are described in Table 21.6.

Solar Farms

One interesting proposal by A. B. Meinel and M. P. Meinel envisions a solar energy "farm."[193] The **solar farm** would be able to provide energy continuously (night as well as day) by using sunlight to heat liquid to about 500 °C. Liquid sodium has a high specific heat and thus can store a lot of energy for each degree rise in temperature. Heat exchangers allow the liquid sodium to cool, giving off the energy absorbed to water. The resulting steam would drive conventional turbines.

The farm would have to be located in a sunny region, such as a desert. It would consist of long rows of energy absorbers painted black to absorb heat, alternating with rows painted white to reflect heat. If this alternation were not done, the thermal balance of the large area would change, which could change the weather. By alternating reflectors and absorbers, the average amount of heat absorbed by the plant would be the same as if it had still been desert. If the solar energy could be captured at 100% efficiency, a land area of only 1300 km^2 could supply all the electric energy currently used in the United States. At 10% efficiency—a more realistic figure—such a farm would still need a net area of only 13,000 km^2 (5000 mi^2).

A "farm" of photovoltaic cells could also be constructed. Central-station photovoltaic systems now cost about \$0.20/kWh. This is expected to drop to \$0.06/kWh by 2005 to 2010. Most improvement in price is expected to come in the cost of the photovoltaic material.[194] A site or sites having a total area of 13,000 to 30,000 km^2, depending on the amount of shading allowed and whether tracking of the Sun is done, and utilizing solar cells operating at 10% efficiency, could supply all annual U.S. electric energy needs, 13 EJ.[110,113] The balance of the system (thermal storage, etc.) is assumed to have an overall efficiency of 81%.[110]

Climatological and Social Effects

All the aforementioned solar energy devices involve no *net* addition of thermal energy to the atmosphere. There is, rather, a redistribution of thermal energy from one place to another. Most conventional energy generation involves net additions of heat, which can affect climate in local areas, raising the possibility that we may be tampering with Earth's overall heat balance (see Chapters 17 and 26). This could, of course, happen even with redistribution, as we shall see in the discussion of ocean thermal gradient power in Chapter 22.

Local climatological effects include increased rainfall and snowfall near generating facilities.[195] Cities, with their high levels of energy use, are consistently warmer (they are referred to by some as "**heat islands**") than the surrounding countryside the year around. The Los Angeles basin ultimately bears a local heat burden from conventional energy generation that is over 5% of the total solar energy absorbed by the basin! This burden will continue to grow unless renewable energy resources are used instead of fossil fuel or nuclear installations. Wind energy and other renewables decrease the temperature difference between city and country and restore the original **heat balance**.

Further, as pointed out in **Extension 21.5**, *Photovoltaic Systems*, in some places, there is no substitute supply of electric energy available. Only wind or solar cells can supply sites far from the grid. In the United States, the cost to extend grid service can be as much as $10 per meter ($15,000 per mile).[196] While extension cost is lower in less-developed countries, it is still substantial. This cost is avoided when such renewable energy is used.

One motivation for use of renewable energy is that the social costs (called externalities in economics) of conventional electricity generation are not included among the costs listed in Chapters 4 through 9. There *are* large costs for society from conventional thermal steam systems, as we discovered in Chapters 12 through 15 and Chapter 20. These costs range from $0.02 to $0.06/kWh for fossil fuels and from $0.06 to $0.13/kWh for nuclear energy.[195] These societal costs are largely avoided with renewables.

SUMMARY

Radiant (light) energy is emitted in the form of electromagnetic waves, which travel at the speed of light, 300,000 km/s. The product of the wavelength and frequency of the wave is the speed of the wave. From longer to shorter wavelengths, there are radio waves, microwaves, infrared light, visible light (wavelengths approximately 700 to 400 nm), ultraviolet light, x rays, and gamma rays.

The Sun is the source of almost all the world's energy, and the source of the Sun's energy is fusion inside the Sun. The Sun radiates energy from its surface characteristic of a blackbody at a temperature of 5800 K (as described in more detail in **Extension 21.1**), the Sun's surface temperature, which produces a lot of radiation in the visible region of the spectrum. Some of this radiation impinges on Earth and is absorbed. Because

Earth is at a nonzero temperature and radiates, the amount of energy it absorbs from the Sun can exactly balance the amount of energy it radiates away.

Wind energy is ubiquitous and available in large amounts. Denmark, Germany, and California lead the world in development of wind power. Windmill blades can capture some of the wind's kinetic energy and produce electricity. The technology is now mature. Windmills of the 1990s and 2000s are reliable, and the cost is near or in some cases slightly under busbar costs of other forms of energy. Extended detail on how windmills work is given in **Extension 21.2** and advantages and disadvantages of wind energy are considered in **Extension 21.3**.

Solar energy can fall on photovoltaic cells or chemical solutions to produce electricity through the photoelectric effect. Solar cells made of semiconductor materials (in a p–n junction) act to force the current to flow through an external circuit to return to neutralize the charge. More detail is found in **Extension 21.4**. This is a technology that is mature in parts, but much work is still needed to allow photovoltaic solar energy to reach its promise of cheap, widely available energy (see also **Extension 21.5**). Photovoltaic electricity is already the most economic source of energy in much of the world not connected to the electric grid. Such solar energy promises individuals eventual freedom from monopoly utilities.

Sunlight can be absorbed in large installations run by utilities. There it can be concentrated to make ultrahot water or oil to generate steam (as in the solar farms, Solar One and Solar Two, parabolic trough collectors and parabolic dishes—see **Extension 21.6**), which can then be run through ordinary turbines.

In Chapter 22 we consider solar energy from water. In Chapters 23 and 24, we consider the one important form of solar energy not yet addressed: biomass. Nature has solved the problem of creating convenient fuels from sunlight.

PROBLEMS AND QUESTIONS

MULTIPLE CHOICE REVIEW

1. To provide all the electricity used in the entire United States using photovoltaic systems, about how much land area would be needed?
   a. About the size of the entire state of Alaska.
   b. About the size of the entire state of Texas.
   c. About the same area as is covered by roads and bridges (0.3% of U.S. area).
   d. About one-fourth the area covered by roads and bridges (0.08% of U.S. area).
   e. About the size of a very large farm or ranch.

2. What advantage(s) do(es) solar towers have as a supplier of electrical energy?
   a. No fuel is needed.
   b. No waste heat escapes to the environment.
   c. There are no moving parts to the system.
   d. All the above reasons listed are advantages.
   e. None of the reasons listed are advantages.

3. What advantage(s) do(es) parabolic troughs have as a supplier of electrical energy?
   a. No fuel is needed.
   b. No waste heat escapes to the environment.
   c. There are no moving parts to the system.
   d. All the above reasons listed are advantages.
   e. None of the reasons listed are advantages.

4. About how much area would an efficient Meinel-type solar farm have to have to supply the United States with all its electricity needs?
   a. About the size of the entire state of Alaska.
   b. About the size of the entire state of Texas.
   c. About the same area as is covered by roads and bridges (0.3% of U.S. area).
   d. About one-fourth the area covered by roads and bridges (0.08% of U.S. area).
   e. About the size of a very large farm or ranch.

5. Typical windmills being built in the 2000s can produce 1.5 MW. About how many would have been needed to supply the entire U.S. electricity needs in 2000 (13 EJ)?
   a. 8×10^{12}
   b. 1.25×10^{11}
   c. 8×10^{6}

   d. 2.5×10^{5}
   e. 1.25×10^{4}

6. If Europe does have 100 GW of wind power installed by 2030, how many 1.5 MW windmills would be needed to supply this power?
   a. 150,000
   b. 66,667
   c. 6667
   d. 1500
   e. 667

7. Parabolic solar troughs are sometimes used, because they
   a. are convenient to construct.
   b. offer the possibility of concentrating sunlight coming from many angles.
   c. are aesthetically more pleasing than circular collectors.
   d. are cutting-edge technology.
   e. can replace flat collector areas that might not "catch" sunlight as effectively.

8. The potential difference delivered by a particular solar cell in sunlight depends on
   a. how many solar cells are arranged in series.
   b. how much sunlight is falling on the cell.
   c. how many devices are hooked up to the cell.
   d. how many solar cells are arranged in parallel.
   e. the material's band gap.

9. The true amortized cost of photovoltaic systems is the total cost spread out over the lifetime of the system (assumed to be 12 yr), divided by the net lifetime energy output. This cost is now around 25 cents/kWh. At this price, more than the 10 cents/kWh of ordinary grid electricity, how far away from the grid do residential customers have to be for photovoltaics to be cost-competitive? Assume it costs $10/m to extend the grid connection and that the household uses 10,000 kWh/yr.(196)
   a. Less than 0.5 km (~0.25 mi)
   b. About 1 to 2 km (~0.5 to 1 mi)
   c. About 5 km (~3 mi)
   d. About 10 km (~6 mi)
   e. About 15 km (~9 mi)

10. Under what circumstances could photovoltaics be used effectively to supplement utilities' other energy sources?

a. Where photovoltaic power correlates well with the utilities' daily load patterns.
b. Where power is available when it is needed most—during daylight hours.
c. Where there is abundant sunlight in the geographical area.
d. When roofs are being replaced for other reasons, and costs can be amortized over 20 yr.
e. Under all the foregoing circumstances.

SHORT ANSWER
Explain whether the statement made is correct, and justify your answer.

11. There are no environmental drawbacks to wind energy.
12. Net metering laws, which pay a customer for energy produced in excess of use and charge the customer for use in excess of production, can help increase the spread of photovoltaics.
13. Ultraviolet radiation cannot be used for solar cells, because its energy is too low.
14. PURPA helped spread renewable energy in the United States.
15. Churches are not interested in energy issues, because these are "of man, not God."
16. Photovoltaic cells will never be much cheaper than now.
17. China can benefit from investments in renewable energy sources.
18. Denmark is the world leader in installed percentage of renewable energy.
19. Offshore wind and solar facilities are difficult to construct because of the water.
20. Solar farms such as the Meinels' are not realistic visions of the energy future.

DISCUSSION QUESTIONS
21. How are solar towers similar to and different from solar trough collectors?
22. Are solar cell efficiencies of over 30% realizable under realistic use conditions (include cost in these conditions)? Explain.
23. Heat exchangers are important. Explain their use in power plants, including large-scale solar power plants.
24. Explain how one could characterize the Dakotas as the "Saudi Arabia of wind energy."
25. How is it that we can "feel" radiation from the Sun that we cannot see?
26. Explain how a black surface such as the Solar Two tower can appear white.
27. Why are inverters necessary for use with photovoltaic cells connected to the grid? Can solar electricity from photovoltaic cells be used *without* using an inverter?

28. What is the basic difference between electrons in a valence and a conduction band in a semiconductor? Explain.
29. Explain why California is so prominent in the use of renewable energy sources.
30. California is ranked high both in total renewable energy (#2) and in renewable energy not including hydroelectricity (#1). Why should there be any difference between these two at all? Explain using specific examples to illustrate your points.

PROBLEMS
31. A wave has a speed of 1000 m/s and a wavelength of 250 m. What are the wave's period and frequency?
32. A wave has a speed of 100.0 m/s and a period of 25.0 s. What are the wave's wavelength and frequency?
33. The speed of light in water is 200,000 km/s. The frequency of light traveling in the water is 5.50×10^{14} Hz. What is the wavelength of this light?
34. Ears of young people can hear sound frequencies as high as 23,000 Hz. What wavelength would this sound wave have if the speed of sound in air is 330 m/s?
35. The amortized cost of a photovoltaic facility is $300/kW$_e$/yr. The cost of the electricity due to maintenance and operations is $0.017/kW$_e$h. The collector array has an area of 10,000 m^2.
    a. What amount of energy does this plant produce per year if the sky is cloudy 20% of the time?
    b. What does it cost the utility to produce that much energy?
    c. Could the utility make money generating this photovoltaic energy off peak? On peak?
36. A solar cell produces electricity with an efficiency of 30%. How much greater area would a collector made of a dye thin-film collector of efficiency 8.8% require to produce the same amount of energy?
37. A photon of ultraviolet electromagnetic radiation of wavelength $\lambda = 200$ nm has how much more energy than a photon of wavelength $\lambda = 0.60$ μm?
38. A photon of blue light has a wavelength $\lambda = 400$ nm. A photon of red light has a wavelength $\lambda = 630$ nm. Which has more energy? By what factor is red larger or smaller?
39. The radius of the Sun is 696 Mm. Determine the Sun's surface area and, given that the radiated power is 3.9×10^{26} W, find the amount of power per unit area radiated by the Sun's surface. How does this compare to the value predicted by the Stefan–Boltzmann law, 64 MW/m^2? (This law is discussed in Extension 21.1; however, it is not necessary to know the law to answer this question.)
40. What area of typical single-crystal silicon photovoltaic cells would be needed to produce 1 GW of electrical power?

CHAPTER 22

Solar Energy and Water

Continuing the consideration of various solar energy resources, this chapter examines several forms of solar energy involving water. There are several major types of stored solar energy involving water considered in this chapter: On land we look at diversion, impoundment, and pumped storage of water as energy resources; all are in use currently. On the oceans, we consider energy from tides, from ocean waves, and from ocean-layer temperature differences. At the ocean–land interface, the mixing of fresh and salt water is considered. All these ocean-linked resources are speculative or under development, but few concrete installations have been built.

KEY TERMS

latent heat of vaporization / head / undershot wheel / breastshot wheel / overshot wheel / schistosomiasis / bilharzia / pumped storage / tidal range / resonance / ocean thermal gradients / salt-gradient solar pond

22.1   ENERGY FROM WATER ON LAND

One of the important uses of water is to supply energy. Water is evaporated, clouds form, the clouds produce rain, and the runoff from the rain may be diverted or captured by dams and used to run water wheels or turbines. This is an "old technology" in common use. We do not know exactly when water power for purposes other than transportation was first used, but it was certainly in use by Roman times.

A CLOSER LOOK

Evaporation in the Water Cycle

The water cycle that allows dams to be built would not run without solar energy. The Sun supplies thermal energy to water on Earth, causing the water to evaporate, gain gravitational potential energy, and form clouds. The clouds then precipitate waterfall all over Earth's surface (except, of course, the interiors of deserts). When rain falls on heights, some gravitational potential energy remains, the water runs downhill, where it (and its gravitational potential energy) may be trapped by judiciously placed dams.

Latent heat of vaporization (see Chapter 7) supplied to the top layer of water in oceans and lakes provides energy from the Sun at that geographical point. The water can then be transformed into vapor and gain potential energy as it forms clouds. As the clouds deliver rain, they also deliver energy—part of the energy absorbed in the original evaporation process, both as kinetic energy carried by raindrops and as gravitational potential energy by virtue of the drop's final position relative to sea level. This energy is delivered to the area in which the rain falls, which is usually different from the place at which it was absorbed. Hence, this latent heat transfer assists the distribution of energy from the tropics to regions nearer the poles, and thus helps drive the weather. This redistribution was treated in more detail in Chapter 16.

Diversion: Millraces and Water Wheels

Water Wheels

The water wheel preceded the windmill, and the water wheel created wealth for the millers who owned them. These wheels operated usually with water diverted from a higher part of the river or backed up behind a small impoundment. Water wheels were used extensively in Europe, and many factories eventually were built on rivers where dams were built to collect water. The height difference between the water level at the top of the dam and the lower level, or height of water drop available, is called the **head**. The head is a direct measure of the gravitational potential energy that can be converted to work (it is the h in mgh). The industrialization of America began in New England, because New England was rich in water resources that could be used to make a head to make a water wheel turn. New England and the Middle Atlantic states boasted 23,000 water wheels in the nineteenth century.[1]

The world's first water wheels were probably the vertical-axis type, in which an influx of water drives into the wheel's blades, causing rotation (Figure 22.1). Such wheels may be used in a streambed, with the minor modification that streamflow be directed onto the blades (in the Middle Ages, these were used for streamflow energy, attached to barges). The horizontal-axis water wheel, in use since Roman times, requires a good deal more engineering. One Roman mill developed the equivalent of 15 kW_e during the time of the Roman Empire.[2]

There are three general types of horizontal-axis water wheels. The **undershot wheel** can be run on streamflow like the vertical-axis wheel: The flowing water pushes at the partially submerged blades, causing the wheel to rotate. The

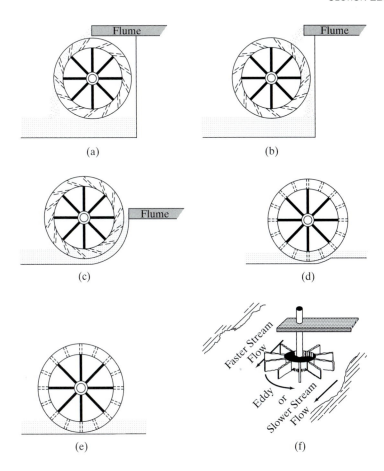

(a)

(b)

(c)

(d)

(e)

(f)

FIGURE 22.1

Types of water wheel. (a) Overshot wheel; the flume can come in at some angle above the horizontal. (b) Overshot wheel with reverse pitch. (c) Breastshot wheel. (d) Undershot wheel. (e) Run-of-stream wheel. (f) Early vertical-axis water wheel. The wheel turns because there is an eddy at the wheel's place in the stream or because the deeper part of the stream carries water faster than the water near the bank.

breastshot wheel has water brought in part way up the wheel so that both the impact of the water and the weight of the water cause rotary motion. In the last of the three, the **overshot wheel**, water is brought to the top of the wheel by a flume, or penstock (Figure 22.2). Here again, the wheel rotates because of the water impact and the weight of the water. Both the breastshot and overshot wheels involve nontrivial engineering.[1]

The overshot wheel is most efficient, delivering about 65% to 80% of the available power. The undershot wheel is the least efficient, delivering about 25% to 30% of the available power.[1,2] Modern hydraulic turbines are considerably more efficient than these, at approximately 90% efficiency. There are three sorts of hydraulic turbine: reaction, impact, and kinetic.[3] A reaction turbine, which is embedded in the water, needs only about 6 m of head. Some reaction turbines have variable pitch blades. In impulse turbines, jets of water hit cup-shaped blades (these are called pelton wheels). They need a head of 300 m or more to function efficiently. Kinetic turbines, which use the force of impact of flowing water to turn the shaft, were actually the first sort used, dropped out of use, and now are again under development.[2] See **Extension 22.1, *Dams*** for more information.

The Dead Sea Project

An interesting project may be nearing reality in Israel. The Dead Sea is 400 m lower than the Mediterranean. Because the water that once flowed to the Dead

FIGURE 22.2

Painting of a grain mill with a water wheel.
(John Constable, *Parham Mill, Gillingham*, Yale Museum, New Haven, used by permission)

Sea from the Jordan River is now being utilized for irrigation, the Dead Sea has shrunk. Water levels are now 8 m or so below those of a half-century ago. There is a proposal to construct a canal to divert water from the Mediterranean, to run south of Beersheba into the Dead Sea.[43] In the course of refilling the Dead Sea to 1930s levels over 10 to 20 years, the 1.6 km^3 of water per year could generate some 1600 GWh (1.6 TWh). Storage facilities would exist so that fluctuations in electricity demand would be met. Even after refilling, about two-thirds as much water will still have to be brought in every year to replace evaporative losses, generating about 1 TWh.[43]

Impoundment: Dams

FIGURE 22.3

Grand Coulee dam, the largest dam in the United States.
(U.S. Department of the Interior, Bureau of Reclamation, Pacific Northwest Region. Photo by H. S. Holmes)

Most modern ways of tapping the stored solar energy of water (see *A Closer Look*: Evaporation in the Water Cycle) involve building dams. Dams increase the available head over that (generally small) head available from stream diversion. The world's dams are getting higher. By 1990, 113 dams exceeded 150 m head.[5] The world has 150 "large" dams that together account for 40% of world capacity. The largest dam in the world in terms of generation of electricity is Itaipu Falls, on the Brazil–Paraguay border, producing 10,500 MW of electricity. The largest dam in the United States is Grand Coulee (Figure 22.3), producing 9780 MW. In Canada, the largest producer is La Grande 2, at 5328 MW.[2] World hydroelectricity in 1999 totaled 17.5% of world electric generation.[44]

The definition of large dams varies. The Department of Energy classifies as "large" dams those producing more than 30 MW. The World Commission on Dams uses the definition of the International Commission on Large Dams (ICOLD), established in 1928, which defines a large dam as having a height of 15 m or more from the foundation or being more than 5 m high while having a reservoir volume of more than 3 million m^3.[4] By the latter definition, there are more than 45,000 large dams worldwide.

Surprisingly for such a well-developed technology, there have been recent gains. The use of roller-compacted concrete involves 40% less water and 30% less cement to accomplish the same task, and the compaction can be done by ordinary construction equipment. The cost of construction is lower by 66%, partly because the pace can be much faster than with conventional construction. There were no dams built using this technique in 1980; now about 50 have been constructed.[2] Typical dam construction times are around 3.5 years for 20 MW units and 7.3 years for 2000 MW units.[2]

Of course, the silt carried along by the stream in its flow is lost after a dam is built, as the silt settles out in the calm reservoir. In the case of the Tarbela Dam in Pakistan, the input silt concentration is so large that the deep reservoir will be entirely filled with silt 40 years after its construction was completed (the late 1970s).[45] The silt that settles in the reservoir cannot fertilize land further downstream. These siltation processes call into question the rationale behind such projects as the Aswan Dam in Egypt. Since construction of the dam, the silt carried down from upriver settles at the upper end of Lake Nasser, where current speed decreases. Erosion in Ethiopia is reducing the life of the Aswan Dam.[46] The Chinese have tackled siltation in their Sanmen Gorge Project by constructing sluices to allow the heavy load of sediment to pass through.[18]

The first Aswan Dam, constructed in 1905, led to many similar complications. It appears that the experience garnered from the smaller dam was ignored by Egyptian planners during construction of the new Aswan High Dam in 1969 in their eagerness to obtain the energy necessary to industrialize.

Other effects of the modern dam have included erosion in the Nile Delta because of loss of sediment from upriver, a loss of fisheries along the Egyptian coast, riverbed erosion, and, of course, the increased salinity of the Nile River water from its impoundment in Lake Nasser.[12] Meanwhile, some land now irrigated in Egypt is becoming salt flat as capillary action brings up salts from the ground underneath.[12,47] This salt problem has led the World Bank to finance a multibillion-dollar installation of drainage tiles on several million hectares in Egypt along the model of the tiles installed in the Imperial Valley,[47] where 14% of the land now in drainage tiles has become productive. Even in Egypt, the gigantic cost of land reclamation has stopped the increase in cultivated area (it has shrunk by 200,000 acres since the mid-1960s).[47]

The increase in stagnant canal waters due to the dam set off an epidemic in Egypt of **schistosomiasis**, or **bilharzia**, which causes blindness. The disease is of parasitic origin, and the stagnation of the water allows the snails harboring the parasite to breed.[47] A similar epidemic of schistosomiasis struck Brazil in the wake of dam construction there.[17] The World Commission on Dams recommended thorough study of possible ecological effects *before* construction decisions are made as a way of preventing some of the problems and alerting local people to others before they find out by experience, or deciding that the problems are serious enough to make construction infeasible.

By now, it is obvious that hydroelectric power is not free of problems. Several other byproducts of dam construction are not at first apparent. Well over a dozen of these problems exist, including neglect of human rights of people displaced by dam projects, drowning of productive agricultural land, sedimentation, evaporative salinization, and changes in species of fish, insects, and plants. A much more complete list is given in the **Extension 22.1**.

Although the list above (and in the extension) is by no means complete, it should convey the idea that the building of a dam should proceed with caution. The insect and parasite problems can be treated if thought is given before construction. Spillways cannot be used during blackfly breeding season. A Venezuelan snail displaces the snails that harbor bilharzia and cattle flukes.[16] Similarly, cutting vegetation before reservoirs are filled minimizes many problems. Today the tendency is to make the reservoir as small as possible. A low power-to-flooded-area ratio (less than about 5 kW/ha) is seen as a cautionary signal to those considering building a dam. This criterion has led to reevaluations of existing plans.[2]

An important report by the World Commission on Dams focused attention on the rationale and human costs of dam construction.[4] The report gathered information from many sources and undertook case studies of the effects of some of the world's largest dam projects. Some results are good, for example, creation of wetlands. Some are bad, for example, displacement of people from their ancestral homes. The Commission emphasized recognition of rights of local people to be consulted in early planning and recompensed for their losses, as well as dispassionate assessment of risks and alternatives as of highest importance. **Extension 22.1** has additional information on the report.

Catastrophic Risks from Dams

Society is not risk-free, and it must decide how to allocate rare resources to manage risks. The refusal on the part of some (those who refuse to recognize TANSTAAFL) to accept that there is no absolute safety may lead the great mass of people astray. For example, dams generate hydroelectricity and have been perceived as benign. However, dams have failed historically at a rate of one failure per 5000 dams per year (2×10^{-4} failures per dam per year). A UCLA study indicated that 250,000 deaths could result from the failure of the largest California dam.[48] These researchers also found that the failure rate was probably closer to 0.01 per dam per year. In the San Fernando earthquake of 1971, the Van Norman Dam almost failed because of soil liquefaction; had it gone, the water from the dam would have caused 50,000 to 100,000 deaths among the people living in suburban communities just below it.[48] Every alternative does have its drawbacks.

While many dam failures have caused little loss of life (for example, the dam in Baldwin Hills, California, the Teton Dam in Idaho, the Malpasset Dam near Frejus, France), the failure of the Vaiont Dam in northern Italy took 1925 lives.[49] The effect of a dam failure can be great if the population density is high near the dam. For example, the tragic 1889 Johnstown flood in Pennsylvania, which killed 2209 people, followed a dam failure.[50] The much smaller, remote Mill River Valley disaster in 1874 in Massachusetts cost the lives of 139 people.[51]

Most dam failures occur in the first year, and over two-thirds of all failures happen within the first decade after construction. Dams built in the last half of the twentieth century fail at a rate of 0.5%, while those built in the first half of the century fail at four times that rate. There has been an average of 10 dam failures per year in the United States over the past 150 years. (The annual dam failure rate varies; for example, it was 29 in 1996 and 1997.) Of the approximately 80,000 small and large dams in the United States, 9326 are rated as "high hazard," meaning serious consequences would follow a failure.[4]

The most common cause of failure of simple gravity dams is sliding along the foundation, but dams are also subject to flood failure, earthquake, human misjudgment, and so on.[49] Gravity dam failure probabilities are typically estimated at about 10^{-5} to 10^{-4} per dam per year (1 for every 10,000 to 1 for every 100,000 dams per year), roughly the same as the historical dam failure rate.[48,52]

Low-Head Hydroelectricity

Already existing small dams may be used to produce electricity. Such low-head facilities were often built for electricity but were abandoned as utilities formed monopolies to gain from economies of scale. More than 770 hydroelectric plants have been abandoned since 1940.[53] Of more than 2800 dams built in New England, only 200 now produce electricity.[54] Towns and individuals are rehabilitating old sites to increase local energy supplies, which is causing a growing market for industrial upgrading and rebuilding efforts (Figure 22.4).[55] Refurbishing of small hydro installations could add 7% to New England's capacity of 1500 MW$_e$.[3]

Even though such efforts make sense and make use of equipment already in place, the total amount of energy supplied is not large. In Michigan, 80 hydro sites produce 2.7% of the state's electric energy total (101 TWh/yr = 0.36 EJ/yr). If the next best 30 sites were added, the total would still be under 3%.[53] In New York State, as of 1981, small hydroelectric plant sites produced about 4.2 MW, another 5.4 MW of capacity was under installation, and 3.4 MW of capacity was in design. Further sites that could produce a total of 12.4 MW were under study.[56] As of 1984, New York had 35 small hydroelectric plants operational, and in 2000, New York hydroelectric facilities supplied 18% (24 TWh/yr of the total electricity sold of 136 TWh/yr), but small hydro totals under 0.2 TWh/yr of that.[57] The developments in New York came about partly because of a 1979 state law setting the price of energy from independently owned plants at 6 cents/kWh.[57] The estimated total potential of the reasonable hydroelectric sites in New York is 1000 MW.[57] New equipment is being designed and installed to take advantage of this new market.[55] For example, Cornell University renovated a small hydroelectric plant on campus. The plant cost $1.25 million to renovate and costs $25,000/yr to run. The power plant produces 1.3 MW and Cornell sells it for $0.25 million to the New York State Power Agency; it paid for itself in just 5 years. The renovation cost was $960/kW, whereas new hydroelectric facilities cost about $3000/kW.[3]

Many developing countries, such as Burundi, Papua-New Guinea, Liberia, and Nepal are developing low-head hydroelectric resources.[58] China has made use of some of its small hydro possibilities. It now has 90,000 small dams to generate electricity and a booming export market in small turbines and generators.[59]

There are several types of turbines that can generate energy from streamflow, just as tethered water wheels once did. The Davis turbine built by Blue Energy[40] and the Gorlov turbine (**Extension 22.1**) promise economic run-of-stream energy.[40] These hold the promise of increased efficiencies and availability of water power.

Pumped Storage

In **pumped storage**, a reservoir is constructed some distance above a source of water. In times of slack demand, electricity is used to pump water from the supply below to the reservoir above. When demand for electricity is high, the water

(a)

(b)

(c)

FIGURE 22.4

(a) A rehabilitated 2.25 MW dam on the Battenkill River in Greenwich, New York, supplies an industrial site and the local utility with energy. (Photo courtesy of New York State Energy Research and Development Administration)
(b) The city of Idaho Falls, Idaho, is supplied by a renovated low-head hydroelectricity system. (U.S. Department of Energy)
(c) A recently refurbished dam in upper New York supplies local energy. (Photo courtesy of New York State Energy Research and Development Administration)

in the higher reservoir returns to the lower level through turbines, adding to the supply of energy available. The idea of pumped storage is that it is most efficient for the utility to increase its base load, evening out the daily fluctuations. Consequently, the utility should try to store the extra energy it produced when demand was low to recover it at times of peak demand. Such pumped storage returns about three-quarters of the energy used to pump the water uphill—not bad by comparison with technologically competing possibilities such as battery storage. About 2% of all U.S. electrical energy cycles through pumped storage load shifting.[3]

A pumped storage plant at Ludington, Michigan, is the world's largest facility. It has an 85 m head and uses six pumps/turbines to pump water up to an artificial lake over Lake Michigan. It can, when needed, run back to produce 2000 MW.[3] The second largest facility is the pumped storage facility in Goldisthal, Thuringia, Germany. [It has an approximate 300 meter head of water going through the four turbines at the rate of about 400 m^3/s (those numbers, if exact, would indicate 1200 MW at 100% efficiency); actual output power is 1076 MW.[3]] Another fairly large facility is the Kinuza Dam in northwest Pennsylvania, which is an 870 MW electricity peaking facility supplying Cleveland. It has a 220 m head and can pump water up for 2 cents/kWh at times of low demand and sell the electricity produced by letting the water fall back down for 5 cents/kWh at times of high demand.[3]

Another lesson in water use is provided by the Storm King Mountain pumped storage facility proposed by Consolidated Edison of New York City in 1962 and finally abandoned in 1981. The Environmental Defense Fund of New York City had calculated that a conventional fossil-fuel system would be a cheaper alternative than Storm King to supply the same amount of energy, but despite its own economic crisis and the prospective cost of $750 million,[60] the utility was convinced that pumped storage was its best alternative. Proponents expected that the mountaintop reservoir could be used for recreation, but one wonders what happens when a small lake is filled or drained at the designed pump rate of 32 million liters (8 million gallons) per minute from or into the Hudson River.[60] Some of the most effective opponents found that the huge fluctuations in river flow rate would destroy 25% to 75% of the striped bass hatch in the Hudson River. Consolidated Edison finally agreed to give up Storm King in return for the dropping of demands by the Natural Resources Defense Council that cooling towers be built onto existing Con Ed thermal plants.[61]

22.2   OCEAN ENERGY

Tides and waves carry kinetic and potential energy, while the ocean has available thermal energy stored in its temperature differences. The first sort of energy we can tap as wave energy; the second by using an engine between the strata at different temperatures.

Energy from the Tides

Tidal energy arises from the interaction of the Moon (and the Sun, whose effect is half as large) with Earth. Many methods that produce energy from flowing water and from waves can also be used to obtain usable energy from the tides. The oscillatory flow of water in filling and emptying a partly enclosed region along a coast can be tapped to produce electricity. This energy may be made available by cutting off the basin with dams to create a water level difference. Additionally, tapered channels ("tapchans") can be used to force the water to rise into a basin. (Water is being pushed from behind, and the taper increases the speed of the water.) Of course, the total amount of energy available depends strongly on the difference between the high and low tide levels, the **tidal range**, because it uses the potential energy generated by the rise and fall of the tides to generate hydroelectricity.

Use of tidal energy is not new. Before 1100, tide mills were operating on the European coast.[62] The total world tidal power available is estimated to be 0.33 TW (2.9 PWh/yr = 10 EJ/yr) and is only about 1% of the world's available water power.[63,64] The largest facility in operation, at 240 MW installed capacity, is at La Rance, France. It has a tidal range of ~8 m and produces about half a billion kWh of electricity yearly (about 20% of the maximum available energy and 90% of the design value).[65] Overall, the world's operating units have capacity factors between 22% to 35%.[65]

The energy could be tapped by taking power only when the basin is emptying, or both as it fills and empties. Two-way generation is not advantageous compared to one-way because of the limitation of available head from obstructing inflow. However, two-way generation operates over a longer time, sending a steadier supply to the grid, and so may be preferable.[62] A 5% to 15% extra energy gain is possible when water is pumped up into the reservoir near high tide (that is, at small head).[62] It is allowed to flow out near low tide, and because it has a large head, there is a large change in gravitational potential energy to supply electrical energy (see *A Closer Look:* Tapping Tidal Energy).

In the Bay of Fundy, which has a 15 m average tidal range, a plant was proposed for the Minas Basin; if built, it would provide about 15 TWh$_e$ yearly.[66] The total potential electric generation capability of the entire Bay of Fundy system is between 30 and 60 GWh/yr.[64] The scheme would consist of 128 turbines with an installed capacity of 5 GW$_e$, and with the intermittent character of the tidal rise and fall, the power output would average ~1.75 GW$_e$.[66] The Bay of Fundy has the largest tidal range in the world, which is due to the resonance between the tidal forcing and the Gulf of Maine, of which Fundy is a part. (**Resonance** refers to what happens to the response amplitude at some particular frequency of forcing—for example, if your tires are out of balance, the imbalance provides a periodic jar; the jarring can cause the car to shake as if it would fall apart at one particular speed. If you speed up or slow down, the shaking vanishes. You have just found a resonant frequency for your car. Bodies of water can resonate, too. Ask any child who used a washcloth to push on water in the bathtub with just the right frequency to make a huge wave go over the edge of the bathtub onto the floor!)

The Gulf of Maine has a natural period of about 13.3 hours, while the Moon-caused tidal period is 12.4 hours.[66] This near-match causes the huge water flow: about $\frac{1}{40}$ (km)^3/s. There is a possible problem with the Minas Basin project. Models indicate that closing off the Minas Basin would decrease the natural period of the Gulf of Maine, causing an increase in the tidal range as a result of the closer match.[66] Therefore, tides would be higher as far south as Cape Cod. This project might even change tides in Britain, although by only a few millimeters.[66]

TABLE 22.1

World tidal power facilities

Location	Power	Year began production
La Rance, France	240 MW$_e$	1966
Annapolis Royal, Nova Scotia	17.8 MW$_e$	1984
Jiangxia, China	3.2 MW$_e$	1980
Kislaya Guba, Russia	0.4 MW$_e$	1969

China has several smaller barrage (moving barrier) schemes that total 1.8 MW$_e$.

Source: References 62 and 65.

A CLOSER LOOK

Tapping Tidal Energy

Energy from the tides is generated in much the same way as hydroelectric energy from dams is generated. Tidal basins (partly enclosed river mouths, bays, and so on) empty and fill with the tides. If the basins are closed off by dams, water level differences will exist between the basin and the ocean that could be used to run turbines.

The total amount of energy available from such an enclosed basin is $\frac{1}{2}Mgh$, where M is the mass of extra water held by the basin at high tide as compared to low tide, g is the acceleration of gravity (9.82 m/s^2), and h (the tidal range) is the difference in heights between high and low tide ($\frac{h}{2}$ is the height of the basin center of mass). The mass M of water stored in the basin is proportional to the surface area of the basin and to h. Thus, the mass of captured water is given by the density of seawater (1030 kg/m^3) multiplied by the volume of water captured by the dam; the volume is (area) $\times$ h; thus,

$$M = (\text{density of water}) \times (\text{area}) \times h.$$

This means that the stored energy grows as the square of h:

$$\text{stored energy} = \tfrac{1}{2}Mgh = (\text{density of water}) \times (\text{area}) \times h \times gh$$
$$= \tfrac{1}{2}(\text{density of water}) \times (\text{area}) \times gh^2.$$

For this reason, regions with high tidal ranges, such as the Bay of Fundy, have the greatest potential for energy production. Clearly, the area is also important, which is why all possibilities considered are large bodies of water.

The energy actually available is generally only a fraction of the maximum potential. The operation at La Rance, one of only two working tidal power installations, produces between 20% and 25% of the theoretical maximum. Still, although tidal energy is only a minor *world* energy resource, locally, it could have a large impact indeed. The closing of the San Jose Bay in Argentina; Mont Saint-Michel in France; Severn in England; Cook Inlet in Alaska; and the White Sea, the Sea of Okhotsk, and the Penzhinsk Gulf in Russia could produce average power in the gigawatt range. Other potential local sources are relatively minor compared to these. Most of these, however, would be too expensive to develop.

The effect on the Georges Bank, a major world fishery, is still to be determined. So far, only a test facility is operating (see Table 22.1).

A tidal barrage is a barrier that can be raised to prevent inrush to or outflow of water from a river connected to the ocean (a tidal barrage was built on the Thames to help prevent floods in London). The costs of tidal barrage energy systems are not going to drop very much because of the extensive construction and operational experience that already exist. There also are fewer surprises with this form of energy, and it is the only one of the alternatives considered in this chapter that can be clearly economic at present.

However, several experimental projects are proceeding that are slightly different from those described above, in that they involve reaping energy from coastal currents. One such project is the tidal fence. In this type of project, water flowing through an underwater valley (which has to move faster because it is constrained) passes through a sunken structure (a caisson) holding vertical turbines.[67,68] The force of the inrushing and outgoing water turn these turbines except at slack water, producing electricity. Submerged horizontal turbines (much like sunken windmills, 20 to 30 m under water) can also be placed to generate electricity. These turbines work best at water speeds of 2 to 3 m/s (7 to 11 km/h) and generate 4 to 13 kW/m^2.[67,68] The greater density of water means that smaller blades can be used at slower water speeds and still equal or beat windmill energy output. Also, the tides are more predictable than the winds.

Promising sites for this type of tidal energy are in Europe and are found near British and Norwegian coasts. Elsewhere, the best sites are at the peripheries of oceans.[67] Projects are under way in Britain (I. T. Power, British) in the Severn,[69] off Devon,[70] and in the Orknies (the "TidEl" system),[71] in the Philippines (Blue Energy, Canadian), in Italy in the Straits of Messina,[72,73] and in the Kvall Sound in Finnmark, Norway (Hammerfest Stroem, Norwegian).[67,74] Two demonstration projects are in the works for the United States: at the Golden Gate in San Francisco Bay (750 kW), and in New York City's East River (200 kW).[75]

Wave Energy

Just as the Sun drives the winds, the winds drive the waves. The United States and European west coasts, and the coasts of New Zealand and Japan, offer the best prospects for utilizing this energy. In some regions, wave power per length of wave can reach a megawatt per meter.[76] The energy could be used to pump water up (controlled by a check valve) to a height at which the water would be allowed to fall through a turbine, or air could be compressed and sent through a turbine. One problem with wave energy is the low frequency of typical wave motion, 0.10 Hz, relative to grid electricity (and especially turbine rotation rate). Shoreline installations and nearshore facilities can offer power in the megawatt range. Deep offshore facilities could give power in the gigawatt range.[62]

A lot of work on ways to extract energy from waves has been done in Europe, especially in the United Kingdom.[62,65] The U.K. National Energy Laboratories evaluated 300 concepts and tested many of them using scale models.[62] Tests are underway or planned in the United Kingdom, Norway, China, India, Japan, and Portugal (see the **Extension 22.2, Wave Energy Devices** for more detail).[62] The cost of electricity in a 1988 analysis of wave energy possibilities was characterized[62] as

$$\text{cost} = \frac{\$1.129/\text{kWh}_e}{(\text{wave "power density" in kW}_e/\text{m})^{0.64}}.$$

An installation in the North Atlantic with 50 kW/m would have produced energy at $0.0923/kWh_e. This is still not quite economically feasible. There are only a few regions with power densities higher than this.

Ocean Thermal Energy Conversion

Temperature gradients in the sea, **ocean thermal gradients**, can be transformed into useful forms of energy: This technology is known as ocean thermal energy conversion (OTEC). It has been estimated that the thermal gradients of the Gulf Stream could produce energy at a rate of 182 PWh_e per day, about 75 times the 1980 U.S. energy use rate.[94] One kilogram of water changing its temperature from 25 °C to 5 °C gives as much energy (~8 kJ) as it could by falling about 820 m. Of course, the efficiency of such a plant is less than the maximum thermodynamic efficiency (see Chapter 7) of $\frac{20}{298} = 0.067$, or 6.7%. In practice, such an OTEC operation is only about 2.5% efficient[65] because of the energy cost of pumping seawater against density differences (the colder, lower water is more dense). Thus, such facilities need a temperature difference of at least 17 °C to work.[65] OTEC facilities must be put near shore if they are to feed electricity to the grid. If not, they could be sited far offshore and perhaps used to generate hydrogen from electrolysis. The gas could be shipped on tankers to shore and used in place of natural gas.[65,95]

The colder water from lower levels has nutrients that could "fertilize" the surface waters and attract fish. OTEC energy farms could also be fish farms.

There are two cycles used to produce electricity, the open (or Claude) cycle and the closed cycle (see accompanying box *A Closer Look*: Open and Closed OTEC Cycles). The first working OTEC facility, built by French physicist Georges Claude, operated in Cuba's Havana harbor in the 1920s for only 2 weeks before being destroyed by a hurricane.

A CLOSER LOOK

Open and Closed OTEC Cycles

OPEN
Warm seawater, about 25 °C to 28 °C from perhaps 12 m down, is brought to an evaporator, a vacuum vessel, that is at low pressure (about 0.03 times the atmospheric pressure). The water boils at such a low pressure. The low-pressure steam runs a turbine for electricity. The spent steam is then condensed to become fresh water by heat exchangers connected to cold (6 °C) seawater. The fresh water is siphoned off for use. See Figure 22.5.

CLOSED
Warm seawater is pumped through a heat exchanger containing, for example, ammonia or some other working fluid. The ammonia vapor released drives turbines to produce electricity; at a second heat exchanger, the ammonia is cooled and condensed. It is then returned to the first container and run through the cycle again. See Figure 22.6.

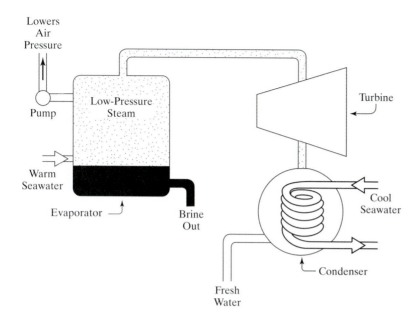

FIGURE 22.5

An open-cycle OTEC facility. The word "open" refers to the spent steam's exhaust to the environment.

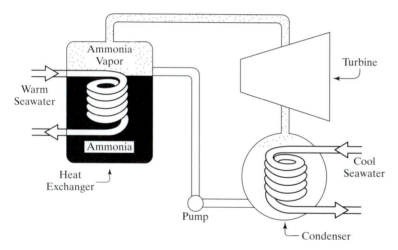

FIGURE 22.6

A closed-cycle OTEC facility. The ammonia cycles through a totally closed loop in the system.

The upwelling cold water could alter the entire world climate if such OTEC plants were built on a large enough scale. Calculations show that if 60 TW$_e$ of OTEC is produced, it would reduce the average surface temperature by 1 °C.[65] Also, there is release of the entrained carbon dioxide from deep water, so OTEC is not entirely free of pollution.[62,96] More on OTEC energy can be found in the **Extension 22.3, *OTEC***.

22.3   MIXING OF FRESH AND SALT WATER

Ocean–Land Interface

A recently recognized energy source is available from the mixing of salt water and fresh water as rivers flow into oceans. Theoretically, it is possible to liberate

$0.65 \text{ kW}_e\text{h/m}^3$ in a seawater–fresh water interface, an energy equivalent to a dam 240 m high.[62,102,103]

Three possible ways to get useful energy from the mixing of salt water and fresh water have been suggested. Semipermeable membranes could generate osmotic pressure (to raise the water around 247 m, in theory). The estimated present costs of $0.22 to $0.30 per kWh_e make this a prohibitive option.[62] It is possible to use reverse dialysis to set up a "salt battery," which would be a very expensive option at around $100,000/$\text{kW}_e$.[62] The different vapor pressures of fresh and salt waters could be exploited; the fresh water would be evaporated, then condensed in seawater after the vapor had gone through the turbine. This consumes fresh water and is expensive, too.[62,103]

The first two methods use membranes to separate the two fluids, but this involves expensive pumping and short membrane life. Because the fresh water is constantly renewed by the Sun, and because the resource potential is so large, such a resource can be used even if the process has a very low efficiency. All alternatives are, however, currently judged too expensive.

Solar Ponds

Salt water is also involved in another type of solar energy storage. Ponds are constructed in which salt water sits covered by fresh water. The water naturally separates into regions of different densities (or salinities). This salt density gradient provides zones in which heat is trapped and stored. The water at the top is fairly fresh. There is a middle region that is very stable, and that lets light through into the bottom but not out (it acts something like a window). The middle zone also prevents the very salty hot water at the bottom from mixing with the rest of the water.

Such **salt-gradient solar ponds** can use the stored heat to heat water or air somewhere else. C. Nielsen of Ohio State University built a pond that demonstrates the economical use of salt-gradient solar ponds for heating barns and swimming pools in the northern United States.[104] The world's largest solar pond has been operating in El Paso, Texas, since 1986. It produces 70 kW_e plus 5000 gallons of fresh water per day.[105]

In most temperate zones, solar ponds can provide 250 $\text{kWh}_e\text{/m}^2$ of heating in the winter, for an efficiency of 18%.[104] Work on such ponds is proceeding in areas with lots of sunlight, especially Israel. Measurements in Beersheva, Israel, indicate that an average of 18.4 MJ of energy reaches each square meter of surface per day.[106] Salt-gradient solar ponds in Israel are used for production of electricity. The two reservoirs for the heat engine are the high-temperature reservoir at the bottom of the pond and the low-temperature reservoir at the top. A turbine using a low-boiling-point fluid operates between the two temperatures. A 1250 m^2 pond built in the early 1970s is still producing 6 kW; a 7000 m^2 pond built in 1979 is producing 35 kW (and can produce a peak of 150 kW).[107] The Israelis were planning to produce around 2000 MW by the year 2000 from such solar ponds, but got only to around 5 MW.[108] Solar ponds are limited by the fierce evaporation experienced, which translates into a necessity for large amounts of water.[105] Also, it costs energy to maintain the salt density gradient.

Solar "Unmixing"

Salty water is much more abundant in parts of the world than the fresh water needed for human use—the 50 liters per day for bathing, drinking, and sanitation.[4]

Solar energy is readily available in many parts of the world where there is a lack of fresh water, and so it is natural to consider solar desalination methods. Recent advances in solar energy and recognition of the needs of fresh-water poor (and economically poor) countries led to more exploitation of solar-energy-based desalination. The three major methods used for solar desalination are multieffect distillation, multistage flash evaporation, reverse osmosis, solar distillation, and rooftop active solar energy.[109]

SUMMARY

Water is necessary for almost all forms of energy generation, from production of coal-fired electricity to production of food. Hydroelectricity is stored solar energy. It may be produced from running water directly or from impoundments or diversions.

The building of dams (see also **Extension 22.1**) for water storage and for electricity is usually considered benign and does confer many benefits. However, dams can cause increased soil salinization; increased disease carried by mosquitoes and waterborne human parasites; drowning of productive land; replacement of fish and insect species; changes in groundwater levels; dislocation of people; and fouling by water plants.

Pumped storage facilities also have environmental side effects. Fewer side effects accompany the refurbishment of old dam sites, as is being done in New England and the upper Midwest.

Solar energy that reaches Earth drives the winds and ocean currents or is used by plants. The winds drive ocean waves, and waves and the tides can be made to produce prodigious amounts of energy (see a more extended discussion in **Extension 22.2**).

The Sun warms the surface of the ocean, and the temperature difference between the surface and the ocean deeps can produce energy and increase yields of seafood (this method also has advantages and disadvantages considered in more detail in **Extension 22.3**). The interface between fresh water and salt water where rivers flow into the ocean can also produce energy.

PROBLEMS AND QUESTIONS

MULTIPLE CHOICE REVIEW

1. Which of the following can be considered stored solar energy?
   a. Diversion of water
   b. Impoundment of water
   c. Tidal energy
   d. Ocean-layer temperature differences
   e. All the other choices are stored solar energy.
2. Which of the following may be used for gaining energy from water in streams?
   a. Millrace
   b. Horizontal-axis breastshot water wheel
   c. Vertical-axis water wheel
   d. Reaction turbine
   e. All of the other choices may be used to obtain energy from a stream.
3. Which is *not* a possible drawback that occurs during or after a dam is built on a river?
   a. Evaporative salinization
   b. Increased median temperature in the locality
   c. Sedimentation of the reservoir
   d. Change in fish species found
   e. Displacement of local population
4. Which of the following may be used for gaining electric energy from water in streams?
   a. Small-head hydro
   b. Horizontal-axis overshot water wheel
   c. Gorlov turbine
   d. Run-of-stream reaction turbine
   e. All of the other choices may be used to obtain energy from a stream.
5. Which of the following is *not* (strictly speaking) using solar energy?
   a. A small-head hydro reservoir
   b. Water in a large dam reservoir
   c. Run-of-stream Davis turbine
   d. Pumped storage in a reservoir
   e. They all use solar energy.
6. Human exploitation of tidal energy began in the
   a. eleventh century.
   b. fifteenth century.
   c. eighteenth century.
   d. nineteenth century.
   e. twentieth century.
7. The typical thermodynamic efficiency of an OTEC installation running between surface water at 22°C and deep water at 4°C is about
   a. 6%.
   b. 18%.
   c. 82%.

d. 86%.

e. 99%.

8. The World Commission on Dams report recommended
   a. equity for the inhabitants of the region in which the dam is built.
   b. consultation with local communities.
   c. careful study of ecological effects before building a dam.
   d. beneficent use of new wetlands.
   e. all of the above measures.

9. Which site for a proposed dam would probably be worse from the point of view of producing increased water salinity? A dam in
   a. the Pacific Northwest, west of the Cascades.
   b. Southern California.
   c. New England.
   d. Labrador.
   e. northeast Russia.

10. Pumped storage is best used to provide
    a. baseload.
    b. cycling load.
    c. peak load.
    d. all of the preceding loads.
    e. none of the other loads because of all the environmental disadvantages of the technique.

SHORT ANSWER

Explain whether the statement made is correct, and justify your answer.

11. Underdeveloped countries can gain by exploiting low-head hydroelectricity.

12. For a tidal energy device, bigger tidal range is better.

13. Solar ponds store solar energy.

14. It is impossible to use appropriate "run-of-stream" turbines to capture tidal energy.

15. Much settlement in early American history was near rivers and streams so people could take advantage of water power to run factories.

16. Water loss through evaporation exceeds input for much of the land area of the United States.

17. The sum total of all the wave energy in the oceans could supply all the world's energy needs.

18. Egyptian dams have brought unmitigated blessings to Egypt.

19. Wave energy is economically feasible at the present time.

20. Tidal energy projects have no environmental drawbacks.

DISCUSSION QUESTIONS

21. How can the mixing of fresh and salt water be considered stored solar energy? Explain.

22. Discuss the economies of scale (if any) that apply to building dams.

23. How could OTEC facilities alter the world's climate? Explain.

24. Are wave-energy machines environmentally benign? Explain.

25. Why is siltation such a pressing question for dam builders? Explain how you decided on your answer.

26. Evaporation losses from impoundments could be countered by release of millions of white Ping-Pong balls to float on the water's surface. The white surface of the balls would reflect most of the incident sunlight and reduce evaporative losses substantially. Assess the feasibility of such a measure in the "real world."

27. How can a tapered channel (tapchan) work to produce energy? Explain.

28. How can OTEC installations provide fresh water as well as electricity? Explain.

29. How can resonance effects in tidal energy systems disrupt the environment?

30. Why are there so few tidal energy installations in existence? Explain.

PROBLEMS

31. Consider the construction of a tidal installation near Passamaquoddy, New Brunswick, Canada. Given that seawater has a density of 1030 kg/m^3, if the engineer says that the impoundment will be 30.3 km^2 in area and have a tidal head of 8.95 m, how much energy can the electric company expect to obtain? How great is the average project power? You may assume that there are two tides per day. Explain.

32. Off the British coast is an inlet with a surface area of 32.1 km^2 that could be closed off to make a tidal basin of head 5.72 m. Given that seawater has a density of 1030 kg/m^3, determine the energy stored in the basin. How much power could a group of investors hope to sell? Explain.

33. In a certain region off the Irish coast, wave "power density" averages 60 kW/m. What will it cost to produce electricity from this region?

34. Earth is about 72% water covered. The latent heat of vaporization of water is 2.25 MJ/kg. The density of seawater is 1030 kg/m^3 and the density of fresh water is 1000 kg/m^3.
    a. How much energy would it take to evaporate the top centimeter of the world's oceans?
    b. How much energy would it take to evaporate the top meter of the world's oceans?

35. By how much does the potential energy of a drop of water of mass 1.5 g change as it falls from a cloud at 2.265 km as rain on ground at sea level? How does this compare to the energy cost of evaporating the water from seawater?

36. You wish to build a solar-powered electric plant that operates by using the temperature difference between the top (25 °C) and bottom (15 °C) waters of a lake. Suppose the Sun deposits 550 W/m^2 into the lake. What is the minimum area of lake you would need for your power plant to generate 500 kW of electric power?

CHAPTER 23

Biomass Energy

In the preceding chapters, we saw several different methods for utilizing solar energy. This chapter and the next focus on solar energy and the production of biological organisms (biomass). This chapter deals mainly with the energy and chemical resources in biomass.

KEY TERMS

anaerobic process / solar window / photosynthesis / chloroplast / chlorophyll / reaction center / adenosine triphosphate (ATP) / C$_3$ plants / C$_4$ plants / ethanol / bagasse / cassava / anaerobic digestion / feedstock / jojoba / guayule

The Sun has shone on Earth for about 5 billion years. The original atmosphere was composed of hydrogen, helium, nitrogen, methane, ammonia, and some carbon dioxide. By 3 billion years ago, the hydrogen and helium had oxidized or escaped, and life was beginning in the oceans. Only because there was available solar energy to keep Earth warm and to supply energy for chemical processes was life able to develop.

It should not be surprising that plants have succeeded at using and storing solar energy over the course of 3 billion years, while humanity has not yet been successful at capturing and storing solar energy for use at later times. Without plants, animals could not exist in their present forms. The original plants were able to use the relatively abundant carbon dioxide (CO_2), methane (CH_4), water (H_2O), and sunlight to produce sugar-like compounds and oxygen. The sugars are food for animals, and without oxygen, animals could not metabolize the sugars. The oxygen atmosphere, with its protective 2-billion-year-old ozone layer, is an artifact of life on Earth. Because oxygen is a byproduct of plant photosynthesis, new forms of plants had to arise as adaptations to new conditions. Many of the old bacteria are still with us. These remnants of the early Earth are responsible for **anaerobic processes** (processes that take place in the absence of oxygen). Bacteria were also mostly responsible for the decrease in atmospheric carbon dioxide, a great amount of which was processed through the oceans and deposited as limestone.

23.1 SOLAR WINDOW

The atmosphere, as we saw in Chapter 16, has a "window" (the **solar window**) allowing solar radiation in a restricted range to arrive on Earth: About 300 to 700 nm is clear for transmission of all solar energy (which is why this is the visible part of the spectrum); there are also some regions in the range from 700 to ~3000 nm that allow some near-infrared from the Sun through the atmosphere. If you were going to design a solar energy receptor, you would probably just want to use the wavelength range from 300 to 700 nm, because longer-wavelength infrared radiation would cause heating, which could interfere with the chemical reactions you would like to use in photosynthesis. A leaf would evolve to avoid wavelengths from 700 to 3000 nm. In fact, real leaves do exactly this:[1] Figure 23.1 shows the absorptance of the leaf of the eastern cottonwood (*Populus deltoides*). The absorptance peaks in the infrared are caused by water in the leaf and correspond to gaps in transmitted solar radiation. Little thermal energy is therefore absorbed in leaves. The broad peak in the visible region is due to absorption by the cells of the leaf.

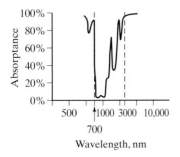

FIGURE 23.1

Absorptance (percentage of incident radiation absorbed) of *Populus deltoides* as a function of wavelength of the incident light. (C. G. Granqvist, "Radiative heating and cooling with spectrally selective surfaces," *Appl. Opt.* **20**, 2606 (1981), © 1981 by Optical Society of America)

23.2 PHOTOSYNTHESIS

Photosynthesis involves the chemical reaction of water and carbon dioxide to produce oxygen and organic materials such as sugars. As emphasized in Chapter 6, photosynthesis still holds a few mysteries for us. Nevertheless, during the course of the activity, three processes are known to occur: Membranes called **chloroplasts**, containing **chlorophyll**, absorb light. Each light photon (γ) boosts an electron in chlorophyll to a higher state. The electron jumps across to the carrier, then again and again as more photons are absorbed. A replacement electron comes from hydroxyl (though it is not known how). There is a separation of charges across the membrane, and electrons are donated or accepted on the sides of the membrane.[2]

Photosynthesis takes place in bacteria as well as in plants and algae, but plants and algae release oxygen and fix carbon, while bacteria do not.[3] Chlorophyll molecules of two different sorts exist in the chloroplast in plants and algae. The chlorophyll molecules can absorb the blue and the red light from incident sunlight and so appear to be green, because the green part of the spectrum is reflected (the color of nonluminous objects depends on the color of the light reflected from them). The energy of the incoming photons is collected in a **reaction** (or "trapping") *center*, in which water is split, oxygen is released, and energy is stored in intermediate compounds. The energy stored in the compound **adenosine triphosphate** (ATP) by the one sort of chlorophyll is used to make sugars; animals also use ATP in their internal energy economy.[2,3] The energy stored in the compound nicotinamide adenine dinucleotide phosphate hydride (NaDPH) by the other sort of chlorophyll is used to store hydrogen and the electrons.[2–4] Energy is exchanged between the two sorts of chlorophyll, as well as electron flow. It is this "light" part of the reaction that is the most mysterious, especially the one involving the splitting of water and the synthesis of ATP. The rest of photosynthesis consists of a "dark" chemical reaction, one not using light energy, in which NaDPH and ATP shuffle energy into the cell to run the Calvin cycle (which converts CO_2 into glucose), returning $NaDP^+$ and ADP (adenosine diphosphate) for reconstitution by the light-absorbing reactions. Each transition between ADP and ATP involves about one-third of an electronvolt per molecule, or 31.8 kJ/mol. Overall,

$$6CO_2 + 12H_2O + \text{energy (sunlight)} \longrightarrow C_6H_{12}O_6 + 6O_2 + 6H_2O,$$

where $C_6H_{12}O_6$ is the sugar glucose. In animals, glucose supplies the energy needed for metabolism, the overall reaction being the reverse of glucose formation:

$$C_6H_{12}O_6 + 6O_2 + 6H_2O \longrightarrow 6CO_2 + 12H_2O$$
$$+ \text{ energy (in the form of ATP)}.$$

Synthetic molecules that will mock natural photosynthesis are being developed. O'Regan and Grätzel[4] created a triruthenium dye molecule "antenna" to focus the solar energy onto tiny particles of semiconductors. They adsorb the light-absorbing photosensitizer molecule onto the surface of polycrystalline electrode. The dye emits electrons, which go to the semiconductor, which sends the current to the negative electrode. It then gets more electrons from a liquid electrolyte containing iodide ions, which give up electrons to the dye and return to the positive electrode.[4] The presence of four manganese atoms appears crucial to the process.[5] The structure of the cytochrome protein complex is important, because that is where the synthesis of ATP takes place, and recent progress has been made in determining that structure.[6]

Photosynthesis is made difficult because of the presence of oxygen (which inhibits photosynthesis), but photosynthesis utilizes about 42% of the incident sunlight at about 700 nm. Only about two-thirds of the sunlight is of the appropriate wavelength, that is, the wavelength corresponding to half the energy necessary to drive the reaction. Sunlight of shorter wavelength (higher energy) is absorbed as well. The excess energy is transferred to the environment as heat. As a result, the efficiency drops to about 27% averaged across the visible region, which, in turn, is only 44% of available sunlight.[3] Hence, plant photosynthesis is only $0.27 \times 0.44 = 0.119$ (12%) efficient *per se* in transforming radiant energy to chemical energy, not counting losses due to plant metabolism (the taking in of carbon dioxide and the emission of oxygen). Assuming a 10% loss to reflection, the overall efficiency cannot exceed 10% for conversion of solar energy to chemical energy.[3] Biomass useful to humans is only a part of the total production of a plant, so the efficiency for harvestable biomass is much lower than 10%. Despite the low efficiency, photosynthesis produces 220 Gt/yr of usable biomass.[7]

Photosynthesis rates are limited by the small proportion of CO_2 in the atmosphere.[3] Increased CO_2 concentration causes more rapid plant growth. Efficiency varies from species to species, being greater for corn (*Zea mays*) or sugar cane (*Saccharum officinarum*) than for wheat (*Tritieum aestivum*) or soybeans (*Glycine max*). (These are representative of species of plants utilizing different types of photosynthesis, called C_4 and C_3, respectively.) The intermediate compounds produced utilize four and three carbons, respectively. **C$_3$ plants** will be more efficient at conversion of carbon to plant matter as the carbon dioxide increases.[7] **C$_4$ plants** have better water-use efficiencies than C$_3$ plants. Most food crops are C$_3$ and produce 1.4 kg of biomass per GJ.[7]

A study of a complete ecosystem in Hubbard Brook, New Hampshire, Experimental Forest reveals that the actual year-round efficiency is only about 0.8%.[8] Of course, plants grow only in the summer months, so actual efficiency is somewhat higher. Sixty years after it was logged over, the forest is still accumulating about 628 kJ/m^2 each year.[8] When the forest finally matures, it is expected that the energy flow should be in balance. Trees lose branches that then rot, and older trees do not add much biomass to their trunks, so the system is in dynamic balance.

For most human purposes, production should be out of balance to enable us to produce food for people and animals. About 1.7 Gt of grain is produced each year, of which 218 Mt is traded (about half is from the United States alone).[9,10] As we saw in Chapter 2 and will investigate further in the next chapter, this energy is vital to our well-being. This chapter focuses on biomass as used for purposes other than food or animal feeds.

23.3 BIOMASS FOR FUEL

Production of biomass is a desirable way to use land. It is one of the few methods of energy production that does not arouse much public opposition. Biomass is renewable. Agricultural fields produce every year, and even trees can grow over periods of 30 to 100 years. Biomass also is well placed to take advantage of the ambient conditions. Plants grow almost everywhere, despite the fact that they spend half their lives in darkness. Plants are automatic energy storage devices, transforming solar energy to plant matter, and are well suited to the intermittent availability of sunlight. Some algae can convert nearly 10% of available light energy to biomass. Sugar cane is about 2% efficient, and corn is about 1% efficient. Most other cultivated plants are less efficient than the Hubbard Brook forest in utilization of solar energy. Plants are not nearly as efficient as the most efficient solar cells.

Given all of these advantages, it is natural to ask whether we could use some of the stored solar energy as a fuel for transportation, heat, or electricity. The answer is clearly yes. A century ago, over 90% of our energy demand was met by wood. Wood currently provides less than half as much energy as is generated in nuclear plants,[11] perhaps as much as 5% of the total U.S. energy supply,[12] and it may be cheaper than alternatives. One factor leading to this rather large amount is PURPA, which led producers to try various sources of renewable energy.[13] A small Kentucky college was able to save some $221,000 a year by burning wood, instead of gas, to heat dormitories.[14] In lumbering areas, such as Oregon, power plants burn "hog fuel," which is wood and bark residue, to produce electricity. The Eugene Water and Electric Board, a municipal utility in Eugene, Oregon, can generate up to 34 MW$_e$ or produce about 1000 tonnes of steam per hour to be used by local industry or for local heating.[15] The McNeil Generating Station, in Burlington, Vermont, produces 50 MW$_e$.[12,16] Bagasse (woody residue) from sugar cane supplies about 7% of Hawaii's electricity needs.[17]

Sugar cane is also involved in an ambitious Brazilian project to produce **ethanol** in order to reduce dependence on imported oil for automobiles and trucks. Brazil has few oil reserves, and in response to the oil price increases of 1973 and 1979, it implemented more stringent energy conservation measures than were taken in the United States. In addition, research had determined that less than 2% of Brazil's land area grown in sugar cane could produce enough fuel to replace all imported petroleum.[18] Between 1973 and 1987, Brazil increased its energy use from 90 million tons of oil equivalent (Mtoe) to 160 Mtoe, but gasoline went from 12% to 4% of total energy use. Brazil's efficiency of ethanol production rose also; between 1977 and 1985, the amount of ethanol produced per area rose from 2663 to 3811 L/ha. Liquid fuel sold in Brazil now is either neat ethyl alcohol (100% pure ethanol, used in 4.2 million cars) or a 78% gasoline, 22% ethanol mixture (5 million cars).[19] No unmixed gasoline is sold.

Many in the U.S. Midwest buy gasoline mixed with 10% ethanol. Ethanol and methanol are easier to handle if oxygen is added, so some is sold as ETBE (ethyl tertiary butyl ether) or MTBE (methyl tertiary butyl ether). MTBE or ETBE can then be transported and handled on the same basis as gasoline. Ethanol engines run 30% more efficiently than gasoline engines (in older cars) according to the EPA, a great advantage (ethanol burns less efficiently in modern cars). There is another reason to use ethanol instead of gasoline: the reduction in pollution. In São Paulo, lead levels dropped 80% after use of ethanol additives began.[19] Ethanol is superior in reducing pollution overall. Methanol or ethanol spills are less threatening than gasoline spills. Alcohol decreases the carbon monoxide concentration, and neat alcohol decreases smog.[19,20] In the United States, 100 areas exceed the national ambient air quality standards (NAAQS; see Chapter 13) for ozone, affecting half the population; and 40 areas exceed the NAAQS for carbon monoxide.[21] These areas could be helped by use of neat alcohol or blends, and the Clean Air Act Amendments of 1991 mandated mixing and oxygenation in them.

Brazil is going forward with a program involving research on producing ethanol from sugar cane, cassava, and sorghum.[19,30] Table 23.1[30] presents its results, showing that cassava (and perhaps grain sorghum) will probably not be useful in the Brazilian context because the energy balance is not so much greater than zero (see box *Applications: Ethanol and Methanol*).

New technologies for distillation are under active investigation. It is of little importance whether distillation shows a net energy gain or loss when the resulting ethanol is marketed as a drink for people, but it is of crucial importance in considering the use of ethanol as a replacement for all or part of our gasoline use. Techniques showing promise include using cellulose to absorb water, using permeable membranes to separate alcohol and water, using CO_2 at high pressure to

TABLE 23.1

Analysis of Brazilian net energy gain for selected crops

Crop	Produced/Consumed	Gross gain	Net gain
Sugar cane	4.53	2.43	1.43
Cassava	1.71	1.16	0.16
Sorghum			
Grain and stems	3.39	1.89	0.89
Stems only	2.27	2.01	1.01

Source: J. Goldemberg, "Brazil: Energy options and current outlook," *Science* **200**, 158 (1978), © 1978 by AAAS.

APPLICATIONS

Ethanol and Methanol

Ethanol can be made by fermentation from almost any crop containing sugars: corn, sugar cane, woody wastes, and so on. In Brazil, the ethanol process is an energy gainer[22] because of the burning of the **bagasse** (solid residue) from sugar cane to supply energy to make the ethanol, and the ratio of energy out as ethanol to fossil fuel used is between 5.9 and 8.2.[19] For the ethanol process to be economic with corn, the lignin must be used as a fuel. The process is

$$C_6H_{12}O_6 \xrightarrow{\text{yeast}} 2C_2H_5OH + 2CO_2$$

About 3% to 12% of the liquid becomes alcohol. This process would be more attractive economically if the feedstocks were cheaper.

Methanol is produced through a process involving both heating and chemical changes. First a feedstock is treated with steam and heat to become synthesis gas, then this is reacted to produce methanol. World demand for methanol is 23 GL/yr.[20] At present, the methanol is mostly produced from natural gas, but methanol is known as wood alcohol, and any carboniferous substance (biomass, coal, and so forth) will produce methanol. The process for making synthesis gas from biomass is

$$\text{dried biomass} + O_2 \longrightarrow CO + H_2 + \text{heat,}$$
$$\text{heat} + \text{biomass} + \text{steam} \longrightarrow CO + H_2.$$

The gas is cleaned and scrubbed, the hydrogen-to-carbon monoxide ratio is adjusted to 2 to 1, and the processes making methanol out of synthesis gas are

$$2H_2 + CO \longrightarrow CH_3OH + \text{heat}$$

and

$$3H_2 + CO_2 \longrightarrow CH_3OH + H_2O.$$

Brazil is considering growing **cassava** (also know as manioc; *Manihot esculeuta*) in addition to huge quantities of sugar cane to supply ethanol. Assuming production of 6 tonnes of cassava per hectare (which is realistic), a 30 km delivery radius could supply 2 Mt of feed per day to a plant that would then produce 15,000 bbl of ethanol per day.[23] It has been estimated that the Brazilian ethanol program could ultimately supply 250,000 to 1,000,000 jobs, mainly in agriculture.[24]

The energy question is more open in the United States because the higher cost of labor invites the substitution of energy for labor.[25] Production of ethanol from wastes has worked in the United States. The notorious Al Capone distilled bootleg liquor from Chicago garbage during Prohibition.[26] One analysis shows that distillation of ethanol from corn (or other grains) can produce disparate results. At best, it can produce a *gain* equivalent to 1 liter of gasoline per liter produced if coal, biomass, or solar energy is used in the distillation process; the ethanol is added to gasoline; and the distiller's grain (leftover mash) is used as animal feed. At worst, it can produce a *loss* of as much as a quarter of a liter of gasoline per liter produced when oil or gas is used in distillation; the ethanol is used straight; and the distiller's grain is not used.[27]

The processes as used currently in the United States to produce ethanol (and methanol) require about as much energy to produce as is produced (an energy balance about zero),[28] though progress has been made. Ethanol from biomass supplies about 800 to 850 Mgal/yr (3 GL/yr)[20] in the United States. The Department of Energy (DOE) estimates that production of 270 Ggal/yr of ethanol is feasible through the normal fermentation process. DOE-sponsored research reduced the cost from $3.60 to $1.35/gal by 1992 (this is $13/GJ, or $0.28/L). Large areas of idled cropland could be used to produce ethanol. The cost has been declining over the past decade. It is expected to reach $0.60/gal from the present $1.35.[21] Currently, 7 Mt of maize is being used to make ethanol each year in the United States.[20] Fermentation of 1 tonne of maize produces 440 L of ethanol, 330 kg of CO_2, 275 kg of corn gluten feed, 70 kg of corn gluten meal, and 37 kg of corn oil. At 18 t/ha/yr, there is an output of 2060 L/ha of ethanol.[29] At a yield of 20 t/ha/yr, there could be 1.5 Gt cellulose-like material collected per year, which could produce 1012 GL/yr of ethanol.[20] Current U.S. gasoline use is 425 GL/yr.[20] Short-rotation crops are being produced at twice the rate and half the cost of a few years ago.

Some Brazilians doubt that the energy gains are so great, but most agree that the situation is much better than in the United States.[30] The experience of Brazil may lead to the adoption of Brazilian agriculture production methods in other Third World countries, to the betterment of everyone concerned.

take up water, and using molecular sieves (zeolites, working by ion exchange) to absorb water from alcohol.[31] Research has been successful at reducing the dollar cost of ethanol,[32] but substantial progress must be made before ethanol could take even 10% of the market.[33] More information is available in **Extension 23.1, *Fuel from Biomass***.

Other Biomass Sources

Given that it is possible to make alcohol from agricultural products, should we do it on a large scale? We need to study the feasibility of moving to a totally solar-derived energy economy. Actually, such studies have been done by the Review Panel on Biomass Energy, chaired by Professor D. Pimentel of Cornell University,[9] by Pimentel and collaborators,[90] and by others.[27,92] They conclude that there is scant chance of biomass supplying total energy demand as solid, liquid, or gaseous fuel in the United States, or even of supplying the transportation sector (about 20% of total energy demand). A target of 5% may be realistic.

In 1979, energy from biomass totaled about 2% of U.S. energy use, roughly the same amount as in 1850, when biomass produced over 90% of our energy.[9] By 1992, the proportion was much greater (see Table 23.2).

While in 1987 biofuels produced 3 EJ of primary energy, people already appropriate another 26 EJ/yr of crops, forage, and forest products.[93] The most important categories of biomass sources are shown in Table 23.3.[9] We consider sources in the order listed in the table.

Wood

Wood is the most common biomass-based fuel. Much is burned inefficiently for cooking (see the discussion in the next section). Half of the wood fuel consumed

TABLE 23.2

Sources of U.S. energy (EJ), 2003

Source	Energy	Corrected[a] energy
Petroleum	40.46	40.46
Natural gas	20.72	20.72
Coal	20.36	20.36
Nuclear	8.41	8.41
Renewables		12.70
Hydro	2.93	8.80
Biomass	3.04	3.04
Geothermal	0.33	0.33
Solar	0.06	0.19
Wind	0.12	0.35
Total energy		95.76

[a] Energies corrected to take into account the distinction between thermal kilowatthours and electric kilowatthours.

is used in the pulp and paper industry, which meets 55% of its own energy needs.[94] Costs of wood collection could exceed energy gained.[9] Biomass growth in forests and growth in commercial forests exceeds 7×10^{18} J/yr.[92] Sustainable yields could be increased to about 1.1% of current fossil-fuel production, but such increases might be accompanied by severe erosion on steep hillsides when residues are taken.[15,95] It would take 635,000 hectares of forest to supply a plant producing 100 MW of electricity.[95] An area of 400 Mha of woodland would be needed to replace all U.S. fossil fuels (for comparison, as of 1988, U.S. cropland totaled 130 Mha).[93] Sustainable biomass production might

TABLE 23.3

Net biomass energy resources

	Potential			Current use
	Biomass total (Mt)	Total available (Mt)	Amount (Mt)	Thermal energy (TWh$_t$)
Wood (total)	651	261		
Mill residues (forest)	135	118	58–75	207–214
Logging residues	164	45	3.3	12
Thinnings	43	43	2.2	9
Residential fuel wood	27	27	27	59
Mortality, excess over harvest	282	27	—	—
Forage	682	118	—	—
Animal wastes	159	45	1	0.3 (biogas)
Grains	321	9–20	1	2.4 (liquid)
Bagasse	3.6	3.6	3.6	12
Food-processing waste	14	14	—	—
Industrial wastes	90	21	7.1	15
Municipal solid wastes	124	60	—	—
Municipal sewage	12	9	—	5.7 (biogas)[a]
Aquatic plants	18	3–12	—	—
Crop residues	430	73	—	—

Source: Reprinted with permission from Review Panel on Biomass Energy, "Biomass Energy," *Solar Energy* **30**, 1 (1983), copyright 1983, Pergamon Journals, Ltd.

[a] Energy is used in sewage digestion so is not net energy.

aid reforestation; but plantations cost more than residues.[13] Because wood is burned at low temperature, emissions of sulfur and nitrogen oxides are low.[15,27] However, woodstoves have caused local pollution problems, especially when many are used in valleys, and a natural inversion layer traps the smoke (see Chapter 13).

Forage

Some land now used for forage could be used for crops instead, or the forage could be used to produce energy by burning, gasification, or conversion to alcohol. It would be necessary to increase the price of forage, which would increase the cost of meat. It would also be necessary to fertilize fields to a greater extent than is now done,[26,95] but the ground cover after harvest will prevent erosion.

Animal Wastes

Much waste (about 60%) is dropped on the range, where it is not readily retrievable for agricultural fertilizer. Only 6% of U.S. manure is being used effectively.[96] However, feedlots are concentrated sources of olfactory pollution and water pollution from runoff. Fermentation of manure to methane and fertilizer by use of anaerobic bacteria, **anaerobic digestion**, could help clean up the environment. A plant in Guymon, Oklahoma, sends methane into the gas pipeline.[97] A proposal has been made to set up a closed-system dairy farm to use gas and produce cattle feed from the digested waste. In California, cattle dung that used to be sent to a landfill supplies gas to 20,000 homes; the 1000 t/d produces 15 MW$_e$.[96,98] A herd of a hundred cattle could get about 66% of their needs recycled in summer (52% in winter).[9] A jokester once proposed tapping the flatulence of cattle; a cow can produce 100 to 500 liters of methane per day.[27]

Grains

In 1980, a million tons of corn supplied 100 million gallons of ethanol. It would take 10% of the yearly corn yield to supply only about 1% of gasoline consumption.[9] Pure replacement seems doubtful. If much corn were used to produce ethanol, the demand would rise, causing corn prices to rise (and the cost of meat as well, because much corn is used as animal feed). This has caused controversy among people concerned about world food shortages. Grain that fuels cars cannot feed people. Historically, however, the United States has had an oversupply of food, and much rots in storage. Fermentation of spoiled stored grains would seem a feasible and useful alternative.

Bagasse

Bagasse is the woody residue of sugar cane. This resource is being used in Hawaii, Brazil, and the Caribbean. The amount is limited, and this is not expected to be a major source of energy elsewhere.

Food-Processing Waste

Very little waste is now being processed. Such processing would be of environmental benefit, as it would reduce the volume of waste products to be disposed of. Treatment is similar to that of animal wastes. A plant to produce gas in Florida proved uneconomic.[99]

Industrial Waste

Obtaining energy from industrial wastes is often problematic because of contamination in the waste by toxic substances, especially "heavy metals." Nevertheless, some energy (mostly in the form of paper) is currently being recovered.

Municipal Solid Wastes

The organic material in garbage is capable of generating methane of heat content from 7450 to 22,350 kJ/m^3.[99] Currently, there are about 25 landfills in the United States tapping this gas.[97] Some municipal power facilities burn wastes (see Chapters 11 and 25).

Municipal Sewage

Sewage treatment plants yield about 1 m^3 of digester gas per kilogram of volatile solids. Many plants utilize this gas in situ, so it makes no net contribution (Table 23.3), though it is an alternative energy source for the treatment plants.

Aquatic Plants

Dead plants matter, made up of carbon compounds, can produce methane when fermented anaerobically (see box *Case Study*: Duckweed). Such plants can be used to concentrate the "heavy metals" in wastewater streams as well (see **Extension 22.3**).[9]

Crop Residues

Removal of crop residues for energy has often proven disastrous.[96] If too much ground cover is gathered, problems can result from fertilizer replacement, increased erosion of soil, and probable net energy cost rather than gain. Research has indicated that some wastes can be gathered with the harvest and used to dry the grain with no adverse effects. In Illinois and Indiana, delivered corn residue is more expensive than coal before processing.[9] The Review Panel on Biomass Energy[9] concluded that potentially adverse environmental consequences of current crop management practices, even under ideal cropping conditions, are so serious as to draw into question the use of residues for biomass energy production until crop production technologies are improved. Sequestration (see Chapter 17) of agricultural wastes could reduce Earth's overall CO_2 emissions, though.

A different overall perspective on the ability of biomass to contribute energy is given in Table 23.4.

TABLE 23.4

Possible additional energy production from biomass in the United States

Category	Current		Additional Potential	
	Area (Mha)	Energy (EJ)	Area (Mha)	Energy (EJ)
Cropland	170	20	40	8–22
Pastureland	50	3	24	5–13
Rangeland	340	11	20	4–11
Forestland	290	12	17	3–10
Others	70	<1		

Source: Reference 100.

Biomass for Feedstocks

Feedstocks are raw materials supplied to industrial processors. Large quantities of the petroleum and natural gas used in the world are consumed as feedstocks—materials used to make other materials (oil into plastic, for example). There are many reasons to believe that biomass is currently better suited to production of industrial chemical feedstocks than to fuels.[101–103] Production of chemical feedstocks provides an alternative for disposal of billions of kilograms of toxic substances released each year by the chemical industry; biomass materials could be part of this process (Figure 23.2).[103] Fuel production is very large, but the chemical production scale matches biomass availability rather well. The problems of fuels are not those of chemicals, and the oil price increases of the 1970s engendered a permanent shift in costs. In 1970, ethylene cost half as much per kilogram as cornstarch; later the cost doubled.[101] Also, biomass can react to market changes by changes in composition, whereas petroleum composition is set by geology. Biomass is flexible as well as being adaptable to market changes.[101] In the fuel market, biomass will certainly have to adapt to conditions set by the greater volume petroleum and natural gas markets. In chemical production, it is possible for biomass to be changed into new chemicals that could build markets for themselves. One example is levulinic acid; it comes from biomass and can be used to make biodegradable strong fibers and transparent fibers.[101]

Industrial processes are often proprietary. However, most processes for fuels necessitate an investment of energy greater than the energy values of the resultant fuel. One that does not was developed at MIT, using specially selected strains of *Clostridium* bacteria.[102] Cellulose is anaerobically transformed into sugars, then it is fermented into ethanol and residues that can be burned to obtain steam or electricity. Corn is currently used to supply over 85% of the ethanol produced in the United States.[103] Switchgrass is an alternative to corn as a source of ethanol, producing 4 to 17 tonnes of carbon per hectare per year; the authors of Reference 103 argue that it is over twice as energy efficient as corn ($E_{out, alcohol}/E_{in, fossil fuel}$ = 20.1 versus 8.8 for corn plus stover or versus 6.7 for corn alone).[103] In addition, growing grasses aids soil conservation and regeneration of fertility. Other promising industrial processes involve production of long cellulose chains for making paper, alcohol, and mixed sugars that may be used for cattle feed.[104] In another example, the British chemical company ICI is growing plastic, polyhydroxybutyrate (PHB) using the bacterium *Alcaligenes eutrophus*.[105] With biopolymers, manufacturers can think of building to order. The bacterium *Rhodospirillum rubrum* can make a polyester out of sunlight.[105]

FIGURE 23.2

Waste wood, bark, and resin are used as feedstocks. Several pyrolysis oils are shown.
(U.S. Department of Energy)

CASE STUDY

Duckweed

Rushes and reeds have been used to clean water for some time. Other plants, such as water hyacinth and duckweed, can remove organic and inorganic materials from treated water. They also concentrate heavy metals. Duckweed can grow very fast. It sends out fronds that produce more duckweed, producing more fronds, and so on.[106] Duckweed is a good bet, because it can grow even in brackish or polluted water, as long as the water is still.[106]

Duckweed flows easily in water, and the plants are small enough to be pumped through pipes.[106] They do not need to be chopped to be fermented as hyacinths do. Duckweed can be mixed with other materials or fermented by itself.

If a farmer used duckweed with a 100-head dairy farm, fermenting the methane for energy and feeding the duckweed on the holding lagoons directly to the cattle, 60% of the protein requirement could be met in summer (15% in winter).

(a)

FIGURE 23.3a

Lemont rice is a high-yielding semidwarf rice variety.
(David Nance, U.S. Department of Agriculture)

(b)

FIGURE 23.3b

Jojoba flowers in Arizona (male plant). These plants produce a high-quality oil.
Photo by Michael Charters

23.4   NEW RESOURCES FROM BIOMASS

Several plants can make contributions to the chemical or fuels industry. Some plants are especially suited to increasing total production because they grow well in dry climates. The buffalo gourd (*Cucurbita foetidissima*) is a dry-climate plant. Seeds produce edible oil as well as protein (each ~35% by weight). The roots can produce industrial starch (food additives), and the vines yield forage.[107–110] The yield is about 2 tonnes of seed and 8 tonnes of roots per hectare.[107]

Humans eat mostly standard familiar plants,[108] only about 50 sorts of which are widely available [with just four—wheat, rice (Figure 23.3a), maize (corn), and potatoes (*solanum tuberasum*), in that order—accounting for the bulk of yearly commercial production], while many different neglected plants (especially tropical fruits) have great potential as foodstuffs. Sales of tropical foods by poorer countries could help slow deforestation by providing an economic base that depends on forest preservation, providing economic incentives to residents of these regions. An example of a recent successful commercial food introduction is the kiwifruit, originally from New Zealand. Below we focus on "industrial" products plants. See **Extension 23.2, *Fuel, Food, and World Politics*** for more information on biomass potential for interesting foods as well as for fuel.

Jojoba (*Simmondsia chinensis*, Figure 23.3b) is already a limited commercial success. The oil is in use in the cosmetics and fashion industry as well as electronics, and it can be used to treat acne and other skin disorders.[109] Jojoba is an evergreen tree that reaches about 5 m height, native to the Sonoran desert. Its oil (~50% by weight of the seeds) is similar to that produced by the sperm whale (it is even superior to sperm oil in some ways).[109] It resists biological degradation, so it is nonrancid and can be used as a replacement oil in foods. It can also be used in cosmetics, hair oil, and transmission fluid.[107,110,111] The price of jojoba is lower than in the past, about $2200/t (yields are around 1 t/ha), and more uses are expected to be found as the price continues to fall. The United States pioneered jojoba oil production in the late 1970s and produced 150 tonnes of jojoba oil as early as 1983, which far exceeded demand at that time. Many farmers lost money and switched crops.[107] Potential world demand is estimated at 200,000 tonnes per year, with current world production only about 1100 t/yr.[109] Jojoba is also grown in Australia, Chile, and Israel. The meal residue currently has no commercial use—the meal contains materials making it toxic and unpalatable as feed, and it is not a good fertilizer.[109,110]

Guayule (*Parthenium argentatum*, Figure 23.3c) produces rubber almost identical to that from the rubber tree. During World War II, guayule was planted in the southwestern United States as a hedge on loss of Malayan rubber. It is still uncertain if crops can be grown economically in the United States.[107,110]

The gumweed (*Grindelia camporum*) produces a rosin-like material. Rosin is a very important naval stores item with a market in excess of 700 kt/yr. Rosins are used for adhesives, tackifiers, paper sizings, and other industrial products. At present, gum rosin is extracted by the slashing of pine trees, which is very labor-intensive; wood rosin is obtained from 300-year-old pine stumps, and an estimated supply of 10 to 15 years is left.[107] The residues of gumweed are high-quality animal feed, and the plant could become a chemical feedstock of importance for pharmaceuticals and for agricultural products.

Crambe (*Crambe abyssinica*) grew originally in countries bordering the Mediterranean and has been grown in many American states.[111] It is a source of erucic acid to replace rapeseed, from which amounts of erucic acid are declining. Erucic acid is used in the production of plasticizers and lubricants.[111] Euricic acid from oil plants such as rapeseed and crambe can be rearranged to approximate crude oil.

Kenaf (*Hibiscus camabinus*, Figure 23.3d) is a bushy plant having fibers ideally suited for making high-quality paper and paperboard,[111–119] and it produces three to five times as much pulp per acre as pine at half the cost.[117] Kenaf grows naturally in Africa, Asia, and Central America.[111,118] Ordinary paper production involves smelly chemicals, but kenaf needs only high temperatures and pressures.[119] About 90% of kenaf's weight becomes fiber as opposed to wood pulp, of which only 50% becomes fiber.[119] It could be a valuable resource in regions such as Scandinavia, which currently must import wood chips.[111]

The gopher plant (*Euphorbia lathyrus*) grows wild in northern California. Its latex is an emulsion of hydrocarbons in water, and it can produce about 25 barrels of crude oil per hectare in California.[110,120] It grows well in both wet and dry regions.

A tree discovered in 1979 in Brazil, *Copaifera multijuga*, produces what seems to resemble pure diesel fuel.[121,122] Local residents use it for medicinal purposes. They drill a 5 cm hole into the trunk, into which a bung is placed. About twice a year they remove the bung to collect 15 to 20 liters of hydrocarbon. A hundred trees could produce 25 barrels of diesel fuel per year and cover only an acre.[121] The millions of barrels used per year would, of course, require immense plantations, and readily available corn oil may also be used directly to fuel diesel engines.

The bladderpod (*Lesquerella fendleri*, Figure 23.4) grows in areas getting only 30 to 40 mm of rain per year, and it produces an inedible oil, about 20% to 30% by weight.[107] The oil can be used as a source of chemicals to make a very strong plastic and high-quality lubricants. *Salicornia bigelovii* is another oilseed crop that can be grown in arid regions and irrigated with seawater. The seeds are 26% to 33% oil.[123]

(c)

FIGURE 23.3c

Guayule plants produce high-quality latex used for making rubber. (Jack Dykinga, U.S. Department of Agriculture)

(d)

FIGURE 23.3d

Kenaf stalks; kenaf is a fast-growing alternative to wood pulp for making paper. (Scott Bauer, U.S. Department of Agriculture)

FIGURE 23.4

A field of bladderpod (*Lesquerella fendleri*). The three hydroxy fatty acids—lesquerolic, densipolic, and auricolic—found in bladderpod seed oil are similar to ricinoleic acid, the main fatty acid in castor oil. Castor oil provides essential lubricants for racing cars and heavy equipment. Lesquerella oil is superior to castor oil for some uses.
(USDA-ARS New Crops Research Unit, National Center for Agricultural Utilization Research)

Biomass Combustion

Generally, biomass-fueled electric generators are planned to be about 100 MW because of the dispersed nature of the supplies.[13] The drawback is that an efficient system would be very expensive, so the compromise is to use lower-quality materials to lower the capital cost that reduces the possible efficiency. California biomass plants average 14% to 18% efficiency, as compared to the 35% typical of fossil-fuel plants.[13] This means that only negative-cost or very low-cost feedstocks can be used. The Electric Power Research Institute suggests that it would be possible to produce a module for $1365/kW that would be 34% efficient.[13]

Steam turbine development ended effectively in 1960, but gas turbines have been supported by military jet research and operate with increasing efficiency. A typical steam-injected gas turbine producing 51 MW at 40% efficiency can be bought skid-delivered for $700/kW.[13] In several years, it is expected that an intercooled steam-injected gas electric turbine system of 114 MW size and 47% efficiency will be deliverable. The GE LM-6000 produces 53 MW at 48% efficiency and is available now.[13]

Fixed-bed gasification is promising for biofuels. Eventually, the fluidized-bed gasifier should be obtainable, 33% to 43% efficient, at $900 to $1800/kW.[13] There can be more NO_x problems with biofuels than fossil fuels because biofuels contain nitrogen. If forest products are burned, the ash should be returned to the forest, because the removal of all residues could acidify the soil.[7]

Biogas

Biogas (methane from fermentation of agricultural waste) gives environmental benefits as compared to fossil fuels. Only 38% of the energy in the sugar cane goes to alcohol; 12% is vinasse (waste water from the sugar cane to ethanol process) and 50% is bagasse.[124] Vinasse is a pollutant when dumped into water-ways, but it is a fertilizer when applied in smaller quantities on land.[124] Vinasse contains 26% of the energy content of the alcohol. Much of this wasted energy resource can be recovered: Vinasse in Brazil could be turned into methane, along with the half of the bagasse that is not burned to supply heat to the process. This can be accomplished by enclosing the material in a container from which air is excluded. The anaerobic bacteria then make methane naturally. The maximal methane production occurs at 35 °C to 37 °C (about body temperature).

After anaerobic digestion, the remaining solids can be used as animal fodder, and the water is of irrigation quality.[124] The combined ethanol and methane approach can liberate 60% of the total cane energy.[124] It is far more efficient to digest anaerobically than to produce alcohol, so that if production becomes widespread, the methane production option may become dominant.

Electricity from biogas generated from open-top wood gasifiers has proved cheaper than electricity from fossil-fueled diesel generators in small Indian villages.[125,126] Commercial biogas generators producing 5 to 100 kW_e are available. The gasifiers supply electricity reliably and are reasonably economic to maintain. The gas produced is 18% to 20% H_2, 18% to 20% CO, 2% to 3% CH_4, and 40% unburnable gases (CO_2, N_2, and other gases).[125]

Politics of Biomass Energy

As an illustration of the political effect of benign appropriate technology, consider the situation in India as described by J. B. Tucker:[127] Distribution of biogas technology (in which methane gas is produced by anaerobic digestion) in India led to redistribution of income to the richer members of society. Ownership of biogas plants is confined to those in the top third of income levels. This occurred because the system requires a substantial capital investment. Capital is in the hands of the rich, and bank loans are available only to people having at least three head of cattle (about 75% of cattle owners in India have only two). Currently, there are 36,000 biogas plants in India, and the government is attempting to deal with the political problems by finding other ways to set up biogas plants. Two general kinds of digesters are in use: the Indian inverted-drum digester, in which the drum gives the seal and is the gas tank; and the Chinese fixed-dome digester.[128] The inverted-drum digester is reliable and robust, but it is also costly and requires a lot of maintenance.[124] The fixed-dome digester is cheaper but prone to gas leaks.[128] Both types of biogas plants require greater discipline of peasants than should be expected and require a cadre of trained maintenance and repair workers.

The plants are economic in the long term, because the dung, night soil, and organic residues produce methane at low pressure for 30 to 40 days. This is piped into the kitchen for heat, light, and cooking. The slurry remaining after fermentation is a fertilizer of high quality. The parasites and bacteria that often infest people are killed during fermentation. For this reason, it would improve public health to have these biogas plants widely available. As more plants are installed, cattle dung, which is now an economic "free good," will, Tucker believes, become a commodity.[127] This could widen the income gap in India and other developing countries still further.

Some of these issues were examined in the village of Pura, Karnataka, India.[128] The system in this village consists of the biogas plant, where mostly bovine waste is anaerobically digested, a sandbox filtration system, an electricity generation system, an electrical distribution system, and a water supply system.

(Detailed descriptions of various types of gasifier may be found in Reference 126.) The water system was added to pump water to the village from wells. The Pura system cost $1207/kW$_e$ and took 6 months to build.[128] It is operating about 4.2 hours per day at present, with 18 hours per day anticipated eventually.[128]

The Pura digester is a wider than normal inverted-drum unit, giving more biogas, cheaper. The daily throughput is 1.25 tonnes of cattle dung plus 1.25 m³ of water to make the input slurry.[128] At 25 °C to 26 °C, 42.5 m³/d of methane is produced. There is an amount of sludge, 1.2 tonnes per day, that can be used as fertilizer. It is more useful than fresh dung, as there are fewer weeds (many seeds are destroyed in the digester). The citizens get 0.6 kg of dried sludge back for every kilogram of dung contributed. The citizens all cooperate, an inverse of the tragedy of the commons.

It is partly for this reason that digesters that produce biogas have been so widely introduced in developing countries. A more direct way to influence wood use may lie in designing and marketing more fuel-efficient stoves for urban and rural cooking, especially those using biogas. Of course, digesters may be used in developed countries, too.[129]

There has been much concern about the denudation of hills in Nepal and the rest of the world. Ninety percent of people in developing countries use firewood or charcoal from wood as a source of fuel, and wood prices are climbing in most of Africa, Asia, and Latin America.[130] People must go farther afield to gather wood for cooking. Cooking with wood accounts for roughly 60% of all energy consumed in the African Sahel, and an Indian study indicated that the figure there is closer to 80%, at least for the six villages studied.[128] Where wood for fuel is part of the market economy, as much as 40% of household expenditures are for wood.[131] The indigenous stoves can lead to acute respiratory and other health problems.[131–150] Locally developed improved cookstoves are more energy efficient than traditional methods (usually involving cooking in a pot resting on stones above an open fire). See **Extension 23.3, *Cookstoves*** for more information on how developing countries can save labor and energy by using local craftspeople to construct safe and efficient stoves that appeal to their countrymen.

Deforestation is proceeding apace in Brazil, despite concern expressed by some experts that the loss of trees in great numbers might cause a decrease in the amount of oxygen in the atmosphere, and an increase in carbon dioxide (see Chapter 14). The Agency for International Development has been trying to persuade Third World countries that it would be useful to save tropical rain forests. Ironically, an ill-conceived operation in Hawaii is chewing up several acres of tropical forest a day to supply wood chips for electricity generation.[162] This certainly provides a poor example for Third World countries. Better forest husbandry and further study of agriculture in the tropics show promise for reducing the risks associated with the opening of new areas to settlement by farmers.[163]

SUMMARY

Biomass energy is a form of solar energy already exploited by people. A century ago, it accounted for over 90% of U.S. energy; as of 2000, energy consumption from other sources had grown so much that biomass totaled a paltry 2%. Plants are as much as 2% efficient in changing sunlight into usable material (for sugar cane, the efficiency is 2%; for corn, 1%; most others are less than 1% efficient). Production of ethanol or methanol from wood is economically and energetically more viable in countries exhibiting less reliance on fossil fuels than the United States. **Extension 23.1** discusses direct use of crops, as well as fuel distilled from crops and produced by microbes. Whether or not such production in the United States breaks even is a tossup.

It is unlikely that our present economy could be run on liquid fuels purely from biomass. The prospects for chemicals from biomass are considerably better.

Recently discovered plants hold some promise of help for fuel and resource shortages. Their discovery underlines the concern for widespread cutting of tropical forests and the attendant loss of species. The tropics also are home to many edible foodstuffs that can supplement the diet of people worldwide (while also allowing income for developing and preserving the forest home of such plants). Many fruits mentioned in **Extension 23.2** could enliven the diet of human beings,

who rely so heavily on just a few of the many plant varieties for their food.

Some forms of biomass energy are well suited to developing countries, especially biogas digesters, which are widespread in China and India. Nevertheless, even the introduction of what seems "benign" appropriate technology may have deleterious effects on society.

Given that so much effort in the undeveloped world goes into gathering firewood, and so many health problems arise from exposure to the pollutants, new methods of cooking can save energy and lives, both. Replacement of open fires by intelligently designed stoves saves lives and saves energy, as discussed in more detail in **Extension 23.3**.

PROBLEMS AND QUESTIONS

MULTIPLE CHOICE REVIEW

1. What gas of the following was *not* in Earth's original atmosphere?
   a. Hydrogen
   b. Helium
   c. Oxygen
   d. Nitrogen
   e. Methane
2. Which of the following has the greatest potential to increase biomass energy production?
   a. Cropland
   b. Pastureland
   c. Rangeland
   d. Forestland
   e. Municipal parks
3. Which of the following plants has the greatest economic potential in the United States?
   a. Kenaf
   b. Crambe
   c. Jojoba
   d. Guayule
   e. Buffalo gourd
4. Use of digesters for treatment of human waste and animal manure to produce energy is a good idea, because
   a. it replaces energy formerly obtained from trees.
   b. gas cookstoves are generally more efficient than those utilizing wood.
   c. the digested manure is a better fertilizer than fresh manure.
   d. less methane, a greenhouse gas, is emitted.
   e. of all of the above reasons.
5. About what proportion of energy consumed in the less-developed countries comes from wood?
   a. 1/10
   b. 1/4
   c. 1/3
   d. 2/3
   e. 95/100

6. Consider a food crop such as maize (corn). About how efficient is maize in converting sunlight into plant material?
   a. <0.1%
   b. 0.5%
   c. 2%
   d. 5%
   e. 30%
7. Consider growing apples. About how efficient is the plant in converting sunlight into apples?
   a. <0.1%
   b. 0.5%
   c. 2%
   d. 5%
   e. 30%
8. How do improved cookstoves *not* help people in developing countries?
   a. They produce less air pollution.
   b. They use less fuel.
   c. They are more easily used to purify local water.
   d. They are less likely to injure children in the vicinity.
   e. Local people can get their fuel for free.
9. Which of the following crops is most efficient at utilizing sunlight for energy?
   a. Cabbage
   b. Wheat
   c. Rice
   d. Millet
   e. Sugar cane
10. Using biomaterials instead of mining for materials is advantageous, because
   a. biomaterials do not get paid, while miners do.
   b. the biomaterials are chemically better than mined materials.
   c. there is much less waste material to dispose of if biomaterials are used.
   d. biomaterials interact more quickly with other biomaterials.
   e. of only (a) and (c) above.

SHORT ANSWER

Explain whether the statement made is correct, and justify your answer.

11. A blue pansy flower reflects red and green light from sunlight.
12. Grain used for producing ethanol cannot be used in any way other than as humus.
13. Not much possibility remains to increase use of mill residues of forest products, but much remains to be gained from use of logging residues.
14. Many facilities that burn wood and hog waste (residues) are currently operating in the United States.
15. C_3 plants are more efficient at converting sunlight to plant matter than C_4 plants.
16. Infrared solar radiation can be harmful to plants.
17. Red asters are red because the red light from sunlight is absorbed by the flowers.
18. Industrial wastes are more difficult to process for energy than municipal wastes.
19. Both C_3 and C_4 plants are equally responsive to increased levels of carbon dioxide.
20. Biomass should not be used to produce feedstocks because of the energy expense involved.

DISCUSSION QUESTIONS

21. Why do you think that more plants per capita in Europe than in the United States burn municipal solid waste? Give your reasons, and explain the differences.
22. Is gasahol a good idea? Explain.
23. How are biomass feedstocks more flexible than petroleum? Explain.
24. Why are leaves green and not yellow, for example? Explain.
25. If the world grows more food than it needs, how can there be famines? Explain.
26. Explain why leaves and plants are the colors they are, not black or white.
27. Which is more energetically favorable, biogas or ethanol? Does your answer differ at different places? Explain.
28. Why is guayule particularly useful? Explain.
29. What possible use could a weed be? Explain how weeds can be useful, starting with your definition of weed.
30. It is energetically feasible to use more logging residue for energy. There is a large amount available. What reasons, if any, are there against using this resource? Explain.

PROBLEMS

31. In the Pura digester, how much methane per kilogram of manure is produced? Explain how you figured out the answer.
32. Estimate the total area necessary to supply the United States with its fossil-fuel needs in 2004, if the diesel fuel tree discovered in Brazil were to be the sole source of energy. (In 2004, the United States used 74 Gbbl of oil; each barrel has a volume of about 160 L.)
33. In a local area, several villages band together to purchase a 25 MW$_e$ electrical generator to be supplied with wood from the forests in the area. How large an area of pine woods would they need to support the generator if the woods can sustain production at
    a. 5.0 t/ha/yr?
    b. 12 t/ha/yr?
    c. 17.5 t/ha/yr?

CHAPTER 24

The Energy Cost of Agriculture— A Case Study

This chapter on solar (biomass) energy emphasizes differing practices between "primitive" and technological agriculture. Energy use in the agricultural system is examined, in particular, in terms of a comparison of energy output to energy input. The agricultural revolution of the developed countries (mechanization, application of chemistry, and application of genetics) and its advantages and disadvantages are discussed. The use of machines and energy for agricultural production is necessary, but not necessarily efficient.

KEY TERMS

swidden agriculture / less-developed countries (LDCs) / ammonia / genetic engineering / genetic modification (GM) / green revolution / productivity / stability

Our remote ancestors found it advantageous to settle down and farm rather than depend on hunting, which their ancestors had done. The concentration of people into large agricultural units in southeast modern-day Turkey around 11,000 years ago allowed the small surplus per family to be gathered to support a local priesthood or gentry.[1] This upper class of nobles was then free from the everyday labor of farming. In return, the nobility and priesthood protected the farmers from evils both natural and spiritual. (The nobles rode out to protect the farmers from marauders. The priests propitiated the gods.)

Thus, the excess productive capacity allowed for the beginnings of culture. Only men of leisure could be scholars, remembrancers, scribes.

24.1 TRADITIONAL AGRICULTURE

One way to estimate the agricultural efficiency of our forebearers is to study contemporary peoples who are using traditional approaches to farming. The Tsembaga tribe of New Guinea has been the subject of such a study.[2] The tribe lives in the interior New Guinea highlands and is relatively unaffected by modern farming ideas. In this chapter, different measures of efficiency are studied: food energy produced per hectare, per hour of human labor, per joule of fossil-fuel input, per dollar, and so on.

Human energy input in the Tsembaga farming method was calculated by closely examining the tribe's practice of **swidden agriculture** (slash-and-burn). After a section of forest is cut down, the residue is burned to fertilize the soil (because tropical soils are thin and do not retain water well). People work at the burning, at erecting fence, at planting, at weeding, and at transporting their produce back to the village. The total energy input per hectare per crop is estimated to be about 5.8 GJ, while the total energy output per hectare is estimated to be about 100 GJ. This is a ratio of 17 energy units of output to one of input. Some of the harvest is fed to pigs, in turn, they provide a rich protein supply in emergencies such as sickness or bereavement.[3] Only about 62 GJ/ha is consumed by people.

Humans need between 6 and 10 MJ/d (1500 to 2500 kcal/d) to live. If we use the higher figure to estimate Tsembaga needs, a hectare supports 17 people for 1 year. The entire Tsembaga population (about 200) requires only about 10 hectares at full productivity (that is, new fields; an old field is nowhere near as productive, and such fields are allowed to revert to forest). About 35 hectares of old and new field are in use each year.[2] Because food is invariably lost to pests, there is some excess production.

Similar results have been found in the slash-and-burn agriculture practiced in Brazil.[4] The Indians harvest the jungle without destroying it; they use the cleared area for 10 to 15 years to cultivate food, then allow the clearing to return to forest for 50 to 60 years.[4] Although this style of agriculture works in remote places with few people, it clearly is insufficient when the population density gets too high. When there are too many people, the method results in permanent degradation of the forest (because the "fallowing" time is not possible).

The overlords in Europe had resources at their command, because their peasants produced more than they ate. We can guess that overlords in the New Guinea highlands probably could not last long with their style of agriculture. The small tribal excess could not support "conspicuous consumption." If customs were altered and wide areas were planted, the soil would soon be exhausted. Tribal behavior seems admirably suited to its clime. We might also guess that these tribes are semidemocratic of necessity.

An important aspect to remember is that this "primitive" agriculture is well suited to its environment and produces about 17 units of energy output for each unit of human energy input. Many so-called primitive agricultural systems are quite productive. This level of productivity held even for the pastoralists of the Sahel (sub-Sahara lands) before the advent of sophisticated European agricultural ideas,[5] which were unsuited to conditions there.

Another example of productive ancient agricultural techniques was found around Lake Titicaca in the South American Andes. The technique began to be used about 1000 B.C. and lasted until about 1400 A.D., when its use mysteriously ended.[6] The local Quechua people used raised platforms—no fertilizer or machinery was needed. The platforms were reconstructed in the late 1980s to test yields. The platforms are 4 to 10 m wide, 10 to 100 m long, and 1 m high.[6,7] Canals run alongside the platforms. The sod is cut out of the canals and put onto the platforms. Potato harvests on demonstration plots were 10 t/ha, whereas nearby conventional fields yielded 1 to 4 t/ha, using fertilizer.[6] The green algae at the canal bottom served as fertilizer. These plots outperform conventional agriculture there and show how great the natural understanding of those ancient farmers was.

Humans increased substantially in numbers only after the beginnings of agriculture in the fertile crescent. There is unresolved controversy over whether agriculture spread by assimilation or by movement of agriculturalists into new territory. In one clear case, that of India, the results favor movement of people into new territory bringing their agricultural practices with them.[8]

24.2   AGRICULTURE IN NORTH AMERICA—THREE REVOLUTIONS

Let us draw the contrast with traditional agriculture clearly by shifting our focus to modern North American agriculture, highly mechanized and utilizing technology. Amazingly, despite low-productivity agriculture in many locations, humans worldwide have already cornered use of 25% to 40% of the world's net primary productivity, which is all the biological material produced by life on Earth.[6]

North American agriculture has seen three revolutionary changes in the past century or so. The first change was the substitution of machines as replacements for human and animal labor. The second revolution involved application of chemistry to farming. The third revolution, currently in its beginnings, is in applying genetics to farming. All three changes have resulted in a decrease in the number of farm workers. With even a 1900 style of agriculture, three-quarters of the population would have to work in agriculture to feed the rest. Today, fewer than one worker in 20 in North America works directly in farming.

North American agricultural practices are apparently so successful that many people enthusiastically advocate exporting these techniques to the poor countries of the world (in euphemistic parlance, **less-developed countries**, or LDCs). Such technique transfer presents a tantalizing picture to those convinced of the superiority of technology, but we must see if it has any basis in reality.

Machines Replace Human Labor

Table 24.1[9] shows how energy efficiency varies for different methods of production. Note that primitive methods are much more energy-efficient than modern

TABLE 24.1

Yields and energetic efficiencies of different crops

Food	Method	Yield	Input	Efficiency[a]
Corn	United States, 1945	1.77	0.58	3.04
Corn	United States, 1975	4.48	1.84	2.44
Corn	Mexico, manpower	1.61	0.16	10.02
Corn	Mexico, oxen	0.78	0.23	3.34
Rice	United States, 1970	5.03	3.67	1.37
Rice	Philippines, 1970	1.43	0.44	3.26
Rice	Philippines, water buffalo	0.36	0.02	16.00
Yams	Tsembaga	0.34	0.02	16.50
Cassava	Africa, manpower	4.59	0.17	26.88

Source: Reference 9.

[a] Efficiency is defined as yield/input; yield and input are given in kilojoules per hectare.

agricultural methods, although they involve an investment of many more hours of human and perhaps animal work. Mexican corn farmers may have to work more than 1100 hours per hectare as compared to 17 hours per hectare in the United States.[9]

Several sources of information[10–12] will be used to place agricultural technology in its proper perspective. Corn is a good crop to study. A crop native to the Americas (it was first farmed about 7000 years ago),[13] it produces more digestible nutrients per acre than any other food crop, and it is the most efficient in the use of technology.[14]

Census Bureau data[15] are shown in Figure 24.1 for the past few decades. The most striking trend is an increase in production per acre and a decrease in the labor component of the total energy cost, which have characterized these years. The farm yield has increased substantially,[16] as Figure 24.2 also shows for several important crops. Grain production rose to 1586 Mt by 1992, near the record set in 1981, but production has been falling in per capita terms since 1984.[17] Grain production dropped because eroded land was being abandoned.[17]

Several examples illustrating the decrease in human labor over four decades are examined in Figure 24.3.[18] The decreases shown are due to increased energy input. For example, cotton was handpicked until the introduction of cotton harvesters in the 1950s. Machines replaced many poor African-Americans who had been employed for farm labor, and the replacement of people by machines was at least partly responsible for the African-American exodus to the northern cities. Tobacco farming has undergone the least improvement because of all the human care required. The leaves are placed by hand and are dried in sheds.

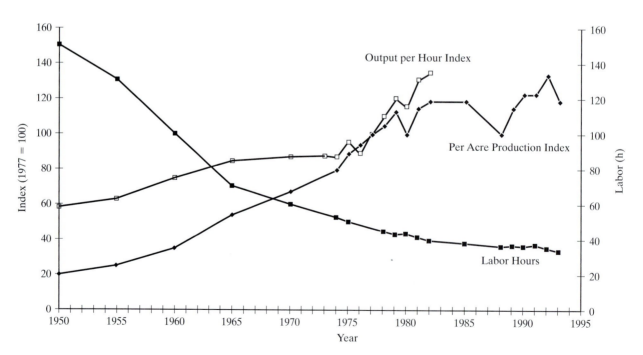

FIGURE 24.1

The growth in output per hour and output per acre between 1950 and 1993 is shown compared to a 1977 value set to 100 (left axis scale). To find the number of person-hours of labor between 1950 and 1982 required by the agricultural sector of the economy, use the right axis scale.
(U.S. Department of Agriculture, Statistical Abstracts of the United States)

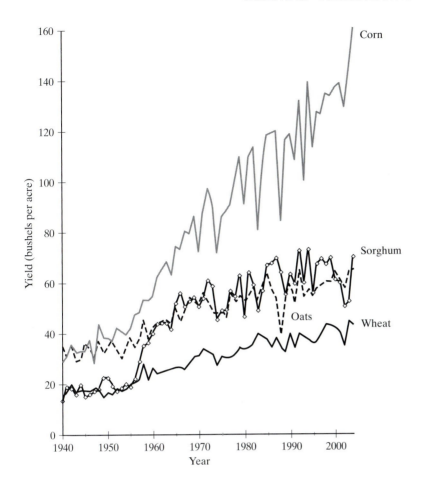

FIGURE 24.2

The yield of selected grains, in bushels per acre, from 1940 to 2004. (U.S. Department of Agriculture, National Agricultural Statistics Service, http://www.usda.gov/nass/, March 2005)

Anyone who has passed through the tobacco-growing areas of Connecticut has seen the fields covered by burlap (to bleach the tobacco for cigar wrappers) strung at great cost in human effort.

Meat is a growing proportion of the protein people consume. Ruminants increased to 4 billion worldwide, doubling since the 1940s as the human population has doubled.[19] Cattle and hog raising show the more recent effect of feedlots (see **Extension 24.1,** *Feedlots*). In the feedlots, many animals are kept close together, and their food and water requirements can be delivered mechanically, cutting down on human labor. Special feed supplements (including antibiotics and hormones) accelerate weight gain.

The number of fowl increased from 3 to 11 billion between 1945 and 1992.[19] The spectacular decrease in human labor in raising poultry can be explained in much the same fashion. "Chickens" are raised on a farm in large buildings but get some exercise walking around. "Broilers" (sometimes called "Arkansas chickens") are chickens raised in tiny cages with no room for exercise. The same holds for egg-laying chickens. They are kept in tiny cages and are fed, watered, and cleaned mechanically.

The trends shown in Figure 24.3 are obviously long-term and substantial. They are mostly due to the spectacular increase in energy input associated with reducing human labor: energy-expensive tractors, herbicides, pesticides, and fertilizers.

FIGURE 24.3

The decrease in labor per unit output (5-yr averages) is shown by decades, from 1935–1939 through 1975–1979, for various agricultural products.
(U.S. Department of Commerce, *Statistical Abstract of the United States* (Washington, DC: GPO,1972, 1984). Data were not published after 1976–1979.)

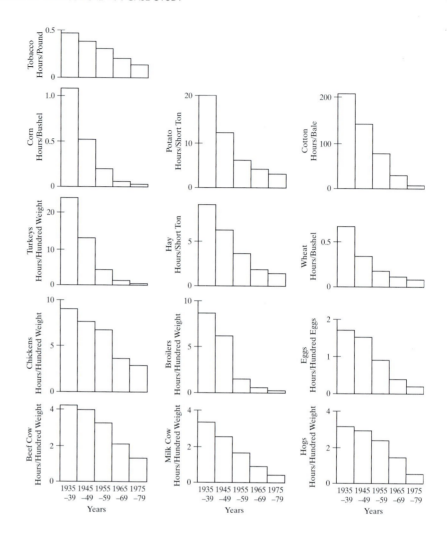

We see in Figure 24.4[11] that between 1930 and 1960, the increase in energy input nearly doubled the output. Table 24.2 shows how the increasing energy input and the increasing crop energy output per acre compare over the period from 1945 through 1970. Note the steady trend of the drop in efficiency, the ratio of output per unit input.

FIGURE 24.4

Farm output (1957–1959 is set to 100) in terms of energy input to the U.S. agricultural sector from 1920 to 1970.
(J. S. Steinhart and C. E. Steinhart, "Energy use in the U.S. food system," *Science* **184**, 307 (1974))

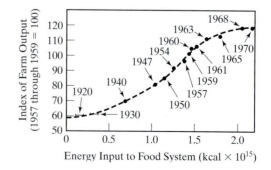

Pesticides Control Losses, Fertilizers Increase Yields

The past half-century has seen the introduction of pesticides such as DDT and many more modern chemical pesticides. This has had at least a short-term effect, reducing pest populations (although pests often develop resistance to the pesticides). Another gift of chemistry is the wider use of fertilizer in crop management. Fertilizer use has leveled off at about 300 kg/ha averaged over all crops.[34] About half of fertilizer need is because of soil loss.[35]

TABLE 24.2

Energy input and output per hectare (in MJ) and ratio, 1945–1970

Year	Energy input	Energy output	Ratio
1945	221.1	818.7	3.70
1950	288.2	915.0	3.18
1955	369.8	987.3	2.67
1960	451.3	1300.3	2.88
1965	535.6	1637.4	3.06
1970	692.0	1950.5	2.82

Source: J. S. Steinhart and C. E. Steinhart, "Energy use in the U.S. food system," *Science* **184**, 307 (1974), copyright 1974 by AAAS.

TABLE 24.3

Energy input to the food system

Crop	Energy input (kJ/kg of production)
Corn	1800
Wheat	143
Average, all U.S. grain production	1430
Broilers	9300
Milk	11,200
Eggs	17,000
Beef	20,800 (feedlot beef 24,000)
Pork	41,999
Average, all U.S. animal husbandry	18,200

Source: D. Pimentel, P. A. Oltenacu, M. C. Nesheim, J. Krummel, M. S. Allen, and S. Chick, "The potential for grass-fed livestock: resource constraints," *Science* **207**, 843 (1980).

Table 24.3 presents energy inputs for selected crops and livestock.[23] There are many ways to minimize the amount of energy put into the food system, some are labor intensive, such as applying pesticides by hand, and some are conservation-minded, such as using chemicals (fertilizers and pesticides) only as necessary. See **Extension 24.2, *Pesticides*** for ideas of how pesticides may be used and misused. Because of the buildup of resistance in the targeted pest, pesticides lose much of their clout over time. They also have deleterious side effects, such as in the appearance of malformed amphibians in wild areas near farm fields. Integrated pest management (relying as much as possible on natural enemies, rotating crops, applying pesticide sparingly at the optimal time, etc.) can lessen these side effects.

The American agricultural system has substituted fossil energy for labor. Labor can then be used for other purposes in the economy. Without the fossil energy input, the United States would still need to have a majority of the population engaged in farming. Figure 24.5 shows how profoundly this input energy changed farming between 1945 and 1970. The human labor component has become all but invisible. The energy cost for combines and tractors (machinery), while rising in absolute numbers from about 760 MJ (210 kWh) to about 1750 MJ (490 kWh), has decreased as a proportion of total energy because of the spectacular growth in fertilizer use. In 1970, about 260 MJ (1300 kWh) of energy per acre was attributable to artificial fertilizer use; fertilizer application reached a peak in 1989 and has declined slightly since.[86] The percentage of direct energy use (gasoline, electricity, drying, transportation) decreased as the energy used on crop care (nitrogen, phosphorus, potassium, insecticides, herbicides) increased faster than direct uses.

Two major fertilizers are used to enhance plant growth: nitrogen and phosphorus (see the extensions). Some nitrogen is fixed by legumes—plants such as alfalfa.

FIGURE 24.5

Comparison of the energy inputs to the agricultural sector, 1945 and 1970.
(J. S. Steinhart and C. E. Steinhart, "Energy use in the U.S. food system," *Science* **184**, 307 (1974))

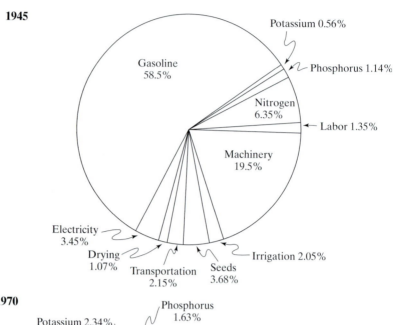

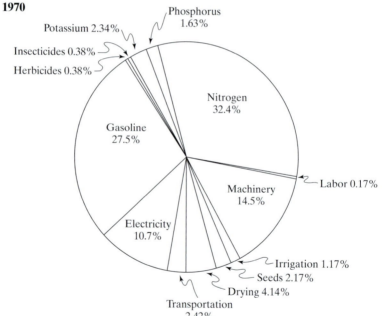

Legumes have bacteria living in a symbiotic relationship with them that put atmospheric nitrogen into the soil. Some nitrogen fertilizer is created by lightning, and this falls on all the world's soils (estimated at 3 to 10 Tg [3000 to 10,000 tonnes] per year).[87,88] Most nitrogen used as fertilizer is in the form of **ammonia** (NH_3), which is easily assimilated by plants. Ammonia is made from methane (CH_4) and air at high pressures. It is convenient to use natural gas, which is mostly CH_4, as the starting point for producing ammonia. The natural gas also supplies the energy needed for ammonia production. The production of ammonia takes a large amount of energy. Because much natural gas is flared anyway, especially in underdeveloped areas of the world, production of fertilizer provides a useful product while decreasing waste.

Even in this country, we must question past practices. For example, fertilizer is apt to get more, not less, expensive. Oil or gas is necessary as both a raw material and an energy source for fertilizer manufacture. The United States uses the equivalent of 12.7 billion m³ (4.5 × 10¹¹ ft³) of natural gas yearly to slake the thirst for fertilizer. Humans provide 170 Tg (170,000 tonnes) of nitrogen in the form of fertilizer each year, dwarfing natural mechanisms.[87]

Another consideration is that overapplication of fertilizer can decrease yield efficiency. Figure 24.6[10] shows that the maximum in energy output/energy input (efficiency) occurs with an application of about 135 kg per hectare (120 lb per acre) of nitrogen. Actual crop yields increase (somewhat) with further fertilizer application. In consequence, if the price of fertilizer is artificially low (because of artificially low energy costs), farmers are forced to increase energy expenditure per acre (decreasing the energy output/energy input ratio) to maximize yield (which is directly proportional to dollar return).

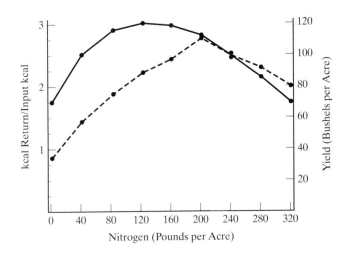

FIGURE 24.6

Ratio of food energy output to farmer energy input as nitrogen fertilizer applied increases (solid line); per-acre yield as nitrogen fertilizer applied increases (broken line). Note that yield actually decreases if too much fertilizer is applied.
(D. Pimentel, L. E. Hurd, A. C. Bellotti, M. J. Forster, I. N. Oka, O. D. Sholes, and R. J. Whitman, "Food production and the energy crisis," *Science* **182**, 443 (1973))

Our final remark on fertilizer is that it can cause pollution problems when it runs off the land. Barry Commoner[89] showed how the nitrate load of the Missouri River depends on farm runoff and vividly pictured the adverse health effects of fertilizers in the groundwater.[90] Anywhere there is agriculture, there is water degradation through runoff.[87,88,91–93] Intensive German agriculture has put nitrates and phosphorus in the rivers in regions having many farms. For example, at the Mosel's mouth in Koblenz, Germany, nitrate concentration averages over 20 mg/L as nitrogen,[94] incredibly high for such a major river.

In any case, we can save energy and help the soil by applying animal manure to the soil or by using "green manure," that is, rotating crops of legumes instead of supplying nitrogen (ammonia).[10] In addition to relieving the health hazard posed by runoff and the enhanced eutrophication that results from the high phosphate levels in lakes, these latter two methods save essentially the entire energy cost of the fertilizer. Crop rotation has the added bonus of being a good way to reduce the need for pesticides. Ridge planting, which also reduces pesticide use, is another technique that has been gaining in popularity.[37]

CASE STUDY

Organic Farming

The use of "organic" farming methods, in which the farmer uses natural fertilizer and rotates crops and uses natural, biological pest control, decreases the energy input, again at an increased labor cost. Organic farms, which do use crop rotation and manure spreading, have been shown to be essentially as efficient, in terms of total dollar return, as more conventional farms.[95] Although the yield used to be somewhat lower on the organic farms, the fertilizer costs were low. The two effects were roughly in balance (economically). A 21-year Swiss study demonstrated a savings of half the energy compared to conventional farms, half the fertilizer, and used almost no pesticides (97% lower).[95] Yields were lower than conventional farms, at 90% for wheat. As organic farmers gain experience, they have been able to increase yields while still not overusing fertilizers. The economic balance should tip further to the organic farmer when energy becomes much more expensive. Organic farmers do not eschew all modern methods. For example, they use grain dryers (the newer hybrid corn is wetter than older varieties, and so takes more energy to dry).

Organic farms now produce a higher yield than in the past, still use less fossil energy, and have lower costs; but they grow less maize per hour of labor input compared to conventional farms. The result for corn is that the ratio of energy output to energy input was 3.21 for the conventional farms, while it was 5.90 for organic farmers.[35] For potatoes, the ratio of energy out to energy in was 1.06 for the conventional farms, while it was 1.08 for the organic farms, essentially a wash.[35] Another study of "biodynamic" farms in New Zealand found similar financial viability with either method, while the soil quality was better on biodynamic farms.[96] This appears to be true on all organic farms. This was part of the reason that the Swiss organic farms needed less fertilizer.[95]

The amount of certified organic cropland more than doubled in the United States in the 1990s.[97] The organic farmers are more focused on no-till agriculture, which has the potential to reduce release of greenhouse gases.[95,98] There are more flavonoids in organically grown produce, and these antioxidant compounds are known to fight cancer and are thought to slow aging.[99] Farmers who sell certified organic produce can charge premium prices, while improving their soil and minimizing environmental damage from pesticides and fertilizers. These are all reasons that many small farmers are "going organic."[100]

The "organic" certification is so important that it led to controversy about what the meaning of the word was, and how much pesticide would be allowed to be used while still retaining the organic designation.[101] The most important factor in organic farming is money. Sales of organic crops are estimated to be $7.76 billion in 2001, an almost eightfold increase from 1990.[102] Many "brand-name" companies are entering the burgeoning organic foods market.[103] Despite this, less than 1% of American farmland is "organic."[103]

Agricultural Advances Due to Use of Genetics

New varieties of standard farm crops and domesticated animals and new types of crops may be cultivated in regions previously unusable for the crop. Both cereal crops and animals have been subject to genetic manipulation since the advent of sedentary agricultural societies. Modern wheat and corn bear scant resemblance to the native grasses they originally were. Modern cattle are bred to gain weight or produce milk with success; humans manipulate animals' environment to make them become more productive of products that humans value.[104] Our ancestors did much of this breeding, but modern agriculture has systematized and accelerated the changes. The field of **genetic engineering**, changing the genes of plants and animals in the laboratory, is invading the farm. For example, 80 genes have been synthesized and put into the potato using the bacterium *Agrobacterium tumorifaciens*.[105] Genetic adaptation accounts for about half the increased yield since 1935.[106]

It is important to protect the genetic blueprints of crop precursor plants. Plant genes contain information that might someday be bred into new strains of plant or identify a weakness that could be strengthened. Seed depositories are important institutions, at once museums and historical record (see also **Extension 26.2**).

Some changes can be made through crop research and genetic alterations, or **genetic modifications (GM)**; for example, by work on increasing the protein content of the crop and on increasing the length of time the crop is harvestable (see **Extension 24.3, *Genetic Modification and Agriculture***). Nitrogen fixation research is promising the possibility of nitrogen-fixing bacterial help for many food crops.[196,197] So far, 11 microbial pest-control agents have been approved by the EPA.[198]

This is very important because of the problem of human-fixed nitrogen entering water supplies. The addition of so much runoff nitrogen is bad for the environment;[199] for example, it is undermining the health of natural forests.[200] Wheat is helped by a natural bacterium that is harvested and applied to the wheat seeds before planting.[198] (In addition, land-use changes are responsible for altering the carbonate loading of the oceans, which measurably changes their alkalinity.[201])

In use or on the horizon are other successes of modern agricultural research. The "**green revolution**" followed development of crops for LDCs with spectacularly higher yields, although the newly developed strains of cereals require substantial amounts of fertilizer during growth. China increased yields by partly restoring private-sector rewards to productive farmers. New developments that bode well for future crop production from LDCs include genetically engineered vaccines, viral insecticides, weeds turned into forage, additives or genetic strains that raise the productivity of dairy animals, and development of new hybrids that extend the geographical growing range of a crop.[202] For more information on the "green revolution" and farming practices, see **Extension 24.4, *The Green Revolution, Fertilizer, and Erosion***.

24.3   FARMS AND ENERGY EFFICIENCY OF THE AGRICULTURAL SYSTEM

Although farms consume only 4% of the U.S. energy supply, about 17% of all the energy use takes place in the food system.[34,269,270] We can compare the distribution of energy to the various stages of the food system in 1940 (Figure 24.7a) and 1970

FIGURE 24.7

(a) Energy in the food system, 1940.
(b) Energy in the food system, 1970.
(J. S. Steinhart and C. E. Steinhart,
"Energy use in the U.S. food
system," *Science* **184**, 307 (1974))

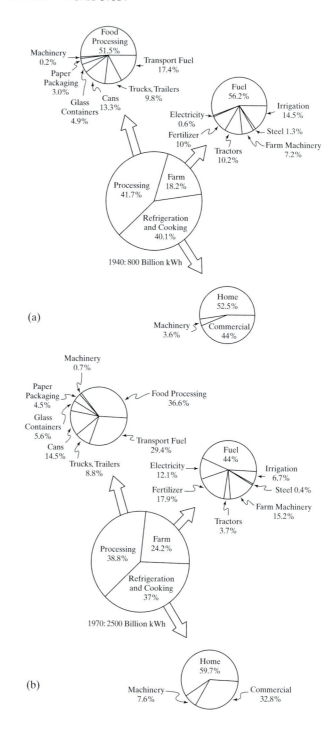

(a)

(b)

(Figure 24.7b).[11] The changes between 1940 and 1970 indicated in the figures took place smoothly, as we see in Figure 24.8.[11] Food energy produced has kept pace with population growth, but energy use has far outstripped the gain in population as well as the food energy ultimately obtained. The difference we see in the ratio compared to Figure 24.4 is that Figure 24.8 considers the total energy input to the system, including the energy use of middle marketers and transport of food from source to warehouse, and the energy at the retail end of the system as well as transport to the store. (The energy used by the customer cannot readily be accounted for, and so is not considered.)

Actually, the estimate of energy use in the food system is probably too low, because we have not charged the energy costs of constructing buildings used for processing, storing, and distributing the food. Also, a part of the fuel production is lost in processing, and a great deal of fuel is used in the production and distribution of food; the proportion representing food production should be charged to the system. Other factors neglected include exports and waste disposal.

Energy gained in North American agriculture has been less than energy spent for quite a long time, as is seen in Figure 24.9.[11] Even 30 years ago, it was estimated[270] that the United States used 7.5 times more energy in the food system than was obtained from it (also see Table 24.4 for the situation as of 1991). Other countries have differing experiences—the British ratio is estimated at 2.5 to 1[271]—but the pace of mechanization continues to increase all over the world.

One-sixth of American energy use is tied to food production and distribution. Much of the actual energy used in the food system is spent in packaging and handling of the food and in transport of the food from farm to distribution facility to markets and stores and from the point of sale to home or restaurant. Another large energy input to the food system is in refrigeration and cooking.

A study of the changes in Turkish agriculture over the period 1975 to 2000 shows a similar trend (obviously happening long after what occurred in the United States). The energy input to agriculture rose staggeringly, from 17.4 GJ/ha to 47.4 GJ/ha between 1975 and 2000. The major underlying change was replacement of human and animal labor by machinery. In 1975, over 60% of the energy input was such labor, but it was under 15% by 2000.[273]

The energy value of the farms' output also rose, from 38.8 to 55.8 GJ/ha in that time. This means the output/input ratio decreased from 2.23 to just 1.18 in 2000.[273]

Similar substitution of machine energy for human and animal labor is occurring across the world. Even in Bangladesh, one of the world's poorest countries, agriculture is becoming less burdensome but also less energy-efficient.[274]

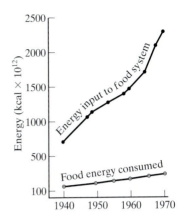

FIGURE 24.8

Energy input to food system and food energy consumed (trillion kilocalories) from 1940 to 1970. The 1970 energy input corresponds to about 10 EJ. (J. S. Steinhart and C. E. Steinhart, "Energy use in the U.S. food system," *Science* **184**, 307 (1974))

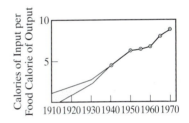

FIGURE 24.9

Ratio of energy input to food energy consumed, 1910 to 1970. Note that Figures 24.8 and 24.9 are consistent: the U.S. agricultural system increasingly subsidizes food production with fossil energy. (J. S. Steinhart and C. E. Steinhart, "Energy use in the U.S. food system," *Science* **184**, 307 (1974))

TABLE 24.4

Energy cost of food production and distribution, as percent of total U.S. energy use

Component of food system	Percent of processing	Percent of total use
Preparation of soil, planting, fertilizing, harvesting		3.0
Processing total		5.4
Food processing	36.6	
Food-processing machinery	0.7	
Paper packaging	4.5	
Glass containers	5.6	
Aluminum/steel cans	14.5	
Fuel	29.3	
Manufacture of trucks, trailers	8.8	
Transportation		0.5
Wholesale/retail		2.6
Preparation		5.0
TOTAL		16.5

Source: J. Dowling, "Agricultural uses of energy," in R. Howes and A. Fainberg, eds., *The Energy Sourcebook* (New York: American Institute of Physics, 1991), 401.

A CLOSER LOOK

Irrigation

The *average* irrigation cost per acre is not large, because only about 4% of all corn is irrigated. If we had chosen a crop such as watermelon, the proportion of irrigated field would have been much higher. Each kilogram of corn grown on irrigation requires use of 1400 liters of fresh water (8 ML/ha).[275] Assuming the irrigation water is pumped up from a depth of 50 meters, it takes around 6 GJ/ha extra to irrigate a field of corn.[276] It takes an average of 9.6 GJ/ha to grow soybeans in Ohio, but 15.5 in Nebraska, where they are irrigated.[239,269] On an irrigated field, over half the energy spent growing the crop is used for fertilizers and irrigating. We might say that the reason the energy return on corn is so high is that only 4% of the fields are irrigated. There are opportunities for saving energy in irrigation by choosing the appropriate method.[277]

Irrigation is a further problem because of the energy cost and the lack of available water: only 30% of potentially irrigable land has enough water available. This arises from the maldistribution of runoff. About a third of the world's total runoff (precipitation minus evaporation) comes from tropical South America (the entire continent is only 15% of the world land area). Southwest Asia, North Africa, Mexico, temperate South America, the southwest United States, and Australia, with 25% of the world land area, have only 5% of the runoff.[278] In Chapter 13, we discussed other problems of irrigation.

DOING YOUR PART

Cosmetics and Packaging

A vast amount of energy could be saved in marketing in the United States by using rail transport in place of truck transport and by using a reduced amount of packaging or substitution of paper packaging for packaging made of petroleum-based products. Germany has blazed the packaging trail by mandating that, starting in 1994, manufacturers take back their packaging materials for ultimate disposal. This mandate should give manufacturers an incentive to reduce their packaging. The German experience could help other countries in their quest to reduce packaging (see Chapter 11).

A change in cosmetic standards of U.S. consumers in buying produce (more as is the case in Europe) could save otherwise unusable fruits and vegetables, thus reducing waste and saving energy.

The ultimate user could help save energy by demanding standardized returnable bottles for products marketed in glass (see Chapter 11), as well as by using more energy-efficient refrigeration (or returning to the sort of refrigerators that must be defrosted by hand) and self-cleaning ovens. Both energy-efficient refrigerators and self-cleaning ovens are much better insulated than the more conventional products, and the additional purchase cost is soon regained from the lower operating cost.

24.4   COULD NORTH AMERICAN AGRICULTURE SUCCEED ELSEWHERE?

The three revolutions in North American agriculture have, basically, substituted energy for brawn. North American agriculture succeeded this way because it had to. We manage to feed our population, and we would not have been able to had we not adopted the three revolutions. That meant, though, that we had to live on past savings of energy, the deposits of fossil fuel accumulated over half a billion years (see Chapter 12). This strategy has succeeded because the cost of fossil energy is so low and the cost of human labor is so high.

With development of the economy in the twentieth century, fewer workers were available, and those who remained could not work more hours. Energy had to be substituted to feed the growing population. The sheer quantity of food that must be produced is staggering, and it has to be produced by fewer and fewer people.

To feed the world at a U.S. level with U.S. techniques would take 80% of the current world energy expenditure; looked at another way, the total known reserves of fossil oil would be exhausted in only 13 years if we used the energy to feed the world on the U.S. standard.[270] Further, LDCs are characterized by low labor costs and high energy costs (the reverse of North American conditions). If so much energy is required, can the benefits seen in North America be replicated? The cost of the energy would surely rise in response to increased demand. How could LDCs pay? How would substitutes for oil be found?

North American agriculture appears to have reached the stage at which further energy input does not substantially increase output (refer to Figure 24.4) unless a new technology is discovered. In the past, there have been examples of such strides: breeding of cereal strains that mature earlier; development of cereals, such as triticale, that have a better amino acid balance;[11,214,279] and so on. The future will probably hold more such jumps.

24.5   MONOCULTURE VERSUS DIVERSITY—STABILITY AND PRODUCTIVITY

The most stable natural ecosystems seem to be the most complex and to have the lowest apparent productivity. This occurs because, in the natural state, everything feeds on or depends on everything else. From a human perspective, this is low productivity. Simplified ecosystems typified by farm fields—monocultures—produce easily accessible excess, which we human beings see as highly productive. Even the output of the best monoculture practice may not be as much as the output of a diverse ecosystem.[139,280]

One of the difficulties we face as agriculturalists is that all ecosystems developed by humans are fragile compared to natural systems. Our aim is maximum **productivity**; nature attempts to maximize **stability**. In biological terms, productivity refers to the total biomass produced by a system. In human terms, productivity means biomass that can be harvested by humans. Stability means that a change induces forces that tend to counteract the change.

The Arctic, with few species, is much less stable than temperate grassland or forest (bulldozer tracks dating to World War II still scar the Arctic tundra). In a complex system of diverse species such as is found in the tropics, virtually any scar would be healed (in Brazil, the forest is already trying to overgrow the recently built Amazonian road system). To place the productivity of agriculture in perspective (even our cultivated lawns are quite productive),[281] an alfalfa field is over seven times as productive as a pine forest; tropical rain forests and the oceans have zero productivity.[282] Economic forces such as economies of scale drive farmers to monoculture—forces made stronger by the three revolutions referred to earlier.[283] Because of the simplification (emphasized in monoculture) we impose on nature, our system is very unstable. However, the lands that become agricultural are generally ones that do not have a high diversity to begin with.[284]

A study of prairie ecosystems verifies that ecosystems having more plant species withstand stresses better.[285] Laboratory experiments show that increases in biodiversity cause analogous increases in productivity.[43,139,280,286–290] (And, contrariwise, fertilization leads to less diversity.[291]) Ecologists have argued about whether it is the diversity itself or the way the different species are represented at differing levels that is the most important. However, it is clear that both are important, just as it has become clear that some species ("keystone species") are more important to the functioning of the ecosystem than others.[43,280,287–290]

An experiment on European grassland plots scattered over the continent showed that diversity itself is a really important variable—a loss in diversity led to a decrease in productivity on the test plots compared to the control plots. Every halving of diversity on a plot led to a 10% to 20% decrease in productivity.[292] The best outcomes were noted for diversity similar to natural diversity, around 100 species per hectare.

According to Reference 289, there appears to be a "single general relationship between species richness and diversity across all sites." Combinations of species on a plot led to a yield greater than any single species' yield on an equivalent plot (this is called "overyielding"). Perhaps this arises because high diversity in ecosystems allows for different paths of energy flow in the ecosystem and enhances stability.[287]

Consider the corn blight epidemic, *Helminthosporium maydis*, that struck the U.S. Midwest in 1970. It was disastrous because of the overwhelming use of one particular strain of hybrid corn by farmers. Were several types of corn planted together, such diseases would have a relatively harder time spreading. In a natural setting, similar plants are not usually close to one another; the distance inhibits the spread of diseases and rots. A human analogy to this might be the danger of flu. In a preschool or retirement community, the effect of flu is much stronger than in a community of mixed ages. Because children and older persons are more susceptible, the disease becomes more contagious. A field of one sort of plant runs a danger similar to these groups. The problem is worse when the same crop is planted in the same fields year after year (some 20% of corn crops are grown in this way).[283] The blight emphasized, for expert and novice alike, the need to keep an inventory of plants of differing genetic background so that we can use the qualities of the "wild" grain to enhance the qualities of the cultivated grain. Monoculture also leads to increased soil erosion and prevents the farm from reaching its maximum agricultural potential.

These experiences have not stopped ordinary farmers from using very high-yielding types of all sorts of grain instead of different varieties. Rice breeding began in 1966 at the International Rice Research Institute in the Philippines and involved the use of semi-dwarf varieties to make "miracle rice."[293] This sort of rice has stems that do not fall over, which helps the farmer harvest the crop more easily. Most of the genetic background of rice used in the LDCs comes from a single Indonesian variety, "Cina." Soon there might be conditions set up that would lead to a "rice blight" of the dimensions of the 1972 American corn blight. Because rice is so essential to nutrition in populous Asia, the result could be famine.

Monoculture seems productive from the human perspective, and for millennia, farmers in the Middle East, Asia, North Africa, and Europe (and later in North America and the colonial parts of Latin America) traditionally planted one crop. What we have learned from recent research is that a farm field might be more productive if other plants are mixed in (see also Chapter 26). Such fields would need less energy-expensive fertilizer. Fields with a single crop are vulnerable to attack in ways that fields with more plants are not. So, the planting of fields with several crops could minimize vulnerability to pests without reliance on the application of so much pesticide. These diverse fields would be more resistant to drought, and so might need less energy-expensive irrigation. Depending on how the intermixing is done, it might or might not require additional energy to sort out the harvest.

24.6   PRESENT AND FUTURE

We have shown that current agricultural practice is not sustainable. Soil is being depleted. More energy is required to grow, market, and eat food than the energy value of the food, 7.3 times as much in the United States in 1995 (10.8 EJ versus 1.5 EJ).[294] In addition to energy cost, many criteria air pollutants are emitted in the food production process—VOCs, particulates, and NO_x. About two-thirds of all ammonia emissions are due to agriculture.[294]

To feed the world, a new and bigger "green revolution" is urgently needed. The greatest promise, and threat, of modern agriculture comes from genetically modified crops. We discuss the possibilities in **Extension 24.3**. Not many other opportunities exist for increasing yield substantially. We will have to deal with the risks as best we can. The moral of these stories is that care and thoughtfulness are requisite in the agricultural system, too. Choices should be made with as much knowledge as possible.

SUMMARY

"Primitive" forms of agriculture, characterized by heavy reliance on human labor, are very energy efficient. North American agriculture uses fossil-fuel energy as a substitute for human energy and is, therefore, much less energy efficient. The change is, however, efficient in terms of human labor. Far fewer hours are needed for each unit of production, as pointed out in **Extension 24.1**. In addition, chemical fertilizers and pesticides have increased yields (see the extended discussion in **Extension 24.2**). This energy-expensive agriculture has succeeded in the developed countries, because the cost of oil is low, and the cost for labor is high.

Genetic modification in agriculture can increase yields, but there are dangers that genetically modified plants could transfer their germ plasm to conventional plants. The genetically modified plants are unpopular in many developed countries, but they are the possible salvation of many LDCs, as explained in **Extension 24.3**. Benefits include enhanced nutrition and longer shelf life. Risks include creation of "superweeds" and possible generation of new diseases.

The "green revolution" has exported high-technology North American-style agriculture around the world. It has been most successful where farmers had sufficient capital available to take advantage of it. Many LDCs are characterized by low labor costs and high energy costs. In a world dogged by hunger, any chance to increase yields looks good. The major costs and benefits of agricultural choices are discussed in **Extension 24.4**. It also has led to enhanced pesticide use, with unintended consequences such as pests' development of resistance and harm to beneficial organisms.

Monoculture is traditional agriculture, but recent research suggests that fields with several crops could reduce energy use and be more productive overall. Such fields would be more stable and resistant to attack by pests. They might be more resistant to droughts, which come regularly and will continue to unless humans learn how to control weather.

In addition to problems associated with high input of fossil energy, the number of plants consumed has decreased, and the number of strains of plants grown has decreased significantly. With this sort of monoculture, one disease can wipe out national grain supplies (as almost happened in the United States in 1972 with corn). Wild strains of domesticated plants should be preserved so that genetic diversity is not lost.

PROBLEMS AND QUESTIONS

MULTIPLE CHOICE REVIEW

1. Which sort of agriculture is most energy efficient?
   a. Swidden agriculture
   b. Terrace agriculture
   c. Paddy agriculture
   d. Mexican agriculture
   e. Pastorialism
2. The number of ruminant livestock has in recent times _____ as the number of people has _____.
   a. tripled, doubled
   b. doubled, tripled
   c. doubled, doubled
   d. quadrupled, doubled
   e. doubled, quadrupled
3. Hogs are
   a. producers.
   b. primary consumers.
   c. secondary consumers.
   d. tertiary consumers.
   e. neither producers nor consumers.
4. Agriculture involves about one in _____ workers in the U.S. economy in the 2000s.
   a. two
   b. four
   c. six
   d. ten
   e. twenty
5. Which farm animals were the least efficient over the period covered by the data, 1935–1979?
   a. Chickens
   b. Broilers
   c. Beef cows
   d. Milk cows
   e. Hogs
6. Which farm animals gained the most in production between 1935 and 1979?
   a. Chickens
   b. Broilers
   c. Beef cows
   d. Milk cows
   e. Hogs
7. What side effect(s) can ensue from the application of pesticides to crops?
   a. Increased crop death
   b. Lower costs
   c. Increased health of farm workers as pests are driven off
   d. Increased resistance to the pesticide
   e. Ability to use less pesticide every year
8. The replacement of hand labor by machine can reduce time needed from a farmer to bring in a crop by
   a. half an order of magnitude.
   b. an order of magnitude.
   c. two orders of magnitude.
   d. three orders of magnitude.
   e. more than three orders of magnitude.
9. Irrigation of cropland is
   a. energetically inefficient.
   b. economically inefficient.
   c. virtually cost-free.
   d. without any deleterious consequences.
   e. ineffective at increasing yields.
10. About how many billion tonnes of topsoil are estimated to be lost from the land area of the United States each year?
   a. 1/2
   b. 2
   c. 8
   d. 22
   e. 45

SHORT ANSWER

Explain whether the statement made is correct, and justify your answer.
11. Swidden agriculture typically gives yields that are much higher than those in developed countries.
12. Pesticides form an integral part of the "green revolution."

13. It takes fewer workers to produce crops in developed countries because of substitution of machines that ultimately run on fossil fuel for hand labor.
14. Techniques for improving yield of livestock involve risks because of the concentration of waste products as the numbers raised in one place increase.
15. "Green manure" is an environmentally friendly way to fertilize fields.
16. Mechanization of southern agriculture was responsible for the increase in northern African-American populations.
17. Some crops are more efficient at utilizing their environments to produce grain growth than others.
18. Diversity of plant species in a plot increases overall yields.
19. Many crops are grown to be bland-tasting but pretty to look at.
20. Farm yield in the United States has continued its rise over many decades.

DISCUSSION QUESTIONS

21. Discuss the advantages and disadvantages of use of herbicides to replace hand weeding.
22. In Bt corn, genetic material from a bacterium has been put into the corn. What consequences might ensue?
23. Some commentators have characterized genetically modified food as "frankenfood." Does such labeling advance real discussions of the value of genetically modified organisms (GMO) or accomplish the reverse? Discuss how this labeling helps or hinders rational discussions.
24. Discuss the pros and cons of organic farming to the best of your knowledge.
25. Much research has taken place on the value of diversity of organisms in a given plot of land. What appears to be the value of such diversity?
26. The "green revolution" appeared to allow the Malthusian vision of widespread famine to be conquered forever.

27. Nitrogen fixation and runoff of nitrogen applied to the soil pose problems for agricultural production and the environment. Why is this so? Explain.
28. How did the corn blight of the 1970s affect agriculture? Was that outcome useful in the long term?
29. Why is the food-for-oil tradeoff possible? Explain how availability or lack of availability of energy influences the agricultural system.
30. Genetic modification has met with much more resistance in Europe than in the United States. Why should the Europeans be so against the use of genetically modified organisms (GMO)? Why are Americans less opposed to use of GMO? Are you personally for or against use of GMO? Explain as best you can with the facts available.
31. Historians claim that mechanization (development of the cotton gin) was responsible for preservation of the institution of slavery in the American South. Discuss the possibility that they could be correct or incorrect, and try to find evidence to support your thesis.

PROBLEMS

32. How long can use of fossil energy to produce food in the ratio of 10 to 1 continue for the United States? Explain your assumptions.
33. How does natural gas use for fertilizer affect use of natural gas for fuel? Explain what proportion of yearly gas use is as fertilizer.
34. What is the relative energy impact of broilers, feedlot cattle, and pork?
35. What would the corn yield be if the United States could operate at the efficiency of Mexican hand labor raising corn? How would that change if it were as efficient as Mexicans using oxen?
36. About how long could the trading off of fossil energy for food energy continue to be done if the entire world did it the North American way? (North Americans make up 1/20th of the world population.)

CHAPTER 25

Energy Storage and Energy Alternatives

In this chapter, we examine many different technologies pertaining to the storage of energy. Most of these are innovative, unconventional technologies, and they give hints of developments to come (even if only a few of these do become viable). In addition, we consider proven alternatives for generating electricity, including geothermal energy sources. We conclude with a discussion of the "hydrogen economy," an alternative system for energy distribution.

KEY TERMS

load management / pumped storage / phase change / latent heat of fusion / heat-of-fusion device / shape-memory devices / nitinol / PACER / thermal regeneration / magnetohydrodynamics / municipal solid waste / energy density

In Chapter 9, we discussed the rise of the utilities as monopolies as an outcome of the difficulty in storing electric energy. In this chapter, we present and evaluate several feasible ideas as well as more speculative energy storage possibilities. The most exciting recent development, which has many implications for the electric distribution system as well as for energy storage, is the discovery of "high-temperature" superconductivity.

25.1   MANAGING LOAD USING ENERGY STORAGE

Load management may be the cheapest way to smooth out demand. This method uses customers' electricity consumption to flatten out the load curve.[1] To help manage load, the utility starts and stops commercial or home electric water heating systems by radio or by timer. In places that have time-of-day pricing, customers have an incentive to do this themselves. When I lived in Germany, I noticed that some people who had electric water heaters used them only at night for hot water because electricity costs were lower then. This self-regulation helps smooth out demand in a small way. I understand such self-regulation is common in rural areas in the United States as well.

Storage water heaters are large energy storage devices. Water has a high specific heat, so a large volume can lose substantial energy and not lower its temperature by much. Typical storage units in use for home heating systems include tanks for pressurized hot water, floor slab heaters, and ceramic units for individual rooms. A comparison of conventional and solar hot water cost with and without storage is given in Figure 25.1.[1] A solar heating system costs only once—for installation.[1] One using electricity can be economical if heating is done when electricity is cheap.

In Chapter 22, we discussed the use of **pumped storage** as a way for utilities to even out hour-by-hour fluctuations in use of electric energy by customers. While pumped storage is only about 70% efficient, cheap base load electricity can be used to pump the water up, and the water can be allowed to return downward during times of high demand. Peak demand is roughly twice base load, and the utilities typically use much more expensive generators to produce it. It is in a utility's economic interest to increase base load and decrease the cost of supplying peak demand, and pumped storage accomplishes this goal. All storage modes involve cost inefficiencies, but even with the extra cost, the total can be less than what the utility would have to pay to obtain that energy at peak demand by other means.

Other strategies for storing energy during off-peak hours are available; some are speculative and others proven. These strategies allow for substitution of stored energy for new energy sources, because they allow an overall increase in efficiency of electricity generation. Some of the devices we discuss can be used on a scale smaller than a large generating facility.

A variant on the pumped storage idea is that of huge underground reservoirs excavated in hard rock at a depth of about a kilometer below ground. (This minimizes the possibility of leakage to groundwater and provides a substantial head).[1] As energy is needed, water can be allowed to fall from a lake into the underground reservoir. Base load energy can then be used to pump it back up.

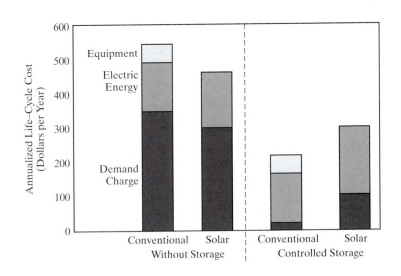

FIGURE 25.1

The cost of heating water for residential use may be decreased by the addition of hot-water storage. The "demand charge" is the amount charged by the utility to assure backup generating capacity. (Adapted from F. R. Kalhammer, "Energy Storage Systems," *Sci. Am.* **241**(6), 56 (1979), Copyright © 1979 by Scientific American. All rights reserved)

Although this system does not eliminate all environmental objections, it is superior in many respects to aboveground pumped storage facilities, which can involve substantial fluctuations in storage reservoir level, because the smaller head would involve a much greater volume of water transfer for a given amount of energy generated. It is also attractive because the excavation and tunneling methods that would have to be used already exist,[1] but it is much more expensive than aboveground reservoirs.

Another possibility is storage of hot water to be used to generate peaking power. This was first used in Berlin–Charlottenburg in the 1930s. The water was stored in steel vessels at high pressures and released as high-pressure steam to run a turbine as electricity was needed.[1] There have even been proposals to use natural aquifers to store hot water in the ground.[2] It has been estimated that such storage facilities would cost only $40 per installed kilowatt.

Superconductors for Energy Storage and Electrical Transmission

In 1911, H. Kammerlingh Onnes discovered that certain materials at very low temperatures—several kelvin—exhibited strange properties. Resistance to the flow of electric currents suddenly disappeared at some "critical" temperature. Onnes found that the resistance of mercury vanished at 4.2 kelvin, for example. As these very cold materials are perfect conductors of electricity, they were called superconductors (as we mentioned in Chapter 8).

Over the years since 1911, progress has been made in understanding the mechanism by which superconductivity works. John Bardeen, Leon Cooper, and J. Richard Schrieffer explicated the mechanism in the 1950s and won the 1972 Nobel Prize for their theory of superconductivity, called the BCS theory, after their names.

A classical model of conduction attributes resistance to collisions of electrons (the normal charge carriers) with atoms, transferring energy to the lattice. When the temperature is high, the atoms in a conductor vibrate from thermal energy about their equilibrium conditions, and collisions are common. When the temperature becomes low enough, this thermal vibration is of very small amplitude, and resistance is lower. In the BCS theory, the electrons in the superconductor are able to conduct electricity with no resistance, because they form "Cooper pairs." The electrons are paired through the lattice of atoms in the very cold superconductor, which supply the conduction electrons in the first place, and they trade momentum with the lattice. As an electron moves between two lattice atoms, it attracts them electrically (the collision, which transfers the electron's energy to the lattice) and causes them to move toward the position the electron had occupied between them. These lattice atoms, in turn, present a region of higher positive charge to the paired electron, attracting it and transferring the energy back to it. The result is that the pair's energy is unaffected as it moves through the lattice. There is no resistance to the *paired* motion.

The temperature of the superconducting transition T_c for the materials investigated by Onnes was very low, a few kelvin. Later research discovered materials of steadily higher transition temperature, until, by the 1960s, it was around 23 K for a niobium–tin alloy (Nb_3Sn). Indeed, the BCS theory seemed to indicate that transition temperature could not be much higher than already found. This material (Nb_3Sn) could be worked, and most superconducting devices made up until 1987 used wires made from it. This remained the highest transition temperature seen until late 1986, when G. Bednorz and A. Müller[3] reported a material—a perovskite, a type of mineral made of barium, copper, lanthanum, and oxygen—with a transition temperature of 36 K. After such a long spell of stagnation in transition temperatures, this publication set off a race to find other materials having higher transition temperature (the highest known T_c at present is around 150 K). Bednorz and Müller won the 1987 Nobel Prize for their discovery. These new ceramic materials had not been previously investigated for superconducting behavior, because the oxygen atoms in the mineral were expected to make it act as an insulator and to prevent superconductivity. It was exactly these defects—oxygen vacancies, or lack of oxygen atoms where they should have been—in the material that were later recognized to be the key to its puzzling properties. See **Extension 25.1,** *More on Superconductors* for a description of uses and information on the newest "low-T_c" superconductor, magnesium bromide (39 K transition temperature).

The discovery of high-T_c materials led physicists to speculate that some material might be found that would be a superconductor even at room temperature (293 K). True, workable room-temperature superconductors would revolutionize technology. They would make superconducting transmission lines and magnetically levitated trains feasible, and they would hold the promise of incredible advances in computing power for the next generations of computers.

Superconductors in Production and Transmission of Electricity

Conventional electric generators are already 98.6% efficient. Superconducting generators will be about 99.5% efficient, but this 1% gain is not generally worth the trouble of needed refrigeration equipment and its maintenance.[27] For room-temperature superconductors, however, this gain will be worthwhile.

The most promising use of the high-T_c superconductors will be in superconducting transmission lines (see Chapter 8). Up to now, such transmission lines were very expensive, because they had to be cooled using liquid helium. The new superconductors exhibit their marvelous behavior above the temperature of liquid nitrogen, which is much cheaper to produce, because its vaporization temperature is so much higher. Cooling to this temperature is much easier and less expensive, making transmission lines feasible. The world's first high-T_c superconducting transmission line was built in Detroit.[18]

The perhaps 10% to 15% of Joule heating losses attendant on short-distance transmission will be saved, as well as the 2% losses suffered in long-distance transmission. A 1% improvement in electricity transmission efficiency translates into $1 billion in savings.[20] This can be an important development because of the recent rise in the amount of "wheeling," or interutility transfers of electricity through intervening transmission lines.[28] As more electricity is transmitted over longer distances, the savings could be large. The most important consequence of the high-temperature superconductors is the fact that the current sent through each transmission line could rise spectacularly—the lines would be able to transfer much more power than current technology allows.

High-T_c Superconducting Magnets

In Chapter 4, we saw that electromagnets could be made by winding wire around an iron core. The greater the current, the stronger the electromagnet. If wires of electromagnets are made of material that can become superconducting, they can carry very high currents, and very strong magnetic fields can result, much stronger than possible with conventional electromagnets. The technology of fabricating ceramic wires for the high-temperature superconductors was daunting, but now superconducting ceramic wire is being made by the American Superconducting Corp., Intermagnetics General Corp., and Sumitomo Electric Industries.[29]

The powerful magnets used in high-energy physics experiments are often low-T_c superconducting magnets. Several large accelerators (the Tevatron at Fermilab, HERA at DESY, RHIC at Brookhaven, the LHC at CERN) utilize such superconducting magnets. A major cost in running current superconductivity magnets is cooling them to low enough temperatures. The development of high-T_c electromagnets should eventually make contstruction of new (or upgraded) particle accelerators easier and less costly (because they do not have to be cooled to as low a temperature).

In the energy storage mode, an electromagnet's current (and so the strength of the magnetic field) would be increased when there was an excess of energy available; current would be withdrawn as demand increased. Given enough sufficiently large magnets, a great amount of energy could be stored. Large normal electromagnets could be used to store energy, but in actuality, the energy cost is substantial—large volumes of cold water must be pumped through the magnets to cool them. With superconducting magnets, at least short-term energy storage has become feasible. Because the materials are superconductors, the stored currents circulate with no loss. Of course, their low temperatures must be maintained at some energy cost. These magnets would be relatively large, perhaps tens of meters in diameter.[20] Room-temperature-T_c superconducting magnetic energy storage would make it nearer 100% efficient than the current 90% to 95% characteristic of low-T_c superconductors (however, there may still be problems with Eddy losses, for example).[27]

The new superconductors raise the promise of larger and much cheaper electromagnets. The technology is developing rapidly, especially since the discovery of perovskite superconductors. The advantage is that the response is virtually instantaneous, because the energy need not be transformed (it is still electrical when it circulates in the magnet), and there is no loss from circulating the current. The magnets will need to store in excess of 10 GWh to be economically feasible.[1,20] As above, superconducting magnetic energy storage at room temperature would make magnetic energy storage nearer 100% efficient rather than the current 90% to 95%.[27] It is also true that with so much energy stored in a superconductor, any abrupt transition back to normal behavior would be catastrophic, because incredible amounts of energy would be transformed into thermal energy in a very short time, leading to a meltdown or even an explosion. Care will have to be taken to prevent this from happening.

Energy Storage in Flywheels

A flywheel is a massive disk that rotates, carrying rotational inertia. Flywheels have been in use for many years. As children, we played with "friction" cars; these may have had a small flywheel to store the energy (though more often it was a spring that stored the energy). In an automobile, a flywheel is attached to the crankshaft to keep the rotation basically steady despite the intermittent nature of the explosions in the cylinders. The stored rotational energy in the flywheel is used to keep the engine running smoothly between cylinder strokes. This rotational energy provides the work to compress the air in the next cylinder in the sequence so that it can fire, and on and on.

What is new about the flywheel is that extremely strong plastics and epoxies have been developed in recent years. They allow for the design of flywheels specifically meant for storing energy, and the total mass need not be so great as in present solid metal flywheels in order to store the same amount of energy. Large amounts of energy may be stored in epoxy flywheels, because they can be spun at much higher speeds than ordinary metal flywheels with no danger of falling apart.[29,30] Composite flywheels have reached edge speeds of 1400 m/s in normal operation at Oak Ridge.[31] Such flywheels (of large size) could be used to store huge amounts of energy from power plants to be drawn on for peak demand. Even in the event of failure of the material, tests at Oak Ridge have shown that composite flywheel disintegrates harmlessly rather than sending pieces ricocheting.[30]

In smaller versions, flywheels could be used to run cars—electricity from a home outlet could be used to start the flywheel rotating rapidly, regenerative braking turns the energy generated in braking into electricity (which is used to speed the flywheel back up and partially reclaim the energy), so the family car could be driven using the rotational energy stored on the flywheel. Instead of the corner gas station or special infrastructure, all travelers would need is a wall outlet to power their travels. Chrysler developed a natural gas, turbine-powered racing car that uses energy from a flywheel in parallel with the turbine while the car accelerates (turbines without assist provide sluggish acceleration).[31] The flywheel car is closer to feasibility because of California rules that mandated that a substantial number of electric cars would have to be sold there (starting in 1998; the date has been delayed several times because of problems with all-electric cars; see **Extension 15.8**). American Flywheel Systems (AFS) has developed a flywheel "concept car," the AFS 20 (which is not yet practical).[31] AFS bought the rights to use the Oak Ridge technology.[32]

The entry of industrial giant United Technologies into the pure electric auto market bodes well for the future. It is an indication of the seriousness with which the zero-emission vehicle is being approached. In contrast to the AFS system, which has no energy source but a flywheel, the United Technologies design is for a flywheel "battery" that supplements electric batteries.[33] The company believes that flywheels can soak up energy better when the car is braking and supply larger amounts of power assist for acceleration than batteries by themselves. United Technologies delivered a 0.55 kWh flywheel "battery" to BMW in summer 1994.[33] The wheel spins at 30,000 revolutions per minute.[33] A Pasadena company, US Flywheel Systems, has a flywheel that rotates at 60,000 revolutions per minute.[34] Flywheels such as these have been subjected to tens of thousands of charge–discharge cycles with no degradation.[34]

Energy Stored in Phase Changes

As we saw in Chapter 7, when materials change from solid phase to liquid phase (as, for example, ice becoming water), they lose some of their internal structure. In a solid, the atoms are fixed in place. They do not move around freely throughout the solid. The attractive forces between atoms prevent this. In a liquid, however, some of this structure is lost. While atoms in a liquid feel attractive forces, they are not bound in one particular place. The atoms can slide around other atoms to a certain extent.

Materials tend to occupy the state of lowest energy; therefore, it takes energy to "unlock" the atoms from their relative stability in the solid and allow them the relative freedom in the liquid. For example, to change ice at 0 °C to water at 0 °C, a **phase change**, an energy input of about 333 kilojoules for each kilogram of water is required. This is called the **latent heat of fusion** of water.

The same kind of considerations hold in analyzing the transition from liquid to gas. In the gas, the interatomic attractive forces are weak. Thus, a certain amount of energy per unit mass (the latent heat of vaporization) must be paid to change liquid to gas. For water at sea level at 100 °C, the latent heat of vaporization is 2260 kilojoules per kilogram.

In energy storage applications, it is useful to use a material that is solid at room temperature and has a melting point slightly above room temperature. Energy can be fed in at a low temperature and be used merely to change the phase of the liquid without requiring the temperature to rise. Recall from Chapter 7 that the temperature is a constant during a phase change. These materials are called **heat-of-fusion** devices. As an example of the large amount of energy that can be stored in a change of phase, we note that it takes as much energy to melt ice at 0 °C as it does afterward to heat the resulting water to 80 °C.

Of course, we are looking for a material that melts at slightly above room temperature. One such material is sodium sulfate,[35] which has a melting point of 31 °C and a latent heat of fusion of about 213 kilojoules per kilogram. Other materials, such as Glauber's salt, are suitable as well. The material to use would vary from application to application. The point is that the latent heat is large, and the thermal energy can be stored even with small temperature changes.

The proposal for the national solar energy farm discussed in Chapter 21 included liquid sodium phase changes for energy storage.[36] Liquid sodium has a rather large heat capacity as well (amount of thermal energy held per unit change in the temperature). Energy generated during off-peak hours could be used to heat a large amount of liquid sodium. The stored thermal energy could later be used to run a generator.[35,36] (Of course, if the sodium encounters water, watch out!)

Pumped Storage Using Compressed Air

One ploy is similar to the pumped storage of water: off-peak power is used to pump air into a sealed underground cavern to a high pressure. This high pressure then drives turbines, as the air is heated and pressure is slowly released; the resulting power produced may be used at peak hours. There are many geological formations that can be used in this scheme, which has the advantage of not involving huge, costly installations. The first working compressed-air storage facility was built in Huntorf, Germany, in 1979 by Nordwestdeutschland Kraftwerke, a local utility.[1,37] The air is stored in two salt caverns with a total volume of 300,000 m³ at pressures of about 70 times atmospheric pressure. It can produce about 290 MW for roughly 2 hours for peak demand.[1,37] For each kilowatthour output, about 0.8 kWh of electrical energy is spent in air compression and cooling (the gas gets hot on compression and must be cooled before storage to prevent heat fracturing of the salt formation), and 1.6 thermal kWh is used for reheating.[1] The compressed-air system overall can be made about 50% efficient.[38]

A system was built by Alabama Electric Cooperative in McIntosh, Alabama, in 1991. The system pressurizes air in an underground chamber hollowed out of salt to about 75 times atmospheric pressure. The storage volume is 538,000 m³, which can supply 26 h of use. The McIntosh plant is 87% efficient compared to water pumped storage, which is 70% to 75% efficient.[39] As with the Huntorf facility, natural gas is burned using compressed air from the cavern to run a turbine, giving energy both from the natural gas and the compression. Exhaust gases from the turbine preheat the air that is used to burn the natural gas.[39] Both the Huntorf and McIntosh systems have operated reliably over long time periods.[40]

In contrast to the two salt mines, a compressed-air facility is being built in an abandoned limestone mine in Norton, Ohio. It was once owned by the Pittsburgh Plate Glass Company, which used the mine to make soda ash for its glass. The mine can hold a volume of 10 Mm^3, far larger than the other two facilities, and the compressed air drives nine 300 MW turbines (2.7 GW overall). Eighteen compressors will be used to pressurize the cavern.[40]

Research in progress is aimed at reducing the energy penalty for storage to acceptable levels. Typical proposed underground pumped storage plants should have a capacity of 1 to 2 GW, with 6 to 10 hours of storage.[41] Schemes with and without intermediate storage reservoirs have been considered; the most economic choice appears to be the system without intermediate storage, but with multistage pumps.[41]

Other Modes of Storage

Storage as Synthesis Gas

Israelis have developed a process to store solar energy using methane as the carrier. Methane is broken down to $CO + H_2$ on exposure to steam (produced by solar energy installations reminiscent of Solar One) to produce synthesis gas; the gas is transported by pipeline, and at the delivery site, it is reacted to regenerate methane.[42] This system is uneconomical for the United States, even in the Southwest, but it might work in Israel or in other countries with higher-intensity solar energy.

Batteries for Energy Storage

The oldest electric energy storage device is the battery (see Chapter 6). The commercial lead-acid battery used in automobiles is capable of producing 50 to 100 W/kg for total stored energy of 25 to 35 Wh/kg. It will withstand about 200 to 300 charge–discharge cycles,[43] although new designs could extend this to 3000 cycles.[44] Lithium-ion and metal hydride batteries are now in common use as well. We have discussed some of these issues in **Extension 15.8** and in **Extension 25.2, *More on Batteries***.

Choices, Choices

Among all these storage alternatives, choices need to be made for particular purposes. Figure 25.2[38] presents the analog of the screening curve (Figure 9.2) for various possible energy storage systems. In some cases, technology has lagged enthusiasm. The figure indicates why compressed-air storage, of all the alternatives considered, has actually advanced to the commercial phase (peak demand is perhaps 6 hours per day).

Still more energy storage alternatives exist to minimize diurnal fluctuations:[1] Refrigerators could run at night to chill water or make ice that could be used to cool the house during the next day; natural ponds could be used to store warm or cool water that could be used when desired. Some home groundwater heat

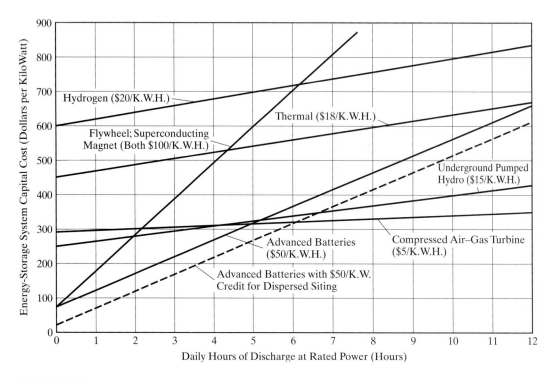

FIGURE 25.2

There is great variation in costs to utilities, as shown in this screening curve.
(F. R. Kalhammer, "Energy Storage Systems," *Sci. Am.* **241**(6), 56 (1979), Copyright © 1979 by Scientific
American. All rights reserved)

pumps (Chapter 9) store cooled and heated water in the ground near the house, hence, increasing efficiency of the system both winter and summer, as the reservoirs are reversed with the season.

25.2 NOVEL ENERGY ALTERNATIVES

Shape-Changing Devices

Even more speculative possibilities for energy storage and transformation are available. Some materials retain an apparently (permanent) plastic deformation strain (that is, appear to retain a new shape) below some critical temperature T_c; above T_c, the materials return to their original shapes (this is called a martensite transition). This phenomenon was discovered in 1932 in Sweden.[62] Machines built with such materials are called **shape-memory devices**. In the return to the original shape, heat is absorbed and converted into mechanical work.[62] There are about 20 elements with alloys that exhibit this martensite transition,[62] one of which is a nickel–titanium alloy called **nitinol**. A motor can be built with nitinol that uses the temperature difference between hot water and air to drive a shaft. Model engines and demonstration kits are available.[63]

Shape-memory materials include plastic as well as metal. Their use in surgery is a novel (non-energy-saving) use of this material.[64] After use, they biodegrade inside the body. Actually, the plastic shape-changing material could also be used for other engineering purposes.

PACER (Peaceful Nuclear Explosion Reactor)

One of the more intriguing proposals for an immediate energy source, although it had a brief existence, was called PACER.[65,66] Scientists from Los Alamos National Laboratory claimed that reactor fuel and electricity could be produced by repeated underground thermonuclear explosions. They proposed detonating at least two devices a day in a mile-deep cavity, most likely in a salt formation. The superheated steam produced from the proposed million tons of water in the cavity would power a regular generating plant on site, and if uranium-238 cladding were added to the bombs, the facility could be used to breed reactor fuel. PACER devices would have yields in the range from 10 to 100 kilotons (kt). One of the problems with the proposal was that it would have invited arguments similar to those over nuclear plants in regard to safety. There was also the fear that the cavity could collapse, releasing the radioactivity inside, contaminating the groundwater, and so on. As a breeder, it could have ended fuel worries forever, if it had not been scuttled. If we assume one 50 kiloton explosion at 10% efficiency per day, we get a total energy output of

$$(50 \text{ kt/d})(4.19 \times 10^{12} \text{ J/kt})(0.1)/(2.6 \times 10^6 \text{ J/kWh})$$
$$= 6 \times 10^6 \text{ kWh/d}$$

for an average power output[67] of

$$6 \times 10^6 \text{ kWh/24 h} = 240 \text{ MW}.$$

At a cost of $40,000 per nuclear device, this would have produced energy at a cost of about 7 cents/kWh, comparable to current prices.

Solar Power Satellite

A proposed 10 GW solar power satellite (SPS, Figure 25.3) would sit in geosynchronous orbit, one in which it would remain stationary over a given location on Earth. It would be shadowed by Earth less than 5% of the time at that distance. In the proposal, a 100 km^2 array would convert solar energy to electricity, then convert the electricity to microwave energy. The microwave energy would be beamed at low intensity down to a 100 km^2 antenna on Earth. This satellite array would take 0.13% of total world refined aluminum, require 500 shuttle launches with concomitant pollution, and could cause health effects from spillover of straying microwaves.[68] The Herendeen, Kary, and Rebitzer analysis showed a ratio of energy delivered over lifetime to primary energy for construction of around 2 for SPS.[68] If one neglects the energy cost of running fossil-fuel plants, then they *appear* superior to the SPS, but SPS is favored if it is considered.

Renewed interest has surfaced in SPS. Japan, in particular, has funded research in beamed microwave energy. The Japanese Ministry of Economy, Trade, and Industry (METI) is doing most of the funding.[69] Tests of microwave energy transmitted between rockets show that the amount of atmospheric scattering for microwaves of frequency under 10 GHz is small.[69]

FIGURE 25.3

Orbiting solar energy satellite. The pieces and fuel would be brought to low-Earth orbit by the space shuttle and later moved out to a geosynchronous orbit for assembly. The satellite at geosynchronous orbit would beam microwave energy back to receivers on Earth.
(U.S. Department of Energy)

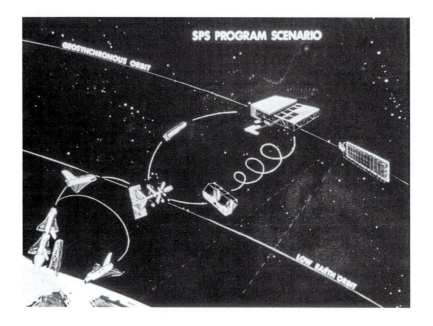

A new idea to be developed is that of putting collectors and transmitters on near-Earth satellites. This would decrease receiving antenna size from 10 km to a few hundred meters across (obviously, the bigger one wouldn't fit on small satellites). The ground rectennas ("rectifying antennas") would also be correspondingly smaller, because the microwave beam could not spread as far in the shorter path it traveled.[70] The ground receivers would receive the microwaves only intermittently, when the satellite would be near enough to send the beam down accurately and without too much scatter. At the rectenna, the power density would be about 100 W/m^2, quite a bit smaller than the original proposals, and far below what is known to cause problems for people. Hoffert and Potter estimated the total amount of power from near-Earth satellites at 10 to 30 TW.[70]

Another similar proposal (originally due to H. Oberth in 1929) is for an orbiting large reflecting sheet that could be used to illuminate cities at night, obviating the need for street lighting (and perhaps reducing crime as well). The Russians successfully tested deployment of such a plastic sheet to redirect sunlight in February 1993.[71] Of course, on a large scale, this could affect Earth's heat balance, but so does any thermal source such as burning coal or oil, or uranium fission.

Thermal Regeneration Conversion

In **thermal regeneration** electrochemical energy systems, reactants (*e.g.*, liquid sodium) are regenerated through flow of thermal energy from a heat source into a heat sink. Such a device is modular, made from a number of small cells (for example, a 100 gram cell can produce 4 to 6 W). It has a membrane (beta-aluminum solid electrolyte, ceramic tubes) that divide a closed cell into two regions. With a way to make one part hotter and the other cooler, these can produce electricity without any moving parts reliably and quietly. The technology is called AMTEC for "alkalai metal thermoelectric converter," and plants have operated at 20% to 40% efficiency for up to 2000 hours.[72] Much of the impetus for development came from NASA's need for electricity on interplanetary space missions. A small reactor could maintain a temperature difference, and because nothing moves but sodium ions when electricity is produced, the spacecraft does not need special stabilization.

In a cell, liquid sodium fills the upper region, maintained at a temperature of 900 to 1300 K. The lower region contains liquid sodium and vapor at 400 to 800 K. Liquid sodium at the hot end moves across the ceramic, turning liquid sodium to vapor. An electric lead inside the tube collects electrons from the vapor. The difference in temperature causes a pressure difference across the membrane, which, in turn, forces sodium ions to the low-pressure surface.[72] At the cool end, the vapor condenses back into liquid sodium and moves across the ceramic, picking up electrons to turn back to sodium atoms.[72] The electrons must flow from one end to the other through an external circuit, that is, they constitute an electric current. The sodium returns to the hot end through a fine-pore wick, retaining the material balance.

Energy from Water

It's not electrolysis or hydroelectricity. It's electrokinetic energy from water. Water running through microchannels (tiny channels in materials, as are found in filters) separates charges. The current and potential difference are microscopic (on the order of nanoamperes), so this is not a replacement for turbines, but the effect could be used to run small machines. It has no moving parts (of course, the water must move through the channels). According to the researchers, any "porous material, such as glass filter, membrane, rock and soil, could be considered as a natural electrokinetic battery."[74]

25.3 PROVEN ENERGY ALTERNATIVES

Geothermal Energy

Earth is the source of geothermal energy. Earth's interior is hot. It became hot originally as Earth formed out of the primordial dust cloud, and radioactive decay of "heavy" elements in Earth's interior keeps it hot. The natural outward heat flow at Earth's surface is estimated at 1.3×10^{11} W,[73] which ranges from place to place between 30 and 500 mW/m^2, with an average of about 60 mW/m^2.[75] A (potential) total energy of 830 EJ (230 PWh) is stored in Earth to a depth of 3 km,[72] and 1.3×10^{27} J (3.5×10^{20} kWh) is stored to a depth of 10 km.[76] In most places, the hot rock is far from the surface; where the heat flow is greatest, there is the potential for exploitation of geothermal resources. There are quite a few geothermal plants operational now: the Geysers, California (2081 MW$_e$); Lardarello, Italy (400 MW$_e$); Cerro Prieto, Mexico (645 MW$_e$);[77] Broadlands, New Zealand (145 MW$_e$);[78] and four fields in the Philippines, among others.[79,80]

In the United States, Alaska, California, and Hawaii have the greatest concentration of exploitable geothermal resources. Figure 25.4 shows geothermal regions in the lower 48 states. (We expect that the geothermal resources of Yellowstone National Park will remain inviolate.) There are four classes of geothermal resources: hydrothermal reservoirs (water trapped in fractured rock formations,

FIGURE 25.4

Geothermal resources in the lower 48 states of the United States. (U.S. Geological Survey)

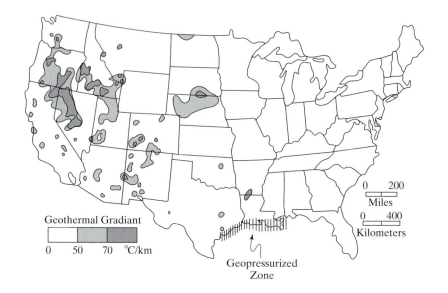

Geothermal Gradient

0   50   70   °C/km

Geopressurized Zone

producing hot water or steam depending on the reservoir); geopressured reserves (methane-bearing water trapped in sand under layers of shale, heated by conduction from underlying rocks); hot dry rock (near-surface rock); and magma (liquid or partially liquid rock at 3 to 10 km depth accessible by drilling). The United States has an estimated 2.6×10^{26} J (7×10^{19} kWh) of thermal energy stored in the upper 10 km of the crust,[76] of which 10% is hydrothermal, 20% is geopressure, 70% is hot dry rock, and approximately 1% is magma.[79] Only a tiny part of this amount is truly exploitable (that is, appears as hot rock near the surface). The most optimistic forecasts envision an eventual domestic geothermal supply of 400 GW.[76]

Hydrothermal energy is the only developed form of geothermal energy at present. Iceland has used hydrothermal heat since 1925 to provide heating for homes in Reykjavik and in other sites having hot magma near the surface. Some 44 states have the potential for direct hot water energy with a total capacity of 1.7 GW_t (at more than 67,000 sites).[79] Boise, Idaho, and Klamath Falls, Oregon, have been using geothermal energy for space heating and hot water for almost a century. For more detail, see **Extension 25.3, *Geothermal Energy*.**

Although use of geothermal energy for electricity began in Italy in 1913, California now leads the world in geothermal energy development. There are 22 plants at the Geysers (Figure 25.5), producing a total of 2081 MW_e.[79] This is the single largest field in the world, and the resource appears here as "dry steam" at the surface (it contains little or no liquid water). Unfortunately, hot dry steam resources total only 0.5% of the total world hydrothermal energy. As of 2000, a total of 3548 MW_e of geothermal electric generation had been installed in the United States.[96]

One of the larger fields, at Broadlands, New Zealand, which produces 161 MW_e, has been studied for effluents.[78] Large amounts of arsenic, mercury, hydrogen sulfide, and carbon dioxide come up along with the hot water and are released by the plant.[97] The temperature in a nearby river used for cooling sometimes rises as much as 6°C. The Broadlands facility has a thermal efficiency of 7.5%—this geothermal plant attains only 25% of its maximum thermodynamic efficiency, while a typical fossil-fuel plant reaches 70% of its maximum thermodynamic efficiency.[78] A second

FIGURE 25.5

Clapine's units 7 and 8 at the Geysers (Sonoma and Lake counties, California) exhaust spent steam. These two units supply 110 MW_e. (U.S. Department of Energy)

geothermal plant was built at Ohaaki, producing 104 MW_e.[98] Geothermal energy represents about 7% of New Zealand's total energy supply. New Zealanders are considering scaling back their exploitation of geothermal energy near Rotorua to encourage increased tourism. They are planning also to reinject the water.[98] It is clear that geothermal energy is not necessarily always the best alternative because of the emissions, the relatively large capital expense, and pipe clogging at some sites.

The exploitation of hot dry rock is closest to being realized on a large scale. One could build a geothermal device in any suitable place—one with a sufficient volume of hot dry rock—by fracturing a large volume of rock, pumping cold water down into the fracture, and drawing heated water up to run a generator (of course, reinjecting the spent water after any further use for space heat or agriculture). Such a system was tested by scientists at Los Alamos National Laboratory for 9 months in a small test at Fenton Hill, New Mexico,[75] and larger-scale tests were run there until the plant was decommissioned in 1996.[99] This approach experiences problems of mineral caking in the pipes in the same fashion that natural geothermal plants do. It produces no effluent, however, because the water is reinjected into the central well for reuse.

The world's resources are located in tectonically active regions such as around the Pacific Rim, the Alps, the Himalayas, and the mountains of North Africa. Most of the potential geothermal resources in the United States are west of the Rockies. These are areas of great surface heat flow, primarily in Idaho and Oregon, with smaller scattered areas in Nebraska, New Mexico, Nevada, and Colorado. Reasonable heat fluxes are available throughout most of the western United States and in small areas of New Jersey, Delaware, Maryland, and Virginia (see Figure 25.4). Three to six kilometers under the coastal rock of Texas and Louisiana is hot water under high pressure. It has been estimated that the deposits could supply methane equivalent to 6 to 55 years use at the current use rate,[100] as well as geothermal energy. Geopressure energy is very dilute, however, and probably will not be a large energy source.

MHD

Magnetohydrodynamics (MHD) refers to interaction of electrically conducting fluids with a magnetic field. MHD provides extra electric energy directly in use with conventional electric generation. Hot gas directly from combustion of coal or oil is "seeded" with a chemical agent that helps ionize it. The hot gas of ions carries charge with it (it is a plasma), and as it passes through a magnet at the high speed of the high-temperature gas, the magnet separates the charges to two terminals of a generator, producing a direct current with no moving parts (the DC is converted to AC in an inverter, see Chapter 4). Electrical energy is generated.

Because MHD generation by itself is not spectacularly efficient (20% to 30%),[101] and because its efficiency increases with temperature, it is probably best used as a topping cycle (that is, as a cycle at the high-temperature end to get some energy there, with subsequent use of the hot gas from the MHD generator to boil water to run a normal power plant to generate electrical energy as usual). In this mode, the net efficiency for the combined cycle can be much greater (~50%) than for either part alone. If a superconducting magnet were used instead of a conventional magnet, efficiency could be even higher, and the amount of water necessary for cooling could be much lower (conventional electromagnets run hot). The sulfur in the gas effluent from combustion of coal or oil combines with the seeding agent (usually potassium) so that the total amount of sulfur emitted is much less in a plant with MHD, although the smallest particles may be increased in number.[101]

Energy from Garbage

Another proven technology is conversion of garbage to energy. It is well known that garbage (often termed **municipal solid waste**, or MSW) produces methane at a rate of 250 m³/t [10⁴ ft³/ton]. The Garrett pyrolysis process (a proprietary process developed by Garrett Research & Development Company; pyrolysis is a process used to convert refuse into a liquid fuel by decomposing organic material at elevated temperatures in the absence of gases such as air) produces about a barrel of oil, 140 lb of ferrous metals, 120 lb of glass, and varying amounts of gas (power gas) per ton of waste.[102] This sort of list makes it clear that garbage can be more than just something to throw away. Italy produces gas from MSW to use in steelmaking and making terra cotta tile, and New York City is considering building a gasifier rather than an incinerator.[103]

Seventy to eighty percent of municipal waste is combustible.[104] Unprocessed solid waste can produce 10 to 17 MJ/kg (compared to 23 MJ/kg for coal).[105] Municipal wastes have been burned to generate steam at least since 1890 and to generate steam and electricity since 1896.[104] It is appropriate that the first plant generating electricity was in Europe (Hamburg, Germany), because Europe has adopted this method of dealing with municipal wastes with much more vigor and interest than the United States. One estimate is that incineration of all U.S. municipal waste could supply 2.5% of the total U.S. energy use.[106]

European countries historically have been short of space for people, let alone for landfills; in the United States, the cost of landfill disposal dropped far below incineration, especially from the early 1950s to the late 1970s.[104] In addition, European municipalities generally supply electricity as well as collect garbage, whereas in the United States, these services are supplied by different entities. As a result, Europeans could more easily see the benefits of burning trash. The situation in Japan is more like that in Europe than in the United States A Japanese study indicates that it is possible to recover 0.3 tonnes of fertilizer, 0.3 tonnes of paper pulp, 30 to 40 kilograms of iron, and over 7 m³ of fuel gas per tonne of waste.[107] The Japanese have made the processing of most garbage compulsory and are actively considering more incineration. Again, it is the Germans who have acted most boldly. Forty-five 200 kt/yr MSW incinerators are being constructed across Germany to reduce the flow into landfills.[108] See **Extension 25.4, *Burning Trash*** for more information on emission of pollutants such as dioxin.

In 1992, there were 145 waste-to-energy plants in the United States,[119] burning 20% of urban garbage.[127] The use of garbage to generate energy is more attractive in densely settled metropolitan areas, and in these locations, this method should probably be pursued. Separation is clearly to the advantage of all facilities, because most recyclables are more valuable before they are burned or do not burn at all, and separation (while expensive when the input is mixed) is not extraordinarily expensive if homeowners separate their materials before collection. Removal of heavy metals before combustion results in reduced production of ash. The Resource Conservation and Recovery Act (RCRA) regulates only hazardous waste disposal plants and hazardous incinerator ash, and much ash is indeed hazardous;[119] in fact, the United States Supreme Court ruled that incinerator ash not specifically measured as nontoxic is hazardous waste.[128] The major hazard is discarded batteries and electronic parts, which contribute lead, cadmium, and mercury to the ash.[129,130]

Waste can be converted to liquid fuel directly. During research on coal lique-faction, researchers found that the addition of 5% to 15% polycarbon waste (rubber, plastic, and so on) ups the yield of the process from 50% to 60%. The cost is competitive with current oil prices (about $45 per barrel).[131] Wastes can also be converted to hydrogen (see below).[132]

Finally, as noted in Chapter 23, microbes can generate current (we focused on seafloor microbes there). In the absence of sulfur, *Chlamydomonas reinhardtii* can be made to give off prodigious amounts of hydrogen.[133] Cycling the microbes between sulfur-rich and sulfur-lean environments could produce hydrogen without paying the energy price directly (see the section on hydrogen below for understanding *C. reinhardtii*'s utility). A Wisconsin group led by J. Dumesic has found ways to convert the sugars in organic wastes at relatively low temperatures (~200 °C) using an expensive platinum as a catalyst,[134] using a much less expensive nickel–tin catalyst,[135] and using gold in nanotubes.[136]

If microbes can generate current directly, and if municipal solid waste is composed of a bountiful feast of carbohydrates, microbes should be able to turn MSW to energy. Chaudhuri and Lovley demonstrated direct conversion of sugars such as those found in MSW directly to electricity. As they say, they've found "a novel microorganism, *Rhodoferax ferrireducens*, that can oxidize glucose to CO_2 and quantitatively transfer electrons to graphite electrodes without the need for an electron-shuttling mediator."[137] While these microbes operate at low current densities at present, there is hope that current density can be increased. Meanwhile, the energy source can be used for small applications; it has no moving parts to wear out.

Microturbines

Microturbines are not so "micro." A typical microturbine, such as produced by Capstone Turbine, might produce 30 kW.[138,139] They run on practically any fuel and are extremely efficient. A Capstone turbine is using waste gas from the Puente Hills landfill of Los Angeles.[138] On oil rigs, the turbine solves two problems—bringing in energy to run the rigs, and eliminating the flaring of gas from the well. The "sour gas" (natural gas containing hydrogen sulfide) produced can run a microturbine and provide the energy to run the rig.[138,139] The novel application here is the success of air bearings at very high revolutions. No lubrication is needed—ever.[138,139] The turbines could capture much flared gas for local use of the grid instead of wasting it and engender less pollution into the deal. Capstone's microturbines current sell for around $27,000, a little under $1000/kW.

25.4   AN ALTERNATIVE ENERGY DISTRIBUTION SYSTEM?

Hydrogen

The *Hindenburg*, a German zeppelin (an aluminum-frame fabric-covered airship using hydrogen to make the ship less dense than air), burned spectacularly upon its landing in Lakehurst, New Jersey, in 1937. Hydrogen in bags was used for buoyancy of the airship. The *Hindenburg* disaster followed from the explosion of the hydrogen in the bags. The live radio report and the films of the disaster effectively ended the era of lighter-than-air ships.

The **energy density**, or energy content per unit volume is low for hydrogen in comparison to the hydrocarbons, but its energy content per unit mass is over twice as large. (Hydrogen has only 10% of the mass of the lightest hydrocarbons.) The comparison of total available energy for various compounds is presented in Table 25.1. The favorable energy-to-mass ratio for hydrogen has led to the proposal that hydrogen replace gasoline as a motor fuel. Because of the large volume of hydrogen in comparison to other fuels, and its extreme flammability, there are storage problems for vehicles that would carry the hydrogen as fuel. Burning hydrogen in internal combustion engines for transportation is not as efficient as using hydrogen for fuel cells (most likely, solid polymer fuel cells, because of their low temperatures and high power densities—see **Extension 15.8** and **Extension 25.5,** *Hydrogen Storage and Fuel Cells*).

TABLE 25.1

Comparison of available energy from various fuels (MJ/kg)

Hydrogen (H_2)	114	Octane (C_8H_{18})	46.1
Methane (CH_4)	50.5	Carbon (C)	32.9
Ethane (C_2H_6)	48.4	Carbon monoxide (CO)	9.23
Propane (C_3H_8)	47.5	Gasoline (C_8H_{18})	2.7
Ethylene (C_2H_9)	47.3		

Source: Reprinted with permission from *Handbook of Chemistry and Physics*, 51st ed. Copyright 1970, CRC Press, Inc., Boca Raton, FL.

One advantage of a hydrogen-powered vehicle or power plant is that it is nearly free of pollution. If hydrogen is burned in pure oxygen, the result is energy plus water. No other chemical is exhausted. However, it is more likely that hydrogen will be burned in air; in this case, nitrogen oxides will also be produced. The combustion temperature can be set low, which minimizes release of this class of pollutants (and also the efficiency). There is no NO_x problem if H_2 is used in fuel cells. Recall from Chapter 6 that fuel cells are porous materials in which hydrogen or methane can combine with oxygen to produce electricity. Hydrogen-powered cars run much cleaner than gasoline-powered cars, and hydrogen-powered airplanes are cleaner than kerosene-powered jets.

Safety Issues

In view of the *Hindenburg* disaster, you may be tempted to think that hydrogen is very unsafe. Despite the fire breaking out when the cabin was 35 m (120 ft) above the ground, 65 of the 97 passengers and crew (and the ground crew) still survived because the flames went upwards. A study determined that the root cause of the fire actually was the flammable paint covering the ship's fabric skin, and not the hydrogen it carried.[145] One promising storage method for hydrogen is use of metal hydrides, solid alloys that can absorb large amounts of gaseous hydrogen and store it inside the metal's atomic lattice. See **Extension 25.5** for more information about hydrogen and the way hydrogen can be stored. A test with armor-piercing bullets of a hydrogen metal hydride storage container resulted in a short-lived flame and a 2% hydrogen loss; in a similar test with gasoline, a flame 2 meters long burned for 5 minutes.[107] All the hydrocarbons in liquid form are much denser than air, remain close to the point of release, sink, burn any object that is nearby, and spread as they burn, which can be dangerous. Hydrogen is much less dense than air and very volatile; it is soon diluted if it escapes. Because of hydrogen's lower density, burning hydrogen rises straight up. It was for this reason that there *were* survivors of the *Hindenburg* and that the crew cabin of the space shuttle *Challenger* survived the hydrogen explosion to plunge into the Atlantic, where it broke apart.

Storage is an important problem for automobiles, because the fuel capacity available is limited, and one would want to have about the same range available for a hydrogen-powered car as for a car with a typical internal combustion engine. At least one gas storage tank for cars has been approved.[148] If cars are to run on hydrogen, current internal combustion engines are capable of burning hydrogen directly, but the carburetors must be replaced and the compression ratio improved.[145,161] This can easily be done by a trained mechanic.

Hydrogen Pipeline Distribution

A further proposal has been made to replace our present electric distribution system with a system of pipeline distribution of hydrogen. For long-distance energy transmission, Gregory and Pangborn[147] calculated that pipeline transmission of hydrogen costs somewhat more than natural gas but is much less expensive than overhead electric transmission lines. A 210 km pipeline in the Ruhr region of Germany has operated reliably since 1938.[147]

We may compare the present cost of natural gas delivery to that of electricity. Typical transmission costs for natural gas run 0.003 cents per kWh per 100 miles of transmission. For electricity, average distribution costs are typically 0.74 cents/kWh out of a total transmission cost of 1.63 cents/kWh.[146] Table 25.2 shows a comparison of distribution costs in various energy distribution systems, including the estimated cost of a hydrogen system.[195]

The most important objection to this scheme is the shortage of hydrogen and the high-energy cost of obtaining it. It would take roughly 3×10^{10} m³ of hydrogen gas to replace totally the electricity used in a given year. About 5.5×10^{11} m³ of hydrogen is already being used per year, 80% for chemical feedstock.[176] Most proposals would use the output of present large power plants (perhaps nuclear, to minimize CO_2 production) to obtain hydrogen in some way (for example, by dissociation of water). An economic analysis based on various strategies[196] found that the proposed hydrogen distribution system was the cheapest (see Table 25.3).

The hydrogen could be supplied to the home using present gas pipelines. The problem is that the present valves would leak hydrogen if they were used. Another problem with hydrogen is that it makes certain (cheaper) steels brittle.[197] The entire system might have to be dug up and replaced. Many home appliances could still operate electrically; fuel cells such as those developed by NASA (see Chapter 6) could burn hydrogen locally to produce electricity. If a hydrogen-powered auto with metal hydride storage tank is used, a fill-up could take place at home. In home space and water heating, hydrogen could be used with flameless catalytic heaters with efficiencies approaching 100%.[176]

TABLE 25.2

Cost of various energy transport modes, in cents per kilowatthour per 100 miles (based on 2000 dollars)

Method	Local	Long distance
Methane by pipeline	0.72	0.04
Hydrogen by pipeline	0.79	0.04
Electricity by high voltage	3.06	0.25
Gasoline by tanker	0.83	0.12

Source: T. H. Maugh II, "Hydrogen: synthetic fuel of the future," *Science* **178**, 849 (1972).

TABLE 25.3

Net annual cost per household of alternatives for supplying residential energy

Alternative	Cost	
	(1973 dollars)	**(2000 dollars)**
Nuclear—electric and gasoline	822	2959
H_2 from nuclear and coal	815	2935
Nuclear to H_2	928	3342
Coal to H_2	486	1750
Coal to methane	515	1855
Nuclear—electric only	815	2935
Coal to electricity and gasoline	764	2752

Note: The costs are roughly one-third greater if the delivery system is buried instead of aboveground.
Source: W. E. Winsche, K. C. Hoffman, and F. J. Salzano, "Hydrogen: its future role in the nation's energy economy," *Science* **180**, 1325 (1973), copyright 1973 by AAAS.

Possibilities are good that a hydrogen economy could help less-developed countries, many of which are in the tropics and have a lot of sunlight available. At present, hydrogen could be produced by electrolysis, but only a few large-scale plants exist, and the process is only 70% to 75% efficient.[197] Of course, if solar energy is the source of the energy for electrolysis, the efficiency is of less concern. A more efficient method employs proton exchange membrane electrolysis, explained in more detail in **Extension 25.5**. This is 80% to 90% efficient at a size less than 110 kW.[197] High-temperature steam electrolysis could be 90% to 95% efficient.[197] Studies are in progress on use of hydrogen from North Africa to be used in Germany,[157,197] and German interests were involved in the construction of the Hysolar solar-to-hydrogen demonstration plant in Saudi Arabia.[157]

Power plants could work more efficiently in the case of a hydrogen system simply because hydrogen can be stored, in contrast to electricity. Thus, the entire load can be taken to be base load, and anything not being used would be stored. Time delays in the distribution system and inconstant demand are then easily handled. Energy expert John O'M. Bockris has estimated that coal energy costs $1800 per person per year because of its pollution, while it would only cost $120 per person per year to build a solar hydrogen economy.[198]

We end on a cautionary note. Air pollution in a fully hydrogen economy would be much lower,[199] if we assume the hydrogen is generated renewably. If it is from coal, pollution could well increase. While hydrogen does not release greenhouse gases per se, it can sop up the hydroxyl radical (OH^-) from the atmosphere, indirectly increasing the atmospheric abundance of methane by an estimated 4%.[200] Reduced nitrogen oxide emissions would at least partially counter the methane increase.[200,201] Modeling also indicates a substantial decrease in stratospheric ozone.[201] Tromp et al.[201] warned that changes in hydrogen concentration could affect other trace gases in unexpected ways, that increases in water concentrations could lead to more cloudiness (and a consequent effect on climate), and that the increase in hydrogen concentration might nourish soil microbes, leading to currently unforeseen effects. Perhaps hydrogen is not quite the magic bullet we all once thought.

SUMMARY

Energy storage can be helpful in leveling off the daily utility demand curve. Among the currently popular methods are load management using home heating systems, pumped water storage, and compressed air storage in underground caverns. Storage methods under discussion include use of superconducting magnets (**Extension 25.1**), novel batteries (**Extension 25.2**), epoxy flywheels, and shape-memory devices.

Speculative energy sources include solar power satellites, geopressure energy, and AMTEC conversion. Proven energy sources are geothermal energy (see also **Extension 25.3**), magnetodynamic topping cycles, and the burning of trash (**Extension 25.4**). The United States leads the world in installed geothermal power stations. Trash-burning power plants (see **Extension 25.4**) are

growing in number because of the increase in price of electric energy and because burning offers reduction in the volume of material that must be taken to landfills, thus prolonging landfill lives by a significant factor.

Hydrogen burns cleanly and is safer than hydrocarbon fuels. The use of an energy distribution system utilizing hydrogen seems feasible for several reasons: the natural gas pipelines could be used to transport hydrogen with relatively minor modifications; storage for hydrogen is cheap, so its use will automatically alleviate the storage problem electricity has; hydrogen can be burned with high efficiency; and hydrogen is cheaper to transport than electricity. Hydrogen fuel cells, as explained in **Extension 25.5**, are used in emergency power plants and transportation.

PROBLEMS AND QUESTIONS

MULTIPLE CHOICE REVIEW

1. Which of the following is FALSE about gasoline vehicle fuel as compared to hydrogen fuel?
   a. Gasoline is more dangerous if the container ruptures.
   b. Gasoline vapor leaks downward from the leak point, hydrogen vapor upward.
   c. The volume of gasoline needed to take a vehicle a given distance is much lower than the volume of hydrogen needed to go that same distance.
   d. The energy per unit mass is greater for hydrogen than for gasoline.
   e. Only gasoline may be stored in metal containers.
2. The energy stored in heat-of-fusion devices is
   a. electrical energy.
   b. heat energy.
   c. chemical energy.
   d. kinetic energy.
   e. thermal energy.
3. Flywheel energy storage in automobiles is possible today, because
   a. metals technology has advanced mightily in the last decade.
   b. batteries have been developed that can hook into the flywheel.
   c. now energy from braking can be stored in them.
   d. new composite plastics allow the flywheels to be fabricated using less material.
   e. unique new flywheel covers have been developed in the last decade.
4. Which of the following is a drawback of a hydrogen economy?
   a. Hydrogen pipelines leak.
   b. Hydrogen contains very little stored chemical energy.

   c. Current methods of generating hydrogen are too dangerous.
   d. Hydrogen pipelines are more costly than natural gas pipelines.
   e. Both (a) and (d) are drawbacks of a hydrogen economy.
5. Superconducting magnets promise advances, because
   a. they are able to circulate currents making magnetic fields with virtually no losses.
   b. the discovery of "super" conductors made of plastic will greatly lower costs.
   c. the magnetic fields can be stored in them, as they cannot in ordinary magnets.
   d. superconductors can withstand more and higher electric fields.
   e. of both (c) and (d).
6. Commercial storage of large amounts of energy using compressed air is possible
   a. only using gigantic steel storage tanks that can withstand the incredible pressure.
   b. only when strong enough turbines are developed.
   c. in Germany, with its relaxed environmental laws.
   d. in underdeveloped countries, where corrupt politicians can be bought easily.
   e. only if it is possible to find a large geological formation that can hold substantial pressure.
7. Stored energy is
   a. always more expensive than direct energy generation.
   b. economically feasible, because peak energy is more expensive than baseload.
   c. possible using a variety of schemes.
   d. feasible in more ways today than previously because of the advance of technology.
   e. correctly described by all the above.

8. Sodium sulfate is to be used in a heat-of-fusion device. How much energy can be stored in 100 kg of sodium sulfate?
   a. 2.13 J
   b. 31 J
   c. 100 J
   d. 213 kJ
   e. 21 MJ

9. A large energy storage facility uses sodium sulfate in its heat-of-fusion energy storage apparatus. At 31 °C. the facility can supply 1.7 MW for 6 h. About how much sodium sulfate was used to construct this facility?
   a. 0.3 kg
   b. 10 kg
   c. 1330 kg
   d. 47,900 kg
   e. 2,830,000 kg

10. Which method below can NOT be used to store energy?
    a. Pumping water up into a reservoir.
    b. Turning a flywheel.
    c. Changing the phase of a material.
    d. Compressing air.
    e. All of the above means could be used to store energy.

11. Which of the following statements is FALSE?
    a. Hydrogen burns cleaner than hydrocarbon fuels.
    b. Natural gas pipelines could be converted to transport hydrogen with minor modifications.
    c. Storage of hydrogen is cheaper than storage of electricity.
    d. Hydrogen can be burned with high efficiency.
    e. There is already a large number of European countries using a hydrogen energy system.

12. Which has the greatest overall cost to deliver electricity to your home, given the table below of transportation costs?

Cost of various transportation modes in cents/kWh/100 mi

Method	Local	Long distance
Methane by pipeline	0.20	0.01
Hydrogen by pipeline	0.21	0.02
Gasoline by tanker	0.23	0.03
Electricity by high voltage	0.85	0.07

   a. Methane (long distance) for 1000 mi, then electricity (local) for 10 mi.
   b. Hydrogen (long distance) for 1000 mi, then electricity (local) for 10 mi.
   c. Gasoline (long distance) for 500 mi, then electricity (local) for 10 mi.
   d. Electricity (long distance) for 200 mi, then local for 5 mi.

13. Currently, the most used form of geothermal energy is
    a. geopressurized reservoirs.
    b. hydrolytic reserves.
    c. magma.
    d. hot steam and hot water.
    e. hot dry rock.

14. Which of the following statements is FALSE?
    a. Energy storage can be used to level the daily utility demand curve.
    b. Pumped water storage is a popular method of load management.

c. Superconducting magnets may be used for short-term energy storage.
    d. Energy cannot be stored by simply changing the phase of a material.
    e. All of the above statements are false.

15. Which country leads the world in installed geothermal power stations?
    a. United States
    b. France
    c. Germany
    d. Russia
    e. Japan

SHORT ANSWER

Explain whether the statement made is correct, and justify your answer.

16. Microturbines will revolutionize the electricity industry.
17. MHD allows the efficiency of conventional power plants to increase.
18. Batteries able to withstand only 200 recharge–discharge cycles are unsuitable for commercial use.
19. The epoxy flywheel is still under development by commercial interests.
20. Magnesium bromide superconductors will revolutionize the electricity distribution system.
21. European trash has a greater thermal energy content than American trash.
22. Production of energy from geothermal sources involves generation of significant pollution.
23. Energy storage is normally used to produce baseload.
24. No geothermal district energy systems yet exist, but some are under development.
25. It was exciting to discover higher-transition-temperature superconductors, because cooling represents a significant energy cost.

DISCUSSION QUESTIONS

26. Explain how it is that hydrogen-powered transportation is cleaner than other alternatives.
27. Discuss your understanding of the author's meaning in the statement, referring to efficiency of a hydrogen economy, that "if solar energy is the source of the energy …, the efficiency" of the hydrogen energy production system "is of less concern."
28. If available energy from hydrogen is so large, explain how there is any controversy about an energy economy based on hydrogen.
29. Is the fact that sodium reacts rapidly with water a concern when building change-of-phase devices with sodium?
30. If long-distance electric transmission lines are over 99% efficient, why should anyone be excited by the prospect of superconducting transmission lines?
31. How could superconducting magnets be used for short-term energy storage?
32. One plan to produce energy (PACER) would detonate nuclear weapons underground. What environmental concerns might a project engineer building the underground cavity need to consider?
33. It is not necessary to separate metals from paper in the feedstock for a MSW energy plant.

34. There are too many practical reasons why pumped water storage cannot be done.
35. The volume of material leaving a waste-burning power plant is smaller than the volume that arrived.

PROBLEMS

36. Stannic bromide is used in a heat-of-fusion device. Its transition temperature is 25.5 °C, and the latent heat of fusion is 106 kJ/kg. How much energy is available if there is a mass of 500 kg of stannic bromide changing phase at the transition temperature?

37. A large energy storage facility uses stannic bromide in its heat-of-fusion energy storage apparatus. At 25.5 °C, the facility can supply 2.0 MW for 6 h. About how much stannic bromide was used to construct this facility?
38. What is the average geothermal power available through the surface area of the United States (9.8×10^6 km^2)?
39. If an underground compressed-air generating plant has a storage capacity of 2 GW for 10 h, how much energy is stored in the underground formation?
40. A flywheel spins at 30,000 rpm. How many revolutions would the device make per second? How fast would the edge of a 10 cm radius epoxy be traveling?

RUNNING ON EMPTY...

TOCSIN

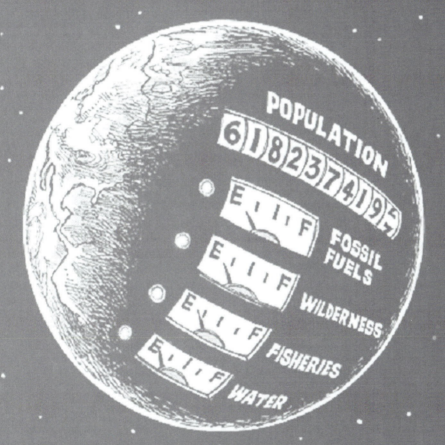

PART VIII

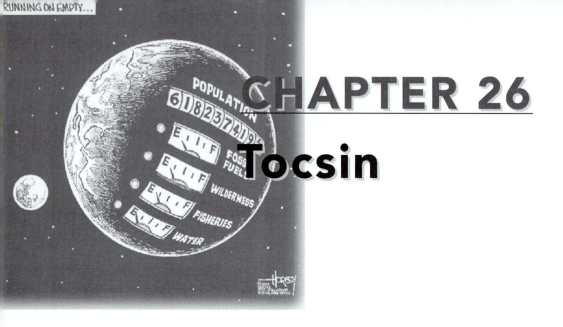

CHAPTER 26
Tocsin

The tocsin, the alarm bell, has been sounded. What we human beings choose to do now will determine our collective future, traveling on Spaceship Earth. Since change will occur, let us agree to make it change for the better!

KEY WORDS

Spaceship Earth / ecological footprint / dust bowl / Sahel / significant risk / microrisk / tocsin / technological fix / appropriate technology

No man is an island, entire of itself,
every man is a piece of the continent, a part of the main;
if a clod be washed away by the sea, Europe is the less;
as well as it a promontory were,
as well as it a manor of thy friends or of thine own were.
Any man's death diminishes me, because I am involved in mankind;
and therefore never send to know for whom the bell tolls;
it tolls for thee.

John Donne, *Devotions XVII* (1632)

26.1   THE FIRST PEAL OF THE BELL: BLINDING OURSELVES TO WHAT *ACTUALLY IS*

Now is the time to listen to the alarm bells that are ringing all around us. Emissions of pollutants and their consequences have reached the state described in this book because, often, people who make the decisions leading to these consequences are not aware of those consequences or of the interrelationship of resource use, energy use, population pressure, and the like. People have beliefs, and facts that do not fit with their beliefs can be ignored—for a time. In a paper entitled "Sorry, wrong number," Koomey et al. point out how numerical distortions filter through the system and lead to bad decisions, because "numbers that prove decisive in policy debates are not always carefully developed, credibly documented, or correct."[1] Eventually, reality bites.

People sometimes turn a blind eye when personal gain is involved. John Stuart Mill[2] characterized Americans (not kindly) as people "devoted to dollar-hunting." Perhaps business people, not only American ones, ignore information that could be used to persuade them that they were wrong. I would prefer to think that we are not so crass in our willful ignorance or even dishonesty. In any case, it is never easy to visualize the action of a complex system when some part of it is changed, even when one tries with the purest of intentions.

For example, I referred in the chapter introduction to a metaphorical "**Spaceship Earth**." John Stuart Mill (as early as 1857) eloquently defended the need of the closed world to move toward a stationary-state economy and a limited population:[2]

> If Earth must lose that great portion of its pleasantness which it owes to things that the unlimited increase of wealth and population would extirpate from it, for the mere purpose of enabling it to support a larger, but not a happier or a better population, I sincerely hope, for the sake of posterity, that they will be content to be stationary, long before necessity compels them to it.

This was an early instance of a knowledgeable person castigating the blindness of economists, a blindness displayed, for example, in their analyses of the greenhouse effect. Modern critics also have ascerbic things to say about economists. Economists' analyses depend on assigning costs where they see them. It is especially hard to assign dollar values to ecosystems. Ethnobiologist E. O. Wilson[3] says that "[n]eoclassical economics is bankrupt.... [Economists] are unable to handle multiple margins outside a narrowly defined market economy." Ecologist D. W. Orr says of economists[4] that their

> unstated assumptions are that we can summon neither the civic and moral wisdom to create a more equitable distribution of wealth nor the wit to redefine well-being in a less stuff-oriented and ecologically destructive manner.

Economists do not want to recognize limits. The sort of untrammeled growth desired by economists has an analog in animals' cells: it is called cancer.

26.2   THE SECOND PEAL OF THE BELL: NEED FOR POPULATION CONTROL

> I think I may fairly make two postulata.
> First, That food is necessary to the existence of man.
> Secondly, That the passion between the sexes is necessary, and will remain nearly in its present state.
>
> Rev. Thomas Malthus, *First Essay on Population* (1798)

Scientists have the reputation of being staid. That cannot be said of von Foerster, Mora, and Amiot, the authors of "Doomsday: Friday, 13 November, A.D. 2026."[5] Talk about a catchy title! It is an extrapolation of population trends as of that time, as found in 24 references to world population at various stages of prehistory and history. The date in the title refers to a least squares fit to the data through 1957.

In spirit, "Doomsday" is close to what we did in Chapter 2 with exponential growth. All humans are passengers on our tiny planet, Spaceship Earth. "Doomsday" notes that productivity is a function of population and carries the logic through to a singularity of infinite population—*something* will have to limit population, soon! It is a delight to read:

> It requires only simple calculations to show ... that Charlemagne ... could have predicted doomsday accurately within 300 years. Elizabeth I of England could have predicted the critical date within 110 years, and Napoleon within 30 years. Today ... we are required to extrapolate our evidence only 4 percent beyond our last point of observation: we can predict doomsday within approximately 10 years.

There was spirited debate through the 1960s on population that involved religious, political, scientific, and demographic argumentation. Serrin's reconsideration of "Doomsday" found the population estimate for 1975 eerily accurate compared to others'.[6] Umpleby in 1990 validated the von Foerster et al. view and reviewed the political implications of the arguments between demographers and scientists.[7] In Umpleby's view, the scientists (von Foerster et al.) had won the debate. As Bartlett has pointed out, "[p]opulation growth in our communities never pays for itself. Taxes and utility costs must escalate in order to pay for the growth. In addition, growth brings increased levels of congestion, frustration, and air pollution."[8]

Palmer et al. observed that because of the crush of people currently alive, "few places on Earth do not bear the stamp of human impacts."[9] Because this huge impact exists, Palmer et al. assert that the future must include humans as part of all ecological considerations.[9] Earth will be composed of human-managed ecosystems, and we know how difficult such management can be.

There are limits to how far humans can push ecosystems, and nature will enforce respect for those limits. That was and may again be the ruthless certainty of the apocryphal Four Horsemen of the Apocalypse: war, pestilence, famine, and death.

26.3 THE THIRD PEAL OF THE BELL: HUMANITY'S ECOLOGICAL FOOTPRINT

In Chapter 2, we analyzed other constraints on humanity's population growth, namely, the problem that human beings cannot appropriate all the solar energy falling on Earth. It has become a useful metaphor to characterize human's extraction of Earth's net primary productivity (NPP; the production of biomass by the biota) as a human "footprint" on the ecology.[10,11,12] Parris and Kates defined the **ecological footprint** as "a global and country-by-country calculation of consumption and waste relative to the Earth's capacity to create new resources and absorb waste."[13] The term was originally created by William Rees, and "grabs attention because it focuses on personal consumption and translates it into a corresponding land area—something else that ordinary citizens can understand."[14] Rees goes on to say in Reference 14 that he

> originally proposed ecological footprinting to counter economists' arguments that the concept of carrying capacity is irrelevant to human beings. Conventional carrying capacity is defined as the population of a given species that could be supported indefinitely by a specified habitat. However, economists argue that trade

and improved resource productivity can raise the carrying capacity of region indefinitely, thus rendering the concept meaningless for humans. Eco-footprint analysis subverts this argument simply by inverting the carrying-capacity ratio. Instead of asking how many people a particular area might support, we ask what area is required to support a particular population.

Figure 26.1 shows aspects of humanity's ecological footprint.

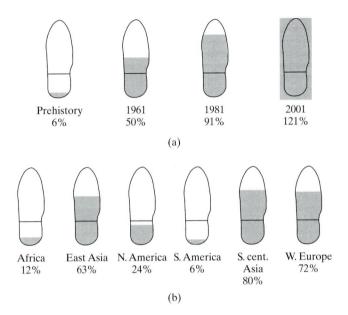

(a)

Prehistory	1961	1981	2001
6%	50%	91%	121%

(b)

Africa	East Asia	N. America	S. America	S. cent. Asia	W. Europe
12%	63%	24%	6%	80%	72%

FIGURE 26.1

(a) Ecological footprint of humanity (the shoe represents 1.9 ha, the world average).
(Prehistory, my estimate; 1961, 1981, 2001 from J. Loh and M. Wackernagel, eds., *Living Planet Report 2004* (Gland, Switzerland: World Wide Fund For Nature, 2004))
(b) Representation of human appropriated NPP in terms of available NPP by world region.
(M. L. Imhoff, L. Bounoua, T. Ricketts, C. Loucks, R. Harriss, and W. T. Lawrence, "Global patterns in human consumption of net primary production," *Nature* **429**, 870 (2004))
One may calculate one's footprint on many websites including http://www.myfootprint.org/ and http://redefiningprogress.org/.

As early as 1986, it was recognized that humans were already using at least 32% of NPP.[15] Field noted that this was a "conservative definition and [NPP appropriation totals] 40% using the most reasonable definition of human appropriation. This is a huge fraction. If we already control two-fifths of the land's productive capacity, then the prospects for future increases are strongly constrained, especially if there is to be anything left for other species."[16] Rojstaczer, Sterling, and Moore go further in estimating human–terrestrial NPP, or HTNPP, as they call it.[17] According to them, prior

global estimates of the human impact on terrestrial photosynthesis products depended heavily on extrapolation from plot-scale measurements. Here, we estimated this impact with the use of recent data, many of which were collected at global and continental scales. Monte Carlo techniques that incorporate known and estimated error in our parameters provided estimates of uncertainty. We estimate that humans appropriate 10 to 55% of terrestrial photosynthesis products. This broad range reflects uncertainty in key parameters and makes it difficult to ascertain whether we are approaching crisis levels in our use of the planet's resources. Improved estimates will require high-resolution global measures within agricultural lands and tropical forests.

A scientific approach can attempt to quantify the carrying capacity of an ecosystem,[18] and how close to "sustainability" we are in that ecosystem. One special ecosystem, the entire Earth, was studied in Reference 10. The authors found that the human footprint was such that "[i]t would require 1.2 Earths, or one Earth for 1.2 years, to regenerate what humanity used in 1999." A reporter quoted the lead author of the study, Mathis Wackernagel, as saying, "You can overdraw on nature's accounts and leave a debt. We are no longer living off nature's interest, but nature's capital. Sustainable economies are not possible if we live beyond the means of nature."[12] Human demand for ecosystem resources has doubled since 1961, and that demand now overstretches Earth's ability to replace those resources by 20%.[10] It takes almost 10 hectares to sustain each American, over 5 hectares for each European, and 2.33 hectares to sustain an average person (there are just 1.9 hectares per person).[10]

The value of natural ecosystems to people is not easy to determine, but as we find in **Extension 26.1, *Sustainability and Ecosystem Value***, it may be about $300 billion per year in the United States and around $30 trillion yearly worldwide (almost twice as great as the value of worldwide human-supplied goods and services).[51,52] Paradoxically, government subsidies of about $1 trillion per year are providing additional impetus to pillage the biomes that provide these services.[52] Nature reserves to safeguard these services, as suggested by Balmford et al. (Reference 52), would cost only about $45 billion per year, and "our reserve costs would have to be off by a factor of 100 for the reserve program envisaged to not make economic sense."[52] Ehrlich has argued strongly for considering the ethical view of ecosystems in addition to the pragmatic one.[53]

All this is evidence of the overwhelming effect of humanity on our environment. We live in the age of "human domination"[65] of Earth. Field wrote that "[h]umans appropriate a large fraction of land NPP, probably more than any other species in Earth history. We do not yet know the maximum limit for a sustainable future or whether it has already been passed. We do know, however, that insuring a sustainable future entails sharing NPP with a great host of other species."[16] Or, as Rojstaczer et al. said, "it is clear that human impact on TNPP is significant. The lower bound on our estimate (12 Pg [dry matter], 6.0 Pg C), although nowhere near total TNPP, indicates that humans have had more impact on biological resources than any single species of megafauna known over the history of Earth."[17] Imhoff et al. find twice that much carbon, 11.5 Pg, and find Asia to have a bigger impact than others find, but they generally agree.[66] This big an overdraft of resources cannot be sustained.

26.4   THE FOURTH PEAL OF THE BELL: NEED FOR RENEWABLE SOURCES OF ENERGY

It is clear that continued human intellectual life and realization of human creativity (not to mention comfort) depend on reliable sources of energy in perpetuity. We have seen in Chapters 7 and 8 that thermodynamics condemns us to reject waste heat into the environment from any energy transformation scheme that is thermal. We have also seen that almost all human energy resources *are* used in thermal systems (in heat engines). We have also seen in Chapters 13 and 16 that cities are "heat islands" that are warmer than their surroundings and, in some cases, are emitting waste heat in amounts that reach more than 10% of the solar energy incident upon their territory.

In a more unremarkable title than von Foerster et al., "A global view of solar energy in rational units," Rose has pointed out an obvious limitation on thermal energy sources.[67] Rose defined a "solar unit" as the average power reaching Earth's surface throughout the day and over the year (about 200 W/m^2). If the world's power output from thermal sources equaled one solar unit, Earth's mean temperature would have to be "close to that of boiling water"[67] to radiate the excess heat (as a blackbody; see Chapter 21). If the world's power totaled one-tenth of a solar unit, it would make Earth's climate "tropical" and, as Rose noted, such a rise in temperature "is intolerable owing, at least, to the melting of the ice caps ... The possible influence on the cloud cover and weather are further major consequences whether or not the specific effects can be detailed." (Recall that many cities' heat islands exceed this!) This is on top of any rise due to other human actions such as the global warming already underway.

Rose characterized a rise of 1 °C in Earth's temperature due to waste heat an "absolute upper limit." At this level, waste heat totals about one one-hundredth of a solar unit. Rose noted that the world's waste heat (in 1979; it is greater now) was around 1% of that, 10^{-4} solar units, while the United States' contribution was around 0.1% of a solar unit.[67] Exceeding 1% of a solar unit, Rose said, is "not the type of experiment the world should choose to explore."[67]

Rose further pointed out that the world is currently committed (through use of "usual" types of energy production, even including nuclear fusion) to about half the "absolute limit." This is due to population increase (to roughly twice the current level), the desire of the rest of the world for an "American style" of life, and the increases in the American standard of living that are pulling those aspirations even higher.[67]

Recall that Malthus (see quote above) had assumed "my postulata as granted, I say, that the power of population is indefinitely greater than the power in the earth to produce subsistence for man."[68] Rose went on to analyze food production (see the analysis of Chapter 2). This is important to Rose's argument, because with a growing population and pure solar energy, the Malthusian tradeoff is extended to be between land for solar collectors or land for food.[67]

Dukes has pointed out how unsustainable our current use of fossil fuels is.[69] Every liter of gasoline required the prior existence of about 24 tonnes of plant matter (and millions of years). Fossil fuels burned in 1997 constituted 400 times current NPP.[69] Were the ecological footprint to include the effect of use of fossil fuels, it would have been much greater than the estimates of References 15 through 66 and shown in Figure 26.1. In fact, Dukes estimated that conversion to biomass energy from fossil-fuel energy would lead to an approximately 50% greater demand by humanity on NPP.[69]

An additional problem is that the less-developed countries are developing, albeit piecemeal. The newest consumers on the globe want access to goods that others have, and this pressure will need to be considered here. Satisfying these new consumers will put more pressure on the environment.[70] The final answer is absolutely clear. Mother Nature cannot be gainsaid. The world must move toward exploitation of truly renewable energy sources. That means wind, tides, river or ocean currents, and technology such as solar photovoltaic energy must be, *MUST BE*, the ultimate future goal of societal energy transformation—and sooner, rather than later.

26.5   THE FIFTH, SIXTH, AND SEVENTH PEALS OF THE BELL: INTERCONNECTIONS CANNOT BE IGNORED

So that we can try to understand how action on one matter might cause some unexpected reaction on some completely different matter, we must try to picture

FIGURE 26.2

Interconnectivity of Earth's support systems.

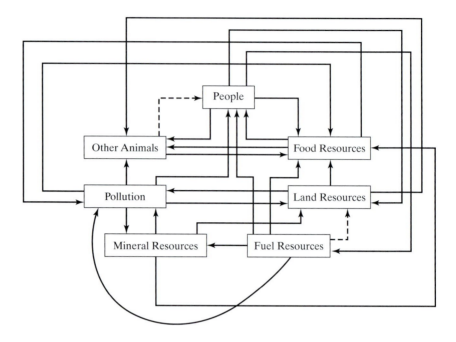

the entire network of relationships in some way. Figure 26.2 is my attempt to indicate the most important reticulations in the closed system making up Spaceship Earth. Notice that everything affects practically everything else. Of course, the categories are grossly simplified in Figure 26.2, but it does give a whiff of the *complexity* of the system. In some sense, the complexity of the system, with its many component pieces that interact, guarantees the stability of the system in the face of perturbations. A change in one variable can be made up by other changes induced in others. On the other hand, complex systems can find many ways to fail because of the sheer complexity. Because our concern in this book has been investigating the interaction of people with everything else, let me briefly present (or remind you of, as the case may be) a few of the interconnections in Figure 26.2.

People and the Land

René Dubos has emphasized that the world around us is very much a human-made world, even our natural surroundings.[71] In Eurasia, with historically high population densities over millennia, the landscape has been shaped by people. (This is less true of Africa south of the Sahara and of the Americas.) The very geography has been in some way domesticated.

There is both good news and bad news on this front. People have been responsible for creation of deserts (desertification) and for soil loss—the notorious American **dust bowl** of the 1930s is one example, and the **Sahel**, a dry area just south of the Sahara desert, is another. Deserts are spreading through what one writer characterized as "Toyota-ization," the breaking of the desert surface crust by off-road vehicles.[72] On the other hand, farmers have made productive and long-lasting fertile soils in many parts of the world. This apparent contradiction is detailed in **Extension 26.2, *People and the Land*.**

People and Plants

Few of us realize that North American agriculture is imported agriculture.[116] Potatoes come from South America, as do sweet potatoes. Corn originated in Central America, wheat in the Middle East, rye in northern Europe. The United States of the late nineteenth century actually had a government agency with the sole purpose of introducing new plant and animal species here. In the year 1916 alone, 2294 species were introduced into the country for the first time.[117]

Invading species outperform many native species all over the world. A species that is benign in its home may turn into a monster when it leaves home. Meanwhile, many invaders have fit in well, and efforts are being made to utilize "good" characteristics of plants that are found in native grounds to improve their offspring worldwide. This dichotomy is further explored in **Extension 26.3, *People and Plants***.

Insects, Other Animals, and People

According to a study by the National Academy of Sciences, "introductions by natural means are uncommon; almost all introductions today are in some way facilitated by human activities."[268] In addition to an imported agriculture, we have imported many nonagricultural animals such as starlings, sparrows, and carp. Some imports have been successful, some have not. For example, the brown trout was beneficial, whereas the carp are destroying aquatic habitats and spreading disease.[117] The huge numbers of animals brought in are truly mind-boggling. In 1968, we imported 120,000 mammals, and in 1972, 100 million fish were brought into the country. No one has any idea of which species will cause permanent trouble in a new home, or where in the world it will find to cause it (obviously, successful invasions depend on hospitable local conditions).[268]

Successful exotic species "eat like whiteflies, breed like rabbits, and colonize like crabgrass."[269] Disturbed areas with few species are the most open to invasion, and humans are ideal disturbers. See **Extension 26.4, *Interspecies Competition*** for more ideas on how one species affects another.

CASE STUDY

What's Happening to Oceanic Life?

Alarming drops in marine populations have been noted globally. The Atlantic has been worst hit. Cod have essentially vanished from the Grand Banks.[309] Drift nets and general overfishing are causing problems. Nets are 10 m high and may be 50 km long. There are 700 or more fishing vessels using the nets, mostly from Japan, South Korea, and Taiwan. Drift-net fishing has been called "biological strip mining."[310]

The nets trap hundreds of porpoises, seals, seabirds, and other creatures. Lost or jettisoned nets also kill large numbers of animals. The Taiwanese especially are implicated in the loss of salmon returning to the United States.[309,310] The U.S. National Marine Fisheries Service found that 45% of fish stocks are overfished. About 60% of stock is fished every year! The population of some species of fish now is at 10% of optimal levels. Fishing alone is not responsible; dam building and clearcut logging also play a role in the decline of Pacific salmon.[311] Furthermore, fishing has been implicated in applying evolutionary pressure to fish stocks, decreasing the size of the average fish.[312] And, of course, one must not forget that by changing the climate, humans are indirectly managing fish stocks in addition to directly managing them.[313]

The contemporary Adam Smith has remarked[314] that "killing whales is very profitable until the day when there are no more whales, because we have only been amortizing the ships and the radar and the depth charges and the harpoons. We haven't amortized the whales, and anyway, how do you replace whales?" The same could be said about salmon or about cod.

Extension 26.5, *The Oceanic Food Web* provides more details on the collapse of cod fisheries, overfishing, and aquaculture.

Progress has been made—or at least accommodation reached—in dealing with some of the invaders and learning what makes invasions successful in Great Lakes fish,[365] plants invading 9.2 Mha of California grassland,[366] and invasive species in general.[196] Genetic research holds out hope that we can come to terms with fire ants and Africanized honeybees. Further detail on the possibilities are detailed in **Extension 26.6,** *Insects, Other Animals, and People*.

26.6   THE EIGHTH PEAL OF THE BELL: TOXIC WASTE VERSUS PEOPLE

The story of industrial wastes has qualities reminiscent of a nightmare. Disease, miscarriage, and birth defects were found among residents in the suburban tracts built over an old toxic chemical waste dump near the Love Canal in Niagara Falls, New York.[495] The publicity served to remind us all that over 250 million tonnes of hazardous industrial waste must be disposed of each year.[496]

The Superfund cleanup, after much controversy, finally seems to be scoring some successes.[497] Some 926 sites have been cleaned up as of the end of 2004. There are 1237 hazardous waste sites that have been identified for action under the Superfund (Figure 26.3),[498] which was created by a tax on chemical production as part of a bill enacted by Congress in 1980 to allow cleanup of contaminated disposal sites in the aftermath of Love Canal, and costs about $1.5 billion per year.[497] About 85% of Superfund money in the past went to fund lawyers instead of funding cleanups.[499]

Superfund proved insufficient for the task, so Congress passed a further $10 billion program in 1985, funded by a tax of $3.1 billion on petroleum companies, $2 billion from chemical companies, and $2 billion on toxic wastes.[500] The Germans are now worrying about PCBs and other chemicals[501] and are contemplating setting up a Superfund of their own.[502]

In the old days, all people could do was remove the contamination from one spot and deposit it in another essentially unchanged. We are learning how to perform remediation that actually cleans up sites and doesn't "rob Peter to pay Paul." See **Extension 26.7,** *Toxic Waste Versus People* for more information on the problems and possible solutions.

FIGURE 26.3

Hazardous waste sites as of October, 2004. The total number of sites in the United States is 1237. These are to be cleaned up under the Comprehensive Environmental Response Compensation and Liability Act of 1980, the "Superfund," which was renewed in 1987. Taxes on polluters expired and Congress has not reinstated them, so Superfund future funding for cleanup is in doubt. (U.S. Environmental Protection Agency, *http://www.epa.gov/superfund/sites/query/queryhtm/nplfin.htm*)

It would certainly be advantageous if the world could clean up after itself. Surprisingly, some materials have been found that do just that. **Extension 26.8, *Interesting Cleanup Ideas*** details the effects of titanium dioxide and ceramic silver coatings for health and plastic coatings for steel in arresting rust.

26.7 THE NINTH PEAL OF THE BELL: RISKS (AND BENEFITS)

Expert Assessment versus Popular Assessment

As we discussed in Chapter 20, there are risks that people enter into voluntarily that are greater than those they will suffer involuntarily. People also respond to the knowledge of a risk differently from "expert" analysts, who presumably know better which risks are more risky. Where does the mismatch between actual and perceived risk arise? The classic studies of Paul Slovic [619] compared the risk assessments of 15 national experts with those of 40 members of the League of Women Voters for 30 activities and technologies. He learned some interesting things about the differences (many already discussed in Chapter 20):

- "Outrage factors" (how could they do that to us?) make people feel that even small risks are unacceptable.
- Imposed risks are more worrisome than assumed risks.
- Unfairly shared risks seem more hazardous to the public.
- Risks that can be controlled to some extent by individuals are more acceptable.
- Natural risks are less threatening than risks caused by humans.
- Risks that have catastrophic outcomes are more unacceptable than those that do not.
- Risks from exotic technologies are more problematic than those from familiar technology.

The public mistrusts the technical assessments of experts. Apparently, people feel that the experts have a self-interest that tends to make them skew their assessments. Negative events seem to carry more weight in the public mind than positive events. People appear to put more emphasis on negative news than is warranted. Just seeing that something is in the news validates its riskiness for many people, regardless of the true element of risk involved.

The public also misses the reverse part of the story—that inaction or lack of exposure to one risk carries risk, too. If you were ill and were told that radiation treatment could save your life, you as an individual can weigh your personal risk of remaining sick and almost certainly dying against the risk of exposure to large doses of radiation. You would probably choose the benefit of exposure. But when a similar calculation is made with exposure to radiation because of the presence of nuclear reactors, the same people will not think about the risks of *not* having the electricity. Or consider the tradeoff between nuclear energy and coal energy. Many studies have shown that coal energy is far more likely to result in monetary loss, injury, or death than nuclear energy (Chapters 12, 13, 15, and 20, and especially, **Extension 20.12**); people still choose the devil they know.

The basic difficulty with regulation is how to allocate limited resources most effectively. To do this well, we need to address social issues: comparability of various risks, comprehension of magnitudes of risk, mental models of risk, and risk literacy.[620] A study of the efficacy of a benzene-removal program was run jointly by Amoco and the EPA. Amoco was going to be required to spend $31 million to rebuild a wastewater treatment plant to reduce evaporation of benzene (178 tons of pollutants, including 3 tons of benzene).[621] But most benzene (784 tons) came from the marine terminal in the York River, where oil and gasoline are transshipped. Control of benzene there would cost only $6 million.[621] To spend money most effectively to reduce risk, one should concentrate on the *goal*, not specify the method by which the risk reduction is attained.

Education about true risk is a start. Rowe calculated that 1 tonne of coal flue gas particulates translates into 100 μg/m^3 of exposure.[622] Cigarette smoke, because of its nearness, amounts to only one-thirtieth the exposure to coal flue gas; however, recalling that Dose = Potency $\times$ Exposure, the dose is 80 times higher.[622] Until recently, the public was not educated enough to know how grave the danger of secondhand smoke was; now, after 30 years, the message has finally penetrated, and people are doing something about it (such as banning smoking in public places). California has chosen a unique way to educate the public. California's toxics law, Proposition 65, forces business to warn the public when it knowingly exposes them to a substance posing a "significant risk" of cancer or birth defects.[623,624] The state has defined **significant risk** as 10^{-5} cancers per person, or 10 "microrisks," in the parlance of one analyst (who defines a **microrisk** as a risk of 10^{-6} cancers per person; see Table 26.1).[624] Presumably, one microrisk is an ignorable risk. (Although one microrisk is small, if an entire population is exposed, the collective risk can be large.)[624] California has listed 250 chemicals and set standards for 50.[623] After a listing, the industry has a year to provide warnings (on labels, for example). Whether or not the idea is scientific, it will provide a reason for removing toxic substances from products by informing the consumer. Consumers who stop buying a product are a powerful force for change, one that could never come directly from the government.

Extension 26.9, *Assessing Risks of Air Pollution and Benefits of Lack of Pollution* tells about the tradeoffs required in this specific case as well as some more general considerations.

The public cannot put the risk numbers in perspective. For example, Chernobyl may cause 40,000 deaths during the next 70 years. This is the number of children that die every day from lack of clean water.[637] Are Americans willing to spend billions to clean up water supplies in the Third World? This is the number of early deaths due to emissions from the British coal industry every 2 1/2 years.[637] Does Britain stop using coal? Do these comparisons change your perception of the risk from Chernobyl? New York City's public schools were closed for weeks in the fall of 1993 at parents' insistence to remove asbestos. Children in schoolrooms containing asbestos are at less risk than pedestrians are of being hit by a car.[638] More children were injured while schools were closed than if they had stayed in school. Does the decision make sense? For the United States, with 280 million people, the whole population being exposed to a microrisk implies 280 deaths. Is that negligible or not? Note again how the way a risk is presented (one in a million chance versus 280 deaths) can have a lot to do with the way the risk is perceived.

It is reasonable for the public to questions the experts' assertions of risk. Consider the cancer risk paradigm of rodent exposure (Chapter 20). Is it reasonable to assume that all animal carcinogens lead to cancer risk in humans? The animals are fed huge doses; the dose is at a level just below fatally poisonous to tweak out of the animal any significant effect.[638] Statistical models are needed as controls, but no knowledge is available to guide us to the correct model.[638] "The proclaimed causes are so small and the statistical associations so weak that the effects become indecipherable," according to one expert, who pointed out, in addition, that comparison among many variables will always turn up associations in any small data set, confounding whether there is any effect at all.[638]

Cleaning Up—How Much Is Appropriate? At What Cost?

People want to clean up toxic Superfund sites. How "clean" should clean be, and who decides? It would cost $71,000 to isolate an abandoned factory in Holden, Missouri, but $3.6 million to clean all residues and bury the remaining traces under a layer of clay.[639] State and federal laws mandate a cleanup that would cost between $13.6 to $41.5 million.[639] That is because the EPA calculates a 10^{-4} chance of a cancer from eating a cow that grazes at the perimeter of the property. The extra money is to restore the land to preindustrial condition.[639]

Government is finally recognizing that the level of cleanup is a decision to be made, at least in part, due to public pressure.[639-641] The Clinton administration

TABLE 26.1

Activities entailing one microrisk (risk of 10^{-6} cancers per person)

2000 km of traveling by plane
2500 km of traveling by train
80 km of traveling by bus
65 km of driving a car
12 km of bicycling
3 km of motorcycling
1.5 cigarettes
Living with a smoker for 2 months
Drinking one-half a liter of wine
Living in a brick house for 10 days
Breathing polluted air for 3 days

Source: Based on G. Marx, "Everyday risks," *Phys. Educ.* **28**, 22 (1993).

proposed different levels of cleanliness for different Superfund sites.[642] The sites need to be most fully cleaned only if the new use will be residential. Lesser restoration would be required for industrial and commercial uses. The assumption that every site would become a housing development would be abandoned. The local community would be involved in the decision of which level of cleanup was finally made. Such regulations could increase the effectiveness of the Superfund.

It is estimated that EPA regulations cost the U.S. economy $115 billion per year, about $450 per person in higher taxes and prices. The 1990 Clean Air Act Amendments might have added another $25 billion per year.[499] On the other hand, the cost of no regulations is estimated to be one death per $12 to $15 billion not spent because of the loss of living standard suffered.[499]

Many observers believe that government regulations are getting out of hand because of politicians' ignorance about what they're voting for, compounded by the ignorance of the public demanding action.[620,639–641] The original impetus came from being conservative—when in doubt, get it out. But economic costs mount, while the benefits are not calculable if they even exist. The money could be used more beneficially in other programs.[639,640] The cost per life saved is becoming incredible. Table 26.2 shows the cost of some recent regulations per life saved.[620,641] While an individual estimate might be argued with, the estimates are internally consistent comparisons. Environmental regulation can spur innovation, but only when the right incentives are provided. The key is to set goals without setting the means. Well-thought-out regulations could end up helping U.S. companies abroad.[643]

TABLE 26.2

Cost of government regulations per life saved (million dollars)

Regulation	Cost
Ban on unvented space heaters	0.1
Aircraft cabin fire protection standards	0.1
Auto passive restraint/seat belt standards	0.1
Trihalomethane drinking water standards	0.2
Aircraft floor emergency lighting standard	0.6
Concrete and masonry construction standards	0.6
Ban on flammable sleepware	0.8
Grain dust explosion standards	2.8
Rear lap/shoulder belts	3.2
Ethylene dibromide drinking water standard	5.7
Asbestos exposure limit for workers	8.3
Benzene exposure limit for workers	8.9
Standards for electrical equipment in coal mines	9.2
Arsenic emission standards for glass plants	13.5
Ethylene oxide exposure limit for workers	20.5
Listing petroleum refining sludge as hazardous	27.6
Acrylonitrile exposure limits for workers	51.5
Asbestos exposure limits for workers	74.9
Arsenic exposure limit for workers	106.9
Asbestos ban	110.7
1,2-Dichloropropane limits in drinking water	653
Hazardous waste land disposal ban	4190.4
Formaldehyde exposure limits for workers	82,201.8
Alachlor/atrazine limits in drinking water	92,069.7
Hazardous waste listing for wood preservatives	5,700,000

Source: J. F. Morrall, III, Office of Management and Budget.

In the end, the principle of TANSTAAFL is at work. If you stop building power plants, you will not have hospital equipment to use when you need it. In assessing risk, it is nice to have certainty, but seldom is certainty found. John Snow used a novel graphic technique to implicate one public pump on Broad Street, London, as the source of a cholera epidemic (Figure 26.4).[644] He acted on his knowledge by removing the pump handle. The end of the cholera epidemic was his proof.[645] It is seldom so easy nowadays. Nevertheless, risk must be dealt with as realistically as possible, so wherever possible, the "pump handle" can be removed. If removal is not possible, at least the most cost-effective use of limited resources can be pursued.

FIGURE 26.4
John Snow's 1854 graph of cholera cases implicated the pump on Broad Street (marked X) as the source.

26.8   PROSPECTS FOR THE FUTURE: STOPPING THE TOLLING OF THE TOCSIN?

It may seem that I am terribly pessimistic about the world and its future from the almost endless list of difficulties I have recounted here and in Chapters 13, 14, 15, 17, and 20. I have proposed few solutions and outlined few courses of action, because an important purpose of this book is to sound the alarm, the **tocsin**, and to arm the reader with the facts about energy and the environment necessary to deal rationally with the myriad of interrelated problems facing the human race. (Of course, you must be paying attention—see *The First Peal*, above.) In this, the end of the book, I want to indulge my exhibitionist tendencies, and share with you, the reader, the reasons that I feel guardedly optimistic about the prospects for the survival and prosperity of the human race.

I do not believe that I am being Panglossian in my vision of the future. Why not? My father always said to me that things I would call problems (whatever they happened to be), weren't *problems*, they were *opportunities*. In some sense, all the problems I have listed here are opportunities waiting for someone to take advantage of them.

The stationary-state economy advocated by Mill (*The First Peal*) has been discussed at some length in a modern, detailed analysis.[646,647] All such ideas are based on the closed aspect of Earth's system. But the Spaceship Earth metaphor is at least inaccurate, if not downright misleading. After all, how did Earth come to possess supplies of fossil fuel? The fossils were living plants, or animals that lived by feeding on plants or feeding on other animals who fed on plants, once long ago. The plants, the ultimate food source, grew because the Sun shone, supplying the energy necessary for photosynthesis. All our fossil fuels are simply stored solar energy. We are not self-sufficient in this sense. It is solar energy that supplies energy sufficient to allow a decrease in the entropy here on Earth (although clearly, the entropy of the universe increases in the process). Earth is simply not a closed system.

It might be more accurate if we were to call it "Spaceship Solar System" instead of Spaceship Earth. If we think of the solar system as a closed system, then some of our problems become opportunities. There are a lot of minerals available in the solar system—those in the asteroid belt are especially easy to send to Earth, because there is no energy penalty for climbing out of a gravitational well. This is so because we do not have to spend energy fighting a planet's gravitational pull as we would if we were transporting something from, say, Mars. Of course, we do not know the mineral assay of the asteroids at present, but we should have one fairly soon (by which I mean the next 25 to 50 years). There must be sufficient minerals in the asteroid belt to supply humanity for eons. By suitably adjusting the speed of an asteroid at an appropriate point in its orbit, we ought to be able to make the asteroid pass through Earth's orbit. We might also process the asteroids on site, and ship the refined materials directly to Earth. Even for refined minerals, the asteroids have a great advantage over the other planets as a source. We can use much less energy to transport the same mass of minerals from the asteroid belt. (Lest you think I am over the top here, recent articles in the popular press—not to mention by science fiction writers—have adopted this same view.)[648,649]

In fact, if we were to establish factories in orbit, perhaps at the gravitationally stable Lagrange points à la O'Neill's proposal to build habitats (see **Extension 2.1**) there,[650] we could run industrial processes in space. It would be impossible to pollute Earth, then, in the processing of materials. Of course, it would be difficult to run industrial processes as they are run today, because in space, there are no free goods—air or water—to be used at will. Industrial design in space would have to be sensible and conservative (in the best sense). This would be possible only because of the huge amounts of solar energy available to orbiting industrial establishments. Gossamer-thin reflective plastic could easily be shaped into an extremely large parabolic mirror that could concentrate solar energy for use by the orbiting factories. This should be economically viable, because only finished products would have to be sent down to the ground. Finished products are worth much more per unit mass than raw ore. This is important because the energy cost of bringing down material depends on the mass involved. A down-to-earth example is that of the ratio of the cost of a computer chip made of silicon and germanium to that of the raw materials—a very large number!

This solution is essentially what is known in the trade as a "**technological fix**." There seems of late a great animosity toward anything hinting of technology. I think that there is nothing intrinsically wrong with using technology to improve human life, and to repair some of the mistakes we have made in applying the fruits of science through technology. As one example of technology applied to problems, electrons from the decay of ^{60}Co can be used to cleanse sewage sludge, and near Munich, Germany, a gamma-ray facility is doing a similar job.[651] Accelerators can be used to get rid of waste from nuclear reactors—an intense neutron bombardment can transmute wastes and reduce necessary storage time from many thousands to several hundred years.[652] Water that was polluted can be cleaned up by concentrated sunlight, which destroys most pollutants.[653] Battelle Memorial Institute has developed a process to vitrify (make glass of) contamination at former weapons plants using 10-m-long electrodes driven into the ground; the resulting solid block immobilizes the radioactive materials.[654]

The irony of the antitechnology stance is that we just have no choice but to embrace technology. I am unwilling to "turn back the clock," and I have no intention of making my children turn it back either. Alvin Toffler, in *Future Shock*,[655] wrote (emphasis in the original):

> To turn our back on technology would not only be stupid but immoral. Given that a majority of men still figuratively live in the twelfth century, who are we even to contemplate throwing away the key to economic advance? Those who prate antitechnological nonsense in the name of some vague "human values" need to be asked "*which* humans?" To deliberately turn back the clock would be to condemn billions to enforced and permanent misery at precisely the moment in history when their liberation is becoming possible. We clearly need not less but *more* technology.

I do not mean to imply that only large-scale technological endeavor is important or even that only huge technological projects are useful. The writings of E. F. Schumacher[656] emphasize the application of **appropriate technology,** technology that is appropriate to the necessary end. Should tractors, say, be exported to Chad to help the country become agriculturally self-sufficient in lieu of encouraging the country toward a quasi-1890s technological base (reapers, improved harnesses, and other such small advances) grown indigenously? This level of technology teaches the user and accustoms people to the use of comprehensible technology. Grafting the complete 2000s technology onto the unsophisticated, unsuspecting, and semifeudal Chadian economic base would not work; the graft would be rejected by a national "immune response system."

Lovins[657] emphasized the desirability of the "soft energy path," one that is dispersed and technically simple, in contrast to the "hard energy path" of centralized, large-scale energy generation (Chapter 6). Lovins sees the two paths as mutually exclusive, nay, contradictory; I am sure that synthesis is possible. Each has its own appropriate place in an organized economy. A recent social trend of the sort to illustrate this complementarity can be observed in the phenomenon of the small, cottage industry, business. No large concern would consider running a business that merely breaks even, that is, provides employees' salaries and covers costs. Grocery chains routinely close profitable stores because their rates of return are not as high as the chain's management desires. Thus, the advent of the small handcraft or exclusive-service business that provides employment but not gigantic profit gives balance to the economy and provides a measure of self-respect to the people involved, people who otherwise might have been reduced to bitter parasitism.

The point is that in diversity lies strength—for energy sources, for plant species, for economic networks, and for application of technology. I have already discussed how plant monoculture can easily be destroyed by small changes in conditions, while natural plant systems recover from even rather large perturbations, how specialized predators can easily be wiped out when specific prey vanish. The same applies to applied technology. No utopia will result from utilizing the soft path or the hard path, from using only small-scale or large-scale technology. (Thank goodness. I have always thought that utopias were especially boring places.) Orbiting factories will not reconstitute the Garden of Eden;[658] they will just serve to increase the resilience of the system. No matter which way we turn, no matter how different the conditions, people will be people in the same old, familiar, endearingly human, frustratingly human ways.

Returning to the problem of vanishing resources, I can see an opportunity for ingenuity closer to home: mine garbage dumps for the mineral resources we need.[659] This is not being done to a great extent, even though it is generally much cheaper energetically to do so (Chapter 25). Our modern midden heaps are an atrocity in conception and in realization.

Of course, even with enhanced recycling of materials, there will still be resources "falling through the cracks." In the computer scenarios of the Club of Rome report,[660] the recycling of materials merely staved off disaster for some short time interval. We shall have to continue exploiting resources whether here on Earth or elsewhere. It is very clear to me that recycling and reuse is a moral necessity. John Donne's words from the beginning of the chapter remind us: "No man is an island, entire of itself." We are obligated not to sacrifice our children's future to prolong our ease right now.

The developed nations of the world consume most of the world's nonrenewable resources to support their living standards. If it is moral to wish everyone to have a standard of living comparable to our own (often seen in the religious notion known as "right sharing"), we must either be prepared to lower our own standard of living to help achieve the balance, or we must find new sources of resources so that there is enough for all to raise standards in the rest of the world. This is the nub of the mistake made by those who argue that it is more immoral for Americans to have children than, say, Indians, because of the much greater impact of Americans on the environment. What is desirable is for all to live in plenty (or at least free of want); the *ultimate* impact of an Indian child is the same as that of an American child. Ultimately, both persons, or their descendants, will have to be supported in a similar style of life—and this is already happening through globalization.[661]

We may try to export population to other planets or to O'Neill's space colonies to raise living standards for those who remain. Europe chose this sort of solution. Over 50 million left to take up new lives in the Americas, of a population varying between 250 million (1850) and 450 million (1950). Despite this emigration, despite the loss of millions killed in each of two world wars, the population doubled over the century. It is unlikely that emigration will, in the short run, ease the pressure of demand for resources.

The idea of lowering my personal living standard is unpleasant but not intolerable. However, I would much prefer to hope that unbridled imagination will lead to a world in which standards of living will level off upwards. Humanity has many talents and many opportunities for exercising them now that the mass of people has begun to awaken to an awareness of the seriousness of the current energy and pollution situation. Will we use our talents and imagination to end the folly of our behavior? This conundrum is expressed by the historian Ortega y Gasset:[662]

> To modern man is happening what was said of the Regent during the minority of Louis XV: he had all the talents except the talent to make use of them.

SUMMARY

Everything is connected to everything else. It is a sometimes sorry history—Earth's exploitation by its inhabitants meddling in ecosystems with disastrous consequences, such as desertification or epidemics of plant diseases; disappearance of species from Earth; throwing away of valuable resources; and careless handling of toxic wastes. The tocsin is sounding for us now. It would be easy to succumb to despair.

Many people worry about small risks while neglecting big ones. To face a life of risk, we must know how to quantify it somehow and come to terms with the tradeoffs needed. TANSTAAFL rules, and we need to decide what we want to get for what we pay.

Nevertheless, where there is life, there is hope. The solar system has resources in plenty that can be used to make life pleasant for all Earthlings. The Sun is an energy source for the next 4 billion years or so. The asteroids contain mineral resources in easily collectable form. Two other planets in the solar system can most likely be made habitable. All this is possible.

Technology misuse has caused problems, but at this stage, technology (applied science) is humanity's hope for the future. *Ad astra per aspera.* (Through work to the stars.)

PROBLEMS AND QUESTIONS

MULTIPLE CHOICE REVIEW

1. Landscapes on Earth are for the most part
   a. wild, at least far from cities.
   b. affected by humans only in cities and on farms.
   c. the same as they were before humans evolved.
   d. affected in some way by human influence.
   e. a human creation.
2. Alien species that travel to new habitats
   a. are always invasive.
   b. are likely to exist in the new habitat for some time before their invasive character is recognized.
   c. sometimes fail to thrive, and sometimes outperform native species.
   d. seldom are recognized as problems.
   e. do not ever adapt to a new habitat.

3. Which ecosystems are most likely to suffer successful invasion by alien species?
   a. Regrown forests
   b. Farm fields
   c. Plantations of trees that are all of the same species
   d. All the above ecosystems suffer successful invasions.
   e. None of the above ecosystems suffer successful invasions.
4. Which of the following is most dangerous in terms of microrisks?
   a. Traveling 100 km by personal automobile
   b. Traveling 1500 km by plane
   c. Traveling 1500 km by train
   d. Smoking one cigarette
   e. All choices are equivalent in risk.

5. Ecosystem services are not always valued in economic analysis. Which of the following ecosystem services is (are) actually valued in at least some economic analyses?
   a. Decomposition of wastes
   b. Renewal of soil fertility
   c. Cleaning of air
   d. Maintenence of biodiversity
   e. All the other choices are found in some economic analyses.

6. How do "amateur" assessments of risk differ from "expert" assessments?
   a. Amateur assessments are more likely to interpret involuntary risk as higher risk.
   b. Risks that are unfairly shared seem riskier to amateurs.
   c. Human-caused risk is interpreted as more risky than natural risk.
   d. Risks are assessed differently by amateurs in all the above ways.
   e. There are no differences between risks as assessed by amateurs and experts.

7. Which of the following people on a two-week vacation is most at risk for cancer?
   a. Hans took a train 200 km from his home to Frankfurt Airport, flew 8000 km from Germany to the United States, rented a car and drove 10,000 km, then went home as he had come.
   b. Allison stayed at home in a brick house and smoked two packs of cigarettes a day.
   c. James and his wife Effie (Effie smokes while James does not) drove 2000 km to and from Provo, Utah, to Yreka, California, to see the redwoods.
   d. Rick flew 2500 km to New York from his farm and its clapboard house and stayed in the brick townhouse home of his friend Sallie. Neither smoke, but they shared 1.5 L of wine at dinner each evening for the two weeks.
   e. Bob and Ann drive 130 km from their home and ride a tandem bike together 60 km each day in a 10-day rally, then return home.

8. Which of the following regulations makes the most economic sense?
   a. Asbestos exposure limits for construction workers
   b. A ban on untreated cotton nightshirts
   c. Exposure limits to dust for silo operators
   d. Mandating seat belts in seats of school buses
   e. Construction workers' exposure to cement

9. Which of the following would NOT represent appropriate technology in Niger?
   a. Better animal harnesses
   b. Improved tractors
   c. Improved seed drillers
   d. Only (b) and (c) represent appropriate technology in Niger.
   e. The choices (a) through (c) represent appropriate technology in Niger.

10. O'Neill habitats
    a. are viable alternatives to living on planets.
    b. can be made with today's technology.
    c. are at orbital locations with high likelihood of stability.
    d. are described by all of the above answers.
    e. are described by none of the above answers.

SHORT ANSWER

Explain whether the statement made is correct, and justify your answer.

11. All insects are harmful to their ecosystems.
12. Small risks may seem more risky than large risks under certain circumstances.
13. All Superfund sites must be cleaned to the same standard.
14. Alien species are usually aggressive invaders.
15. Ecosystem services cannot be quantified, so they should be totally left out of economic analyses.
16. Spaceship Earth can be self-sufficient in all resources.
17. Trash may be considered a resource.
18. Desertification has geographical effects beyond the local ones.
19. Sometimes a company spending money to control pollution can reap income that more than compensates for the expense.
20. Asbestos is so risky that it should always be removed from public buildings.

DISCUSSION QUESTIONS

21. Why is off-planet migration unlikely to help Earth deal with overpopulation? Explain using examples.
22. List reasons that might explain
    a. the success of some invading species.
    b. the failure of other invading species.
23. Ehrenfeld[119] argued that nonresources should be conserved. By this, he means that conservation cannot and should not rely on economic or ecological justifications. He thinks that each species on the Earth, evolved over millions of years, is natural art, and that a species lost is permanently gone. He asks: "What sort of change in the world view would favor the conservation of nonresources? Nothing less than a rejection of the heroic, Western ethic with its implicit denial of man's biological roots and evolved structure." Comment on Ehrenfeld's ideas.
24. Why might some choose a "soft" energy path, while others choose a "hard" path? Explain their respective reasons.
25. What are appropriate measures of risk? Judge how they might be used.
26. Global climate change has been suggested as a reason for societal changes or collapses in several instances. Why should drought or other climate-induced change cause social change? Explain how this could possibly occur.
27. Contrast "good" anthropogenic change with "bad" anthropogenic change by citing examples of each.
28. In recent reports,[315–320,324] overfishing is cited as a cause for species' numerical collapse and possible forthcoming extinction. In (b) and (c), explain the parallelism or lack of parallelism with fishery management.
    a. What does this imply about human management of fishery resources? Cite specific instances and consequences.
    b. Consider, for example, politicians dealing with mining.
    c. Consider, for example, politicians dealing with forestry.
29. Brown[155] reported that much toxic material is illegally dumped. Violators have loosened tank-truck valves and dumped contaminants along the road (a New York company spilled out polychlorinated biphenyls—PCBs—from a truck onto 270 mi of North Carolina highway).

What are the attitudes of company officials that could tolerate this atrocity? Is indiscriminate dumping a national pastime? Where do you dispose of trash and garbage on long car trips?

30. What is a species worth? How do we balance economics and ethics? Are there values other than economic that are important? Explain.

31. It is thought that the ancestors of most Europeans originally shared the Mongolian plateau with the mongolian peoples of Asia. It is also known that the Hittites, the Scythians, and the various Aryan invaders of India came from that region over quite a long time span. This area is now the Gobi desert. Advance scenarios for the historical migrations of these various peoples.

APPENDIX 1
Scientific Notation

KEY TERMS

multiplication / commutativity / division / equation / powers of ten / scientific notation / exa- / peta- / tera- / giga- / mega- / kilo- / milli- / micro- / nano- / pico- / femto- / atto- / centi- / hecto- / order of magnitude

A1.1 THE LANGUAGE OF SCIENCE

The language of science is mathematics. Most scientific ideas are presented most succinctly as equations. To talk about science, we have to recall some mathematical ideas and agree on a way of talking that will reduce confusion. In this section we briefly review multiplication, division, the meaning of an equation, and the manipulation of equations. We then continue to the main business at hand, scientific notation.

Multiplication

When we take two numbers and multiply them together, for example 3×6, we are forming the number that results from taking the second number and adding it to itself the first number of times, as in the example $6 + 6 + 6 = 18$. Alternatively, in **multiplication** we take the first number and add it to itself the second number of times, in the example $3 + 3 + 3 + 3 + 3 + 3 = 18$. The operation is more complex when the numbers are not integer, but the concept itself is unchanged. Note that it is a property of the real numbers (all the infinite number of numbers running in the continuum from $-|$ to $+|$) that they are **commutative;**

this means that the positions of the first and second and so on numbers do not matter to the result of the operation. Specifically, the operation of multiplication is commutative.

Division

Division occurs when we divide one number by another. We are asking how many times the denominator (the number on the bottom) fits into the numerator (the number on the top). For example, 18/3 is a shorthand for the number which results when 18 is broken up in units of 3; the result is 6. The operation might work this way: we begin with 18. Now, $18 - 3$ is 15, and 3 has fit once so far. $15 - 3$ is 12, and 3 fits another time, that is, twice so far. $12 - 3$ is 9, and 3 has so far fit $2 + 1 = 3$ times. In a similar way, $9 - 3$ is 6, with 4 threes so far; $6 - 3$ is 3. At last we find that 4 plus 2 is 6, the total number of 3s in 18.

Although a division such as 10.8/2.42 is mechanically more complicated, the procedure is the same; we ask how many times the number 2.42 fits into the number 10.8 or, how big 10.8 is when it is broken into units of length 2.42.

There is no restriction whatsoever on the numbers allowed in either position. For example, 2 divided by 5 (2/5) asks how big 2 is in units 5 big. This is, of course, a number smaller than one and we have a way of expressing it as a decimal. In the example, 2/5 is 0.4. The extra zero in front of the decimal point does not matter. It was simply put in that position for clarity.

Meaning and Manipulation of Equations

An **equation** is a statement. It says that one collection of numbers (or symbols) has the same value as another collection of numbers (or symbols). An example of an equation containing numbers is $18/3 = 6$. This equation states that the numerical value on the left-hand side of the equal sign is the same as the numerical value on the right-hand side of the equal sign.

An equation is invariably such a statement of numerical equivalence. However, it may appear to be more subtle in some cases because of unfamiliarity with the content of the equation, when that equation contains symbols rather than numbers. An example of a simple equation containing both symbols and numbers is $2.5z = 10$. In this example, the numerical value of the left-hand side of the equation must be the same as the numerical value of the right-hand side of the equation, just as before. In this particular example, the numerical value of z is fixed; only one number when substituted for z will leave the numerical value of the right hand side the same as that of the left hand side. This number is 4, because 4×2.5 is $2.5 + 2.5 + 2.5 + 2.5$, or 10.

There is another way to achieve the same result. It may occur to us that if we divide both sides of the equation $2.5z = 10$ by the same number, the resulting statement is still an equation. By that I mean that the number on the right-hand side and the number on the left-hand side have the same numerical value. The number will in general be different from the number that was previously the equivalent on both sides of the equal sign. Suppose we divide both sides of $2.5z = 10$ by the number 2.5. On the left-hand side, 2.5 $z/2.5$ asks how many times 2.5 fits into 2.5 z. By our definitions, this is simply z. On the right-hand side 10/2.5 is just 4 by our previous discussion of division. We may then restate the equation $2.5z = 10$ as $z = 4$.

When an equation has mixed numbers and symbols on both sides, it is a general statement of relation. For example, the equation $v^2 = 2gx$ states that the product of v with itself is the same as 2 times g times x. If we are given that g is some number (say, 10), we can simplify the right hand side: $2gx = 2(10)x = 20x$. Here we have said that if g is equal to 10, t'e equation is the same whether the symbol or the number be used. Our example equation now takes the form $v^2 = 20x$. Since we do not know a specific value for v or for x, we cannot specifically solve the equation numerically, that is, state the numerical values of v and x the way we did the value for z in the previous example. What we can say is that there is a relation between the *possible*; values of v and the *possible* values of x. If we knew x were 20, for example, we would know v to be 20; if x were zero, v would be zero; and so on. There are an infinite number of possible values of x for the infinite possible values of v, two for each value of x. There are two v's for each x because the sign of v is not determined.

The most concise way to present this infinite amount of information is in a graph, which can show some sample of the behavior of v as x is chosen differently. The graph for our example is shown in Figure A1.1. Note that both positive and negative values of v give the same value of x.

Equations in physics are often shorthand ways of presenting just such an infinitude of information. If x is the symbol for the height in meters above the ground, and v is the symbol for speed in meters per second, then the equation $v^2 = 2(10 \text{ m/s}^2)\, x$ describes their relation to one another. The relation is more important than the numerical values because the equation is a completely general statement about any body's speed when that body is any arbitrary distance x above the ground.

We also may solve equations which are general relations to isolate the dependence of one of the unknown quantities in terms of the others. For example, we may have an equation $ab = cd/e$. Suppose we wish to solve for c in terms of the other unknown quantities. This involves manipulations of the equation given in ways that do not change its identity as an equation. We multiply both sides of the equation by e: $abe = cd$. Then we divide both sides of the

FIGURE A1.1

The graph of a quadratic equation.

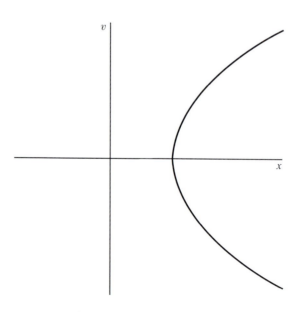

equation by d: $abc/d = c$. By this process, the desired unknown, c, has been isolated. We often go through such manipulations in this book.

Powers of Ten

Numbers are expressible as powers of other numbers. By the power of a number we mean the number of times it multiplies itself; 10 to the power 2 (10^2) means 10 times 10 ($10^2 = 10 \times 10$). An interesting property of such numbers is that two numbers that are expressed this way are easily multiplied. As an example, $10^2 \times 10^3$ is a shorthand for $((10 \times 10) \times (10 \times 10 \times 10))$ which is otherwise known as 10^5. That is, we may multiply two numbers of this form by simply adding the exponents: $5 = 2 + 3$. The word exponent is a synonym for the power of a number; in 10^A, A is the power or exponent of 10. Although we have so far restricted ourselves to integer exponents, there is nothing to prevent noninteger exponents. These are discussed in Chapter 8 of this book and in Appendix 3. In general,

$$10^A \times 10^B \text{ is } 10^{A+B}:$$

$$\underbrace{(10 \times 10 \times \cdots \times 10)}_{A \text{ tens}} \times \underbrace{(10 \times \cdots \times 10)}_{B \text{ tens}}$$

$$= \underbrace{10 \times 10 \times \cdots \times 10}_{A + B \text{ tens}}$$

These numbers, 10 and its powers, arise naturally in our number system. Because we have ten fingers, and our ancestors used them to keep track of numbers of things, we have ten unit numbers, ten units of tens of units, ten units of hundreds of units, and so forth. Our numbers are implicitly expressed in terms of **powers of ten**.

Imagine drawing a line across the page and dividing it into columns under the heading units (1), tens (10), hundreds (10^2), thousands (10^3), and so on. The year of the first notice of the energy crises by the general public, 1973, is, in terms of entries under this column, 1 in the thousands column, 9 in the hundreds column, 7 in the decades column, and 3 in the units column:

10^5	10^4	10^3	10^2	10	1
		1	9	7	3

Note that this just reproduces the number 1973 as the column entries. In fact, the number 1973 is just the column-entry shorthand for one thousand, nine hundreds, seven decades, and three units. We could also write 1973 as $1 \times 10^3 + 9 \times 10^2 + 7 \times 10 + 3$. Any number in our number system can be expressed in one of these alternative ways.

So far we have only discussed multiplication of these numbers. Division is also simple. Consider the division of 10^3 by 10, which we will write as the multiplication $10^3 \times (1/10)$. The result with this rewriting is easily obtained:

$$(10 \times 10 \times 10) \times (1/10) = (10 \times 10) \times (10 \times 1/10)$$
$$= (10 \times 10) \times 1 = 10^2.$$

We may follow this pattern for any number of the form 10^A:

$$10^A/10 = 10^A \times (1/10) = 10^{A-1+1} \times (1/10)$$
$$= 10^{A-1} \times 10 \times 1/10 = 10^{A-1}.$$

It is clear that if we wish to follow the pattern $10^A \times 10^B = 10^{A+B}$, we must write $1/10$ as 10^{-1}. It follows that $1/100 = 1/(10 \times 10)$ is 10^{-2}, $1/10^3 = 10^{-3}$, and so forth.

Now suppose we were to multiply 100 by $1/100$. The result is of course 1. If we use our notation, though, $10^2 \times 10^{-2} = 10^{2-2} = 10^0$. In order to continue to use the rule, we must choose $10^0 = 1$. With this completion, we can write our column headings more fully;

$$\ldots \quad 10^5 \quad 10^4 \quad 10^3 \quad 10^2 \quad 10^1 \quad 10^0 \quad 10^{-1} \quad 10^{-2} \quad \ldots$$

Suppose we filled in a number in our columns

	10^1	10^0	10^{-1}	10^{-2}	
$\ldots$	1	5	2	5	$\ldots$

Had we filled only the two columns under 10^1 and 10^0, we would know how to write the number: 15. We can't write the number 1525, because that would mean 1×10^3 plus $\ldots$ instead of 1×10^1 plus. $\ldots$ In the scheme we have adopted, position matters. To supplement the original scheme and allow translation of the column headings (to eliminate some of the clumsiness), we add a positional indicator: the decimal point. The decimal point is the indicator of the shift from numbers whose size is greater than 1, to numbers whose size is smaller than 1. In the case of the example given above, the number is written 15.25. The decimal point thus marks the switch from powers greater than zero to powers less than zero.

Scientific Notation

In an effort to preserve clarity in the transmission of scientific information, scientists have made an agreement to use a special way of writing numbers. A number such as the speed of light, 300,000,000 m/s, is so large it boggles the mind. It would be very easy to add or lose a zero in writing such a number. To eliminate such possible confusion as could result, all numbers of interest are written in **scientific notation** as $M \times 10^N$, where M is a number between 1.0 and 9.999 $\ldots$, and N is an integer exponent.

The speed of light in this notation is 3.0×10^8 m/s. There is a greatly reduced possibility of error here.

Unit Prefixes

In ordinary speech we have a shorthand for large numbers. They are grouped and given names such as thousands, billions, and so on. As in common speech, it is usual to give special names to factors 10^3 apart in magnitude. Large international conventions of scientists agree on the names to be given to multipliers, and assign symbols to stand for the size. We shall here illustrate the most common names by their use as prefixes for lengths.

The meter is the standard of length for every country in the world, and we shall explore its prefixes:

10^{18} m	Em	**exa**meter
10^{15} m	Pm	**peta**meter
10^{12} m	Tm	**tera**meter
10^9 m	Gm	**giga**meter

10^6 m	Mm	**mega**meter
10^3 m	km	**kilo**meter
10^0 m	m	meter
10^{-3} m	mm	**milli**meter
10^{-6} m	μm	**micro**meter
10^{-9} m	nm	**nano**meter
10^{-12} m	pm	**pico**meter
10^{-15} m	fm	**femto**meter
10^{-18} m	am	**atto**meter

Two other prefixes are in common use. **Centi-** means one-one hundredth, and is commonly used with meter: a centimeter is slightly less than a 1/2 inch. **Hect(o)-** means one hundred. Areas are commonly given in hectares (10^4 m²). A hectare is slightly less than 2 1/2 acres (2.471 acres). Occasionally, volumes are given in hectoliters.

Order of Magnitude

An **order of magnitude** is a factor of 10^1. The important unit names are separated by three orders of magnitude, that is by a factor of $(10^1)^3 = 10^3$.

A human ovum may be about 1 μm in diameter. The human adult has a typical height of 1.5 to 2 m. Thus, we say that the size of an adult human is about 6 orders of magnitude larger than the ovum from which it developed. A newborn child is, say, about 50 cm [20 in.] long. Thus an adult is only a factor of 3 or 4 longer than an infant. Since this is not a factor of 10, we may say that the adult and the baby have lengths of the about the same order of magnitude.

A child at birth typically weights 5 to 9 lb and an adult 100 to 200 lb (this depends on the adult's sex). Since the ratio of weights is 10 to 40, we may say that an adult's weight is about an order of magnitude greater than a child's.

Beware the common incorrect usage of order of magnitude for a factor of 2.

PROBLEMS AND QUESTIONS

SHORT ANSWER
(Explain whether the statement made is correct and justify your answer.)
1. A meter is an order of magnitude longer than a centimeter.
2. A gallon of milk contains an order of magnitude more milk than a pint.
3. The number 13×10^{-7} is expressed in scientific notation.
4. Multiplication is just repeated addition.

MULTIPLE CHOICE REVIEW
5. The population of Mauritius is 900,000. If infant mortality is 70 per 1000 infants born, and the birthrate is 31 per 1000 population, the yearly number of dead infants is
   a. 1.95×10^2.
   b. 2.03×10^7.
   c. 3.99×10^4.
   d. 4.15×10^8.
   e. none of the above.

PROBLEMS
6. $10^2 \times 10^5 =$
7. $10^{17} \times 10^{-24} =$
8. $10^{-5} \times 10^{-10} =$
9. $10^4/10^{-5} =$
10. $10^6 \times 10^{-6} =$
11. The Republic of South Africa had an estimated 1990 population of 38,880,000. Its people belong to the following "racial" groups: blacks, 67%; whites, 20%; "colored," 10%; "Asians," 3%. In addition, 60% of the whites speak Afrikaans and the rest speak English. Express all these numbers in scientific notation and calculate the number of people in each subgroup (expressed in scientific notation).

12. Write in scientific notation:
    a. 250,000,000 (U.S. population)
    b. 1/10,000
    c. 3,617,204 (U.S. area, in mi^2)
    d. 96,981 (Oregon area, in mi^2)
    e. 668,700 (number of American Indians in United States in 1970)

13. A person can run 100 m in about 10 s. A cheetah can run about 300 m; a jet plane can go about 2.7 km; and an Apollo spacecraft may go 20 km in about the same time. Express the respective speeds in meters per second in scientific notation. Express the speed of light in the same units in scientific notation. Calculate the ratios of the speeds to the speed of light and express in scientific notation. The speed of light is 300,000,000 m/s.

14. Solve the equation $F = qvB$ for v.

15. Given the equation: $mgh_1 + 1/2 mv_1^2 = mgh_2$, solve for v_1.

DISCUSSION QUESTIONS

16. For what reasons do you think it would be useful to develop scientific notation? Need it have been in powers of ten?

17. Of what utility is the idea of "order of magnitude?"

18. Use the ideas introduced in the section on division to show that 1 over a fraction is the reciprocal of the fraction: $1/(a/b) = b/a$.

APPENDIX 2

Logarithms

When noninteger real numbers appear as powers of ten, the resulting number can be *any* number. As we vary the exponent of 10 between 0 and 1, all numbers between 1 and 10 result. There is a one-to-one correspondence between the numbers in the interval 0 to 1 and the numbers in the interval 1 to 10.

Such a table is clearly useful if only because of the difficulties of translation of numbers such as 10^{ct}. In fact such tables were constructed long ago: they are called tables of logarithms. The logarithm (log for short) of a number is the power of some base number, which corresponds one-to-one with that number. The number ct is the log to the base 10 of 10^{ct} (we write $ct = \log_{10} 10^{ct}$). Since $2^3 = 8$, we could say that the log of 8 to the base 2 is 3. Any number whatsoever can be used as a base. The most commonly used bases are 10 and $e = 2.718\ldots$ The former base we have considered in some detail; the latter base is noninteger, and logarithms expressed to this base are known as natural logarithms. In this natural base, equations of the form $\Delta N = aN \, \Delta t$ have the simple solution $N(t) = N_0 \, e^{at}$.

A table of logarithms to the base 10 may be found at the end of this appendix. In Table A2.1, we present selected integer logarithms. The log tables contain only logs of numbers between 1 and 10. It is easy to understand why these are all that are necessary. Consider the multiplication of two numbers of the form 10^A and 10^B. The resultant number, R, is

$$R = 10^A \times 10^B = 10^{A+B}.$$

By our definition of $\log_{10}$, then

$$\log_{10} R = \log_{10} 10^A + \log_{10} 10^B = A + B.$$

This is simply a reexpression of the rule discussed above: the log of the product of two numbers is the sum of their respective logs. Thus, the $\log_{10}$ of a number such as 62 is

$$\log_{10} 62 = \log_{10}(6.2 \times 10) = \log_{10} 6.2 + \log_{10} 10$$

$$= \log_{10} 6.2 + 1.$$

We need only look up the number representing 6.2 in the log table.

TABLE A2.1

Selected integer logarithms

$10^{-4} = 1/10{,}000$	$\log_{10} 10^{-4} = -4$
$10^{-3} = 1/1000$	$\log_{10} 10^{-3} = -3$
$10^{-2} = 1/100$	$\log_{10} 10^{-2} = -2$
$10^{-1} = 1/10$	$\log_{10} 10^{-1} = -1$
$10^{0} = 1$	$\log_{10} 10^{0} = 0$
$10^{1} = 10$	$\log_{10} 10^{1} = 1$
$10^{2} = 100$	$\log_{10} 10^{2} = 2$
$10^{3} = 1000$	$\log_{10} 10^{3} = 3$
and so on.	

EXPONENTIAL GROWTH

How is it possible that $N(t) = N_0 \, 10^{ct}$ is a solution to $N = aN \, \Delta t$? Let us suppose that it is a solution, and see if we can calculate ΔN. We write

$$\Delta N = N(t + \Delta t) - N(t)$$
$$= N_0 \, 10^{c(t+\Delta t)} - N_0 \, 10^{ct}.$$

Using the fact that $10^{ct+c\Delta t} = 10^{ct} \times 10^{c\,\Delta t}$, we gather common factors:

$$\Delta N = N_0 \, 10^{ct} (10^{c\,\Delta t} - 1).$$

If Δt were ~zero (that is, the time interval is small), the factor in parentheses would be near zero because $10^0 = 1$. So $\Delta N = 0$ if $\Delta t = 0$. In fact, for small Δt, ΔN is proportional to Δt, as seems plausible from our argument:

$$\Delta N = N_0 \, 10^{ct}(c[log_e 10] \, \Delta t).$$

Since $N_0 \, 10^{ct}$ is just $N(t)$,

$$\Delta N = (c \, log_e 10)N(t) \, \Delta t.$$

Therefore, $c = a/log_e 10$ and

$$\Delta N(t) = aN_0 \, e^{at}.$$

Here, e represents a special base in which the solution to $\Delta N = aN\Delta t$ is particularly simple.

EXPONENTIAL DECAY—RADIOACTIVE DECAY

Radioactive decay in nuclei is a random occurrence; whether a particular nucleus will undergo decay is an unanswerable question. However, if a substantial number of the same sort of radioactive nuclei are gathered together, we can predict that a certain percentage will decay spontaneously in a given time. That is, ΔN, the number of nuclei that have decayed in time Δt, is proportional to N. Again, we get a geometrical (or exponential) change. Such a situation is described by

$$N(t) = N_0 \, e^{-\lambda t};$$

because of the minus sign, N decreases with time instead of growing, as in the examples in the text. The time required for half the nuclei in a group to decay is called the half-life of the nucleus. A free neutron, for example, has a half-life of about ten minutes. If we begin with 1,000,000 free neutrons when the clock starts ticking, we will have only 500,000 ten minutes later. In a further ten minutes this is halved to 250,000; and in the next ten minutes it is halved again to 125,000. This will continue until there are too few neutrons left for us to predict decay with any certainty.

Many nuclei have half-lives less than ten minutes and many others have very long half-lives. Particles involved in so-called strong nuclear interactions may have lifetimes as short as 10^{-21} s. The proton has been predicted by Grand Unified Theories to have a lifetime in excess of 10^{31} years, and recent experiments have shown that it must actually be greater than 10^{32} years.

To find the $\log_{10}$ of a number such as 0.037, we rewrite 0.037 as 3.7×10^{-2}:

$$\log_{10} 0.037 = \log_{10}(3.7 \times 10^{-2}) = \log_{10} 3.7 + \log_{10}(10^{-2})$$
$$= \log_{10} 3.7 + (-2) = \log_{10} 3.7 - 2.$$

Some further examples are presented in the box.

Logarithms are useful because they provide a way for us to graph exponential, or geometrical, growth easily. Since the logarithm of a number changes from 0 to 1 as the number changes from 1 to 10, we may choose to make an axis of our graph paper proportional on a linear scale to the log of the number. This provides a nonlinear, logarithmic, scale for the number itself.

EXAMPLES OF CALCULATIONS WITH LOGARITHMS

1. $\log_{10}(10^{-3} \times 10^{-4}) = \log_{10} 10^{-3} + \log_{10} 10^{-4}$
$$= (-3) + (-4) = -7$$

2. $32 \times 98 = ?$
$$\log_{10}(32 \times 98) = \log_{10} 32 + \log_{10} 98$$
$$= \log_{10}(3.2 \times 10^1 + \log_{10}(9.8 \times 10^1)$$
$$= \log_{10} 3.2 + 1 + \log_{10} 9.8 + 1$$

Now read off the logs from Table A2.2:

$$\log_{10} 3.2 = 0.5051$$

$$\log_{10} 9.8 = 0.9912$$

Then, substituting,

$$\log_{10}(32 \times 98) = 0.5051 + 1 + 0.9912 + 1 = 3.4963.$$

Thus the answer to 32×98 is $10^{3.4963} = 10^3 \times 10^{0.4963} = 1000 \times 10^{0.4963}$. Looking up 0.4963 in the table of logarithms, Table A2.2, we find t is between 3.13 and 3.14, closer to 3.14. For most purposes, this answer, 3.140 is good enough. Note that the answer resulting from multiplication is 3.136. We could actually approximate this rather well by interpolation: 0.4963 is 8/14 of the way from the log of 3.13 to the log of 3.14; $8/14 = 0.57 \sim 0.6$. Thus, the answer is about $1000 \times 3.136 = 3136$.

3. $32/98 = ?$
Since division is the inverse of multiplication (see Appendix 1), the rule is that the log of one number dividing another is the respective difference of the logs:

$$\log_{10}(32/98) = \log_{10} 32 - \log_{10} 98$$
$$= (1.5051) - (1.9912) = -0.4861,$$

from the previous example. It is not clear at first how to handle this number until we recognize that, as with the last example, we eventually wish to have the number in the form of scientific notation before we can decipher it. Thus we write $10^{-0.4861}$ as $10^{-1+0.5139} = 10^{-1} \times 10^{0.5139}$. In this form, we see that the number having the log of 0.5139 is between 3.26 and 3.27, closer to 3.27 ($7/13 = 0.54$ of the way). Hence, we get 3.265×10^{-1}. By long division we get 0.32653...

TABLE A2.2

Table of logarithms—numbers from 1.00 to 9.99

1.0	0	1	2	3	4	5	6	7	8	9
.0	.000	.004	.009	.013	.017	.021	.025	.029	.033	.037
.1	.041	.045	.049	.053	.057	.061	.064	.068	.072	.076
.2	.079	.083	.086	.090	.093	.097	.100	.104	.107	.111
.3	.114	.117	.121	.124	.127	.130	.134	.137	.140	.143
.4	.146	.149	.152	.155	.158	.161	.164	.167	.170	.173
.5	.176	.179	.182	.185	.188	.190	.193	.196	.199	.201
.6	.204	.207	.210	.212	.215	.217	.220	.223	.225	.228
.7	.230	.233	.236	.238	.241	.243	.246	.248	.250	.253
.8	.255	.258	.260	.262	.265	.267	.270	.272	.274	.276
.9	.279	.281	.283	.286	.288	.290	.292	.294	.297	.299

2.0	0	1	2	3	4	5	6	7	8	9
.0	.301	.303	.305	.307	.310	.312	.314	.316	.318	.320
.1	.322	.324	.326	.328	.330	.332	.334	.336	.338	.340
.2	.342	.344	.346	.348	.350	.352	.354	.356	.358	.360
.3	.362	.364	.365	.367	.369	.371	.373	.375	.377	.378
.4	.380	.382	.384	.386	.387	.389	.391	.393	.394	.396
.5	.398	.400	.401	.403	.405	.407	.408	.410	.412	.413
.6	.415	.417	.418	.420	.422	.423	.425	.427	.428	.430
.7	.431	.433	.435	.436	.438	.439	.441	.442	.444	.446
.8	.447	.449	.450	.452	.453	.455	.456	.458	.459	.461
.9	.462	.464	.465	.467	.468	.470	.471	.473	.474	.476

3.0	0	1	2	3	4	5	6	7	8	9
.0	.477	.479	.480	.307	.483	.484	.486	.487	.489	.490
.1	.491	.493	.494	.496	.497	.498	.500	.501	.502	.504
.2	.505	.507	.508	.509	.511	.512	.513	.515	.516	.517
.3	.519	.520	.521	.522	.524	.525	.526	.528	.529	.530
.4	.531	.533	.534	.535	.537	.538	.539	.540	.542	.543
.5	.544	.545	.547	.548	.549	.550	.551	.553	.554	.555
.6	.556	.558	.559	.560	.561	.562	.563	.565	.566	.567
.7	.568	.569	.571	.572	.573	.574	.575	.576	.577	.579
.8	.580	.581	.582	.583	.584	.585	.587	.588	.589	.590
.9	.591	.592	.593	.594	.595	.597	.598	.599	.600	.601

4.0	0	1	2	3	4	5	6	7	8	9
.0	.602	.603	.604	.605	.606	.607	.609	.610	.611	.612
.1	.613	.614	.615	.616	.617	.618	.619	.620	.621	.622
.2	.623	.624	.625	.626	.627	.628	.629	.630	.631	.632
.3	.633	.634	.635	.636	.637	.638	.639	.640	.641	.642
.4	.643	.644	.645	.646	.647	.648	.649	.650	.651	.652
.5	.6532	.6542	.6551	.6561	.6571	.6580	.6590	.6599	.6609	.6618
.6	.6628	.6637	.6646	.6656	.6665	.6675	.6684	.6693	.6702	.6712
.7	.6721	.6730	.6739	.6749	.6758	.6767	.6776	.6785	.6794	.6803
.8	.6812	.6821	.6830	.6839	.6848	.6857	.6866	.6875	.6885	.6893
.9	.6902	.6911	.6920	.6928	.6937	.6946	.6955	.6964	.6972	.6981

5.0	0	1	2	3	4	5	6	7	8	9
.0	.6990	.6998	.7007	.7016	.7024	.7033	.7042	.7050	.7059	.7067
.1	.7076	.7084	.7093	.7101	.7110	.7118	.7126	.7135	.7143	.7152
.2	.7160	.7168	.7177	.7185	.7193	.7202	.7210	.7218	.7226	.7235
.3	.7243	.7251	.7259	.7267	.7275	.7284	.7292	.7300	.7308	.7316
.4	.7324	.7332	.7340	.7355	.7356	.7364	.7372	.7380	.7388	.7396
.5	.7404	.7412	.7419	.7427	.7435	.7443	.7451	.7459	.7466	.7474
.6	.7482	.7490	.7497	.7505	.7513	.7520	.7528	.7536	.7543	.7551
.7	.7559	.7566	.7574	.7582	.7589	.7597	.7604	.7612	.7619	.7627
.8	.7634	.7642	.7649	.7657	.7664	.7672	.7679	.7686	.7694	.7701
.9	.7709	.7716	.7723	.7731	.7738	.7745	.7752	.7760	.7767	.7774

(Continued)

TABLE A2.2

(Continued)

6.0	0	1	2	3	4	5	6	7	8	9
.0	.7782	.7789	.7796	.7803	.7810	.7818	.7825	.7832	.7839	.7846
.1	.7853	.7860	.7868	.7875	.7882	.7889	.7896	.7903	.7910	.7917
.2	.7924	.7931	.7938	.7945	.7952	.7959	.7966	.7973	.7980	.7987
.3	.7993	.8000	.8007	.8014	.8021	.8028	.8035	.8041	.8048	.8055
.4	.8062	.8069	.8075	.8082	.8089	.8096	.8102	.8109	.8116	.8122
.5	.8129	.8136	.8142	.8149	.8156	.8162	.8169	.8176	.8182	.8189
.6	.8195	.8202	.8209	.8215	.8222	.8228	.8235	.8241	.8248	.8254
.7	.8261	.8267	.8274	.8274	.8287	.8293	.8299	.8306	.8312	.8319
.8	.8325	.8331	.8338	.8344	.8351	.8357	.8363	.8370	.8376	.8382
.9	.8388	.8395	.8401	.8407	.8414	.8420	.8426	.8432	.8439	.8445

7.0	0	1	2	3	4	5	6	7	8	9
.0	.8451	.8457	.8463	.8470	.8476	.8482	.8488	.8494	.8500	.8506
.1	.8513	.8519	.8525	.8531	.8537	.8543	.8549	.8555	.8561	.8567
.2	.8573	.8579	.8585	.8591	.8597	.8603	.8609	.8615	.8621	.8627
.3	.8633	.8639	.8645	.8651	.8657	.8663	.8669	.8675	.8681	.8686
.4	.8692	.8698	.8704	.8710	.8716	.8722	.8727	.8733	.8739	.8745
.5	.8751	.8766	.8762	.8768	.8774	.8779	.8785	.8791	.8797	.8802
.6	.8808	.8814	.8820	.8825	.8831	.8837	.8842	.8848	.8854	.8859
.7	.8865	.8871	.8876	.8882	.8887	.8893	.8899	.8904	.8910	.8915
.8	.8921	.8932	.8932	.8938	.8943	.8949	.8954	.8960	.8965	.8971
.9	.8976	.8982	.8987	.8993	.8998	.9004	.9009	.9015	.9020	.9025

8.0	0	1	2	3	4	5	6	7	8	9
.0	.9031	.9036	.9042	.9047	.9053	.8058	.9063	.9069	.9074	.9079
.1	.9085	.9090	.9096	.9101	.9106	.9112	.9117	.9122	.9128	.9133
.2	.9138	.9143	.9149	.9154	.9159	.9165	.9170	.9175	.9180	.9186
.3	.9191	.9196	.9201	.9206	.9212	.9217	.9222	.9227	.9232	.9238
.4	.9243	.9248	.9253	.9258	.9263	.9269	.9274	.9279	.9284	.9289
.5	.9294	.9299	.9304	.9309	.9315	.9320	.9325	.9330	.9335	.9340
.6	.9345	.9350	.9355	.9360	.9365	.9370	.9375	.9380	.9385	.9390
.7	.9395	.9400	.9405	.9410	.9415	.9420	.9425	.9430	.9435	.9440
.8	.9445	.9450	.9455	.9460	.9465	.9369	.9375	.9479	.9484	.9489
.9	.9494	.9499	.9504	.9509	.9513	.9518	.9523	.9528	.9533	.9538

9.0	0	1	2	3	4	5	6	7	8	9
.0	.9542	.9547	.9552	.9557	.9562	.9566	.9571	.9576	.9581	.9586
.1	.9590	.9595	.9600	.9605	.9609	.9614	.9619	.9624	.9628	.9633
.2	.9638	.9643	.9647	.9652	.9657	.9661	.9666	.9671	.9675	.9680
.3	.9685	.9689	.9694	.9699	.9703	.9708	.9713	.9717	.9722	.9727
.4	.9731	.9736	.9741	.9745	.9750	.9754	.9759	.9763	.9768	.9773
.5	.9777	.9782	.9786	.9791	.9795	.9800	.9805	.9809	.9814	.9818
.6	.9823	.9827	.9832	.9836	.9841	.9845	.9850	.9854	.9859	.9863
.7	.9868	.9872	.9877	.9881	.9886	.9890	.9894	.9899	.9903	.9908
.8	.9912	.9917	.9921	.9926	.9930	.9934	.9939	.9943	.9948	.9952
.9	.9956	.9961	.9965	.9969	.9974	.9978	.9983	.9987	.9991	.9996

PROBLEMS AND QUESTIONS

MULTIPLE CHOICE REVIEW

1. The logarithm of 4.2 is 0.6232. The logarithm of 3 is 0.4771. The logarithm of $420 \times (3 \times 10^{-4})$ is
   a. -0.8997
   b. 0.1260
   c. 0.7148
   d. 3.1003
   e. 7.1003

PROBLEMS
(Use the logarithm tables, Table A2.2, to solve)

2. $12 \times 20 =$

3. $(6 \times 10^8) \times (1.6 \times 10^{27}) =$

4. $(7 \times 10^{-8}) \times (7 \times 10^{20}) =$

5. $(9 \times 10^{-17}) \times (8 \times 10^{+21}) =$

6. $(800 \text{ million}) \times (0.1\%) =$

APPENDIX 3

Understanding Tabular Data

Often the information a person needs is in tables full of numbers that seem so overwhelming that the search for the information is given up in disgust. Organizing data into tabular form is certainly one way to try to put information into a form suitable for interpretation. Such attempts usually are successful only when the table is sufficiently short or when only modest correlations are presented.

It is often possible to obtain relevant information from government sources in tabular form. In this section, we tackle the question of how to interpret and use a barrage of tables to examine questions of public interest.

In this appendix Tables 20, 21, 29, 30, 31, and 32 are reproduced from the Government-commissioned report *Patterns of Energy Consumption in the United States*.[1] The tables indicate the extent of saturation (percentage of American households owning a particular item) and of energy use for the years 1960 to 1968. The amount of information in these tables is overwhelming. How can any sense be made of the data?

Perhaps the first thing to do is to try to understand the trend. We can move our eyes down the columns one at a time. Looking at Table 20 for example, we see that the number of households increased between 1960 and 1968 steadily. So too did the number of gas and electric water heaters. By looking at the saturation column, we discover that both gas and electric water heating use grew faster than the number of households. In addition, we see that gas water heaters are growing in use more rapidly than electric water heaters. Rapid scanning of the other tables gives much the same trend.

You now have some information. Is there more information in the tables? The answer is yes. If you were an appliance manufacturer, these tables could tell you which items enjoy the biggest possibilities for growth—namely, those with the smallest market saturation. You would compare and then make a decision on the basis of that comparison. Such work is tedious, indeed, even with the number of tables to be considered here. In a real-life situation, more alternatives would probably be considered. The problems is one of *organization* and *presentation*.

By constructing comparison tables from the original tables, we can make the requisite comparisons, but this also is a tedious task and the comparisons are still difficult to make and interpret.

The simplest solution is to use graphs to organize the data in the tables. We might plot the saturation of many items on one sheet of graph paper. This allows the observers to use their eyes to make visual comparisons more easily than if they were reading the same data from tables. With this method we can compare saturation and observe the trends in saturation with time.

TABLE A3.1

(Table 20) Saturation of water heaters in residences, 1960–68

	Numbers of households[*] (millions)	Percentage saturation[**]		Number of water heaters in use (millions)	
		Gas	Electric	Gas	Electric
1960	53.0	54%	20%	28.6	10.6
1961	53.3	56	21	29.8	11.2
1962	54.7	58	22	31.7	12.0
1963	55.2	59	22	32.7	12.2
1964	56.0	61	23	34.2	12.9
1965	57.3	63	23	36.1	13.2
1966	58.1	65	23	37.8	13.7
1967	58.8	66	24	38.8	14.1
1968	60.4	68	24	41.2	14.5

[*] *Statistical Abstract of the United States*, various years; household count as of March of years shown; mobile homes are included.

[**] *Merchandising Week*; 1960 Census, Gas Appliance Manufacturers Association, and SRI estimates.

TABLE A3.2

(Table 21) Residential energy consumption for water heating

	Electricity				Gas			
			Total consumption			Unit consumption[**] (millions of Btu)	Total consumption[†] (trillions of Btu)	Total energy consumption (trillions of Btu)
Year	Units (millions)	Consumption[*] (kWh)	(billions of kWh)	(trillions of Btu)	Units (millions)			
1960	10.6	4272	45.3	429	28.6	25.5	730	1159
1961	11.2	4272	47.8	444	29.8	25.5	761	1205
1962	12.0	4290	51.6	474	31.7	25.5	808	1282
1963	12.2	4300	51.7	480	32.7	25.8	846	1326
1964	12.9	4320	55.6	512	34.2	26.0	889	1401
1965	13.2	4390	57.8	532	36.1	26.4	952	1484
1966	13.7	4400	60.1	563	37.8	26.8	1013	1576
1967	14.1	4420	62.4	579	38.8	27.0	1049	1628
1968	14.5	4490	65.4	613	41.2	27.2	1125	1738
		12,816 (1960)				7508 (1960)		

[*] Source is Edison Electric Institute with SRI estimates.

[**] Source is American Gas Association with SRI estimates.

[†] Both natural and liquefied petroleum gas.

TABLE A3.3

(Table 29) Saturation of television sets in residences (thousands of units)

	Sales[*]		Replacements[**]		Sets in use	
	Monochrome	**Color**	**Monochrome**	**Color**	**Monochrome**	**Color**
Base †					44,924	1389
1960	5605	117	2242	47	48,287	1459
1961	6047	144	2419	58	51,915	1545
1962	6460	438	2584	175	55,791	1808
1963	7141	749	2856	300	60,076	2257
1964	8542	1541	3417	616	65,201	3182
1965	8954	2827	3582	1131	70,573	4878
1966	7904	5549	3162	2220	75,315	8207
1967	5384	6496	2154	2598	78,545	12,105
1968	6296	7865	2518	3146	82,323	16,824
1969	7270	6607	2908	2643	86,685	20,788

[*] Source is Current Industrial Reports, Bureau of the Census.

[**] Assumed at 40% of sales.

† Source is National Survey of Television Sets in U.S. households, June 1967, Advertising Research Foundation.

TABLE A3.4

(Table 30) Residential energy consumption for television

	Monochrome sets			Color sets				
							Total energy consumption	
Year	**Number of units (millions)**	**Unit consumption[*] (kWh)**	**Total consumption (billions of kWh)**	**Number of units (millions)**	**Unit consumption (kWh)**	**Total consumption (billions of kWh)**	**(billions of kWh)**	**(trillions of Btu)**
1960	48.2	345	16.6	1.5	450	0.7	17.3	163
1961	51.9	345	17.9	1.5	450	0.7	18.6	172
1962	55.8	346	19.3	1.8	455	0.8	20.1	186
1963	60.1	349	21.0	2.3	460	1.1	22.1	203
1964	65.2	350	22.8	3.2	465	1.5	24.3	225
1965	70.6	352	24.9	4.9	470	2.3	27.2	251
1966	75.3	356	26.8	8.2	475	3.9	30.7	288
1967	78.5	359	28.2	12.1	482	5.8	34.0	315
1968	82.3	360	29.6	16.8	490	8.2	37.8	352

[*] Source is Edison Electric Institute with SRI estimates.

It is clear from Figure A3.1 that there is really no possibility of great sales of refrigerators, ranges, or washing machines. You sell only to new families or to replace old units. Color TV, air conditioners, electric dryers, and dishwashers offered (as of 1968) considerable scope for increasing market penetration. There were great possibilities for sales to families not owning these items. Notice that many households had two black and white TVs or one color and one black and white TV.

TABLE A3.5

(Table 31) Saturation of air-conditioning units in residences, (thousands of units) 1960–68

	Electric room air conditioners				Electric central air conditioners				Gas central air conditioners	
Base‡	Sales*	Replace-ment units (50%)	Incre-mental units (50%)	Units in use	Ship-ments**	Replace-ment units (30%)	Incre-mental units (70%)	Units in use	Sales†	Units in use
				7126				996		
1960	1402	701	701	7827	312	94	218	1214		
1961	1327	663	664	8491	366	110	256	1470		
1962	1445	722	723	9214	467	140	327	1797		15***
1963	1868	934	934	10,148	580	174	406	2203	11	26
1964	2565	1282	1283	11,431	701	210	491	2694	17	43
1965	2755	1377	1378	12,809	828	248	580	3274	20	63
1966	3101	1550	1551	14,360	960	288	672	3946	29	92
1967	3839	1919	1920	16,280	1047	314	733	4679	29	121
1968	3747	1873	1874	18,154	1165	350	815	5494	39	160

*Source is Current Industrial Reports, Bureau of the Census.

**Source is Current Statistical Review, Metal Products Manufacturing.

†Source is H. R. Linden, "Current Trends in U.S. Gas Demand and Supply," *Public Utilities Fortnightly*.

‡Source is 1960 Census of Housing.

***SRI estimate.

TABLE A3.6

(Table 32) Residential energy consumption for air conditioning

	Electric							Gas			
	Room air conditioning		Central air conditioning			Total consumption		Central air conditioning			Total energy con sumption (trillions of Btu)
Year	Units (millions)	Unit con-sumption* (kWh)	Total con-sumption (billions of kWh)	Units (thou-sands)	Unit con-sumption** (kWh)	Total con-sumption (billions of kWh)	(billions of kWh)	(trillions of Btu)	Units (thou-sands)	Unit con-sumption (millions of Btu)	Total con sumption (trillions of Btu)
1960	7.8	1250	10	1214	3600	4	14	133	—	—	—	133
1961	8.5	1265	11	1470	3600	5	16	150	—	—	—	150
1962	9.2	1280	12	1797	3600	6	18	167	15	20	—	167
1963	10.1	1295	13	2203	3600	8	21	195	26	20	1	196
1964	11.4	1310	15	2694	3600	10	25	230	43	20	1	231
1965	12.8	1325	17	3274	3600	12	29	272	63	20	1	273
1966	14.4	1310	19	3916	3600	14	33	310	92	20	2	312
1967	16.3	1360	22	4679	3600	17	39	361	121	20	2	363
1968	18.2	1375	25	5494	3600	20	45	423	160	20	3	426

*Source is Edison Electric Institute with SRI estimates.

**SRI estimate.

If you sell electricity, the saturation curves in Figure A3.1 are interesting but insufficient. In order to supply electricity, you also must know how much energy is used per item as well as how fast the saturation changes. Note the energy use curves in Figure A3.2, in which all energies are given in kWh_e. This is partly due to population growth, but it also depends on efficiency.

FIGURE A3.1

Saturation curves for selected
appliances. Source: Ref. 1

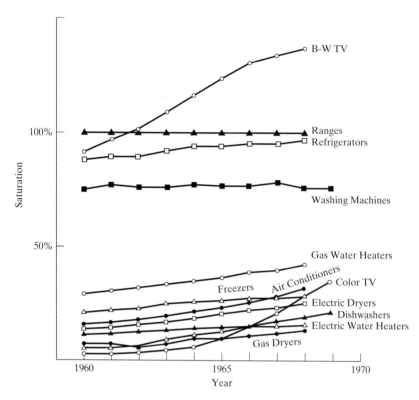

FIGURE A3.2

Energy use for specific purposes (thermal
kilowatthours).

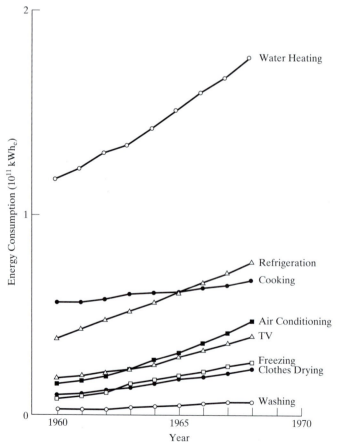

FIGURE A3.3

Per household energy consumption.

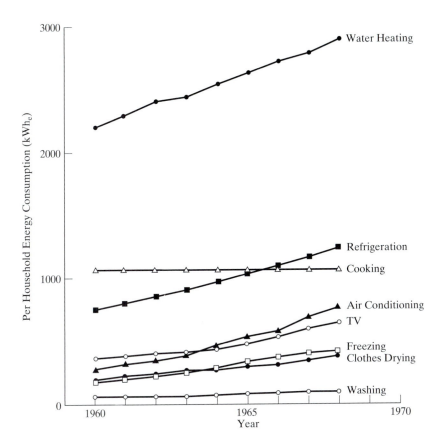

We can see about how population increases by looking at the cooking or washing curves for energy consumption, since from Figure A3.1 these are about constant in saturation. The refrigeration curve is growing faster because people were replacing older refrigerators with newer frost-free models, which are less efficient. Another way to see the same result is to present the energy consumption per household as in Figure A3.3.

These three curves give us a pretty complete idea of all the information contained in the tables. The information is more readily used in graphic form and the comparisons are much easier. The graph is one of the most important tools of the informed person.

REFERENCE

1. Stanford Research Institute, *Patterns of Energy Consumption in the United States* (Washington, D.C.: GPO, 1972).

APPENDIX 4
Vector Addition

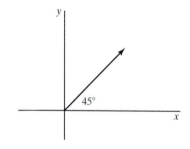

FIGURE A4.1

A vector at a 45° angle.

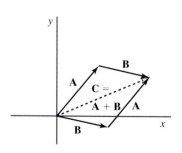

FIGURE A4.2

Vector addition.

Any physical quantity that has a direction associated with it is called a vector. We exist in three dimensions [length, width, and height, for example]. Most vectors of our experience exist in these three directions. We usually think of vectors in terms of the three linear dimensions of our world.

The size of a vector is usually referred to as the magnitude of the vector. Of course, a quantity must have a direction associated with it to be a vector. Vectors thus are said to have both magnitude and direction. The zero vector has magnitude zero and any direction.

In an arbitrary one-dimensional coordinate system, there is only one axis. Suppose we call it the x-axis. The one-dimensional vector is characterized by giving its size and its direction. It would be denoted by an arrow lying along the axis and pointing in the appropriate direction (toward the plus or minus end of the number line).

In a two-dimensional coordinate system such as this page, a vector can point in any direction. Again, we show it as an arrow. Shown in Figure A4.1 is a coordinate system with the vector pointing at an angle of 45° from the x-axis.

If you have two arrows in space representing vectors there, you add the two vectors graphically by placing the tail of the first arrow [whichever of the two that you choose] at the origin (the point the y- and x-axes cross). The tip of that first arrow is where to put the tail end of the second vector, the one you want to add to the first vector. Then the vector that is the sum of the two vectors is the vector you would draw directly that connects the tail of the first arrow to the tip of the second arrow.

In Figure A4.2, the vectors **A** and **B** are added to make vector **C**. We write **C** = **A** + **B** = **B** + **A**. The tail of **C** starts at the origin, the tip goes to the point at the end of the second arrow.

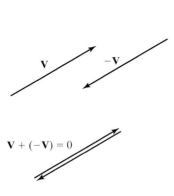

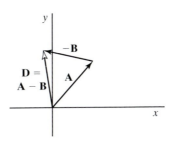

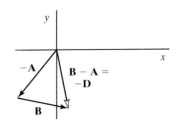

FIGURE A4.3

Vector subtraction of a vector with itself.

FIGURE A4.4

Vector subtraction with two different vectors: $\mathbf{A} - \mathbf{B}$.

FIGURE A4.5

Vector subtraction with two different vectors: $\mathbf{B} - \mathbf{A}$.

Subtraction of vectors is accomplished in a similar way. First, the negative of a vector $\mathbf{V}$, $-\mathbf{V}$, is defined as that vector that produces a vector 0 when added to $\mathbf{V}$ (see Figure A4.3).

We would write vector subtraction $\mathbf{A} - \mathbf{B}$ as vector addition of $\mathbf{A} + (-\mathbf{B})$, which is added by the tail-to-tip method discussed above. So, suppose we use the same vectors $\mathbf{A}$ and $\mathbf{B}$ for which addition was illustrated in Figure A4.3 and find $\mathbf{D} = \mathbf{A} - \mathbf{B} = \mathbf{A} + (-\mathbf{B})$.

Of course, this time changing the order matters, in that it must reverse the sign. In other words, $\mathbf{B} - \mathbf{A} = -\mathbf{D}$. It may be seen from Figures A4.4 and A4.5 that this is indeed the case.

REFERENCES

A longer list with all authors' names and article titles is to be found on the Companion Web Site.
If the number of authors is greater than two, et al. designates the rest of the author names.
Abbreviations used in the references are:

Agence France-Presse, AFP
Associated Press, AP
Die Welt, DW
Die Zeit, DZ
Government Printing Office, GPO
The Akron Beacon Journal, ABJ
The Baltimore Sun, BS
The Boston Globe, BG
The Charleston Gazette, CG
The Chicago Tribune, CT
The Christian Science Monitor, CSM
The Cleveland Plain Dealer, CPD
The Columbus Dispatch, CD
The Contra Costa Times, CCT

The Dallas Morning News, DMN
The Guardian (UK), TG
The Fresno Bee, FB
The Houston Chronicle, HC
The Independent (UK), TI
The Irish Times, IT
The Lansing State Journal, LSJ
The Las Vegas Review-Journal, LVRJ
The Los Angeles Times, LAT
The Milwaukee Journal Sentinel, MJS
The Modesto Bee, MB
The New York Times, NYT
The Orlando Sentinel, OS
The Palm Beach Post, PBP

The Portland Oregonian, PO
The Raleigh News and Observer, RNO
The Sacramento Bee, SB
The San Francisco Chronicle, SFC
The San Francisco Examiner, SFE
The San Jose Mercury News, SJMN
The Seattle Post-Intelligencer, SPI
The Seattle Times, ST
The Times of London, TL
The Toledo Blade, TB
The Toronto Star, TS
The Wall Street Journal, WSJ
The Washington Post, WP

CHAPTER 1

1. J. C. Fisher, *Energy Crises in Perspective* (New York: John Wiley & Sons, 1974).
2. Department of Energy, *Annual Energy Review 2000* (Washington, DC: GPO, 2000).
3. Anonymous, *Phys. Today* **27**(9), 77 (1974).
4. B. J. Finlayson-Pitts and J. N. Pitts, Jr., *Chemistry of the upper and lower atmosphere* (San Diego and London: Academic Press, 2000), Ch. 12.
5. P. J. Crutzen, *Q. J. Meterol. Soc.* **96**, 320 (1970). P. J. Crutzen, *J. Geophys. Res.* **76**, 7311 (1971).
6. R. Simó and C. Pedros-Allo, *Nature* **402**, 396 (1999).
7. Y. Yokouchi et al., *Nature* **403**, 295 (2000). Y. Yokouchi et al., *Nature* **416**, 163 (2002).
8. F. Keppler et al., *Nature* **403**, 298 (2000).
9. R. G. Prinn, *Annu. Rev. Environ. Resour.* **28**, 29 (2003).
10. World Meterological Organization, "Scientific Assessment of Ozone Depletion: 1994," Global Ozone Research and Monitoring Project, Report 37 (1995); "Scientific Assessment of Ozone Depletion: 1998," Global Ozone Research and Monitoring Project, Report 44 (1999).
11. K. Yagi et al., *Proc. Natl. Acad. Sci.* **90**, 8420 (1993). K. Yagi et al., *Science* **267**, 1979 (1995). See, A. C. Revkin, *NYT*, 10 November 2003; this proposal ultimately failed to get support, as discussed by Anonymous, AP dispatch, 15 November 2003. The U.S. is continuing to press for an exemption, as reported

in A. C. Revkin, *NYT*, 27 March 2004 and J. Eilperin, *WP*, 13 July 2004
12. S. A. Montzka et al., *Nature* **398**, 690 (1999).
13. R. C. Rhew et al., *Nature* **403**, 292 (2000).
14. K. R. Redeker et al., *Science* **290**, 966 (2000).
15. D. W. Fahey et al., *Science* **291**, 1026 (2001). R. A. Kerr, *Science* **291**, 962 (2001).
16. A. E. Waibel et al., *Science* **283**, 2064 (1999); A. Tabazadeh et al., *Science* **288**, 1407 (2000); A. Tabazadeh et al., *Science* **291**, 2591 (2001).
17. M. R. Schoeberl and D. L. Hartmann, *Science* **251**, 46 (1991). O. B. Toon and R. P. Turco, *Sci. Am.* **264**(6), 68 (1991); and P. Hamill and O. B. Toon, *Phys. Today* **44**(12), 34 (1991).
18. R. Derwent and R. Friedl, eds., I. L. Karol et al. (Main authors), in J. E. Penner et al., eds., *Aviation and the Global Atmosphere* (New York and London: Cambridge University Press, 1999). This is a special report of IPCC Working Groups I and III in collaboration with the Scientific Panel to the Montreal Protocol on Substances that Deplete the Ozone Layer.
19. K. L. Foster et al., *Science* **291**, 471 (2001).
20. P. O. Wennberg et al., *Science* **279**, 49 (1998).
21. J. G. Anderson et al., *Science* **251**, 39 (1991).
22. R. Cooke, *Newsday*, 30 January 2001.
23. S. Solomon, *Nature* **427**, 289 (2004).
24. S. Solomon, *Nature* **427**, 291 (2004).
25. About 8,000 deaths per year in the U.S. are due to melanoma. For predictions,

see C. P. Shea, in L. Starke, ed., *State of the World 1989* (New York: W. W. Norton, 1989), 77.
26. K. D. Malloy et al., *Proc. Natl. Acad. Sci.* **94**, 1258 (1997). Also M. C. Rousseaux et al., *Proc. Natl. Acad. Sci.* **96**, 15310 (1999) and G. Ries et al., *Nature* **406**, 98 (2000).
27. R. McKenzie et al., *Science* **285**, 1709 (1999).
28. V. Smil, *Sci. Am.* **277**(1), 76 (1997).
29. H. Johnston, *Science* **173**, 517 (1971).
30. F. S. Rowland, *New Scientist* **64**, 717 (1974); A. L. Hammond and T. H. Maugh, *Science* **186**, 335 (1974).
31. L. Zuckerman, *NYT*, 30 March 2001; J. Wallace, *SPI*, 31 March 2001.
32. M. J. Prather and R. T. Watson, *Nature* **344**, 455 (1990).
33. M. J. Molina and F. S. Rowland, *Nature* **249**, 810 (1974); for a more up-to-date review see also F. S. Rowland, *Am. Sci.* **77**, 36 (1989).
34. P. M. Solomon, *Science* **224**, 1210 (1984).
35. A. M. Hough and R. G. Derwent, *Nature* **344**, 645 (1990).
36. T. H. Maugh, *Science* **223**, 1051 (1984).
37. See the special ozone issues of *Geophys. Res.*: *Geophys. Res. Lett.* **13** (#12) (1986); *Geophys. Res.* **94** (D9 & D14) (1989); *Geophys. Res. Lett.* **17** (#4) (1990); *Geophys. Res.* **97** (D8) (1992), all devoted to papers about the Arctic and Antarctic ozone holes.
38. G. H. Mount et al., *Science* **242**, 555 (1988) and also see S. Solomon et al., *Science* **242**, 550 (1988).
39. D. J. Hofmann et al., *Nature* **340**, 117 (1989).

40. J. F. Gleason et al., *Science* **260**, 523 (1993).

41. See in the popular press, for example, W. Sullivan, *NYT*, 7 November 1985; J. Gleick, *NYT*, 29 July 1986; R. A. Kerr, *Science* **232**, 1602 (1986); R. A. Kerr, *Science* **246**, 324 (1989); R. A. Kerr, *Science* **252**, 204 (1991); J. Horgan, *Sci. Am.* **266**(3), 28 (1992); R. A. Kerr, *Science* **262**, 501 (1993); and I. H. Rowlands, *Environment* **35**(6), 25 (1993).

42. J. C. Farman et al., *Nature* **315**, 207 (1985).

43. S. Solomon et al., *Nature* **321**, 755 (1986).

44. R. A. Kerr, *Science* **239**, 1489 (1988). See also J. Gleick, *NYT*, 20 March 1988.

45. M. R. Schoeberl and A. J. Krueger, *Geophys. Res. Lett.* **13**, 119 (1986).

46. R. S. Stolarski et al., *Geophys. Res. Lett.* **18**, 1015 (1991).

47. S. Madronich, *Geophys. Res. Lett.* **19**, 37 (1992).

48. M. R. Schoeberl and D. L. Hartmann, *Science* **251**, 46 (1991). M. R. Schoeberl et al., *J. Geophys. Res.* **97**, 7859 (1992).

49. W. J. Randel et al., *Science* **285**, 1689 (1999). See also G. P. Collins, *Phys. Today* **51**(1), 19 (1998) and the articles in issue 22 of *Geophys. Res. Lett.* **24**, (1997).

50. D. L. Albritton, in R. A. Anthes, ed., *Ozone Depletion, Greenhouse Gases, and Climate Change* (Washington, DC: National Research Council, 1989), 10.

51. G. Brasseur and C. Granier, *Science* **257**, 1239 (1992); A. Tabazadeh and R. P. Turco, *Science* **260**, 1082 (1993).

52. D. W. Fahey et al., *Nature* **363**, 509 (1993) show the effect of Mount Pinatubo by direct measurement comparisons of ozone loss before and after the eruption. See also the review by G. Brasseur, *Nature* **359**, 275 (1992).

53. W. Hively, *Am. Sci.* **77**, 219 (1989).

54. W. K. Stevens, *NYT*, 9 April 1991.

55. H. F. French, in L. Starke, ed., *State of the World, 1997* (New York: Norton, 1997).

56. W. K. Stevens, *NYT*, 26 November 1992. See, however, about later developments as given in, for example, T. Maliti, AP dispatch, 11 November 2003.

57. See, for example, for the Northern Hemisphere, M. Blumthaler and W. Ambach, *Science* **248**, 206 (1990); M. Blumthaler et al., *Theor. Appl. Climatol.* **46**, 39 (1992) and J. B. Kerr and C. T. McElroy, *Science* **262**, 1032 (1993).

58. See, for example, for the Southern Hemisphere, R. W. Atkinson et al., *Nature* **340**, 290 (1989).

59. G. Seckmeyer and R. L. McKenzie, *Nature* **359**, 135 (1992).

60. J. E. Frederick and A. D. Alberts, *Geophys. Res. Lett.* **18**, 1869 (1991).

61. A. Pollack, *NYT*, 15 May 1991; J. Holusha, *NYT*, 18 December 1991.

62. M. Henderson, *TL*, 4 December 2000.

63. R. A. Kerr, *Science* **271**, 32a (1996).

64. D. A. Fisher et al., *Nature* **344**, 508 (1990).

65. A. Smith, *Wealth of Nations* (New York: Oxford University Press, 1976), 423.

66. G. Hardin, *Science* **162**, 1243 (1968). Hardin attributes the idea to W. F. Lloyd, whose *Two Lectures on the Checks to Population* (New York: Oxford University Press, 1833), is reprinted in G. Hardin, ed., *Population, Evolution, and Birth Control* (San Francisco: Freeman, 1964).

67. G. C. Daily and P. R. Ehrlich, *BioScience* **42**, 761 (1992).

68. D. R. Klein, *J. Wildl. Manage.* **32**, 350 (1968).

69. Dept. of Energy, *Coal Industry Annual 1999* DOE/EIA-0584(99) (Washington, DC: GPO, 2000). Surface mined coal cost $12.37/ton, while underground coal cost $24.33/ton. Wyoming coal cost $5.38/ton. Delivered coal averaged $23.63/ton in 1999. For historical perspective, see H. Perry, *Science* **222**, 377 (1983). Strip mine reclamation costs were estimated to run between $1 and $5 per ton of coal strip mined. The delivered cost of coal in Illinois was $38.94/ton for Wyoming coal.

70. Allete, Inc., "2000 Annual Report," Duluth, Minnesota (www.allete.com).

71. Bureau of Mines, *Mineral Facts and Problems*, 1980 ed., Bu. Mines Bull. 650, U.S. Dept. of the Interior (Washington, DC: GPO, 1980).

72. M. Milinski et al., *Nature* **415**, 424 (2002).

73. J. Pretty, *Science* **302**, 1912 (2003).

74. T. Dietz et al., *Science* **302**, 1907 (2003). W. M. Adams et al., *Science* **302**, 1915 (2003).

75. A. C. Revkin, *NYT*, 10 October 2000.

76. T. G. Slanger et al., *Science* **241**, 945 (1988).

77. D. Johnson et al., *Nature* **415**, 82 (2002).

CHAPTER 2

1. R. Freedman and B. Berelson, *Sci. Am.* **231**(3), 31 (1974).

2. F. Daniels, *Direct Use of the Sun's Energy* (New York: Ballantine, 1974); *Handbook of Physical Constants*, 51st ed. (Cleveland: C. R. C. Publ. Co.).

3. *Encyclopedia Americana*. The ratio is 28.3%, including Antarctica.

4. J. P. Holdren and P. R. Ehrlich, *Am. Sci.* **62**, 282 (1974).

5. I. Asimov, *Life and Energy* (New York: Avon, 1972), 362.

6. S. Rojstaczer et al., *Science* **294**, 2549 (2001). C. B. Field, *Science* **294**, 2490 (2001).

7. P. M. Vitousek et al., *BioScience* **36**, 368 (1986).

8. R. Revelle, *Sci. Am.* **231**(3), 161 (1974).

9. D. H. Meadows et al., *The Limits to Growth* (New York: Signet, 1972), 60.

10. R. Dubos, *The God Within* (New York: Charles Scribner's Sons, 1972).

11. Many of these issues are discussed in A. A. Bartlett, *Am. J. Phys.* **46**, 876 (1978). He speculates further in A. A. Bartlett, *Population and Environment* **16**, 5 (1994) and *Renew. Res. J.* **15**(4), 6 (1998). See also N. Keyfitz, *Deutschland* #2, 44 (1994). In addition, a National Academy report (Committee on Earth Resources (C. Otte et al.), National Academy of Sciences, *Mineral Resources and Sustainability: Challenges for Earth Scientists*, URL: http://www.nap.edu/catalog/9077.html) suggests that "activities that deplete renewable resources or exhaust nonrenewable resources are consistent with sustainability as long as an appropriate portion of the proceeds of these activities is invested in reproducible, man-made resources such as education and technology." They define sustainability with Solow (R. Solow, *Resources Policy* **19**, 162 (1993)) as that which "allows every future generation the option of being as well off as its predecessors."

12. J. Brunner, *Stand on Zanzibar* (New York: Ballantine, 1976); H. Harrison, *Make Room, Make Room* (New York: Ace, 1984); and C. M. Kornbluth, *The Marching Morons* (New York: Ballantine, 1959).

13. F. S. Roberts, in H. Marcus-Roberts and M. Thompson, eds., *Modules in Applied Mathematics*, V. 4, *Life Science Models*, (New York: Springer-Verlag, 1984), 250 (see especially pages 250-258).

14. V. Smil, *Sci. Am.* **277**(1), 76 (1997).

15. A. Trewavas, *Nature* **402**, 231 (1999) and A. Trewavas, *Nature* **418**, 668 (2002).

16. D. Tilman et al., *Nature* **418**, 671 (2002); R. S. Hails, *Nature* **418**, 685 (2002)

17. D. Pimentel et al., *Environment, Development, and Sustainability* **1**, 19 (1999).

18. J. C. J. M. van den Bergh and P. Rietveld, *BioScience* **54**, 195 (2004). See from among earlier analyses H. von Foerster et al., *Science* **132**, 1291 (1960) and S. Umpleby, *Population and Environment* **11**, 159 (1990). For more up-to-date analysis, see, for example, J. E. Cohen, *Science* **302**, 1172 (2003).

19. Population Division, Department of Economic and Social Affairs, United Nations, ESA/P/WP.165, 28 February 2001. Hopwever et al., *Nature* **412**, 543 (2001) consider that there is a 60%

probability that the population will be lower than 10×10^9 people in 2100.

20. G. K. O'Neill, *The High Frontier: Human Colonies in Space* 3rd. Ed. (Burlington, Ontario: Collectors Guide Publishing, Inc., 2000). G. K. O'Neill, *Phys. Today* **27**(9), 32 (1974). See also G. K. O'Neill, *NYT Magazine*, 18 January 1976, 10 and J. I. Merritt, *Princeton Alumni Weekly* **90**(2), 22 (1989).

21. J. Pournelle, *Galaxy* **36**(8), 65 (1975) [September 1975] and S. L. Gillett, *Analog* **104**(12), 64 (1984) [December 1984].

22. A. L. Robins, *Science* **195**, 668 (1977).

23. C. P. McKay et al., *Nature* **352**, 489 (1991).

24. M. C. Malin and K. S. Edge, *Science* **288**, 2330 (2000). H. Y. McSween, T. L. Grove, R. C. F. Lentz, J. C. Dann et al., *Nature* **409**, 487 (2001). J. F. Mustard et al., *Nature* **412**, 411 (2001). F. Costard et al., *Science* **295**, 110 (2001). More indirect older evidence is found in D. A. Paige, *Nature* **356**, 43 (1992) and V. R. Baker et al., *Nature* **352**, 589 (1991).

25. J. Simon, *Cato Policy Report* **17**(5), 131 (1995). See also R. Bailey, *Futurist* **29**(1), 14 (1995) and J. L. Simon, *Futurist* **29**(1), 19 (1995).

26. G. Easterbrook, *Science Friday*, PBS, 27 December 1996.

CHAPTER 3

1. I. Asimov, *Words of Science* (New York: Mentor, 1969), 110. *The American Heritage Dictionary of the English Language* points out that Aristotle used the word *energia* (*en*-at + *ergon*, work).

2. E. M. Rogers, *Physics for the Enquiring Mind* (Princeton, New Jersey: Princeton University Press, 1960).

3. M. J. Cormier, *Natl. Hist.* **83**(3), 26 (1974).

4. I. Asimov, *Life and Energy* (New York: Avon, 1972), 182.

CHAPTER 4

1. L. Ruby, *Phys. Teach.* **40,** 272 (2002).

2. M. Gottlieb and J. Glanz, *NYT*, 15 August 2003.

3. Federal Power Commission, *Report to the president by the Federal Power Commission on the power failure in the northeastern United States and the province of Ontario on November* 9-10, 1965 (Washington, DC: U.S. GPO, 9 December 1965).

4. S. J. Blum et al., eds., *The Night the Lights Went Out* (New York: Signet Books, 1965).

5. G. D. Friedlander, *IEEE Spectrum* **13**(10), 83 (1976).

6. Consolidated Edison, "Lessons learned from the 1977 blackout," undated.

7. R. Booth et al., *Newsweek*, 25 July 1977, p. 23.

8. See, for example, G. Lucas, *SFC*, 13 October 2002.

9. C. P. Dahlberg, *SB*, 16 May 2002.

10. U.S.-Canada Power System Outage Task Force, *Final Report on the August 14, 2003 Blackout in the United States and Canada: Causes and Recommendations*, 5 April 2004.

11. A. C. Revkin and M. L. Wald, *NYT*, 18 August 2003. P. Behr, *WP*, 15 August 2003. Anonymous, *IEEE Spectrum* **40**(9), 9 (2003).

12. A. Clark and D. Gow, *TG*, 30 August 2003. H. Timmons, *NYT*, 2 September 2003.

13. National Grid Company PLC, 28 August 2003, blackout report 10 September 2003.

14. G. Schneider and K. Bredemeier, *WP*, 20 August 2003.

15. T. J. Overbye, *Am. Sci.* **88**, 220 (2000). W. Sweet, *IEEE Spectrum* **37**(6), 43 (2000).

16. W. Sweet, *IEEE Spectrum* **40**(10), 12 (2003).

17. Ref. 10, p. 8; Ref. 10, p. 26.

18. T. Robertson, *BG*, 12 July 2004. J. Blum, *WP*, 10 August 2004. J. Apt and L. B. Lave, *WP*, 10 August 2004.

19. J. Funk, *CPD*, 8 August 2004. M. L. Wald, *NYT*, 10 August 2004.

CHAPTER 5

1. E. Cook, in G. Piel, D. Flanagan, F. Bello, P. Morrison, J. B. Piel et al., eds., *Energy and Power* (San Francisco: Freeman, 1971), 83.

2. C. M. Summers, in G. Piel et al., eds., *Energy and Power* (San Francisco: Freeman, 1971), 95.

3. Dept. of Commerce, *Statistical Abstract of the United States*, 2000 (Washington, DC: GPO, 2001). Department of Energy, *Annual Energy Review 2000* (Washington, DC: GPO, 2001), DOE/EIA-0384(2000).

4. Dept. of Energy, *Improving Technology: Modeling Energy Futures for the National Energy Strategy*, SR/NES/90-01 and *Energy Consumption and Conservation Potential: Supporting Analysis for the National Energy Strategy*, SR/NES/90-02 (Washington, DC: GPO, 1991).

5. Office of Emergency Preparedness, *The Potential for Energy Conservation: A Staff Study* (Washington, DC: GPO, 1972).

6. An interesting discussion of the movement of the center of population in the United States over the years is found in A. A. Bartlett and R. L. Conklin, *Am. J. Phys.* **53**, 242 (1985).

7. N. Keyfitz, in P. Cloud et al., eds., *Resources and Man* (San Francisco: Freeman, 1969), 43. N. Keyfitz, *Sci. Am.* **261**(3), 118 (1989).

8. D. Chapman et al., *Science* **178**, 703 (1972).

9. L. Uchitelle, *NYT*, 3 October 1993.

10. C. Steinhart and J. Steinhart, *The Fires of Culture* (N. Scituate, MA: Duxbury, 1974).

11. R. Jones, Columbus and Southern Ohio Electric Company, private communication. The prices have been adjusted for the rise in the CPI. Also, Dept. of Energy, *Annual Energy Review 1992* (Washington, DC: GPO, 1993), DOE/EIA-0384(92), Table 8.11.

12. Economic Council of Canada, *Connections* (Ottawa: Canadian Government Publishing Centre, 1985).

13. W. E. Mooz and C. C. Mow, as quoted in A. Hammond, *Science* **178**, 1079 (1972).

14. *A Time to Choose: America's Energy Future* (Cambridge, MA: Ballinger, 1974).

15. J. Fisher, *Energy Crises in Perspective* (New York: John Wiley & Sons, 1974).

16. E. Marshall, *Science* **208**, 1353 (1980).

17. Members of the National Energy Policy Development Group (R. Cheney et al.), *National Energy Policy: Report of the National Energy Policy Development Group*, (Washington, DC: GPO, May 2001).

18. A. L. Hammond, ed., *World Resources 1992-93* (New York and Oxford: Oxford University Press, 1992), Table 21-2. See also W. Sassin, *Sci. Am.* **243**(3), 118 (1980).

19. L. Carroll, *Through the Looking Glass* (New York: Clarkson N. Potter, 1960).

20. W. Häfele, *Science* **209**, 174 (1980).

21. C. J. Cleveland et al., *Science* **225**, 890 (1984).

22. L. D. Roper, *Am. J. Phys.* **47**, 467 (1979). See also T. Modis, *Futurist* **28**(5), 20 (1994).

23. J. P. Cohn, *BioScience* **40**, 10 (1990).

CHAPTER 6

1. V. A. Szalai and G. W. Brudvig, *Am. Sci.* **86**, 542 (1998).

2. G. C. Dismukes, *Science* **292**, 447 (2001).

3. K. N.Ferreira et al., *Science* **303**, 1831 (2004).

4. E. Marshall, *Science* **224**, 268 (1984).

CHAPTER 7

1. E. F. Hammel et al., *Science* **223**, 789 (1984).

2. C. A. Price, in J. A. Harte and R. H. Socolow, eds., *Patient Earth*, (New York: Holt, Rinehart & Winston, 1971), 70; and J. A. Harte and R. H. Socolow, in *Patient Earth*, 84.

3. M. W. Browne, *NYT*, 27 October 1987.

4. S. A. Verhovek, *NYT*, 21 September 1993.

5. M. O. Scully et al., *Science* **299**, 862 (2003). H. Linke, *Science* **299**, 841 (2003).

6. W. Carnahan et al., eds., *Efficient Use of Energy, The APS Studies on the Technical Aspects of the More Efficient Use of Energy* (New York: American Institute of Physics, 1975).

7. R. U. Ayres and I. Nair, *Phys. Today* **37**(11), 62 (1984).

8. W. D. Metz, *Science* **188**, 820 (1975).

9. Dept. of Energy, *Annual Energy Review 2002* (Washington, DC: GPO, 2003), DOE/EIA-0384(2002), Table A6.

CHAPTER 8

1. Dept. of Energy, *Annual Energy Review 2000* (Washington DC: GPO. 2001), DOE/EIA-0384(00).

2. Stanford Research Institute, *Patterns of Energy Consumption in the United States* (Washington, DC: GPO. 1972).

3. S. Novick, *Environment* **17**(8), 7 (1975) and *Environment* **17**(8), 35 (1975).

4. E. Nalder and M. Gladstone, *SJMN*, 3 June 2001.

5. Office of Coal, Nuclear, Electric and Alternate Fuels, U.S. Department of Energy, *Electric Power Annual 2000* (Washington, DC: GPO. 2001), DOE/EIA-0348(2000)/1.

6. C. Berthelsen and S. Winokur, *SFC*, 11 March 2001

7. N. Vogel, *LAT*, 9 December 2000.

8. J. Jelter, Reuters dispatch, 7 January 2001.

9. W. Booth, *WP*, 28 January 2001.

10. N. Vogel, *LAT*, 13 April 2001.

11. T. Monmaney, *LAT*, 7 January 2001.

12. S. Stanton, *SB*, 6 May 2001.

13. P. H. King, *LAT*, 7 February 2001.

14. D. Kasler, J. Hill, and S. Stanton, *SB*, 23 September 2001.

15. C. Squatriglia, *SFC*, 8 April 2001.

16. J. Jelter, Reuters dispatch, 28 January 2001.

17. C. Said, *SFC*, 17 June 2001

18. P. Rogers, *SJMN*, 25 February 2001.

19. M. Landsberg et al., *LAT*, 20 January 2001.

20. C. Berthelsen and S. Winokur, *SFC*, 20 May 2001.

21. B. Bailey and C. O'Brien, *SJMN*, 25 May 2001.

22. V. Laroi, Reuters dispatch, 31 March 2001; R. Scherer, *CSM*, 20 February 2001.

23. A. Gibbs, *The Tacoma News Tribune*, 4 February 2001.

24. J. Bjorhus, *SJMN*, 6 July 2001.

25. N. Riccardi and S. Berry, *LAT*, 7 February 2001.

26. D. R. Francis, *CSM*, 2 February 2001.

27. P. H. King, *LAT*, 15 April 2001.

28. L. M. Holson with R. A. Oppel, Jr., *NYT*, 16 June 2001; B. Tansey, *SFC*, 17 June 2001.

29. B. Klein, *IEEE Spectrum* **38**(3), 31 (2001).

30. D. Lazarus, *SFC*, 10 June 2001.

31. J. Gerth, *NYT*, 19 June 2001.

32. E. Lau, *SB*, 1 January 2002.

33. R. Jurgens, *CCT*, 19 August 2001.

34. C. Peyton, *SB*, 23 February 2002.

35. Office of Integrated Analysis and Forecasting, EIA. Department of Energy, *Annual Energy Outlook 2002 With Projections to 2020*, DOE/EIA-0383(2002), See also C. Peyton, *SB*, 20 August 2001,

36. R. Salladay, *SFC*, 24 June 2001, L. Gledhill, *SFC*, 4 October 2001, K. Russell and P. Felsenfeld, *CCT*, 4 February 2002, M. Z. Barabak and N. Riccardi, *LAT*, 7 February 2002.

37. L. Gledhill, *SFC*, 31 January 2002.

38. S. Bandrapalli, *CSM*, 22 February 2001.

39. D. Lazarus, *SFC*, 18 February 2001.

40. S. H. Verhovek, *NYT*, 26 January 2001.

41. S. Leavenworth, *SB*, 10 June 2001, E. Shogren, *LAT*, 8 February 2001.

42. M. Janofsky, *NYT*, 4 June 2001.

43. N. Banerjee, *NYT*, 2 May 2001.

44. R. Burnett, *OS*, 13 May 2001.

45. N. Banerjee and R. Pérez-Peña, *NYT*, 1 June 2001.

46. A. Berenson, *NYT*, 25 February 2001.

47. D. Whitney, *SB*, 26 February 2001.

48. Economic Council of Canada, *Connections* (Ottawa: Canadian Government Publishing Centre, 1985).

49. C. A. Berg, *Science* **184**, 264 (1974).

50. J. A. Fisher, *Energy Crises in Perspective* (New York: John Wiley & Sons, 1974).

51. R. Stuart, *NYT*, 10 August 1975; also data on time-of-day pricing from Consolidated Edison of New York, quoted in M. Leebaw and H. Heyman, *NYT*, 12 September 1976.

52. R. D. Hershey, Jr., *NYT*, 8 March 1981.

53. M. L. Wald, *NYT*, 1 January 1984 and *NYT*, 20 January 1985.

54. P. Lewis, *NYT*, 11 December 1983.

55. I. M. Torrens et al., *Annu. Rev. Energy Environ.* **17**, 211 (1992); Dept, of Energy, *The U.S. Petroleum Refining Industry in the 1980's* (Washington, DC: GPO. 1992), DOE/EIA-0536; Dept, of Energy, *Natural Gas Annual 1991* (Washington, DC: GPO. 1992), DOE/EIA-0131(91); and Dept, of Energy, *Crude Oil, Natural Gas, and Natural Gas Liquids Reserves* (Washington, DC: GPO. 1992), DOE/EIA-0216(91).

56. D. S. Johnson, senior planning engineer, Southern California Edison, private communication, June 1985.

57. C. F. Fisher, Jr. and W. R. Schriver, *Power Plant Cost Study* (Washington, DC: Edison Electric Institute, October 1984).

58. For a modern combined-cycle plant (as of 2002), burning natural gas at $5 per 1000 ft^3 at 50% efficiency, the operating cost is 3. 4 cents/kWh, An earlier publication: Dept, of Energy, *Electric Plant Cost and Power Production Expense 1991* (Washington, DC: GPO. 1991), DOE/EIA-0545(91), quotes 3. 1 cents/kWh, so the cost has not changed much over the years.

59. D. McInnis, *NYT*, 4 October 1981; J. M. Fowler et al., *Environment* **20**(3), 25 (1978); and R. L. Goble and C. Hohenemser, *Environment* **21**(8), 32 (1979).

60. N. Vogel, *LAT*, 9 October 2001.

61. C. Said, *SFC*, 15 June 2001; Anonymous, AP dispatch, 15 June 2001.

62. N. R. Brooks, *LAT*, 5 June 2002, M. Martin, *SFC*, 16 June 2002.

63. R. Salladay, *SFC*, 15 June 2001.

64. R. J. Lopez and R. Connell, *LAT*, 19 May 2001.

65. C. Devall et al., *SJMN*, 19 May 2001.

66. C. P. Dahlberg, *SB*, 18 September 2002, N. R. Brooks, *LAT*, 31 January 2003.

67. S. Berry and N. Riccardi, *LAT*, 14 January 2001, D. Lazarus, *SFC*, 19 May 2001.

68. R. Salladay, *SFC*, 17 May 2001.

69. D. Nissenbaum, *SJMN*, 22 June 2001, K. Yamamura and E. Bazar, *SB*, 22 June 2001, D. Thompson, AP dispatch, 6 July 2001, K. Yamamura, *SB*, 6 July 2001.

70. M. Richtel, *NYT*, 23 June 2001.

71. L. Gledhill, *SFC*, 23 June 2001.

72. Anonymous, AP dispatch, 18 September 2001.

73. N. Vogel and N. Riccardi, *LAT*, 19 January 2001.

74. Non-public Appendix to order directing Williams Energy Marketing & Trading Company and AES Southland, Inc., to show cause, Docket No, IN01-3-000, See also N. R. Brooks, *LAT*, 16 November 2002.

75. S. Johnson and M. A. Ostrom, *SJMN*, 18 February 2001. The accusations of deliberate misrepresentation were confirmed when the FERC made public the document of Ref. 127.

76. P. Behr and J. Eilperin, *WP*, 21 June 2001.

77. M. Garvey and R. Simon, *LAT*, 28 March 2001

78. S. Watson, *PO*, 27 December 2000.

79. T. Detzel, *PO*, 17 February 2001.

80. M. Martin, *SFC*, 12 March 2002.

81. J. Peterson, *LAT*, 28 April 2003, and *LAT*, 25 June 2003, J. Peterson and E. Douglass, *LAT*, 26 June 2003, P. Krugman, *NYT*, 2 September 2003, D. Kasler, *SB*, 17 September 2003, D. Whitney, *SB*, 18 October 2003, and J. Peterson, *LAT*, 31 March 2004.

82. Z. Coile, *SFC*, 7 July 2001. D. Whitney, *SB*, 8 December 2001.

83. D. Whitney, *SB*, 26 June 2003.

84. N. R. Brooks, *LAT*, 7 January 2003.

85. L. Williams and C. Marinucci, *SFC*, 16 December 2001, E. Bazar, *SB*, 28 June 2002.

86. D. Whitney, *SB*, 17 June 2004, C. Lochhead, *SFC*, 17 June 2004.

87. See, for example, J. Peterson, *LAT*, 25 April 2003, J. Peterson, *LAT*, 5 July 2004, and *LAT*, 29 July 2004.

88. L. Goldberg, *HC*, 29 August 2003 (Reliant), D. Kasler, *SB*, 1 November 2003 (Sempra), E. Douglass, *LAT*, 26 February 2004 (Williams Cos.), J. Peterson and E. Douglass, *LAT*, 27 April 2004 (Dynergy), D. Kasler, *SB*, 14 July 2004 (Duke).

89. F. P. Sioshansi, *Energy Policy* **30**, 245 (2002).

90. B. Tansey, *SFC*, 11 February 2001.

91. Anonymous, AP dispatch, 12 December 2003.

92. R. A. Oppel, Jr. and L. Bergman, *NYT*, 26 March 2001, M. Bustillo and J. Tamaki, *LAT*, 19 April 2001, R. Alonso-Zaldivar, *LAT*, 14 May 2001, B. Tansey, *SFC*, 15 May 2001.

93. J. Bjorhus, *SJMN*, 10 October 2001.

94. Anonymous, Reuters dispatch, 6 December 2003.

95. R. Alonso-Zaldivar, *LAT*, 19 May 2001.

96. R. A. Oppel Jr., *NYT*, 5 June 2001, R. Alonso-Zaldivar, *LAT*, 9 June 2001.

97. J. Peterson, *LAT*, 30 January 2003.

98. L. Goldberg, *HC*, 21 April 2004. Anonymous, Reuters dispatch, 12 November 2003.

99. N. Banerjee, *NYT*, 2 June 2002.

100. FERC. "Initial Report on Company-Specific Separate Proceedings and Generic Reevaluations; Published Natural Gas Price Data; and Enron Trading Strategies, Fact-Finding Investigation of Potential Manipulation of Electric and Natural Gas Prices," Docket No, PA02-2-000, August 2002.

101. A. LaMar, *CCT*, 13 October 2001, V. Ellis and N. Vogel, *LAT*, 18 October 2001, A. Chance, *SB*, 24 February 2002, J. Sterngold, *NYT*, 25 February 2002, T. Reiterman, *LAT*, 25 February 2002, M. Martin et al., *SFC*, 25 February 2002.

102. M. Martin, *SFC*, 26 February 2002.

103. N. Vogel, *LAT*, 1 February 2002.

104. J. Woolfolk and S. Johnson, *SJMN*, 8 July 2001.

105. C. Peyton, *SB*, 22 March 2002, *Sacramento Bee* Editorial, *SB*, 26 March 2002.

106. L. Gledhill and M. Martin, *SFC*, 10 July 2001, J. Rosen, *SB*, 26 March 2002.

107. R. Connell, R. J. Lopez, and D. Smiths, *LAT*, 10 July 2001.

108. M. Martin, *SFC*, 13 February 2002.

109. R. W. Stevenson, *NYT*, 12 April 2002; C. Lochhead, *SFC*, 12 April 2002; D. Whitney, *SB*, 12 April 2002.

110. R. A. Oppel, Jr., *NYT*, 11 April 2002, R. A. Oppel, Jr. and J. Gerth, *NYT*, 7 May 2002, N. R. Brooks et al., *LAT*, 7 May 2002, M. Martin, *SFC*, 7 May 2002, R. A. Oppel, Jr., *NYT*, 8 May 2002.

111. Editorial, *LAT*, 19 June 2004.

112. L. M. O'Rourke, *SB*, 23 February 2002.

113. Z. Coile, *SFC*, 18 January 2002.

114. J. D. Glater, *NYT*, 20 April 2002, See also Anonymous, Reuters dispatch, 19 April 2002, K. Eichenwald, *NYT*, 7 May 2002, and K. Eichenwald, *NYT*, 8 May 2002, and the *Chicago Tribune* series "A final accounting: The fall of Andersen," a series written by F. McRoberts and reported by D. Alexander et al.: *CT*, 1 September 2002, *CT*, 2 September 2002, *CT*, 3 September 2002, and *CT*, 4 September 2002.

115. K. Eichenwald, *NYT*, 10 April 2002.

116. K. Eichenwald, *NYT*, 22 April 2002.

117. C. Lochhead, *SFC*, 23 February 2002, C. Lochhead, *SFC*, 12 February 2002.

118. L. Bergman and J. Gerth, *NYT*, 25 May 2001, See also D. Van Natta, Jr., and N. Banerjee, *NYT*, 28 March 2002.

119. C. Lochhead, *SFC*, 14 February 2002, M. Martin, *SFC*, 30 January 2002.

120. N. Vogel, *LAT*, 28 January 2002, M. Martin, *SFC*, 3 February 2002.

121. Anonymous, Bloomberg News, 23 November 2002, C. Berthelsen and M. Martin, *SFC*, 23 November 2002, R. A. Oppel, Jr., *NYT*, 4 March 2003, R. A. Oppel, Jr., *NYT*, 27 March 2003.

122. J. Peterson and N. R. Brooks, *LAT*, 4 March 2003, D. Whitney, *SB*, 4 March 2003.

123. R. Alonso-Zaldivar and N. R. Brooks, *LAT*, 21 November 2002.

124. L. Gledhill, *SFC*, 13 December 2002.

125. As quoted in M. Bazeley, *SJMN*, 4 March 2003.

126. J. Peterson, *LAT*, 23 July 2004.

127. Memorandum from attorneys C. Yoder and S. Hall of Stoel Reeves PLC to R. Sanders of Enron, "Traders strategies in the California wholesale power markets/ ISO sanctions," 6 December and 8 December 2000, (Evidence released to the public.)

128. H. Rice, *HC*, 12 December 2002.

129. C. Schwarzen, *ST*, 3 June 2004.

130. Anonymous, AP dispatch, 4 August 2004.

131. R. A. Oppel, Jr., *NYT*, 13 June 2004.

132. D. Kasler, *SB*, 15 June 2004, Z. Coile, *SFC*, 15 June 2004.

133. D. Richman, *SPI*, 26 August 2004.

134. Supplemental Testimony of Dr. Carl Pechman on behalf of Public Utility District No, 1 of Snohomish County, Washington, FERC. Ex, SNO-160, This FERC document contains the transcripts prepared by the Snohomish County Public Utility District of telephone conversations among Enron traders, Personal information was edited out from the transcripts.

135. J. Peterson and D. Wotapka, *LAT*, 18 June 2004.

136. D. Barboza, *NYT*, 10 April 2002.

137. K. Eichenwald, *NYT*, 13 January 2002.

138. G. Kessler and M. Allen, *WP*, 13 January 2002.

139. K. Eichenwald, *NYT*, 15 April 2002.

140. S. Labaton and R. A. Oppel, Jr., *NYT*, 8 February 2002, C. Lochhead, *SFC*, 8 February 2002.

141. K. Eichenwald, *NYT*, 27 February 2002.

142. L. Wayne, *NYT*, 13 January 2002.

143. T. Fowler, *HC*, 20 October 2002.

144. E. A. Hirs, III and P. W. MacAvoy, *HC*, 25 August 2003.

145. D. C. Johnston, *NYT*, 17 January 2002.

146. B. Tammeus, *The Kansas City Star*, 26 March 2002.

147. K. Eichenwald, *NYT*, 28 January 2002, C. Lochhead and Z. Coile, *SFC*, 25 January 2002.

148. K. Eichenwald, *NYT*, 15 March 2002, K. Eichenwald, *NYT*, 18 March 2002. See also C. Swett, *SB*, 29 January 2002, A. Berenson with J. D. Glater, *NYT*, 13 January 2002, K. Eichenwald, *NYT*, 16 June 2002.

149. K. Eichenwald, *NYT*, 12 April 2002.

150. R. A. Oppel, Jr., *NYT*, 19 April 2002.

151. L. Wayne, *NYT*, 20 April 2002.

152. Atomic Energy Commission, *The Safety of Nuclear Power Reactors and Related Facilities*, WASH-1250 (Washington, DC: AEC. 1973), Members of the National Energy Policy Development Group (R. Cheney et al.), *National Energy Policy: Report of the National Energy Policy Development Group*, (Washington, DC: GPO, May 2001).

153. J. A. Fisher, *Energy Crises in Perspective* (New York: John Wiley & Sons, 1974).

154. C. A. Berg, *Science* **181**, 128 (1973), In 1973, cost of maintenance was 0.03 cents/kWh, This is 0. 095 cents/kWh in 2000 dollars.

155. A. Lovins, *Foreign Affairs* **55**, 65 (1976)

156. S. Diamond, *NYT*, 24 June 1984.

157. J. Karkheck et al., *Science* **195**, 948 (1977); W. Leonhardt et al., eds., *Kommunales Energie-Handbuch* (Karlsruhe: Verlag C. F. Müller, 1991).

158. C. Stein, *ASHRAE J.* **27**(2), 36 (1985).

159. J. S. Feher et al., *Environment* **26**(10), 12 (1984).

160. P. Navarro, *NYT*, 7 July 1985.

161. M. Crawford, *Science* **229**, 248 (1985).

162. Office of Coal, Nuclear, Electric and Alternate Fuels, U.S. Department of Energy, *Electric Power Annual 2000* (Washington, DC: GPO. 2001), DOE/EIA-0348(2000)/1, Table 2.

163. S. D. Freeman, *Environment* **27**(3), 6 (1985).

164. U. Columbo, *Science* **217**, 705 (1982).

165. *Environment Policies for the 1980s* (Paris: Organization for Economic Cooperation and Development, 1980).

166. L. B. Young, *Environment* **20**(4), 16 (1978).

167. M. W. Miller and G. E. Kaufman, *Environment* **20**(1), 6 (1978); A. A. Marino and R. O. Becker, *Environment* **20**(9), 6 (1978).

168. K. J. McLeod et al., *Science* **236**, 1465 (1987).

169. R. Adair, *Phys, Rev, A* **43**, 1039 (1991), W. R. Bennett, Jr., *Phys. Today* **47**(4), 23 (1994).

170. See, for example, the discussions of the experiments in H. K. Florig, *Science* **257**, 468 (1992) and R. Stone, *Science* **258**, 1724 (1992), The first intimations of possible effects are found in N. Wertheimer and E. Leeper, *Am. J. Epidem.* **109**, 273 (1979) and *Int. J. Epidem.* **11**, 345 (1982).

171. J. L. Kirschvink, *Phys, Rev, A* **46**, 2178 (1992) and R. K. Adair, *Phys, Rev, A* **46**, 2185 (1992).

172. J. L. Kirschvink et al., *Proc. Natl. Acad. Sci.* **89**, 7683 (1992), See also L. E. Fernenden and H. Mühlensiegen, *Endeavour* **12**, 119 (1988).

173. D. A. Savitz et al., *Am. J. Epidem.* **131**, 763 (1990) and S. J. London et al., *Am. J. Epidem.* **134**, 923 (1991).

174. I. Nair et al., *Biological Effects of Power Frequency Electric and Magnetic Fields—Background Paper*, OTA-BP-E-53 (Washington, DC: GPO. 1989).

175. B. Shi et al., *Environ. Health Perspect.* **111**, 281 (2003).

176. W. Grundler et al., *Naturw.* **79**, 551 (1992).

177. J. D. Brain et al., *Environ. Health Perspect.* **111**, 962 (2003), Another report, by the ICNIRP (International Commission for Non-Ionizing Radiation Protection, Standing Committee on Epidemiology) A. Ahlbom et al., *Environ. Health Perspect.* **109**(S6), 911 (2001), is more circumspect but agrees that there is no clear connection between health and low-frequency fields in most (if not all) cases, These results agree with the conclusion of the National Academy panel of Ref 178.

178. National Academy of Sciences: Committee on the Possible Effects of Electromagnetic Fields on Biologic Systems (C. F. Stevens et al.), *Possible Health Effects of Exposure to Residential Electric and Magnetic Fields* (Washington, DC: National Academy Press, 1997) that "the current body of evidence does not show that exposure to these fields presents a human-health hazard. "

179. E. R. Schoenfeld et al. (Long Island Breast Cancer Study Project), *Am. J. Epidemiol.* **158**, 47 (2003), See also S. Davis et al., *Am. J. Epidemiol.* **155**, 446 (2002) nnd S. J. London et al., *Am. J. Epidemiol.* **158**, 969 (2003),

180. N. Håkansson et al., *Am. J. Epidemiol.* **158**, 534 (2003),

181. J. Sahl et al., *Am. J. Epidemiol.* **156**, 913 (2002).

182. See, for example, the article by B. Gillette, *E/The Environmental Magazine* **XII**(6), 40 (2001).

183. G. J. Aubrecht et al., in M. A. Moreira, ed., *Proceedings of the Seventh InterAmerican Conference in Physics Education* (Porto Alegre, Brasil: IAC. 2000).

184. R. Stuart, *NYT*, 25 May 1976; P. Khiss, *NYT*, 18 February 1979; and M. L. Wald, *NYT*, 11 November 1984.

185. M. L. Wald, *NYT*, 10 May 1992.

186. G. Laroque, *Bull. Am. Phys. Soc.* **30**, 32 (1985).

187. A. McGowan, *Environment* **19**(6), 48 (1977).

188. I. Urbina, *NYT*, 6 February 2004.

189. T. H. Maugh, II. *Science* **178**, 849 (1972).

190. S. Linke and R. E. Schuler, *Annu. Rev. Energy Environ.* **13**, 23 (1988).

191. M. K. Wu et al., *Phys. Rev. Lett.* **58**, 908 (1987); S. R. Ovshinsky et al., *Phys. Rev. Lett.* **58**, 2579 (1987), See also T. H. Geballe and J. K. Hulm, *Sci. Am.* **243**(5), 138 (1980); and J. K. Hulm and B. T. Matthias, *Science* **208**, 881 (1980),

CHAPTER 9

1. M. Allen, *WP*, 1 May 2001.

2. National Energy Policy Development Group (R. Cheney et al.), *National Energy Policy: Report of the National Energy Policy Development Group*, (Washington, DC: GPO, 2001).

3. R. Pool, *LAT*, 12 May 2001.

4. Interlaboratory Working Group, *Scenarios for a Clean Energy Future* (Oak Ridge, TN: Oak Ridge National Laboratory; Berkeley, CA: Lawrence Berkeley National Laboratory; and Golden, CO: National Renewable Energy Laboratory), ORNL/CON-476, LBNL-44029, and NREL/TP-620-29379, November 2000.

5. See the Federal Energy Management Program website at URL: http://www.eren.doe.gov/femp/prodtech/awards/winners01.html#reneworg.

6. See the Federal Energy Management Program website at URL: http://www.eren.doe.gov/femp/prodtech/awards/winners00.html.

7. Department of Energy, FEMP brochure, December 1999, PNNL-SA-32209.

8. Department of Energy, FEMP brochure, December 1999, PNNL-SA-32207.

9. K. Riesselmann, *Fermi News*, 9 November 2001.

10. A. Rosenfeld, *Annu. Rev. Energy Environ.* **24**, 33 (1999).

11. Department of Energy, *1995 Commercial Buildings Energy Consumption Survey* (Washington, DC: GPO, 1997) DOE-0318(95).

12. T. Schevitz, *SFC*, 5 March 2001.

13. E. Mills and M. A. Piette, *Energy* **18**, 75 (1993).

14. J. Schuman et al., *Technology Reviews: Lighting Systems*, LBL-33200 (September 1992).

15. A. B. Lovins and L. H. Lovins, *Annu. Rev. Energy Environ.* **16**, 433 (1991).

16. Dept. of Energy, *Lighting in Commercial Buildings* (Washington, DC: GPO, 1992), DOE/EIA-0555(92)/1.

17. S. M. Nadel et al., *Energy* **18**, 145 (1993).

18. Department of Energy, *Buildings and Energy in the 1980s*, (Washington, DC: GPO, 1995), DOE/EIA-0555(95)/1.

19. K. Fagan, *SFC*, 30 January 2001.

20. See the EnergyStar compact fluorescent website at URL http://www.energystar.gov/products/cfls/#design.

21. Anonymous, AP dispatch, 10 August 2001.

22. T. P. Khánh, *SJMN*, 23 February 2001.

23. K. Gaudette, AP dispatch, 28 July 2001.

24. T. Moran, *MB*, 10 June 2001.

25. H. Chiang, *SFC*, 3 September 2001.

26. C. Burress, *SFC*, 9 May 2001.

27. M. L. Wald, *NYT*, 1 June 1992.

28. D. Hsu, *CCT*, 19 June 2001. The lamp may be found at URL: http://www.lightcorp.com.

29. J. Kahn, *NYT*, 6 May 2001.

30. K. Davidson, *SFC*, 21 May 2001.

31. L. Brezosky, AP dispatch, 25 September 2000.

32. Anonymous, *FEMP Focus Newsletter*, July/August 19179. URL: http://www.eren.doe.gov/femp/newsevents/femp_focus/aug98_effsulfr.html.

33. D. Ranii, *RNO*, 9 February 2001.

34. M. G. Craford et al., *Sci. Am.* **284**(2), 62 (2001). G. Zorpette, *Sci. Am.* **283**(2), 30 (2000) and *IEEE Spectrum* **39**(9), 70 (2002). See also N. Savage, *Tech. Rev.* **103**(5), 38 (2000).

35. M. A. Baldo et al., *Nature* **403**, 750 (2000).

36. A. Bergh et al., *Phys. Today* **54**(12), 42 (2001). P. Ball, Nature Science Update, URL http://www.nature.com/nsu/000217/000217-11.html. See also I. Austen, *NYT*, 31 May 2001.

37. M. Suozzo, *A Market Transformation Opportunity Assessment for LED Traffic Signals*, (Washington, DC: American Council for an Energy-Efficient Economy, 1998).

38. M. Barrows, *SB*, 7 March 2001.

39. *EPRI Journal*, August 2001. See the article online at URL http://www.epri.com/journal/details.asp?doctype=features&id=213.

40. C. Said, *SFC*, 10 March 2001.

41. Environmental Protection Agency and Department of Energy, Energy Star buildings brochure.

42. Consortium for Energy Efficiency website, URL http://www.cee1.org/resrc/facts/led-fx.php3.

43. M. Barrows, *SB*, 15 April 2003.

44. Committee on Benefits of DOE R&D on Energy Efficiency and Fossil Energy Board on Energy and Environmental Systems, Commission on Engineering and Technical Systems (R. W. Fri et al.), National Academy of Sciences, *Energy Research at DOE: Was It Worth It? Energy Efficiency and Fossil Energy Research 1978 to 2000* (Washington, DC: National Academy Press, 2001).

45. J. Kahn, *NYT*, 5 April 2001.

46. Panel on Policy Implications of Greenhouse Warming (D. J. Evans et al.), National Academy of Sciences, *Policy Implications of Greenhouse Warming: Mitigation, Adaptation, and the Science Base* (Washington, DC: National Academy Press, 1992).

47. California Energy Commission, Webpage URL: http://www.consumerenergycenter.org/homeandwork/homes/inside/lighting/bulbs.html.

48. E. Hirst and M. Brown, *Res. Cons. Recycling* **3**, 267 (1990).

49. D. Wooley and R. Cavanaugh, *NYT*, 3 January 1993.

50. W. Carnahan et al., eds., *Efficient Use of Energy, The APS Studies on the Technical Aspects of the More Efficient Use of Energy* (The New York: American Institute of Physics, 1975).

51. Dept. of Energy, *Household Energy Consumption and Expenditures 1990* (Washington, DC: GPO, 1993), DOE/EIA-0321(90).

52. This number is attributed to personnel at Consolidated Edison of New York in Anonymous, *NYT*, 3 February 1974.

53. R. A. Masullo, *SB*, 14 January 2001.

54. P. Hammond, *Science* **178**, 1079 (1972).

55. C. M. Summers, in G. Piel et al., eds., *Energy and Power* (San Francisco: Freeman, 1971), 95.

56. A. J. Wilkins, *Energy* **18**, 123 (1993).

57. Dept. of Energy, *Energy Consumption & Conservation Potential: Supporting Analysis for the National Energy Strategy* (Washington, DC: GPO, 1990), SR/NES 90-02.

58. Dept. of Energy, *State Energy Data Report 1991* (Washington, DC: GPO, 1993), DOE/EIA-0214(91).

59. M. deC. Hinds, *NYT*, 27 January 1980.

60. J. Holusha, *NYT*, 23 November 1980.

61. R. H. Williams et al., *Annu. Rev. Energy* **8** (1983), 269.

62. S. Berman, in D. Hafemeister, W. Kelly, and B. Levi, eds., *Energy Sources: Conservation and Renewables* (The New York: American Institute of Physics, 1985), 247.

63. P.-E. Nilsson and S. Aronsson, *Energy* **18**, 115 (1993).

64. K. Rave, *Energy* **18**, 159 (1993).

65. B. Nielsen, *Energy* **18**, 211 (1993).

66. E. Mills, *Energy* **18**, 131 (1993).

67. J. Cherfis, *Science* **251**, 154 (1991).

68. J. Holusha, *NYT*, 26 October 1994.

69. D. H. W. Li and J. C. Lam, *Energy & Buildings* **35**, 365 (2003). M. R. Atif and A.D. Galasiu, *Energy & Buildings* **35**, 441 (2003).

70. J. Schuman et al., Technology Reviews: Glazing Systems, LBL-33204 (September 1992). See also S. Chirarattananon, J. Nooritanon, and R. Balaka, *Energy* **21**, 505 (1996).

71. G. Sweitzer, *Energy* **18**, 107 (1993). See also Anonymous, *Tech. Rev.* **103**(7), 24 (2000).

72. C. A. Berg, *Science* **181**, 128 (1973).

73. The New York City Board of Education Manual of School Planning, as quoted in Reference 3.

74. S. Bartlett, *Energy* **18**, 171 (1993).

75. H. S. Geller, in D. Hafemeister, W. Kelly, and B. Levi, eds., *Energy Sources: Conservation and Renewables* (The New York: American Institute of Physics, 1985), 270.

76. A. H. Rosenfeld, in D. Hafemeister, W. Kelly, and B. Levi, eds., *Energy Sources: Conservation and Renewables* (The New York: American Institute of Physics, 1985), 92. R. Bevington and A. H. Rosenfeld, *Sci. Am.* **263**(3), 76 (1990).

77. R. L. Rose, *WSJ*, 19 August 1994.

78. S. Prokesch, *NYT*, 12 February 1986.

79. J. Holusha, *NYT*, 30 August 1992; M. L. Wald, *NYT*, 8 July 1992.

80. J. Holusha, *NYT*, 30 June 1993.

81. National Renewable Energy Laboratory, *Energy Savers: Tips on Saving Energy & Money at Home* (Washington, DC: GPO, 2000), DOE/GO-102000-1211.

82. Department of Energy, *Residential Lighting: Use and Potential Savings* (Washington, DC: GPO, 1996), DOE/EIA-0555(96)/2.

83. C. Calwell et al., *Lighting the Way to Energy Savings* (The New York: Natural Resources Defense Council, 1999).

84. Appliance Committee, Consortium for Energy Efficiency, "National Residential Home Appliance Market Transformation Strategic Plan," Consortium for Energy Efficiency report, December 2000.

85. M. L. Willman, *LAT*, 6 April 2001.

86. American Council on an Energy Efficient Economy, URL http://www.aceee.org/consumerguide/topfridge.htm87. Consortium for Energy Efficiency, *Super-Efficient, Apartment-Sized Refrigerator Initiative* (June 2001).

88. J. McMahon and S. Pickle, Center for Building Science News #7, Summer 1995, Lawrence Berkeley National Laboratory, URL http://eetd.lbl.gov/cbs/newsletter/NL7/APS.html.

89. A. Meier, URL: http://ducts.lbl.gov/.

90. EnergyStar website, URL http://www.energystar.gov/ products/, There are separate web pages for refrigerators, dishwashers, washing machines, and air conditioners.

91. M. L. Wald, *NYT*, 19 January 2001.

92. M. L. Wald, *NYT*, 31 March 2001.

93. M. L. Wald, *NYT*, 13 April 2001.

94. M. L. Wald, *NYT*, 14 April 2001.

95. M. L. Wald, *NYT*, 19 June 2001.

96. J. Jardine, *MB*, 8 February 2001.

97. M. L. Wald, *NYT*, 11 April 2001.

98. H. J. Hebert, AP dispatch, 25 October 2001.

99. Shel Feldman Management Consulting, Research Into Action, Inc., and XENERGY, Inc., *The Residential Clothes Washer Initiative: A Case Study of the Contributions of a Collaborative Effort to Transform a Market* Consortium for Energy Efficiency report, June 2001.

100. L. Guernsey, *NYT*, 13 November 2003.

101. M. L. Vellinga, *SB*, 21 May 2001.

102. J. Garofoli, *SFC*, 22 April 2001.

103. Department of Energy, DOE/GO-10095-064, January 1995.

104. P. duPont, *Energy Auditor and Retrofitter* **3**(3), 31 (1986). A. Meier, *Energy Auditor and Retrofitter* **3**(3), 37 (1986).

105. Department of Energy, "Solar water heating: Well-proven technology pays off in several situations," Federal Technology Alert (May 1996).

106. Department of Energy, brochure, DOE/GO-10096-051 March 1996.

107. Department of Energy, brochure, DOE/GO-10098-557, October 1999.

108. Dethman & Associates, *Dishwasher Survey Report*, report submitted to The Northwest Energy Efficiency Alliance and The Consortium for Energy Efficiency, April 1999.

109. K. Rosen and A. Meier, *Energy* **25**, 219 (2000).

110. A. Chen, *CCT*, 2 July 2001.

111. H. Fairfield, *NYT*, 27 February 2001.

112. G. Gugliotta, *WP*, 24 August 2001. P. Wayner, *NYT*, 1 May 2003.

113. P. Wayner, *NYT*, 1 May 2003.

114. A. S. Fung et al., *Energy and Buildings* **35**, 217 (2003).

115. K. Kawamoto et al., *Energy* **27**, 255 (2002).

116. International Energy Agency, *Things That Go Blip in the Night: Standby Power and How to Limit it* (Paris: Organization for Economic Cooperation & Development, 2001).

117. A. LaMar, *CCT*, 24 April 2001. See also S. Leavenworth and D. Kasler, *SB*, 22 February 2001.

118. A. Bridges, AP dispatch, 5 March 2001.

119. J. Tamaki, *LAT*, 6 March 2001.

120. R. Hartway et al., *Energy* **24**, 895 (1999).

121. P. Rogers, *SJMN*, 22 March 2001.

122. L. J. Magid, *SFC*, 29 January 2001.

123. A. Banerjee and B. D. Solomon, *Energy Policy* **31**, 109 (2003).

124. E. Vine et al., *Energy* **26**, 1041 (2001). S. Nadel, *Annu. Rev. Energy Environ.* **27**, 159 (2002).

125. H. J. Hebert, AP dispatch, 30 August 2003.

126. J. Holusha, *NYT*, 27 June 1993.

127. D. Mackenzie, *Science* **278**, 2060 (1997). M. W. Browne, *NYT*, 25 February 1992. S. Garfinkel, *Tech. Rev.* **97**(8), 17 (1994). A. Witze, *DMN*, 9 December 2002.

128. O. Tegus et al., *Nature* **415**, 150 (2002).

129. Anonymous, AP dispatch, 31 December 2001. K. Chang, *NYT*, 19 February 2002.

130. E. Hirst, *Environment* **32**(1), 4 (1990).

131. M. F. Fels and K. M. Keating, *Annu. Rev. Energy Environ.* **18**, 57 (1993).

132. A. H. Rosenfeld and D. Hafemeister, *Sci. Am.* **258**(4), 78 (1988).

133. W. Kempton, *Annu. Rev. Energy Environ.* **18**, 217 (1993).

134. L. Lutzenhisen, *Annu. Rev. Energy Environ.* **18**, 247 (1993).

135. A. Dütz, in W. Leonhardt et al., eds., *Kommunales Energie-Handbuch* (Karlsruhe: Verlag C. F. Müller, 1991), 57.

136. E. Hirst and M. Schweitzer, *Risk Analysis* **10**, 137 (1990).

137. AP dispatch, *NYT*, 16 September 1984; R. Smiley, *NYT*, 11 November 1984; and D. Roe, *NYT*, 11 November 1984.

138. M. L. Wald, *NYT*, 11 September 1992.

139. P. Asmus, *Amicus J.* **14**(4), 38 (1993).

140. M. Kushler et al., *Energy* **28**, 303 (2003).

141. K. N. Lee, *Annu. Rev. Energy Environ.* **16**, 337 (1991).

142. C. Starr, in G. Piel et al., eds., *Energy and Power* (San Francisco: Freeman, 1971), 3.

143. E. Hirst and B. Hannon, *Science* **205**, 656 (1979); E. Hirst and J. Carney, *Science* **199**, 845 (1978).

144. R. L. Lehman and H. E. Warren, *Science* **199**, 879 (1978).

145. L. Hodges, *Phys. Teach.* **22**, 576 (1984).

146. A. M. Al-Turki and G. M. Zaki, *Energy and Buildings* **17**, 35 (1991). N. Klitsikas et al., *Energy and Buildings* **21**, 755 (1994).

147. J. R. Simpson, *Energy and Buildings* **34**, 1067 (2002). I. G. Capeluto, *Energy and Buildings* **35**, 367 (2003).

148. N. Cardinale et al., *Energy and Buildings* **35**, 153 (2003).

149. E. Hirst and J. C. Moyers, *Science* **179**, 1299 (1973).

150. G. A. Lincoln, *Science* **180**, 155 (1973).

151. B. L. Cohen, *Am. J. Phys.* **52**, 614 (1984).

152. Dept. of Energy, *Commercial Buildings Energy Consumption & Expenditures 1989* (Washington, DC: GPO, 1992), DOE/EIA-0318(89).

153. M. F. Fels, *Energy and Buildings* **9**, 5 (1986). M. F. Fels, in D. Hafemeister, W. Kelly, and B. Levi, eds., *Energy Sources: Conservation and Renewables* (The New York: American Institute of Physics, 1985), 169. See also S. Selkowitz, in D. Hafemeister, W. Kelly, and B. Levi, eds., *Energy Sources: Conservation and Renewables* (The New York: American Institute of Physics, 1985), 258. A. Rabl and A. Rialhe, *Energy and Buildings* **19**, 143 (1992).

154. G. S. Dutt et al., *Energy and Buildings* **9**, 21 (1986).

155. M. L. Goldberg, *Energy and Buildings* **9**, 37 (1986).

156. E. Hirst, *Energy and Buildings* **9**, 45 (1986); L. S. Rodberg, *Energy and Buildings* **9**, 55 (1986); M. J. Hewett, T. S. Dunsworthy et al., *Energy and Buildings* **9**, 65 (1986); and C. A. Goldman and R. L. Ritschord, *Energy and Buildings* **9**, 89 (1986).

157. E. Hirst et al., *Energy and Buildings* **8**, 83 (1985).

158. M. D. Levine et al., in D. Hafemeister, W. Kelly, and B. Levi, eds., *Energy Sources: Conservation and Renewables* (The New York: American Institute of Physics, 1985), 299.

159. D. E. Claridge and R. J. Mowris, in D. Hafemeister, W. Kelly, and B. Levi, eds., *Energy Sources: Conservation and Renewables* (The New York: American Institute of Physics, 1985), 201. For an implementation, see M. Shuman, *NYT*, 14 August 1994.

160. T. R. Johansson et al., *Science* **219**, 355 (1983).

161. M. Cunningham, *Insight*, 29 September 1986.

162. E. L. McFarland et al., *Energy, Physics, and the Environment* (Winnipeg, Canada: Wuertz Publ. Ltd., 1994).

163. J. Bidder, *DW*, 28 April 2001.

164. T. D. Manning and I. P. Parkin, *J. Mater. Chem.* **14**, 2554 (2004). For a popular description, see M. Peplow, Nature Science Update, 10 August 2004, URL http://www.nature.com/news/2004/04080 9/pf/040809-2_pf.html. The work is based on U. Qureshi et al., *J. Mater. Chem.* **14**, 1190 (2004), which found a transition temperature of 51 °C.

165. G. T. Armstrong, *Bull. Am. Phys. Soc.* **30**, 32 (1985).

166. J. Amos, *Fine Homebuilding* #24, 65 (1985).

167. G. E. Buchanan and B. G. Honey, *Energy and Buildings* **20**, 205 (1994). Carbon is emitted during housing construction (46 kg/m^2 of carbon) and commercial construction (64 kg/m^2 for timber frame construction and 31 kg/m^2 for steel frame construction).

168. T. Oka et al., *Energy and Buildings* **19**, 303 (1993). They estimate carbon dioxide emission for commercial buildings at 750 to 1140 kg/m^2 (carbon emission 200 to 310 kg/m^2), higher than in Ref. 167, and SO and SO$_2$ emissions of 0.72 to 1.43 kg/m^2, NO and NO$_2$ emissions of 0.70 to 1.14 kg/m^2, and dust generated at 0.07 to 0.13 kg/m^2.

169. S. Meyers and L. Schipper, *Annu. Rev. Energy Environ.* **17**, 463 (1992).

170. C. M. Anders, *SFE*, 23 August 1992.

171. D. Hafemeister, *Am. J. Phys.* **55**, 307 (1987).

172. H. Akbari and O. Sezgea, *Energy and Buildings* **19**, 133 (1992).

173. D. Arasteh and S. Selkowitz, *A Superwindow Field Demonstration Program in Northwest Montana*, LBL-26069 (September 1989).

174. D. Hafemeister and L. Wall, in R. Howes and A. Fainberg, eds., *The Energy Sourcebook* (The New York: American Institute of Physics, 1991), 441.

175. G. S. Dutt, in D. Hafemeister et al., eds., *Energy Sources: Conservation and Renewables* (The New York: American Institute of Physics, 1985), 122; M. Sherman, in D. Hafemeister et al., eds., *Energy Sources: Conservation and Renewables*, 655.

176. M. L. Savitz, *Energy and Buildings* **8**, 93 (1985).

177. P. E. McNall, Jr., *ASHRAE J.* **28**(6), 39 (1986) and **28**(7), 37 (1986). See also J. E. Snell et al., *Science* **192**, 1305 (1976).

178. W. J. Fisk et al., Lawrence Berkeley Laboratory, Berkeley, CA, Report LBL-16493, 1985. See also R. G. Sextro et al., in D. Hafemeister, W. Kelly, and B. Levi, eds., *Energy Sources: Conservation and Renewables* (The New York: American Institute of Physics, 1985), 229.

179. J. D. Spengler and K. Sexton, *Science* **221**, 9 (1983). The authors discuss the Canadian ban; the Consumer Product Safety Commission banned urea formaldehyde insulation in September 1982.

180. D. T. Harrje and K. J. Gadsby, *ASHRAE J.* **28**(7), 32 (1986). See also

A. V. Nero, Jr., *Sci. Am.* **258**(5), 42 (1988).

181. L. C. Oliver and B. W. Shackleton, *Public Health Rep.* **113**, 398 (1998).

182. P. Shabecoff, *NYT*, 14 July 1985.

183. W. R. Ott and J. W. Roberts, *Sci. Am.* **278**(2), 86 (1998).

184. V. Cocheo et al., *Nature* **404**, 141 (2000).

185. L. A. Wallace, *Annu. Rev. Energy Environ.* **26**, 269 (2001).

186. J. D. Spengler and Q. Chen, *Annu. Rev. Energy Environ.* **25**, 567 (2000).

187. W. J. Fisk, *Annu. Rev. Energy Environ.* **25**, 537 (2000).

188. P. Harrison and P. Holmes (U. K. Institute for Environment and Health), *Indoor Air Quality in the Home*, Final Report on DETR Contract EPG 1/5/12 (2001). See a popular explanation of the report in R. Edwards, New Scientist Online News, 17 December 2001. See also L. A. Wallace, Ref. 185.

189. W. J. Fisk and A. H. Rosenfeld, *Indoor Air* **7**, 158 (1997).

190. J. N. Wilford, *NYT*, 26 July 1991. See also Anonymous, *NYT*, 30 July 1991.

191. R. H. Bedzak et al., *Science* **203**, 1214 (1979).

192. J. G. Asbury and R. O. Mueller, *Science* **195**, 445 (1977).

193. E. Kahn, *Annu. Rev. Energy* **4**, 313 (1979).

194. G. Cook, in R. Howes and A. Fainberg, eds., *The Energy Sourcebook* (The New York: American Institute of Physics, 1991), 175.

195. A. L. Hammond and W. D. Metz, *Science* **201**, 36 (1978).

196. Dept. of Energy, *Heating Your Home with an Active Solar Energy System* (Washington, DC: Advanced Sciences, 1990), FSS 177.

197. C. Whipple, *Science* **208**, 262 (1980).

198. W. Schulz, in W. Leonhardt et al., eds., *Kommunales Energie-Handbuch* (Karlsruhe: Verlag C. F. Müller, 1991), 279.

199. G. Cook, in R. Howes and A. Fainberg, eds., *The Energy Sourcebook* (The New York: American Institute of Physics, 1991), 175. See also M. L. Wald, *NYT*, 30 March 1991.

200. S. Cravotta, in T. N. Veziroglu, ed., *Alternative Energy Sources III*, V. 2 (Washington, DC: Hemisphere Publ., 1984), 47.

201. Dept. of Energy, *Passive Cooling* (Washington, DC: Advanced Sciences, 1987), FSS 186.

202. J. G. Ingersoll, in R. Howes and A. Fainberg, eds., *The Energy Sourcebook* (The New York: American Institute of Physics, 1991), 207.

203. K. Frost et al., *Savings from Energy Efficient Windows*, LBL-33956 (April 1993).

204. B. Feder, *NYT*, 8 April 1990.

205. M. L. Wald, *NYT*, 16 August 1992; Anonymous, *NYT*, 29 September 1992.

206. H. H. Wiechmann and Z. Varsek, *Ziegelindustrie International* **5**, 307 (1982). D. W. Hawes et al., *Energy and Buildings* **20**, 77 (1993). See also O. A. Aboloru and K. S. Murali, *Energy and Buildings* **21**, 111 (1993).

207. K. Talib, in T. N. Veziroglu, ed., *Alternative Energy Sources III*, V. 2 (Washington, DC: Hemisphere Publ., 1984), 57.

208. J. K. Nayak et al., *Solar Energy* **30**, 51 (1983).

209. Department of Energy, brochure, DOE/CH10093-350, November 1994.

210. A. N. Tombazis and S. A. Preuss, *Solar Energy* **70**, 311 (2001).

211. National Renewable Energy Laboratory, *Clean Energy Choices* (Washington, DC: GPO, 2000), DOE/GO-102000-1012.

212. D. McLellan, *LAT*, 15 July 2001. A. H. Rosenfeld et al., *Tech. Rev.* **100**(2), 52 (1997).

213. S. H. Samuels, *NYT*, 16 January 2001.

214. G. Kats et al. (A Report to California's Sustainable Building Task Force, by Capital E, the California Department of Health Services, and Lawrence Berkeley Laboratory), *The Costs and Financial Benefits of Green Buildings*, October 2003. For a popular description, see Anonymous, GreenBiz.com, 28 October 2003.

215. R. Vincent, *LAT*, 30 March 2003.

216. Anonymous, GreenBiz.com, 7 May 2003.

217. D. Snoonian, *IEEE Spectrum* **40**(8), 18 (2003).

218. R. Brown , C. Webber, and J.G. Koomey, *Energy* **27**, 505 (2002).

219. Anonymous, GreenBiz.com, 17 March 2004.

220. T. Pristin, *NYT*, 4 June 2003.

221. J. Finer, *WP*, 24 April 2004.

222. C. Hawthorne, *NYT*, 13 November 2003. D. Whitcomb, Reuters dispatch, 14 November 2003.

223. W. Potter, *The Chronicle of Higher Education*, 26 March 2004.

224. D. Wedner, *LAT*, 21 March 2004. M. Rich, *NYT*, 6 May 2004.

225. M. Holtz et al., *Solar Age* **10**(10), 49 (1985).

226. C. A. Goldman, *Energy and Buildings* **8**, 137 (1985).

227. J. D. Balcomb, *Energy and Buildings* **7**, 281 (1984).

228. L. R. Glicksman et al., *Annu. Rev. Energy Environ.* **26**, 83 (2001).

229. E. A. MacDougall, in T. N. Veziroglu, ed., *Alternative Energy Sources II, V. 2* (Washington, DC: Hemisphere Publ., 1982), 931 and in T. N. Veziroglu, ed., *Alternative Energy Sources III, V. 2* (1984), 44. See also N. E. Collins and W. E. Handy, in T. N. Veziroglu, ed., *Alternative Energy Sources II, V. 2* (Washington, DC: Hemisphere Publ., 1982), 937, C. Benard and D. Gobin, in T. N. Veziroglu, ed., *Alternative Energy Sources II* V. 2 (Washington, DC: Hemisphere Publ., 1982), 949, and E. M. Abu El-salam, in T. N. Veziroglu, ed., *Alternative Energy Sources III, V. 2* (1984), 447.

230. M. N. Öziik et al., in T. N. Veziroglu, ed., *Alternative Energy Sources II* (Washington, DC: Hemisphere Publ., 1982), 891.

231. J.-L. Bourgeois, *Natl. Hist.* **89**(11), 70 (1980). M. N. Bahadori, *Sci. Am.* **238**(2), 144 (1978).

232. R. V. Pound, *Science*, **208**, 494 (1980). S. W. Jacobson, *NYT*, 23 December 1982.

233. K. R. Foster and A. W. Gray, *Sci. Am.* **255**(3), 32 (1986).

234. D. Adelman, *Fine Homebuilding* #22, 68 (1984).

235. M. P. Yazigi, *NYT*, 8 November 1992.

236. R. Marshall, *Phys. Educ.* **37**, 158 (2002).

237. H. Hottel and J. Howard, *New Energy Technology—Some Facts and Assessments* (Boston: MIT Press, 1970).

238. R. A. Macriss, *Annu. Rev. Energy* **8**, 247 (1983).

239. R. Idler, in W. Leonhardt et al., eds., *Kommunales Energie-Handbuch* (Karlsruhe: Verlag C. F. Müller, 1991), 207.

240. C. E. Hewett et al., *Annu. Rev. Energy* **6**, 139 (1981).

241. B. Gladstone, *NYT*, 22 December 1977.

242. L. Cranberg, *Am. J. Phys.* **49**, 596 (1981).

243. B. Franklin, *An Account of the New Invented Pennsylvania Fireplaces*, Philadelphia, 1744 (as quoted in E. R. Berndt, *Annu. Rev. Energy* **3**, 225 (1978)). Quotations are from pages 1 and 32 of the original.

244. J. F. Kowalczyk and B. J. Tombleson, *J. Air Poll. Control Assoc.* **35**, 619 (1985).

245. R. H. Socolow, *Annu. Rev. Energy* **2**, 239 (1977); and R. H. Socolow, in D. Hafemeister et al., eds., *Energy Sources: Conservation and Renewables* (The New York: American Institute of Physics, 1985), 15.

246. National Renewable Energy Laboratory, *Geothermal Today: Clean energy for the 21st century* (Washington, DC: GPO, 2000), DOE/GO-102000-1066.

247. S. Martínez, *SB*, 19 February 2001.

248. J. V. Iovine, *NYT*, 4 January 2001.

249. GeoExchange manufacturer website URL: http://www.geoexchange.org/residential/residential.htm.

250. J. Hagnegruber, AP dispatch, 23 March 2001.

251. Stanford Research Institute, *Patterns of Energy Consumption in the United States* (Washington, DC: GPO, 1972).

252. G. J. Aubrecht, II, *Phys. Teach.* **21**, 30 (1981).

253. R. Klopfleisch, in W. Leonhardt et al., eds., *Kommunales Energie-Handbuch* (Karlsruhe: Verlag C. F. Müller, 1991), 15.

254. A. B. Peters, in W. Leonhardt et al., eds., *Kommunales Energie-Handbuch* (Karlsruhe: Verlag C. F. Müller, 1991), 189.

255. G. Luther et al., in W. Leonhardt et al., eds., *Kommunales Energie-Handbuch* (Karlsruhe: Verlag C. F. Müller, 1991), 231.

256. C. Zuleg, Diplomarbeit (thesis) submitted to Fachhochschule Offenburg, 2001.

257. J. Schilling, *DW*, 5 January 2000. M. Ronzheimer, *DW*, 29 May 2001.

258. H. Keßler, *DZ*, 3 August 1996.

259. A. J. Parisi, *NYT*, 29 October 1977.

260. P. A. A. Berle, *NYT*, 6 June 1993.

261. T. Kelly, *Canada Today*, March 1982, 4.

262. A. H. Rosenfeld and D. Hafemeister, in D. Hafemeister et al., eds., *Energy Sources: Conservation and Renewables* (The New York: American Institute of Physics, 1985), 148.

263. N. J. Numack and A. A. Bartlett, *Am. J. Phys.* **50**, 329 (1982).

264. K. Yamanouchi, *LAT*, 12 July 2002

265. Anonymous, AP dispatch, 2 June 2003.

266. See D. B. Crawley and R. S. Briggs, *ASHRAE J.* **27**(11), 18 (1985) for a discussion of ASHRAE Standard 90.

267. S. Meyers et al., *Energy* **28**, 755 (2003).

268. J. G. Koomey et al., *Energy* **24**, 69 (1999).

269. M. Ross, *Science* **244**, 311 (1989).

270. R. C. Marlay, *Science* **226**, 1277 (1984). R. A. Frosch and N. E. Gallopoulos, *Sci. Am.* **261**(3), 144 (1989).

271. L. Schipper et al., *Annu. Rev. Energy* **15**, 455 (1990).

272. M. Ross, *Annu. Rev. Energy* **6**, 379 (1981). M. H. Ross and D. Steinmeyer, *Sci. Am.* **263**(3), 88 (1990).

273. Dept. of Energy, *Energy Use and Carbon Emissions: Some International Comparisons* (Washington, DC: GPO, 1994), DOE/EIA-0579.

274. M. Ross, in D. Hafemeister, W. Kelly, and B. Levi, eds., *Energy Sources: Conservation and Renewables* (The New York: American Institute of Physics, 1985), 347.

275. D. J. BenDaniel and E. E. David, Jr., *Science* **206**, 773 (1979).

276. S. F. Baldwin, *Annu. Rev. Energy* **13**, 67 (1988). See also S. F. Baldwin, in R. Howes and A. Fainberg, eds., *The Energy Sourcebook* (The New York: American Institute of Physics, 1991), 475.

277. J. P. Hicks, *NYT*, 10 October 1990.

278. Consortium for Energy Efficiency, Efficient motors brochure, 1999.

279. J. R. Miller, *Sci. Am.* **250**(5), 32 (1984).

CHAPTER 10

1. E. D. Larson et al., *Sci. Am.* **254**(6), 34 (1986).

2. C. Mayer, *Science* **227**, 1421 (1985).

3. G. P. Eaton, *Am. Sci.* **72**, 368 (1984).

4. C. Holden, *Science* **212**, 305 (1981). See also *Mining and Minerals Policy*, a report by the Secretary of the Interior to the Congress, 1973.

5. Bureau of Mines, *Mineral Facts and Problems*, Bulletin 617 (Washington, DC: GPO, 1980).

6. U.S. Department of the Interior, U.S. Geological Survey, *Mineral Commodity Summaries 2001* (Washington, DC: GPO, 2001) [online at URL http://minerals.usgs.gov/minerals/pubs/mcs/2001/mcs2001.pdf]. See also R. C. Kirby and A. S. Prokopovitsh, *Science* **191**, 713 (1976).

7. Dept. of the Interior, *First Annual Report of the Secretary of the Interior under the Mining and Minerals Policy Act of 1970*; this report to Congress was delivered in March 1972.

8. P. E. Cloud, Jr., *Texas Quarterly* **11**(2), 103 (1968). See also P. Cloud, *Ecologist* **7**, 273 (1977).

9. J. Diamond, *Guns, Germs, and Steel* (New York and London: W. W. Norton & Co., 1999).

10. D. R. Francis, *CSM*, 5 December 2002.

11. D. A. Taylor, *Environ. Health Perspect.* **112**, A168 (2004). P. McLoughlin, Reuters dispatch, 24 August 2004.

12. D. Ludwig et al., *Science* **260**, 17 (1993).

13. P. Sampat, in L. Starke, ed. *State of the World 2003* (New York and London: W. W. Norton & Co., 2003), 110.

14. J. E. Young, in L. Starke, ed., *State of the World 1992* (New York and London: W. W. Norton, 1992), 100.

15. R. Cockle, *PO*, 26 January 2001.

16. Anonymous, AP dispatch, 22 October 2000.

17. S. Fields, *Environ. Health Perspect.* **109**, A474 (2001). S. Fields, *Environ. Health Perspect.* **111**, A154 (2003).

18. D. K. Nordstrom and C. N. Alpers, *Proc. Natl. Acad. Sci.* **96**, 3455 (1999).

19. J. Pasternak, *LAT*, 3 May 2001.

20. J. Kay, *SFE*, 26 September 2000. See also S. Leavenworth, *SB*, 19 October 2003, S. Leavenworth, *SB*, 3 September 2004.

21. D. Thompson, AP dispatch, 11 April 2003. J. Cart, *LAT*, 15 April 2003.

22. P. Grandjean et al., *Environ. Health Perspect.* **107**, 587 (1999).

23. E. M. Yokoo et al., *Environ. Health.* **2**(1), 8 (2003).

24. Anonymous, AP dispatch, 30 August 2000.

25. J. Kay, *SFC*, 13 April 2001. R. A. Saar, *NYT*, 2 November 1999.

26. H. H. Harris et al., *Science* **301**, 1203 (2003).

27. H. G. McCann, AP dispatch, 15 September 1999.

28. M. Janofsky, *NYT*, 4 August 2004. M. Janofsky, *NYT*, 25 August 2004.

29. H. J. Hebert, AP dispatch, 14 December 2000.

30. J. 8. Lee, *NYT*, 3 December 2003.

31. Anonymous, *LAT*, 14 May 2004.

32. M. Cone, *LAT*, 12 July 2000; A. C. Revkin, *NYT*, 12 July 2000.

33. J. Sandin, *MJS*, 24 June 2001.

36. E. Pianin, *WP*, 10 December 2003.

37. E. Pianin, *WP*, 6 November 2003.

34. T. Hosaka, *PO*, 20 February 2001. B. Daley, *BG*, 28 August 2001.

35. M. Cole, *PO*, 1 April 2004.

38. W. H. Schroeder et al. *Nature* **394**, 331 (1998). J. Y. Lu et al., *Geophys. Res. Lett.* **28**, 3219 (2001).

39. A. Martínez-Cortizas et al., *Science* **284**, 939 (1999).

40. H. Brim et al., *Nature Biotechnology* **18**, 85 (2000).

41. See, for example, S. P. Bizily et al., *Nature Biotechnology* **18**, 213 (2000). See also A. S. Moffat, *Science* **285**, 369 (1999).

42. U. Noikorn, AP dispatch, 16 July 2001. For an American experience, see C. Savoye, *CSM*, 4 March 2002; J. Stange, AP dispatch, 17 January 2002; J. Wilgoren, *NYT*, 19 January 2002; J. Stange, AP dispatch, 28 February 2002; M. Harden, *CD*, 23 June 2002.

43. D. E. Jacobs et al., *Environ. Health Perspect.* **110**, A599 (2002).

44. H. W. Mielke et al., *Environ. Health Perspect.* **109**, 973 (2001).

45. D. E. Jacobs et al., *Environ. Health Perspect.* **111**, 185 (2003).

46. D. Q. Rich et al., *Environ. Health Perspect.* **110**, 889 (2002). L.-M. Yiin et al., *Environ. Health Perspect.* **110**, 1233 (2002).

47. J. Wakefield, *Environ. Health Perspect.* **110**, A574 (2002). E. Agnvall, *WP*, 30 March 2004.

48. M. Scott, *BG*, 24 October 2000.

49. L. Tanner, AP dispatch, 15 May 2000.

50. S. Pistoi, Scientific American Online, 17 May 2001.

51. H.-Y. Chuang et al., *Environ. Health Perspect.* **109**, 527 (2001).

52. B. P. Lanphear, *Science* **281**, 1617 (1998).

53. R. L. Canfield et al., *N. Engl. J. Med.* **348**, 1517 (2003). Only 30 years ago, 30 µg per deciliter was considered a "safe" level; the accompanying commentary, W. J. Rogan and J. H. Ware, *N. Engl. J. Med.* **348**, 1515 (2003),

suggests that no level is "safe." The result of Canfield et al. was confirmed by DC. Bellinger and H. L. Needleman, *N. Engl. J. Med.* **349**, 500 (2003).

54. P. J. Landrigan et al., *Environ. Health Perspect.* **110**, 721 (2002).

55. S. D. Grosse et al., *Environ. Health Perspect.* **110**, 563 (2002).

56. P. B. Stretesky and M. J. Lynch, *Arch. Pediatr. Adolesc. Med.* **155**, 579 (2001).

57. R. Kaiser et al., *Environ. Health Perspect.* **109**, 563 (2001).

58. D. J. Steding et al., *Proc. Natl. Acad. Sci.* **97**, 11181 (2000).

59. I. Renberg and M. W. Persson, *Nature* **368**, 323 (1994). W. Shotyk et al., *Science* **281**, 1635 (1998). See also J. O. Nriagu, *Science* **272**, 223 (1996) and J. O. Nriagu, *Science* **281**, 1622 (1998).

60. B. Nowack et al., *J. Environ. Quality* **30**, 919 (2001).

61. D.Gleba et al., *Proc. Natl. Acad. Sci.* **96**, 5973 (1999).

62. L. Q. Ma et al., *Nature* **409**, 579 (2001) and Å. Olson and J.Stenlid, *Nature* **411**, 438 (2001). P. N. Spotts, *CSM*, 1 February 2001.

63. National Academy of Sciences, Committee on Earth Resources (C. Otte et al.) *Mineral Resources and Sustainability: Challenges for Earth Scientists* (Washington, DC: National Academy Press, 1996).

64. D. B. Brooks and P. W. Andrews, *Science* **185**, 13 (1974).

65. W. Greene, *NYT Magazine*, 5 January 1975, 9.

66. G. Hardin, *BioScience* **24**, 561 (1974); G. Hardin, *Psychology Today* **8**(4) 38 (1974).

67. G. Matos and L. Wagner, *Annu. Rev. Energy Environ.* **23**, 107 (1998).

68. R. M. Emrick, *Phys. Teach.* **31**, 242 (1993).

69. V. E. McKelvey, *Am. Sci.* **60**, 32 (1972).

70. P. Flawn, *Mineral Resources* (New York: Rand McNally & Co., 1966), 14.

71. C. F. Westoff, *Sci. Am.* **231**(3), 108 (1974).

72. T. S. Lovering, in P. Cloud et al., eds., *Resources and Man* (San Francisco: Freeman, 1969), 109. See also T. S. Lovering, *Texas Quarterly* **11**(2), 127 (1968).

73. E. Cook, *Science* **191**, 677 (1976).

74. S. G. Lasky, *Eng. Mining J.* **151**(4), 81 (1950).

75. J. Harte and R. Socolow, in J. Harte and R. Socolow, eds., *Patient Earth* (New York: Holt, Rinehart & Winston, 1971), 84; W. D. Metz, *Science* **183**, 59 (1974). Metz points out that each billion cubic feet of helium retrieved from air costs 10% of the U.S. annual energy use. For a more recent analysis, see E. F. Hammel et al., *Science* **223**, 789 (1984).

76. Energy Information Administration, *Largest U.S. Oil and Gas Fields*, DOE/EIA-TR-0567 (Washington, DC: GPO, 1993).

77. D. H. Meadows et al., *The Limits to Growth* (New York: Universe Publ., 1972). For a later reconsideration of these issues, see D. H. Meadows et al., *Beyond the Limits: Confronting Global Collapse, Envisioning a Sustainable Future* (Post Mills, VT: Chelsea Green Publ. Co., 1992). See also D. H. Meadows, *Amicus J.* **14**(1), 27 (1992).

78. K. Schneider, *NYT*, 27 April 1993.

79. I. Mayer-List, *DZ*, 13 August 1984.

80. H. E. Goeller and A. M. Weinberg, *Science* **191**, 683 (1976).

81. A. G. Chynoweth, *Science* **191**, 725 (1976).

82. W. B. Hillig, *Science* **191**, 733 (1976).

83. M. W. Barsoum and T. El-Raghy, *Am. Sci.* **89**, 334 (2001). R. Cooke, *Newsday*, 17 July 2001.

84. P. E. Cloud, Jr., in P. Cloud et al., eds., *Resources and Man* (San Francisco: Freeman, 1969), 135.

85. A. L. Hammond, *Science* **183**, 502, 644 (1974).

86. R. A. Kerr, *Science* **223**, 576 (1984).

87. F. T. Manheim, *Science* **232**, 600 (1986).

88. K. S. Jayaraman, *Nature* **417**, 212 (2002).

89. P. A. Rona, *Science* **299**, 673 (2003).

90. A. A. Bartlett, *Am. J. Phys.* **54**, 398 (1986).

CHAPTER 11

1. U.S EPA, Office of Solid Waste and Emergency Response, Municipal Solid Waste in The United States: 2001 Facts and Figures (Washington, DC: GPO, 2003), EPA530-R-03-011. N. Shute, *Amicus J.* **12**(3), 44 (1990).

2. Umweltbundesamt, *Daten zur Umwelt* (Berlin: Erich Schmidt Verlag, 1992).

3. Anonymous, *DZ*, 22 November 1991.

4. J. R. Luoma, *Audubon* **92**(2), 86 (1990).

5. A glimpse into the lives of a rag-picker's family may be had in the all-of-a-kind family books by Sidney Taylor, originally published by Follette Publishing, Chicago, and in paperback by the Dell Yearling Series.

6. See, for example, Stewart Brand, ed., *The Last Whole Earth Catalog; Access to Tools* (New York: Random House, 1980).

7. Environmental Protection Agency, *Environmental Quality-1970*, (Washington, DC: GPO, 1970).

8. Dept. of Commerce, *Statistical Abstracts of the United States, 2000* (Washington, DC: GPO, 2000), Table 396. Dept. of Commerce, *Statistical Abstracts of the United States, 1992* (Washington, DC: GPO, 1992), Table 373.

9. R. U. Ayres et al., *Energy* **23**, 355 (1998).

10. Dept. of Commerce, *Statistical Abstracts of the United States, 1992* (Washington, DC: GPO, 1992), Table 915.

11. Dept. of Energy, *Annual Energy Review 2000* (Washington, DC: GPO, 2001), DOE/EIA-0384(00).

12. C. H. Deutsch, *NYT*, 17 March 1999.

13. A. P. Carter, *Science* **184**, 325 (1974).

14. Office of Emergency Preparedness, *The Potential for Energy Conservation* (Washington, DC: GPO, 1972).

15. M. Ross, *Annu. Rev. Energy* **6**, 379 (1981).

16. J. P. Hicks, *NYT*, 27 February 1991.

17. J. R. Miller, *Sci. Am.* **250**(5), 32 (1984).

18. E. T. Hayes, *Science* **191**, 661 (1976).

19. R. Iranpour et al., *Science* **285**, 706 (1999). See also L. Martin, *Environment* **28**(3), 35 (1986).

20. J. E. Young, in L. Starke, ed., *State of the World 1991* (New York: W. W. Norton, 1991), 39.

21. A. Modl and F. Hermann, *Annu. Rev. Energy Environ.* **20**, 233 (1995).

22. R. M. Abbott, "Eco-Labeling and the Green Economy: Strategic Options for British Columbia," report to the British Columbia Green Economy Secretariat, November 2000.

23. Umweltbundesamt, *Daten zur Umwelt* (Berlin: Erich Schmidt Verlag, 1997), Sec. 1.

24. Duales System Deutschland, AG, website URL: http://www.gruener-punkt.de. See the section "Unternehmen," with data available in English at URL: http://www.gruener-punkt.de/en/frames.php3?choice1=bilanz&choice2=mengenstrom.

25. J. Zurheide, *Der Tagesspiegel*, 14 January 1992 (translated in the *German Tribune*).

26. J. Holusha, *NYT*, 9 August 1992.

27. J. Sandin, *MJS*, 29 April 2001. W. Bergstrom, AP dispatch, 7 May 2000.

28. K. Ueta and H. Koizumi, *Environment* **43**(9), 20 (2001).

29. Anonymous, AP dispatch, 13 July 2000.

30. Anonymous, Reuters dispatch, 23 December 2002.

31. J. Mozingo, *LAT*, 12 May 2001.

32. K.-H. Joepen, *DZ*, 17 February 1984.

33. C. J. Williams, *LAT*, 14 July 2001. This epitaph was a bit premature (see Anonymous, *Stuttgarter Zeitung*, 2 July 2002), but it is true that the 0.25 Euro deposit bill is facing uncommonly tough sledding. As of this writing, a judge has ruled that the deposit is not allowed (see Anonymous, Deutsche Press Agentur dispatch, 10 September 2002, Anonymous, AP dispatch, 11 September 2002, and D. Dehmer, *Der Tagesspiegel* [Berlin], 12 September 2002) but the government planned to go ahead with the deposit beginning 1 January 2003 with the blessing of the

European Union (see Anonymous, Deutsche Press Agentur dispatch, 23 August 2002). M. Landler, *NYT*, 1 January 2003.

34. S. M. Gupta and P. Veerakamolmal, *Disassembly of Products*. Final grant report, NIST Systems Integration for Manufacturing Application Program Grant No. 60NANB5D0112, December 1996.

35. Anonymous, AP dispatch, 4 June 2003; Anonymous, AP dispatch, 2 October 2003.

36. J. D. Underwood and B. Fishbein, *NYT*, 4 April 1993.

37. F. Protzman, *NYT*, 12 July 1992.

38. F. Protzman, *NYT*, 4 July 1993.

39. C. J. Williams, *LAT*, 11 August 2002.

40. S. P. Reynolds, *Villanova Environmental Law J.* **6**, 1 (1995).

41. Duales System Deutschland, AG, *The Closed-Cycle Economy in Figures*, brochure, June 2002.

42. Press Office, Ireland Department of the Environment and Local Government, *What is the plastic shopping bag levy?*, brochure, January 2002.

43. S. Pogatchnik, AP dispatch, 5 March 2002. See also Anonymous, *CSM*, 17 April 2002.

44. M. Kuhl, *CSM* , 17 April 2002.

45. D. Kriebel et al., *Environ. Health Perspect.* **109**, 871 (2001).

46. K. R. Foster et al., *Science* **288**, 97 (2000).

47. M. MacGarvin et al., *Late lessons from early warnings: the Precautionary Principle 1896–2000* (Copenhagen: European Environment Agency, 2001).

48. T. O'Riordan and J. Cameron, eds., *Interpreting the Precautionary Principle* (London: Earthscan Publications, 1994).

49. J. Tickner et al., *The Precautionary Principle in Action: A Handbook* (Windsor ND and Lowell, MA: The Science and Environmental Health Network and Lowell Center for Sustainable Production, 1999).

50. C. Raffensperger and J. Tickner, eds., *Protecting Public Health and the Environment: Implementing the Precautionary Principle* (Washington, DC: Island Press, 1999).

51. J. Cairns, Jr., *Environ. Health Perspect.* **111**, 877 (2003).

52. R. Costanza and C. Perrings, *Ecological Economics* **2**, 57 (1990). R. Costanza and L. Cornwell, *Environment* **34**(9) 12 (1992).

53. J. D. Graham, Heritage Lecture #818, 15 January 2004. Available at URL http://www.heritage.org/Research/Regulation/hl818.cfm. Graham was at the time the Administrator of the Office of Information and Regulatory Affairs at the Office of Management and Budget.

54. J. Holusha, *NYT*, 28 May 1991.

55. R. Schwalbe, *DW*, 20 October 1990 (translated in the *German Tribune*). See also R. Praetorius, *DZ*, 9 October 1987.

56. G. Küffner, *Frankfurter Allgemeine Zeitung*, 24 July 1990.

57. D. Oehler, *Stuttgarter Zeitung*, 18 August 1990.

58. Anonymous, *Allgemeine Zeitung* (Mainz), 14 July 1990.

59. B. Daviss, *Futurist* **23**(3), 28 (1999).

60. M. Krebs, *NYT*, 27 April 1999.

61. European Union Directorate, "Proposal for a directive on waste from electrical and electronic equipment," 1999. See also F. Cosentino and I. Clasper, *Measurement + Control* **32**(12), 1 (1999).

62. J. Holusha, *NYT*, 25 May 1993.

63. R. Ferrone, *IEEE Spectrum* **35**(1), 95 (1998).

64. K. S. Betts, *IEEE Spectrum* **36**(1), 104 (1999).

65. W. McCartney, *IEEE Spectrum* **36**(1), 106 (1999).

66. C. Frankel, *IEEE Spectrum* **33**(1), 76 (1996).

67. N. Heinze, *IEEE Spectrum* **35**(1), 98 (1998).

68. S. Lohr, *NYT*, 14 April 1993.

69. D. F. Gallagher, *NYT*, 18 July 2002.

70. Y. M. Ibrahim, *NYT*, 14 December 1977.

71. I. Meyer-List, *DZ*, 13 August 1984.

72. A. Pollack and K. Bradsher, *NYT*, 13 March 2004.

73. C. Fagan, AP dispatch, 2 March 2003. E. L. Andrews and S. Sachs, *NYT*, 18 May 2003.

74. Office of Solid Waste and Emergency Response, Environmental Protection Agency, *Municipal Solid Waste in the United States: 2000 Facts and Figures* (Washington, DC: GPO, 2002), EPA530-R-02-001, June 2002. See also R. A. Denison, *Annu. Rev. Energy Environ.* **21**, 191 (1996).

75. U.S. Department of the Interior, U.S. Geological Survey, *Mineral Commodity Summaries 2005* (Washington, DC: GPO, 2005).

76. Container Recycling Institute, Container Recycling Institute website, URL http://www.container-recycling.org/. P. Franklin, Container Recycling Institute press release, 19 April 2000.

77. J. Gitlitz, Container Recycling Institute (2004) finds a 44% aluminum recycling rate, private communication, 25 February 2005.

78. K. Engstrom, *Endeavour* **16**, 117 (1992). See also J. Fußer, *DZ*, 23 September 1988, who discusses implementation of the German law mandating deposits for plastic beverage containers.

79. Anonymous, Environmental News Network dispatch, 4 December 2000.

80. H. F. Kurtz and L. H. Baumgardner, in *Mineral Facts and Problems*, 1980 ed., Bu. Mines Bull. 650, U.S. Dept. of the Interior (Washington, DC: GPO, 1980), 9.

81. Stanford Research Institute, *Patterns of Energy Consumption in the United States* (Washington, DC: GPO, 1972).

82. S. Bronstein, *NYT*, 14 September 1986.

83. W. U. Chandler, *Conservation and Recycling* **9**, 87 (1986).

84. J. de Beer et al., *Annu. Rev. Energy Environ.* **23**, 123 (1998).

85. I. K. Wernick and N. J. Themelis, *Annu. Rev. Energy Environ.* **23**, 465 (1998).

86. J. B. Rosenbaum, *Science* **191**, 720 (1976).

87. J. D. Good, *Endeavour* **10**, 150 (1986).

88. C. H. Deutsch, *NYT*, 18 March 1990.

89. J. Holusha, *NYT*, 7 November 1990.

90. T. Geisen, *General Anzeiger* (Bonn), 23 June 1990.

91. H. Keßler, *DZ*, 10 December 1993.

92. J. Holusha, *NYT*, 19 August 1993.

93. J. Holusha, *NYT*, 10 September 1989.

94. J. Holusha, *NYT*, 21 October 1993.

95. J. Holusha, *NYT*, 6 January 1991.

96. J. Holusha, *NYT*, 20 October 1993.

97. R. A. Denison, *EDF Letter* **24**(6), 4 (1993).

98. H. Keßler, *DZ*, 28 January 1994.

99. B. R. Hook, American Recycler 7 (9), 1 (September 2004).

100. B. J. Feder, *NYT*, 8 January 1991.

101. N. Wolf and E. Feldman, *Plastics: America's Packaging Dilemma* (Washington, DC and Covelo, CA: Island Press, 1991).

102. S. Diesenhouse, *NYT*, 20 February 1994.

103. L. Jacobson, *WP*, 8 December 2003.

104. M. B. Hocking, *Science* **251**, 504 (1991).

105. D. W. Laist, in J. M. Coe and D. B. Rogers, eds., *Marine Debris: Sources, Impacts and Solutions* (New York: Springer-Verlag, 1997). J. G. B. Derraik, *Marine Pollution Bulletin* **44**, 842 (2002). Y. Dandonneau et al., *Science* **302**, 1548 (2003). R. C. Thompson et al., *Science* **304**, 838 (2004). Preliminary Report of the U.S. Commission on Ocean Policy, Governors' Draft, Washington, DC: April 2004, Ch. 18.

106. A. C. Palmisano and C. A. Pettigrew, *BioScience* **42**, 680 (1992).

107. S. J. Ainsworth and A. M. Thayer, *Chem. & Engg. News* **72**(42), 10 (1994).

108. T. E. Graedel, *Annu. Rev. Energy Environ.* **21**, 69 (1996). See also L. W. Jelinski et al., *Proc. Natl. Acad. Sci.* **89**, 793 (1992), R. A. Frosch, *Phys. Today* **47**(11), 63 (1994), R. A. Frosch, *Sci. Am.* **273**(3), 178 (1995), and N. Gertler and J. R. Ehrenfeld, *Tech. Rev.* **99**(2), 48 (1996).

109. B. R. Allenby, *IEEE Spectrum* **33**(1), 81 (1996). H. Kane, in L. Starke, ed. *State of the World 1996* (New York & London: W. W. Norton, 1996), 152.

110. R. A. Denison and J. F. Ruston, *Tech. Rev.* **100**(7), 55 (1997).

111. C. H. Deutsch, *NYT*, 14 July 1998.

112. C. H. Deutsch, *NYT*, 9 September 2001.

113. United States Environmental Protection Agency, *Integrated Environmental Management Systems* (Washington, DC: GPR, 2000), EPA 744-R-00-011.

114. J. Kaiser, *Science* **285**, 685 (1999).

115. G. Matos and L. Wagner, *Annu. Rev. Energy Environ.* **23**, 107 (1998).

116. J. Kaiser, *Science* **285**, 686 (1999).

117. C. Woodard, *CSM*, 19 July 2001.

118. M. R. Chertow, *Annu. Rev. Energy Environ.* **25**, 313 (2000).

119. B. Piasecki, *Proc. Natl. Acad. Sci.* **89**, 873 (1992).

120. P. Franklin, talk presented at the Take it Back! '97 Producer Responsibility Forum, 18 November 1997.

121. A. Cortese, *NYT*, 6 May 2001.

122. S. Kinsella and D. Knapp, URL http://www.grrn.org/zerowaste/grrn1.html.

123. European Environmental Bureau, EEB Document no 2001/007, Brussels, April 2001.

124. E. Lymberidi, *Towards Waste-Free Electrical and Electronic Equipment*, European Environmental Bureau argumentation paper concerning the proposals for Directives on Waste Electrical and Electronic Equipment and on the Restriction of the use of certain Hazardous Substances in Electrical and Eletronical Equipment, March 2001.

125. N. Grove, *Natl. Geog.* **186**(1), 92 (1994).

126. U.S. EPA, Office of Solid Waste and Emergency Response, Municipal Solid Waste in The United States: 2001 Facts and Figures (Washington, DC: GPO, 2003), EPA530-R-03-011.

127. J. Holusha, *NYT*, 23 January 1994. D. Jackson, *Waste Age*, February 2002.

128. N. S. Glance and B. A. Huberman, *Sci. Am.* **270**(3), 76 (1994).

129. M. Magnier, *LAT*, 13 May 2001.

130. M. Clayton, *CSM*, 2 January 2004.

131. C. Swett, *SB*, 14 July 2004.

132. J. Holusha, *NYT*, 23 January 1994. D. Jackson, Waste Age 33(2), 16 (2002).

133. B. Burmeier, *CSM*, 21 April 2003.

134. A. O'Connor, *NYT*, 8 October 2002.

135. C. W. Schmidt, *Environ. Health Perspect.* **110**, A188 (2002).

136. Anonymous, report on electric appliance recycling Japan Foreign Press Center press release, April 2001.

137. S. Elston, Environmental News Network dispatch, 2 January 2001.

138. J. Motavalli , *E/The Environmental Magazine* **12**(2), 26 (2001). A description of the Minnesota Sony program may be found at the Sony website URL http://www.sel.sony.com/SEL/esh/mnproj/wpaper.html.

139. Anonymous, GreenBiz.com dispatch, 1 May 2002.

140. E. de Bony, IDG dispatch, 15 June 2000.

141. M. Brady, *The Institute* **23**(5), 1 (1999).

142. J. E. Boon et al. Paper presented at the 2000 Environmentally Conscious Manufacturing Conference, Boston, Massachusetts, November 6-8, 2000.

143. J. Pepi, *University of Massachusetts Amherst Scrap Electronics Processing*, Chelsea Center for Recycling and Economic Development, University of Massachusetts, Chelsea, MA, August 1998.

144. M. J. Coren, *SJMN*, 19 June 2001.

145. Anonymous, AP dispatch, 22 May 2001.

146. S. Deveaux, IDG dispatch, 3 April 2000.

147. M. Cole, *PO*, 5 June 2002.

148. See, for example, M. Bustillo, *LAT*, 26 June 2002 and Anonymous, *LAT*, 26 June 2002. M. Bustillo, *Tribune News Service*, 10 September 2002.

149. M. Pflum, Cable News Network dispatch, 8 December 2000.

150. D. Prial, AP dispatch, 14 November 2000.

151. J. Markoff, *NYT*, 27 June 2003.

152. Anonymous, CNET News.com dispatch, 14 July 2004.

153. M. Skomial, *SB*, 22 July 2004.

154. Anonymous, *CSM*, 18 December 2000.

155. J. deBlanc-Knowles, *E/The Environmental Magazine*, 22 August 2003.

156. The Massachusetts law is 310 CMR 19.006. State of Massachusetts, URL http://www.state.ma.us/ dep/recycle/crt/CRTRSBZ.htm discusses options.

157. J. Puckett et al., *Exporting Harm: The High-Tech Trashing of Asia* (The Basel Action Network and the Silicon Valley Toxics Coalition, 25 February 2002). The report is available at URL http://www.ban.org/. Several articles were based on the report, including J. Markoff, *NYT*, 25 February 2002 and P. J. Huffstutter, *LAT*, 26 February 2002.

158. Richard Dahl, *Environ. Health Perspect.* **110**, A196 (2002).

159. C.-C. Ni, *LAT*, 6 April 2004.

160. J. deBlanc-Knowles, E/The Environment Magazine XIV(3), 12 (2003).

161. I. Peterson, *NYT*, 16 April 1987.

162. P. Passell, *NYT*, 26 February 1991.

163. A. R. Good, *NYT*, 18 July 1990.

164. I. Peterson, *NYT*, 1 May 1987.

165. T. Henry, *TB*, 7 July 2002.

166. League of Women Voters Education Fund, *The Garbage Primer* (New York: Lyons and Burfort, 1993).

167. B. Meier, *NYT*, 7 September 1993.

168. D. C. Wilson and W. L. Rathje, *Natl. Hist.* **99**(5), 54 (1990).

169. R. E. Fahey, *Natl. Hist.* **99**(5), 58 (1990).

170. See, for example, P. S. Gutis, *NYT*, 24 April 1987, 25 April 1987; 4 May 1987; and 2 September 1987 chronicles the voyages of the *Mobro*. See also G. B. Ward, *NYT*, 23 October 1988 and L. Goff, *The Queens Tribune*, 8 February 2001.

171. AP dispatch, *NYT*, 28 February 1988; K. Meadows, AP dispatch, 2 May 2000. Anonymous, Reuters dispatch, 26 January 2001. H. Reeves, *NYT*, 18 February 2001. M. Levy, AP dispatches, 28 June 2002.

172. C. Pollock, in L. Starke, ed., *State of the World 1987* (New York: W. W. Norton, 1987), 101. See also W. U. Chandler, *Futurist* **18**(1), 89 (1984).

173. Anonymous, *NYT*, 21 April 1987; J. F. Sullivan, *NYT*, 11 January 1987.

174. S. Bronstein, *NYT*, 12 February 1987.

175. A Rozens, Reuters dispatch, 14 July 2004. Mongo describes materials found as discards that can be reused. For an exegesis on mongo, see T. Botha, *Mongo: Adventures In Trash* (New York: Bloomsbury USA, 2004).

176. J. Tierney, *NYT Magazine*, 30 June 1996, 24.

177. E. S. Wood, *Resource Recycling* 15(1), 44 (1996).

178. Dept. of Commerce, *Statistical Abstracts of the United States, 2000* (Washington, DC: GPO, 2000), Table 398.

179. J. Gitlitz, *Trashed Cans: The Global Environmental Impacts of Aluminum Can Wasting in America* (Arlington, VA: Container Recycling Institute, 2002).

180. "The trillionth wasted can: background on aluminum can recycling and wasting in the United States," Container Recycling Institute fact sheet, 2004.

181. Container Recycling Institute, "Report shows plastic bottle waste doubled since 1995," news release, 12 September 2003.

182. California Department of Conservation, "Billions of plastic water bottles tossed in trash," press release, 29 May 2003.

183. J. Gitlitz and P. Franklin, "The 10% incentive to recycle," Container Recycling Institute, 2004.

184. J. C. Kuehner, *CPD*, 1 September 2003.

185. L. P. Vellequette, *TB*, 30 November 2003.

186. J. Barron, *NYT*, 10 December 1989.

187. A. R. Gold, *NYT*, 12 October 1990.

188. A. R. Good, *NYT*, 17 March 1990.

189. A. R. Gold, *NYT*, 30 December 1990.

190. A. Rinard, *MJS*, 29 August 2003.

191. C. Kahn, AP dispatch, 12 May 2003.

192. J. S. Manuel, *Environ. Health Perspect.* **111**, A880 (2003). Anonymous, GreenBiz.com dispatch, 20 February 2003.

193. J.I. Miller, American Recycler **7**(12), 1 (December 2004).

194. R. R. Grinstead, *Environment*, **14**(3), 2 (1972); G. Hill, *NYT*, 14 October 1976.

195. Container Recycling Institute, 2004.

196. Container Recycling Institute, "The can costs more than the cola," brochure, 2003.

197. P. Sarnoff, ed., *NYT Encyclopedic Dictionary of the Environment*, (New York: Arno Press, 1975), Table 13.

198. J. C. Kuehner, *CPD*, 14 July 2004. L. Mergener, *TB*, 15 July 2004. Amazingly, nearly a million bottles of urine are dumped on Ohio's roadsides each year, and many bags of feces were also found. In a year, an average of 475 pounds of litter is collected per mile of highway in Ohio.

199. State of California, Department of Transportation, District 11, *Fact Sheet, District 11 Litter Program*, January 2000. This may be downloaded from URL http://www.dot.ca.gov/dist11/facts/litter.htm

200. State of Texas, *2004 Litter Attitudes & Behavior Study—Fact Sheet*. This may be downloaded from URL http://www.dontmesswithtexas.org/research_detail.php?report_id=8. Rugs, stoves, refrigerators, matresses, and even ladders are recovered in great numbers.

201. J. McCaull, *Environment* **16**(1), 6 (1974). A similar effect was seen in Vermont following passage of the bottle bill there. See J. M. Jeffords and D. M. Webster, *Vermont 5 cent Deposit*, U.S. House of Representatives and Northeast Recycling Council, *Recycling and the Environment: Facts about Recycling in Vermont*, fact sheet, August 1992. For Connecticut, Northeast Recycling Council, *Recycling and the Environment: Facts about Recycling in Connecticut*, fact sheet, August 1992. For Oregon, see State of Oregon, Dept. of Environmental Quality, *Oregon's Bottle Bill*, the 1977 report and Oregon League of Conservation Voters Education Fund, *The 2001 Environmental Handbook for the Oregon Legislature* (Salem, OR: Oregon League of Conservation Voters, 2002), Ch. 8. For Iowa, R. W. Beck, Inc., *Economic Impacts of Recycling in Iowa* (Des Moines: Recycle Iowa, August 2001) claims almost 10,000 jobs are created by the state's recycling programs. For certain New Jersey communities, J. F. Sullivan, *NYT*, 4 December 1983; also see Northeast Recycling Council, *Recycling and the Environment: Facts about Recycling in New Jersey*, fact sheet, August 1992. See also M. Sullivan, *The ABC's About Beverage Containers*, National Wildlife Federation, 1977 and the rest of the Northeast Recycling Council's fact sheets.

202. B. Platt and D. Rowe, Reduce, Reuse, Refill! (Washington, DC: Institute for Local Self-Reliance, April 2002),

produced under a joint project with the GrassRoots Recycling Network.

203. B. M. Hannon, *Environment* **14**(2), 11 (1972).

204. L. D. Orr, *Environment*, **18**(10), 33 (1976).

205. C. Morawski, Solid Waste & Recycling #6, 44 (December/January 2004).

206. C. Morawski, Solid Waste & Recycling #5, 16 (October/November 2004).

207. Gesellschaft für Umfassende Analysen, GmbH, Volkswirtschaftlicher Vergleich von Einweg- und Mehrwegsystemen (Vienna: Austrian Ministry of Environment, 2000).

208. Editorial, PO, 16 April 2001.

209. C. Holden, Science 202, 34 (1978)

210. S. H. Verhovek, *NYT*, 28 December 1986.

211. D. Gonzalez, *NYT*, 22 March 2005.

212. D. Hajela, AP dispatch, 15 February 2002. K. Johnson, *NYT*, 25 February 2002. See also E. Lipton, *NYT*, 2 February 2004; as usual in these analyses, this report counts only direct costs.

213. M. Cooper, *NYT*, 17 June 2002. J. Steinhauer, *NYT*, 29 June 2002. K. Roth, AP dispatch, 2 July 2002. J. J. Goldman, *LAT*, 2 July 2002. D. Cardwell, *NYT*, 12 August 2002.

214. M. Bloomberg, testimony on bill 201-A, "Temporary Emergency Recycling Requirements," 1 July 2002.

215. D. Cardwell, *NYT*, 12 August 2002.

216. M. Cooper, *NYT*, 14 January 2003.

217. A. DePalma, *NYT*, 4 March 2004.

218. E. Daley, *BG*, 25 November 2001; E. Daley, *BG*, 31 January 2002.

219. Anonymous, *The Chicago Tribune News Services*, 14 February 2001.

220. Environmental Protection Agency and Department of Energy, "Waste sector: source reduction and recycling: Iowa," brochure, 1996.

221. J. DeFao, *SFC*, 27 December 2001.

222. E. Wilson et al., California Department of Conservation press release NR 2004-38, 20 December 2004.

223. P. Franklin, "Americans break record for beverage containers wasted; 129 billion bottles and cans trashed in 2004," Container Recycling Institute press release, 30 December 2004.

224. C. Morawski, Solid Waste & Recycling, December/January 1999. C. Morawski, Solid Waste & Recycling, December/January 2000.

225. J. Gitlitz, personal communication, 25 February 2005.

226. Businesses and Environmentalists Allied for Recycling (BEAR), a Project of Global Green USA, 16 January 2002.

CHAPTER 12

1. C. Hall, P. Tharakan, J. Hallock, C. Cleveland, and M. Jefferson, *Nature* **426**, 318 (2003).

2. M. K. Hubbert, *U.S. Energy Resources, a Review as of 1972*, U.S. Senate Committee on Interior and Insular Affairs report, GPO, Washington, DC: 1974; M. K. Hubbert, *Am. J. Phys.* **49**, 1007 (1981); M. K. Hubbert, in P. Cloud et al., eds., *Resources and Man*, (San Francisco: Freeman, 1969), 157.

3. M. K. Hubbert, in G. Piel et al., eds., *Energy and Power* (San Francisco: Freeman, 1971), 31.

4. G. Ourisson et al., *Sci. Am.* **251**(2), 44 (1984). M. A. Gough and S. J. Rowland, *Nature* **344**, 648 (1990). See also K. S. Deffeyes, *Hubbert's Peak* (Princeton and Oxford: Princeton University Press, 2001).

5. M. Jones and S. R. Larter, *Nature* **426**, 344 (2003).

6. J. S. Dukes, *Climatic Change* **61**, 31 (2003). Dukes quotes G. Demaison, in M. L. Bordenave, ed., *Applied Petroleum Geochemistry*, (Paris: Éditions Technip, 1993), 4812, in support of the proportion of oil from coast, deltas, and lakes.

7. Department of Energy website, URL: http://www.fe.doe.gov/education/oil_history.html.

8. Department of Energy website, URL: http://www.fe.doe.gov/education/gas_history.html.

9. J. Z. de Boer et al., *Geology* **29**, 707 (2001). For a more popular view, see J. Roach, *National Geographic News*, 14 August 2001, URL: http://news.nationalgeographic.com/news/2001/08/0814_delphioracle.html and W. J. Broad, *NYT*, 19 March 2002.

10. B. Clark and R. Kleinberg, *Phys. Today* **55**(4), 48 (2002).

11. Dept. of Energy, *Annual Energy Review 2000* (Washington, DC: GPO, 2001), DOE/EIA-0384(00), Section 5.

12. R. D. Hershey, Jr., *NYT*, 25 September 1983.

13. H. Herberg, *Reports of the DFG*(2-3), 35 (1984).

14. S. F. Singer, *Annu. Rev. Energy* **8**, 451 (1983).

15. Federal Energy Administration, *Project Independence Report* (Washington DC: GPO, 1974) (as quoted in R. Gillette, *Science* **186**, 718 (1974)). See also R. Stuart, *NYT*, 1 September 1974.

16. R. N. Kaufmann and C. J. Cleveland, *Annu. Rev. Energy Environ.* **16**, 379 (1991).

17. Members of the National Energy Policy Development Group (R. Cheney et al.), *National Energy Policy: Report of the National Energy Policy Development Group*, (Washington, DC: GPO, 2001).

18. C. J. Cleveland and R. K. Kaufmann, *Energy Policy* **31**, 485 (2003).

19. D. Van Natta, Jr. and N. Banerjee, *NYT*, 28 March 2002. D. Van Natta, Jr., *NYT*, 4 April 2002.

20. K. Q. Seelye, *NYT*, 28 April 2002.
21. C. Lochhead, *SFC*, 26 January 2002. D. Van Natta, Jr., *NYT*, 31 January 2002. M. Allen, *WP*, 31 January 2002.
22. D. Lazarus, *SFC*, 30 January 2002. D. Van Natta Jr. and N. Banerjee, *NYT*, 21 April 2002. D. Van Natta, Jr., *NYT*, 27 April 2002.
23. D. Van Natta, Jr., *NYT*, 11 April 2002.
24. H. J. Hebert, AP dispatch, 18 March 2004.
25. L. Alvarez, *NYT*, 18 July 2001.
26. M. Kilian, *CT*, 1 March 2001.
27. E. Pianin, *WP*, 19 April 2002; D. E. Rosenbaum, *NYT*, 19 April 2002.
28. "Anonymous" (M. Scheuer), *Imperial Hubris* (New York: , 2004). See also B. Woodward, *Plan of Attack* (New York: Simon & Schuster, 2004).
29. N. Banerjee, *NYT*, 26 November 2002.
30. T. Detzel, *PO*, 14 February 2001.
31. W. W. Gibbs, *Sci. Am.* **284**(5), 62 (2001). See also C. W. Schmidt, *Environ. Health Perspect.* **110**, A 22 (2002).
32. R. N. Anderson, *Sci. Am.* **278**(3), 86 (1998). See also J. Pelley, *Environ. Sci. & Technology* **35**, 240A (2001).
33. C. Shenot, *OS*, 3 September 2001.
34. A. C. Revkin, *NYT*, 30 January 2001.
35. ANWR Assessment Team, U.S. Geological Survey, (Washington, DC: GPO, 1999).
36. N. Banerjee, *NYT*, 12 November 2000.
37. R. Walker, *CSM*, 10 May 2001.
38. G. Martin, *SFC*, 19 August 2001.
39. J. Heilprin, AP dispatch, 18 January 2002. M. Grunwald, *WP*, 18 January 2002.
40. K. Murphy, *LAT*, 30 March 2002. S. H. Verhovek, *NYT*, 30 March 2002.
41. K. Murphy, *LAT*, 8 April 2002. J. Kaiser, *Science* **296**, 444 (2002).
42. J. G. Mitchell, *Natl. Geog.* **200**(2), 46 (2001).
43. National Resources Defense Council, URL: http://www.nrdc.org/land/ wilderness/anwr/anwrinx.asp. T. Detzel, *PO*, 14 February 2001.
44. M. Grunwald, *WP*, 6 March 2002.
45. J. G. Koomey et al., *Annu. Rev. Energy Environ.* **27**, 119 2002.
46. S. H. Verhovek, *NYT*, 16 March 2001.
47. J. Landers, *DMN*, 15 April 2001.
48. S. H. Verhovek, *NYT*, 10 June 2001.
49. S. H. Verhovek, *NYT*, 30 May 2001.
50. S. Greenhouse, *NYT*, 2 September 2001.
51. R. Simon, *LAT*, 22 April 2003.
52. M. Chambers, AP dispatch, 2 April 2004.
53. Anonymous, Reuters dispatch, 13 June 2003. J. Heilprin, AP dispatch, 21 November 2003. E. Pianin, *WP*, 23 January 2004.
54. T. Doggett, Reuters dispatch, 18 February 2004.
55. D. Jehl, *NYT*, 6 April 2001. D. Schoch, *LAT*, 5 May 2001.

56. Y. Rosen, *CSM*, 7 October 2003. A. C. Revkin, *NYT*, 13 January 2004. C. Krauss, *NYT*, 6 September 2004.
57. F. T. Keimig and R. S. Bradley, *Geophys. Res. Lett.* **29**, 1163 (2002).
58. J. Kahn, *NYT*, 14 July 2001.
59. N. Banerjee and B. Spiess, *NYT*, 6 September 2001.
60. D. Martin, *NYT*, 29 August 1985.
61. L. Schipper and A. N. Ketoff, *Science* **230**, 1118 (1985).
62. N. D. Kristof, *NYT*, 8 April 1986; R. W. Stevenson, D. F. Cuff et al., *NYT*, 22 January 1986.
63. R. L. Hirsch, *Science* **235**, 1467 (1987). See also M. Crawford, *Science* **235**, 626 (1987).
64. K. Axtman and R. Scherer, *CSM*, 2 August 2001.
65. T. Doggett, Reuters dispatch, 29 March 2001.
66. Dept. of Energy, *Annual Energy Outlook 1993* (Washington, DC: GPO, 1993), DOE/EIA-0383(93).
67. Dept. of Energy, *Improving Technology: Modeling Energy Futures for the National Energy Strategy* (Washington, DC: GPO, 1991), SR/NES/90-01.
68. Dept. of Energy, *Annual Outlook for Oil & Gas 1991* (Washington, DC: GPO, 1991), DOE/EIA-0517(91).
69. H. M. Hubbard, *Sci. Am.* **264**(4), 36 (1991).
70. P. Shabecoff, *NYT*, 5 March 1989.
71. R. Suro, *NYT*, 3 May 1992.
72. V. E. McKelvey, *Am. Sci.* **60**, 32 (1972).
73. A. A. Bartlett, *Math. Geol.* **32**, 1 (2000).
74. C. J. Campbell and J. H. Laherrère, *Sci. Am.* **278**(3), 78 (1998).
75. Table 11.3, Ref. 3.
76. BP, *BP Statistical Review of World Energy* (52nd ed.) (London: BP, 2003).
77. J. Hakes, presentation made to the American Association of Petroleum Geochemists, New Orleans, Louisiana, URL http:www.eia.doe.gov/pub/oil_gas/ petroleum/presentations/2000/ long_term_supply/index.htm (2000), as quoted in Ref. 1a.
78. Dept. of Energy, *International Energy Outlook 2001* (Washington, DC: GPO, 2001).
79. R. A. Kerr, *Science* **281**, 1128 (1998).
80. H. W. Menard and G. Sharman, *Science* **190**, 337 (1975).
81. D. H. Root and C. J. Drew, *Am. Sci.* **67**, 648 (1979).
82. Dept. of Energy, *Geologic Distribution of U.S. Oil & Gas* (Washington, DC: GPO, 1992), DOE/EIA-0557.
83. R. R. Berg, J. C. Calhoun, Jr., and R. L. Whiting, *Science* **184**, 331 (1974).
84. R. A. Kerr, *Science* **212**, 427 (1981).
85. T. C. Hayes, *NYT*, 4 July 1990.
86. H. W. Menard, *Sci. Am.* **244**(1), 55 (1981).

87. W. L. Fisher, *Science* **236**, 1631 (1987).
88. J. Spears, *TS*, 25 July 2004.
89. See, for example just in 2004, M. Meacher, *The Financial Times* (London), 5 January 2004; D. R. Francis, *CSM*, 29 January 2004; P. Raeburn, *NYT*, 8 February 2004; P. Roberts, *LAT*, 7 March 2004; P. Roberts, *LAT*, 25 March 2004; A. E. Moreno, *SFC*, 2 April 2004; J. Jordan and J. R. Powell, *WP*, 6 June 2004; N. Banerjee, *NYT*, 20 June 2004.
90. N. Antosh, *HC*, 5 February 2004. D. Olive, *TS*, 24 July 2004.
91. J. Fuerbringer, *NYT*, 10 January 2004.
92. J. Gerth and S. Labaton, *NYT*, 8 April 2004.
93. S. Labaton and J. Gerth, *NYT*, 9 March 2004.
94. J. Gerth and H. Timmons, *NYT*, 16 April 2004.
95. S. Labaton and J. Gerth, *NYT*, 17 March 2004.
96. S. Labaton and H. Timmons, *NYT*, 20 April 2004.
97. B. Gardiner, AP dispatch, 24 May 2004.
98. D. Kohn, *Newsday*, 6 June 2000. K. Reich, *LAT*, 11 June 2001.
99. C. Norman, *Science* **228**, 974 (1985).
100. F. M. Orr and J. J. Taber, *Science* **224**, 563 (1984).
101. T. M. Doesher, *Am. Sci.* **69**, 193 (1981).
102. G. W. Hinman, in R. Howes and A. Fainberg, eds., *The Energy Sourcebook* (New York: American Institute of Physics, 1991), 99.
103. A. Wilhelms et al., *Nature* **411**, 1034 (2001).
104. R. W. Fisher, *Futurist* **31**(5), 43 (1997).
105. F. R. Stabler, *Futurist* **32**(8), 31 (1998).
106. K. Spear, *OS*, 2 September 2001.
107. K. Spear and C. Shenot, *OS*, 2 September 2001.
108. Dept. of Energy, *Oil and gas development in the United States in the early 1990s: An expanded role for independent producers*, (Washington, DC: GPO, 1995), DOE/EIA-03600.
109. D. Fleshler, *OS*, 3 September 2001.
110. J. Brooke, *NYT*, 7 September 2000.
111. T. Lytle, *OS*, 9 May 2001.
112. D. Shapley, *Science* **187**, 1064 (1975).
113. A. R. Fowler, *Sci. Am.* **238**(3), 42 (1978).
114. R. Gillette, *Science* **187**, 723 (1975).
115. M. Kenward, *New Scientist* **65**, 142 (1975).
116. S. F. Singer, *Science* **188**, 401 (1975).
117. National Petroleum Council, *U.S. Energy Outlook* (Washington, DC: 1972); R. H. Cram, ed., *Future Petroleum Provinces of the United States(their Geological Potential*, V. 1, A. A. P. G. Memoir 15; P. K. Theobald, S. P. Schweinfurth, and D.C. Duncan, *Energy Resources of the United States* U.S. Geological Survey Circular 650, 1972.

118. R. A. Kerr, *Science* **223**, 382 (1984).
119. Dept. of Energy, *Annual Energy Review 1986* (Washington DC: GPO, 1987).
120. Anonymous, National Geographic Special Report, *Energy*, February 1981.
121. C. D. Masters et al., *Annu. Rev. Energy Environ.* **15**, 23 (1990).
122. Dept. of Energy, *U.S. Crude Oil, Natural Gas, and Natural Gas Liquids Reserves* (Washington, DC: GPO, 1992), DOE/EIA-0216(91).
123. Dept. of Energy, *The Domestic Oil & Gas Recoverable Resource Base: Supporting Analysis for the National Energy Strategy* (Washington, DC: GPO, 1990), SR/NES/90-05.
124. T. Scanlan, *Science* **217**, 325 (1982).
125. P. Belluck, *NYT*, 14 June 2000.
126. T. Pack, *OS*, 24 June 2001.
127. J. Dunkerley and W. Ramsay, *Science* **216**, 590 (1982).
128. Dept. of Energy, *Petroleum, An Energy Profile* (Washington, DC: GPO, 1991), DOE/EIA-0545(91); Ref. 3, Table 10.16.
129. See, for example, P. W. MacAvoy, *NYT*; 14 March 1982; E. J. Oppenheimer, *NYT*, 1 April 1984.
130. G. Szegö, *The U.S. Energy Problem*, RANN Report, Appendix S (Washington DC: GPO, 1972).
131. W. D. Metz, *Science* **184**, 1271 (1974).
132. R. A. Dick and S. P. Wimpfen, *Sci. Am.* **243**(4), 182 (1980).
133. J. P. Sterba, *NYT*, 3 November 1974. E. Cowan, *NYT*, 26 January 1975.
134. A. Schriesheim and I. Kirschenbaum, *Am. Sci.* **69**, 536 (1981).
135. R. L. George, *Sci. Am.* **278**(3), 84 (1998).
136. E. Marshall, *Science* **204**, 1283 (1979).
137. G. D. Mossop, *Science* **207**, 145 (1980).
138. J. Brooke, *NYT*, 23 January 2001.
139. N. Antosh, *HC*, 6 December 2003.
140. W. M. Burnett and S. D. Ban, *Science* **244**, 305 (1989).
141. Dept. of Energy, *Natural Gas Monthly*, May 2001, DOE/EIA-0130(05/01).
142. Table 9.1, Ref. 3.
143. G. J. MacDonald, *Annu. Rev. Energy* **15**, 53 (1990).
144. Verbandes der Gas- und Wasserwerke, Baden-Württemberg e. V., *Haushalten mit Energie* (Karlsruhe, Germany: Stadtwerk Karlsruhe, 1980).
145. Dept. of Energy, *Natural Gas Monthly*, October 2000, DOE/EIA-0130(10/00).
146. D. Jehl, *NYT*, 4 February 2001; M. Janofsky, *NYT*, 8 April 2000.
147. J. Landers, *DMN*, 16 April 2001.
148. E. Nakashima, *WP*, 17 May 2002. E. Shogren, *LAT*, 19 May 2002. J. Lloyd, *CSM*, 14 August 2002. Anonymous, AP dispatch, 15 November 2002. Blaine Harden, WP, 4 November 2003. S. Romero, *NYT*, 5 February 2004. E. Shogren, *LAT*,

29 February 2004. J. Eilperin, *WP*, 1 March 2004. C. Roberts, AP dispatch, 19 March 2004. B. Harden, *WP*, 5 July 2004.
149. D. R. Francis, *CSM*, 6 March 2001. B. Tansey, *SFC*, 15 July 2001.
150. D. Jehl, *NYT*, 8 March 2001.
151. D. Adam, *Nature* **415**, 913 (2002).
152. Q. Schiermeier, *Nature* **423**, 681(2003).
153. E. Suess et al., *Sci. Am.* **281**(5), 76 (1999).
154. K. A. Kvenvolden, *Proc. Natl. Acad. Sci.* **96**, 3420 (1999).
155. R. L. Kleinberg and P. G. Brewer, *Am. Sci.* **89**, 244 (2001).
156. B. A. Buffett, *Annu. Rev. Earth Planet. Sci.* **28**, 477 (2000). See also D. Graham, *Tech. Rev.* **98**(7), 17 (1995). See also G. R. Dickens and M. S. Quinby-Hunt, *Geophys. Res. Lett.* **21**, 2115 (1994).
157. J. Barnes, *Mother of Storms* (New York: Tor, 1995).
158. M. J. Hornbach, D. M. Saffer, and W. S. Holbrook, *Nature* **427**, 142 (2004).
159. R. Kerr, *Science* **287**, 576 (2000). G. Ryskin, *Geology* **31**, 737 (2003).
160. M. H. Nederlof, *Energy* **13**, 95 (1988).
161. T. C. Hayes, *NYT*, 26 December 1990.
162. H. Clifford, *CSM*, 3 January 2002. S. Swartz, Reuters dispatch, 8 May 2002. B. Harden and D. Jehl, *NYT*, 29 December 2002. V. Klinkenborg, *NYT*, 1 December 2003.
163. M. L. Wald, *NYT*, 10 January 1993.
164. Dept. of Energy, *Coal Industry Annual 1999*, (Washington, DC: GPO, 2000), DOE/EIA-0584(99).
165. J. Walsh, *Science* **184**, 336 (1974).
166. E. D. Griffeth and A. W. Clarke, *Sci. Am.* **240**(1), 38 (1979).
167. Dept. of Energy, *Clean Coal Technology* (Washington, DC: GPO, 1992), DOE/FE-0217P.
168. Dept. of Energy, *Coal Production 1992* (Washington, DC: GPO, 1992), DOE/EIA-0118(92).
169. M. L. Wald, *NYT*, 8 February 1990; E. S. Tavoularas, *Annu. Rev. Energy Environ.* **16**, 25 (1991).
170. F. X. Clines, *NYT*, 4 May 2001.
171. M. E. Dry, *Endeavour* **8**, 2 (1984).
172. K. Brown, *Science* **299**, 1177 (2003).
173. F. X. Clines, *NYT*, 23 July 2002.
174. R. Wingfield-Hayes, BBC News Online, 3 August, 2000.
175. B. Sloat, *CPD*, 12 May 2003.
176. S. Bhattacharya, New Scientist Online, 14 February 2003. J. Amos, BBC News Online, 14 February, 2003.
177. H. Maass and D. Tatlow, *SFC*, 3 April 2003.
178. H. Perry, *Science* **222**, 377 (1983).
179. R. L. Gordon, *Science* **200**, 153 (1978).
180. W. Berry, AP dispatch, 8 December 2002.
181. A. Rose, T. Torries, and W. Labys, *Annu. Rev. Energy Environ.* **16**, 59 (1991).

182. I. M. Torrens, *Environment* **32**(6), 10 (1990).
183. S. B. Alpert, *Annu. Rev. Energy Environ.* **16**, 1 (1991).
184. E. Corcoran, *Sci. Am.* **264**(5), 106 (1991).
185. S. C. Morris et al., *Science* **206**, 654 (1979).
186. A. M. Squires et al., *Science* **230**, 1329 (1985).
187. M. L. Wald, *NYT*, 26 July 1992.
188. Lynda V. Mapes, *ST*, 26 March 2001.
189. M. Newkirk and R Downing, *ABJ*, 7 January 2001.
190. M. Newkirk, *ABJ*, 7 January 2001.
191. A. M. Squires, as quoted by W. D. Metz, *Science* **179**, 54 (1973).
192. E. F. Osborn, *Science* **183**, 477 (1974).
193. J. T. Dunham, C. Rampacek, T. A. Henrie, *Science* **184**, 346 (1974).
194. A. M. Squires, "The fossil fuel development gap," (unpublished).
195. A. L. Hammond, *Science* **193**, 750 (1976).
196. A. M. Squires, *Science* **184**, 340 (1974).
197. H. R. Linden et al., *Annu. Rev. Energy* **1**, 65 (1976).
198. H. C. Hottel and J. B. Howard, *New Energy Technologies* (Cambridge, MA: MIT Press, 1971).
199. W. R. Long, *NYT*, 21 February 1999.
200. G. K. Vick and W. R. Epperly, *Science* **217**, 311 (1982).
201. H. E. Swift, *Am. Sci.* **71**, 616 (1983).
202. R. L. Hirsch et al., *Science* **215**, 121 (1982).
203. R. E. Lumpkin, *Science* **239**, 873 (1989).
204. Dept. of Energy, *The Potential for Coal Liquefaction* (Washington, DC: GPO, 1990), SR/NES/90-07.
205. C. Starr et al., *Science* **256**, 981 (1992).
206. D. Jehl, *NYT*, 4 August 2001.
207. D. F. Spencer et al., *Science* **215**, 1571 (1982).
208. P. F. Fennelly, *Am. Sci.* **72**, 254 (1984).
209. M. Crawford, *Science* **230**, 1022 (1985).
210. S. A. Fouda, *Sci. Am.* **278**(3), 78 (1998).
211. A. Salpukas, *NYT*, 2 January 1998.
212. D. F. Spencer et al., *Science* **232**, 609 (1986).
213. V. Shorter, head engineer, Cool Water Gasification Plant, private communication.
214. W. N. Clark, project manager, Cool Water Coal Gasification Plant, news release, 25 June 1985.
215. M. L. Wald, *NYT*, 10 June 1992.
216. R. E. Balzhiser and K. E. Yeager, *Sci. Am.* **257**(3), 100 (1987).
217. M. Crawford, *Science* **228**, 565 (1985).
218. M. Crawford, *Science* **228**, 1410 (1985).
219. S. Borenstein, *SJMN*, 18 April 2001.
220. T. Williams, *Audubon* **103**(3), 36 (2001). See also J. Goodell, *NYT Magazine*, 22 July 2001, 30.
221. U.S. Environmental Protection Agency, with the U.S. Army Corps of Engineers, the U.S. Fish and Wildlife Service,

Office of Surface Mining, and the West Virginia Department of Environmental Protection, *Mountaintop Mining/Valley Fill, Draft Programmatic Environmental Impact Statement*, May 2003.

222. E. Shogren, 26 April 2002. K. Q. Seelye, *NYT*, 26 April 2002. E. Pianin, *WP*, 26 April 2002. Anonymous, *NYT*, 4 May 2002.

223. E. Shogren, *LAT*, 14 April 2002.

224. See as an example of the news coverage E. Shogren, *LAT*, 30 May 2003.

225. J. Heilprin, AP dispatch, 31 March 2004.

226. K. Ward, Jr., *CG*, 25 July 2003.

227. E. Shogren, *LAT*, 9 May 2002 and E. Pianin, *WP*, 10 May 2002. F. X. Clines, *NYT*, 19 May 2002. P. Ramsey, AP dispatch, 18 June 2002.

228. E. Pianin, *WP*, 31 January 2003.

229. C. Haden, decision in Civil Action No. 2:01-0770: "Kentuckians for the Commonwealth, Inc., Plaintiff, v. Colonel John Rivenburgh, Colonel, District Engineer; Robert B. Flowers, Lieutenant General, Chief of Engineers and Commander of the U.S. Army Corps of Engineers; and Michael D. Gheen, Chief of the Regulatory Branch, Operations and Readiness Division, U.S. Army Corps of Engineers, Huntington District, Defendants," 17 June 2002.

230. N. Zuckerbrod, AP dispatch, 8 January 2004.

231. J. Warrick, *WP*, 17 August 2004.

232. R. Weiss, *WP*, 16 August 2004.

233. C. Drew and R. A. Oppel, Jr., *NYT*, 9 August 2004.

234. A. Goldstein and S. Cohen, *WP*, 15 August 2004.

235. E. Shogren, *LAT*, 18 January 2004. J. Donn, AP dispatch, 6 March 2004.

236. K. Ward, Jr., *CG*, 6 September 2003.

237. K. Ward, Jr., *CG*, 5 September 2003.

238. K. Ward, Jr., *CG*, 31 March 2004.

239. P. Slavin, *WP*, 9 June 2003.

240. V. Smith, AP dispatch, 18 July 2004.

241. E. Shogren, *LAT*, 7 January 2004.

242. F. Barringer, *NYT*, 10 July 2004.

243. J. R. Godwin, decision in Civil Action No. 3:03-2281: "Ohio Valley Environmental Coalition, Coal River Mountain Watch, and Natural Resources Defense Council, Plaintiffs, v. William Bulen, Colonel, District Engineer, U.S. Army Corps of Engineers, Huntington District, and Robert B. Flowers, Lieutenant General, Chief of Engineers and Commander of the U.S. Army Corps of Engineers, Defendants," 8 July 2004.

244. Editorial, *CG*, 16 July 2004.

245. T. Appenzeller, *Natl. Geog.* **205**(6), 80 (2004).

CHAPTER 13

1. M. B. Ibrahim and R. M. Kovach, *Energy* **18**, 961 (1993). See also C. Sims, *NYT*, 4 February 1986.

2. An exhaustive analysis of the effects of pollution on a river is found in J. M. Fallows, *The Water Lords* (New York: Bantam, 1971).

3. H. McGee, J. McInerney, and A. Harrus, *Phys. Today* **52**(11), 30 (1999).

4. J. W. Gibbons and R. R. Sharitz, *Am. Sci.* **62**, 660 (1974).

5. W. A. Brungs, in H. Foreman, ed., *Nuclear Power and the Public* (Minneapolis: University of Minnesota Press, 1970), 52.

6. See, for example, D. G. J. Larsson and L. Förlin, *Environ. Health Perspect.* **110**, 739 (2002). This study was unable to identify whether the temperature or the effluent was responsible, however.

7. A. W. Eipper, in *Patient Earth*, ed. J. Harte and R. Socolow (New York: Holt, Rinehart & Winston, 1971), 112.

8. T. C. Elliot, *Power* (December 1985), S1.

9. M. L. Kramer et al., *Science* **193**, 1239 (1976).

10. J. S. Steinhart and C. E. Steinhart, *Fires of Culture* (N. Scituate, MA: Duxbury, 1974), Table 8-2; or *Energy Sources* (N. Scituate, MA: Duxbury, 1974), Table 10-3. These are based on data taken from *Electric Power and Thermal Discharges*, M. Eisenbud and G. Gleason, eds. (New York: Gordon and Breach, 1969), 372.

11. W. Heller, Chairman, Council of Economic Advisers in the Kennedy-Johnson Administration, in Testimony before the Committee on Interior and Insular Affairs (U.S. Senate) appearing in Committee on Interior and Insular Affairs, *Conservation of Energy* (Washington, DC: U.S. Senate, 1972). See also the news report by A. L. Hammond, *Science* **178**, 1186 (1972).

12. C. Bowden, *Killing the Hidden Waters* (Austin: University of Texas Press, 1977).

13. P. H. Gleick, *Annu. Rev. Environ. Resour.* **28**, 275 (2003).

14. P. Ball, *Nature Science Update*, 27 January 2000 (http://www.nature.com/nsu/000127/000127-11.html).

15. Department of Commerce, *Statistical Abstract of the United States, 2000* (Washington, DC: GPO, 2001), Table 387.

16. S. L. Dingman, in *Academic American Encyclopedia* (Danbury, CT: Grolier, 1992). For the situation in 1981, see R. Reinhold, *NYT*, 9 August 1981.

17. Alex Kirby, BBC News Online dispatch, 27 January 2003.

18. H. L. Penman, in *The Biosphere* (San Francisco: Freeman, 1970), 37

19. F. H. Forrester, *Weatherwise* **38**, 82 (1985).

20. R. L. Nace, U.S. Geological Survey, Circular 536, 1967.

21. A. M. Piper, U.S. Geological Survey Water Supply Paper 1797 (Washington, DC: GPO, 1973).

22. C. J. Vörösmarty et al., *Science* **289**, 284 (2000).

23. S. Postel, *Sci. Am.* **284**(2), 46 (2001).

24. R. P. Ambroggi, *Sci. Am.* **243**(3), 100 (1980).

25. P. H. Gleick, *Sci. Am.* **284**(2), 40 (2001).

26. P. H. Gleick, *Environment* **43**(2), 18 (2001). See also the overview by F. Montaigne, *Natl. Geog.* **202**(3), 2 (2002).

27. N. Johnson et al., *Science* **292**, 1071 (2001).

28. P. Ball, *Nature Science Update*, 27 January 2000 (http://www.nature.com/nsu/000127/000127-12.html).

29. J. S. Meyers, "Evaporation from the 17 western states," with a section on evapotranspiration rates by T. J. Nordenson, U.S. Weather Bureau, 1962.

30. E. Groth, III, *Environment* **17**(1), 28 (1975).

31. D. Sheridan, *Environment* **23**(3), 6 (1981).

32. A. F. Pillsbury, *Sci. Am.* **245**(1), 54 (1981).

33. A. L. Dellon, in T. N. Veziroglu, ed., *Alternative Energy Sources II* (Washington, DC: Hemisphere Publ., 1984), 3715. See also W. Robbins, *NYT*, 25 August 1974.

34. R. Reinhold, *NYT*, 15 February 1991. See also R. Lindsey, *NYT*, 16 August 1987.

35. S. Postel, *NYT*, 18 February 1990. See also S. Postel, in L. Starke, ed., *State of the World 1990* (New York and London: W. W. Norton, 1990), 39.

36. Anonymous, BBC News Online dispatch, 19 November 2002.

37. M. Rosenblum, AP dispatch, 20 August 2002; AP dispatch, 25 August 2002. J. Tagliabue, *NYT*, 26 August 2002. K. Hall, AP dispatch, 18 March 2003.

38. D. Jehl, *NYT*, 10 February 2003.

39. J. Langman, *SFC*, 2 February 2002.

40. P. H. Gleick, *Science* **302**, 1524 (2003).

41. P. H. Gleick, *Nature* **418**, 373 (2002).

42. P. Aldhous, *Nature* **422**, 251 (2003). Anonymous, *Nature* **422**, 252 (2003). R. Weiss, *WP*, 5 March 2003. J. Jowit, *The Observer* (UK), 15 February 2004.

43. T. Weiner, *NYT*, 14 April 2001.

44. W. Marston, *NYT*, 8 December 1999.

45. A. Blankstein, *LAT*, 15 February 2001. J. Tamaki, *LAT*, 15 February 2003. M. Lifsher, *LAT*, 6 March 2004.

46. A. Blankstein, *LAT*, 22 August 2002.

47. E. Gaggelli et al., *Environ. Health Perspect.* **110**(Suppl 5), 733 (2002).

48. D. Blowes, *Science* **295**, 2024 (2002).

49. A. S. Ellis et al., *Science* **295**, 2060 (2002).

50. T. Egan, *NYT*, 14 April 2001. On fighting the federal standards, see for example E. Shogren, *LAT*, 18 March 2002.

51. See, for example, D. Jehl, *NYT*, 21 March 2001; J. Kaiser, *Science* **291**, 2533 (2001); and D. Jehl, *NYT*, 19 April 2001. On finally embracing the reductions, see, for example, E. Pianin, *WP*, 11 September 2001 and K. Q. Seelye, *NYT*, 11 September 2001. On costs, see for example D. Behm, *MJS*, 3 November 2001.

52. B. Bearak, *NYT*, 8 December 1999. U. K. Chowdhury et al., *Environ. Health Perspect.* **108**, 393 (2000). B. Bearak, *NYT*, 14 July 2002.

53. S. Postel, in L. Starke, ed., *State of the World 1996* (New York: Norton, 1996), 40, P. Sampat, Worldwatch Paper 154 (December, 2000), and also P. Sampat, in L. Starke, ed., *State of the World 2001* (New York: Norton, 2001), 21.

54. T. Clarke, *Nature* **422**, 254 (2003).

55. J. A. Centeno et al., *Environ. Health Perspect.* **110**(Suppl 5), 883 (2002).

56. International Consortium of Investigative Journalists, Center for Public Integrity release, 4 February 2003; Center for Public Integrity release, 6 February 2003; Center for Public Integrity release, 12 February 2003; Center for Public Integrity release, 13 February 2003; Center for Public Integrity release, 14 February 2003; Center for Public Integrity release, 19 February 2003; Center for Public Integrity release, 20 February 2003.

57. S. Leavenworth, *SB*, 30 January 2002. E. Bailey, *LAT*, 14 December 2002.

58. T. Wilkinson, *CSM*, 3 May 2001.

59. W. M. Alley et al., *Science* **296**, 1985 (2002).

60. B. Alexander, *NYT*, 8 December 1999. See also S. Postel, in L. Starke, ed., *State of the World 2000* (New York: Norton, 2000), 39-58,

61. Anonymous, AP dispatch, 5 April 2001.

62. R. Sanchez, *WP*, 1 June 2004.

63. K. F. Sherry, *LAT*, 19 January 2002.

64. D. Kasler, *SB*, 9 February 2002.

65. J. Yardley, *NYT*, 16 April 2001.

66. Glen Martin, *SFC*, 9 April 2001.

67. Anonymous, AP dispatch, 3 March 2001.

68. K. Horan, *The Bradenton Herald*, 6 May 2001.

69. L. Clark, *The Miami Herald*, 2 May 2001.

70. D. Behm, *MJS*, 4 March 2002.

71. S. Leavenworth, *SB*, 3 September 2002.

72. R. Fausset, *LAT*, 8 November 2002.

73. M. A. Hiltzik, *LAT*, 19 May 2002.

74. M. A. Hiltzik, *LAT*, 22 August 2002, *LAT*, 6 September 2002.

75. T. Perry and M. Hiltzik, *LAT*, 30 August 2002.

76. M. A. Hiltzik, *LAT*, 9 October 2002.

77. D. Kasler, *SB*, 9 April 2003.

78. V. Lazarova, S. Hills, and R. Birks, *Wat. Sci. Techn. Wat. Supply* 3, 69 (2003).

79. D. Martindale, *Sci. Am.* **284**(2), 54 (2001).

80. M. Hiltzik, *LAT*, 28 April 2003.

81. T. A. Larsen et al., *Environ. Sci. Technol.* **35**, 192 A (2001).

82. C. Pahl-Wostl et al., *Wat. Sci. Techn.* **48**, 57 (2003). T. A. Larsen and J. Lienert, *Wat. Intel. Online* (2003).

83. T. Asano et al., *Wat. Sci. Techn.* **34**, 11 (1996).

84. C. H. Deutsch, *NYT*, 16 May 1999.

85. W. M. Orme, Jr., *NYT*, 23 June 2001. B. Mason, *Nature* **426**, 110 (2003).

86. Anonymous, AFP dispatch, 23 March 2003.

87. C. Arthur, *TI*, 10 June 2004.

88. R. La Corte, AP dispatch, 26 March 2003.

89. L. Wides, AP dispatch, 14 January 2003 (Metropolitan Water District of Southern California, considering five desalination plants). P. Fimrite, *SFC*, 30 December 2002 (Marin Municipal Water District, considering one plant). D. Glaister, *TG*, 13 April 2004 (San Diego County, one operating demonstration plant in Carlsbad).

90. W. Booth, *WP*, 30 March 2003. B. Boxall, *LAT*, 20 March 2004.

91. S. Allison, *LAT*, 8 August 2003.

92. D. Martindale, *Sci. Am.* **284**(2), 53 (2001).

93. D. Martindale, *Sci. Am.* **284**(2), 54 (2001).

94. R. Hanley, *NYT*, 23 March 1989; Anonymous, *NYT*, 13 June 1990.

95. P. H. Gleick, *Sci. Am.* **284**(2), 53 (2001). For an earlier idea, see Anonymous, *NYT*, 3 November 1976.

96. A. H. Rosenfeld, *Annu. Rev. Energy Environ.* **24**, 33 (1999); A. Gadgil, *Annu. Rev. Energy Environ.* **23**, 253 (1998).

97. S. Rattner, *NYT*, 26 November 1976.

98. J. S. Gilmore, *Science* **191**, 535 (1976). D. Champion and A. Ford, *Environment* **22**(5), 25 (1980).

99. Dept. of the Interior, *Kaiparowits, Environmental Impact Statement*, (Washington, DC: U.S.D.I., 1975).

100. *Kaiparowits EIS*, Figure III-68, page III-323.

101. G. Lichtenstein, *NYT*, 3 August 1975; G. Hill, *NYT*, 15 April 1976; and G. Lichtenstein, *NYT*, 16 April 1976.

102. Kaiparowits EIS, pages VI-32, 33.

103. P. Chapman, *New Scientist* **64**, 866 (1974); J. Wright and J. Syrett, *New Scientist* **65**, 66 (1975).

104. C. Whipple, *Science* **208**, 262 (1980).

105. J. L. Marx, *Science* **186**, 809 (1974); R. D. Lyons, *NYT*, 20 November 1974; C. W. Krusé, in J. N. Pitts, Jr., and R. L. Metcalf, eds., *Advances in environmental sciences*, Vol. 1 (New York: John Wiley & Sons, 1969), 60. A. H. Smith et al., *Science* **296**, 2145 (2002). D. K. Nordstrom, *Science* **296**, 2143 (2002).

106. D. Zwick and M. Benstock, *Water Wasteland* (New York, Bantam, 1972).

107. J. Tinker, *New Scientist* **65**, 551 (1975).

108. K. Johnson, *NYT*, 16 January 2001. P. Hellman, *NYT Magazine*, 24 October 1976, 16.

109. A. C. Revkin, *NYT*, 6 June 2000.

110. A. C. Revkin, *NYT*, 22 February 1997.

111. K. Johnson, *NYT*, 7 December 2000.

112. D. Campagna, *BG*, 22 April 2001.

113. K. Johnson, *NYT*, 2 December 2000; K. Johnson, *NYT*, 13 December 2000.

114. A. C. Revkin, *NYT*, 7 December 2000.

115. K. Johnson, *NYT*, 6 December 2000.

116. A. C. Revkin, *NYT*, 5 January 2001.

117. K. Johnson, *NYT*, 3 January 2001.

118. K. Johnson, *NYT*, 7 December 2000.

119. R. Perez-Peña, *NYT*, 5 August 1999.

120. Committee on Remediation of PCB-Contaminated Sediments (J. W. Farrington et al.), National Academy of Sciences, *A Risk Management Strategy for PCB-Contaminated Sediments* (Washington, DC: National Academy Press, 2001). For a report in the popular press, see K. Johnson, *NYT*, 5 January 2001.

121. M. Hill, AP dispatch, 5 December 2000.

122. J. Gorman, *NYT*, 25 March 2003.

123. E. Pianin and M. Powell, *WP*, 5 December 2001. K. Johnson, *NYT*, 5 December 2001.

124. D. Barrett, AP dispatch, 5 March 2004.

125. E. Pianin, *WP*, 30 July 2003.

126. M. Grunwald, *WP*, 23 February 2002.

127. M. Grunwald, *WP*, 24 March 2002.

128. Anonymous, *St. Louis Business Journal*, 21 March 2003, *St. Louis Business Journal*, 9 April 2003.

129. C. Houge, *Chem. & Engineering News* **81**(34), 10 (2003). J. M. Taylor, *Environment & Climate News*, 1 October 2003.

130. M. Reisch, *Chem. & Engineering News* **81**(24), 6 (2003).

131. G. V. Jarnes, *Intern. J. Environmental Studies* **1**, 47 (1970).

132. E.-M. Thoms and C. Uebel, *DZ*, 9 October 1992.

133. B. Boxall, *LAT*, 22 June 2004.

134. D. Jehl, Scientific American Online, 16 December 2002 (the states are Florida, Nebraska, Kansas and Mississippi). R. Blumenthal, *NYT*, 11 December 2003.

135. I. Dror et al., *Science* **300**, 950 (2003). See also K. Spear, *OS*, 9 May 2004.

136. D. Kelley, *LAT*, 15 February 2003.

137. B. Webster, *NYT*, 7 March 1975. See also W. J. Lewell, *Am. Sci.* **82**, 366 (1994) for more information on bulrushes as filters.

138. D. Jehl, *NYT*, 13 April 2001; J. Ash, *PBP*, 21 April 2001.

139. H. Tanner, *NYT*, 4 May 1975.
140. J. R. Luoma, *NYT*, 28 March 1989.
141. W. K. Stevens, *NYT*, 13 March 1990.
142. R. Gillette, *Science* **182**, 456 (1973).
143. B. A. Franklin, *NYT Magazine*, 29 September 1974, 22.
144. W. D. Metz, *Science* **184**, 1271 (1974).
145. Committee on Mineral Resources and the Environment (B. J. Skinner et al.), National Academy of Sciences, *Mineral Resources and the Environment* (Washington, DC: National Academy of Sciences, 1975). See also Commission on Accessory Elements, *Redistribution of Accessory Elements in Mining and Mineral Processing. Part I: Coal and Oil Shale* (Washington, DC: National Academy of Sciences, 1979).

CHAPTER 14

1. Committee on Mineral Resources and the Environment (B. J. Skinner et al.), National Academy of Sciences, *Mineral Resources and the Environment* (Washington, DC: National Academy of Sciences, 1975). See also Committee on Earth Resources (C. Otte et al.), National Academy of Sciences, *Mineral Resources and Sustainability: Challenges for Earth Scientists*, 1996, URL: http://www.nap.edu/catalog/9077.html).
2. J. M. Mitchell, Jr., in W. H. Matthews, W. W. Kellogg, and G. D. Robinson, eds., *Man's Impact on the Climate* (Cambridge, MA: MIT Press, 1971), 167; Table 8.1, 168.
3. Dept. of Energy, *Saving Energy by Managing Urban Heat Islands* (Washington, DC: GPO, unknown date), DOE/CE-0361P.
4. W. P. Lowry, *Sci. Am.* **217**(2), 15 (1967).
5. P. V. Hobbs et al., *Science* **183**, 909 (1974).
6. V. Brodine, *Environment* **13**(1), 2 (1971). See also V. Brodine, *Environment* **14**(1), 2 (1972).
7. W. Sullivan, *NYT*, 28 February 1977; B. A. Franklin, *NYT*, 13 July 1986.
8. J. Doyle, *SFC*, 6 March 2000
9. J. M. Shepherd et al., *J. Appl. Meterology* **41**, 689 (2002). See also K. Chang, *NYT*, 17 August 2000 and S. Goforth, *DMN*, 8 July 2002.
10. A. H. Rosenfeld et al., *Tech. Rev.* **100**(2), 52 (1997).
11. H. Akbari et al., *ASHRAE Trans.* **96**, 1381 (1990). H. Akbari and H. Taha, *Energy* **17**, 141 (1992).
12. J. Brooke, *NYT*, 13 August 2002.
12. W. K. Stevens, *NYT*, 31 July 1990.
14. J. C. Semenza et al., *N. Engl. J. Med.* **335**, 84 (1996).
15. L. Bowman, Scripps Howard News Service dispatch, 9 June 1999.

16. A. H. Rosenfeld, *Annu. Rev. Energy Environ.* **24**, :33-82 (1999). H. Akbari et al., *Energy* **24**, 391 (1999).
17. H. Taha et al., *Building and Environment* **23**, 271 (1988). A. Rosenfeld et al., *Energy and Buildings* **22**, 255 (1995). H. Taha et al., *Theor. Appl. Climatology* **62**, 175 (1999).
18. T. B. Smith, in W. H. Matthews, W. W. Kellogg, and G. D. Robinson, eds., *Man's Impact on the Climate* (Cambridge, MA: MIT Press, 1971), 400.
19. D. Rosenfeld, *Science* **287**, 1793 (2000). See also O. B. Toon, *Science* **287**, 1763 (2000).
20. P. H. Fischer, B. Brunekreef, and E. Lebret, *Atmos. Environ.* **38**, 1083 (2004). J. R. Stedman, *Atmos. Environ.* **38**, 1087 (2004). J. Larsen, Earth Policy Institute release, 9 October 2003, URL http://www.earth-policy.org/Updates/Update29.htm.
21. T. Bahrampour, *NYT*, 13 August 2002.
22. E. A. Goldstein and M. A. Izeman, *Amicus J.* **12**(3), 22 (1990).
23. S. C. Davis, *Transportation Energy Data Book: Edition 20*, Oak Ridge National Laboratory, ORNL-6959, Tables 4.3, 4.5, 4.7, 4.9, 6.3, and 6.5.
24. Volkswagen website (URL: http://www.volkswagen. de/lupo_3l_tdi/index_. htm).
25. For more information on combustion and pollution control, visit the OL-2000 website, J. R. Richards et al., URL: http://registrar. ies. ncsu. edu/ol_2000/index. htm.
27. Dept. of Energy, *Clean Coal Technology* (Washington, DC: GPO, 1992), DOE/FE-0217P.
28. I. M. Torrens, *Environment* **32**(6), 10 (1990).
29. M. L. Wald, *NYT*, 30 September 1992.
26. R. Swanekamp, *Power* **146**(3), 28 (2002).
30. L. F. Diaz et al., *CRC Critical Reviews in Environmental Control* **14**, 251 (1985).
31. S. B. Alpert, *Annu. Rev. Energy Environ.* **16**, 1 (1991).
32. Office of Fossil Energy, Department of Energy, URL: http://www. lanl. gov/projects/cctc/factsheets.
33. C. H. Deutsch, *NYT*, 9 September 2001.
34. D. Altman, *NYT*, 31 May 2002.
35. D. Morgan, *WP*, 2 June 2002.
36. R. R. Gould, *Annu. Rev. Energy* **9**, 529 (1984).
37. R. D. Doctor et al., *J. Air Pollut. Control Assoc.* **35**, 331 (1985).
38. J. G. McCarty, *Nature* **403**, 35 (2000).
39. C. Hagner, *Water, Air, & Soil Pollution* **134**, 1 (2002).
40. D. M. Settle and C. C. Patterson, *Science* **207**, 1167 (1980).
41. D. Ferber, Science OnLine, URL: http://sciencenow. sciencemag. org/cgi/content/full/2001/514/2. See also J. Kaiser, *Science* **293**, 1567 (2001).

42. M. Scott, *BG*, 24 October 2000.
43. R. McEnery, *CPD*, 2 September 2001. H W. Mielke et al., *Environ. Health Perspect.* **109**, 973 (2001). R. Chepesiuk, *Environ. Health Perspect.* **109**, A 434 (2001).
44. H. L. Needleman et al., *J. Am. Med. Assn.* **275**, 363 (1996). L. Tanner, AP dispatch, 15 May 2000.
45. R. Kaiser et al., *Environ. Health Perspect.* **109**, 563 (2001).
46. W. J. Rogan et al. (The Treatment of Lead-Exposed Children Trial Group), *N. Engl. J. Med.* **344**, 1421 (2001). H. L. Needleman et al., *N. Engl. J. Med.* **322**, 83 (1990) and P. A. Baghurst et al. (The Port Pirie Cohort Study), *N. Engl. J. Med.* **327**, 1279 (1992). The cost to children is estimated to be $43. 3 billion per year in P. J. Landrigan et al., *Environ. Health Perspect.* **110**, 721 (2002).
47. H. -Y. Chuang et al., *Environ. Health Perspect.* **109**, 527 (2001).
48. P. B. Stretesky and M. J. Lynch, *Arch. Pediatr. Adolesc. Med.* **155**, 579 (2001).
49. P. B. Stretesky and M. J. Lynch, *J. Health Social Behav.* **45**, 214 (2004).
50. M. Allen, *WP*, 18 April 2001.
51. U.S. Environmental Protection Agency, *Regulatory Finding on the Emissions of Hazardous Air Pollutants from Electric Utility Steam Generating Units*, December 2000.
52. Board on Environmental Studies and Toxicology, Commission on Life Sciences, National Resource Council, *Toxicological Effects of Methylmercury* (Washington, DC: NRC, 2000).
53. The National Energy Policy Development Group (R. Cheney et al.), *National Energy Policy: Report of the National Energy Policy Development Group*, (Washington, DC: GPO, 2001).
54. Environmental Protection Agency, *Regulatory Finding on the Emissions of Hazardous Air Pollutants From Electric Utility Steam Generating Units* [AD–FRL–6919–6] 2060–ZA10, *Federal Register* **65**, 79825 (2000).
55. J. O. Löfken, *DW*, 16 February 2001.
56. W. K. Stevens, *NYT*, 17 July 1990.
57. L. T. Marufu et al., *Geophys. Res. Lett.* **31**, L13106 (2004).
58. R. R. Dickerson et al., *Science* **278**, 827 (1997).
59. M. Petras et al., *Environ. Health Perspect.* **111**, 1175 (2003). D. M. Guvenius et al., *Environ. Health Perspect.* **111**, 1235 (2003). A. Mazdai et al., *Environ. Health Perspect.* **111**, 1249 (2003). C. Washam, *Environ. Health Perspect.* **111**, A480 (2003).
60. A. Schecter et al., *Environ. Health Perspect.* **111**, 1723 (2003). C. W. Schmidt, *Environ. Health Perspect.* **111**, A774 (2003).

61. M. G. Ikonomou et al., *Environ. Sci. Technol.* **36**,1886 (2002). See also H. Wolkers et al., *Environ. Sci. Technol.* **38**, 1667 (2004).

62. M. Cone, *LAT*, 19 June 2003.

63. E. Den Hond et al., *Environ. Health Perspect.* **110**, 771 (2002).

64. J. Giles, *Nature* **427**, 768 (2004).

65. A. C. Revkin, *NYT*, 27 July 2004.

66. D. W. Griffin et al., *Am. Sci.* **90**, 228 (2002). M. Roig-Franzia, *WP*, 21 January 2004.

67. S. A. Mims and F. M. Mims, III, *Atmospheric Environ.* **38**, 651 (2004).

68. R. Dalton, *Nature* **416**, 668 (2002). G. Polakovic, *LAT*, 26 April 2002. J. P. Miller, *CT*, 3 May 2004. Asia now emits in toto more NOx and aerosols than North America, as discussed in H. Akimoto, *Science* **302**, 1716 (2003).

69. M. A. Lev, *CT*, 2 February 2002. H. Chu, *LAT*, 19 June 2002. E. Cody, *WP*, 25 May 2004.

70. K. Bradsher, *NYT*, 26 June 2004.

71. Anonymous, AP dispatch, 9 October 2003.

72. Global Air Monitoring System (UNEP, WHO), *Environment* **36**(2), 4 (1994).

73. T. S. Purdom, *NYT*, 21 April 1998.

74. G. Polakovic, *LAT*, 14 January 2001.

75. M. O. Andreae and P. Crutzen, *Science* **276**, 1052 (1997).

76. R. J. Charlson et al., *Nature* **326**, 655 (1987).

77. D. Kley, *Science* **276**, 1043 (1997).

78. R. K. Monson and E. A. Holland, *Annu. Rev. Ecol. Syst.* **32**, 547 (2001).

79. B. J. Finlayson-Pitts and J. N. Pitts, Jr., *Science* **276**, 1045 (1997).

80. A. R. Ravishankara et al., *Science* **280**, 60 (1998).

81. R. G. Prinn et al., *Science* **292**, 1882 (2001).

82. S. A. Montzka et al., *Science* **288**, 500 (2000).

83. J. Lelieveld et al., *Science* **291**, 1031 (2001). J. Lelieveld et al., *IEEE Spectrum* **35**(12), 50 (1999). T. Nakajima et al., *Geophys. Res. Lett.* **26**, 2421 (1999). An early indication of the trouble with Indonesian forest fires is found in R. Willemsen, *DZ*, 10 June 1988.

84. M. Z. Jacobson, *Nature* **409**, 695 (2001).

85. E. W. Kenworthy, *NYT*, 2 May 1976; and Dept. of the Interior, *Kaiparowits, Environmental Impact Statement*, Sec. III (Washington, DC: GPO, 1975).

86. D. V. Bates, *A Citizen's Guide to Air Pollution* (Montreal: McGill-Queens University Press, 1972).

87. M. L. Bell and D. L. Davis, *Environ. Health Perspect.* **109** (Supplement 3), 389 (2001).

88. A. Hunt et al., *Environ. Health Perspect.* **111**, 1209 (2003). R. Stone, *Science* **298**, 2106 (2002). See also E. Nagourney, *NYT*, 12 August 2003.

89. K. Johnson, *NYT*, 29 September 2002.

90. K. L. Demerjian et al., in J.N. Pitts, Jr., and R.L. Metcalf, eds., *Advances in Environmental Sciences and Technology*, Vol. 4, 1-262, 1974.

91. J. R. Odum et al., *Science* **276**, 96 (1997).

92. A. C. Lewis et al., *Nature* **405**, 778 (2000).

93. B. A. Robertson, *SB*, 30 January 2004.

94. P. F. Fennelly, *Am. Sci.* **64**, 46 (1976).

95. J. L. Adgate et al., *Environ. Health Perspect.* **112**, 1386 (2004).

96. J. Kay, *SFC*, 16 April 2004.

97. G. Polakovic, *WP*, 24 September 2003. R. Sanchez, *WP*, 16 February 2003. M. Arax and G. Polakovic, *LAT*, 8 December 2002.

98. D. B. Wood, *CSM*, 28 October 2003.

99. J. O'Dell, *LAT*, 23 September 2003.

100. M. Bustillo, *LAT*, 3 August 2003.

101. M. Bustillo, *LAT*, 24 May 2004. G. Polakovic, *LAT*, 16 January 2003.

102. M. J. Campbella and A. Tobías, *Int. J. Epidemiol.* **29**, 271 (2000).

103. E. R. Stephens, in J. N. Pitts, Jr., and R. L. Metcalf, eds., *Advances in Environmental Sciences* Vol. 1 (New York: John Wiley and Sons, 1969), 119.

104. R. D. Brook et al., *Circulation* **105**, 1534 (2002).

105. J. Kay, *SFC*, 1 May 2001. Times Staff and Wire Reports, *LAT*, 7 February 2001.

106. M. S. Goldberg et al., *Am. J. Epidemiol.* **154**, 817 (2001).

107. R. Zhang et al., *Proc. Natl. Acad. Sci.* **101**, 6346 (2004).

108. J. M. Samet et al., *Am. J. Respir. Crit. Care Med.* **164**, 819 (2001).

109. B. B. Jalaludin et al., *Int. J. Epidemiol.* **29**, 549 (2000).

110. D. E. Abbey et al., *Am. J. Respir. Crit. Care Med.* **159**, 373 (1999). J. M. Lents and W. J. Kelly, *Sci. Am.* **269**(4), 32 (1993).

111. M. Grossi et al., *FB*,15 December 2002.

112. J. L. Marx, *Science* **187**, 731 (1975).

113. P. H. Abelson, *Science* **226**, 1263 (1984).

114. J. N. Pitts, Jr., in J. N. Pitts, Jr., and R. L. Metcalf, eds., *Advances in Environmental Sciences* Vol. 1 (New York: John Wiley and Sons, 1969), 289.

115. W. H. Smith, *Environ. Pollut.* **6**, 111 (1974).

116. D. L. Mauzerall and X. Wang, *Annu. Rev. Energy Environ.* **26**, 9002 (2001).

117. J. Cardoso-Vilhena et al., *J. Experimental Botany* **55**, 771 (2004).

118. A. S. Katzenstein et al., *Proc. Natl. Acad. Sci.* **100**, 11975 (2003).

119. R. G. Prinn, *Annu. Rev. Environ. Resour.* **28**, 29 (2003).

120. J. Sandin, *MJS*, 22 June 2001.

121. G. Schueller, Environmental News Network dispatch, 3 January 2001.

122. H. J. Hebert, AP dispatch, 13 January 2000.

123. J. Yardley, *NYT*, 24 September 2000.

124. N. Gott, AP dispatch, 6 December 2000.

125. M. L. Wald, *NYT*, 10 September 1999.

126. D. Mansfield, AP dispatch, 9 August 1999.

127. M. L. Wald, *NYT*, 6 August 1999.

128. D. K. Taylor et al., *Annu. Rev. Energy Environ.* **25**, 115 (2000). J. M. DeSimone, *Science* **297**, 799 (2002). For a popular view, see M. Waggoner, AP dispatch, 6 February 2000; M. Waggoner, AP dispatch, 13 March 2000.

129. P. Di Carlo et al., *Science* **304**, 722 (2004).

130. J. M. Shepherd et al., *J. Appl. Meterology* **41**, 689 (2002). A. Bridges, AP dispatch, 13 December 2003.

131. S. M. Steiger et al., *J. Geophys. Res. - Atmospheres* **107**, D11, 4117 (2002).

132. D. Thompson, AP dispatch, 11 June 2004. A. Hammond, Reuters dispatch, 24 February 2004.

133. L. Rotstayn and U. Lohmann, *J. Climate* **15**, 2103 (2002).

134. J. E. Penner et al., *Nature* **427**, 231 (2004).

135. D. Rosenfeld et al., *Science* **297**, 1667 (2002).

136. P. M. Solomon et al., *Science* **224**, 1210 (1984); T. H. Maugh, *Science* **223**, 1051 (1984).

137. A. M. Hough and R. G. Derwent, *Nature* **344**, 645 (1990). P. J. Crutzen and J. Lelieveld, *Annu. Rev. Earth Planet. Sci.* **29**, 17 (2001). Crutzen and Lelievoeld also discuss many other atmospheric gases.

138. J. Lilieveld and P. J. Crutzen, *Nature* **343**, 227 (1990).

139. J. 8. Lee, *NYT*, 16 May 2003.

140. W. R. Moomaw, *Ambio* **31**, 184 (2002).

141. M. J. Bradley and B. M. Jones, *Ambio* **31**, 141 (2002).

142. M. E. Fenn et al., *BioScience* **53**, 391 (2003).

143. T. B. Ryerson et al., *Science* **292**, 719 (2001).

144. M. Lerdau et al., *BioScience* **47**, 373 (1997).

145. J. H. Cushman, Jr., *NYT*, 16 March 1997.

146. D. Stout, *NYT*, 24 June 2000; L. Greenhouse, *NYT*, 28 February 2001.

147. T. Kenworthy, *WP*, 15 June 1999; M. L. Wald, *NYT*, 15 June 1999.

148. M. L. Wald, *NYT*, 18 December 1999.

149. M. L. Wald, *NYT*, 16 May 2001.

150. M. Preusch, *NYT*, 20 March 2004.

151. M. E. Kahn, *J. Urban Econ.* **56**, 51 (2004).

152. W. L. Chameides et al., *Science* **276**, 916 (1997).

153. C. -Y. C. Lin et al., *Geophys. Res. Lett.* **27**, 3465 (2000).

154. G. Wotawa and M. Trainer, *Science* **288**, 324 (2000).

155. P. E. Biscaye, *Science* **290**, 2258 (2000) (letter to editor).

156. K. E. Wilkening et al., *Science* **290**, 65 (2000).

157. M. O. Andreae et al., *Geophys. Res. Lett.* **28**, 951 (2001).

158. P. A. Newman et al., *Geophys. Res. Lett.* **28**, 959 (2001).

159. J. Lelieveld et al., *Science* **298**, 794 (2002).

160. J. Sturcke, *TI*, 12 July 2004. P. N. Spotts, *CSM*, 5 August 2004. N. Lossau, *DW*, 19 August 2004.

161. R. D. Borys et al., *Geophys. Res. Lett.* **30** 1538 (2003).

162. A. Givati and D. Rosenfeld, *J. Appl. Meteor.* **43**, 1038 (2004).

163. J. Schwartz et al., *Environ. Health Perspect.* **107**, 339 (1999).

164. D. W. Dockery et al., *N. Engl. J. Med.* **329**, 1753 (1993).

165. B. Urch et al., *Inhalation Toxicology* **16**, 345 (2004).

166. C. A. Pope et al., *J. Am. Med. Assn.* **287**, 1132 (2002).

167. W. J. Gauderman et al., *Am. J. Respir. Crit. Care Med.* **162**, 1383 (2000).

168. E. L. Avol et al., *Am. J. Respir. Crit. Care Med.* **164**, 2067 (2001).

169. J. M. Peters et al., *Am. J. Respir. Crit. Care Med.* **159**, 760 (1999).

170. J. M. Peters et al., *Am. J. Respir. Crit. Care Med.* **159**, 768 (1999).

171. M. Brauer et al., *Am. J. Respir. Crit. Care Med.* **166**, 1092 (2002).

172. W. J. Gauderman et al., *Am. J. Respir. Crit. Care Med.* **166**, 76 (2002). W. J. Gauderman et al., *N. Engl. J. Med.* **351**, 1057 (2004).

173. K. Leutwyler, *Sci. Am.* **269**(5), 23 (1993); P. J. Hilts, *NYT*, 19 July 1993.

174. R. Severo, *NYT*, 23 July 1989.

175. D. E. Abbey et al., Ref. 75. R. C. Gwynn et al., *Environ. Health Perspect.* **108**, 125 (2000). T. F. Mar et al., *Environ. Health Perspect.* **108**, 347 (2000). B. Ritz et al., *Am. J. Epidemiol.* **155**, 17 (2002). J. Schwartz, *Environ. Health Perspect.* **112**, 557 (2004).

176. J. Schwartz, *Epidemiology* **10**, 17 (1999). See also J. Pekkanen et al., *Circulation* **106**, 933 (2002), which details how ischemia, or decreased blood flow toward the heart results from air pollution, which provides the mechanism for the observed effect.

177. N. Künzli et al., *Lancet* **356**, 795 (2000). P. Penttinen et al., *Environ. Health Perspect.* **109**, 319 (2001). S. Ebelt et al., *Environ. Health Perspect.* **109**, 325 (2001). J. Schwartz et al., *Environ. Health Perspect.* **109**, 1001 (2001). A. D. Kappos et al., *Intl. J. Hygiene Environ. Health* **207**, 399 (2004).

178. A. Peters et al., *Environ. Health Perspect.* **108**, 283 (2000).

179. M. Ezzati and D. M. Kammen, *Environ. Health Perspect.* **109**, 481 (2001).

180. Environmental Protection Agency, *Air Quality Criteria for Particulate Matter*, EPA 600/P-99/002aB (2001).

181. A. C. Revkin, *NYT*, 11 September 2001.

182. I. Romieu et al., *Am. J. Respir. Crit. Care Med.* **154**, 300 (1996). A. Peters et al., *Am. J. Respir. Crit. Care Med.* **155**, 1376 (1997). A. Churg and M. Brauer, *Am. J. Respir. Crit. Care Med.* **155**, 2109 (1997).

183. J. M. Samet et al., *N. Engl. J. Med.* **329**, 1753 (2000).

184. J. Schwartz, *Environ. Health Perspect.* **112**, 557 (2004).

185. W. Jedrychowski et al., *Environ. Health Perspect.* **112**, 1398 (2004).

186. C. Braun-Fahrlander et al. (SCARPOL Team. Swiss Study on Childhood Allergy and Respiratory Symptoms with Respect to Air Pollution, Climate and Pollen), *Am. J. Respir. Crit. Care Med.* **155**, 1042 (1997). K. L. Timonen and J. Pekkanen, *Am. J. Respir. Crit. Care Med.* **156**, 546 (1997). S. V. Glinianaia et al., *Environ. Health Perspect.* **112**, 1365 (2004).

187. J. Schwartz and D. W. Dockery, *Am. Rev. Respir. Dis.* **145**, 600 (1992). J. Schwartz, *Am. J. Respir. Crit. Care Med.* **150**, 648 (1994). R. J. Delfino et al., *Am. J. Respir. Crit. Care Med.* **155**, 568 (1997).

188. A. Peters et al., *Circulation* **103**, 2819 (2001). C. H. Holloman et al., *Environ. Health Perspect.* **112**, 1282 (2004).

189. T. S. Fitz-Simons et al., U.S. Environmental Protection Agency, Office of Air Quality Planning and Standards (29 September 2000).

190. J. A. Cushman, Jr., *NYT*, 21 January 1997.

191. Committee of the Environmental and Occupational Health Assembly of the American Thoracic Society, *Am. J. Respir. Crit. Care Med.* **153**, 3 (2001).

192. R. Burnett et al., *Environ. Health Perspect.* **109** (Supplement 3), 375 (2001).

193. Health Effects Institute (A Special Report of the Institute's Particle Epidemiology Reanalysis Project), *Reanalysis of the Harvard Six Cities Study and the American Cancer Society Study of Particulate Air Pollution and Mortality* (Cambridge MA: Health Effects Institute, 2000). J. M. Samet et al., *The National Morbidity, Mortality and Air Pollution Study, Part I: Methods and Methodologic Issues, Research Report 94, Part I* (Cambridge MA: Health Effects Institute, 2000). J. M. Samet et al., *The National Morbidity, Mortality and Air Pollution Study, Part II: Morbidity and Mortality from Air Pollution in the United States, Research Report 94, Part II* (Cambridge MA: Health Effects Institute, 2000).

194. B. Downing, *ABJ*, 22 March 2002. B. Downing, *ABJ*, 30 May 2002.

B. Downing, *ABJ*, 4 October 2002. T. Henry, *TB*, 2 July 2003. J. C. Kuehner, *CPD*, 24 September 2004.

195. D. Krewski et al., *J. Toxicol. Environ. Health* Part A **66**, 1507 (2003).

196. K. Hoover et al., *J. Toxicol. Environ. Health* Part A **66**, 1553 (2003).

197. C. A. Pope, *Am. J. Pub. Health* **79**, 623 (1989). C. A. Pope, *Arch. Environ. Health* **46**, 90 (1991).

198. A. J. Ghio and R. B. Devlin, *Am. J. Respir. Crit. Care Med.* **164**, 704 (2001).

199. A. C. Revkin, *NYT*, 9 September 2001. J. Glanz, *NYT*, 30 October 2001.

200. A. J. Ghio, *J. Aerosol. Med.* **17**, 157 (2004).

201. D. Zmirou et al., *J. Occupational Environ. Med.* **41**, 847 (1999).

202. E. Shogren, *LAT*, 24 April 2003.

203. J. Eilperin, *WP*, 30 June 2004. H. J. Hebert, AP dispatch, 30 June 2004.

204. J. C. Kuehner, *CPD*, 30 June 2004.

205. Z. Coile, *SFC*, 15 April 2004.

206. E. G. McPherson et al., *Chicago's Urban Forest Ecosystem: Results of the Chicago Urban Forest Climate Project* USDA Forest Service Northeastern Forest Experiment Station, General Technical Report NE-186, 1994.

207. P. H. Freer-Smith et al., *Water, Air, and Soil Poll.* **155**, 173 (2004).

208. K. Q. Seelye, *NYT*, 22 June 2001.

209. Duncan Mansfield, AP dispatch, 24 September 2002.

210. W. Rawlins, *RNO*, 26 April 2002. E. Pianin, *WP*, 21 June 2002. A. Bethea, *RNO*, 21 June 2002.

211. G. Polakovic, *LAT*, 26 April 2002. Anonymous, Reuters dispatch, 23 May 2003.

212. E. Shogren, *LAT*, 17 April 2004.

213. National Center for Environmental Assessment, Office of Research and Development, U.S. Environmental Protection Agency, *Health Assessment Document For Diesel Engine Exhaust* EPA/600/8-90/057F, May 2002.

214. M. L. Wald, *NYT*, 20 November 2000.

215. G. Polakovic, *LAT*, 20 November 2000.

216. H. J. Hebert, AP dispatch, 20 November 2000.

217. J. Kay, *SFC*, 15 July 2004.

218. Office of Air Quality Planning and Standards, Environmental Protection Agency, *National Air Pollutant Emission Trends, 1900 - 1998* EPA-454/R-00-002, March 2000.

219. J. D. McDonald et al., *Environ. Health Perspect.* **112**, 1307 (2004).

220. M. Z. Jacobson et al., *Geophys. Res. Lett.* **31**, L02116 (2004).

221. C. Baltimore, Reuters dispatch, 10 May 2004. M. Janofsky, *NYT*, 11 May 2004. J. Eilperin, *WP*, 11 May 2004.

222. Ceres, the Natural Resources Defense Council, and Public Service Electric & Gas Company, *Benchmarking air*

emissions of the 100 largest electric power producers in the United States - 2002, April 2004.

223. M. R. Taylor et al., *Environ. Sci. & Technol.* **37**, 4627 (2003).

224. E. Pianin and D. Morgan, *WP*, 18 March 2002.

225. Natural Resources Defense Council, NRDC press release, 20 March 2002. See also N. Banerjee, *NYT*, 21 March 2002.

226. J. Walke, NRDC press release, 13 June 2002.

227. J. 8. Lee, *NYT*, 19 April 2003. M. Kilian, *CT*, 22 April 2003.

228. J. Williams, AP dispatch, 30 April 2003.

229. J. 8. Lee, *NYT*, 9 April 2003. Anonymous, AP dispatch, 10 April 2003.

230. J. Eilperin, *WP*, 7 October 2004.

231. M. Janofsky, *NYT*, 10 October 2004.

232. M. L. Wald, *NYT*, 16 July 1999. J. Carter, AP dispatch, 3 November 1999. D. Stout, *NYT,* 3 November 1999. M. L. Wald, *NYT*, 26 December 1999. B. Barcoli, *NYT Magazine*, 4 April 2004, 38.

233. This is discussed at length in an investigative series "Power to pollute," which ran in *ABJ*. The articles are: M. Newkirk and R. Downing, *ABJ*, 7 January 2001; M. Newkirk, *ABJ*, 7 January 2001; R. Downing, *ABJ*, 7 January 2001; R. Downing, *ABJ*, 7 January 2001; M. Newkirk, *ABJ*, 8 January 2001; R. Downing, *ABJ*, 8 January 2001; J. Craig, *ABJ*, 8 January 2001; Beacon Journal staff report, *ABJ*, 8 January 2001; M. Newkirk, *ABJ*, 9 January 2001; M. Newkirk, *ABJ*, 9 January 2001; R. Downing, *ABJ*, 9 January 2001; R. Downing, *ABJ*, 9 January 2001.

234. M. Hawthorne, *CD*, 17 April 2002. C. Powell, *ABJ*, 17 April 2002. K. Q. Seelye, *NYT*, 13 May 2002. Read about the Cheshire "blue cloud" in R. Downing, *ABJ*, 2 August 2001. Personal stories of residents are discussed in M. B. Lane, *CD*, 7 July 2002, and Cheshire neighbors express upset at lack of compensation in R. Price, *CD*, 20 May 2002 and S. Simon, *LAT*, 25 August 2002. See also A. Goodheart, *NYT Magazine*, 8 February 2004, 38.

235. R. Downing, *ABJ*, 9 May 2002.

236. R. C. Archibold, *NYT*, 22 December 2000. AP dispatch, 22 December 2000.

237. M. Hawthorne, *CD*, 17 March 2002.

238. M. Newkirk and R. Downing, *ABJ*, 18 May 2001.

239. N. Banerjee, *NYT*, 13 April 2002. E. Pianin, *WP*, 23 May 2002. K. Q. Seelye, *NYT*, 14 June 2002 and G. Polakovic, *LAT*, 14 June 2002.

240. M. L. Wald, *NYT*, 27 August 2002.

241. C. Baltimore, Reuters dispatch, 15 March 2002. D. Morgan, *WP*, 15 April 2002.

242. J. Heilprin, AP dispatch, 26 June 2003.

243. M. Williams, AP dispatch, 4 February 2003. S. Eaton, *CPD*, 6 May 2004.

244. K. Q. Seelye, *NYT*, 28 April 2002.

245. K. Q. Seelye, *NYT*, 22 August 2003.

246. J. Eilperin, *WP*, 15 July 2004.

247. G. Gugliotta and E. Pianin, *WP*, 2 July 2003. E. Pianin, *WP*, 10 October 2003.

248. J. Heilprin, AP dispatch, 26 August 2003.

249. J. 8. Lee, *NYT*, 15 December 2003.

250. J. 8. Lee, *NYT*, 14 July 2003.

251. D. Kocieniewski, *NYT*, 29 August 2003. D. Barrett, AP dispatch, 28 October 2003. Anonymous, *LAT*, 7 November 2003. R. A. Oppel Jr. and C. Drew, *NYT*, 9 November 2003. D. Barrett, AP dispatch, 18 November 2003.

252. E. Pianin, *WP*, 6 October 2003. R. A. Oppel, Jr., *NYT*, 25 October 2003.

253. E. Shogren, *LAT*, 31 August 2003.

254. A. DePalma, *NYT*, 5 June 2004.

255. J. 8. Lee, *NYT*, 8 August 2003 (Ohio). J. Eilperin, *WP*, 21 May 2004 (West Virginia).

256. E. Lau, *SB*, 23 August 2003.

257. C. Marquis, *NYT*, 30 July 2002.

258. R. Kirbach, *DZ*, 1 February 1985.

259. J. J. Romm and C. A. Ervin, *Pub. Health Rep.* **111**, 390 (1996).

260. C. -M. Wong et al., *Environ. Health Perspect.* **109**, 335 (2001).

261. M. W. Thring, *Int. J. Environmental Studies* **5**, 251 (1974); J. W. Sawyer, *Environment* **20**(2), 25 (1978); and R. Frank, in L. C. Ruedisili and M. W. Firebaugh, eds., *Perspectives on Energy* (New York: Oxford University Press, 1975), 166.

262. E. N. Allred et al., *N. Engl. J. Med.* **321**, 1426 (1989).

263. M. Simons, *NYT*, 8 April 1990.

264. M. Bobak, *Environ. Health Perspect.* **108**, 173 (2000). M. Maisonet et al., *Environ. Health Perspect.* **109** (Supplement 3), 351 (2001).

265. S. Cohanim et al., *Summary of Air Quality in California's South Coast and Southeast Desert Air Basins* 1987-1990 (Los Angeles: South Coast Air Quality Management District, 1991).

266. C. Holden, *Science* **187**, 818 (1975).

267. P. A. Bromberg, in A. Y. Watson et al., eds., *Air Pollution, the Automobile, and Public Health* (Washington, DC: National Academy of Sciences, 1988), 465.

268. E. Goldsmith, *Ecologist* **7**, 160 (1977).

269. L. B. Lave and E. P. Seskin, *Science* **169**, 723 (1970); and L. B. Lave and L. P. Silverman, *Annu. Rev. Energy* **1**, 601 (1976).

270. P. Nafstad et al., *Environ. Health Perspect.* **112**, 610 (2004).

271. R. S. Gao et al., *Science* **303**, 516 (2004). A. M. Fridlind et al., *Science* **304**, 718 (2004).

272. K. Sexton and P. B. Ryan, in A. Y. Watson et al., eds., *Air Pollution, the Automobile, and Public Health* (Washington, DC: National Academy of Sciences, 1988), 207.

273. M. Ash and T. R. Fetter, *Social Science Quarterly* **85**, 441 (2004).

274. W. Bach, *Atmospheric Pollution* (New York: McGraw-Hill Book Co., 1972).

275. J. V. Hall et al., *Science* **255**, 812 (1992).

276. Landesanstalt für Umweltschutz Baden-Württemberg, "Die Luft in Baden-Württemberg," 1993.

277. Y. Wang and K. He, *IEEE Spectrum* **35**(12), 55 (1999).

278. R. L. Babcock, Jr., *J. Air Pollut. Control Assoc.* **20**, 653 (1970).

279. R. B. Schlesinger, in A. Y. Watson et al., eds., *Air Pollution, the Automobile, and Public Health* (Washington, DC: National Academy of Sciences, 1988), 239. See also Y. Mamane et al., *Atmos. Environ.* **11**, 2125 (1986).

280. D. F. S. Nautsch and J. R. Wallace, *Science* **186**, 695 (1974).

281. R. Atkinson, in A. Y. Watson, R. R. Bates, and D. Kennedy, eds., *Air Pollution, the Automobile, and Public Health* (Washington, DC: National Academy of Sciences, 1988), 99.

282. A. J. Haagen-Smit, in R. J. Seale and R. A. Sierka, eds., *Energy Needs and the Environment* (Tucson: University of Arizona Press, 1973), 53

283. J. A. Last, in A. Y. Watson, R. R. Bates, and D. Kennedy, eds., *Air Pollution, the Automobile, and Public Health* (Washington, DC: National Academy of Sciences, 1988), 415.

284. M. Lippmann, *J. Air Waste Manage. Assoc.* **39**, 672 (1989).

285. L. Mansnerus, *NYT*, 21 August 1988.

286. J. L. Wright, in A. Y. Watson et al., eds., *Air Pollution, the Automobile, and Public Health* (Washington, DC: National Academy of Sciences, 1988), 441.

287. J. D. Sun et al., in A. Y. Watson et al., eds., *Air Pollution, the Automobile, and Public Health* (Washington, DC: National Academy of Sciences, 1988), 299; S. S. Hecht, in A. Y. Watson et al., eds., *Air Pollution, the Automobile, and Public Health* (Washington, DC: National Academy of Sciences, 1988), 555.

288. D. G. Kaufman, in A. Y. Watson et al., eds., *Air Pollution, the Automobile, and Public Health* (Washington, DC: National Academy of Sciences, 1988), 519.

289. SCEP (Study of Critical Environmental Problems) report, W. W. Kellogg et al.,

in W. H. Matthews et al., eds., *Man's Impact on the Climate* (Cambridge, Mass.: MIT Press, 1971).

290. F. P. Prera and A. K. Ahmed, *Respirable Particles* (New York: Natural Resources Defense Council, 1978).

291. C. L. Comar and L. A. Sagan, *Annu. Rev. Energy* **1**, 581 (1976).

292. M. W. McElroy et al., *Science* **215**, 13 (1982).

293. R. W. Shaw, *Sci. Am.* **257**(2), 96 (1987).

294. A. R. Gold, *NYT*, 18 April 1990.

295. M. Hoggan et al., *Air Quality Trends in California's South Coast and Southeast Desert Air Basins 1976-1990* (Los Angeles: South Coast Air Quality Management District, 1991).

296. D. Hafemeister, *Am. J. Phys.* **50**, 713 (1982).

297. Anonymous, *Air Conservation Newsletter* #92 (1977).

298. Abt Associates, Inc., *The Particulate-Related Health Benefits of Reducing Power Plant Emissions* (October 2000). The report is to be found at URL: http://www. abtassociates. com/html/ reports/reports-index. html.

299. Abt Associates, Inc., *Particulate-Related Health Impacts of Eight Electric Utility Systems* (April 2002). See also Office of Air and Radiation, U.S. EPA, *Regulatory Impact Analysis for the NO_x SIP Call, FIP, and Section 126 Petitions* EPA-452/R-98-003, December, 1998.

300. C. L. Blanchard, *Annu. Rev. Energy Environ.* **24**, 329 (1999).

301. W. S. Cleveland et al., *Science* **191**, 179 (1976).

302. D. D. Davis et al., *Science* **186**, 733 (1974).

303. W. S. Cleveland and T. E. Graedel, *Science* **204**, 1273 (1979).

304. K. A. Rahn and D. H. Lowenthal, *Science* **228**, 275 (1985); *Natl. Hist.* **95**(7), 62 (1986).

305. J. O. Nriagu, *Nature* **338**, 47 (1989). For an example of unnatural concentrations, see S. Hong et al., *Science* **272**, 246 (1996).

306. S. Shulman, *Tech. Rev.* **100**(7), 11 (1997).

307. J. O. Nriagu, *Nature* **349**, 142 (1991).

308. R. A. Kerr, *Science* **212**, 1013 (1987) reports on a special issue of *Atmospheric Environment* that has many articles on this subject; K. A. Rahn, *Natl. Hist.* **93**(5), 30 (1984). According to more recent research, the answer to Rahn's question is both Europe and North America: D. H. Lowenthal et al., *Atmos. Environ.* **31**, 3707 (1997).

309. C. Holden, *Science* **260**, 1586 (1993); J. R. Luoma, *NYT*, 1 June 1993.

310. X.-H. Song et al., *Atmos. Environ.* **35**, 5277 (2001).

311. S. E. Schwartz, *Science* **243**, 753 (1989).

312. Anonymous, *Der Spiegel* (No. 47), 96 (1981); Anonymous, *Der Spiegel* (No. 48), 188 (1981); and Anonymous, *Der Spiegel* (No. 49), 174 (1981). For more information on crossborder transport, see Umweltbundesamt, *Daten zur Umwelt* (Berlin: Erich Schmidt Verlag, 1998), 180-181; *Daten zur Umwelt* (Berlin: Erich Schmidt Verlag, 1992), 212-219; and *Daten zur Umwelt* (Berlin: Erich Schmidt Verlag, 1998), 180-186.

313. Air Pollution and Health: A European Information System, *Health Impact Assessment of Air Pollution In 26 European Cities* (Paris: Institut de Veille Sanitaire, 2002). Air Pollution and Health: A European Information System, *Health Impact Assessment of Air Pollution and Communication Strategy* (Paris: Institut de Veille Sanitaire, 2004). S. Medina et al., *Bundesgesundheitsbl.- Gesundheitsforsch.- Gesundheitsschutz* **44**, 537 (2001). S. Medina et al., *Pollution Atmosphérique* **176**, 499 (2002).

314. J. T. Peterson and C. E. Junge, in W. H. Matthews et al., eds., *Man's Impact on the Climate* (Cambridge, MA: MIT Press, 1971), 310.

315. J. E. Jiusto, *The Physics of Weather Modification* (College Park, MD: AAPT, 1986).

316. J. Shulevitz, *NYT*, 6 August 1989.

317. L. B. Lave and L. C. Freeburg, *Nuc. Safety* **14**, 409 (1973).

318. J. P. McBride et al., *Science* **202**, 1045 (1978).

319. F. X. Clines, *NYT*, 23 July 2002.

320. A. A. Cook and J. D. Rosenberg, *Amicus J.* **8**(2), 5 (1986).

321. T. E. Graedel, in A. Y. Watson et al., eds., *Air Pollution, the Automobile, and Public Health* (Washington, DC: National Academy of Sciences, 1988), 133.

322. Umweltbundesamt, *Daten zur Umwelt* (Berlin: Erich Schmidt Verlag, 1992), 196-211.

323. R. W. Lindsay et al., *J. Air Waste Manage. Assoc.* **39**, 40 (1989); D. R. Lawson, *Air and Waste* **43**, 1567 (1993).

324. A. M. Hough and R. G. Derwent, *Nature* **344**, 645 (1990); A. M. Terry, *Science* **256**, 1157 (1992); and A. S. Lefohn et al., *J. Air Waste Manage. Assoc.* **42**, 156 (1992).

325. S. J. Oltmans and H. Levy, II, *Nature* **358**, 392 (1992).

326. B. J. Finlayson-Pitts and J. N. Pitts, Jr., *Air and Waste* **43**, 1091 (1993).

327. W. H. Brune, *Science* **256**, 1154 (1992).

328. G. J. McRae and A. G. Russell, *Comp. in Phys.* **4**, 227 (1990); R. Pool, *Science* **244**, 1438 (1989).

329. Committee on Tropospheric Ozone Formation and Measurement, National Research Council (J. H. Seinfeld et al.), *Rethinking the Ozone Problem in Urban and Regional Air Pollution* (Washington, DC: National Academy Press, 1991).

330. S. Sillman, *Annu. Rev. Energy Environ.* **18**, 31 (1993).

331. J. H. Seinfeld, *Science* **243**, 745 (1989).

332. A. G. Russell et al., *Science* **247**, 201 (1990).

333. M. L. Wald, *NYT*, 7 April 1989.

334. M. L. Wald, *NYT*, 30 October 1991 and *NYT*, 8 November 1991, and also the common action *NYT*, 3 April 1992.

335. Dept. of Energy, *Petroleum Supply Annual 1992* (Washington, DC: GPO, 1992), V. 1, DOE-EIA-0340 (92)/1.

336. M. Bustillo, *LAT*, 29 July 2004.

337. E. Pianin, *WP*, 19 September 2003. G. Polakovic, *LAT*, 26 September 2003. G. Polakovic, *LAT*, 24 October 2003.

338. S. Allison, *LAT*, 24 May 2004.

339. Dept. of Energy, *Acid Rain Compliance Strategies for the Clean Air Act Amendments of 1990* (Washington, DC: GPO, 1994), DOE/EIA-0582.

340. K. R. Smith, *Annu. Rev. Energy Environ.* **18**, 529 (1993).

341. I. M. Torrens et al., *Annu. Rev. Energy Environ.* **17**, 187 (1992).

342. Dept. of Energy, *Improving Technology: Modeling Energy Futures for the National Energy Strategy* (Washington, DC: GPO, 1990), SR/NES/90-01

343. A. J. Krupnick and P. R. Portney, *Science* **252**, 522 (1991).

344. R. A. Kerr, *Science* **282**, 1024 (1998).

345. B. J. Finlayson and J. N. Pitts, *Science* **192**, 111 (1976).

346. Environmental Protection Agency, *National Air Quality And Emissions Trends Report, 1998* (2000).

347. Office of Air and Radiation, Environmental Protection Agency, *The Benefits and Costs of the Clean Air Act 1990 to 2010*, EPA-410-R-99-001 (1999).

348. D. Munton, *Environment* **40**(6), 4 (1998).

349. M. L. Wald, *NYT*, 12 May 1992; *NYT*, 13 May 1992.

350. R. Reinhold, *NYT*, 26 November 1993.

351. M. Simons, *NYT*, 8 February 1992.

352. M. Hawthorne, *CD*, 3 August 2001. See also M. Newkirk, Ref. 139.

353. A. Jenkins, *Nature* **401**, 537 (1999).

354. T. H. Maugh, II, *Science* **193**, 871 (1976).

355. P. Liss, *Science* **285**, 1217 (1999).

356. G. E. Likens et al., *Sci. Am.* **241**(4), 43 (1979).

357. R. Partrick et al., *Science* **211**, 446 (1981).

358. J. M. Miller and A. Yoshinaga, *Geophys. Res. Lett.* **8**, 779 (1981); see also R. A. Kerr, *Science* **212**, 1014 (1981).

359. S. J. Nagourney and DC. Bogen, in C. M. Bhumralkar, ed., *Meteorological Aspects of Acid Rain* (Boston: Butterworth, 1984), 145.

360. Umweltbundesamt, *Daten zur Umwelt* (Berlin: Erich Schmidt Verlag, 1998), Fig. 3.1.

361. H.-G. Kemmer, *DZ*, 3 August 1984; H. Bieber et al., *DZ*, 10 August 1984;

J. Nawrocki, *DZ*, 30 November 1984; G. Lütge, *DZ*, 20 May 1988 all discuss the controversy over the Buschhaus coal-fired facility, which by itself poured 35 kt of SO_2 into the atmosphere each year.

362. D. Spurgeon, *Nature* **390**, 6 (1997).
363. R. Triendl, *Nature* **392,** 426 (1998).
364. R. E. Newell, *Sci. Am.* **224**(1), 32 (1971). E. Marshall, *Science* **229**, 1070 (1985).
365. M. Oppenheimer et al., *Science* **229**, 859 (1985).
366. M. Sun, *Science* **228**, 34 (1985).
367. J. Aber et al., *BioScience* **48**, 921 (1998).
368. T. E. Jordan and D. E. Weller, *BioScience* **46**, 655 (1996).
369. G. P. Asner et al., *BioScience* **47**, 226 (1997).
370. C. T. Driscoll et al., *BioScience* **51**, 180 (2001).
371. Dept. of Commerce, *Statistical Abstract of the United States 1990* (Washington, DC: GPO, 1990), Table 349; *Statistical Abstract of the United States 1992* (Washington, DC: GPO, 1992), Table 352 *Statistical Abstract of the United States 2002* (Washington, DC: GPO, 1992), Table 392.
372. D. H. DeHayes et al., *BioScience* **49**, 789 (1999).
373. J. N. Cape, *Environ. Pollut.* **82**, 167 (1993).
374. B. I. Chevone et al., *J. Air Poll. Control Assoc.* **36**, 813 (1986). H. Blaschke, *Environ. Pollut.* **68**, 409 (1990). S. V. Krupa and A. H. Legge, in R. S. Ambasht, ed., *Modern trends in ecology and environment* (Leiden, The Netherlands: Backhuys Publishers, 1998) p. 285–306.
375. S. Huttunen, in K. E. Percy et al., eds., *Air pollutants and the leaf cuticle* (New York, NY: Springer-Verlag, 1994), p. 81–96. (NATO ASI series)
376. L. O. Hedin and G. E. Likens, *Sci. Am.* **275**(6), 88 (1996). G. E. Likens et al., *Science* **272**, 244 (1996).
377. L. O. Hedin et al., *Nature* **367**, 351 (1994).
378. R. S. Hames et al., *Proc. Natl. Acad. Sci.* **99**, 11235 (2002).
379. T. J. Sullivan et al., *Nature* **345**, 54 (1990).
380. W. M. Stigliani and R. W. Shaw, *Annu. Rev. Energy* **15**, 201 (1990).
381. Environment Canada, *Downwind: The Acid Rain Story* (Ottawa: Government Printing Centre, 1981).
382. M. Uman, in D. Hafemeister, ed., *Acid Rain: How Serious and What to Do* (College Park, MD: AAPT, 1986), 1.
383. D. Martin, *NYT*, 26 February 1984; E. Marshall, *Science* **217**, 1118 (1982).
384. Environmental Protection Agency website, URL http://www. epa. gov/ airmarkets/acidrain/effects/ surfacewater. html.

385. J. Dao, *NYT*, 27 March 2000.
386. U. Schnabel, *DZ*, 2 October 1992.
387. Environment Canada website, URL http://www. ec. gc. ca/science/sandesept01/ article2_e. html.
388. J. N. Galloway and G. E. Likens, *Atmos. Environ.* **15**, 1081 (1981).
389. S. E. W. Bonga and L. H. T. Dederson, *Endeavour* **10**, 198 (1986).
390. R. Howard and M. Perley, *Acid Rain* (Toronto: Anansi, 1980).
391. E. Gorham, in C. M. Bhumralkar, ed., *Meteorological Aspects of Acid Rain* (Boston: Butterworth, 1984), 1.
392. T. H. Maugh, *Science* **226**, 1408 (1984).
393. D. W. Schindler, *Science* **239**, 149 (1988). J. R. Luoma, *NYT*, 13 September 1988.
394. D. W. Schindler et al., *Science* **228**, 1396 (1985).
395. J. M. W. Rudd et al., *Science* 240, 1515 (1988).
396. T. K. Morris and G. L. Krueger, in C. M. Bhumralkar, ed., *Meteorological Aspects of Acid Rain* (Boston: Butterworth, 1984), 231.
397. R. F. Wright et al., *Nature* **334**, 670 (1988). See more about the Norwegian acid rain work in R. F. Wright et al., *Nature* **343**, 63 (1990).
398. R. Shaw, Environmental News Network dispatch, 25 October 2000.
399. J. L. Stoddard et al., *Nature* **401**, 575 (1999).
400. W. K. Stevens, *NYT*, 31 January 1989.
401. W. K. Stevens, *NYT*, 16 January 1990.
402. Environmental Protection Agency website, URL http://www. epa. gov/EPA_particulates/state_data/ State_Emission_Index_Page.
403. J. Stashenko, AP dispatch, 2 May 2000.
404. Environment Canada website, URL http://www. ec. gc. ca/science/ sandesept01/article1_e. html.
405. R. W. Shaw, in C. M. Bhumralkar, ed., *Meteorological Aspects of Acid Rain* (Boston: Butterworth, 1984).
406. M. Sun, *Science* **229**, 949 (1985).
407. R. A. Kerr, *Science* **221**, 254 (1983).
408. J. V. Galloway et al., *Science* **226**, 829 (1984).
409. T. Hilchey, *NYT*, 7 September 1993.
410. J. M. Prospero and D. L. Savoie, *Nature* **339**, 687 (1989).
411. Umweltbundesamt, *Daten zur Umwelt* (Berlin: Erich Schmidt Verlag, 1994), Table III. 1, p. 262-263; Umweltbundesamt, *Daten zur Umwelt* (Berlin: Erich Schmidt Verlag, 1998), Table III. 2, p. 135-136.
412. E. Marshall, *Science* **221**, 241 (1983).
413. Anonymous, *OMRA Newsletter* (Ohio Mining and Reclamation Association), 1978.
414. R. A. Smith, *Air and Rain* (London: Longmans, Green, 1872).

415. B. L. Niemann, in C. M. Bhumralkar, ed., *Meteorological Aspects of Acid Rain* (Boston: Butterworth, 1984), 57.
416. S. Postel, *Futurist* **18**(4), 39 (1984).
417. S. B. McLaughlin, *J. Air Poll. Control Assoc.* **35**, 512 (1985).
418. P. Shabecoff, *NYT*, 26 February 1984; H. W. Vogelman, *Natl. Hist.* **91**(11), 8 (1982).
419. Environmental Resources Ltd., *Acid Rain* (New York: Unipub, 1983).
420. P. Winkler, in Jack L. Durham, ed., *Chemistry of Particles, Fogs, and Rain* (Boston: Butterworth, 1984), 161.
421. E. C. Krug and C. R. Frink, *Science* **221**, 520 (1983).
422. C. Alewell et al., *Nature* **407**, 856 (2000).
423. R. A. Kerr, *Science* **211**, 692 (1981).
424. T. Paces, *Nature* **315**, 31 (1985).
425. P. Lewis, *NYT*, 3 August 1984.
426. W. K. Stevens, *NYT*, 20 February 1990.
427. B. D. Ayres, *NYT*, 20 June 1989.
428. M. del Monte and O. Vittori, *Endeavour* **9**, 117 (1985).
429. J. N. Wilford, *NYT*, 8 August 1989.
430. C. Rodriguez-Navarro et al., *Appl. Environmental Microbiol.* **69**, 2182 (2003).
431. L. A. Baker et al., *Science* **252**, 1151 (1991).
432. M. Oppenheimer, in D. Hafemeister, ed., *Acid Rain: How Serious and What to Do* (College Park, MD: AAPT, 1986), 27.

CHAPTER 15

1. Dept. of Energy, *Annual Energy Review 2000* (Washington, DC: GPO, 2001), DOE/EIA-0384(01).
2. S. C. Davis, *Transportation Energy Data Book: Edition 20*, Oak Ridge National Laboratory, ORNL-6959, Table 2.7 (November, 2000).
3. Dept. of Energy, *Energy Consumption & Conservation Potential: Supporting Analysis for the National Energy Strategy* (Washington, DC: GPO, 1990), SR/NES 90-02.
4. U.S. Departement of Transportation, *Our Nation's Travel: 1995 NPTS Early Results Report* (2000).
5. E. A. Bretz, *IEEE Spectrum* **37**(1), 91 (2001).
6. Dept. of Commerce, *Statistical Abstract of the United States, 1984* (Washington, DC: GPO, 1984), Table 1048. Dept. of Commerce, *Statistical Abstract of the United States, 2000* (Washington, DC: GPO, 2001), Table 1067.
7. E. E. Covert, *Sci. Am.* **273**(3), 110 (1995). For a novel look at drag, see S. Alben, M. Shelley, and J. Zhang, *Nature* **420**, 479 (2002) and the accompanying commentary by V. Steinberg, *Nature* **420**, 473 (2002).

8. Dept. of Commerce, *Statistical Abstract of the United States, 1984* (Washington, DC: GPO, 1984), Tables 1086 and 1087. Dept. of Commerce, *Statistical Abstract of the United States, 2000* (Washington, DC: GPO, 2001), Tables 1007, 1008, and 1051.

9. R. Bezdek and B. Hannon, *Science* **185**, 669 (1974).

10. C. Shirley and C. Winston, *J. Urban Econ.* **55**, 398 (2004).

11. T. R. Eastham, *Sci. Am.* **273**(3), 100 (1995).

12. Office of Emergency Preparedness, *The Potential for Energy Conservation* (Washington, DC: GPO, 1972), Appendix C.

13. Dept. of Transportation, *Compendium of National Urban Mass Transportation Statistics for the 1984 Report Year* (Washington, DC: GPO, 1987).

14. S. Meyers and L. Schipper, *Annu. Rev. Energy Environ.* **17**, 463 (1992).

15. W. Brown, *WP*, 9 March 2000.

16. C. Sheehan, AP dispatch, 24 December 2002.

17. See, for example, P. Hergersberg, *DW*, 29 September 2002.

18. D. L. Greene, *Annu. Rev. Energy Environ.* **17**, 537 (1992). A more recent discussion is found in J. J. Lee et al., *Annu. Rev. Energy Environ.* **26**, 167 (2001).

19. A. B. Lovins and L. H. Lovins, *Annu. Rev. Energy Environ.* **16**, 433 (1991).

20. M. Rauner, *DZ*, 26 December 2002.

21. R. Witkin, *NYT*, 20 July 1986.

22. E. Weiner, *NYT*, 29 August 1990.

23. W. E. Leary, *NYT*, 22 May 2001. E. Guizzo, *IEEE Spectrum* **41**(1), 66 (2004).

24. M. W. Browne, *NYT*, 11 September 1990. See also Ref. 7.

25. P. Pae, *LAT*, 1 September 2003.

26. E. Wong, *NYT*, 21 December 2002.

27. E. A. Bretz, *IEEE Spectrum* **38**(1), 97 (2001).

28. D. L. Giles, *Sci. Am.* **277**(4), 126 (2001).

29. T. Jüngling, *DW*, 2 December 2001. A. Johansen, *DW*, 2 December 2001.

30. H. Y. Erbil et al., *Science* **299**, 1377 (2003).

31. C. L. Dybas, *WP*, March 31, 2003.

32. E. J. Wasp, *Sci. Am.* **249**(5), 48 (1983).

33. Dept. of Energy, *Annual Prospects for World Coal Trade 1991* (Washington, DC: GPO, 1991), DOE/EIA-0363(91).

34. Dept. of Commerce, *Statistical Abstract of the United States, 2000* (Washington, DC: GPO, 2001), Table 1096.

35. J. B. C. Jackson et al., *Science* **243**, 37 (1989).

36. J. W. Farrington, *Oceanus*, **27**(3), 3 (1985) summarizes the National Research Council report *Oil in the Sea: Inputs, Fates, and Effects*.

37. B. J. Feder, *NYT*, 27 June 1990.

38. R. J. Seymour and R. A. Geyser, *Annu. Rev. Energy Environ.* **17**, 261 (1992). See also J. R. Bragg et al., *Nature* **368**, 413 (1994) and K. Schneider, *NYT*, 7 July 1994.

39. E. A. Bretz, *IEEE Spectrum* **41**(1), 60 (2004).

40. *NYT Encyclopedic Almanac*, 1971 ed. (New York: New York Times, 1971), 191. Most moves are short distances. The 18% figure eliminates very short moves (for example, within the same block).

41. E. Hirst and J. C. Moyers, *Science* **179**, 1299 (1973).

42. S. C. Davis and L. F. Truett, Oak Ridge report ORNL/TM-2000/147, August 2000.

43. Dept. of Commerce, *Statistical Abstract of the United States, 2000* (Washington, DC: GPO, 2001), Table 1048.

44. L. Hodges, *Am. J. Phys.* **42**, 456 (1974).

45. J. Dillin, *CSM*, 18 October 2000.

46. J. B. Treaster and K. Bradsher, *NYT*, 2 December 2000.

47. M. L. Wald, *NYT*, 16 April 2002. R. Alonso-Zaldivar, *LAT*, 16 April 2002.

48. K. Bradsher, *NYT*, 18 May 2001.

49. D. Greene and S. Plotkin, Ch. 6 in Interlaboratory Working Group 2000, *Scenarios for a Clean Energy Future* (Oak Ridge, TN: Oak Ridge National Laboratory and Berkeley, CA: Lawrence Berkeley National Laboratory), ORNL/CON-476 and LBNL-44029, November 2000.

50. K. Bradsher, *NYT*, 19 May 2001.

51. G. Miller, *LAT*, 21 May 2001. R. Simon and T. Y. Jones, *LAT*, 26 August 2001.

52. N. Pickler, AP dispatch, 20 December 2000.

53. K. Bradsher, *NYT*, 7 April 2001.

54. C. Lochhead, *SFC*, 14 March 2002. D. E. Rosenbaum, *NYT*, 14 March 2002.

55. D. Hakim, *NYT*, 2 October 2004.

56. Committee on the Effectiveness and Impact of Corporate Average Fuel Economy (CAFE) Standards (P. R. Portney et al.), *Effectiveness and impact of Corporate Average Fuel Economy (CAFE) standards* (Washington, DC: National Academy Press, 2001).

57. J. O'Dell, *LAT*, 2 January 2002. Anonymous, Reuters dispatch, 6 September 2002. D. Hakim, *NYT*, 31 July 2003. V. Klinkenborg, *NYT*, 2 August 2003.

58. Anonymous, Reuters dispatch, 5 April 2002.

59. H. Stoffer, *Automotive News*, No. 5921:2, 2001.

60. K. Bradsher, *NYT*, 23 September 2004.

61. D. Weikel, *LAT*, 8 April 2004.

62. J. Hyde, Reuters dispatch, 2 April 2003. D. Hakim, *NYT*, 23 December 2003. Anonymous, Reuters dispatch, 26 March 2004.

63. J. O'Dell, *LAT*, 2 October 2002. D. Hakim, *NYT*, 22 December 2002.

64. P. F. Waller, *Annu. Rev. Public Health* **23**, 93 (2002). D. Hakim, *NYT*, 16 January 2003. D. Hakim, *NYT*, 30 January 2003. R. Alonso-Zaldivar, *LAT*, 5 February 2003. D. Hakim, *NYT*, 28 June 2003.

65. R. Alonso-Zaldivar, *LAT*, 29 April 2004. D. Hakim, *NYT*, 17 August 2004.

66. D. Hakim, *NYT*, 10 August 2004.

67. D.-A. Durbin, AP dispatch, 7 June 2004. D. Hakim, *NYT*, 8 June 2004.

68. D. Hakim, *NYT*, 14 February 2003. D. Hakim, *NYT*, 4 December 2003.

69. Insurance Institute for Highway Safety, press release, 25 August 2003. One article based on the IIHS press release is D. Hakim, *NYT*, 26 August 2003. The government's response is delineated in G. Schneider, *WP*, 12 May 2004.

70. R. Alonso-Zaldivar, *LAT*, 7 September 2003. See also R. Alonso-Zaldivar, *LAT*, 22 June 2003.

71. D. Hakim, *NYT*, 30 October 2003. J. O'Dell, *LAT*, 14 March 2004.

72. D. Hakim, *NYT*, 15 October 2003.

73. J. B. Treaster, *NYT*, 24 August 2002.

74. M. Gardner, *CSM*, 4 June 2001. K. Ellison, *WP*, 8 November 2002. E. P. Dalesio, AP dispatch, 14 November 2002.

75. A. Donovan, *NYT*, 23 August 2002.

76. K. Edds, *WP*, 23 August 2003. N. Madigan, *NYT*, 31 August 2003. M. Martinez, *CT*, 28 September 2003. Anonymous, Reuters dispatch, 10 March 2004.

77. R. Elvik, *Accident Analysis & Prevention* **31**, 265 (1999). Elvik refers to the article by C. Tingvall, in H. von Holst et al., eds., *Transportation, Traffic Safety and Health* (Berlin: Springer Verlag, 1997), 37.

78. I. W. H. Parry, *J. Urban Econ.* **56**, 346 (2004).

79. Anonymous, Deutsche Press Agentur dispatch, 9 December 2002.

80. S. Etzold, *DZ*, 12 July 2002.

81. T. Moran, *NYT*, 22 October 2003.

82. Anonymous, Reuters dispatch, 29 July 2002.

83. J. Selingo, *NYT*, 22 May 2003.

84. R. F. Raney, *NYT*, 22 October 2003.

85. J. Motavalli, *NYT*, 12 September 2004.

86. J. Motavalli, *NYT*, 12 April 2002.

87. K. Bradsher, *NYT*, 21 February 2001.

88. K. Bradsher, *NYT*, 1 March 2001.

89. Freightliner press release, 21 February 2001. (URL at http://www.freightlinertrucks.com/about_us/press_room/press_release.asp?id=261)

90. F. Protzman, *NYT*, 5 March 1994.

91. S. C. Davis, *Transportation Energy Data Book: Edition 20*, Oak Ridge National Laboratory, ORNL-6959, Table 7.21.

92. C. L. Gray, Jr. and F. von Hippel, *Sci. Am.* **224**(5), 48 (1981).

93. J. Grey et al., *Science* **200**, 135 (1978).

94. R. K. Whitford, *Annu. Rev. Energy* **9**, 375 (1984).

95. Dept. of Commerce, *Statistical Abstract of the United States*, 2000 (Washington, DC: GPO, 2001), Table 1050.

96. Dept. of Energy, *Household Vehicles Energy Consumption 1988* (Washington, DC: GPO, 1991), DOE/EIA-0464(88).

97. Dept. of Commerce, *Statistical Abstract of the United States*, 2000 (Washington, DC: GPO, 2001), Tables 1042 to 1044.

98. P. Zielbauer, *NYT*, 27 December 1999.

99. S. M. Rock, *Accident Analysis & Prevention* **27**, 207 (1995).

100. D. Hakim, *NYT*, 24 November 2003.

101. M. Bernstein, *Invention & Technology* **17**(4), 38 (2002).

102. D. J. Steding et al., *Proc. Natl. Acad. Sci.* **97**, 11181 (2000).

103. J. V. Spadero and A. Rabl, *Atmos. Environ.* **35**, 4763 (2001).

104. Dept. of Commerce, *Statistical Abstract of the United States, 2000* (Washington, DC: GPO, 2001), Table 394. Environmental Protection Agency, *Environmental Quality 1975* (Washington, DC: GPO, 1975).

105. South Coast Air Quality Management District, *The Path to Clean Air: Attainment Strategies* (El Monte, Calif., 1987); *Air Quality Management Plan* (1988 revision), App. III-A (El Monte, Calif., 1988).

106. S. C. Davis, *Transportation Energy Data Book: Edition 20*, Oak Ridge National Laboratory, ORNL-6959, Tables 4.1 to 4.7.

107. P. V. Hobbs et al., *Science* **183**, 909 (1974).

108. R. A. Bryson, *Science* **184**, 753 (1974).

109. J. Holusha, *NYT*, 4 March 1990.

110. M. L. Wald, *NYT*, 27 March 1991; J. Holusha, *NYT*, 28 July 1991; and D. F. Cuff, *NYT*, 12 March 1992.

111. M. L. Wald, *NYT*, 26 January 1994.

112. D. McCosh, *Pop. Sci.* **257**(3), 84 (2000). D. Stover, *Pop. Sci.* **256**(6), 33 (2000).

113. J.-D. Eckhardt, *UniKaTh* 1/2001.

114. K. Bradsher, *NYT*, 22 March 2001.

115. D. R. Lawson, *Air and Waste* **43**, 1567 (1993).

116. D. Yourk, *The Toronto Globe and Mail*, 2 January 2003.

117. J. G. Calvert et al., *Science* **261**, 37 (1993).

118. R. Weinhold, *Environ. Health Perspect.* **109**, A 422 (2001).

119. M. Bustillo, *LAT*, 4 May 2004.

120. Committee on Tropospheric Ozone Formation and Measurement, National Research Council (J. H. Seinfeld et al.), *Rethinking the Ozone Problem in Urban and Regional Air Pollution* (Washington, DC: National Academy Press, 1991). See also Committee to Assess the North American Research Strategy for Tropospheric Ozone (NARSTO) Program (M. Russell et al.), National Academy of Sciences, *Review of the NARSTO Draft Report: An Assessment of Tropospheric Ozone Pollution—A North American Perspective* (Washington, DC: National Academy Press, 2000).

121. Anonymous, *NYT*, 13 January 1991.

122. B. G. Wicke et al., *Nature* **338**, 492 (1989).

123. B. Weinhold, *Environ. Health Perspect.* **110**, A458 (2002).

124. B. Commoner, *J. Am. Inst. Planners* **39**, 147 (1973).

125. M. Specter, *NYT*, 21 October 1992.

126. E. Caplun et al., *Endeavour* **8**, 135 (1984).

127. K. B. Noble, *NYT*, 5 August 1984. R. A. Searls, *Endeavour* **13**, 2 (1989).

128. J. H. Trefry et al., *Science* **230**, 439 (1985).

129. K. J. R. Rosman et al., *Nature* **362**, 333 (1993). Anonymous, *NYT*, 17 August 1993.

130. G. Hill, *NYT*, 13 February 1979.

131. P. Christ and R. Gaul, *DZ*, 27 April 1984.

132. D. Asendorpf, *DZ*, 15 May 2002 documents the success of the German program in eliminating lead is in the environment from gasoline.

133. D. Dickson, *Science* **228**, 159 (1985).

134. Various editions of Umweltbundesamt (German Ministry for the Environment), *Daten zur Umwelt* (Erich Schmidt Verlag, 1989, 1992, 1995, 1998) were used, as not all data is reported continually.

136. J. Godoy, IPS News Agency dispatch, 1 June 2004.

137. P. Brown, *TG*, 27 March 2004.

135. D. M. Settle and C. C. Patterson, *Science* **207**, 1167 (1980).

138. M. L. Wald, *NYT*, 13 August 1989.

139. M. L. Wald, *NYT*, 8 November 1991.

140. M. L. Wald, *NYT*, 15 September 1989.

141. R. Reinhold, *NYT*, 16 December 1989.

142. G. Rifkin, *NYT*, 11 November 1992.

143. A. Pollack, *NYT*, 9 September 2000. G. Polakovic, *LAT*, 26 January 2001.

144. R. Snow et al., *J. Air Poll. Control Assoc.* **39**, 48 (1989).

145. K. Wright, *Sci. Am.* **262**(5), 92 (1990).

146. A. C. Lloyd et al., *J. Air Poll. Control Assoc.* **39**, 696 (1989); W. K. Stevens, *NYT*, 28 August 1990; M. L. Wald, *NYT*, 14 April 1993; and M. L. Wald, *NYT*, 28 April 1993.

147. P. C. Judge, *NYT*, 18 July 1990.

148. G. Polakovic, *LAT*, 8 December 2000.

149. Robert Salladay, *SFC*, 26 January 2001.

150. M. J. Rienzenman, *IEEE Spectrum* **35**(11), 18 (1998). For a more popular description of the future of electric vehicles, see S. D. Freeman, *NYT*, 28 August 1994. Also see M. L. Wald, *NYT*, 2 December 1994.

151. B. Fowler, AP dispatch, 21 March 2002.

152. G. Polakovic and J. O'Dell, *LAT*, 16 June 2002.

153. See J. Kay, *SFC*, 11 June 2002 and E. Pianin, *WP*, 11 June 2002.

154. Dept. of Commerce, *Statistical Abstract of the United States*, 2000 (Washington, DC: GPO, 2001), Table 394.

155. D. Jehl, *NYT*, 21 December 2000. These rules were ultimately adopted by the Bush administration after initial hesitation. A most persuasive report by the National Center for Environmental Assessment, U.S.Environmental Protection Agency, *Health Assessment Document for Diesel Engine Exhaust* (Washington, DC: GPO, 2002) played a significant role by indicating clearly the health cost of diesels.

156. Volkswagen website, URL: http://www.volkswagen.de/lupo_3l_tdi/index_.htm.

157. E. L. Andrews with K. Bradsher, *NYT*, 27 May 2001.

158. M. Landler, *NYT*, 8 March 2003.

159. Anonymous, AP dispatch, 15 January 2003.

160. Staff of the Air Resources Board and the Office of Environmental Health Hazard Assessment, *Proposed Identification of Diesel Exhaust as a Toxic Air Contaminant*, California Air Resources Board, 22 April 1998.

161. H. J. Hebert, AP dispatch, 20 December 2000. R. Weinhold, *Environ. Health Perspect.* **110**, A 458 (2002).

162. M. L. Wald, *NYT*, 20 November 2000.

163. National Center for Environmental Assessment, *Health Assessment Document For Diesel Engine Exhaust*, U.S. Environmental Protection Agency, EPA/600/8-90/057F, May 2002.

164. E. Pianin, *WP*, 11 June 2002.

165. J. Eilperin, *WP*, 11 May 2004.

166. E. Shogren, *LAT*, 20 September 2001.

167. J. Cart, *LAT*, 4 September 2003.

168. F. Barringer, *NYT*, 12 February 2004. T. R. Reid, *WP*, 15 March 2004. J. Krist, Environmental News Network dispatch, 26 March 2004. A. Fram, AP dispatch, 18 June 2004. E. Shogren, *LAT*, 20 August 2004.

169. E. J. Hutchinson and P. J. G. Pearson, *Environ. Health Perspect.* **112**, 132 (2004).

170. Y. Nishihata et al., *Nature* **418**, 164 (2002).

171. W. G. Phillips, *Pop. Sci.* **256**(3), 38 (2000).

172. J. O'Dell, *LAT*, 10 February 2004.

173. D. Hakim, *NYT*, 22 October 2003.

174. D. E. Sanger, *NYT*, 31 July 1991; D. P. Levin, *NYT*, 20 September 1991.

175. D. L. Chandler, *BG*, 31 October 2000.

176. A. J. Haagen-Schmidt, in *Energy Needs and the Environment*, ed. R. L. Seale and R. A. Sierka (Tucson: University of Arizona Press, 1973), Figure 3.13, 80.

177. M. Walsh, *Annu. Rev. Energy* **15**, 217 (1990).

178. G. J. McRae's data were published in a diagram in *LAT* on 22 July 1979. See the description of this work in G. McRae et al., California Air Resources Board A5-046-87 and A7-187-30 (1982) and *Atmos. Environ.* **16**, 679 (1982). An updated ozone discussion is found in R. Lu and R. P. Turco, *Atmos. Environ.* **30**, 4155 (1996).

179. M. L. Wald, *NYT*, 26 January 1994.

180. P. Shabecoff, *NYT*, 17 December 1985.

181. T. B. Reed and R. M. Lerner, *Science* **182**, 1299 (1973); E. E. Wigg, *Science* **186**, 785 (1974); E. R. Holles, *NYT*, 1 December 1974; and R. K. Mullen and E. E. Wigg, *Science* **188**, 209 (1975).

182. A. L. Hammond, *Science* **195**, 564 (1977).

183. C. L. Gray, Jr. and J. A. Alson, *Sci. Am.* **261**(5), 108 (1989).

184. R. K. Perfley et al., in *Hydrogen and Other Synthetic Fuels A Summary of the Work of the Synthetic Fuels Panel*, AEC report TID-26136, paper 719008.

185. A. Gabele, *J. Air Waste Manage. Assoc.* **40**, 296 (1990).

186. B. Larsen, *Logos* **7**(2), 2 (1989).

187. R. S. Chambers et al., *Science* **206**, 789 (1979).

189. Z. Hu et al., *Renewable Energy* **29**, 2183 (2004).

188. D. Pimentel, *J. Agr. Environmental Ethics* **4**, 1 (1991).

190. K. Schneider, *NYT*, 15 December 1993.

191. D. A. Hickman and L. D. Schmidt, *Science* **259**, 343 (1993); R. Periana et al., *Science* **259**, 340 (1993). See also I. Amato, *Science* **259**, 311 (1993).

192. S. Mehta, *LAT*, 14 October 2001.

193. D. Whitney, *SB*, 8 October 2002.

194. P. Yost, AP dispatch, 17 February 2004.

195. E. Douglass, *LAT*, 4 December 2003.

196. M. Schuon, *NYT*, 3 December 1989.

197. R. Severo, *NYT*, 20 April 1975.

198. R. Haitch, *NYT*, 2 June 1985.

199. M. Janofsky, *NYT*, 1 September 2001. J. Brooke, *NYT*, 21 April 1998.

200. W. M. Thring, *Intern. J. Environmental Studies* **5**, 251 (1974); D. G. Wilson, *Intern. J. Environmental Studies* **6**, 35 (1974).

201. M. Guido, *SJMN*, 7 May 2001.

202. D. Lawson, California Air Quality Board, private communication, June 1988; M. Hoggan et al., *1985 Summary of Air Quality in California's South Coast Air Basin* (El Monte, CA: South Coast Air Quality Management District, 1986); Air Quality Data, 1982, 1986 (South Coast Air Quality Management District); *Air Quality Management Plan*, 1988 revision, App. III-A (South Coast Air Quality Management District, 1988). For earlier data, see R. D. Cadle and E. R. Allen, *Science* **167**, 243 (1970).

203. I. Meyer-List, *Scala*, August 1984. See also D. Dickson, *Science* **228**, 37 (1985).

204. G. Haaf et al., *DZ*, 26 October 1984; R. Klingholz, *DZ*, 8 November 1985.

205. S. B. McLaughlin, *J. Air Poll. Control Assoc.* **35**, 512, 923 (1985).

206. H. Hatzfeldt, *DZ*, 9 March 1984.

207. F. Haemmerli, *DZ*, 24 August 1984.

208. B. Prinz, *J. Air Poll. Control Assoc.* **35**, 913 (1985); B. Prinz, *Environment* **29**(9), 10 (1987).

209. R. W. Leonhardt, *DZ*, 9 March 1984.

210. L. B. Lave, *Science* **212**, 893 (1981).

211. H. Blüthmann, *DZ*, 7 December 1984.

212. R. Gaul, *DZ*, 14 September 1984.

213. H. Blüthmann, *DZ*, 1 February 1985.

214. F. von Hippel, *Science* **228**, 263 (1985).

215. H. Schuh, *DZ*, 6 December 1991; H. Schuh, *DZ*, 4 December 1993.

216. L. W. Blank et al., *Nature* **336**, 27 (1988).

217. E.-D. Schultze, *Science* **244**, 776 (1989).

218. R. Klingholz, *DZ*, 21 June 1985; and Anonymous, *NYT*, 9 December 1985.

219. D. Denniston, in L. R. Brown et al., eds., *Vital Signs 1993* (New York: W. W. Norton & Co., 1993), 108.

220. H. W. Vogelman, *Natl. Hist.* **91**(11), 8 (1982).

221. R. W. Shaw, *Sci. Am.* **257**(2), 96 (1987).

222. P. Shabecoff, *NYT*, 26 February 1984.

223. S. E. Plotkin, *Environment* **31**(6), 25 (1989); M. L. Wald, *NYT*, 21 November 1990; and M. L. Wald, *NYT*, 17 October 1993.

224. L. Schipper et al., *Annu. Rev. Energy* **14**, 273 (1989).

225. D. L. Greene, *Energy J.* **13**, 117 (1992).

226. M. Woods, *TB*, 17 March 2003.

227. M. DeLuchi, Q. Wang, and D. L. Greene, ORNL-6715, 1992 (unpublished).

228. L. M. Fisher, *NYT*, 7 May 1990. See also M. Åhman, *Energy* **26**, 973 (2001).

229. J. Cherfis, *Science* **251**, 154 (1991).

230. M. L. Wald, *NYT*, 16 August 1990.

231. F. R. Field, III and J. P. Clark, *Tech. Rev.* **100**(1), 28 (1997).

232. M. L. Wald, *NYT*, 6 October 1993.

233. C. Holden, *Science* **284**, 583 (1999).

234. A. Lovins, *Amicus J.* **21**(3), 25 (1999). See the URL: http://www.hypercar-center.org.

235. G. C. Eads, *Environment* **39**(1), 28 (1997).

236. G. Stix, *Sci. Am.* **284**(5), 28 (2001).

237. N. Lossau, *DW*, 20 January 2004. K. Jopp, *DW*, 16 March 2003 .

238. C. Greenman, *NYT*, 18 March 1999.

239. D. P. Levin, *NYT*, 26 November 1989.

240. A. Pollack, *NYT*, 18 November 1992.

241. J. Holusha, *NYT*, 12 August 1992.

242. P. E. Ross, *Sci. Am.* **267**(4), 112 (1992); L. M. Fisher, *NYT*, 8 July 1990.

243. D. G. Wilson, *Tech. Rev.* **98**(2), 50 (1995). P. Baldus et al., *Science* **285**, 699 (1999).

244. M. L. Wald, *NYT*, 29 May 2001.

245. J. O'Dell, *LAT*, 6 April 2003. See also Anonymous (Tech Notes), *NYT*, 25 February 1990.

246. A. Eisenberg, *NYT*, 6 April 2000.

247. J. Schilling, *DW*, 6 February 2001.

248. C. Heredia, *SFC*, 4 March 2002.

249. D. L. Greene and K. G. Duleep, *Costs and Benefits of Automotive Fuel Economy Improvement: A Partial Analysis* (Oak Ridge, TN: Oak Ridge National Laboratory, 1992), ORNL-6704.

250. P. Patton, *NYT*, 16 September 2001.

251. Opel website (URL: http://www.opel.de/showroom/corsa/).

252. D. Talbot, *Tech. Rev.* **105**(2), 26 (2002).

253. G. L. Hunt, *IEEE Spectrum* **35**(11), 21 (1998).

254. R. C. Stempel et al., *IEEE Spectrum* **35**(11), 29 (1998).

255. B. Phillips, *NYT*, 22 October 2003. P. Rogers, *SJMN*, 26 June 2004.

256. L. M. Fisher, *NYT*, 25 February 1990.

257. M. L. Wald, *NYT*, 28 January 1994.

258. G. Schneider, *WP*, 22 October 2003.

259. B. J. Feder, *NYT*, 2 October 2000.

260. G. Polakovic and J. O'Dell, *LAT*, 16 June 2002. K. Q. Seelye, *NYT*, 10 October 2002. P. Rogers, *SJMN*, 4 March 2003. B. Melley, AP dispatch, 28 March 2003. G. Polakovic, *LAT*, 25 April 2003.

261. J. Motavalli, *NYT*, 20 January 2001.

262. M. Walton, CNN dispatch, 9 March 2001.

263. M. Maynard, *NYT*, 31 August 2002.

264. Anonymous, Bloomberg News dispatch, 17 September 2004.

265. K. Hafner, *NYT*, 21 January 2001.

266. J. K. Morita, *SB*, 19 May 2003. J. O'Dell, *LAT*, 26 December 2002.

267. B. Wade, *NYT*, 24 December 2000.

268. M. L. Wald, *NYT*, 11 July 1993. For more information on the development of various types of battery, see F. R. McLennon and E. J. Cairns, *Annu. Rev. Energy* **14**, 241 (1989).

269. S. R. Ovshinsky et al., *Science* **260**, 176 (1993).

270. M. Clayton, *CSM*, 29 July 2004.

271. C. B. Toepfer, *IEEE Spectrum* **35**(11), 41 (1998).

272. L. B. Lave and H. L. MacLean, *IEEE Spectrum* **38**(3), 47 (2001). See also R. De Neufville et al., *Tech. Rev.* **99**(1), 30 (1996).

273. L.B. Lave et al., *Science* **268**, 992 (1995).

274. A. Cocconi, with T. Gage, *IEEE Spectrum* **39**(4), 14 (2002).

275. D. Hakim, *NYT*, 10 January 2002. N. Banerjee with D. Hakim, *NYT*, 9 January 2002. P. Grier and L. Belsie, *CSM*, 10 January 2002.

276. T. Gilchrist, *IEEE Spectrum* **35**(11), 35 (1998).

277. A. J. Appleby, *Sci. Am.* **281**(1), 74 (1999).

278. M. J. Rienzenman, *IEEE Spectrum* **38**(1), 95 (2001). M. W. Jensen and M. Ross, *Environment* **42**(7), 10 (2000). J. DeCicco and M. Ross, *Sci. Am.* **217**(6), 52 (1994).

279. J. Yip, *LAT*, 29 November 2000.

280. D. H. Freedman, *Tech. Rev.* **105**(1), 40 (2002).

281. E. C. Evarts, *CSM*, 31 January 2002.

282. M. L. Wald, *NYT*, 14 March 2002.

283. J. M. Ogden, *Phys. Today* **55**(4), 69 (2002).

284. DaimlerChrysler website, URL: http://ww.daimlerchrysler.com/ research/research_e.htm.

285. E. Garsten, AP dispatch, 20 September 2001.

286. J. O'Dell, *LAT*, 8 October 2002.

287. J. O'Dell, *LAT*, 3 December 2002. Times Staff and Wire Reports, *LAT*, 3 July 2002. J. O'Dell, *LAT*, 28 April 2004.

288. D. Hakim, *NYT Magazine*, 29 September 2002, 100. J. O'Dell, *LAT*, 19 February 2003. E. C. Evarts, *CSM*, 13 May 2004.

289. V. Wouk, *Sci. Am.* **277**(4), 70 (1997).

290. J. O'Dell, *LAT*, 17 September 2004.

291. D. Hermance and S. Sasaki, *IEEE Spectrum* **35**(11), 48 (1998).

292. K. Bradsher, *NYT*, 20 February 2001.

293. H. A. Rosen and D. R. Castleman, *Sci. Am.* **277**(4), 75 (1997).

294. Anonymous, Reuters dispatch, 2 October 2003.

295. E. C. Evarts, *CSM*, 8 December 2003.

296. C. Canfield, AP dispatch, 15 May 2002. K. Bradsher, *NYT*, 1 January 2000.

297. J. O'Dell, *LAT*, 7 April 2004. K. Bradsher, *NYT*, 29 January 2001.

298. J. O'Dell and T. Y. Jones, *LAT*, 19 May 2001. M. Maynard, *NYT*, 9 September 2001.

299. J. Garofoli, *SFC*, 9 September 2001.

300. J. Garofoli, *SFC*, 30 December 2001.

301. Anonymous, Reuters dispatch, 15 June 2001.

302. M. L. Wald, *NYT*, 5 April 2002.

303. D. Smith, *SB*, 7 May 2004.

304. C. Jensen, *NYT*, 14 April 2002.

305. G. Schneider, *WP*, 23 August 2004; J. F. Peltz, *LAT*, 5 August 2004.

306. C. Jensen, *NYT*, 14 April 2002.

307. D. Hakim, *NYT*, 21 October 2003.

308. Early discussions of catalytic converters include R. H. Ebel, in J. N. Pitts, Jr., and R. L. Metcalf, eds., *Advances in Environmental Sciences* Vol. 1 (New York: John Wiley and Sons, 1969), 237; R. W. Irwin, *NYT*, 13 October 1974 (as we know, it did work!); C. Holden, *Science* **187**, 818 (1975).

309. J. H. Johnson, in A. Y. Watson et al., eds., *Air Pollution, the Automobile, and Public Health* (Washington, DC: National Academy of Sciences, 1988), 39.

310. B. S. Feder, *NYT*, 9 May 1990.

311. J. P. Hicks, *NYT*, 17 November 1991.

312. G. Stapley and J. Holland, *MB*, 17 September 2000. G. Stapley and J. Holland, *MB*, 17 September 2000. J. Holland, *MB*, 19 September 2000. J. Miller, *MB*, 31 May 2001.

313. See Dept. of Commerce, *Statistical Abstract of the United States, 1972* (Washington, DC: GPO, 1972), Table 83 and Dept. of Commerce, *Statistical Abstract of the United States, 2000* (Washington, DC: GPO, 2001), Table 1071.

314. L. Frazier, *Environ. Health Perspect.* **109**, A 430 (2001). The basic research is found in W. S. Gutowski and E. R. Pankevicius, *Composite Interfaces* **1**, 141 (1993) and W. S. Gutowski, D. Y. Wu, and S. Li, *J. Adhesion* **43**, 139 (1993).

315. P. M. Salzberg and J. M. Moffat, *J. Safety Res.* **35**, 215 (2004).

316. E. A. Bretz, *IEEE Spectrum* **38**(1), 92 (2001).

317. W. D. Jones, *IEEE Spectrum* **38**(9), 40 (2001). T. Jüngling, *DW*, 10 November 2002.

318. A. Eisenberg, *NYT*, 7 June 2001.

319. E. A. Bretz, *IEEE Spectrum* **38**(4), 68 (2001).

320. W. D. Jones, *IEEE Spectrum* **39**(1), 82 (2002) and I. Berger, *IEEE Spectrum* **39**(4), 40 (2002).

321. T. Moran, *NYT*, 19 May 1999.

322. D.-A. Durbin, AP dispatch, 21 July 2003.

323. P. Patton, *NYT*, 26 December 2003.

324. D. Hakim, *NYT*, 4 April 2004.

325. M. Stabaty, *DW*, 13 July 2004.

326. D. Zetsche, *Sci. Am.* **273**(3), 102 (1995).

327. M. L. Wald, *NYT*, 19 April 1990.

328. P. C. Judge, *NYT*, 19 April 1990.

329. M. Bertozzi et al., *Robotics and Autonomous Systems* **32**, 1 (2000). See also 313. M. P. Yazigi, *NYT*, 13 September 1992, Y. Deguilhem, *Le magazine de l'innovation—Cyberthèque R&D*, April 1996, URL: www.retdmag.renault.com; Department of Computing, Imperial College of Science, Technology and Medicine, University of London, website, undated, URL: http://www.doc.ic.ac.uk/~op98/ report/report.htm, and Ref. 82.

330. J. R. Quain, *NYT*, 7 November 2003.

331. E. A. Bretz, *IEEE Spectrum* **39**(4), 9 (2002).

332. D. Haldane, *LAT*, 15 July 2001.

333. Anonymous, Environmental News Network dispatch, URL: http://ww.enn.com/news/ enn-stories/2001/09/09182001/ concrete_44989.asp.

334. J. Hecht, *Tech. Rev.* **101**(6), 78 (1998). S. Shulman, *Tech. Rev.* **98**(2), 18 (1995).

CHAPTER 16

1. P. Cloud, *Oasis in Space* (New York: W. W. Norton & Co., Inc., 1988); D. M. Hunten, *Science* **259**, 915 (1993); J. M. Kasting, *Science* **259**, 920 (1993); C. J. Allègre and S. H. Schneider, *Sci. Am.* **271**(4), 66 (1994), and H. J. Melosh, *Nature* **424**, 22 (2003).

2. J. Xiong et al., *Science* **289**, 1724 (2000). See also the commentary by D. J. Des Marais, *Science* **289**, 1703 (2000).

3. D. E. Canfield et al., *Science* **288**, 658 (2000). See also the commentary by A. Paytan, *Science* **288**, 626 (2000).

4. D. C. Catling et al., *Science* **293**, 839 (2001). See also the commentary by J. F. Casting, *Science* **293**, 819 (2001) and U. H. Wiechert, *Science* **298**, 2341 (2002).

5. D. S. Heckman et al., *Science* **293**, 1129 (2001).

6. Anonymous, *NYT*, 11 January 1981; *NYT*, 5 January 1986. These figures are usually to be found in the first Sunday peper of the new year.

7. T. H. Maugh, II, *Science* **220**, 39 (1983).

8. J. J. Tribbia and R. A. Anthes, *Science* **237**, 493 (1987); R. A. Kerr, *Science* **228**, 40 (1985).

9. R. A. Kerr, *Science* **244**, 30 (1989); *Science* **244**, 1137 (1989).

10. B. J. Mason, *Contemp. Phys.* **27**, 463 (1986).

11. B. Reinhold, *Science* **235**, 437 (1987).

12. T. N. Krishnamurthi et al., *Science* **285**, 1548 (1999).

13. A. L. F. Braga et al., *Environ. Health Perspect.* **110**, 859 (2002).

14. J. E. Walsh, *Am. Sci.* **72**, 50 (1984); R. A. Kerr, *Science* **216**, 608 (1982).

15. W. M. Gray, *Science* **249**, 1251 (1990).

16. R. Reiter and R. Sladkovic, *Arch. Met. Geoph. Biocl.* **A33**, 297 (1985).

17. Dept. of Commerce, *Statistical Abstract of the United States, 1993* (Washington, DC: GPO, 1993).

18. R. A. Bryson, *Science* **184**, 753 (1974).

19. S. H. Schneider, *The Genesis Strategy* (New York: Plenum, 1976).

20. R. A. Bryson, *Ecologist* **6**, 205 (1976). A. L. Hammond, *Science* **191**, 455 (1976); and J. W. Sawyer, *Environment* **20**(2), 25 (1978).

21. J. P. Peixóto and A. H. Oort, *Physics of Climate* (New York: American Institute of Physics, 1992); J. P. Peixoto and A. H. Oort, *Rev. Mod. Phys.* **56**, 365 (1984); W. L. Gates, in M. C. MacCracken and F. M. Luther, eds., *Projecting the Climatic Effects of Increasing Carbon Dioxide* (Washington, DC: GPO, 1986), DOE/ER-0237, 57; and M. E. Schlesinger, ibid., 281.

22. W. S. Broecker, in B. G. Levi, D. Hafemeister, and R. Scribner, eds., *Global Warming: Physics and Facts* (New York: American Institute of Physics, 1992), 129. See also W. S.

Broecker, *Natl. Hist.* **96**(10), 74 (1987) and W. S. Broecker, *Annu. Rev. Energy Environ.* **25**, 1 (2000).

23. U. von Zahn, *Alexander von Humboldt-Stiftung Mitteilungen* #39, 15 (1981).

24. R. E. Newell, *Am. Sci.* **67**, 405 (1979).

25. E. M. Rasmussen, *Am. Sci.* **73**, 168 (1985). See also G. A. Jacobs et al., *Nature* **370**, 360 (1994), the commenary by M. J. McPhaden, *Nature* **370**, 326 (1994) and the popular article by W. K. Stevens, *NYT*, 9 August 1994.

26. G. T. Walker and E. W. Bliss, *Mem. Royal Meteorol. Soc.* **4**, 53 (1932).

27. R. T. Barber and F. P. Chavez, *Science* **222**, 1203 (1983).

28. R. A. Kerr, *Science* **249**, 1246 (1990).

29. C. S. Ramage, *Sci. Am.* **254**(6), 77 (1986); R. A. Kerr, *Science* **216**, 608 (1982) and *Science* **221**, 940 (1983).

30. M. A. Cane and S. E. Zebiak, *Science* **228**, 1085 (1985).

31. A. V. Fedorov and S. G. Philander, *Science* **288**, 1997 (2000).

32. J. D. Neelin and M. Latif, *Phys. Today* **51**(12), 32 (1998).

33. A. Upgren and J. Stock, *Weather* (Cambridge, MA: Perseus Publishing, 2000).

34. B. Fagan, *The Little Ice Age* (New York: Basic Books, 2000) explains the effects of the North Atlantic Oscillation and its effects on European weather clearly.

35. J. Uppenbrink, *Science* **283**, 948 (1999).

36. C. Appenzeller et al., *Science* **282**, 446 (1998).

37. J. W. Hurrell et al., *Science* **291**, 603 (2001).

38. D. W. J. Thompson and J. M. Wallace, *Science* **293**, 85 (2001). See also the news story by R. Kerr, *Science* **284**, 241 (1999).

39. M. P. Baldwin and T. J. Dunkerton, *Science* **294**, 581 (2001). See also the news story by R. A. Kerr, *Science* **294**, 494 (2001).

40. B. Naujokat, *J. Atmos. Sci.* **43**, 1873 (1986).

41. Joint Institute for the Study of the Atmosphere and Ocean website, URL http://www.jisao.washington.edu/pdo/.

42. R. A. Madden and P. R. Julian, *J. Atmos. Sci.* **28**, 702 (1971); R. A. Madden and P. R. Julian, *J. Atmos. Sci.* **29**, 1109 (1972).

43. M. J. McPhaden, *Science* **283**, 950 (1999). See also R. A. Kerr, *Science* **285**, 322 (1999).

44. Y. N. Takayabu et al., *Nature* **402**, 279 (1999).

45. R. A. Kerr, *Science* **238**, 1507 (1987).

46. T. Y. Canby, *Natl. Geog.* **165**, 144 (1984).

47. R. T. Barber and F. P. Chávez, *Nature* **319**, 279 (1986).

48. A. Thayer, *Chem. Engg. News* **75**(45), 32 (1997).

49. L. L. Ely et al., *Science* **262**, 410 (1993).

50. F. P. Chavez et al., *Science* **286**, 2126 (1999). M. J. Behrenfeld et al., *Science* **291**, 2594 (2001).

51. X. Rodó et al., *Proc. Natl. Acad. Sci.* **99**, 12901 (2002).

52. R. A. Kerr, *Science* **262**, 656 (1993); C. Holden, *Science* **263**, 607 (1994).

53. R. A. Kerr, *Science* **290**, 257 (2000).

54. B. S. Orlove et al., *Nature* **403**, 68 (2000).

55. J. Angell, *Geophys. Res. Lett.* **19**, 285 (1992).

56. H. F. Diaz and V, Markgraf, eds., *Historical and Paleoclimatic Aspects of the Southern Oscillation* (Cambridge, UK and New York: Cambridge University Press, 1992).

57. J. Cole, *Science* **291**, 1496 (2001).

58. D. T. Rodbell et al., *Science* **283**, 516 (1999).

59. D. H. Sandweiss et al., *Science* **273**, 1531 (1996).

60. C. D. Charles et al., *Science* **277**, 925 (1997). A. W. Tudhope et al., *Science* **291**, 1511 (2001).

61. T. P. Guilderson and D. P. Schrag, *Science* **281**, 240 (1998).

62. J. Overpeck and R. Webb, *Proc. Natl. Acad. Sci.* **97**, 1335 (2000).

63. K. K. Kumar et al., *Science* **284**, 2156 (1999).

64. M. J. McPhaden and D. Zhang, *Nature* **415**, 603 (2002).

65. A. J. Miller et al., *J. Oceanogr.* **60**, 163 (2004).

66. K. M. Cobb et al., *Nature* **424**, 271 (2003). See also the accompanying commentary by S. Tudhope and M. Collins, *Nature* **424**, 261 (2003).

67. E. J. Hendy et al., *Science* **295**, 1511 (2002).

68. C. M. Moy et al., *Nature* **420**, 162 (2002).

69. D. W. Lea, *Science* **297**, 202 (2002).

70. Y. Rosenthal and A. J. Broccoli, *Science* **304**, 219 (2004).

71. A. Koutavas et al., *Science* **297**, 226 (2002).

72. L. Stott et al., *Science* **297**, 222 (2002).

73. C. S. M. Turney et al., *Nature* **428**, 306 (2004).

74. J. B. Adams et al., *Nature* **426**, 274 (2003). See also the accompanying commentary by S. de Silva, *Nature* **426**, 239 (2003) and R. A. Kerr, *Science* **299**, 337 (2003).

75. D. Chen et al., *Nature* **428**, 733 (2004). See also the accompanying commentary by D. Anderson, *Nature* **428**, 709 (2004).

76. K. E. Trenberth et al., *Science* **242**, 1640 (1988); for a popular view, see W. K. Stevens, *NYT*, 3 January 1989.

77. J. Shukla, in B. G. Levi et al., eds., *Global Warming: Physics and Facts* (New York: American Institute of Physics, 1992), 113.

78. W. M. Washington and T. W. Bettge, *Comp. in Phys.* **4**, 240 (1990).

79. R. D. Cess et al., *Science* **245**, 513 (1989). These models stem from the original model of S. Manabe and

R. T. Wetherald, *J. Atmos. Sci.* **24**, 241 (1967). See also R. D. Cess and G. L. Potter, *Clim. Change* **6**, 365 (1984).

80. R. D. Cess et al., *Science* **262**, 1252 (1993).

81. B. J. Mason, *Contemp. Phys.* **28**, 49 (1987).

82. S. H. Schneider, *Sci. Am.* **256**(5), 72 (1987).

83. J. Maddox, *Nature* **346**, 605 (1990); J. L. Le Mouël et al., *Nature* **355**, 26 (1992); R. Morrow et al., *Nature* **357**, 482 (1992); and J. O. Dickey et al., *Nature* **357**, 484 (1992).

84. L. O'Hanlon, *Nature* **415**, 360 (2002). See also R. A. Kerr, *Science* **300**, 1081 (2003).

85. M. P. Baldwin et al., *Science* **301**, 317 (2003).

86. M. P. Baldwin et al., *Science* **301**, 636 (2003).

87. See, for example, B. Bova, *The Weathermakers* (New York: Ace Books, 1979). The book is based on a series of short stories Bova published in *Analog Science Fiction Magazine* in the mid-1960s.

88. S. Lewis, *Elmer Gantry* (New York: Signet Classics, 1976).

89. P. M. de F. Forster and S. Solomon, *Proc. Natl. Acad. Sci.* **100**, 11225 (2003). R. S. Cerveny and K. J. Coakley, *Geophys. Res. Lett.* **29**, 1028 (2002).

90. R. B. Standler, URL http://www.rbs2.com/w2.htm.

91. P. N. Spotts, *CSM*, 2 January 2003. W. Weir, *The Hartford Courant*, 4 February 2004.

92. Council of the American Meteorological Society, Policy Statement as adopted on 5 January 1992.

93. The GLACE Team: R. D. Koster et al., *Science* **305**, 1138 (2004).

94. T. P. Barnett, in J. Gribben, ed., *Climatic Change* (New York: Cambridge University Press, 1978), 157.

95. J. Gribben and H. H. Lamb, in J. Gribben, ed., *Climatic Change* (New York: Cambridge University Press, 1978), 68.

96. J. J. Walsh, *BioScience* **34**, 499 (1984).

97. S. W. Matthews, *Natl. Geog.* **150**, 576 (1976).

98. J. Esper et al., *Science* **295**, 2250 (2002). See also K. R. Briffa and T. J. Osborn, *Science* **295**, 2227 (2002).

99. W. W. Kellogg and S. H. Schneider, *Science* **186**, 1163 (1974).

100. K. R. Briffa et al., *Nature* **346**, 434 (1990).

101. J. Norwine, *Environment* **19**(8), 6 (1977).

102. C. B. Beaty, *Am. Sci.* **66**, 452 (1978).

103. R. A. Bryson, *Natl. Hist.* **89**(6), 65 (1980).

104. V. A. Todorov, *J. Clim. Appl. Met.* **24**, 97 (1985); D. Winstanley, *Weatherwise* 38, 74 (1985).

105. S. Huang et al., *Nature* **403**, 756 (2000). See also H. N. Pollack and D. S. Chapman, *Sci. Am.* **268**(6), 44 (1993).

106. C. Covey, *Ann. Glaciology* **5**, 43 (1984).

107. B. McGowran, *Am. Sci.* **78**, 30 (1990).

108. A. Gerdes, *DZ*, 16 July 1993.

109. Panel on Climatic Variation [W. L. Gates et al.], U.S. Committee for the Global Atmospheric Research Program, National Research Council, *Understanding Climatic Change* (Detroit: Grand River Books, 1980).

110. P. L. Koch et al., *Nature* **358**, 319 (1992).

111. P. U. Clark et al., *Science* **293**, 283 (2001).

112. H. N. Pollack and S. Huang, *Annu. Rev. Earth Planet. Sci.* **28**, 339 (2000).

113. H. Beltrami, *Science* **297**, 206 (2002).

114. M. E. Mann, *Science* **297**, 1481 (2002). K. E. Trenberth and B. L. Otto-Bliesner, *Science* **300**, 589 (2003).

115. See, for example, C. J. Proctor et al., *Climate Dyn.* **19**, 449 (2002).

116. N. Ohkouchi et al., *Science* **298**, 1224 (2002).

117. H. Kitagawa and J. van der Plicht, *Science* **279**, 1187 (1998) and H. Kitagawa and J. van der Plicht, *Radiocarbon* **42**, 369 (2000) [Japan]; J. W. Beck et al., *Science* **292**, 2453 (2001) [Bahamas].

118. E. Bard, F. Rostek, and G. Ménot-Combes, *Science* **303**, 178 (2004).

119. S. J. Hovan et al., *Nature* **340**, 296 (1989).

120. M Rossignol-Strick and N. Planchais, *Nature* **342**, 413 (1989).

121. G. Eglinton et al., *Nature* **356**, 423 (1992); M. Lyle, *Nature* **356**, 385 (1992); and F. Rostek, G. Ruhland et al., *Nature* **364**, 319 (1993).

122. R. G. Fairbanks, *Nature* **342**, 637 (1989); see also the commentary by N. J. Shackleton, *Nature* **342**, 616 (1989); C. G. H. Rooth, (letter) *Nature* **343**, 702 (1990); J. W. Beck et al., *Science* **257**, 644 (1992); R. L. Edwards et al., *Science* **260**, 962 (1993); T. P. Guilderson et al., *Science* **263**, 663 (1994); C. D. Gallup et al., *Science* **263**, 796 (1994); and D. A. Richards et al., *Nature* **367**, 260 (1994).

123. R. A. Kerr, *Science* **263**, 173 (1994).

124. W.-X. Li, J. Lundberg et al., *Nature* **339**, 534 (1989); J. A. Dorale et al., *Science* **258**, 1626 (1992).

125. I. J. Winograd et al., *Science* **258**, 255 (1992); J. Imbrie et al., *Nature* **363**, 531 (1993); see also R. A. Kerr, *Science* **248**, 31 (1990).

126. J. N. Wilford, *NYT,* 1 December 1992.

127. S. Manabe and A. J. Broccoli, *Ann. Glaciology* **5**, 100 (1984).

128. D. Dahl-Jensen and S. J. Johnson, *Nature* **320**, 250 (1986).

129. E. Bard, *Phys. Today* **55**(12), 32 (2002).

130. P. U. Clark et al., *Science* **295**, 2438 (2002). See also the commentary by

R. Sabadini, *Science* **295**, 2376 (2002). A. J. Weaver et al., *Science* **299**, 1709 (2003).

131. S. M. Colman, *Science* **296**, 1251 (2002).

132. J. F. Kasting and J. C. G. Walker, in B. G. Levi et al., eds., *Global Warming: Physics and Facts* (New York: American Institute of Physics, 1992), 175; M. A. Arthur et al., *Nature* **335**, 714 (1988); and R. A. Berner and A. C. Lasaga, *Sci. Am.* **260**(3), 74 (1989).

133. J. C. G. Walters, *BioScience* **34**, 486 (1984).

134. K. Caldeira and J. F. Kasting, *Nature* **360**, 721 (1992); T. Volk, *Nature* **360**, 707 (1992); and E. T. Sundquist, *Science* **259**, 934 (1993).

135. C. C. Mann, *Science* **283**, 314 (1999).

136. S. A. Cowling, *Science* **285**, 1500 (1999).

137. F. A. Street-Perrott et al., *Science* **278**, 1422 (1997). See also G. D. Farquhar, *Science* **278**, 1411 (1997).

138. M. Pagani et al., *Science* **285**, 876 (1999). See also M. E. Morgan et al., *Nature* **367**, 162 (1994).

139. Y. Huang et al., *Science* **293**, 1647 (2001). See also R. A. Kerr, *Science* **293**, 1572 (2001). See also D. R. Cole and H. C. Monger, *Nature* **368**, 533 (1994).

140. S. D. Smith et al., *Nature* **408**, 79 (2000).

141. W. S. Broecker et al., *Science* **206**, 409 (1979).

142. N. J. Shackleton et al., *Nature* **306**, 319 (1983).

143. D. B. Botkin et al., *BioScience* **34**, 508 (1984); and R. A. Houghton et al., in J. R. Trabalka, ed., *Atmospheric Carbon Dioxide and the Global Carbon Cycle* (Washington, DC: GPO, 1986), DOE/ER-0239, 113.

144. G. M. Woodwell et al., *Science* **222**, 1081 (1983).

145. R. Revelle, *Sci. Am.* **247**(2), 35 (1982).

146. B. Moore, III and B. Bolin, *Oceanus* **29**(4), 9 (1986).

147. J. J. Walsh, *BioScience* **34,** 499 (1984).

148. C. S. Wong, *Science* **200**, 197 (1978).

149. J. J. McCarthy et al., *Oceanus* **29**(4), 16 (1986).

150. H. Oeschger and B. Stauffer, *Science* **199**, 388 (1978); B. Stauffer et al., *Ann. Glaciology* **5**, 160 (1984); and G. I. Pearman et al., *Nature* **320**, 248 (1986).

151. R. A. Berner, *Science* **249**, 1382 (1990); W. M. Post et al., *Am. Sci.* **78**, 310 (1990).

152. Y. Sugimura and Y. Suzuki, *Mar. Chem.* **24**, 105 (1988); J. R. Toggweiler, *Nature* **356**, 665 (1992); H. Ogawa and N. Ogura, *Nature* **356**, 696 (1992); J. H. Martin and S. E. Fitzwater, *Nature* **356**, 699 (1992); and J. I. Hedges and B. A. Bergamaschl, *Nature* **359**, 202 (1992).

153. W. S. Broecker et al., *Science* **206**, 409 (1979).

154. C. F. Baes et al., in J. R. Trabalka, ed., *Atmospheric Carbon Dioxide and the*

Global Carbon Cycle (Washington, DC: GPO, 1986), DOE/ER-0239, 81.

155. R. A. Kerr, *Science* **223**, 1053 (1984).

156. W. S. Bracken et al., *Nature* **315**, 21 (1985); T. P. Barnett, in M. C. MacCracken and F. M. Luther, eds., *Detecting the Climatic Effects of Increasing Carbon Dioxide* (Washington, DC: GPO, 1986), DOE/ER-0235, 149; and W. S. Broecker, *Nature* **328**, 123 (1987). See also S. H. Schneider *Science* **263**, 341 (1994).

157. M. Leuenberger et al., *Nature* **357**, 488 (1992); see also the commentary by N. J. Shackleton, *Nature* **347**, 427 (1990). J. W. C. White et al., *Nature* **367**, 153 (1994) (see, however, S. K. Rice and L. Giles, [letter] *Nature* **371**, 111 (1994)).

158. N. G. Pisias and J. Imbrie, *Oceanus* **29**(4), 43 (1986).

159. J. J. McCarthy, P. G. Brewer, and G. Feldman, *Oceanus* **29**(4), 16 (1986).

160. G. M. Woodwell, *Oceanus* **29**(4), 71 (1986).

161. C. P. Rinsland and J. S. Levine, *Nature* **318**, 250 (1985).

162. CO_2/Climate Review Panel (J. Smagorinsky et al.), National Research Council, *Carbon Dioxide and Climate: A Second Assessment* (Washington, DC: National Academy Press, 1982).

163. G. M. Woodwell, *Sci. Am.* **238**(1), 34 (1978); G. M. Woodwell et al., *Science* **199**, 141 (1978).

164. J. Hobbie et al., *BioScience* **34**, 492 (1984).

165. P. P. Tans et al., *Science* **247**, 1431 (1990); P. Wallich, *Sci. Am.* **262**(5), 25 (1990); and R. S. Webb and J. T. Overpeck, *Nature* **361**, 497 (1993).

166. W. S. Broecker and T.-H. Peng, *Nature* **356**, 587 (1992); J. L. Sarmiento and E. J. Sundquist, *Nature* **356**, 589 (1992); U. Siegenthaler and J. L. Sarmiento, *Nature* **365**, 119 (1993); P. D. Quay et al., *Science* **256**, 74 (1992); and a commentary on the preceding article, R. A. Kerr, *Science* **256**, 35 (1992).

167. U. Siegenthaler and J. L. Sarmiento, *Nature* **365**, 119 (1993); R. K. Dixon et al., *Science* **263**, 185 (1994).

168. M. E. Raymo and W. F. Ruddiman, *Nature* **359**, 117 (1992).

169. S. Manabe and A. J. Broccoli, *Science* **247**, 192 (1990); R. A. Kerr, *Science* **244**, 1441 (1989).

170. G. J. Kukla and H. J. Kukla, *Science* **183**, 709 (1974); W. Sullivan, *NYT,* 22 February 1976.

171. P. Wadhams, *Nature* **345**, 795 (1990); see also the commentary by A. S. McLaren et al., *Nature* **345**, 762 (1990).

172. W. Sullivan, *NYT,* 14 August 1990; W. K. Stevens, *NYT,* 30 October 1990; and P. Ya. Groisman et al., *Science* **263**, 198 (1994).

173. W. L. Gates, *Science* **191**, 1138 (1976).
174. G. J. Kukla, in J. Gribben, ed. *Climate Change* (New York: Cambridge University Press, 1978), 114.
175. H. J. Zwally, *Ann. Glaciology* **5**, 191 (1984).
176. E. Raschke and J. Schmetz, *Rep. Deutsche Forschunggemeinshaft* **1**, 10 (1984).
177. J. M. Mitchell, Jr., in F. Singer, ed., *Global Effects of Environmental Pollution* (Dodrecht, Neth.: D. Reidel Publ. Co., 1970); and W. H. Matthews, *Intern. J.Environmental Studies* **4**, 283 (1973).
178. R. A. Kerr, *Science* **260**, 1724 (1993).
179. H. H. Lamb, *Phil. Trans. Roy. Soc. London A* **266**, 425 (1970); H. H. Lamb, *Ecologist* **4**, 10 (1974).
180. H. Stommel and E. Stommel, *Sci. Am.* **240**(6), 176 (1979).
181. K. Sassen et al., *Geophys. Res. Lett.* **30**, 1633 (2003). P. J. DeMott et al., *Geophys. Res. Lett.* **30**, 1732 (2003). See also O. B. Toon, *Nature* **424**, 623 (2003).
182. K. Sassen, *Geophys. Res. Lett.* **29**, 1465 (2002).
183. D. Cyranoski, *Nature* **421**, 101 (2003).
184. R. Marquand, *CSM*, 22 March 2002.
185. O. B. Toon and J. B. Pollack, *Am. Sci.* **68**, 268 (1980). See also F. M. Luther and R. G. Ellingson, n M. C. MacCracken and F. M. Luther, ed., *Projecting the Climatic Effects of Increasing Carbon Dioxide* (Washington, DC: GPO, 1986), DOE/ER-0237, 25.
186. A. D. Clarke and R. J. Charlson, *Science* **229**, 263 (1985).
187. J. E. Hansen et al., *Proc. Natl. Acad. Sci.* **95**, 12753 (1998).
188. D. Rosenfeld et al., *Proc. Natl. Acad. Sci.* **98**, 5975 (2001).
189. The INDOEX website is at URL http://www-indoex.ucsd.edu/index.html
190. J. M. Haywood et al., *Science* **283**, 1299 (1999). J. T. Kiehl, *Science* **283**, 1273 (1999).
191. F.-M. Bréon et al., *Science* **295**, 834 (2002).
192. S. E. Schwartz et al., *Proc. Natl. Acad. Sci.* **99**, 1784 (2002).
193. S. K. Satheesh and V. Ramanathan, *Nature* **405**, 60 (2000).
194. A. S. Ackerman et al., *Science* **288**, 1042 (2001). S. E. Schwartz and P. R. Buseck, *Science* **288**, 989 (2000).
195. V. Ramanathan et al., *Science* **294**, 2119 (2001).
196. R. A. Kerr, *Science* **217**, 1023 (1982). M. R. Rampino and S. Self, *Sci. Am* **250**(1), 48 (1984).
197. R. Reiter and H. Jäger, *Meteorol. Atmos. Phys.* **35**, 19 (1986).
198. R. A. Kerr, *Science* **259**, 594 (1993).
199. R. J. Charlson et al., *Tellus* **43AB**, 152 (1991); J. Lelieveld and J. Heintzenberg, *Science* **258**, 117 (1992); J. T. Kiehl and

B. P. Briegleb, *Science* **260**, 311 (1993); and J. Hansen et al., *Natl. Geog. Res. Exploration* **9**, 142 (1993).
200. P. Minnis et al., *Science* **259**, 1411 (1993).
201. L. Gu et al., *Science* **299**, 2035 (2003). G. D. Farquhar and M. L. Roderick, *Science* **299**, 1997 (2003).
202. W. Lucht et al., *Science* **296**, 1687 (2002).
203. A. J. Broccoli et al., *J. Geophys. Res.* **108**(D24), 4798 (2003).
204. N. P. Gillett et al., *Geophys. Res. Lett.* **31**, L12217 (2004).
205. R. J. Charlson and T. M. L. Wigley, *Sci. Am.* **270**(2), 48 (1994).
206. C. D. O'Dowd et al., *Nature* **417**, 632 (2002).
207. M. Z. Jacobson, *Nature* **409**, 695 (2001).
208. R. J. Charlson et al., *Science* **292**, 202 (2001).
209. M. O. Andreae, *Nature* **409**, 671 (2001).
210. V. M. H. Lavanchy et al., *J. Geophys. Res.* **104** (D17), 21,227 (1999).
211. Y. Liu and P. H. Daum, *Nature* **419**, 580 (2002).
212. W. R. Cotton, *Am. Sci.* **73**, 275 (1985).
213. R. P. Turco et al., *Science* **222**, 1283 (1983) and *Sci. Am.* **251**(2), 33 (1984). This study is known as TTAPS after the first initials of the authors' last names.
214. S. L. Stevens and J. W. Birks, *BioScience* **35**, 557 (1985); C. Covey, *BioScience* **35**, 563 (1985).
215. J. E. Penner, *Nature* **324**, 222 (1986).
216. R. P. Turco et al., *Science* **247**, 166 (1990).
217. J. P. Kennett and R. C. Thunell, *Science* **187**, 497 (1975).
218. W. Sullivan, *NYT*, 9 August 1981.
219. W. Sullivan, *NYT*, 20 July 1993.
220. R. A. Bryson and B. M. Goodman, *Science* **207**, 1041 (1980). See also A. Robock, in M. E. Schlesinger, ed., *Greenhouse–Gas–Induced Climatic Change: A Critical Appraisal of Simulations and Observations* (Amsterdam: Elsevier, 1991), 429.
221. C. Covey, *Sci. Am.* **250**(2), 38 (1984).
222. M. Milankovitch, in W. Koppen and R. Geiger, eds., *Handbuch der Klimatologie* (Berlin: Borntraeger, 1930).
223. M. Stuiver and P. D. Quay, *Science* **207**, 11 (1980).
224. J. D. Hays et al., *Science* **194**, 1121 (1976).
225. R. A. Kerr, *Science* **201**, 144 (1978).
226. R. C. Balling, Jr., and R. S. Cerveny, *Geophys. Res. Lett.* **22**, 3199 (1995). See also the commentary by M. Szpir, *Am. Sci.* **84**, 119 (1996).
227. J. R. Herman and R. A. Goldberg, *Sun, Weather, and Climate* (Detroit: Grand River Books, 1980).
228. M. Lyle, *Nature* **335**, 529 (1988); J. R. Petit, et al., *Nature* **343**, 56 (1990);

T. J. Crowley et al., *Science* **255**, 705 (1992); H.-S. Liu, *Nature* **358**, 397 (1992); and D. Raynaud et al., *Science* **259**, 926 (1993).
229. P. E. Olsen, *Science* **234**, 842 (1986).
230. N. G. Pisias and J. Imbrie, *Oceanus* **29**(4), 43 (1986).
231. J. Imbrie and J. Z. Imbrie, *Science* **207**, 943 (1980).
232. R. A. Kerr, *Science* **233**, 1053 (1984); *Science* **235**, 973 (1987).
233. M. Khodri et al., *Nature* **410**, 570 (2001).
234. W. K. Stevens, *NYT*, 4 February 1992; E. Culotta, *Science* **259**, 906 (1993).
235. T. J. Crowley, *Science* **295**, 1473 (2002).
236. T. D. Herbert and A. G. Fisher, *Nature* **321**, 739 (1986).
237. W. L. Gates and M. C. MacCracken, in M. C. MacCracken and F. M. Luther, eds., *Detecting the Climatic Effects of Increasing Carbon Dioxide* (Washington, DC: GPO, 1986), DOE/ER-0235, 1.
238. A. McIntyre and B. Molfino, *Science* **274**, 1867 (1996).
239. S. C. Clemens et al., *Science* **274**, 943 (1996). F. Sirocko, *Science* **274**, 937 (1996).
240. C. H. Stirling et al., *Science* **291**, 290 (2001).
241. D. F. Williams et al., *Science* **278**, 1114 (1997).
242. K. Kashiwaya et al., *Nature* **410**, 71 (2001).
243. J. C. Zachos et al.,*Science* **292**, 686 (2001).
244. N. J. Shakleton, *Science* **289**, 1897 (2000). R. A. Kerr, *Science* **289**, 1868 (2000).
245. J. C. Zachos et al., *Science* **292**, 274 (2001). R. A. Kerr, *Science* **292**, 191 (2001).
246. T. R. Naish et al., *Nature* **413**, 719 (2001). See also the Ohio State University Byrd Center press release, 14 October 2001.
247. K. A. Maasch and B. Salzman, *J. Geophys. Res.* **95**, 11641 (1990).
248. R. A. Muller and G. J. MacDonald, *Science* **277**, 215 (1997). S. J. Kortenkamp and S. F. Dermott, *Science* **280**, 874 (1998).
249. J. A. Rial, *Science* **285**, 564 (1999).
250. D. Paillard, *Nature* **391**, 378 (1998). M. E. Raymo, *Science* **281**, 146 (1998).
251. K. J. Willis et al., *Science* **285**, 568 (1999).
252. I. J. Winograd et al., *Science* **258**, 255 (1992). K. R. Ludwig et al., *Science* **258**, 284 (1992). T. B. Coplen et al., *Science* **263**, 361 (1994).
253. D. B. Karner and R. A. Muller, *Science* **288**, 2143 (2000).
254. R. L. Edwards et al., *Science* **276**, 782 (1997). R. A. Kerr, *Science* **276**, 680 (1997). See earlier correlations of climate change in J. Guiot et al., *Nature* **338**, 309 (1989).

255. T. D. Herbert et al., *Science* **293**, 71 (2001). D. W. Lea, *Science* **293**, 59 (2001).

256. H. M. Perks and R. F. Keeling, *Paleoceanography* **13**, 63 (1998). H. M. Perks et al., *Paleoceanography* **17**, 1037 (2002).

257. W. F. Ruddiman and A. McIntyre, *Bull. Geol. Soc. Am.* **95**, 381 (1984).

258. Z. Liu and T. D. Herbert, *Nature* **427**, 721 (2004). K. Billups, *Nature* **427**, 686 (2004).

259. A. Hall et al., "Atmospheric dynamics govern northern hemisphere wintertime climate variations forced by changes in earth's orbit," Preprint, 2003.

260. R. A. Kerr, *Science* **234**, 283 (1986).

261. D. A. Short and J. C. Mengel, *Nature* **323**, 48 (1986).

262. J. T. Overpeck et al., *Nature* **338**, 553 (1989). See also W. K. Stevens, *NYT*, 4 September 1990 for an explanation of a similar conclusion in G. S. Boulton and C. D. Clark, *Nature* **346**, 813 (1990).

263. S. J. Lehman and L. D. Keigwin, *Nature* **356**, 757 (1992); G. Schaffer and J. Bendtsen, *Nature* **367**, 354 (1994).

264. A. J. Weaver and T. M. C. Hughes, *Nature* **367**, 447 (1994).

265. D. H. Tarling, in J. Gribben, ed. *Climatic Change* (New York: Cambridge University Press, 1978).

266. S. Sofia, P. Demarque, and A. Endal, *Am. Sci.* **73**, 326 (1985).

267. E. Friis-Christensen and K. Lassen, *Science* **254**, 698 (1991). P. Foukal et al., *Science* **306**, 68 (2004).

268. E. Friis-Christensen, *Energy* **18**, 1273 (1993); K. Labitzke, *Geophys. Res. Lett.* **14**, 535 (1987).

269. W. K. Stevens, *NYT*, 13 June 1989.

270. D. A. Hodell et al., *Science* **292**, 1367 (2001). R. A. Kerr, *Science* **292**, 1293 (2001).

271. D. Shindell et al., *Science* **284**, 305 (1999). R. A. Kerr, *Science* **284**, 234 (1999). This explains the connection noted by J. D. Haigh, *Nature* **370**, 544 (1994). See also M. I. Hoffert, in M. E. Schlesinger, ed., *Greenhouse-Gas-Induced Climatic Change: A Critical Appraisal of Simulations and Observations* (Amsterdam: Elsevier, 1991), 413.

272. U. Neff et al., *Nature* **411**, 290 (2001).

273. M. Lockwood et al., *Nature* **399**, 437 (1999); E. N. Parker, *Nature* **399**, 416 (1999).

274. R. C. Willson, *Science* **277**, 1963 (1997). R. A. Kerr, *Science* **277**, 1923 (1997).

275. G. Bond et al., *Science* **294**, 2130 (2001). J. D. Haigh, *Science* **294**, 2109 (2001) and R. A. Kerr, *Science* **294**, 143 (2001).

276. J. Lean and D. Rind, *Science* **292**, 234 (2001). D. Rind, *Science* **296**, 673 (2002). J. D. Haigh, *Phil. Trans. R. Soc. Lond. A* **361**, 95 (2003).

277. J. Lean, plenary talk presented at the National Astronomy Meeting for 2003, Dublin, Ireland, 8 April 2003. H. Briggs, BBC News Online, 8 April, 2003.

278. J. A. Eddy, *Science* **192**, 1189 (1976) and *Sci. Am.* **236**(5), 80 (1977); see also R. A. Kerr, *Science* **231**, 339 (1986).

279. S. H. Schneider and C. Maas, *Science* **190**, 741 (1975); W. W. Kellogg, in W. H. Matthews et al., eds., *Man's Impact on the Climate* (Cambridge, MA: MIT Press, 1971), 123.

280. J. Lean, *Annu. Rev. Astron. Astrophys.* **35**, 33 (1997).

281. S. Baliunas and R. Jastrow, *Energy* **18**, 1285 (1993).

282. K. Caldeira and J. F. Kasting, *Nature* **359**, 226 (1992); J. F. Kasting, *Nature* **364**, 759 (1993).

283. T. E. Graedel et al., *Geophys. Res. Lett.* **18**, 188 (1991); W. R. Kuhn, *Nature* **359**, 196 (1992). See also J. F. Kasting, *Sci. Am.* **291**(1), 78 (2004).

284. G. E. Williams, *Sci. Am.* **255**(2), 88 (1986).

285. L. A. Pustilnik and G. Yom Din, preprint (2003), http://xxx.lanl.gov/abs/astro-ph/0312244.

286. D. Shindell et al., *Science* **294**, 2149 (2001).

287. P. Foukal, *Science* **264**, 238 (1994).

288. J. Lean and D. Rind, *Science* **292**, 234 (2001).

289. R. A. Kerr, *Science* **262**, 1370 (1993).

290. L. W. Alvarez et al., *Science* **208**, 1095 (1980). See also L. W. Alvarez, *Phys. Today* **40**(7), 24 (1987) and W. Sullivan, *NYT*, 1 November 1988.

291. M. M. Waldrop, *Science* **239**, 977 (1988); I. Venkatesan and J. Duhl, *Nature* **338**, 57 (1989).

292. J. Smit, *Annu. Rev. Earth Planet. Sci.* **27**, 75 (1999).

293. D. R. Prothero, *Science* **229**, 550 (1985).

294. C. C. Swisher et al., *Science* **257**, 954 (1992).

295. R. Grieve and A. Therriault, *Annu. Rev. Earth Planet. Sci.* **28**, 305 (2000).

296. L. C. Ivany et al., *Nature* **407**, 887 (2000).

297. P. D. Ward et al., *Science* **289**, 1740 (2000); Y. G. Jin et al., *Science* **289**, 432 (2000). See also the news article by R. A. Kerr, *Science* **289**, 1666 (2000).

298. S. P. Hesselbo et al., *Nature* **406**, 392 (2000).

299. J. W. Kirchner and A. Weil, *Nature* **404**, 177 (2000). D. Erwin, *Nature* **404**, 129 (2000).

300. C. K. Yoon, *NYT*, 16 May 2000.

301. J. E. Houlihan et al., *Nature* **404**, 752 (2000); J. M. Kiesecker et al., *Nature* **410**, 681 (2001). See also the discussion in Ch. 24.

302. N. Shaviv and J. Veizer, *GSA Today* **13**(7), 4 (2003).

CHAPTER 17

1. W. W. Kellogg and S. H. Schneider, *Science* **186,** 1163 (1974).

2. C. B. Beaty, *Am. Sci.* **66,** 452 (1978).

3. A. L. Hammond, *Science* **191,** 455 (1976); J. W. Sawyer, *Environment* **20**(2), 25 (1978).

4. W. L. Gates, *Science* **191**, 1138 (1976).

5. W. Dansgaard et al., *Nature* **339**, 532 (1989). S. J. Johnsen et al., *Nature* **359**, 311 (1992). K. C. Taylor et al., *Nature* **361**, 432 (1993). R. G. Fairbanks, *Nature* **362**, 495 (1993). M. Anklin et al., (Greenland Ice-core Project members), *Nature* **364**, 203 (1993). A. Landair et al., *J. Geophys. Res.* **108** (D18), 4563, (2003). W. Dansgaard et al., *Nature* **364**, 218 (1993). D. Grossman, *NYT*, 18 July 2003. D. Dahl-Jensen et al., *Ann. Glaciol.* **35**, 1 (2002).

6. W. Sullivan, *NYT*, 30 June 1993; *NYT*, 15 July 1993; *NYT*, 20 July 1993; R. A. Kerr, *Science* **260**, 890 (1993); S. Lehman, *Nature* **361**, 404 (1993); M. Holloway, *Sci. Am.* **269**(6), 34 (1993); and R. A. Kerr, *Science* **261**, 292 (1993).

7. R. B. Alley, *Quaternary Sci. Rev.* **19**, 213 (2000).

8. J. Oerlemans, *Nature* **364**, 783 (1993); R. A. Kerr, *Science* **262**, 1972 (1993); W. Sullivan, *NYT*, 7 September 1993.

9. C. D. Charles and R. G. Fairbanks, *Nature* **355**, 416 (1992). W. S. Broecker, in B. G. Levi, D. Hafemeister, and R. Scribner, eds., *Global Warming: Physics and Facts* (New York: American Institute of Physics, 1992), 148. R. Zahn, *Nature* **356**, 744 (1992); K. S. Thomson, *Am. Sci.* **81**, 522 (1993). C. Wunsch, *Science* **298**, 1179 (2002). R. A. Wood et al., *Phil. Trans. R. Soc. Lond. A* **361**, 1961 (2003).

10. E. Bard et al., *Science* **289**, 1321 (2000).

11. F. M. Phillips et al., *Science* **274**, 749 (1996).

12. P. U. Clark et al., *Science* **293**, 283 (2001).

13. K. K. Andersen et al., (North Greenland Ice Core Project members), *Nature* **431**, 147 (2004).

14. R. A. Kerr, *Science* **305**, 1693 (2004).

15. R. A. Houghton and G. M. Woodwell, *Sci. Am.* **260**(4), 36 (1989). D. Raynaud et al., *Science* **259**, 926 (1993) and T. Appenzeller, *Science* **259**, 908 (1993). J. Jouzel, et al., *Nature* **364**, 407 (1993). J. R. Petit et al., *Nature* **399**, 429 (1999).

16. L. Augustin et al., (EPICA community members), *Nature* **429**, 623 (2004). J. F. McManus, *Nature* **429**, 611 (2004) and J. W. C. White, *Science* **304**, 1609 (2004). G. Walker, *Nature* **429**, 596 (2004).

17. O. Watanabe et al., *Nature* **422**, 509 (2003).

18. C. Lorius et al., *Nature* **347**, 139 (1990); B. Saltzman and M. Verbitsky, *Nature* **367**, 419 (1994).

14b. D. Genty et al., *Nature* **421**, 833 (2003).

19. L. G. Thompson et al., *Science* **282**, 1858 (1998).

20. T. F. Stocker and D. G. Wright, *Nature* **351**, 729 (1991), A. J. Weaver and T. M. C. Hughes, *Nature* **367**, 447 (1994), L. D. Keigwin et al., *Nature* **371**, 323 (1994), R. Zahn, *Nature* **371**, 289 (1994), and W. S. Broecker et al., *Nature* **315**, 21 (1985).

21. R. Curry et al., *Nature* **426**, 826 (2003).

22. W. Munk, *Science* **300**, 2041 (2003). P. Wadhams and W. Munk, *Geophys. Res. Lett.* **31**, L11311 (2004). S. Häkkinen and P. B. Rhines, *Science* **304**, 555 (2004).

23. P. J. Crutzen and E. F. Stoermer, *IGBP Newsletter* No 41, 17 (2002).

24. W. F. Ruddiman, *Clim. Change* **61**, 261 (2003). T. J. Crowley, *Clim. Change* **61**, 259 (2003) and B. Mason, *Nature* **427**, 582 (2004).

25. S. H. Schneider, *The Genesis Strategy* (New York: Plenum, 1976).

26. H. Cachier et al., *Nature* **340**, 371 (1989).

27. J. M. Lobert et al., *Nature* **346**, 552 (1990).

28. S. E. Page et al., *Nature* **420**, 61 (2002). D. Schimel and D. Baker, *Nature* **420**, 29 (2002). R. L. Langenfelds et al., *Glob. Biogeochem. Cycles* **16**, 1048 (2002). R. C. Cowen, *CSM*, 7 November 2002 and L. Rohter, *NYT*, 4 November 2003.

29. P. J. Crutzen and J. Lelieveld, *Annu. Rev. Earth Planet. Sci.* **29**, 17 (2001). F. Achard et al., *Science* **297**, 999 (2002). P. M. Fearnside and W. F. Laurance, *Ecol. Appl.* **14**, 982 (2004).

30. J. J. Hidore, *J. Geog.* **77**, 214 (1978).

31. J. M. Melillo et al., in J. T. Houghton et al., eds., *Climate Change, The IPCC Scientific Assessment* (Cambridge: Cambridge University Press, 1990), 283; D. Kupfer and P. Karimanzera, in F. M. Bernthal, ed., *Climate Change: the IPCC Response Strategies* (Washington, DC. & Covela, CA: Island Press, 1991), 73; and J. S. Levine et al., in R. Bras, ed., *The World at Risk: Natural Hazards and Climate Change* (New York: American Institute of Physics, 1993), 131 support the tropical estimate. D. S. Schimel et al., *Nature* **414**, 169 (2001) supports the higher worldwide estimate (but includes cement manufacturing).

32. R. S. DeFries et al., *Proc. Natl. Acad. Sci.* **99**, 14256 (2002).

33. J. E. Richey et al., *Nature* **416**, 617 (2002).

34. D. A. Clark et al., *Proc. Natl. Acad. Sci.* **100**, 5852 (2003). J. Kaiser, *Science* **300**, 566 (2003) and C. Choi, United Press International dispatch, 21 April 2003.

35. E. Berg et al., talk presented at the 88th Annual Meeting of The Ecological Society of America held jointly with the International Society for Ecological Modeling - North American Chapter, 6 August 2004. See also T. Egan, *NYT*, 25 June 2003. N. Rozell, Geophysical Institute, University of Alaska Fairbanks, Alaska Science Forum Article #1688, 26 February 2004.

36. K. E. Percy et al., *Nature* **420**, 403 (2002). J. A. Logan et al., *Frontiers in Ecology and the Environment* **1**, 130 (2003).

37. A. P. Kershaw, *Nature* **322**, 47 (1986).

38. S. Senkowsky, *BioScience* **51**, 916 (2001).

39. A. C. Kurtz et al., *Paleoceanography* **18**, 1090 (2003). H. Weissert and S. M. Bernasconi, *Nature* **428**, 130 (2004).

40. A. Grübler et al., *Energy* **18**, 499 (1993).

41. H. Herzog et al., *Sci. Am.* **282**(2), 72 (2000).

42. W. F. Laurance et al., *Science* **291**, 438 (2001).

43. T. Smith, *NYT*, 28 June 2003 and Anonymous, Reuters dispatch, 27 June 2003.

44. F. Achard et al., *Science* **297**, 999 (2002).

45. M. Keller et al., *Ecol. Appl.* **14**, S3 (2004).

46. C. Sagan et al., *Science* **206**, 1363 (1979).

47. L. I. Wilder, *Little House on the Prairie* (New York: Harper & Row, 1970); S. Postel, *Futurist* **18**(4), 39 (1984).

48. J. U. Nef, *Sci. Am.* **237**(5), 140 (1977).

49. D. A. Robinson and G. Kukla, *J. Clim. Appl. Met.* **23**, 1626 (1984).

50. J. Charney et al., *Science* **187**, 434 (1974). See also J. Osterman, *Science* **186**, 531 (1974).

51. R. C. Balling, Jr., *Bull. Am. Meterolog. Soc.* **72**, 232 (1991).

52. N. Wade, *Science* **185**, 234 (1974).

53. M. Hulme and M. Kelly, *Environment* **35**(6), 4 (1993).

54. S. L. O'Hara et al., *Nature* **362**, 48 (1993); and *Nature* **364**, 197 (1993).

55. J. J. Walsh, *BioScience* **34**, 499 (1984).

56. H. Brückner, *Geog. J.* **13**, 7 (1986).

57. J. P. Peixoto and A. H. Oort, *Rev. Mod. Phys.* **56**, 365 (1984); W. L. Gates, in M. C. MacCracken and F. M. Luther, eds., *Projecting the Climatic Effects of Increasing Carbon Dioxide* (Washington, DC: GPO, 1986), DOE/ER-0237, 57; and M. E. Schlesinger, ibid., 280.

58. J. P. Peixóto and A. H. Oort, *Physics of Climate* (New York: American Institute of Physics, 1992).

59. E. Raschke and J. Schmetz, *Rep. Deutsche Forschunggemeinshaft* **1**, 10 (1984).

60. W. S. Smith, *Natl. Hist.* **92**(11), 36 (1983); W. Sullivan, *NYT*, 31 December 1985.

61. R. Caputo, *Natl. Geog.* **167**, 577 (1985); A. F. Pillsbury, *Sci. Am.* **245**(1), 54 (1981).

62. E. Pallé et al., *Science* **304**, 1299 (2004). K. Chang, *NYT*, 28 May 2005. D. Perlman, *SFC*, 28 May 2004.

63. J. M. Mitchell, Jr., in F. Singer, ed., *Global Effects of Environmental Pollution* (Dodrecht, Neth.: D. Reidel Publ. Co., 1970); and W. H. Matthews, *Intern. J.Environmental Studies* **4**, 283 (1973).

64. H. H. Lamb, *Ecologist* **4**, 10 (1974).

65. R. H. Gammon et al., in J. R. Trabalka, ed., *Atmospheric Carbon Dioxide and the Global Carbon Cycle* (Washington, DC: GPO, 1986), DOE/ER-0239, 25.

66. A. J. Kaufman and S. Xiao, *Nature* **425**, 379 (2003). S. J. Mojzsis, *Nature* **425**, 249 (2003).

67. See, for example, M. L. Hoke and J. H. Shaw, *Appl. Opt.* **21**, 929 (1982), M. L. Hoke and J. H. Shaw, *Appl. Opt.* **22**, 328-332 (1983), and M. S. Abubakar and J. H. Shaw, *Appl. Opt.* **25**, 1196 (1986).

68. U. von Zahn, *Alexander von Humboldt-Stiftung Mitteilungen* #39, 15 (1981).

69. Energy Committee of the German Physical Society, *Phys. Bl.* **39**, 320 (1983).

70. R. T. Watson et al., in J. T. Houghton, G. J. Jenkins, and J. J. Ephraums, eds., *Climate Change, The IPCC Scientific Assessment* (Cambridge: Cambridge University Press, 1990), 1.

71. R. T. Watson et al., eds., *Climate Change 1992* (Cambridge: Cambridge University Press, 1992), 25.

72. R. A. Kerr, *Science* **222**, 1107 (1983).

73. R. T. Watson et al. (Authors) in D. J. Verardo et al., eds., *Land Use, Land-Use Change, and Forestry* (Cambridge and New York: Cambridge University Press, 2001).

74. World Resources Institute, *World Resources 1990—91* (Oxford University Press, New York, 1990). Table 21.1 (pp. 316-317) discusses fossil fuel use.

75. Dept. of Energy, *Energy Use and Carbon Emissions: Some International Comparisons* (Washington, DC: GPO, 1994), DOE/EIA-0579.

76. G. M. Woodwell, *Oceanus* **29**(4), 71 (1986).

77. U. Siegenthaler and H. Oeschger, *Science* **199**, 388 (1978); U. Siegenthaler and H. Oeschger, *Ann. Glaciology* **5**, 153 (1984); and G. I. Pearman et al., *Nature* **320**, 248 (1986).

78. A. Neftel et al., *Nature* **315**, 45 (1985); D. Raynaud and J. M. Barnola, *Nature* **315**, 309 (1985); and H. Friedli et al., *Nature* **324**, 237 (1987).

79. R. E. Dickinson and R. J. Cicerone, *Nature* **319**, 109 (1986).

80. J. J. McCarthy et al., eds., *Climate Change 2001: Impacts, Adaptation, and Vulnerability* (Cambridge and New York: Cambridge University Press, 2001)

81. C. D. Keeling and T. P. Whorf, in *Trends: A Compendium of Data on Global Change*, Carbon Dioxide Information Analysis Center, Oak Ridge National Laboratory, U.S. Department of Energy, Oak Ridge, Tennessee (2000). The figure is based on work previously published, for example, C. D. Keeling, et al., *Aspects of Climate Variability in the Pacific and the Western Americas* (Washington, DC: American Geophysical Union, 1989), App. A. For earlier versions, see R. B. Barcastow et al., *J. Geophys. Res.* **90,** 10529 (1985); and C. D. Keeling et al., *J. Geophys. Res.* **90,** 10511 (1985).

82. G. M. Woodwell, *Sci. Am.* **238**(1), 34 (1978); G. M. Woodwell et al., *Science* **199**, 141 (1978).

83. J. Hansen et al., *Science* **213,** 957 (1981).

84. Panel on Climatic Variation [W. L. Gates et al.], U.S. Committee for the Global Atmospheric Research Program, National Research Council, *Understanding Climatic Change* (Detroit: Grand River Books, 1980).

85. P. P. Tans and J. W. C. White, *Science* **281**, 183 (1998).

86. G. J. Retallack, *Phil. Trans. R. Soc. Lond. A* **360**, 659 (2002). D. H. Rothman, *Proc. Natl. Acad. Sci.* **99**, 4167 (2002). K. Pahnke et al., *Science* **301**, 948 (2003).

87. J. P. Sachs et al., *Geochemistry Geophysics Geosystems* **1**, 200GC000059 (2000) (e-journal of the AGU and the Geochemical Society). M. Pagani, *Phil. Trans. R. Soc. Lond. A* **360**, 609 (2002).

88. B. K. Linsley et al., *Science* **290**, 1145 (2000). M. A. Cane and M. Evans, *Science* **290**, 1107 (2000).

89. D. Dahl-Jensen et al., *Science* **282**, 268 (1998).

90. L. G. Thompson et al., *Science* **276**, 1821 (1997).

91. P. D. Jones et al., *Science* **292**, 662 (2001).

92. M. E. Mann et al., *Geophys. Res. Lett.* **26**, 759 (1999). K. R. Briffa and T. J. Osborn, *Science* **284**, 926 (1999). M. E. Mann, R. S. Bradley, and M. K. Hughes, *Nature* **392**, 779 (1998) and P. Jones, *Science* **280**, 544 (1998). S. McIntyre and R. McKitrick, *Energy & Environment* **14**, 751 (2003) and M. E. Mann et al., *Nature* **430**, 105 (2004). P. D. Jones, *Endeavour* **14**, 129 (1990) also discusses this topic for a more popular audience.

93. T. J. Crowley, *Science* **289**, 270 (2000). See also the commentary by Mann, Ref. 103.

94. W. S. Broecker, *Science* **291**, 1497 (2001).

95. R. Bradley, *Science* **288**, 1353 (2000). R. S. Bradley, in M. E. Schlesinger, ed., *Greenhouse−Gas−Induced Climatic Change: A Critical Appraisal of Simulations and Observations* (Amsterdam: Elsevier, 1991), 301.

96. J. J. Magnuson et al., *Science* **289**, 1743 (2000).

97. L. D. Keigwin, *Science* **274**, 1504 (1996).

98. K. J. Kreutz et al., *Science* **277**, 1294 (1997).

99. D. Verschuren et al., *Nature* **403**, 410 (2000).

100. S. Huang et al., *Nature* **403**, 756 (2000). J. T. Overpeck, *Nature* **403**, 714 (2000). H. N. Pollack, S. Huang, and P.-Y. Shen, *Science* **282**, 279 (1998). H. N. Pollack and D. S. Chapman, *Sci. Am.* **268**(6), 44 (1993).

101. As an example of such a local reconstruction, see data from Utah in R. N. Harris and D. S. Chapman, *Science* **275**, 1618 (1997). Cooling before 1900 is clearly visible, but little is clear about the time before ~1750.

102. J. Overpeck et al., *Science* **278**, 1251 (1997).

103. G. C. Jacoby et al., *Science* **273**, 771 (1996)

104. M. E. Mann, *Science* **289**, 253 (2000).

105. E. A. Parson, Ch. 10 in J. Mellilo et al., eds., (National Assessment Synthesis Team) *Climate Change Impacts on the United States: The Potential Consequences of Climate Variability and Change* (Cambridge and New York: Cambridge University Press, 2001). See also R. Sagarin and F. Micheli, *Science* **294**, 811 (2001).

106. D. G. Vaughan et al., *Science* **293**, 1777 (2001).

107. C. Barbraud and H. Weimerskirch, *Nature* **411**, 183 (2001).

108. T. L. Delworth and T. R. Knutson, *Science* **287**, 2246 (2000). T. M. L. Wigley et al., *Science* **282**, 1676 (1998).

109. P. A. Stott et al., *Science* **290**, 2133 (2000). S. F. B. Tett et al., *Nature* **399**, 569 (1999). F. W. Zwiers and A. J. Weaver, *Science* **290**, 2081 (2000).

110. J. Hansen et al., *Science* **295**, 275 (2002).

111. B. Wuethrich, *Science* **285**, 37 (1999).

112. D. A. Rothrock et al., *Geophys. Res. Lett.* **26**, 3469 (1999). P. Wadhams and N. R. Davis, *Geophys. Res. Lett.* **27**, 3973 (2000). K. Krajick, *Science* **291**, 424 (2001). S. Laxon et al., *Nature* **425**, 947 (2003).

113. B. G. Levi, *Phys. Today* **53**(1), 19 (2000).

114. B. J. Peterson et al., *Science* **298**, 2171 (2002).

115. R. Thomas et al., *Science* **289**, 426 (2000). P. Chylek, *Clim. Change* **63**, 201 (2004).

116. W. Krabill et al., *Science* **289**, 428 (2000). W. Krabill et al., *Science* **283**, 1522 (1999). D. Dahl-Jensen, *Science* **289**, 404 (2000).

117. A. B. Paterson and N. Reeh, *Nature* **414**, 60 (2001).

118. A. A. Arendt et al., *Science* **297**, 382 (2002); M. F. Meier and M. B. Dyurgerov, *Science* **297**, 350 (2002):

119. A. Shepherd et al., *Science* **291**, 862 (2001).

110. E. J. Rignot, *Science* **281**, 549 (1998).

121. S. S. Jacobs et al., *Science* **297**, 386 (2002).

122. I. Joughin and S. Tulaczyk, *Science* **295**, 476 (2002). R. B. Alley, *Science* **295**, 451 (2002).

123. L. G. Thompson et al., *Science* **289**, 1916 (2000).

124. J. Oerlemans, *Nature* **320,** 607 (1986). J. Oerlemans, *Science* **264**, 243 (1994).

125. P. Recer, AP dispatch, 19 February 2001; J. Mangels, *CPD*, 19 February 2001; A. C. Revkin, *NYT*, 19 February 2001.

126. K. Alverson et al., *Science* **293**, 47 (2001).

127. L. G. Thompson, *Quat. Sci. Rev.* **19**, 19 (2000).

128. M. Bowen, *Natl. Hist.* **107**(1), 28 (1998).

129. Anonymous, Environmental News Network dispatch, 13 June 2001, URL: http://www.enn.com/news/enn-stories/2001/06/06132001/glaciers_43943.asp.

130. P. Recer, AP dispatch, 14 September 2000; M. Henderson, *TL*, 15 September 2000.

131. J. E. Hansen et al., *Science* **281**, 930 (1998). R. A. Kerr, *Science* **281**, 1948 (1998).

132. B. D. Santer et al., *Science* **287**, 1227 (2000). D. E. Parker, *Science* **287**, 1216 (2000).

133. C. A. Mears et al., *J. Climate* **16**, 3650 (2003). Q. Fu et al., *Nature* **429**, 55 (2004). Q. Schiermeier, *Nature* **429**, 7 (2004).

134. R. A. Kerr, *Science* **304**, 805 (2004). K. Y. Vinnikov and N. C. Grody, *Science* **302**, 269 (2003).

135. D. J. Seidel et al., *J. Climate* **17**, 2225 (2004). R. W. Reynolds et al., *J. Climate* **17**, 2938 (2004). P. W. Thorne et al., *Climate Dyn.* **21**, 573 (2003).

136. P. B. Duffy et al., *J. Climate* **14**, 2809 (2001).

137. S. Levitus et al., *Science* **287**, 2225 (2000). R. A. Kerr, *Science* **287**, 2126 (2000).

138. S. Levitus et al., *Science* **292**, 267 (2001). R. A. Kerr, *Science* **292**, 193 (2001).

139. T. P. Barnett et al., *Science* **292,** 270 (2001).

140. S. T. Gille, *Science* **295**, 1275 (2002). D. Perlman, *SFC*, 25 February 2002.

141. B. G. Levi, *Phys. Today* **54**(6), 19 (2001).

142. W. C. Wang et al., *Science* **194,** 685 (1976).

143. R. A. Kerr, *Science* **220**, 1364 (1983); G. Marland and R. M. Rotty, *J. Air Poll. Control Assoc.* **35,** 1033 (1985).

144. L. S. Rothman et al., *Appl. Opt.* **22**(11), 1616 (1983).

145. A. Mosier et al., *Nutr. Cyc. Agroecosyst.* **52**, 225 (1998).

146. E. J. Dlugokencky et al., *Geophys. Res. Lett.* **28**, 499 (2001).

147. E. J. Dlugokencky et al., *Geophys. Res. Lett.* **30**, 1992 (2003).

148. Enquete-Kommission "Vorsorge zum Schutz der Erdatmosphäre" der Deutschen Bundestages, *Schutz der Erde*, V. 1 (Bonn: Economia Verlag and Karlsruhe: Verlag C. F. Müller, 1991).

149. World Meteorological Organization, *WMO Global Ozone Research and Monitoring Project, Rep. No. 14 of the Meeting of Experts on Potential Effects of Ozone and Other Minor Trace Gases* (World Meteorological Organization, Geneva, 1982).

150. S. Seidel and D. Keyes, *Can We Delay a Greenhouse Warming?* (Washington, DC: Office of Policy and Resources Management, EPA, 1983).

151. D. A. Lashof and D. R. Ahuja, *Nature* **344**, 529 (1990).

152. J. E. Hansen and M Sato, *Proc. Natl. Acad. Sci.* **98**, 14778 (2001).

153. M. A. K. Khalil and R. A. Rasmussen, *Ann. Glaciology* **10**, 160 (1988).

154. D. R. Blake and F. S. Rowland, *Science* **239**, 1129 (1989). H.-U. Neue, *BioScience* **43**, 466 (1996).

155. D. H. Ehhalt, *Environment* **30**(10), 6 (1985).

156. P. A. Matson and P. M. Vitousek, *BioScience* **40**, 667 (1990).

157. M. H. Thiemens and W. C. Trogler, *Science* **251**, 932 (1991).

158. M. A. K. Khalil, *Annu. Rev. Energy Environ.* **24**, 645 (1999).

159. E. J. Brook et al., *Science* **273**, 1087 (1996).

160. S. K. Kaharabata et al., *Environ. Sci. Technol.* **34**, 3296 (2000).

161. Anonymous, BBC dispatch, 7 June 2001.

162. A. B. Lovins and L. H. Lovins, *Annu. Rev. Energy Environ.* **16**, 433 (1991).

163. V. L. St. Louis et al., *BioScience* **50**, 766 (2000).

164. B. Stauffer et al., *Science* **229**, 1386 (1985). R. A. Kerr, *Science* **226**, 954 (1984).

165. C. P. Rinsland et al., *Nature* **318**, 245 (1985); R. A. Rasmussen and M. A. K. Khalil, *Science* **232**, 1623 (1986).

166. U.S. Environmental Protection Agency, URL: http://www.epa.gov/ghginfo/ topics/topic1.htm.

167. L. P. Steele et al., *Nature* **358**, 313 (1993).

168. D. I. Stern and R. K. Kaufmann, n *Trends Online: A Compendium of Data on Global Change* (1998). Carbon Dioxide Information Analysis Center, Oak Ridge National Laboratory, U.S. Department of Energy, Oak Ridge, Tenn., U.S.A.

169. J. E. Hansen et al., *Proc. Natl. Acad. Sci.* **97**, 9875 (2000). R. J. Cicerone, *Proc. Natl. Acad. Sci.* **97**, 10304 (2000). F. de la Chesnaye et al., *Energy Policy* **29**, 1325 (2001).

170. K. A. Kvenvolden, *Global Biogeochem. Cycles* **2**, 221 (1988). E. G. Nisbet, *Phil Trans. R. Soc. Lond. A* **360**, 581 (2002).

171. E. Nisbet, *Nature* **347**, 23 (1990); S. C. Whalen and W. S. Reeburgh, *Nature* **346**, 160 (1990); W. K. Stevens, *NYT*, 19 February 1991; and R. A. Sommerfeld et al., *Nature* **361**, 140 (1993).

172. S. D. Bridgeham et al., *BioScience* **45**, 262 (1995). J. W. H. Dacey et al., *Nature* **370**, 47 (1994).

173. M. L. Goulden et al., *Science* **279**, 214 (1998).

174. M. A. K. Khalil and R. A. Rasmussen, *Science* **232**, 56 (1986).

175. A, McCulloch and P. M. Midgley, *Atmos. Environ.* **35**, 5311 (2001).

176. M. Parry and T. Carter, *Climate Impact and Adaptation Assessment* (London: Earthscan Publications, Ltd., 1998).

177. Panel on Policy Implications of Greenhouse Warming (D. J. Evans et al.), *Policy Implications of Greenhouse Warming: Mitigation, adaptation, and the science base* (Washington, D.C: National Academy Press, 1992); E. S. Rubin et al., *Science* **257**, 148 (1992).

178. S. Schneider, *Sci. Am.* **286**(1), 62 (2002).

179. J. P. Holdren, *Sci. Am.* **286**(1), 65 (2002).

180. J. Tyndall, *Philos. Mag.* (Series 4) **22**, 169, 273 (1861).

181. S. Arrhenius, *Philos. Mag.* (Series 5) **41**, 237 (1896).

182. R. Revelle and H. E. Suess, *Tellus* **9**, 18 (1957).

183. R. M. Garrels et al., *Am. Sci.* **64**, 306 (1976); J. C. G. Walker, *BioScience* **34**, 486 (1984).

184. N. J. Shackleton et al., *Nature* **306**, 319 (1983); J. M. Palais, *Oceanus* **29**(4), 55 (1986).

185. J. M. Barnola, et al., *Nature* **329**, 410 (1987).

186. N. Petit-Maire et al., *Paleogeogr. Paleoclimat. Paleoecol.* **86**, 197 (1991).

187. V. Ramanathan, *Science* **240**, 293 (1988); V. Ramanathan et al., *Science* **243**, 57 (1989); V. Ramanathan et al., *Phys. Today* **42** (5), 22 (1989). M. La Brecque, *Mosaic* **21**(2), 2 (1990) and J. T. Kiehl, *Phys. Today* **47**(11), 36 (1994).

188. H. Oeschger and B. Stauffer, *Science* **199**, 388 (1978); B. Stauffer et al., *Ann. Glaciology* **5**, 160 (1984).

189. G. M. MacDonald et al., *Nature* **361**, 243 (1993). J. Pastor, *Nature* **361**, 208 (1993).

190. P. V. Hobbs et al., *Science* **183**, 909 (1974).

191. J. Hansen et al., *Atmos. Res.* **37**, 175 (1995).

192. C.-D. Schönwiese, *Arch. Met. Geophy. Biocl.* **B35**, 155 (1984).

193. P. D. Jones et al., *Nature* **322**, 430 (1986).

194. G. A. Maul, in B. G. Levi et al., eds., *Global Warming: Physics and Facts* (New York: American Institute of Physics, 1992), 78.

195. J. Hansen and S. Lebedeff, *Geophys. Res. Lett.* **15**, 323 (1988); and *J. Geophys. Res.* **92**, 13345 (1987).

196. P. D. Jones and T. M. L. Wigley, *Sci. Am.* **263**(2), 84 (1990).

197. G. Parrilla et al., *Nature* **369**, 48 (1994).

198. CO_2/Climate Review Panel (J. Smagorinsky et al.), National Research Council, *Carbon Dioxide and Climate: A Second Assessment* (Washington, DC: National Academy Press, 1982).

199. J. T. Houghton et al., *Climate Change, The IPCC Scientific Assessment* (Cambridge: Cambridge University Press, 1990).

200. J. T. Houghton et al., eds., *Climate Change 1992* (Cambridge: Cambridge University Press, 1992).

201. R. T. Watson et al., eds., *Technologies, Policies and Measures for Mitigating Climate Change* (Geneva, Switzerland: IPCC, 1996).

202. J. T. Houghton et al., eds., *Stabilization of Atmospheric Greenhouse Gases: Physical, Biological and Socio-economic Implications* (Geneva, Switzerland: IPCC, 1997).

203. T. M. L. Wigley et al., eds., *Implications of Proposed CO_2 Emissions Limitations* (Geneva, Switzerland: IPCC, 1997).

204. J. T. Houghton et al., eds., *Climate Change 1995: The Science of Climate Change* (Contribution of Working Group I to the Second Assessment of the Intergovernmental Panel on Climate Change) (Cambridge and New York: Cambridge University Press, 1996).

205. B. Bolin (Chair of the IPCC, Chairman of the Drafting Team); J. T. Houghton et al., *IPCC Second Assessment Synthesis of Scientific-Technical Information relevant to interpreting Article 2 of the UN Framework Convention on Climate Change* (Geneva, Switzerland: ICPP, 1996).

206. R. T. Watson et al., eds., *Climate Change 1995: Impacts, Adaptations and Mitigation of Climate Change: Scientific-Technical Analyses* (Contribution of Working Group II to the Second Assessment of the Intergovernmental Panel on Climate Change) (Cambridge and New York: Cambridge University Press, 1996).

207. J. P. Bruce et al., eds., *Climate Change 1995: Economic and Social Dimensions of Climate Change* (Contribution of Working Group III to the Second Assessment of the Intergovernmental Panel on Climate Change) (Cambridge and New York: Cambridge University Press, 1996).

208. R. S. Cerveny and R. C. Balling, Jr., *Geophys. Res. Lett.* **26**, 1605 (1999). H. Gee, Nature Science Update, 24 June 1999, URL: http://www.nature.com/nsu/990624/990624-9.html. See also R. C. Balling, Jr. and R. S. Cerveny, *Science* **267**, 1481 (1995).

209. C. P. Sonett and L. A. Smith, *Geophys. Res. Lett.* **26**, 1569 (1999).

210. P. R. Goode, et al., *Geophys. Res. Lett.* **28**, 1671 (2001). G. Taubes, *Science* **264**, 1529 (1994).

211. J. T. Houghton et al., eds., *Climate Change 2001: The Scientific Basis* (Cambridge and New York: Cambridge University Press, 2001).

212. D. L. Albritton et al., "Report of Working Group I," IPCC Report (2001).

213. K. S. White et al., "Report of Working Group II," IPCC Report (2001).

214. T. Banuri et al., "Report of Working Group III," IPCC Report (2001).

215. B. Metz et al., eds., *Climate Change 2001: Mitigation* (Cambridge and New York: Cambridge University Press, 2001)

216. Committee on the Science of Climate Change (R. J. Cicerone et al.), National Academy of Sciences, *Climate Change Sciences* (Washington, DC: National Academy Press, 2001).

217. K. R. Gurney, *Nature* **353**, 23 (1991).

218. W. T. Sturges et al., *Science* **289**, 611 (2000).

219. Energy Information Administration, *Emissions of Greenhouse Gases in the United States 2000* (Washington, DC: GPO, 2001), DOE/EIA-0573(2000).

220. J. E. Hansen et al., *Proc. Natl. Acad. Sci.* **95**, 12753 (1998).

221. D. J. Karoly et al., *Science* **302**, 1200 (2003).

222. N. P. Gillett et al., *Nature* **422**, 29 (2003).

223. T. L. Root et al., *Nature* **421**, 87 (2003). J. C. Pugh, *BioScience* **53**, 542 (2003).

224. C. Parmesan and G. Yohe, *Nature* **421**, 37 (2003).

225. L. F. Richardson, *Weather Prediction by Numerical Process* (Cambridge: Cambridge University Press, 1922).

226. J. G. Charney et al., *Tellus* **2**, 237 (1950).

227. N. A. Phillips, *Quarterly J. Royal Meteorological Soc.* **82**, 123 (1956).

228. R. S. Keir, in T. M. L. Wigley and D. S. Schmiel, eds., *The Carbon Cycle* (Cambridge and New York: Cambridge University Press, 2000). See also J. R. Toggweiler, *Phys. Today* **47**(11), 45 (1994).

229. L. D. D. Harvey, in T. M. L. Wigley and D. S. Schmiel, eds., *The Carbon Cycle* (Cambridge and New York: Cambridge University Press, 2000).

230. L. D. D. Harvey, *Global Biogeochem. Cycles* **3**, 137 (1989); L. D. D. Harvey, *Clim. Change* **15**, 343 (1989).

231. W. L. Gates et al., eds., *Climate Change, The IPCC Scientific Assessment* (Cambridge: Cambridge University Press, 1990), 93.

232. R. D. Cess et al., *Science* **245**, 513 (1989).

233. R. D. Cess et al., *Science* **262**, 1252 (1993).

234. E. A. Graham et al., *Proc. Natl. Acad. Sci.* **100**, 572 (2003).

235. W. L. Gates et al., in J. T. Houghton et al., eds., *Climate Change 1992* (Cambridge: Cambridge University Press, 1992), 97.

236. T. C. Johns et al., *Climate Dyn.* **13**, 103 (1997).

237. G. J. Boer et al., *Atmosphere-Ocean* **22**, 397 (1984). G. J. Boer et al., *Climate Dyn.* **16**, 405 (2000).

238. GISS refers to the model developed at the NASA Goddard Institute of Space Science in New York.

239. GFDL refers to the model developed at the NOAA Geophysical Fluid Dynamics laboratory at Princeton University.

240. M. MacCracken et al., Ch. 1 in J. Mellilo, A, Janetos, and T. Karl, eds., (National Assessment Synthesis Team) *Climate Change Impacts on the United States: The Potential Consequences of Climate Variability and Change* (Cambridge and New York: Cambridge University Press, 2001).

241. J. T. Houghton et al., eds., *An Introduction to Simple Climate Models used in the IPCC Second Assessment Report* (Report by IPCC Working Group I) (Geneva, Switzerland: IPCC, 1997).

242. N. Nakić enović et al., *Emissions Scenarios* (Geneva, Switzerland: ICPP, 2000). N. Nakić enović and R. Swart, eds., *Special Report on Emissions Scenarios* (Geneva, Switzerland: IPCC, 1997) (ICPP website http://www.grida.no/climate/ipcc/emission/).

243. R. G. Prinn et al., *Science* **292**, 1882 (2001).

244. E. B. Rastetter, *BioScience* **46**, 190 (1996).

245. G. R. Shaver et al., *BioScience* **50**, 871 (2000).

246. F. S. Chapin et al., *Ecology* **76**, 694 (1996). See also W. C. Oechel et al., *Nature* **361**, 520 (1993).

247. J. Mellilo et al., eds., (National Assessment Synthesis Team) *Climate Change Impacts on the United States: The Potential Consequences of Climate Variability and Change* (Cambridge and New York: Cambridge University Press, 2001).

248. Committee on Abrupt Climate Change (R. Alley et al.), National Academy of Sciences, *Abrupt Climate Change: Inevitable Surprises* (Washington, DC: National Academy Press, 2002).

249. R. B. Alley and P. U. Clark, *Annu. Rev. Earth Planet. Sci.* **27**, 149 (1999).

250. C. Lang et al., *Science* **286**, 934 (1999). J. Jouzel, *Science* **286**, 910 (1999).

251. K. Taylor, *Am. Sci.* **87**, 320 (1999).

252. P. deMenocal et al., *Science* **288**, 2198 (2000).

253. G. Bond et al., *Science* **278**, 1257 (1997). G. G. Bianchi and I. N. McCave, *Nature* **397**, 515 (1999).

254. D. A. Meese et al., *Science* **266**, 1680 (1994). S. Björck et al., *Science* **274**, 1155 (1996). K. C. Taylor et al., *Science* **278**, 825 (1997). Z. Yu and U. Eicher, *Science* **282**, 2235 (1998). J. P. Severinghaus and E. J. Brook, *Science* **286**, 930 (1999). S. J. Lehman and L. D. Keigwin, *Nature* **356**, 757 (1992).

255. Y. G. Jin et al., *Science* **289**, 432 (2000).

256. G. R. Dickens et al., *Paleoceanography* **10**, 965 (1995).

257. S. Bains et al., *Science* **285**, 724 (1999). M. E. Katz et al., *Science* **286**, 1531 (1999). S. Simpson, *Sci. Am.* **282**(2), 24 (2000); R. A. Kerr, *Science* **286**, 1465 (1999). R. D. Norris and U. Röhl, ;*Nature* **401**, 775 (1999). G. R. Dickens, *Nature* **401**, 752 (1999).

258. J. P. Kennett et al., *Science* **288**, 128 (2000). T. Blunier, *Science* **288**, 68 (2000). J. P. Kennett et al., *Methane Hydrates in Quaternary Climate Change: The Clathrate Gun Hypothesis*, (Washington, DC: American Geophysical Union, 2002).

259. F. P. Bretherton et al., in J. T. Houghton et al., eds., *Climate Change, The IPCC Scientific Assessment* (Cambridge: Cambridge University Press, 1990), 173; D. A. Randall, in B. G. Levi et al., eds., *Global Warming: Physics and Facts* (New York: American Institute of Physics, 1992), 24; S. Manabe and R. J. Stouffer, *Nature* **364**, 215 (1993); and J. Marotzke, in R. Bras, ed., *The World at Risk: Natural Hazards and Climate Change* (New York: American Institute of Physics, 1993), 150.

260. W. B. Harland, *Geol. Rundsch.* **54**, 45 (1964); J. L. Kirschvink, in J. W. Schopf and C. Klein, eds., *The Proterozoic Biosphere* (Cambridge and New York: Cambridge University Press, 1992).

261. P. F. Hoffman et al., *Science* **281**, 1342 (1998). See also P. F. Hoffman and D. P. Schrag, *Sci. Am.* **282**(1), 68 (2000) and R. A. Kerr, *Science* **287**, 1734 (2000).

262. W. T. Hyde et al., *Nature* **405**, 425 (2000). B. Runnegar, *Nature* **405**, 403 (2000) and R. A. Kerr, *Science* **288**, 1316 (2000).

263. P. J. Sellers et al., *Science* **275**, 502 (1997).

264. H. Grassl, *Science* **288**, 1991 (2000).

265. S. Manabe and R. T. Wetherald, *J. Atmos. Sci.* **24**, 241 (1967); S. Manabe, in W. H. Matthews, et al., eds., *Man's Impact on the Climate* (Cambridge, MA: MIT Press, 1971).

266. S. Manabe and R. T. Wetherald, *J. Atmos. Sci.* **32**, 3 (1975).

267. R. A. Kerr, *Science* **305**, 932 (2004).

268. C. K. Folland et al., in J. T. Houghton et al., eds., *Climate Change 1992* (Cambridge: Cambridge University Press, 1992), 135.

269. C. K. Folland et al., in J. T. Houghton et al., eds., *Climate Change, The IPCC Scientific Assessment* (Cambridge: Cambridge University Press, 1990), 195; T. M. L. Wigley and T. P. Barnett, ibid., 239.; T. R. Karl, in L. Rosen and R. Glasser, eds., *Climate Change and Energy Policy* (New York: American Institute of Physics, 1992), 40.

270. T. M. L. Wigley and S. C. B. Raper, *Science* **293**, 451 (2001). An even wider statement about uncertainty is made in C. E. Forest et al., *Science* **295**, 113 (2002), that their uncertainty is smaller than that of IPCC, Ref. 196, but they find a temperature range of 1.4 °C to 7.7 °C, an even greater upper 95% confidence limit than in Ref. 197. R. A. Kerr, *Science* **295**, 29 (2002).

271. J. Hansen et al., *Natl. Geog. Res. Exploration* **9**, 142 (1993).

272. M. I. Hoffert and C. Covey, *Nature* **360**, 573 (1992); G. Walker, *Nature* **362**, 110 (1993).

273. T. P. Ackerman and G. M. Stokes, *Phys. Today* **56**(1), 38 (2003).

274. N. Gruber et al., *Science* **298**, 2374 (2002). P. Quay, *Science* **298**, 2344 (2002).

275. K. R. Gurney et al., *Nature* **415**, 626 (2002).

276. K. Heki, *Science* **293**, 89 (2001). K. Heki, *Earth Planetary Sci. Lett.* **207**, 159 (2003). T. Clarke, Nature Science Update, 15 February 2003.

277. J. J. McCarthy et al., *Oceanus* **29**(4), 16 (1986).

278. M. C. Facchini et al., *Nature* **401**, 257 (1999). H. Rodhe, *Nature* **401**, 223 (1999).

279. R. J. Charlson et al., *Science* **292**, 202 (2001).

280. R. S. Gao et al., *Science* **303**, 516 (2004).

281. J. H. Seinfeld, *Nature* **391**, 837 (1998).

282. R. A. Kerr, *Science* **276**, 1649 (1997). D. Rind, *Science* **281**, 1152 (1998).

283. D. J. Travis et al., *Nature* **418**, 601 (2002).

284. J. E. Penner et al., eds., *Aviation and the Global Atmosphere* (Cambridge and New York: Cambridge University Press, 1999).

285. K. H. Rosenlof et al., *Geophys. Res. Lett.* **28**, 1195 (2001).

286. H. Douville et al., *Climate Dyn.* **20**, 45 (2002).

287. J. F. B. Mitchell et al., *Nature* **341**, 132 (1989); A. Slingo, *Nature* **343**, 49 (1990).

288. D. Entekhabi and P. S. Eagleson, in R. Bras, ed., *The World at Risk: Natural Hazards and Climate Change* (New York: American Institute of Physics, 1993), 168.

289. I. M. Held and B. J. Soden, *Annu. Rev. Energy Environ.* **25**, 441 (2000).

290. NASA, NASA website URL: http://www.gsfc.nasa.gov/gsfc/service/gallery/fact_sheets/earthsci/terra/aerosols.htm.

291. J. M. Haywood et al., *Science* **283**, 1299 (1999). J. T. Kiehl, *Science* **283**, 1273 (1999). T. L. Anderson et al., *Science* **300**, 1103 (2003).

292. C. D. Jones et al., *Geophys. Res. Lett.* **30**, 1479 (2003). They predict the land carbon source to release carbon at a rate of 7 Gt/yr by 2100, to exceed the ocean carbon sink beyond about 2080 (the sink is predicted to saturate at a rate of 5 Gt/yr by 2100).

293. J. E. Penner et al., *Science* **256**, 1432 (1992) and R. J. Charlson et al., *Science* **255**, 423 (1992) find a sulfate forcing of −1 to −2 W/m²; J. T. Kiehl and B. P. Briegleb, *Science* **260**, 311 (1993) find -0.3; R. A. Kerr, *Science* **255**, 682 (1992).

294. R. C. Bay et al., *Proc. Natl. Acad. Sci.* **101**, 6341 (2004).

295. J. Hansen et al., *Geophys. Res. Lett.* **19**, 215 (1992); A. M. Vogelmann et al., *Nature* **359**, 47 (1992).

296. P. C. Novelli et al., *Science* **263**, 1587 (1994). M. A. K. Khalil and R. A. Rasmussen, *Nature* **370**, 639 (1994)

297. J. E. Penner, *J. Air Waste Manage. Assoc.* **40**, 456 (1990); R. J. Charlson and T. M. L. Wigley, *Sci. Am.* **270**(2), 48 (1994).

298. G. Stanhill and S. Cohen, *Agric. For. Meteorol.* **107**, 255 (2001). K. Chang, *NYT*, 13 May 2005. R. S. Boyd, Knight Ridder Newspapers dispatch, 7 May 2004.

299. M. L. Roderick and G. D. Farquhar, *Science* **298**, 1410 (2002). S. Cohen et al., *Agric. For. Meteorol.* **111**, 83 (2002). A. Ohmura and M. Wild, *Science* **298**, 1345 (2002).

300. B. A. Wielicki et al., *Science* **295**, 841 (2002).

301. Y. Liu and P. H. Daum, *Nature* **419**, 580 (2002).

302. T. Nakajima et al., in T. Matsuno and H. Kida, eds. *Present and Future of Modeling Global Environmental Change: Toward Integrated Modeling*, (Tokyo: Terrapub, 2001), 77.

303. S. E. Schwartz, *Energy* **18**, 1229 (1993).

304. J. T. Houghton et al., eds., *Climate Change 1994: Radiative Forcing of Climate Change and an Evaluation of the IPCC IS92 Emission Scenarios* (Cambridge and New York: Cambridge University Press, 1995). See Ch. 3.

305. D. Rosenfeld et al., *Proc. Natl. Acad. Sci.* **98**, 5975 (2001).

306. S. K. Satheesh and V. Ramanathan, *Nature* **405**, 60 (2000).

307. V. Ramanathan et al., *Science* **294**, 2119 (2001). A. S. Ackerman, et al., *Science* **288**, 1042 (2001). S. E. Schwartz and P. R. Buseck, *Science* **288**, 989 (2000).

308. V. M. H. Lavanchy et al., *J. Geophys. Res.-Atmos.* **104D**, 21227 (1999).

309. P. V. Hobbs et al., *Science* **275**, 1777 (1997).

310. P. J. Crutzen and M. O. Andreae, *Science* **250**, 1669 (1990).

311. M. Z. Jacobson, *Nature* **409**, 695 (2001). M. O. Andreae, *Nature* **409**, 671 (2001).

312. M. Sato et al., *Proc. Natl. Acad. Sci.* **100**, 6319 (2003).

313. J. M. Prospero and P. J. Lamb, *Science* **302**, 1024 (2003).

314. N. P. Gillett and D. W. J. Thompson, *Science* **302**, 273 (2003). D. J. Karoly, *Science* **302**, 236 (2003).

315. I. A. Janssens et al., *Science* **300**, 1538 (2003).

316. T. M. L. Wigley and S. C. B. Raper, *Nature* **357**, 293 (1992).

317. G. R. Shaver et al., *BioScience* **42**(6), 433 (1992).

318. M. C. Mack et al., *Nature* **431**, 440 (2004). W. M. Loya and P. Grogan, *Nature* **431**, 406 (2004).

319. Y. Malhi, *Phil. Trans. R. Soc. Lond. A* **360**, 2925 (2002).

320. See the articles in T. M. L. Wigley and D. S. Schmiel, eds., *The Carbon Cycle* (Cambridge and New York: Cambridge University Press, 2000). Of particular interest in assessing the uncertainties in carbon storage are Chs. 3 through 11.

321. V. Ramaswamy et al., Ch. 6 in J. T. Houghton et al., eds., *Climate Change 2001: The Scientific Basis* (Cambridge and New York: Cambridge University Press, 2001)

322. R. A. Kerr, *Science* **292**, 192 (2001).

323. A. G. Patt and D. P. Schrag, *Clim. Change* **61**, 17 (2003).

324. S. C. Zehr, *Public Understand. Sci.* **9**, 85 (2000).

325. R. J. Bord et al., *Public Understand. Sci.* **9**, 205 (2000).

326. J. Giles, *Nature* **418**, 476 (2002).

327. M. E. Mann, *Science* **297**, 1481 (2002).

328. M. E. Mann and P. D. Jones, *Geophys. Res. Lett.* **30**, 1820 (2003).

329. H. von Storch et al., *Science* **306**, 679 (2004). T. J. Osborn and K. R. Briffa, *Science* **306**, 621 (2004).

330. R. B. Alley, *Phil. Trans. R. Soc. Lond. A* **361**, 1831 (2003).

331. R. S. J. Tol, *Clim. Change* **56**, 265 (2003).

332. R. B. Govindan et al., *Phys. Rev. Lett.* **89**, 028501 (2002).

333. D. P. C. Peters et al., *Proc. Natl. Acad. Sci.* **101**, 15130 (2004).

334. K. Visser et al., *Nature* **421**, 152 (2003). R. B. Dunbar, *Nature* **421**, 121 (2003).

335. W. S. Broecker, *Science* **300**, 1519 (2003). See also R. B. Alley, *Sci. Am.* **291**(5), 62 (2004), W. Steffen et al.,

Environment **46**(3), 8 (2004), and V. Morell, *Natl. Geog.* **206**(3), 56 (2004).

336. M. V. Shabalova and A. F. V. Van Engelen, *Clim. Change* **58**, 219 (2003).

337. J. Luterbacher et al., *Science* **303**, 1499 (2004).

338. J. M. Murphy et al., *Nature* **430**, 768 (2004). T. F. Stocker, *Nature* **430**, 737 (2004). See also D. Osumi-Sutherland, Nature Science Update, 11 August 2004.

339. R. Knutti et al., *Nature* **416**, 719 (2002).

340. P. A. Stott and J. A. Kettleborough, *Nature* **416**, 723 (2002).

341. F. W. Zwiers, *Nature* **416**, 690 (2002).

342. S. Hitz and J. Smith, *Global Environmental Change* **14**, 201 (2004).

343. T. P. Bennett, in M. C. MacCracken and F. M. Luther, eds., *Detecting the Climatic Effects of Increasing Carbon Dioxide* (Washington, DC: GPO, 1986), DOE/ER-0235, 149.

344. W. W. Kellogg, in J. Gribben, ed., *Climate Change* (New York: Cambridge University Press, 1978), 124. See also M. E. Schlesinger and J. F. B. Mitchell, in M. C. MacCracken and F. M. Luther, eds., *Projecting the Climatic Effects of Increasing Carbon Dioxide* (Washington, DC: GPO, 1986), DOE/ER-0237, 81.

345. T. R. Karl et al., *Geophys. Res. Lett.* **18**, 2253 (1992).

346. U. Cubasch and R. D. Cess, in J. T. Houghton et al., eds., *Climate Change, The IPCC Scientific Assessment* (Cambridge: Cambridge University Press, 1990), 69; J. F. B. Mitchell et al., in J. T. Houghton et al., eds., *Climate Change, The IPCC Scientific Assessment* (Cambridge: Cambridge University Press, 1990), 131.

347. R. J. Charlson et al., *Nature* **340**, 438 (1989).

348. R. J. Stouffer et al., *Nature* **342**, 660 (1989).

349. H. S. Sahsamanoglou and T. J. Makrogiannis, *Theor. Appl. Climatol.* **45**, 183 (1992); T. J. Makrogiannis and H. S. Sahsamanoglou, *Theor. Appl. Climatol.* **45**, 193 (1992).

350. A. Henderson-Sellers and K. McGuffie, *Nature* **340**, 436 (1989); S. J. Ghan et al., *Nature* **340**, 438 (1989).

351. P. H. Whelton and A. B. Pittock, in L. Rosen and R. Glasser, eds., *Climate Change and Energy Policy* (New York: American Institute of Physics, 1992), 109.

352. R. Revelle, *Sci. Am.* **247**(2), 35 (1982); B. Moore, III and B. Bolin, *Oceanus* **29**(4), 9 (1986). S. Manabe and R. J. Stouffer, Ref. 259.

353. J. Hansen, et al., in M. E. Schlesinger, ed., *Greenhouse−Gas−Induced Climatic Change: A Critical Appraisal of Simulations and Observations* (Amsterdam: Elsevier, 1991), 211.

354. W. K. Stevens, *NYT,* 5 March 1990.

355. J. A. Lowe et al., *Climate Dyn.* **18**, 179 (2001). R. E. Peterson and T. E. Warner, in R. Bras, ed., *The World at Risk: Natural Hazards and Climate Change* (New York: American Institute of Physics, 1993), 34.

356. T. R. Knutson and R. E. Tuleya, *J. Climate* **17**, 3477 (2004). A. C. Revkin, *NYT*, 30 September 2004.

357. K. A. Emanuel, in R. Bras, ed., *The World at Risk: Natural Hazards and Climate Change* (New York: American Institute of Physics, 1993), 25.

358. E. Williams and S. Heckman, in R. Bras, ed., *The World at Risk: Natural Hazards and Climate Change* (New York: American Institute of Physics, 1993), 77.

359. C. Price and D. Rind, in R. Bras, ed., *The World at Risk: Natural Hazards and Climate Change* (New York: American Institute of Physics, 1993), 68.

360. L. M. Thompson, *Science* **188,** 535 (1975). D. Pimentel and J. Krummel, *Ecologist* **7,** 254 (1977). R. A. Bryson, *Science* **184,** 753 (1974). S. W. Matthews, *Natl. Geog.* **150,** 576 (1976).

361. P. E. Waggoner, *Am. Sci.* **72,** 179 (1984).

362. H. Weiss and R. S. Bradley, *Science* **291**, 609 (2001). See also a popular account by L. W. Milbrath, *Futurist* **28**(3), 26 (1994).

363. P. B. deMenocal, *Science* **292**, 667 (2001). See also W. K. Stevens, *NYT*, 2 April 1991.

364. D. H. Sandweiss et al., *Science* **283**, 499 (1999).

365. H. M. Cullen et al., *Geology* **28**, 379 (2000). A. S. Issar, *Am. Sci.* **83**, 350 (1995).

366. V. J. Polyak and Y. Asmerom, *Science* **294**, 148 (2001).

367. R. A. Bryson, *Natl. Hist.* **89**(6), 65 (1980).

368. S. D. Schubert et al., *Science* **303**, 1855 (2004).

369. D. E. Stahle et al., *Science* **280**, 564 (1998). For more information, see S. Stine, *Nature* **369**, 546 (1994) as well as F. A. Street-Perrrott, *Nature* **369**, 519 (1994), and also see E. R. Cook et al., *Science* **306**, 1015 (2004).

370. D. R. Easterling et al., *Science* **289**, 2068 (2000).

371. Board on Natural Disasters, *Science* **284**, 1943 (1999).

372. S. Fankhauser, eds.,Ch. 19 in J. J. McCarthy et al., eds., *Climate Change 2001: Impacts, Adaptation, and Vulnerability* (Cambridge and New York: Cambridge University Press, 2001)

373. K. Jacobs et al., Ch. 14 in J. Mellilo et al., eds., (National Assessment Synthesis Team) *Climate Change Impacts on the United States: The Potential Consequences of Climate*

Variability and Change (Cambridge and New York: Cambridge University Press, 2001).

374. C. J. Vörösmarty et al., *Science* **289**, 284 (2000).

375. R. Aktar et al., in A. McMichael and A. Githenko, eds., Ch. 10 in J. J. McCarthy et al., eds., *Climate Change 2001: Impacts, Adaptation, and Vulnerability* (Cambridge and New York: Cambridge University Press, 2001).

376. L. M. Carter et al., Ch. 11 in J. Mellilo et al., eds., (National Assessment Synthesis Team) *Climate Change Impacts on the United States: The Potential Consequences of Climate Variability and Change* (Cambridge and New York: Cambridge University Press, 2001).

377. K. Broad and S. Agrawala, *Science* **289**, 1693 (2000).

378. Committee on Climate, Ecosystems, Infectious Diseases, and Human Health, Board on Atmospheric Sciences and Climate (D. Burke et al.), National Research Council., *Under the Weather: Climate, Ecosystems, and Infectious Disease* (Washington, DC: National Academy Press, 2002).

379. P. R. Epstein, *Sci. Am.* **236**(2), 50 (2000).

380. C. D. Harvell et al., *Science* **296**, 2158 (2002).

381. J. A. Patz et al., Ch. 15 in J. Mellilo et al., eds., (National Assessment Synthesis Team) *Climate Change Impacts on the United States: The Potential Consequences of Climate Variability and Change* (Cambridge and New York: Cambridge University Press, 2001). S. M. Bernard et al., *Environ. Health Perspect.* **109**(Suppl 2), 199 (2001).

382. P. Martens, *Am. Sci.* **87**, 534 (1999). See also L. S. Kalkstein and J. S. Greene, *Environ. Health Perspect.* **105**, 84 (1997).

383. D. L. Davis et al., (Working Group on Public Health and Fossil-Fuel Combustion), *The Lancet* **350**, 1341 (1997).

384. K. J. Linthicum et al., *Science* **285**, 397 (1999). P. R. Epstein, *Science* **285**, 347 (1999).

385. R. R. Colwell, *Science* **274**, 2025 (1996). S. Hales et al., *Environ. Health Perspect.* **107**, 99 (1999). X. Rodó et al., *Proc. Natl. Acad. Sci.* **99**, 12901 (2002). G. Taubes, *Science* **278**, 1004 (1997).

386. C. D. Harvell et al., *Science* **285**, 1505 (1999).

387. D. J. Rogers et al., *Nature* **415**,710 (2002).

388. D. J. Rogers and S. E. Randolph, *Science* **289**, 1763 (2000). C. Dye and P. Reiter, *Science* **289**, 1697 (2000). B. Greenwood and T. Mutabingwa, *Nature* **415**, 670 (2002).

389. D. J. Gubler et al., *Environ. Health Perspect.* **109**(Suppl 2):223 (2001).
390. K. G. Kuhn et al., *Proc. Natl. Acad. Sci.* **100**, 9997 (2003).
391. J. Small et al., *Proc. Natl. Acad. Sci.* **100**, 15341 (2003). C. Thomas, *Nature* **427**, 690 (2004).
392. S. I. Hay et al., *Nature* **415**, 905 (2002). J. A. Patz et al., *Nature* **420**, 627 (2002) and S. I. Hay et al., *Nature* **420**, 628 (2002).
393. S. Hales et al., *Environ. Health Perspect.* **107**, 99 (1999).
394. L. K. Altman and G. Kolata, *NYT*, 6 January 2002.
395. L. Berkowitz, *Psych. Bull.* **95**, 410 (1984).
396. C. A. Anderson, *Currrent Directions Psych. Sci.* **10**, 33 (2001).
397. W. Behringer, *Climatic Change* **43** 335 (1999).
398. G.-R. Walther et al., *Nature* **416**, 389 (2002).
399. J. Harte and R. Shaw, *Science* **267**, 876 (1995). C. K. Yoon, *NYT*, 21 June 1994.
400. K. Krajick, *Science* **303**, 1600 (2004).
401. C. D. Thomas et al., *Nature* **427**, 145 (2004). J. A. Pounds and R. Puschendorf, *Nature* **427**,107 (2004).
402. O. E. Sala et al., *Science* **287**, 1770 (2000).
403. J. A. Morgan et al., *Ecol. Appl.* **14**, 208 (2004).
404. M. B. Davis and R. G. Shaw, *Science* **292**, 673 (2001).
405. C. P. Osborne et al., *Proc. Natl. Acad. Sci.* **101**, 10360 (2004).
406. J. M. Melillo et al., *Nature* **363**, 234 (1993).
407. W. C. Oechel et al., *Nature* **406**, 978 (2000).
408. B. A. Hungate et al., *Science* **304**, 1291 (2004).
409. E. S. Zavaleta et al., *Proc. Natl. Acad. Sci.* **100**, 7650 (2003). The conclusion that a single-factor analysis is appropriate is not supported by the same group's paper, M. R. Shaw et al., *Science* **298**, 1987 (2002), which shows that single-factor analysis is inadequate. See also J. A. Morgan, *Science* **298**, 1903 (2002).
410. G. Hoch et al., *Plant, Cell and Environment* **26**, 1067 (2003).
411. R. Oren et al., *Nature* **411**, 469 (2001).
412. R. D. Alward et al., *Science* **283**, 229 (1998).
413. J. M. Melillo, *Science* **283**, 183 (1999).
414. J. Peñuelas and I. Filella, *Science* **294**, 793 (2001).
415. A. H. Fitter and R. S. R. Fitter, *Science* **296**, 1689 (2002).
416. D. Primack et al., *Am. J. Bot.* **91**, 1260 (2004).
417. A. R. Aston, *J. Hydrol.* **67**, 273 (1984); V. C. LaMarche et al., *Science* **225**, 1019 (1984); and S. B. Idso and J. Brazel, *Nature* **312,** 51 (1984).
418. H. A. Mooney et al., *BioScience* **41**, 96 (1991).
419. W. J. Arp et al., in J. Fozema et al., eds., *CO₂ and Biosphere* (Dodrecht: Kluwer Academic, 1993), 133.
420. R. L. Graham et al., *BioScience* **40**, 575 (1990).
421. W. K. Stevens, *NYT,* 18 September 1990.
422. P. E. Waggoner, in *Changing Climate: Report of the Carbon Dioxide Assessment Committee* (Washington, DC: National Academy of Sciences Press, 1983), 383 (The Changing Climate Carbon Dioxide Assessment team was W. A. Nierenberg et al.); N. J. Rosenberg, (p. 324) in response to C. F. Cooper, (p. 297) in W. C. Clark, ed., *Carbon Dioxide Review: 1982* (New York: Oxford University Press, 1982); W. C. Clark, *Clim. Change* **7**, 5 (1985); and J. M. Callaway and J. W. Currie, in M. R. White, ed., *Characterization of Information Requirements for Studies of Carbon Dioxide Effects: Water Resources, Agriculture, Fisheries, Forests, and Human Health* (Washington, DC: GPO, 1986), DOE/ER-0236, 23.
423. C. Körner and J. A. Arnone, *Science* **257**, 1672 (1992); M. W. Browne,*NYT,* 18 September 1992.
424. S. D. Smith et al., *Nature* **408**, 79 (2000).
425. S. H. Schneider and R. L. Temkin, in J. Gribben, ed., *Climate Change* (New York: Cambridge University Press, 1978), 228.
426. T. H. Jones et al., *Science* **280**, 441 (1998).
427. G. Fischer et al., *Global Agroecological Assessment for Agriculture in the 21st Century* (Laxenburg, Austria: International Institute for Applied Systems Analysis, 2001).
428. G. Daily et al., *Science* **281**, 1291 (1998).
429. S. O. Andersen et al., in S. N. van Rooijen et al., eds., *Methodological and Technological Issues in Technology Transfer* (Cambridge and New York: Cambridge University Press, 2001).
430. B. H. Braswell et al., *Science* **278**, 870 (1997).
431. C. Freeman et al., *Nature* **412**, 785 (2001).
432. W. W. Gregg et al., *Geophys. Res. Lett.* **30**, 1809 (2003).
433. J. Norwine, *Environment* **19**(8), 7 (1977).
434. R. A. Kerr, *Science* **226**, 326 (1984).
435. S. Manabe and R. T. Wetherald, *Science* **232,** 626 (1986).
436. R. O. Lawton et al., *Science* **294**, 584 (2001).
437. A. D. Richardson et al., *J. Climate* **16**, 2093 (2003).
438. T. G. Siccama, *Ecol. Monogr.* **44**, 325 (1974). A. Ananthaswamy, New Scientist Online, 26 March 2003.
439. D. J. Karoly et al., *Science* **302**, 1200 (2003).
440. E. Kalnay and M. Cai, *Nature* **423**, 528 (2003).
441. K. Hayhoe et al., *Proc. Natl. Acad. Sci.* **101**, 12422 (2004).
442. E. Marris, *Nature* **430**, 818 (2004).
443. R. Basu and J. M. Samet, *Environ. Health Perspect.* **110**, 1219 (2002).
444. G. A.Meehl and C. Tebaldi, *Science* **305**, 994 (2004).
445. M. A. McGeehin and M. Mirabelli, *Environ. Health Perspect.* **109**(Suppl 2), 185 (2001).
446. J. M. Lenihan et al., *Ecol. Appl.* **13**, 1667 (2003).
447. Union of Concerned Scientists, *Climate Change in California: Choosing Our Future*, May 2004.
448. E. Asimov, *NYT*, 6 August 2003. J. Gaffney, *The Wine Spectator*, 14 November 2003.
449. B. Mason, Nature Science Update, 4 November 2003.
450. V. Fievez et al., *Animal Feed Sci. Technol.* **104**, 41 (2003).
451. See, for example, M. R. Allen, *Nature* **425**, 242 (2003), Anonymous, *TI*, 3 July 2003, and F. Bruni, *NYT*, 6 August 2003.
452. C. Schär et al., *Nature* **427**, 332 (2004). F. Guterl, *Newsweek*, 12 July 2004.
453. M. Beniston, *Geophys. Res. Lett.* **31**, L02202 (2004).
454. J. H. Christensen and O. B. Christensen, *Nature* **421**, 805 (2002).
455. Anonymous, *The Observer* (UK), 19 September 2004.
456. S. Connor, *TI*, 3 December 2003.
457. G. Seenan, *TG*, 14 February 2004.
458. N. Poisson et al., *Chemosphere—Global Change Sci.* **3**, 353 (2001). Anonymous, BBC News Online, 10 May 2004.
459. N. Carslaw et al., *Atmos. Environ.* **34**, 2827 (2002).
460. G. Greenough et al., *Environ. Health Perspect.* **109**(Suppl 2), 191 (2001).
461. A. Becker and U. Grünewald, *Science* **300**, 1099 (2003).
462. Anonymous, BBC News Online, 22 April 2004.
463. T. Radford, *TG*, 14 June 2004.
464. Q. Schiermeier, *Nature* **428**, 114 (2004). J. M. Gregory et al., *Nature* **428**, 616 (2004).
465. J. C. R. Hunt, *Phil. Trans. R. Soc. Lond. A* **360**, 1531 (2002). N. L. Poff, *Phil. Trans. R. Soc. Lond. A* **360**, 1497 (2002).
466. B. P. Finney et al., *Science* **290**, 795 (2000). L. Busch, *Science* **269**, 1507 (1994).
467. E. Sanford, *Science* **283**, 2095 (1999).
468. M. M. Hefting et al., *J. Environ. Qual.* **32**, 1194 (2003).
469. N. Nosengo, *Nature* **425**, 894 (2003).
470. D. S. Reay, *Nature* **427**, 485 (2004).
471. D. E. Schindler et al., *Science* **277**, 248 (1997).

472. J. A. Thomas et al., *Science* **303**, 1879 (2004). E. Pennisi, *Science* **303**,1747 (2004).

473. W. E. Bradshaw and C. M. Holzapfel, *Proc. Natl. Acad. Sci.* **97**, 14509 (2001).

474. M. S. Warren et al., *Nature* **414**, 65 (2001).

475. M. E. Visser and L. J. M. Holleman, *Proc. Royal Soc. Lond. B* **268**, 289 (2001).

476. B. W. Alto and S. A. Juliano, *J. Med. Entomol.* **38**, 548 (2001).

477. B.-E. Saether, *Science* **288**, 1975 (2000).

478. T. S. Sillett et al., *Science* **288**, 2040 (2000).

479. S. L. Olson and P. J. Hearty, *Proc. Natl. Acad. Sci.* **100**, 12825 (2003). J. Whitfield, Nature Science Update, 14 October 2003.

480. C. Gjerdrum et al., *Proc. Natl. Acad. Sci.* **100**, 9377 (2003).

481. T. S. Sillett et al., *Science* **288**, 2040 (2000).

482. J. Price and P. Glick, *The Birdwatcher's Guide to Global Warming* (Reston and The Plains, VA: National Wildlife Federation and American Bird Conservancy, 2002). L. Line, *Natl. Wildlife* **41**(1), 20 (2003).

483. P. A. Cotton, *Proc. Natl. Acad. Sci.* **100**, 12219 (2003).

484. O. Hüppop and K. Hüppop, *Proc. R. Soc. Lond. B* **270**, 233 (2003). A. J. van Noordwijk, *Nature* **422**, 29 (2003).

485. B.-E. Sæther et al., *Science* **287**, 854 (2000).

486. P. O. Dunn and D. W. Winkler, *Proc. R. Soc. Lond. B* **266**, 2487 (1999). D. W. Winkler et al., *Proc. Natl. Acad. Sci.* **99**, 13595 (2002).

487. L. G. Sorenson et al., *Clim. Change* **40,** 343 (1998).

488. D. W. Thomas et al., *Science* **291**, 2598 (2001). E. Pennisi, *Science* **291**, 2532 (2001).

489. M. E. Visser et al., *Proc. Royal Soc. Lond. B* **265**, 1867 (1998). M. E. Visser et al., *Proc. R. Soc. Lond. B* **270**, 367 (2003).

490. B. Wuethrich, *Science* **287**, 795 (2000).

491. K. Schneider, *NYT,* 13 August 1991.

492. D. W. Schindler et al., *Science* **250**, 967 (1990). J. A. Foley et al., *Nature* **371**, 52 (1994).

493. C. E. Burns et al., *Proc. Natl. Acad. Sci.* **100**, 11474 (2003).

494. N. Leader-Williams, *Phil. Trans. R. Soc. Lond. A* **360**, 1787 (2002).

495. L. Roberts, *Science* **242**, 1010 (1988).

496. J. B. Smith and D. A. Tirpak, *The Potential Effects of Global Climate Change on the United States* (Washington, DC: Congressional Report, U.S. Environmental Protection Agency, June 1989).

497. E. Post and N. Chr. Stenseth, *Ecology* **80**, 1322 (1999):

498. A. Mysterud et al., *Nature* **410**, 1096 (2001).

499. E. Post and M. C. Forchhammer, *Nature* **420**, 168 (2002). E. Post and M. C. Forchhammer, *Proc. Natl. Acad. Sci.* **101**, 9286 (2004).

500. N. C. Stenseth et al., *Proc. Natl. Acad. Sci.* **101**, 6056 (2004).

501. D. Réale et al., *Proc. R. Soc. Lond. B* **270**, 591 (2003).

502. J. A. Pounds et al., *Nature* **398**, 611 (1999).

503. E. A. Beever et al., *J. Mammalogy* **84**, 37 (2003). See also E. A. Beever, *Sierra Nature Notes* **3**, December 2003, URL http://www.yosemite.org/naturenotes/Pika1.htm

504. K. Reid and J. P. Croxall, *Proc. Royal Soc. B* **268**, 377 (2001).

505. O. J. Schmitz et al., *BioScience* **53**, 1199 (2003).

506. M. B. Bush et al., *Science* **303**, 827 (2004).

507. A. T. Peterson et al., *Nature* **416**, 626 (2002).

508. A. D. Barnosky et al., *Proc. Natl. Acad. Sci.* **101**, 9297 (2004).

509. D. R. Schiel et al., *Ecology* **85**, 1833 (2004).

510. D. Arnold, *BG*, 2 July 2002.

511. J. J. Stachowicz et al., *Proc. Natl. Acad. Sci.* **99**, 15497 (2002).

512. A. J. Richardson and D. S. Schoeman, *Science* **305**, 1609 (2004). E. Stokstad, *Science* **305**, 1548 (2004).

513. M. Edwards and A. J. Richardson, *Nature* **430**, 881 (2004).

514. J. L. Sarmiento et al., *Global Biogeochem. Cycles* **18**, GB3003 (2004).

515. G. Beaugrand et al., *Nature* **426**, 661 (2003). Q. Schiermeier, *Nature* **428**, 4 (2004).

516. K. Johnson, *NYT*, 9 November 2002. J. Donn, AP dispatch, 21 August 2004.

517. C. M. Wapnick et al., *Ecology Lett.* **7**, 354 (2004).

518. R. B. Aronson et al., *Ecological Monographs* **72**, 233 (2002).

519. J. M. Pandol et al., *Science* **301**, 955 (2003).

520. A. C. Baker et al., *Nature* **430**, 741 (2004).

521. T. A. Gardner et al., *Science* **301**, 958 (2003).

522. W. F. Precht and R. B. Aronson, *Front. Ecol. Environ.* **2**, 307 (2004;). B. Mason, Nature Science Update, 4 November 2003.

523. D. R. Bellwood et al., *Nature* **429**, 827 (2004). See also V. H. Garrison et al., *BioScience* **53**, 469 (2003).

524. P. Verburg et al., *Science* **301**, 505 (2003). D. A. Livingstone, *Science* **301**, 468 (2003).

525. C. M. O'Reilly et al., *Nature* **424**, 766 (2003). D. Verschuren, *Nature* **424**, 731 (2003).

526. J. Reilly et al., *Clim. Change* **57**, 43 (2003).

527. D. B. Lobell and G. P. Asner, *Science* **299**, 1032 (2003). E. Stokstad, *Science* **299**, 997 (2003).

528. P. G. Jones and P. K.Thornton, *Global Environ. Change* **13**, 51 (2003).

529. R. Jansson, *Proc. R. Soc. Lond. B* **270**, 583 (2003).

530. S. H. Millspaugh et al., *Geology* **28**, 211 (2000).

531. P. Wilf and C. C. Labandeira, *Science* **284**, 2153 (1999).

532. R. G. Barry, in M. C. MacCracken and F. M. Luther, eds., *Detecting the Climatic Effects of Increasing Carbon Dioxide* (Washington, DC: Dept. of Energy, DOE/ER-0235, 1986), 109.

533. C. Rosenzweig, *Clim. Change* **7**, 367 (1985).

534. C. Rosenzweig and M. L. Parry, *Nature* **367**, 133 (1994). J. Reilly, *Nature* **367**, 118 (1994).

535. W. L. Chameides et al., *Science* **264**, 74 (1994).

536. J. E. Newman, *Biometeorol.* **7,** 128 (1980). See also W. L. Dicker et al., in M. R. White, ed., *Characterization of Information Requirements for Studies of Carbon Dioxide Effects: Water Resources, Agriculture, Fisheries, Forests, and Human Health* (Washington, DC: GPO, 1986), DOE/ER-0236, 69.

537. R. Lewin, *Science* **228,** 165 (1985). K. J. Willis, *Endeavour* **20**, 110 (1996) gives a decription of how plants found refuges in the south during previous glaciations.

538. K. Powell, Nature Science Update, 6 January 2003. M. N. Jensen, *Science* **299**, 38 (2003).

539. P. C. Tzedakis et al., *Science* **297**, 2044 (2002). P. Taberlet and R. Cheddadi, *Science* **297**, 2009 (2002). K. C. Rowe et al., *Proc. Natl. Acad. Sci.* **101**, 10355 (2004).

540. R. L. Peters and J. D. S. Darling, *BioScience* **35**, 707 (1985). L. Roberts, *Science* **243**, 735 (1989).

541. W. F. Laurance et al., *Nature* **428**, 171 (2004). T. Radford, *TG*, 11 March 2004.

542. O. L. Phillips et al., *Nature* **418**, 770 (2002).

543. O. L. Phillips and A. H. Gentry, *Science* **263**, 954 (1994).

544. T. L. Root and S. H. Schneider, *Cons. Biol.* **7**, 256 (1993).

545. S. A. Elias, *BioScience* **41**, 552 (1991).

546. R. Lewin, *Science* **244**, 527 (1989).

547. J. Halfpenny and J. Clark, *BioScience* **38**, 399 (1988).

548. D. Roemmich, *Science* **257**, 373 (1992).

549. J. G. Titus and V. K. Narayanan, *The Probability of Sea Level Rise*, Environmental Protection Planning, and Evaluation Agency, United States

Office of Policy, EPA 230-R-95-008, September 1995.

550. B. C. Douglas and W. R. Peltier, *Phys. Today* **55**(3), 35 (2002).

551. L. Miller and B. C. Douglas, *Nature* **428**, 406 (2004). R. Warrick and J. Oerlemans, in J. T. Houghton et al., eds., *Climate Change, The IPCC Scientific Assessment* (Cambridge: Cambridge University Press, 1990), 257; A. R. Solow, in R. Bras, ed., *The World at Risk: Natural Hazards and Climate Change* (New York: American Institute of Physics, 1993), 57. P. N. Spotts, *CSM*, 25 March 2004.

552. J. Oerlemans and J. P. F. Fortuin, *Science* **258**, 115 (1992).

553. S. Pelto, in R. Bras, ed., *The World at Risk: Natural Hazards and Climate Change* (New York: American Institute of Physics, 1993), 61.

554. L. G. Thompson et al., *Science* **298**, 589 (2002). T. Mölg et al., *J. Geophys. Res.* **108**, D4731 (2003). T. Mölg and D. R. Hardy, *J. Geophys. Res.* **109**, D16104 (2004). A. C. Revkin, *NYT*, 23 March 2004.

555. L. G. Thompson et al., *Clim. Change* **59**, 137 (2003).

556. J. Oerlemans, *Science* **264**, 243 (1994). The rate of thinning corresponds to a warming of 0.66 °C since about 1890, consistent with other estimates of global warming in the twentieth century. M. F. Meier et al., *Clim. Change* **59**, 123 (2003). D. Glick, *Natl. Geog.* **206**(3), 12 (2004).

557. E. Berthier et al., *Geophys. Res. Lett.* **31**, L17401 (2004).

558. G. W. K. Moore et al., *Nature* **420**, 401 (2002). J. Graham, *CT*, 17 July 2003. U. Lee McFarling, *LAT*, 12 October 2003.

559. M. H. P. Hall and D. B. Fagre, *BioScience* **53**, 131 (2003). U. L. McFarling, *LAT*, 18 November 2002.

560. G. Hoffmann, *Science* **301**, 776 (2003).

561. S. G. Evans and J. J. Clague, in R. Bras, ed., *The World at Risk: Natural Hazards and Climate Change* (New York: American Institute of Physics, 1993), 69. U. L. McFarling, *LAT*, 25 September 2002 and B. I. Konviser, *BG*, 16 July 2002.

562. K. M. Cuffey and S. J. Marshall, *Nature* **404**, 591 (2000). C. S. Hvidberg, *Nature* **404**, 551 (2000).

563. E. Rignot et al., *Science* **302**, 434 (2003).

564. R. M. DeConto and D. Pollard, *Nature* **421**, 245 (2003). P. Barrett, *Nature* **421**, 221 (2003).

565. R. H. Gammon et al., in J. R. Trabalka, ed., *Atmospheric Carbon Dioxide and the Global Carbon Cycle* (Washington, DC: GPO, 1986), DOE/ER-0239, 25.

566. R. Thomas et al., *Science* **306**, 255 (2004). E. Rignot et al., *Geophys. Res.*

Lett. **31**, L18401 (2004). T. A. Scambos et al., *Geophys. Res. Lett.* **31**, L18402 (2004).

567. M. A. J. Curran et al., *Science* **302**, 1203 (2003). E. W. Wolff, *Science* **302**, 1164 (2003).

568. D. G. Vaughan et al., *Clim. Change* **60**, 243 (2003).

569. K. R. Arrigo and G. L. van Dijken, *Geophys. Res. Lett.* **30**, 1836 (2003).

570. H. De Angelis and P. Skvarca, *Science* **299**, 1560 (2003). T. Lincoln, *Nature* **431**, 519 (2004).

571. A. Shepherd et al., *Science* **302**, 856 (2003).

572. C. R. Bentley, *Science* **275**, 1077 (1997).

573. S. L. Kanfoush et al., *Science* **288**, 1815 (2000).

574. J. L. Bamber et al., *Science* **287**, 1248 (2000).

575. D. Schneider, *Sci. Am.* **276**(3), 112 (1997).

576. B. U. Hag et al., *Science* **235**, 1156 (1987). R. A. Kerr, *Science* **235**, 1141 (1987).

577. L. Polyak et al., *Nature* **410**, 453 (2001). R. Spielhagen, *Nature* **410**, 427 (2001).

578. E. A. Colhoun et al., *Nature* **358**, 316 (1992).

579. J. O. Stone et al., *Science* **299**, 99 (2003). R. P. Ackert, Jr., *Science* **299**, 57 (2003).

580. D. A. Robinson and G. Kukla, *J. Clim. Appl. Met.* **23**, 1626 (1984). J. C. Knox, *Nature* **361**, 430 (1993).

581. J. A. Church et al., Ch. 11 in J. T. Houghton et al., eds., *Climate Change 2001: The Scientific Basis* (Cambridge and New York: Cambridge University Press, 2001).

582. J. R. De Wolde et al., *Climate Dyn.* **11**, 2881 (1995). J. R. De Wolde, P. Huybrechts, J. Oerlemans, and R. S. W. van de Wal, *Tellus* **49A**, 486 (1997).

583. J. A. Church et al., *J. Clim.* **4**, 438 (1991).

584. D. R. Jackett et al., *J. Clim.* **13**, 1384 (2000).

585. C. Cabanes et al., *Science* **294**, 840 (2001). J. A. Church, *Science* **294**, 802 (2001).

586. J. E. Hansen, *Nature* **313**, 349 (1985).

587. W. R. Peltier and A. M. Tushingham, *Science* **244**, 806 (1989).

588. U. Mikolajewicz et al., *Nature* **345**, 589 (1990).

589. D. K. Jacobs and D. L. Sahagian, *Nature* **361**, 710 (1993); D. L. Sahagian et al., *Nature* **367**, 54 (1994).

590. W. S. Newman and R. W. Fairbridge, *Nature* **320**, 319 (1986).

591. A. Henderson-Sellers and K. McGuffie, *New Scientist* **110**(1512), 24 (1986).

592. K. Zhang et al., *Clim. Change* **64**, 41 (2004). C. Day, *Phys. Today* **57**(2), 24 (2004).

593. U.S. Environmental Protection Agency, URL http://www. epa. gov/globalwarming/climate/future/index. html

594. O. H. Pilkey and J. A. G.Cooper, *Science* **303**, 1781 (2004).

595. S. H. Schneider and R. S. Chen, *Annu. Rev. Energy* **5**, 107 (1980).

596. J. Gilbert and P. Vellinga, in F. M. Bernthal, ed., *Climate Change: the IPCC Response Strategies* (Washington, D.C. & Covela, CA: Island Press, 1991), 148. J. G.Titus et al., *Coastal Manage.* **19**, 171 (1991).

597. J. P. Donnelly and M. D. Bertness, *Proc. Natl. Acad. Sci.* **98**, 14218 (2001).

598. Anonymous, *NYT*, 20 December 1989.

599. R. C. Paddock, *LAT*, 4 October 2002. C. J. Hanley, AP dispatch, 30 May 2004.

600. C. Arthur, *The New Zealand Herald*, 1 November 2004.

601. T. Sterling, *LAT*, 23 February 2003.

602. A. Jha, *TG*, 12 June 2003.

603. G. Lean, *TI*, 7 December 2003.

604. J. Barnett and W. N. Adger, *Clim. Change* **61**, 321 (2003).

605. W. K. Nuttle, in R. Bras, ed., *The World at Risk: Natural Hazards and Climate Change* (New York: American Institute of Physics, 1993), 43. C. A. Senior et al., *Phil. Trans. R. Soc. Lond. A* **360**, 1301 (2002).

606. R. Potts, *Science* **273**, 922 (1996).

607. K. Caldiera and J. F. Kasting, *Nature* **366**, 251 (1993).

608. N. Myers, *BioScience* **43**, 752 (1993).

609. W. D. Nordhaus, *Am. Sci.* **82**, 45 (1994).

610. R. A. Bryson and B. M. Goodman, *Science* **207**, 1041 (1980).

611. J. G. Titus, in S. K. Majumdar et al., eds., *Global Climate Change: Implications, Challenges, and Mitigation Measures* (Philadelphia: Pennsylvania Academy of Science, 1992).

612. G. Berz, *Our Planet* (1/2001). United Nations Environment Programme, United Nations Environment Programme press release (2001/11), 5 February 2001.

613. A. Cortese, *NYT*, 18 August 2002. B. J. Feder, *NYT*, 10 July 2003.

614. Anonymous, GreenBiz.com dispatch, 16 April 2003. D. Hakim, *NYT*, 25 July 2004.

615. M. Allen, *Nature* **421**, 891 (2003).

616. C. C. Jaeger et al., *DZ*, 13. February 2003. T. Atkins, Reuters dispatch, 3 March 2004. M. Lubber, *BG*, 4 May 2004. Anonymous, GreenBiz.com dispatch, 11 November 2003.

617. J. P. Bruce, in R. Bras, ed., *The World at Risk: Natural Hazards and Climate Change* (New York: American Institute of Physics, 1993), 3; G. A. Berz, ibid., 217.

618. J. M. Nigg, in R. Bras, ed., *The World at Risk: Natural Hazards and Climate Change* (New York: American Institute of Physics, 1993), 289.

619. W. K. Stevens, *NYT*, 14 November 1989.

620. P. Shabecoff, *NYT*, 26 June 1988.

621. M. L. Wald, *NYT*, 28 August 1988.

622. P. Shabecoff, *NYT*, 19 July 1988.

623. R. M. Bierbaum et al., in B. G. Levi et al., eds., *Global Warming: Physics and Facts* (New York: American Institute of Physics, 1992), 237.

624. W. Fulkerson et al., *Science* **246**, 868 (1989); H. Akbari and A. Rosenfeld, testimony before the California Energy Commission, LBL (25 January 1990); K. Yokobon and S.-X. Xie, in F. M. Bernthal, ed., *Climate Change: the IPCC Response Strategies* (Washington, DC: & Covela, CA: Island Press, 1991), 45; Enquete-Kommission "Vorsorge zum Schutz der Erdatmosphäre" der Deutschen Bundestages, *Schutz der Erde*, V. 2 (Bonn: Economia Verlag and Karlsruhe: Verlag C. F. Müller, 1991); A. H. Rosenfeld and L. Price, in B. G. Levi, D. Hafemeister, and R. Scribner, eds., *Global Warming: Physics and Facts* (New York: American Institute of Physics, 1992), 261; J. A. Laurmann, in L. Rosen and R. Glasser, eds., *Climate Change and Energy Policy* (New York: American Institute of Physics, 1992), 341; A. Grübler et al., *Energy* **18**, 461 (1993).

625. D. F. Spencer, *Annu. Rev. Energy Environ.* **16**, 259 (1991); D. F. Spencer, in L. Rosen and R. Glasser, eds., *Climate Change and Energy Policy* (New York: American Institute of Physics, 1992), 176.

626. J. F. Kasting and J. C. G. Walker, in B. G. Levi, D. Hafemeister, and R. Scribner, eds., *Global Warming: Physics and Facts* (New York: American Institute of Physics, 1992), 175.

627. K. Caldeira et al., *Science* **299**, 2052 (2003).

628. M. I. Hoffert et al., *Science* **298**, 981 (2002).

629. D.C. White, in R. Bras, ed., *The World at Risk: Natural Hazards and Climate Change* (New York: American Institute of Physics, 1993), 224.

630. S. Pacala and R. Socolow, *Science* **305**, 968 (2004).

631. Dept. of Energy, *Annual Energy Outlook 1994—with Projections to 2010* (Washington, DC: GPO, 1994), DOE/EIA-0384(94).

632. R. Lal et al., *Science* **304**, 393 (2004).

633. L. J. Tranvik and M. Jansson, *Nature* **415**, 861 (2002).

634. Q. Schiermeier, *Nature* **423**, 681 (2003).

635. M. Fox, Reuters dispatch, 21 October 2003.

636. N. Nakić enović and S. Messner, *Energy* **18**, 485 (1993).

637. J. Makansi, *Power* **137**(8), 60 (1993).

638. S. Holloway, *Annu. Rev. Energy Environ.* **26**, 145 (2001). E. A. Parson and D. W. Keith, *Science* **282**, 1053 (1998).

639. K. Jopp, *DW*, 29 December 2003.

640. I. García and J. V. M. Zorraquino, *Energy* **27**, 607 (2002).

641. J. David and H. Herzog, talk presented at the Fifth International Conference on Greenhouse Gas Control Technologies, Cairns, Australia, August 13-16 (2000). J. David, *Economic Evaluation of Leading Technology Options for Sequestration of Carbon Dioxide* M.I.T. Masters Thesis, (2000).

642. Interlaboratory Working Group 2000, *Scenarios for a Clean Energy Future* (Oak Ridge, TN: Oak Ridge National Laboratory and Berkeley, CA: Lawrence Berkeley National Laboratory), ORNL/CON-476 and LBNL-44029, November 2000.

643. A. H. Rosenfeld et al., *Phys. Today* **53**(11), 29 (2000).

644. Interlaboratory Working Group, M. A. Brown and M. D. Levine, eds., *Potential Impacts of Energy-Efficient and Low-Carbon Technologies by 2010 and Beyond*, Report number LBNL-40533 or ORNL/CON-444, September 1997.

645. T. Sarbu et al., *Nature* **405**, 165 (2000). W. Leitner, *Nature* **405**, 129 (2000).

646. D. K. Taylor et al., *Annu. Rev. Energy Environ.* **25**: 115 (2000).

647. I. R. Summerfield et al., *Proc. Instn. Mech. Engrs.* **207**, 81 (1993); E. Mot et al., *Air and Waste* **43**, 835 (1993).

648. H. J. Herzog, *Environ. Sci. Technol.* **35**, 148A (2001).

649. A. C. Revkin, *NYT*, 25 June 2003. A. Doyle, Reuters dispatch, 1 July 2003.

650. H. Kroker, *DW*, 12 May 2004.

651. H. Hoag, *Nature* **422**, 7 (2003). U.S. Department of Energy, at URL http://www.fossil.energy.gov/programs/powersystems/futuregen/.

652. S. Cooke, AP dispatch, 29 January 2004.

653. S. von der Weiden, *DW*, 15 April 2003.

654. T. A. Torp and J. Gale, *Energy* **29**, 1361 (2004). J. Gale, *Energy* **29**, 1329 (2004).

655. I. Sethov, Reuters dispatch, 21 November 2003.

656. D. W. Keith and E. A. Parson, *Sci. Am.* **282**(2), 78 (2000).

657. R. W. Klusman, *Energy Conversion and Management* **44**, 1921 (2003).

658. P. N. Spotts, *CSM*, 24 June 2003.

659. S. Elliott et al., *Geophys. Res. Lett.* **28**, 1235 (2001).

660. K. Chang, *NYT*, 17 June 2001.

661. M. Kakizawa et al., *Energy* **26**, 341 (2001).

662. P. M. Haugen and H. Drange, *Nature* **357**, 318 (1992); J. C. Orr, *Nature* **357**, 283 (1992). The original ocean storage idea came from C. Marchetti, *Clim. Change* **1**, 59 (1977).

663. P. G. Brewer et al., *Science* **284**, 943 (1999).

664. P. G.Brewer et al., *Marine Chemistry* **72**, 83 (2000).

665. M. N. Tamburri et al., *Marine Chemistry* **72**, 95 (2000). R. A. Kerr, *Science* **293**, 790 (2001).

666. J. Giles, *Nature* **431**, 115 (2004).

667. V. Gewin, *Nature* **417**, 888 (2002).

668. J. Giles, *Nature* **419**, 6 (2002).

669. B. K. Mignone et al., *Energy* **29**, 1467 (2004).

670. D. Adam, *TG*, 5 September 2003.

671. A. Yamasaki et al., *Energy* **25**, 85 (2000).

672. K. Caldeira and G. H. Rau, *Geophys. Res. Lett.* **27**, 225 (2000).

673. H. Herzog et al., *Clim. Change* **59**, 293 (2003).

674. J. L. Sarmiento and N. Gruber, *Phys. Today* **55**(8), 30 (2002).

675. C. Gough, I. Taylor and S. Shackley, *Energy & Environment* **13**, 883 (2002).

676. K. Caldeira and M. E. Wickett, *Nature* **425**, 365 (2003). Q. Schiermeier, *Nature* 430, 820 (2004).

677. B. A. Seibel and P. J. Walsh, *Science* **294**, 319 (2001). G. H. Rau and K. Caldeira, *Science* **295**, 275 (2002) and B. A. Seibel and P. J. Walsh, *Science* **295**, 275 (2002).

678. J. Arístegui et al., *Science* **298**, 1967 (2002).

679. P. N. Spotts, *CSM*, 9 September 2004.

680. B. I. McNeil et al., *Science* **299**, 235 (2003).

681. C. L. Sabine et al., *Science* **305**, 367 (2004).

682. H. Thomas et al., *Science* **304**, 1005 (2004).

683. K. D. Bidle et al., *Science* **298**, 1980 (2002).

684. R. A. Feely et al., *Science* **305**, 362 (2004). S. Barker et al., *Phil. Trans. R. Soc. Lond. A* **361**, 1977 (2003).

685. K. S. Lackner, *Science* **300**, 1677 (2003).

686. P. Falkowski et al., *Science* **290**, 291 (2000).

687. P. G. Falkowski et al., *Science* **281**, 200 (1998).

688. P. G. Falkowski and C. S. Davis, *Nature* **431**, 131 (2004).

689. A. J. Ridgwell, *Phil. Trans. R. Soc. Lond. A* **360**, 2905 (2002).

690. R. A. Kerr, *Science* **263**, 1090 (1994).

691. R. Röthlisberger et al., *Geophys. Res. Lett.* **31**, L16207 (2004).

692. See, for example, W. J. Broad, *NYT*, 12 November 1996, A. C. Revkin, *NYT*, 12 October 2000, and A. Wellmann, Nature Science Update, 26 January 2004.

693. M. Pahlow and U. Riebesell, *Science* **287**, 831 (2000).

694. J. Wu et al., *Science* **289**, 759 (2000).

695. M. M. Mills et al., *Nature* **429**, 292 (2004).

696. C. B. Field et al., *Science* **281**, 237 (1998).

697. J. Gillon and D. Yakir, *Science* **291**, 2584 (2001). F. I. Woodward, *Science* **291**, 2562 (2001).

698. Q. Schiermeier, *Nature* **421**, 109 (2003).
699. P. W. Boyd et al., *Nature* **407**, 695 (2000).
700. P. W. Boyd et al., *Nature* **428**, 549 (2004).
701. S. M. Turner et al., *Geophys. Res. Lett.* **31**, L14307 (2004).
702. K. O. Buesseler and P. W. Boyd, *Science* **300**, 67 (2003).
703. V. Smetacek et al., *Geophys. Res. Abstracts* **5**, 14444 (2003).
704. A. Tsuda et al., *Science* **300**, 958 (2003).
705. J. K. B. Bishop et al., *Science* **304**, 417 (2004). P. Boyd, *Science* **304**, 396 (2004).
706. K. H. Coale et al., *Science* **304**, 408 (2004).
707. K. O. Buesseler et al., *Science* **304**, 414 (2004). Q. Schiermeier, *Nature* **428**, 788 (2004). R. Dalton, *Nature* **420**, 722 (2002).
708. S. Nadis, *Sci. Am.* **278**(4), 33 (1998). The original idea was due to J. H. Martin and S. E. Fitzwater, *Nature* **331**, 341 (1988) and found support in Z. S. Kolber et al., *Nature* **371**, 145 (1994).
709. S. W. Chisholm et al., *Science* **294**, 309 (2001).
710. Z. S. Kolber et al., *Science* **292**, 2492 (2001). T. Fenchel, *Science* **292**, 2444 (2001).
711. R. Amundson, *Annu. Rev. Earth Planet. Sci.* **29**, 535 (2001).
712. P. Bousquet et al., *Science* **290**, 1342 (2000).
713. R. A. Houghton, *Glob. Chang. Biol.* **1**, 275 (1995).
714. G. P. Robertson et al., *Science* **289**, 1922 (2000).
715. W. H. Schlesinger, *Science* **284**, 2095 (1999).
716. C. D. Keeling et al., in D. H. Peterson, ed., *Aspects of Climate Variability in the Pacific and the Western Americas* (AGU Monograph 55) (Washington DC: American Geophysical Union, 1989), 305.
717. P. P. Tans et al., *Science* **247**, 1431 (1990).
718. G. C. Hurtt et al., *Proc. Natl. Acad. Sci.* **99**, 1389 (2002). R. A. Gill et al., *Nature* **417**, 279 (2002).
719. E. Marshall, *Science* **239**, 973 (1988).
720. D. Schimel et al., in T. M. L. Wigley and D. S. Schmiel, eds., *The Carbon Cycle* (Cambridge and New York: Cambridge University Press, 2000).
721. R. J. Scholes and J. R. Noble, *Science* **294**, 1012 (2001).
722. M. Battle et al., *Science* **287**, 2467 (2000).
723. P. H. Martin et al., *Annu. Rev. Energy Environ.* **26**, 435 (2001).
724. C. S. Potter, *BioScience* **49**, 769 (1999).
725. W. B. Cohen et al., *BioScience* **46**, 836 (1996).
726. C. C. Barford et al., *Science* **294**, 1688 (2001). M. L. Goulden et al., *Science* **271**, 1576 (1996).

727. J. Fang et al., *Science* **292**, 2320 (2001).
728. O. L. Phillips et al., *Science* **282**, 439 (1998).
729. R. A. Houghton et al., *Nature* **403**, 301 (2000).
730. J. Q. Chambers et al., *Nature* **410**, 429 (2001).
731. S. Payette et al., *Geophys. Res. Lett.* **31**, L18208 (2004).
732. M. Sturm et al., *Nature* **411**, 546 (2001).
733. V. A. Barber et al., *Nature* **405**, 668 (2000).
734. R. Valentini et al., *Nature* **404**, 861 (2001).
735. R. A. Betts, *Nature* **408**, 187 (2000).
736. W. C. Oechel et al., *Nature* **361**, 520 (1993).
737. W. C. Oechel et al., *Nature* **406**, 978 (2000).
738. Free Air CO_2 Enrichment website, URL: http://www.face.bnl.gov/.
739. P. B. Reich et al., *Nature* **410**, 809 (2001).
740. J. P. Caspersen et al., *Science* **290**, 1148 (2000).
741. P. M. Cox et al., *Nature* **408**, 184 (2000). This is supported by the new Hadley Centre model described in C. D. Jones et al., *Geophys. Res. Lett.* **30**, 1479 (2003).
742. S. L. LaDeau and J. S. Clark, *Science* **292**, 95 (2001). L. Tangley, *Science* **292**, 36 (2001).
743. E. H. DeLucia et al., *Science* **284**, 1177 (1999).
744. D. A. Clark et al., *Ecol. Appl.* **11**, 371 (2001).
745. R. R. Nemani, et al., *Science* **300**, 1560 (2003).
746. W. Schlesinger and J. Lichter, *Nature* **411**, 466 (2001).
747. R. Lal, *Science* **304**, 1623 (2004).
748. B. A. Hungate et al., *Science* **302**, 1512 (2003).
749. R. B. Jackson et al.,*Nature* **418**, 623 (2002).
750. J. M. Melillo et al., *Science* **298**, 2173 (2002).
751. E. S. Zavaleta et al., *Proc. Natl. Acad. Sci.* **100**, 9892 (2003).
752. J. O. Niles et al., *Phil. Trans. R. Soc. Lond. A* **360**, 1621 (2002).
753. R. Nemani et al., *Geophys. Res. Lett.* **29**, 1468 (2002).
754. R. A. Lovett, *Science* **296**, 1787 (2002). P. N. Spotts, *CSM*, 6 June 2003.
755. D. A. Clark, *Frontiers in Ecology and the Environment* **2**, 73 (2004). N. C. Stenseth et al., *Science* **297**, 1292 (2002).
756. N. Buchmann and E.-D. Schulze, *Global Biogeochem. Cycles* **13**, 751 (1999). E.-D. Schulze et al., *Science* **289**, 2058 (2000).
757. R. J. Norby et al., *Proc. Natl. Acad. Sci.* **101**, 9689 (2004). R. Matamala et al., *Science* **302**, 1385 (2003) find that sequestration potential is overstated.
758. G. Marland, in T. M. L. Wigley and D. S. Schmiel, eds., *The Carbon Cycle*

(Cambridge and New York: Cambridge University Press, 2000).
759. G. Marland, as quoted by W. Booth, *Science* **242**, 19 (1988).
760. F. Myers, as quoted by W. Booth, *Science* **242**, 19 (1988).
761. A. S. Moffat, *Science* **277**, 315 (1997).
762. R. A. Sedjo, *Environment* **31**(1), 15 (1989); W. K. Stevens, *NYT*, 18 July 1989; Anonymous, *NYT*, 12 June 1992
763. D. Kupfer and P. Karimanzera, in F. M. Bernthal, ed., *Climate Change: the IPCC Response Strategies* (Washington, D.C. & Covela, CA: Island Press, 1991), 73.
764. M. U. F. Kirschbaum, *Clim. Change* **58**, 47 (2003).
765. K. R. Richards and C. Stokes, *Clim. Change* **63**, 1 (2004).
766. M. C. Trexler, in B. G. Levi et al., eds., *Global Warming: Physics and Facts* (New York: American Institute of Physics, 1992), 201.
767. M. E. Harmon et al., *Science* **247**, 699 (1990).
768. C. P. Giardina and M. G. Ryan, *Nature* **404**, 858 (2000). J. Grace and M. Raymont, *Nature* **404**, 819 (2000).
769. Y. Luo et al., *Nature* **413**, 622 (2001). L. Rustad, *Nature* **413**, 578 (2001).
770. P. Jarvis and S. Linder, *Nature* **405**, 904 (2000).
771. S. Fan et al., *Science* **282**, 442 (1998). J. Kaiser, *Science* **282**, 386 (1998).
772. R. A. Houghton et al., *Science* **285**, 574 (1999). C. B. Field and I. Y. Fung, *Science* **285**, 544 (1999).
773. D. Schimel et al., *Science* **287**, 2004 (2000).
774. T. R. Christensen et al., *Geophys. Res. Lett.* **30**, 1414 (2003). R. C. Cowen, *CSM*, 11 March 2004.
775. V. Gauci et al., *Proc. Natl. Acad. Sci.* **101**, 12583 (2004).
776. G. B. Bonan, *J. Climate* **14**, 2430 (2001).
777. H. A. C. Denier van der Gon et al., *Proc. Natl. Acad. Sci.* **99**, 12021 (2002).
778. J. P. K Aye et al., *Ecol. Appl.* **14**, 975 (2004).
779. E. Nemitz et al., *Environ. Sci. Technol.* **36**, 3139 (2002).
780. C. S. B. Grimmond et al., *Environ. Poll.* **116**, S243 (2002).
781. D. S. Guertin et al., *BioScience* **47**, 287 (1997).
782. G. J.Collatz et al., *Geophys. Res. Lett.* **27**, 3381 (2000).
783. S. W. Pacala et al., *Science* **292**, 2316 (2001). S. C. Wofsy, *Science* **292**, 2261 (2001).
784. D. S. Schimel et al., *Nature* **414**, 169 (2001).
785. D. A. Clark, *Ecol. Appl.* **12**, 3 (2002).
786. D. Adam, *Nature* **412**, 108 (2001).
787. J. Maddox, *Nature* **346**, 311 (1990).
788. S. S. Penner and J. Haraden, *Energy* **18**, 1087 (1993). The original idea is found in S. S. Penner et al., *Acta Astronautica* **11**, 345 (1984).

789. S. J. Smith et al., *Science* **290**, 1109 (2000).
790. K. Hayhoe et al., *Science* **286**, 905 (1999).
791. J. Reilly et al., *Nature* **401**, 549 (1999).
792. G. A. De Leo et al., *Nature* **413**, 478 (2001).
793. M. G. Morgan, *Science* **289**, 2285 (2000).
794. L. Cifuentes et al., *Science* **293**, 1257 (2001). See also M. L. Bell et al., *Environ. Health Perspect.* **110**, 1163 (2002).
795. J. Palca, *Science* **247**, 520 (1990).
796. P. Sudha and N. H. Ravindranath, Ch. 16, Case Study 26 in B. L. Metzger et al., eds., *Methodological and Technological issues in Technology Transfer* (Cambridge, U.K. and New York: Cambridge University Press, 2001). This is a Special Report of Working Group III of the Intergovernmental Panel on Climate Change. See also Ch. 2 in M. Dutschke, *Sustainable Forestry Investment under the Clean Development Mechanism: The Malaysian Case* (Hamburg, Germany: Hamburgisches Welt-Wirtschafts-Archiv (HWWA), 2002).
797. W. D. Nordhaus, *Science* **258**, 1315 (1992).
798. D. A. Lashof and D. A. Tirpak, *Policy Options for Stabilizing Global Climate* (Washington, DC: Congressional Report, U.S. Environmental Protection Agency, February 1989).
799. P. Baer et al., *Science* **289**, 2287 (2000). J. Lubchenco, *Science* **279**, 491 (1998).
800. D. Koplow and J. Dernbach, *Annu. Rev. Energy Environ.* **26**, 361 (2001).
801. W. D. Nordhaus, *Managing the Global Commons: the Economics of Climate Change* (Cambridge, MA: MIT Press, 1994). W. D. Nordhaus and J. Boyer, *Roll the Dice again: Economic Models of Global Warming* (Cambridge, MA: MIT Press, 2000).
802. B. Lomborg, *The Skeptical Environmentalist* (Cambridge, U.K. and New York: Cambridge University Press, 2001).
803. M. A. Brown et al., *Energy Policy* **29**, 1179 (2001).
804. E. Worrell and L. Price, *Energy Policy* **29**, 1223 (2001).
805. D. S. Reay, *Phil. Trans. R. Soc. Lond. A* **360**, 2947 (2002).
806. A. Doyle, Reuters dispatch, 30 March 2004.
807. D. Frey, *NYT Magazine*, 8 December 2002, 98.
808. J. H. Cushman, Jr., *NYT*, 2 July 2002.
809. M. Sappenfield, *CSM*, 14 June 2004. J. O'Dell, *LAT*, 21 June 2004. D. Hakim, *NYT*, 7 August 2004. D. Hakim, *NYT*, 23 September 2004. M. Bustillo, *LAT*, 25 September 2004.
810. D. Hakim, *NYT*, 11 June 2004.

811. A. C. Revkin, *NYT*, 21 July 2004. B. Egelko, *SFC*, 22 July 2004.
812. K. Johnson, *NYT*, 5 June 2003.
813. P. C. D. Milly et al., *Nature* **415**, 514 (2002). J. A. Lowe et al., *Climate Dyn.* **18**, 179 (2001).
814. R. Schnur, *Nature* **415**, 483 (2002). T. N. Palmer and J. Räisänen, *Nature* **415**, 512 (2002).
815. J. J. Stachowicz et al., *Proc. Natl. Acad. Sci.* **99**, 15497 (2002).
816. C. D. Harvell et al., *Science* **296**, 2158 (2002).
817. S. Abdulla, Nature Science Update, 2 February 2002, http://www.nature.com/nsu/020218/0202 18-5.html.
818. J. G. Titus and C. Richman, *Climate Res.* **18**, 205 (2001).
819. P. N. Schweitzer and R. S. Thompson, U.S. Geological Survey, Open-File Report 96-000.
820. T. Egan, *NYT*, 16 June 2002.
821. R. K. Monson and E. A. Holland, *Annu. Rev. Ecol. Syst.* **32**, 547 (2001).
822. D. Schimel et al., in J. T. Houghton et al., eds., *Climate change 1995. The science of climate change* (Cambridge and New York: Cambridge University Press, 1995), 65.
823. A. Fangmeier and H.-J. Jäger, in R. Guderian, ed., *Handbuch der Umweltveränderungen und Öko-toxikologie, Vol. 2a: Terrestrische Ökosysteme* (Berlin: Springer, 2001).
824. M. Hulme, *Phil. Trans. R. Soc. Lond. A* **361**, 2001 (2003).
825. B. C. O'Neill and M. Oppenheimer, *Science* **296**, 1971 (2002). H. Graßl et al., *Climate Protection Strategies for the 21st Century. Kyoto and beyond* (Berlin: German Advisory Council on Global Change [WBGU], 2003).
826. A. Cortese, *NYT*, 18 August 2002.
827. See, for example, Anonymous, Reuters dispatch, 21 February 2001, J. Yardley, *NYT*, 16 April 2001, B. Stewart, *NYT*, 24 February 2002, W. Booth, *WP*, 3 July 2002. In regions, for New York, I. Peterson, *NYT*, 1 March 2002 and P. Bodo, *NYT*, 28 April 2002; for Georgia, D. Jehl, *NYT*, 27 May 2002; for Florida, M. Grunwald, *WP*, 23 June 2002, ibid., 23 June 2002, ibid., 24 June 2002, ibid., 24 June 2002, ibid., 25 June 2002, ibid., 25 June 2002, and ibid., 26 June 2002; for the Midwest, T. Egan, *NYT*, 3 May 2002 and K. Axtman et al., *CSM*, 1 June 2001; for the West J. Cart, *LAT*, 25 April 2002, J. Wilson, *LAT*, 23 June 2002, and W. Booth, *WP*, 12 August 2002. For the effect on American farmers and ranchers, see M. Janofsky *NYT*, 23 May 2002, P. Henetz, AP dispatch, 18 June 2002, and J. R. Moehringer, *LAT*, 23 June 2002.

828. P. Ball, Nature Science Update, 27 January 2000, URL http://www.nature.com/nsu/000127/0001 27-11.html.
829. See K. Bradsher, *NYT*, 8 May 2002 and *NYT*, 7 May 2002. For a view of what happened to the Aral Sea of Asia in I. Small et al., *Environ. Health Perspect.* **109**, 547 (2001) and B. Keller, *NYT*, 20 December 1988.
830. See also African history given in Ref. 98 and what's happened to Lake Chad in N. Pawelski, CNN dispatch, 27 February 2001.
831. B. G. Hunt and T. I. Elliott, *Climate Dyn.* **20**, 1 (2002).
832. D. A. King, *Science* **303**, 176 (2004).
833. K. Kübler, *Energy & Environment* **13**, 423 (2002).
834. T. P. Tomich et al., *Am. J. Alternative Agriculture* **17**, 125 (2002). L. S. Saunders et al., *Phil. Trans. R. Soc. Lond. A* **360**, 1763 (2002). N. McDowell, *Nature* **420**, 4 (2002).
835. G. Brumfiel, *Nature* **419**, 869 (2002).
836. E.-D. Schulze et al., *Science* **299**, 1669 (2003).
837. P. K. Dutta and R. Radner, *Proc. Natl. Acad. Sci.* **101**, 5174 (2004).
838. A. S. Manne and R. G. Richels, *Nature* **410**, 675 (2001). J. Reilly et al., *Nature* **401**, S49 (1999).
839. K. Hasselmann et al., *Science* **302**, 1923 (2003).
840. A. F. Massardo et al., *Energy* **28**, 607 (2003). R. Walz, *Energy & Environment* **10**, 169 (1999).
841. S. Mydans and A. C. Revkin, *NYT*, 1 October 2004. P. N. Spotts, *CSM*, 7 October 2004.
842. J. Eilperin, *WP*, 3 October 2004. J. Flanigan, *LAT*, 10 October 2004. E. Shogren, *LAT*, 30 July 2003.
843. E. Bard, *Phys. Today* **55**(12), 32 (2002).
844. R. B. Alley et al., *Science* **299**, 2003 (2003). T. R. Karl and K. E. Trenberth, *Science* **302**, 1719 (2003). J. E. Hansen, *Sci. Am.* **290**(3), 68 (2004). B. Hansen et al., *Science* **305**, 953 (2004).
845. H. Gildor and E. Tziperman, *Phil. Trans. R. Soc. Lond. A* **361**, 1935 (2003).
846. P. Schwartz and D. Randall, Report to the Pentagon, October 2003.
847. R. S. Bradley et al., *Science* **302**, 404 (2003).
848. E. Claussen, *Science* **306**, 816 (2004).
849. P. A. Stott et al., *Nature* **432**, 610 (2004).

CHAPTER 18
1. F. J. Shore, *Phys. Teach.* **12**, 327 (1974).
2. S. Glasstone, *Sourcebook on Atomic Energy,* 3rd ed. (New York: Van Nostrand Reinhold Co., 1967).
3. R. F. Post and P. L. Ribe, *Science* **186**, 397 (1974). W. Meier et al., *Energy & Environment* **13**, 647 (2002).

4. See the website of the Contemporary Physics Education Project, URL http://cpepweb.org for more information on quarks, on fusion, and on radioactive decay.

5. R. M. Barnett et al., *The Charm of Strange Quarks* (New York: Springer-Verlag, 2000).

6. G. J. Aubrecht and H. S. Matis, eds., *A Teacher's Guide to the Nuclear Science Wall Chart or, You do not have to be a Nuclear Physicist to Understand Nuclear Science*, Lawrence Berkeley Laboratory, 1998.

CHAPTER 19

1. S. Glasstone, *Sourcebook on Atomic Energy*, 3rd ed. (New York: Van Nostrand Reinhold Co., 1967).

2. S. Glasstone and W. H. Jordan, *Nuclear Power and Its Environmental Effects* (Chicago: American Nuclear Society, 1980).

3. United States Nuclear Regulatory Commission, *Producing Nuclear Fuel* NUREG/BR-0280. United States Nuclear Regulatory Commission, *Gaseous Diffusion* (Fact Sheet) (2000).

4. C. Norman, *Science* **228**, 1407 (1985).

5. C. Norman, *Science* **221**, 730 (1983).

6. Department of Energy, *Commercial Nuclear Fuel from U.S. and Russian Surplus Defense Inventories: Materials, Policies, and Market Effects* (Washington, DC: GPO, 1998), DOE/EIA-0619.

7. S. W. Kidd, *Energy & Environment* **13**, 901, 2002.

8. Atomic Energy Commission, *The Safety of Nuclear Power Reactors and Related Facilities*, WASH-1250 (Washington, DC: AEC, 1973).

9. United States Nuclear Regulatory Commission, *Reactor Safety Study* (Rasmussen report), WASH-1400 (NUREG-75/014), October 1975.

10. D. H. Worledge, *Annu. Rev. Energy Environ.* **17**, 285 (1992).

11. R. P. Hammond, *Am. Sci.* **62**, 146 (1979).

12. C. Norman, *Science* **228**, 31 (1985).

13. Dept. of Energy, *Annual Energy Review 2000* (Washington, DC: GPO, 2001), DOE/EIA-0384(2000).

14. Department of Energy, *Annual Energy Outlook 2001 With Projections to 2020* (Washington, DC: GPO, 2000), DOE/EIA-0383(2001).

15. Department of Energy, *Electric Power Annual 2000* (Washington, DC: GPO, 2001), DOE/EIA-0348(2000).

16. Dept. of Energy, *Nuclear Power Generation and Fuel Cycle Report 1997* (Washington, DC: GPO, 1997), DOE/EIA-0436(97).

17. Energy Information Administration, Department of Energy, *Electric Power Annual, Volume II* (Washington, DC: GPO, 2000) DOE/EIA-0348(99)/2, Table 13.

18. W. Sweet, *IEEE Spectrum* **34**(11), 23 (1997). R. E. Löfstedt, *Environment* **43**(4), 20 (2001).

19. A. Salpukas, *NYT*, 6 March 1999. A. Salpukas, *NYT*, 24 September 1999. D. W. Chen, *NYT*, 29 March 2000.

20. M. L. Wald, *NYT*, 20 December 2000.

21. G. Szegö, *The U.S. Energy Problem*, RANN report, Appendix E (Washington, DC: GPO 1972); also Atomic Energy Commission, Annual report to Congress, 1973 (Washington, DC: GPO, 1974).

22. E. Cowan, *NYT*, 18 July 1976; D. Burnham, *NYT*, 16 November 1975.

23. M. L. Wald, *NYT*, 24 February 1984.

24. M. R. Copulos, *Natl. Rev.* **31**, 156 (1979). A. J. Parisi, *NYT*, 8 April 1979.

25. K. S. Deffeyes and I. D. MacGregor, *Sci. Am.* **242**(1), 66 (1980).

26. M. K. Hubbert, in P. Cloud et al., eds., *Resources and Man* (San Francisco: Freeman, 1969), 157.

27. Dept. of Energy, *Uranium Industry Annual 1991* (Washington, DC: GPO, 1992), DOE/EIA-0478 (91).

28. Economic Council of Canada, *Connections* (Ottawa: Canadian Government Publishing Centre, 1985).

29. G. A. Cowan, *Sci. Am.* **239**(1), 36 (1978).

30. J. M. Herndon, *Proc. Natl. Acad. Sci.* **100**, 3047 (2003).

31. N. A. Dollezhal', *Nucl. Energy* **20**, 385 (1981). Also, Annex 2, Report of the Commission on the Causes of the Accident at the Fourth Unit of the Chernobyl Nuclear Power Plant, 1986. See also B. G. Levi, *Phys. Today* **39**(7), 17 (1986) and R. Wilson, *Nature* **323**, 29 (1986).

32. M. L. Wald, *NYT*, 20 January 1985.

33. Dept. of Energy, *Commercial Nuclear Power 1991* (Washington, DC: GPO, 1991), DOE/EIA-0438(91).

34. E. Marshall, *Science* **219**, 265 (1983).

35. K. Davidson, *SFC*, 24 May 2001.

36. S. Novick, *Environment* **16**(10), 6 (1974); D. Burnham, *NYT*, 9 March 1975.

37. D. B. Myers, *The Nuclear Power Debate* (New York: Praeger, 1977).

38. A. D. Rosser and T. A. Rieck, *Science* **201**, 582 (1978).

39. J. M. Fowler et al., *Environment* **20**(6), 25 (1978); R. L. Goble and C. Hohenemser, *Environment* **21**(3), 31 (1979).

40. M. L. Wald, *NYT*, 31 March 1993.

41. M. L. Wald, *NYT*, 8 December 1985.

42. M. L. Wald, *NYT*, 27 February 1990.

43. M. L. Wald, *NYT*, 26 November 1989; M. L. Wald, *NYT*, 12 July 1992.

44. J. F. Aherne, *Am. Sci.* **81**, 24 (1993).

45. J. J. Taylor, *Science* **244**, 318 (1989). R. K. Lester, *Sci. Am.* **254**(3), 31 (1986).

46. L. Mez and A. Piening, *Energy & Environment* **13**, 161 (2002). Anonymous, BBC News Online, 6 December 2002.

47. M. W. Golay and N. E. Todreas, *Sci. Am.* **262**(11), 82 (1990).

48. C. W. Forsberg and A. M. Weinberg, *Annu. Rev. Energy* **15**, 133 (1990).

49. H. M. Agnew, *Sci. Am.* **224**(13), 55 (1981).

50. E. Marshall, *Science* **224**, 699 (1984). A. C. Kadak et al., MIT-ANP-PR-075, July 2000.

51. Anonymous, *NYT*, 31 August 1989.

52. A. D. Rossin, *Annu. Rev. Energy* **15**, 153 (1990).

53. G. von Randow, *DZ*, 12 March 1993.

54. K. Krüger and J. Cleveland, *Trans. Am. Nucl. Soc.* **60**, 735 (1989).

55. Department of Energy, URL: http://www.eia.doe.gov/oiaf/ieo/nuclear.html.

56. J. Johnson, *Chem. & Eng. News* **79**(36), 29 (2001).

57. W. Sweet, *IEEE Spectrum* **34**(11), 41 (1997).

58. The Royal Society and The Royal Academy of Engineering Nuclear Energy Group (E. Ash et al.), *Nuclear energy—the future climate* (Portsmouth, UK: Holbrooks Printers Ltd., 1999).

59. G. Gugliotta, *WP*, 10 June 2001.

60. United States Nuclear Regulatory Commission, *Next-Generation Reactors* (Fact Sheet) (2001).

61. Office of Nuclear Regulatory Research, United States Nuclear Regulatory Commission, *Nuclear Research Programs To Ensure Public Health and Safety* NUREG/BR-0282.

62. World Resources Institute, *World Resources 1994-1995* (Oxford & New York: Oxford University Press, 1994), Table 21.1.

63. A. M. Weinberg, *Science* **232**, 695 (1986). A. Camplani and A. Zambelli, *Endeavour* **10**, 132 (1986). G. A. Vendryes, *Sci. Am.* **236**(3), 26 (1977).

64. H.-P. Arndt et al., *DZ*, 31 July 1992.

65. J. Hénard, *DZ*, No. 31, 1 August 1997.

66. F. Venneri et al., talk presented at the Uranium Institute's twenty-fourth annual symposium, September 1999.

67. C. Smith, LLNL Press Release, URL http://www.llnl.gov/str/JulAug04/Smith.html.

68. R. Klapisch, *Europhysics News* **31**(6), 26 (2000).

69. M. S. Kazimi, *Am. Sci.* **91**, 408 (2003).

70. C. Rubbia, in E.Arthur et al., eds., *Proceedings of the First International Conference on Accelerator-Driven Transmutation Technologies and Applications, Las Vegas, 25—29 July 1994* (Woodbury, NY: AIP Press, 1995).

C. Rubbia et al., CERN/LHC/97-01, 1997. R. Matthews, *The Sunday Telegraph* (UK), 11 July 2004.

71. D. Normile, *Science* **302**, 379 (2003).
72. H. Bieber, *DZ*, 26 February 1993.
73. M. L. Wald, *NYT*, 22 June 1989.
74. R. Gillette, *Science* **177**, 330 (1972); Atomic Energy Commission, Annual Report to Congress, 1973.
75. E. Marshall, *Science* **215**, 1596 (1982).
76. K. E. Sickafus et al., *Science* **289**, 748 (2000).
77. F. L. Vook et al., *Rev. Mod. Phys.* **47**, Supplement 3, S1 (1975).
78. R. Gillette, *Science* **179**, 360 (1973).
79. M. L. Wald, *NYT*, 2 September 1998.
80. M.L. Wald, *NYT*, 29 July 1997 ibid., 2 August 1997, ibid., 2 October 1997. I Goodwin, *Phys. Today* **50**(9), 54 (1997).
81. J. Johnson, *Chem. & Eng. News* **78**(6), 10 (2000).
82. M. L. Wald, *NYT*, 23 March 1998. M. L. Wald, *NYT*, 29 January 2000. Anonymous, AP dispatch, 6 February 2000.
83. R. E. Schmid, AP dispatch, 21 October 1999.
84. R. Gillette, *Science* **181**, 728 (1973); D. Burnham, *NYT*, 21 October 1976; *NYT*, Anonymous, AP dispatch, 15 April 1979; D. Burnham, *NYT*, 6 May 1979; T. Lash, *Amicus J.* **1**(2), 24 (1979); E. Marshall, *Science* **236**, 1616 (1987); and M. L. Wald, *NYT,* 11 July 1993.
85. D. Burnham, *NYT*, 10 November 1974.
86. R. Gillette, *Science* **177**, 771 (1972); ibid., *Science* **177**, 867 (1972); ibid., *Science* **177**, 970 (1972); *Science* **177**, 1080 (1972).
87. D. Burnham, *NYT*, 25 August 1974.
88. M. L. Wald, *NYT*, 21 October 2000.
89. D. Burnham, *NYT*, 29 December 1974.
90. K. Schneider, *NYT*, 13 July 1990.
91. M. L. Wald, *NYT*, 31 July 1990.
92. Anonymous, *NYT*, 21 June 1993.
93. M. L. Wald, *NYT*, 21 June 1993.
94. L. Ashton, AP dispatch, 11 March 2001.
95. F. Munger, *The Knoxville News-Sentinel*, 29 August 2000.
96. R. M. Anjos et al., *Am. J. Phys.* **69**, 377 (2001).
97. Anonymous, AP dispatch, 22 February 2000.
98. D. Burnham, *NYT*, 25 August 1974.
99. S. R. Weisman, *NYT*, 21 March 1979.
100. A. C. Revkin, *NYT*, 24 February 2000.
101. F. R. McCoy et al., Department of Energy, October 1999.
102. Division of Fuel Cycle Safety and Safeguards, Office of Nuclear Material Safety and Safeguards, United States Nuclear Regulatory Commission, April 2000.

103. IAEA, *Report on the preliminary fact finding mission following the accident at the nuclear fuel processing facility in Tokaimura, Japan* (Vienna, Austria: IAEA, 1999).
104. H. W. French, *NYT*, 1 October 1999.
105. The Nuclear Safety Commission of Japan, 24 December 1999.
106. H. W. French, *NYT*, 16 September 2002.
107. A. Faiola, *WP*, 9 August 2004.
108. R. Downing, *ABJ*, 28 March 2004.
109. U.S. Nuclear Regulatory Commission, NRC Update, September 2002.
110. T. Henry, *TB*, 19 February 2003.
111. J. Mangels and J. Funk, *CPD*, 20 June 2004.
112. J. Mangels and J. Funk, *CPD*, 29 December 2002.
113. K. Lecker, *TB*, 14 July 2002.
114. J. Mangels and J. Funk, *CPD*, 30 August 2002. J. Funk, J. Mangels, and S. Koff, *CPD*, 14 January 2003.
115. J. Mangels, *CPD*, 22 October 2003.
116. J. Mangels and J. Funk, *CPD*, 4 October 2002. M. L. Wald, *NYT*, 4 January 2003. T. Henry, *TB*, 20 February 2003. T. Henry, *TB*, 15 February 2004.
117. J. Funk and J. Mangels, *CPD*, 5 February 2003.
118. J. Funk and J. Mangels, *CPD*, 6 February 2003.
119. U.S. Nuclear Regulatory Commission, NRC News No. 04-053, 4 May 2004. See also T. Henry, *TB*, 5 May 2004.
120. U.S. Nuclear Regulatory Commission, NRC Information Notice 2003-02, January 16, 2003.
121. T. Henry, *TB*, 11 September 2002. T. Henry, *TB*, 10 October 2003. T. Henry, *TB*, 23 October 2003. J. Mangels and J. Funk, *CPD*, 24 October 2003. T. Henry, *TB*, 19 November 2003. M. L. Wald, *NYT*, 9 March 2004. U.S. Nuclear Regulatory Commission, NRC News No. 04-117, 20 September 2004. J. Funk and J. Mangels, *CPD*, 21 September 2004.
122. T. Henry, *TB*, 12 December 2002.
123. J. Mackinnon, *ABJ*, 16 July 2003.
124. Performance, Safety, and Health Associates, Inc., *Safety Culture Evaluation of the Davis-Besse Nuclear Power Station*, 14 April 2003.
125. J. Funk, *CPD*, 11 September 2003. J. Funk and J. Mangels, *CPD*, 2 October 2003. J. Mangels, *CPD*, 30 December 2003. J. Funk, *CPD*, 20 December 2003. A. McFeatters, *TB*, 21 May 2004.
126. J. Funk and J. Mangels, *CPD*, 12 February 2004.
127. Anonymous, *TB*, 17 March 2004.
128. J. Mangels and J. Funk, *CPD*, 17 March 2004.
129. Anonymous, *TB*, 5 April 2004.
130. Anonymous, *TB*, 17 July 2003.
131. J. Funk and J. Mangels, *CPD*, 4 April 2003. T. Henry, *TB*, 22 May 2003.

132. J. Mackinnon, *ABJ*, 12 February 2003. J. Mangels and J. Funk, *CPD*, 14 February 2003. T. Henry, *TB*, 1 August 2003.
133. Anonyymous, *TB*, 12 December 2003. J. Funk, *CPD*, 25 May 2004. J. Funk, *CPD*, 22 June 2004.
134. Anonymous, AP dispatch, 17 November 2002.
135. Anonymous, *TB*, 15 March 2003. Anonymous, *TB*, 3 October 2003. Anonymous, *TB*, 5 September 2004.
136. J. Mangels, *CPD*, 27 June 2003. M. L. Wald, *NYT*, 8 8eptember 2003.
137. Anonymous, Reuters dispatch, 5 May 2002.
138. J. Funk and J. Mangels, *CPD*, 18 April 2003. M. L. Wald, *NYT*, 19 April 2003.
139. J. Mangels, *CPD*, 23 May 2003.
140. M. L. Wald, *NYT*, 1 May 2003. M. L. Wald, *NYT*, 6 June 2003.
141. W. J. Broad, *NYT*, 19 March 2004.
142. D. M. Chapin et al., *Science* **297**, 1997 (2002).
143. H. J. Hebert, AP dispatch, 23 December 2002. T. Henry, *TB*, 26 December 2002.
144. F. J. Shore, *Phys. Teach.* **12**, 327 (1974).
145. G. Gugliotta, *WP*, 5 December 2001. P. Richter, *LAT*, 24 April 2002. D. Oldenburg, *WP*, 13 June 2002. T. Siegfried, *DMN*, 9 December 2002.
146. R. F. Post and P. L. Ribe, *Science* **186**, 397 (1974).
147. W. D. Metz, *Science* **178**, 291 (1972); V. L. Parsegian et al., *Science* **187**, 213 (1975).
148. J. D. Lawson, *Proc. Phys. Soc.* (London) **B70**, 6 (1957).
149. M. M. Waldrop, *Science* **222**, 1002 (1983).
150. K. I. Thomassen, *Annu. Rev. Energy* **9**, 281 (1984).
151. M. W. Browne, *NYT*, 11 December 1993.
152. J. D. Strachan et al., *Phys. Rev. Lett.* **72**, 3526 (1994). (TFTR)
153. R. J. Hawryluk et al., Princeton Plasma Physics Laboratory preprint PPPL-3318 (1998).
154. U. Columbo and U. Farinelli, *Annu. Rev. Energy Environ.* **17**, 123 (1992). The JET first achieved fusion in November 1991.
155. A. Gibson et al. (the JET Team), *Phys. Plasmas* **5**, 1839 (1998).
156. L. M. Ariza, *Sci. Am.* **282**(3), 19 (2000).
157. R. W. Conn et al., *Sci. Am.* **266**(4), 102 (1992).
158. T. A. Friedrich, *DW*, 24 September 2004. D. Clery, *Science* **306**, 26 (2004). See also G. Brumfiel, *Nature* **425**, 887 (2003).
159. W. Sweet, *IEEE Spectrum* **42**(1), 41 (2004). G. Brumfiel, *Nature* **421**, 563 (2003).
160. M. K. Matzen et al., *Plasma Phys. Control. Fusion* **41**, A175 (1999). The

Z-beamlet project, a collaboration of Sandia and LLL, is described in J. A. Caird et al., UCRL-ID-134409 (1999).

161. T. A. Mehlhorn et al., *Plasma Phys. Control. Fusion* **45**, A325 (2003). W. A. Stygar et al., *Phys. Rev.* **E69**, 046403 (2004). G. Brumfiel, *Nature* **422**, 549 (2003). W. Triplett, *Nature* **413**, 338 (2001).

162. R. S. Craxton et al., *Sci. Am.* **255**(2), 68 (1986). W. D. Metz, *Science* **186**, 519 and 1194 (1974).

163. J. P. VanDevender and D. L. Cook, *Science* **232**, 831 (1986).

164. J. P. Holdren, *Annu. Rev. Energy Environ.* **16**, 235 (1991).

165. N. Rostocker et al., *Science* **278**, 1419 (1997).

166. W. D. Metz, *Science* **194**, 307 (1976).

CHAPTER 20

1. R. Jungck, *Brighter than a Thousand Suns* (New York: Harcourt, Brace, 1958).

2. G. Aubrecht and D. Torick, in R. Pintó and S. Surinach, eds., *International Conference Physics Teacher Education Beyond 2000. Selected Contributions* (Elsevier Editions: Paris, 2001), 209. G. Aubrecht and D. Torick, in M. A. Moreira, Ed., *Proceedings of the Seventh InterAmerican Conference in Physics Education* (Porto Alegre, Brasil: IAC, 2000). G. Aubrecht, *AURCO J.* **7**, 4 (2001).

3. U.S. Atomic Energy Commission, *Theoretical Possibilities and Consequences of Major Accidents in Large Nuclear Power Plants*, WASH-740, March 1957.

4. U.S. Atomic Energy Commission, *The Safety of Nuclear Power Reactors and Related Facilities*, WASH-1250 (Washington, DC: AEC, 1973).

5. U.S. Nuclear Regulatory Commission, *Reactor Safety Study* (Rasmussen report), WASH-1400 (NUREG-75/014), October 1975.

6. Anonymous, *Die Deutsche Universitäts-Zeitung* **33**, 530 (1979), describes the Birkhofer report (Reaktor Sicherheits Kommission, SNR-300 Accident Risk Study) to the German Ministry for Research and Technology.

7. D. Burnham, *NYT*, 10 November 1974.

8. United Nations Scientific Committee on the Effects of Atomic Radiation (UNSCEAR), *Sources and Effects of Ionizing Radiation* (New York: United Nations Publ., 2000).

9. S. Glasstone and W. H. Jordan, *Nuclear Power and its Environmental Effects* (LaGrange Park, IL: American Nuclear Society, 1980).

10. A. C. Upton, *Sci. Am.* **246**(2), 41 (1982).

11. A. C. Upton, *Phys. Today* **44**(8), 34 (1991).

12. Office of Civilian Radioactive Waste Management, Department of Energy, DOE/YMP-0403 (June 2000).

13. M. L. Wald, *NYT*, 12 June 2001. M. L. Wald, *NYT*, 14 February 1990, ibid., 19 February 1990, and ibid., 1 March 1990.

14. M. L. Wald, *NYT*, 1 February 1990.

15. W. J. Fisk et al., *Indoor Air Quality Control Techniques: A Critical Review*, LBL-16493 (1983).

16. D. J. Crawford and R. W. Leggett, *Am. Sci.* **68**, 524 (1980).

17. Committee on the Biological Effects of Ionizing Radiations (BEIR V: A. C. Upton et al.), National Academy of Sciences, *Health Effects of Exposure to Low Levels of Ionizing Radiation* (Washington, DC: National Academy Press, 1990). Committee on Health Effects of Exposure to Low Levels of Ionizing Radiations (BEIR VII, Phase I: R. B. Setlow et al.), National Academy of Sciences, *Health Effects of Exposure to Low Levels of Ionizing Radiations: Time for Reassessment?* (Washington, DC: National Academy Press, 1998).

18. Committee on the Biological Effects of Ionizing Radiations (BEIR IV: J. I. Fabrikant et al.), National Academy of Sciences, *Health Risks of Radon and other Internally Deposited Alpha-Emitters* (Washington, DC: National Academy Press, 1988). Committee on Health Risks of Exposure to Radon (BEIR VI: J. M. Samet et al.,), National Academy of Sciences, *Health Effects of Exposure to Radon* (Washington, DC: National Academy Press, 1999).

19. H. Wesch et al., *Radiat. Res.* **152**, S48 (1999); T. Wiethege et al., *Radiat. Res.* **152**, S52 (1999); M. Kreuzer et al., *Radiat. Res.* **152**, S56 (1999); and L. Tomásek and V. Placek, *Radiat. Res.* **152**, S59 (1999).

20. H. W. Tso, *AAPT Announcer* **15**(2), 45 (1985).

21. Anonymous, *NYT*, 16 August 1986.

22. G. Charpak and R. Garwin, *Megawatts and Megatons: A Turning Point in the Nuclear Age?* (New York: A. A. Knopf, 2002).

23. G. Charpak and R. Garwin, *Europhysics News* **33**(1), 14 (2002). See also T. Feder, *Phys. Today* **55**(6), 21 (2002).

24. H. Karagiannis, and D. A. Hagemeyer, *Occupational Radiation Exposure at Commercial Nuclear Power Reactors and Other Facilities 1999* (Washington, DC: GPO, 1999) Division of Regulatory Applications, U.S. Nuclear Regulatory Commission, NUREG-0713, Vol. 21.

25. E. E. Prather and R. R. Harrington, *J. Coll. Sci. Teach.* **31**, 89 (2001). See also E. E. Prather, *An Investigation into what students think and how they learn about ionizing radiation and radioactivity*, Ph. D. Thesis, University of Maine (2000).

26. G. Aubrecht, *AURCO J.* **7**, 4 (2001).

27. R. Millar, *Phys. Educ.* **25**, 338 (1990). R. Millar, *Public Understanding Sci.* **3**, 53 (1994).

28. J. B. Allen, *Pop. Sci.* **256**(5), 43 (2000).

29. B. M. Coursey and R. Nath, *Phys. Today* **53**(4), 25 (2000).

30. G. J. Aubrecht and H. S. Matis, eds., *A Teacher's Guide to the Nuclear Science Wall Chart or, You do not have to be a Nuclear Physicist to Understand Nuclear Science*, Lawrence Berkeley Laboratory, 1998.

31. U.S. Department of Agriculture, in E. Golan et al., *Food Safety Innovation in the United States*, AER-831 (2004), 10.

32. R. Lutter, *Science* **286**, 2275 (1999). R. J. Woods, *Endeavour* **18**, 104 (1994).

33. E. Becker, *NYT*, 5 March 2002. M. Burros, *NYT*, 28 February 2001.

34. Anonymous, Reuters duspatch, 5 February 2003.

35. H. Goldstein, *IEEE Spectrum* **40**(8), 24 (2003).

36. C. Ness, *SFC*, 22 October 2002.

37. E. Meyer-Miebach, *Lebenesm.-Wiss. u.-Technol.* **26**, 493 (1993).

38. M. Fulmer, *LAT*, 1 November 2002. M. Fulmer, *LAT*, 9 September 2003.

39. J. Han et al., *Lebensm.-Wiss. u.-Technol.* **37**, 705 (2004).

40. R. S. Farag and K. H. A. M. El-Khawas, *Intern. J. Food Sci. Nutr.* **49**, 109 (1998).

41. B. Bhushan and P. Thomas, *Intern. J. Food Sci. Nutr.* **49**, 485 (1998).

42. C. Starr, *Science* **165**, 1232 (1969); C. Starr et al., *Nucl. News* **15**(10), 37 (1972); C. Starr and C. Whipple, *Science* **208**, 1114 (1980); and G. Marx, *Phys. Educ.* **28**, 170 (1993).

43. P. Slovic, *Science* **236**, 280 (1987). P. Slovic et al., *Psychology Today* **14**(8), 44 (1980) and M. G. Morgan, *Sci. Am.* **269**(1), 32 (1993).

44. J. F. Aherne, *Environment* **35**(2), 16 (1993).

45. M. L. Wald, *NYT*, 15 June 1997.

46. G. Apostolakis, *Science* **250**, 1359 (1990).

47. U.S. Nuclear Regulatory Commission, *Risk Assessment Review Group Report to the United States Nuclear Regulatory Commission*, NUREG/CR-0400, September 1978.

48. H. W. Lewis et al., *Rev. Mod. Phys.* **47**, Supp. 1, S1 (1975). (This is usually referred to as the "Lewis report.")

49. E. Marshall, *Science* **204**, 152 (1979); C. R. Herron and D. Lewis, *NYT*, 15 April 1979.

50. E. Marshall, *Science* **204**, 280 (1979); *Science* **204**, 594 (1979). B. G. Levi, *Phys. Today* **32**(6), 77 (1979).

51. Anonymous, *NYT*, 21 June 1993, sidebar that accompanies M. L. Wald, *NYT*, 21 June 1993.

52. W. Booth, *Science* **238**, 1342 (1987).

53. M. L. Wald, *NYT*, 24 April 1990.

54. AP dispatch, *NYT*, 15 August 1993.

55. J. W. Senders, *Psychology Today*, **13**(11), 52 (1980).

56. A. O. Sulzberger, Jr., *NYT*, 22 April 1979; C. Mohr, *NYT*, 13 May 1979.

57. M. L. Wald, *NYT*, 22 June 1985.

58. C. Norman, *Science* **228**, 31 (1985).

59. B. G. Levi, *Phys. Today* **38**(5), 67 (1985).

60. R. Wilson et al., *Rev. Mod. Phys.* **57**, S1 (1985).

61. A. A. Abagyan et al., *The Accident at the Chernobyl Nuclear Power Plant and its Consequences, Commission on the Causes of the Accident at the Fourth Unit of the Chernobyl Nuclear Power Plant*, appointed by the USSR State Committee on the Utilization of Atomic Energy, 1986.

62. J. F. Ahearne, *Science* **236**, 673 (1987).

63. Annex 2, Ref. 61.

64. D. Bodansky, *Nuclear Energy: Principles, Practices, and Prospects* (Woodbury, New York: AIP Press, 1996).

65. B. L. Cohen, *Am. J. Phys.* **55**, 1076 (1987).

66. C. Hohenemser, *Annu. Rev. Energy* **13**, 383 (1988).

67. B. G. Levi, *Phys. Today* **39**(7), 17 (1986); F. von Hippel and T. B. Cochran, *Bull. At. Sci.* **43**(1), 18 (1986).

68. C. Norman, *Science* **232**, 1331 (1986); C. Norman and D. Dickson, *Science* **233**, 1141 (1986); and E. Marshall, *Science* **233**, 1375 (1986). S. Diamond, *NYT*, 16 August 1986.

69. R. Wilson, *Science* **236**, 1636 (1987).

70. S. Diamond, *NYT*, 18 August 1986.

71. C. Hohenemser et al., *Environment* **28**(5), 6 (1986).

72. S. Schmeman, *NYT*, 6 May 1986; S. Diamond, *NYT*, 13 May 1986.

73. H. Salisbury, *NYT Magazine*, 27 July 1986, 18.

74. L. Devell et al., *Nature* **321**, 192 (1986).

75. F. A. Fry et al., *Nature* **321**, 569 (1986); C. R. Hill et al., *Nature* **321**, 655 (1986); C. Hohenemser et al., *Nature* **321**, 817 (1986); A. J. Thomas and J. M. Martin, *Nature* **321**, 818 (1986); B. Holliday et al., *Nature* **321**, 821 (1986); F. B. Smith and M. J. Clark, *Nature* **327**, 690 (1986); I. R. Falconer, *Nature* **322**, 692 (1986); N. G. Alexandropoulos et al., *Nature* **322**, 779 (1986); and K. Bangert et al., *Naturw.* **73**, 495 (1986). A retrospective look is found in R. Pöllännen et al., *Atmos. Environ.* **31**, 3575 (1997). This work found that transport was enhanced because warm air currents lifted the contaminated air (they found an effective release height greater than 2 km, which was not characteristic of the fire plume itself).

76. M. Aoyama et al., *Nature* **321**, 819 (1986).

77. R. H. Clarke, *Ann. Nucl. Sci. Engg.* **1**, 73 (1974).

78. Committee on the Possible Effects of Electromagnetic Fields on Biologic Systems (C. F. Stevens et al.,), *Possible Health Effects of Exposure to Residential Electric and Magnetic Fields* (Washington, DC: National Academy Press, 1997). S. F. Cleary, *BioScience* **33**, 269 (1983). D. I. McRee, *J. Air Poll. Contr. Assn.* **24**, 122 (1974).

79. See *IEEE Engg. in Med. Biol. Mag.* **14**, 336 (1995). This reference states "There is no reliable scientific evidence that continuous exposure to low intensity RFEM fields with whole-body averaged energy absorption rates (i.e., specific absorption rates or SAR) less than 0.4 W/kg results in damage, irreversible or otherwise, to biological molecules and tissues."

80. B. E. Taylor, *AAPT Announcer* **23**(2), 94 (1993) and L. Austin, *Phys. Ed.* **32**, 108 (1997).

81. D. Vergano, *Science* **285**, 23 (1999).

82. K. R. Foster and J. E. Moulder, *IEEE Spectrum* **37**(8), 23 (2000).

83. S. K. Moore, *IEEE Spectrum* **38**(2), 30 (2001).

84. Report of the 1978 Interagency Task Force on Ionizing Radiation, as quoted in J. L. Marx, *Science* **204**, 160 (1979).

85. A. Nero, *Phys. Today* **42**(4), 32 (1989).

86. K. Z. Morgan, in A. B. Kline, Jr., ed., *The Environmental and Ecological Forum*, 1970-1971, (Washington, DC: GPO, 1972), AEC publication.

87. D. J. Steck, et al., *Environ Health Perspect* **107**, 123 (1999) and R. W. Field et al., *Am. J. Epidemiol.* **151**, 1092 (2000). R. W. Field, Department of Epidemiology, University of Iowa Hospitals and Clinics, June 1999, URL http://www.vh.org/Providers/Textbooks/Radon/HealthRisk.html.

88. H. Raines, *NYT*, 16 March 1979; D. Henry, *NYT*, 13 May 1979.

89. P. Shabecoff, *NYT*, 19 May 1985; W. Greer, *NYT*, 28 October 1988. P. Shabecoff, *NYT*, 14 July 1985; R. Hanley, *NYT*, 3 April 1986 and *NYT*, 20 April 1986; A. A. Narvaez, *NYT*, 13 August 1986; and R. Pear, *NYT*, 15 August 1986.

90. M. Eisenbud, *Environment* **26**(10), 6 (1984).

91. N. Misra, AP dispatch, 11 May 1999.

92. M. K. Nair et al., *Radiat. Res.* **152**, S145 (1999). G. Jaikrishan et al., *Radiat. Res.* **152**, S149 (1999). V. D. Cheriyan et al., *Radiat. Res.* **152**, S154 (1999).

93. J. de Oliveira et al., *J. Environ. Radioact.* **53**, 99 (2001); J. M. Godoy et al., *J. Environ. Radioact.* **53**, 175 (2001).

94. R. Stone, *Science* **261**, 1514 (1993); R. A. Kerr, *Science* **240**, 606 (1988).

95. Committee on Risk Assessment of Exposure to Radon in Drinking Water (J. Doull et al.), National Academy of Sciences, *Risk Assessment of Radon in Drinking Water* (Washington, DC: National Academy Press, 1999).

96. J. E. Brody, *NYT*, 8 January 1991.

97. National Radiological Protection Board, as quoted in I.G. Draganiet al., *Radiation and Radioactivity on Earth and Beyond* (CRC Press, Boca Raton, FL, 1990).

98. D. Bruce Henschel (U.S. Environmental Protection Agency), *Radon Reduction Techniques for Detached Houses*, 2nd. Ed. (GPO, Washington, DC: 1988), EPA-625-5-87-019.

99. Bundesministerium des Innern, *Umweltradioactivität und Strahlenbelastung*, West Germany, 1983.

100. D. Marples, *Bull. At. Sci.* **49**(7), 38 (1993).

101. J. P. Holdren, *Annu. Rev. Energy Environ.* **17**, 235 (1992).

102. T. Feder, *Phys. Today* **50**(4), 55 (1997).

103. A. Lawler, *Science* **275**, 1730 (1997).

104. M. L. Wald, *NYT*, 27 September 1999.

105. M. L. Wald, *NYT*, 2 June 1999.

106. Anonymous, AP dispatch, 26 October 2003.

107. A. Dworkin, *PO*, 21 March 2004.

108. M. L. Wald, *NYT*, 27 July 2004.

109. J. Rojas-Burke, *PO*, 16 September 2003. Anonymous, *SPI*, 21 February 2004. M. L. Wald and S. Kershaw, *NYT*, 26 February 2004. H. Bernton, *ST*, 17 July 2004. L. Stiffler, *SPI*, 17 July 2004.

110. R. A. Kerr, *Science* **283**, 1626 (1999).

111. R. A. Kerr, *Science* **283**, 1627 (1999).

112. T. Gorman, *LAT*, 18 March 2002.

113. J. Beard, *IEEE Spectrum* **34**(11), 33 (1997).

114. K. Rogers, *LVRJ*, 3 December 2002. E. Werner, AP dispatch, 19 December 2003.

115. Office of Civilian Radioactive Waste Management, Department of Energy, DOE/YMP-0402 (June 2000).

116. Office of Civilian Radioactive Waste Management, Department of Energy, *Yucca Mountain Preliminary Site Suitability Evaluation* (2000) DOE/RW-0540.

117. R. A. Kerr, *Science* **274**, 913 (1996).

118. Office of Civilian Radioactive Waste Management, Department of Energy, DOE/YMP-0026 (May 2000).

119. K. Rogers, *LVRJ*, 22 September 2002. H. J. Hebert, AP dispatch, 17 May 2002.

120. T. Gorman, *LAT*, 29 May 2002.

121. D. Egan, *MJS*, 28 September 2002.

122. J. Fahys, *The Salt Lake Tribune*, 11 March 2003.

123. Office of Civilian Radioactive Waste Management, Department of Energy, DOE/YMP-0338 (September 2000).

124. J. Christensen, *NYT*, 10 August 19102. W. J. Broad, *NYT Magazine*, 18 November 1990, 36.

125. B.L. Cohen, *Phys. Today* **29**(1), 9 (1976); *Rev. Mod. Phys.* **49**, 1 (1977); *Sci. Am.* **236**(6), 21 (1977); *Phys. Teach.* **18**, 526 (1978); *Am. J. Phys.* **54**, 38 (1986).

126. R. C. Ewing, *Science* **286**, 415 (1999).

127. K. Davidson, *SFC*, 17 November 2003.

128. S. Sonner, AP dispatch, 19 February 2004.

129. M. L. Wald, *NYT*, 12 June 2004. The reason may be recent advances in the ability to sequester radioactive materials not discussed in the text, for example, A. J. Lupinetti et al., in K. K. S. Pillay and K. C. Kim, *Plutonium Futures* —The Science, Topical Conference on Plutonium and Actinides Santa Fe, New Mexico, USA July 10–13, 2000, 127 (capture in boron compounds); K.-A. H. Kubatko et al., *Science* **302**, 1191 (2003) (capture in peroxides); T. M. Nenoff and M. D. Nyman, United States Patent 6,596,254, awarded 22 July 2003 (molecular sieve); J. Li et al., *Science* **295**, 2242 (2002) (argon containment); S. Komarneni et al., Ref. 143 (clay storage); M. D. Tucker et al., *Biotech. Bioengg.* **60**, 88 (2000), F. Malekzadeh et al., *World J.Microbiol. Biotech.* **18**, 295 (2002), and R. B. Payne et al., *Appl. Environ. Microbiol.* **68**, 3129 (2002) (bacterial remediation).

130. M. L. Wald, *NYT*, 10 July 2004. C. Baltimore, Reuters dispatch, 14 July 2004.

131. M. L. Wald, *NYT*, 2 March 1999.

132. R. Stone, *Science* **297**, 1801 (2002).

133. J. M. Haschke, T. H. Allen, and L. A. Morales, *Science* **287**, 285 (2000). C. Madic, *Science* **287**, 243 (2000).

134. E. L. Christensen et al., *Rev. Mod. Phys.* **50**, Suppl. 1, S1 (1978).

135. L. J. Carter, *Science* **202**, 191 (1978); G. Lichtenstein, *NYT*, 22 May 1976.

136. Anonymous, AP dispatch, *NYT*, 22 February 1979. Anonymous, AP dispatch, *NYT*, 9 March 1979.

137. K. Schneider, *NYT*, 30 December 1990.

138. K. Schneider, *NYT*, 3 May 1993.

139. United States Nuclear Regulatory Commission, *Final Generic Environmental Impact Statement of Uranium Milling* (Washington, DC: GPO, 1980) NUREG-0706.

140. M. Crawford, *Science* **229**, 537 (1985).

141. F. G. F. Gibb and P. G. Atrill,*Geology* **31**, 657 (2003).

142. Office of Civilian Radioactive Waste Management, Department of Energy, DOE/YMP-0017 (December 2000).

143. C. D. Hollister and S. Nadia, *Sci. Am.* **278**(1), 60 (1998). There is strong evidence that clays can trap radium, as shown in S. Komarneni et al., *Nature* **410**, 771 (2001); see also R. Dahl, *Environ. Health Perspect.* **110**, A528 (2002).

144. J. M. Harrison, *Science* **226**, 11 (1984).

145. Council on Environmental Quality, *Environmental Quality* (11th Ed.) (Washington, DC: GPO, 1980).

146. W. Sullivan, *NYT*, 24 November 1992; W. J. Broad, *NYT*, 27 April 1993.

147. R. A. Kerr, *Science* **204**, 603 (1979).

148. R. P. Hammond, *Am. Sci.* **67**, 146 (1979).

149. I. J. Winograd, *Science* **212**, 1457 (1981).

150. P. A. Witherspoon et al., *Science* **211**, 894 (1981).

151. A. Barbreu et al., *Science* **197**, 519 (1977).

152. G. von Randow, *DZ*, 14 March 1997.

153. E. Grimmel, *Hamburger Geogr. Studien* **47**, 79 (1995).

154. B. Müller, *Bild d. Wissenschaft* (May 2002), 88.

155. German Federal Ministry for the Environment, *Nuclear Safety in Germany* (Report under the Convention on Nuclear Safety by the Government of the Federal Republic of Germany for the First Review Meeting in April 1999), September 1998, Ch. 19.

156. Select Committee on Science and Technology, House of Lords, HL Paper 41, 10 March 1999.

157. The Royal Society and The Royal Academy of Engineering Nuclear Energy Group (E. Ash et al.), *Nuclear energy—the future climate* (Portsmouth, UK: Holbrooks Printers Ltd., 1999).

158. Office of Civilian Radioactive Waste Management, Department of Energy, DOE/YMP-0405 (June 2001).

159. G. Lucas, *SFC*, 19 March 2002. K. Yamamura, *SB*, 24 June 2002.

160. M. Bustillo, *LAT*, 8 December 2002.

161. A. Rojas, *SB*, 7 April 2003.

162. U.S. Encironmental Protection Agency, *Fed. Reg.* **68**, 65119 (2003).

163. J. Lochard and C. Schreiber, *J. Radiol. Prot.* **20**, 101 (2000).

164. T. H. Maugh, *Science* **202**, 37 (1978).

165. G. B. Gori, *Science* **208**, 256 (1980).

166. M. Crawford and R. Wilson, *Human and Ecological Risk Assessment.* **2**, 305 (1996); M. Heitzman and R. Wilson, *BELLE Newsletter* **6**, No. 1 (1997). BELLE stands for biological effects of low level exposures, http://www.belleonline.com/.

167. L. B. Lave, *Science* **236**, 291 (1987).

168. J. Brinkley, *NYT*, 23 March 1993.

169. E. Marshall, *Science* **212**, 900 (1981); and L. Roberts, *Science* **238**, 1649 (1987).

170. C. Haberman, *NYT*, 4 August 1985.

171. K. Z. Morgan, in M. Kaku and J. Trainer, eds., *Nuclear Power: Both Sides* (New York: W. W. Norton, 1982), 35.

172. T. Straume et al., *Health Phys.* **63**, 421 (1992); W. J. Broad, *NYT*, 13 October 1992.

173. T. Straume et al., *Nature* **424**, 539 (2003). M. P. Little, *Nature* **424**, 495 (2003). D. Normale, *Science* **301**, 742 (2003).

174. A. Brodsky, in M. Kaku and J. Trainer, eds., *Nuclear Power: Both Sides* (New York: W. W. Norton, 1982), 46. The data are based on R. L. Gotchy, *Health Phys.* **35**, 563 (1978).

175. D. Chemelevsky et al., in W. Gössner et al., eds., *The Radiobiology of Radium and Thorotrast* (Urban and Schwarzenberg, Munich, W. Germany, 1986), p. 32.

176. R. L. Garwin, *Phys. Today* **53**(5), 12 (2000).

177. B. L. Cohen, *Technology* **6**, 43 (1999).

178. Z. Jaworowski, *Phys. Today* **52**(9), 24 (1999). J. F. Ward, *Rad. Res.* **142**, 362 (1995). J. F. Ward, *Prog, Nucleic Acids Mol. Biol.* **35**, 95 (1988).

179. H. O. Dickinson and L. Parker, *Intern. J. Cancer* **99**, 437 (2002).

180. Z. Jaworowski, *Phys. Today* **53**(4), 15 (2000).

181. Yu. E. Dubrova et al., *Science* **295**, 1037 (2003).

182. B. L. Cohen, *Health Phys.* **68**, 157 (1995).

183. E. J. Calabrese and L. A. Baldwin, *BELLE Newsletter* **8**, No. 2 (1999).

184. E. J. Calabrese and L. A. Baldwin, *BELLE Newsletter* **8**, No. 2 (1999).

185. J. Kaiser, *Science* **302**, 378 (2003).

186. K. Rothkamm and M. Löbrich, *Proc. Natl. Acad. Sci.* **100**, 5057 (2003). W. M. Bonner, *Proc. Natl. Acad. Sci.* **100**, 4973 (2003).

187. M. Goldberg et al., *Nature* **421**, 952 (2003).

188. G. S. Stewart et al., *Nature* **421**, 961 (2003).

189. J. Kaiser, *Science* **285**, 177 (1999).

190. R. Sielken, *BELLE Newsletter* **6**, No. 1 (1997).

191. D. Hoel, *BELLE Newsletter* **6**, No. 1 (1997).

192. D. J. Brenner et al., *Proc. Natl. Acad. Sci.* **100**, 13761 (2003).

193. R. Wilson, *Phys. Today* **53**(5), 11 (2000).

194. C. N. Coleman et al., *Science* **304**, 693 (2004).

195. J. Mintz, *WP*, 19 May 2003.

196. E. Marshall, *Science* **204**, 711 (1979); M. E. Jacobs, *Phys. Today*, **32**(7), 78 (1979).

197. I. Asimov and T. Dobzhansky, *The Genetic Effects of Radiation* (Washington, DC: GPO, 1966), AEC publication.

198. J. W. Gofman, fact sheet, Committee for Nuclear Responsibility, Inc., November 1993. J. W. Gofman and A. Tamplin, *Environment* **12**(3), 12 (1970). A. M. Stewart, *Bull. At. Sci.* **46**(9), 15 (September 1990).

199. Committee on Biological Exposure to Ionizing Radiation (C. L. Comar et al.), National Academy of Sciences, *The Effects on Populations of Exposure to*

Low Levels of Ionizing Radiation (BEIR III), (Washington, DC: National Academy Press, 1972).

200. E. Marshall, *Science* **236**, 658 (1987).

201. A. Berrington de González and S. Darby, *The Lancet* **363**, 345 (2003).

202. R. Wilson and E. A. C. Crouch, *Science* **236**, 267 (1987).

203. F. Barringer, *NYT*, 6 September 1986.

204. M. C. Hatch et al., *Am. J. Epidemiol.* **132**, 397 (1990), and E. O. Talbott et al., *Environ. Health Perspect.* **108**, 545 (2000).

205. UNSCEAR, 2001 Report.

206. S. Wing et al., *Environ. Health Perspect.* **105**, 52 (1997).

207. E. O. Talbott et al., *Environ. Health Perspect.* **111**, 341 (2003). D. C. Holzman, *Environ. Health Perspect.* **111**, A166 (2003) and M. L. Wald, *NYT*, 1 November 2002.

208. Annex 4, Ref. 61.

209. Annex J, Ref. 8.

210. D. R. Francis, *CSM*, 29 May 2001.

211. Annex 7, Ref. 61.

212. L. R. Anspaugh et al., *Science* **242**, 1513 (1988).

213. Annex D, UNSCEAR, *Sources and Effects of Ionizing Radiation* (New York: United Nations Publ., 1988).

214. P. Taylor, *New Scientist* **110**(#1508), 24 (1986).

215. V. K. Savchenko, *The Ecology of the Chernobyl Catastrophe* (Carnforth, UK: Parthenon Publ. Group, 1995).

216. G. Gugliotta, *WP*, 29 December 2002. R. Stone, *Science* **292**, 422 (2001). H. Kroker, *DW*, 26 April 2004.

217. A. K. Guskova et al., Appendix to Annex G, UNSCEAR, *Sources and Effects of Ionizing Radiation* (New York: United Nations Publ., 1988).

218. C. Norman and D. Dickson, *Science* **233**, 1141 (1986).

219. V. S. Kazakov, E. P. Dimidchik, L. N. Astrakhova, *Nature* **359**, 21 (1992).

220. K. Baverstock et al., *Nature* **359**, 21 (1992).

221. L. N. Astakhova et al., *Radiat. Res.* **150**, 349 (1998).

222. Y. Shirahige et al., *Endocrin. J.* **45**, 203 (1998).

223. M. D. Tronko et al., *Cancer* **86**, 149 (1999).

224. R. Stone, *Science* **292**, 420 (2001).

225. Der Bundesminister für Umwelt, Naturschutz, und Reaktorsicherheit, *Bericht über den Reaktorunfall in Tschernobyl, seine Auswirkungen und die getroffenen bzw. zu treffenden Vorkehrungen*, West Germany, 1986.

226. J. W. Gofman, talk delivered to the American Chemical Society meeting, September 1986.

227. O. Kovalchuk et al., *Nature* **407**, 583 (2000).

228. R. Stone, *Science* **292**, 421 (2001).

229. D. Röhrlich, *DW*, 16 April 1996

230. R. Stone, *Science* **292**, 420 (2001).

231. K. Z. Morgan, *Environment* **13**(1), 28 (1971).

232. J. Gofman and J. R. Tamplin, *Poisoned Power* (New York: Signet, 1974), Appendix I; J. Gofman, in A. B. Kline, ed., *The Environmental and Ecological Forum, 1970-1971* (Washington, DC: GPO, 1972), 72, AEC publication, TID 25857.

233. F. G. Laoman et al., in *Radioactivity in the Marine Environment* (Washington, DC: National Academy Press, 1971), Ch. 7, 161; M. E. Eisenbud, *Environment* **20**(8), 6 (1978).

234. D. P. Burkitt, *Intern. J. Environmental Studies* **1**, 275 (1971); R. J. C. Harris, *Intern. J. Environmental Studies* **1**, 59 (1971).

235. M. Woods, *TB*, 1 May 2001.

236. J. Lederberg, *WP*, 19 July 1970.

237. H. L. Beck and P. W. Krey, *Science* **220**, 18 (1983).

238. S. A. Fetter and K. Tsipis, *Sci. Am.* **244**(4), 41 (1981).

239. C. Norman, *Science* **223**, 258 (1984); *Science* **229**, 448 (1985).

240. M. L. Wald, *NYT*, 27 August 1993.

241. L. J. Carter, *Science* **200**, 1135 (1978).

242. M. Sun, *Science* **215**, 1483 (1982); L. J. Carter, *Science* **222**, 1104 (1983).

243. F. Donath, in M. Kaku and T. Trainer, eds., *Nuclear Power: Both Sides* (New York: W. W. Norton, 1982), 115.

244. R. A. Kerr, *Science* **204**, 289 (1979); R. O. Pohl, in M. Kaku and J. Trainer, eds., *Nuclear Power: Both Sides* (New York: W. W. Norton, 1982), 123.

245. A. S. Kubo and D. J. Rose, *Science* **182**, 1205 (1973); D. J. Rose, *Science* **184**, 351 (1974).

246. E. J. Moniz and T. L. Neff, *Phys. Today* **31**(4), 42 (1978); T. B. Taylor, *Annu. Rev. Nucl. Sci.* **25**, 407 (1975).

247. D. J. Rose and R. K. Lester, *Sci. Am.* **238**(4), 46 (1978).

248. C. L. Rickard and R. C. Dahlberg, *Science* **202**, 581 (1978).

249. H. A. Ferverson, et al., *Science* **203**, 330 (1979).

250. V. K. McElheney, *NYT*, 28 February 1978.

251. S. Rattner, *NYT*, 16 March 1978. As we now know, it will not happen before about 2012.

252. L. A. Sagan, *Science* **177**, 487 (1972). Committee on Biological Exposure to Ionizing Radiation (G. W. Casarett et al.), National Academy of Sciences, *Considerations of Health Benefit-Cost Analysis for Activities Involving Ionizoing Radiation Exposure and Alternatives* EPA 520/4-77-003, 1977; this gives some indications of the trade-offs needed.

253. Mine Safety and Health Administration, U.S. Department of Labor, *Coal Fatalities for 1900 Through 2001* website URL: http://www.msha.gov/stats/charts/chartshome.htm.

254. H. W. Lorber and A. Ford, *Environment* **22**(4), 25 (1980).

255. J. E. Martin et al., in D. A. Berkowitz and A. M. Squires, eds., *Power Generation and Environmental Change* (Cambridge, MA: MIT Press, 1971), 107.

256. J. P. McBride et al., *Science* **202**, 1045 (1978).

257. B. L. Cohen, in D. J. Paustenbach, ed., *The Risk Assessment of Environmental and Human Health Hazards* (New York: Wiley, 1989).

258. B. L. Cohen, colloquium, Department of Physics, Ohio State University, 22 February 2000; B. L. Cohen, private communication.

259. D. R. Francis, *CSM*, 29 May 2001.

260. D. R. Roeck et al., *Health Phys.* **52**, 311 (1987).

261. J. O. Corbett, *Rad. Protect. Dosimetry* **4**, 5 (1983).

262. A. Gabbard, *ORNL Review* **26** http://www.ornl.gov/ORNLReview/rev26-34/text/colmain.html.

263. J. Tadmor, *J. Environ. Rad.* **4**, 177 (1986).

264. National Council on Radiation Protection and Measurements, Report No. 92 (1987); Report No. 95 (1987)

265. Annex B, Ref. 8.

266. Environmental Protection Agency, *Study of Hazardous Air Pollutant Emissions from Electric Utility Steam Generating Units—Final Report to Congress*, V. I (Washington, DC: GPO, 1998), EPA-453/R-98-004a.

267. H. Inhaber, *Science* **203**, 718 (1979).

268. J. Lovelock, *TI*, 24 May 2004. See also D. Butler, *Nature* **429**, 238 (2004).

269. J. Lovelock, *TI*, 28 August 2004.

270. C. Arthur, *TI*, 25 May 2004, T. Davenport, *TI*, 4 June 2004, and Z. Goldsmith, *TI*, 28 August 2004.

271. K. Schneider, *NYT*, 26 December 1990.

CHAPTER 21

1. S. Wieder and E. Jacobi, *Am. J. Phys.* **45**, 981 (1977).

2. A. L. Hammond, *Science* **179**, 1116 (1973).

3. D. L. Klass, *Energy Policy* **31**, 353 (2003). The author asserts that the decline in oil must inevitably bring America back to renewables.

4. M. L. Wald, *NYT*, 8 December 2002.

5. K. Porter, et al., NREL/TP-620-28674 (1999 Ed., August 2000).

6. S. Sklar, *Annu. Rev. Energy* **15**, 121 (1990).

7. D. Pimentel et al., *BioScience* **52**, 1111 (2002).
8. Department of Energy, *Photovoltaic Energy Program Overview Fiscal Year 2000* (Washington, DC: GPO, 2001).
9. U.S. Department of Energy, (2000). Million Solar Roofs Website, URL: http://www.MillionSolarRoofs.org.
10. See, for example, C. P. Dahlberg, *SB*, 10 September 2002; D. Kelly, *LAT*, 23 September 2002; D. E. Murphy, *NYT*, 24 November 2002; C. P. Dahlberg, *SB*, 7 July 2003; N. R. Brooks, *LAT*, 5 December 2003; E. Fletcher, *SB*, 24 May 2004; D. M. Kammen, *SFC*, 18 July 2004; and R. Redford, *SFC*, 9 August 2004.
11. T. R. Reid, *WP*, 25 July 2004.
12. T. Nugent, *CT*, 3 March 2003.
13. Virginia Department of Mines, Minerals, and Energy, URLhttp://www.mme.state.va.us/De/chap7c.html.
14. J. Rendon, *NYT*, 12 October 2003. M. Clayton, *CSM*, 12 February 2004 and also R. Atkin, *CSM*, 3 September 2003.
15. Dept. of Energy, *Annual Energy Review 2000* (Washington, DC: GPO, 2001), DOE/EIA-0384(00), Ch. 10.
16. UK Department of Trade and Industry, *New & Renewable Energy: Prospects for the 21st Century* (London: Department of Trade and Industry, 1999).
17. H. J. Koch, *Energy & Environment* **13**, 673 (2002). A. Uzun, *Energy Policy* **30**, 131 (2002). A. C. Christiansen, *Energy Policy* **30**, 235 (2002). T. Czuczka, AP dispatch, 6 September 2002. P. Meller, *NYT*, 16 October 2002. N. I. Meyer and A. L. Koefoed, *Energy Policy* **31**, 597 (2003). K. Belson, *NYT*, 29 July 2003. T. A. Friedrich, *DW*, 10. April 2004. T. Hamilton, *TS*, 24 July 2004.
18. Deutsche Bundestag, *Solar Energy* **70**, 489 (2001) (Renewable Energy Sources Act, Germany, 2000). See also W. Krewitt and J. Nitsch, *Renewable Energy* **28**, 533 (2003).
19. Anonymous, BBC News Online dispatch, 23 May 2002.
20. S. Blakeslee, *NYT*, 4 December 1983.
21. G. W. Braun and D. R. Smith, *Annu. Rev. Energy Environ.* **17**, 97 (1992).
22. Department of Energy, *Renewable Resources in the U.S. Electricity Supply* (Washington, DC: GPO, 1993) DOE/EIA-0562.
23. R. L. San Martin, *Futurist* **23**(3), 37 (1989).
24. L. Guey-Lee, in Energy Information Administration, Departement of Energy, *Renewable Energy 1998:Issues and Trends* (Washington, DC: GPO, 1999) DOE/EIA-0628(98), p. 1.
25. A. J. Cavallo et al., in T. B. Johansson et al., eds., *Renewable Energy Sources for Fuels and Electricity* (Washington, D.C. and Covelo, CA: Island Press, 1993), 121.
26. M. Simons, *NYT*, 7 December 1997.
27. L. Fisher, *NYT*, 26 October 1985.
28. S. Bronstein, *NYT*, 23 June 1985.
29. Anonymous, *CD* (*SFE* story), 13 September 1992.
30. National Renewable Energy Laboratory, *Wind Power Today* DOE/GO-102001-1325 (2001).
31. J. Chapman, in R. Howes and A. Fainberg, eds., *The Energy Sourcebook* (New York: American Institute of Physics, 1991), 313.
32. M. R. Gustavson, *Science* **204**, 13 (1979).
33. H. M. Drees, 2nd Interanational Symposium on Wind Energy Systems, Oct. 3-6, 1978, pp. E7-82 to E7-129. See also D. W. Lobitz and T. D. Ashwill, Sandia Report SAND85–0957, 1986.
34. Department of Energy website, http://www.eren.doe.gov/RE/wind_technologies.html.
35. D. M. Dodge, *Illustrated History of Wind Power Development* http://telosnet.com/wind/index.html.
36. P. Fairley, *IEEE Spectrum* **40**(8), 35 (2003).
37. B. Sørensen, *Am. Sci.* **69**, 500 (1981).
38. M. F. Merriam, *Annu. Rev. Energy* **3**, 29 (1978). See also C. Garrett, *Endeavour* **8**, 58 (1984).
39. P. M. Moretti and L. V. Divone, *Sci. Am.* **254**(6), 110 (1986).
40. M. J. Grubb and N. I. Meyer, in T. B. Johansson et al., eds., *Renewable Energy Sources for Fuels and Electricity* (Washington, D.C. and Covelo, CA: Island Press, 1993), 157.
41. L. Guey-Lee, in Energy Information Administration, Departement of Energy, *Renewable Energy 1998: Issues and Trends* (Washington, DC: GPO, 1999) DOE/EIA-0628(98), p. 79.
42. R. W. Righter, *Wind Energy in America: A history* (Norman, OK and London: Oklahoma University Press, 1996).
43. R. J. Smith, *Science* **207**, 739 (1980). See also J. W. Shupe, *Science* **216**, 1193 (1982). Since 1984, wind energy has supplied about 10% of Hawaii's energy needs.
44. M. Parrish, *LAT*, 14 February 1994.
45. G. P. Corten and H. F. Veldkamp, *Nature* **412**, 41 (2001).
46. I. Sample, *New Scientist* **166** (2243), 14 (2000).
47. B. Ernst et al., NREL/CP-500-26722 (1999).
48. D. Graham-Rowe, *New Scientist* **171** (2299), 20 (2001). D. Sasse, *DW*, 19 July 2001.
49. International Energy Agency, *World Energy Outlook 2000—Highlights* (Paris: OECD, 2001).
50. M. Franken, *DW*, 7 October 2002.
51. D. Asendorpf and M. Rauner, *DZ*, 6 May 2004. D. Asendorpf, *DZ*, 19 June 2003.
52. R. Ringer, *NYT*, 18 May 1993.
53. S. N. Paleocrassas, *Solar Energy* **16**, 45 (1974).
54. D. L. Elliott et al., *Wind Energy Resource Atlas of the United States*, DOE/CH10093-4, Solar Energy Research Institute, Golden, Colorado (1987). See also D.L. Elliott and M.N. Schwartz, PNL-SA-23109 (September 1993).
55. D. Jehl, *NYT*, 26 November 2000.
56. Sr. P. Larson, *The Benedictine Witness* **21** (#2), 1998 and *The Benedictine Witness* **22** (#3), 1999; Sr. B. Bodine, http://www.rc.net/bismarck/shm/windturbine.htm.
57. Citizens Action Coalition of Indiana, Iowa RENEW, Izaak Walton League of America, RENEW Wisconsin, Union of Concerned Scientists, Dakota Resource Council, Minnesotans for an Energy-Efficient Economy, and the Environmental Law & Policy Center, *Repowering the Midwest The Clean Energy Development Plan for the Heartland* (Chicago: Environmental Law & Policy Center, 2001).
58. National Renewable Energy Laboratory, *A Consumer's Guide to Buying a Solar Electric System* (Department of Energy, 1999).
59. International Energy Agency, *Benign Energy? The Environmental Implications of Renewables* (Paris: OECD, 1997), Appendix 3d. See also M. L. Morrison et al., National Renewable Energy Laboratory, NREL/CP-500-25009 (1998).
60. J. Donn, AP dispatch, 19 June 2004.
61. M. Perry, Reuters dispatch, 14 February 2001.
62. S. Krohn, http://www.windpower.dk/articles/success.htm.
63. W. Hoge, *NYT*, 9 October 1999.
64. Anonymous, AP dispatch, 31 January 2003.
65. M. Simons, *NYT*, 8 December 2002.
66. U. L. K., *DW*, 3 September 2002.
67. W. E. Heronemus, in L. C. Ruedisili and M. W. Firebaugh, eds. *Perspectives on Energy: Issues, Ideas, and Environmental Dilemmas* (New York: Oxford University Press, 1975), 364.
68. D. R. Inglis, *Environment* **20**(8), 17 (1978).
69. J. Constans, *Marine Sources of Energy* (New York: Pergamon Press, 1979). This is the report of the EUROCEAN group, presented to the United Nations Office of Science and Technology. It investigates the engineering feasibility of alternative sources of energy.

70. S. Krohn, http://www.windpower.dk/articles/offshore.htm.

71. B. W. Byrne and G. T. Houlsby, *Phil. Trans. R. Soc. Lond. A* **361**, 2909 (2003).

72. J. G. McGowan and S. R. Connors, *Annu. Rev. Energy Environ.* **25**, 147 (2000).

73. C. Woodard, *SFC*, 23 April 2001.

74. D. Graham-Rowe, *New Scientist* **171** (2310), 29 (2001).

75. M. Hamer, *New Scientist* **142** (1921), 21 (1994).

76. Anonymous, Sapa-AFP dispatch, 7 July 2004.

77. M. Z. Jacobson and G. M. Masters, *Science* **293**, 1438 (2001).

78. S. Torok, *New Scientist* **167**(2257), 38 (2000).

79. L. Hattersley, *New Scientist* **161**(2176), 30 (1999).

80. K. Edds, *WP*, 24 December 2003. M. Kolber, *SB*, 14 August 2004.

81. R. Tempest, *LAT*, 8 December 2003.

82. S. Orloff and A. Flannery, Report to the California Energy Commission and Planning Departments of Alameda, Contra Costa and Solano Counties by BioSystems Analysis, Inc. (1992).

83. N. R. Brooks, *LAT*, 11 February 2003. See also M. Wiltenburg, *CSM*, 28 August 2003.

84. K. L. Ziner, *NYT*, 16 April 2002. B. Daley, *BG*, 25 July 2002. Anonymous, AP dispatch, 22 January 2003. S. Ebbert, *BG*, 11 November 2003.

85. B. Daley, *BG*, 20 April 2002. E. Mehren, *LAT*, 11 August 2002. P. Ferdinand, *WP*, 20 August 2002. E. Burkett, *NYT Magazine*, 15 June 2003, 48. B. Daley, *BG*, 25 July 2003. J. Peter, AP dispatch, 12 August 2003. J. Haughton and D. G. Tuerck, *BG*, 3 July 2004.

86. B. Ray, AP dispatch, 12 February 2003. Anonymous, AP dispatch, 22 August 2003.

87. G. Collins, *NYT*, 28 August 2002.

88. B. Jespersen, *The Waterville (Maine) Morning Sentinel*, 9 June 2004.

89. A. Crawford, *The Scottish Sunday Herald* (Glasgow, UK), 18 July 2004. R. Syal, *The Sunday Telegraph* (UK), 20 June 2004.

90. Anonymous, BBC News Online, 23 January 2002.

91. Anonymous, BBC News Online, 8 July 2002.

92. H. Timmons, *NYT*, 19 December 2003.

93. Anonymous, *TB*, 4 July 2002. Editorial, *TB*, 17 November 2003. Anonymous, *TB*, 29 February 2004. J. Feehan, *TB*, 19 April 2004. Anonymous, *TB*, 28 April 2004.

94. J. Newman, *The Wisconsin State Journal*, 16 July 2004.

95. J. Kohler, AP dispatch, 2 July 2004.

96. Anonymous, *New Scientist* **165** (2224), 23 (2000).

97. W. Zank, *DZ*, 3 January 1986.

98. M. Rink, private communication, 25 July 2001.

99. G. G. Piepers, *Alt. Sources Energy* **81**, 40 (1986).

100. J. Reeves, Alternative Energy Project Manager, Southern California Edison, private communication.

101. R. Kahn, *Alt. Sources Energy* **81**, 44 (1986).

102. D. Kasler, *SB*, 1 April 2001. M. Liedtke, AP dispatch, 19 April 2001.

103. V. Balzani et al., *Science* **189**, 852 (1975).

104. J. R. Bolton, *Science* **202**, 705 (1978).

105. S. Licht and D. Peramunage, *Nature* **345**, 330 (1990) and *Nature* **354**, 440 (1991).

106. T. H. Maugh, *Science* **222**, 151 (1983).

107. A. Heller, *Science* **223**, 1141 (1984).

108. E. Becquerel, *Comptes Rendus* **9**, 145 (1839) and *Comptes Rendus* **9**, 561 (1839).

109. A. Wilson, *Alt. Sources Energy* **81**, 8 (1986).

110. J. L Stone, *Phys. Today* **46**(9), 22 (1993).

111. E. A. Perez-Albuerne and Y.-S. Tan, *Science* **208**, 901 (1980).

112. E. M. Hubbard, *Science* **244**, 297 (1989).

113. K. Zweibel, *Am. Sci.* **81**, 362 (1993). T. Beardsley *Sci. Am.* **270**(4), 115 (1994) and M. L. Wald, *NYT*, 14 June 1994; thin 30 μm cells allow higher concentrations of impurities and so cheaper cell modules.

114. E. A. DeMeo and R. W. Taylor, *Science* **224**, 245 (1984).

115. M. A. Green, in T. B. Johansson et al., eds., *Renewable Energy Sources for Fuels and Electricity* (Washington, D.C. and Covelo, CA: Island Press, 1993), 337.

116. D. M. Chapin et al., *J. Appl. Phys.* **25**, 686 (1954).

117. A. Shah et al., *Science* **285**, 692 (1999). See also M. Grätzel, *Nature* **414**, 338 (2001) for a review of such cells.

118. J. P. Benner and L. Kazmerski, *IEEE Spectrum* **36**(9), 34 (1999).

119. H. Kelly, in T. B. Johansson et al., eds., *Renewable Energy Sources for Fuels and Electricity* (Washington, D.C. and Covelo, CA: Island Press, 1993), 297.

120. National Renewable Energy Laboratory, *Photovoltaic Energy Program Overview, Fiscal year 2000* DOE/GO-102001-1168 (2001).

121. E. W. McFarland and J. Tang, *Nature* **421**, 616 (2003). M. Grätzel, *Nature* **421**, 586 (2003).

122. P. Wang et al., *Nature Materials* **2**, 402 (2003).

123. W. U. Huynh et al., *Science* **295**, 2425 (2002).

124. J. Wu et al., *Appl. Phys. Lett.* **80**, 3967 (2002). J. Wu et al., *Appl. Phys. Lett.* **80**, 4741 (2002). J. Wu et al., *Phys. Rev.* **B66**, 201403 (2002).

125. J. Wu et al., *J. Appl. Phys.* **94**, 6477 (2003).

126. J. Nelson, *Current Opinion in Solid State and Materials Science* **6**, 87 (2002).

127. C. Norman, *Science* **226**, 319 (1984).

128. K. Firor et al., in T. B. Johansson et al., eds., *Renewable Energy Sources for Fuels and Electricity* (Washington, D.C. and Covelo, CA: Island Press, 1993), 483.

129. D. Carlson and S. Wagner, in T. B. Johansson et al., eds., *Renewable Energy Sources for Fuels and Electricity* (Washington, D.C. and Covelo, CA: Island Press, 1993), 403.

130. T. H. Maugh, *Science* **221**, 1358 (1983).

131. F. Frisch, *DZ*, 10 May 1991.

132. SMUD website on PV Pioneers, http://www.smud.org/pv/pv_pioneer1.html. See also the related sites at the www.smud.org web address.

133. T. Erge et al., *Solar Energy* **70**, 479 (2001).

134. E. C. Kern, Jr. and D. L. Greenberg, EPA/600/SR-96/130 (1996).

135. J. Johnson, *Chem. & Engg. News* **76**(13), 28 (1998).

136. D. E. Osborn, *Solar Today* p. 30 (May/June 2000).

137. Sacramento Municipal Utility District, SMUD newsletter (August/September 2001).

138. Sacramento Municipal Utility District, Ref. 137.

139. D. Hull, *SJMN*, 19 January 2001. G. Chan, *SB*, 13 February 2001. H. Martin, *LAT*, 6 March 2001.

140. E. Bailey, *LAT*, 16 July 2001. J. K. Morita, *SB*, 23 June 2001. M. Morgante, AP dispatch, 18 February 2001.

141. K. Gaudette, AP dispatch, 15 October 2001.

142. E. Banducci, *SJMN*, 5 April 2001.

143. W. Lindelof, *SB*, 29 March 2001.

144. R. Alessi, Scripps Howard News Service dispatch, 29 August 2000.

145. B. Pimentel, *SFC*, 13 June 2001. C. Burress, *SFC*, 25 February 2001.

146. Sacramento Municipal Utility District, Ref. 137.

147. S. Herel, *SFC*, 22 February 2001.

148. Anonymous, Environmental News Network dispatch, 26 March 2001, URL: http://www.enn.com/news/enn-stories/2001/03/03262001/solar_42674.asp.

149. Sacramento Municipal Utility District, SMUD report (2001).

150. Department of Energy, *Renewable energy annual 1998 with data for 1997* DOE/EIA-0603(98)/1 (1998).

151. National Renewable Energy Laboratory, *Photovoltaics: the power of choice* DOE/GO-10096-017 (1996).

152. J. Johnson, *Chem. & Engg. News* **76**(13), 24 (1998).

153. M. L. Wald, *NYT*, 16 August 1997.

155. Department of Energy, *Annual Energy Review 1992* (Washington, DC: GPO, 1993), DOE/EIA-0384(92).

154. G. T. Pakulski, *TB*, 16 February 2003.

156. T. Meredith, AP dispatch, 16 April 2001.

157. Department of Energy, URL: http://www.eren.doe.gov/pv/ pvbatcase.html. Also see URL: http://www.eren.doe.gov/pv/ pvutcase.html.

158. WisconSUN, URL: http://www.wiscon-sun.org/learn/cs_davenport.shtml.

159. C. P. Dahlberg, *SB*, 25 June 2002. C. P. Dahlberg, *SB*, 6 September 2002. SMUD did emerge, as we see in G. Crump, *SB*, 18 February 2003 and in G. Burke, *SB*, 10 June 2004.

160. Wisconsin Focus on Energy, brochure (2000).

161. National Tour of Solar Homes, see the URL: http://www.solarhouse.com.

162. R. Klüting, *DZ*, 16 October 1992.

163. D. Schwab, *Nürnberger Nachrichten*, 24 October 1992.

164. Solardyne Corporation, press release, 13 March 2001.

165. Anonymous, Environmental News Network dispatch, 2 July 2001, URL: http://www.enn.com/ enn-news-archive/2001/03/03142001/ backpack_42460.asp.

166. U.S. Department of Energy, *Clean Energy Partnerships: A Decade of Success* DOE/EE-0213 March 2000.

167. T. Riordan, *NYT*, 1 September 2003.

168. M. L. Wald, *NYT*, 12 August 2004.

169. D. Lipschultz, *NYT*, 9 September 2001.

170. D. M. Kammen, *Environment* **41**(5), 10 (1999).

171. E. F. Schumaker, *Small is beautiful* (London: Hartley & Marks Publishers, Incorporated, 1998).

172. K. Knapp and T. Jester, *Solar Energy* **71**, 165 (2001).

173. T. J. Lueck, *NYT*, 16 October 1983.

174. L. Peck, *Amicus J.* **12**(2), 27 (1990). For a more extended discussion of the ARCO Solar sale, see M. L. Wald, *NYT*, 7 March 1989 and M. Crawford, *Science* **244**, 918 (1989).

175. M. L. Kramer et al., *Science* **193**, 1239 (1976).

176. S. von der Weiden, *DW*, 25 August 2002.

177. J. Zhao et al., *Prog. Photovolt: Res. Appl.* **7**, 471 (1999).

178. S. Bronstein, *NYT*, 23 June 1985.

179. C. Lopes, Research and Development site manager, Solar One, private communication.

180. I. R. Straughan, *Middle Management Perspectives Program*, Southern California Edison, 1985.

181. J. J. Bartel, (New Mexico: Sandia Labs, 1984).

182. F. Kreith and R. T. Meyer, *Am. Sci.* **71**, 598 (1983).

183. J. G. Ingersoll, in R. Howes and A. Fainberg, eds., *The Energy Sourcebook* (New York: American Institute of Physics, 1991), 207.

184. C. Macilwain, *Nature* **362**, 778 (1993).

185. M. Karus, *DZ*, 12 January 1990.

186. Department of Energy, *Solar Collector Manufacturing Activity 1991* (Washington, DC: GPO, 1992), DOE/EIA-0174(91)

187. J. Thornton, presentation delivered at the Southern Ohio AAPT Section Meeting, Columbus, Ohio, 4 May 1984.

188. P. De Laquil et al., in T. B. Johansson et al., eds., *Renewable Energy Sources for Fuels and Electricity* (Washington, D.C. and Covelo, CA: Island Press, 1993), 213.

189. Department of Energy, SAND2000-0613 (March 2000).

190. Department of Energy, DOE/GO-10097-406 (1999).

191. M. Nichols, Reuters dispatch, 7 January 2003.

192. Department of Energy, "Solar dish engine," Ch. 5.

193. A. B. Meinel and M. P. Meinel, *Phys. Today* **25**(2), 44 (1972).

194. K. Zweibel and A. M. Barnett, in T. B. Johansson et al., eds., *Renewable Energy Sources for Fuels and Electricity* (Washington, DC and Covelo, CA: Island Press, 1993), 437.

195. O. Hohmeyer, *Social Costs of Energy Consumption* (Berlin: Springer Verlag, 1988).

196. Depending on the sort of photovoltaic system, the actual cost is between 25¢ to 50¢ per kWh. According to, for example, the Southwest Public Power District, 221 N. Main Street, Palisade NE 69040, it costs between $12,000 and $15,000 per mile to extend a grid connection. This is $7.50 to $9.40 per meter of extension. See their URL: http://www.swppd.com/pv_system.htm. See also Virginia Department of Mines, Minerals, and Energy, t URL: http://www.mme.state.va.us/De/chap7c. html. According to this latter reference, "For demand less than 20 kilowatt-hours (kWh) per month, PV will cost less than extending lines more than 100 feet; for 500 kWh/mo, PV will compete with utility line extensions of more than 1112 miles."
A good source for general information on solar energy alternatives discussed in this chapter is L. Hodges, ed., *Solar Energy, Book II* (College Park, Md.: AAPT 1986).

CHAPTER 22

1. D. McGuigan, *Harnessing Water Power for Home Energy* (Charlotte, VT: Garden Way, 1978); R. Wolfe and P. Clegg, *Home Energy for the 'Eighties* (Charlotte, VT: Garden Way, 1979), 167-81.

2. J. R. Moreira and A. D. Poole, in T. B. Johansson et al., eds., *Renewable Energy Sources for Fuels and Electricity* (Washington, D.C. and Covelo, CA: Island Press, 1993), 73.

3. J. Dowling, in R. Howes and A. Fainberg, eds., *The Energy Sourcebook* (New York: American Institute of Physics, 1991), 225. D. Asendorpf, *DZ*, 9 October 2002. VEB means volkseigener Betrieb, or nationally owned enterprise. The plans were developed when Thuringia was part of communist East Germany.

4. The World Commission on Dams (K. Asmal et al.), *Dams and development: A new framework for decision-making* (London and Sterling, VA: Earthscan Publications, Ltd., 2000).

5. C. P. Shea, in L. Starke, ed., *State of the World 1988* (New York and London: W. W. Norton, 1988), 62.

6. J. Chao, *PBP*, 6 May 2001. See also discussions of Chinese resettlement in A. Zich, *Natl. Geog.* **192**(3), 2 (1997); W. Sweet, *IEEE Spectrum* **38**(9), 46 (2001); T. Anthony, AP dispatch, 6 November 2002; Anonymous, Xinhua News Agency dispatch, 18 May 2003; E. Eckholm, *NYT*, 9 June 2003; and G. Baker, AP dispatch, 10 June 2003

7. P. Nixon, Reuters dispatch, 22 July 2003.

8. G. Long, Reuters dispatch, 4 December 2003.

9. C. Kraul, *LAT*, 9 May 2004.

10. L. Pendharkar, *SFC*, 17 January 2002.

11. D. E. Murphy, *NYT*, 14 September 2004.

12. E. Groth, III, *Environment* **17**(1), 28 (1975).

13. S. Trussart et al., *Energy Policy* **30**, 1251 (2002).

14. A. M. Josephy, Jr., *Audubon* **77**(2), 77 (1975).

15. D. Zimmerman, *Smithsonian* **14**(3), 28 (1983).

16. W. R. Jobin, *Intl. Water Power and Dam Construction* **38**(11), 19 (1986).

17. C. Canfield, *Natl. Hist.* **92**(7), 60 (1983).

18. W. Xiutao, *Intl. Water Power and Dam Construction* **38**(11), 23 (1986). The sedimentation problems of Chinese dams are also discussed in X. Lei, *Science* **280**, 24 (1998), in W. Sweet, *IEEE Spectrum* **38**(9), 46 (2001), and a more general discussion of sedimentation problems overall in C. J. Vörösmarty and D. Sahagian, *BioScience* **50**, 753 (2000), in F. K. Ligon et al., *BioScience* **45**, 183 (1995), and in K. F. Lagler, in D. A. Berkowitz and A. M. Squires, eds., *Power Generation and Environmental Change* (Cambridge, MA: MIT Press, 1971), 133.

19. R. Goodland, *Intl. Water Power and Dam Construction* **38**(11), 25 (1986).
20. C. Bodeen, AP dispatch, 13 June 2003. Anonymous, Reuters dispatch, 12 September 2002.
21. P. M. Fearnside, *Environ. Conserv.* **22**, 7 (1995).
22. G. H. Schueller, Environmental News Network dispatch, 10 January 2001. URL: http://www.enn.com/news/enn-stories/2001/01/01102001/hydropower_41184.asp
23. J. Yardley, *NYT*, 10 March 2004.
24. Anonymous, AP dispatch, 28 April 2004.
25. Anonymous, Reuters dispatch, 16 September 2003.
26. J. Rojas-Burke, *PO*, 18 May 2004.
27. G. W. Frey and D. M. Linke, *Energy Policy* **30**, 1261 (2002).
28. A. Weiner, *Am. Sci.* **60**, 466 (1972).
29. P. Jonsson, *CSM*, 30 January 2001. B. Harden, *WP*, 16 April 2003. J. Rojas-Burke, *PO*, 7 July 2003. S. Leavenworth, *SB*, 6 August 2003. P. Belluck, *NYT*, 7 October 2003. E. Shogren, *LAT*, 24 February 2004.
30. C. Saillant, *LAT*, 10 February 2004.
31. G. W. Griggs, *LAT*, 27 July 2004.
32. T. Philp, *SB*, 21 April 2002. H. A. Sample, *SB*, 4 August 2002. W. Booth, *WP*, 27 April 2003.
33. J. Brinckman, *PO*, 19 October 2002. J. Rojas-Burke, *PO*, 2 May 2003. J. Rojas-Burke, *PO*. 12 December 2003. Anonymous, AP Dispatch, 29 July 2004.
34. M. Higgins, Environmental News Network dispatch, 16 January 2001. URL: http://www.enn.com/enn-news-archive/2001/01/01162001/energydam_41320.asp
35. B. Harden, *WP*, 1 September 2004.
36. E. Whitelaw and E. MacMullan, *BioScience* **52**, 724 (2002).
37. N. L. Poff and D. D. Hart, *BioScience* **52**, 659 (2002).
38. D. D. Hart et al., *BioScience* **52**, 669 (2002). J. Pizzuto, *BioScience* **52**, 683 (2002).
39. R. E. Schmid, AP dispatch, 6 June 2000.
40. D. L. Chandler, *BG*, 24 April 2001. S. Steindorf and T. Regan, *CSM*, 17 May 2001.
41. B. V. Davis, Nova Energy, Ltd. report NEL-002 to the Canadian National Research Council, March 1980.
42. H.N. Halvorson Consultants Ltd., Report to Ministry of Employment and Investment, Government of British Columbia, 1994. Nova Energy changed its name to Blue Energy Canada, Inc.
43. I. Steinhorn and J. R. Gat, *Sci. Am.* **249**(4), 102 (1983); and Y. Ne'eman and I. Schul, *Annu. Rev. Energy* **8**, 113 (1983).
44. F. H. Koch, *Energy Policy* **30**, 1207 (2002).
45. R. Revelle, *Science* **209**, 164 (1980).
46. H. Tanner, *NYT*, 4 May 1975.
47. D. Deudney, *Environment* **23**(7), 16 (1981).
48. D. Okrent, *Science* **208**, 372 (1980).
49. Anonymous, *Water, Power, and Dam Construction* **37**(11), 33 (1985). This is a report on the International Workshop on Dam Failures from 1985.
50. The Johnstown Flood is commemorated by the National Park Service at the Johnstown Flood National Memorial. Refer to URL http://www.nps.gov/jofl/.
51. R. E. Schmid, AP dispatch, 13 July 2004.
52. K.V. Bury and H. Kreuzer, *Water, Power, and Dam Construction* **37**(11), 46 (1985).
53. P. Kakela, G. Chilson, and W. Patric, *Environment* **27**(1), 31 (1985).
54. B. P. Smith, *Environment* **20**(9), 16 (1978).
55. C. A. Neumann, *Alt. Sources Energy* **82**, 24 (1986); and D. Marier, *Alt. Sources Energy* **82**, 10 (1986).
56. E. J. Dionne, *NYT*, 4 October 1981.
57. New York State Energy Research and Development Authority, *Patterns And Trends: New York State Energy Profiles: 1986-2000* (December 2001). See also E. A. Gargan, *NYT*, 11 March 1984 for the hydroelectric situation as of 1984.
58. M. Johnson, *Alt. Sources Energy* **78**, 23 (1986); A. R. Inversin, *Alt. Sources Energy* **82**, 33 (1986).
59. Anonymous, *Futurist* **19**(3), 50 (1985).
60. L. J. Carter, *Science* **184**, 1353 (1974).
61. P. Hine, *NRDC News* **1**(2), 1 (1981).
62. J. E. Cavanagh et al., in T. B. Johansson et al., eds., *Renewable Energy Sources for Fuels and Electricity* (Washington, D.C. and Covelo, CA: Island Press, 1993), 513.
63. S. N. Paleocrassas, *Solar Energy* **16**, 45 (1974).
64. M. F. Merriam, *Annu. Rev. Energy* **3**, 29 (1978).
65. M. M. Sanders, in R. Howes and A. Fainberg, eds., *The Energy Sourcebook* (New York: American Institute of Physics, 1991), 257.
66. A. B. Meinel and M. P. Meinel, *Phys. Today* **25**(2), 44 (1972).
67. F. Pearce, *New Scientist* **158** (2139), 38 (1998).
68. K. O'Mara et al., URL: http://www.acre.murdoch.edu.au/refiles/tidal/text.html, 1999.
69. Anonymous, BBC News Online, 9 February 2003.
70. Anonymous, BBC News Online, 16 June 2003.
71. M. Peplow, Nature Science Update, 24 March 2004. T. Jüngling, *DW*, 16 May 2004.
72. T. Migge, *DW*, 28 February 2003.
73. House of Commons, Select Committee on Science and Technology, *Wave and Tidal Energy*, seventh report, February 2001.
74. Anonymous, *New Scientist* **156** (2108), 15 (1997). Anonymous, Reuters dispatch, 28 August 2001. A. Doyle, Reuters dispatch, 7 November 2002. R. Stone, *Science* **299**, 339 (2003). A. Doyle, Reuters dispatch, 23 September 2003.
75. Anonymous, Reuters dispatch, 7 May 2003. H. Pearson, Nature Science Update, 13 August 2004; |doi:10.1038/news040809-17.
76. D. F. Othmer and O. A. Roels, *Science* **182**, 121 (1973); C. Zener, *Phys. Today* **26**(1), 48 (1973).
77. UK Department of Trade and Industry, *New & Renewable Energy: Prospects for the 21st Century* (London: Department of Trade and Industry, 1999). Anonymous, Reuters dispatch, 21 November 2000. Anonymous, *Futurist* **35**(3), 2 (2001).
78. M. Knott, *New Scientist* **167** (2257), 16 (2000).
79. R. Edwards, *New Scientist* **160** (2154), 30 (1998).
80. Paul Brown, *TG*, 10 February 2004.
81. J. O. Löfken, *DW*, 28 May 2004. S. von der Weiden, *DW*, 22 March 2003.
82. B. Daley, *BG*, 6 June 2002.
83. T. W. Thorpe, *A Brief Review of Wave Energy, A report produced for The UK Department of Trade and Industry*, ETSU-R120, May 1999.
84. L.Claeson, *Energi från havets vågor* (Stockholm, Sweden: Energiforskningsnämnden, 1987).
85. T. Thorpe, in World Energy Organization, *Survey of Energy Resources* (London:WEC, 1998), Chapter 15; ibid., "Country notes."
86. J. Lear, *Saturday Review*, 5 December 1970, 53.
87. J. Constans, *Marine Sources of Energy* (New York: Pergamon Press, 1979). This is the report of the EUROCEAN group, presented to the United Nations Office of Science and Technology. It investigates the engineering feasibility of alternative sources of energy.
88. M. W. Browne, *NYT*, 10 February 1987.
89. J. Kurtz, *NYT*, 25 August 1991.
90. M. Locke, AP dispatch, 17 June 2001.
91. S. Endo, *New Scientist* **148** (2005), 42 (1995).
92. K. J. Tjugen, in G. Elliot and K. Diamantaras, eds., *Proceedings of the Second European Wave Power Conference* (European Commission: Brussels, 1995), pp 42-32.
93. C. Mitchell, *Annu. Rev. Energy Environ.* **25**, 285 (2000).

94. J. H. Anderson and J. H. Anderson, Jr., *Mech. Engg.* **88**(4), 41 (1966).

95. T. H. Daniel, Sustainable Development International, Sept. 2000.

96. L.A.Vega, in *Encyclopedia of Energy Technology and the Environment* (New York: John Wiley & Sons, Inc., 1995), pp. 2104.

97. R. Halloran, *NYT*, 22 May 1990.

98. R. J. Smith, *Science* **207**, 739 (1980). See also J. W. Shupe, *Science* **216**, 1193 (1982). Since 1984, wind energy has supplied about 10% of Hawaii's energy needs.

99. W. J. Broad, *NYT*, 13 July 1993.

100. Sea Solar Power website, URL: http://www.seasolarpower.com/.

101. B. Command, *West Hawaii Today*, 26 May 2001.

102. R. S. Norman, *Science* **186**, 350 (1974).

103. J. D. Isaacs and W. R. Schmitt, *Science* **207**, 265 (1980).

104. C. Nielsen, private communication. Professor Nielsen was a consultant for the first commercial use of a salt-gradient solar pondæfor heating a public swimming pool in Miamisburg, Ohio.

105. P. De Laquil et al., in T. B. Johansson et al.s, eds., *Renewable Energy Sources for Fuels and Electricity* (Washington, D.C. and Covelo, CA: Island Press, 1993), 213.

106. A. I. Kudish, D. Wolf, and Y. Machlav, *Solar Energy* **30**, 33 (1983).

107. T. H. Maugh, *Science* **216**, 1213 (1982).

108. D. Faiman, Israel Ministry of Foreign Affairs website, URL http://www.mfa.gov.il/MFA/Facts About Israel/Science—Technology/Solar Energy in Israel, 22 November 2002.

109. G. S. Virka et al., *Desalination* **137**, 149 (2001).

CHAPTER 23

1. D. M. Gates and W. Tantraporn, *Science* **115**, 613 (1952); C. G. Granqvist, *Appl. Opt.* **20**, 2606 (1981); and C. G. Granqvist, *Phys. Teach.* **24**, 372 (1984).

2. J. R. Bolton and D. O. Hall, *Annu. Rev. Energy* **4**, 353 (1979); G. R. Fleming and R. van Grondelle, *Phys. Today* **47**(2), 46 (1994). See also W. K. Stevens, *NYT*, 19 February 1991.

3. R. Radmer and B. Kok, *BioScience* **27**, 599 (1977).

4. B. O'Regan and M. Grätzel, *Nature* **353**, 737 (1991). T. E. Mallouk, *Nature* **353**, 698 (1991) and M. W. Browne, *NYT*, 24 October 1991.

5. K. N. Ferreira et al., *Science* **303**, 1831 (2004). A. W. Rutherford and A. Boussac, *Science* **303**, 1782 (2004).

6. G. Kurisu et al., *Science* **302**,1009 (2003).

7. D. O. Hall et al., in T. B. Johansson, H. Kelly et al., eds., *Renewable Energy Sources for Fuels and Electricity* (Washington, D.C. and Covelo, CA: Island Press, 1993), 593.

8. J. R. Gosz et al., *Sci. Am.* **238**(3), 93 (1978).

9. D. Pimentel et al. (Review Panel on Biomass Energy), *Solar Energy* **30**, 1 (1983).

10. L. R. Brown, in L. R. Brown, H. Kane, and E. Ayres, eds., *Vital Signs 1993* (New York: W. W. Norton, 1993), 26. G. Gardner, in L. Starke, ed. *State of the World 1997* (New York & London: W. W. Norton, 1997), 42; L. R. Brown, in L. Starke, ed. *State of the World 1998* (New York & London: W. W. Norton, 1998), 79; and L. R. Brown, in L. Starke, ed. *State of the World 1999* (New York & London: W. W. Norton, 1999), 115.

11. M. M. Waldrop, *Science* **211**, 914 (1981). A. I. Fraser, *Endeavour* **17**, 168 (1993); according to Fraser, about 1 Gt/yr of wood is burned, primarily as fuel for cooking in undeveloped countries.

12. S. Bronstein, *NYT*, 23 June 1985.

13. R. H. Williams and E. D. Larson, in T. B. Johansson et al., eds., *Renewable Energy Sources for Fuels and Electricity* (Washington, D.C. and Covelo, CA: Island Press, 1993), 729.

14. Anonymous, *NYT*, 16 December 1984.

15. C. E. Hewett et al., *Annu. Rev. Energy* **6**, 139 (1981).

16. AP dispatch, *NYT*, 13 October 1985.

17. J. W. Shupe, *Science* **216**, 1193 (1982).

18. J. Goldemberg, *Science* **200**, 158 (1978); A. L. Hammond, *Science* **195**, 564 (1977) and **195**, 566 (1977).

19. J. Goldemberg et al., in T. B. Johansson et al., eds., *Renewable Energy Sources for Fuels and Electricity* (Washington, D.C. and Covelo, CA: Island Press, 1993), 841.

20. C. E. Wyman et al., in T. B. Johansson et al., eds., *Renewable Energy Sources for Fuels and Electricity* (Washington, D.C. and Covelo, CA: Island Press, 1993), 865.

21. L. R. Lynd et al., *Science* **251**, 1318 (1991).

22. J. Goldemberg, *Science* **223**, 1357 (1984).

23. W. G. Pollard, *Am. Sci.* **64**, 509 (1976).

24. A. L. Hammond, *Science* **195**, 564 (1977).

25. C. C. Burwell, *Science* **199**, 1041 (1978).

26. N. Wade, *Science* **204**, 928 (1979).

27. S. E. Plotkin, *Environment* **22**(9), 6 (1980).

28. R. S. Chambers et al., *Science* **206**, 789 (1979). L. P. Rosa et al., *Energy* **23**, 987 (1998).

29. J. W. Ranney and J. H. Cushman, in R. Howes and A. Fainberg, eds., *The Energy Sourcebook* (New York: American Institute of Physics, 1991), 299.

30. J. Gomes Da Silva et al., *Science* **201**, 903 (1978).

31. F. F. Hartline, *Science* **206**, 41 (1979).

32. C. E. Wyman, *Annu. Rev. Energy Environ.* **24**, 189 (1999).

33. H. S. Kheshgi et al., *Annu. Rev. Energy Environ.* **25**, 199 (2000).

34. M. S. Sheya and S. J. S. Mushi, *Appl. Energy* **65**, 257 (2000).

35. S. Yokoyama et al., *Biomass and Bioenergy* **18**, 405 (2000).

36. K. Porter, D. Trickett, and L. Bird, NREL/TP-620-28674 (1999 Ed., August 2000).

37. M. L. Wald, *NYT*, 12 August 1999.

38. R. Fields, *LAT*, 23 January 2001. M. Taugher, *CCT*, 13 April 2001. T. Reiterman and D. Morain, *LAT*, 14 June 2001. But see also E. Bailey, *LAT*, 2 January 2001.

39. D. Pitt, AP dispatch, 15 January 2001.

40. G. Antolín et al., *Energy* **21**, 165 (1996).

41. N. Nowak, *DW*, 26 April 2000.

42. D. Lewerenz, AP dispatch, 22 February 2001. See also S. Levine, *WP*, 19 November 2002 and R. Weiss, *WP*, 27 January 2003 about corn-fed stoves in Takoma Park, Maryland.

43. N. Nowak, *DW*, 29 March 2001.

44. A. Melis et al., *Plant Physiol.* **122**, 127 (2000).

45. L. Dye, *LAT*, 28 February 2000.

46. C. E. Reimers et al., *Environ. Sci. Technol.* **35**, 192 (2001).

47. D. R. Bond et al., *Science* **295**, 483 (2002). J. Mullins, *New Scientist* **172** (2371), 24 (2001).

48. U. Schröder et al., *Angew. Chem.* **115**, 2986 (2003), *Angew. Chem. Intl. Edn.* **42**, 2880 (2003).

49. I. Ieropoulos et al., in W. Banzhaf et al., eds., *Advances in Artificial Life* (New York and Heidelberg: Springer-Verlag, 2004).

50. N. Mano et al., *J. Am. Chem. Soc.* **125**, 6588 (2003).

51. V. Soukharev et al., *J. Am. Chem. Soc.* **126**, 8368 (2004).

52. J. Howard, AP dispatch, 23 August 1999. T. E. Bull, *Science* **285**, 120 (1999).

53. M. L. Wald, *NYT*, 25 October 1997.

54. L. R. Lynd, *Annu. Rev. Energy Environ.* **21**, 403 (1996).

55. M. Giampietro et al., *BioScience* **47**, 587 (1997).

56. S. T.Bagley et al., *Environ. Sci. Technol.* **32**, 1183 (1998).

57. H. Özkaynak et al., Harvard University School of Public Health, Department of Environmental Health; Boston, MA (1996).

58. V. Singh, REPP Research Report number 12, Renewable Energy Policy Project, Winter 2001. T. Hendricks, *SFC*, 24 April 2001 and J. Kay, *SFC*, 23 May 2001.

59. C. Dixon, *NYT*, 22 April 2003.

60. Lovelace Respiratory Research Institute, Study Report Number FY98-056, submitted to the National Biodiesel Board, 1998.

61. N. Y. Kado et al.,Report to the Montana State Department of Environmental Quality and U.S. Department of Energy, November 1996.

62. National Renewable Energy Laboratory, Fact sheet, DOE/GO-102000-1048, May 2000. J. Duffield et al., *U.S. Biodiesel Development: New Markets for Conventional and Genetically Modified Agricultural Fats and Oils* U.S. Department of Agriculture Agricultural Economic Report No. 770 (1998).

63. The Veggivan website, URL: http://www.veggivan.org/. See also J.Tickell and K. Tickell, *From the Fryer to the Fuel Tank: The Complete Guide to Using Vegetable Oil as an Alternative Fuel* (third edition) (Sarasota, FL: Greenteach Publications, 2000).

64. S. Linden, Reuters dispatch, 16 March 2004. R. Haglund, Newhouse News Service dispatch, 10 April 2004.

65. T. Shore, *San Francisco Bay Crossings* **2**(4), (2001). For more about the Bay Area approach to biodiesel, see also T. Hendricks, *SFC*, 24 April 2001 and J. Kay, *SFC*, 23 May 2001.

66. S. Chawkins, *LAT*, 3 May 2002.

67. J. Higgins, National Biodiesel Board Press Release, www.biodiesel.org/Press releases/DESC11012001-1.htm, 1 November 2001.

68. R. Chase, AP dispatch, 6 May 2003.

69. C. Gammelin and F. Vorholz, *DZ*, 16 2002.

70. M. Y. E. Selim et al., *Renewable Energy* **28**, 1401 (2003).

71. Y. He and Y. D. Bao, *Renewable Energy* **28**, 1447 (2003), D. Özçimen and F. Karaosmanoğlu, *Renewable Energy* **29**, 779 (2004), and P. Janulis, *Renewable Energy* **29**, 861 (2004). (Rapeseed oil); S. Bari, T. H. Lim, and C. W. Yu, *Renewable Energy* **27**, 339 (2002) (Palm oil); K. Pramanik, *Renewable Energy* **28**, 239 (2003), F. K. Forson et al., *Renewable Energy* **29**, 1135 (2004) (Jatropha oil); and M. A. Kalam et al., *Renewable Energy* **28**, 2405 (2003) (Coconut oil).

72. A. S. Ramadhas et al., *Renewable Energy* **29**, 727 (2004).

73. M. Higgins, Environmental News Network dispatch, 2 March 2001, URL: http://www.enn.com/enn-news-archive/2001/03/03022001/biodiesel_42246.asp.

74. E. Baard, *NYT*, 12 May 2002. Anonymous, GreenBiz.com release, 16 October 2002. K. Gaudette, AP dispatch 30 December 2002. C. Doering, Reuters dispatch, 25 March 2003.

75. S. Lyall, *NYT*, 15 October 2002.

76. Anonymous, Reuters dispatch, 23 May 2003.

77. Anonymous, BBC News Online, 27 August 2004.

78. Anonymous, AP dispatch, 10 May 2004. This discusses the BioMax 15, a device for turning waste into electricity.

79. M. C. Heller et al., *Renewable Energy* **29**, 1023 (2004).

80. Anonymous, Reuters dispatch, 27 February 2003.

81. B. Bergstrom, AP dispatch, 16 May 2003. L. Belsie and M. Wiltenburg, *CSM*, 25 September 2003.

82. M. Smith, Bloomberg Business News, 30 August 2003.

83. Anonymous, Reuters dispatch, 22 February 2003. H. Fountain, *NYT*, 20 April 2004 for a new and different way to make pig manure into oil with an energy gain of 3 to 1.

84. S. Srinivasan, AP dispatch, 15 October 2003.

85. J. O'Dell, *LAT*, 18 October 2003.

86. L. Young and C. C.P. Pian, *Energy* **28**, 655 (2003).

87. G. Chellam, *The Jakarta Post*, 9 July 2001.

88. O. Khaselev and J. A. Turner, *Science* **280**, 425 (1998). R. F. Service, *Science* **280**, 382 (1998).

89. J. H. Schön et al., *Appl. Phys. Lett.* **77**, 2473 (2000). Pentacene-based printed thin-film solar cells have been made that can exhibit 4.3% efficiency, S. E. Shaheen et al., *Appl. Phys. Lett.* **79**, 2996 (2001).

90. M. Machida et al., *Chem. Commun.* **19**, 1939 (1999).

91. A. F. Heyduk and D. G. Nocera, *Science* **293**, 1639 (2001). J. K. McCusker, *Science* **293**, 1599 (2001).

92. C. C. Burwell, *Science* **199**, 1041 (1978).

93. J. H. Cook et al., *Annu. Rev. Energy Environ.* **16**, 401 (1991). D. Pimentel et al., *BioScience* **44**, 536 (1994).

94. C. P. Shea, in L. Starke, ed., State of the World 1988 (New York and London: W. W. Norton, 1988), 62.

95. D. Pimentel et al., *Science* **212**, 1110 (1981).

96. D. Pimentel et al., *BioScience* **42**, 354 (1992).

97. D. L. Klass, *Science* **223**, 1021 (1984).

98. Anonymous, *Futurist* **24**(1), 6 (1990).

99. L. F. Diaz et al., *CRC Critical Reviews in Environmental Control* **14**, 251 (1984).

100. D. Pimentel et al., *Adv. Food Res.* **32**, 185 (1988).

101. E. S. Lipinsky, *Science* **212**, 1465 (1981).

102. H. R. Bungay, *Science* **218**, 643 (1982).

103. M. Downing et al., in *Proceedings, Second Biomass Conference of the Americas: Energy, Environment,* *Agriculture, and Industry* NREL/CP-200-8098 (Golden, CO: National Renewable Energy Laboratory, 1996), 288.

104. G. Stix, *Sci. Am.* **269**(5), 104 (1993).

105. R. Pool, *Science* **245**, 1187 (1989). J. Holusha, *NYT*, 21 October 1990.

106. W. S. Hillman and D. D. Culley, *Am. Sci.* **66**, 442 (1978).

107. C. W. Hinman, *Science* **225**, 1445 (1984). L. Icerman and C. W. Hinman, *Science* **227**, 466 (1985).

108. R. Prescott-Allen and C. Prescott-Allen, *Cons. Biol.* **4**, 365 (1990).

109. Recent information on jojoba may be found in P. Milthorpe, in K. Hyde, ed., *The New Rural Industries: A handbook for Farmers and Investors* (Canberra: Rural Industries Research and Development Corporation, 1997), A. Benzioni, New Crop FactSHEET, Purdue University Center for New Crops and Plant Products, 1997, and D. J. Undersander et al., in *Alternative Field Crops Manual*, University of Wisconsin-University of Minnesota Cooperative Extension, 1992.

110. J. D. Johnson and C. W. Hinman, *Science* **208**, 460 (1980).

111. C. W. Hinman, *Sci. Am.* **255**(1), 32 (1986). J. Ralston, *Audubon* **96**(2), 20 (1994).

112. H. Cleveland and A. King, *Futurist* **14**(3), 47 (1980).

113. L. Brown, *Environment* **22**(4), 32 (1980).

114. J. H. Cock, *Science* **218**, 755 (1982).

115. M. A. El-Sharkawy, *BioScience* **43**, 441 (1993).

116. N. D. Vietmeyer, *Science* **232**, 1379 (1986). S. Lyall, *NYT*, 17 May 1987.

117. C. R. Clement, *BioScience* **39**, 624 (1989).

118. K. Schneider, *NYT*, 28 March 1986.

119. J. E. Brody, *NYT*, 13 December 1988.

120. T. H. Maugh, *Science* **194**, 46 (1976).

121. T. H. Maugh, *Science* **206**, 436 (1979).

122. M. Calvin, *J. Chem. Ed.* **64**, 335 (1987).

123. E. P. Glenn et al., *Science* **251**, 1065 (1991).

124. G. Lattinga and A. C. Van Haandel, in T. B. Johansson et al., eds., *Renewable Energy Sources for Fuels and Electricity* (Washington, D.C. and Covelo, CA: Island Press, 1993), 817.

125. H. S. Makunda et al., in T. B. Johansson et al., eds., *Renewable Energy Sources for Fuels and Electricity* (Washington, D.C. and Covelo, CA: Island Press, 1993), 699.

126. S. Kartha and E. D. Larson, *Bioenergy Primer: Modernised Biomass Energy for Sustainable Development* (: United Nations Development Programme, 2000).

127. J. B. Tucker, *Environment* **24**(8), 13 (1982).

128. P. Rajabapaiah et al., in T. B. Johansson et al., eds., *Renewable Energy Sources*

for Fuels and Electricity (Washington, D.C. and Covelo, CA: Island Press, 1993), 787. S. Kartha and E. D. Larson, *Bioenergy Primer* (New York: United Nations Development Programme, 2000), Chapters 5 and 7.

129. Anonymous, Reuters dispatch, 19 July 2002. T. King, *CSM*, 21 November 2002. A. Scholzen, *DW*, 28 March 2003. Anonymous, *NYT*, 11 May 2003. L Rathke, AP dispatch, 19 September 2003. F. Alvarez, *LAT*, 8 December 2003.

130. E. P. Eckholm, *Losing Ground* (New York: W. W. Norton & Co., 1976); S. Postel, *Natl. Hist.* **94**(4), 58 (1985).

131. F. R. Manibog, *Annu. Rev. Energy* **9**, 199 (1984).

132. M. R. Dove and D. M. Kammen, *Environment* **39**(6), 10 (1997).

133. J. Manuel, *Environ. Health Perspect.* **111**, A28 (2003).

134. M. Ezzati and D. M. Kammen, *The Lancet* **358**, 619 (2001).

135. D. M. Kammen, *Sci. Am.* **273**(1), 72 (1995).

136. J. Sundell, *Indoor Air* **14** (Suppl 7), 51 (2004).

137. M. Ezzati and D. M. Kammen, *Environ. Health Perspect.* 110, 1057 (2002).

138. R. D. Edwards et al., *Energy Policy* **32**, 395 (2004).

139. M. Ezzati et al., *Environ. Sci. Technol.* **34**, 578 (2000).

140. V. V. N. Kishore and P. V. Ramana, *Energy* **27**, 47 (2002).

141. M. S. Musthapa et al., *Environ. Molec. Mutagenesis* **43**, 243 (2004).

142. E. E. Dooley, *Environ. Health Perspect.* **111**, A31 (2003).

143. E. Park and K. Lee, *Indoor Air* **13**, 253 (2003).

144. K. Balakrishnan et al., *Environ. Health Perspect.* **110**, 1069 (2002).

145. K. R. Weiss, *LAT*, 29 August 2002.

146. E. Hood, *Environ. Health Perspect.* **110**, A691 (2002).

147. V. Mishra, *Environ. Health Perspect.* **111**, 71 (2003).

148. M. Ezzati and D. M. Kammen, *Energy Policy* **30**, 815 (2002).

149. D. M. Kammen, *Environment* **41**(5), 10 (1999).

150. A. H. Rosenfeld, *Annu. Rev. Energy Environ.* **24**, 33 (1999).

151. K. R. Smith et al., *Annu. Rev. Energy Environ.* **25**, 741 (2000). R. Bailis et al., *Environ. Sci. Technol.* **37**, 2051 (2003).

152. G. S. Dutt and N. H. Ravindranath, in T. B. Johansson et al., eds., *Renewable Energy Sources for Fuels and Electricity* (Washington, D.C. and Covelo, CA: Island Press, 1993), 653.

153. J. Zhang et al., *Atm. Environ.* **34**, 4537 (2000).

154. J. Ludwig et al., *J. Atm. Chem.* **44**, 23 (2003). J. S. Levine, *Nature* **423**, 28 (2003).

155. S. C. Bhattacharya et al., *Energy* **25**, 169 (2000).

156. C. K. Yoon, *NYT*, 9 July 2002.

157. R. A. Lucky and I. Hossain, *Energy* **26**, 221 (2001).

158. N. V. Patel and S. K. Philip, *Renewable Energy* **20**, 347 (2000).

159. A. V. Sonune and S. K. Philip, *Renewable Energy* **28**, 1225 (2003).

160. Anonymous, *NYT*, 11 December 1990.

161. A. H. Algifri and H. A. Al-Towaie, *Solar Energy* **70**, 165 (2001).

162. C. Holden, *Science* **228**, 1073 (1985).

163. J. J. Nicholaides et al., *BioScience* **35**, 279 (1985).

CHAPTER 24

1. M. A. Zeder and B. Hesse, *Science* **287**, 2254 (2000). J. N. Wilford, *NYT*, 18 November 1997.

2. R. Rappaport, in G. Piel et al., eds., *Energy and Power* (San Francisco: Freeman, 1972), p. 69. For further information on energy costs in agriculture, see also F. S. Roberts, in H. Marcus-Roberts and M. Thompson, eds., *Modules in Applied Mathematics*, V. 4, *Life Science Models*, (New York: Springer-Verlag, 1984), 250 and F. S. Roberts and H. Marcus-Roberts, in H. Marcus-Roberts and M. Thompson, eds., *Modules in Applied Mathematics*, V. 4, *Life Science Models*, (New York: Springer-Verlag, 1984), 287.

3. R. Allen, *Ecologist* **5**, 4 (1975).

4. W. K. Stevens, *NYT*, 3 April 1990.

5. R. Baker, *Ecologist* **4**, 170 (1974); M. B. Coughenour et al., *Science* **230**, 619 (1985).

6. S. Rojstaczer et al., *Science* **294**, 2549 (2001). They determine 32%. C. B. Field, *Science* **294**, 2490 (2001). P. M. Vitousek et al., *BioScience* **36**, 368 (1986).

7. W. K. Stevens, *NYT*, 22 November 1988.

8. R. Cordaux et al., *Science* **304**, 1125 (2004).

9. D. Pimentel and E. L. Terhune, *Annu. Rev. Energy* **2**, 171 (1977), Tables 2, 3, 4, 5, and 7; F. S. Roberts, in H. Marcus-Roberts and M. Thompson, eds., *Modules in Applied Mathematics*, V. 4, *Life Science Models*, (New York: Springer-Verlag, 1984), 250 (see page 269).

10. D. Pimentel et al., *Science* **182**, 443 (1973); and V. W. Ruttan and D. Pimentel, *Science* **187**, 560 (1975).

11. J. S. Steinhart and C. E. Steinhart, *Science* **184**, 307 (1974).

12. Dept. of Commerce, *Statistical Abstract of the United States*, 1972 (Washington, DC: GPO, 1972). Dept. of Commerce, *Statistical Abstract of the United States*, 1984 (Washington, DC: GPO, 1984).

Dept. of Commerce, *Statistical Abstract of the United States*, 1993 (Washington, DC: GPO, 1993).

13. K. O. Pope et al., *Science* **292**, 1370 (2001).

14. S. H. Wittwer, *BioScience* **24**, 216 (1974).

15. Reference 12, Table 1183, 670 (1983).

16. Reference 12, Table 1010, 608 (1974); Reference 12, Table 1193, 676 (1974).

17. L. R. Brown, in L. R. Brown et al., eds., *Vital Signs 1993* (New York: W. W. Norton, 1993), 26, and ibid. 41. See also G. Gardner, in L. R. Brown et al., eds., *Vital Signs 1997* (New York: W. W. Norton, 1997), 40.

18. Reference 12, Table 1002, 604 (1974); Reference 13, Table 1185, 672 (1983).

19. A. T. Dunning and H. B. Brough, in L. Starke, ed., *State of the World 1992* (New York: W. W. Norton, 1992), 66. See also the discussions on livestock husbandry in G. Gardner, in L. Starke, ed., *State of the World 1998* (New York: W. W. Norton, 1998), 96 and L. R. Brown, in L. Starke, ed. *State of the World 2001* (New York & London: W. W. Norton, 2001), 43.

20. D. Pimentel, *J. Agr. Environmental Ethics* **6**, 53 (1993).

21. B. Melley, *LAT*, 10 September 2004.

22. G. Polakovic, *LAT*, 6 May 2003.

23. D. Pimentel et al., *Science* **207**, 843 (1980).

24. J. 8. Lee, *NYT*, 11 May 2003.

25. K. A. Johnson and D. E. Johnson, *J. Anim. Sci.* **73**, 2483 (1995).

26. R. Sanchez, *WP*, 11 July 2004.

27. J. Sterngold, *NYT*, 15 May 2002. S. Gold, *LAT*, 20 November 2002. M. Grossi, *FB*, 23 December 2002. E. Shogren, *LAT*, 25 September 2003. C. Pogash, *NYT*, 1 July 2004.

28. M. Janofsky, *NYT*, 3 June 2004.

29. A. D. G. Wright et al., *Vaccine* **22**, 3976 (2004). C. Dennis, *Nature* **429**, 119 (2004).

30. Y. S. Do et al., *Appl Environ Microbiol.* **69**, 1710 (2003).

31. J. C. Kuehner, *CPD*, 25 February 2003.

32. M. McCulloch et al., *Nature* **421**, 727 (2003). J. Cole, *Nature* **421**, 705 (2003).

33. D. L. Beck, *SJMN*, 10 July 2004.

34. D. Pimentel et al., in S. R. Gleissman, ed., *Agroecology* (Berlin and New York: Springer-Verlag, 1990), 305. See also D. Pimentel and W. Dazhong, in C. R. Carroll et al., eds., *Agroecology* (New York: McGraw Hill, 1990).

35. D. Pimentel et al., *BioScience* **37**, 277 (1987).

36. Office of Pesticide Programs, United States Environmental Protection Agency, *The Role of Use-Related Information in Pesticide Risk Assessment and Risk Management* (Washington, DC: GPO, 2000).

37. D. Pimentel et al., *BioScience* **41**, 402 (1991).

38. D. Pimentel et al., *BioScience* **42**, 750 (1992). D. Pimentel et al., *BioScience* **28**, 376 (1978).
39. A. DePalma, *NYT*, 1 September 2004.
40. C. Norman, *Science* **242**, 366 (1988).
41. D. Pimentel and L. Levitan, *BioScience* **36**, 86 (1986).
42. Umweltbundesamt (German Environment Ministry), *Daten zur Umwelt 1990/91* (Berlin: Erich Schmidt Verlag, 1992), 334.
43. P. A. Matson et al., *Science* **277**, 504 (1997).
44. U.S. Environmental Protection Agency, *Implementing the Food Quality Protection Act* (Washington, DC: GPO, 1999), EPA 735-R-99001.
45. Committee on the Future Role of Pesticides in US Agriculture (M. Berenbaum et al.), National Academy of Sciences, *The Future Role of Pesticides in US Agriculture* (Washington, DC: National Academy Press, 2002).
46. D. Hanson, *Chem. & Engg. News* **76**(18), 38 (1998).
47. S. Fulwood, *LAT*, 9 June 2000. A. C. Revkin, *NYT*, 9 June 2000.
48. Anonymous, *NYT*, 11 December 2000.
49. C. Bowman, *SB*, 27 March 2001.
50. Committee on Pesticides in the Diets of Infants and Children (P. I. Landrigan et al.), National Academy of Sciences, *Pesticides in the Diets of Infants and Children* (Washington, DC: National Academy Press, 1993).
51. L. Bergquist, *MJS*, 10 December 2002. Anonymous, AP dispatch, 19 March 2003.
52. M. Lee, *SB*, 12 November 2003.
53. S. Lee, R. McLaughlin et al., *Environ. Health Perspect.* **110**, 1175 (2002). S. Kegley, A. Katten, and M. Moses, *Secondhand Pesticides: Airborne Pesticide Drift in California* (San Francisco: Pesticide Action Network North America, 2003).
54. A. Jennings, in the conference Changing Expectations, Columbus Ohio, 24 February 1993.
55. Umweltbundesamt (German Environment Ministry), *Daten zur Umwelt 1990/91* (Berlin: Erich Schmidt Verlag, 1992), 158.
56. D. Pimentel et al., *BioScience* **39**, 606 (1989).
57. A. S. Moffat, *Science* **261**, 550 (1993).
58. Food and Agricultural Organization/World Health Organization, Press release, 5 January 2001, URL: http://www.fao.org/ WAICENT/OIS/PRESS_NE/ PRESSENG/2001/pren0105.htm.
59. T. B. Hayes et al., *Proc. Natl. Acad. Sci.* **99**, 5476 (2002). T. Hayes et al., *Nature* **419**, 895 (2002).

60. S. I. Storrs and J. M. Kiesecker, *Environ. Health Perspect.* **112**, 1054 (2004).
61. E. Walsh, *WP*, 1 February 2003. M. Cone, *LAT*, 27 July 2004.
62. T. Henry, *TB*, 22 February 2004.
63. R. Renner, *Science* **298**, 938 (2002). R. Dalton, *Nature* **420**, 256 (2002).
64. J. N. Perry et al., *Nature* **428**, 313 (2004).
65. R. A. Relyea, *Ecol. Appl.* **13**, 1515 (2003).
66. M. D. Boone et al., *Ecol. Appl.* **14**, 685 (2004).
67. E. Shogren, *LAT*, 4 June 2003.
68. J. Kaiser, *Science* **299**, 327 (2003).
69. See Ref. 60. T. Harlow, U.S. Geological Survey Press Release, 7 December 2000. C. T. Hall, *SFC*, 8 December 2000 and M. Cone, *LAT*, 8 December 2000. A. Mattoon, in L. Starke, ed., *State of the World 2001* (New York & London: Norton, 2001), 63.
70. P. Frost, Ohio State University Press Release, 15 November 1999, URL: http://www.acs.ohio-state.edu/units/ research/archive/evrgreen.htm.
71. C. De Moraes, ScienceUpdate, USDA Agricultural Research (July 2000).
72. W. K. Stevens, *NYT*, 5 July 1991.
73. D. Kelly, *LAT*, 21 July 2003.
74. C. S. Gasser and R. T. Fraley, *Science* **244**, 1293 (1989).
75. R. Ahlstrom, *IT*, 17 April 2000.
76. T. S. Walker et al., *Plant Physiol.* **132**, 44 (2003).
77. S.-W. Park et al., *Plant Physiol.* **130**, 164 (2002).
78. M. Nishimura and S. Somerville, *Science* **295**, 2032 (2002). C. Azevedo et al., *Science* **295**, 2073 (2002). M. J. Austin et al., *Science* **295**, 2077 (2002).
79. F. Alvarez, *LAT*, 26 August 2001.
80. M. L. Henneman and J. Memmott, *Science* **293**, 1314 (2001).
81. R. Nowak, New Scientist Online, 5 March 2003. S. Whyard, talk presented at the Conference Horizons in Livestock Sciences 2004: Gene Silencing and Therapeutic Innovations, Gold Coast International Hotel, Surfers Paradise, Queensland, Australia, 15 September 2004.
82. D. P. Collins et al., *Biol. Control* **26**, 224 (2003).
83. B. J. Jacobsen, CSREES Program Support for Pest Management Project Details, 2002.
84. J. Huang et al., *Nature* **418**, 678 (2002).
85. W. Witte, *Science* **279**, 996 (1998).
86. L. R. Brown, in L. R. Brown, H. Kane, and E. Ayres, eds., *Vital Signs 1993* (New York: W. W. Norton, 1993), 40.
87. J. N. Galloway et al., *BioScience* **53**, 341 (2003).
88. S. Fields, *Environ. Health Perspect.* **112**, A 556 (2004).

89. B. Commoner, in F. Singer, ed., *Global Effects of Environmental Pollution* (Dodrecht, Neth.: D. Reidel Publ. Co., 1970), 70.
90. B. Commoner, *The Closing Circle* (New York: Alfred A. Knopf, 1971).
91. M. E. Fenn et al., *BioScience* **53**, 404 (2003).
92. C. T. Driscoll et al., *BioScience* **53**, 357 (2003).
93. J. D. Aber et al., *BioScience* **53**, 375 (2003).
94. Umweltbundesamt (German Environment Ministry), *Daten zur Umwelt 1990/91* (Berlin: Erich Schmidt Verlag, 1992), 277-378.
95. P. Mäder et al., *Science* **296**, 1694 (2002). E. Stokstad, *Science* **296**, 1589 (2002) and C. Macilwain, *Nature* **428**, 792 (2004). E. Green, *LAT*, 31 May 2002 and R. Reed, *NYT*, 20 July 1975.
96. J. P. Reganold et al., *Science* **260**, 344 (1993).
97. C. R. Greene, Economic Research Service/USDA AIB-770 (June 2001). C. R. Greene, Economic Research Service/USDA, Agricultural Outlook/April 2000, 9.
98. K. Lydersen, *WP*, 26 August 2002. D. Barboza, *NYT*, 25 November 2003.
99. D. K Asami et al., *J. Agric. Food Chem.* **51**, 1237 (2003). E. Lau, *SB*, 7 March 2003.
100. L. Belsie, *CSM*, 5 July 2002.
101. C. Ness, *SFC*, 27 May 2004.
102. R. T. Estrada, *MB*, 16 January 2001.
103. K. Severson, *SFC*, 13 October 2002.
104. M. Oesterheld et al., *Nature* **356**, 234 (1992).
105. J. E. Brody, *NYT*, 9 October 1990.
106. J. S. Boyer, *Science* **218**, 443 (1982).
107. J. Harwood et al., in R. Shoemaker, ed., *Economic Issues in Agricultural Biotechnology* Resource Economics Division, Economic Research Service, U.S. Department of Agriculture, Agriculture Information Bulletin 762 (2001).
108. L. Mitchell, U.S. Department of Agriculture Economic Research Service, Agriculture Information Bulletin 765-11 (June 2001).
109. P. W. Heisey et al., U.S. Department of Agriculture, Agriculture Information Bulletin 762 (2001).
110. I. Sarageldin, *Science* **285**, 387 (1999).
111. J. Harwood et al., in R. Shoemaker, ed., *Economic Issues in Agricultural Biotechnology* Resource Economics Division, Economic Research Service, U.S. Department of Agriculture, Agriculture Information Bulletin 762 (2001).
112. J. Harwood et al., in R. Shoemaker, ed., *Economic Issues in Agricultural Biotechnology* Resource Economics Division, Economic Research Service,

U.S. Department of Agriculture, Agriculture Information Bulletin 762 (2001).

113. D. Normile, *Science* **283**, 313 (1999).

114. C. Sommerville and S. Sommerville, *Science* **285**, 380 (1999).

115. C. Mann, *Tech. Rev.* **102**(4), 36 (1999).

116. G. Conway and G. Toenniessen, *Nature* **402**, C55 (1999).

117. K. M. Leisinger, in P. Pinstrup-Andersen, ed. *Biotechnology for Developing-country Agriculture: Problems and Opportunities* (Washington, DC: International Food Policy Research Institute, 1999).

118. M. W. Rosegrant et al., International Food Policy Research Institute paper, Washington, D.C. (August 2001).

119. M. D. Gale and K. M. Devos, *Science* **282,** 656 (1998).

120. C. Mann, *Science* **283**, 310 (1999).

121. C. Mann, *Science* **277**, 1040 (1997).

122. A. S. Moffat, *Science* **285**, 369 (1999).

123. L. L. Wolfenbarger and P. R. Phifer, *Science* **290**, 2088 (2000).

124. Y. Carrière et al., *Proc. Natl. Acad. Sci.* **100**, 1519 (2003).

125. Y. Ismael et al., *AgBioForum* **5**, 1 (2003). R. Bennett et al., *Outlook on Agriculture* **32**, 123 (2003). S., Morse et al., *Nature Biotechnology* **22**, 379 (2004).

126. M. Qaim and D. Zilberman, *Science* **299**, 900 (2003).

127. J. Whitfield, Nature Science Update, 7 February 2003.

128. A. Pollack, *NYT*, 15 May 2001.

129. A. S. Moffat, *Science* **296**, 1226 (2002).

130. J. Yu et al. *Science* **296**, 79 (2002); S. A. Goff et al., *Science* **296**, 92 (2002); J. Bennetzen, *Science* **296**, 60 (2002); T. Sasaki et al., *Nature* **420**, 312 (2002); Q. Feng et al., *Nature* **420**, 316 (2002); E. Check, *Nature* **420**, 259 (2002); and Y. Yu et al., *Science* **300**, 1566 (2003).

131. K. S. Fischer et al., *Science* **290**, 279 (2000).

132. P. L. E. Bodeller et al., *Nature* **403**, 421 (2000). J. Schmiel, *Nature* **403**, 375 (2000).

133. H. A. C. Denier van der Gon et al., *Proc. Natl. Acad. Sci.* **99**, 12021 (2002).

134. X. Ye et al., *Science* **287**, 303 (2000). M. L. Guerinot, *Science* **287**, 241 (2000).

135. A. Pollack, *NYT*, 7 March 2001.

136. A. S. Moffat, *Science* **282**, 2176 (1998).

137. R. Palmer, Reuters dispatch, 20 January 2004.

138. B. Simon, *NYT*, 22 May 2004. S. Leahy, IPS News Agency dispatch, 5 October 2004.

139. M. A. Rieger et al., *Science* **296**, 2386 (2002).

140. E. Stokstad, *Science* **296**, 2314 (2002).

141. C. Y. Huang et al., *Nature* **422**, 72 (2003).

142. C. L. Cummings et al., *Ecol. Appl.* **12**, 1661 (2002).

143. A. A. Snow et al., *Ecol. Appl.* **13**, 279 (2003).

144. R. Dalton, *Nature* **419**, 655 (2002).

145. M. D. Halfhill et al., *Environ. Biosafety Res.* **1**, 19 (2002).

146. M. J. Wilkinson et al., *Science* **302**, 457 (2003).

147. V. Jaenicke-Després et al., *Science* **302**, 1206 (2003). N. V. Fedoroff, *Science* **302**, 1158 (2003).

148. D. Quist and I. H. Chapela, *Nature* **414**, 541 (2001). See, however, M. Metz and J. Fütterer, *Nature* **416**, 600 (2002), N. Kaplinsky et al., *Nature* **416**, 601 (2002), and D. Quist and I. H. Chapela, *Nature* **416**, 602 (2002). D. Butler, *Nature* **415**, 948 (2002) and C. C. Mann, *Science* **296**, 236 (2002).

149. C. L. Carpentier and H. Herrmann, Report by the Secretariat of the Commission for Environmental Cooperation of North America as part of the Article 13 initative on Maize and Biodiversity: the Effects of Transgenic Maize in Mexico, 2004.

150. J. P. Schernthaner et al., *Proc. Natl. Acad. Sci.* **100**, 6855 (2003).

151. J.-F.Arnaud et al., *Proc. R. Soc. Lond. B* **270**, 1565 (2003).

152. M. J May, *Ann. Appl. Biology* **142**, 41 (2003).

153. J. Harwood et al., in R. Shoemaker, ed., *Economic Issues in Agricultural Biotechnology* Resource Economics Division, Economic Research Service, U.S. Department of Agriculture, Agriculture Information Bulletin 762 (2001).

154. G. Gaskell et al., *Science* **285**, 384 (1999).

155. C. Tilstone, *Nature* **423**, 672 (2003).

156. M. McCarthy, *TI*, 25 September 2003.

157. J. Giles, *Nature* **428**, 107 (2004).

158. E. Masood, *Nature* **426**, 224 (2003). A. Coghlan, New Scientist Online, 29 January 2003.

159. W. S. A., *DW*, 6 January 2003. J. Gillis, *WP*, 17 May 2004. P. Murphy, *TG*, 17 May 2004.

160. S. R. Owens, *EMBO Reports* **4**, 229 (2003).

161. N. C. Ellstrand, *Phil. Trans. R. Soc. Lond. B* **358**, 1163 (2003).

162. J. F. Hancock, *BioScience* **53**, 512 (2003).

163. R. S. Hails, *Nature* **418**, 685 (2002).

164. J. E. Losey et al., *Nature* **399**, 214 (1999).

165. D. S. Pimentel and P. H. Raven, *Proc. Natl. Acad. Sci.* **97**, 8198 (2000).

166. C. L. Wraight et al., *Proc. Natl. Acad. Sci.* **97**, 7700 (2000).

167. R. P. Freckleton et al., *Science* **302**, 994 (2003).

168. A. Pollack, *NYT*, 15 May 2001.

169. C. Mann, *Science* **277**, 1038 (1997).

170. B. Mazur et al., *Science* **285**, 372 (1999).

171. S. D. Tanksley and S. R. McCouch, *Science* **277**, 1063 (1997).

172. C. C. Mann, *Science* **283**, 314 (1999).

173. C. H. Foyer and G. Noctor, *Science* **284**, 599 (1999).

174. T. Gura, *Science* **287**, 412 (2000).

175. J. Cohen, *Science* **276**, 1960 (1997).

176. D. DellaPenna, *Science* **285**, 375 (1999).

177. W. M. Gelbart, *Science* **282**, 659 (1998).

178. A. S. Moffat, *Science* **249**, 630 (1990).

179. J. Smith, Reuters dispatch, 17 November 2001.

180. R. J. Davenport, *Science* **291**, 807 (2001).

181. J. Harwood et al., in R. Shoemaker, ed., *Economic Issues in Agricultural Biotechnology* Resource Economics Division, Economic Research Service, U.S. Department of Agriculture, Agriculture Information Bulletin 762 (2001).

182. A. Pollack, *NYT*, 15 May 2001.

183. G. Gardner and B. Halweil, in L. Starke, ed., *State of the World 2000* (New York & London: Norton, 2000), 59.

184. J. Bergelson et al., *Nature* **395**, 25 (1998).

185. B. Halweil, *World Watch*, 21 (July August 1999).

186. A. S. Moffat, *Science* **257**, 482 (1992).

187. A. S. Moffat, *Science* **252**, 211 (1991).

188. B. E. Tabashnik et al., *Proc. Natl. Acad. Sci.* **97**, 12980 (2000). There is concern also about new diseases, as in B. W. Falk and G. Bruening, *Science* **263**, 1395 (1994).

189. J. J. Obrycki et al., *BioScience* **51**, 353 (2001).

190. E. E. Ortman et al., *BioScience* **51**, 900 (2001).

191. L. J. Gahan et al., *Science* **293**, 857 (2001).

192. J. S. Griffitts et al.,*Science* **293**: 860 (2001).

193. S. H. Strauss, *Science* **300**, 61 (2003).

194. D. Charles, *Science* **294**, 772 (2001).

195. D. Charles, *Science* **294**, 1263 (2001).

196. A. S. Moffat, *Science* **250**, 910 (1990). R. N. F. Thorneley, *Nature* **360**, 532 (1992).

197. S. E. Lindow et al., *Science* **244**, 1300 (1989).

198. W. K. Stevens, *NYT*, 10 July 1990.

199. I. C. Burke et al., *BioScience* **52**, 813 (2002). J. C. Neff et al., *Nature* **419**, 915 (2002). S. V. Smith et al., *BioScience* **53**, 235 (2003).

200. N. Nosengo, *Nature* **425**, 894 (2003).

201. P. A. Raymond and J. J. Cole, *Science* **301**, 88 (2003). V. Ittekkot, *Science* **301**, 56 (2003).

202. D. Avery, *Science* **230**, 408 (1985).

203. G. Daily et al., *Science* **281**, 1291 (1998).

204. G. Lean, *TI*, 29 August 2004.

205. L. Belsie, *CSM*, 20 February 2003.

206. D. Brough, Reuters dispatch, 7 March 2003.

207. M. W. Rosegrant and S. A. Cline, *Science* **302**, 1917 (2003).

208. M. A. Stocking, *Science* **302**, 1356 (2003).

209. H. Tiessen, *AvH-Mitteilungen* #74, 33 (1999).

210. D. Gonzalez, *NYT*, 15 January 2001.

211. P.E. Rasmussen et al., *Science* **282**, 893 (1998).

212. V. Smil, *Ambio* **31**, 126 (2002).

213. P. E. Fixen and F. B. West, *Ambio* **31**, 169 (2002).

214. N. C. Brady, *Science* **218**, 847 (1982).

215. L. R. Brown, *Natl. Hist.* **94**(4), 63 (1985).

216. R. Revelle, *Science* **192**, 969 (1976).

217. G. Conway and G. Toenniessen, *Science* **299**, 1187 (2003).

218. "Prester John," *Ecologist* **4**, 304 (1974); P. R. Kann, *WSJ*, 18 November 1974; P. R. Jennings, *Science* **186**, 1085 (1974); N. Wade, *Science* **186**, 1093 and *Science* **186**, 1186 (1974); T. T. Poleman, *Science* **188**, 510 (1975); and R. B. Trenbath, *Ecologist* **5**, 76 (1975).

219. V.K. McElheny, *NYT*, 1 September 1974.

220. United Nations Environment Programme, R. Clarke, ed., *Global Environmental Outlook 2000* (London: Earthscan Publications, Ltd, 1999). The report is also available on the web at URL http://www.grida.no/geo2000/.

221. D. A. Wedin and D. Tilman, *Science* **274**, 1720 (1996).

222. A. J. Symstad et al., *BioScience* **53**, 89 (2003).

223. D. Tilman et al., *Nature* **418**, 671 (2002)

224. B. J. Peterson et al., *Science* **292**, 86 (2001); R. B. Alexander et al., *Nature* **401**, 756 (2000).

225. E. M. Bennett et al., *BioScience* **51**, 227 (2001).

226. D. Malakoff, *Science* **279**, 1308.(1998).

227. A. P. Kinzig and R. H. Socolow, *Phys. Today* **47**(11), 24 (1994). A. S. Moffat, *Science* **279**, 988.(1998).

228. G. P. Asner et al., *BioScience* **47**, 226 (1997).

229. D. A. Wedin and D. Tilman, *Science* **274**, 1720 (1996).

230. R. W. Howarth et al., *Ambio* **31**, 88 (2002).

231. R. H. Socolow, *Proc. Natl. Acad. Sci.* **96**, 6001 (1999).

232. J. Kaiser, *Science* **294**, 1268 (2001).

233. S. Seitzinger and C.Kroeze, *Nutrient Cycling in Agroecosystems* **52**, 195 (1998).

234. K. G. Cassman et al., *Ambio* **31**, 132 (2002).

235. O. Oenema and S. Pietrzak, *Ambio* **31**, 59 (2002).

236. R. N. Roy, R. V. Misra, and A. Montanez, *Ambio* **31**, 77 (2002).

237. K. N. Chaudhuri, *Endeavour* **15**, 68 (1991) discusses qanats in detail. A. Smith, *Blind White Fish in Persia* (London: Penguin Travel Library,

1953). H. E. Wulff, *Sci. Am.* **218**(4), 95 (1968).

238. S. L. Postel, *BioScience* **48**, 629 (1998).

239. D. Pimentel et al., *BioScience* **47**, 97 (1997).

240. L. R. Brown and J. E. Young, in L. R. Brown, ed., *State of the World 1990* (New York: W. W. Norton, 1990), 59.

241. R. Lal, in K. Wiebe, N. Ballenger, and P. Pinstrup-Andersen, eds., *Who Will Be Fed in the 21st Century?* (Washington, DC: International Food Policy Research Institute/Johns Hopkins University Press, 2001), 17.

242. M. Byrnes, Reuters dispatch, 28 March 2001.

243. L. Parks, FoxNews dispatch, 4 April 2000.

244. A. J. Chepstow-Lusty et al., *Mountain Res. Devel.* **18**, 159 (2000). J. L. Brown, *Civil Engg.* **71**, 32 (2001) gives information on Inca water engineering at Machu Picchu.

245. Anonymous, BBC News Online, 22 May 2003. This details the rebuilding of the water system in Pampachiri, a very poor area in Peru, as restored to its original configuration using only local materials by the Cusichaca Trust NGO.

246. R. Dubos, *The God Within* (New York: Charles Scribners Sons, 1972). C. N. Runnels, *Sci. Am.* **272**(3), 96 (1995).

247. R. M. Klein, in A. Ternes, ed., *Ants, Indians, and Little Dinosaurs* (New York: Charles Scribners Sons, 1975), 275.

248. E. Furness, *NYT*, 26 January 1975.

249. R. E. Peterson and J. M. Gregory, in R. Bras, ed., *The World at Risk: Natural Hazards and Climate Change* (New York: American Institute of Physics, 1993), 125. See also how agricultural practices can lead to generation of dust in N. A. Gordon and P. E. Todhunter, *At. Env.* **32**, 1587 (1998).

250. M. Klockenbrink, *NYT*, 20 August 1991.

251. N. C. Paul, *CSM*, 1 November 2001. G. B. Triplett, Jr. and D. M. Van Doren, Jr., *Sci. Am.* **236**(1), 28 (1977).

252. L. R. Oldeman, in D. J. Greenland and I. Szabolcs, eds., *Soil resilience and sustainable land use* (Wallingford, U.K.: CAB International, 1994).

253. E. A. Shinn et al., *Geophys. Res. Lett.* **27**, 3029 (2000).

254. M. W. Rosegrant et al., *Global Food Projections to 2020: Emerging Trends and Alternative Futures* (Washington, DC: International Food Policy Research Institute, 2001).

255. M. R. Gebhandt et al., *Science* **230**, 625 (1985).

256. E. C. Wolf, *Natl. Hist.* **94**(4), 53 (1985); H. M. Peskin, *Environment* **28**(4), 30 (1986).

257. W. E. Larson et al., *Science* **219**, 458 (1983).

258. S. Postel, in L. Starke, ed., *State of the World 1989* (New York: W. W. Norton, 1989), 21.

259. S. W. Trimble, *Science* **285**, 1244.(1999). J. Glanz, *Science* **285**, 1188 (1999).

260. S. W. Trimble and P. Crosson, *Science* **289**, 253.(2000).

261. Anonymous, AP dispatch, 4 April 1998.

262. U.S. Department of Energy, case study (1999).

263. D.L. Forster, *J. Soil Water Conserv.* **55**, 309 (2000).

264. Anonymous, Environmental News Network dispatch, 7 June 2001; URL: http://www.enn.com/news/ enn-stories/2001/06/06072001/ brownback_43896.asp?site=email

265. J. Pretty, in P. Pinstrup-Andersen, ed., *Appropriate Technology for Sustainable Food Security* (Washington, DC: International Food Policy Research Institute, 2001), p. 5.

266. P. L.Pingali, in P. Pinstrup-Andersen, ed., *Appropriate Technology for Sustainable Food Security* (Washington, DC: International Food Policy Research Institute, 2001), 7.

267. D. Kleijn et al., *Nature* **413**, 723 (2001).

268. R. Ahlstrom, *IT*, 11 September 2000.

269. J. Dowling, in R. Howes and A. Fainberg, eds., *The Energy Sourcebook* (New York: American Institute of Physics, 1991), 401.

270. D. Pimentel et al., *Science* **190**, 754 (1975).

271. E. Hirst, *Natl. Hist.* **82**(10), 21 (1973). E. Hirst, *Science* **184**, 134 (1974).

272. K. Blaxter, *New Scientist* **65**, 697 (1975).

273. B. Ozkan et al., *Renewable Energy* **29**, 39 (2004).

274. M. S. Alam et al., *Energy* **24**, 537 (1999).

275. D. Pimentel and M. Giampietro, Carrying Capacity Network paper, November 1994, URL: http://dieoff.org/page55.htm.

276. According to figures pertaining to use fo an electric water pump found at the University of Arkansas Agricultural Engineering Department site, URL http://www.aragriculture.org/agengineering/irrigation/crop/rice/pumpingcost.asp, it costs 18.85 joules of electricity per liter per meter of water table depth to pump up water for irrigation. Since 5 to 8 ML/ha of water is used for irrigating the typical cornfield (D. Pimentel et al., *Frontiers* **3**, 105 (1997)), it takes 4.71 to 7.54 GJ/ha $\approx$ 6 GJ/ha to pump this water from a 50-m deep water table. Water twice as deep would cost twice as much energy.

277. W. L. Kranz et al., *Ag. Wat. Mgmt.* **22**, 325 (1992).

278. R. Revelle, *Sci. Am.* **231**(3), 161 (1974).

279. L. R. Brown, in *The Biosphere* (San Francisco: Freeman, 1970), 93; and J. H. Hulse and D. Spurgen, *Sci. Am.* **231**(2), 72 (1974).

280. D. U. Hooper and P. M. Vitousek, *Science* **277**, 1302 (1997).

281. J. H. Falk, *Ecology* **57**, 141 (1976). B. Webster, *NYT*, 2 May 1976.

282. J. P. Holdren and P. R. Ehrlich, *Am. Sci.* **62**, 282 (1974).

283. J. F. Power and R. F. Follett, *Sci. Am.* **256**(3), 78 (1987).

284. M. Huston, *Science* **262**, 1676 (1993).

285. D. Tilman and J. D. Downing, *Nature* **367**, 363 (1994). W. K. Stevens, *NYT*, 1 February 1994.

286. S. Naeem et al., *Nature* **368**, 734 (1994). P. Kareiva, *Nature* **368**, 687 (1994).

287. F. S. Chapin et al., *Science* **277**, 500 (1997).

288. J. P. Grime, *Science* **277**, 1260 (1997).

289. D. A. Wardle et al., *Science* **277**, 1296 (1997).

290. D. Tilman et al., *Science* **277**, 1300 (1997).

291. P. D. Moore, *Nature* **424**, 26 (2003).

292. A. Hector et al., *Science* **286**, 1123 (1999). D. Tilman, *Science* **286**, 1099 (1999).

293. T. R. Hargrove et al., *BioScience* **38**, 675 (1988).

294. M. C. Heller and G. A. Keoleian, *Life Cycle-Based Sustainability Indicators for Assessment of the U.S. Food System* University of Michigan, Center for Sustainable Systems Report No. CSS00-04 (6 December 2000).

CHAPTER 25

1. F. R. Kalhammer, *Sci. Am.* **241**(6), 56 (1979); F. R. Kahammer and T. R. Scheider, *Annu. Rev. Energy* **1**, 311 (1976).

2. W. J. Schaetzle et al., in T. N. Veziroglu, ed., *Alternative Energy Sources II* (Washington, DC: Hemisphere Publ., 1981), 285.

3. J. G. Bednorz and K. A. Müller, *Z. Phys. B* **64**, 189 (1986).

4. M. K. Wu et al., *Phys. Rev. Lett.* **58**, 908 (1987). A.L. Robinson, *Science* **235**, 1137 (1987).

5. C. W. Chu et al., *Phys. Rev. Lett.* **60**, 941 (1988); Z. Z. Sheng and A. M. Hermann, *Nature* **332**, 138 (1988); R. Pool, *Science* **240**, 25 (1988) and *Science* **240**, 146 (1988); and A. Khurana, *Phys. Today* **41**(4), 21 (1988). The highest critical temperature achieved with the new materials stalled out at about 130 K, see A. Schilling et al., *Nature* **363**, 56 (1993).

6. C. C. Torardi et al., *Science* **240**, 631 (1988).

7. P. Yam, *Sci. Am.* **270**(1), 18 (1994).

8. M. Laguës et al., *Science* **262**, 1850 (1993). R. Pool, *Science* **262**, 1817 (1993).

9. J. Nagamatsu et al., *Nature* **410**, 63 (2001).

10. S. Jin et al., *Nature* **411**, 563 (2001). K. Chang, *NYT*, 11 July 2001.

11. T. He et al., *Nature* **411**, 54 (2001).

12. J. H. Schön et al., *Nature* **408**, 549 (2000).

13. K. Chang, *NYT*, 11 September 2001.

14. J. H. Schön et al., *Science* **293**, 2432 (2001).

15. J. L. Sarrao et al., *Nature* **420**, 297 (2002).

16. B. J. Feder, *NYT*, 3 January 1994.

17. S. F. Baldwin, in R. Howes and A. Fainberg, eds., *The Energy Sourcebook* (The New York: American Institute of Physics, 1991), 475.

18. G. Gugliotta, *WP*, 20 May 2001. K. Chang, *NYT*, 29 May 2001. W. S. Browne, *NYT*, 3 November 1998.

19. D. Larbalestier et al., *Nature* **414**, 368 (2001).

20. T. R. Schneider, *Annu. Rev. Energy Environ.* **16**, 533 (1991).

21. R. A. Uher, *Futurist* **24**(5), 28 (1990).

22. E. H. Brandt, *Science* **243**, 349 (1989).

23. B. J. Feder, *NYT*, 27 April 2004.

24. M. L. Wald, *NYT*, 14 August 2003.

25. H. J. Choi et al., *Nature* **418**, 758 (2002).

26. R. F.Service, *Science* **295**, 786 (2002).

27. A. M. Wolsky et al., *Sci. Am.* **260**(2), 61 (1989).

28. A. Pollack, *NYT,* 11 August 1987, *NYT,* 12 August 1987, and *NYT,* 13 August 1987.

29. R. F. Post and S. F. Post, *Sci. Am.* **229**(6), 17 (1973).

30. M. Olszewski and D. U. O'Kain, *Endeavour* **11**, 58 (1987).

31. M. L. Wald, *NYT*, 3 January 1994 (AFS); 6 January 1994 (racer).

32. M. L. Wald, *NYT*, 22 June 1994.

33. M. L. Wald, *NYT*, 29 June 1994. For a lengthier discussion of hybrid car power sources, see J. J. MacKenzie, *The Keys to the Car* (Washington, DC: World Resources Institute, 1994).

34. R. Hebner and J. Beno, *IEEE Spectrum* **39**(4), 46 (2002).

35. A. L. Robinson, *Science* **184**, 884 (1974).

36. A. B. Meinel and M. P. Meinel, *Phys. Today* **25**(2), 44 (1972).

37. J. Ingersoll, in R. Howes and A. Fainberg, eds., *The Energy Sourcebook* (New York: American Institute of Physics, 1991), 325.

38. F. R. McLennon and E. J. Cairns, *Annu. Rev. Energy* **14**, 241 (1989).

39. M. L. Wald, *NYT*, 29 September 1991.

40. S. Miller, *IEEE Spectrum* **38**(8), 27 (2001).

41. S. W. Tam et al., in T. N. Veziroglu, ed., *Alternative Energy Sources II* (Washington DC: Hemisphere Publ., 1981), 459.

42. M. L. Wald, *NYT*, 31 October 1990.

43. A. Chilenskas and R. K. Steinenberg, *Logos* **23**, 10 (1985).

44. B. Sorenson, *Annu. Rev. Energy* **9**, 1 (1984).

45. J. Holusha, *NYT*, 16 March 1994.

46. D. R. MacFarlane et al., *Nature* **402**, 792 (1999).

47. S. L. Wilkinson, *Chem. & Engg. News* **75**(41) 18 (1997). J. Giles, Nature Science Update, 17 February 2004 and R. F.Service, *Science* **303**, 1122 (2004).

48. L. Frazer, *Environ. Health Perspect.* **110**, A201 (2002).

49. O. Chusid et al., *Adv. Mater.* **15**, 627 (2003). D. Aurbach et al., *Nature* **407**, 724 (2000).

50. A. K. Pahdi et al., *J. Electrochem. Soc.* **144**, 1188 (1997). J.-M. Tarascon and M. Armand, *Nature* **414**, 359 (2001). S.-Y. Chung et al., *Nature Materials* **1**, 123 (2002). M. Thackeray, *Nature Materials* **1**, 81 (2002).

51. W. Li et al., *Science* **264**, 1115 (1994).

52. M. L. Wald, *NYT*, 16 March 1994.

53. M. L. Wald, *NYT*, 6 March 1994.

54. M. L. Wald, *NYT*, 16 February 1994.

55. M. L. Wald, *NYT*, 27 May 1994.

56. T. Jüngling, *DW*, 27 October 2002.

57. M. L. Wald, *NYT*, 10 March 1994.

58. B. D. Ayres, Jr., *NYT*, 6 December 1996.

59. R. Froböse, *DW*, 24 May 2001.

60. M. Reisch, *Chem. & Engg. News* **79**(43) 22 (2001).

61. D. Malakoff, *Science* **285**, 680 (1999).

62. A. A. Golestaneh, *Phys. Today* **37**(4), 62 (1984); R. D. Spence and M. J. Harrison, *Am. J. Phys.* **52**, 1144 (1984); and A. D. Johnson and J. L. McNichols, Jr., *Am. J. Phys.* **54**, 745 (1986). J. Van Humpbeck et al., *Endeavour* **15**, 148 (1991). G. Cook, *BG*, 26 April 2002 about A. Lendlein and R. Langer, *Science* **296**, 1673 (2002) and B. J. Feder, *NYT*, 22 July 2002.

63. Model heat engines: TINI Sales, Box 1431, Lafayette, CA 94549; demonstration kits: Nitinol Devices, 1436 View Point, Escondido, CA 92027.

64. A. Lendlein and R. Langer, *Science* **296**, 1673 (2002).

65. R&D Associates, *Project PACER Final Report* RDA-TR-4100-003 (July 1974). For a newer reworking of the idea, see C. J. Call and R. W. Moir, *Nucl. Sci. & Engineering* **104**, 364 (1990).

66. W. D. Metz, *Science* **188**, 136 (1975) and H. W. Hubbard et al., *Science* **188**, 780 (1975); and W. D. Metz, *Science* **189**, 440 (1975).

67. D. Hafmeister, *Am. J. Phys.* **47**, 671 (1979).

68. R. A. Herendeen et al., *Science* **205**, 451 (1979). The original idea is due to P. E. Glaser, *Science* **162**, 957 (1968). It would be more convenient if there were widespread photovoltaic solar

available for the satellite to send energy onto Earth's night side. This idea has been discussed by G. A. Landis, *Proceedings SPS '97 Conference* (Montreal, Canada: Canadian Aeronautics and Space Institute, Canada, and the Société des Électriciens et Électroniciens, France, 1997), 327.

69. D. Normile, *Science* **294**, 1273 (2001).
70. M. I. Hoffert and S. D. Potter, *Tech. Rev.* **100**(7), 30 (1997).
71. W. E. Leary, *NYT*, 12 January 1993.
72. T. Cole, *Science* **221**, 915 (1983). See also Anonymous, *Am. Ceramic Soc. Bull.* **80**(4), 43 (2001).
73. P. E. White, *Geothermal Energy*, U.S. Geological Survey Circular 519, 1965.
74. J. Yang et al., *J. Micromech. Microeng.* **13**, 963 (2003). I. Austen, *NYT*, 1 January 2004.
75. C. G. Palmierini, in T. B. Johansson et al., eds., *Renewable Energy Sources for Fuels and Electricity* (Washington, D.C. and Covelo, CA: Island Press, 1993), 549.
76. P. Kreiger, *Annu. Rev. Energy* **1**, 159 (1976).
77. J. Lear, *Saturday Review*, 5 December 1970, 53.
78. R. C. Axtmann, *Science* **187**, 795 (1975).
79. Department of Energy, *Geothermal Energy* (Washington, DC: Advanced Services, 1989), FS188, 2nd ed. See also R. A. Wright, *NYT*, 13 October 1974 for some interesting facts about Klamath Falls and some of its residents.
80. C. Flavin, *Futurist* **19**(2), 36 (1985); K. Collins et al., *USA: Living with the Sun* National Center for Appropriate Technology, DOE circular CE-0017 (1985).
81. M. Milstein, *PO*, 13 June 2001. P. Lineau and K. Rafferty, *Geo-Heat Center Q. Bull.* **13**(4), 8 (1991) and B. Brown, *Geo-Heat Center Q. Bull.* **16**(4), 23 (1995).
82. National Renewable Energy Laboratory, Fact Sheet (1999).
83. J. W. Shupe, *Science* **216**, 1193 (1982).
84. R. H. Howes, in R. Howes and A. Fainberg, eds., *The Energy Sourcebook* (New York: American Institute of Physics, 1991), 239.
85. I. R. Straughan, *Middle Management Perspectives Program* (Rosemead, CA: Southern California Edison, 1985).
86. Anonymous, Environmental News Network dispatch, 7 May 2001; URL: http://www.enn.com/news/enn-stories/2001/05/05072001/geothermal_43413.asp?site=email
87. National Renewable Energy Laboratory, *Geothermal Today: Clean energy for the 21st century* (Washington, DC: GPO, 2000), DOE/GO-102000-1066.
88. P. J. Podger, *SFC*, 10 July 2001.

89. J. E. Mock et al., *Annu. Rev. Energy Environ.* **22**, 305 (1997).
90. H. Kroker, *DW*, 15 July 2004.
91. E. Bailey, *LAT*, 27 November 2002.
92. E. Bailey, *LAT*, 17 July 2002.
93. M. McInerney, Reuters dispatch, 22 August 2003.
94. R. H. K., *DW*, 16 May 2003 .
95. Anonymous, Reuters dispatch, 26 April 2001.
96. K. Porter et al., NREL/TP-620-28674 (1999 Ed., August 2000).
97. Dept. of Energy, *Geothermal Energy in the Western United States & Hawaii: Resources & Projected Electricity Generation Supplies* (Washington, DC: GPO, 1991), DOE/EIA-0544.
98. Contact Energy, Ltd., *Power from the Earth* (Taupo, New Zealand, Contact Energy, Ltd., unknown). Contact Energy, Ltd., Fact Sheet.
99. D. V. Duchane, in L. Rosen and R. Glasser, eds., *Climate Change and Energy Policy* (New York: American Institute of Physics, 1992), 369. See also M. L. Wald, *NYT*, 3 November 1991.
100. R. A. Kerr, *Science* **207**, 1455 (1980).
101. J. Melcher, *Environment* **20**(2), 20 (1978).
102. T. H. Maugh, *Science* **178**, 599 (1972). See also Anonymous, *NYT*, 1 September 1974.
103. M. L. Wald, *NYT*, 4 May 1994.
104. L. F. Diaz et al., *CRC Critical Reviews in Environmental Control* **14**, 251 (1984).
105. P. R. O'Leary et al., *Sci. Am.* **259**(6), 36 (1988).
106. C. F. Golueke and P. H. McGauhey, *Annu. Rev. Energy* **1**, 257 (1976). A. Hirschkowitz, *Tech. Rev.* **90**(5), 26 (1987) points out that the Japanese recycle about 65% of their MSW.
107. Anonymous, *NYT*, 3 November 1976.
108. H. Keßler, *DZ*, 19 February 1993.
109. D. Morris, *Alt. Sources Energy* **80**, 14 (1986).
110. W. K. Stevens, *NYT,* 9 March 1986.
111. A. A. Narvaez, *NYT*, 4 August 1987.
112. M. Bergmeier, *Energy* **28**, 1359 (2003).
113. J. Barron, *NYT*, 11 December 1983.
114. J. T. McQuiston, *NYT,* 10 January 1982.
115. J. Mayer-List, *DZ*, 18 June 1982.
116. S. Bronstein, *NYT*, 23 June 1985.
117. R. A. Denison and J. Ruston, *Recycling and Incineration, Evaluating the Choices* (Washington, D.C. and Covelo, CA: Island Press, 1990).
118. D. Marier, *Alt. Sources Energy* **80**, 19 (1986).
119. League of Women Voters Education Fund, *The Garbage Primer* (New York: Lyons and Burfort, 1993).
120. R. Smothers, *NYT*, 14 January 1979.
121. J. R. Luoma, *NYT*, 2 August 1988.
122. P. Shabecoff, *NYT*, 26 November 1987.
123. W. E. Schmidt, *NYT*, 27 May 1990.

124. H. Levinson, *Environment* **32**(2), 10 (1990); D. Erickson, *Sci. Am.* **264**(5), 122 (1991).
125. G. Bishop, *Newark Star Ledger* (Newark, NJ), 24 December 1984.
126. B. S. Feder, *NYT*, 9 May 1990.
127. J. P. Hicks, *NYT*, 17 November 1991.
128. L. Greenhouse, *NYT*, 3 May 1994.
129. K. Schneider, *NYT*, 3 May 1994. J. Connor, *WSJ*, 19 August 1994 and K. Schneider, *NYT*, 11 October 1994.
130. K. Schneider, *NYT*, 18 March 1994.
131. I. Amato, *Science* **261**, 1388 (1993).
132. P. H. Wallman et al., *Energy* **23**, 271 (1998).
133. A. Melis et al., *Plant Physiol.* **122**, 127 (2000).
134. R. D. Cortright et al., *Nature* **418**, 964 (2002). S. Fields, *Environ. Health Perspect.* **111**, A 38 (2003).
135. G. W. Huber et al., *Science* **300**, 2075 (2003).
136. W. B. Kim et al., *Science* **305**, 1280 (2004).
137. S. K. Chaudhuri and D. R. Lovley, *Nature Biotechnology* **21**, 1229 (2003).
138. L. Zuckerman, *NYT*, 2 December 1997.
139. P. Fairley, *Tech. Rev.* **103**(6), 54 (2000). P. Fairley, *Tech. Rev.* **104**(4), 70 (2001).
140. N. Banerjee, *NYT*, 7 March 2003.
141. R. F. Service, *Science* **305**, 958 (2004).
142. P. Meller, *NYT*, 17 June 2003.
143. G. Vogel, *Science* **305**, 967 (2004).
144. C. Sims, *NYT*, 27 December 1992.
145. L. Schlapbach and A. Züttel, *Nature* **414**, 353 (2001).
146. R. S. El-Mallakh, *Environment* **23**(3), 30 (1981).
147. D. P. Gregory and J. B. Pangborn, *Annu. Rev. Energy* **1**, 279 (1976).
148. E. Garsten, AP dispatch, 31 July 2002.
149. P. Chen et al., *Science* **285**, 91 (1999).
150. J. J. Reilly and G. D. Sandrock, *Sci. Am.* **242**(2), 118 (1980). W. Sullivan, *NYT,* 29 March 1978.
151. P. Chen et al., *Nature* **420**, 302 (2002).
152. M. D. Ward, *Science* **300**, 1104 (2003).
153. O. M. Yaghi et al., *Nature* **423**, 705 (2003).
154. N. L. Rosi et al., *Science* **300**, 1127 (2003).
155. H. K. Chae et al., *Nature* **427**, 523 (2004).
156. G. A. Deluga et al., *Science* **303**, 993 (2004). A. Cho, *Science* **303**, 942 (2004).
157. E. Greenbaum, *Science* **230**, 1373 (1985).
158. G. Kurisu et al., *Science* **302**,1009 (2003).
159. S. U. M. Khan et al., *Science* **297**, 2243 (2002). R. F. Service, *Science* **297**, 2189 (2002).
160. M. Arndt, *Hamburger Abendblatt*, 10 June 1992 (as translated in *the German Tribune*).
161. C. H. Deutsch, *NYT*, 26 August 1990.
162. P. C. Cruver, *Futurist* **23**(6), 24 (1989).
163. H. Michaels, *DZ*, 24 December 1993.
164. G. Vogel, *Science* **305**, 966 (2004). N. Lossau, *DW*, 27 April 2003.

165. A. Cho, *Science* **305**, 964 (2004).
166. B. C. H. Steele and A. Heinzel, *Nature* **414**, 345 (2001).
167. N. V. Skorodumova et al., *Phys. Rev. Lett.* **89**, 166601 (2002).
168. H. Rogner et al., *Energy* **18**, 461 (1993).
169. D. Voss, *Science* **285**, 683 (1999).
170. C. Greenman, *NYT*, 10 May 2001.
171. R. F. Service, *Science* **285**, 682 (1999).
172. J. Naar, *OnEarth* **23**(3), 34 (2001).
173. M. L. Wald, *NYT*, 17 August 1999.
174. National Renewable Energy Laboratory, Wastewater to energy Fact sheet.
175. R. H. Wolk, *IEEE Spectrum* **36**(5), 45 (1999).
176. J. Mandak, AP dispatch, 26 September 2001.
177. M. L. Wald, *NYT*, 29 May 1994. J. M. Ogden, *Annu. Rev. Energy Environ.* **24**, 227 (1999).
178. D. L. Chandler, *BG*, 10 April 2001.
179. S. M. Haile et al., *Nature* **410**, 910 (2001). R. Fitzgerald, *Phys. Today* **54**(7), 22 (2001).
180. C. T. Hall, *SFC*, 2 April 2001.
181. BMW, BMW press release, 5 February 2001, URL: http://www.bmw.com/bmwe/intro/news/index.html. A. Bridges, AP dispatch, 13 July 2001.
182. S. von der Weiden, *DW*, 8 February 2001.
183. N. Demirdöven and J. Deutch, *Science* **305**, 974 (2004).
184. E. P. Murray et al., *Nature* **400**, 649 (1999).
185. T. Monmaney, *LAT*, 19 November 2001.
186. A. Eisenberg, *NYT*, 21 October 1999.
187. W. D. Jones, *IEEE Spectrum* **38**(3), 25 (2001).
188. A. M. Hayashi, *Sci. Am.* **278**(4), 32 (1998).
189. I. Austen, *NYT*, 26 August 2004.
190. R. F. Service, *Science* **296**, 1222 (2002).
191. B. J. Feder, *NYT*, 16 March 2003.
192. M. J. Rienzenman, *IEEE Spectrum* **38**(1), 95 (2001).
193. D. Voss, *Tech. Rev.* **104**(9), 68 (2001).
194. S. Srinivasan et al., *Annu. Rev. Energy Environ.* **24**, 281 (1999).
195. T. H. Maugh, *Science* **178**, 849 (1972).
196. W. E. Winsche et al., *Science* **180**, 1325 (1973).
197. J. M. Ogden and J. Nitch, in T. B. Johansson et al., eds., *Renewable Energy Sources for Fuels and Electricity* (Washington, D.C. and Covelo, CA: Island Press, 1993), 925.
198. J. Stein, *Amicus J.* **12**(2), 33 (1990).
199. M. G. Schultz et al., *Science* **302**, 624 (2003).
200. M. J. Prather, *Science* **302**, 581 (2003).
201. T. K. Tromp et al., *Science* **300**, 1740 (2003).

CHAPTER 26
1. J. G. Koomey et al., *Annu. Rev. Energy Environ.* **27**, 119 (2002).

2. J. S. Mill, *Principles of Political Economy* (Toronto: University of Toronto Press, 1965).
3. E. O. Wilson, in D. Western and M. Pearl, eds., *Conservation for the Twenty-First Century* (New York: Oxford University Press, 1989), 3.
4. D. W. Orr, *Cons. Biol.* **5**, 439 (1991).
5. H. von Foerster et al., *Science* **132**, 1291 (1960).
6. J. Serrin, *Science* **189**, 86 (1975).
7. S. Umpleby, *Population and Environment* **11**, 159 (1990).
8. A. A. Bartlett, *Focus* **9**, 49 (1999).
9. M. Palmer et al., *Science* **304**, 1251 (2004). A. J. McMichael et al., *Science* **302**, 1919 (2003).
10. M. Wackernagel et al. (Redefining Progress Group), *Proc. Natl. Acad. Sci.* **99**, 9266 (2002) says humans use the resources of 1.2 Earths every year.
11. E. W. Sanderson et al., *BioScience* **52**, 891 (2002) claims that humans affect 83% of Earth's surface.
12. G Polakovic, *LAT*, 25 June 2002.
13. T. M. Parris and R. W. Kates, *Annu. Rev. Environ. Resour.* **28**, 559 (2003).
14. W. E. Rees, *Nature* **421**, 898 (2003).
15. P. M. Vitousek et al., *BioScience* **36**, 368 (1986).
16. C. Field, *Science* **294**, 2490 (2001). R. S. DeFries et al., *Global Biogeochem. Cycles* **13**, 803 (1999) [world NPP] and J. A. Hicke et al., *Global Biogeochem. Cycles* **16**, 1018 (2002).
17. S. Rojstaczer et al., *Science* **294**, 2549 (2001).
18. M. Giampietro et al., *Population and Environment* **14**, 1091 (1992).
19. G. C. Daily and B. H. Walker, *Nature* **403**, 243 (2000).
20. J. D. Parrish et al., *BioScience* **53**, 851 (2003).
21. A. Cortese, *NYT*, 6 May 2001.
22. P. R. Ehrlich et al., *The Atlantic Monthly* **280**(6), 98 (1997).
23. D. Pauly et al., *Science* **302**, 1359 (2003).
24. J. Tibbetts, *Environ. Health Perspect.* **112**, A282 (2004).
25. N. Sizer, Forest Notes, World Resources Institute, June 2000.
26. A. Balmford et al., *Proc. Natl. Acad. Sci.* **100**, 1046 (2003).
27. Committee on Trade and Environment, World Trade Organization, WT/CTE/W/134, 23 February 2000.
28. N. Mabey et al. (UK Fisheries Committee), Prime Minister's Strategy Unit Report, March 2004.
29. G. Wright, *TG*, 25 March 2004. M. Woolf, *TI*, 26 March 2004.
30. Q. Schiermeier, *Nature* **419**, 866 (2002).
31. R. Hilborn et al., *Annu. Rev. Environ. Resour.* **28**, 359 (2003).
32. M. Schrope, *Nature* **418**, 718 (2002).
33. I. R. Noble and R. Dirzo, *Science* **277**, 622 (1997).

34. R. Bonnie et al., *Science* **288**, 1763 (2000).
35. C. Kremen et al., *Science* **288**, 1828 (2000). E. Niesten et al., *Phil. Trans. R. Soc. Lond. A* **360**, 1787 (2002).
36. L. Helmuth, *Science* **286**, 1283 (1999). P. Zhang et al., *Science* **288**, 2135 (2000).
37. D. A. Taylor, *Environment* **39**(1), 6 (1997).
38. M. Murray, *Science* **299**, 1851 (2003).
39. R. W. Kates et al., *Science* **292**, 641 (2001).
40. E. Ayensu et al., *Science* **286**, 685 (1999).
41. W. W. Hargrove and F. M. Hoffman, *Comp. Sci Engg.* **1**(4), 18 (1999).
42. D. M. Olson et al., *BioScience* **51**, 933 (2001). See the map on the World Wildlife Fund's Web site, URL: www. wwf. org/.
43. D. Schoch, *LAT*, 5 November 2001.
44. J. Henrich et al., *Am. Economic Rev.* **91**, 73 (2001).
45. E. Fehr and U. Fischbacher, *Nature* **425**, 785 (2003).
46. E. Fehr and B. Rockenbach, *Nature* **422**, 137 (2003). T. Bewley, *Nature* **422**, 125 (2003).
47. R. Boyd et al., *Proc. Natl. Acad. Sci.* **100**, 3531 (2003). A. G. Sanfey et al., *Science* **300**, 1755 (2003). D. Semmann et al., *Nature* **425**, 390 (2003). T. Singer et al., *Neuron* **41**, 653 (2004). J. M. McNamara et al., *Nature* **428**, 745 (2004).
48. C. W. Johnson, preprint, 2003 (unpublished). See also J. Whitfield, Nature Science Update, 15 August 2003.
49. G. Vogel, *Science* **303**, 1128 (2004).
50. P. J. Mumby et al., *Nature* **427**, 533 (2004).
51. G. C. Daily et al., *Science* **289**, 395 (2000).
52. R. Costanza et al., *Nature* **387**, 253 (1997). A. Balmford et al., *Science* **297**, 950 (2002).
53. See, for example, P. Ehrlich, *BioScience* **52**, 31 (2002) or N. Leader-Williams, *Phil. Trans. R. Soc. Lond. A* **360**, 1787 (2002).
54. D. Pimentel et al., *BioScience* **47**, 747 (1997). S. L. Pimmand and A. M. Sugden, *Science* **263**, 933 (1994) and S. L. Pimm et al., *Science* **269**, 347 (1995).
55. E. Zavaleta, *Ambio* **29**, 462 (2000).
56. E. S. Zavaleta et al., *Trends in Ecology & Evolution* **16**, 454 (2001).
57. S. Nee and R. M. May, *Science* **278**, 692 (1997).
58. K. V. Flannery, in P. J. Ucko, and G. W. Dimbleby, eds. *The Domestication and Exploitation of Plants and Animals* (London: Duckworth, 1969), 73.
59. E. Weiss et al., *Proc. Natl. Acad. Sci.* **101**, 9551 (2004).
60. A. Beja-Pereira et al., *Science* **304**, 1781 (2004). J. Hecht, New Scientist Online, 17 June 2004.

61. B. Fowler, *NYT*, 27 July 2004.
62. J. Diamond, *Guns, Germs, and Steel* (New York and London: W. W. Norton & Co., 1999).
63. J. Diamond, *Nature* **429**, 616 (2004).
64. J. Diamond, *Nature* **418**, 700 (2002).
65. P. M. Vitousek et al., *Science* **277**, 494 (1997).
66. M. L. Imhoff et al., *Nature* **429**, 870 (2004).
67. A. Rose, *Physica Status Solidi* **A 56**,11 (1979).
68. T. Malthus, *First Essay on Population* (Bristol, U.K.: Thoemmes Press, 1999).
69. J. S. Dukes, *Clim. Change* **61**, 31 (2003).
70. N. Myers and J. Kent, *Proc. Natl. Acad. Sci.* **100**, 4963 (2003).
71. R. Dubos, *The God Within* (New York: Charles Scribners Sons, 1972).
72. M. McCarthy, *TI*, 20 August 2004. C. Hawley, AP dispatch, 16 June 2004.
73. G. T. T. Molitor, *Futurist* **8**, 169 (1974); P. Sears, in J. Harte and R. E. Socolow, eds., *Patient Earth* (New York: Holt, Rinehart, & Winston, 1971), 2.
74. S. S. King, *NYT*, 10 December 1978.
75. W. Robbins, *NYT*, 19 March 1989.
76. D. Pimentel et al., *Science* **267**, 1117 (1995). D. Pimentel, in G. A. Aistars, ed., *A Life Cycle Approach to Sustainable Agriculture Indicators Proceedings* (Ann Arbor, MI: The Center for Sustainable Systems, 1999), 8.
77. R. E. Peterson and J. M. Gregory, in R. Bras, ed., *The World at Risk: Natural Hazards and Climate Change* (New York: American Institute of Physics, 1993), 125.
78. M. Klockenbrink, *NYT*, 20 August 1991.
79. J. R. McNeill and V. Winiwarter, *Science* **304**, 1627 (2004).
80. H. J. Geist and E. F. Lambin, *BioScience* **54**, 817 (2004).
81. I. Campbell, *Ecologist* **4**, 164 (1974).
82. R. Baker, *Ecologist* **4**, 170 (1974).
83. J. Charney et al., *Science* **187**, 434 (1974).
84. L. D. Rotstayn and U. Lohmann, *J. Climate* **15**, 2103 (2002).
85. W. H. Matthews, *Intern. J.Environmental Studies* **4**, 283 (1973); W. W. Kellogg and S. H. Schneider, *Science* **186**, 1163 (1974); C. Sagan et al., *Science* **206**, 1363 (1979).
86. E. P. Eckholm, *Losing Ground* (New York: W. W. Norton & Co., 1976).
87. Anonymous, Environmental News Network dispatch, 19 January 2001.
88. A. J. Belsky et al., *J. Soil and Water Cons.* **54**, 419 (1999).
89. E. Pianin, *WP*, 29 April 2001.
90. J. Barnard, AP dispatch, 28 January 2001. D. Hogan, *PO*, 7 February 2001. D. Hogan, *PO*, 12 February 2001. D. Hogan, *PO*, 14 February 2001.
91. D. Hogan, *PO*, 7 March 2001.

92. P. R. Krausman, in K. L. Launchbaugh et al., eds., *Grazing Behavior of Livestock and Wildlife*, Idaho Forest, Wildlife & Range Exp. Sta. Bull. #70 (Moscow, ID: University of Idaho, 1999), 85.
93. W. K. Stevens, *NYT*, 19 March 1991.
94. B. Kadik, Algerian Technical Bulletin No. 175, 1978.
95. A. J. Belsky and C. D. Canham, *BioScience* **44**, 77 (1994).
96. W. Campbell-Purdie, *Ecologist* **4**, 300 (1974).
97. Algerian Ministry of Water, Environment, and Forests, *La bande forestiere de protection en zone aride et semi-aride* (*The Battle Against Desertification in Algeria: Experience of the Green Line*), Algerian Technical Bulletin No. 50, 1984.
98. B. Kadik, Algerian Technical Bulletin No. 56, 1982.
99. P. J. Vesilind, *Natl. Geog.* **168**, 2 (1985).
100. C. Haberman, *NYT*, 7 July 1985.
101. S. J. Milton et al., *BioScience* **44**, 70 (1994).
102. R. F. Dasman, *BioScience* **38**, 487 (1988); C. M. Peters et al., *Nature* **339**, 655 (1989); Anonymous, *NYT*, 4 July 1989; J. O. Browder, *BioScience* **42**, 174 (1992); C. Dold, *NYT*, 28 April 1992; M. J. Balick and R. Mendelsohn, *Cons. Biol.* **6**, 128 (1992); N. Salafsky et al., *Cons. Biol.* **7**, 39 (1993); T. A. Carr et al., *Environment* **35**(7), 12 (1993); J. Alper, *Science* **260**, 1895 (1993); and J. N. Abramovitz, in L. Starke, ed. *State of the World 1997* (New York & London: W. W. Norton, 1997), 95.
103. C. Burrow, *NYT*, 13 December 1992.
104. T. E. E. Oldfield et al., *Nature* **423**, 531 (2003). See also M. Hopkin, Nature Science Update, 29 May 2003.
105. T. DelCurto et al., in K. L. Launchbaugh et al., eds., *Grazing Behavior of Livestock and Wildlife*, Idaho Forest, Wildlife, & Range Expimental Station Bulletin #70 (Moscow, ID: University of Idaho, 1999), 119.
106. F. D. Provenza and K. L. Launchbaugh, in K. L. Launchbaugh et al., eds., *Grazing Behavior of Livestock and Wildlife*, Idaho Forest, Wildlife & Range Exp. Sta. Bull. #70 (Moscow, ID: University of Idaho, 1999), 1.
107. S. Blakeslee, *NYT*, 26 December 2000.
108. M. Higgins, Environmental News Network dispatch, 12 February 2001. URL: http://www. enn. com/news/enn-stories/2001/02/02122001/neastcons_41896. asp.
109. T. Knudson, *SB*, 26 April 2001.
110. New England Forestry Foundation website, URL: http://www. neforestry. org/Pages/listings. htm. (According to the site, "NEFF was founded in 1944 as

a non-profit corporation for the purpose of protecting New England's working forests. NEFF owns and manages 21, 000 acres in 120 demonstration forests and holds over 50 easements across New England.")
111. R. Channell and M. V. Lomolino, *Nature* **403**, 84 (2000). T. Brooks, *Nature* **403**, 26 (2000).
112. K. R. Weiss and E. Shogren, *LAT*, 12 September 2001.
113. R. E. Gullison et al., *Nature* **404**, 923 (2000).
114. W. F. Laurance et al., *Science* **291**, 438 (2001).
115. G. Martin, *SFC*, 22 April 2001. T. Egan, *NYT*, 27 May 2001.
116. J. R. Harlan, *Science* **188**, 618 (1975).
117. W. R. Courtenay, Jr. and C. R. Robins, *BioScience* **25**, 306 (1975).
118. R. M. Klein, in A. Ternes, ed., *Ants, Indians, and Little Dinosaurs* (New York: Charles Scribners Sons, 1975), 275.
119. E. Furness, *NYT*, 26 January 1975.
120. J. Brooke, *NYT*, 14 November 1989; D. Skole and C. Tucker, *Science* **260**, 1905 (1993). W. K. Stevens, *NYT*, 29 June 1993.
121. M. Simons, *NYT*, 11 October 1988 (Brazil); M. Simons, *NYT*, 19 February 1989 (Brazil); L. Rohter, *NYT*, 10 July 1990 (Mexico); G. M. Green and R. W. Sussman, *Science* **248**, 212 (1990) (Madagascar); P. Shabecoff, *NYT*, 8 June 1990; S. Christian, *NYT*, 3 April 1990 (Chile); N. C. Nash, *NYT*, 3 June 1991 (Chile); N. C. Nash, *NYT*, 21 June 1993 (Bolivia); and S. B. Hecht, *BioScience* **43**, 687 (1993).
122. T. Egan, *NYT*, 11 June 1992
123. L. Roberts, *Science* **242**, 1508 (1988); P. Shabecoff, *NYT*, 6 December 1988; E. Norse, *Amicus J.* **12**(1), 42 (1990).
124. D. C. Duffy and A. J. Meier, *Cons. Biol.* **6**, 196 (1992); C. Dold, *NYT*, 1 September 1992. L. R. Boring et al., *Ecol.* **62**, 1244 (1981) gives a more optimistic view.
125. R. O. Lawton et al., *Science* **294**, 584 (2001).
126. C. J. Still et al., *Nature* **398**, 608 (1999); J. A. Pounds et al., *Nature* **398**, 611 (1999).
127. R. Revelle, *Science* **209**, 164 (1980).
128. Quote from United Nations journal *Ceres*, as given in Reference 86, 22.
129. M. Lindow, *CSM*, 8 January 2004.
130. J. A. McNeely, *Environment* **46**(6), 16 (2004).
131. J. Withgott, *Science* **295**, 2201 (2002).
132. W. Kaczor, AP dispatch, 21 October 2003.
133. T. Martin, *LSJ*, 24 August 2003.
134. D. Normile, *Science* **306**, 968 (2004).
135. J. Randerson, New Scientist Online, 12 May 2003.

136. J. L. Gelbard and J. Belnap, *Cons. Biol.* **17**, 420 (2003). E. Lau, *SB*, 4 May 2003.

137. S. C. H. Barrett, *Sci. Am.* **261**(4), 90 (1989).

138. W. K. Stevens, *NYT*, 18 December 1990.

139. M. Scheffer et al., *Proc. Natl. Acad. Sci.* **100**, 4040 (2003).

140. H. Faber, *NYT*, 20 April 1975.

141. S. H. Wittwer, *Science* **188**, 579 (1975); J. E. Brody, *NYT*, 26 May 1987.

142. M. E. Soulé, *Science* **253**, 744 (1991).

143. W. Sullivan, *NYT*, 10 January 1978.

144. H. Faber, *NYT*, 12 March 1989.

145. W. Sullivan, *NYT*, 28 September 1993.

146. G. Wilkes, *Ecologist* **7**, 312 (1977).

147. P. R. Ehrlich and E. O. Wilson, *Science* **253**, 758 (1991). P. Shabecoff, *NYT*, 30 July 1989 and W. K. Stevens, *NYT*, 20 August 1991.

148. J. B. Hughes et al., *Science* **278**, 689 (1997).

149. D. Bryant et al., *The Last Frontier Forests: Ecosystems and Economies on the Edge* (Washington, DC: World Resources Institute, 1997).

150. G. Ceballos and P. R. Ehrlich, *Science* **296**, 904 (2002).

151. F. S. Chapin et al., *Science* **277**, 500 (1997).

152. D. J. Schoen and A. H. D. Brown, *BioScience* **51**, 960 (2001).

153. S. L. Pimm and P. Raven, *Nature* **403**, 843 (2000).

154. F. S. Chapin et al., *Nature* **405**, 234 (2000).

155. C. Gascon et al., *Science* **288**, 1356 (2000).

156. W. F. Laurance et al., *Science* **278**, 1117 (1997). W. F. Laurance et al., *Nature* **404**, 836 (2000).

157. A. P. Dobson et al., *Science* **277**, 515 (1997).

158. J. L. Pierce et al., *Nature* **432**, 87 (2004). C. Whitlock, *Nature* **432**, 28 (2004).

159. N. Myers, *Environmentalist* **8**, 178 (1988).

160. N. Myers et al., *Nature* **403**, 853 (2000). J. R. Prendergast et al., *Nature* **365**, 335 (1993) discusses how hotspots are different for different species in Britain.

161. R. P. Cincotta et al., *Nature* **404**, 990 (2000).

162. World Resources Institute, *World Resources 2000-2001* (Washington, DC; World Resources Institute, 2000). See also N. Williams, *Science* **281**, 1426 (1998).

163. R. Doyle, *Sci. Am.* **277**(2), 26 (1997). A. P. Dobson et al., *Science* **275**, 550 (1997).

164. W. K. Stevens, *NYT*, 216 March 2000.

165. E. W. Seabloom et al., *Proc. Natl. Acad. Sci.* **99**, 11229 (2002).

166. L. M. Curran et al., *Science* **286**, 2184 (1999). G. Hartshorn and N. Bynum, *Science* **286**, 2093 (1999).

167. S. E. Page et al., *Nature* **420**, 61 (2002).

168. P. Aldhous, *Nature* **432**, 144 (2004).

169. P. N. Spotts, *CSM*, 6 November 2003.

170. A. Downie, *CSM*, 22 April 2004.

171. S. Connor, *TI*, 10 July 2004.

172. I. Koren et al., *Science* **303**, 1342 (2004).

173. M. O. Andreae et al., *Science* **303**, 1337 (2004).

174. D. A. Wedin and D. Tilman, *Science* **274**, 1720 (1996).

175. D. Kleijn et al., *Nature* **413**, 723 (2001).

176. A. Anderson, Jr., *NYT Magazine*, 27 April 1975, 15; D. W. Ehrenfeld, *Am. Sci.* **64**, 648 (1976); M. Reisner, *NRDC Newsletter* **6**(1), 1 (1977); and E. Eckholm, *Futurist* **12** 289 (1978).

177. B. Webster, *NYT*, 28 March 1976.

178. S. Seagrave, *BioScience* **26**, 153 (1976).

179. B. Rensberger, *NYT*, 14 April 1977; Anonymous, *NYT*, 14 January 1979.

180. F. T. Ledig, *BioScience* **38**, 471 (1988).

181. A. R. Hughes and J. J. Stachowicz, *Proc. Natl. Acad. Sci.* **101**, 8998 (2004).

182. A. P. Dobson et al., *Science* **275**, 550 (1997). H. R. Pulliam and B. Babbitt, *Science* **275**, 499 (1997).

183. R. Stone, *Science* **285**, 817 (1999). R. Stone, *Science* **285**, 1837 (1999).

184. D. Tilman and J. A. Downing, *Nature* **367**, 363 (1994). Y. Baskin, *Science* **264**, 202 (1994); A. Kreuss and T. Tscharntke, *Science* **264**, 1581 (1994).

185. S. Naeem et al., *Nature* **403**, 762 (2000). P. J. Morin, *Nature* **403**, 718 (2000).

186. W. Sullivan, *NYT*, 15 November 1988 and *NYT*, 5 December 1989; and Anonymous, *Futurist* **27**(5), 6 (1993). R. Callahan, AP dispatch, 28 May 2004.

187. W. K. Stevens, *NYT*, 2 March 1993.

188. T. Egan, *NYT*, 25 June 1989; J. Brooke, *NYT*, 30 March 1993.

189. E. Stokstad, *Science* **291**, 2294 (2001). However, see E. Matthews, WRI Forest Briefing paper No. 1, March, 2001.

190. A. Gómez-Pompa and A. Kaus, *Proc. Natl. Acad. Sci.* **96**, 5982 (1999).

191. A. Packer and K. Clay, *Nature* **404**, 278 (2000). W. H. van der Putten, *Nature* **404**, 232 (2000).

192. M. Loreau and A. Hector, *Nature* **412**, 72 (2001).

193. K. J. Willis and R. J. Whattaker, *Science* **287**, 1406 (2000).

194. D. Tilman and C. Lehman, *Proc. Natl. Acad. Sci.* **98**, 5433 (2001).

195. P. B. Reich et al., *Proc. Natl. Acad. Sci.* **101**, 10101 (2004).

196. I. M. Côté and J. D. Reynolds, *Science* **298**, 1181 (2002). J. A. Lockwood, *BioScience* **53**, 99 (2003). N. J. Sanders et al., *Proc. Natl. Acad. Sci.* **100**, 2474 (2003). C. J. Costello and A. R. Solow, *Proc. Natl. Acad. Sci.* **100**, 3321 (2003). J. Withgott, *Science* **305**, 1100 (2004).

197. E. S. Zavaleta and K. B. Hulvey, *Science* **306**, 1175 (2004).

198. C. C. Mann, *Science* **297**, 920 (2002).

199. J. Lehmann, Terra Preta de Indio Soil Biogeochemistry webpage, URL http://www.css.cornell.edu/faculty/lehmann/terra_preta/TerraPretahome.htm.

200. J. Lehmann et al., *Plant and Soil* **249**, 343 (2003).

201. C. L. Erickson, *Nature* **408**, 190 (2000).

202. M. Higgins, Environmental News Network dispatch, 16 March 2000; http://www. enn. com/enn-news-archive/2000/03/03162000/ecolog_11067. asp.

203. R. Shaw, Environmental News Network dispatch, 22 November 1999; URL http://www. enn. com/enn-news-archive/1999/11/112299/forestplan_75113. asp. M. Higgins, Environmental News Network dispatch, 10 March 2000; URL. http://www. enn. com/enn-news-archive/2000/03/03102000/endold_10897. asp.

204. S. H. Verhovek, *NYT*, 5 August 1999.

205. T. Wilkinson, *CSM*, 15 November 2000. D. Jehl, *NYT*, 5 January 2001. R. A. Rosenblatt and R. Trounson, *LAT*, 5 January 2001. M. Milstein, *PO*, 5 January 2001. G. Martin, *SFC*, 9 January 2001. D. Jehl, *NYT*, 9 January 2001.

206. K. Pfleger, AP dispatch, 4 May 2001.

207. S. Borenstein, Knight-Ridder Washington Bureau dispatch, 2 May 2001.

208. The John Muir Project, The John Muir Project website URL: http://ww.john-muirproject.org/. See also C. McKinney, Rep. McKinney website URL: http://www. house. gov/mckinney/hl/nfppa. htm.

209. D. Jehl, *NYT*, 4 May 2001. E. Pianin and M. Allen, *WP*, 4 May 2001. E. Pianin, *WP*, 5 May 2001.

210. B. Knickerbocker, *CSM*, 7 May 2001.

211. D. Jehl, *NYT*, 5 May 2001. K. Pfleger, AP dispatch, 6 May 2001.

212. E. Brazil, *SFC*, 11 May 2001. E. Pianin, *WP*, 11 May 2001. D. Jehl, *NYT*, 11 May 2001.

213. M. Milstein, *PO*, 28 June 2001.

214. C. Doering, Reuters dispatch, 8 June 2001.

215. K. Pfleger, AP dispatch, 6 July 2001. G. Martin, *SFC*, 7 July 2001.

216. K. Pfleger, AP dispatch, 28 September 2001.

217. G. R. Flematti et al., *Science* **305**, 977 (2004).

218. A. S. Moffat, *Science* **282**, 1253 (1999). J. Gerstenzang et al., *LAT*, 21 August 2002. M. Daly, AP dispatch, 22 August 2002 and AP dispatch, 30 August 2002. Z. Coile, *SFC*, 6 September 2002. E. Shogren, *LAT*, 6 September 2002. E. Pianin, *WP*, 8 September 2002. D. Malakoff, *Science* **297**, 2194 (2002).

219. T. D. Hooker and J. E. Compton, *Ecol. Appl.* **13**, 299 (2003).

220. S. Mehta, *LAT*, 11 December 2001.

221. K. S. Powell, *LAT*, 28 September 2000.

222. M. M. Woodsen, *NYT*, 4 September 2001.

223. P. Fimrite, *SFC*, 7 March 2001. P. Fimrite, *SFC*, 2 July 2001.

224. C. K. Yoon, *NYT*, 10 January 2002. P. Fimrite, *SFC*, 11 January 2001. B. Boxall, *LAT*, 16 January 2001. C. K. Yoon, *NYT*, 5 September 2002. D. E. Murphy, *NYT*, 8 September 2002.

225. A. Kirby, BBC News Online dispatch, 4 December 2003. S. Connor, *TI*, 5 December 2003.

226. P. Fimrite, *SFC*, 11 August 2001.

227. P. Fimrite, *SFC*, 28 July 2004.

228. P. Fimrite, *SFC*, 11 March 2004.

229. P. Fimrite, *SFC*, 18 May 2001.

230. M. Jones, *SB*, 10 December 2001. S. Leavenworth, *SB*, 2 March 2002.

231. P. Fimrite, *SFC*, 18 December 2002.

232. S. Leavenworth, *SB*, 21 December 2002.

233. L. M. Krieger, *SJMN*, 28 April 2004.

234. P. Fimrite and G. Raine, *SFC*, 27 March 2004.

235. P. Fimrite, *SFC*, 9 March 2001. C. K. Yoon, *NYT*, 1 October 2002.

236. M. Org, AP dispatch, 3 October 2003.

237. G. Burke, *WP*, 19 January 2004.

238. B. Mason, *CCT*, 11 June 2004.

239. G. Garelik, *Science* **298**, 1702 (2002). G. Gugliotta, *WP*, 18 March 2002.

240. Anonymous, *TG*, 13 November 2003. G. Collins, Press Agency News dispatch, 13 November 2003.

241. S. M. Louda et al., *Science* **277**, 1088 (1997). D. R. Strong, *Science* **277**, 1058 (1997).

242. D. H. Strong and R. W. Pemberton, *Science* **288**, 1969 (2000).

243. U. Schaffner, *BioScience* **51**, 951 (2001).

244. S. Mirsky, *Audubon* **101**(3), 71 (1999).

245. J. E. Brody, *NYT*, 9 June 1998.

246. D. Tenenbaum, *Tech. Rev.* **99**(6), 32 (1996).

247. M. Koidin, AP dispatch, 28 February 2000.

248. D. Malakoff, *Science* **285**, 1255 (1999). See also popular accounts of the actual release of the beetles in S. M. Bryan, AP dispatch, 14 October 2003 and N. Lofholm, *The Denver Post*, 23 October 2003.

249. J. O. Luken and J. W. Thieret, *BioScience* **46**, 18 (1996).

250. R. A. Malecki et al., *BioScience* **43**, 680 (1993).

251. S. Swanson, *CT*, 15 April 2001. S. Swanson, *CT*, 16 April 2001.

252. G. Maranto, *NYT*, 27 April 1997.

253. W. Owen, *PO*, 29 April 2001.

254. J. Christensen, *NYT*, 1 February 2000w.

255. M. Enserink, *Science* **285**, 1834 (1999). As an example, see W. K. Stevens, *NYT*, 16 August 1994.

256. Anonymous, Environmental News Network dispatch, 8 December 2000. D. Lewerenz, AP dispatch, 1 July 2002.

257. American Chestnut Cooperators' Foundation webpage, URL http://www. ppws. vt. edu/griffin/blight. html.

258. I. A. Bowles et al., *Science* **280**, 1899 (1998).

259. L. Stiffler, *SPI*, 9 June 2001.

260. M. Milstein, *PO*, 5 April 2001.

261. W. M. Getz et al., *Science* **283**, 1855 (1999).

262. R. L. Chazdon, *Science* **281**, 1295 (1998).

263. W. Berry, *Solving for Pattern* (1980), quoted in G. A. Aistars, in G. A. Aistars, ed., *A Life Cycle Approach to Sustainable Agriculture Indicators Proceedings* (Ann Arbor, MI: The Center for Sustainable Systems, 1999)

264. R. Weiss, *WP*, 25 December 2000. J. Robbins, *NYT*, 10 July 2001.

265. K. Cornish, *Science* **274**, 924 (1996).

266. T. U. Gerngross and S. C. Slater, *Sci. Am.* **283**(2), 36 (2000). E. S. Stevens, *Green Plastics* (Princeton & Oxford: Princeton University Press, 2002).

267. S. L. Wilkinson, *Chem. & Engg. News* **75**(31), 35 (1997).

268. Committee on the Scientific Basis for Predicting the Invasive Potential of Nonindigenous Plants and Plant Pests in the United States (R. N. Mack et al.), National Academy of Sciences, *Predicting invasions of nonindigenous plants and plant pests* (Washington, DC: National Academy Press, 2002).

269. E. Culotta, *Science* **254**, 1444 (1991).

270. T. W. Schoener, *Science* **185**, 27 (1974); R. S. Miller and D. B. Botkin, *Am. Sci.* **62**, 172 (1974).

271. J. R. Luoma, *NYT*, 14 March 1989.

272. J. Terborgh et al., *Science* **294**, 1923 (2001).

273. H. P. Bais et al., *Science* **301**, 1377 (2003).

274. B. W. Brook et al., *Nature* **404**, 365 (2000).

275. P. Wilf and C. C. Labandeira, *Science* **284**, 2153 (1999). P. D. Coley, *Science* **284**, 2098 (1999).

276. J. G. Robinson et al., *Science* **284**, 595 (1999).

277. D. A. Wardle et al., *Science* **277**, 1296 (1997).

278. M. Sankaran and S. J. McNaughton, *Nature* **404**, 691 (1999).

279. M. J. Crawley and J. E. Harral, *Science* **291**, 864 (2001).

280. R. M. May and M. P. H. Stumpf, *Science* **290**, 2084 (2000).

281. J. B. Plotkin et al., *Proc. Natl. Acad. Sci.* **97**, 10850 (2000).

282. S. P. Hubbell et al., *Science* **283**, 554 (1999). D. Tilman, *Science* **283**, 495 (1999).

283. J. -F. Molino and D. Sabatier, *Science* **294**, 1702 (2001).

284. S. K. M. Ernest and J. H. Brown, *Science* **292**, 101 (2001). W. Bond, *Science* **292**, 63 (2001) and C. Bright, in L. Starke, ed. *State of the World 1996* (New York & London: W. W. Norton, 1996), 95.

285. T. A. Kennedy et al., *Nature* **417**, 636 (2002).

286. A. R. Watkinson et al., *Science* **289**, 1554 (2000).

287. J. N. Klironomos, *Nature* **417**, 67 (2002).

288. A. K. Knapp, *Science* **291**, 481 (2001). J. Kaiser, *Science* **291**, 413 (2001).

289. R. J. Wilson et al., *Nature* **432**, 393 (2004).

290. L. Lens et al., *Science* **298**, 1236 (2002).

291. B. Van Valkenburgh et al., *Science* **306**, 101 (2004).

292. R. A. Kerr, *Science* **300**, 885 (2003).

293. C. N. Johnson, *Proc. R. Soc. Lond. B* **269**, 2221 (2002) M. Cardillo and A. Lister, *Nature* **419**, 440 (2002). A. M. Lister, *Phil. Trans. R. Soc. Lond. B* **359**, 221 (2004).

294. A. J. Stuart et al., *Nature* **431**, 684 (2004). G. Gugliotta, *WP*, 11 October 2004.

295. C. J. Stevens et al., *Science* **303**, 1876 (2004).

296. T. M. Blackburn et al., *Science* **305**, 1955 (2004).

297. D. Jablonski, *Nature* **427**, 589 (2004).

298. International Union for Conservation of Nature and Natural Resources, *The IUCN Red List of threatened species 2003* URL http://www.redlist.org/.

299. B. W. Brook et al., *Nature* **424**, 420 (2003). M. Hopkin, Nature Science Update, 24 July 2003.

300. J. A. Thomas et al., *Science* **303**, 1879 (2004).

301. A. C. Revkin and C. K. Yoon, *NYT*, 20 August 2002.

302. M. A. Davis, *BioScience* **53**, 481 (2003).

303. E. W. Seabloom et al., *Proc. Natl. Acad. Sci.* **99**, 11229 (2002).

304. D. J. D. Earn et al., *Science* **290**, 1360 (2000).

305. M. E. Soulé and M. A. Sanjayan, *Science* **279**, 2060 (1998).

306. G. Ferraz et al., *Proc. Natl. Acad. Sci.* **100**, 14069 (2003).

307. A. Inamdar et al., *Science* **283**, 1856 (1999).

308. R. Hilborn et al., *Proc. Natl. Acad. Sci.* **100**, 6564 (2003).

309. D. E. Pitt, *NYT*, 3 August 1993.

310. P. Shabecoff, *NYT* 21 March 1989.

311. B. Holmes, *Science* **264**, 1252 (1994). R. A. Myers et al., *Science* **269**, 1106 (1995).

312. D. O. Conover and S. B. Munch, *Science* **297**, 31 (2002). D. Malakoff, *Science* **297**, 31 (2002).

313. B. P. Finney et al., *Nature* **416**, 729 (2002). M. J. Attrill and M. Power, *Nature* **417**, 275 (2002).

314. As quoted in D. Hayes, *Natl. Hist.* **99**(4), 55 (1990).

315. P. Recer, AP dispatch, 26 July 2001.
316. J. B. C. Jackson et al., *Science* **293**, 629 (2001).
317. L. W. Botsford et al., *Science* **277**, 509 (1997).
318. D. Pauly et al., *Science* **279**, 860 (1998).
319. N. Williams, *Science* **279**, 809 (1998). A. P. McGinn, in L. Starke, ed. *State of the World 1998* (New York & London: W. W. Norton, 1998), 59 and in L. Starke, ed. *State of the World 1999* (New York & London: W. W. Norton, 1999), 82. As a sad example of how human meddling, introduction of Nile perch into Lake Victoria, had "disastrous consequences," as pointed out in C. D. N. Barel et al., *Nature* **315**, 19 (1985).
320. P. K. Dayton, *Science* **279**, 821 (1998).
321. A. Kirby, BBC dispatch, 11 November 2000. The situation continues to worsen; as of the end of 2001, twelve species of North Atlantic fish were at a level near collapse. See R. Casert, AP dispatch, 18 December 2001.
322. R. K. O'Dor, *The Unknown Ocean: Baseline Report of the Census of Marine Life 2003* (Washington, DC: Consortium for Oceanographic Research and Education, 2003).
323. S. Connor, *TI*, 24 October 2003. D. Perlman, *SFC*, 24 October 2003.
324. R. Watson and D. Pauly, *Nature* **414**, 534 (2001).
325. D. Malakoff, *Science* **277**, 486 (1997).
326. B. A. Block et al., *Science* **293**, 1310 (2001). J. J. Magnuson, *Science* **293**, 1267 (2001).
327. G. J. Herbert, *Marine Policy* **19**, 301 (1995).
328. D. Paul et al., *Nature* **418**, 689 (2002).
329. D. MacKenzie, New Scientist Online, 13 June 2003. A. Kirby, BBC News Online, 28 August, 2003. International Council for the Exploration of the Sea, press release, 18 October 2004.
330. Q. Schiermeier, *Nature* **423**, 212 (2003).
331. R. A. Myers and B. Worm, *Nature* **423**, 280 (2003).
332. J. K. Baum et al., *Science* **299**, 389 (2003).
333. A. M. Springer et al., *Proc. Natl. Acad. Sci.* **100**, 12223 (2003).
334. J. Roman and S. R. Palumbi, *Science* **301**, 508 (2003).
335. E. M. Olsen et al., *Nature* **428**, 932 (2004).
336. J. A. Hutchings, *Nature* **428**, 899 (2004).
337. F. C. Coleman et al., *Science* **305**, 1958 (2004). D. Grimm, *Science* **305**, 1235 (2004).
338. S. J. Cooke and I. G. Cowx, *BioScience* **54**, 857 (2004).
339. B. P. Finney et al., *Nature* **416**, 729 (2002).
340. R. A. Myers et al., *Ecol. Appl.* **7**, 91 (1997).
341. A. Atkinson et al., *Nature* **432**, 100 (2004). E. Marris, *Nature* **432**, 4 (2004).

342. B. S. Halpern, *Ecol. Appl.* **13**, S117 (2003).
343. D. O. Conover and S. B. Munch, *Science* **297**, 94 (2002).
344. J. E. Neigel, *Ecol. Appl.* **13**, S138 (2003).
345. A. Balmford et al., *Proc. Natl. Acad. Sci.* **101**, 9694 (2004).
346. K. F. Schmidt, *Science* **277**, 49 (1997).
347. C. M. Roberts, *Science* **278**, 1454 (1997). J. C. Ogden, *Science* **278**, 1414 (1997).
348. K. R. Weiss, *LAT*, 10 July 2001.
349. L. Rohter, *NYT*, 27 December 2000.
350. R. A. Myers et al., *Science* **303**, 1980 (2004).
351. R. Dalton, *Nature* **420**, 451 (2002).
352. R. Dalton, *Nature* **502**, 431 (2004).
353. K. Powell, *Nature* **426**, 378 (2003).
354. R. A. Hites et al., *Science* **303**, 226 (2004).
355. R. A. Hites et al., *Environ. Sci. Technol.* **38**, 4945 (2004).
356. R. A. Hites, *Environ. Sci. Technol.* **38**, 945 (2004).
357. S. Rayne et al., *Environ. Sci. Technol.* **38**, 4293 (2004).
358. Q. Schiermeier, *Nature* **425**, 753 (2003).
359. I. A. Fleming et al., *Proc. R. Soc. Lond. B* **267**, 1517 (2000).
360. P. McGinnity et al., *Proc. R. Soc. Lond. B* **270**, 2443 (2003).
361. B. Daley, *BG*, 6 September 2001. R. L. Naylor et al., *Science* **282**, 883 (1998).
362. K. R. Weiss, *LAT*, 9 December 2002.
363. R. L. Naylor et al., *Science* **294**, 1655 (2001).
364. E. Stokstad, *Science* **297**, 1797 (2002).
365. C. S. Kolar and D. M. Lodge, *Science* **298**, 1233 (2002).
366. E. W. Seabloom et al., *Proc. Natl. Acad. Sci.* **100**, 13384 (2003).
367. S. R. Palumbi, *Science* **293**, 1786 (2001). K. V. Sykora, *Endeavour* **8**, 118 (1984) and F. di Castri, in J. A. Drake et al., eds., *Biological Invasions: A Global Perspective* (New York: John Wiley & Sons, 1989).
368. J. J. Ewel et al., *BioScience* **49**, 619 (1999).
369. R. M. Callaway and E. T. Aschehoug, *Science* **290**, 521 (2000).
370. P. Niemelä and W. J. Mattson, *BioScience* **46**, 741 (1996).
371. D. Pimentel et al., *BioScience* **50**, 53 (2000).
372. N. D. Tsutsui et al., *Proc. Natl. Acad. Sci.* **100**, 1078 (2003).
373. J. A. Grob, Scientific American Online, 24 February 2003.
374. R. E. Smith, *TB*, 1 May 2004.
375. Government of New Zealand, URL http://www.maf.govt.nz/biosecurity/pests-diseases/plants/giant-african-snail/
377. U.S. Department of Agriculture, Press release 0085.00, March 17, 2000; URL: http://ww.usda.gov/news/releases/2000/03/0085.
378. G. Martin, *SFC*, 7 May 2004.

379. D. Haddix, *CD*, 29 October 2003.
380. P. B. McEvoy, *BioScience* **46**, 401 (1996).
381. M. Derr, *NYT*, 4 September 2001.
382. C. E. Mitchell and A. G. Power, *Nature* **421**, 625 (2003).
383. M. E. Torchin et al., *Nature* **421**, 628 (2003).
384. K. Clay, *Nature* **421**, 585 (2003).
385. P. D. Moore, *Nature* **403**, 492 (2000).
386. J. Kaiser and R. Gallagher, *Science* **277**, 1204 (1997).
387. E. A. Francis, *CD*, 28 October 2003.
388. J. B. Little, *SB*, 6 September 2003. D. Thompson, AP dispatch, 24 May 2004.
376. C. Locke, *SB*, 16 November 2003.
389. W. Wiley, *SB*, 29 August 2003.
390. D. Haddix, *CD*, 29 October 2003.
391. D. Haddix, *CD*, 28 October 2003.
392. J. Robbins, *NYT*, 8 February 2000.
393. P. Gorner, *CT*, 29 August 2001.
394. R. W. Stack et al., brochure PP-324 (Revised), North Dakota State University Extension Service website, URL: http://www. ext. nodak. edu/extpubs/plantsci/trees/pp324. htm.
395. T. Robertson, *BG*, 25 February 2003.
396. B. J. Cabrera et al., University of Florida, Institute of Food and Agricultural Sciences, URL http://edis.ifas.ufl.edu/MG064.
397. D. L. Roberts, Michigan State University Extension Service, November 2004 version, URL http://www.msue.msu.edu/reg_se/roberts/ash/.
398. E. Minor, AP dispatch, 5 July 2002. E. Slater, *LAT*, 19 July 2002. P. Recer, AP dispatch, 7 November 2002. R. Beamish, *WP*, 5 July 2004. C. K. Yoon, *NYT*, 9 September 2003. H. Bäsemann, *DW*, 16 December 2003. A. Doyle, Reuters dispatch, 10 February 2004. T. Henry, *TB*, 3 May 2004. D. Simberloff, *BioScience* **54**, 247 (2004).
399. M. Jones, *TB*, 3 September 2004. J. Feehan, *TB*, 2 September 2004. J. Seewer, AP dispatch, 8 February 2004. S. Murphy, *TB*, 2 December 2003. S. Murphy, *TB*, 17 September 2003. T. Vezner, *TB*, 2 September 2003. W. Sloat, *CPD*, 25 May 2003. T. Henry, *TB*, 30 October 2003. J. Sielicki, *TB*, 6 May 2003. T. Henry, *TB*, 30 March 2003. L. P. Vellequette, *TB*, 6 February 2003.
400. D. Behm and A. Johnson, *MJS*, 23 November 2002.
401. W. B. Ennis et al., *Science* **188** 593 (1975).
402. P. M. Vitousek et al., *Am. Sci.* **84**, 468 (1996).
403. H. R. Smith and C. L. Remington, *BioScience* **46**, 436 (1996).
404. J. L. Ruesink et al., *BioScience* **45**, 465 (1995).

405. T. Henry, *TB*, 29 September 2003.

406. T. Martin, *LSJ*, 24 August 2003.

407. J. Hansen, *MJS*, 1 September 2001.

408. J. J. Stachowicz et al., *Science* **286**, 1577 (1999).

409. A. N. Cohen and J. T. Carlton, *Science* **279**, 555 (1998).

410. Committee on Nonnative Oysters in the Chesapeake Bay (J. Anderson et al.), Ocean Studies Board, National Academy of Sciences, *Nonnative Oysters in the Chesapeake Bay* (Washington, DC: National Academy Press, 2003). See also J. Dao, *NYT*, 15 August 2003.

411. J. E. Brody, *NYT*, 20 April 1975.

412. C. E. Christian, *Nature* **413**, 635 (2001).

413. D. Fickling, *TG*, 12 August 2004.

414. M. Beroza, *Am. Sci.* **59**, 320 (1971).

415. E. Marshall, *Science* **213**, 417 (1981); 849 (1981).

416. M. R. Berenbaum, *Audubon* **102**(1), 74 (2000).

417. J. Kaiser, *Science* **285**, 1836 (1999).

418. T. McCann and R. E. Igoe, *CT*, 12 April 2001.

419. Anonymous, Environmental News Network dispatch, 28 March 2001.

420. C. L. Dybas, *WP*, 9 July 2001.

421. F. Pearce, *BG*, 23 March 2003.

422. T. Oguz et al., *Global Biogeochem. Cycles* **17**, 1088 (2003).

423. D. Ferber, *Science* **292**, 203 (2001).

424. E. Stokstad, *Science* **301**, 157 (2003). S. Ziemba, *CT*, 13 July 2004. T. Henry, *TB*, 8 October 2004.

425. D. Haddix, *CD*, 29 October 2003.

426. C. K. Yoon, *NYT*, 8 April 2003.

427. State of New Jersey, Department of Environmental Protection, press release, 25 August 2004, URL http://www.state.nj.us/dep/newsrel/2004/04_0099.htm.

428. J. Brown, Delaware River Fisheries Coordinator, U.S. Fish & Wildlife Service, Spring 2004 field notes.

429. S. Borenstein, Knight Ridder Newspapers, 19 August 2004.

430. L. O'Hanlon, Discovery News Online dispatch, 12 February 2002.

431. K. Ringle, *WP*, 4 July 2002.

432. F. X. Clines, *NYT*, 13 July 2002. C. Thomson, *BS*, 5 September 2002.

433. W. R. Courtenay, Jr., and J. D. Williams, U.S. Department of the Interior, U.S. Geological Survey Circular 1251, 1 April, 2004.

434. J. Powder, *BS*, 18 September 2002. Anonymous, *WP*, 3 October 2002. S. Braun, *LAT*, 30 November 2002.

435. D. A. Fahrenthold and J. Partlow, *WP*, 14 May 2004. Anonymous, *BS*, 28 May 2004. D. O'Brien, *BS*, 8 June 2004.

436. Editorial, *WP*, 23 May 2004.

437. Anonymous, CBS News story, 7 July 2004. J. L. Yang, *LAT*, 25 July 2004.

438. S. Lin, *LAT*, 23 May 2004.

439. U.S. Fish and Wildlife Service, Invasive Species Program circular, July 2002.

440. F. Barringer, *NYT*, 3 November 2004.

441. P. F. Hendrix and P. J. Bohlen, *BioScience* **52**, 1 (2002).

442. E. Byron, *WSJ*, 7 July 2003. E. Slater, *LAT*, 18 September 2003. A. Minard, *NYT*, 28 October 2003.

443. R. Koenig, *Science* **287**, 1737 (2000).

444. D. Jehl, *NYT*, 20 February 2001.

445. M. Kelley, AP dispatch, 22 November 2000.

446. T. Wilkinson, *CSM*, 16 February 2001. D. Jehl, *NYT*, 12 April 2001.

447. Anonymous, AP dispatch, 22 November 2000.

448. R. N. Holdaway and C. Jacomb, *Science* **287**, 2250 (2000). J. Diamond, *Science* **287**, 2170 (2000).

449. E. O. Wilson, *Consilience: The unity of knowledge* (New York: Knopf, 1998), 292-297. See also the interview with Wilson in B. Rensberger, *Audubon* **101**(6), 64 (1999).

450. A. Hastings and L. W. Botford, *Science* **284**, 1537 (1999).

451. C. M. Roberts et al., *Science* **294**, 1920 (2001). D. Malakoff, *Science* **294**, 1807 (2001).

452. K. Wallace, CNN dispatch, 4 December 2000. Anonymous, AP dispatch, 4 December 2000. Anonymous, AP dispatch, 5 December 2000.

453. E. Stokstad, *Science* **285**, 1838 (1999).

454. D. Malakoff, *Science* **285**, 1841 (1999).

455. C. Seabrook, *The Atlanta Journal-Constitution*, 15 March 2004.

456. M. Zaloudek, *The Sarasota Herald-Tribune*, 17 November 2002.

457. L. Siggins, *IT*, 2 May 2002.

458. H. De Groote et al., *Ecological Economics* **45**, 105 (2003). Not all the stories are of total success. While the beetles did control the hyacinth in Malawi, in east Africa, there a different weed is taking the place of hyacinth; see R. Carroll, *TG*, 24 May 2004.

459. C. K. Yoon, *NYT*, 6 March 2001.

460. A. E. Kideys et al., "Laboratory studies on *Beroe Ovata* and *Mnemiopsis Leidyi* in the Caspian Sea Water (Preliminary Report)," URL http://www.caspianenvironment.org/mnemiopsis/mnemmenu6.htm.

461. D. W. Onstad and M. L. McManus, *BioScience* **46**, 430 (1996).

462. J. Brooke, *NYT*, 6 September 1988.

463. H. Fountain, *NYT*, 25 May 1997.

464. K. Schneider, *NYT*, 28 February 1988; 448. K. Schneider, *NYT*, 11 June 1989.

465. Anonymous, *NYT*, 23 October 1988; A. Raver, *NYT*, 5 June 1994.

466. K. Schneider, *NYT*, 23 August 1987

467. D. D. Thomas et al., *Science* **287**, 2474 (2000).

468. R. Cheverton, *LAT*, 2 September 2001.

469. N. C. Veitch et al., *J. Natl. Prod.* **62**, 1260 (1999).

470. J. Kathirithamby et al., *Proc. Natl. Acad. Sci.* **100**, 7655 (2003). M. Watanabe, *BioScience* **54**, 383 (2004).

471. J. Kathirithamby et al., *Tijdschr. Entomol.* **144**, 187 (2001).

472. J. Kathirithamby and S. J. Johnston, *Proc. R. Soc. Lond. B* (Biology Letters Suppl.) **271**, S5 (2004).

473. W. E. Leary, *NYT*, 4 January 1994.

474. J. D. Sachs, *Science* **298**, 122 (2002).

475. D. F. Wirth, *Nature* **419**, 495 (2002).

476. D. Cyranoski, *Nature* **432**, 259 (2004).

477. M. D'Antonio, *LAT*, 2 September 2001. T. Clarke, *Nature* **419**, 429 (2002).

478. S. L. Hoffman et al., *J. Infect. Dis.* **185**, 1155 (2002). T. C. Luke and S. L. Hoffman, *J. Exptl. Biol.* **206**, 3803 (2003). D. Butler, *Nature* **425**, 437 (2003).

479. F. Catteruccia et al., *Nature* **405**, 959 (2000).

480. F. Catteruccia et al., *Science* **299**, 1225 (2003). T. Clarke, Nature Science Update, 21 February 2003.

481. H. Ranson et al., *Science* **298**, 179 (2002).

482. J. Hemingway et al., *Science* **298**, 96 (2002).

483. R. A. Holt et al., *Science* **298**, 129 (2002). C. M. Morel et al., *Science* **298**, 79 (2002), L. H. Miller and B. Greenwood, *Science* **298**, 121 (2002), and E. De Gregorio and B. Lemaitre, *Nature* **419**, 496 (2002).

484. M. Weill et al., *Nature* **423**, 136 (2003).

485. M. J. Gardner et al., *Nature* **419**, 498 (2002). J. M. Carlton et al., *Nature* **419**, 512 (2002). L. Florens et al., *Nature* **419**, 520 (2002). N. Hall et al., *Nature* **419**, 527 (2002). M. J. Gardner et al., *Nature* **419**, 531 (2002). R. W. Hyman et al., *Nature* **419**, 534 (2002). E. Lasonder et al., *Nature* **419**, 537 (2002).

486. A. S. Moffat, *Science* **284**, 1249 (1999).

487. W. D. Hamilton and S. P. Brown, *Proc. Royal Soc. B* **268**, 1489 (2001).

488. A. Kessler and I. T. Baldwin, *Science* **291**, 2141 (2001). M. W. Sabelis, A. Janssen, and M. R. Kant, *Science* **291**, 2104 (2001).

489. J. Long, *CT*, 29 August 2001.

490. M. R. Strand and J. J. Obrycki, *BioScience* **46**, 422 (1996).

491. B. A. Federici and J. V. Maddox, *BioScience* **46**, 410 (1996).

492. A. T. Peterson and D. A. Vieglais, *BioScience* **51**, 363 (2001).

493. C. Mlot, *Science* **293**, 1238 (2001).

494. J. Thomas, *NYT*, 20 October 1998.

495. J. Crossland, *Environment* **19**(5), 6 (1977); P. G. Hollie, *NYT*, 22 March 1978; M. H. Brown, *NYT Magazine*, 21 January 1979, 23; and S. H. Verhovek, *NYT*, 7 August 1988.

496. C. Norman, *Science* **220**, 34 (1983); P. Shabecoff, *NYT*, 18 November 1984.
497. E. Pianin, *WP*, 10 July 2001.
498. Dept. of Commerce, *Statistical Abstracts of the United States, 1992* (Washington, DC: GPO, 1992), Table 357.
499. P. Brimelow and L. Spencer, *Forbes* (6 July 1992), 59. See also P. Passel, *NYT,* 10 November 1994.
500. P. Shabecoff, *NYT*, 11 December 1985. Of course, attention comes in cycles to this, as other, problems.
501. W. Gehrmann, *DZ*, 26 July 1985.
502. I. Meyer-List, *DZ*, 5 July 1985.
503. J. Kay, *SFC*, 1 February 2003.
504. J. Eilperin, *WP*, 23 June 2004.
505. R. Smothers, *NYT*, 17 December 1978.
506. M. Waldron, *NYT*, 17 December 1978.
507. M. L. Wald, *NYT*, 17 December 1978.
508. J. Miller, *NYT*, 27 April 1979.
509. B. A. Franklin, *NYT*, 27 April 1979.
510. F. F. Marcus, *NYT*, 27 April 1979.
511. G. Hill, *NYT*, 27 April 1979.
512. W. Robbins, *NYT*, 27 April 1979.
513. A. A. Boraiko, *Natl. Geog.* **167**, 318 (1985).
514. See, for example, M. Sun, *Science* **219**, 468 (1983).
515. W. Williams, *NYT*, 13 March 1983.
516. I. Peterson, *NYT*, 11 December 1983.
517. K. Q. Seelye, *NYT*, 1 July 2002. E. Shogren, *LAT*, 5 August 2002. J. Heilprin, AP dispatch, 16 October 2002. E. Pianin, *WP*, 31 October 2002. E. Pianin, *WP*, 3 September 2003. J. 8. Lee, *NYT*, 9 March 2004. J. Eilperin, *WP*, 28 July 2004.
518. C. W. Schmidt, *Environ. Health Perspect.* **111**, A 162 (2003).
519. E. Pianin, *WP*, 14 September 2003.
520. D. Dickson, *Science* **220**, 1362 (1983).
521. M. Dowling, *Environment* **27**(3), 18 (1985).
522. B. Meier, *NYT*, 6 April 1990; K. Schneider, *NYT*, 26 August 1991.
523. J. O. Nrigau, *Nature* **363**, 589 (1993).
524. E. B. Swain et al., *Science* **257**, 784 (1992).
525. I. Renberg et al., *Nature* **368**, 323 (1994); J. Holusha, *NYT*, 13 May 1990 and *NYT*, 20 December 1992.
526. P. J. Hilts, *NYT*, 17 October 2000. K. L. Capozza, *SFC*, 11 June 2001.
527. R. Blumenthal, *NYT*, 25 November 1984.
528. G. Polakovic, *LAT*, 11 February 2003.
529. Centers for Disease Control and Prevention, U.S. Department of Health and Human Services, *Second National Report on Human Exposure to Environmental Chemicals* January 2003, NCEH Pub. No. 02-0716, Revised 26 August 2003.
530. A. Rome, *SFC*, 28 March 2004.
531. M. Hawthorne, *CT*, 27 July 2004.
532. J. Kaiser and M. Enserink, *Science* **290**, 2053 (2000).

533. M. S. Reich, *Chem. & Engg. News* **79**(36), 17 (2001).
534. T. H. Maugh, *Science* **215**, 490 (1982); R. J. Smith, *Science* **217**, 714 (1982); R. J. Smith, *Science* **217**, 808 (1982).
535. T. H. Maugh, *Science* **204**, 1295 (1979).
536. T. H. Maugh, *Science* **204**, 1188 (1979).
537. S. J. Marcus, *NYT*, 8 January 1984.
538. P. Shabecoff, *NYT*, 20 November 1988; *NYT*, 23 March 1989.
539. K. Schneider, *NYT,* 17 May 1991; J. Holusha, *NYT,* 13 October 1991; and K. Schneider, *NYT,* 26 March 1993.
540. R. Suro, *NYT*, 2 July 1989; D. A. Sheiman, *Amicus J.* **12**(3), 19 (1990).
541. M. W. Browne, *NYT*, 26 December 1989.
542. L. A. Daniels, *NYT*, 3 November 1985.
543. J. R. Luoma, *NYT*, 6 December 1988; Anonymous, *NYT*, 27 January 1991; Anonymous, *Scala* (No. 6), 8 (1991); S. Wöhler, *Scala* (No. 6), 8 (1991); and R. F. Probstein and R. E. Hicks, *Science* **260**, 498 (1993).
544. Anonymous, *NYT*, 8 December 1985.
545. J. Holusha, *NYT,* 15 September 1991.
546. J. Holusha, *NYT,* 13 March 1991.
547. J. McCullough et al., *Bioremediation: what it is and how it works*, Department of Energy, Natural and Accelerated Bioremediation Research Office, January 1999
548. J. Holusha, *NYT*, 9 March 1994.
549. U.S. Department of Energy, *Linking Legacies* (Washington, DC: GPO, 1997), DOE/EM-319.
550. R. G. Riley et al., *Chemical Contaminants on DOE Lands and Selection of Contaminant Mixtures for Subsurface Research* (Washington, DC: GPO, 1992), DOE-0547T.
551. J. Daniszewski, *LAT*, 30 March 2001.
552. J. Holusha, *NYT*, 21 April 1991; A. L. Penenberg, *NYT,* 15 August 1993.
553. J. A. Bumpus, M. Tien, D. Wright, and S. D. Aust, *Science* **228**, 1434 (1985).
554. P. Pauls, *Der Tagesspiegel*, 8 December 1990.
555. L. Roberts, *Science* **237**, 975 (1987); Anonymous, *NYT,* 8 November 1988.
556. J. D. Coates et al., *Nature* **411,** 1039 (2001).
557. D. R. Lovley, *Science* **293**, 1444 (2001).
558. Anonymous, *NYT*, 9 April 1991; Anonymous, *NYT,* 8 November 1992. J. Holusha, *NYT*, 22 July 1990.
559. A. S. Moffat, *Science* **264**, 778 (1994).
560. A. C. Revkin, *NYT*, 6 March 2001.
561. S. L. Doty et al., *Proc. Natl. Acad. Sci.* **97**, 6287 (2000).
562. E. Pianin, *WP*, 26 April 2001.
563. M. Hawthorne, *CD*, 25 February 2001.
564. A. Appel, *NYT*, 25 June 2000.
565. R. A. Oppel, Jr., *NYT*, 22 September 2000.
566. F. B. Pyatt and J. P. Grattan, *J. Public Health Med.* **23**, 235 (2001).

567. N. T. Basta et al., *J. Environ. Qual.* **30**, 1222 (2001).
568. G. M. Hettiarachchi et al., *J. Environ. Qual.* **30**, 1214 (2001).
569. D. Gleba et al., *Proc. Natl. Acad. Sci.* **96**, 5973 (1999).
570. L. Tanner, AP dispatch, 8 October 2000.
571. C. C., *DW*, 15 April 2001.
572. N. S. Pence et al., *Proc. Natl. Acad. Sci.* **97**, 4956 (2000).
573. L. Q. Ma et al., *Nature* **409**, 579 (2001); ibid., *Nature* **411**, 438 (2001).
574. R. C. Kaltreider et al., *Environ. Health Perspect.* **109**, 245 (2001).
575. J. W. Huang et al., *Environ. Sci. Technol.* **38**, 3412 (2004).
576. S. D. Ebbs and L. V. Kochian, *Environ. Sci. Technol.* **32**, 802 (1998).
577. D. Kantachote et al., *J. Chem. Technol. Biotechnol.* **79**, 6328 (2004).
578. S. P. Bizily et al., *Nature Biotechnology* **18**, 213 (2000).
579. A. S. Moffat, *Science* **285**, 369 (2000). E. M. Bernstein, *NYT*, 8 September 1992.
580. C. L. Rugh et al., *Proc. Natl. Acad. Sci.* **93**, 3182 (1996). S. P. Bizily et al., *Proc. Natl. Acad. Sci.* **96**, 6808 (1999).
581. D. Biller, Policy Research Working Paper 1304, The World Bank, May 1994. A. J. Gunson and M. M. Veiga, *Environmental Practice* **6**, 109 (2004).
582. F. N. Moreno et al., *Environmental Practice* **6**, 165 (2004).
583. J. Ross, *CSM*, 15 April 2004.
584. D. Johnson, *Futurist* **33**(4), 6 (1999).
585. M. Poliakoff et al., *Science* **297**, 807 (2002).
586. H. E. Schoemaker et al., *Science* **299**, 1694 (2003).
587. J. M. DeSimone, *Science* **297**, 799 (2002).
588. D. J. Cole-Hamilton, *Science* **299**, 1702 (2003).
589. D. Bradley, *Science* **300**, 2022 (2003).
590. S. S. Gupta et al., *Science* **296**, 326 (2002).
591. R. A. Gross and B. Kalra, *Science* **297**, 803 (2002).
592. S. Lyon, *Nature* **427**, 406 (2004).
593. H. Ashassi-Sorkhabi et al., *Appl. Surface Sci.* **225**, 176 (2004).
594. Anonymous, BBC News dispatch, 20 August 2000.
595. J. Robins, *NYT*, 11 December 2001. See Lotusan described on the Ispo website at URL www. ispo-online. de/prod04a. htm.
596. Fraunhofer Institutes, *Fraunhofer Cleaning Technology* (München: Fraunhofer-Gesellschaft, 2003).
597. E. Jelen et al., *Pesticide Outlook* **14**, 7 (2003).
598. J. Jakob and J. Danzig, Final Report, July 2003. V. Heil et al., *GIT Fachzeitschrift für das Laboratorium* **7**, 748 (2003). Fraunhofer-Institut für Umwelt-, Sicherheits- und Energietechnik UMSICHT, *Jahresbericht 2003* (2004).

599. W. J. Blanford et al., *Ground Water Monitoring and Remediation* **21**, 58 (2001).

600. C. E. Divine et al., presented at the 4th International Conference on Remediation of Chlorinated and Recalcitrant Compounds, Monterey, California, May 2004.

601. T. B. Boving and M. L. Brusseau, *J. Contam. Hydrol.* **42**, 51 (2000).

602. T. Boving, Department of Defense grant, URL http://www.estcp.org/projects/cleanup/cu-0113.cfm. See also T. Boving et al., *Geological Society of America Abstracts with Programs* **35**, 371 (2003).

603. C. Bowman, *SB*, 9 January 2002.

604. W. Guarini, Powerpoint presentation August 23-24, 2000, URL https://www.denix.osd.mil/denix/Public/Library/Water/Perchlorate/Technology/USFilter_FBR_Bio_P2.ppt.

605. E. E. Cox et al., presentation at the Annual International Conference on Soils, Sediments, and Water, 21 October 2003.

606. B. Sun, et al., *Science* **298**, 1023 (2002). J. M. Gossett, *Science* **298**, 974 (2002).

607. J. He et al., *Nature* **424**, 62 (2003).

608. M. Bunge et al., *Nature* **421**, 357 (2003).

609. D. R. Bond et al., *Science* **295**, 483 (2002).

610. B. A. Methé et al., *Science* **302**, 1967 (2003).

611. J. S. Levinton et al., *Proc. Natl. Acad. Sci.* **100**, 9889 (2003).

612. A. Gadgil, *Annu. Rev. Energy Environ.* **23**, 253 (1998). A. Rosenfeld, *Annu. Rev. Energy Environ.* **24**, 33 (1999).

613. S. Barisic, AP dispatch, 1 January 2001.

614. S. Strauss, *Tech. Rev.* **99**(2), 22 (1996).

615. L. Frazer, *Environ. Health Perspect.* **109**, A174 (2001).

616. J. C. Tiller et al., *Proc. Natl. Acad. Sci.* **98**, 5981 (2001).

617. L. Cornwell, AP dispatch, 10 June 2001.

618. A. Riding, *NYT*, 10 December 2001.

619. D. Goleman, *NYT*, 1 February 1994.

620. J. F. Aherne, *Environment* **35**(2), 16 (1993). M. G. Morgan, *Sci. Am.* **269**(1), 32 (1993).

621. K. Schneider, *NYT*, 29 November 1993.

622. K. R. Smith, *Annu. Rev. Energy Environ.* **18**, 529 (1993).

623. L. Roberts, *Science* **243**, 306 (1989).

624. G. Marx, *Phys. Educ.* **28**, 22 (1993).

625. L. A. Cienfuentes and L. B. Lave, *Annu. Rev. Energy Environ.* **18**, 319 (1993).

626. J. V. Hall et al., *Science* **255**, 812 (1992).

627. U. K. Ministry for Trade and Industry, *Our energy future—creating a low carbon economy* Presented to Parliament by the Secretary of State for Trade and Industry by Command of Her Majesty, February 2003.

628. R. U. Ayres et al., *Energy* **28**, 219 (2003).

629. E. Worrell and L. Price, *Energy Policy* **29**, 1223 (2001).

630. V. Smil, *Annu. Rev. Energy Environ.* **25**, 21 (2000).

631. J. Chow et al., *Science* **302**, 1528 (2003).

632. S. W. Pacala et al., *Science* **301**, 1187 (2003).

633. J. Holt, *NYT Magazine*, 28 March 2004, 7.

634. C. W. Schmidt, *Environ. Health Perspect.* **111**, A530 (2003).

635. J. Graham, in *U.S. Budget FY 2003, Analytical Perspectives* (Washington, DC: GPO, 2002), 419.

636. J. Kaiser, *Science* **299**, 1836 (2003)z. M. Bustillo, *LAT*, 30 April 2003. K. Q. Seelye and J. Tierney, *NYT*, 8 May 2003.

637. M. C. Grimston, *Phys. Educ.* **27**, 202 (1992).

638. A. Wildavsky, *Nature* **367**, 227 (1994). For a slightly different (and somewhat humorous) view, see J. Paling and S. Paling, *Up to Your Armpits in Alligators? How to Sort out What Risks Are Worth Worrying About!* (Gainesville, FL: John Paling & Company, 1994).

639. P. Passell, *NYT*, 1 September 1991. See also M. Harden, *CD*, 23 June 2002 about the *human* cost of the Times Beach, MO debacle.

640. J. Alper, *Science* **260**, 1895 (1993).

641. K. Schneider, *NYT*, 24 March 1993.

642. K. Schneider, *NYT*, 21 March 1993; 371; J. A. Cushman, Jr., *NYT*, 31 January 1994.

643. W. K. Stevens, *NYT*, 8 September 1992.

644. E. R. Tutte, *The Visual Display of Quantitative Information* (Cheshire, CT: Graphics Press, 1983), 24.

645. E. Gorham, in C. M. Bhumralkar, ed., *Meteorological Aspects of Acid Rain* (Boston: Butterworth, 1984), 1.

646. H. E. Daly, in J. Harte, and R. E. Socolow, eds., *Patient Earth* (New York: Holt, Rinehart, & Winston, 1971), 226.

647. D. L. Meadows, ed., *Alternatives to Growth A Search for Sustainable Futures* (Cambridge, MA: Ballinger, 1977).

648. M. Ingebretsen, *IEEE Spectrum* **38**(8), 34 (2001).

649. J. Smith, Reuters dispatch, 26 November 2001.

650. G. K. O'Neill, *NYT Magazine*, 18 January 1976, 10; G. K. O'Neill, *The High Frontier* (New York: Morrow, 1977).

651. J. G. Trump, *Am. Sci.* **69**, 276 (1981).

652. E. D. Arthur, in L. Rosen and R. Glasser, eds., *Climate Change and Energy Policy* (New York: American Institute of Physics, 1992), 2179. See also D. Gibson, *Bull. At. Sci.* **47**(6), 12 (1991); M. Salvatores and C. Prunier, *Endeavour* **17**, 116 (1993); and R. A. Jameson et al., *Alexander von Humboldt Magazin* #66, 13 (1995).

653. J. G. Ingersoll, in R. Howes and A. Fainberg, eds., *The Energy Sourcebook* (New York: American Institute of Physics, 1991), 207; R. Pool, *Science* **245**, 130 (1989).

654. J. R. Luoma, *NYT*, 3 January 1989.

655. A. Toffler, *Future Shock* (New York: Random House, 1970).

656. E. F. Schumacher, *Small Is Beautiful* (New York: Perennial Library, 1975).

657. A. Lovins, *Foreign Affairs* **55**, 65 (1976); A. Lovins, *Soft Energy Paths* (Cambridge, MA: Friends of the Earth—Ballinger, 1977).

658. P. Csonka, *Futurist* **11**, 285 (1977).

659. R. R. Grinstead, *Environment* **14**(3), 2 (1972) and *Environment* **14**(4), 34 (1972); J. G. Abert et al., *Science* **183**, 1052 (1974); J. McCaull, *Environment* **16**(1), 6 (1974); L. D. Orr, *Environment* **18**(10), 33 (1976); P. K. DeJoie, *Environment* **19**(7), 32 (1977); L. Hastings, *Environment* **19**(7), 38 (1977); A. Van Dam, *New Ecologist* **1**, 20 (1978); J. P. Sterba, *NYT*, 16 May 1978; and R. Smothers, *NYT*, 14 January 1979.

660. D. H. Meadows et al., *The Limits to Growth* (New York: Signet, 1972); M. Mesarovic and E. Pestel, *Mankind at the Turning Point* (New York: Dutton, 1974); D. H. Meadows et al., *Beyond the Limits: Confronting Global Collapse, Envisioning a Sustainable Future* (Post Mills, VT: Chelsea Green Publ., 1992). D. H. Meadows, *Amicus J.* **14**(1), 27 (1992).

661. M. Crenson, AP dispatch, 16 January 2001.

662. J. Ortega y Gasset, *The Revolt of the Masses* (New York: W. W. Norton, 1974).

INDEX

SELECTED UNITS AND THEIR MEANINGS

Units' names are written out with small letters (with the exception of degree Celsius and degree Fahrenheit, which were established before the convention). Abbreviations for the unit names use small letters unless the unit is named for a person; in this case, the abbreviation begins with a capital letter. As examples: the unit of energy in the International System (SI, from Système Internationale) is the joule, abbreviated J, named for James Prescott Joule; the unit of frequency is the hertz, abbreviated Hz, named for Heinrich Hertz.

Units shown in **bold italic** are accepted units in the International System. Note that the Companion Web Site has spreadsheets available that will convert from any set of units to SI units.

ampere (A): unit of electric current, one of the base units of SI; a flow of charge of one coulomb per second constitutes a current of 1 ampere.

atomic mass unit (*unified*) (u): unit of mass defined by the convention that the atom $^{12}_{6}C$ has a mass of exactly 12 u; the u represents a mass of $1.6605402 \times 10^{-27}$ kg.

barrel of oil (bbl): a volume of oil containing about 160 liters [or 42 gallons]. One *barrel of oil equivalent* is the amount of energy contained in one barrel of oil, 6.12 GJ.

becquerel (Bq): unit of activity (radioactive disintegration); one becquerel is one disintegration per second. (1 Bq = 27 pCi)

British thermal unit (Btu): the amount of thermal energy needed to raise the temperature of one pound of water one Fahrenheit degree. One Btu is equivalent to 1055 J.

Celsius degree (°C): 1/100 of the temperature interval between the freezing point of water, set to 0 °C, and the boiling point of water, set to 100 °C, at sea level. Body temperature is 37 °C, and typical room temperature is around 20 °C.

coulomb (C): unit of electric charge, equivalent to that produced by 6.28×10^{18} electrons or protons.

curie (Ci): unit of activity (radioactive disintegration); one curie is 3.7×10^{10} disintegrations per second (1 Ci = 37 GBq).

degree: measure of 1/360 of the full circle; the *minute of arc* is 1/60 of one degree, and the *second of arc* is 1/60 of one minute.

electronvolt (eV): unit of energy used as the basis of measurement for atomic (eV), electronic (keV), nuclear (MeV), and subnuclear processes (GeV or TeV). One electronvolt is equal to the amount of energy gained by an electron dropping through a potential difference of one volt, 1.6×10^{-19} J.

Fahrenheit degree (°F): 1/180 of the temperature interval between the freezing point of water, set to 32 °F, and the boiling point of water, set to 212 °F, at sea level. The temperature 0 °F was supposed to have been set by the lowest temperature at which a water and salt mixture could be liquid; the temperature 96 °F was supposed to have been set by body temperature, now given as 98.6 °F.

foot pound (ft-lb): unit of work in the English system, equivalent to that done in lifting a weight of one pound a distance of one foot.

gray (Gy): unit of exposure (technically, "absorbed dose") from γ-radiation losing one joule per kilogram of material (such as tissue).

hectare (ha): subsidiary unit of area in the international system, equal to 10,000 m^2; areas are commonly measured in hectares for agricultural uses, and yields are measured in kilograms per hectare. The hectare is 2.47 acres.

hertz (Hz): unit of frequency, equivalent to one complete cycle every second.

horsepower (hp): unit of power in the English system, equivalent to 550 foot-pounds per second, or about 750 watts.

joule (J): unit of work and energy, equivalent to the work done in lifting a one newton weight a distance of one meter. This is the basic energy unit in the international system; all other units are defined in terms of joules.

kelvin (K): unit of temperature equal in size to the Celsius degree, but with the zero set by the absolute zero of temperature, 273.15 °C below 0° Celsius. Ice freezes at 273 K, room temperature is about 293 K, and water boils at 373 K, at sea level.

kilocalorie (kcal): Amount of thermal energy needed to raise the temperature of one kilogram of water one Celsius degree (strictly, to raise the temperature from 14.5 °C to 15.5 °C). Typical Americans require 2500 kcal of food per day. A kilocalorie is 4186 joules. The kilocalorie is sometimes referred to as the "food calorie."

kilogram (kg): unit of mass, or quantity of inertia; originally set as the mass of one liter of water.

kilowatthour (kWh): an amount of energy supplied to a device drawing one kilowatt of power for one hour; equal to 3.6 megajoules.

liter (L): subsidiary unit of volume in the international system equal to 1/1000 of a cubic meter (10^{-3} m^3).

A *milliliter* is a volume of 1/1,000,000 of a cubic meter (10^{-6} m^3), and is the same as the volume of a cube one centimeter on a side.

meter (m): unit of length in the international system, originally set as 1/10,000,000 of the distance from Earth's equator to the pole.

mole (gram molecular mass, or sometimes, incorrectly, "gram molecular weight") (mol): the amount of material having a mass in grams equal to the mass of one *average* atom or molecule in u; the number known as Avogadro's number, 6.02×10^{23}, is the number of atoms or molecules in one mole.

newton (N): unit of force (and weight) in the international system; a mass of 1 kg will have an acceleration of 1 m/s^2 if acted on by a force of 1 N.

ohm (Ω): unit of electrical resistance in the international system; a potential difference of one volt will cause a current of one ampere to flow through a resistance of one ohm.

pascal (Pa): unit of pressure; equal to one newton per square meter. Atmospheric pressure at sea level is 1.013×10^5 Pa.

person-sievert: a collective measure of dose, based on SI. If the dose to a large number of people adds to one sievert, we would call that a person-sievert (this implies that the dose is spread among more than one person). It is assumed that radioactivity doses in terms of the harm that is done are additive in all senses—for example, that the dose to a person now adds to what dose the person had gotten before, or that the effect of one thousand small doses to one thousand people has the same effect as the sum of all these doses acting on one person. If thresholds exist, or if small doses confer protection but large doses harm, this additivity would not be the case.

quad (Q): unit of energy equivalent to 10^{15} Btu (one quadrillion Btu). The quad is equivalent to 1.055 EJ.

rad: unit of dose representing the exposure to one röntgen in soft body tissue; 100 rad equals 1 gray (1 rad = 0.01 Gy).

radian: angular measure, defined by the ratio of arc length to radius; since the arc length of a full circle is a circumference, 2π times the radius, a full circle has an angular measure of 2π radians. A right angle has a measure of $\pi/2$ radians.

rem (**r**öntgen **e**quivalent, **m**an): a measure of dose deposited in body tissue, averaged over the body. One rem is the dose corresponding to exposure to one röntgen; 100 rem is equal to 1 Sv.

röntgen or *roentgen* (R): unit of exposure measuring the ionizing ability of γ-radiation; one röntgen produces one electric charge (1.6×10^{-19} C) per 10^{-6} m^3 of dry air at 0 °C and atmospheric pressure. This corresponds to an energy loss of 0.0877 joule per kilogram of air.

second (s): unit of time. Originally 1/86,400 of a mean solar day. A more accurate standard (1967) is the time taken for 9,192,631,770 vibrations of a cesium-133 atom. The 1985 definition is the time required for light to travel 1/299,792,458 m. Derived units are the *minute* (min), 60 seconds, the *hour* (h), 60 minutes or 3600 seconds, the *day* (d), 86,400 seconds, and the *year* (yr or a), 3.156×10^7 s.

sievert (Sv): a measure of dose (technically, "dose equivalent") deposited in body tissue, averaged over the body. Such a dose would be caused by an exposure to ionizing radiation undergoing an energy loss of 1 joule per kilogram of body tissue (1 gray). One sievert is 100 rem or 1 mSv = 100 mrem.

tesla (T): unit of magnetic field (magnetic flux density). A mass carrying a charge of 1 C, moving at a speed of 1 m/s at a right angle to a 1 T magnetic field will experience a force of 1 N. The *gauss* is 1/10,000 of a tesla.

ton (ton): unit of weight in the English system, 2000 lb.

tonne (t): derived unit of mass equal to 1000 kilograms, a metric ton. The weight of a tonne, a tonne times g, is equivalent to 1.1 tons.

tonne of coal equivalent (tce): amount of energy contained in an average tonne of coal, 29.29 gigajoules.

tonne of oil equivalent (toe): the amount of energy contained in an average tonne of oil, 44.67 gigajoules.

volt (V): unit of electrical potential energy difference per unit charge; an electrical potential energy difference of one joule between two points in space, for an object having a charge of one coulomb, will mean that the electrical potential energy difference between the two points is one volt.

watt (W): unit of power, or work per unit time; doing work at a rate of one joule each second requires one watt of power.

working level (WL): A unit of radioactive exposure to 1.3×10^5 MeV per liter of air (or about 3.7 disintegrations per liter of air, or about 3700 disintegrations per cubic meter).

working level month (WLM): This is the dose that results from exposure to one working level for one month. The dose is 4.9 mSv.